BIG IDEAS MATH®

A Common Core Curriculum

RED ACCELERATED

Ron Larson
Laurie Boswell

Erie, Pennsylvania
BigIdeasLearning.com

Big Ideas Learning, LLC
1762 Norcross Road
Erie, PA 16510-3838
USA

For product information and customer support, contact Big Ideas Learning
at **1-877-552-7766** or visit us at ***BigIdeasLearning.com***.

Printed in the U.S.A.

ISBN 13: 978-1-60840-299-1
ISBN 10: 1-60840-299-1

6 7 8 9 10 WEB 16 15 14 13

AUTHORS

Ron Larson is a professor of mathematics at Penn State Erie, The Behrend College, where he has taught since receiving his Ph.D. in mathematics from the University of Colorado. Dr. Larson is well known as the lead author of a comprehensive program for mathematics that spans middle school, high school, and college courses. His high school and Advanced Placement books are published by Holt McDougal. Ron's numerous professional activities keep him in constant touch with the needs of students, teachers, and supervisors. Ron and Laurie Boswell began writing together in 1992. Since that time, they have authored over two dozen textbooks. In their collaboration, Ron is primarily responsible for the pupil edition and Laurie is primarily responsible for the teaching edition of the text.

Laurie Boswell is the Head of School and a mathematics teacher at the Riverside School in Lyndonville, Vermont. Dr. Boswell received her Ed.D. from the University of Vermont in 2010. She is a recipient of the Presidential Award for Excellence in Mathematics Teaching. Laurie has taught math to students at all levels, elementary through college. In addition, Laurie was a Tandy Technology Scholar, and served on the NCTM Board of Directors from 2002 to 2005. She currently serves on the board of NCSM, and is a popular national speaker. Along with Ron, Laurie has co-authored numerous math programs.

ABOUT THE BOOK

The Big Ideas Math Red Accelerated book is the newest book in the Big Ideas Math series. The program uses the same research-based strategy of a balanced approach to instruction that made the Big Ideas Math series so successful. This approach opens doors to abstract thought, reasoning, and inquiry as students persevere to answer the Essential Questions that drive instruction. The foundation of the program is the Common Core Standards for Mathematical Content and Standards for Mathematical Practice. This series exposes students to highly motivating and relevant problems that offer the depth and rigor needed to prepare them for Calculus and other college-level courses that they will study during their senior year in high school. The Big Ideas Math Red Accelerated book, along with the Algebra 1 book, completes the compacted pathway for middle school students.

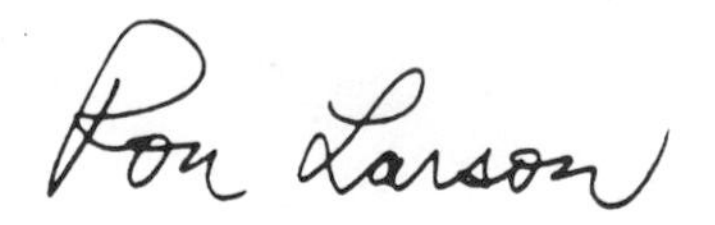

Laurie Boswell

TEACHER REVIEWERS

- Aaron Eisberg
  Napa Valley Unified School District
  Napa, CA
- Gail Englert
  Norfolk Public Schools
  Norfolk, VA
- Alexis Kaplan
  Lindenwold Public Schools
  Lindenwold, NJ
- Lou Kwiatkowski
  Millcreek Township School District
  Erie, PA
- Marcela Mansur
  Broward County Public Schools
  Fort Lauderdale, FL
- Bonnie Pendergast
  Tolleson Union High School District
  Tolleson, AZ
- Tammy Rush
  Hillsborough County Public Schools
  Tampa, FL
- Patricia D. Seger
  Polk County Public Schools
  Bartow, FL
- Denise Walston
  Norfolk Public Schools
  Norfolk, VA

STUDENT REVIEWERS

- Ashley Benovic
- Vanessa Bowser
- Sara Chinsky
- Kaitlyn Grimm
- Lakota Noble
- Norhan Omar
- Jack Puckett
- Abby Quinn
- Victoria Royal
- Madeline Su
- Lance Williams

CONSULTANTS

- Patsy Davis
  Educational Consultant
  Knoxville, Tennessee
- Bob Fulenwider
  Mathematics Consultant
  Bakersfield, California
- Deb Johnson
  Differentiated Instruction Consultant
  Missoula, Montana
- Mark Johnson
  Mathematics Assessment Consultant
  Raymond, New Hampshire
- Ryan Keating
  Special Education Advisor
  Gilbert, Arizona
- Michael McDowell
  Project-Based Instruction Specialist
  Tahoe City, California
- Sean McKeighan
  Interdisciplinary Advisor
  Norman, Oklahoma
- Bonnie Spence
  Differentiated Instruction Consultant
  Missoula, Montana

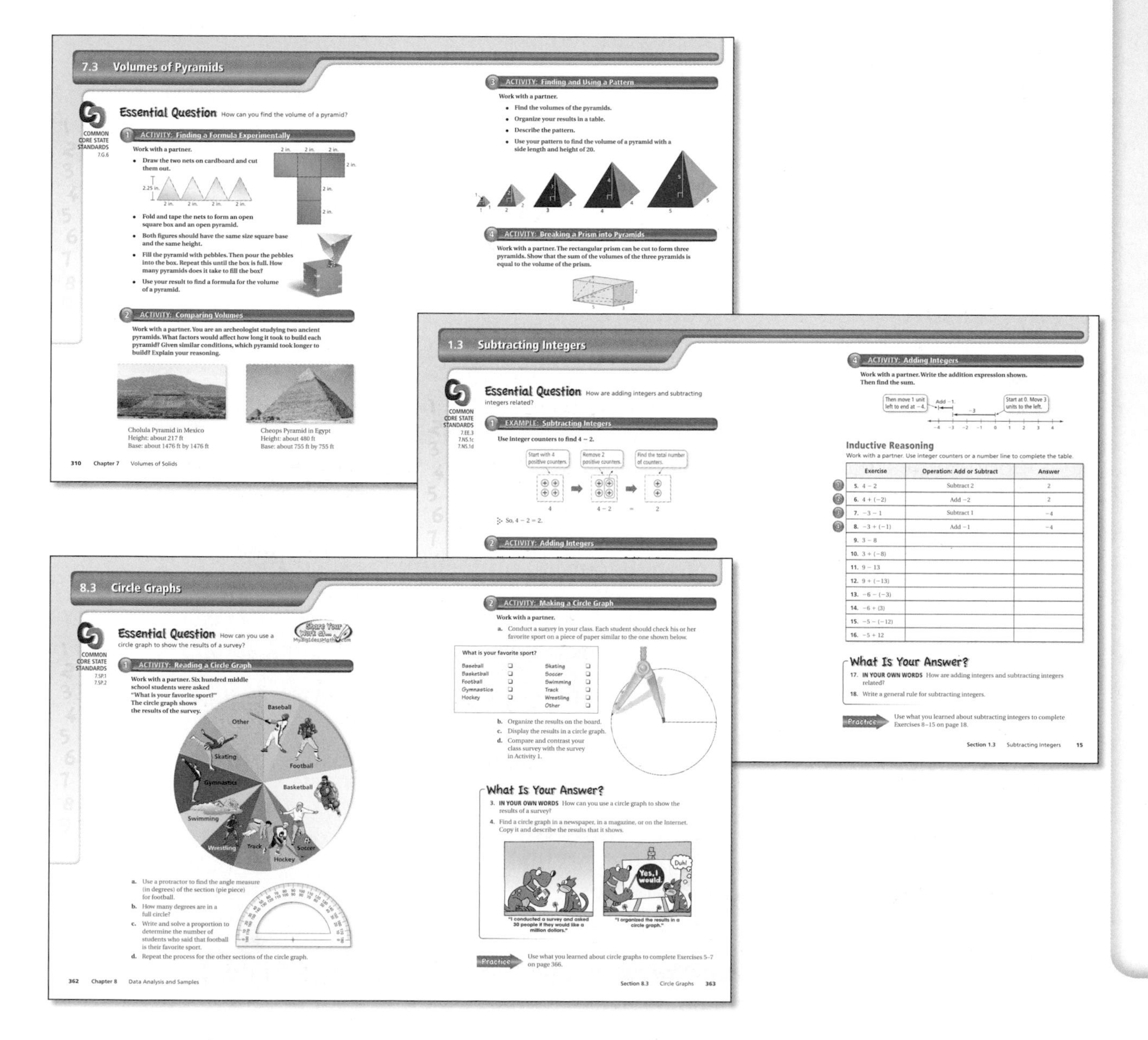

Common Core State Standards for Mathematical Practice

Make sense of problems and persevere in solving them.

- Multiple representations are presented to help students move from concrete to representative and into abstract thinking
- *Essential Questions* help students focus and analyze
- *In Your Own Words* provide opportunities for students to look for meaning and entry points to a problem

Reason abstractly and quantitatively.

- Visual problem solving models help students create a coherent representation of the problem
- Opportunities for students to decontextualize and contextualize problems are presented in every lesson

Construct viable arguments and critique the reasoning of others.

- *Error Analysis*; *Different Words, Same Question*; and *Which One Doesn't Belong* features provide students the opportunity to construct arguments and critique the reasoning of others
- *Inductive Reasoning* activities help students make conjectures and build a logical progression of statements to explore their conjecture

Model with mathematics.

- Real-life situations are translated into diagrams, tables, equations, and graphs to help students analyze relations and to draw conclusions
- Real-life problems are provided to help students learn to apply the mathematics that they are learning to everyday life

Use appropriate tools strategically.

- *Graphic Organizers* support the thought process of what, when, and how to solve problems
- A variety of tool papers, such as graph paper, number lines, and manipulatives, are available as students consider how to approach a problem
- Opportunities to use the web, graphing calculators, and spreadsheets support student learning

Attend to precision.

- *On Your Own* questions encourage students to formulate consistent and appropriate reasoning
- Cooperative learning opportunities support precise communication

Look for and make use of structure.

- *Inductive Reasoning* activities provide students the opportunity to see patterns and structure in mathematics
- Real-world problems help students use the structure of mathematics to break down and solve more difficult problems

Look for and express regularity in repeated reasoning.

- Opportunities are provided to help students make generalizations
- Students are continually encouraged to check for reasonableness in their solutions

Go to *BigIdeasMath.com* for more information on the Common Core State Standards for Mathematical Practice.

Common Core State Standards for Mathematical Content for Grade 7 Accelerated

Chapter Coverage for Standards

Conceptual Category Number and Quantity

- The Real Number System
- Quantities

1 2 3 4 5 6 7 8 9 10 11 12 13 AT

Conceptual Category Algebra

- Seeing Structure in Expressions
- Creating Equations
- Reasoning with Equations and Inequalities

1 2 3 4 5 6 7 8 9 10 11 12 13 AT

Conceptual Category Geometry

- Congruence
- Similarity, Right Triangles, and Trigonometry
- Geometric Measurement and Dimension

1 2 3 4 5 6 7 8 9 10 11 12 13 AT

Conceptual Category Statistics and Probability

- Interpreting Categorical and Quantitative Data
- Making Inferences and Justifying Conclusions
- Conditional Probability and the Rules of Probability

Go to *BigIdeasMath.com* for more information on the Common Core State Standards for Mathematical Content.

Operations with Integers

"I love my math book. It has so many interesting examples and homework problems. I have always liked math, but I didn't know how it could be used. Now I have lots of ideas."

Rational Numbers and Equations

"I like starting each new lesson with a partner activity. I just moved to this school and the activities helped me make friends."

Proportions and Variation

"I like having the book on the Internet. The online tutorials help me with my homework when I get stuck on a problem."

Percents

"I love the cartoons. They are funny and they help me remember the math. I want to be a cartoonist some day."

Similarity and Transformations

"I like how I can click on the words in the book that is online and hear them read to me. I like to pronounce words correctly, but sometimes I don't know how to do that by just reading the words."

← how old is this?

Surface Areas of Solids

"I really liked the projects at the end of the book. The history project on ancient Egypt was my favorite. Someday I would like to visit Egypt and go to the pyramids."

Volumes of Solids

"I like how the glossary in the book is part of the index. When I couldn't remember how a vocabulary word was defined, I could go to the index and find where the word was defined in the book."

Data Analysis and Samples

"I like the practice tests in the book. I get really nervous on tests. So, having a practice test to work on at home helped me to chill out when the real test came."

Probability

"I like the review at the beginning of each chapter. This book has examples to help me remember things from last year. I don't like it when the review is just a list of questions."

Linear Equations and Functions

"I like that the student book is available on my tablet. Now, I always have my book."

Angles and Similarity

"I like that the book teaches me to think. At first I just wanted someone to tell me how. Now I discover it!"

Radicals and the Pythagorean Theorem

"The website support is really helpful. I use it a lot to help me do my homework and study for tests."

Exponents and Scientific Notation

"I like the graphic organizers because they help me organize my math notes."

Additional Topics

Appendix A: My Big Ideas Projects

"I like the workbook (Record and Practice Journal). It saved me a lot of work to not have to copy all the questions and graphs."

How to Use Your Math Book

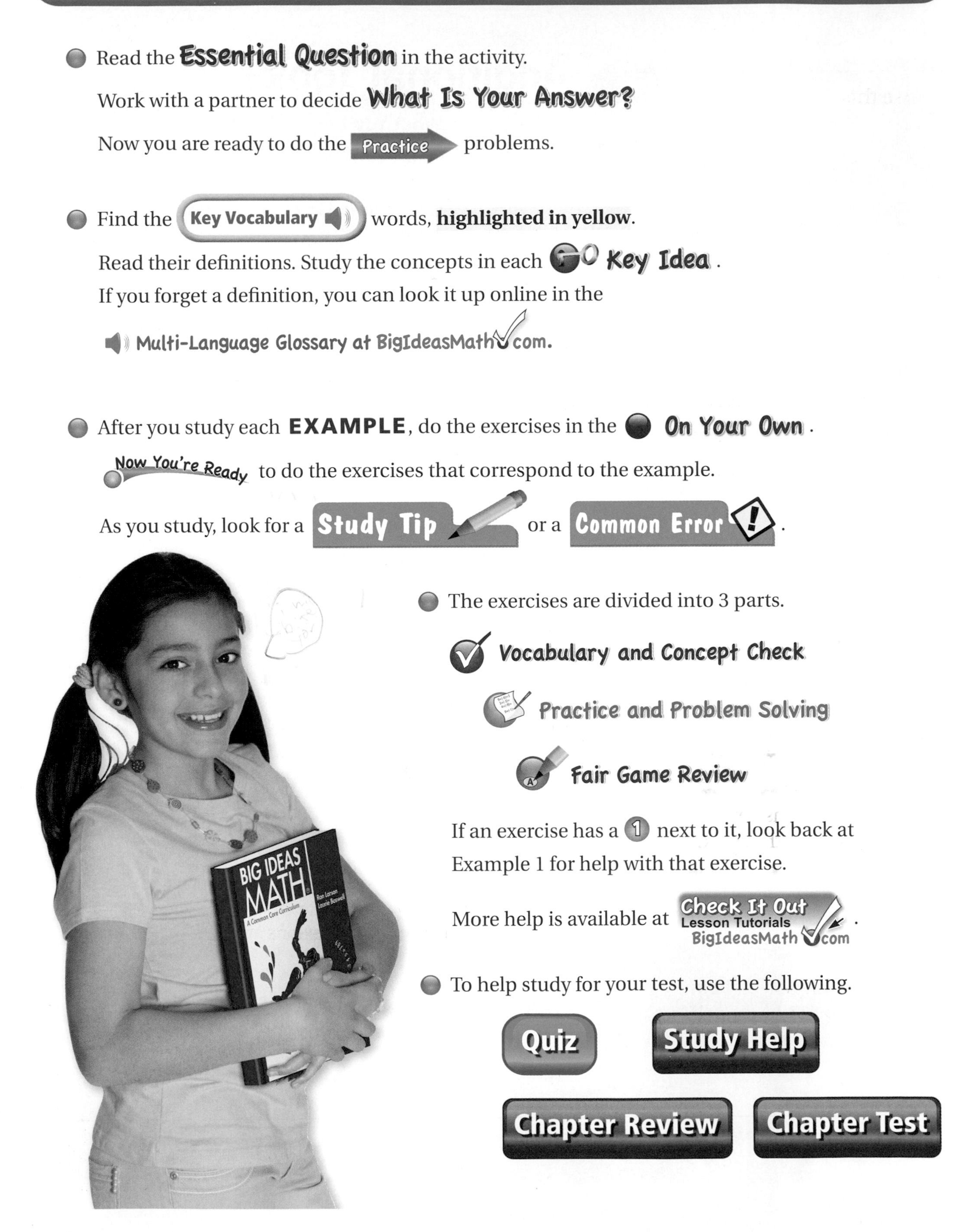

- Read the **Essential Question** in the activity.

  Work with a partner to decide **What Is Your Answer?**

  Now you are ready to do the **Practice** problems.

- Find the **Key Vocabulary** words, **highlighted in yellow**.

  Read their definitions. Study the concepts in each **Key Idea**.

  If you forget a definition, you can look it up online in the

  Multi-Language Glossary at BigIdeasMath.com.

- After you study each **EXAMPLE**, do the exercises in the **On Your Own**.

  Now You're Ready to do the exercises that correspond to the example.

  As you study, look for a **Study Tip** or a **Common Error**.

- The exercises are divided into 3 parts.

  Vocabulary and Concept Check

  Practice and Problem Solving

  Fair Game Review

  If an exercise has a 1 next to it, look back at Example 1 for help with that exercise.

  More help is available at **Check It Out Lesson Tutorials BigIdeasMath.com**.

- To help study for your test, use the following.

  Quiz **Study Help**

  Chapter Review **Chapter Test**

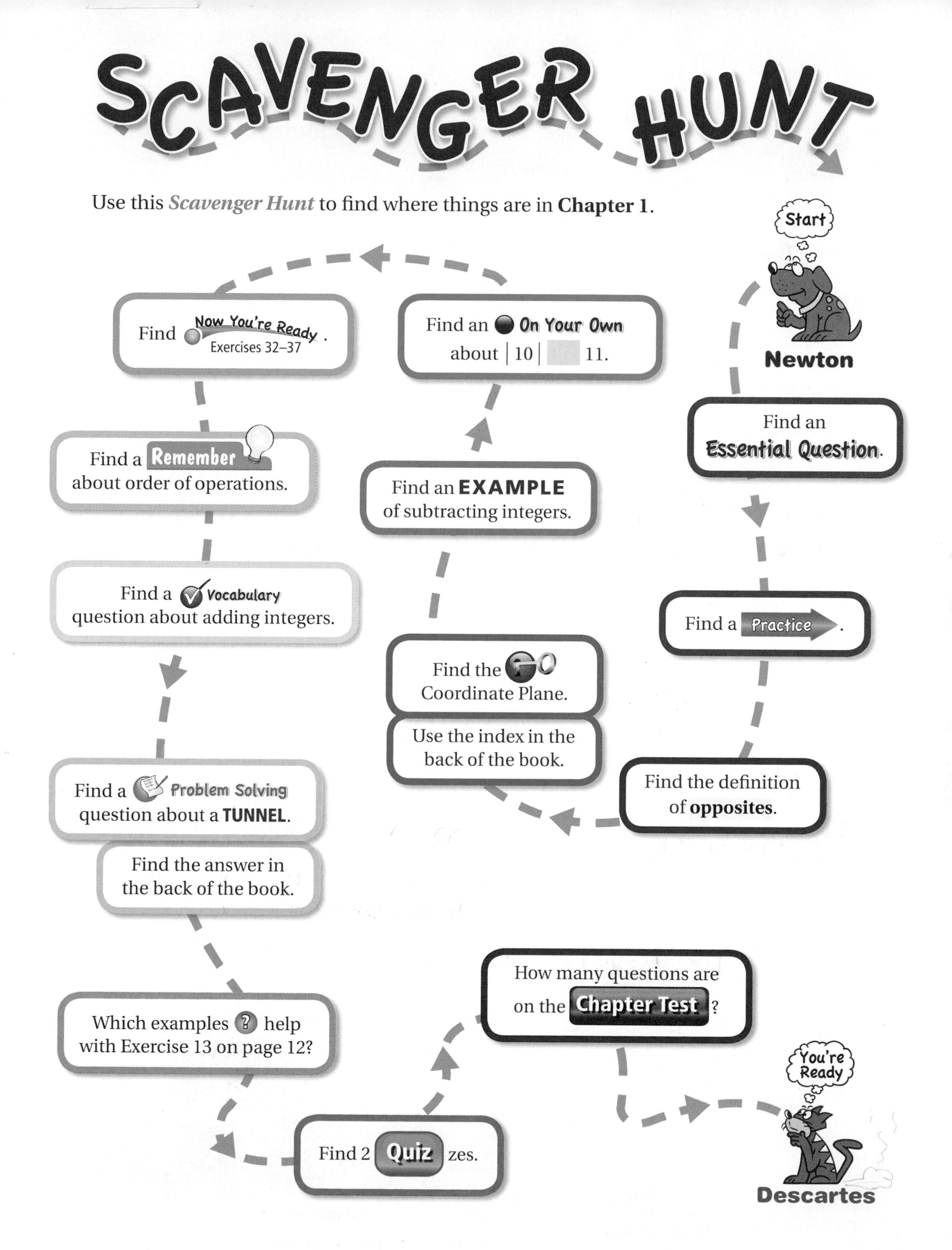
SCAVENGER HUNT
Use this Scavenger Hunt to find where things are in Chapter 1.
Start
Newton
Find an Essential Question.
Find a Practice.
Find the definition of opposites.
Find the Coordinate Plane.
Use the index in the back of the book.
Find an EXAMPLE of subtracting integers.
Find an On Your Own about 10 11.
Find Now You're Ready Exercises 32–37.
Find a Remember about order of operations.
Find a Vocabulary question about adding integers.
Find a Problem Solving question about a TUNNEL.
Find the answer in the back of the book.
Which examples help with Exercise 13 on page 12?
Find 2 Quizzes.
How many questions are on the Chapter Test?
You're Ready
Descartes

1 Operations with Integers

"Look, subtraction is not that difficult. Imagine that you have five squeaky mouse toys."

"After your friend Fluffy comes over for a visit, you notice that one of the squeaky toys is missing."

"Now, you go over to Fluffy's and retrieve the missing squeaky mouse toy. It's easy."

"Dear Sir: You asked me to 'find' the opposite of -1."

That's not the only thing that is missing.

"I didn't know it was missing."

What You Learned Before

"I liked it because it is the opposite of the freezing point on the Fahrenheit temperature scale."

Ordering Integers (6.NS.6c)

Example 1 Order 0, −1, 2, 5, and −6 from least to greatest.

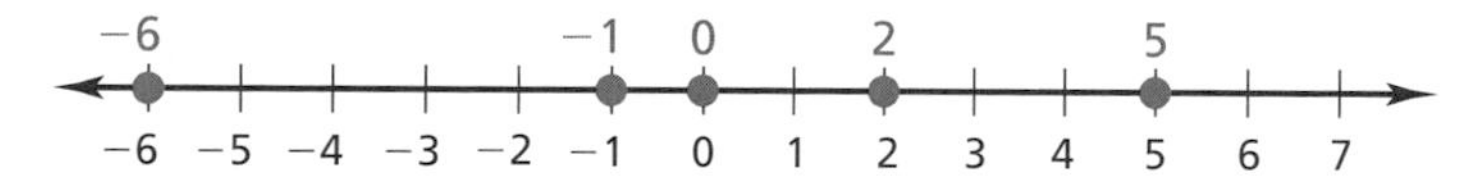

Try It Yourself

Order the integers from least to greatest.

1. −10, 15, 4, −2, −12

2. 7, −5, 3, −3, 1

Plotting Points (5.G.1)

Example 2 Plot the point (2, 3).

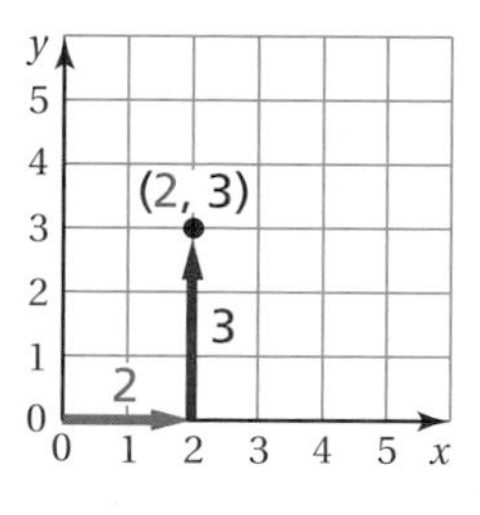

Example 3 Write an ordered pair corresponding to Point Q.

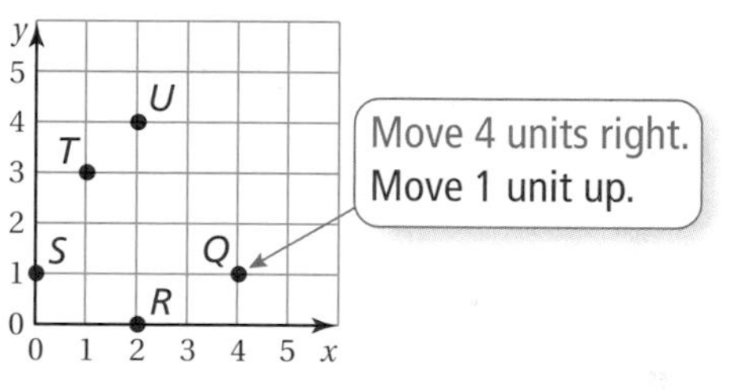

∴ The ordered pair (4, 1) corresponds to Point Q.

Try It Yourself

Use the graph in Example 3 to write an ordered pair corresponding to the point.

3. Point S

4. Point T

5. Point U

6. Point R

Using Order of Operations (6.EE.1)

Example 4 Evaluate $6^2 \div 4 - 2(9 - 5)$.

First:	Parentheses	$6^2 \div 4 - 2(9 - 5) = 6^2 \div 4 - 2 \cdot 4$
Second:	Exponents	$= 36 \div 4 - 2 \cdot 4$
Third:	Multiplication and Division (from left to right)	$= 9 - 8$
Fourth:	Addition and Subtraction (from left to right)	$= 1$

Try It Yourself

Evaluate the expression.

7. $15\left(\frac{8}{4}\right) + 2^2 - 3 \cdot 7$

8. $5^2 \cdot 2 \div 10 + 3 \cdot 2 - 1$

9. $3^2 - 1 + 2(4(3 + 2))$

1.1 Integers and Absolute Value

COMMON CORE STATE STANDARDS
7.EE.3

Essential Question How are velocity and speed related?

On these two pages, you will investigate vertical motion (up or down).

- Speed tells how fast an object is moving, but does not tell the direction.
- Velocity tells how fast an object is moving and also tells the direction.
  If velocity is positive, the object is moving up.
  If velocity is negative, the object is moving down.

1 EXAMPLE: Falling Parachute

You are gliding to the ground wearing a parachute. The table shows your height at different times.

Time (seconds)	0	1	2	3
Height (feet)	45	30	15	0

a. How many feet do you move each second?

b. What is your speed? Give the units.

c. Is your velocity positive or negative?

d. What is your velocity? Give the units.

a. For each 1 second of time, your height is 15 feet less.

b. You are moving at 15 feet per second.

c. Because you are moving down, your velocity is negative.

d. Your velocity is -15 feet per second. This can be written as -15 ft/sec.

2 ACTIVITY: Rising Balloons

Work with a partner. The table shows the height of a group of balloons.

Time (seconds)	0	1	2	3
Height (feet)	0	4	8	12

a. How many feet do the balloons move each second?

b. What is the speed of the balloons? Give the units.

c. Is the velocity positive or negative?

d. What is the velocity? Give the units.

3 ACTIVITY: Finding Speed and Velocity

Work with a partner. The table shows the height of a firework's parachute.

Time (seconds)	Height (feet)
0	480
1	360
2	240
3	120
4	0

a. How many feet does the parachute move each second?

b. What is the speed of the parachute? Give the units.

c. Is the velocity positive or negative?

d. What is the velocity? Give the units.

Inductive Reasoning

4. Copy and complete the table.

Velocity (feet per second)	−14	20	−2	0	25	−15
Speed (feet per second)						

5. Find two different velocities for which the speed is 16 feet per second.

6. Which number is greater: −4 or 3? Use a number line to explain your reasoning.

7. One object has a velocity of −4 feet per second. Another object has a velocity of 3 feet per second. Which object has the greater speed? Explain your answer.

What Is Your Answer?

In this lesson, you will study **absolute value**. Here are some examples:

Absolute value of −16 = 16　　　Absolute value of 16 = 16

Absolute value of 0 = 0　　　Absolute value of −2 = 2

8. **IN YOUR OWN WORDS** How are velocity and speed related?

9. Which of the following is a true statement? Explain your reasoning.

a. Absolute value of velocity = speed

b. Absolute value of speed = velocity

Practice Use what you learned about absolute value to complete Exercises 4–11 on page 6.

1.1 Lesson

The following numbers are **integers.**

$\ldots, -3, -2, -1, 0, 1, 2, 3, \ldots$

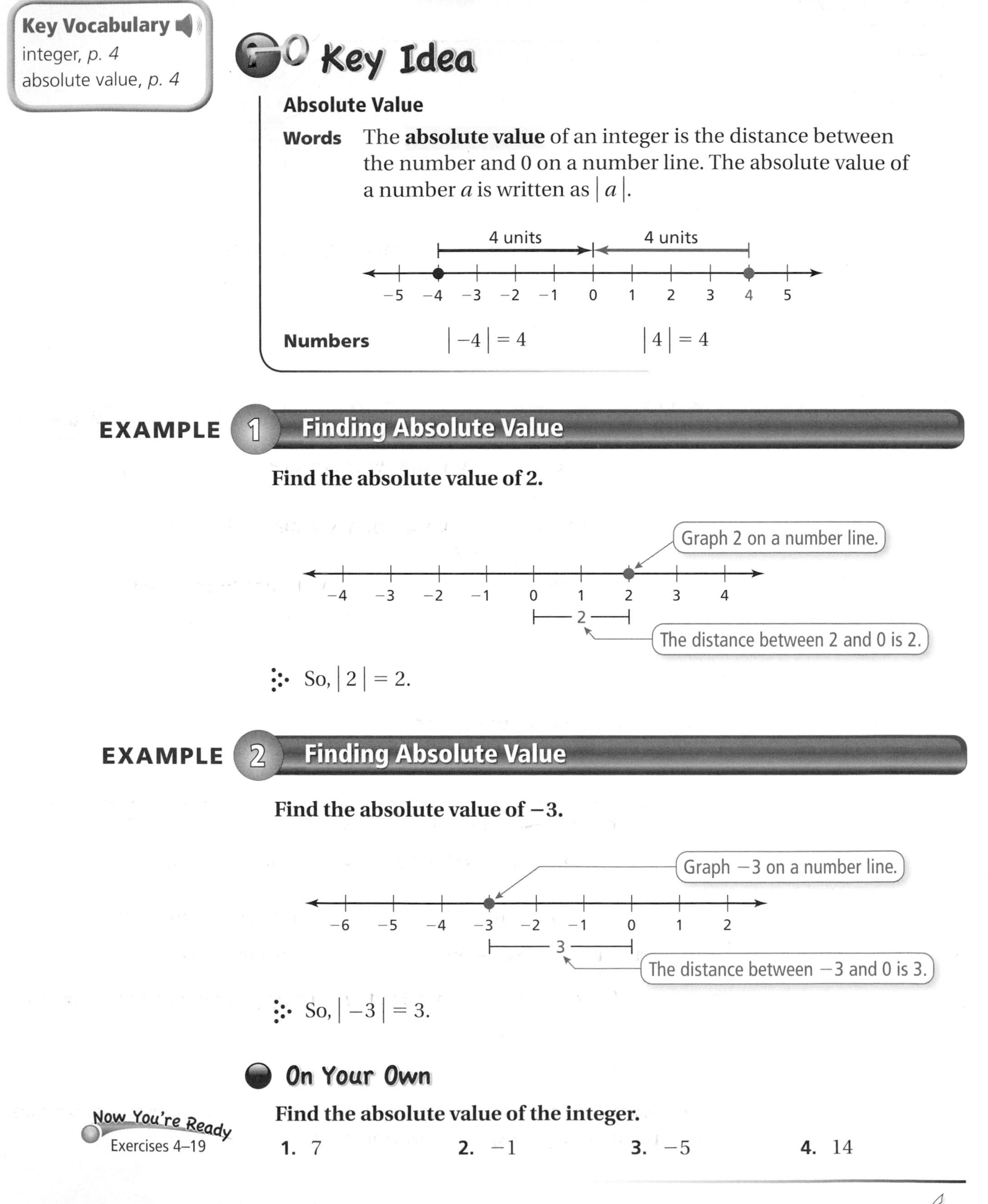

Key Vocabulary

integer, *p. 4*

absolute value, *p. 4*

Key Idea

Absolute Value

Words The **absolute value** of an integer is the distance between the number and 0 on a number line. The absolute value of a number a is written as $|a|$.

Numbers $|-4| = 4$ $\quad |4| = 4$

EXAMPLE 1 Finding Absolute Value

Find the absolute value of 2.

So, $|2| = 2$.

EXAMPLE 2 Finding Absolute Value

Find the absolute value of −3.

So, $|-3| = 3$.

On Your Own

Now You're Ready
Exercises 4–19

Find the absolute value of the integer.

1. 7 **2.** -1 **3.** -5 **4.** 14

EXAMPLE 3 Comparing Values

Compare 1 and $|-4|$.

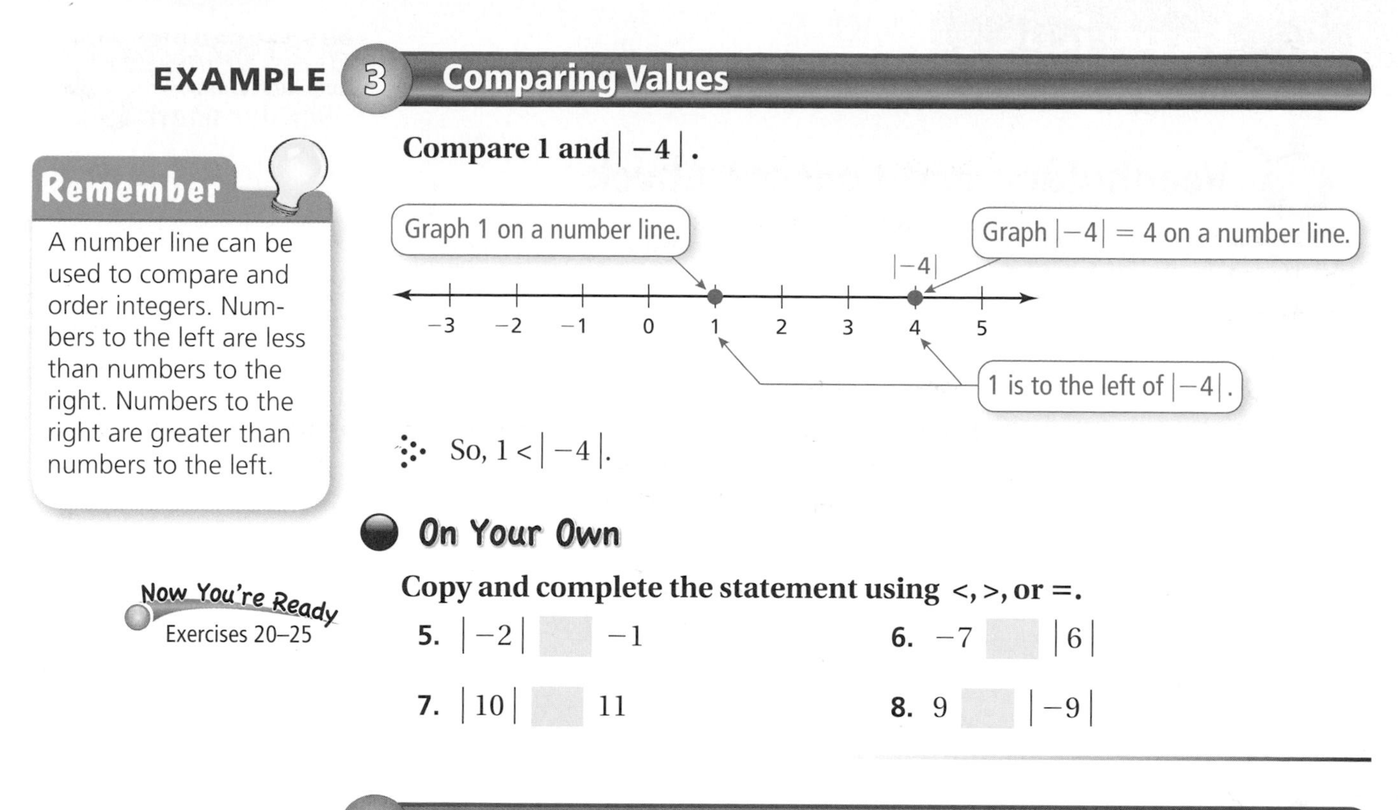

Remember

A number line can be used to compare and order integers. Numbers to the left are less than numbers to the right. Numbers to the right are greater than numbers to the left.

So, $1 < |-4|$.

On Your Own

Now You're Ready
Exercises 20–25

Copy and complete the statement using <, >, or =.

5. $|-2|$ ▢ -1
6. -7 ▢ $|6|$
7. $|10|$ ▢ 11
8. 9 ▢ $|-9|$

EXAMPLE 4 Real-Life Application

Substance	Freezing Point (°C)
Butter	35
Airplane fuel	−53
Honey	−3
Mercury	−39
Candle wax	55

The *freezing point* is the temperature at which a liquid becomes a solid.

a. Which substance in the table has the lowest freezing point?

b. Is the freezing point of mercury or butter closer to the freezing point of water, 0°C?

a. Graph each freezing point.

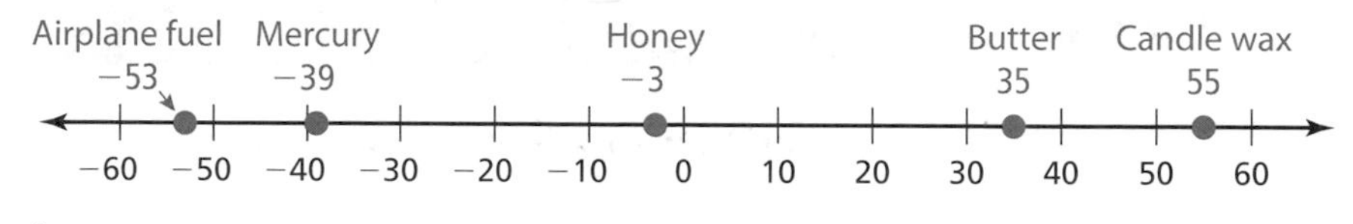

Airplane fuel has the lowest freezing point, −53°C.

b. The freezing point of water is 0°C, so you can use absolute values.

Mercury: $|-39| = 39$ **Butter:** $|35| = 35$

Because 35 is less than 39, the freezing point of butter is closer to the freezing point of water.

On Your Own

9. Is the freezing point of airplane fuel or candle wax closer to the freezing point of water? Explain your reasoning.

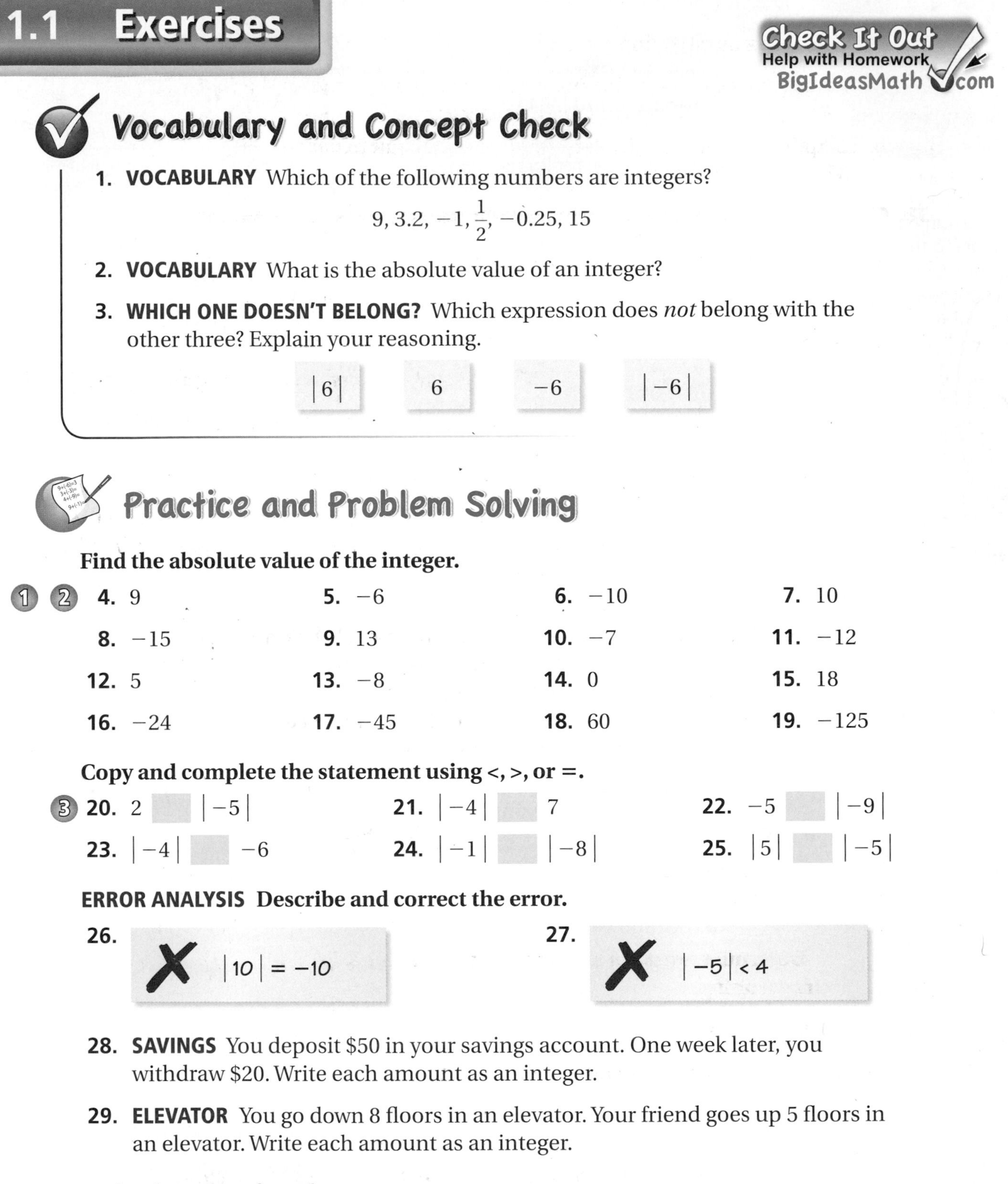

1.1 Exercises

Check It Out
Help with Homework
BigIdeasMath.com

Vocabulary and Concept Check

1. **VOCABULARY** Which of the following numbers are integers?

$$9,\ 3.2,\ -1,\ \frac{1}{2},\ -0.25,\ 15$$

2. **VOCABULARY** What is the absolute value of an integer?
3. **WHICH ONE DOESN'T BELONG?** Which expression does *not* belong with the other three? Explain your reasoning.

$|6|$ 6 -6 $|-6|$

Practice and Problem Solving

Find the absolute value of the integer.

(1) (2)

4. 9
5. -6
6. -10
7. 10
8. -15
9. 13
10. -7
11. -12
12. 5
13. -8
14. 0
15. 18
16. -24
17. -45
18. 60
19. -125

Copy and complete the statement using <, >, or =.

(3)

20. $2 \ \square \ |-5|$
21. $|-4| \ \square \ 7$
22. $-5 \ \square \ |-9|$
23. $|-4| \ \square \ -6$
24. $|-1| \ \square \ |-8|$
25. $|5| \ \square \ |-5|$

ERROR ANALYSIS **Describe and correct the error.**

26. ✗ $|10| = -10$
27. ✗ $|-5| < 4$

28. **SAVINGS** You deposit \$50 in your savings account. One week later, you withdraw \$20. Write each amount as an integer.
29. **ELEVATOR** You go down 8 floors in an elevator. Your friend goes up 5 floors in an elevator. Write each amount as an integer.

Order the values from least to greatest.

30. $8,\ |3|,\ -5,\ |-2|,\ -2$
31. $|-6|,\ -7,\ 8,\ |5|,\ -6$
32. $-12,\ |-26|,\ -15,\ |-12|,\ |10|$
33. $|-34|,\ 21,\ -17,\ |20|,\ |-11|$

Simplify the expression.

34. $|-30|$
35. $-|4|$
36. $-|-15|$

37. PUZZLE Use a number line.

a. Graph and label the following points on a number line: $A = -3, E = 2, M = -6, T = 0$. What word do the letters spell?

b. Graph and label the absolute value of each point in part (a). What word do the letters spell now?

38. OPEN-ENDED Write a negative integer whose absolute value is greater than 3.

REASONING Determine whether $n \geq 0$ or $n \leq 0$.

39. $n + |-n| = 2n$

40. $n + |-n| = 0$

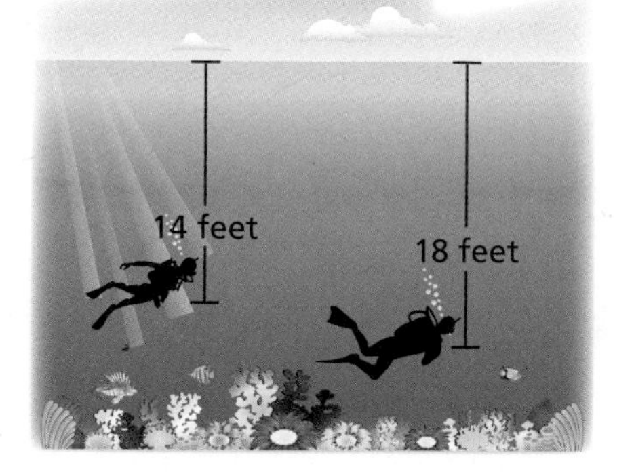

41. CORAL REEF Two scuba divers are exploring a living coral reef.

a. Write an integer for the position of each diver relative to sea level.

b. Which integer in part (a) is greater?

c. Which integer in part (a) has the greater absolute value? Compare this with the position of the diver farther from sea level.

42. VOLCANOES The *summit elevation* of a volcano is the elevation of the top of the volcano relative to sea level. The summit elevation of the volcano Kilauea in Hawaii is 1277 meters. The summit elevation of the underwater volcano Loihi in the Pacific Ocean is −969 meters. Which summit is closer to sea level?

43. MINIATURE GOLF The table shows golf scores, relative to *par*.

a. The player with the lowest score wins. Which player wins?

b. Which player is at par?

c. Which player is farthest from par?

Player	Score
1	+5
2	0
3	−4
4	−1
5	+2

True or False? **Determine whether the statement is *true* or *false*. Explain your reasoning.**

44. If $x < 0$, then $|x| = -x$.

45. The absolute value of every integer is positive.

Fair Game Review What you learned in previous grades & lessons

Add. *(Skills Review Handbook)*

46. 19 + 32 **47.** 50 + 94 **48.** 181 + 217 **49.** 1149 + 2021

50. MULTIPLE CHOICE Which value is *not* a whole number? *(Skills Review Handbook)*

Ⓐ −5 Ⓑ 0 Ⓒ 4 Ⓓ 113

1.2 Adding Integers

COMMON CORE STATE STANDARDS
7.EE.3
7.NS.1b
7.NS.1d

Essential Question Is the sum of two integers *positive*, *negative*, or *zero*? How can you tell?

1 EXAMPLE: Adding Integers with the Same Sign

Use integer counters to find $-4 + (-3)$.

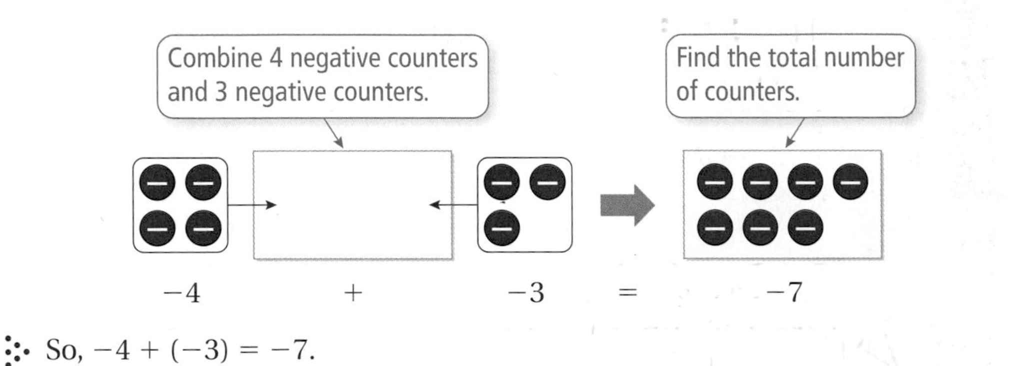

Combine 4 negative counters and 3 negative counters.
Find the total number of counters.
−4
+
−3
=
−7

So, $-4 + (-3) = -7$.

2 ACTIVITY: Adding Integers with Different Signs

Work with a partner. Use integer counters to find $-3 + 2$.

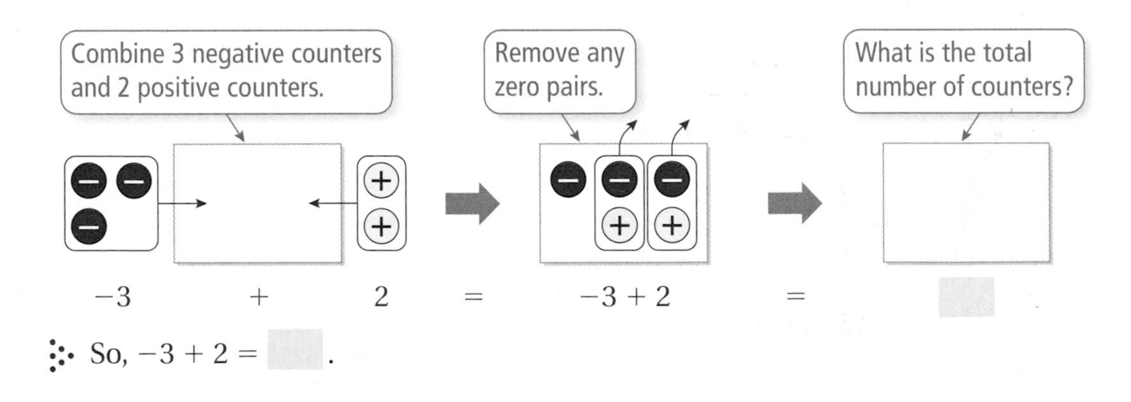

Combine 3 negative counters and 2 positive counters.
Remove any zero pairs.
What is the total number of counters?
−3
+
2
=
−3 + 2
=

So, $-3 + 2 =$ ______.

3 EXAMPLE: Adding Integers with Different Signs

Use a number line to find $5 + (-3)$.

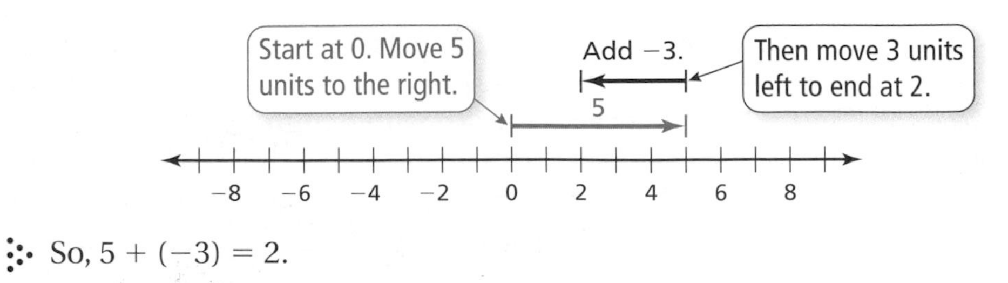

Start at 0. Move 5 units to the right.
Add −3.
Then move 3 units left to end at 2.
5
−8
−6
−4
−2
0
2
4
6
8

So, $5 + (-3) = 2$.

4 ACTIVITY: Adding Integers with Different Signs

Work with a partner. Write the addition expression shown. Then find the sum.

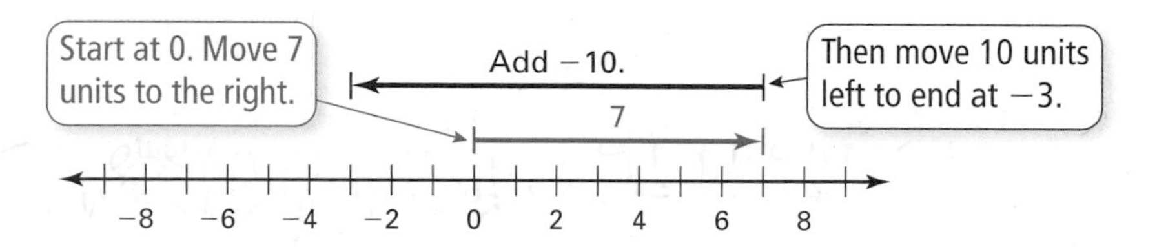
Start at 0. Move 7 units to the right.
Add −10.
7
Then move 10 units left to end at −3.
−8 −6 −4 −2 0 2 4 6 8

Inductive Reasoning

Work with a partner. Use integer counters or a number line to complete the table.

Exercise	Type of Sum	Sum	Sum: Positive, Negative, or Zero
(1) **5.** $-4 + (-3)$	Integers with the same sign		
(2) **6.** $-3 + 2$			Negative
(3) **7.** $5 + (-3)$		2	
(4) **8.** $7 + (-10)$	Integers with different signs		
9. $2 + 4$			
10. $-6 + (-2)$			
11. $-5 + 9$			
12. $15 + (-9)$			
13. $-10 + 10$			
14. $-6 + (-6)$			
15. $12 + (-12)$			

What Is Your Answer?

16. IN YOUR OWN WORDS Is the sum of two integers *positive, negative,* or *zero*? How can you tell?

17. Write general rules for adding (a) two integers with the same sign, (b) two integers with different signs, and (c) an integer and its opposite.

Practice Use what you learned about adding integers to complete Exercises 8–15 on page 12.

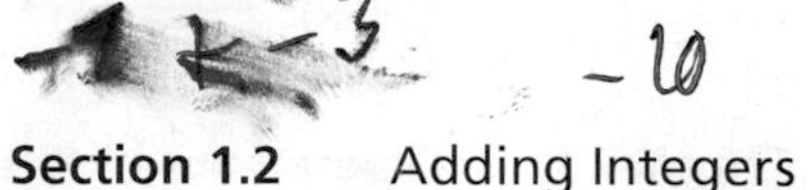

1.2 Lesson

Check It Out
Lesson Tutorials
BigIdeasMath.com

Key Idea

Key Vocabulary
opposites, *p. 10*
additive inverse, *p. 10*

Adding Integers with the Same Sign

Words Add the absolute values of the integers. Then use the common sign.

Numbers $2 + 5 = 7$ $\quad -2 + (-5) = -7$

EXAMPLE 1 Adding Integers with the Same Sign

Find $-2 + (-4)$. Use a number line to check your answer.

$-2 + (-4) = -6$ Add $|-2|$ and $|-4|$.

Use the common sign.

The sum is -6.

Check

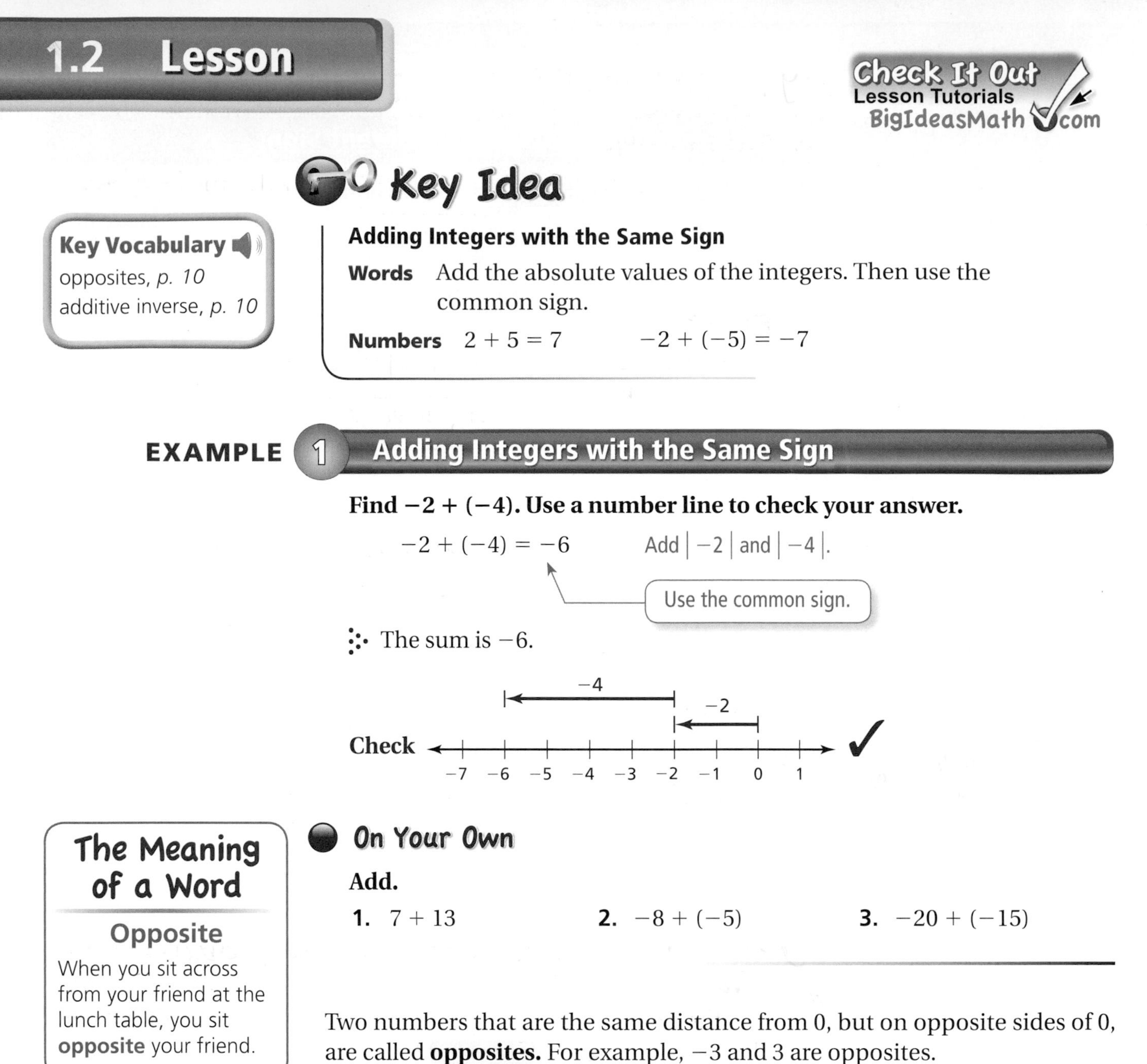

The Meaning of a Word

Opposite

When you sit across from your friend at the lunch table, you sit **opposite** your friend.

On Your Own

Add.

1. $7 + 13$ **2.** $-8 + (-5)$ **3.** $-20 + (-15)$

Two numbers that are the same distance from 0, but on opposite sides of 0, are called **opposites.** For example, -3 and 3 are opposites.

Key Ideas

Adding Integers with Different Signs

Words Subtract the lesser absolute value from the greater absolute value. Then use the sign of the integer with the greater absolute value.

Numbers $8 + (-10) = -2$ $\quad -13 + 17 = 4$

Additive Inverse Property

Words The sum of an integer and its **additive inverse,** or opposite, is 0.

Numbers $6 + (-6) = 0$ $\quad -25 + 25 = 0$

Multi-Language Glossary at BigIdeasMath.com.

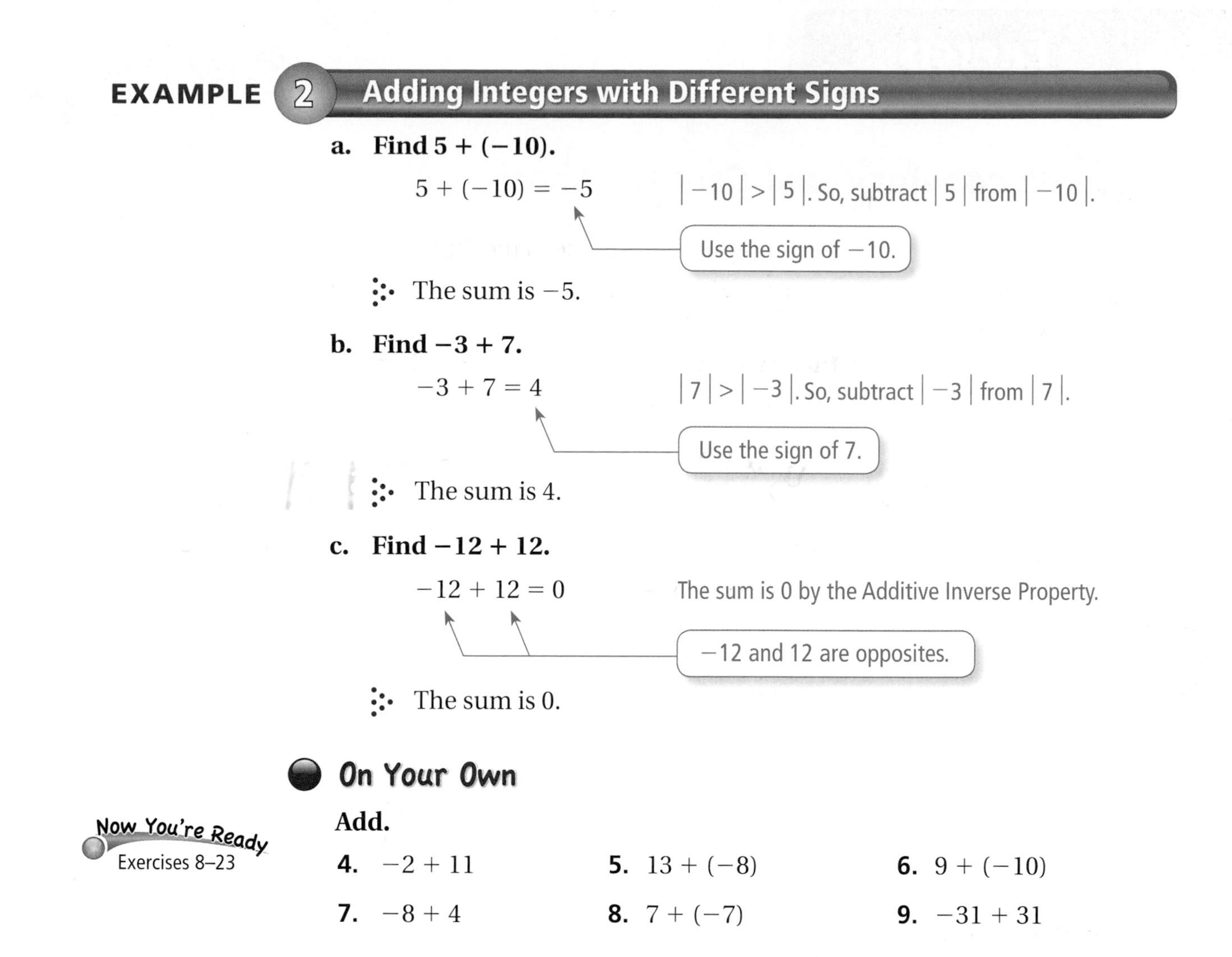

EXAMPLE 2 Adding Integers with Different Signs

a. Find $5 + (-10)$.

$5 + (-10) = -5$ $\quad$ $|-10| > |5|$. So, subtract $|5|$ from $|-10|$.

Use the sign of -10.

The sum is -5.

b. Find $-3 + 7$.

$-3 + 7 = 4$ $\quad$ $|7| > |-3|$. So, subtract $|-3|$ from $|7|$.

Use the sign of 7.

The sum is 4.

c. Find $-12 + 12$.

$-12 + 12 = 0$ $\quad$ The sum is 0 by the Additive Inverse Property.

-12 and 12 are opposites.

The sum is 0.

On Your Own

Now You're Ready
Exercises 8–23

Add.

4. $-2 + 11$ $\quad$ **5.** $13 + (-8)$ $\quad$ **6.** $9 + (-10)$

7. $-8 + 4$ $\quad$ **8.** $7 + (-7)$ $\quad$ **9.** $-31 + 31$

EXAMPLE 3 Adding More than Two Integers

The list shows four bank account transactions in July. Find the change C in the account balance.

JULY TRANSACTIONS	
Deposit	\$50
Withdrawal	-\$40
Deposit	\$75
Withdrawal	-\$50

Find the sum of the four transactions.

$C = 50 + (-40) + 75 + (-50)$ $\quad$ Write the sum.

$= 10 + 75 + (-50)$ $\quad$ Add 50 and -40.

$= 85 + (-50)$ $\quad$ Add 10 and 75.

$= 35$ $\quad$ Add 85 and -50.

Because $C = 35$, the account balance increased \$35 in July.

On Your Own

Now You're Ready
Exercises 28–33

10. WHAT IF? In Example 3, the deposit amounts are \$30 and \$55. Find the change C in the account balance.

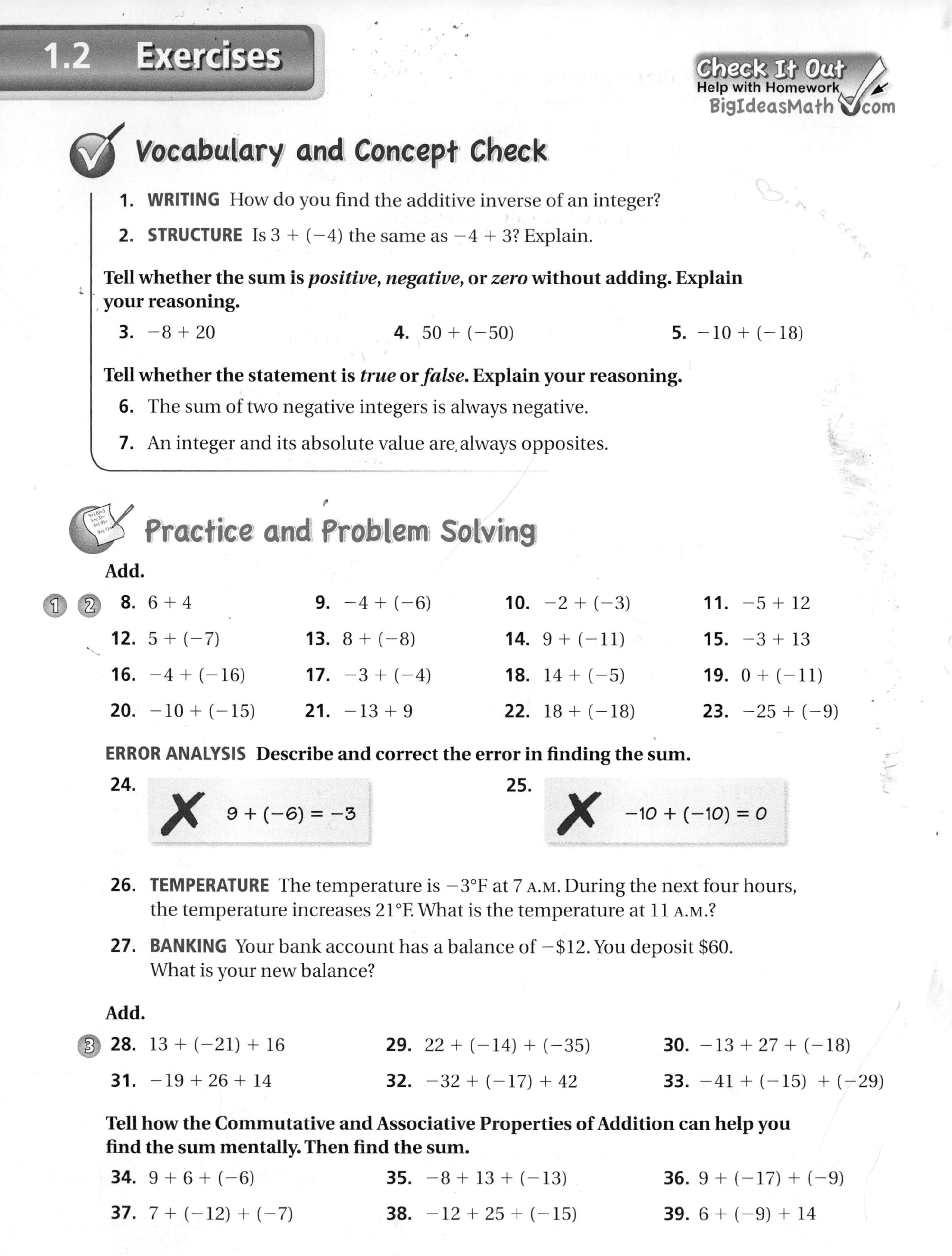

1.2 Exercises

Check It Out
Help with Homework
BigIdeasMath.com

Vocabulary and Concept Check

1. **WRITING** How do you find the additive inverse of an integer?
2. **STRUCTURE** Is $3 + (-4)$ the same as $-4 + 3$? Explain.

Tell whether the sum is *positive, negative,* or *zero* without adding. Explain your reasoning.

3. $-8 + 20$
4. $50 + (-50)$
5. $-10 + (-18)$

Tell whether the statement is *true* or *false*. Explain your reasoning.

6. The sum of two negative integers is always negative.
7. An integer and its absolute value are always opposites.

Practice and Problem Solving

Add.

(1) (2)

8. $6 + 4$
9. $-4 + (-6)$
10. $-2 + (-3)$
11. $-5 + 12$
12. $5 + (-7)$
13. $8 + (-8)$
14. $9 + (-11)$
15. $-3 + 13$
16. $-4 + (-16)$
17. $-3 + (-4)$
18. $14 + (-5)$
19. $0 + (-11)$
20. $-10 + (-15)$
21. $-13 + 9$
22. $18 + (-18)$
23. $-25 + (-9)$

ERROR ANALYSIS **Describe and correct the error in finding the sum.**

24. ✗ $9 + (-6) = -3$

25. ✗ $-10 + (-10) = 0$

26. **TEMPERATURE** The temperature is -3°F at 7 A.M. During the next four hours, the temperature increases 21°F. What is the temperature at 11 A.M.?

27. **BANKING** Your bank account has a balance of $-$\$12. You deposit \$60. What is your new balance?

Add.

(3)

28. $13 + (-21) + 16$
29. $22 + (-14) + (-35)$
30. $-13 + 27 + (-18)$
31. $-19 + 26 + 14$
32. $-32 + (-17) + 42$
33. $-41 + (-15) + (-29)$

Tell how the Commutative and Associative Properties of Addition can help you find the sum mentally. Then find the sum.

34. $9 + 6 + (-6)$
35. $-8 + 13 + (-13)$
36. $9 + (-17) + (-9)$
37. $7 + (-12) + (-7)$
38. $-12 + 25 + (-15)$
39. $6 + (-9) + 14$

ALGEBRA **Evaluate the expression when $a = 4$, $b = -5$, and $c = -8$.**

40. $a + b$ **41.** $b + c$ **42.** $|a + b + c|$

43. **OPEN-ENDED** Write two integers with different signs that have a sum of -25. Write two integers with the same sign that have a sum of -25.

MENTAL MATH **Use mental math to solve the equation.**

44. $d + 12 = 2$ **45.** $b + (-2) = 0$ **46.** $-8 + m = -15$

47. **FIRST DOWN** In football, a team must gain 10 yards to get a first down. The team gains 6 yards on the first play, loses 3 yards on the second play, and gains 8 yards on the third play. Which expression can be used to decide whether the team gets a first down?

$10 + 6 - 3 + 8$ $6 + (-3) + 8$ $6 + (-3) + (-8)$

48. **DOLPHIN** Starting at point A, the path of a dolphin jumping out of the water is shown.

a. Is the dolphin deeper at point C or point E? Explain your reasoning.

b. Is the dolphin higher at point B or point D? Explain your reasoning.

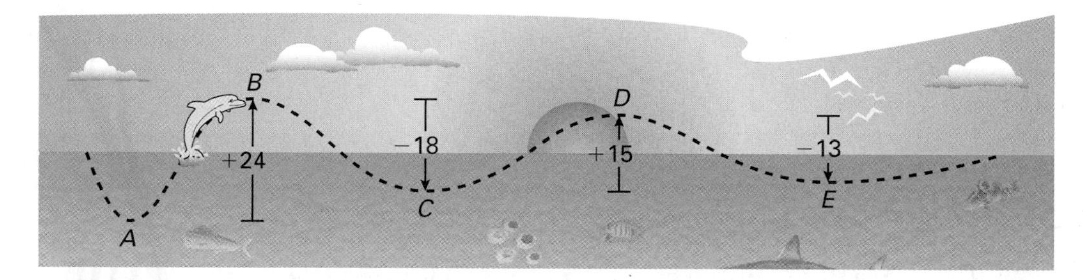

49. Puzzle According to a legend, the Chinese Emperor Yu-Huang saw a magic square on the back of a turtle. In a *magic square*, the numbers in each row and in each column have the same sum. This sum is called the magic sum.

Copy and complete the magic square so that each row and each column has a magic sum of 0. Use each integer from -4 to 4 exactly once.

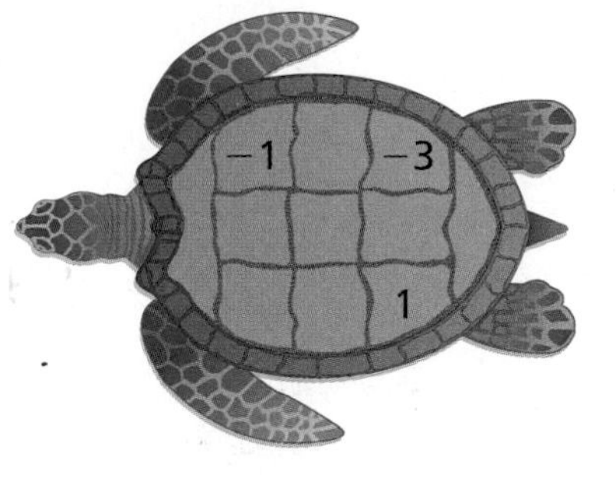

Fair Game Review What you learned in previous grades & lessons

Subtract. *(Skills Review Handbook)*

50. $69 - 38$ **51.** $82 - 74$ **52.** $177 - 63$ **53.** $451 - 268$

54. **MULTIPLE CHOICE** What is the range of the numbers below? *(Skills Review Handbook)*

12, 8, 17, 12, 15, 18, 30

Ⓐ 12 Ⓑ 15 Ⓒ 18 Ⓓ 22

1.3 Subtracting Integers

COMMON CORE STATE STANDARDS
7.EE.3
7.NS.1c
7.NS.1d

Essential Question How are adding integers and subtracting integers related?

1 EXAMPLE: Subtracting Integers

Use integer counters to find $4 - 2$.

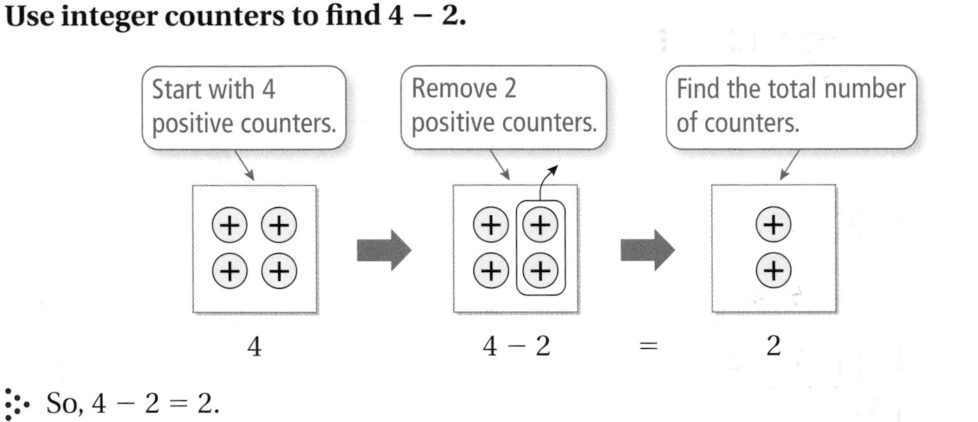

Start with 4 positive counters.
Remove 2 positive counters.
Find the total number of counters.
4
4 − 2
=
2

So, $4 - 2 = 2$.

2 ACTIVITY: Adding Integers

Work with a partner. Use integer counters to find $4 + (-2)$.

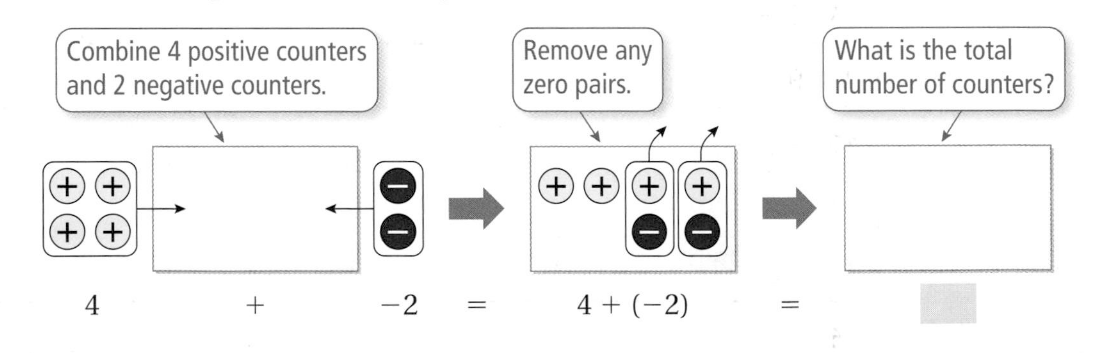

Combine 4 positive counters and 2 negative counters.
Remove any zero pairs.
What is the total number of counters?
4
+
−2
=
4 + (−2)
=

3 EXAMPLE: Subtracting Integers

Use a number line to find $-3 - 1$.

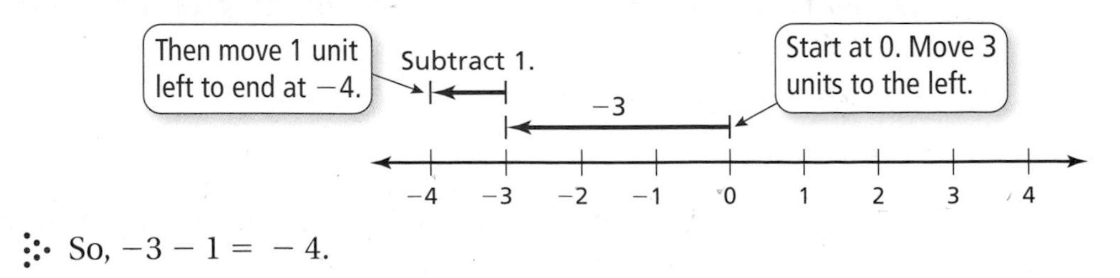

Then move 1 unit left to end at −4.
Subtract 1.
−3
Start at 0. Move 3 units to the left.
−4
−3
−2
−1
0
1
2
3
4

So, $-3 - 1 = -4$.

4 ACTIVITY: Adding Integers

Work with a partner. Write the addition expression shown. Then find the sum.

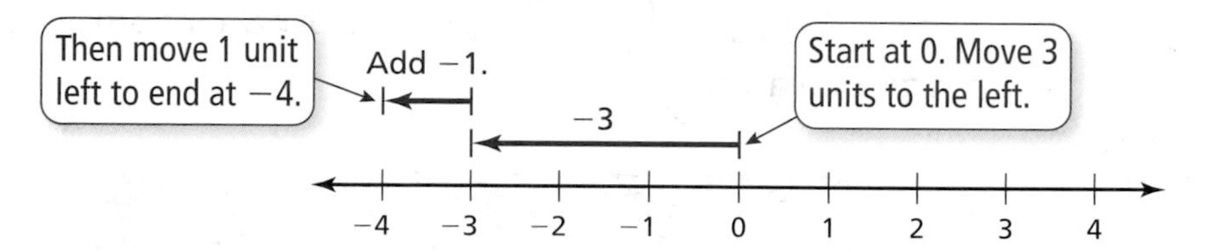
Then move 1 unit left to end at −4.
Add −1.
−3
Start at 0. Move 3 units to the left.
−4 −3 −2 −1 0 1 2 3 4

Inductive Reasoning

Work with a partner. Use integer counters or a number line to complete the table.

Exercise	Operation: Add or Subtract	Answer
5. $4 - 2$	Subtract 2	2
6. $4 + (-2)$	Add −2	2
7. $-3 - 1$	Subtract 1	−4
8. $-3 + (-1)$	Add −1	−4
9. $3 - 8$		
10. $3 + (-8)$		
11. $9 - 13$		
12. $9 + (-13)$		
13. $-6 - (-3)$		
14. $-6 + (3)$		
15. $-5 - (-12)$		
16. $-5 + 12$		

What Is Your Answer?

17. IN YOUR OWN WORDS How are adding integers and subtracting integers related?

18. Write a general rule for subtracting integers.

Practice Use what you learned about subtracting integers to complete Exercises 8–15 on page 18.

1.3 Lesson

Key Idea

Subtracting Integers

Words To subtract an integer, add its opposite.

Numbers $3 - 4 = 3 + (-4) = -1$

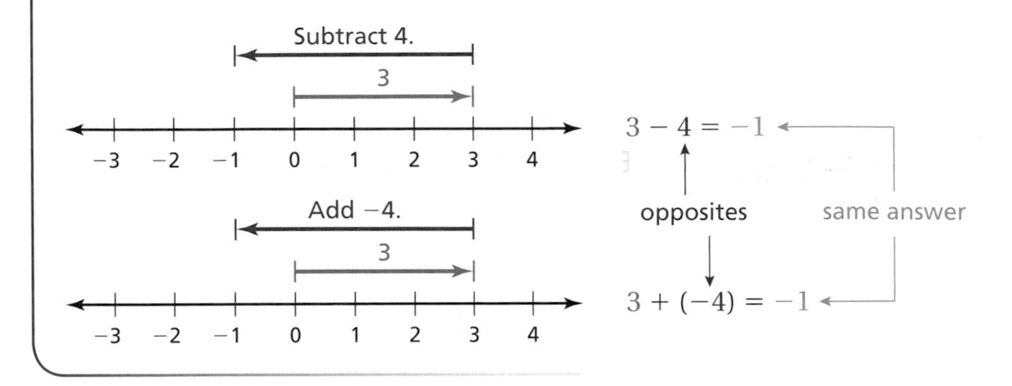

EXAMPLE 1 Subtracting Integers

a. Find 3 − 12.

$$3 - 12 = 3 + (-12) \quad \text{Add the opposite of 12.}$$
$$= -9 \quad \text{Add.}$$

The difference is −9.

b. Find −8 − (−13).

$$-8 - (-13) = -8 + 13 \quad \text{Add the opposite of } -13.$$
$$= 5 \quad \text{Add.}$$

The difference is 5.

c. Find 5 − (−4).

$$5 - (-4) = 5 + 4 \quad \text{Add the opposite of } -4.$$
$$= 9 \quad \text{Add.}$$

The difference is 9.

On Your Own

Now You're Ready
Exercises 8–23

Subtract.

1. $8 - 3$ **2.** $9 - 17$ **3.** $-3 - 3$

4. $-14 - 9$ **5.** $9 - (-8)$ **6.** $-12 - (-12)$

EXAMPLE 2 Subtracting Integers

Evaluate $-7 - (-12) - 14$.

$$
\begin{aligned}
-7 - (-12) - 14 &= -7 + 12 - 14 && \text{Add the opposite of } -12. \\
&= 5 - 14 && \text{Add } -7 \text{ and } 12. \\
&= 5 + (-14) && \text{Add the opposite of } 14. \\
&= -9 && \text{Add.}
\end{aligned}
$$

So, $-7 - (-12) - 14 = -9$.

On Your Own

Now You're Ready
Exercises 27–32

Evaluate the expression.

7. $-9 - 16 - 8$ **8.** $-4 - 20 - 9$

9. $0 - 9 - (-5)$ **10.** $0 - (-6) - 8$

11. $15 - (-20) - 20$ **12.** $13 - 18 - (-18)$

EXAMPLE 3 Real-Life Application

Which continent has the greater range of elevations?

	North America	Africa
Highest Elevation	6198 m	5895 m
Lowest Elevation	−86 m	−155 m

To find the range of elevations for each continent, subtract the lowest elevation from the highest elevation.

North America

$$
\begin{aligned}
\text{range} &= 6198 - (-86) \\
&= 6198 + 86 \\
&= 6284 \text{ m}
\end{aligned}
$$

Africa

$$
\begin{aligned}
\text{range} &= 5895 - (-155) \\
&= 5895 + 155 \\
&= 6050 \text{ m}
\end{aligned}
$$

Because 6284 is greater than 6050, North America has the greater range of elevations.

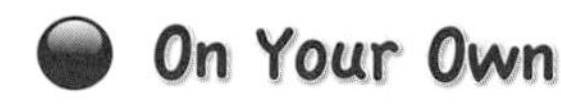

On Your Own

13. The highest elevation in Mexico is 5700 meters, on Pico de Orizaba. The lowest elevation in Mexico is −10 meters, in Laguna Salada. Find the range of elevations in Mexico.

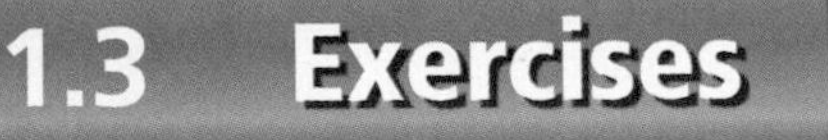

1. **WRITING** How do you subtract one integer from another?
2. **OPEN-ENDED** Write two integers that are opposites.
3. **DIFFERENT WORDS, SAME QUESTION** Which is different? Find "both" answers.

Find the difference of 3 and -2.	What is 3 less than -2?
How much less is -2 than 3?	Subtract -2 from 3.

MATCHING Match the subtraction expression with the corresponding addition expression.

4. $9 - (-5)$
5. $-9 - 5$
6. $-9 - (-5)$
7. $9 - 5$

A. $-9 + 5$ B. $9 + (-5)$ C. $-9 + (-5)$ D. $9 + 5$

Practice and Problem Solving

Subtract.

(1)
8. $4 - 7$
9. $8 - (-5)$
10. $-6 - (-7)$
11. $-2 - 3$
12. $5 - 8$
13. $-4 - 6$
14. $-8 - (-3)$
15. $10 - 7$
16. $-8 - 13$
17. $15 - (-2)$
18. $-9 - (-13)$
19. $-7 - (-8)$
20. $-6 - (-6)$
21. $-10 - 12$
22. $32 - (-6)$
23. $0 - (20)$
24. **ERROR ANALYSIS** Describe and correct the error in finding the difference $7 - (-12)$.

    ✗ $7 - (-12) = 7 + (-12) = -5$

25. **SWIMMING POOL** The floor of the shallow end of a swimming pool is at -3 feet. The floor of the deep end is 9 feet deeper. Which expression can be used to find the depth of the deep end?

    $-3 + 9$ $-3 - 9$ $9 - 3$

26. **SHARKS** A shark is at -80 feet. It swims up and jumps out of the water to a height of 15 feet. Write a subtraction expression for the vertical distance the shark travels.

Evaluate the expression.

(2)
27. $-2 - 7 + 15$
28. $-9 + 6 - (-2)$
29. $12 - (-5) - 8$
30. $8 + 14 - (-4)$
31. $-6 - (-8) + 5$
32. $-15 - 7 - (-11)$

MENTAL MATH **Use mental math to solve the equation.**

33. $m - 5 = 9$

34. $w - (-3) = 7$

35. $6 - c = -9$

ALGEBRA **Evaluate the expression when $k = -3$, $m = -6$, and $n = 9$.**

36. $4 - n$

37. $m - (-8)$

38. $-5 + k - n$

39. $|m - k|$

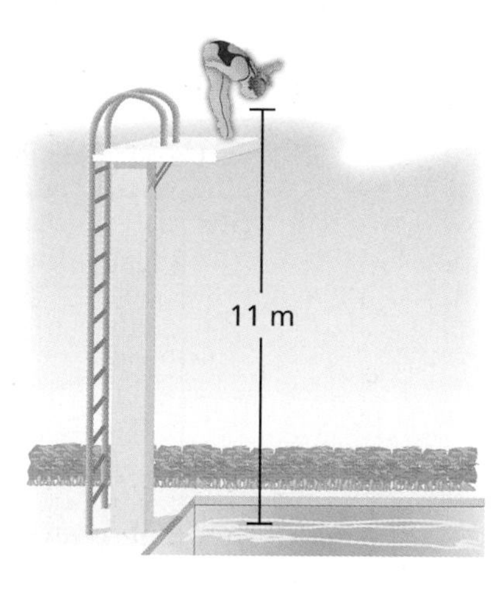

40. **PLATFORM DIVING** The figure shows a diver diving from a platform. The diver reaches a depth of 4 meters. What is the change in elevation of the dive?

41. **OPEN-ENDED** Write two different pairs of negative integers, x and y, that make the statement $x - y = -1$ true.

42. **TEMPERATURE** The table shows the record monthly high and low temperatures in Anchorage, AK.

	Jan	Feb	Mar	Apr	May	Jun	Jul	Aug	Sep	Oct	Nov	Dec
High (°F)	56	57	56	72	82	92	84	85	73	64	62	53
Low (°F)	−35	−38	−24	−15	1	29	34	31	19	−6	−21	−36

a. Find the range of temperatures for each month.

b. What are the all-time high and all-time low temperatures?

c. What is the range of the temperatures in part (b)?

REASONING **Tell whether the difference between the two integers is *always*, *sometimes*, or *never* positive. Explain your reasoning.**

43. Two positive integers

44. Two negative integers

45. A positive integer and a negative integer

46. A negative integer and a positive integer

Number Sense **For what values of a and b is the statement true?**

47. $|a - b| = |b - a|$

48. $|a + b| = |a| + |b|$

49. $|a - b| = |a| - |b|$

Fair Game Review What you learned in previous grades & lessons

Add. *(Section 1.2)*

50. $-5 + (-5) + (-5) + (-5)$

51. $-9 + (-9) + (-9) + (-9) + (-9)$

Multiply. *(Skills Review Handbook)*

52. 8×5

53. 6×78

54. 36×41

55. 82×29

56. **MULTIPLE CHOICE** Which value of n makes the value of the expression $4n + 3$ a composite number? *(Skills Review Handbook)*

Ⓐ 1 Ⓑ 2 Ⓒ 3 Ⓓ 4

1 Study Help

You can use an **idea and examples chart** to organize information about a concept. Here is an example of an idea and examples chart for absolute value.

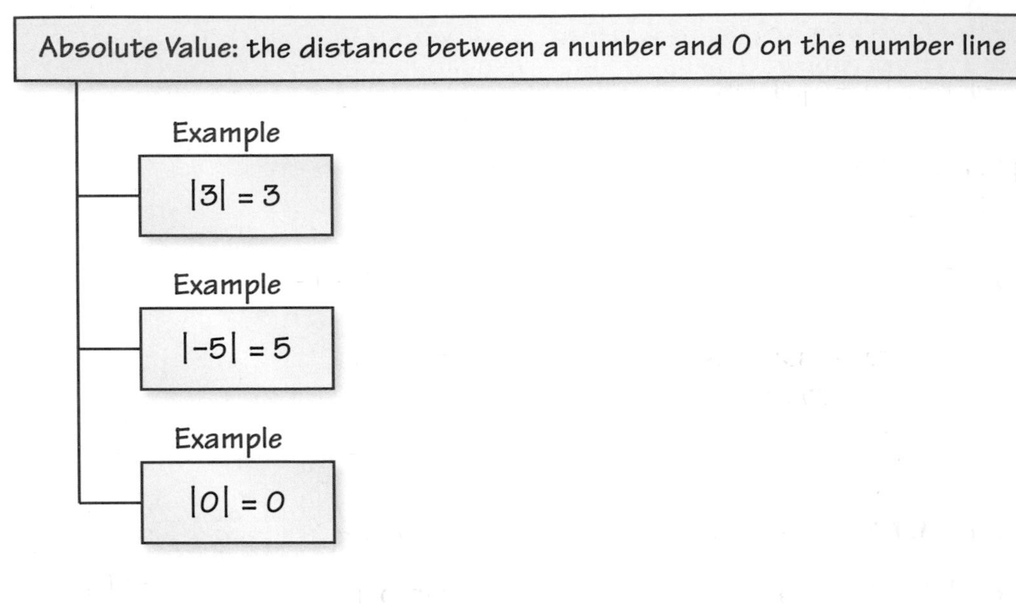

On Your Own

Make an idea and examples chart to help you study these topics.

1. integers
2. adding integers
   a. with the same sign
   b. with different signs
3. Additive Inverse Property
4. subtracting integers

"I made an idea and examples chart to give my owner ideas for my birthday next week."

After you complete this chapter, make idea and examples charts for the following topics.

5. multiplying integers
   a. with the same sign
   b. with different signs
6. dividing integers
   a. with the same sign
   b. with different signs
7. quadrants
8. plotting ordered pairs

1.1–1.3 Quiz

Copy and complete the statement using <, >, or =. *(Section 1.1)*

1. $|-8|$ ☐ 3

2. 7 ☐ $|-7|$

Order the values from least to greatest. *(Section 1.1)*

3. $-4, |-5|, |-4|, 3, -6$

4. $12, -8, |-15|, -10, |-9|$

Simplify the expression. *(Section 1.2 and Section 1.3)*

5. $-3 + (-8)$

6. $-4 + 16$

7. $3 - 9$

8. $-5 - (-5)$

Evaluate the expression when $a = -2$, $b = -8$, and $c = 5$.
(Section 1.2 and Section 1.3)

9. $4 - a - c$

10. $|b - c|$

11. EXPLORING Two climbers explore a cave. *(Section 1.1)*

a. Write an integer for the depth of each climber relative to the surface.

b. Which integer in part (a) is greater?

c. Which integer in part (a) has the greater absolute value?

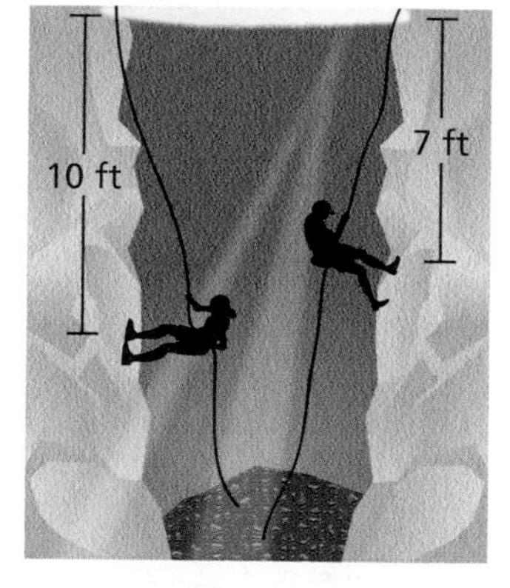

12. SCHOOL CARNIVAL The table shows the income and expenses for a school carnival. The school's goal was to raise $1100. Did the school reach its goal? Explain. *(Section 1.2)*

Games	Concessions	Donations	Flyers	Decorations
$650	$530	$52	−$28	−$75

13. TEMPERATURE Temperatures in the Gobi Desert reach $-40°\text{F}$ in the winter and $90°\text{F}$ in the summer. Find the range of the temperatures. *(Section 1.3)*

1.4 Multiplying Integers

COMMON CORE STATE STANDARDS
7.EE.3
7.NS.2a
7.NS.2c

Essential Question Is the product of two integers *positive*, *negative*, or *zero*? How can you tell?

1 EXAMPLE: Multiplying Integers with the Same Sign

Use repeated addition to find $3 \cdot 2$.

Recall that multiplication is repeated addition. $3 \cdot 2$ means to add 3 groups of 2.

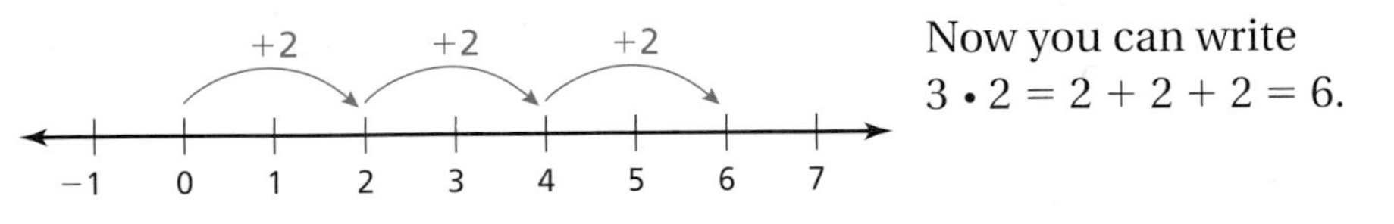

Now you can write
$3 \cdot 2 = 2 + 2 + 2 = 6$.

So, $3 \cdot 2 = 6$.

2 EXAMPLE: Multiplying Integers with Different Signs

Use repeated addition to find $3 \cdot (-2)$.

$3 \cdot (-2)$ means to add 3 groups of -2.

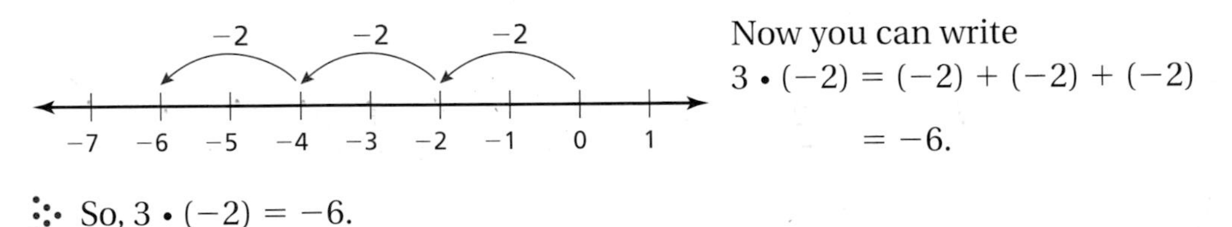

Now you can write
$3 \cdot (-2) = (-2) + (-2) + (-2)$
$= -6$.

So, $3 \cdot (-2) = -6$.

3 ACTIVITY: Multiplying Integers with Different Signs

Work with a partner. Use a table to find $-3 \cdot 2$.

Describe the pattern in the table. Use the pattern to complete the table.

2	•	2	=	4
1	•	2	=	2
0	•	2	=	0
−1	•	2	=	
−2	•	2	=	
−3	•	2	=	

Notice the products decrease by 2 in each row.

So, continue the pattern.

$-1 \cdot 2$: $0 - 2 =$ ___

$-2 \cdot 2$: $-2 - 2 =$ ___

$-3 \cdot 2$: $-4 - 2 =$ ___

So, $-3 \cdot 2 =$ ___.

4 ACTIVITY: Multiplying Integers with the Same Sign

Work with a partner. Use a table to find $-3 \cdot (-2)$.

Describe the pattern in the table. Use the pattern to complete the table.

-3	$\cdot$	3	$=$	-9
-3	$\cdot$	2	$=$	-6
-3	$\cdot$	1	$=$	-3
-3	$\cdot$	0	$=$	
-3	$\cdot$	-1	$=$	
-3	$\cdot$	-2	$=$	

Notice the products increase by 3 in each row.

So, continue the pattern.

$-3 \cdot 0$: $-3 + 3 =$ ___

$-3 \cdot -1$: $0 + 3 =$ ___

$-3 \cdot -2$: $3 + 3 =$ ___

So, $-3 \cdot (-2) =$ ___.

Inductive Reasoning

Work with a partner. Complete the table.

	Exercise	Type of Product	Product	Product: Positive or Negative
1	**5.** $3 \cdot 2$	Integers with the same sign		
2	**6.** $3 \cdot (-2)$	Integers with different signs		
3	**7.** $-3 \cdot 2$	Integers with different signs		
4	**8.** $-3 \cdot (-2)$	Integers with the same sign		
	9. $6 \cdot 3$			
	10. $2 \cdot (-5)$			
	11. $-6 \cdot 5$			
	12. $-5 \cdot (-3)$			

13. Write two integers whose product is 0.

What Is Your Answer?

14. IN YOUR OWN WORDS Is the product of two integers *positive, negative,* or *zero*? How can you tell?

15. Write general rules for multiplying (a) two integers with the same sign and (b) two integers with different signs.

Practice Use what you learned about multiplying integers to complete Exercises 8–15 on page 26.

1.4 Lesson

Check It Out
Lesson Tutorials
BigIdeasMath.com

Key Ideas

Multiplying Integers with the Same Sign

Words The product of two integers with the same sign is positive.

Numbers $2 \cdot 3 = 6$ $\quad -2 \cdot (-3) = 6$

Multiplying Integers with Different Signs

Words The product of two integers with different signs is negative.

Numbers $2 \cdot (-3) = -6$ $\quad -2 \cdot 3 = -6$

EXAMPLE 1 Multiplying Integers with the Same Sign

Find $-5 \cdot (-6)$.

The integers have the same sign.

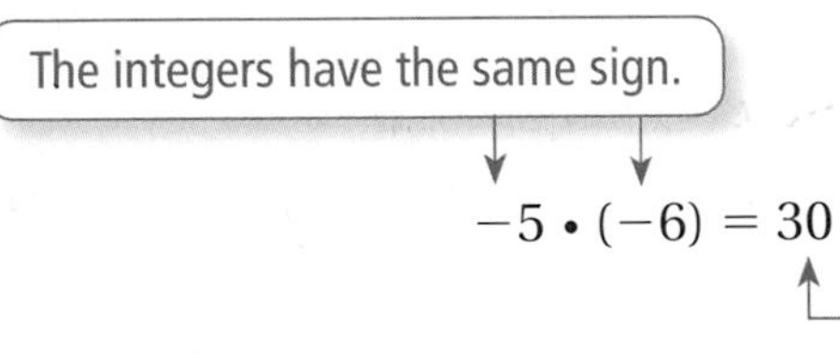
−5 • (−6) = 30
The product is positive.

∴ The product is 30.

EXAMPLE 2 Multiplying Integers with Different Signs

Multiply.

a. $3(-4)$ **b.** $-7 \cdot 4$

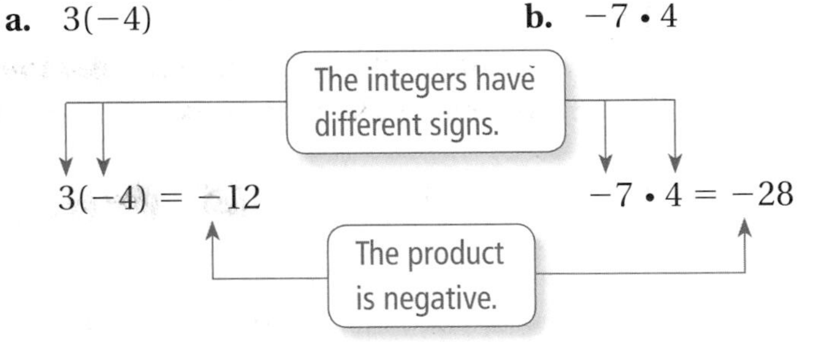
The integers have different signs.
3(−4) = −12
−7 • 4 = −28
The product is negative.

∴ The product is -12. ∴ The product is -28.

On Your Own

Now You're Ready
Exercises 8–23

Multiply.

1. $5 \cdot 5$ **2.** $4(11)$

3. $-1(-9)$ **4.** $-7 \cdot (-8)$

5. $12 \cdot (-2)$ **6.** $4(-6)$

7. $-10(6)$ **8.** $-5 \cdot 7$

EXAMPLE 3 Using Exponents

Study Tip

Place parentheses around a negative number to raise it to a power.

a. Evaluate $(-2)^2$.

$(-2)^2 = (-2) \cdot (-2)$ Write $(-2)^2$ as repeated multiplication.

$= 4$ Multiply.

b. Evaluate -5^2.

$-5^2 = -(5 \cdot 5)$ Write 5^2 as repeated multiplication.

$= -25$ Multiply.

c. Evaluate $(-4)^3$.

$(-4)^3 = (-4) \cdot (-4) \cdot (-4)$ Write $(-4)^3$ as repeated multiplication.

$= 16 \cdot (-4)$ Multiply.

$= -64$ Multiply.

On Your Own

Now You're Ready
Exercises 32–37

Evaluate the expression.

9. $(-3)^2$ 10. $(-2)^3$ 11. -7^2 12. -6^3

EXAMPLE 4 Real-Life Application

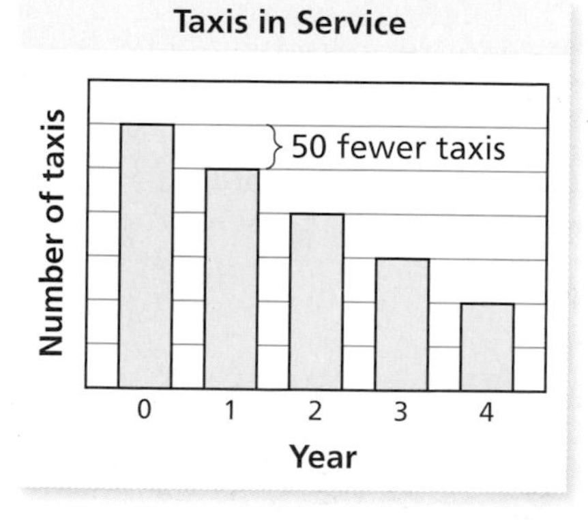

The bar graph shows the number of taxis a company has in service. The number of taxis decreases by the same amount each year for four years. Find the total change in the number of taxis.

The bar graph shows that the number of taxis in service decreases by 50 each year. Use a model to solve the problem.

Total change = Change per year • Number of years

$= -50 \cdot 4$

$= -200$

Use -50 for the change per year because the number *decreases* each year.

∴ The total change in the number of taxis is -200.

On Your Own

13. A manatee population decreases by 15 manatees each year for 3 years. Find the total change in the manatee population.

1.4 Exercises

Vocabulary and Concept Check

1. **WRITING** What do you know about the signs of two integers whose product is (a) positive and (b) negative?
2. **WRITING** How is $(-2)^2$ different from -2^2?

Tell whether the product is *positive* or *negative* without multiplying. Explain your reasoning.

3. $4(-8)$
4. $-5(-7)$
5. $-3 \cdot (12)$

Tell whether the statement is *true* or *false*. Explain your reasoning.

6. The product of three positive integers is positive.
7. The product of three negative integers is positive.

Practice and Problem Solving

Multiply.

(1) (2)

8. $6 \cdot 4$
9. $7(-3)$
10. $-2(8)$
11. $-3(-4)$
12. $-6 \cdot 7$
13. $3 \cdot 9$
14. $8 \cdot (-5)$
15. $-1 \cdot (-12)$
16. $-5(10)$
17. $-13(0)$
18. $-9 \cdot 9$
19. $15(-2)$
20. $-10 \cdot 11$
21. $-6 \cdot (-13)$
22. $7(-14)$
23. $-11 \cdot (-11)$
24. **JOGGING** You burn 10 calories each minute you jog. What integer represents the change in your calories after you jog for 20 minutes?
25. **WETLANDS** About 60,000 acres of wetlands are lost each year in the United States. What integer represents the change in wetlands after 4 years?

Multiply.

26. $3 \cdot (-8) \cdot (-2)$
27. $6(-9)(-1)$
28. $-3(-5)(-4)$
29. $-7(-3)(-5)$
30. $-6 \cdot 3 \cdot (-6)$
31. $3 \cdot (-12) \cdot 0$

Evaluate the expression.

(3)

32. $(-4)^2$
33. $(-1)^3$
34. -8^2
35. -6^2
36. $-5^2 \cdot 4$
37. $-2 \cdot (-3)^3$

ERROR ANALYSIS Describe and correct the error in evaluating the expression.

38. ✗ $-2(-7) = -14$
39. ✗ $-10^2 = 100$

ALGEBRA **Evaluate the expression when $a = -2$, $b = 3$, and $c = -8$.**

40. ab

41. $|a^2c|$

42. $ab^3 - ac$

STRUCTURE **Find the next two numbers in the pattern.**

43. $-12, 60, -300, 1500, \ldots$

44. $7, -28, 112, -448, \ldots$

45. GYM CLASS You lose four points each time you attend gym class without sneakers. You forget your sneakers three times. What integer represents the change in your points?

46. AIRPLANE The height of an airplane during a landing is given by $22{,}000 + (-480t)$, where t is the time in minutes.

a. Copy and complete the table.

b. Estimate how many minutes it takes the plane to land. Explain your reasoning.

Time	5 min	10 min	15 min	20 min
Height				

47. INLINE SKATES In June, the price of a pair of inline skates is \$165. The price changes each of the next three months.

a. Copy and complete the table.

Month	Price of Skates
June	$165 = \$165$
July	$165 + (-12) = \$\_\_\_\_$
August	$165 + 2(-12) = \$\_\_\_\_$
September	$165 + 3(-12) = \$\_\_\_\_$

b. Describe the change in the price of the inline skates for each month.

c. The table at the right shows the amount of money you save each month to buy the inline skates. Do you have enough money saved to buy the inline skates in August? September? Explain your reasoning.

Amount Saved	
June	\$35
July	\$55
August	\$45
September	\$18

48. **Reasoning** Two integers, a and b, have a product of 24. What is the least possible sum of a and b?

Fair Game Review What you learned in previous grades & lessons

Divide. *(Skills Review Handbook)*

49. $27 \div 9$

50. $48 \div 6$

51. $56 \div 4$

52. $153 \div 9$

53. MULTIPLE CHOICE What is the prime factorization of 84? *(Skills Review Handbook)*

(A) $2^2 \times 3^2$ (B) $2^3 \times 7$ (C) $3^3 \times 7$ (D) $2^2 \times 3 \times 7$

1.5 Dividing Integers

COMMON CORE STATE STANDARDS
7.NS.2b
7.NS.2c
7.NS.3

Essential Question Is the quotient of two integers *positive*, *negative*, or *zero*? How can you tell?

1 EXAMPLE: Dividing Integers with Different Signs

Use integer counters to find $-15 \div 3$.

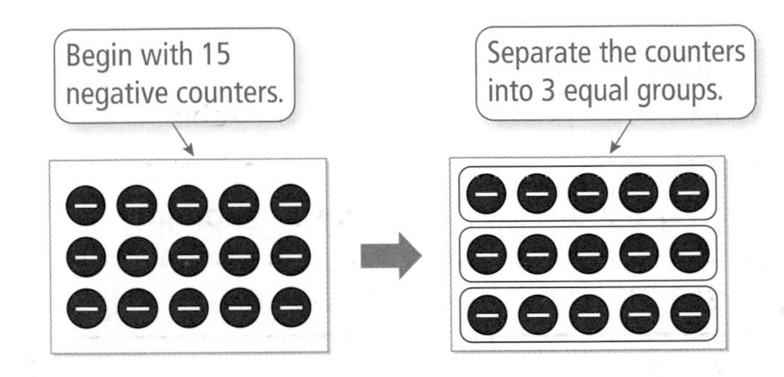
Begin with 15 negative counters.
Separate the counters into 3 equal groups.

∴ Because there are five negative counters in each group, $-15 \div 3 = -5$.

2 ACTIVITY: Rewriting a Product as a Quotient

Work with a partner. Rewrite the product $3 \cdot 4 = 12$ as a quotient in two different ways.

First Way

12 is equal to 3 groups of ___.

∴ So, $12 \div 3 =$ ___.

Second Way

12 is equal to 4 groups of ___.

∴ So, $12 \div 4 =$ ___.

3 EXAMPLE: Dividing Integers with Different Signs

Rewrite the product $-3 \cdot (-4) = 12$ as a quotient in two different ways. What can you conclude?

First Way

$12 \div (-3) = -4$

Second Way

$12 \div (-4) = -3$

∴ In each case, when you divide a positive integer by a negative integer, you get a negative integer.

4 EXAMPLE: Dividing Negative Integers

Rewrite the product $3 \cdot (-4) = -12$ as a quotient in two different ways. What can you conclude?

First Way

$-12 \div (-4) = 3$

Second Way

$-12 \div (3) = -4$

When you divide a negative integer by a negative integer, you get a positive integer. When you divide a negative integer by a positive integer, you get a negative integer.

Inductive Reasoning

Work with a partner. Complete the table.

Exercise	Type of Quotient	Quotient	Quotient: Positive, Negative, or Zero
1 **5.** $-15 \div 3$	Integers with different signs		
2 **6.** $12 \div 4$			Positive
3 **7.** $12 \div (-3)$		-4	
4 **8.** $-12 \div (-4)$	Integers with the same sign		Positive
9. $-6 \div 2$			
10. $-21 \div (-7)$			
11. $10 \div (-2)$			
12. $12 \div (-6)$			
13. $0 \div (-15)$			
14. $0 \div 4$			

What Is Your Answer?

15. **IN YOUR OWN WORDS** Is the quotient of two integers *positive, negative,* or *zero*? How can you tell?

16. Write general rules for dividing (a) two integers with the same sign and (b) two integers with different signs.

Use what you learned about dividing integers to complete Exercises 8–15 on page 32.

1.5 Lesson

Check It Out
Lesson Tutorials
BigIdeasMath.com

Key Ideas

Dividing Integers with the Same Sign

Words The quotient of two integers with the same sign is positive.

Numbers $8 \div 2 = 4$ $\quad -8 \div (-2) = 4$

Dividing Integers with Different Signs

Words The quotient of two integers with different signs is negative.

Numbers $8 \div (-2) = -4$ $\quad -8 \div 2 = -4$

EXAMPLE 1 Dividing Integers with the Same Sign

Find $-18 \div (-6)$.

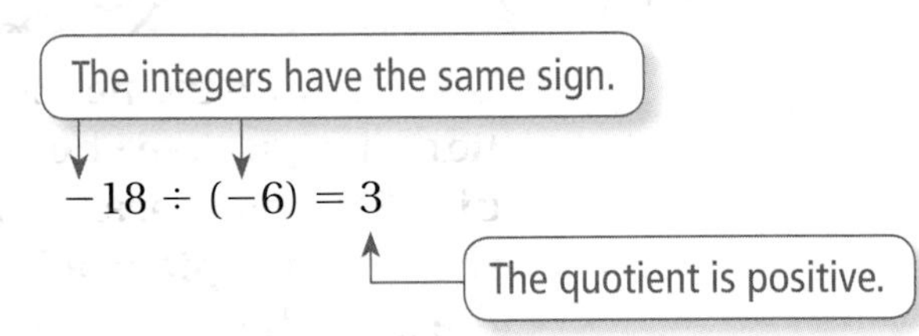
The integers have the same sign.
−18 ÷ (−6) = 3
The quotient is positive.

∴ The quotient is 3.

EXAMPLE 2 Dividing Integers with Different Signs

Divide.

a. $75 \div (-25)$ **b.** $\frac{-54}{6}$

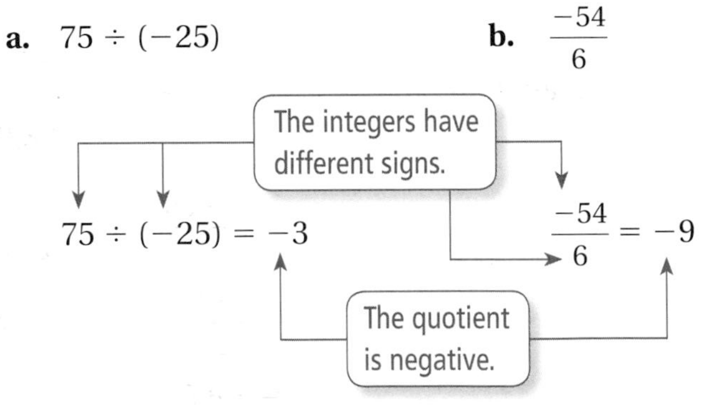
The integers have different signs.
75 ÷ (−25) = −3
−54/6 = −9
The quotient is negative.

∴ The quotient is -3. ∴ The quotient is -9.

On Your Own

Now You're Ready
Exercises 8–23

Divide.

1. $14 \div 2$
2. $-32 \div (-4)$
3. $-40 \div (-8)$
4. $0 \div (-6)$
5. $\frac{-49}{7}$
6. $\frac{21}{-3}$

EXAMPLE 3 Evaluating Expressions

Evaluate $10 - x^2 \div y$ when $x = 8$ and $y = -4$.

Remember

Use order of operations when evaluating an expression.

$10 - x^2 \div y = 10 - 8^2 \div (-4)$	Substitute 8 for x and -4 for y.
$= 10 - 8 \cdot 8 \div (-4)$	Write 8^2 as repeated multiplication.
$= 10 - 64 \div (-4)$	Multiply 8 and 8.
$= 10 - (-16)$	Divide 64 and -4.
$= 26$	Subtract.

On Your Own

Now You're Ready
Exercises 28–31

Evaluate the expression when $a = -18$ and $b = -6$.

7. $a \div b$ **8.** $\dfrac{a + 6}{3}$ **9.** $\dfrac{b^2}{a} + 4$

EXAMPLE 4 Real-Life Application

You measure the height of the tide using support beams of a pier. Your measurements are shown in the picture. What is the mean hourly change in the height?

Use a model to solve the problem.

$$\text{Mean hourly change} = \frac{\text{Final height} - \text{Initial height}}{\text{Elapsed Time}}$$

$= \dfrac{8 - 59}{6}$	Substitute. The elapsed time from 2 P.M. to 8 P.M. is 6 hours.
$= \dfrac{-51}{6}$	Subtract.
$= -8.5$	Divide.

∴ The mean change in the height of the tide is -8.5 inches per hour.

On Your Own

10. The height of the tide at the Bay of Fundy in New Brunswick decreases 36 feet in 6 hours. What is the mean hourly change in the height?

1.5 Exercises

Vocabulary and Concept Check

1. **WRITING** What can you tell about two integers when their quotient is positive? negative? zero?
2. **VOCABULARY** A quotient is undefined. What does this mean?
3. **OPEN-ENDED** Write two integers whose quotient is negative.
4. **WHICH ONE DOESN'T BELONG?** Which expression does *not* belong with the other three? Explain your reasoning.

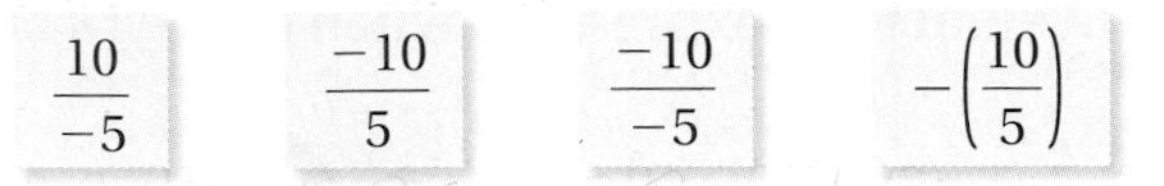

$\frac{10}{-5}$ $\quad \frac{-10}{5}$ $\quad \frac{-10}{-5}$ $\quad -\left(\frac{10}{5}\right)$

Tell whether the quotient is *positive* or *negative* without dividing.

5. $-12 \div 4$
6. $\frac{-6}{-2}$
7. $15 \div (-3)$

Practice and Problem Solving

Divide, if possible.

① ② 8. $4 \div (-2)$
9. $21 \div (-7)$
10. $-20 \div 4$
11. $-18 \div (-6)$
12. $\frac{-14}{7}$
13. $\frac{0}{6}$
14. $\frac{-15}{-5}$
15. $\frac{54}{-9}$
16. $-33 \div 11$
17. $-49 \div (-7)$
18. $0 \div (-2)$
19. $60 \div (-6)$
20. $\frac{-56}{14}$
21. $\frac{18}{0}$
22. $\frac{65}{-5}$
23. $\frac{-84}{-7}$

ERROR ANALYSIS Describe and correct the error in finding the quotient.

24.

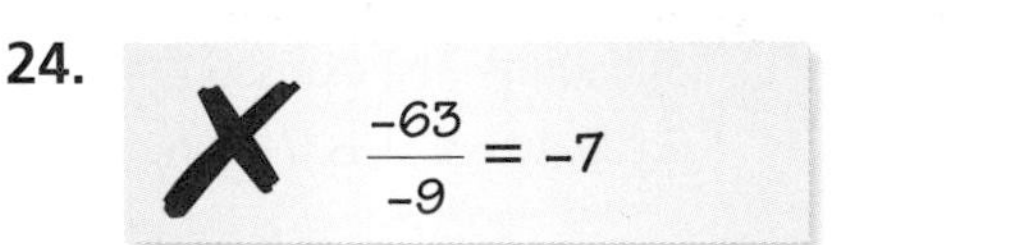

25.

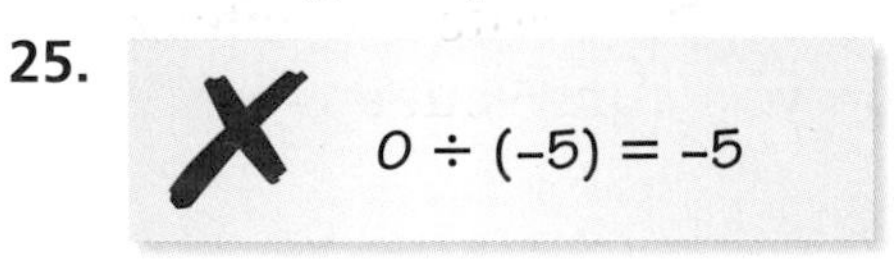

26. **ALLIGATORS** An alligator population in a nature preserve in the Everglades decreases by 60 alligators over 5 years. What is the mean yearly change in the alligator population?
27. **READING** You read 105 pages of a novel over 7 days. What is the mean number of pages you read each day?

ALGEBRA Evaluate the expression when $x = 10$, $y = -2$, and $z = -5$.

③ 28. $x \div y$
29. $\frac{10y^2}{z}$
30. $\left|\frac{xz}{-y}\right|$
31. $\frac{-x^2 + 6z}{y}$

Find the mean of the integers.

32. 3, −10, −2, 13, 11

33. −26, 39, −10, −16, 12, 31

Evaluate the expression.

34. $-8 - 14 \div 2 + 5$

35. $24 \div (-4) + (-2) \cdot (-5)$

36. PATTERN Find the next two numbers in the pattern −128, 64, −32, 16, ... Explain your reasoning.

37. SNOWBOARDING A snowboarder descends a 1200-foot hill in 3 minutes. What is the mean change in elevation per minute?

38. THE MASTERS In 1997, at the age of 21, Tiger Woods became the youngest golfer to win the Masters Tournament. The table shows his score for each round.

Scorecard	
Round 1	−2
Round 2	−6
Round 3	−7
Round 4	−3

a. Tiger set the tournament record with the lowest total score. What was his total score?

b. What was his mean score per round?

39. TUNNEL The Detroit-Windsor Tunnel is an underwater highway that connects the cities of Detroit, Michigan, and Windsor, Ontario. How many times deeper is the roadway than the bottom of the ship?

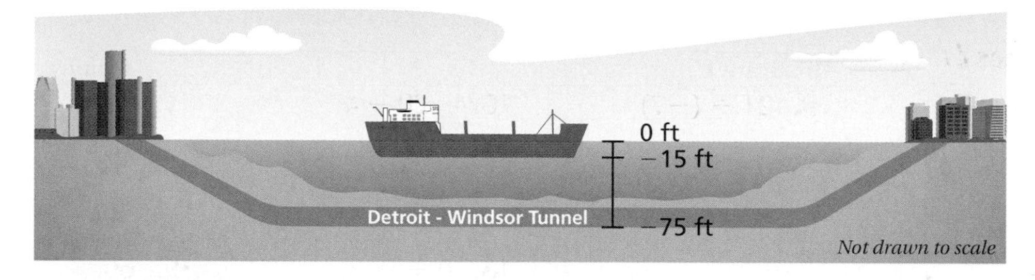

40. AMUSEMENT PARK The regular admission price for an amusement park is $72. For a group of 15 or more, the admission price is reduced by $25. How many people need to be in a group to save $500?

41. **Number Sense** Write five different integers that have a mean of −10. Explain how you found your answer.

Fair Game Review What you learned in previous grades & lessons

Graph the values on a number line. Then order the values from least to greatest. *(Section 1.1)*

42. $-6, 4, |2|, -1, |-10|$

43. $3, |0|, |-4|, -3, -8$

44. $|5|, -2, -5, |-2|, -7$

45. MULTIPLE CHOICE What is the value of $4 \cdot 3 + (12 \div 2)^2$? *(Skills Review Handbook)*

Ⓐ 15 Ⓑ 48 Ⓒ 156 Ⓓ 324

1.6 The Coordinate Plane

COMMON CORE STATE STANDARDS
7.NS.1b

Essential Question How can you use ordered pairs to locate points in a coordinate plane?

1 EXAMPLE: Plotting Points in a Coordinate Plane

Plot the ordered pairs. Connect the points to make a picture. Color the picture when you are done.

1(4, 12)	2(9, 9)	3(12, 4)	4(12, −3)	5(10, −9)
6(9, −10)	7(7, −9)	8(2, −11)	9(−1, −11)	10(−3, −10)
11(−4, −8)	12(−11, −10)	13(−12, −9)	14(−11, −8)	15(−11, −6)
16(−12, −5)	17(−11, −4)	18(−4, −6)	19(−3, −3)	20(−4, 0)
21(−8, 2)	22(−8, 3)	23(−5, 8)	24(−1, 11)	

Wildcats

Chiefs

Bulldogs

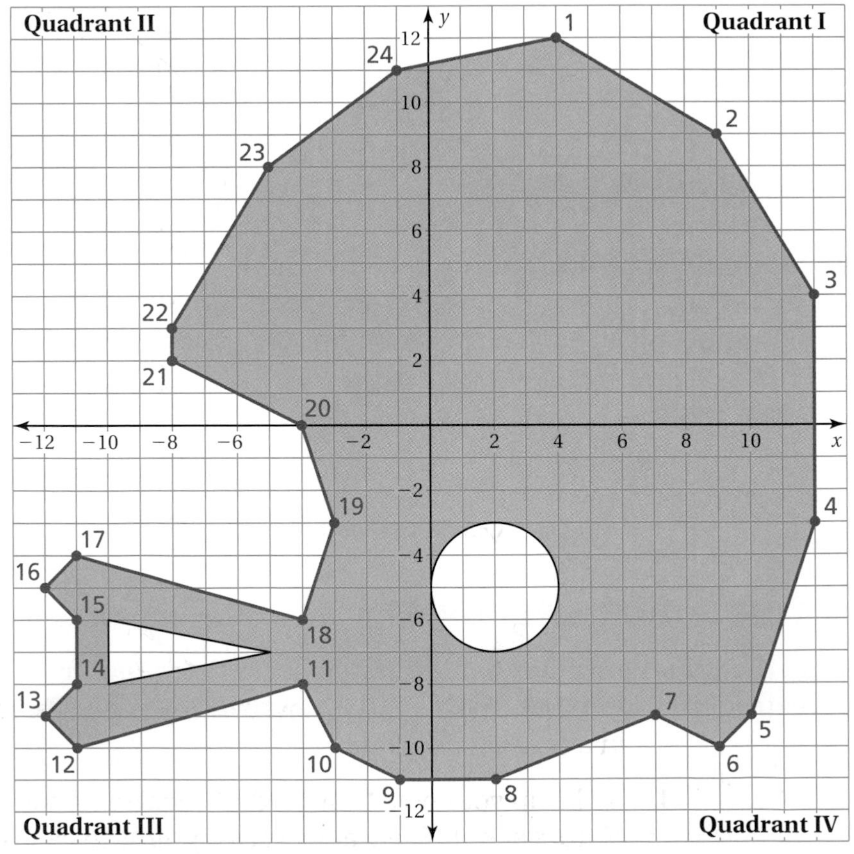

2 ACTIVITY: Plotting Points in a Coordinate Plane

Work with a partner.

Plot the ordered pairs. Connect the points to make a picture. Describe and color the picture when you are done.

1(6, 9)	2(4, 11)	3(2, 12)	4(0, 11)	5(−2, 9)
6(−6, 2)	7(−9, 1)	8(−11, −3)	9(−7, 0)	10(−5, −1)
11(−5, −5)	12(−4, −8)	13(−6, −10)	14(−3, −9)	15(−3, −10)
16(−4, −11)	17(−4, −12)	18(−3, −11)	19(−2, −12)	20(−2, −11)
21(−1, −12)	22(−1, −11)	23(−2, −10)	24(−2, −9)	25(1, −9)
26(2, −8)	27(2, −10)	28(1, −11)	29(1, −12)	30(2, −11)
31(3, −12)	32(3, −11)	33(4, −12)	34(4, −11)	35(3, −10)
36(3, −8)	37(4, −6)	38(6, 0)	39(9, −3)	40(9, −1)
41(8, 1)	42(5, 3)	43(3, 6)	44(3, 7)	45(4, 8)

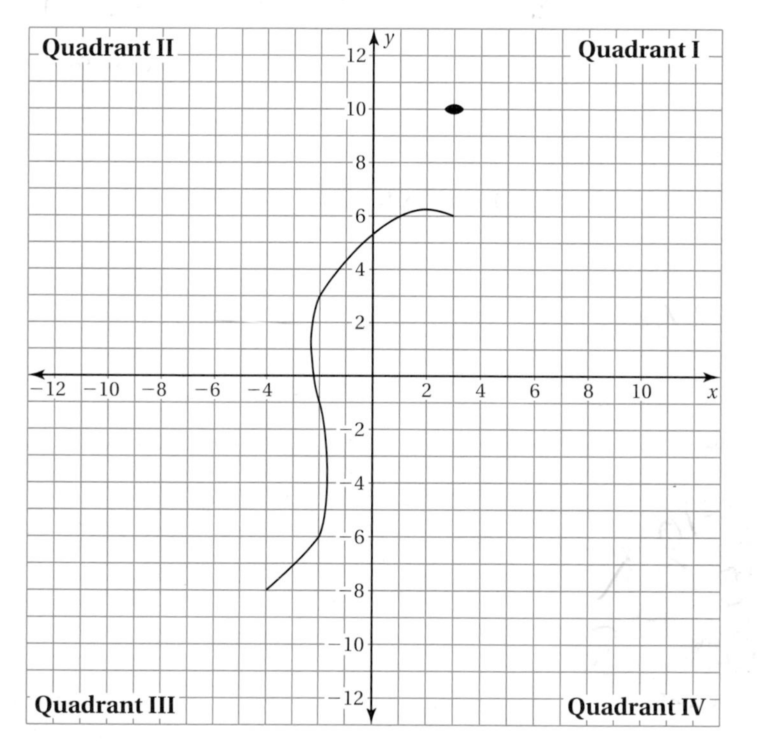

What Is Your Answer?

3. **IN YOUR OWN WORDS** How can you use ordered pairs to locate points in a coordinate plane?

4. Make up your own "dot-to-dot" picture. Use at least 20 points. Your picture should have at least two points in each quadrant.

Practice Use what you learned about the coordinate plane to complete Exercises 15–18 on page 38.

1.6 Lesson

Check It Out
Lesson Tutorials
BigIdeasMath.com

Key Idea

Key Vocabulary
coordinate plane, *p. 36*
origin, *p. 36*
quadrant, *p. 36*
x-axis, *p. 36*
y-axis, *p. 36*

The Coordinate Plane

A **coordinate plane** is formed by the intersection of a horizontal number line and a vertical number line. The number lines intersect at the **origin** and separate the coordinate plane into four regions called **quadrants.**

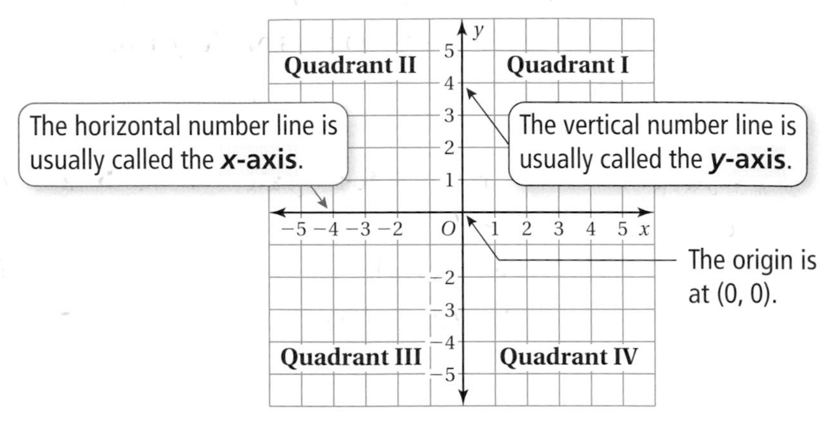

Quadrant II
Quadrant I
Quadrant III
Quadrant IV
The horizontal number line is usually called the **x-axis**.
The vertical number line is usually called the **y-axis**.
The origin is at (0, 0).

An *ordered pair* is a pair of numbers that is used to locate a point in a coordinate plane.

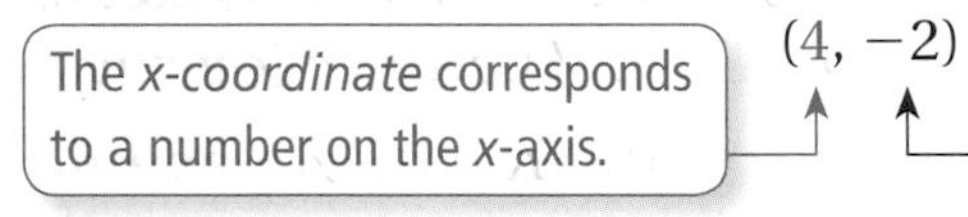

ordered pair
(4, −2)
The *x-coordinate* corresponds to a number on the *x*-axis.
The *y-coordinate* corresponds to a number on the *y*-axis.

EXAMPLE 1 Standardized Test Practice

Which ordered pair corresponds to point T?

(A) $(-3, -3)$ (B) $(-3, 3)$
(C) $(3, -3)$ (D) $(3, 3)$

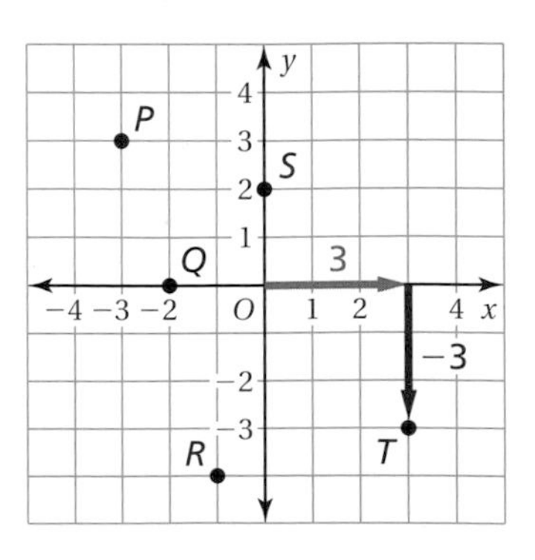

P
S
Q
R
T
3
−3

Point T is 3 units to the right of the origin and 3 units down. So, the x-coordinate is 3 and the y-coordinate is -3.

∴ The ordered pair $(3, -3)$ corresponds to point T. The correct answer is (C).

On Your Own

Now You're Ready
Exercises 5–14

Use the graph in Example 1 to write an ordered pair corresponding to the point.

1. Point P
2. Point Q
3. Point R
4. Point S

EXAMPLE 2 Plotting Ordered Pairs

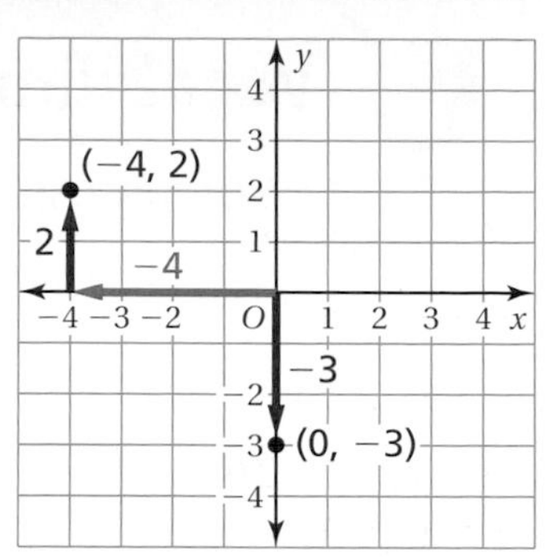

Plot (a) $(-4, 2)$ and (b) $(0, -3)$ in a coordinate plane. Describe the location of each point.

a. Start at the origin. Move 4 units left and 2 units up. Then plot the point.

∴ The point is in Quadrant II.

b. Start at the origin. Move 3 units down. Then plot the point.

∴ The point is on the y-axis.

On Your Own

Now You're Ready
Exercises 15–26

Plot the ordered pair in a coordinate plane. Describe the location of the point.

5. $A(2, 3)$ **6.** $B(-1, 0)$

7. $C(-5, -1)$ **8.** $D(3, -6)$

EXAMPLE 3 Real-Life Application

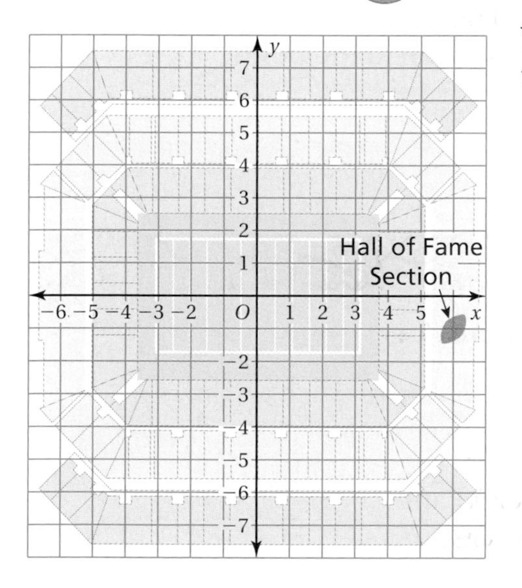

You and a friend have tickets to see a football game. You sit in the Hall of Fame Section and your friend sits at $(-4, -2)$.

a. Write an ordered pair for your location. In which quadrant are you seated?

b. In which quadrant is your friend seated?

c. A fan in Quadrant II is chosen to win a prize. Do you or your friend have a chance to win the prize?

a. The Hall of Fame Section is 6 units to the right of the origin and 1 unit down. So, your seat is located at $(6, -1)$. You are seated in Quadrant IV.

b. Move 4 units to the left of the origin and 2 units down. Your friend is seated in Quadrant III.

c. You are seated in Quadrant IV and your friend is seated in Quadrant III. So, you and your friend do not have a chance to win the prize.

On Your Own

9. WHAT IF? In Example 3, a fan sitting in the level closest to the playing field is chosen to win a prize. Do you or your friend have a chance to win the prize?

1.6 Exercises

Vocabulary and Concept Check

1. **VOCABULARY** How many quadrants are in a coordinate plane?
2. **VOCABULARY** Is the point $(0, -7)$ on the x-axis or the y-axis?
3. **WRITING** How are the locations of the points $(2, -2)$ and $(-2, 2)$ different?
4. **WRITING** Describe the characteristics of ordered pairs in each of the four quadrants.

Practice and Problem Solving

Write an ordered pair corresponding to the point.

① 5. Point A 6. Point B

7. Point C 8. Point D

9. Point E 10. Point F

11. Point G 12. Point H

13. Point I 14. Point J

Plot the ordered pair in a coordinate plane. Describe the location of the point.

② 15. $K(4, 3)$ 16. $L(-1, 2)$ 17. $M(0, -6)$ 18. $N(3, -2)$

19. $P(2, -4)$ 20. $Q(-2, 4)$ 21. $R(-4, 1)$ 22. $S(7, 0)$

23. $T(-4, -5)$ 24. $U(-2, 5)$ 25. $V(-3, 8)$ 26. $W(-5, -1)$

ERROR ANALYSIS Describe and correct the error in the solution.

27. ✗ To plot (4, 5), start at (0, 0) and move 5 units right and 4 units up.

28. ✗ To plot (−6, 3), start at (0, 0) and move 6 units right and 3 units down.

29. **REASONING** The coordinates of three vertices of a square are shown in the figure. What are the coordinates of the fourth vertex?

30. **GEOMETRY** The points $D(1, 1)$, $E(1, -2)$, $F(-2, -2)$, and $G(-2, 1)$ are vertices of a figure.

    a. Draw the figure in a coordinate plane.

    b. Find the perimeter of the figure.

    c. Find the area of the figure.

Tell whether the statement is *sometimes*, *always*, or *never* true. Explain your reasoning.

31. The x-coordinate of a point on the x-axis is zero.

32. The y-coordinate of points in Quadrant III are positive.

33. The x-coordinate of a point in Quadrant II has the same sign as the y-coordinate of a point in Quadrant IV.

ZOO In Exercises 34–38, use the map of the zoo.

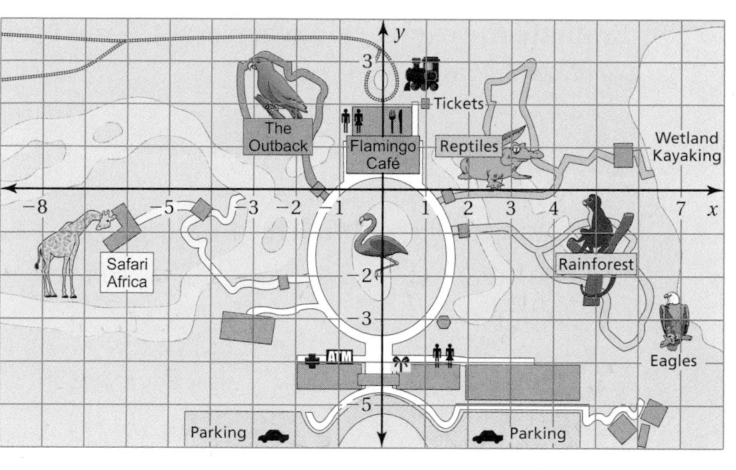

34. Which exhibit is located at (2, 1)?

35. Name an attraction on the positive y-axis.

36. Is parking available in Quadrant II? If not, name a quadrant in which you can park.

37. Write two different ordered pairs that represent the location of the Rainforest.

38. Which exhibit is closest to $(-8, -3)$?

39. **NUMBER SENSE** Name the ordered pair that is 5 units right and 2 units down from $(-3, 4)$.

Plot the ordered pair in a coordinate plane. Describe the location of the point.

40. $A\left(3, -\frac{3}{2}\right)$ **41.** $B\left(-\frac{5}{2}, \frac{10}{3}\right)$ **42.** $C(-5.25, -3.5)$ **43.** $D(-4.75, 0)$

44. **Reasoning** Your school is located at $(2, -1)$, which is 2 blocks east and 1 block south of the center of town. To get from your house to the school, you walk 5 blocks west and 2 blocks north.

a. What ordered pair corresponds to the location of your house?

b. Is your house or your school closer to the center of town? Explain.

Fair Game Review What you learned in previous grades & lessons

Copy and complete the statement using <, >, or =. *(Section 1.5 and Skills Review Handbook)*

45. $\frac{-16}{2}$ ▢ $\frac{-12}{3}$ **46.** $2\frac{2}{5}$ ▢ $\frac{24}{10}$ **47.** 3.45 ▢ $3\frac{3}{8}$

48. **MULTIPLE CHOICE** What is $\frac{1}{3}$ of $3\frac{1}{2}$? *(Skills Review Handbook)*

Ⓐ $\frac{1}{2}$ Ⓑ $1\frac{1}{6}$ Ⓒ $1\frac{1}{2}$ Ⓓ $10\frac{1}{2}$

1.4–1.6 Quiz

Simplify the expression. *(Section 1.4 and Section 1.5)*

1. $-7(6)$

2. $-1(-9)$

3. $\dfrac{-72}{-9}$

4. $-24 \div 3$

Evaluate the expression when $a = 4$, $b = -6$, and $c = -12$.
(Section 1.4 and Section 1.5)

5. c^2

6. $\dfrac{|c-b|}{a}$

Write an ordered pair corresponding to the point.
(Section 1.6)

7. Point A

8. Point B

9. Point C

10. Point D

11. SPEECH In speech class, you lose 3 points for every 30 seconds you go over the time limit. Your speech is 90 seconds over the time limit. What integer represents the change in your points? *(Section 1.4)*

12. MOUNTAIN CLIMBING On a mountain, the temperature decreases by 18°F every 5000 feet. What integer represents the change in temperature at 20,000 feet? *(Section 1.4)*

13. GAMING You play a video game for 15 minutes. You lose 165 points. What integer represents the average change in points per minute? *(Section 1.5)*

14. GEOMETRY The points $A(-4, 2)$, $B(1, -1)$, $C(1, 2)$, and $D(-4, -1)$ are the vertices of a figure. *(Section 1.6)*

a. Draw the figure in a coordinate plane.

b. Find the perimeter of the figure.

c. Find the area of the figure.

1 Chapter Review

Review Key Vocabulary

integer, *p. 4*	additive inverse, *p. 10*	quadrant, *p. 36*
absolute value, *p. 4*	coordinate plane, *p. 36*	*x*-axis, *p. 36*
opposites, *p. 10*	origin, *p. 36*	*y*-axis, *p. 36*

Review Examples and Exercises

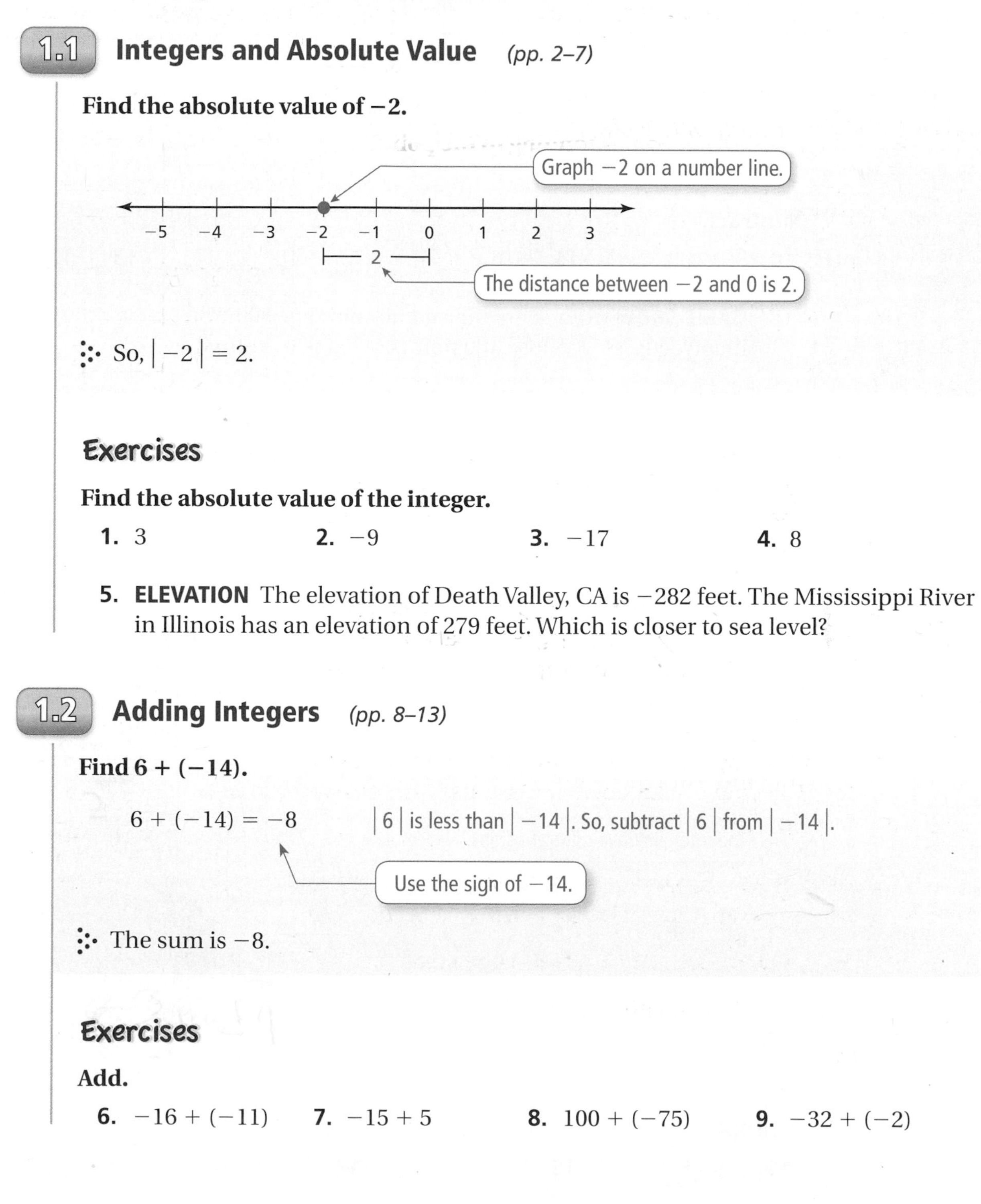

1.1 Integers and Absolute Value *(pp. 2–7)*

Find the absolute value of −2.

So, $|-2| = 2$.

Exercises

Find the absolute value of the integer.

1. 3 **2.** −9 **3.** −17 **4.** 8

5. **ELEVATION** The elevation of Death Valley, CA is −282 feet. The Mississippi River in Illinois has an elevation of 279 feet. Which is closer to sea level?

1.2 Adding Integers *(pp. 8–13)*

Find 6 + (−14).

$6 + (-14) = -8$ $|6|$ is less than $|-14|$. So, subtract $|6|$ from $|-14|$.

Use the sign of −14.

The sum is −8.

Exercises

Add.

6. $-16 + (-11)$ **7.** $-15 + 5$ **8.** $100 + (-75)$ **9.** $-32 + (-2)$

1.3 Subtracting Integers *(pp. 14–19)*

Subtract.

a. $7 - 19 = 7 + (-19)$ Add the opposite of 19.

$= -12$ Add.

The difference is -12.

b. $-6 - (-10) = -6 + 10$ Add the opposite of -10.

$= 4$ Add.

The difference is 4.

Exercises

Subtract.

10. $8 - 18$ **11.** $-16 - (-5)$ **12.** $-18 - 7$ **13.** $-12 - (-27)$

14. GAME SHOW Your score on a game show is -300. You answer the final question incorrectly, so you lose 400 points. What is your final score?

1.4 Multiplying Integers *(pp. 22–27)*

a. Find $-7 \cdot (-9)$.

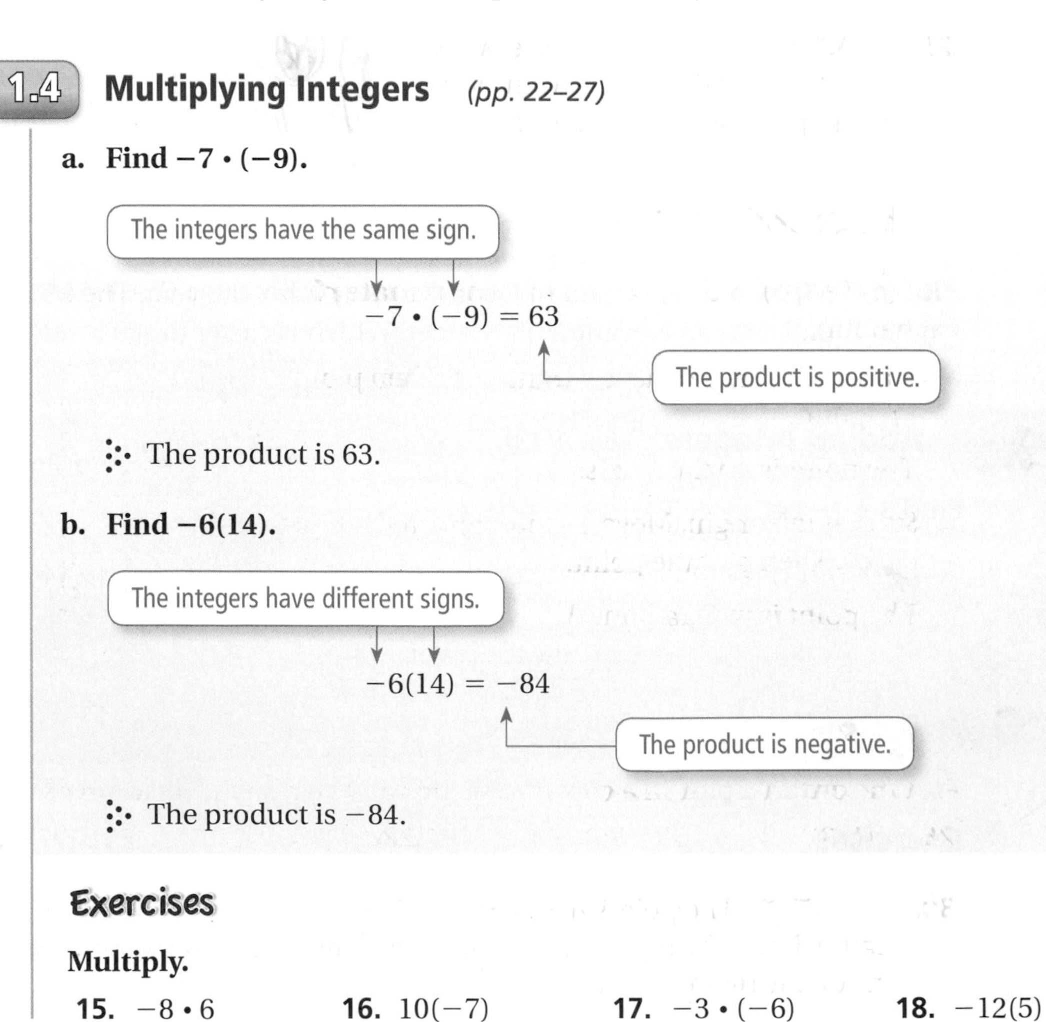

The integers have the same sign.
$-7 \cdot (-9) = 63$
The product is positive.

The product is 63.

b. Find $-6(14)$.

The integers have different signs.
$-6(14) = -84$
The product is negative.

The product is -84.

Exercises

Multiply.

15. $-8 \cdot 6$ **16.** $10(-7)$ **17.** $-3 \cdot (-6)$ **18.** $-12(5)$

1.5 Dividing Integers (pp. 28–33)

Find $30 \div (-10)$.

The integers have different signs.

$$30 \div (-10) = -3$$

The quotient is negative.

∴ The quotient is -3.

Exercises

Divide.

19. $-18 \div 9$ **20.** $\frac{-42}{-6}$ **21.** $\frac{-30}{6}$ **22.** $84 \div (-7)$

Find the mean of the integers.

23. $-3, -8, 12, -15, 9$ **24.** $-54, -32, -70, -25, -65, -42$

25. PROFITS The table shows the weekly profits of a fruit vendor. What is the mean profit for these weeks?

Week	1	2	3	4
Profit	−$125	−$86	$54	−$35

1.6 The Coordinate Plane (pp. 34–39)

Plot (a) $(-3, 0)$ and (b) $(4, -4)$ in a coordinate plane. Describe the location of each point.

a. Start at the origin. Move 3 units left. Then plot the point.

The point is on the x-axis.

b. Start at the origin. Move 4 units right and 4 units down. Then plot the point.

The point is in Quadrant IV.

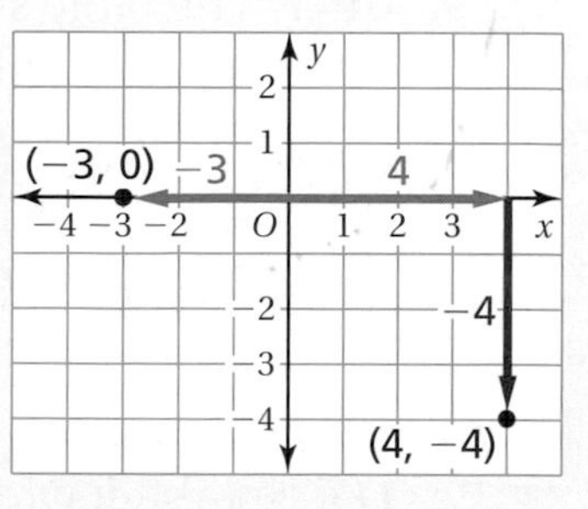

Exercises

Plot the ordered pair in a coordinate plane. Describe the location of the point.

26. $A(1, 3)$ **27.** $B(0, -3)$ **28.** $C(-4, -2)$ **29.** $D(-1, 2)$

30. GEOMETRY The points $A(-3, 5)$, $B(-3, -3)$, and $C(4, -3)$ are vertices of a figure. Draw the figure in a coordinate plane. Name three points that lie inside the figure.

1 Chapter Test

Check It Out
Test Practice
BigIdeasMath.com

Find the absolute value of the integer.

1. -9
2. 64
3. -22

Copy and complete the statement using <, >, or =.

4. $4 \ \square \ |-8|$
5. $-12 \ \square \ |-7|$
6. $-7 \ \square \ |3|$

Simplify the expression.

7. $-6 + (-11)$
8. $2 - (-9)$
9. $-9 \cdot 2$
10. $-72 \div (-3)$
11. $-5 + 17$
12. $-14(21)$

Plot the ordered pair in a coordinate plane. Describe the location of the point.

13. $K(1, 3)$
14. $L(-3, 0)$
15. $M(-4, 5)$
16. $N(2, -1)$

17. **BANKING** The balance of your checking account is \$86. You withdraw \$98. What is your new balance?

18. **NASCAR** A driver receives -25 points for each rule violation. What integer represents the change in points after four rule violations?

19. **GOLF** The table shows your scores, relative to *par*, for nine holes of golf. What is your total score for the nine holes?

Hole	1	2	3	4	5	6	7	8	9	Total
Score	+1	−2	−1	0	−1	+3	−1	−3	+1	?

20. **VISITORS** In a recent 10-year period, the change in the number of visitors to U.S. National Parks was about $-11{,}150{,}000$ visitors.

    a. What was the mean yearly change in the number of visitors?

    b. During the seventh year, the change in the number of visitors was about 10,800,000. Explain how the change for the 10-year period can be negative.

1 Standardized Test Practice

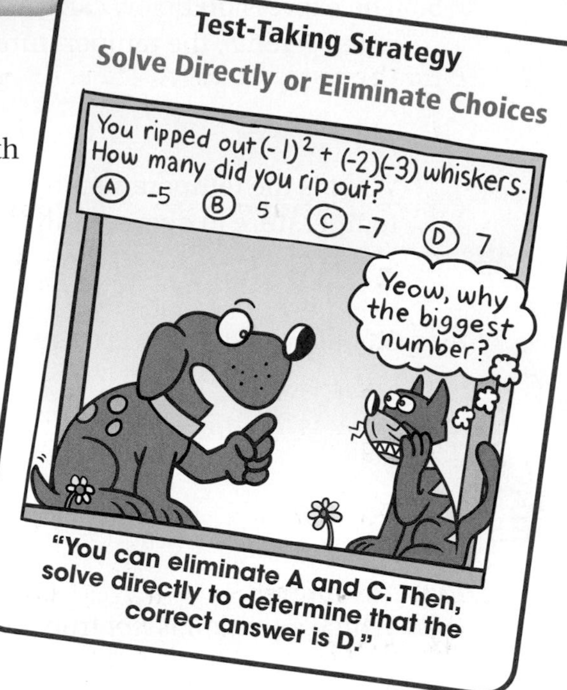

1. A football team gains 2 yards on the first play, loses 5 yards on the second play, loses 3 yards on the third play, and gains 4 yards on the fourth play. What is the team's overall gain or loss for all four plays? *(7.NS.1b)*

A. a gain of 14 yards **C.** a loss of 2 yards

B. a gain of 2 yards **D.** a loss of 14 yards

2. Point P is plotted in the coordinate plane below.

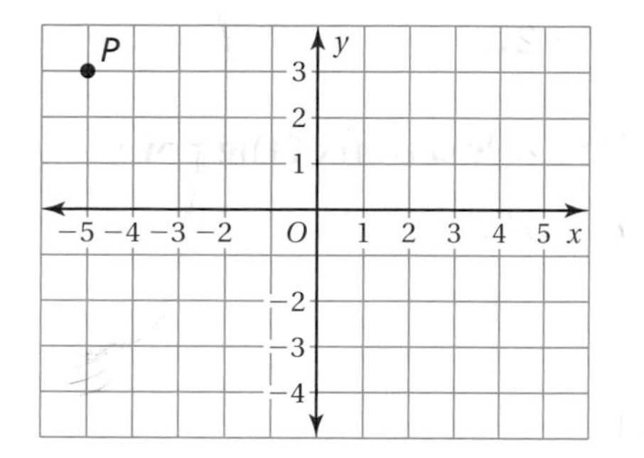

What are the coordinates of point P? *(7.NS.1b)*

F. $(-5, -3)$ **H.** $(-3, -5)$

G. $(-5, 3)$ **I.** $(3, -5)$

3. What is the value of the expression below? *(7.NS.1c)*

$$17 - (-8)$$

4. Sam was simplifying an expression in the box below.

$$|-8 + 6 + (-3)| = |-8| + |6| + |-3|$$
$$= 8 + 6 + 3$$
$$= 17$$

What should Sam do to correct the error that he made? *(7.EE.3)*

A. Find the absolute value of the sum of 8, 6, and 3 and make that the final answer.

B. Find the sum of −8, −6, and −3 and make that the final answer.

C. Find the sum of −8, 6, and −3 and make that the final answer.

D. Find the absolute value of the sum of −8, 6, and −3 and make that the final answer.

5. The expression below can be used to find the temperature in degrees Celsius when given F, the temperature in degrees Fahrenheit.

$$\frac{5}{9}(F - 32)$$

What is the temperature in degrees Celsius, to the nearest degree, when the temperature in degrees Fahrenheit is 27°? *(7.NS.3)*

F. $-33°$

G. $-17°$

H. $-5°$

I. $-3°$

6. What is the missing number in the sequence below? *(7.NS.1d)*

39, 24, 9, ___, −21

7. Which equation is *not* true for all numbers n? *(7.NS.3)*

A. $-n + 0 = -n$

B. $n \cdot (-1) = -n$

C. $n - 0 = -n$

D. $-n \cdot 1 = -n$

8. What is the area of the semicircle below? (Use 3.14 for π.) *(7.NS.2c)*

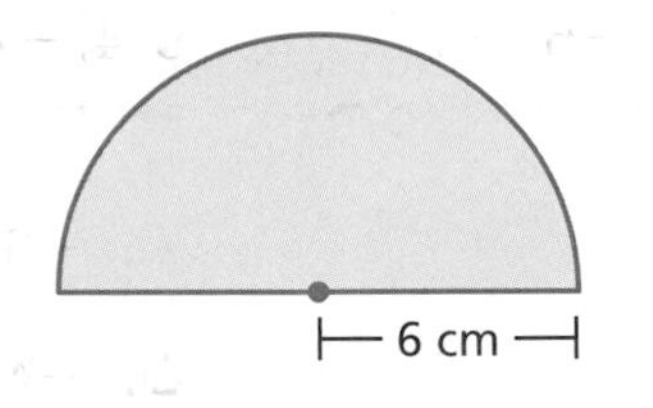

F. 18.84 cm^2

G. 37.68 cm^2

H. 56.52 cm^2

I. 226.08 cm^2

9. The campers at a summer camp held a contest in which they had to run across a field carrying buckets of water that were full at the beginning. The team who lost the least water from its bucket was the winner.

- Team A *lost* 40% of the water from its bucket.
- Team B *lost* 0.3 of the water from its bucket.
- Team C *kept* $\frac{5}{8}$ of the water in its bucket.
- Team D *kept* 67% of the water in its bucket.

Which team was the winner? *(7.EE.3)*

A. Team A

B. Team B

C. Team C

D. Team D

10. Which integer is closest to the value of the expression below? *(7.NS.3)*

$$-5.04 \cdot (16.89 - 20.1)$$

F. -105

G. -15

H. 15

I. 105

11. Answer the following questions in the coordinate plane. *(7.NS.1b)*

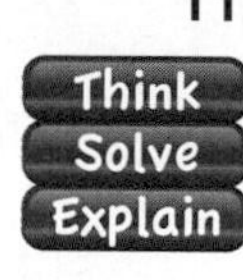

Part A Draw an x-axis and y-axis in the coordinate plane. Then plot and label the point $(2, -3)$.

Part B Plot and label *four* points that are 3 units away from $(2, -3)$.

12. What is the mean of the data set in the box below? *(7.NS.3)*

$$-8, -6, -2, 0, -6, -8, 4, -7, -8, 1$$

A. -8

B. -7

C. -6

D. -4

13. Jane and Manuel measured the lengths of their pet rats.

- Jane's pet rat
  - Body length: $10\frac{1}{2}$ inches
  - Tail length: $7\frac{3}{4}$ inches
- Manuel's pet rat
  - Body length: $9\frac{5}{8}$ inches
  - Tail length: $8\frac{1}{4}$ inches

The total length of each rat is determined by the sum of its body length and its tail length. Whose rat has the longer total length and by how much? *(7.NS.1d)*

F. Jane's rat is longer by $\frac{3}{8}$ inch.

G. Jane's rat is longer by $\frac{1}{6}$ inch.

H. Manuel's rat is longer by $\frac{5}{8}$ inch.

I. Manuel's rat is longer by $\frac{1}{4}$ inch.

2 Rational Numbers and Equations

"I can't find my algebra tiles, so I am painting some of my dog biscuits."

"Now I will be able to solve the equation $2x + (-2) = 2$."

"On the count of 5, I'm going to give you half of my dog biscuits."

"1, 2, 3, 4, $4\frac{1}{2}$, $4\frac{3}{4}$, $4\frac{7}{8}$,..."

What You Learned Before

"Let's play a game. The goal is to say a positive rational number that is less than the other pet's number... You go first."

Writing Decimals and Fractions (4.NF.6)

Example 1 Write 0.37 as a fraction.

$$0.37 = \frac{37}{100}$$

Example 2 Write $\frac{2}{5}$ as a decimal.

$$\frac{2}{5} = \frac{2 \cdot 2}{5 \cdot 2} = \frac{4}{10} = 0.4$$

Try It Yourself

Write the decimal as a fraction or the fraction as a decimal.

1. 0.51 **2.** 0.731 **3.** $\frac{3}{5}$ **4.** $\frac{7}{8}$

Adding and Subtracting Fractions (5.NF.1)

Example 3 Find $\frac{1}{3} + \frac{1}{5}$.

$$\frac{1}{3} + \frac{1}{5} = \frac{1 \cdot 5}{3 \cdot 5} + \frac{1 \cdot 3}{5 \cdot 3}$$
$$= \frac{5}{15} + \frac{3}{15}$$
$$= \frac{8}{15}$$

Example 4 Find $\frac{1}{4} - \frac{2}{9}$.

$$\frac{1}{4} - \frac{2}{9} = \frac{1 \cdot 9}{4 \cdot 9} - \frac{2 \cdot 4}{9 \cdot 4}$$
$$= \frac{9}{36} - \frac{8}{36}$$
$$= \frac{1}{36}$$

Multiplying and Dividing Fractions (6.NS.1)

Example 5 Find $\frac{5}{6} \cdot \frac{3}{4}$.

$$\frac{5}{6} \cdot \frac{3}{4} = \frac{5 \cdot \cancel{3}^{1}}{{}_{2}\cancel{6} \cdot 4}$$
$$= \frac{5}{8}$$

Example 6 Find $\frac{2}{3} \div \frac{9}{10}$.

$$\frac{2}{3} \div \frac{9}{10} = \frac{2}{3} \cdot \frac{10}{9}$$

Multiply by the reciprocal of the divisor.

$$= \frac{2 \cdot 10}{3 \cdot 9}$$
$$= \frac{20}{27}$$

Try It Yourself

Evaluate the expression.

5. $\frac{1}{4} + \frac{13}{20}$ **6.** $\frac{14}{15} - \frac{1}{3}$ **7.** $\frac{3}{7} \cdot \frac{9}{10}$ **8.** $\frac{4}{5} \div \frac{16}{17}$

2.1 Rational Numbers

COMMON CORE STATE STANDARDS
7.EE.3
7.NS.2b
7.NS.2d

Essential Question How can you use a number line to order rational numbers?

The Meaning of a Word ● Rational

The word **rational** comes from the word *ratio.*

If you sleep for 8 hours in a day, then the *ratio* of your sleeping time to the total hours in a day can be written as $\frac{8\text{ h}}{24\text{ h}}$.

A **rational number** is a number that can be written as the ratio of two integers.

$2 = \frac{2}{1}$ $\quad -3 = \frac{-3}{1}$ $\quad -\frac{1}{2} = \frac{-1}{2}$ $\quad 0.25 = \frac{1}{4}$

1 ACTIVITY: Ordering Rational Numbers

Work in groups of five. Order the numbers from least to greatest.

a. Sample: $-0.5, 1.25, -\frac{1}{3}, 0.5, -\frac{5}{3}$

- Make a number line on the floor using masking tape and a marker.

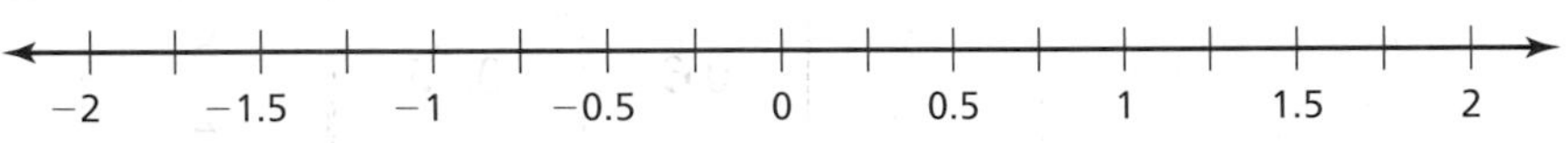

- Write the numbers on pieces of paper. Then each person should choose one.
- Stand on the location of your number on the number line.

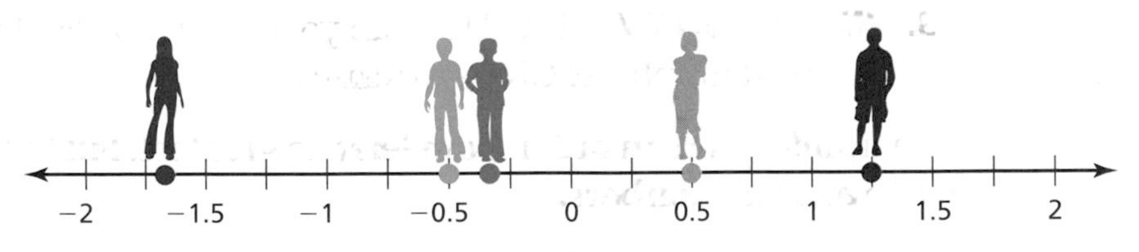

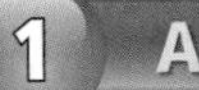

- Use your positions to order the numbers from least to greatest.

∴ So, the numbers from least to greatest are $-\frac{5}{3}, -0.5, -\frac{1}{3}$, 0.5, and 1.25.

b. $-\frac{7}{4}, 1.1, \frac{1}{2}, -\frac{1}{10}, -1.3$

c. $-\frac{1}{4}, 2.5, \frac{3}{4}, -1.7, -0.3$

d. $-1.4, -\frac{3}{5}, \frac{9}{2}, \frac{1}{4}, 0.9$

e. $\frac{9}{4}, 0.75, -\frac{5}{4}, -0.8, -1.1$

2 ACTIVITY: The Game of Math Card War

Preparation:

- Cut index cards to make 40 playing cards.
- Write each number in the table on a card.

To Play:

- Play with a partner.
- Deal 20 cards to each player face-down.
- Each player turns one card face-up. The player with the greater number wins. The winner collects both cards and places them at the bottom of his or her cards.
- Suppose there is a tie. Each player lays three cards face-down, then a new card face-up. The player with the greater of these new cards wins. The winner collects all ten cards and places them at the bottom of his or her cards.
- Continue playing until one player has all the cards. This player wins the game.

$-\frac{3}{2}$	$\frac{3}{10}$	$-\frac{3}{4}$	-0.6	1.25	-0.15	$\frac{5}{4}$	$\frac{3}{5}$	-1.6	-0.3
$\frac{3}{20}$	$\frac{8}{5}$	-1.2	$\frac{19}{10}$	0.75	-1.5	$-\frac{6}{5}$	$-\frac{3}{5}$	1.2	0.3
1.5	1.9	-0.75	-0.4	$\frac{3}{4}$	$-\frac{5}{4}$	-1.9	$\frac{2}{5}$	$-\frac{3}{20}$	$-\frac{19}{10}$
$\frac{6}{5}$	$-\frac{3}{10}$	1.6	$-\frac{2}{5}$	0.6	0.15	$\frac{3}{2}$	-1.25	0.4	$-\frac{8}{5}$

What Is Your Answer?

3. **IN YOUR OWN WORDS** How can you use a number line to order rational numbers? Give an example.

The numbers are in order from least to greatest. Fill in the blank spaces with rational numbers.

4. $-\frac{1}{2}$, ___, $\frac{1}{3}$, ___, $\frac{7}{5}$, ___

5. $-\frac{5}{2}$, ___, -1.9, ___, $-\frac{2}{3}$, ___

6. $-\frac{1}{3}$, ___, -0.1, ___, $\frac{4}{5}$, ___

7. -3.4, ___, -1.5, ___, 2.2, ___

Practice Use what you learned about ordering rational numbers to complete Exercises 28–30 on page 54.

2.1 Lesson

Key Vocabulary
terminating decimal, *p. 52*
repeating decimal, *p. 52*
rational number, *p. 52*

A **terminating decimal** is a decimal that ends.

1.5, –0.25, 10.625

A **repeating decimal** is a decimal that has a pattern that repeats.

$-1.333\ldots = -1.\overline{3}$

$0.151515\ldots = 0.\overline{15}$

Use *bar notation* to show which of the digits repeat.

Terminating and repeating decimals are examples of *rational numbers.*

Key Idea

Rational Numbers

A **rational number** is a number that can be written as $\frac{a}{b}$ where a and b are integers and $b \neq 0$.

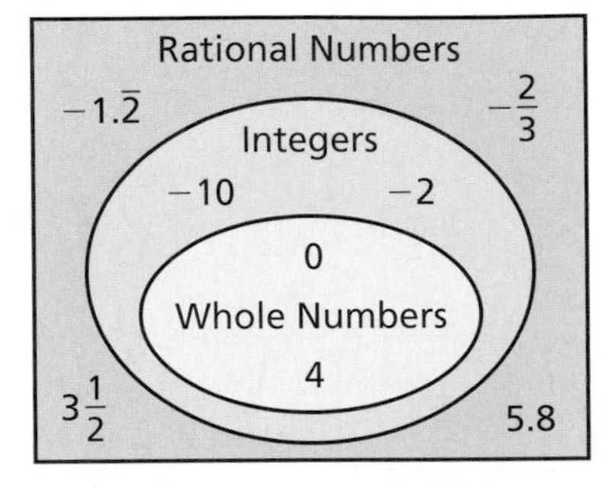

EXAMPLE 1 Writing Rational Numbers as Decimals

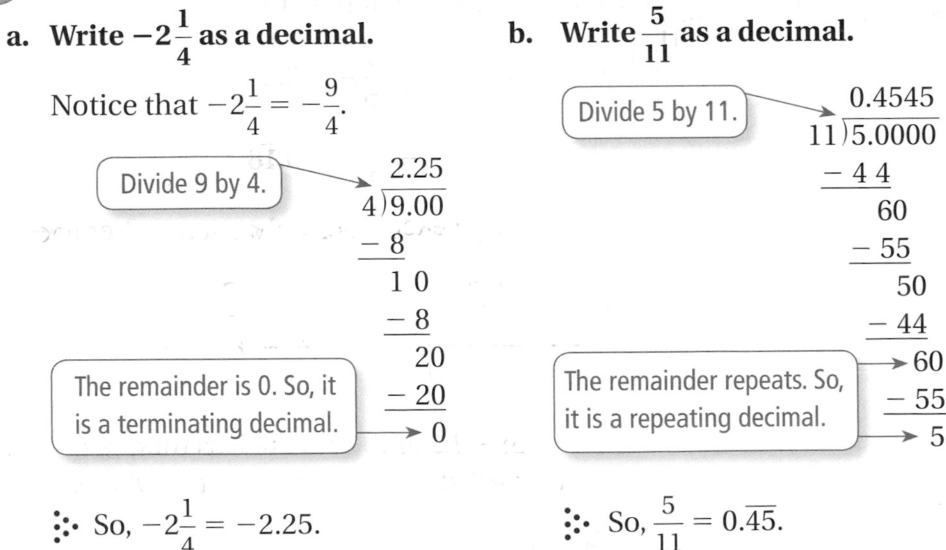

a. Write $-2\frac{1}{4}$ as a decimal.

Notice that $-2\frac{1}{4} = -\frac{9}{4}$.

Divide 9 by 4.

$$\begin{array}{r} 2.25 \\ 4\overline{)9.00} \\ -8 \\ \hline 1\,0 \\ -8 \\ \hline 20 \\ -20 \\ \hline 0 \end{array}$$

The remainder is 0. So, it is a terminating decimal.

So, $-2\frac{1}{4} = -2.25$.

b. Write $\frac{5}{11}$ as a decimal.

Divide 5 by 11.

$$\begin{array}{r} 0.4545 \\ 11\overline{)5.0000} \\ -4\,4 \\ \hline 60 \\ -55 \\ \hline 50 \\ -44 \\ \hline 60 \\ -55 \\ \hline 5 \end{array}$$

The remainder repeats. So, it is a repeating decimal.

So, $\frac{5}{11} = 0.\overline{45}$.

On Your Own

Now You're Ready
Exercises 11–18

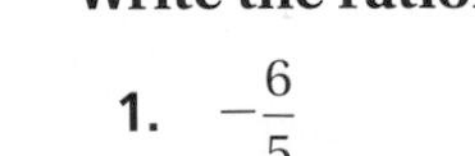

Write the rational number as a decimal.

1. $-\frac{6}{5}$
2. $-7\frac{3}{8}$
3. $-\frac{3}{11}$
4. $1\frac{5}{27}$

Multi-Language Glossary at BigIdeasMath.com.

EXAMPLE 2 Writing a Decimal as a Fraction

Write −0.26 as a fraction in simplest form.

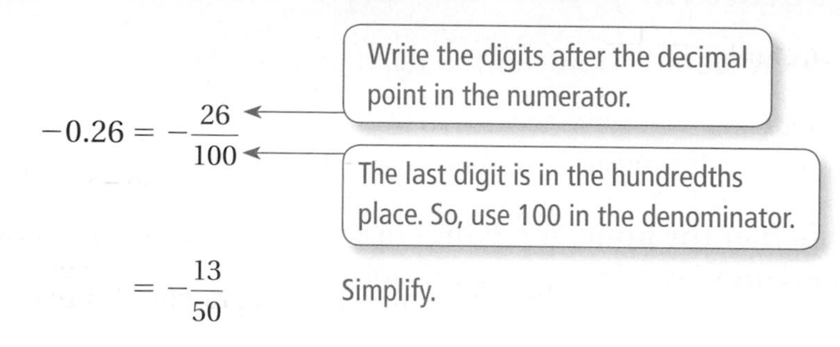

$$-0.26 = -\frac{26}{100}$$

$$= -\frac{13}{50} \quad \text{Simplify.}$$

On Your Own

Now You're Ready
Exercises 20–27

Write the decimal as a fraction or mixed number in simplest form.

5. -0.7 **6.** 0.125 **7.** -3.1 **8.** -10.25

EXAMPLE 3 Ordering Rational Numbers

Creature	Elevations (km)
Anglerfish	$-\frac{13}{10}$
Squid	$-2\frac{1}{5}$
Shark	$-\frac{2}{11}$
Whale	-0.8

The table shows the elevations of four sea creatures relative to sea level. Which of the sea creatures are deeper than the whale? Explain.

Write each rational number as a decimal.

$$-\frac{13}{10} = -1.3$$

$$-2\frac{1}{5} = -2.2$$

$$-\frac{2}{11} = -0.\overline{18}$$

Then graph each decimal on a number line.

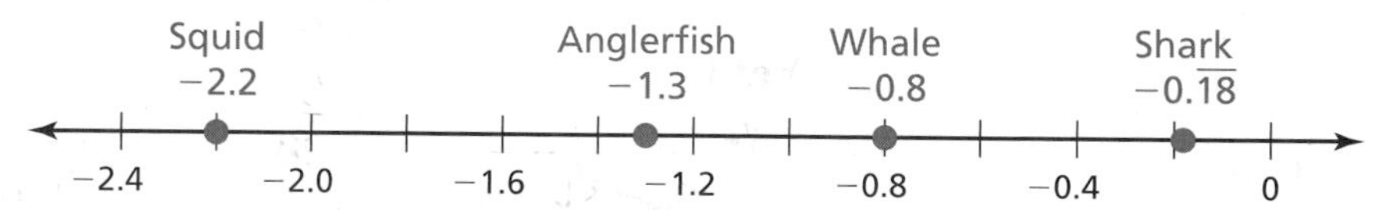

∴ Both −2.2 and −1.3 are less than −0.8. So, the squid and the anglerfish are deeper than the whale.

On Your Own

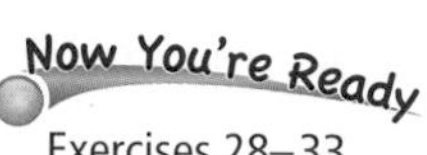

9. **WHAT IF?** The elevation of a dolphin is $-\frac{1}{10}$ kilometer. Which of the sea creatures in Example 3 are deeper than the dolphin? Explain.

2.1 Exercises

Check It Out
Help with Homework
BigIdeasMath.com

Vocabulary and Concept Check

1. **VOCABULARY** How can you tell that a number is rational?
2. **WRITING** You have to write 0.63 as a fraction. How do you choose the denominator?

Tell whether the number belongs to each of the following number sets: *rational numbers, integers, whole numbers.*

3. -5
4. $-2.1\overline{6}$
5. 12
6. 0

Tell whether the decimal is *terminating* or *repeating*.

7. $-0.4848\ldots$
8. -0.151
9. 72.72
10. $-5.2\overline{36}$

Practice and Problem Solving

Write the rational number as a decimal.

1
11. $\frac{7}{8}$
12. $\frac{5}{11}$
13. $-\frac{7}{9}$
14. $-\frac{17}{40}$
15. $1\frac{5}{6}$
16. $-2\frac{17}{18}$
17. $-5\frac{7}{12}$
18. $8\frac{15}{22}$

19. **ERROR ANALYSIS** Describe and correct the error in writing the rational number as a decimal.

✗ $-\frac{7}{11} = -0.6\overline{3}$

Write the decimal as a fraction or mixed number in simplest form.

2
20. -0.9
21. 0.45
22. -0.258
23. -0.312
24. -2.32
25. -1.64
26. 6.012
27. -12.405

Order the numbers from least to greatest.

3
28. $-\frac{3}{4}, 0.5, \frac{2}{3}, -\frac{7}{3}, 1.2$
29. $\frac{9}{5}, -2.5, -1.1, -\frac{4}{5}, 0.8$
30. $-1.4, -\frac{8}{5}, 0.6, -0.9, \frac{1}{4}$
31. $2.1, -\frac{6}{10}, -\frac{9}{4}, -0.75, \frac{5}{3}$
32. $-\frac{7}{2}, -2.8, -\frac{5}{4}, \frac{4}{3}, 1.3$
33. $-\frac{11}{5}, -2.4, 1.6, \frac{15}{10}, -2.25$

34. **COINS** You lose one quarter, two dimes and two nickels.
    a. Write the amount as a decimal.
    b. Write the amount as a fraction in simplest form.

35. **HIBERNATION** A box turtle hibernates in sand at $-1\frac{5}{8}$ feet. A spotted turtle hibernates at $-1\frac{16}{25}$ feet. Which turtle is deeper?

Copy and complete the statement using <, >, or =.

36. -2.2 ☐ -2.42 **37.** -1.82 ☐ -1.81 **38.** $\frac{15}{8}$ ☐ $1\frac{7}{8}$

39. $-4\frac{6}{10}$ ☐ -4.65 **40.** $-5\frac{3}{11}$ ☐ $-5.\overline{2}$ **41.** $-2\frac{13}{16}$ ☐ $-2\frac{11}{14}$

42. OPEN-ENDED Find one terminating decimal and one repeating decimal between $-\frac{1}{2}$ and $-\frac{1}{3}$.

Player	Hits	At Bats
Eva	42	90
Michelle	38	80

43. SOFTBALL In softball, a batting average is the number of hits divided by the number of times at bat. Does Eva or Michelle have the higher batting average?

44. QUIZ You miss 3 out of 10 questions on a science quiz and 4 out of 15 questions on a math quiz. Which quiz has a higher percent of correct answers?

45. SKATING Is the half pipe deeper than the skating pool? Explain.

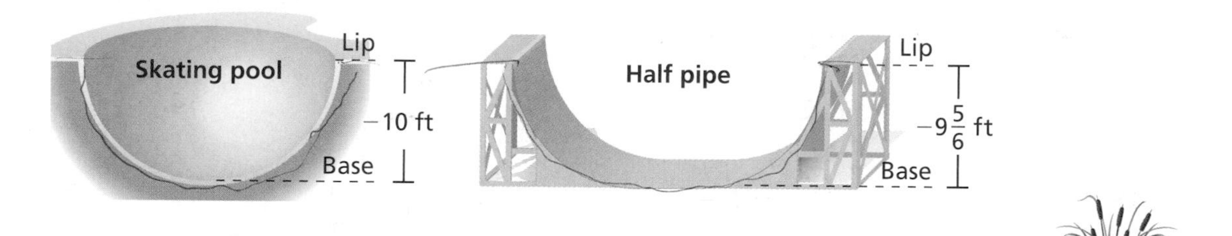

46. ENVIRONMENT The table shows the changes from the average water level of a pond over several weeks. Order the numbers from least to greatest.

Week	1	2	3	4
Change (inches)	$-\frac{7}{5}$	$-1\frac{5}{11}$	-1.45	$-1\frac{91}{200}$

47. Critical Thinking Given: a and b are integers.

a. When is $-\frac{1}{a}$ positive? **b.** When is $\frac{1}{ab}$ positive?

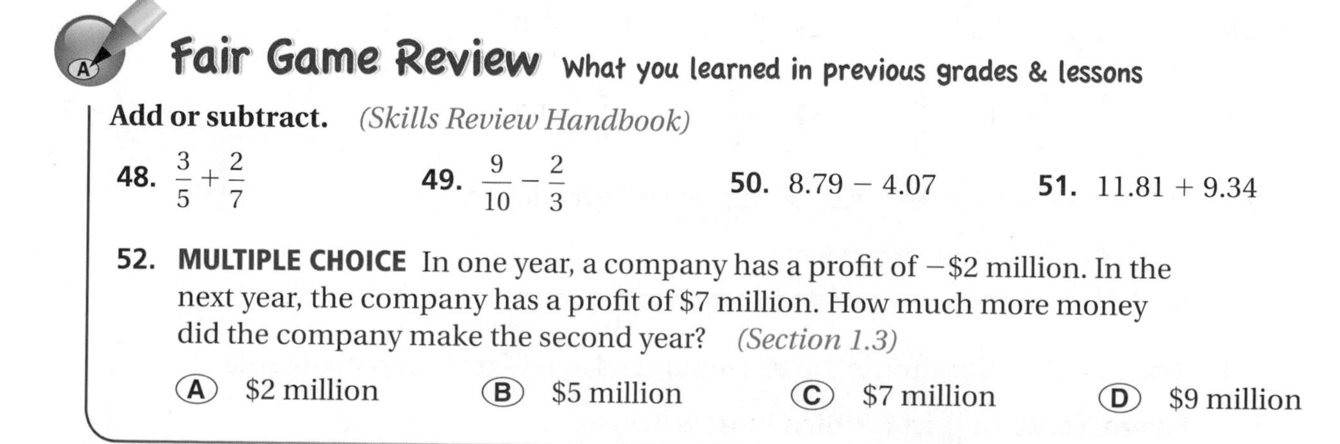

Fair Game Review What you learned in previous grades & lessons

Add or subtract. *(Skills Review Handbook)*

48. $\frac{3}{5} + \frac{2}{7}$ **49.** $\frac{9}{10} - \frac{2}{3}$ **50.** $8.79 - 4.07$ **51.** $11.81 + 9.34$

52. MULTIPLE CHOICE In one year, a company has a profit of −\$2 million. In the next year, the company has a profit of \$7 million. How much more money did the company make the second year? *(Section 1.3)*

Ⓐ \$2 million Ⓑ \$5 million Ⓒ \$7 million Ⓓ \$9 million

2.2 Adding and Subtracting Rational Numbers

Essential Question How does adding and subtracting rational numbers compare with adding and subtracting integers?

COMMON CORE STATE STANDARDS
7.NS.1b
7.NS.1c
7.NS.1d

1 ACTIVITY: Adding and Subtracting Rational Numbers

Work with a partner. Use a number line to find the sum or difference.

a. Sample: $2.7 + (-3.4)$

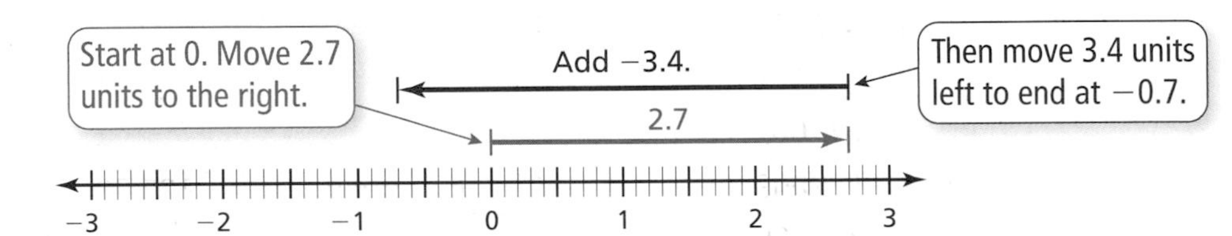
Start at 0. Move 2.7 units to the right.
Add −3.4.
Then move 3.4 units left to end at −0.7.
2.7
−3
−2
−1
0
1
2
3

So, $2.7 + (-3.4) = -0.7$.

b. $\frac{3}{10} + \left(-\frac{9}{10}\right)$

c. $-\frac{6}{10} - 1\frac{3}{10}$

d. $1.3 + (-3.4)$

e. $-1.9 - 0.8$

2 ACTIVITY: Adding and Subtracting Rational Numbers

Work with a partner. Write the numerical expression shown on the number line. Then find the sum or difference.

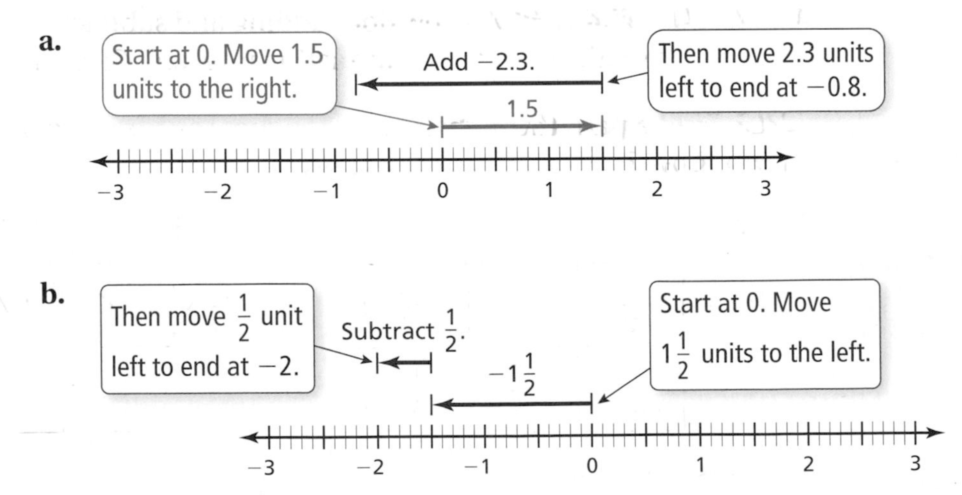
a.
Start at 0. Move 1.5 units to the right.
Add −2.3.
Then move 2.3 units left to end at −0.8.
1.5
−3
−2
−1
0
1
2
3
b.
Then move $\frac{1}{2}$ unit left to end at −2.
Subtract $\frac{1}{2}$.
$-1\frac{1}{2}$
Start at 0. Move $1\frac{1}{2}$ units to the left.
−3
−2
−1
0
1
2
3

3 ACTIVITY: Financial Literacy

Work with a partner. The table shows the balance in a checkbook.

- Black numbers are amounts added to the account.
- Red numbers are amounts taken from the account.

Date	Check #	Transaction	Amount	Balance
--	--	Previous balance	--	100.00
1/02/2009	124	Groceries	34.57	
1/06/2009		Check deposit	875.50	
1/11/2009		ATM withdrawal	40.00	
1/14/2009	125	Electric company	78.43	
1/17/2009		Music store	10.55	
1/18/2009	126	Shoes	47.21	
1/20/2009		Check deposit	125.00	
1/21/2009		Interest	2.12	
1/22/2009	127	Cell phone	59.99	

You can find the balance in the **second row** two different ways.

$100.00 - 34.57 = 65.43$ Subtract 34.57 from 100.00.

$100.00 + (-34.57) = 65.43$ Add −34.57 to 100.00.

a. Copy the table. Then complete the balance column.

b. How did you find the balance in the **tenth row**?

c. Use a different way to find the balance in part (b).

What Is Your Answer?

4. IN YOUR OWN WORDS How does adding and subtracting rational numbers compare with adding and subtracting integers? Give an example.

PUZZLE Find a path through the table so that the numbers add up to the sum. You can move horizontally or vertically.

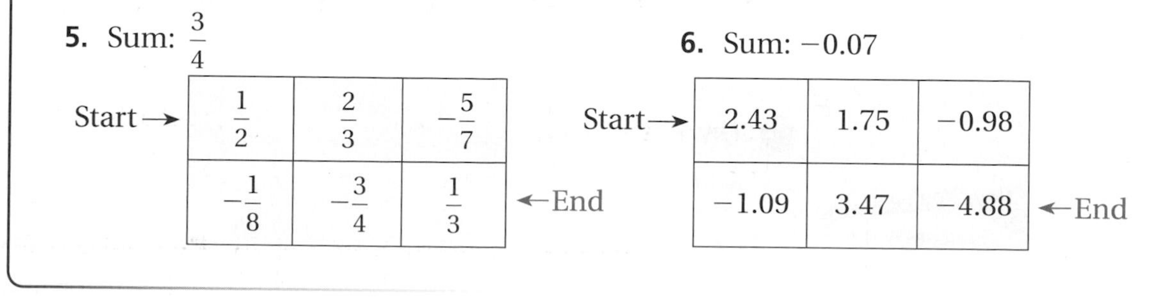

5. Sum: $\frac{3}{4}$

Start →

$\frac{1}{2}$	$\frac{2}{3}$	$-\frac{5}{7}$
$-\frac{1}{8}$	$-\frac{3}{4}$	$\frac{1}{3}$

← End

6. Sum: -0.07

Start →

2.43	1.75	−0.98
−1.09	3.47	−4.88

← End

Practice

Use what you learned about adding and subtracting rational numbers to complete Exercises 7–9 and 16–18 on page 60.

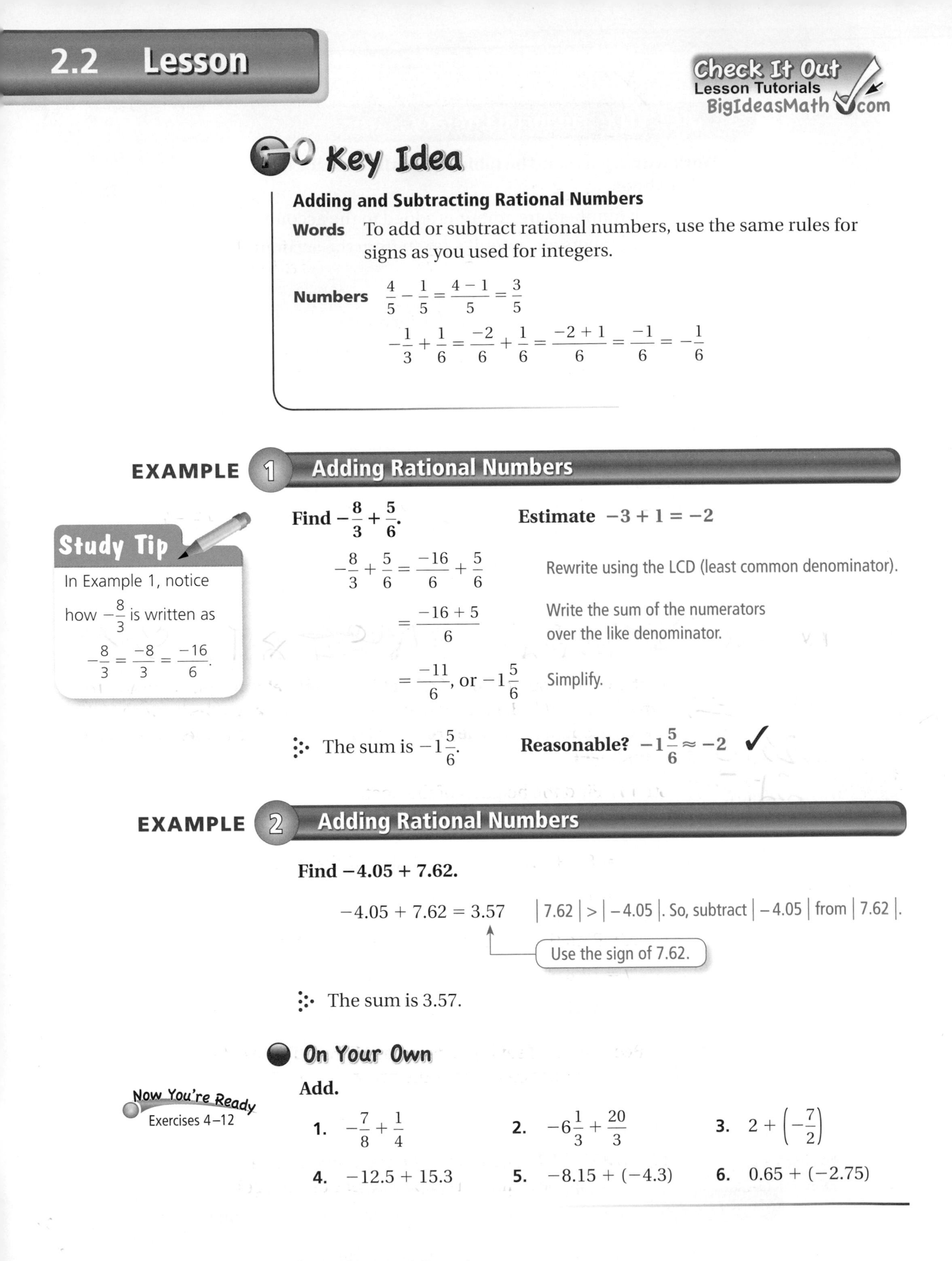

2.2 Lesson

Check It Out
Lesson Tutorials
BigIdeasMath.com

Key Idea

Adding and Subtracting Rational Numbers

Words To add or subtract rational numbers, use the same rules for signs as you used for integers.

Numbers $\frac{4}{5} - \frac{1}{5} = \frac{4-1}{5} = \frac{3}{5}$

$-\frac{1}{3} + \frac{1}{6} = \frac{-2}{6} + \frac{1}{6} = \frac{-2+1}{6} = \frac{-1}{6} = -\frac{1}{6}$

EXAMPLE 1 Adding Rational Numbers

Find $-\frac{8}{3} + \frac{5}{6}$. **Estimate $-3 + 1 = -2$**

$-\frac{8}{3} + \frac{5}{6} = \frac{-16}{6} + \frac{5}{6}$ Rewrite using the LCD (least common denominator).

$= \frac{-16+5}{6}$ Write the sum of the numerators over the like denominator.

$= \frac{-11}{6}$, or $-1\frac{5}{6}$ Simplify.

The sum is $-1\frac{5}{6}$. **Reasonable? $-1\frac{5}{6} \approx -2$** ✓

Study Tip

In Example 1, notice how $-\frac{8}{3}$ is written as $-\frac{8}{3} = \frac{-8}{3} = \frac{-16}{6}$.

EXAMPLE 2 Adding Rational Numbers

Find −4.05 + 7.62.

$-4.05 + 7.62 = 3.57$ $|7.62| > |-4.05|$. So, subtract $|-4.05|$ from $|7.62|$.

Use the sign of 7.62.

The sum is 3.57.

On Your Own

Now You're Ready
Exercises 4–12

Add.

1. $-\frac{7}{8} + \frac{1}{4}$
2. $-6\frac{1}{3} + \frac{20}{3}$
3. $2 + \left(-\frac{7}{2}\right)$
4. $-12.5 + 15.3$
5. $-8.15 + (-4.3)$
6. $0.65 + (-2.75)$

EXAMPLE 3 Subtracting Rational Numbers

Find $-4\frac{1}{7} - \left(-\frac{6}{7}\right)$. **Estimate** $-4 - (-1) = -3$

$$-4\frac{1}{7} - \left(-\frac{6}{7}\right) = -4\frac{1}{7} + \frac{6}{7}$$ Add the opposite of $-\frac{6}{7}$.

$$= -\frac{29}{7} + \frac{6}{7}$$ Write the mixed number as an improper fraction.

$$= \frac{-23}{7}, \text{ or } -3\frac{2}{7}$$ Simplify.

The difference is $-3\frac{2}{7}$. **Reasonable?** $-3\frac{2}{7} \approx -3$ ✓

On Your Own

Subtract.

7. $\frac{1}{3} - \left(-\frac{1}{3}\right)$

8. $-3\frac{1}{3} - \frac{5}{6}$

9. $4\frac{1}{2} - 5\frac{1}{4}$

EXAMPLE 4 Real-Life Application

Clearance: 11 ft 8 in.

In the water, the bottom of a boat is 2.1 feet below the surface and the top of the boat is 8.7 feet above it. Towed on a trailer, the bottom of the boat is 1.3 feet above the ground. Can the boat and trailer pass under the bridge?

Step 1: Find the height h of the boat.

$h = 8.7 - (-2.1)$ Subtract the lowest point from the highest point.

$= 8.7 + 2.1$ Add the opposite of -2.1.

$= 10.8$ Add.

Step 2: Find the height t of the boat and trailer.

$t = 10.8 + 1.3$ Add the trailer height to the boat height.

$= 12.1$ Add.

Because 12.1 feet is greater than 11 feet 8 inches, the boat and trailer cannot pass under the bridge.

On Your Own

Now You're Ready
Exercises 13–21

10. WHAT IF? In Example 4, the clearance is 12 feet 1 inch. Can the boat and trailer pass under the bridge?

2.2 Exercises

Check It Out
Help with Homework
BigIdeasMath.com

Vocabulary and Concept Check

1. **WRITING** Explain how to find the sum $-8.46 + 5.31$.
2. **OPEN-ENDED** Write an addition expression using fractions that equals $-\frac{1}{2}$.
3. **DIFFERENT WORDS, SAME QUESTION** Which is different? Find "both" answers.

Add -4.8 and 3.9.	What is 3.9 less than -4.8?
What is -4.8 increased by 3.9?	Find the sum of -4.8 and 3.9.

Practice and Problem Solving

Add. Write fractions in simplest form.

① ② 4. $\frac{11}{12} + \left(-\frac{7}{12}\right)$

5. $-\frac{9}{14} + \frac{2}{7}$

6. $\frac{15}{4} + \left(-4\frac{1}{3}\right)$

7. $2\frac{5}{6} + \left(-\frac{8}{15}\right)$

8. $4 + \left(-1\frac{2}{3}\right)$

9. $-4.2 + 3.3$

10. $-3.1 + (-0.35)$

11. $12.48 + (-10.636)$

12. $20.25 + (-15.711)$

Subtract. Write fractions in simplest form.

③ ④ 13. $\frac{5}{8} - \left(-\frac{7}{8}\right)$

14. $\frac{1}{4} - \frac{11}{16}$

15. $-\frac{1}{2} - \left(-\frac{5}{9}\right)$

16. $-5 - \frac{5}{3}$

17. $-8\frac{3}{8} - 10\frac{1}{6}$

18. $-1 - 2.5$

19. $5.5 - 8.1$

20. $-7.34 - (-5.51)$

21. $6.673 - (-8.29)$

22. **ERROR ANALYSIS** Describe and correct the error in finding the difference.

✗ $\frac{3}{4} - \frac{9}{2} = \frac{3-9}{4-2} = \frac{-6}{2} = -3$

23. **SPORTS DRINK** Your sports drink bottle is $\frac{5}{6}$ full. After practice the bottle is $\frac{3}{8}$ full. Write the difference of the amounts after practice and before practice.

24. **BANKING** Your bank account balance is –\$20.85. You deposit \$15.50. What is your new balance?

Evaluate.

25. $2\frac{1}{6} - \left(-\frac{8}{3}\right) + \left(-4\frac{7}{9}\right)$

26. $6.3 + (-7.8) - (-2.41)$

27. $-\frac{12}{5} + \left|-\frac{13}{6}\right| + \left(-3\frac{2}{3}\right)$

28. **REASONING** When is the difference of two decimals an integer? Explain.

29. **RECIPE** A cook has $2\frac{2}{3}$ cups of flour. A recipe calls for $2\frac{3}{4}$ cups of flour. Does the cook have enough flour? If not, how much more flour is needed?

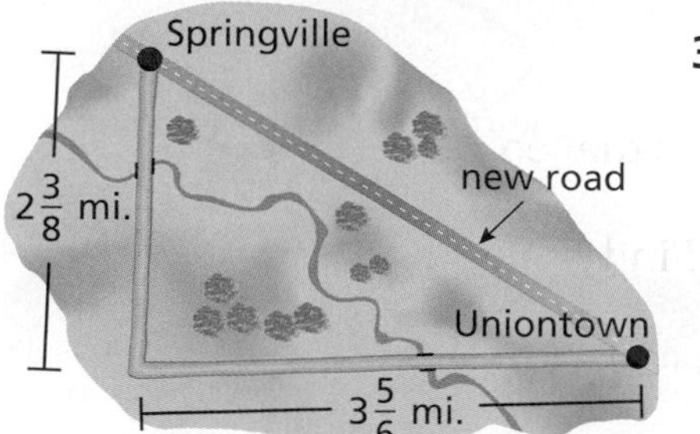

30. **ROADWAY** A new road that connects Uniontown to Springville is $4\frac{1}{3}$ miles long. What is the change in distance when using the new road instead of the dirt roads?

RAINFALL In Exercises 31–33, the bar graph shows the differences in a city's rainfall from the historical average.

31. What is the difference in rainfall between the wettest and driest months?

32. Find the sum of the differences for the year.

33. What does the sum in Exercise 32 tell you about the rainfall for the year?

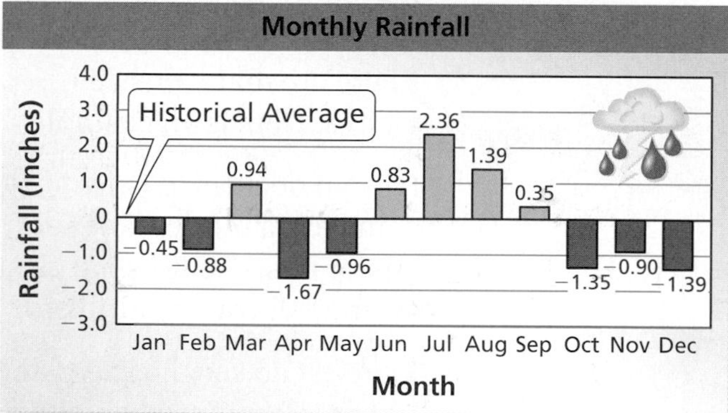

ALGEBRA Add or subtract. Write the answer in simplest form.

34. $-4x + 8x - 6x$

35. $-\frac{3n}{8} + \frac{2n}{8} - \frac{n}{8}$

36. $-4a - \frac{a}{3}$

37. $\frac{5b}{8} + \left(-\frac{2b}{3}\right)$

38. **Puzzle** Fill in the blanks to make the solution correct.

$$5.\square 4 - (\square.8\square) = -3.61$$

Fair Game Review What you learned in previous grades & lessons

Evaluate. *(Skills Review Handbook)*

39. 5.2×6.9

40. $7.2 \div 2.4$

41. $2\frac{2}{3} \times 3\frac{1}{4}$

42. $9\frac{4}{5} \div 3\frac{1}{2}$

43. **MULTIPLE CHOICE** A sports store has 116 soccer balls. Over 6 months, it sells eight soccer balls per month. How many soccer balls are in inventory at the end of the 6 months? *(Section 1.3 and Section 1.4)*

Ⓐ −48 Ⓑ 48 Ⓒ 68 Ⓓ 108

2.3 Multiplying and Dividing Rational Numbers

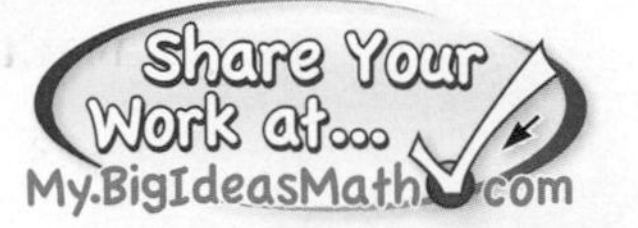

COMMON CORE STATE STANDARDS
7.NS.2a
7.NS.2c
7.NS.3

Essential Question How can you use operations with rational numbers in a story?

1 EXAMPLE: Writing a Story

Write a story that uses addition, subtraction multiplication, or division of rational numbers. Draw pictures for your story.

24 Lemons	-\$11.75
5 cups sugar	-\$1.50
30 plastic glasses	-\$1.50
18 sales (\$0.50 each)	\$9.00
PROFIT	-\$5.75

LEMONADE

All You Can Drink For 50¢!

There are many possible stories. Here is an example.

Lauryn decides to earn some extra money. She sets up a lemonade stand. To get customers, she uses big plastic glasses and makes a sign saying "All you can drink for 50¢!"

Lauryn can see that her daily profit is negative. But, she decides to keep trying. After one week, she has the same profit each day.

Sunday	Monday	Tuesday	Wednesday	Thursday	Friday	Saturday
-\$5.75	-\$5.75	-\$5.75	-\$5.75	-\$5.75	-\$5.75	-\$5.75

Lauryn is frustrated. Her profit for the first week is

$$7(-5.75) = (-5.75) + (-5.75) + (-5.75) + (-5.75) + (-5.75) + (-5.75) + (-5.75)$$
$$= -40.25.$$

She realizes that she has too many customers who are drinking a second and even a third glass of lemonade. So, she decides to try a new strategy. Soon, she has a customer. He buys a glass of lemonade and drinks it.

He hands the empty glass to Lauryn and says "*That was great. I'll have another glass.*" Today, Lauryn says "*That will be 50¢ more, please.*" The man says "*But, you only gave me one glass and the sign says 'All you can drink for 50¢!'*" Lauryn replies, "*One glass IS all you can drink for 50¢.*"

With her new sales strategy, Lauryn starts making a profit of \$8.25 per day. Her profit for the second week is

$$7(8.25) = (8.25) + (8.25) + (8.25) + (8.25) + (8.25) + (8.25) + (8.25) = 57.75.$$

Her profit for the two weeks is $-40.25 + 57.75 = \$17.50$. So, Lauryn has made some money. She decides that she is on the right track.

2 ACTIVITY: Writing a Story

Work with a partner. Write a story that uses addition, subtraction, multiplication, or division of rational numbers.

- At least one of the numbers in the story has to be negative and *not* an integer.
- Draw pictures to help illustrate what is happening in the story.
- Include the solution of the problem in the story.

If you are having trouble thinking of a story, here are some common uses of negative numbers.

- A profit of $-\$15$ is a loss of \$15.
- An elevation of -100 feet is a depth of 100 feet below sea level.
- A gain of -5 yards in football is a loss of 5 yards.
- A score of -4 in golf is 4 strokes under par.
- A balance of $-\$25$ in your checking account means the account is overdrawn by \$25.

What Is Your Answer?

3. **IN YOUR OWN WORDS** How can you use operations with rational numbers in a story? You already used rational numbers in your story. Describe another use of a negative rational number in a story.

PUZZLE Read the cartoon. Fill in the blanks using 4s or 8s to make the equation true.

"Dear Mom, I'm in a hurry. To save time I won't be typing any 4's or 8's."

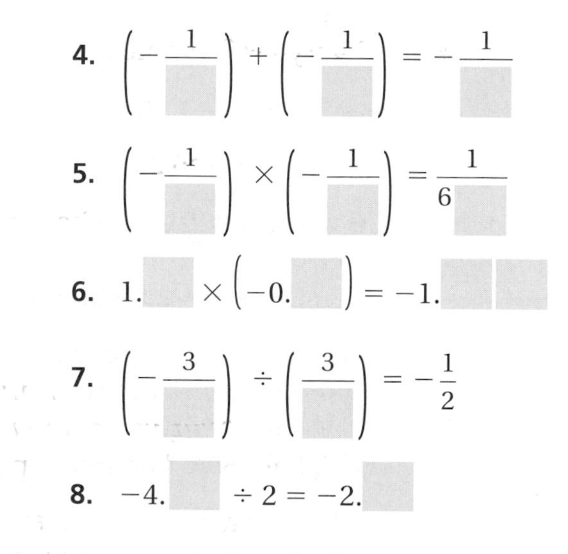

4. $\left(-\dfrac{1}{\square}\right) + \left(-\dfrac{1}{\square}\right) = -\dfrac{1}{\square}$

5. $\left(-\dfrac{1}{\square}\right) \times \left(-\dfrac{1}{\square}\right) = \dfrac{1}{6\square}$

6. $1.\square \times \left(-0.\square\right) = -1.\square\square$

7. $\left(-\dfrac{3}{\square}\right) \div \left(\dfrac{3}{\square}\right) = -\dfrac{1}{2}$

8. $-4.\square \div 2 = -2.\square$

2.3 Lesson

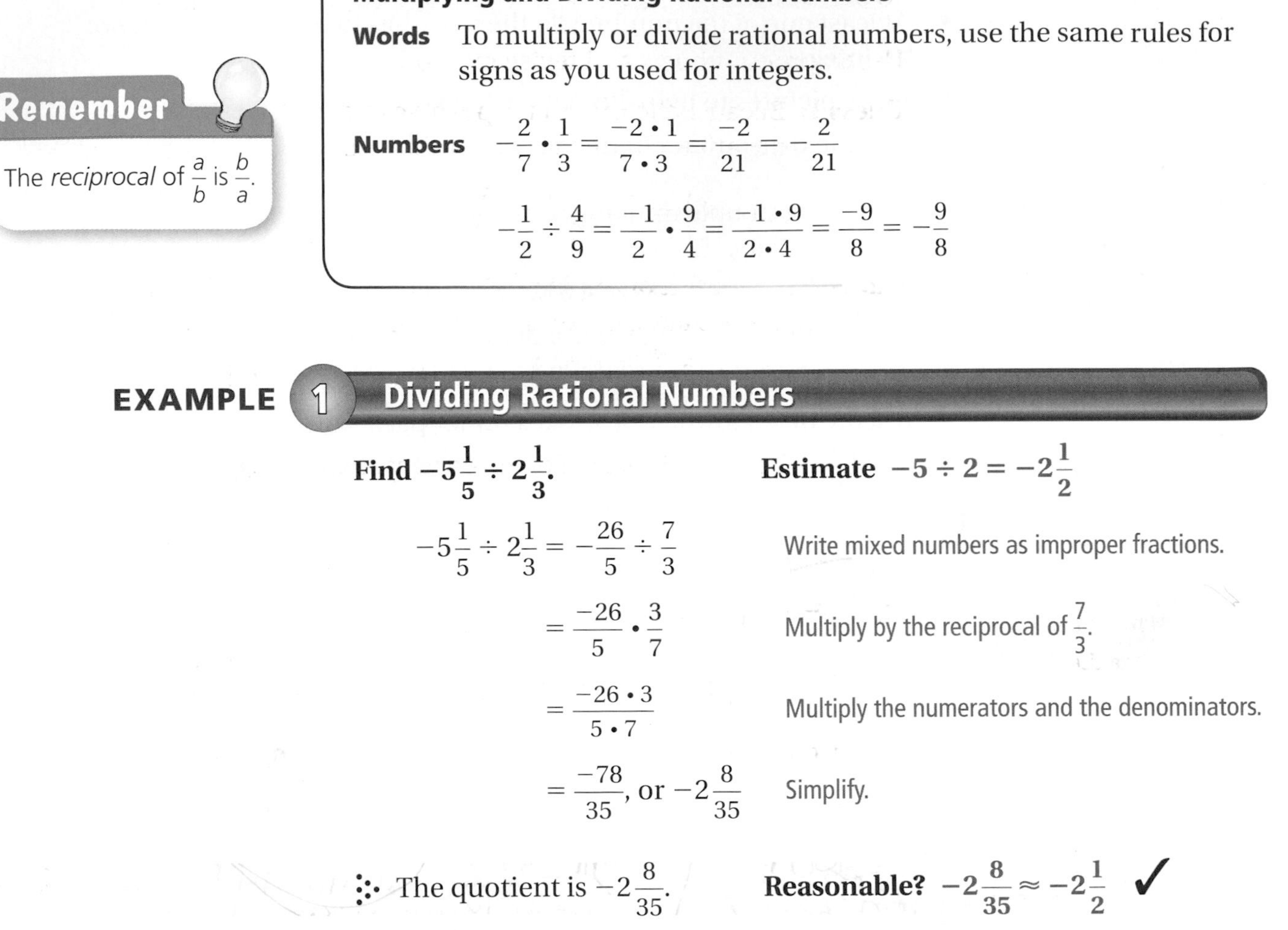

Multiplying and Dividing Rational Numbers

Words To multiply or divide rational numbers, use the same rules for signs as you used for integers.

Numbers $-\frac{2}{7} \cdot \frac{1}{3} = \frac{-2 \cdot 1}{7 \cdot 3} = \frac{-2}{21} = -\frac{2}{21}$

$-\frac{1}{2} \div \frac{4}{9} = \frac{-1}{2} \cdot \frac{9}{4} = \frac{-1 \cdot 9}{2 \cdot 4} = \frac{-9}{8} = -\frac{9}{8}$

Remember

The *reciprocal* of $\frac{a}{b}$ is $\frac{b}{a}$.

EXAMPLE 1 Dividing Rational Numbers

Find $-5\frac{1}{5} \div 2\frac{1}{3}$. **Estimate** $-5 \div 2 = -2\frac{1}{2}$

$-5\frac{1}{5} \div 2\frac{1}{3} = -\frac{26}{5} \div \frac{7}{3}$ Write mixed numbers as improper fractions.

$= \frac{-26}{5} \cdot \frac{3}{7}$ Multiply by the reciprocal of $\frac{7}{3}$.

$= \frac{-26 \cdot 3}{5 \cdot 7}$ Multiply the numerators and the denominators.

$= \frac{-78}{35}$, or $-2\frac{8}{35}$ Simplify.

∴ The quotient is $-2\frac{8}{35}$. **Reasonable?** $-2\frac{8}{35} \approx -2\frac{1}{2}$ ✓

EXAMPLE 2 Multiplying Rational Numbers

Find $-2.5 \cdot 3.6$.

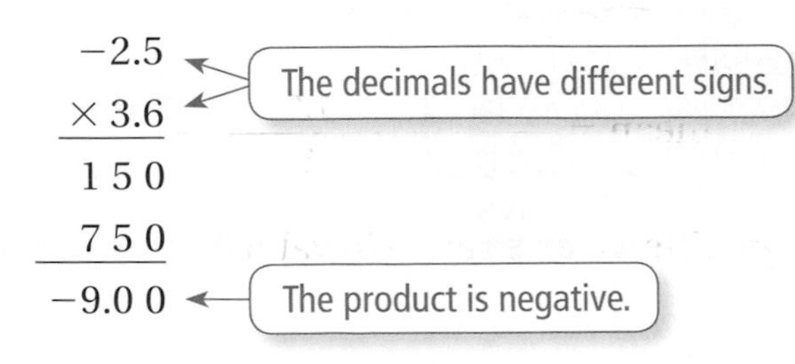

$$\begin{array}{r} -2.5 \\ \times\ 3.6 \\ \hline 1\,5\,0 \\ 7\,5\,0 \\ \hline -9.0\,0 \end{array}$$

The decimals have different signs.

The product is negative.

∴ The product is -9.

EXAMPLE 3 Standardized Test Practice

Which number, when multiplied by $-\frac{5}{3}$, gives a product between 5 and 6?

Ⓐ -6 Ⓑ $-3\frac{1}{4}$ Ⓒ $-\frac{1}{4}$ Ⓓ 3

Use the guess, check, and revise method.

Guess 1: Because the product is positive and the known factor is negative, choose a number that is negative. Try Choice Ⓒ.

$$-\frac{1}{4}\left(-\frac{5}{3}\right) = \frac{-1 \cdot (-5)}{4 \cdot 3} = \frac{5}{12}$$

Guess 2: The result of Choice Ⓒ is not between 5 and 6. So, choose another number that is negative. Try Choice Ⓑ.

$$-3\frac{1}{4}\left(-\frac{5}{3}\right) = -\frac{13}{4}\left(-\frac{5}{3}\right) = \frac{-13 \cdot (-5)}{4 \cdot 3} = \frac{65}{12} = 5\frac{5}{12}$$

∴ $5\frac{5}{12}$ is between 5 and 6. So, the correct answer is Ⓑ.

On Your Own

Now You're Ready
Exercises 10–33

Multiply or divide.

1. $-\frac{6}{5} \div \left(-\frac{1}{2}\right)$
2. $\frac{1}{3} \div \left(-2\frac{2}{3}\right)$
3. $\left(-\frac{1}{2}\right)^3$
4. $1.8(-5.1)$
5. $-6.3(-0.6)$
6. $(-1.3)^2$

EXAMPLE 4 Real-Life Application

Account Positions

Stock	Original Value	Current Value	Change
A	600.54	420.15	−180.39
B	391.10	518.38	127.28
C	380.22	99.70	−280.52

An investor owns stocks A, B, and C. What is the mean change in value of the stocks?

$$\text{mean} = \frac{-180.39 + 127.28 + (-280.52)}{3} = \frac{-333.63}{3} = -111.21$$

∴ The mean change in value of the stocks is −\$111.21.

On Your Own

7. In Example 4, the change in value of stock D is \$568.23. What is the mean change in value of the four stocks?

2.3 Exercises

Vocabulary and Concept Check

1. **WRITING** How is multiplying and dividing rational numbers similar to multiplying and dividing integers?

Find the reciprocal.

2. $-\frac{2}{5}$
3. -3
4. $\frac{16}{9}$
5. $-2\frac{1}{3}$

Tell whether the expression is *positive* or *negative* without evaluating.

6. $-\frac{3}{10} \times \left(-\frac{8}{15}\right)$
7. $1\frac{1}{2} \div \left(-\frac{1}{4}\right)$
8. -6.2×8.18
9. $\frac{-8.16}{-2.72}$

Practice and Problem Solving

Divide. Write fractions in simplest form.

(1) 10. $-\frac{7}{10} \div \frac{2}{5}$
11. $\frac{1}{4} \div \left(-\frac{3}{8}\right)$
12. $-\frac{8}{9} \div \left(-\frac{8}{9}\right)$
13. $-\frac{1}{5} \div 20$
14. $-2\frac{4}{5} \div (-7)$
15. $-10\frac{2}{7} \div \left(-4\frac{4}{11}\right)$
16. $-9 \div 7.2$
17. $8 \div 2.2$
18. $-3.45 \div (-15)$
19. $-0.18 \div 0.03$
20. $8.722 \div (-3.56)$
21. $12.42 \div (-4.8)$

Multiply. Write fractions in simplest form.

(2) (3) 22. $-\frac{2}{3} \times \frac{2}{9}$
23. $-\frac{1}{4} \times \left(-\frac{4}{3}\right)$
24. $\frac{5}{6}\left(-\frac{8}{15}\right)$
25. $-2\left(-1\frac{1}{4}\right)$
26. $-3\frac{1}{3} \cdot \left(-2\frac{7}{10}\right)$
27. $\left(-1\frac{2}{3}\right)^3$
28. $0.4 \times (-0.03)$
29. $-0.05 \times (-0.5)$
30. $-8(0.09)$
31. $-9.3 \cdot (-5.1)$
32. $-95.2 \cdot (-0.12)$
33. $(-0.4)^3$

ERROR ANALYSIS Describe and correct the error.

34. ✗ $-2.2 \times 3.7 = 8.14$

35. ✗ $-\frac{1}{4} \div \frac{3}{2} = -\frac{4}{1} \times \frac{3}{2} = -\frac{12}{2} = -6$

36. **HOUR HAND** The hour hand of a clock moves $-30°$ every hour. How many degrees does it move in $2\frac{1}{5}$ hours?

37. **SUNFLOWER SEEDS** How many 0.75-pound packages can be made with 6 pounds of sunflower seeds?

Evaluate.

38. $-4.2 + 8.1 \times (-1.9)$

39. $2.85 - 6.2 \div 2^2$

40. $-3.64 \cdot |-5.3| - 1.5^3$

41. $1\frac{5}{9} \div \left(-\frac{2}{3}\right) + \left(-2\frac{3}{5}\right)$

42. $-3\frac{3}{4} \times \frac{5}{6} - 2\frac{1}{3}$

43. $\left(-\frac{2}{3}\right)^2 - \frac{3}{4}\left(2\frac{1}{3}\right)$

44. OPEN-ENDED Write two fractions whose product is $-\frac{3}{5}$.

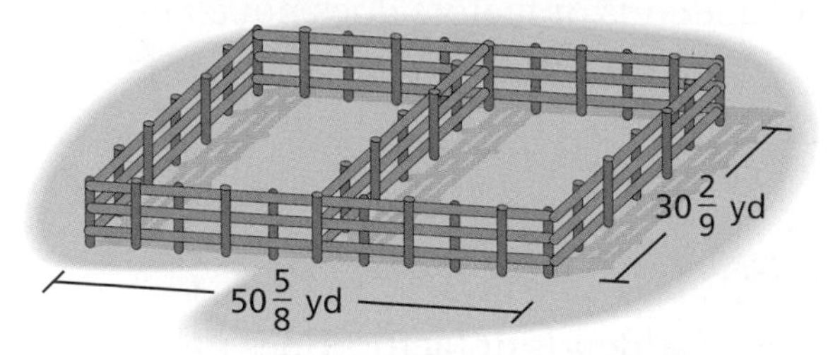

45. FENCING A farmer needs to enclose two adjacent rectangular pastures. How much fencing does the farmer need?

46. GASOLINE A 14.5-gallon gasoline tank is $\frac{3}{4}$ full. How many gallons will it take to fill the tank?

47. BOARDWALK A section of a boardwalk is made using 15 boards. Each board is $9\frac{1}{4}$ inches wide. The total width of the section is 144 inches. The spacing between each board is equal. What is the width of the spacing between each board?

48. RUNNING The table shows the changes in the times (in seconds) of four teammates. What is the mean change?

Teammate	Change
1	−2.43
2	−1.85
3	0.61
4	−1.45

49. Repeated Reasoning Consider $(-2)^1, (-2)^2, (-2)^3, (-2)^4, (-2)^5$, and $(-2)^6$.

a. Evaluate each expression.

b. What pattern do you notice?

c. What is the sign of $(-2)^{49}$?

Fair Game Review What you learned in previous grades & lessons

Add or subtract. *(Section 2.2)*

50. $-6.2 + 4.7$

51. $-8.1 - (-2.7)$

52. $\frac{9}{5} - \left(-2\frac{7}{10}\right)$

53. $-4\frac{5}{6} + \left(-3\frac{4}{9}\right)$

54. MULTIPLE CHOICE What are the coordinates of the point in quadrant IV? *(Section 1.6)*

Ⓐ $(-4, 1)$ Ⓑ $(-3, -3)$

Ⓒ $(0, -2)$ Ⓓ $(3, -3)$

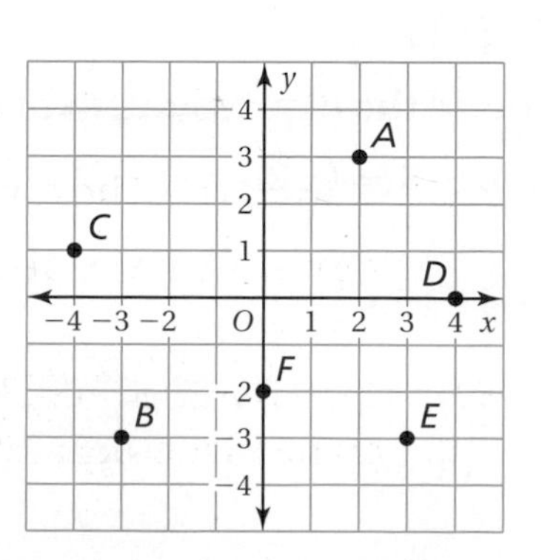

2.3b Number Properties

Key Ideas

Commutative Properties

Words Changing the order of addends or factors does not change the sum or product.

Numbers $-5 + 8 = 8 + (-5)$

$-5 \cdot 8 = 8 \cdot (-5)$

Associative Properties

Words Changing the grouping of addends or factors does not change the sum or product.

Numbers $(7 + 4) + 2 = 7 + (4 + 2)$

$(7 \cdot 4) \cdot 2 = 7 \cdot (4 \cdot 2)$

EXAMPLE 1 Using Commutative and Associative Properties

Study Tip

Use number properties to group numbers that are easy to add or multiply.

a. Evaluate 4.7 + 6 + 3.3.

$4.7 + 6 + 3.3 = 6 + 4.7 + 3.3$	Commutative Property of Addition
$= 6 + (4.7 + 3.3)$	Associative Property of Addition
$= 6 + 8$	Add 4.7 and 3.3.
$= 14$	Add 6 and 8.

b. Evaluate $(7 \cdot 4)\frac{1}{4}$.

$(7 \cdot 4)\frac{1}{4} = 7\left(4 \cdot \frac{1}{4}\right)$	Associative Property of Multiplication
$= 7(1)$	Multiply 4 and $\frac{1}{4}$.
$= 7$	Multiplication Property of One

Practice

Evaluate the expression. Explain each step.

1. $2 + 3 + (-2)$
2. $12 + 6.1 + 5.9$
3. $4 \cdot 19 \cdot \frac{1}{2}$
4. $\frac{1}{3} \cdot 2 \cdot \frac{1}{2}$
5. $5\left(\frac{7}{8} \cdot \frac{2}{5}\right)$
6. $-1.45 + (-8.55 + 2.7)$

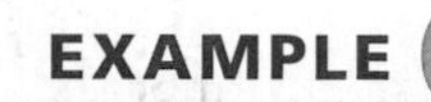

EXAMPLE 2 Real-Life Application

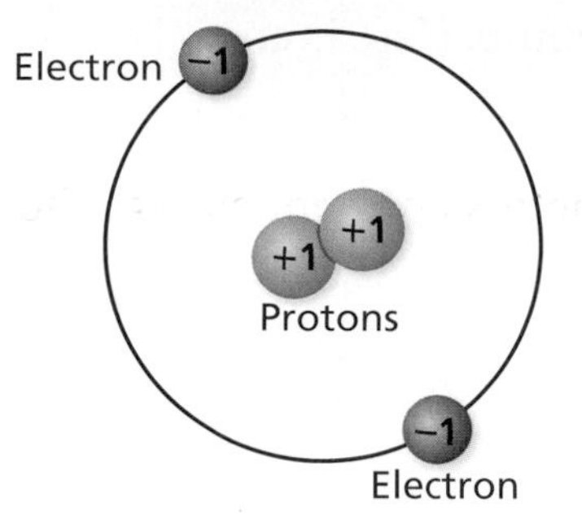

A helium atom has positively-charged protons and negatively-charged electrons, as shown in the diagram. The sum of the charges gives the charge of the helium atom. Find the charge of the atom.

Charge of atom = Charge of protons + Charge of electrons

$= (+1) + (+1) + (-1) + (-1)$ Substitute.

$= (+2) + (-2)$ Simplify.

$= 0$ Additive Inverse Property

∴ The protons and the electrons are oppositely charged. So, the helium atom has a charge of 0.

Practice

7. **SCIENCE** A lithium atom has positively-charged protons and negatively-charged electrons. The sum of the charges gives the charge of the lithium atom. Find the charge of the atom.

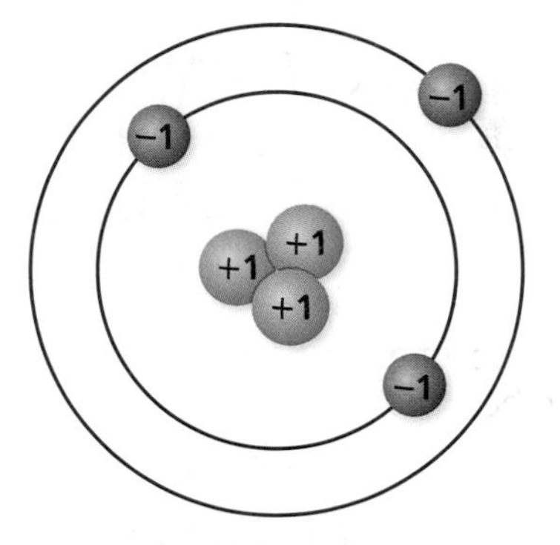

8. **BANK ACCOUNT** The table shows four bank account transactions. Find the change in the account balance.

Transactions	
Deposit	$75
Groceries	−$33.50
Withdrawal	−$75
Deposit	$63.50

9. **VIDEO GAME** The table shows ways to gain and lose points in a video game. List a series of at least three events that results in 0 points.

Event	Points	Event	Points
Find treasure	+100	Find a map	+50
Find a gold coin	+10	Lose a coin	−10
Lose a map	−50	Find a silver coin	+5
Lose a compass	−25	Lose treasure	−100

2 Study Help

Check It Out
Graphic Organizer
BigIdeasMath.com

You can use a **process diagram** to show the steps involved in a procedure. Here is an example of a process diagram for adding rational numbers.

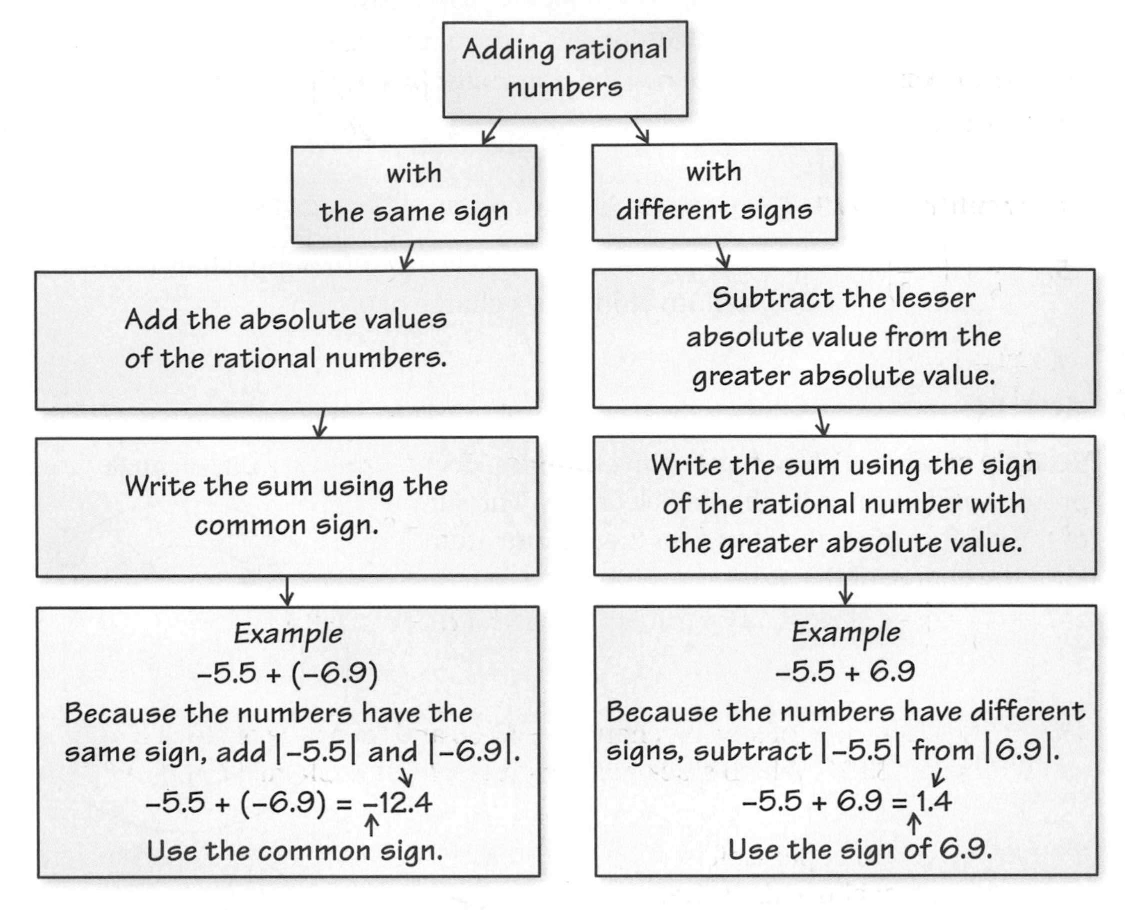

Adding rational numbers
with the same sign
with different signs
Add the absolute values of the rational numbers.
Subtract the lesser absolute value from the greater absolute value.
Write the sum using the common sign.
Write the sum using the sign of the rational number with the greater absolute value.
Example
−5.5 + (−6.9)
Because the numbers have the same sign, add |−5.5| and |−6.9|.
−5.5 + (−6.9) = −12.4
Use the common sign.
Example
−5.5 + 6.9
Because the numbers have different signs, subtract |−5.5| from |6.9|.
−5.5 + 6.9 = 1.4
Use the sign of 6.9.

On Your Own

Make a process diagram with examples to help you study these topics. Your process diagram can have one or more branches.

1. writing rational numbers as decimals
2. subtracting rational numbers
3. dividing rational numbers

After you complete this chapter, make process diagrams with examples for the following topics.

4. solving equations using addition or subtraction
5. solving equations using multiplication or division
6. solving two-step equations

"Does this process diagram accurately show how a cat claws furniture?"

2.1–2.3 Quiz

Write the rational number as a decimal. *(Section 2.1)*

1. $-\frac{3}{20}$

2. $-\frac{11}{6}$

Write the decimal as a fraction or mixed number in simplest form. *(Section 2.1)*

3. -0.325

4. -1.28

Add or subtract. Write fractions in simplest form. *(Section 2.2)*

5. $-\frac{4}{5} + \left(-\frac{3}{8}\right)$

6. $-5.8 + 2.6$

7. $\frac{12}{7} - \left(-\frac{2}{9}\right)$

8. $9.1 - 12.9$

Multiply or divide. Write fractions in simplest form. *(Section 2.3)*

9. $-2\frac{3}{8} \times \frac{8}{5}$

10. $-9.4 \times (-4.7)$

11. $-8\frac{5}{9} \div \left(-1\frac{4}{7}\right)$

12. $-8.4 \div 2.1$

13. STOCK The value of stock A changes $-\$3.68$ and the value of stock B changes $-\$3.72$. Which stock has the greater loss? Explain. *(Section 2.1)*

14. PARASAILING A parasail is at 200.6 feet above the water. After five minutes, the parasail is at 120.8 feet above the water. What is the change in height of the parasail? *(Section 2.2)*

15. FOOTBALL The table shows the statistics of a running back in a football game. How many total yards did he gain? *(Section 2.2)*

Quarter	1	2	3	4	Total
Yards	$-8\frac{1}{2}$	23	$42\frac{1}{2}$	$-2\frac{1}{4}$	?

16. LATE FEES You were overcharged \$4.52 on your cell phone bill three months in a row. The cell phone company will add $-\$4.52$ to your next bill for each month you were overcharged. How much will be added to your next bill? *(Section 2.3)*

2.4 Solving Equations Using Addition or Subtraction

Essential Question How can you use inverse operations to solve an equation?

COMMON CORE STATE STANDARDS
7.EE.4a
8.EE.7a
8.EE.7b

Key: [variable tile] = Variable   [+] = 1   [−] = −1   [+][−] = Zero Pair

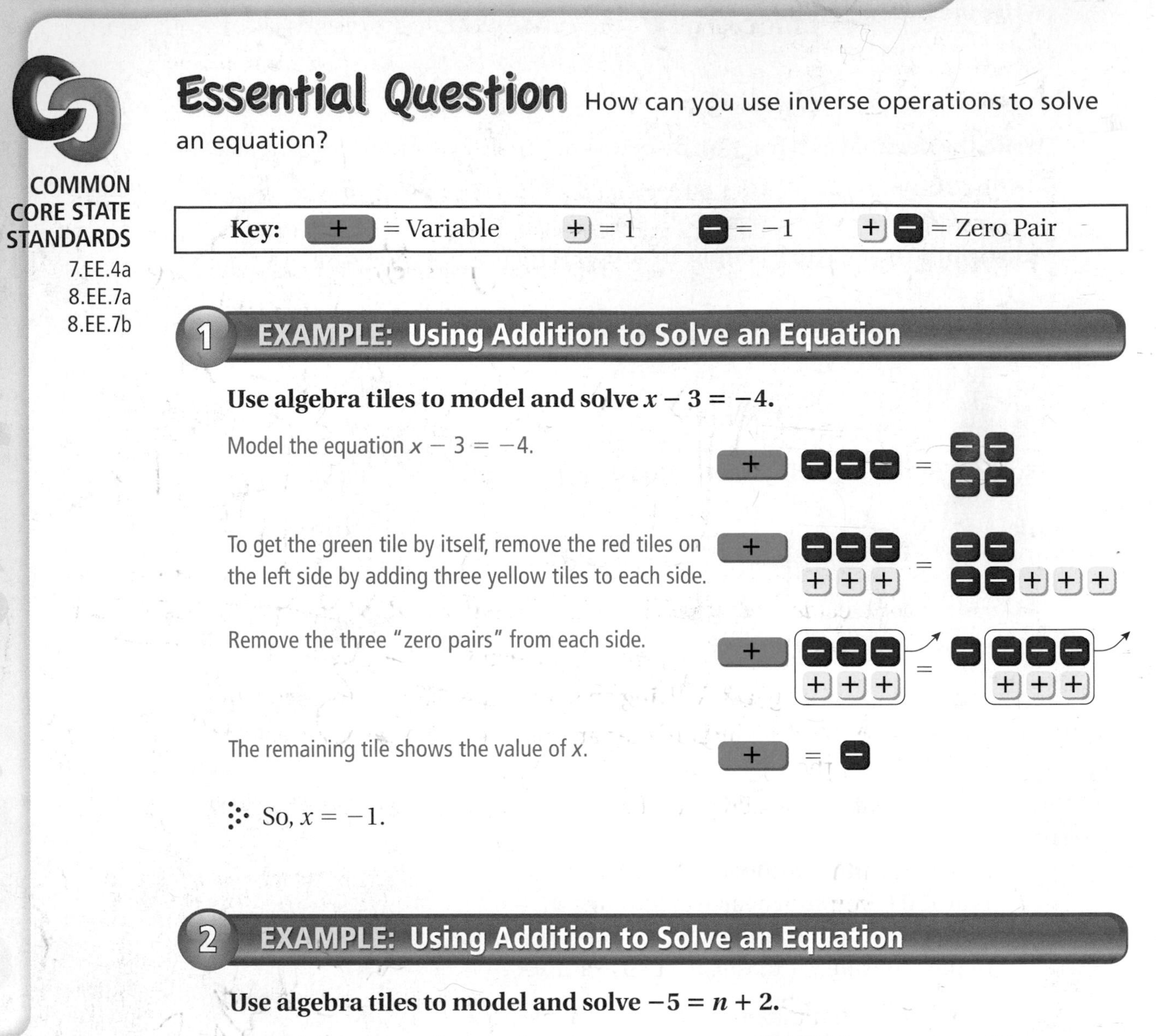

1 EXAMPLE: Using Addition to Solve an Equation

Use algebra tiles to model and solve $x - 3 = -4$.

Model the equation $x - 3 = -4$.

To get the green tile by itself, remove the red tiles on the left side by adding three yellow tiles to each side.

Remove the three "zero pairs" from each side.

The remaining tile shows the value of x.

So, $x = -1$.

2 EXAMPLE: Using Addition to Solve an Equation

Use algebra tiles to model and solve $-5 = n + 2$.

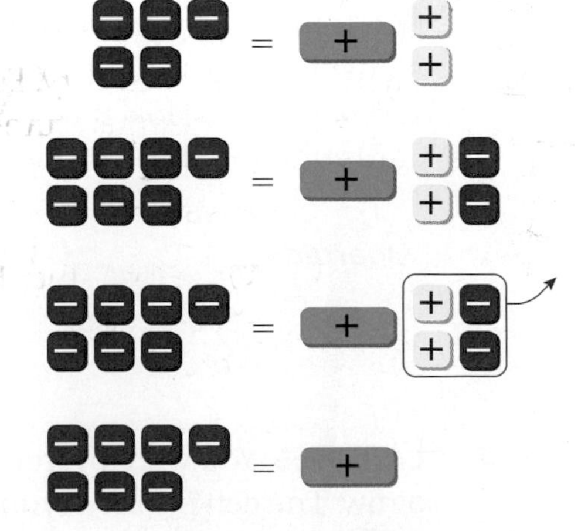

Model the equation $-5 = n + 2$.

Remove the yellow tiles on the right side by adding two red tiles to each side.

Remove the two "zero pairs" from the right side.

The remaining tiles show the value of n.

So, $-7 = n$ or $n = -7$.

3 ACTIVITY: Solving Equations Using Algebra Tiles

Work with a partner. Use algebra tiles to model and solve the equation.

a. $y + 10 = -5$

b. $p - 7 = -3$

c. $-15 = t - 5$

d. $8 = 12 + z$

4 ACTIVITY: Writing and Solving Equations

Work with a partner. Write an equation shown by the algebra tiles. Then solve.

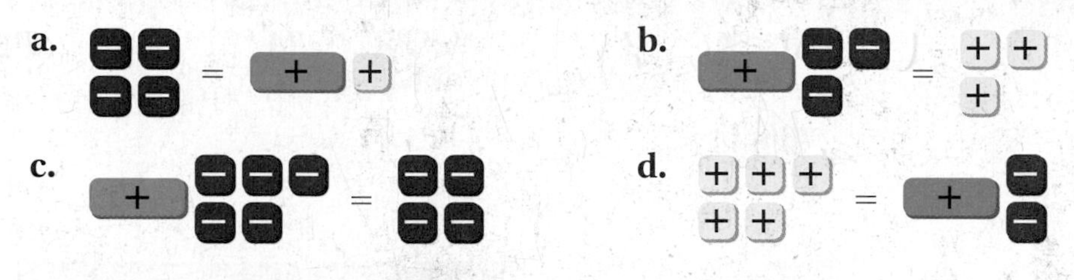

What Is Your Answer?

5. Decide whether the statement is *true* or *false*. Explain your reasoning.

a. In an equation, any letter can be used as a variable.

b. The goal in solving an equation is to get the variable by itself.

c. In the solution, the variable always has to be on the left side of the equal sign.

d. If you add a number to one side, you should add it to the other side.

6. **IN YOUR OWN WORDS** How can you use inverse operations to solve an equation without algebra tiles? Give two examples.

7. What makes the cartoon funny?

8. The word *variable* comes from the word *vary*. For example, the temperature in Maine varies a lot from winter to summer.

Write two other English sentences that use the word *vary*.

"To vary or not to vary." That is the question.

"Dear Sir: Yesterday you said $x = 2$. Today you are saying $x = 3$. Please make up your mind."

Practice Use what you learned about solving equations using inverse operations to complete Exercises 5–8 on page 74.

Check It Out
Lesson Tutorials
BigIdeasMath.com

Key Vocabulary
equivalent equations, *p. 72*

Key Ideas

Addition Property of Equality

Words Two equations are **equivalent equations** if they have the same solutions. Adding the same number to each side of an equation produces an equivalent equation.

Algebra If $a = b$, then $a + c = b + c$.

Subtraction Property of Equality

Words Subtracting the same number from each side of an equation produces an equivalent equation.

Algebra If $a = b$, then $a - c = b - c$.

EXAMPLE 1 Solving Equations

Remember

To solve equations, use *inverse operations* that "undo" each other. For example, use addition to solve an equation with subtraction.

a. Solve $x - 5 = -1$.

$x - 5 = -1$ Write the equation.

$\underline{+5} \quad \underline{+5}$ Add 5 to each side.

$x = 4$ Simplify.

So, the solution is $x = 4$.

Check

$x - 5 = -1$

$4 - 5 \stackrel{?}{=} -1$

$-1 = -1$ ✓

b. Solve $z + \frac{3}{2} = \frac{1}{2}$.

$z + \frac{3}{2} = \frac{1}{2}$ Write the equation.

$\underline{-\frac{3}{2}} \quad \underline{-\frac{3}{2}}$ Subtract $\frac{3}{2}$ from each side.

$z = -1$ Simplify.

So, the solution is $z = -1$.

On Your Own

Now You're Ready
Exercises 5–20

Solve the equation. Check your solution.

1. $p - 5 = -2$
2. $w + 13.2 = 10.4$
3. $x - \frac{5}{6} = -\frac{1}{6}$

Multi-Language Glossary at BigIdeasMath.com.

EXAMPLE 2 Standardized Test Practice

A company has a profit of $750 this week. This profit is $900 more than the profit *P* last week. Which equation can be used to find *P*?

(A) $750 = 900 - P$ (B) $750 = P + 900$

(C) $900 = P - 750$ (D) $900 = P + 750$

Words	The profit this week	is	$900 more than		the profit last week.
Equation	750	=	P	+	900

The equation is $750 = P + 900$. The correct answer is (B).

On Your Own

Now You're Ready
Exercises 22–25

4. A company has a profit of $120.50 today. This profit is $145.25 less than the profit *P* yesterday. Write an equation that can be used to find *P*.

EXAMPLE 3 Real-Life Application

The line graph shows the scoring while you and your friend played a video game. Write and solve an equation to find your score after Level 4.

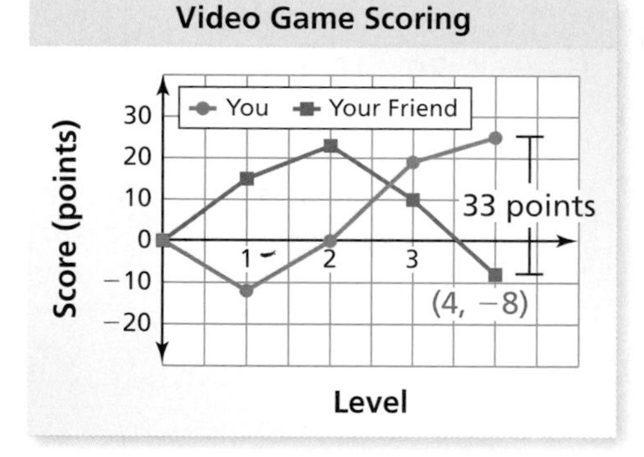

You can determine the following from the graph.

Words	Your friend's score	is	33 points	less than	your score.
Variable	Let s be your score after Level 4.				
Equation	-8	=	s	−	33

$$
\begin{aligned}
-8 &= s - 33 && \text{Write equation.}\\
+33 &\quad\ \ +33 && \text{Add 33 to each side.}\\
25 &= s && \text{Simplify.}
\end{aligned}
$$

Your score after Level 4 is 25 points.

Reasonable? From the graph, your score after Level 4 is between 20 points and 30 points. So, 25 points is a reasonable answer.

On Your Own

5. **WHAT IF?** In Example 3, you have −12 points after Level 1. Your score is 27 points less than your friend's score. What is your friend's score?

2.4 Exercises

Vocabulary and Concept Check

1. **VOCABULARY** What property would you use to solve $m + 6 = -4$?
2. **VOCABULARY** Name two inverse operations.
3. **WRITING** Are the equations $m + 3 = -5$ and $m = -2$ equivalent? Explain.
4. **WHICH ONE DOESN'T BELONG?** Which equation does *not* belong with the other three? Explain your reasoning.

$x + 3 = -1$ $\qquad$ $x + 1 = -5$ $\qquad$ $x - 2 = -6$ $\qquad$ $x - 9 = -13$

Practice and Problem Solving

Solve the equation. Check your solution.

(1)

5. $a - 6 = 13$
6. $-3 = z - 8$
7. $-14 = k + 6$
8. $x + 4 = -14$
9. $c - 7.6 = -4$
10. $-10.1 = w + 5.3$
11. $\frac{1}{2} = q + \frac{2}{3}$
12. $p - 3\frac{1}{6} = -2\frac{1}{2}$
13. $g - 9 = -19$
14. $-9.3 = d - 3.4$
15. $4.58 + y = 2.5$
16. $x - 5.2 = -18.73$
17. $q + \frac{5}{9} = \frac{1}{6}$
18. $-2\frac{1}{4} = r - \frac{4}{5}$
19. $w + 3\frac{3}{8} = 1\frac{5}{6}$
20. $4\frac{2}{5} + k = -3\frac{2}{11}$

21. **ERROR ANALYSIS** Describe and correct the error in finding the solution.

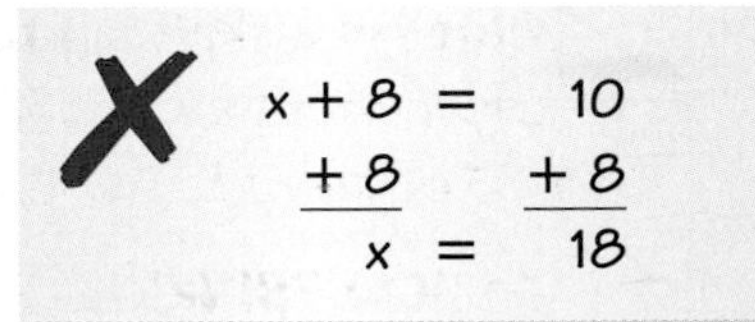

Write the verbal sentence as an equation. Then solve.

(2)

22. 4 less than a number n is -15.
23. 10 more than a number c is 3.
24. The sum of a number y and -3 is -8.
25. The difference between a number p and 6 is -14.

In Exercises 26–28, write an equation. Then solve.

26. **DRY ICE** The temperature of dry ice is -109.3°F. This is 184.9°F less than the outside temperature. What is the outside temperature?
27. **PROFIT** A company makes a profit of \$1.38 million. This is \$2.54 million more than last year. What was the profit last year?
28. **PIER** The difference between the lengths of a paddle boat and a pier is $-7\frac{3}{4}$ feet. The pier is $18\frac{1}{2}$ feet long. How long is the paddle boat?

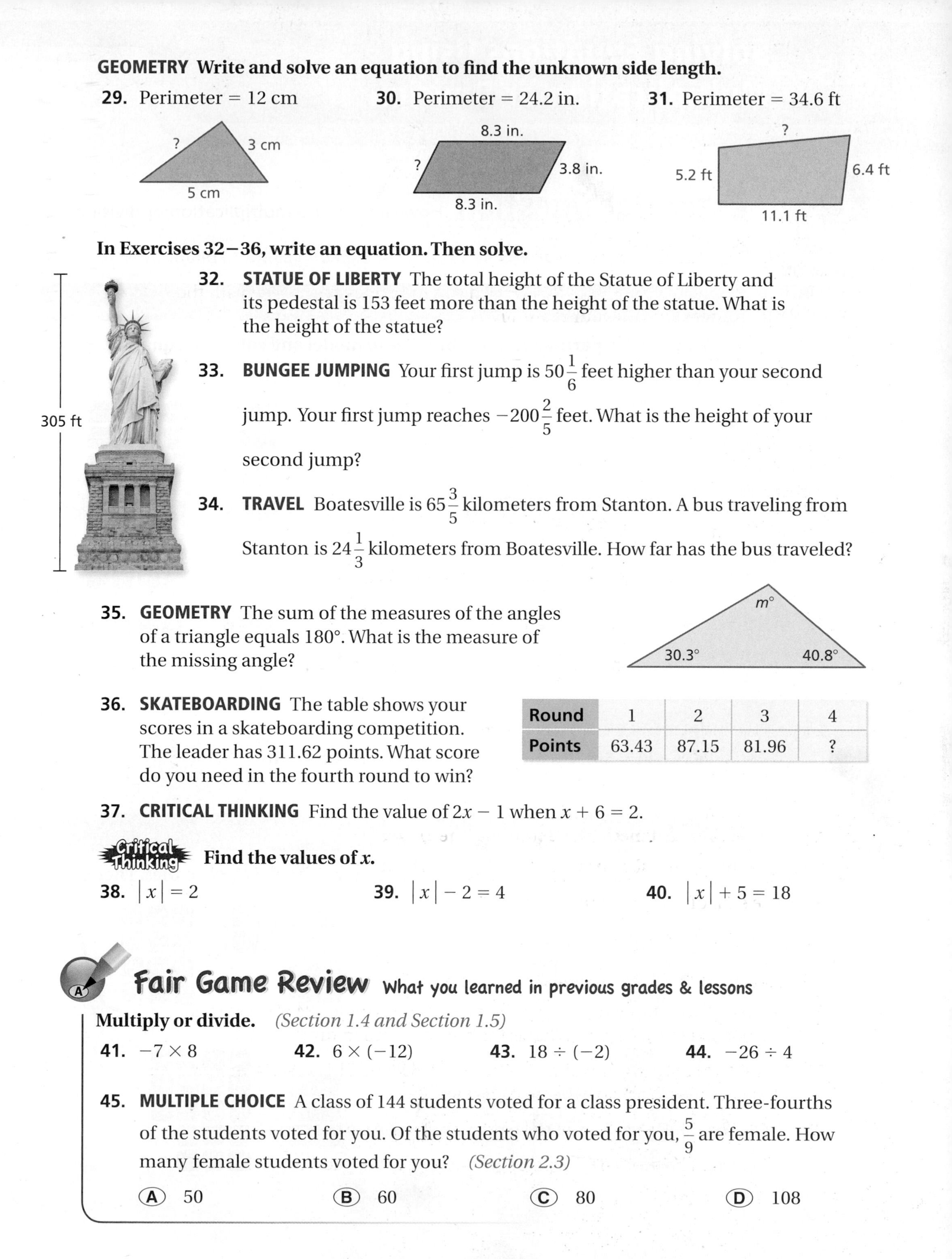

GEOMETRY Write and solve an equation to find the unknown side length.

29. Perimeter = 12 cm

30. Perimeter = 24.2 in.

31. Perimeter = 34.6 ft

In Exercises 32–36, write an equation. Then solve.

32. STATUE OF LIBERTY The total height of the Statue of Liberty and its pedestal is 153 feet more than the height of the statue. What is the height of the statue?

33. BUNGEE JUMPING Your first jump is $50\frac{1}{6}$ feet higher than your second jump. Your first jump reaches $-200\frac{2}{5}$ feet. What is the height of your second jump?

34. TRAVEL Boatesville is $65\frac{3}{5}$ kilometers from Stanton. A bus traveling from Stanton is $24\frac{1}{3}$ kilometers from Boatesville. How far has the bus traveled?

35. GEOMETRY The sum of the measures of the angles of a triangle equals 180°. What is the measure of the missing angle?

36. SKATEBOARDING The table shows your scores in a skateboarding competition. The leader has 311.62 points. What score do you need in the fourth round to win?

Round	1	2	3	4
Points	63.43	87.15	81.96	?

37. CRITICAL THINKING Find the value of $2x - 1$ when $x + 6 = 2$.

Critical Thinking **Find the values of x.**

38. $|x| = 2$

39. $|x| - 2 = 4$

40. $|x| + 5 = 18$

Fair Game Review What you learned in previous grades & lessons

Multiply or divide. *(Section 1.4 and Section 1.5)*

41. -7×8

42. $6 \times (-12)$

43. $18 \div (-2)$

44. $-26 \div 4$

45. MULTIPLE CHOICE A class of 144 students voted for a class president. Three-fourths of the students voted for you. Of the students who voted for you, $\frac{5}{9}$ are female. How many female students voted for you? *(Section 2.3)*

Ⓐ 50 Ⓑ 60 Ⓒ 80 Ⓓ 108

2.5 Solving Equations Using Multiplication or Division

COMMON CORE STATE STANDARDS
7.EE.1
7.EE.2
7.EE4a
8.EE.7a
8.EE.7b

Essential Question How can you use multiplication or division to solve an equation?

1 ACTIVITY: Using Division to Solve an Equation

Work with a partner. Use algebra tiles to model and solve the equation.

a. Sample: $3x = -12$

Model the equation $3x = -12$.

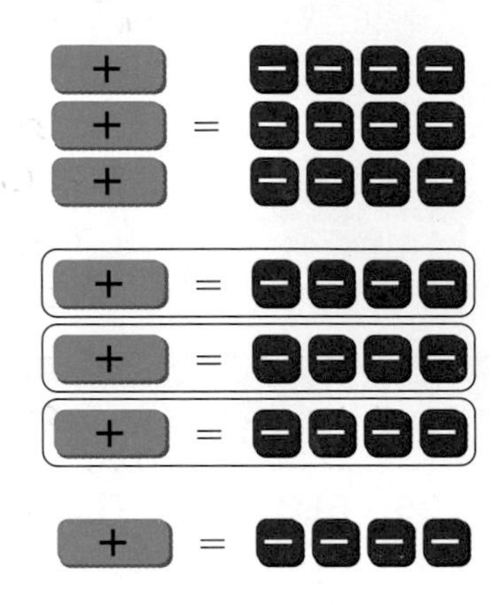

Your goal is to get one green tile by itself. Because there are three green tiles, divide the red tiles into three equal groups.

Keep one of the groups. This shows the value of x.

So, $x = -4$.

b. $2k = -8$

c. $-15 = 3t$

d. $-20 = 5m$

e. $4h = -16$

2 ACTIVITY: Writing and Solving Equations

Work with a partner. Write an equation shown by the algebra tiles. Then solve.

a. **b.** **c.** **d.**

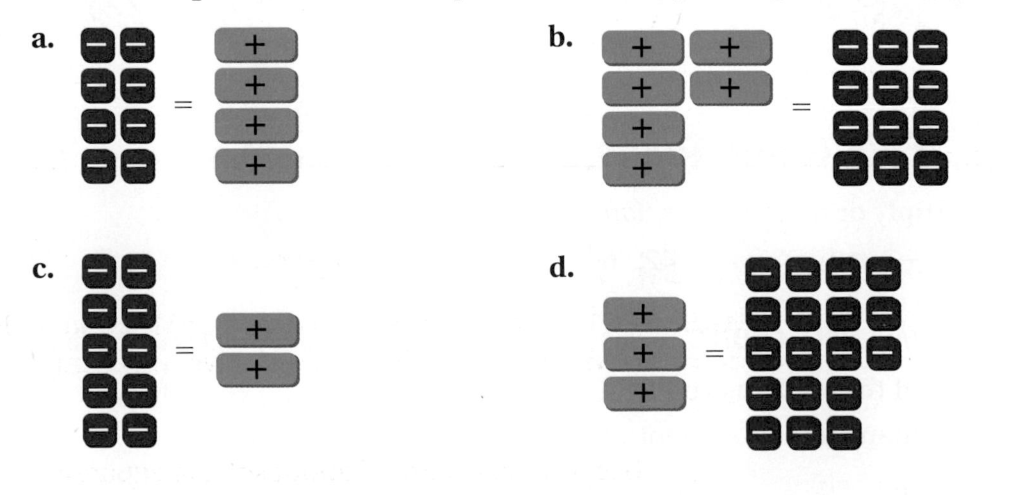

3 ACTIVITY: The Game of Math Card War

Preparation:

- Cut index cards to make 40 playing cards.
- Write each equation in the table on a card.

To Play:

- Play with a partner. Deal 20 cards to each player face-down.
- Each player turns one card face-up. The player with the greater solution wins. The winner collects both cards and places them at the bottom of his or her cards.
- Suppose there is a tie. Each player lays three cards face-down, then a new card face-up. The player with the greater solution of these new cards wins. The winner collects all ten cards, and places them at the bottom of his or her cards.
- Continue playing until one player has all the cards. This player wins the game.

$-4x = -12$	$x - 1 = 1$	$x - 3 = 1$	$2x = -10$	$-9 = 9x$
$3 + x = -2$	$x = -2$	$-3x = -3$	$\frac{x}{-2} = -2$	$x = -6$
$6x = -36$	$-3x = -9$	$-7x = -14$	$x - 2 = 1$	$-1 = x + 5$
$x = -1$	$9x = -27$	$\frac{x}{3} = -1$	$-8 = -2x$	$x = 3$
$-7 = -1 + x$	$x = -5$	$-10 = 10x$	$x = -4$	$-2 = -3 + x$
$-20 = 10x$	$x + 9 = 8$	$-16 = 8x$	$x = 2$	$x + 13 = 11$
$x = -3$	$-8 = 2x$	$x = 1$	$\frac{x}{2} = -2$	$-4 + x = -2$
$\frac{x}{5} = -1$	$-6 = x - 3$	$x = 4$	$x + 6 = 2$	$x - 5 = -4$

What Is Your Answer?

4. **IN YOUR OWN WORDS** How can you use multiplication or division to solve an equation without using algebra tiles? Give two examples.

Practice Use what you learned about solving equations to complete Exercises 7–10 on page 80.

2.5 Lesson

Multiplication Property of Equality

Words Multiplying each side of an equation by the same number produces an equivalent equation.

Algebra If $a = b$, then $a \cdot c = b \cdot c$.

Division Property of Equality

Words Dividing each side of an equation by the same number produces an equivalent equation.

Algebra If $a = b$, then $a \div c = b \div c$, $c \neq 0$.

EXAMPLE 1 Solving Equations

a. Solve $\frac{x}{3} = -6$.

$\frac{x}{3} = -6$ Write the equation.

$3 \cdot \frac{x}{3} = 3 \cdot (-6)$ Multiply each side by 3.

$x = -18$ Simplify.

∴ So, the solution is $x = -18$.

b. Solve $18 = -4y$.

$18 = -4y$ Write the equation.

$\frac{18}{-4} = \frac{-4y}{-4}$ Divide each side by −4.

$-4.5 = y$ Simplify.

∴ So, the solution is $y = -4.5$.

Check

$18 = -4y$

$18 \stackrel{?}{=} -4(-4.5)$

$18 = 18$ ✓

On Your Own

Now You're Ready
Exercises 7–18

Solve the equation. Check your solution.

1. $\frac{x}{5} = -2$ **2.** $-a = -24$ **3.** $3 = -1.5n$

EXAMPLE 2 Solving an Equation Using a Reciprocal

Solve $-\frac{4}{5}x = -8$.

$-\frac{4}{5}x = -8$	Write the equation.
$-\frac{5}{4} \bullet \left(-\frac{4}{5}x\right) = -\frac{5}{4} \bullet (-8)$	Multiply each side by $-\frac{5}{4}$, the reciprocal of $-\frac{4}{5}$.
$x = 10$	Simplify.

So, the solution is $x = 10$.

On Your Own

Now You're Ready
Exercises 19–22

Solve the equation. Check your solution.

4. $-14 = \frac{2}{3}x$ **5.** $-\frac{8}{5}b = 5$ **6.** $\frac{3}{8}h = -9$

EXAMPLE 3 Real-Life Application

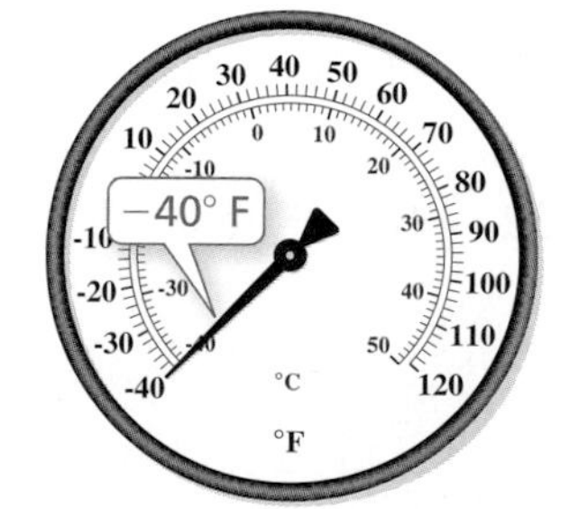

Record low temperature in Arizona

The record low temperature in Arizona is 1.6 times the record low temperature in Rhode Island. What is the record low temperature in Rhode Island?

Words The record low in Arizona is 1.6 times the record low in Rhode Island.

Variable Let t be the record low in Rhode Island.

Equation $-40 \quad = \quad 1.6 \quad \times \quad t$

$-40 = 1.6t$	Write equation.
$-\frac{40}{1.6} = \frac{1.6t}{1.6}$	Divide each side by 1.6.
$-25 = t$	Simplify.

The record low temperature in Rhode Island is −25°F.

On Your Own

7. The record low temperature in Hawaii is −0.15 times the record low temperature in Alaska. The record low temperature in Hawaii is 12°F. What is the record low temperature in Alaska?

2.5 Exercises

Vocabulary and Concept Check

1. **WRITING** Explain why multiplication can be used to solve equations involving division.

2. **OPEN-ENDED** Turning a light on and then turning the light off are considered to be inverse operations. Describe two other real-life situations that can be thought of as inverse operations.

Describe the inverse operation that will undo the given operation.

3. Multiplying by 5
4. Subtracting 12
5. Dividing by -8
6. Adding -6

Practice and Problem Solving

Solve the equation. Check your solution.

1

7. $3h = 15$
8. $-5t = -45$
9. $\frac{n}{2} = -7$
10. $\frac{k}{-3} = 9$
11. $5m = -10$
12. $8t = -32$
13. $-0.2x = 1.6$
14. $-10 = -\frac{b}{4}$
15. $-6p = 48$
16. $-72 = 8d$
17. $\frac{n}{1.6} = 5$
18. $-14.4 = -0.6p$

2

19. $\frac{3}{4}g = -12$
20. $8 = -\frac{2}{5}c$
21. $-\frac{4}{9}f = -3$
22. $26 = -\frac{8}{5}y$

23. **ERROR ANALYSIS** Describe and correct the error in finding the solution.

✗

$$-4.2x = 21$$

$$\frac{-4.2x}{4.2} = \frac{21}{4.2}$$

$$x = 5$$

Write the verbal sentence as an equation. Then solve.

24. A number divided by -9 is -16.
25. A number multiplied by $\frac{2}{5}$ is $\frac{3}{20}$.
26. The product of 15 and a number is -75.
27. The quotient of a number and -1.5 is 21.

In Exercises 28 and 29, write an equation. Then solve.

28. **NEWSPAPERS** You make \$0.75 for every newspaper you sell. How many newspapers do you have to sell to buy the soccer cleats?

29. **ROCK CLIMBING** A rock climber averages $12\frac{3}{5}$ feet per minute. How many feet does the rock climber climb in 30 minutes?

OPEN-ENDED **(a) Write a multiplication equation that has the given solution. (b) Write a division equation that has the same solution.**

30. -3 **31.** -2.2 **32.** $-\frac{1}{2}$ **33.** $-1\frac{1}{4}$

34. REASONING Which of the methods can you use to solve $-\frac{2}{3}c = 16$?

Multiply each side by $-\frac{2}{3}$.	Multiply each side by $-\frac{3}{2}$.
Divide each side by $-\frac{2}{3}$.	Multiply each side by 3, then divide each side by -2.

35. STOCK A stock has a return of $-\$1.26$ per day. Write and solve an equation to find the number of days until the total return is $-\$10.08$.

36. ELECTION In a school election, $\frac{3}{4}$ of the students vote. There are 1464 ballots. Write and solve an equation to find the number of students.

37. OCEANOGRAPHY Aquarius is an underwater ocean laboratory located in the Florida Keys National Marine Sanctuary. Solve the equation $\frac{31}{25}x = -62$ to find the value of x.

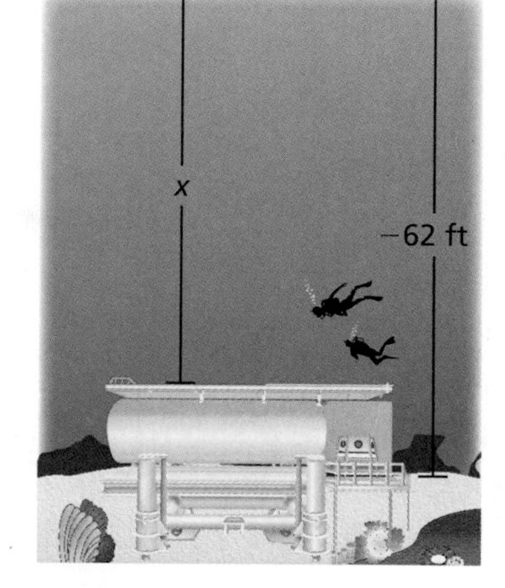

38. SHOPPING The price of a bike at store A is $\frac{5}{6}$ the price at store B. The price at store A is \$150.60. Write and solve an equation to find how much you save by buying the bike at store A.

39. CRITICAL THINKING Solve $-2|m| = -10$.

40. Number Sense In four days, your family drives $\frac{5}{7}$ of a trip. Your rate of travel is the same throughout the trip. The total trip is 1250 miles. How many more days until you reach your destination?

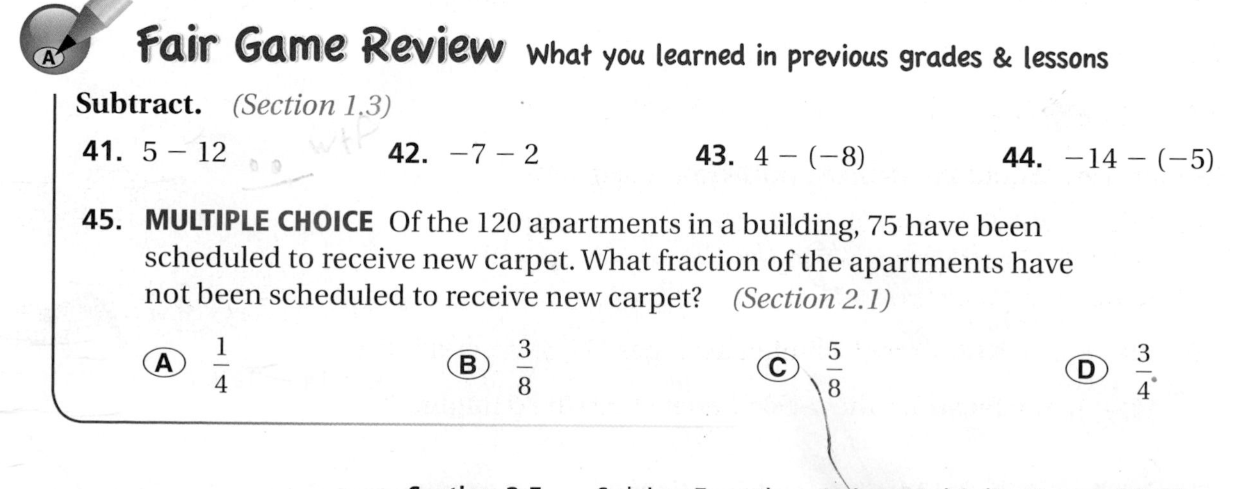

Fair Game Review What you learned in previous grades & lessons

Subtract. *(Section 1.3)*

41. $5 - 12$ **42.** $-7 - 2$ **43.** $4 - (-8)$ **44.** $-14 - (-5)$

45. MULTIPLE CHOICE Of the 120 apartments in a building, 75 have been scheduled to receive new carpet. What fraction of the apartments have not been scheduled to receive new carpet? *(Section 2.1)*

Ⓐ $\frac{1}{4}$ Ⓑ $\frac{3}{8}$ Ⓒ $\frac{5}{8}$ Ⓓ $\frac{3}{4}$

2.5b Algebraic Expressions

Parts of an algebraic expression are called terms. **Like terms** are terms that have the same variables raised to the same exponents. A term without a variable, such as 4, is called a *constant*. Constant terms are also like terms.

Like Terms	Unlike Terms
3 and -4	x and 5
$-2x$ and $7x$	$2x$ and $-6y$

EXAMPLE 1 **Identifying Terms and Like Terms**

Identify the terms and like terms in each expression.

a. $9x - 2 + 7 - x$

$9x - 2 + 7 - x$

Terms: $9x,\ -2,\ 7,\ -x$

Like terms: $9x$ and $-x$, -2 and 7

Same variable raised to same exponent

b. $6 + 5z - 3z + z$

$6 + 5z - 3z + z$

Terms: $6,\ 5z,\ -3z,\ z$

Like terms: $5z$, $-3z$, and z

Remember

The numerical factor of a term that contains a variable is a *coefficient*.

An algebraic expression is in **simplest form** if it has no like terms and no parentheses. To *combine* like terms that have variables, use the Distributive Property to add or subtract the coefficients.

EXAMPLE 2 **Simplifying Algebraic Expressions**

Simplify $\frac{3}{4}y + 12 - \frac{1}{2}y - 6$.

$\frac{3}{4}y$ and $-\frac{1}{2}y$ are like terms. 12 and -6 are also like terms.

$$\frac{3}{4}y + 12 - \frac{1}{2}y - 6 = \frac{3}{4}y - \frac{1}{2}y + 12 - 6 \quad \text{Commutative Property of Addition}$$

$$= \left(\frac{3}{4} - \frac{1}{2}\right)y + 12 - 6 \quad \text{Distributive Property}$$

$$= \frac{1}{4}y + 6 \quad \text{Simplify.}$$

Practice

Identify the terms and like terms in the expression.

1. $y + 10 - \frac{3}{2}y$
2. $2r + 7r - r - 9$
3. $7 + 4p - 5 + p + 2q$

Simplify the expression.

4. $2.5x + 4.3x - 5$
5. $\frac{3}{8}b - \frac{3}{4}b$
6. $14 - 3z + 8 + z$

 Multi-Language Glossary at BigIdeasMath.com.

EXAMPLE 3 Standardized Test Practice

Which expression is equivalent to $5(n - 8) + 4n$?

Ⓐ $49n$　Ⓑ $9n + 40$　Ⓒ $9n - 40$　Ⓓ $5n - 40$

$5(n - 8) + 4n$	$= 5(n) - 5(8) + 4n$	Distributive Property
	$= 5n - 40 + 4n$	Multiply.
	$= 5n + 4n - 40$	Commutative Property of Addition
	$= (5 + 4)n - 40$	Distributive Property
	$= 9n - 40$	Add coefficients.

∴ The correct answer is Ⓒ.

EXAMPLE 4 Real-Life Application

Evening Tickets	$7.50
REFRESHMENTS	
Drinks	
Small	$1.75
Medium	$2.75
Large	$3.50
Popcorn	
Small	$3.00
Large	$4.00

Each person in a group buys a ticket, a medium drink, and a large popcorn. Write an expression in simplest form that represents the amount of money the group spends at the movies.

Words Each ticket is $7.50, each medium drink is $2.75, and each large popcorn is $4.

Variable The same number of each item is purchased. So, x can represent the number of tickets, the number of medium drinks, and the number of large popcorns.

Expression $7.50x$ + $2.75x$ + $4x$

$7.50x + 2.75x + 4x$	$= (7.50 + 2.75 + 4)x$	Distributive Property
	$= 14.25x$	Add coefficients.

∴ The expression $14.25x$ represents the amount of money the group spends at the movies.

Study Tip

In Example 4, rewriting $7.50x + 2.75x + 4x$ as $14.25x$ helps you conclude that the total cost per person is $14.25.

Practice

Simplify the expression.

7. $3(q + 1) - 1$

8. $7x + 4\left(\frac{3}{4}x - \frac{1}{4}\right)$

9. $2(g + 4) + 5(g - 1)$

10. **WHAT IF?** In Example 4, each person buys a ticket, a large drink, and a small popcorn. How does the expression change? Explain.

2.6 Solving Two-Step Equations

Essential Question In a two-step equation, which step should you do first?

COMMON CORE STATE STANDARDS
7.EE.4a
7.EE.4b
8.EE.7a
8.EE.7b

1 EXAMPLE: Solving a Two-Step Equation

Use algebra tiles to model and solve $2x - 3 = -5$.

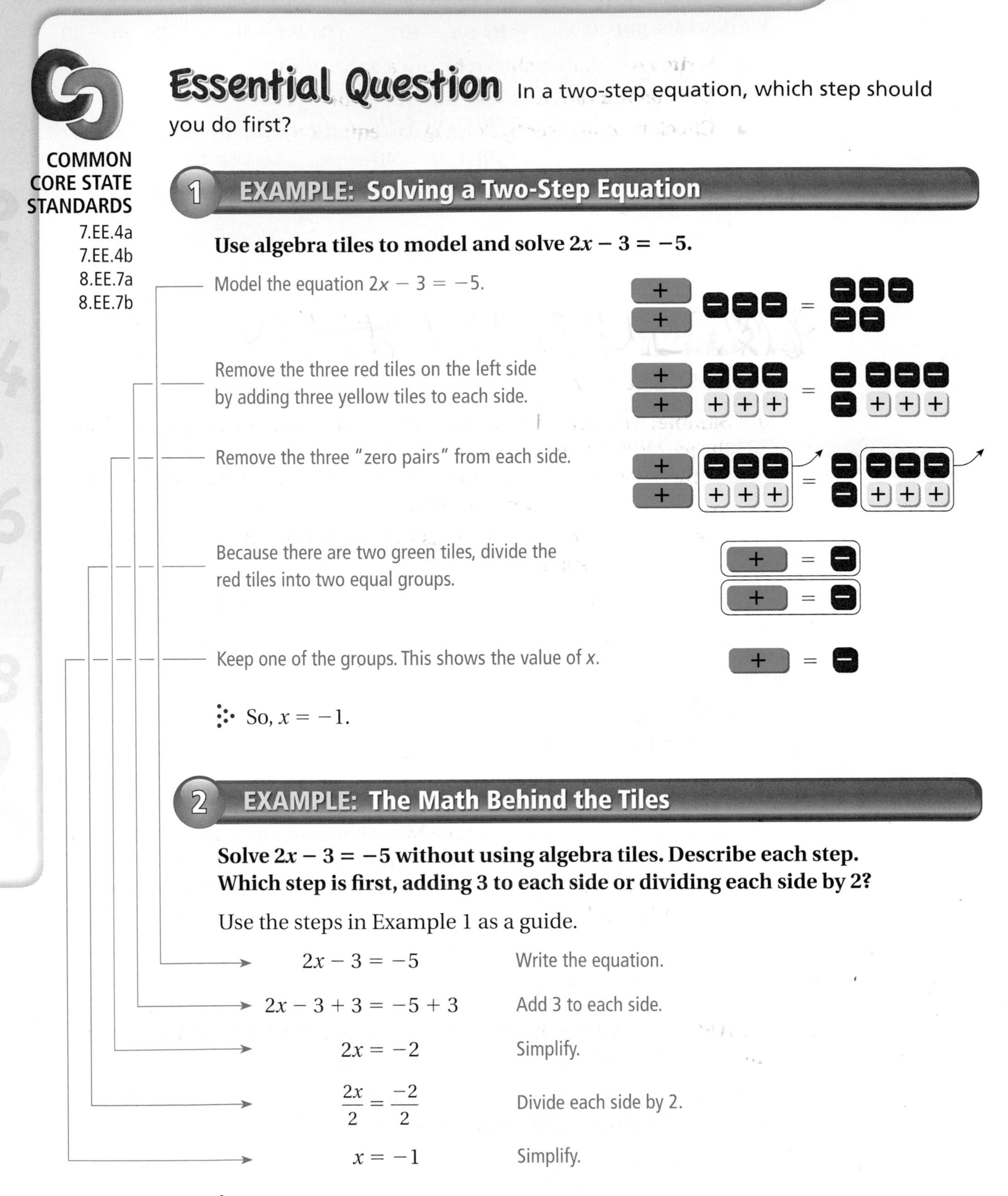

Model the equation $2x - 3 = -5$.

Remove the three red tiles on the left side by adding three yellow tiles to each side.

Remove the three "zero pairs" from each side.

Because there are two green tiles, divide the red tiles into two equal groups.

Keep one of the groups. This shows the value of x.

So, $x = -1$.

2 EXAMPLE: The Math Behind the Tiles

Solve $2x - 3 = -5$ without using algebra tiles. Describe each step. Which step is first, adding 3 to each side or dividing each side by 2?

Use the steps in Example 1 as a guide.

$2x - 3 = -5$	Write the equation.
$2x - 3 + 3 = -5 + 3$	Add 3 to each side.
$2x = -2$	Simplify.
$\frac{2x}{2} = \frac{-2}{2}$	Divide each side by 2.
$x = -1$	Simplify.

So, $x = -1$. Adding 3 to each side is the first step.

3 ACTIVITY: Solving Equations Using Algebra Tiles

Work with a partner.

- **Write an equation shown by the algebra tiles.**
- **Use algebra tiles to model and solve the equation.**
- **Check your answer by solving the equation without using algebra tiles.**

a. **b.**

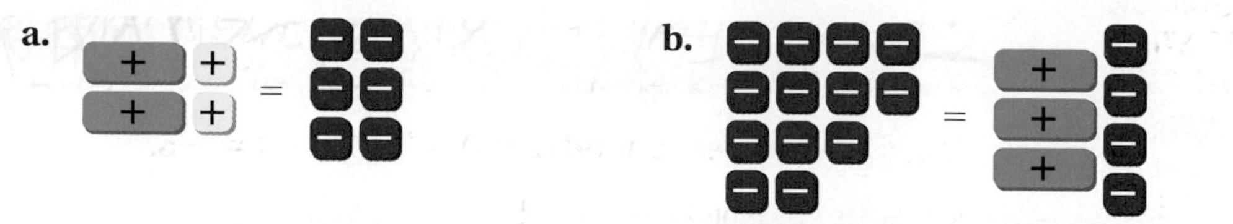

4 ACTIVITY: Working Backwards

Work with a partner.

a. Sample: Your friend pauses a video game to get a drink. You continue the game. You double the score by saving a princess. Then you lose 75 points because you do not collect the treasure. You finish the game with -25 points. How many points did you start with?

One way to solve the problem is to work backwards. To do this, start with the end result and retrace the events.

You have -25 points at the end of the game.	$\mathbf{-25}$
You lost 75 points for not collecting the treasure, so add 75 to -25.	$\mathbf{-25 + 75 = 50}$
You doubled your score for saving the princess, so find half of 50.	$\mathbf{50 \div 2 = 25}$

So, you started the game with 25 points.

b. You triple your account balance by making a deposit. Then you withdraw \$127.32 to buy groceries. Your account is now overdrawn by \$10.56. By working backwards, find your account balance before you made the deposit.

What Is Your Answer?

5. **IN YOUR OWN WORDS** In a two-step equation, which step should you do first? Give four examples.

6. Solve the equation $2x - 75 = -25$. How do your steps compare with the strategy of working backwards in Activity 4?

Use what you learned about solving two-step equations to complete Exercises 6–11 on page 86.

2.6 Lesson

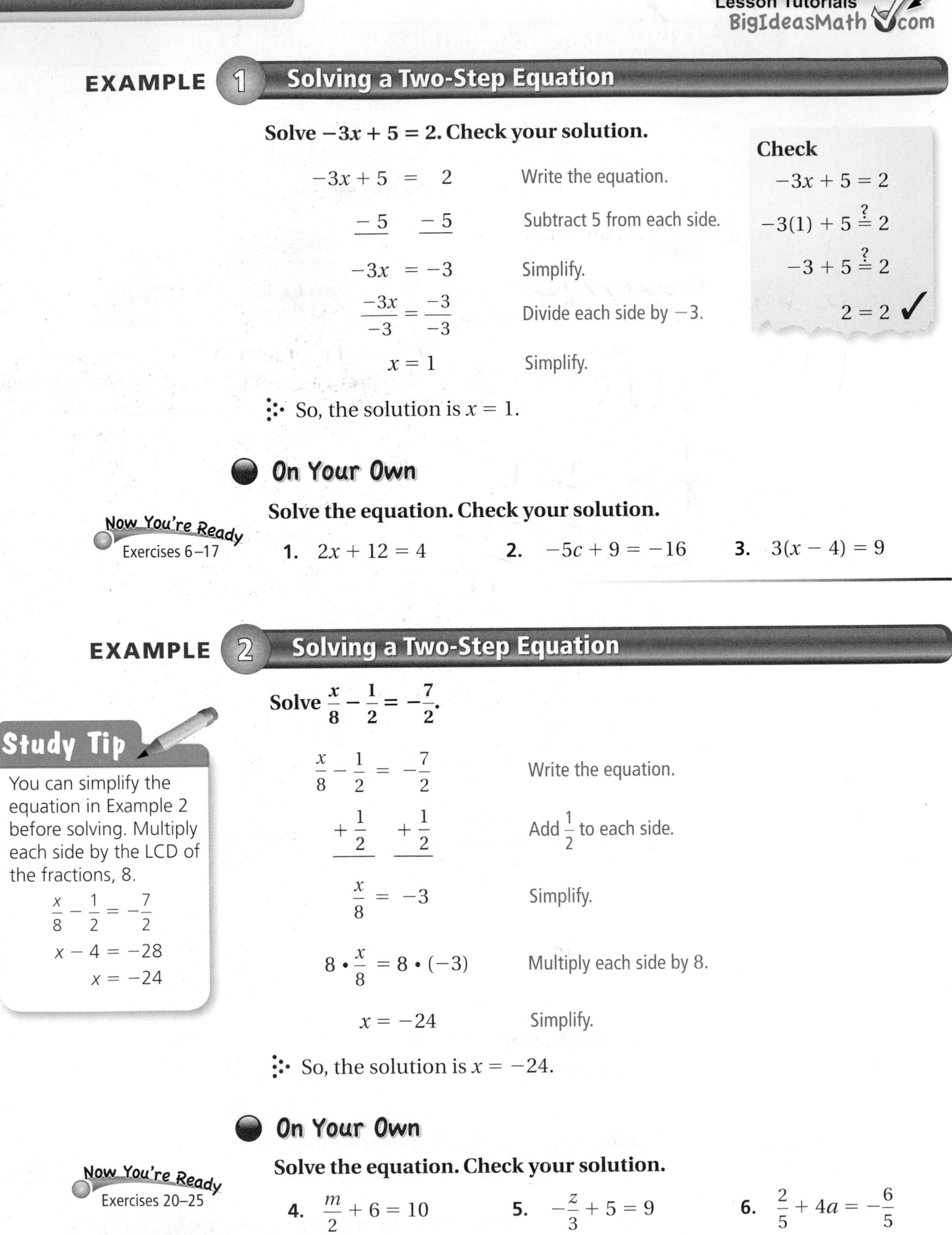

EXAMPLE 1 Solving a Two-Step Equation

Solve $-3x + 5 = 2$. Check your solution.

$$-3x + 5 = 2$$ Write the equation.

$$\underline{-5} \quad \underline{-5}$$ Subtract 5 from each side.

$$-3x = -3$$ Simplify.

$$\frac{-3x}{-3} = \frac{-3}{-3}$$ Divide each side by -3.

$$x = 1$$ Simplify.

So, the solution is $x = 1$.

Check

$-3x + 5 = 2$

$-3(1) + 5 \stackrel{?}{=} 2$

$-3 + 5 \stackrel{?}{=} 2$

$2 = 2$ ✓

On Your Own

Now You're Ready Exercises 6–17

Solve the equation. Check your solution.

1. $2x + 12 = 4$
2. $-5c + 9 = -16$
3. $3(x - 4) = 9$

EXAMPLE 2 Solving a Two-Step Equation

Solve $\frac{x}{8} - \frac{1}{2} = -\frac{7}{2}$.

$$\frac{x}{8} - \frac{1}{2} = -\frac{7}{2}$$ Write the equation.

$$\underline{+\frac{1}{2}} \quad \underline{+\frac{1}{2}}$$ Add $\frac{1}{2}$ to each side.

$$\frac{x}{8} = -3$$ Simplify.

$$8 \cdot \frac{x}{8} = 8 \cdot (-3)$$ Multiply each side by 8.

$$x = -24$$ Simplify.

So, the solution is $x = -24$.

Study Tip

You can simplify the equation in Example 2 before solving. Multiply each side by the LCD of the fractions, 8.

$$\frac{x}{8} - \frac{1}{2} = -\frac{7}{2}$$

$$x - 4 = -28$$

$$x = -24$$

On Your Own

Now You're Ready Exercises 20–25

Solve the equation. Check your solution.

4. $\frac{m}{2} + 6 = 10$
5. $-\frac{z}{3} + 5 = 9$
6. $\frac{2}{5} + 4a = -\frac{6}{5}$

EXAMPLE 3 Combining Like Terms Before Solving

Solve $3y - 8y = 25$.

$3y - 8y = 25$ Write the equation.

$-5y = 25$ Combine like terms.

$y = -5$ Divide each side by -5.

So, the solution is $y = -5$.

EXAMPLE 4 Real-Life Application

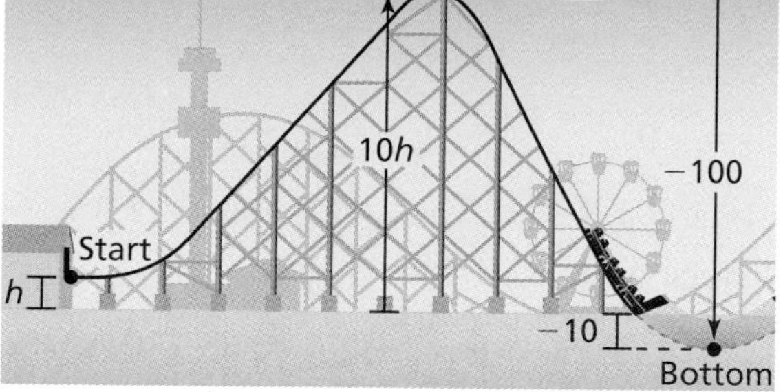

The height at the top of a roller coaster hill is 10 times the height h of the starting point. The height decreases 100 feet from the top to the bottom of the hill. The height at the bottom of the hill is -10 feet. Find h.

Location	Verbal Description	Expression
Start	The height at the start is h.	h
Top of hill	The height at the top of the hill is 10 times the starting height h.	$10h$
Bottom of hill	Height decreases by 100 feet. So, subtract 100.	$10h - 100$

The height at the bottom of the hill is -10 feet. Solve $10h - 100 = -10$ to find h.

$10h - 100 = -10$ Write equation.

$10h = 90$ Add 100 to each side.

$h = 9$ Divide each side by 10.

The height at the start is 9 feet.

On Your Own

Now You're Ready
Exercises 29–34

Solve the equation. Check your solution.

7. $4 - 2y + 3 = -9$ 8. $7x - 10x = 15$ 9. $-8 = 1.3m - 2.1m$

10. **WHAT IF?** In Example 4, the height at the bottom of the hill is -5 feet. Find the height h.

2.6 Exercises

Vocabulary and Concept Check

1. **WRITING** How do you solve two-step equations?

Match the equation with the first step to solve it.

2. $4 + 4n = -12$
3. $4n = -12$
4. $\frac{n}{4} = -12$
5. $\frac{n}{4} - 4 = -12$

A. Add 4.
B. Subtract 4.
C. Multiply by 4.
D. Divide by 4.

Practice and Problem Solving

Solve the equation. Check your solution.

(1)

6. $2v + 7 = 3$
7. $4b + 3 = -9$
8. $17 = 5k - 2$
9. $-6t - 7 = 17$
10. $8n + 16.2 = 1.6$
11. $-5g + 2.3 = -18.8$
12. $2t - 5 = -10$
13. $-4p + 9 = -5$
14. $11 = -5x - 2$
15. $4 + 2.2h = -3.7$
16. $-4.8f + 6.4 = -8.48$
17. $7.3y - 5.18 = -51.9$

ERROR ANALYSIS **Describe and correct the error in finding the solution.**

18. ✗

$$-6 + 2x = -10$$
$$-6 + \frac{2x}{2} = -\frac{10}{2}$$
$$-6 + x = -5$$
$$x = 1$$

19. ✗

$$-3x + 2 = -7$$
$$-3x = -9$$
$$-\frac{3x}{3} = \frac{-9}{3}$$
$$x = -3$$

Solve the equation. Check your solution.

(2)

20. $\frac{3}{5}g - \frac{1}{3} = -\frac{10}{3}$
21. $\frac{a}{4} - \frac{5}{6} = -\frac{1}{2}$
22. $-\frac{1}{3} + 2z = -\frac{5}{6}$
23. $2 - \frac{b}{3} = -\frac{5}{2}$
24. $-\frac{2}{3}x + \frac{3}{7} = \frac{1}{2}$
25. $-\frac{9}{4}v + \frac{4}{5} = \frac{7}{8}$

In Exercises 26–28, write an equation. Then solve.

26. **WEATHER** Starting at 1:00 P.M., the temperature changes -4 degrees per hour. How long will it take to reach -1°?

27. **BOWLING** It costs $2.50 to rent bowling shoes. Each game costs $2.25. You have $9.25. How many games can you bowl?

28. **CELL PHONES** A cell phone company charges a monthly fee plus $0.25 for each text message. The monthly fee is $30.00 and you owe $59.50. How many text messages did you have?

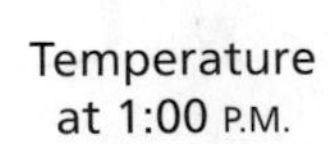

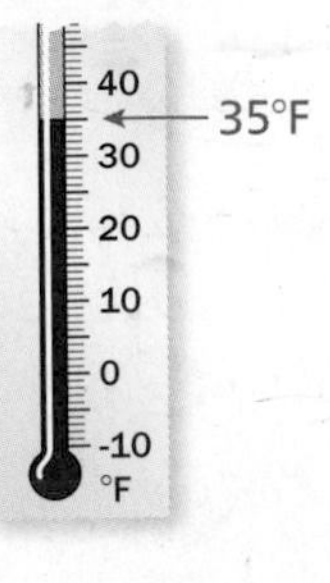

Solve the equation. Check your solution.

③ **29.** $3v - 9v = 30$

30. $12t - 8t = -52$

31. $-8d - 5d + 7d = 72$

32. $6(x - 2) = -18$

33. $-4(m + 3) = 24$

34. $-8(y + 9) = -40$

35. MODELING Write a real-world problem that can be modeled by $\frac{1}{2}x - 2 = 8$. Then solve the equation.

36. GEOMETRY The perimeter of the parallelogram is 102 feet. Find m.

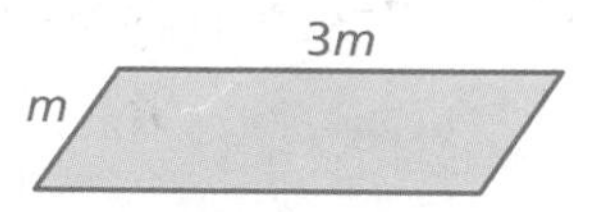

REASONING **Exercises 37 and 38 are missing information. Tell what information is needed to solve the problem.**

37. TAXI A taxi service charges an initial fee plus \$1.80 per mile. How far can you travel for \$12?

38. EARTH The coldest surface temperature on the moon is 57 degrees colder than twice the coldest surface temperature on Earth. What is the coldest surface temperature on Earth?

39. SCIENCE On Saturday, you catch insects for your science class. Five of the insects escape. The remaining insects are divided into three groups to share in class. Each group has nine insects. How many insects did you catch on Saturday?

a. Solve the problem by working backwards.

b. Solve the equation $\frac{x - 5}{3} = 9$. How does the answer compare with the answer to part (a)?

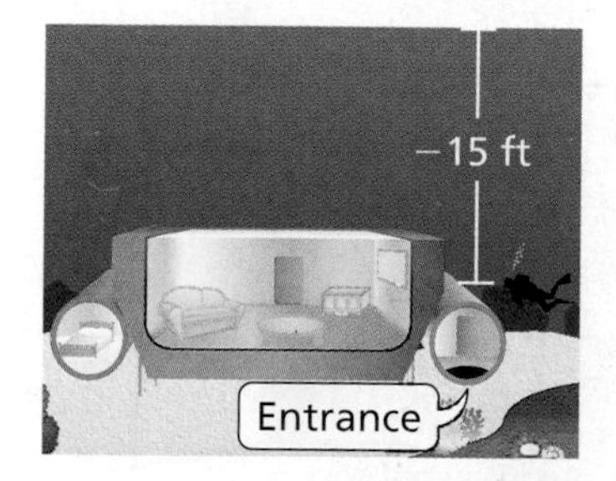

40. UNDERWATER HOTEL You must scuba dive to the entrance of your room at Jule's Undersea Lodge in Key Largo, Florida. The diver is 1 foot deeper than $\frac{2}{3}$ of the elevation of the entrance. What is the elevation of the entrance?

41. Geometry How much should you change the length of the rectangle so that the perimeter is 54 centimeters? Write an equation that shows how you found your answer.

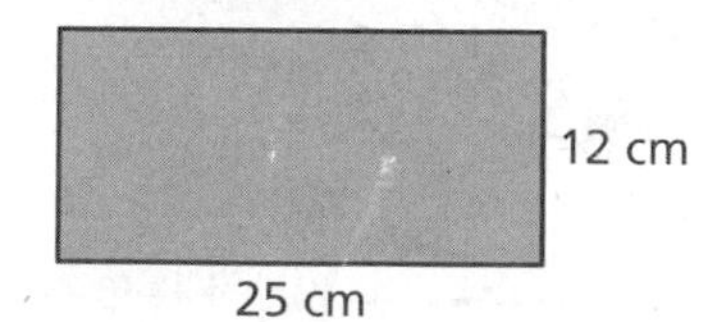

Fair Game Review What you learned in previous grades & lessons

Multiply or divide. *(Section 2.3)*

42. -6.2×5.6

43. $\frac{8}{3} \times \left(-2\frac{1}{2}\right)$

44. $\frac{5}{2} \div \left(-\frac{4}{5}\right)$

45. $-18.6 \div (-3)$

46. MULTIPLE CHOICE Which fraction is *not* equivalent to 0.75? *(Skills Review Handbook)*

Ⓐ $\frac{15}{20}$ Ⓑ $\frac{9}{12}$ Ⓒ $\frac{6}{9}$ Ⓓ $\frac{3}{4}$

2.6b Solving Inequalities

Key Ideas

Addition Property of Inequality

Words If you add the same number to each side of an inequality, the inequality remains true.

Algebra If $a < b$, then $a + c < b + c$.

Subtraction Property of Inequality

Words If you subtract the same number from each side of an inequality, the inequality remains true.

Algebra If $a < b$, then $a - c < b - c$.

These properties are true for $<$, $>$, $\leq$, and $\geq$.

Study Tip

You can solve inequalities in much the same way you solve equations. Use inverse operations to get the variable by itself.

EXAMPLE 1 Solving Inequalities Using Addition or Subtraction

a. Solve $x - 5 < -3$. Graph the solution.

	$x - 5 < -3$	Write the inequality.
Undo the subtraction. →	$\underline{+5} \quad \underline{+5}$	Add 5 to each side.
	$x < 2$	Simplify.

The solution is $x < 2$.

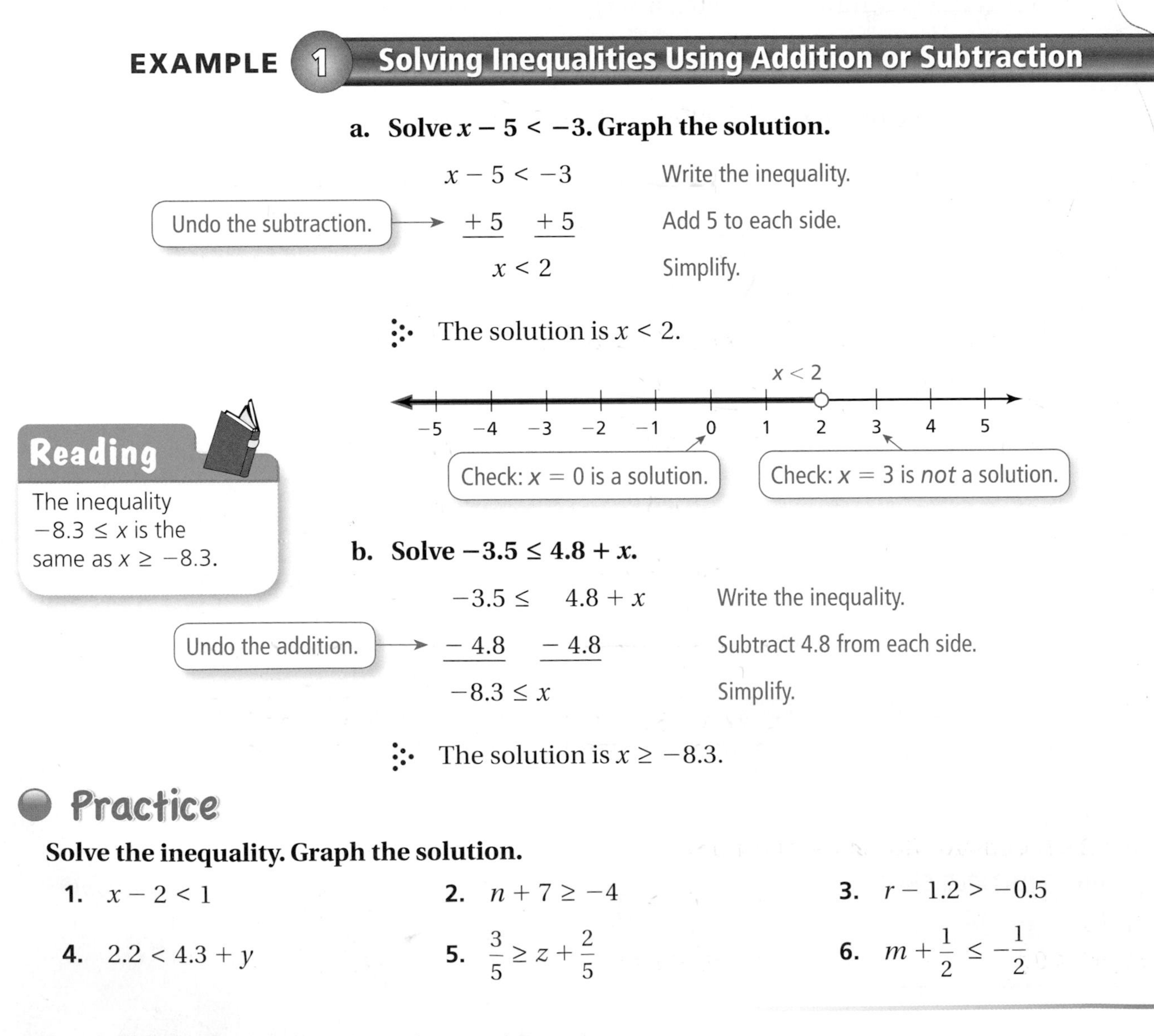

Reading

The inequality $-8.3 \leq x$ is the same as $x \geq -8.3$.

b. Solve $-3.5 \leq 4.8 + x$.

	$-3.5 \leq \quad 4.8 + x$	Write the inequality.
Undo the addition. →	$\underline{-4.8} \quad \underline{-4.8}$	Subtract 4.8 from each side.
	$-8.3 \leq x$	Simplify.

The solution is $x \geq -8.3$.

Practice

Solve the inequality. Graph the solution.

1. $x - 2 < 1$
2. $n + 7 \geq -4$
3. $r - 1.2 > -0.5$
4. $2.2 < 4.3 + y$
5. $\frac{3}{5} \geq z + \frac{2}{5}$
6. $m + \frac{1}{2} \leq -\frac{1}{2}$

Key Idea

Multiplication and Division Properties of Inequality (Case 1)

Words If you multiply or divide each side of an inequality by the same *positive* number, the inequality remains true.

Algebra If $a < b$, then $a \cdot c < b \cdot c$ for a positive number c.

If $a < b$, then $\frac{a}{c} < \frac{b}{c}$ for a positive number c.

EXAMPLE 2 Solving Inequalities Using Multiplication or Division

a. Solve $\frac{x}{10} \le -2$. Graph the solution.

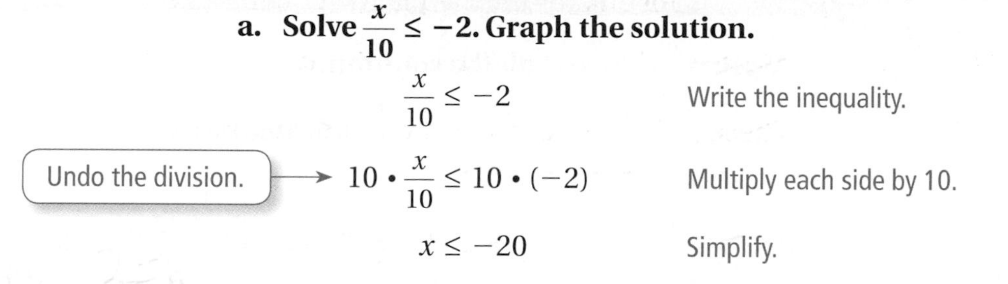

$\frac{x}{10} \le -2$ Write the inequality.
Undo the division.
$10 \cdot \frac{x}{10} \le 10 \cdot (-2)$ Multiply each side by 10.
$x \le -20$ Simplify.

The solution is $x \le -20$.

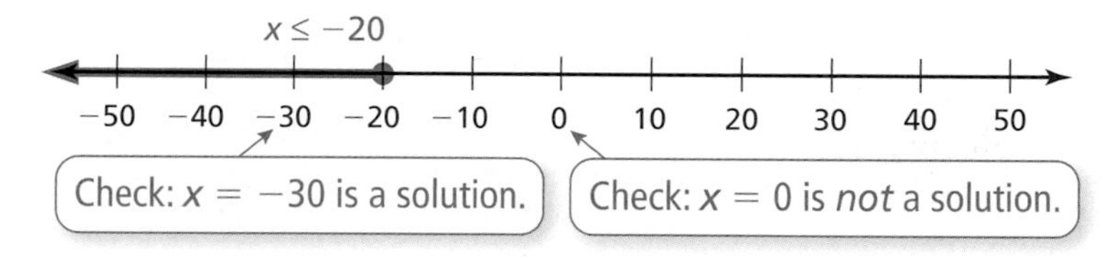

Check: x = −30 is a solution.
Check: x = 0 is not a solution.

b. Solve $2.5x > 11.25$. Graph the solution.

$2.5x > 11.25$ Write the inequality.

Undo the multiplication. → $\frac{2.5x}{2.5} > \frac{11.25}{2.5}$ Divide each side by 2.5.

$x > 4.5$ Simplify.

The solution is $x > 4.5$.

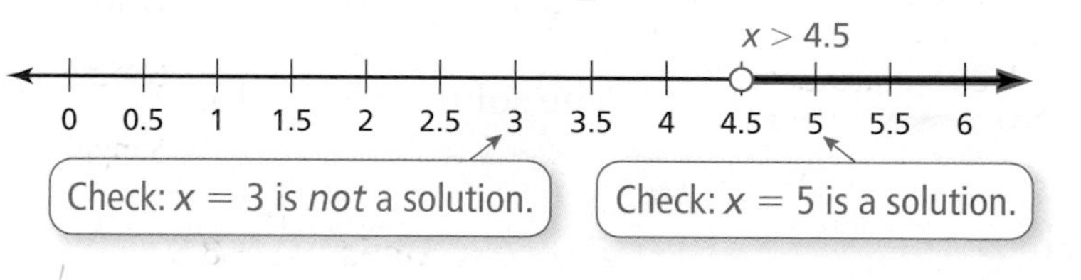

Check: x = 3 is not a solution.
Check: x = 5 is a solution.

Practice

Solve the inequality. Graph the solution.

7. $\frac{b}{8} \ge -5$

8. $-0.4 > \frac{g}{15}$

9. $\frac{2}{3}m \le \frac{8}{9}$

10. $63 < 9q$

11. $60 \le 2.4x$

12. $1.6u > -19.2$

Common Error

A negative sign in an inequality does not necessarily mean you must reverse the inequality symbol.

Only reverse the inequality symbol when you multiply or divide each side by a negative number.

Key Idea

Multiplication and Division Properties of Inequality (Case 2)

Words If you multiply or divide each side of an inequality by the same *negative* number, the direction of the inequality symbol must be reversed for the inequality to remain true.

Algebra If $a < b$, then $a \cdot c > b \cdot c$ for a negative number c.

If $a < b$, then $\frac{a}{c} > \frac{b}{c}$ for a negative number c.

EXAMPLE 3 Solving Inequalities Using Multiplication or Division

a. Solve $\frac{y}{-4} > 6$. Graph the solution.

$\frac{y}{-4} > 6$	Write the inequality.
Undo the division. → $-4 \cdot \frac{y}{-4} < -4 \cdot 6$	Multiply each side by -4. Reverse the inequality symbol.
$y < -24$	Simplify.

The solution is $y < -24$.

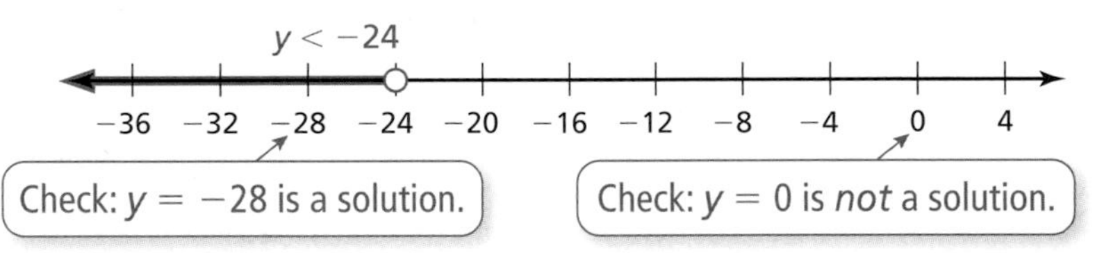

b. Solve $-21 \geq -1.4y$.

$-21 \geq -1.4y$	Write the inequality.
Undo the multiplication. → $\frac{-21}{-1.4} \leq \frac{-1.4y}{-1.4}$	Divide each side by -1.4. Reverse the inequality symbol.
$15 \leq y$	Simplify.

The solution is $y \geq 15$.

Practice

Solve the inequality. Graph the solution.

13. $7 > \frac{j}{-1.5}$

14. $\frac{a}{-3} \leq -2$

15. $-2.5 < k \div (-4.8)$

16. $-2s < 24$

17. $-3.1z \geq 62$

18. $-3.9 \geq -0.6d$

EXAMPLE 4 Solving a Two-Step Inequality

Solve $-3x + 2 > 11$.

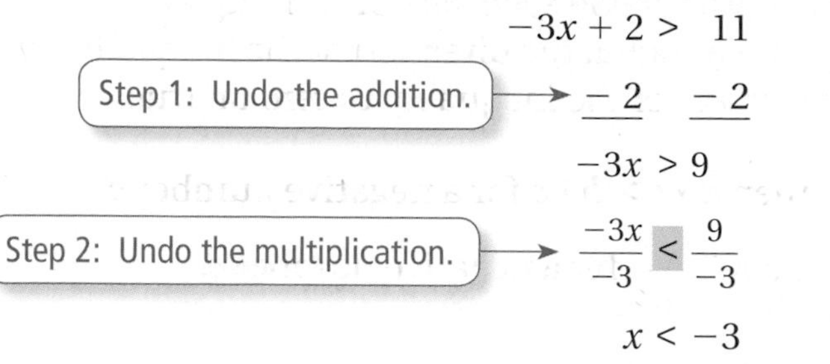
Step 1: Undo the addition.
Step 2: Undo the multiplication.

$-3x + 2 > 11$	Write the inequality.
$\underline{-2} \quad \underline{-2}$	Subtract 2 from each side.
$-3x > 9$	Simplify.
$\dfrac{-3x}{-3} < \dfrac{9}{-3}$	Divide each side by −3. Reverse the inequality symbol.
$x < -3$	Simplify.

∴ The solution is $x < -3$.

EXAMPLE 5 Real-Life Application

Progress Report	
Month	**Pounds Lost**
1	9
2	5
3	x
4	x

A contestant in a weight loss competition wants to lose at least 30 pounds in 4 months. Write and solve an inequality to find the average number x of pounds the contestant must lose in each of the last 2 months to meet the goal.

Use the progress report to write an expression for the number of pounds lost.

Pounds lost: $9 + 5 + x + x = 14 + 2x$

Because the contestant wants to lose *at least* 30 pounds, use the symbol $\geq$.

$14 + 2x \geq 30$	Write an inequality.
$\underline{-14} \quad \underline{-14}$	Subtract 14 from each side.
$2x \geq 16$	Simplify.
$\dfrac{2x}{2} \geq \dfrac{16}{2}$	Divide each side by 2.
$x \geq 8$	Simplify.

∴ The contestant must lose an average of at least 8 pounds in each of the last 2 months to meet the goal.

Practice

Solve the inequality. Graph the solution.

19. $5n - 3 < 12$ **20.** $-3(w - 10) > 27$ **21.** $-7 \geq \dfrac{c}{-2} + 2$

22. BICYCLE You want to purchase a bicycle that costs \$265. So far, you have saved \$128 and you plan to save an additional \$20 per week.

a. Write and solve an inequality to find the number of weeks it will take to save at least \$265.

b. Graph the solution in part (a). Will you have saved enough money after 6 weeks? 8 weeks? Explain.

2.4–2.6 Quiz

Solve the equation. Check your solution. *(Section 2.4 and Section 2.5)*

1. $-6.5 + x = -4.12$

2. $4\frac{1}{2} + p = -5\frac{3}{4}$

3. $-\frac{b}{7} = 4$

4. $2h = -57$

Write the verbal sentence as an equation. Then solve. *(Section 2.4 and Section 2.5)*

5. The difference between a number b and 7.4 is -6.8.

6. $5\frac{2}{5}$ more than a number a is $7\frac{1}{2}$.

7. A number x multiplied by $\frac{3}{8}$ is $-\frac{15}{32}$.

8. The quotient of two times a number k and -2.6 is 12.

Write and solve an equation to find the value of x. *(Section 2.4 and Section 2.6)*

9. Perimeter = 26

10. Perimeter = 23.59

11. Perimeter = 33

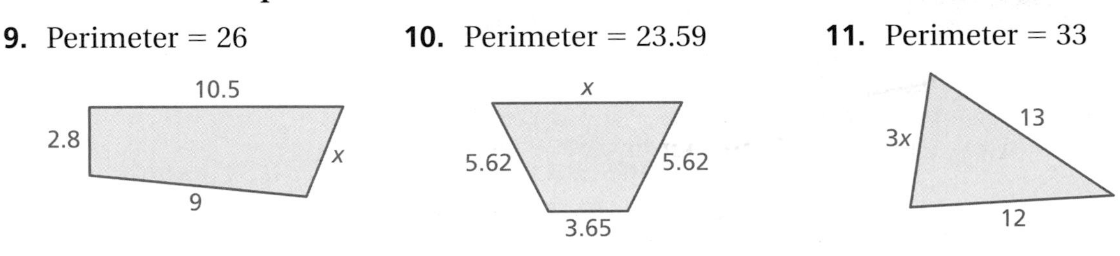

12. BANKING You withdraw \$29.79 from your bank account. Now, your balance is $-\$20.51$. Write and solve an equation to find the amount of money in your bank account before you withdrew the money. *(Section 2.4)*

13. WATER LEVEL During a drought, the water level of a lake changes $-3\frac{1}{5}$ feet per day. Write and solve an equation to find how long it takes for the water level to change -16 feet. *(Section 2.5)*

14. BASKETBALL A basketball game has four quarters. The length of a game is 32 minutes. You play the entire game except $4\frac{1}{2}$ minutes. Write and solve an equation to find the mean time you play per quarter. *(Section 2.6)*

15. SCRAPBOOKING The mat needs to be cut to have a 0.5-inch border on all four sides. *(Section 2.6)*

a. How much should you cut from the left and right sides?

b. How much should you cut from the top and bottom?

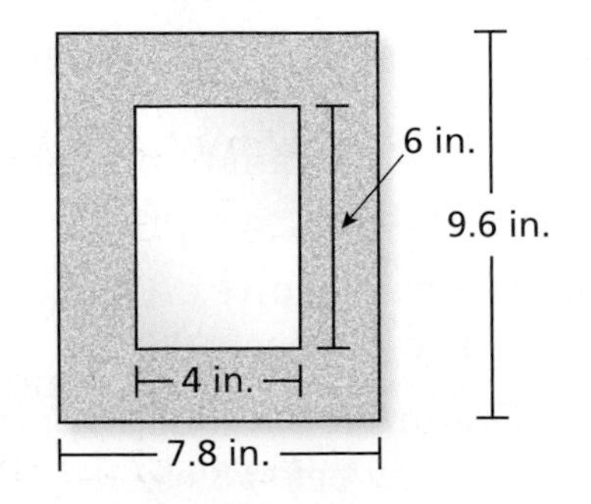

2 Chapter Review

Review Key Vocabulary

terminating decimal, *p. 52*
repeating decimal, *p. 52*
rational number, *p. 52*
equivalent equations, *p. 72*

Review Examples and Exercises

2.1 Rational Numbers *(pp. 50–55)*

Write −0.14 as a fraction in simplest form.

$$-0.14 = -\frac{14}{100}$$

Write the digits after the decimal point in the numerator.

The last digit is in the hundredths place. So, use 100 in the denominator.

$$= -\frac{7}{50}$$ Simplify.

Exercises

Write the rational number as a decimal.

1. $-\frac{8}{15}$ **2.** $\frac{5}{8}$ **3.** $-\frac{13}{6}$ **4.** $1\frac{7}{16}$

Write the decimal as a fraction or mixed number in simplest form.

5. -0.6 **6.** -0.35 **7.** -5.8 **8.** 24.23

2.2 Adding and Subtracting Rational Numbers *(pp. 56–61)*

Find −8.18 + 3.64.

$-8.18 + 3.64 = -4.54$ $|-8.18| > |3.64|$. So, subtract $|3.64|$ from $|-8.18|$.

Use the sign of −8.18.

Exercises

Add or subtract. Write fractions in simplest form.

9. $-4\frac{5}{9} + \frac{8}{9}$ **10.** $-\frac{5}{12} - \frac{3}{10}$ **11.** $-2.53 + 4.75$ **12.** $3.8 - (-7.45)$

13. TURTLES A turtle is $20\frac{5}{6}$ inches below the surface of a pond. It dives to a depth of $32\frac{1}{4}$ inches. How far did it dive?

2.3 Multiplying and Dividing Rational Numbers (pp. 62–67)

Find $-4\frac{1}{6} \div 1\frac{1}{3}$.

$-4\frac{1}{6} \div 1\frac{1}{3} = -\frac{25}{6} \div \frac{4}{3}$ — Write mixed numbers as improper fractions.

$= \frac{-25}{6} \cdot \frac{3}{4}$ — Multiply by the reciprocal of $\frac{4}{3}$.

$= \frac{-25 \cdot 3}{6 \cdot 4}$ — Multiply the numerators and the denominators.

$= \frac{-25}{8}$, or $-3\frac{1}{8}$ — Simplify.

Exercises

Multiply or divide. Write fractions in simplest form.

14. $-\frac{4}{9}\left(-\frac{7}{9}\right)$ **15.** $\frac{9}{10} \div \left(-\frac{6}{5}\right)$ **16.** $\frac{8}{15}\left(-\frac{2}{3}\right)$ **17.** $-\frac{4}{11} \div \frac{2}{7}$

18. $-5.9(-9.7)$ **19.** $6.4 \div (-3.2)$ **20.** $4.5(-5.26)$ **21.** $-15.4 \div (-2.5)$

22. **SUNKEN SHIP** The elevation of a sunken ship is -120 feet. Your elevation is $\frac{5}{8}$ of the ship's elevation. What is your elevation?

2.4 Solving Equations Using Addition or Subtraction (pp. 70–75)

Solve $x - 9 = -6$.

$x - 9 = -6$ — Write the equation.

$\underline{+9} \quad \underline{+9}$ — Add 9 to each side.

$x = 3$ — Simplify.

Exercises

Solve the equation. Check your solution.

23. $p - 3 = -4$ **24.** $6 + q = 1$ **25.** $-2 + j = -22$ **26.** $b - 19 = -11$

27. $n + \frac{3}{4} = \frac{1}{4}$ **28.** $v - \frac{5}{6} = -\frac{7}{8}$ **29.** $t - 3.7 = 1.2$ **30.** $\ell + 15.2 = -4.5$

31. **GIFT CARD** A shirt costs \$24.99. After using a gift card as a partial payment, you still owe \$9.99. What is the value of the gift card?

2.5 Solving Equations Using Multiplication or Division (pp. 76–81)

Solve $\frac{x}{5} = -7$.

$\frac{x}{5} = -7$	Write the equation.
$5 \cdot \frac{x}{5} = 5 \cdot (-7)$	Multiply each side by 5.
$x = -35$	Simplify.

Exercises

Solve the equation. Check your solution.

32. $\frac{x}{3} = -8$

33. $-7 = \frac{y}{7}$

34. $-\frac{z}{4} = -\frac{3}{4}$

35. $-\frac{w}{20} = -2.5$

36. $4x = -8$

37. $-10 = 2y$

38. $-5.4z = -32.4$

39. $-6.8w = 3.4$

40. TEMPERATURE The mean temperature change is -3.2°F per day for five days. What is the total change over the five-day period?

2.6 Solving Two-Step Equations (pp. 82–87)

Solve $\frac{x}{5} + \frac{7}{10} = -\frac{3}{10}$.

$\frac{x}{5} + \frac{7}{10} = -\frac{3}{10}$	Write the equation.
$\frac{x}{5} = -1$	Subtract $\frac{7}{10}$ from each side.
$x = -5$	Multiply each side by 5.

Exercises

Solve the equation. Check your solution.

41. $-2c + 6 = -8$

42. $3(3w - 4) = -20$

43. $\frac{w}{6} + \frac{5}{8} = -1\frac{3}{8}$

44. $-3x - 4.6 = 5.9$

45. EROSION The floor of a canyon has an elevation of -14.5 feet. Erosion causes the elevation to change by -1.5 feet per year. How many years will it take for the canyon floor to have an elevation of -31 feet?

2 Chapter Test

Check It Out
Test Practice
BigIdeasMath.com

Write the rational number as a decimal.

1. $\frac{7}{40}$ **2.** $-\frac{1}{9}$ **3.** $-\frac{21}{16}$ **4.** $\frac{36}{5}$

Write the decimal as a fraction or mixed number in simplest form.

5. -0.122 **6.** 0.33 **7.** -4.45 **8.** -7.09

Add or subtract. Write fractions in simplest form.

9. $-\frac{4}{9}+\left(-\frac{23}{18}\right)$ **10.** $\frac{17}{12}-\left(-\frac{1}{8}\right)$ **11.** $9.2+(-2.8)$ **12.** $2.86-12.1$

Multiply or divide. Write fractions in simplest form.

13. $3\frac{9}{10}\times\left(-\frac{8}{3}\right)$ **14.** $-1\frac{5}{6}\div 4\frac{1}{6}$ **15.** $-4.4\times(-6.02)$ **16.** $-5\div 1.5$

Solve the equation. Check your solution.

17. $7x=-3$ **18.** $2(x+1)=-2$ **19.** $\frac{2}{9}g=-8$

20. $z+14.5=5.4$ **21.** $-14=6c$ **22.** $\frac{2}{7}k-\frac{3}{8}=-\frac{19}{8}$

23. MARATHON A marathon is a 26.2-mile race. You run three marathons in one year. How many miles do you run?

24. RECORD A runner is compared with the world record holder during a race. A negative number means the runner is ahead of the time of the world record holder, and a positive number means that the runner is behind the time of the world record holder. The table shows the time difference between the runner and the world record holder for each lap. What time difference does the runner need for the fourth lap to match the world record?

Lap	Time Difference
1	−1.23
2	0.45
3	0.18
4	?

25. GYMNASTICS You lose 0.3 point for stepping out of bounds during a floor routine. Your final score is 9.124. Write and solve an equation to find your score before the penalty.

26. PERIMETER The perimeter of the triangle is 45. Find the value of x.

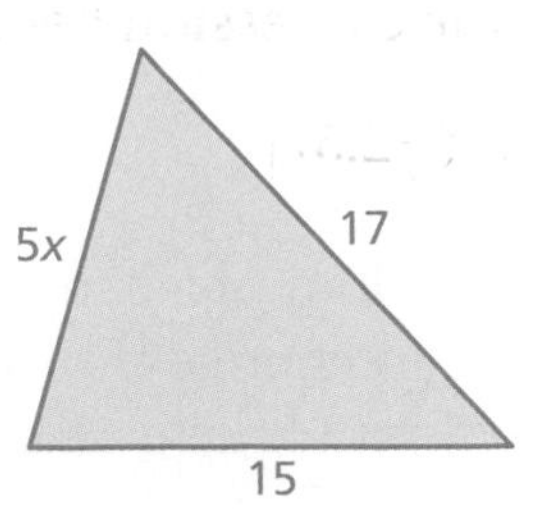

2 Standardized Test Practice

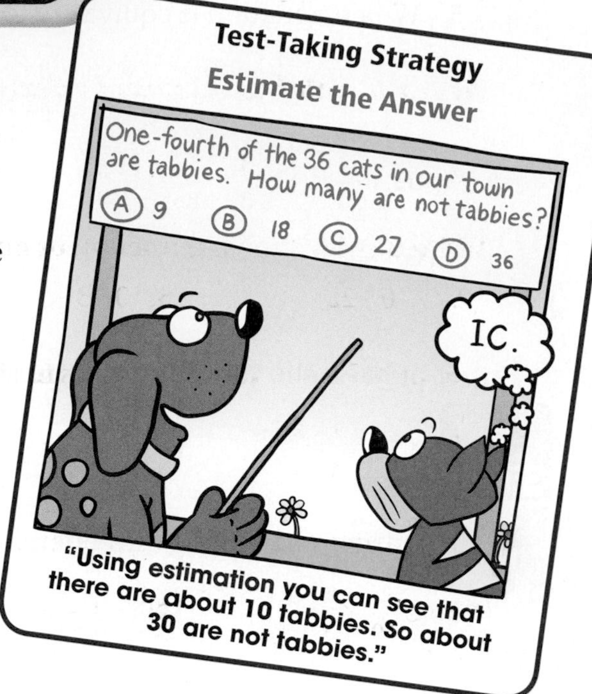

1. When José and Sean were each 5 years old, José was $1\frac{1}{2}$ inches taller than Sean. José grew at an average rate of $2\frac{3}{4}$ inches per year from the time that he was 5 years old until the time he was 13 years old. José was 63 inches tall when he was 13 years old. How tall was Sean when he was 5 years old? *(7.EE.4a)*

   A. $39\frac{1}{2}$ in. **C.** $44\frac{3}{4}$ in.

   B. $42\frac{1}{2}$ in. **D.** $47\frac{3}{4}$ in.

2. A line is graphed in the coordinate plane below.

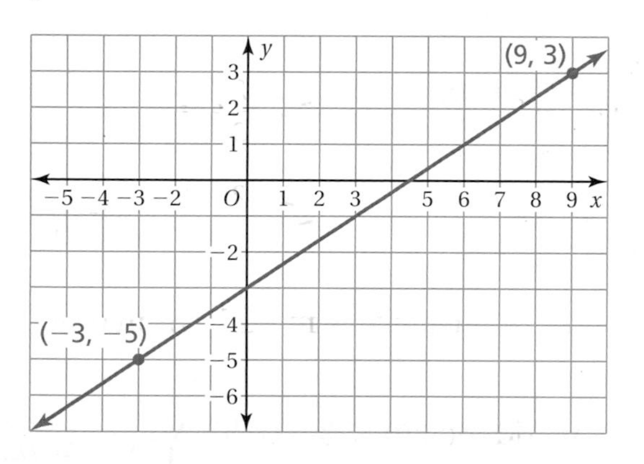

   Which point is *not* on the line? *(7.NS.1b)*

   F. $(-3, 0)$ **H.** $(3, -1)$

   G. $(0, -3)$ **I.** $(6, 1)$

3. What is the missing number in the sequence below? *(7.NS.2c)*

$$\frac{9}{16}, -\frac{9}{8}, \frac{9}{4}, -\frac{9}{2}, 9, \_\_\_\_$$

4. What is the value of the expression below? *(7.EE.3)*

$$|-2 - (-2.5)|$$

   A. -4.5 **C.** 0.5

   B. -0.5 **D.** 4.5

5. Which equation is equivalent to the equation shown below? *(7.EE.4a)*

$$-\frac{3}{4}x + \frac{1}{8} = -\frac{3}{8}$$

F. $-\frac{3}{4}x = -\frac{3}{8} - \frac{1}{8}$

G. $-\frac{3}{4}x = -\frac{3}{8} + \frac{1}{8}$

H. $x + \frac{1}{8} = -\frac{3}{8} \cdot \left(-\frac{4}{3}\right)$

I. $x + \frac{1}{8} = -\frac{3}{8} \cdot \left(-\frac{3}{4}\right)$

6. What is the value of the expression below? *(7.NS.2c)*

$$-5 \div 20$$

7. Karina was solving the equation in the box below.

$$-96 = -6(15 - 2x)$$
$$-96 = -90 - 12x$$
$$-96 + 90 = -90 + 90 - 12x$$
$$-6 = -12x$$
$$\frac{-6}{-12} = \frac{-12x}{-12}$$
$$\frac{1}{2} = x$$

What should Karina do to correct the error that she made? *(7.EE.4a)*

A. First add 6 to both sides of the equation.

B. First add $2x$ to both sides of the equation.

C. Distribute the -6 to get $90 - 12x$.

D. Distribute the -6 to get $-90 + 12x$.

8. Current, voltage, and resistance are related according to the formula below, where I represents the current, in amperes, V represents the voltage, in volts, and R represents the resistance, in ohms.

$$I = \frac{V}{R}$$

What is the voltage when the current is 0.5 ampere and the resistance is 0.8 ohm? *(7.EE.4a)*

F. 4.0 volts

G. 1.3 volts

H. 0.4 volt

I. 0.3 volt

9. What is the area of a triangle with a base length of $2\frac{1}{2}$ inches and a height of 3 inches? *(7.NS.2c)*

A. $2\frac{3}{4}$ in.^{2}

B. $3\frac{3}{4}$ in.^{2}

C. $5\frac{1}{2}$ in.^{2}

D. $7\frac{1}{2}$ in.^{2}

10. What is the circumference of the circle below? (Use 3.14 for π.) *(7.NS.2c)*

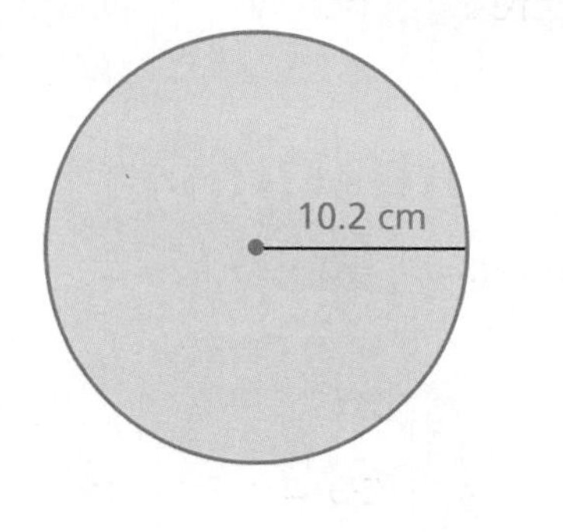

F. 64.056 cm

G. 60.028 cm

H. 32.028 cm

I. 30.028 cm

11. Four points are graphed on the number line below. *(7.EE.3)*

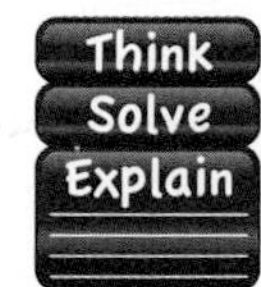

Part A Choose the two points whose values have the greatest sum. Approximate this sum. Explain your reasoning.

Part B Choose the two points whose values have the greatest difference. Approximate this difference. Explain your reasoning.

Part C Choose the two points whose values have the greatest product. Approximate this product. Explain your reasoning.

Part D Choose the two points whose values have the greatest quotient. Approximate this quotient. Explain your reasoning.

12. What number belongs in the box to make the equation true? *(7.NS.3)*

$$\frac{-0.4}{\square} + 0.8 = -1.2$$

A. 1

B. 0.2

C. -0.2

D. -1

3 Proportions and Variation

"I am doing an experiment with slope. I want you to run up and down the board 10 times."

"Now with 2 more dog biscuits, do it again and we'll compare your rates."

"Dear Sir: I counted the number of bacon, cheese, and chicken dog biscuits in the box I bought."

"There were 16 bacon, 12 cheese, and only 8 chicken. That's a ratio of 4:3:2. Please go back to the original ratio of 1:1:1."

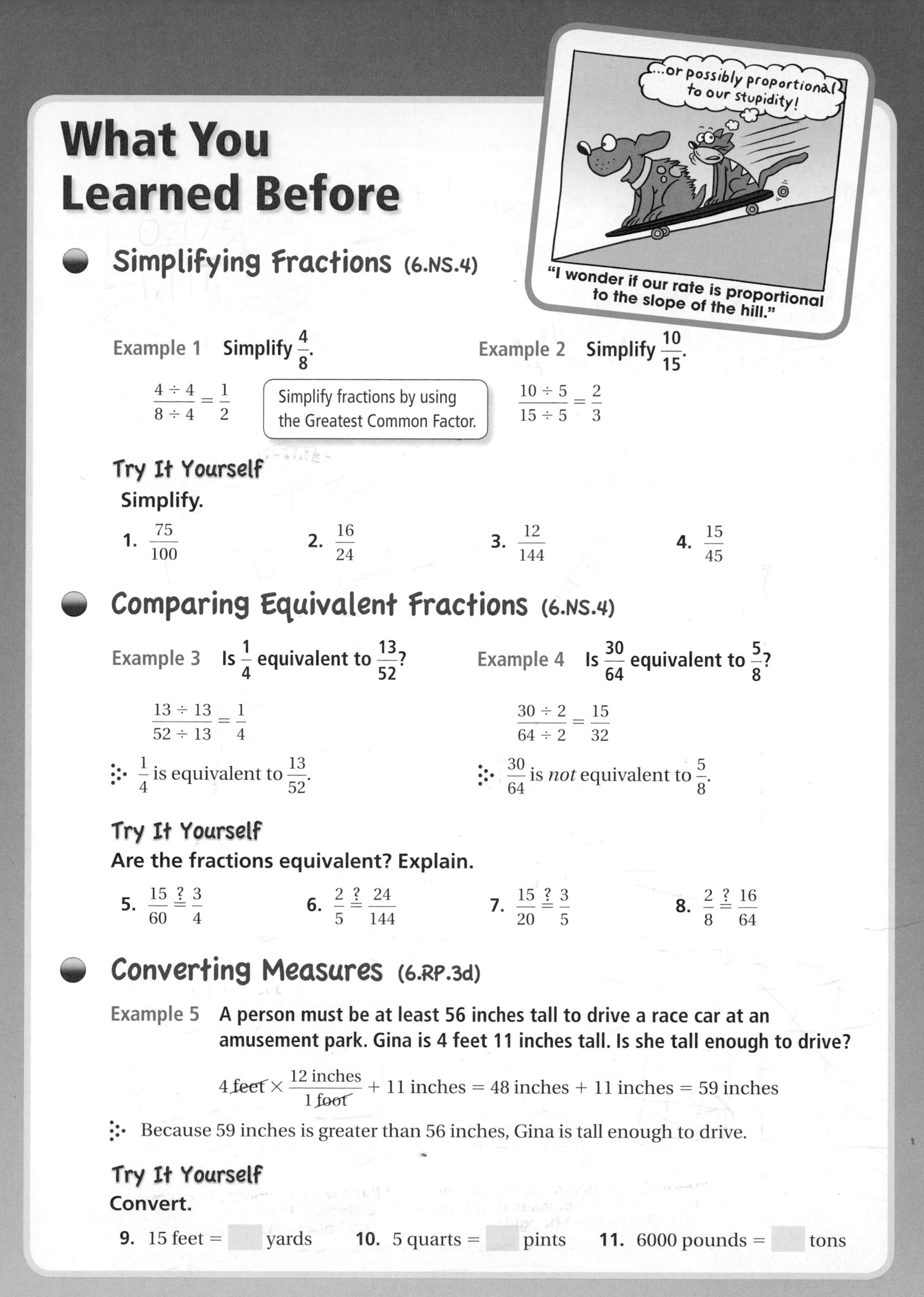

...or possibly proportional to our stupidity!
"I wonder if our rate is proportional to the slope of the hill."

What You Learned Before

Simplifying Fractions (6.NS.4)

Example 1 **Simplify $\frac{4}{8}$.**

$$\frac{4 \div 4}{8 \div 4} = \frac{1}{2}$$

Simplify fractions by using the Greatest Common Factor.

Example 2 **Simplify $\frac{10}{15}$.**

$$\frac{10 \div 5}{15 \div 5} = \frac{2}{3}$$

Try It Yourself

Simplify.

1. $\frac{75}{100}$ **2.** $\frac{16}{24}$ **3.** $\frac{12}{144}$ **4.** $\frac{15}{45}$

Comparing Equivalent Fractions (6.NS.4)

Example 3 **Is $\frac{1}{4}$ equivalent to $\frac{13}{52}$?**

$$\frac{13 \div 13}{52 \div 13} = \frac{1}{4}$$

∴ $\frac{1}{4}$ is equivalent to $\frac{13}{52}$.

Example 4 **Is $\frac{30}{64}$ equivalent to $\frac{5}{8}$?**

$$\frac{30 \div 2}{64 \div 2} = \frac{15}{32}$$

∴ $\frac{30}{64}$ is *not* equivalent to $\frac{5}{8}$.

Try It Yourself

Are the fractions equivalent? Explain.

5. $\frac{15}{60} \stackrel{?}{=} \frac{3}{4}$ **6.** $\frac{2}{5} \stackrel{?}{=} \frac{24}{144}$ **7.** $\frac{15}{20} \stackrel{?}{=} \frac{3}{5}$ **8.** $\frac{2}{8} \stackrel{?}{=} \frac{16}{64}$

Converting Measures (6.RP.3d)

Example 5 **A person must be at least 56 inches tall to drive a race car at an amusement park. Gina is 4 feet 11 inches tall. Is she tall enough to drive?**

$$4 \cancel{\text{feet}} \times \frac{12 \text{ inches}}{1 \cancel{\text{foot}}} + 11 \text{ inches} = 48 \text{ inches} + 11 \text{ inches} = 59 \text{ inches}$$

∴ Because 59 inches is greater than 56 inches, Gina is tall enough to drive.

Try It Yourself

Convert.

9. 15 feet = ☐ yards **10.** 5 quarts = ☐ pints **11.** 6000 pounds = ☐ tons

3.1 Ratios and Rates

COMMON CORE STATE STANDARDS
7.RP.1
7.RP.2b

Essential Question How do rates help you describe real-life problems?

The Meaning of a Word ● Rate

When you rent snorkel gear at the beach, you should pay attention to the rental **rate**. The rental rate is in dollars per hour.

1 ACTIVITY: Finding Reasonable Rates

Work with a partner.

a. Match each description with a verbal rate.

b. Match each verbal rate with a numerical rate.

c. Give a reasonable numerical rate for each description. Then give an unreasonable rate.

Description	*Verbal Rate*	*Numerical Rate*
Your pay rate for washing cars	inches per month	$\frac{\square\ \text{m}}{\text{sec}}$
The average rainfall rate in a rain forest	pounds per acre	$\frac{\square\ \text{people}}{\text{yr}}$
Your average driving rate along an interstate	meters per second	$\frac{\square\ \text{lb}}{\text{acre}}$
The growth rate for the length of a baby alligator	people per year	$\frac{\square\ \text{mi}}{\text{h}}$
Your running rate in a 100-meter dash	dollars per hour	$\frac{\square\ \text{in.}}{\text{yr}}$
The population growth rate of a large city	dollars per year	$\frac{\square\ \text{in.}}{\text{mo}}$
The average pay rate for a professional athlete	miles per hour	$\frac{\$\square}{\text{h}}$
The fertilization rate for an apple orchard	inches per year	$\frac{\$\square}{\text{yr}}$

2 ACTIVITY: Unit Analysis

Work with a partner. Some real-life problems involve the product of an amount and a rate. Find each product. List the units.

a. **Sample:** $6\text{ h} \times \frac{\$12}{\text{h}} = 6\cancel{\text{h}} \times \frac{\$12}{\cancel{\text{h}}}$ Divide out "hours."

$= \$72$ Multiply. Answer is in dollars.

b. $6\text{ mo} \times \frac{\$700}{\text{mo}}$

c. $10\text{ gal} \times \frac{22\text{ mi}}{\text{gal}}$

d. $9\text{ lb} \times \frac{\$3}{\text{lb}}$

e. $13\text{ min} \times \frac{60\text{ sec}}{\text{min}}$

3 ACTIVITY: Writing a Story

Work with a partner.

- **Think of a story that compares two different rates.**
- **Write the story.**
- **Draw pictures for the story.**

What Is Your Answer?

4. **RESEARCH** Use newspapers, the Internet, or magazines to find examples of salaries. Try to find examples of each of the following ways to write salaries.

   a. dollars per hour **b.** dollars per month **c.** dollars per year

5. **IN YOUR OWN WORDS** How do rates help you describe real-life problems? Give two examples.

6. To estimate the annual salary for a given hourly pay rate, multiply by 2 and insert "000" at the end.

   Sample: $10 per hour is about $20,000 per year.

   a. Explain why this works. Assume the person is working 40 hours a week.

   b. Estimate the annual salary for an hourly pay rate of $8 per hour.

   c. You earn $1 million per month. What is your annual salary?

   d. Why is the cartoon funny?

"We had someone apply for the job. He says he would like $1 million a month, but will settle for $8 an hour."

Practice Use what you discovered about ratios and rates to complete Exercises 7–10 on page 102.

3.1 Lesson

Key Vocabulary
ratio, *p. 100*
rate, *p. 100*
unit rate, *p. 100*

A **ratio** is a comparison of two quantities using division.

$\frac{3}{4}$, 3 to 4, 3 : 4

A **rate** is a ratio of two quantities with different units.

$\frac{60 \text{ miles}}{2 \text{ hours}}$

A rate with a denominator of 1 is called a **unit rate**.

$\frac{30 \text{ miles}}{1 \text{ hour}}$

EXAMPLE 1 Finding Ratios and Rates

There are 45 males and 60 females in a subway car. The subway car travels 2.5 miles in 5 minutes.

a. Find the ratio of males to females.

b. Find the speed of the subway car.

a. $\frac{\text{males}}{\text{females}} = \frac{45}{60} = \frac{3}{4}$

∴ The ratio of males to females is $\frac{3}{4}$.

b. 2.5 miles in 5 minutes $= \frac{2.5 \text{ mi}}{5 \text{ min}} = \frac{2.5 \text{ mi} \div 5}{5 \text{ min} \div 5} = \frac{0.5 \text{ mi}}{1 \text{ min}}$

∴ The speed is 0.5 mile per minute.

EXAMPLE 2 Finding a Rate from a Table

The table shows the amount of money you can raise by walking for a charity. Find your unit rate in dollars per mile.

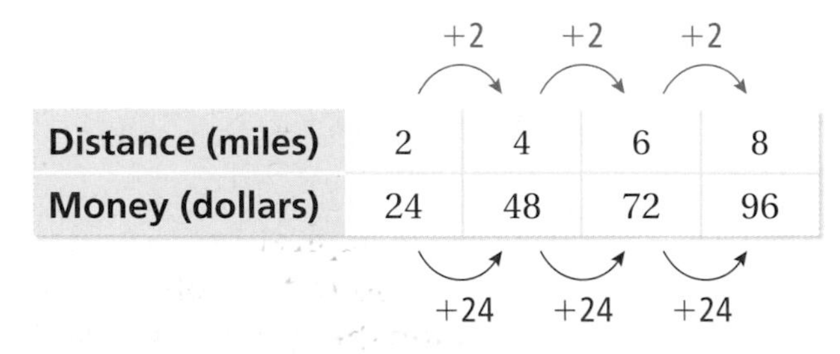

Distance (miles)	2	4	6	8
Money (dollars)	24	48	72	96

Use the table to find the unit rate.

$$\frac{\text{change in money}}{\text{change in distance}} = \frac{\$24}{2 \text{ mi}}$$ The money raised increases by $24 every 2 miles.

$$= \frac{\$12}{1 \text{ mi}}$$ Simplify.

∴ Your unit rate is $12 per mile.

Multi-Language Glossary at BigIdeasMath.com.

On Your Own

Now You're Ready
Exercises 11–26

1. In Example 1, find the ratio of females to males.
2. In Example 1, find the ratio of females to total passengers.
3. The table shows the distance that the International Space Station travels while orbiting Earth. Find the speed in miles per second.

Time (seconds)	3	6	9	12
Distance (miles)	14.4	28.8	43.2	57.6

EXAMPLE 3 Finding a Rate from a Line Graph

The graph shows the distance that sound travels through water. Find the speed of sound in kilometers per second.

Sound through Water

Step 1: Choose a point on the line.

The point (2, 3) shows you that sound travels 3 kilometers in 2 seconds.

Step 2: Find the speed.

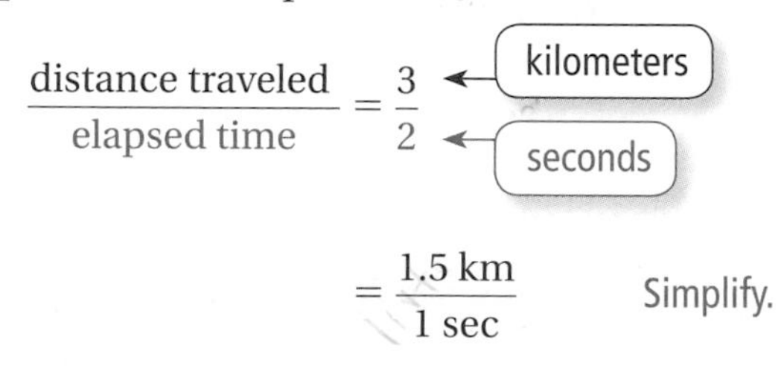

$$\frac{\text{distance traveled}}{\text{elapsed time}} = \frac{3}{2}$$

$$= \frac{1.5 \text{ km}}{1 \text{ sec}} \quad \text{Simplify.}$$

The speed is 1.5 kilometers per second.

On Your Own

4. **WHAT IF?** In Example 3, you use the point (4, 6) to find the speed. Does your answer change? Why or why not?
5. The graph shows the distance that sound travels through air. Find the speed of sound in kilometers per second.
6. Does sound travel faster in water or in air? Explain.

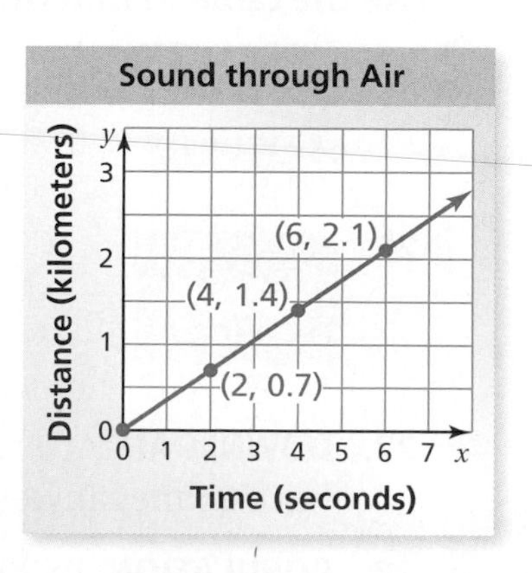

3.1 Exercises

Vocabulary and Concept Check

1. **VOCABULARY** How can you tell when a rate is a unit rate?
2. **WRITING** Why do you think rates are usually written as unit rates?
3. **OPEN-ENDED** Write a real-life rate that applies to you.

Estimate the unit rate.

4. \$74.75   5. \$1.19   6. \$2.35

Practice and Problem Solving

Find the product. List the units.

7. $8 \text{ h} \times \frac{\$9}{\text{h}}$ 8. $8 \text{ lb} \times \frac{\$3.50}{\text{lb}}$ 9. $14 \text{ sec} \times \frac{60 \text{ MB}}{\text{sec}}$ 10. $6 \text{ h} \times \frac{19 \text{ mi}}{\text{h}}$

Write the ratio as a fraction in simplest form.

① 11. 25 to 45   12. 63 : 28   13. 35 girls : 15 boys

14. 2 feet : 8 feet   15. 16 dogs to 12 cats   16. 51 correct : 9 incorrect

Find the unit rate.

17. 180 miles in 3 hours   18. 256 miles per 8 gallons   19. \$9.60 for 4 pounds

20. \$4.80 for 6 cans   21. 297 words in 5.5 minutes   22. 54 meters in 2.5 hours

Use the table to find the rate.

② 23.

Servings	0	1	2	3
Calories	0	90	180	270

24.

Days	0	1	2	3
Liters	0	1.6	3.2	4.8

25.

Packages	3	6	9	12
Servings	13.5	27	40.5	54

26.

Years	2	6	10	14
Feet	7.2	21.6	36	50.4

27. **DOWNLOAD** At 1 P.M., you have 24 megabytes of a movie. At 1:15 P.M., you have 96 megabytes. What is the download rate in megabytes per minute?

28. **POPULATION** In 2000, the U.S. population was 281 million people. In 2008, it was 305 million. What was the rate of population change per year?

29. **TICKETS** The graph shows the cost of buying tickets to a concert.

a. What does the point (4, 122) represent?

b. What is the unit rate?

c. What is the cost of buying 10 tickets?

Concert

Cost (dollars)

Tickets

(2, 61)

(4, 122)

(6, 183)

ADMIT ONE

30. **CRITICAL THINKING** Are the two statements equivalent? Explain your reasoning.

- The ratio of boys to girls is 2 to 3.
- The ratio of girls to boys is 3 to 2.

31. **TENNIS** A sports store sells three different packs of tennis balls. Which pack is the best buy? Explain.

Beverage	Serving Size	Calories	Sodium
Whole milk	1 cup	146	98 mg
Orange juice	1 pt	210	10 mg
Apple juice	24 fl oz	351	21 mg

32. **NUTRITION** The table shows nutritional information for three beverages.

a. Which has the most calories per fluid ounce?

b. Which has the least sodium per fluid ounce?

33. **Open-Ended** Fire hydrants are painted four different colors to indicate the rate at which water comes from the hydrant.

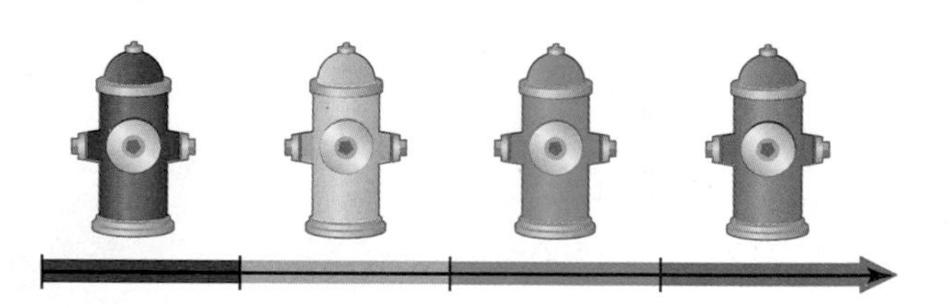

a. **RESEARCH** Use the Internet to find the ranges of the rates for each color.

b. Research why a firefighter needs to know the rate at which water comes out of the hydrant.

Fair Game Review What you learned in previous grades & lessons

Plot the ordered pair in a coordinate plane. *(Section 1.6)*

34. $A(-5, -2)$ 35. $B(-3, 0)$ 36. $C(-1, 2)$ 37. $D(1, 4)$

38. **MULTIPLE CHOICE** Which fraction is greater than $-\frac{2}{3}$ and less than $-\frac{1}{2}$? *(Section 2.1)*

Ⓐ $-\frac{3}{4}$ Ⓑ $-\frac{7}{12}$ Ⓒ $-\frac{5}{12}$ Ⓓ $-\frac{3}{8}$

3.2 Slope

Essential Question How can you compare two rates graphically?

COMMON CORE STATE STANDARDS
7.RP.2b
8.EE.5

1 ACTIVITY: Comparing Unit Rates

Work with a partner. The table shows the maximum speeds of several animals.

a. Find the missing speeds. Round your answers to the nearest tenth.

b. Which animal is fastest? Which animal is slowest?

c. Explain how you convert between the two units of speed.

Animal	Speed (miles per hour)	Speed (feet per second)
Antelope	61.0	
Black Mamba Snake		29.3
Cheetah		102.6
Chicken		13.2
Coyote	43.0	
Domestic Pig		16.0
Elephant		36.6
Elk		66.0
Giant Tortoise	0.2	
Giraffe	32.0	
Gray Fox		61.6
Greyhound	39.4	
Grizzly Bear		44.0
Human		41.0
Hyena	40.0	
Jackal	35.0	
Lion		73.3
Peregrine Falcon	200.0	
Quarter Horse	47.5	
Spider		1.76
Squirrel	12.0	
Thomson's Gazelle	50.0	
Three-Toed Sloth		0.2
Tuna	47.0	

2 ACTIVITY: Comparing Two Rates Graphically

Work with a partner. A cheetah and a Thomson's gazelle are running at constant speeds.

a. Find the missing distances.

	Cheetah	Gazelle
Time (seconds)	Distance (feet)	Distance (feet)
0	0	0
1	102.6	
2		
3		
4		
5		
6		
7		

b. Use the table to complete the line graph for each animal.

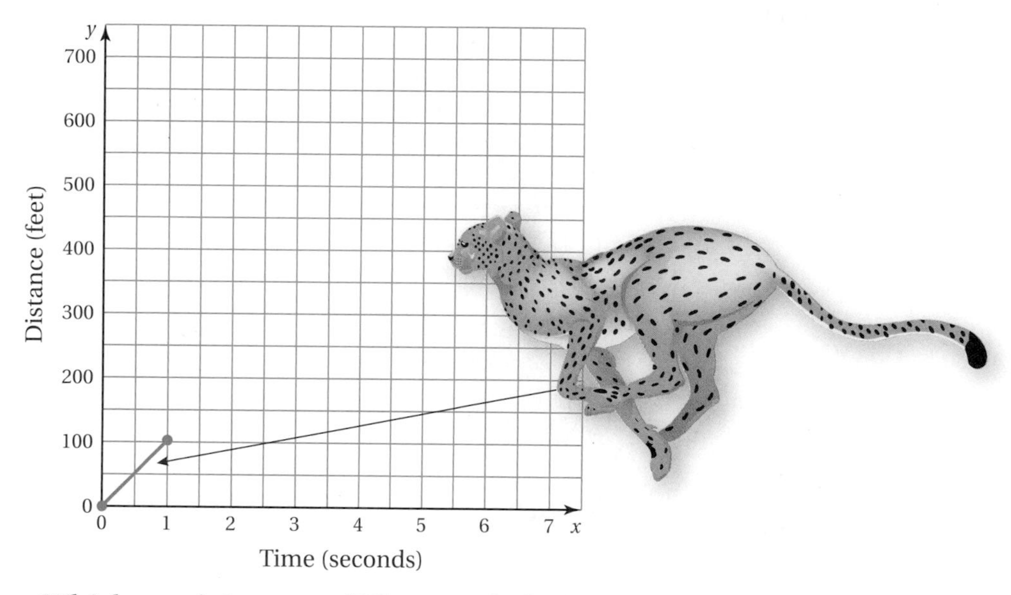

c. Which graph is steeper? The speed of which animal is greater?

What Is Your Answer?

3. **IN YOUR OWN WORDS** How can you compare two rates graphically? Explain your reasoning. Give some examples with your answer.

4. Choose 10 animals from Activity 1.

   a. Make a table for each animal similar to the table in Activity 2.

   b. Sketch a graph of the distances for each animal.

   c. Compare the steepness of the 10 graphs. What can you conclude?

3.2 Lesson

Check It Out
Lesson Tutorials
BigIdeasMath.com

Key Vocabulary
slope, *p. 106*

Key Idea

Slope

Slope is the rate of change between any two points on a line. It is a measure of the *steepness* of a line.

To find the slope of a line, find the ratio of the change in y (vertical change) to the change in x (horizontal change).

$$\text{slope} = \frac{\text{change in } y}{\text{change in } x}$$

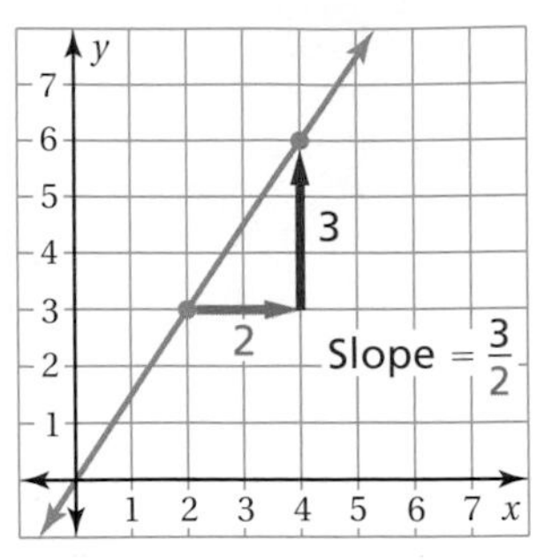

EXAMPLE 1 Finding Slopes

Find the slope of each line.

a.

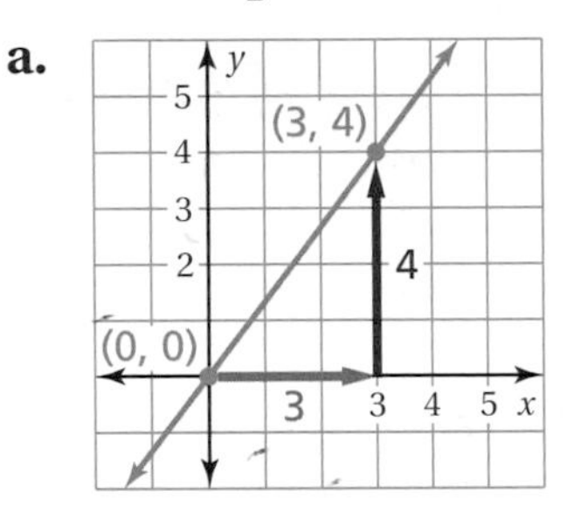

$$\text{slope} = \frac{\text{change in } y}{\text{change in } x}$$

$$= \frac{4}{3}$$

The slope of the line is $\frac{4}{3}$.

b.

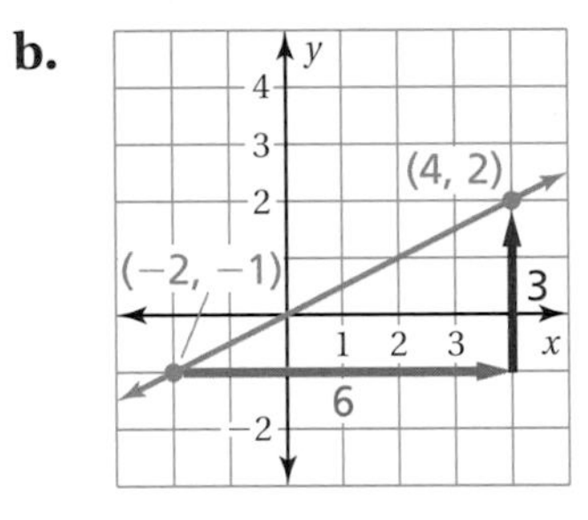

$$\text{slope} = \frac{\text{change in } y}{\text{change in } x}$$

$$= \frac{3}{6} = \frac{1}{2}$$

The slope of the line is $\frac{1}{2}$.

On Your Own

Now You're Ready
Exercises 4–9

Find the slope of the line.

1.

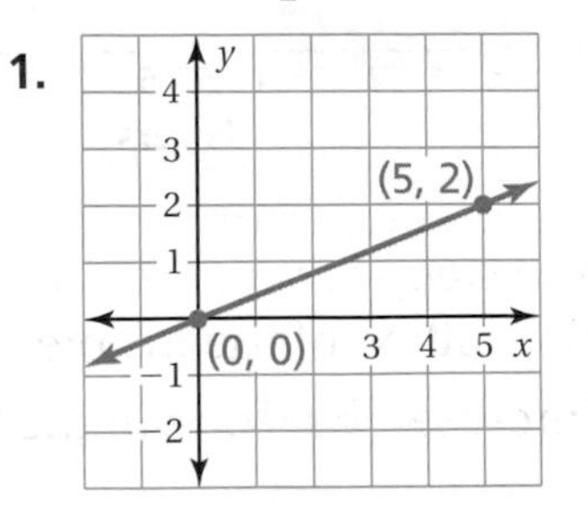

2.

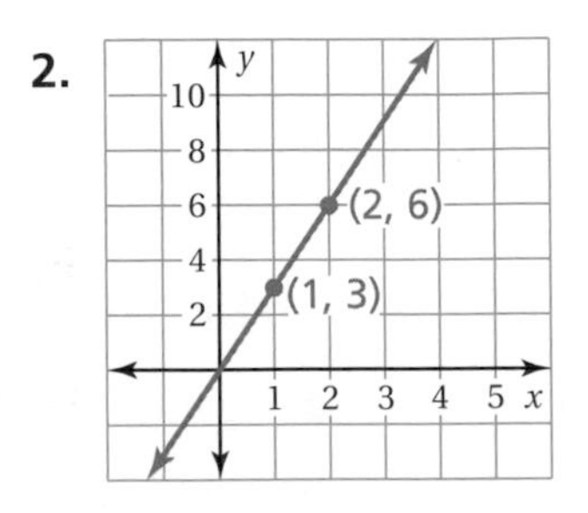

Multi-Language Glossary at BigIdeasMath.com.

EXAMPLE 2 Finding a Slope

The table shows your earnings for babysitting.

a. Graph the data.

b. Find and interpret the slope of the line through the points.

Hours, x	0	2	4	6	8	10
Earnings, y (dollars)	0	10	20	30	40	50

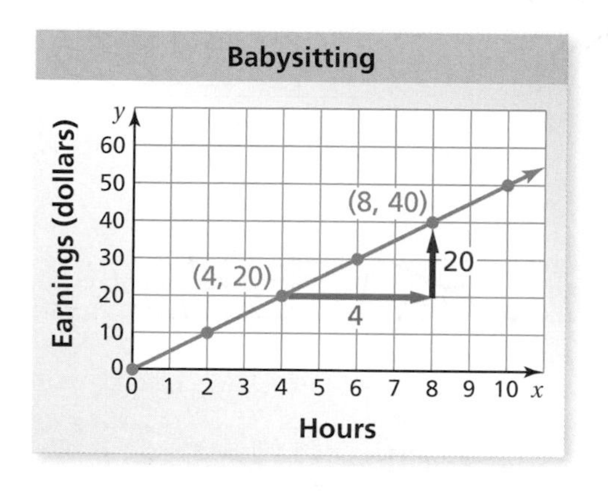

a. Graph the data. Draw a line through the points.

b. Choose any two points to find the slope of the line.

$$\text{slope} = \frac{\text{change in } y}{\text{change in } x}$$

$$= \frac{20}{4} \leftarrow \text{dollars}, \leftarrow \text{hours}$$

$$= 5$$

∴ The slope of the line is 5. So, you earn $5 per hour babysitting.

On Your Own

3. In Example 2, use two other points to find the slope. Does the slope change?

4. The graph shows the earnings of you and your friend for babysitting.

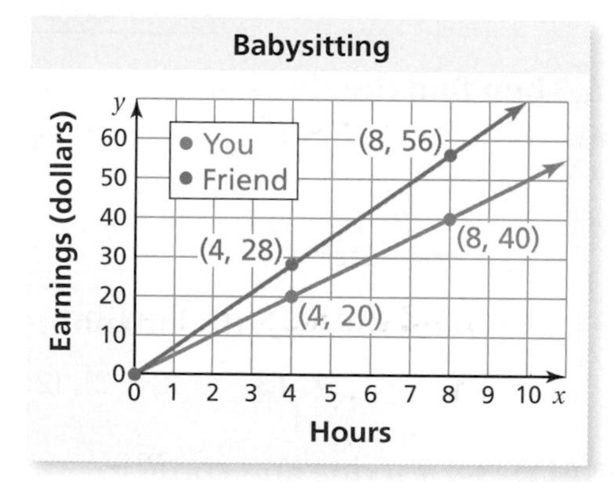

a. Compare the steepness of the lines. What does this mean in the context of the problem?

b. Find and interpret the slope of the blue line.

3.2 Exercises

Check It Out
Help with Homework
BigIdeasMath.com

Vocabulary and Concept Check

1. **VOCABULARY** Is there a connection between rate and slope? Explain.
2. **REASONING** Which line has the greatest slope?
3. **REASONING** Is it more difficult to run up a ramp with a slope of $\frac{1}{5}$ or a ramp with a slope of 5? Explain.

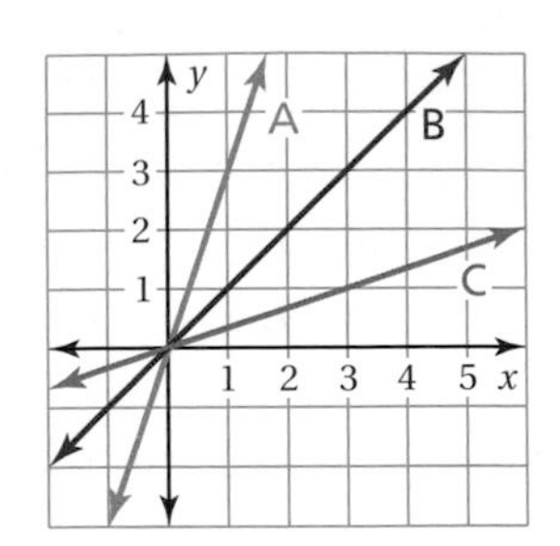

Practice and Problem Solving

Find the slope of the line.

① **4.** **7.**

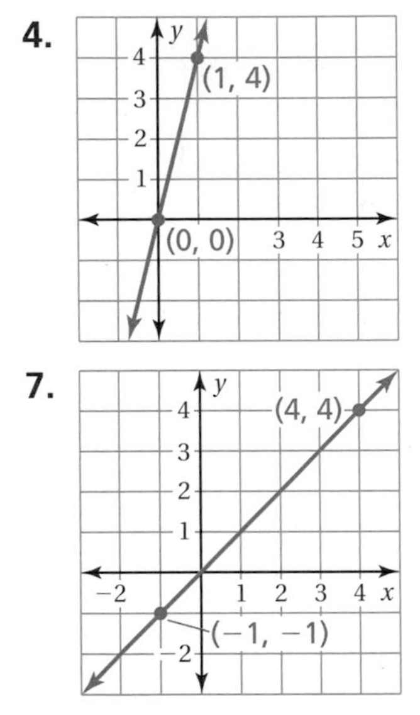

5. **8.**

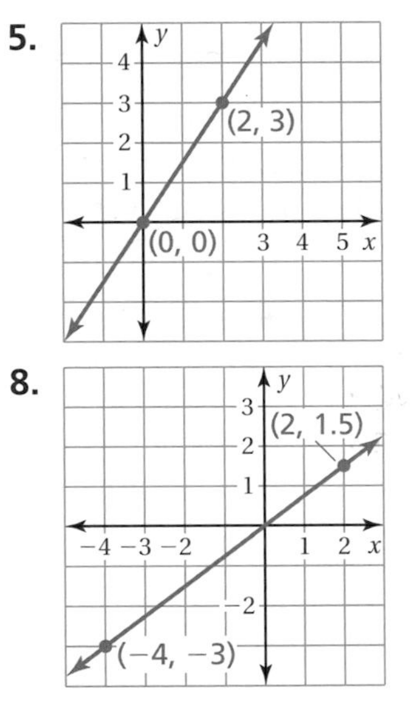

6. **9.**

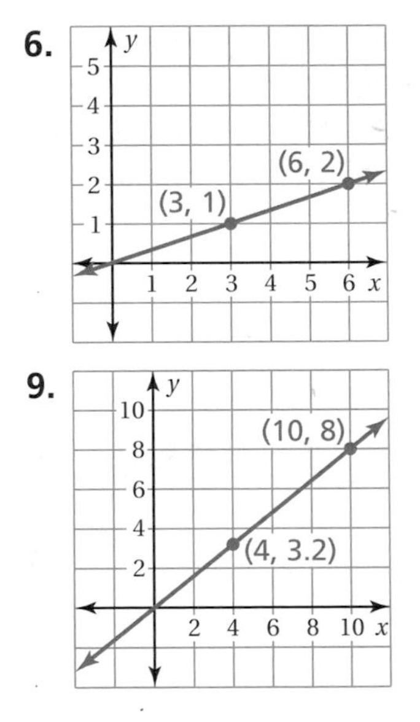

Graph the data. Then find the slope of the line through the points.

② **10.**

Minutes, x	3	5	7	9
Words, y	135	225	315	405

11.

Gallons, x	5	10	15	20
Miles, y	162.5	325	487.5	650

Graph the line that passes through the two points. Then find the slope of the line.

12. (0, 0), (5, 8)

13. (−2, −2), (2, 2)

14. (10, 4), (−5, −2)

15. ERROR ANALYSIS Describe and correct the error in finding the slope of the line passing through (0, 0) and (4, 5).

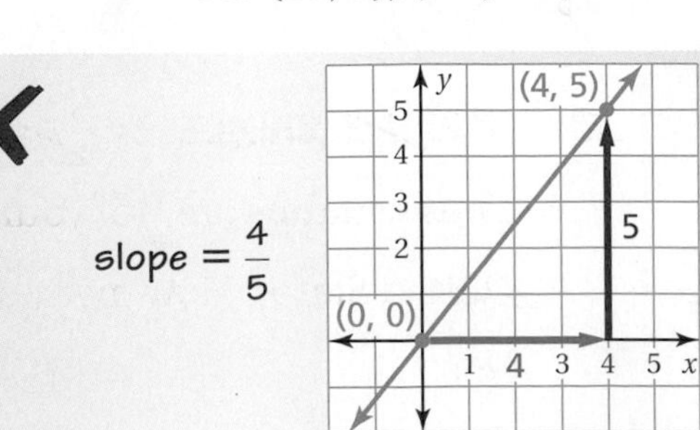

16. **CAMPING** The graph shows the amount of money you and a friend are saving for a camping trip.

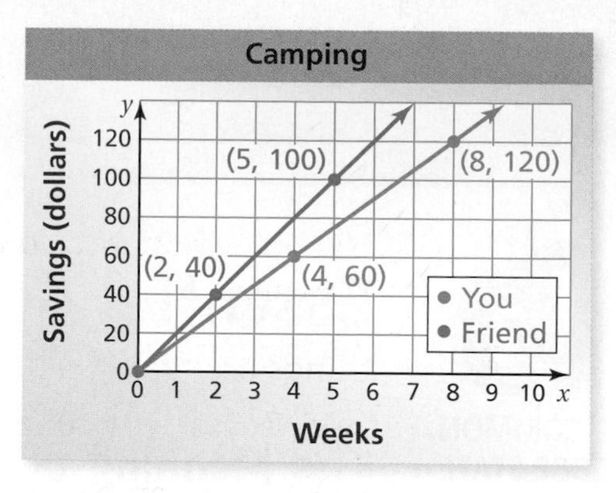

a. Compare the steepness of the lines. What does this mean in the context of the problem?

b. Find the slope of each line.

c. How much more money does your friend save each week than you?

d. The camping trip costs $165. How long will it take you to save enough money?

17. **MAPS** The table shows data from a key to a map of Ohio.

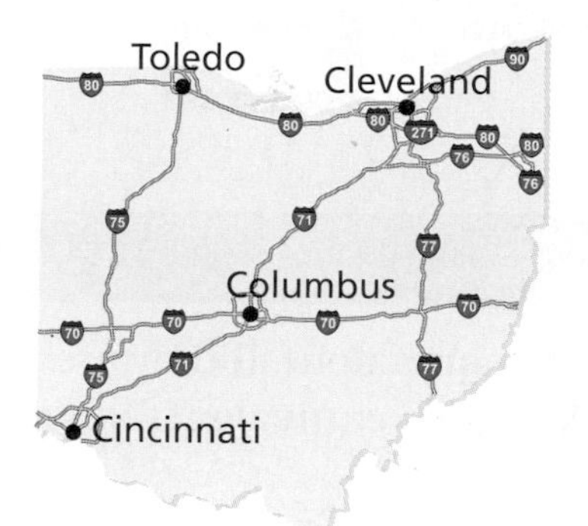

Distance on Map (mm), x	10	20	30	40
Actual Distance (mi), y	25	50	75	100

a. Graph the data.

b. Find the slope of the line. What does this mean in the context of the problem?

c. The map distance between Toledo and Columbus is 48 millimeters. What is the actual distance?

d. Cincinnati is about 225 miles from Cleveland. What is the distance between these cities on the map?

18. **CRITICAL THINKING** What is the slope of a line that passes through the points (2, 0) and (5, 0)? Explain.

19. Number Sense A line has a slope of 2. It passes through the points (1, 2) and (3, y). What is the value of y?

Fair Game Review What you learned in previous grades & lessons

Copy and complete the statement using <, >, or =. *(Section 2.1)*

20. $\frac{9}{2}$ ☐ $\frac{8}{3}$

21. $-\frac{8}{15}$ ☐ $\frac{10}{18}$

22. $\frac{-6}{24}$ ☐ $\frac{-2}{8}$

Multiply. *(Section 2.3)*

23. $-\frac{3}{5} \times \frac{8}{6}$

24. $1\frac{1}{2} \times \left(-\frac{6}{15}\right)$

25. $-2\frac{1}{4} \times -1\frac{1}{3}$

26. **MULTIPLE CHOICE** You have 18 stamps from Mexico in your stamp collection. These stamps are $\frac{3}{8}$ of your collection. The rest of the stamps are from the United States. How many stamps are from the United States? *(Section 2.5)*

Ⓐ 12 Ⓑ 24 Ⓒ 30 Ⓓ 48

3.3 Proportions

COMMON CORE STATE STANDARDS
7.RP.2a

Essential Question How can proportions help you decide when things are "fair?"

The Meaning of a Word ● Proportional

When you work toward a goal, your success is usually **proportional** to the amount of work you put in.

An equation stating that two ratios are equal is a **proportion**.

1 ACTIVITY: Determining Proportions

Work with a partner. Tell whether the two ratios are equivalent. If they are not equivalent, change the second day to make the ratios equivalent. Explain your reasoning.

a. On the first day, you pay $5 for 2 boxes of popcorn. The next day, you pay $7.50 for 3 boxes.

First Day

$$\frac{\$5.00}{\$7.50} \stackrel{?}{=} \frac{2 \text{ boxes}}{3 \text{ boxes}}$$

Next Day

b. On the first day, it takes you 3 hours to drive 135 miles. The next day, it takes you 5 hours to drive 200 miles.

First Day

$$\frac{3 \text{ h}}{5 \text{ h}} \stackrel{?}{=} \frac{135 \text{ mi}}{200 \text{ mi}}$$

Next Day

c. On the first day, you walk 4 miles and burn 300 calories. The next day, you walk 3 miles and burn 225 calories.

First Day

$$\frac{4 \text{ mi}}{3 \text{ mi}} \stackrel{?}{=} \frac{300 \text{ cal}}{225 \text{ cal}}$$

Next Day

d. On the first day, you download 5 songs and pay $2.25. The next day, you download 4 songs and pay $2.00.

First Day

$$\frac{5 \text{ songs}}{4 \text{ songs}} \stackrel{?}{=} \frac{\$2.25}{\$2.00}$$

Next Day

2 ACTIVITY: Checking a Proportion

Work with a partner.

a. It is said that "one year in a dog's life is equivalent to seven years in a human's life." Explain why Newton thinks he has a score of 105 points. Did he solve the proportion correctly?

$$\frac{1\text{ year}}{7\text{ years}} \stackrel{?}{=} \frac{15\text{ points}}{105\text{ points}}$$

b. If Newton thinks his score is 98 points, how many points does he actually have? Explain your reasoning.

"I got 15 on my online test. That's 105 in dog points! Isn't that an A+?"

3 ACTIVITY: Determining Fairness

Work with a partner. Write a ratio for each sentence. If they are equal, then the answer is "It is fair." If they are not equal, then the answer is "It is not fair." Explain your reasoning.

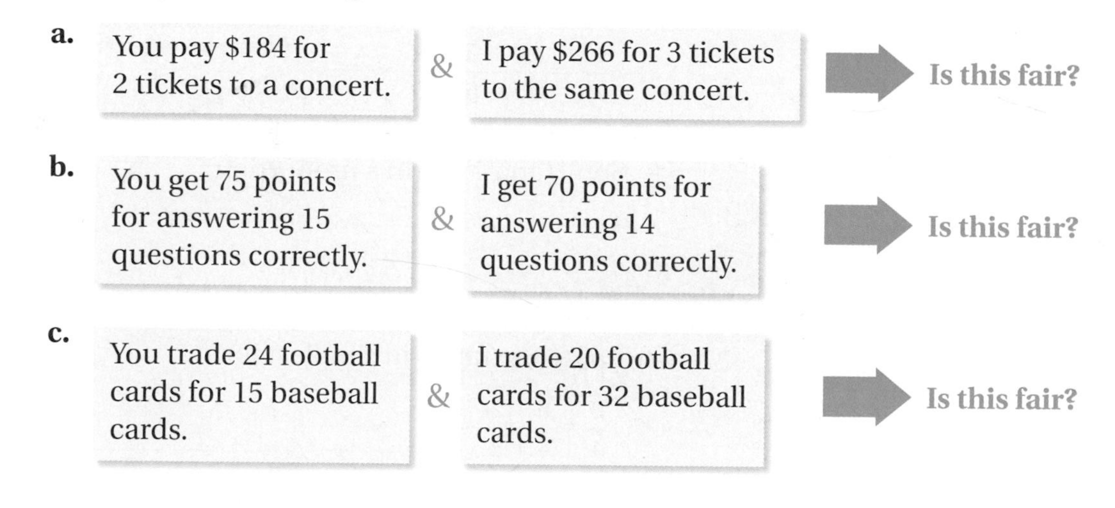

a. You pay $184 for 2 tickets to a concert. & I pay $266 for 3 tickets to the same concert. → Is this fair?

b. You get 75 points for answering 15 questions correctly. & I get 70 points for answering 14 questions correctly. → Is this fair?

c. You trade 24 football cards for 15 baseball cards. & I trade 20 football cards for 32 baseball cards. → Is this fair?

What Is Your Answer?

4. Find a recipe for something you like to eat. Then show how two of the ingredient amounts are proportional when you double or triple the recipe.

5. **IN YOUR OWN WORDS** How can proportions help you decide when things are "fair?" Give an example.

Use what you discovered about proportions to complete Exercises 17–22 on page 114.

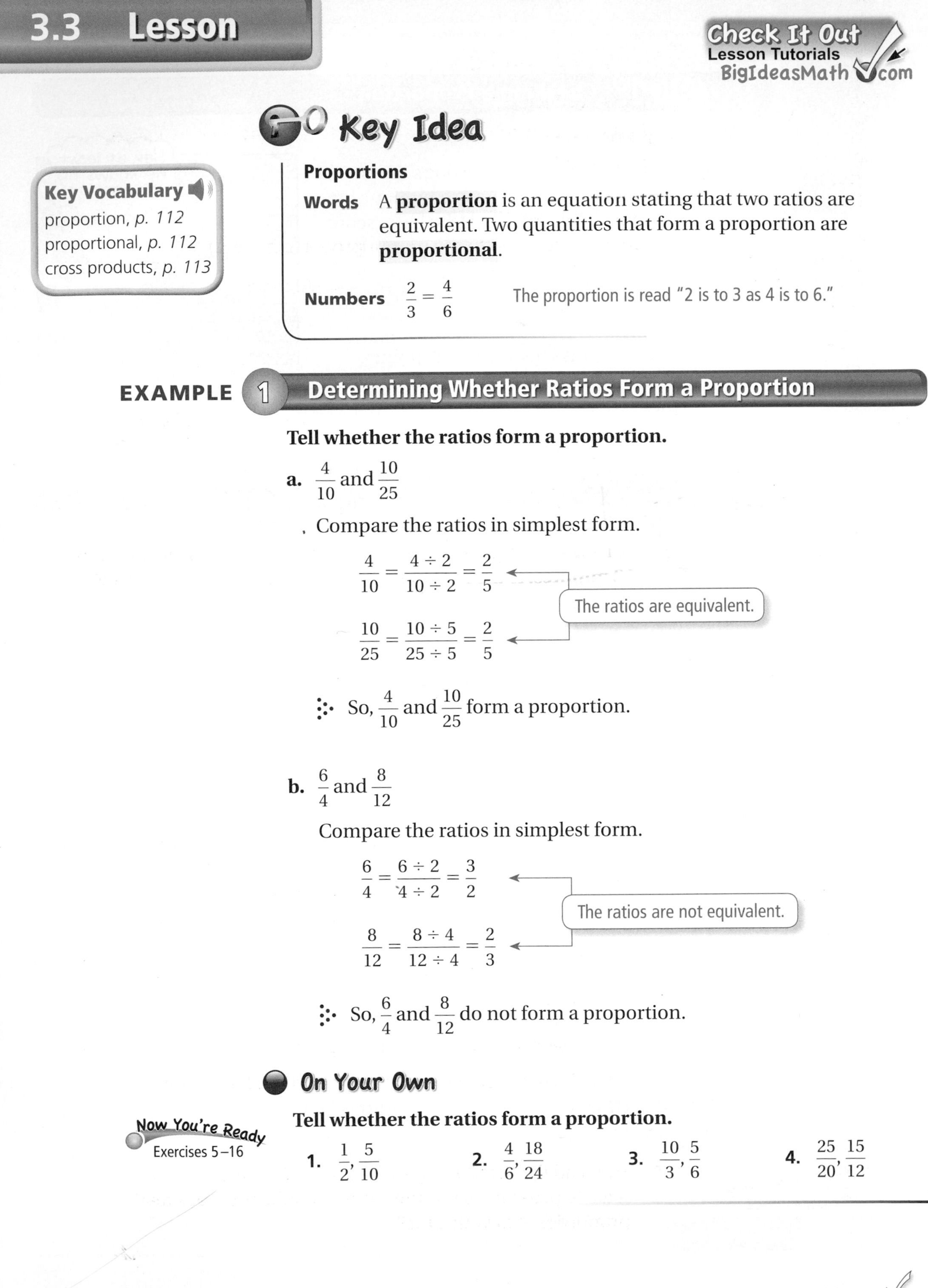

3.3 Lesson

Check It Out
Lesson Tutorials
BigIdeasMath.com

Key Idea

Key Vocabulary
proportion, *p. 112*
proportional, *p. 112*
cross products, *p. 113*

Proportions

Words A **proportion** is an equation stating that two ratios are equivalent. Two quantities that form a proportion are **proportional**.

Numbers $\frac{2}{3} = \frac{4}{6}$ The proportion is read "2 is to 3 as 4 is to 6."

EXAMPLE 1 Determining Whether Ratios Form a Proportion

Tell whether the ratios form a proportion.

a. $\frac{4}{10}$ and $\frac{10}{25}$

Compare the ratios in simplest form.

$$\frac{4}{10} = \frac{4 \div 2}{10 \div 2} = \frac{2}{5}$$

$$\frac{10}{25} = \frac{10 \div 5}{25 \div 5} = \frac{2}{5}$$

The ratios are equivalent.

So, $\frac{4}{10}$ and $\frac{10}{25}$ form a proportion.

b. $\frac{6}{4}$ and $\frac{8}{12}$

Compare the ratios in simplest form.

$$\frac{6}{4} = \frac{6 \div 2}{4 \div 2} = \frac{3}{2}$$

$$\frac{8}{12} = \frac{8 \div 4}{12 \div 4} = \frac{2}{3}$$

The ratios are not equivalent.

So, $\frac{6}{4}$ and $\frac{8}{12}$ do not form a proportion.

On Your Own

Now You're Ready
Exercises 5–16

Tell whether the ratios form a proportion.

1. $\frac{1}{2}, \frac{5}{10}$ **2.** $\frac{4}{6}, \frac{18}{24}$ **3.** $\frac{10}{3}, \frac{5}{6}$ **4.** $\frac{25}{20}, \frac{15}{12}$

Multi-Language Glossary at BigIdeasMath.com.

Key Ideas

Cross Products

In the proportion $\frac{a}{b} = \frac{c}{d}$, the products $a \cdot d$ and $b \cdot c$ are called **cross products**.

Cross Products Property

Words The cross products of a proportion are equal.

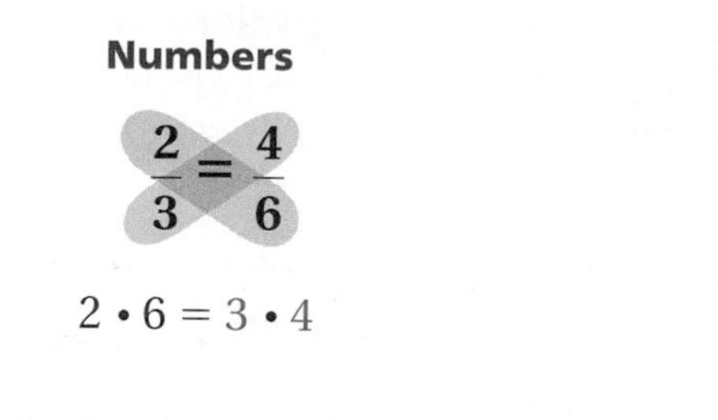

Numbers

$2 \cdot 6 = 3 \cdot 4$

Algebra

$ad = bc$,
where $b \neq 0$ and $d \neq 0$

EXAMPLE 2 Identifying Proportional Relationships

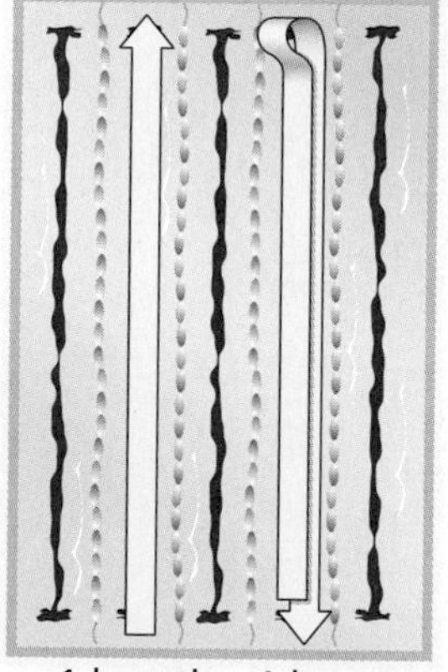

You swim your first 4 laps in 2.4 minutes. You complete 16 laps in 12 minutes. Is the number of laps proportional to your time?

Method 1: Compare unit rates.

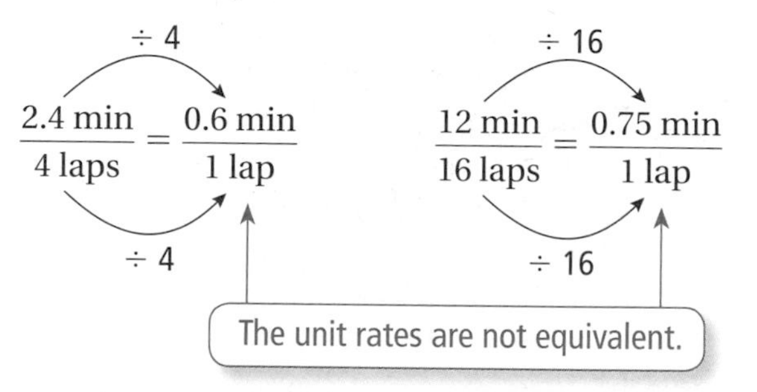

÷ 4
2.4 min / 4 laps = 0.6 min / 1 lap
÷ 4
÷ 16
12 min / 16 laps = 0.75 min / 1 lap
÷ 16
The unit rates are not equivalent.

So, the number of laps is not proportional to the time.

Method 2: Use the Cross Products Property.

$\frac{2.4 \text{ min}}{4 \text{ laps}} \stackrel{?}{=} \frac{12 \text{ min}}{16 \text{ laps}}$	Test to see if the rates are equivalent.
$2.4 \cdot 16 \stackrel{?}{=} 4 \cdot 12$	Find the cross products.
$38.4 \neq 48$	The cross products are not equal.

So, the number of laps is not proportional to the time.

On Your Own

5. You read the first 20 pages of a book in 25 minutes. You read 36 pages in 45 minutes. Is the number of pages read proportional to your time?

3.3 Exercises

Vocabulary and Concept Check

1. **VOCABULARY** What does it mean for two ratios to form a proportion?
2. **VOCABULARY** What are two ways you can tell that two ratios form a proportion?
3. **OPEN-ENDED** Write two ratios that are equivalent to $\frac{3}{5}$.
4. **WHICH ONE DOESN'T BELONG?** Which ratio does *not* belong with the other three? Explain your reasoning.

$\frac{4}{10}$ $\frac{2}{5}$ $\frac{3}{5}$ $\frac{6}{15}$

Practice and Problem Solving

Tell whether the ratios form a proportion.

(1) 5. $\frac{1}{3}, \frac{7}{21}$

6. $\frac{1}{5}, \frac{6}{30}$

7. $\frac{3}{4}, \frac{24}{18}$

8. $\frac{2}{5}, \frac{40}{16}$

9. $\frac{48}{9}, \frac{16}{3}$

10. $\frac{18}{27}, \frac{33}{44}$

11. $\frac{7}{2}, \frac{16}{6}$

12. $\frac{12}{10}, \frac{14}{12}$

13. $\frac{27}{15}, \frac{18}{10}$

14. $\frac{4}{15}, \frac{15}{42}$

15. $\frac{76}{36}, \frac{19}{9}$

16. $\frac{49}{77}, \frac{38}{57}$

Tell whether the two rates form a proportion.

(2) 17. 7 inches in 9 hours; 42 inches in 54 hours

18. 12 players from 21 teams; 15 players from 24 teams

19. 440 calories in 4 servings; 300 calories in 3 servings

20. 120 units made in 5 days; 88 units made in 4 days

21. 66 wins in 82 games; 99 wins in 123 games

22. 68 hits in 172 at bats; 43 hits in 123 at bats

23. **FITNESS** You can do 90 sit-ups in 2 minutes. Your friend can do 135 sit-ups in 3 minutes. Are these rates proportional? Explain.

24. **HEARTBEAT** Find the heartbeat rates of you and your friend. Do these rates form a proportion? Explain.

	Heartbeats	Seconds
You	22	20
Friend	18	15

Tell whether the ratios form a proportion.

25. $\frac{3}{8}, \frac{31.5}{84}$

26. $\frac{14}{30}, \frac{75.6}{180}$

27. $\frac{2.5}{4}, \frac{7}{11.2}$

28. PAY RATE You earn $56 walking your neighbor's dog for 8 hours. Your friend earns $36 painting your neighbor's fence for 4 hours.

- **a.** What is your pay rate?
- **b.** What is your friend's pay rate?
- **c.** Are the pay rates equivalent? Explain.

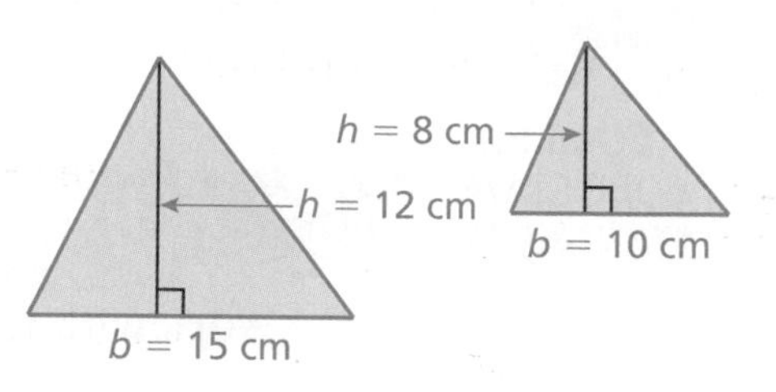

29. GEOMETRY Are the ratios of h to b in the two triangles proportional? Explain.

30. MUSIC You can buy 3 CDs for $52.20 or 5 CDs for $62.45. Are the rates proportional? Explain.

31. BASEBALL The table shows pitching statistics for four pitchers during the 2008 season.

- **a.** Which pitcher has the highest ratio of strikeouts to walks?
- **b.** Which of the pitchers have equivalent strikeout to walk ratios?

2008 Season		
Pitcher	Strikeouts	Walks
Pitcher 1	6	8
Pitcher 2	8	4
Pitcher 3	10	1
Pitcher 4	10	5

32. NAIL POLISH A specific shade of red nail polish requires 7 parts red to 2 parts yellow. A mixture contains 35 quarts of red and 8 quarts of yellow. How can you fix the mixture to make the correct shade of red?

33. LOGIC The ratio of quarters to dimes in a coin collection is 5 : 3. The same number of new quarters and dimes are added to the collection.

- **a.** Is the ratio of quarters to dimes still 5 : 3?
- **b.** If so, illustrate your answer with an example. If not, show why with a "counterexample."

34. Critical Thinking Ratio A is equivalent to ratio B. Ratio B is equivalent to ratio C. Is ratio A equivalent to ratio C? Explain.

Fair Game Review What you learned in previous grades & lessons

Add or subtract. *(Sections 1.2 and 1.3)*

35. $-28 + 15$

36. $-6 + (-11)$

37. $-10 - 8$

38. $-17 - (-14)$

39. MULTIPLE CHOICE Which fraction is not equivalent to $\frac{2}{6}$? *(Skills Review Handbook)*

Ⓐ $\frac{1}{3}$ Ⓑ $\frac{12}{36}$ Ⓒ $\frac{4}{12}$ Ⓓ $\frac{6}{9}$

3.4 Writing Proportions

COMMON CORE STATE STANDARDS
7.RP.2a

Essential Question How can you write a proportion that solves a problem in real life?

1 ACTIVITY: Writing Proportions

Work with a partner. A rough rule for finding the correct bat length is "The bat length should be half of the batter's height." So, a 62-inch-tall batter uses a bat that is 31 inches long. Write a proportion to find the bat length for each given batter height.

a. 58 inches

b. 60 inches

c. 64 inches

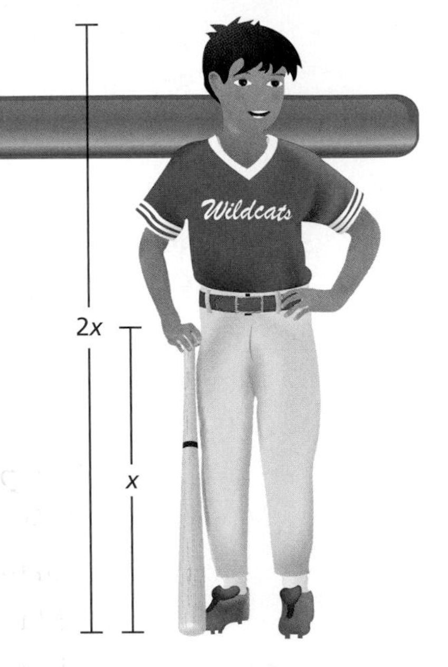

2 ACTIVITY: Bat Lengths

Work with a partner. Here is a more accurate table for determining the bat length for a batter. Find all of the batter heights for which the rough rule in Activity 1 is exact.

	Height of Batter (inches)							
Weight of Batter (pounds)	**45–48**	**49–52**	**53–56**	**57–60**	**61–64**	**65–68**	**69–72**	**Over 72**
Under 61	28	29	29					
61–70	28	29	30	30				
71–80	28	29	30	30	31			
81–90	29	29	30	30	31	32		
91–100	29	30	30	31	31	32		
101–110	29	30	30	31	31	32		
111–120	29	30	30	31	31	32		
121–130	29	30	30	31	32	33	33	
131–140	30	30	31	31	32	33	33	
141–150	30	30	31	31	32	33	33	
151–160	30	31	31	32	32	33	33	33
161–170		31	31	32	32	33	33	34
171–180				32	33	33	34	34
Over 180					33	33	34	34

3 ACTIVITY: Writing Proportions

Work with a partner. The batting average of a baseball player is the number of "hits" divided by the number of "at bats."

$$\text{Batting average} = \frac{\text{Hits } (H)}{\text{At bats } (A)}$$

A player whose batting average is 0.250 is said to be "batting 250."

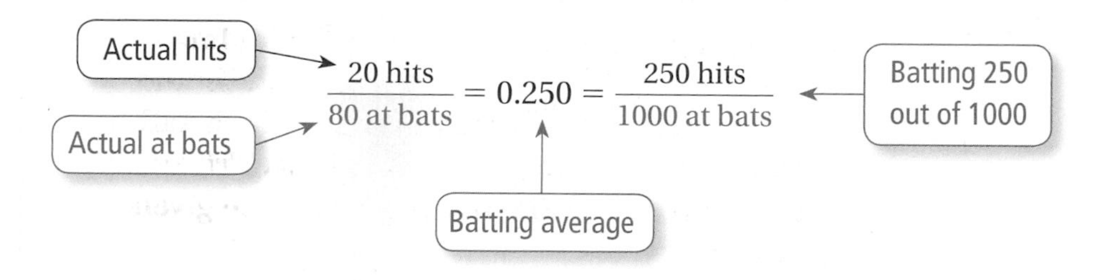

Write a proportion to find how many hits H a player needs to achieve the given batting average. Then solve the proportion.

a. 50 times at bat; batting average is 0.200.

b. 84 times at bat; batting average is 0.250.

c. 80 times at bat; batting average is 0.350.

d. 1 time at bat; batting average is 1.000.

What Is Your Answer?

4. **IN YOUR OWN WORDS** How can you write a proportion that solves a problem in real life?

5. Two players have the same batting average.

	At Bats	Hits	Batting Average
Player 1	132	45	
Player 2	132	45	

Player 1 gets four hits in the next five at bats. Player 2 gets three hits in the next three at bats.

a. Who has the higher batting average?

b. Does this seem fair? Explain your reasoning.

Practice Use what you discovered about proportions to complete Exercises 4–7 on page 120.

3.4 Lesson

One way to write a proportion is to use a table.

	Last Month	This Month
Purchase	2 ringtones	3 ringtones
Total Cost	6 dollars	x dollars

Use the columns or the rows to write a proportion.

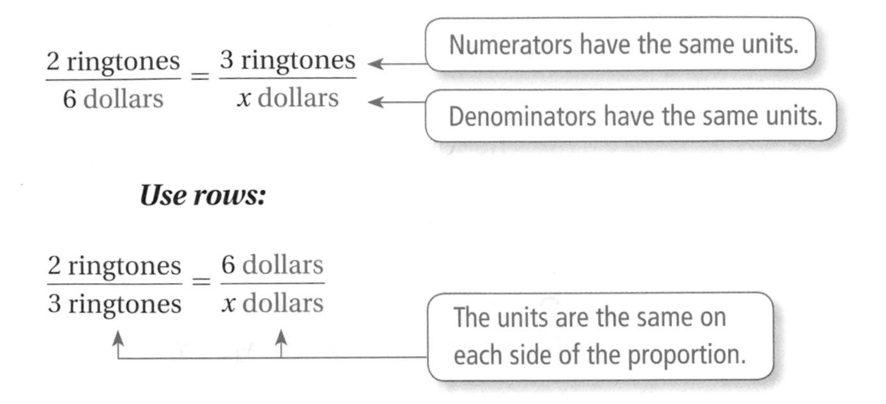

Use columns:
$\frac{2 \text{ ringtones}}{6 \text{ dollars}} = \frac{3 \text{ ringtones}}{x \text{ dollars}}$
Numerators have the same units.
Denominators have the same units.
Use rows:
$\frac{2 \text{ ringtones}}{3 \text{ ringtones}} = \frac{6 \text{ dollars}}{x \text{ dollars}}$
The units are the same on each side of the proportion.

EXAMPLE 1 Writing a Proportion

Black Bean Soup

1.5 cups black beans
0.5 cup salsa
2 cups water
1 tomato
2 teaspoons seasoning

A chef increases the amounts of ingredients in a recipe to make a proportional recipe. The new recipe has 6 cups of black beans. Write a proportion that gives the number x of tomatoes in the new recipe.

Organize the information in a table.

	Original Recipe	New Recipe
Black Beans	1.5 cups	6 cups
Tomatoes	1 tomato	x tomatoes

∴ One proportion is $\frac{1.5 \text{ cups beans}}{1 \text{ tomato}} = \frac{6 \text{ cups beans}}{x \text{ tomatoes}}$.

On Your Own

Now You're Ready
Exercises 8–11

1. In Example 1, write a different proportion that gives the number x of tomatoes in the new recipe.
2. In Example 1, write a proportion that gives the amount y of water in the new recipe.

EXAMPLE 2 Solving Proportions Using Mental Math

Solve $\frac{3}{2} = \frac{x}{8}$.

Step 1: Think: The product of 2 and what number is 8?

$$\frac{3}{2} = \frac{x}{8}$$

$2 \times ? = 8$

Step 2: Because the product of 2 and 4 is 8, multiply the numerator by 4 to find x.

$3 \times 4 = 12$

$$\frac{3}{2} = \frac{x}{8}$$

$2 \times 4 = 8$

∴ The solution is $x = 12$.

EXAMPLE 3 Solving Proportions Using Mental Math

In Example 1, how many tomatoes are in the new recipe?

Solve the proportion $\frac{1.5}{1} = \frac{6}{x}$.

cups black beans

tomatoes

Step 1: Think: The product of 1.5 and what number is 6?

$1.5 \times ? = 6$

$$\frac{1.5}{1} = \frac{6}{x}$$

Step 2: Because the product of 1.5 and 4 is 6, multiply the denominator by 4 to find x.

$1.5 \times 4 = 6$

$$\frac{1.5}{1} = \frac{6}{x}$$

$1 \times 4 = 4$

∴ So, there are 4 tomatoes in the new recipe.

On Your Own

Now You're Ready
Exercises 16–21

Solve the proportion.

3. $\frac{5}{8} = \frac{20}{d}$

4. $\frac{7}{z} = \frac{14}{10}$

5. $\frac{21}{24} = \frac{x}{8}$

6. A school has 950 students. The ratio of female students to all students is $\frac{48}{95}$. Write and solve a proportion to find the number f of students that are female.

3.4 Exercises

Vocabulary and Concept Check

1. **WRITING** Describe two ways you can use a table to write a proportion.

2. **WRITING** What is your first step when solving $\frac{x}{15} = \frac{3}{5}$? Explain.

3. **OPEN-ENDED** Write a proportion using an unknown value x and the ratio 5 : 6. Then solve it.

Practice and Problem Solving

Write a proportion to find how many points a student needs to score on the test to get the given score.

4. Test worth 50 points; test score of 40%

5. Test worth 50 points; test score of 78%

6. Test worth 80 points; test score of 80%

7. Test worth 150 points; test score of 96%

Use the table to write a proportion.

① 8.

	Game 1	Game 2
Points	12	18
Shots	14	w

9.

	May	June
Winners	n	34
Entries	85	170

10.

	Today	Yesterday
Miles	15	m
Hours	2.5	4

11.

	Race 1	Race 2
Meters	100	200
Seconds	x	22.4

12. **ERROR ANALYSIS** Describe and correct the error in writing the proportion.

✗

	Monday	Tuesday
Dollars	2.08	d
Ounces	8	16

$\frac{2.08}{16} = \frac{d}{8}$

13. **T-SHIRTS** You can buy three T-shirts for \$24. Write a proportion that gives the cost c of buying seven T-shirts.

14. **COMPUTERS** A school requires two computers for every five students. Write a proportion that gives the number c of computers needed for 145 students.

15. **SWIM TEAM** The school team has 80 swimmers. The ratio of 6th grade swimmers to all swimmers is 5 : 16. Write a proportion that gives the number s of 6th grade swimmers.

Solve the proportion.

16. $\frac{1}{4} = \frac{z}{20}$

17. $\frac{3}{4} = \frac{12}{y}$

18. $\frac{35}{k} = \frac{7}{3}$

19. $\frac{15}{8} = \frac{45}{c}$

20. $\frac{b}{36} = \frac{5}{9}$

21. $\frac{1.4}{2.5} = \frac{g}{25}$

22. **ORCHESTRA** In an orchestra, the ratio of trombones to violas is 1 to 3.

    a. There are nine violas. Write a proportion that gives the number t of trombones in the orchestra.

    b. How many trombones are in the orchestra?

23. **ATLANTIS** Your science teacher has a 1 : 200 scale model of the Space Shuttle Atlantis. Which of the proportions can be used to find the actual length x of Atlantis? Explain.

$\frac{1}{200} = \frac{19.5}{x}$ $\quad\frac{1}{200} = \frac{x}{19.5}$ $\quad\frac{200}{19.5} = \frac{x}{1}$ $\quad\frac{x}{200} = \frac{1}{19.5}$

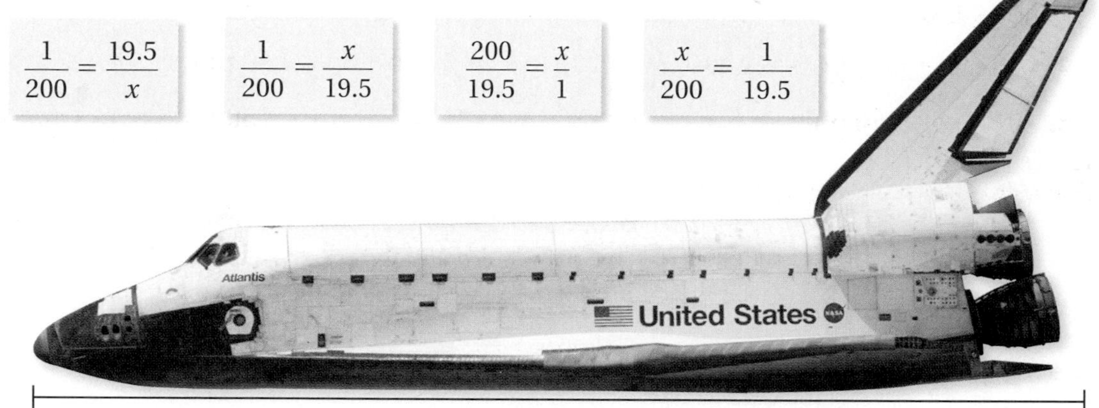

24. **YOU BE THE TEACHER** Your friend says "$48x = 6 \cdot 12$." Is your friend right? Explain.

Solve $\frac{6}{x} = \frac{12}{48}$.

25. **Reasoning** There are 180 white lockers in the school. There are 3 white lockers for every 5 blue lockers. How many lockers are in the school?

Fair Game Review What you learned in previous grades & lessons

Solve the equation. *(Section 2.5)*

26. $\frac{x}{6} = 25$

27. $8x = 72$

28. $150 = 2x$

29. $35 = \frac{x}{4}$

30. **MULTIPLE CHOICE** Which is the slope of a line? *(Section 3.2)*

Ⓐ $\frac{\text{change in } y}{1}$ Ⓑ $\frac{\text{change in } x}{1}$ Ⓒ $\frac{\text{change in } x}{\text{change in } y}$ Ⓓ $\frac{\text{change in } y}{\text{change in } x}$

3.5 Solving Proportions

COMMON CORE STATE STANDARDS
7.RP.2a
7.RP.2b
7.RP.2d

Essential Question

How can you use ratio tables and cross products to solve proportions in science?

1 ACTIVITY: Solving a Proportion in Science

SCIENCE Scientists use *ratio tables* to determine the amount of a compound (like salt) that is dissolved in a solution. Work with a partner to show how scientists use cross products to determine the unknown quantity in a ratio.

a. Sample: Salt Water

Salt Water	1 L	3 L
Salt	250 g	x g

1 liter　　3 liter

$\frac{3 \text{ L}}{1 \text{ L}} = \frac{x \text{ g}}{250 \text{ g}}$　　Write proportion.

$3 \cdot 250 = 1 \cdot x$　　Set cross products equal.

$750 = x$　　Simplify.

So, there are 750 grams of salt in the 3-liter solution.

b. White Glue Solution

Water	½ cup	1 cup
White Glue	½ cup	x cups

c. Borax Solution

Borax	1 tsp	2 tsp
Water	1 cup	x cups

d. Slime (see recipe)

Borax Solution	½ cup	1 cup
White Glue Solution	y cups	x cups

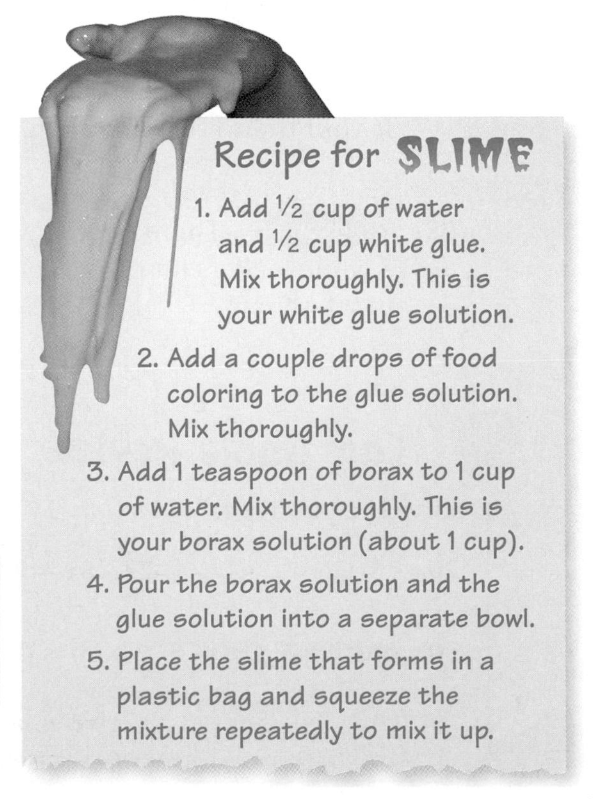

Recipe for SLIME
1. Add ½ cup of water and ½ cup white glue. Mix thoroughly. This is your white glue solution.
2. Add a couple drops of food coloring to the glue solution. Mix thoroughly.
3. Add 1 teaspoon of borax to 1 cup of water. Mix thoroughly. This is your borax solution (about 1 cup).
4. Pour the borax solution and the glue solution into a separate bowl.
5. Place the slime that forms in a plastic bag and squeeze the mixture repeatedly to mix it up.

2 ACTIVITY: The Game of Criss Cross

CRISS CROSS

Preparation:

- Cut index cards to make 48 playing cards.
- Write each number on a card.

  1, 1, 1, 2, 2, 2, 3, 3, 3, 4, 4, 4, 5, 5, 5, 6, 6, 6, 7, 7, 7, 8, 8, 8, 9, 9, 9, 10, 10, 10, 12, 12, 12, 13, 13, 13, 14, 14, 14, 15, 15, 15, 16, 16, 16, 18, 20, 25
- Make a copy of the game board.

To Play:

- Play with a partner.
- Deal 8 cards to each player.
- Begin by drawing a card from the remaining cards. Use four of your cards to try to form a proportion.
- Lay the four cards on the game board. If you form a proportion, say "Criss Cross" and you earn 4 points. Place the four cards in a discard pile. Now it is your partner's turn.
- If you cannot form a proportion, then it is your partner's turn.
- When the original pile of cards is empty, shuffle the cards in the discard pile and start again.
- The first player to reach 20 points wins.

What Is Your Answer?

3. **IN YOUR OWN WORDS** How can you use ratio tables and cross products to solve proportions in science? Give an example.

4. **PUZZLE** Use each number once to form three proportions.

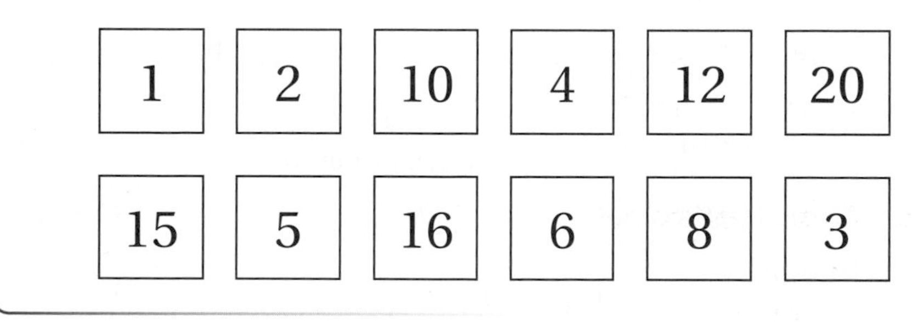

Practice Use what you discovered about solving proportions to complete Exercises 10–13 on page 126.

3.5 Lesson

Solving Proportions

Method 1 Use mental math. *(Section 3.4)*

Method 2 Use the Multiplication Property of Equality. *(Section 3.5)*

Method 3 Use the Cross Products Property. *(Section 3.5)*

EXAMPLE 1 Solving Proportions Using Multiplication

Solve $\frac{5}{7} = \frac{x}{21}$.

$\frac{5}{7} = \frac{x}{21}$ Write the proportion.

$21 \cdot \frac{5}{7} = 21 \cdot \frac{x}{21}$ Multiply each side by 21.

$15 = x$ Simplify.

∴ The solution is 15.

On Your Own

Now You're Ready
Exercises 4–9

Solve the proportion using multiplication.

1. $\frac{w}{6} = \frac{6}{9}$ **2.** $\frac{12}{10} = \frac{a}{15}$ **3.** $\frac{y}{6} = \frac{2}{4}$

EXAMPLE 2 Solving Proportions Using the Cross Products Property

Solve each proportion.

a. $\frac{x}{8} = \frac{7}{10}$

$x \cdot 10 = 8 \cdot 7$ Use the Cross Products Property.

$10x = 56$ Multiply.

$x = 5.6$ Divide.

∴ The solution is 5.6.

b. $\frac{9}{y} = \frac{3}{17}$

$9 \cdot 17 = y \cdot 3$

$153 = 3y$

$51 = y$

∴ The solution is 51.

On Your Own

Now You're Ready Exercises 10–21

Solve the proportion using the Cross Products Property.

4. $\dfrac{2}{7} = \dfrac{x}{28}$ **5.** $\dfrac{12}{5} = \dfrac{6}{y}$ **6.** $\dfrac{40}{z+1} = \dfrac{15}{6}$

EXAMPLE 3 Real-Life Application

The toll due on a turnpike is proportional to the number of miles driven. How much does it cost to drive 150 miles?

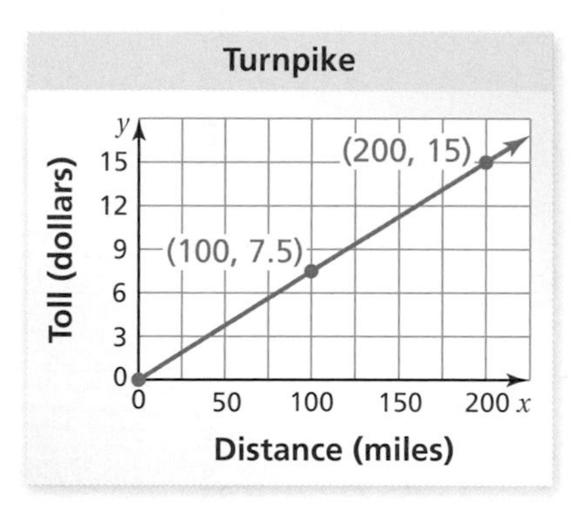

Method 1: Interpret the slope as a unit rate.

$$\text{slope} = \frac{\text{change in } y}{\text{change in } x}$$

$$= \frac{7.5}{100} \qquad \text{Substitute.}$$

$$= 0.075 \qquad \text{Divide.}$$

The unit rate is \$0.075 per mile. Multiply to find the total cost.

$$150 \cancel{\text{mi}} \cdot \frac{\$0.075}{1 \cancel{\text{mi}}} = \$11.25$$

∴ It costs \$11.25 to drive 150 miles on the turnpike.

Method 2: Write and solve a proportion.

$$\frac{7.5}{100} = \frac{y}{150} \qquad \text{Use (100, 7.5) to write a proportion.}$$

(dollars ← y; miles ← 150)

$$150 \cdot \frac{7.5}{100} = 150 \cdot \frac{y}{150} \qquad \text{Multiply each side by 150.}$$

$$11.25 = y \qquad \text{Simplify.}$$

∴ It costs \$11.25 to drive 150 miles on the turnpike.

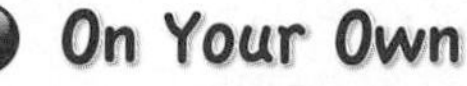

7. WHAT IF? In Example 3, how much does it cost to drive 75 miles on the turnpike?

3.5 Exercises

Vocabulary and Concept Check

1. **WRITING** What are three ways you can solve a proportion?

2. **OPEN-ENDED** Which way would you choose to solve $\frac{3}{x} = \frac{6}{14}$? Explain your reasoning.

3. **NUMBER SENSE** Does $\frac{x}{4} = \frac{15}{3}$ have the same solution as $\frac{x}{15} = \frac{4}{3}$? Use the Cross Products Property to explain your answer.

Practice and Problem Solving

Solve the proportion using multiplication.

1 4. $\frac{9}{5} = \frac{z}{20}$

5. $\frac{h}{15} = \frac{16}{3}$

6. $\frac{w}{4} = \frac{42}{24}$

7. $\frac{35}{28} = \frac{n}{12}$

8. $\frac{7}{16} = \frac{x}{4}$

9. $\frac{y}{9} = \frac{44}{54}$

Solve the proportion using the Cross Products Property.

2 10. $\frac{a}{6} = \frac{15}{2}$

11. $\frac{10}{7} = \frac{8}{k}$

12. $\frac{3}{4} = \frac{v}{14}$

13. $\frac{5}{n} = \frac{16}{32}$

14. $\frac{36}{42} = \frac{24}{r}$

15. $\frac{9}{10} = \frac{d}{6.4}$

16. $\frac{x}{8} = \frac{3}{12}$

17. $\frac{8}{m} = \frac{6}{15}$

18. $\frac{4}{24} = \frac{c}{36}$

19. $\frac{20}{16} = \frac{d}{12}$

20. $\frac{30}{20} = \frac{w}{14}$

21. $\frac{2.4}{1.8} = \frac{7.2}{k}$

22. **ERROR ANALYSIS** Describe and correct the error in solving the proportion $\frac{m}{8} = \frac{15}{24}$.

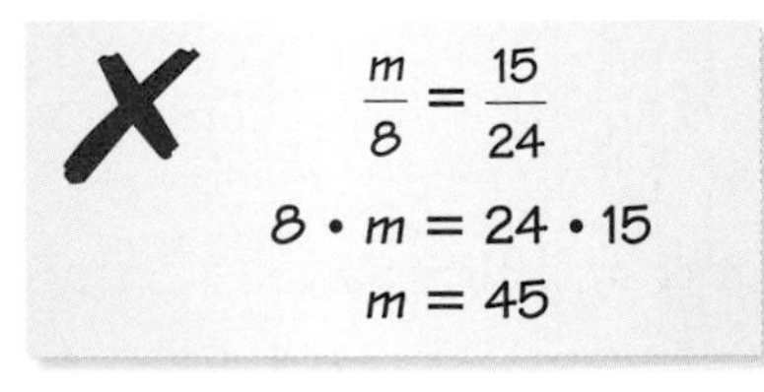

23. **PENS** Forty-eight pens are packaged in four boxes. How many pens are packaged in nine boxes?

24. **PIZZA PARTY** How much does it cost to buy 10 medium pizzas?

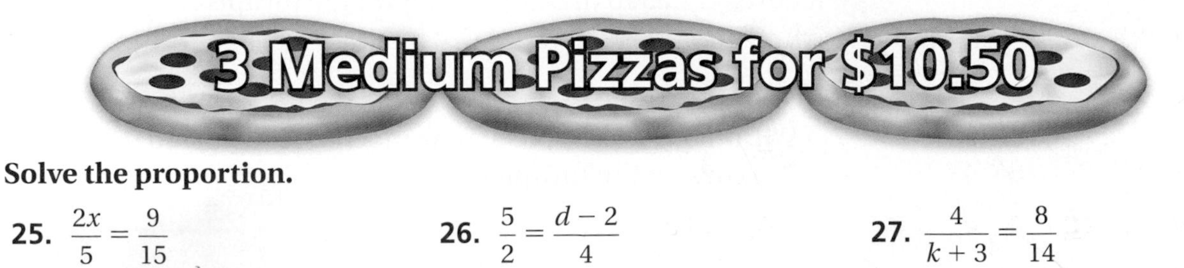

Solve the proportion.

25. $\frac{2x}{5} = \frac{9}{15}$

26. $\frac{5}{2} = \frac{d-2}{4}$

27. $\frac{4}{k+3} = \frac{8}{14}$

28. TRUE OR FALSE? Tell whether the statement is *true* or *false*. Explain.

$$\text{If } \frac{a}{b} = \frac{2}{3}, \text{ then } \frac{3}{2} = \frac{b}{a}.$$

29. CLASS TRIP It costs \$95 for 20 students to visit an aquarium. How much does it cost for 162 students?

30. GRAVITY A person who weighs 120 pounds on Earth weighs 20 pounds on the moon. How much does a 93-pound person weigh on the moon?

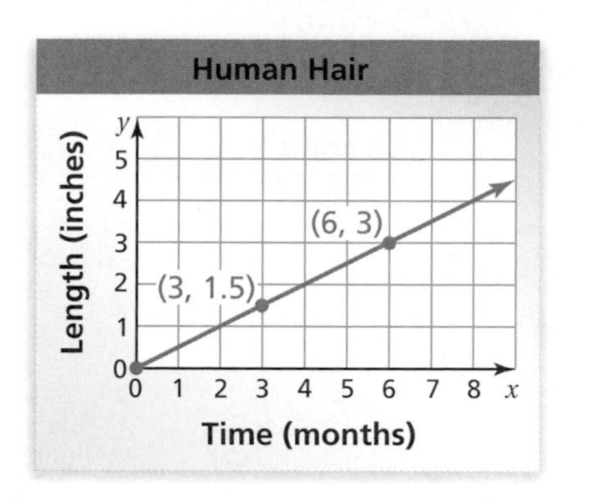

31. HAIR The length of human hair is proportional to the number of months it has grown.

a. How long does it take hair to grow 8 inches?

b. Use a different method than the one in part (a) to find how long it takes hair to grow 20 inches.

32. CHEETAH Cheetahs are the fastest mammals in the world. They can reach speeds of 70 miles per hour.

a. At this speed, how long would it take a cheetah to run 17 miles?

b. RESEARCH Use the Internet or library to find how long a cheetah can maintain a speed of 70 miles per hour.

33. AUDIENCE There are 144 people in an audience. The ratio of adults to children is 5 to 3. How many are adults?

34. LAWN SEED Three pounds of lawn seed covers 1800 square feet. How many bags are needed to cover 8400 square feet?

35. Critical Thinking Consider the proportions $m = \frac{1}{2}$ and $k = \frac{1}{4}$. What is the ratio $\frac{m}{k}$? Explain your reasoning.

Fair Game Review What you learned in previous grades & lessons

Copy and complete. *(Skills Review Handbook)*

36. 530 cm = ▢ m

37. 6.4 kg = ▢ g

38. 56 oz = ▢ lb

39. $1\frac{1}{2}$ mi = ▢ ft

40. MULTIPLE CHOICE How many cups of milk are shown? *(Skills Review Handbook)*

Ⓐ $\frac{7}{10}$ c

Ⓑ $\frac{7}{8}$ c

Ⓒ $1\frac{3}{4}$ c

Ⓓ 14 c

3 Study Help

You can use an **information wheel** to organize information about a concept. Here is an example of an information wheel for slope.

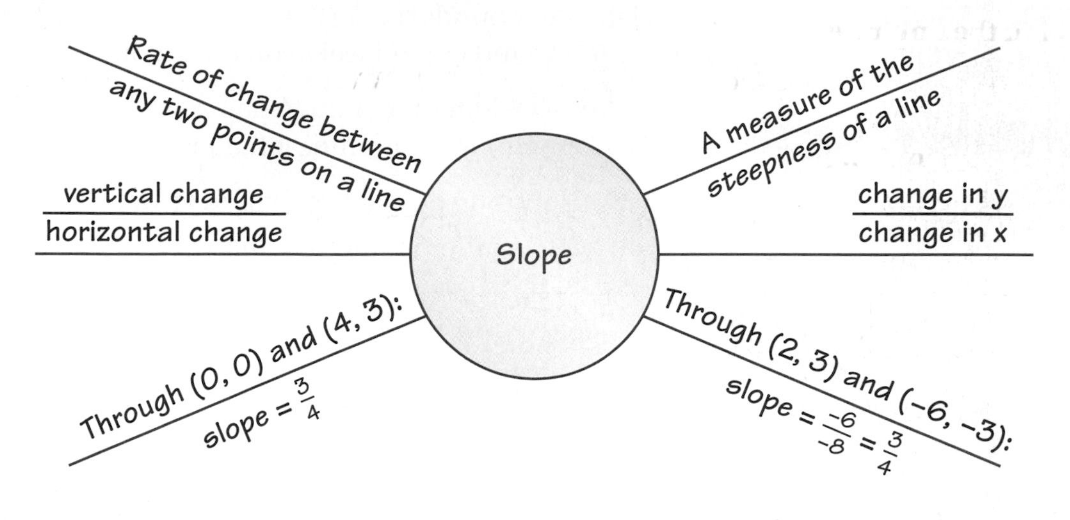

On Your Own

Make an information wheel to help you study these topics.

1. ratio
2. rate
3. unit rate
4. proportion
5. cross products
6. solving proportions

After you complete this chapter, make information wheels for the following topics.

7. U.S. customary system
8. metric system
9. converting units
10. direct variation
11. inverse variation

"My information wheel summarizes how cats act when they get baths."

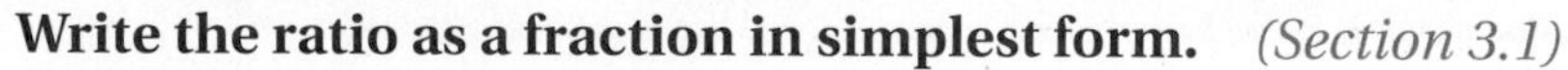

3.1–3.5 Quiz

Check It Out
Progress Check
BigIdeasMath.com

Write the ratio as a fraction in simplest form. *(Section 3.1)*

1. 18 red buttons : 12 blue buttons

2. 30 inches to 3 inches

Find the unit rate. *(Section 3.1)*

3. 108 feet per 3 seconds

4. 360 miles per 12 gallons

Find the slope of the line. *(Section 3.2)*

5.

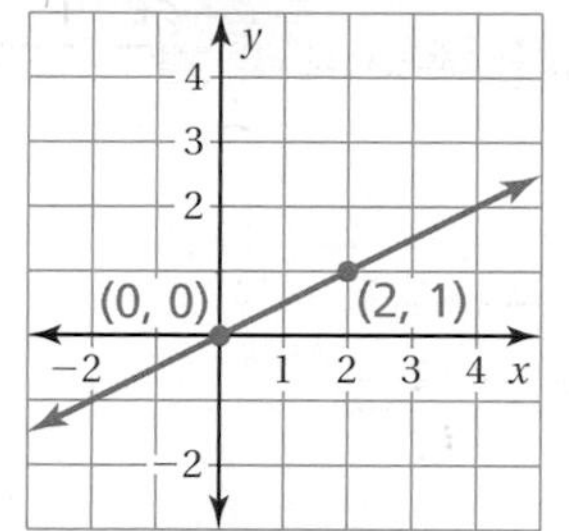

6.

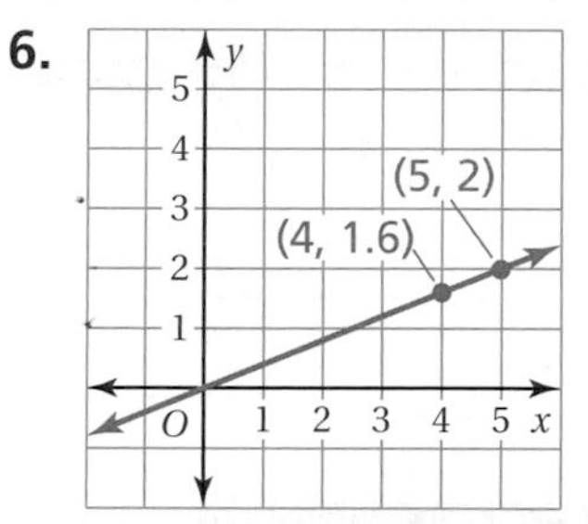

Tell whether the ratios form a proportion. *(Section 3.3)*

7. $\frac{1}{8}, \frac{4}{32}$

8. $\frac{2}{3}, \frac{10}{30}$

Use the table to write a proportion. *(Section 3.4)*

9.

	Monday	Tuesday
Dollars	42	56
Hours	6	h

10.

	Series 1	Series 2
Games	g	6
Wins	4	3

Solve the proportion. *(Section 3.5)*

11. $\frac{x}{2} = \frac{40}{16}$

12. $\frac{3}{11} = \frac{27}{z}$

13. MUSIC DOWNLOAD The amount of time needed to download music is shown in the table. Find the rate in megabytes per second. *(Section 3.1)*

Megabytes	Seconds
2	6
4	12
6	18
8	24

14. GAMING You advance 3 levels in 15 minutes. Your friend advances 5 levels in 20 minutes. Are these rates proportional? Explain. *(Section 3.3)*

15. CLASS TIME You spend 150 minutes in three classes. Write and solve a proportion to find how many minutes you spend in five classes. *(Section 3.4)*

16. CONCERT A benefit concert with three performers lasts 8 hours. At this rate, how many hours is a concert with four performers? *(Section 3.5)*

3.6 Converting Measures Between Systems

COMMON CORE STATE STANDARDS
7.RP.3

Essential Question How can you compare lengths between the customary and metric systems?

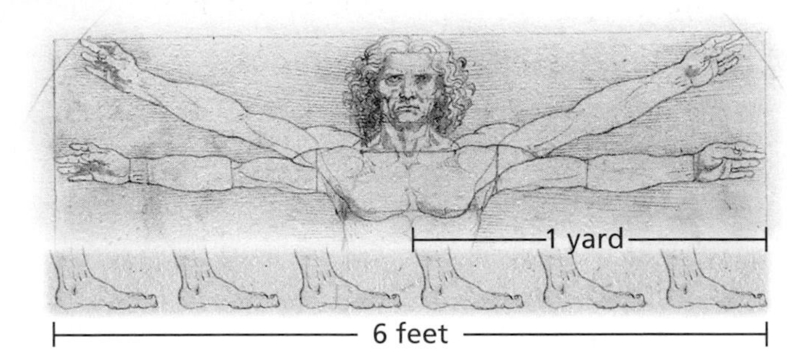

1 ACTIVITY: Customary Measure History

Work with a partner.

a. Match the measure of length with its historical beginning.

Length	*Historical Beginning*
Inch	The length of a human foot.
Foot	The width of a human thumb.
Yard	The distance a human can walk in 1000 paces (two steps).
Mile	The distance from a human nose to the end of an outstretched human arm.

b. Use a ruler to measure your thumb, arm, and foot. How do your measurements compare to your answers from part (a)? Are they close to the historical measures?

You know how to convert measures within the customary and metric systems.

Equivalent Customary Lengths

1 ft = 12 in. 1 yd = 3 ft 1 mi = 5280 ft

Equivalent Metric Lengths

1 m = 1000 mm 1 m = 100 cm 1 km = 1000 m

You will learn how to convert between the two systems.

Converting Between Systems

1 in. ≈ 2.54 cm

1 mi ≈ 1.6 km

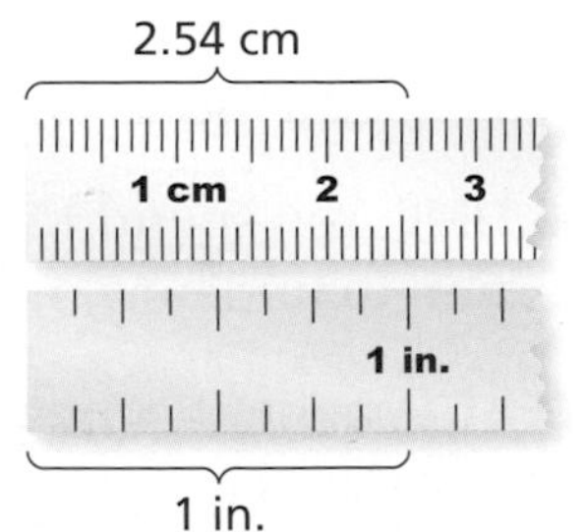

2 ACTIVITY: Comparing Measures

Work with a partner. Answer each question. Explain your answer. Use a diagram in your explanation.

		Metric	*Customary*
a.	**Car Speed:** Which is faster?	80 km/h	60 mi/h
b.	**Trip Distance:** Which is farther?	200 km	200 mi
c.	**Human Height:** Who is taller?	180 cm	5 ft 8 in.
d.	**Wrench Width:** Which is wider?	8 mm	5/16 in.
e.	**Swimming Pool Depth:** Which is deeper?	1.4 m	4 ft
f.	**Mountain Elevation:** Which is higher?	2000 m	7000 ft
g.	**Room Width:** Which is wider?	3.5 m	12 ft

What Is Your Answer?

3. **IN YOUR OWN WORDS** How can you compare lengths between the customary and metric systems? Give examples with your description.

4. **HISTORY** The meter and the metric system originated in France. In 1791, the French Academy of Sciences was instructed to create a new system of measurement. This new system would be based on powers of 10.

The fundamental units of this system would be based on natural values that were unchanging. The French Academy of Sciences decided to find the length of an imaginary arc that began at the North Pole and ended at the equator.

They would then divide this arc into exactly ten million identical pieces. The length of one of these pieces would be the base unit of length for the new system of measurement.

a. Find the distance around Earth in meters.

b. Find the distance around Earth in kilometers.

5. Find the distance around Earth in miles.

Practice Use what you learned about converting measures between systems to complete Exercises 4–9 on page 134.

3.6 Lesson

Key Vocabulary
U.S. customary system, *p. 132*
metric system, *p. 132*

The **U.S. customary system** is a system of measurement that contains units for length, capacity, and weight. The **metric system** is a decimal system of measurement, based on powers of 10, that contains units for length, capacity, and mass.

Use the relationships below to convert units *between* systems.

Length	Capacity	Weight and Mass
1 in. ≈ 2.54 cm 1 mi ≈ 1.6 km	1 qt ≈ 0.95 L	1 lb ≈ 0.45 kg

EXAMPLE 1 Converting Units

Convert 5 liters to quarts.

Method 1: Convert using a ratio.

1 qt ≈ 0.95 L

$$5 \cancel{L} \times \frac{1 \text{ qt}}{0.95 \cancel{L}} \approx 5.26 \text{ qt}$$

So, 5 liters is about 5.26 quarts.

Method 2: Convert using a proportion.

Let x be the number of quarts equivalent to 5 liters.

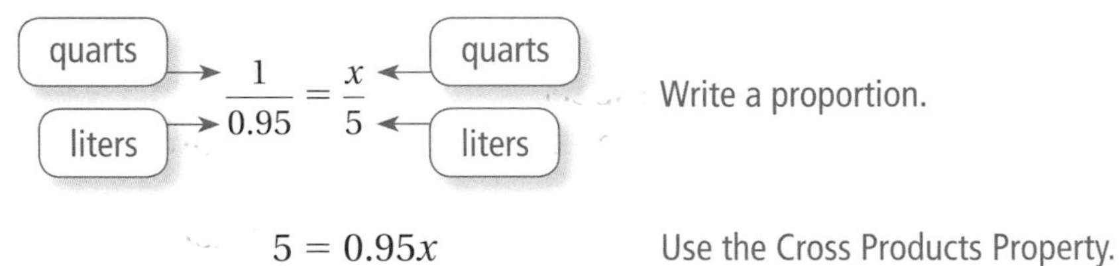

$$\frac{1}{0.95} = \frac{x}{5}$$ Write a proportion.

$$5 = 0.95x$$ Use the Cross Products Property.

$$5.26 \approx x$$ Divide each side by 0.95.

So, 5 liters is about 5.26 quarts.

On Your Own

Now You're Ready
Exercises 10–22

Copy and complete the statement. Round to the nearest hundredth, if necessary.

1. 7 mi ≈ ___ km
2. 12 qt ≈ ___ L
3. 25 kg ≈ ___ lb
4. 8 cm ≈ ___ in.

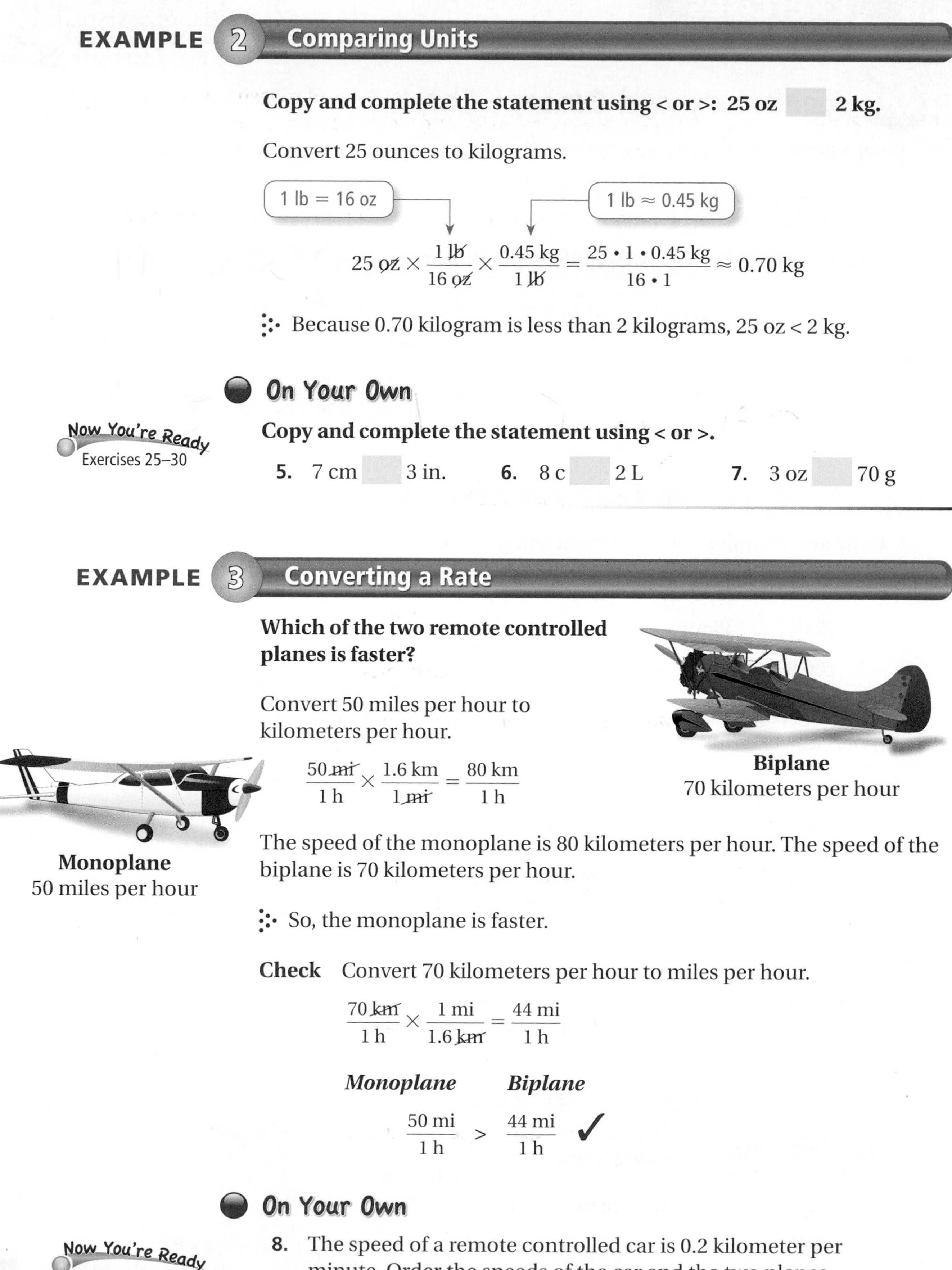

EXAMPLE 2 Comparing Units

Copy and complete the statement using $<$ or $>$: 25 oz ☐ 2 kg.

Convert 25 ounces to kilograms.

1 lb = 16 oz

1 lb ≈ 0.45 kg

$$25 \text{ oz} \times \frac{1 \text{ lb}}{16 \text{ oz}} \times \frac{0.45 \text{ kg}}{1 \text{ lb}} = \frac{25 \cdot 1 \cdot 0.45 \text{ kg}}{16 \cdot 1} \approx 0.70 \text{ kg}$$

∴ Because 0.70 kilogram is less than 2 kilograms, 25 oz $<$ 2 kg.

On Your Own

Now You're Ready
Exercises 25–30

Copy and complete the statement using $<$ or $>$.

5. 7 cm ☐ 3 in. **6.** 8 c ☐ 2 L **7.** 3 oz ☐ 70 g

EXAMPLE 3 Converting a Rate

Which of the two remote controlled planes is faster?

Biplane
70 kilometers per hour

Convert 50 miles per hour to kilometers per hour.

$$\frac{50 \text{ mi}}{1 \text{ h}} \times \frac{1.6 \text{ km}}{1 \text{ mi}} = \frac{80 \text{ km}}{1 \text{ h}}$$

Monoplane
50 miles per hour

The speed of the monoplane is 80 kilometers per hour. The speed of the biplane is 70 kilometers per hour.

∴ So, the monoplane is faster.

Check Convert 70 kilometers per hour to miles per hour.

$$\frac{70 \text{ km}}{1 \text{ h}} \times \frac{1 \text{ mi}}{1.6 \text{ km}} = \frac{44 \text{ mi}}{1 \text{ h}}$$

Monoplane *Biplane*

$$\frac{50 \text{ mi}}{1 \text{ h}} > \frac{44 \text{ mi}}{1 \text{ h}} \checkmark$$

On Your Own

Now You're Ready
Exercises 31–34

8. The speed of a remote controlled car is 0.2 kilometer per minute. Order the speeds of the car and the two planes in Example 3 from least to greatest.

3.6 Exercises

Vocabulary and Concept Check

1. **WRITING** Describe two methods you can use to convert measurements.
2. **OPEN-ENDED** Which method would you use to convert 10 miles to kilometers? Explain your reasoning.
3. **DIFFERENT WORDS, SAME QUESTION** Which is different? Find "both" answers.

Convert 5 inches to centimeters.	Find the number of inches in 5 centimeters.
How many centimeters are in 5 inches?	Five inches equals how many centimeters?

Practice and Problem Solving

Copy and complete the statement using < or >.

4. 1 ft ____ 1 cm
5. 450 yd ____ 450 cm
6. 30 in. ____ 30 mm
7. 125 in. ____ 125 cm
8. 100 ft/h ____ 100 km/h
9. 10 L ____ 10 gal

Copy and complete the statement using a ratio. Round to the nearest hundredth, if necessary.

(1) 10. 3 mi ≈ ____ km
11. 10 qt ≈ ____ L
12. 68 kg ≈ ____ lb
13. 8.3 in. ≈ ____ cm
14. 25.5 lb ≈ ____ kg
15. 5 km ≈ ____ mi

16. **ERROR ANALYSIS** Describe and correct the error in using a ratio to convert 12 kilometers to miles.

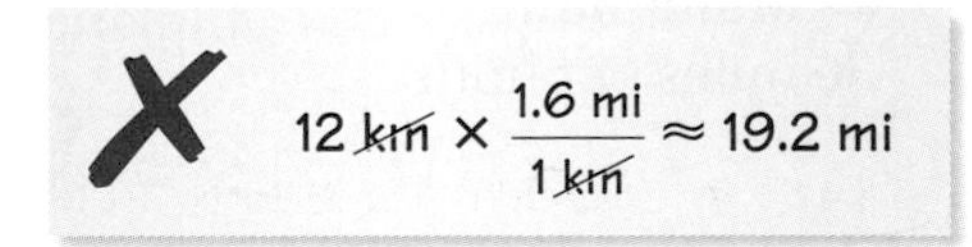

$12 \text{ km} \times \frac{1.6 \text{ mi}}{1 \text{ km}} \approx 19.2 \text{ mi}$

Copy and complete the statement using a proportion. Round to the nearest hundredth, if necessary.

17. 48 in. ≈ ____ cm
18. 2 km ≈ ____ mi
19. 165 cm ≈ ____ in.
20. 85 lb ≈ ____ kg
21. 2.5 qt ≈ ____ L
22. 14.2 L ≈ ____ qt

23. **CAVES** Mammoth Cave is the longest cave system in the world. So far, 365 miles of the cave have been explored. What is this distance in kilometers?
24. **IGUANA** How long is the iguana in inches?

Length: 24.7 cm

Copy and complete the statement using $<$ or $>$.

(2) **25.** 8 kg ____ 30 oz

26. 6 ft ____ 300 cm

27. 3 gal ____ 6 L

28. 10 in. ____ 200 mm

29. 1200 g ____ 5 lb

30. 1500 m ____ 3000 ft

Copy and complete the statement. Round to the nearest hundredth, if necessary.

(3) **31.** 45 mi/h ≈ ____ km/h

32. 5 gal/min ≈ ____ L/min

33. 120 mm/sec ≈ ____ in./sec

34. 900 g/day ≈ ____ lb/day

35. BRACHIOSAURUS One of the largest dinosaurs was the brachiosaurus. How much did it weigh in kilograms?

Weight: 172,000 lb

36. BOTTLE Can you pour the water from a full 2-liter bottle into a 2-quart pitcher without spilling any? Explain.

37. AUTOBAHN Germany suggests a speed limit of 130 kilometers per hour on highways.

a. Is the speed shown greater than the suggested limit?

b. Suppose the speed drops 30 miles per hour. Is the new speed below the suggested limit?

38. SOCCER The size of a soccer field is 50 yards wide by 80 yards long. What is the size in meters?

39. PAINT One liter of paint covers 100 square feet. How many gallons does it take to cover 1400 square feet?

40. **Critical Thinking** The speed of light is about 300,000 kilometers per second. Convert the speed to miles per hour.

Fair Game Review What you learned in previous grades & lessons

Graph the data. Then find the slope of the line through the points. *(Section 3.2)*

41.

Months, x	Height, y
2	3
4	6
6	9

42.

Hours, x	Units, y
4	60
8	120
12	180

43. MULTIPLE CHOICE Which equation has a solution of 4? *(Section 2.6)*

(A) $2x + 7 = -1$ (B) $-3 + 2x = -11$ (C) $2x - 11 = -3$ (D) $11 + 2x = 3$

3.7 Direct Variation

COMMON CORE STATE STANDARDS
7.RP.2a
7.RP.2c
7.RP.3
8.EE.5

Essential Question How can you use a graph to show the relationship between two variables that vary directly? How can you use an equation?

1 ACTIVITY: Math in Literature

Gulliver's Travels was written by Jonathan Swift and published in 1725. Gulliver was shipwrecked on the island Lilliput, where the people were only 6 inches tall. When the Lilliputians decided to make a shirt for Gulliver, a Lilliputian tailor stated that he could determine Gulliver's measurements by simply measuring the distance around Gulliver's thumb. He said "Twice around the thumb equals once around the wrist. Twice around the wrist is once around the neck. Twice around the neck is once around the waist."

Work with a partner. Use the tailor's statement to complete the table.

Thumb, t	Wrist, w	Neck, n	Waist, x
0 in.	0 in.		
1 in.	2 in.		
2 in.	4 in.		
3 in.	6 in.		
4 in.	8 in.		
5 in.	10 in.		

2 EXAMPLE: Drawing a Graph

Use the information from Activity 1 to draw a graph of the relationship between the distance around the thumb t and the distance around the wrist w.

Use the table to write ordered pairs. Then plot the ordered pairs.

(0, 0), (1, 2), (2, 4), (3, 6), (4, 8), (5, 10)

Notice the following about the graph:

1. All the points lie on a line.
2. The line passes through the origin.

This type of relationship is called **direct variation**. You can write an equation to describe the relationship between t and w.

$w = 2t$ Wrist is twice thumb.

3 ACTIVITY: Drawing a Graph

Work with a partner. Use the information from Activity 1 to draw a graph of the relationship. Write an equation that describes the relationship between the two variables.

a. Thumb t and neck n ($n =$ ___ t)

b. Wrist w and waist x ($x =$ ___ w)

c. Wrist w and thumb t ($t =$ ___ w)

d. Waist x and wrist w ($w =$ ___ x)

What Is Your Answer?

4. **IN YOUR OWN WORDS** How can you use a graph to show the relationship between two variables that vary directly? How can you use an equation?
5. Give a real-life example of two variables that vary directly.
6. Work with a partner. Use string to find the distance around your thumb, wrist, and neck. Do your measurements agree with those of the tailor in *Gulliver's Travels*? Explain your reasoning.

Use what you learned about direct variation to complete Exercises 4–7 on page 140.

3.7 Lesson

Check It Out
Lesson Tutorials
BigIdeasMath.com

Key Vocabulary
direct variation, *p. 138*

Key Idea

Direct Variation

Words Two quantities x and y show **direct variation** when $y = kx$, where k is a number and $k \neq 0$.

Graph The graph of $y = kx$ is a line that passes through the origin.

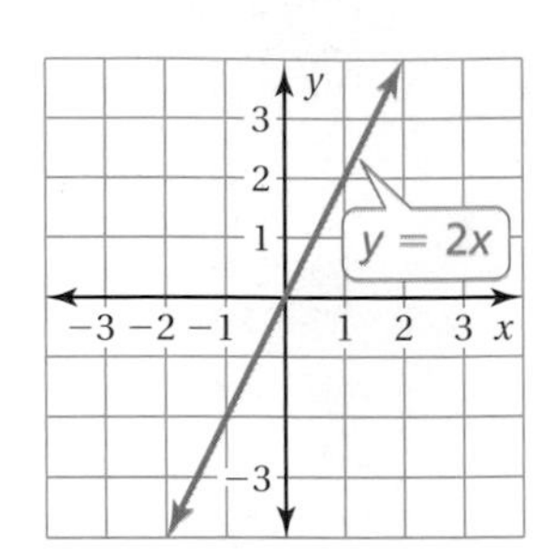

EXAMPLE 1 Identifying Direct Variation

Tell whether x and y show direct variation. Explain your reasoning.

a.

x	1	2	3	4
y	−2	0	2	4

Plot the points. Draw a line through the points.

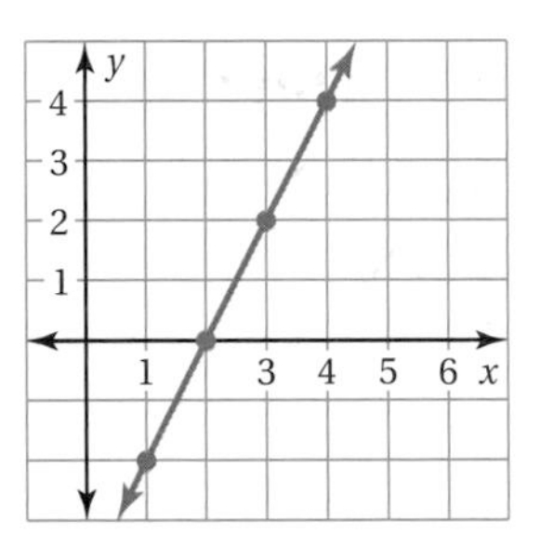

The line does not pass through the origin. So, x and y do *not* show direct variation.

b.

x	0	2	4	6
y	0	2	4	6

Plot the points. Draw a line through the points.

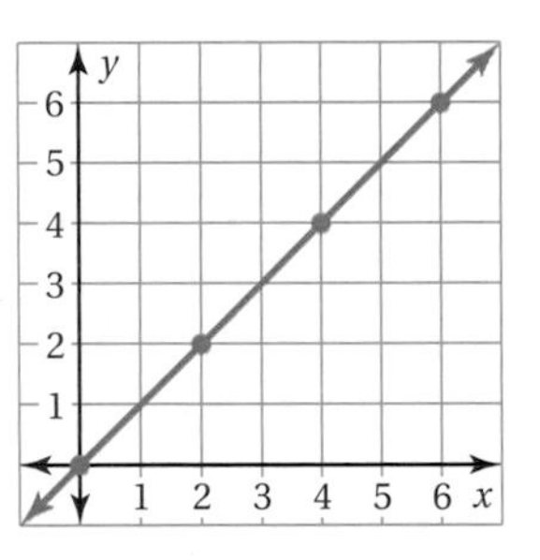

The line passes through the origin. So, x and y show direct variation.

Study Tip

Other ways to say that x and y show direct variation are "y varies directly with x" and "x and y are directly proportional."

EXAMPLE 2 Identifying Direct Variation

Tell whether x and y show direct variation. Explain your reasoning.

a. $y + 1 = 2x$

$y = 2x - 1$ Solve for y.

The equation *cannot* be written as $y = kx$. So, x and y do *not* show direct variation.

b. $\frac{1}{2}y = x$

$y = 2x$ Solve for y.

The equation can be written as $y = kx$. So, x and y show direct variation.

On Your Own

Now You're Ready
Exercises 8–21

Tell whether x and y show direct variation. Explain your reasoning.

1.

x	y
0	−2
1	1
2	4
3	7

2.

x	y
1	4
2	8
3	12
4	16

3.

x	y
−2	4
−1	2
0	0
1	2

4. $xy = 3$

5. $x = \frac{1}{3}y$

6. $y + 1 = x$

EXAMPLE 3 Using a Direct Variation Model

The height y of a television screen varies directly with its width x.

a. Find the height when the width is 48 inches.

b. Sketch the graph of the equation.

a. Use the equation to find the height when $x = 48$ inches.

$$y = \frac{9}{16}(48) \quad \text{Substitute 48 for } x.$$

$$= 27 \quad \text{Simplify.}$$

So, when the width is 48 inches, the height is 27 inches.

b. To sketch a graph, plot the point (48, 27). Then draw the line that passes through this point and the origin.

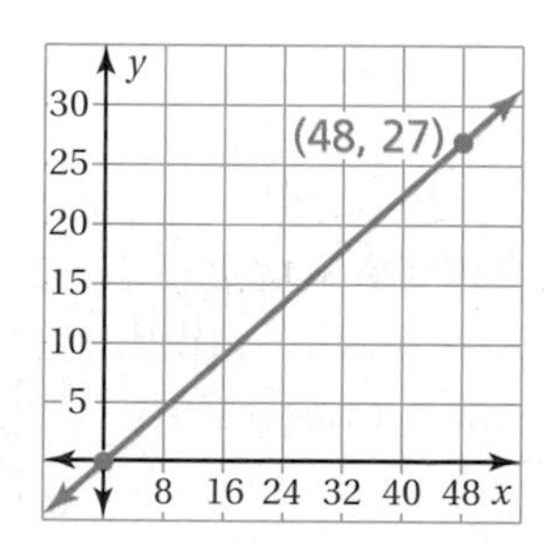

On Your Own

Now You're Ready
Exercises 24–29

7. Your earnings y (in dollars) vary directly with the number x of lawns you mow. Use the equation $y = 7.5x$ to find how much you earn when you mow 5 lawns.

3.7 Exercises

Vocabulary and Concept Check

1. **VOCABULARY** What does it mean for x and y to vary directly?
2. **WRITING** What point is on the graph of every direct variation equation?
3. **WHICH ONE DOESN'T BELONG?** Which graph does *not* belong with the other three? Explain your reasoning.

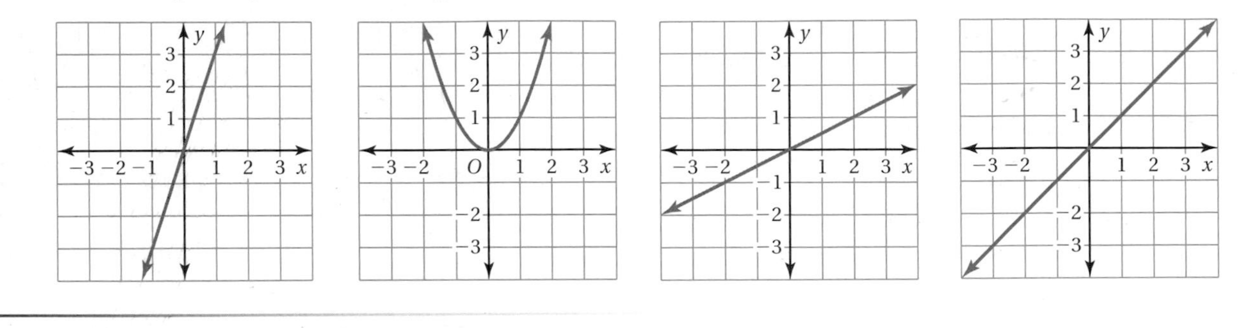

Practice and Problem Solving

Tell whether x and y show direct variation. Explain your reasoning.

4. $(-1, -1), (0, 0), (1, 1), (2, 2)$
5. $(-4, -2), (-2, 0), (0, 2), (2, 4)$
6. $(1, 2), (1, 4), (1, 6), (1, 8)$
7. $(2, 1), (6, 3), (10, 5)\ (14, 7)$

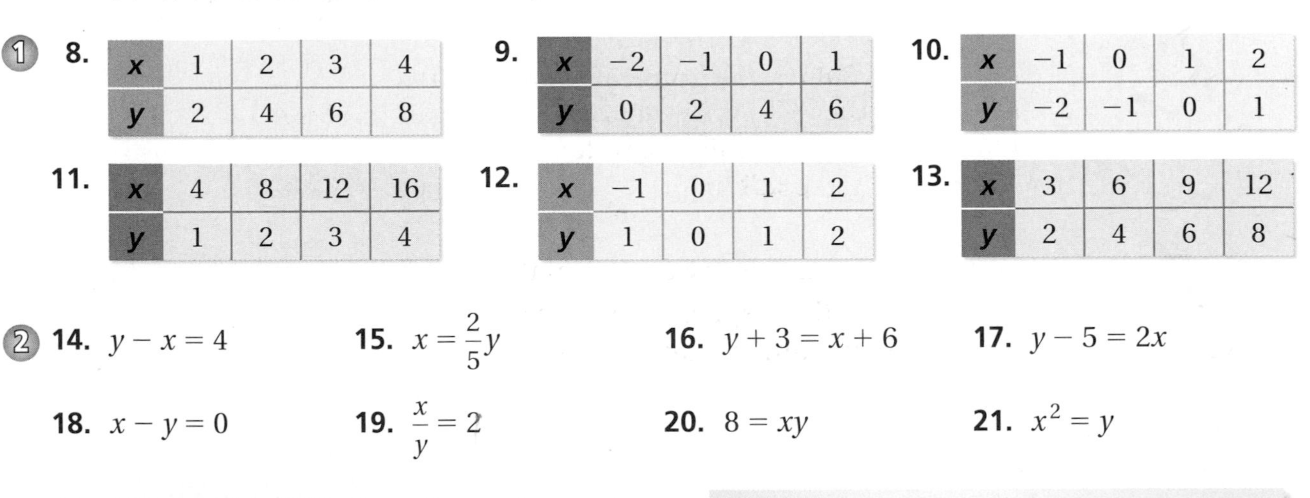

① 8.

x	1	2	3	4
y	2	4	6	8

9.

x	−2	−1	0	1
y	0	2	4	6

10.

x	−1	0	1	2
y	−2	−1	0	1

11.

x	4	8	12	16
y	1	2	3	4

12.

x	−1	0	1	2
y	1	0	1	2

13.

x	3	6	9	12
y	2	4	6	8

② 14. $y - x = 4$

15. $x = \frac{2}{5}y$

16. $y + 3 = x + 6$

17. $y - 5 = 2x$

18. $x - y = 0$

19. $\frac{x}{y} = 2$

20. $8 = xy$

21. $x^2 = y$

22. **ERROR ANALYSIS** Describe and correct the error in telling whether x and y show direct variation.

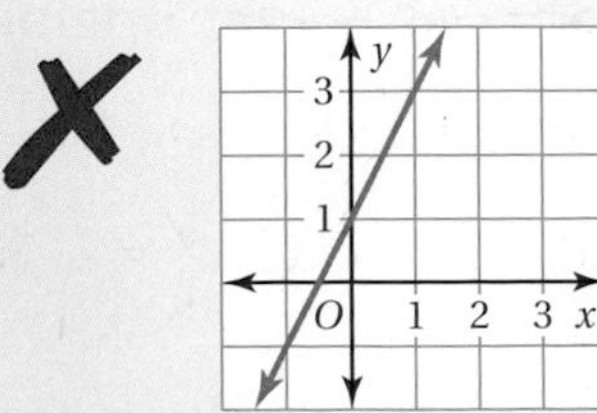

The graph is a line, so it shows direct variation.

23. **RECYCLING** The table shows the profit y for recycling x pounds of aluminum. Tell whether x and y show direct variation.

Aluminum, x	10	20	30	40
Profit, y	\$4.50	\$9.00	\$13.50	\$18.00

The variables x and y vary directly. Use the values to write an equation that relates x and y.

3 **24.** $y = 4; x = 2$ **25.** $y = 25; x = 5$ **26.** $y = 60; x = 15$

27. $y = 72; x = 3$ **28.** $y = 20; x = 12$ **29.** $y = 45; x = 40$

2.54 cm

1 in.

30. MEASUREMENT Write a direct variation equation that relates x inches to y centimeters.

31. MODELING Design a jet ski ramp. Show how you can use direct variation to plan the heights of the vertical supports.

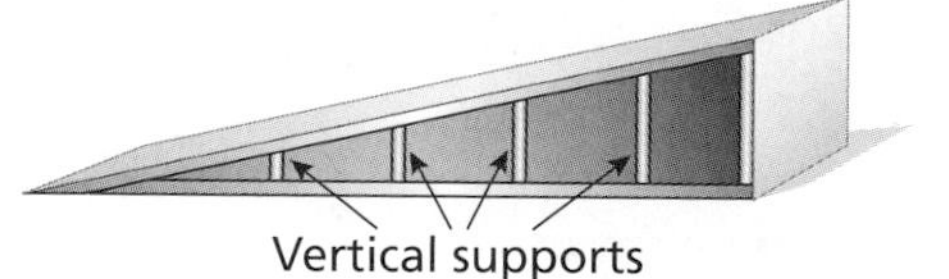

32. JUPITER The weight of an object in our solar system varies directly with the weight of the object on Earth.

Location	Earth	Jupiter	Moon
Weight (lb)	100	214	
Weight (lb)	120		20

a. Copy and complete the table.

b. RESEARCH Why does weight vary throughout our solar system?

Minutes, x	500	700	900	1200
Cost, y	\$40	\$50	\$60	\$75

33. CELL PHONE PLANS Tell whether x and y show direct variation. If so, write an equation of direct variation.

34. CHLORINE The amount of chlorine in a swimming pool varies directly with the volume of water. The pool has 2.5 milligrams of chlorine per liter of water. How much chlorine is in the pool?

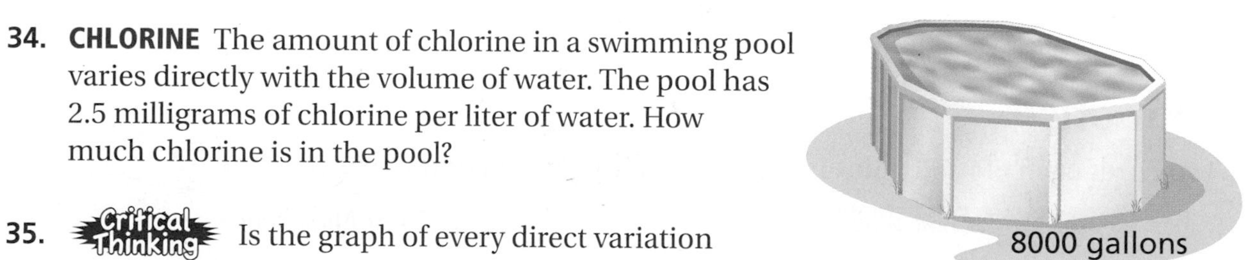

35. Critical Thinking Is the graph of every direct variation equation a line? Does the graph of every line represent a direct variation equation? Explain your reasoning.

Fair Game Review What you learned in previous grades & lessons

Solve the equation. *(Section 2.5)*

36. $-4x = 36$ **37.** $\frac{y}{6} = -10$ **38.** $-\frac{3}{4}m = 24$ **39.** $-17 = \frac{2}{7}d$

40. MULTIPLE CHOICE Which rate is *not* equivalent to 180 feet per 8 seconds? *(Section 3.1)*

Ⓐ $\frac{225 \text{ ft}}{10 \text{ sec}}$ Ⓑ $\frac{45 \text{ ft}}{2 \text{ sec}}$ Ⓒ $\frac{135 \text{ ft}}{6 \text{ sec}}$ Ⓓ $\frac{180 \text{ ft}}{1 \text{ sec}}$

3.7b Proportional Relationships

EXAMPLE 1 Interpreting a Proportional Relationship

The distance traveled by a high speed train is proportional to the number of hours traveled. Interpret each plotted point in the graph.

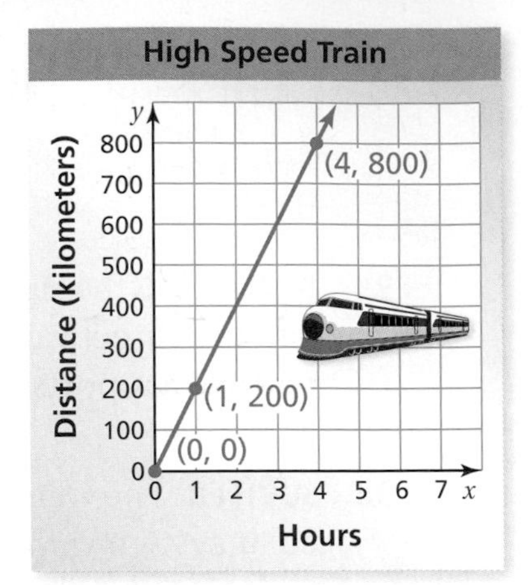

(0, 0): The train travels 0 kilometers in 0 hours.

(1, 200): The train travels 200 kilometers in 1 hour.

This point represents the unit rate, $\frac{200 \text{ km}}{1 \text{ h}}$, or 200 kilometers per hour.

(4, 800): The train travels 800 kilometers in 4 hours.

Because the relationship is proportional, you can also use this point to find the unit rate.

$\frac{800 \text{ km}}{4 \text{ h}} = \frac{200 \text{ km}}{1 \text{ h}}$, or 200 kilometers per hour

Study Tip

In the graph of a proportional relationship, you can find the unit rate from the point $(1, y)$.

Practice

Interpret each plotted point in the graph of the proportional relationship.

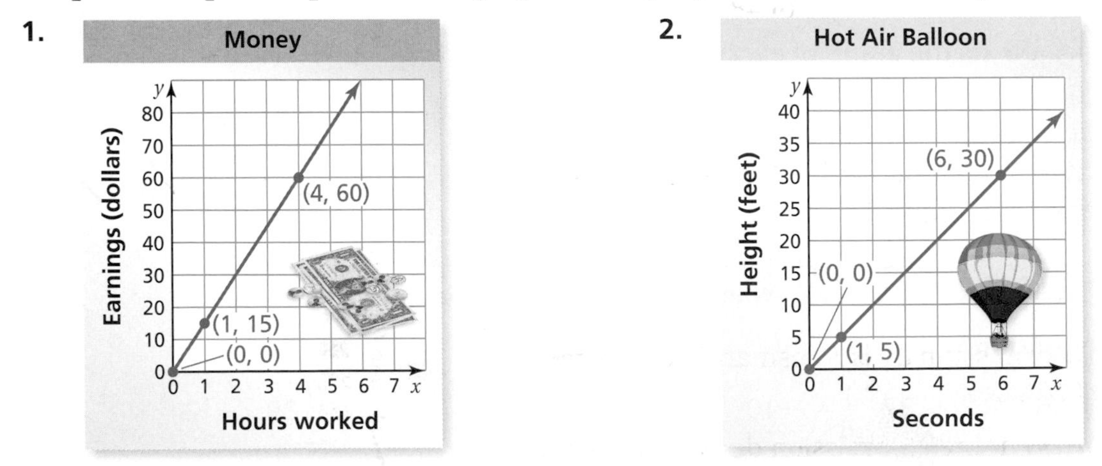

3. **REASONING** The graph of a proportional relationship passes through (3, 4.5) and $(1, y)$. Find y.

EXAMPLE 2 Comparing Graphs of Proportional Relationships

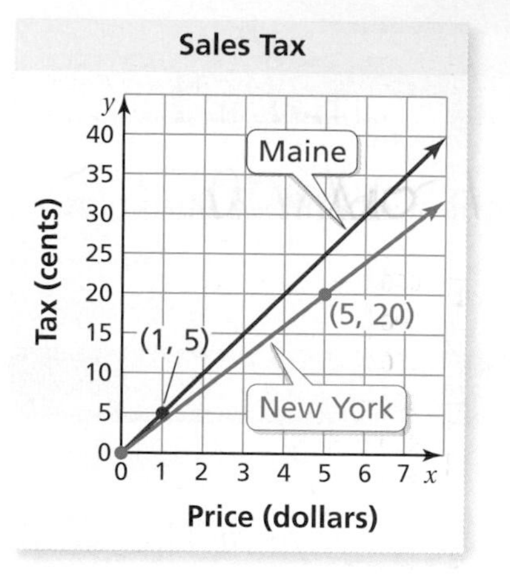

The state sales taxes in Maine and New York are proportional to the price of a purchase, as shown in the graph.

a. Express the sales tax rate for each state as a percent.

Maine: (1, 5) indicates that the tax is 5 cents per dollar.

$$\frac{5 \text{ cents}}{1 \text{ dollar}} = \frac{5 \text{ cents}}{100 \text{ cents}} = 5\%$$

New York: (5, 20) indicates that the tax is 20 cents per 5 dollars.

$$\frac{20 \text{ cents}}{5 \text{ dollars}} = \frac{4 \text{ cents}}{1 \text{ dollar}} = \frac{4 \text{ cents}}{100 \text{ cents}} = 4\%$$

∴ Maine has a 5% sales tax and New York has a 4% sales tax.

b. What is the sales tax on a $12 purchase in New York?

Method 1: Write and solve a proportion to find the sales tax.

$$\frac{20}{5} = \frac{y}{12}$$ ← cents / ← dollars — Use (5, 20) to write a proportion.

$$48 = y$$ Multiply each side by 12.

∴ The sales tax on a $12 purchase in New York is 48 cents.

Method 2: Find 4% of $12.

$$4\% \text{ of } \$12 = 0.04 \cdot 12$$ Write 4% as a decimal.

$$= 0.48$$ Multiply.

∴ The sales tax on a $12 purchase in New York is $0.48, or 48 cents.

Practice

4. **COMMISSION** The graph shows that the commissions of two salesmen are proportional to the amounts of sales.

   a. Express the commission rate for each salesman as a percent.

   b. What commission does Salesman A receive for a $10,000 sale?

   c. How much more commission does Salesman A receive than Salesman B for a $10,000 sale?

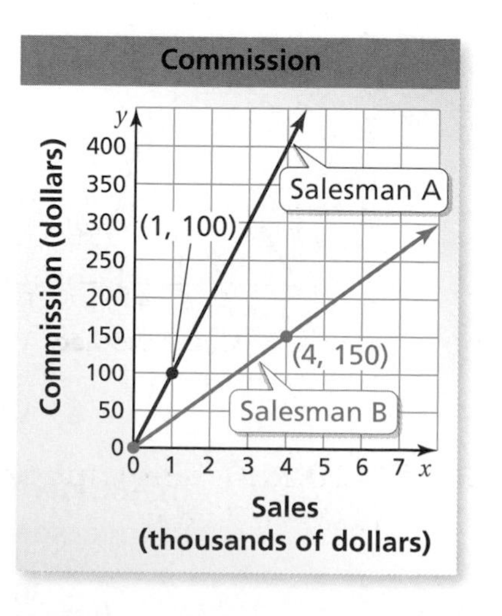

3.8 Inverse Variation

COMMON CORE STATE STANDARDS
7.RP.2a

Essential Question

How can you recognize when two variables are inversely proportional?

1 ACTIVITY: Comparing the Height and the Base

Work with a partner.

a. There are nine ways to arrange 36 square blocks to form a rectangle. Here are two ways. Find the other seven ways.

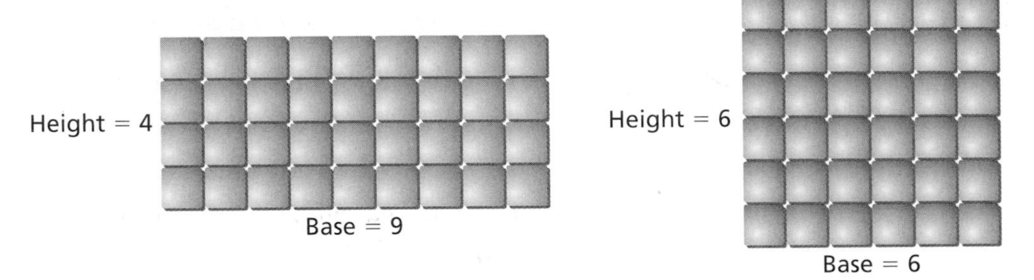

b. Order the nine ways according to height. Record your results in a table.

Height, h	Base, b	Area, A
4	9	$A = 9 \cdot 4 = 36$
6	6	$A = 6 \cdot 6 = 36$

c. Look at the first and second columns. Complete each sentence.

- When the height increases, the base ______.
- When the height decreases, the base ______.

In Activity 1, the relationship between the height and the base is an example of **inverse variation**. You can describe the relationship with an equation.

$$h = \frac{36}{b}$$ h and b are inversely proportional.

2 ACTIVITY: Comparing Direct and Inverse Variation

Work with a partner. Discuss each description. Tell whether the two variables are examples of *direct variation* or *inverse variation*. Use a table to explain your reasoning. Write an equation that relates the variables.

a. You bring 200 cookies to a party. Let n represent the number of people at the party and c represent the number of cookies each person receives.

b. You work at a restaurant for 20 hours. Let r represent your hourly pay rate and p represent the total amount you earn.

c. You are going on a 240-mile trip. Let t represent the number of hours driving and s represent the speed of the car.

What Is Your Answer?

3. **IN YOUR OWN WORDS** How can you recognize when two variables are inversely proportional? Explain how a table can help you recognize inverse variation.

4. **SCIENCE** The *wing beat frequency* of a bird is the number of times per second the bird flaps its wings.

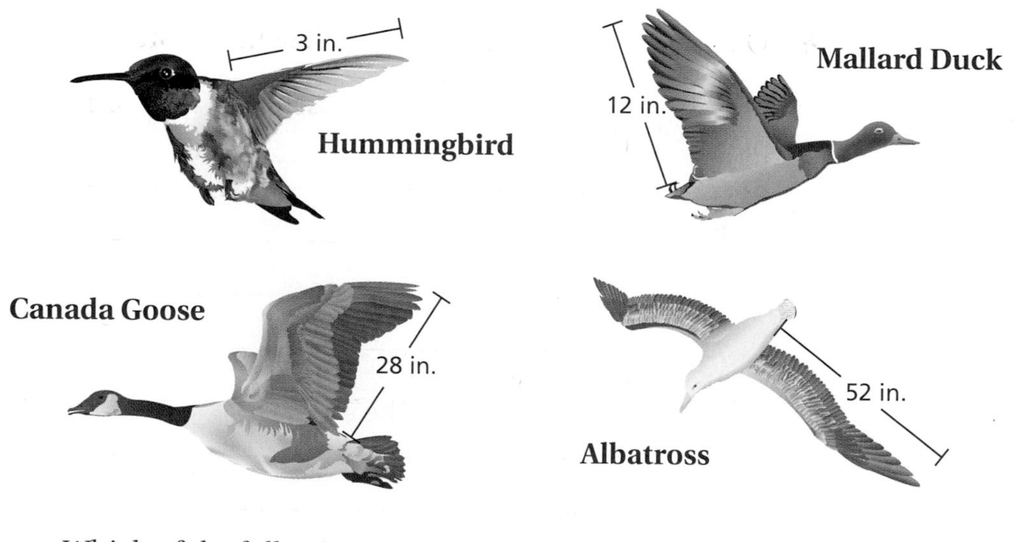

Which of the following seems true? Explain your reasoning.

- Wing length and wing beat frequency are directly proportional.
- Wing length and wing beat frequency are inversely proportional.
- Wing length and wing beat frequency are unrelated.

5. **SCIENCE** Think of an example in science where two variables are inversely proportional.

Practice Use what you learned about inverse variation to complete Exercises 4–7 on page 146.

3.8 Lesson

Check It Out
Lesson Tutorials
BigIdeasMath.com

Key Vocabulary
inverse variation, *p. 144*

Key Idea

Inverse Variation

Words

Two quantities x and y show **inverse variation** when $y = \frac{k}{x}$, where k is a number and $k \neq 0$.

Graph

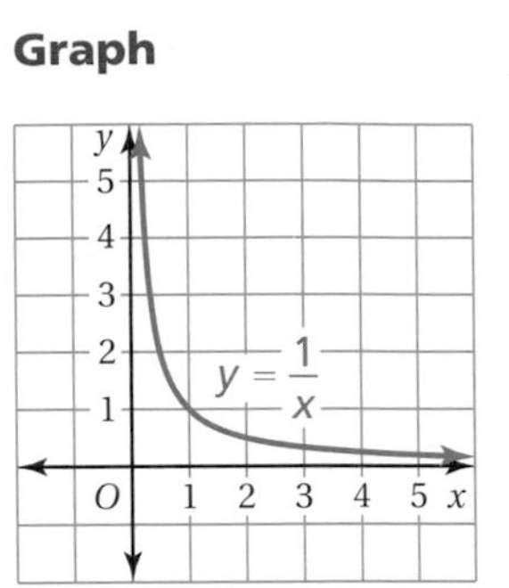

EXAMPLE 1 Identifying Direct and Inverse Variation

Tell whether x and y show *direct variation, inverse variation,* or *neither*. Explain your reasoning.

Study Tip

Other ways to say that x and y show inverse variation are "y varies inversely with x" and "x and y are inversely proportional."

a. $5y = x$

$y = \frac{1}{5}x$ Solve for y.

The equation can be written as $y = kx$. So, x and y show direct variation.

b. $\frac{1}{3}y = \frac{1}{x}$

$y = \frac{3}{x}$ Solve for y.

The equation can be written as $y = \frac{k}{x}$. So, x and y show inverse variation.

c. $-x = y + 7$

$-x - 7 = y$ Solve for y.

The equation cannot be written as $y = kx$ *or* $y = \frac{k}{x}$. So, x and y do *not* show direct or inverse variation.

On Your Own

Now You're Ready
Exercises 4–15

Tell whether x and y show *direct variation, inverse variation, or neither*. Explain your reasoning.

1. $y - 1 = 2x$ **2.** $\frac{1}{5}y = x$ **3.** $2y = \frac{1}{x}$

EXAMPLE 2 Standardized Test Practice

In the graph, x and y show inverse variation. Which equation relates x and y?

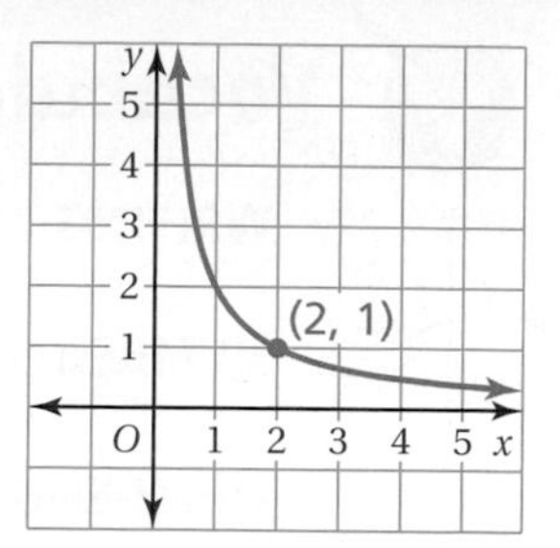

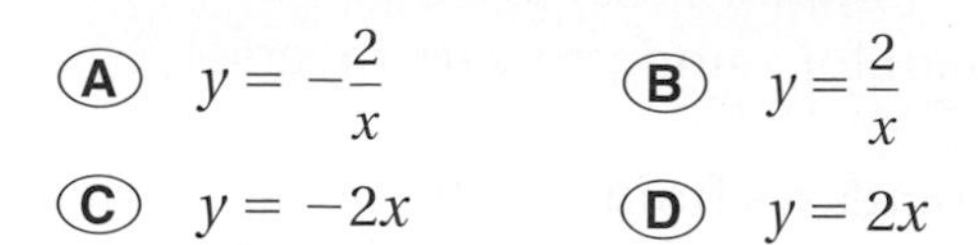

Ⓐ $y = -\frac{2}{x}$ Ⓑ $y = \frac{2}{x}$

Ⓒ $y = -2x$ Ⓓ $y = 2x$

The graph passes through (2, 1). Substitute to find k.

$y = \frac{k}{x}$ Write inverse variation equation.

$1 = \frac{k}{2}$ Substitute 2 for x and 1 for y.

$2 = k$ Solve for k.

∴ So, the equation $y = \frac{2}{x}$ relates x and y. The correct answer is Ⓑ.

EXAMPLE 3 Real-Life Application

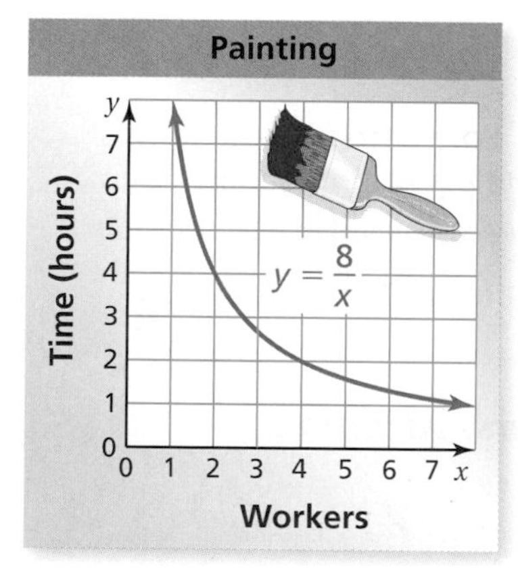

The graph shows the number of hours y it takes x workers to paint a room. (a) How does y change as x increases? (b) Do x and y show direct or inverse variation? (c) How many hours does it take five workers to paint the room?

a. From the graph, you can see that y decreases as x increases. So, as the number of workers increases, the time to paint the room decreases.

b. The equation is written as $y = \frac{k}{x}$. So, x and y show inverse variation.

c. Use the equation to find y when $x = 5$.

$y = \frac{8}{x}$ Write equation.

$= \frac{8}{5} = 1.6$ Substitute. Then simplify.

∴ It takes 1.6 hours for five workers to paint the room.

On Your Own

Now You're Ready
Exercises 23 and 24

4. Suppose y varies inversely with x and $y = 3$ when $x = 1$. Write an equation that relates x and y.
5. **WHAT IF?** In Example 3, how many hours does it take three workers to paint the room?

3.8 Exercises

Vocabulary and Concept Check

1. **WRITING** What does it mean for x and y to vary inversely?
2. **NUMBER SENSE** When x increases from 1 to 10, does $\frac{1}{x}$ increase or decrease?
3. **OPEN-ENDED** Describe a real-life situation that shows inverse variation.

Practice and Problem Solving

Tell whether x and y show *direct variation, inverse variation,* or *neither*. Explain your reasoning.

1 4. $y = \frac{1}{x}$ 5. $xy = 8$ 6. $y - x = 0$ 7. $\frac{1}{2}y = 2x$

8. $\frac{y}{3} = \frac{2}{x}$ 9. $y - 2 = \frac{7}{x}$ 10. $x = y + 9$ 11. $x = 4y$

12. $y = \frac{5}{2x}$ 13. $2y = \frac{6}{x}$ 14. $\frac{5x}{3} = \frac{y}{4}$ 15. $x = \frac{7 + y}{2}$

16. **ERROR ANALYSIS** Describe and correct the error in telling whether x and y show inverse variation.

> ✗ $\frac{y}{2} = \frac{8}{x}$
> The equation does not show inverse variation because it is not of the form $y = \frac{k}{x}$.

Graph the data. Tell whether x and y show *direct variation* or *inverse variation*.

17.

x	−2	2	4	6
y	−1	1	2	3

18.

x	0.5	1	3	6
y	6	3	1	0.5

19.

x	2	5	8	20
y	10	4	2.5	1

20.

x	2	4	8	11
y	1.5	3	6	8.25

Tell whether x and y show *direct variation* or *inverse variation*. Explain.

21. **STADIUM** The time y it takes to empty a stadium and the number x of open exits are related by the equation $y = \frac{0.8}{x}$.
22. **TRAVEL** The number y of miles driven and the number x of gallons of gas used are related by the equation $y = 28.5x$.

The variables x and y vary inversely. Write an equation relating x and y.

23.

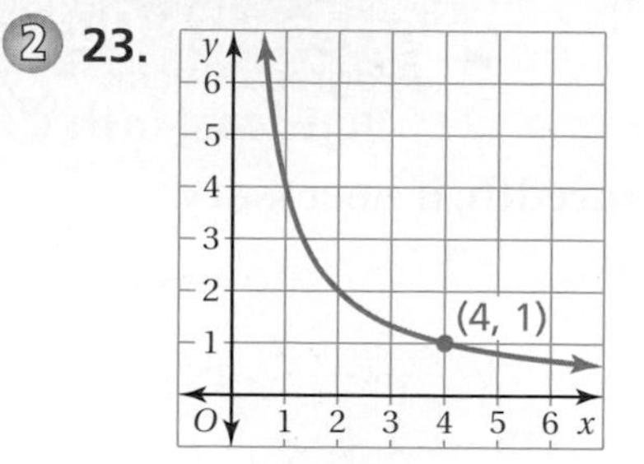

24.

y
12
10
8
6
4
2
O
2 4 6 8 10 12 x
(2, 6)

25. BICYCLING The table shows the times it takes to bicycle 12 miles at various speeds.

Speed (mi/h)	12	6	3	2
Time (h)	1	2	4	6

a. Does the time t vary inversely with the speed s? If so, write an equation relating t and s.

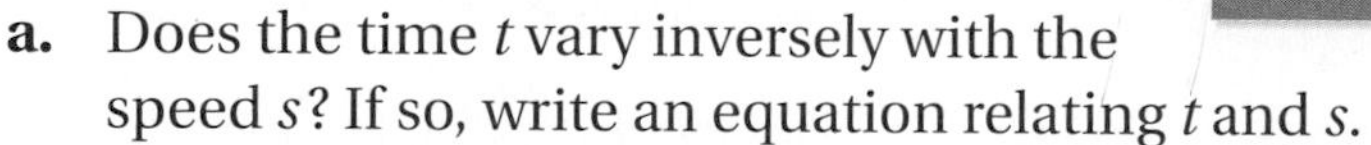

b. What time corresponds to a speed of 4 miles per hour?

26. MARTIAL ARTS It takes 3.6 pounds of force to break a 5-foot board.

a. You remember from science that force and board length vary directly or inversely, but you've forgotten which. How can you use reason to remember?

b. How much force does it take to break the board shown?

27. SALARY A salesperson has a fixed weekly salary. The person works twice as many hours this week as last week. What happens to the person's hourly rate?

28. Reasoning The price per person to rent a limousine varies inversely with the number of passengers. It costs $90 each for five people. How many people are renting the limousine when the cost per person is $56.25?

Fair Game Review What you learned in previous grades & lessons

Find the percent of the number. *(Skills Review Handbook)*

29. 40% of 220 **30.** 32% of 275 **31.** 84% of 75 **32.** 21% of 300

Tell whether the ratios form a proportion. *(Section 3.3)*

33. $\frac{9}{15}, \frac{18}{30}$ **34.** $\frac{21}{9}, \frac{18}{8}$ **35.** $\frac{42}{91}, \frac{24}{52}$ **36.** $\frac{24}{38}, \frac{36}{57}$

37. MULTIPLE CHOICE A gumball machine contains 1000 gumballs. The ratio of red gumballs to the total number of gumballs is 1 : 4. How many red gumballs are in the machine? *(Section 3.5)*

Ⓐ 150 Ⓑ 250 Ⓒ 400 Ⓓ 750

3.6–3.8 Quiz

Check It Out
Progress Check
BigIdeasMath.com

Copy and complete the statement. Round to the nearest hundredth, if necessary. *(Section 3.6)*

1. 10 mi ≈ ___ km

2. 3 qt ≈ ___ L

3. 29 kg ≈ ___ lb

4. 6.8 in. ≈ ___ cm

Tell whether *x* and *y* show direct variation. Explain your reasoning. *(Section 3.7)*

5.

x	*y*
−3	0
−1	1
1	2
3	3

6.

x	*y*
−1	−2
0	0
1	2
2	4

7. $y - 9 = 6 + x$

8. $x = \frac{5}{8}y$

Tell whether *x* and *y* show *direct variation*, *inverse variation*, or *neither*. Explain your reasoning. *(Section 3.8)*

9. $y = \frac{12}{x}$

10. $y - x = 9$

11.

x	*y*
−3	−2
−1	0
1	0
3	2

12.

x	*y*
1	2
2	1
4	0.5
8	0.25

13. HEIGHT The tallest player in Euroleague Basketball is 229 centimeters. The tallest player in the National Basketball Association is 90 inches. Which league has the tallest player? *(Section 3.6)*

14. PIE SALE The table shows the profit of a pie sale. Tell whether there is direct variation between the two data sets. If so, write the equation of direct variation. *(Section 3.7)*

Pies Sold	10	12	14	16
Profit	\$79.50	\$95.40	\$111.30	\$127.20

15. JEWELRY The number of beads on a bracelet varies inversely with the length of the beads. You use 8-millimeter beads to make a bracelet with 25 beads. How many 10-millimeter beads would you need to make a bracelet? *(Section 3.8)*

3 Chapter Review

Check It Out
Vocabulary Help
BigIdeasMath.com

Review Key Vocabulary

ratio, *p. 100*
rate, *p. 100*
unit rate, *p. 100*
slope, *p. 106*
proportion, *p. 112*
proportional, *p. 112*
cross products, *p. 113*
U.S. customary system, *p. 132*
metric system, *p. 132*
direct variation, *p. 138*
inverse variation, *p. 144*

Review Examples and Exercises

3.1 Ratios and Rates *(pp. 98–103)*

Find the unit rate of calories per serving.

Servings	2	4	6	8
Calories	240	480	720	960

The calories increase by 240 for every 2 servings.

$$\frac{\text{change in calories}}{\text{change in servings}} = \frac{240}{2} = \frac{120}{1}$$

∴ The rate is 120 calories per serving.

Exercises

Find the unit rate.

1. 289 miles on 10 gallons
2. 975 revolutions in 3 minutes

3.2 Slope *(pp. 104–109)*

Find the slope of the line.

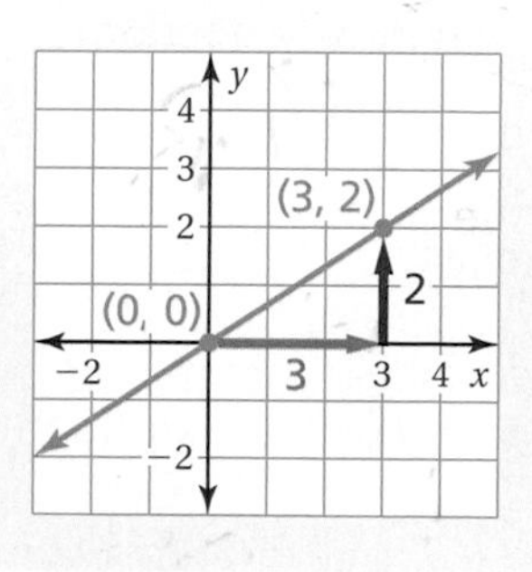

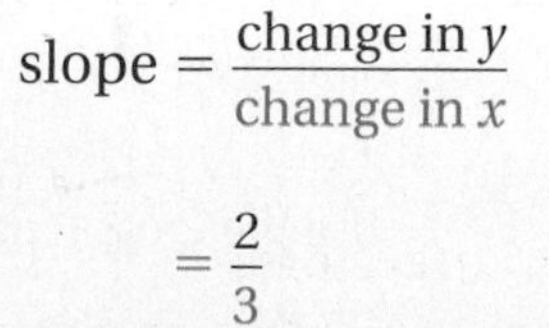

$$\text{slope} = \frac{\text{change in } y}{\text{change in } x}$$

$$= \frac{2}{3}$$

∴ The slope of the line is $\frac{2}{3}$.

Exercises

Graph the line that passes through the two points. Then find the slope of the line.

3. (−3, −3), (1, 1)
4. (−3, −2), (3, 2)
5. (3, 9), (−2, −6)

3.3 Proportions (pp. 110–115)

Tell whether the ratios $\frac{9}{12}$ and $\frac{6}{8}$ form a proportion.

$\frac{9}{12} = \frac{9 \div 3}{12 \div 3} = \frac{3}{4}$ $\frac{6}{8} = \frac{6 \div 2}{8 \div 2} = \frac{3}{4}$

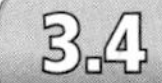

The ratios are equivalent. So, the ratios form a proportion.

Exercises

Tell whether the ratios form a proportion.

6. $\frac{4}{9}, \frac{2}{3}$ **7.** $\frac{12}{22}, \frac{18}{33}$ **8.** $\frac{8}{50}, \frac{4}{10}$ **9.** $\frac{32}{40}, \frac{12}{15}$

3.4 Writing Proportions (pp. 116–121)

Write a proportion that gives the number r of returns on Saturday.

	Friday	Saturday
Sales	40	85
Returns	32	r

$$\frac{40 \text{ sales}}{32 \text{ returns}} = \frac{85 \text{ sales}}{r \text{ returns}}$$

Exercises

Use the table to write a proportion.

10.

	Game 1	Game 2
Penalties	6	8
Minutes	16	m

11.

	Concert 1	Concert 2
Songs	15	18
Hours	2.5	h

3.5 Solving Proportions (pp. 122–127)

Solve $\frac{x}{9} = \frac{4}{5}$.

$x \cdot 5 = 9 \cdot 4$ Use the Cross Products Property.

$5x = 36$ Multiply.

$x = 7.2$ Divide.

Exercises

Solve the proportion.

12. $\frac{x}{4} = \frac{2}{5}$ **13.** $\frac{5}{12} = \frac{y}{15}$ **14.** $\frac{z}{7} = \frac{3}{16}$ **15.** $\frac{8}{20} = \frac{6}{w}$

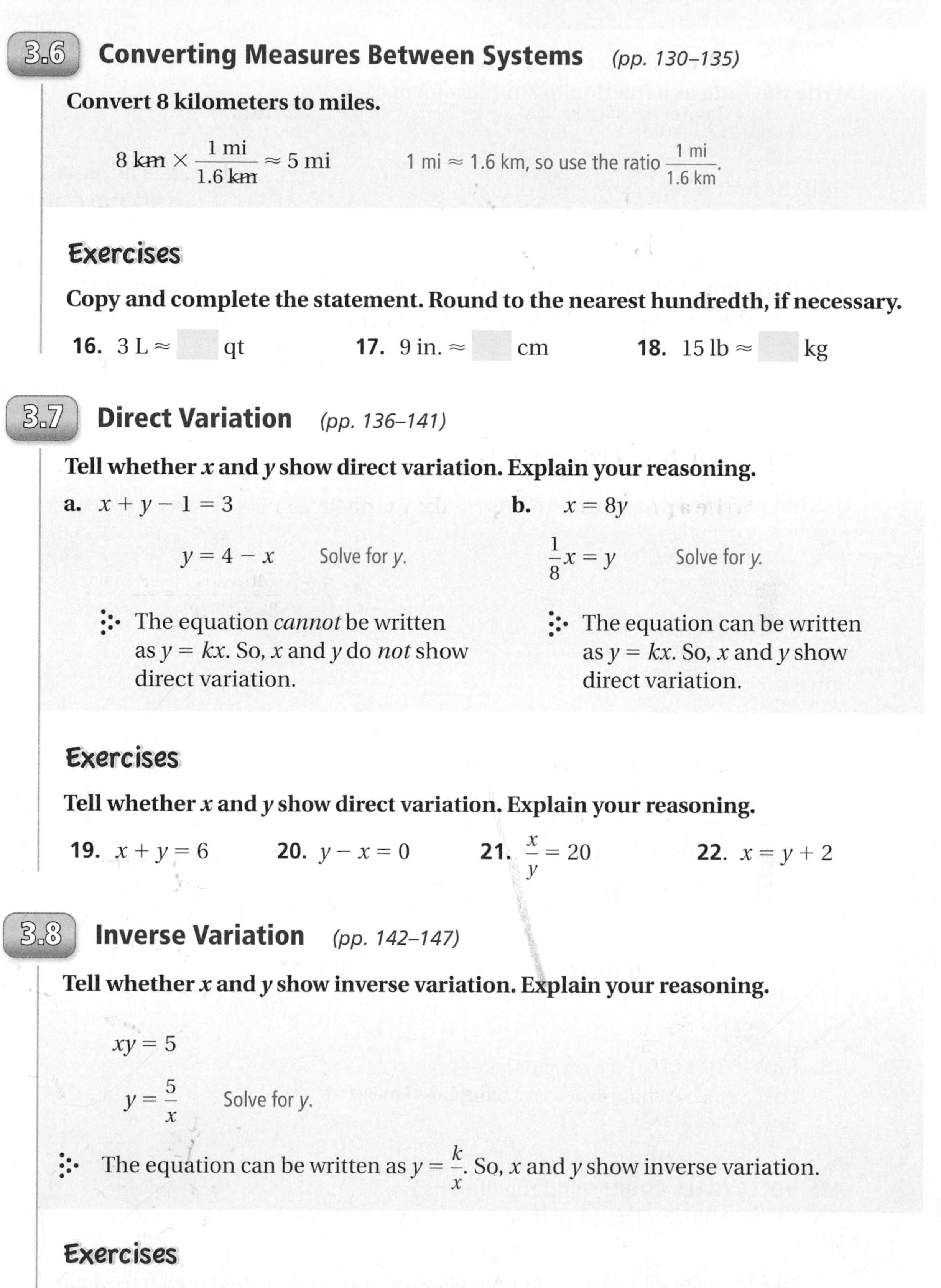

3.6 Converting Measures Between Systems (pp. 130–135)

Convert 8 kilometers to miles.

$8\ \cancel{\text{km}} \times \frac{1\text{ mi}}{1.6\ \cancel{\text{km}}} \approx 5\text{ mi}$ 1 mi ≈ 1.6 km, so use the ratio $\frac{1\text{ mi}}{1.6\text{ km}}$.

Exercises

Copy and complete the statement. Round to the nearest hundredth, if necessary.

16. 3 L ≈ ___ qt **17.** 9 in. ≈ ___ cm **18.** 15 lb ≈ ___ kg

3.7 Direct Variation (pp. 136–141)

Tell whether x and y show direct variation. Explain your reasoning.

a. $x + y - 1 = 3$

$y = 4 - x$ Solve for y.

∴ The equation *cannot* be written as $y = kx$. So, x and y do *not* show direct variation.

b. $x = 8y$

$\frac{1}{8}x = y$ Solve for y.

∴ The equation can be written as $y = kx$. So, x and y show direct variation.

Exercises

Tell whether x and y show direct variation. Explain your reasoning.

19. $x + y = 6$ **20.** $y - x = 0$ **21.** $\frac{x}{y} = 20$ **22.** $x = y + 2$

3.8 Inverse Variation (pp. 142–147)

Tell whether x and y show inverse variation. Explain your reasoning.

$xy = 5$

$y = \frac{5}{x}$ Solve for y.

∴ The equation can be written as $y = \frac{k}{x}$. So, x and y show inverse variation.

Exercises

Tell whether x and y show *direct variation*, *inverse variation*, or *neither*. Explain your reasoning.

23. $\frac{x}{y} = 6$ **24.** $3x + y = 7$ **25.** $8y = 4x$ **26.** $xy = 12$

3 Chapter Test

Check It Out
Test Practice
BigIdeasMath.com

Write the ratio as a fraction in simplest form.

1. 34 cars : 26 trucks
2. 3 feet to 9 feet

Find the unit rate.

3. 84 miles in 12 days
4. $3.20 for 8 ounces

Graph the line that passes through the two points. Then find the slope of the line.

5. (15, 9), (−5, −3)
6. (2, 9), (4, 18)

Tell whether the ratios form a proportion.

7. $\frac{1}{9}, \frac{6}{54}$
8. $\frac{9}{12}, \frac{8}{72}$

Use the table to write a proportion.

9.

	Monday	Tuesday
Gallons	6	8
Miles	180	m

10.

	Thursday	Friday
Classes	6	c
Hours	8	4

Solve the proportion.

11. $\frac{x}{8} = \frac{9}{4}$
12. $\frac{17}{3} = \frac{y}{6}$

Copy and complete the statement. Round to the nearest hundredth, if necessary.

13. 5 L ≈ ▢ qt
14. 56 lb ≈ ▢ kg

Tell whether x and y show *direct variation, inverse variation,* or *neither*. Explain your reasoning.

15. $xy - 11 = 5$
16. $x = \frac{3}{y}$
17. $\frac{y}{x} = 8$

18. **MOVIE TICKETS** Five movie tickets cost $36.25. What is the cost of eight movie tickets?

19. **VOLLEYBALL COURT** Find the dimensions of the volleyball court in feet.

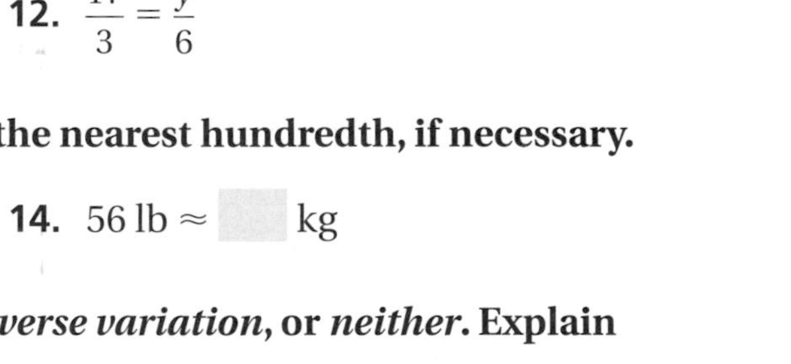

20. **GLAZE** A specific shade of green glaze requires 5 parts blue to 3 parts yellow. A glaze mixture contains 25 quarts of blue and 9 quarts of yellow. How can you fix the mixture to make the specific shade of green glaze?

3 Standardized Test Practice

1. The school store sells 4 pencils for \$0.50. At that rate, what would be the cost of 10 pencils? *(7.RP.2a)*

A. \$1.10 **C.** \$2.00

B. \$1.25 **D.** \$5.00

2. Which expressions do *not* have a value of 3? *(7.EE.3)*

I. $|3|$ II. $|-3|$

III. $-|3|$ IV. $-|-3|$

F. I and II **H.** II and IV

G. I and III **I.** III and IV

3. What is the value of y in the equation below when $x = 12$ and $k = 3$? *(7.NS.2c)*

$$xy = k$$

4. Use the coordinate plane to answer the question below.

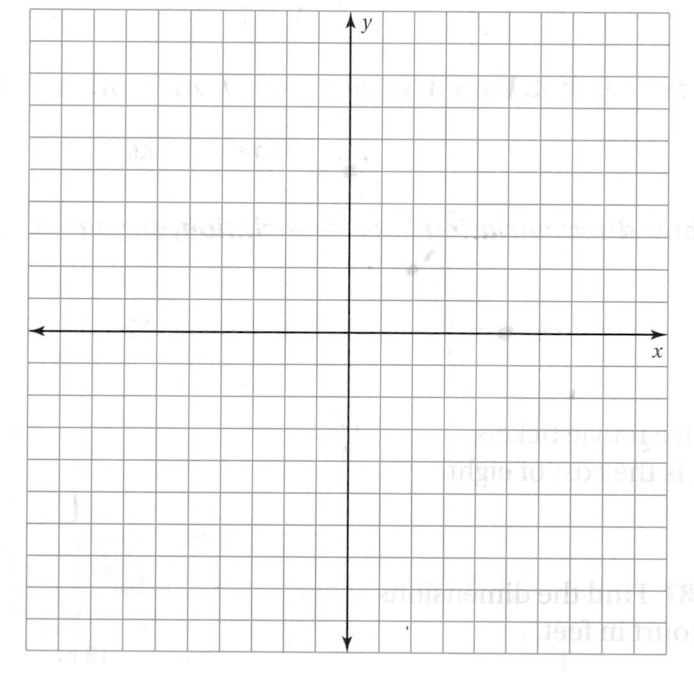

A line contains both the point (0, 5) and the point (5, 0). Which of the following points is also on this line? *(7.NS.1b)*

A. (0, −5) **C.** (5, 5)

B. (3, 3) **D.** (7, −2)

5. The scores from a diving competition are shown in the line plot below.

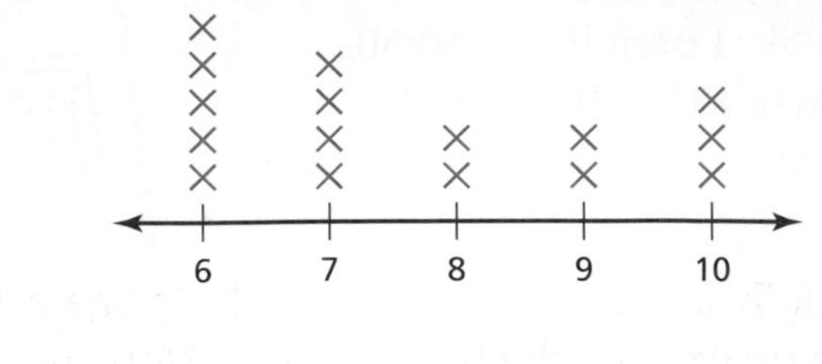

What is the median score? *(7.NS.3)*

F. 7

G. 6

H. 5

I. 4

6. Meli was solving the equation in the box below.

$$-\frac{1}{2}(4x - 10) = -16$$

$$-2x - 5 = -16$$

$$-2x - 5 + 5 = -16 + 5$$

$$-2x = -11$$

$$\frac{-2x}{-2} = \frac{-11}{-2}$$

$$x = \frac{11}{2}$$

What should Meli do to correct the error that she made? *(7.EE.4a)*

A. Distribute the $-\frac{1}{2}$ to get $-2x + 5$.

B. Distribute the $-\frac{1}{2}$ to get $2x - 5$.

C. Divide -11 by -2 to get $-\frac{11}{2}$.

D. Add 2 to -11 to get -9.

7. What is the value of the expression below when $n = -8$ and $p = -4$? *(7.NS.3)*

$$-9n - p$$

F. 76

G. 68

H. -68

I. -76

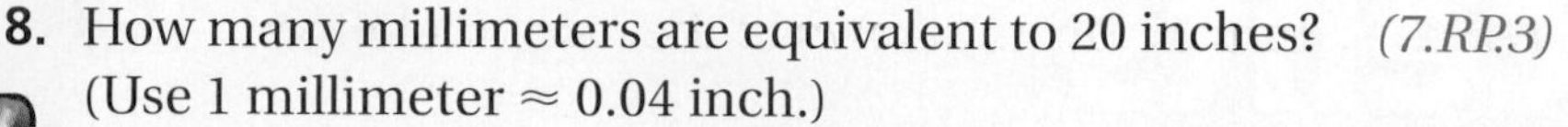

8. How many millimeters are equivalent to 20 inches? *(7.RP.3)*
(Use 1 millimeter $\approx$ 0.04 inch.)

9. If 5 dogs share equally a bag of dog treats, each dog gets 24 treats. Suppose 8 dogs share equally the bag of treats. How many treats does each dog get? *(7.RP.2a)*

A. 3

B. 15

C. 21

D. 38

10. The figure below consists of a rectangle and a right triangle.

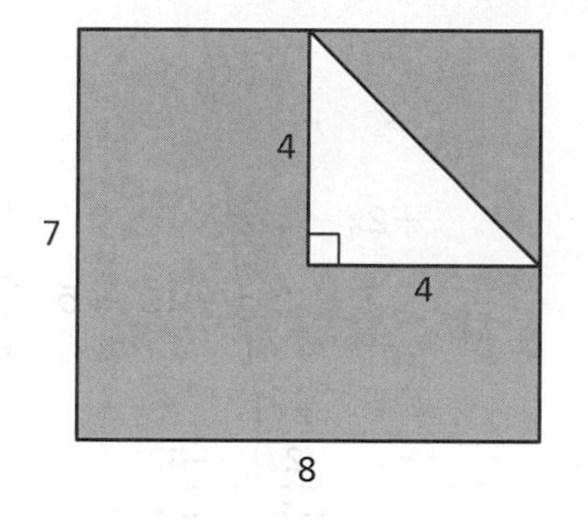

What is the area of the shaded region? *(7.NS.3)*

F. 23 units^2

G. 40 units^2

H. 48 units^2

I. 60 units^2

11. You can mow 800 square feet of lawn in 15 minutes. At this rate, how many minutes will you take to mow a lawn that measures 6000 square feet? *(7.RP.2c)*

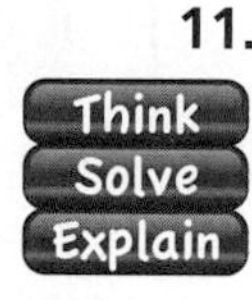

Part A Write a proportion to represent the problem. Use m to represent the number of minutes. Explain your reasoning.

Part B Solve the proportion you wrote in Part A and use it to answer the problem. Show your work.

12. What number belongs in the box to make the equation true? *(7.NS.3)*

$$5\frac{7}{8} = \square \cdot \left(-\frac{1}{4}\right)$$

A. $-23\frac{1}{2}$

B. $-1\frac{15}{32}$

C. $6\frac{1}{8}$

D. $23\frac{1}{2}$

4 Percents

4.1 The Percent Equation

4.2 Percents of Increase and Decrease

4.3 Discounts and Markups

4.4 Simple Interest

"Here's my sales strategy. I buy each dog bone for $0.05."

"Then I mark each one up to $1. Then, I have a 75% off sale. Cool, huh?"

"Dear Vet: I have this strange feeling that I am wagging my tail 15% fewer times than I used to wag it."

"Oh look. He already answered me."

"Dear Newton, I only practice general vet work. I need to refer you to a dog tail specialist."

What You Learned Before

"The fact that these two percents do not total 100 is a sad commentary on humans."

Writing Percents Using Models (6.RP.3c)

What percent of the model is shaded?

Example 1

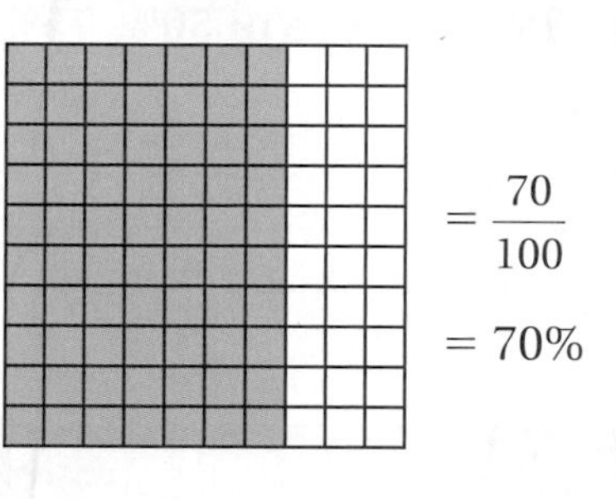

$= \frac{70}{100}$

$= 70\%$

Example 2

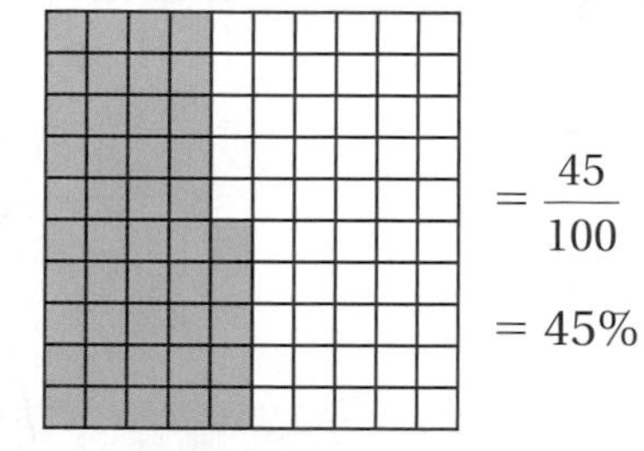

$= \frac{45}{100}$

$= 45\%$

Try It Yourself

What percent of the model is shaded?

1.

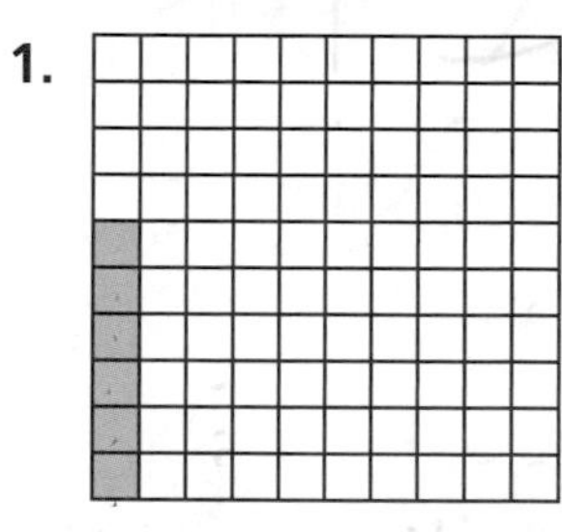

2.

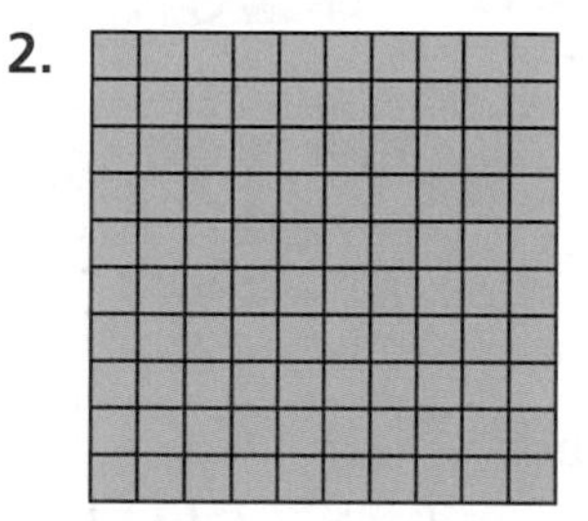

3.

Writing Decimals, Percents, and Fractions (6.RP.3c)

Example 3 **Write $\frac{3}{5}$ as a decimal.**

$$\frac{3}{5} = \frac{3 \cdot 2}{5 \cdot 2} = \frac{6}{10} = 0.6$$

Example 4 **Write $\frac{3}{5}$ as a percent.**

$$\frac{3}{5} = \frac{3 \cdot 20}{5 \cdot 20} = \frac{60}{100} = 60\%$$

Multiply to make the denominator 100.

Try It Yourself

Copy and complete the table.

	Percent	Decimal	Fraction
4.	35%		
5.		0.6	
6.			$\frac{13}{25}$

	Percent	Decimal	Fraction
7.	10%		
8.		0.85	
9.			$\frac{1}{5}$

4.1 The Percent Equation

COMMON CORE STATE STANDARDS
7.EE.3
7.RP.3

Essential Question How can you use models to estimate percent questions?

1 ACTIVITY: Estimating a Percent

Work with a partner. Estimate the locations of 50%, 75%, 40%, 6%, and 65% on the model. 50% is done for you.

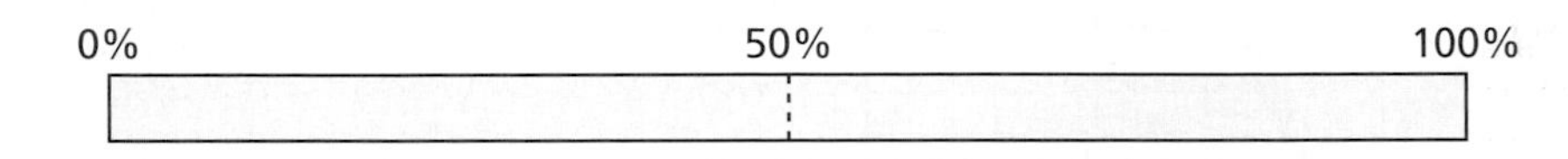

2 ACTIVITY: Estimating a Part of a Number

The statement "25% of 12 is 3" has three numbers. In real-life problems, any one of these numbers can be unknown.

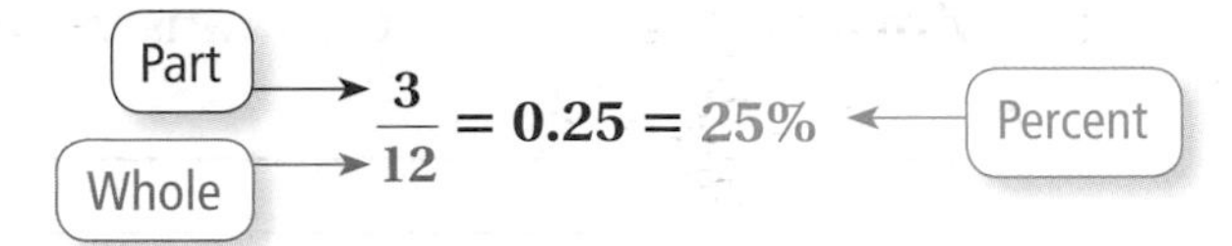

Which number is missing?	Question	Type of Question
3	What is 25% of 12?	Find a part of a number.
25%	3 is what percent of 12?	Find a percent.
12	3 is 25% of what?	Find the whole.

Work with a partner. Estimate the answer to each question using a model.

a. Sample: What number is 50% of 30?

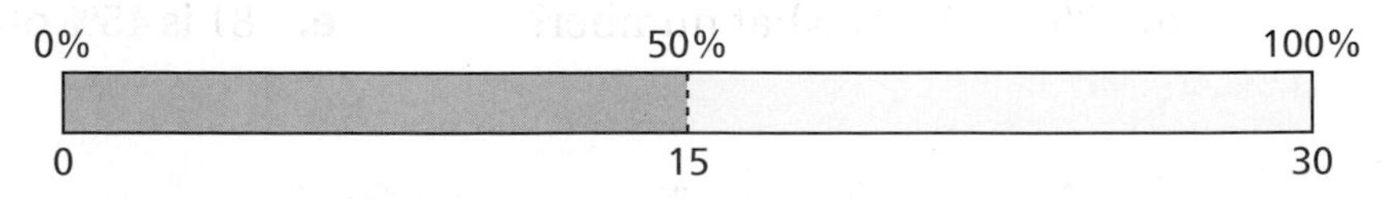

∴ So, from the model, 15 is 50% of 30.

b. What number is 75% of 30?

c. What number is 40% of 30?

d. What number is 6% of 30?

e. What number is 65% of 30?

3 ACTIVITY: Estimating a Percent

Work with a partner. Estimate the answer to the question using a model.

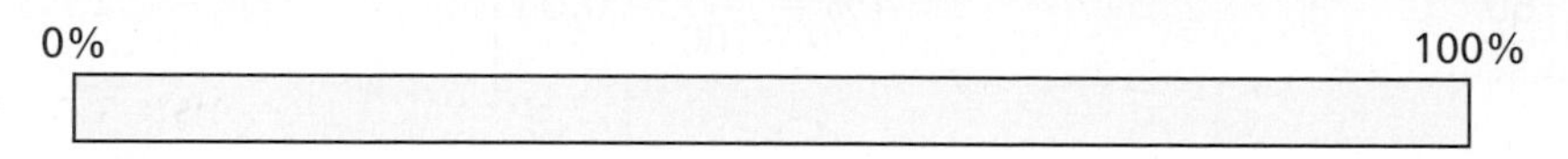

a. Sample: 15 is what percent of 75?

0% 20% 40% 60% 80% 100%

0 15 30 45 60 75

So, 15 is 20% of 75.

b. 5 is what percent of 20?

c. 18 is what percent of 40?

d. 50 is what percent of 80?

e. 75 is what percent of 50?

4 ACTIVITY: Estimating a Whole

Work with a partner. Estimate the answer to the question using a model.

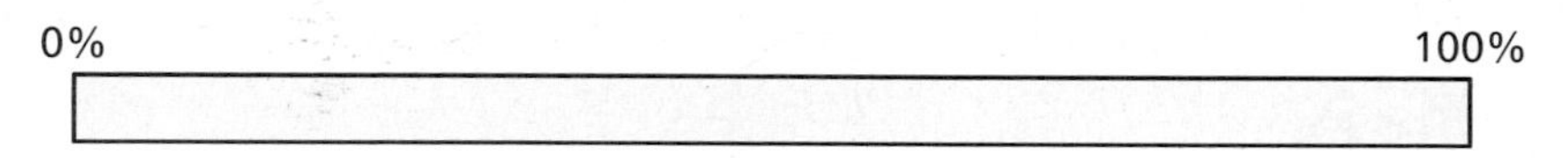

a. Sample: 24 is $33\frac{1}{3}\%$ of what number?

0% $33\frac{1}{3}\%$ $66\frac{2}{3}\%$ 100%

0 24 48 72

So, 24 is $33\frac{1}{3}\%$ of 72.

b. 13 is 25% of what number?

c. 110 is 20% of what number?

d. 75 is 75% of what number?

e. 81 is 45% of what number?

What Is Your Answer?

5. **IN YOUR OWN WORDS** How can you use models to estimate percent questions? Give examples to support your answer.

Practice Use what you learned about estimating percent questions to complete Exercises 4–9 on page 162.

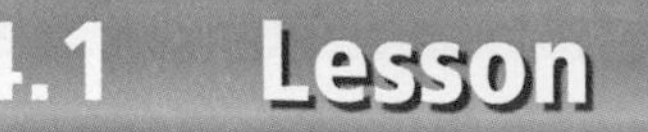

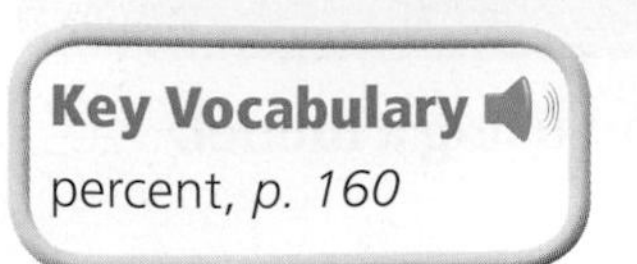

A **percent** is a ratio whose denominator is 100. Here are two examples.

$$4\% = \frac{4}{100} = 0.04 \qquad 25\% = \frac{25}{100} = 0.25$$

Key Idea

The Percent Equation

Words To represent "a is p percent of w," use an equation.

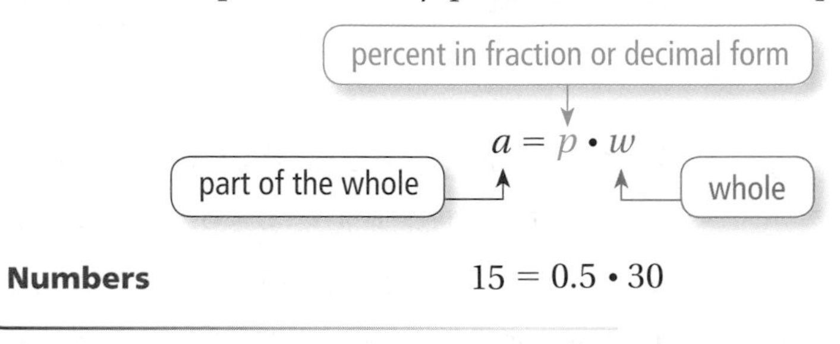

Numbers $15 = 0.5 \cdot 30$

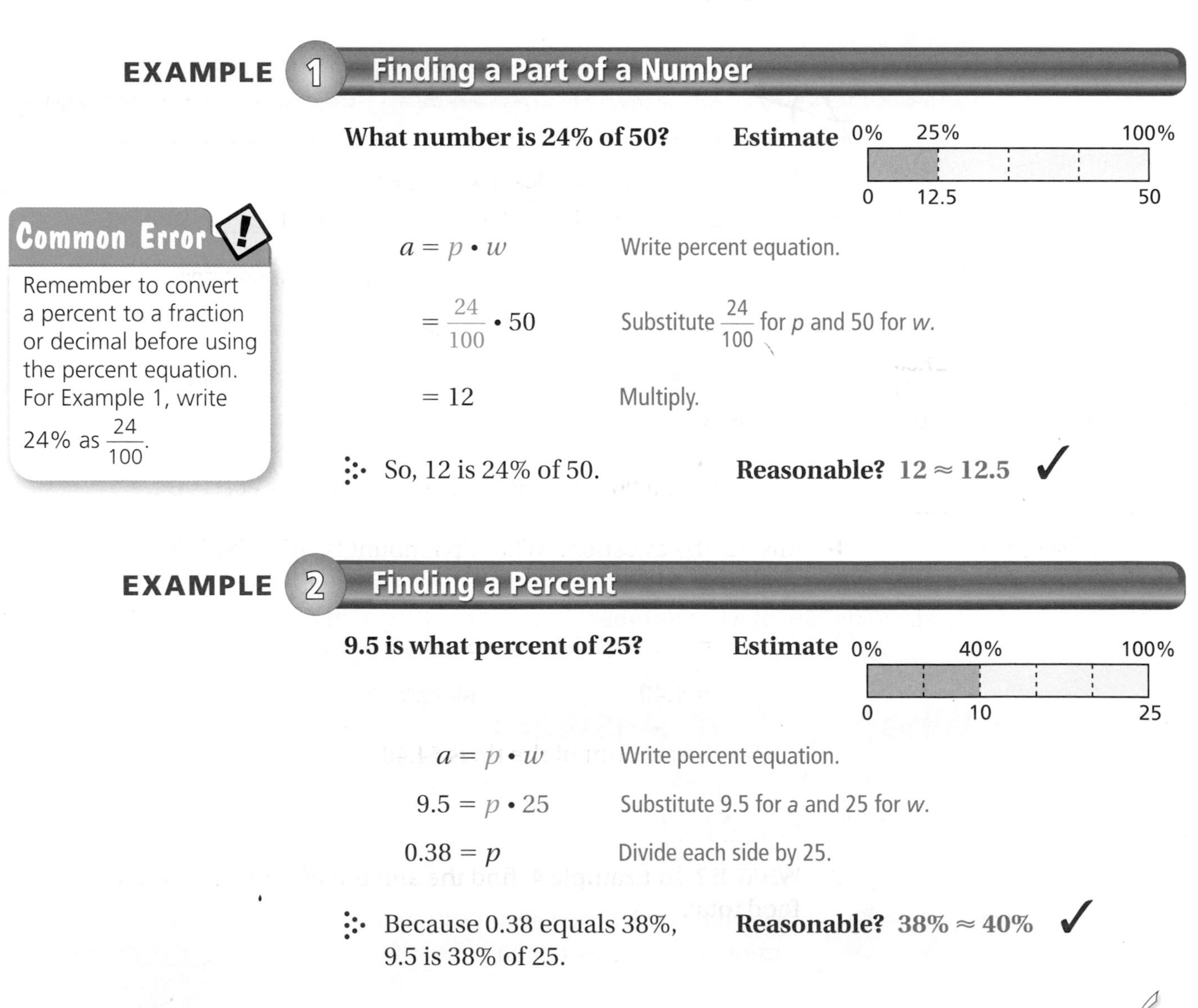

EXAMPLE 1 Finding a Part of a Number

What number is 24% of 50?

Estimate

$a = p \cdot w$ Write percent equation.

$= \frac{24}{100} \cdot 50$ Substitute $\frac{24}{100}$ for p and 50 for w.

$= 12$ Multiply.

∴ So, 12 is 24% of 50. **Reasonable?** $12 \approx 12.5$ ✓

Common Error

Remember to convert a percent to a fraction or decimal before using the percent equation. For Example 1, write 24% as $\frac{24}{100}$.

EXAMPLE 2 Finding a Percent

9.5 is what percent of 25?

Estimate

$a = p \cdot w$ Write percent equation.

$9.5 = p \cdot 25$ Substitute 9.5 for a and 25 for w.

$0.38 = p$ Divide each side by 25.

∴ Because 0.38 equals 38%, 9.5 is 38% of 25. **Reasonable?** $38\% \approx 40\%$ ✓

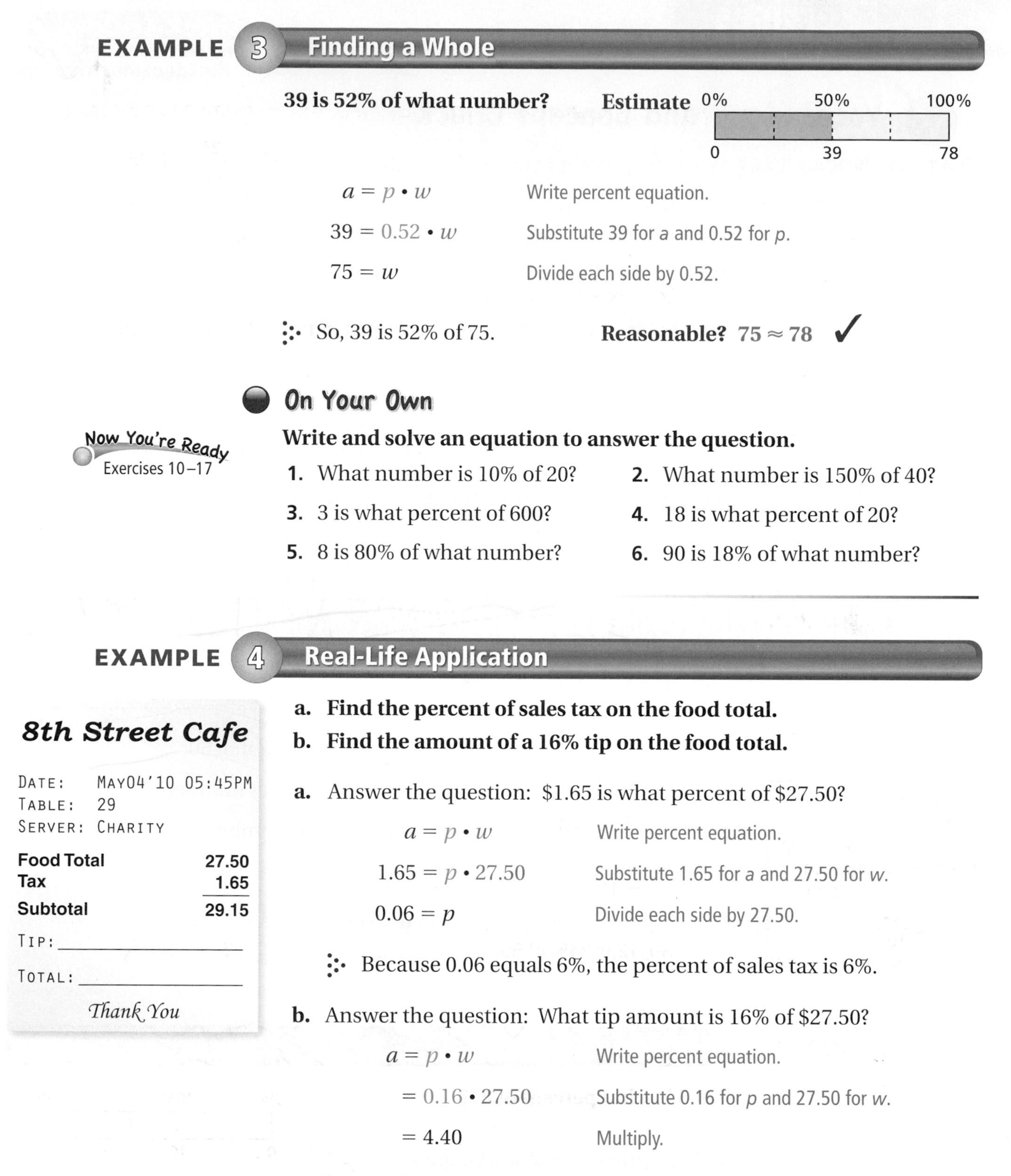

EXAMPLE 3 Finding a Whole

39 is 52% of what number?

Estimate 0% 50% 100%; 0 39 78

$a = p \cdot w$	Write percent equation.
$39 = 0.52 \cdot w$	Substitute 39 for a and 0.52 for p.
$75 = w$	Divide each side by 0.52.

∴ So, 39 is 52% of 75.

Reasonable? $75 \approx 78$ ✓

On Your Own

Now You're Ready
Exercises 10–17

Write and solve an equation to answer the question.

1. What number is 10% of 20?
2. What number is 150% of 40?
3. 3 is what percent of 600?
4. 18 is what percent of 20?
5. 8 is 80% of what number?
6. 90 is 18% of what number?

EXAMPLE 4 Real-Life Application

8th Street Cafe

DATE: MAY04'10 05:45PM
TABLE: 29
SERVER: CHARITY

Food Total	**27.50**
Tax	**1.65**
Subtotal	**29.15**

TIP: ____________

TOTAL: ____________

Thank You

a. Find the percent of sales tax on the food total.

b. Find the amount of a 16% tip on the food total.

a. Answer the question: \$1.65 is what percent of \$27.50?

$a = p \cdot w$	Write percent equation.
$1.65 = p \cdot 27.50$	Substitute 1.65 for a and 27.50 for w.
$0.06 = p$	Divide each side by 27.50.

∴ Because 0.06 equals 6%, the percent of sales tax is 6%.

b. Answer the question: What tip amount is 16% of \$27.50?

$a = p \cdot w$	Write percent equation.
$= 0.16 \cdot 27.50$	Substitute 0.16 for p and 27.50 for w.
$= 4.40$	Multiply.

∴ So, the amount of the tip is \$4.40.

On Your Own

7. **WHAT IF?** In Example 4, find the amount of a 20% tip on the food total.

4.1 Exercises

Check It Out
Help with Homework
BigIdeasMath.com

Vocabulary and Concept Check

1. **VOCABULARY** Write the percent equation in words.
2. **REASONING** A number n is 150% of number m. Is n *greater than, less than,* or *equal to* m? Explain your reasoning.
3. **DIFFERENT WORDS, SAME QUESTION** Which is different? Find "both" answers.

What number is 20% of 55?	55 is 20% of what number?
20% of 55 is what number?	0.2 • 55 is what number?

Practice and Problem Solving

Estimate the answer to the question using a model.

4. What number is 24% of 80?
5. 15 is what percent of 40?
6. 15 is 30% of what number?
7. What number is 120% of 70?
8. 20 is what percent of 52?
9. 48 is 75% of what number?

Write and solve an equation to answer the question.

10. 20% of 150 is what number?
11. 45 is what percent of 60?
12. 35% of what number is 35?
13. 32% of 25 is what number?
14. 29 is what percent of 20?
15. 0.5% of what number is 12?
16. What percent of 300 is 51?
17. 120% of what number is 102?

ERROR ANALYSIS **Describe and correct the error in using the percent equation.**

18. What number is 35% of 20?

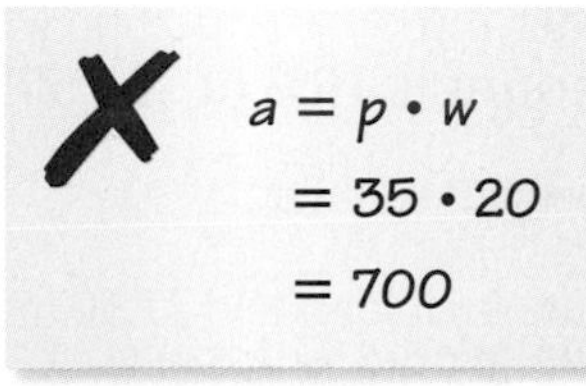

19. 30 is 60% of what number?

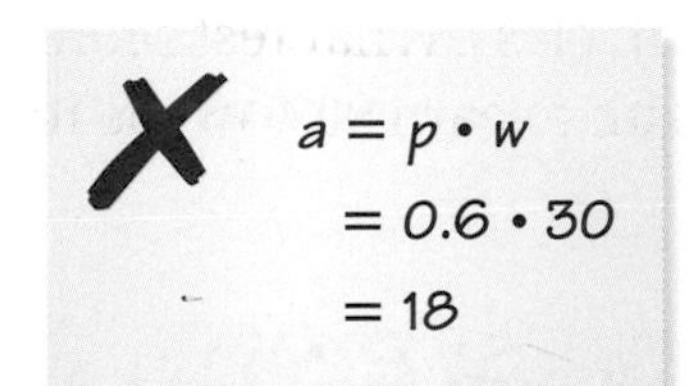

20. **BASEBALL** A pitcher throws 75 pitches. Of these, 72% were strikes. How many strikes did the pitcher throw?
21. **FUNDRAISING** Your school raised 125% of its fundraising goal. The school raised $6750. What was the goal?
22. **SURFBOARD** The sales tax on a surfboard is $12. What is the percent of sales tax?

PUZZLE There were w signers of the Declaration of Independence. The youngest was Edward Rutledge, who was x years old. The oldest was Benjamin Franklin, who was y years old.

23. x is 25% of 104. What was Rutledge's age?
24. 7 is 10% of y. What was Franklin's age?
25. w is 80% of y. How many signers were there?
26. y is what percent of $(w + y - x)$?

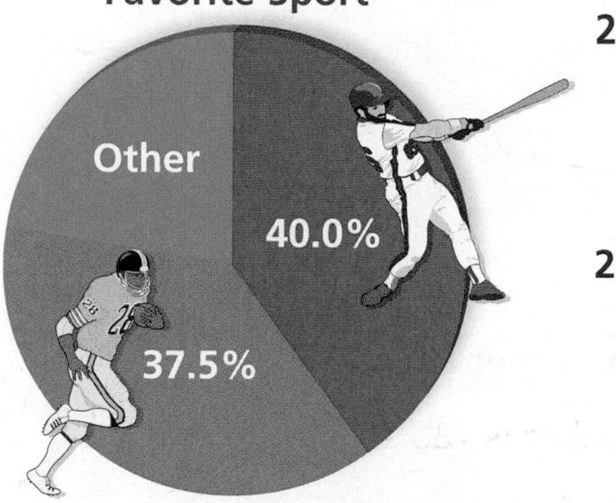

27. **REASONING** How can you tell whether the percent of a number will be *greater than, less than,* or *equal to* the number?

28. **SURVEY** In a survey, a group of students were asked their favorite sport. "Other" sports were chosen by 18 people.
    a. How many students participated?
    b. How many chose football?

29. **PROBLEM SOLVING** Water tank A has a capacity of 550 gallons and is 66% full. Water tank B is 53% full. The ratio of the capacity of tank A to tank B is 11 : 15.
    a. How much water is in tank A?
    b. What is the capacity of tank B?
    c. How much water is in tank B?

30. **TRUE OR FALSE?** Tell whether the statement is *true* or *false*. Explain your reasoning.

    If W is 25% of Z, then $Z : W$ is 75 : 25.

31. **Reasoning** The table shows your test results for math class. What test score is needed on the last exam to earn 90% of the total points?

Test Score	Point Value
83%	100
91.6%	250
88%	150
?	300

Fair Game Review What you learned in previous grades & lessons

Simplify. Write as a decimal. *(Skills Review Handbook)*

32. $\frac{10 - 4}{10}$
33. $\frac{25 - 3}{25}$
34. $\frac{105 - 84}{84}$
35. $\frac{170 - 125}{125}$

36. **MULTIPLE CHOICE** There are 160 people in a grade. The ratio of boys to girls is 3 to 5. Which proportion can you use to find the number x of boys? *(Section 3.4)*

Ⓐ $\frac{3}{8} = \frac{x}{160}$ Ⓑ $\frac{3}{5} = \frac{x}{160}$ Ⓒ $\frac{5}{8} = \frac{x}{160}$ Ⓓ $\frac{3}{5} = \frac{160}{x}$

4.2 Percents of Increase and Decrease

Essential Question What is a percent of decrease? What is a percent of increase?

COMMON CORE STATE STANDARDS
7.EE.3
7.RP.3

1 ACTIVITY: Percent of Decrease

Each year in the Columbia River Basin, adult salmon swim up river to streams to lay eggs and hatch their young.

To go up the river, the adult salmon use fish ladders. But, to go down the river, the young salmon must pass through several dams.

There are electric turbines at each of the eight dams on the main stem of the Columbia and Snake Rivers. About 88% of the young salmon pass through these turbines unharmed.

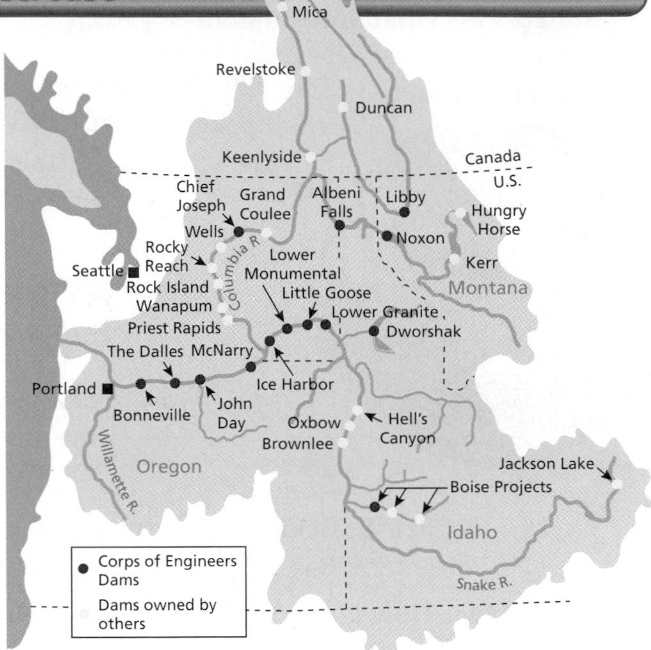

Copy and complete the table and the bar graph to show the number of young salmon that make it through the dams.

Dam	0	1	2	3	4	5	6	7	8
Salmon	1000	880	774						

88% of 1000 $= 0.88 \cdot 1000$
$= 880$

88% of 880 $= 0.88 \cdot 880$
$= 774.4 \approx 774$

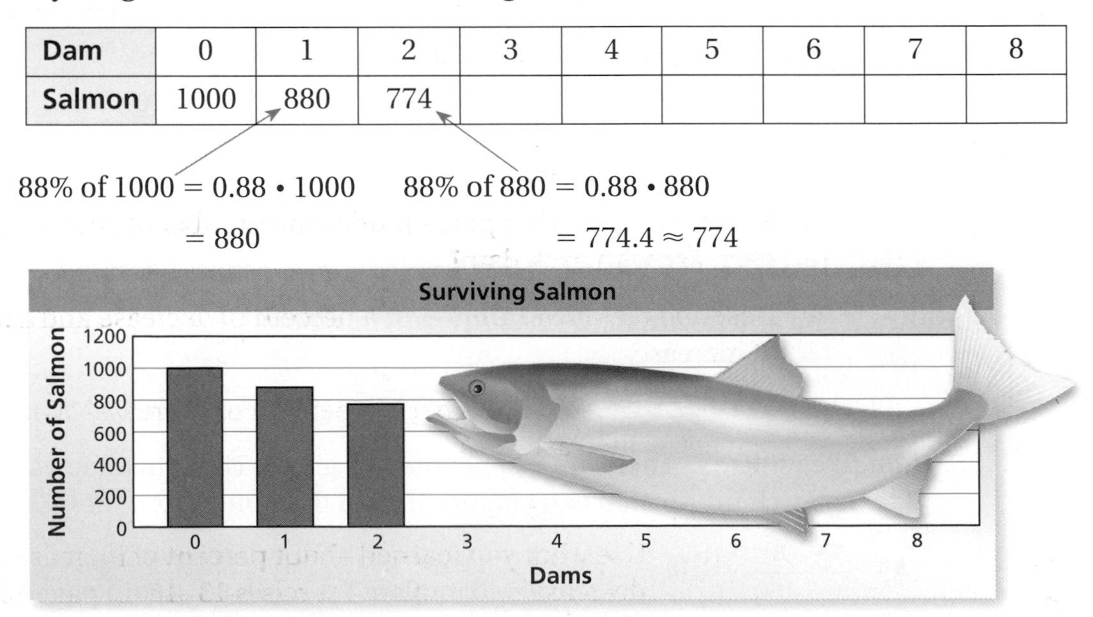

2 ACTIVITY: Percent of Increase

From 2000 to 2006, the population of Florida increased about 2% each year. Copy and complete the table and the bar graph using this pattern. Predict the population in 2015.

For 2007:

$$2\% \text{ of } 18{,}000{,}000 = 0.02 \cdot 18{,}000{,}000$$
$$= 360{,}000$$

$$18{,}000{,}000 + 360{,}000 = 18{,}360{,}000$$

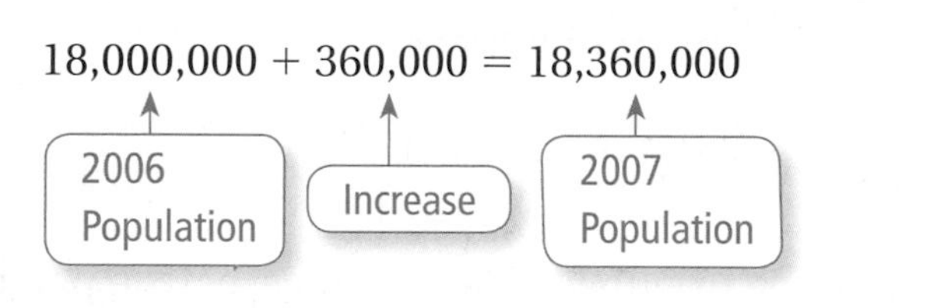

2006 Population 18,000,000

Year	Population
2006	18,000,000
2007	18,360,000
2008	
2009	
2010	
2011	
2012	
2013	
2014	
2015	

Population of Florida

Population: 0, 2,000,000, 4,000,000, 6,000,000, 8,000,000, 10,000,000, 12,000,000, 14,000,000, 16,000,000, 18,000,000, 20,000,000, 22,000,000, 24,000,000, 26,000,000

Year: 2006, 2007, 2008, 2009, 2010, 2011, 2012, 2013, 2014, 2015

What Is Your Answer?

3. In Activity 1, by what percent does the number of young salmon decrease with each dam?

4. Describe real-life examples of a percent of decrease and a percent of increase.

5. **IN YOUR OWN WORDS** What is a percent of decrease? What is a percent of increase?

Practice Use what you learned about percent of increase and percent of decrease to complete Exercises 13–18 on page 168.

4.2 Lesson

Key Vocabulary
percent of change, p. 166
percent of increase, p. 166
percent of decrease, p. 166

A **percent of change** is the percent that a quantity changes from the original amount.

$$\text{percent of change} = \frac{\text{amount of change}}{\text{original amount}}$$

Key Idea

Percents of Increase and Decrease

When the original amount increases, the percent of change is called a **percent of increase**.

$$\text{percent of increase} = \frac{\text{new amount} - \text{original amount}}{\text{original amount}}$$

When the original amount decreases, the percent of change is called a **percent of decrease**.

$$\text{percent of decrease} = \frac{\text{original amount} - \text{new amount}}{\text{original amount}}$$

EXAMPLE 1 Finding a Percent of Increase

The table shows the number of hours you spent online last weekend. What is the percent of change in your online time from Saturday to Sunday?

Day	Hours Online
Saturday	2
Sunday	4.5

The number of hours on Sunday is greater than the number of hours on Saturday. So, the percent of change is a percent of increase.

$$\text{percent of increase} = \frac{\text{new amount} - \text{original amount}}{\text{original amount}}$$

$= \frac{4.5 - 2}{2}$ Substitute.

$= \frac{2.5}{2}$ Subtract.

$= 1.25$, or 125% Write as a percent.

∴ Your online time increased 125% from Saturday to Sunday.

On Your Own

Find the percent of change. Round to the nearest tenth of a percent, if necessary.

1. 10 inches to 25 inches
2. 57 people to 65 people

Multi-Language Glossary at BigIdeasMath.com.

EXAMPLE 2 Finding a Percent of Decrease

The bar graph shows a softball player's home run totals. What was the percent of change from 2007 to 2008?

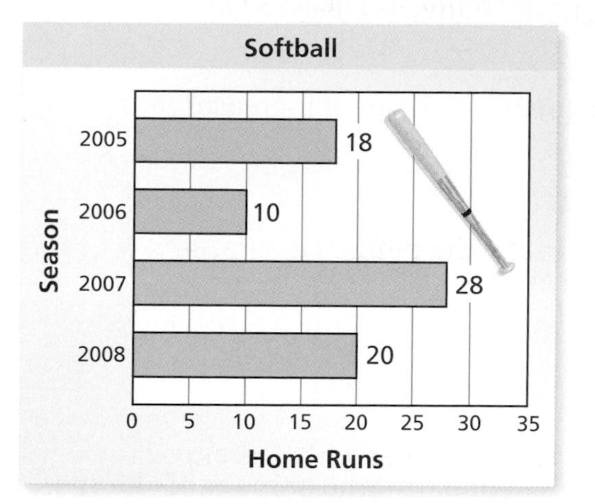

The number of home runs decreased from 2007 to 2008. So, the percent of change is a percent of decrease.

$$\text{percent of decrease} = \frac{\text{original amount} - \text{new amount}}{\text{original amount}}$$

$$= \frac{28 - 20}{28} \qquad \text{Substitute.}$$

$$= \frac{8}{28} \qquad \text{Subtract.}$$

$$\approx 0.286, \text{ or } 28.6\% \qquad \text{Write as a percent.}$$

∴ The number of home runs decreased about 28.6%.

On Your Own

3. What was the percent of change from 2005 to 2006?

EXAMPLE 3 Standardized Test Practice

You have 250 songs on your MP3 player. You delete 20% of the songs. How many songs are left?

Ⓐ 50 Ⓑ 150 Ⓒ 200 Ⓓ 300

Find the amount of decrease.

$$20\% \text{ of } 250 = 0.2 \cdot 250 \qquad \text{Write as multiplication.}$$

$$= 50 \qquad \text{Multiply.}$$

The decrease is 50 songs. So, there are $250 - 50 = 200$ songs left.

∴ The correct answer is Ⓒ.

On Your Own

Now You're Ready
Exercises 13–22

4. **WHAT IF?** After deleting the 50 songs in Example 3, you add 10% more songs. How many songs are on the MP3 player?

4.2 Exercises

Check It Out
Help with Homework
BigIdeasMath.com

Vocabulary and Concept Check

1. **VOCABULARY** How do you know whether a percent of change is a *percent of increase* or a *percent of decrease*?
2. **NUMBER SENSE** Without calculating, which has a greater percent of increase?
   - 5 bonus points on a 50-point exam
   - 5 bonus points on a 100-point exam
3. **WRITING** What does it mean to have a 100% decrease?

Practice and Problem Solving

Identify the percent of change as an *increase* or *decrease*. Then find the percent of change. Round to the nearest tenth of a percent, if necessary.

① ② 4. 12 inches to 36 inches

5. 75 people to 25 people
6. 50 pounds to 35 pounds
7. 24 songs to 78 songs
8. 10 gallons to 24 gallons
9. 72 paper clips to 63 paper clips
10. 16 centimeters to 44.2 centimeters
11. 68 miles to 42.5 miles
12. **ERROR ANALYSIS** Describe and correct the error in finding the percent increase from 18 to 26.

$$\frac{26 - 18}{26} \approx 0.31 = 31\%$$

Find the new amount.

③ 13. 8 meters increased by 25%

14. 15 liters increased by 60%
15. 50 points decreased by 26%
16. 25 penalties decreased by 32%
17. 68 students increased by 125%
18. 1000 grams decreased by 94%
19. 62 kilograms decreased by 32%
20. 124 ounces decreased by 67%
21. **ERROR ANALYSIS** Describe and correct the error in using the percent of change to find a new amount.

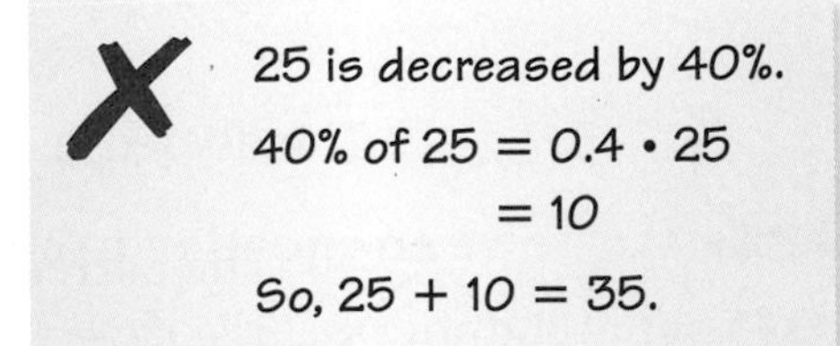
25 is decreased by 40%.
40% of 25 = 0.4 • 25
= 10
So, 25 + 10 = 35.

22. **VIDEO GAME** Last week, you finished Level 2 of a video game in 32 minutes. Today, you finish Level 2 in 28 minutes. What is your percent of change?

Identify the percent of change as an *increase* or *decrease*. Then find the percent of change. Round to the nearest tenth of a percent, if necessary.

23. $\frac{1}{4}$ to $\frac{1}{2}$

24. $\frac{4}{5}$ to $\frac{3}{5}$

25. $\frac{3}{8}$ to $\frac{7}{8}$

26. $\frac{5}{4}$ to $\frac{3}{8}$

27. CRITICAL THINKING Explain why a change from 20 to 40 is a 100% increase, but a change from 40 to 20 is a 50% decrease.

28. POPULATION The table shows population data for a community.

Year	Population
2000	118,000
2006	138,000

a. What is the percent of change from 2000 to 2006?

b. Use this percent of change to predict the population in 2012.

29. GEOMETRY Suppose the length and width of the sandbox are doubled.

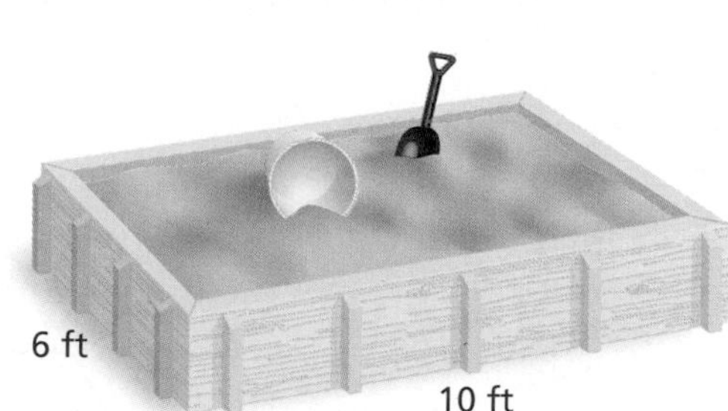

a. Find the percent of change in the perimeter.

b. Find the percent of change in the area.

30. RUNNING Find the percent of change in the time to run a mile from June to September.

31. CRITICAL THINKING A number increases by 10% and then decreases by 10%. Will the result be *greater than*, *less than*, or *equal to* the original number? Explain.

32. DONATIONS Donations to an annual fundraiser are 15% greater this year than last year. Last year, donations were 10% greater than the year before. The amount raised this year is $10,120. How much was raised 2 years ago?

33. Reasoning Forty students are in the science club. Of those, 45% are girls. This percent increases to 56% after new girls join the club. How many new girls join?

Fair Game Review What you learned in previous grades & lessons

Write and solve an equation to answer the question. *(Section 4.1)*

34. What number is 25% of 64?

35. 39.2 is what percent of 112?

36. 5 is 5% of what number?

37. 18 is 32% of what number?

38. MULTIPLE CHOICE Which equation shows direct variation? *(Section 3.7)*

Ⓐ $y - x = 1$ Ⓑ $\frac{y}{x} = 10$ Ⓒ $y = \frac{4}{x}$ Ⓓ $xy = 5$

Check It Out
Graphic Organizer
BigIdeasMath.com

You can use a **summary triangle** to explain a concept. Here is an example of a summary triangle for finding a percent of a number.

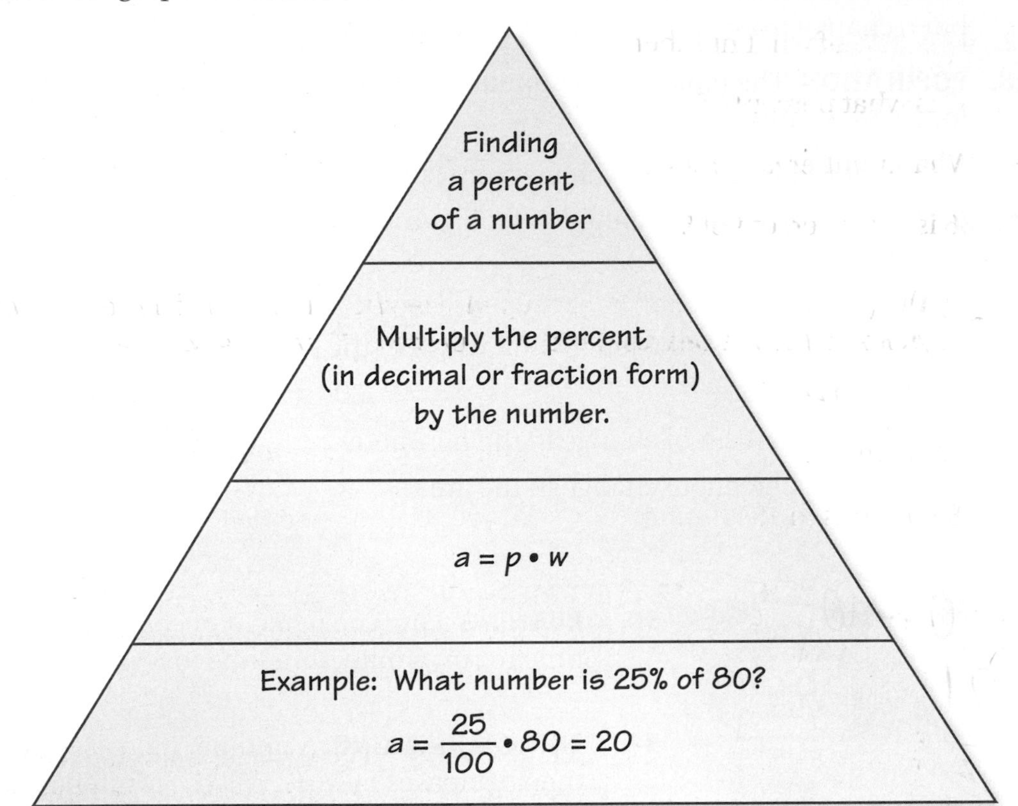

Finding
a percent
of a number
Multiply the percent
(in decimal or fraction form)
by the number.
$a = p \cdot w$
Example: What number is 25% of 80?
$a = \frac{25}{100} \cdot 80 = 20$

On Your Own

Make a summary triangle to help you study these topics.

1. finding the percent given a number and a part of the number
2. finding the number given a part of the number and a percent
3. percent of increase
4. percent of decrease

After you complete this chapter, make summary triangles for the following topics.

5. discount
6. markup
7. simple interest

"I hope my owner sees my summary triangle. I just can't seem to learn 'roll over'."

4.1–4.2 Quiz

Write and solve an equation to answer the question. *(Section 4.1)*

1. What number is 28% of 75?
2. 42 is 21% of what number?
3. 36 is what percent of 45?
4. What number is 68% of 12?
5. 66 is what percent of 55?

Identify the percent of change as an *increase* or *decrease*. Then find the percent of change. Round to the nearest tenth of a percent, if necessary. *(Section 4.2)*

6. 8 inches to 24 inches
7. 300 miles to 210 miles
8. \$42.00 to \$16.00
9. 32 points to 46 points
10. 185 pounds to 153 pounds
11. 35 people to 70 people

12. **TEXT MESSAGES** You have 44 text messages in your inbox. How many messages can your cell phone hold? *(Section 4.1)*

13. **COMPLETIONS** A quarterback completed 68% of his passes in a game. He threw 25 passes. How many passes did the quarterback complete? *(Section 4.1)*

14. **QUIZ** You answered 14 questions correctly on a 15-question quiz. What percent did you receive on the quiz? Round to the nearest hundredth. *(Section 4.1)*

15. **FRUIT JUICE** The graph shows the amount of fruit juice available per person in the United States during a six-year period. *(Section 4.2)*

    a. What is the percent of change from 2002 to 2005?

    b. What is the percent of change from 2002 to 2003?

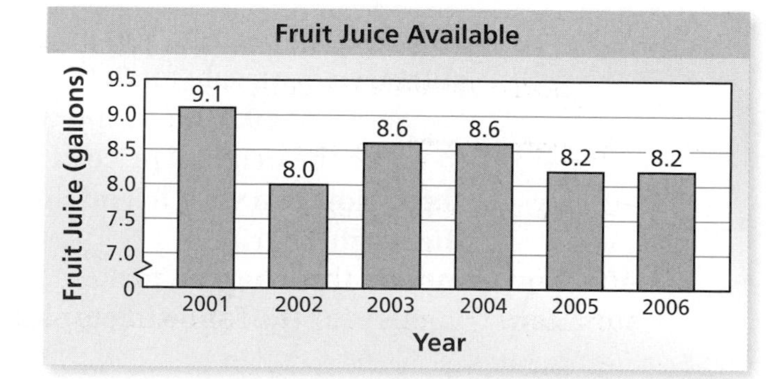

16. **CAR** A car loses 15% of its original value each year. After one year, a car has a value of \$13,600. What is the original value of the car? *(Section 4.2)*

4.3 Discounts and Markups

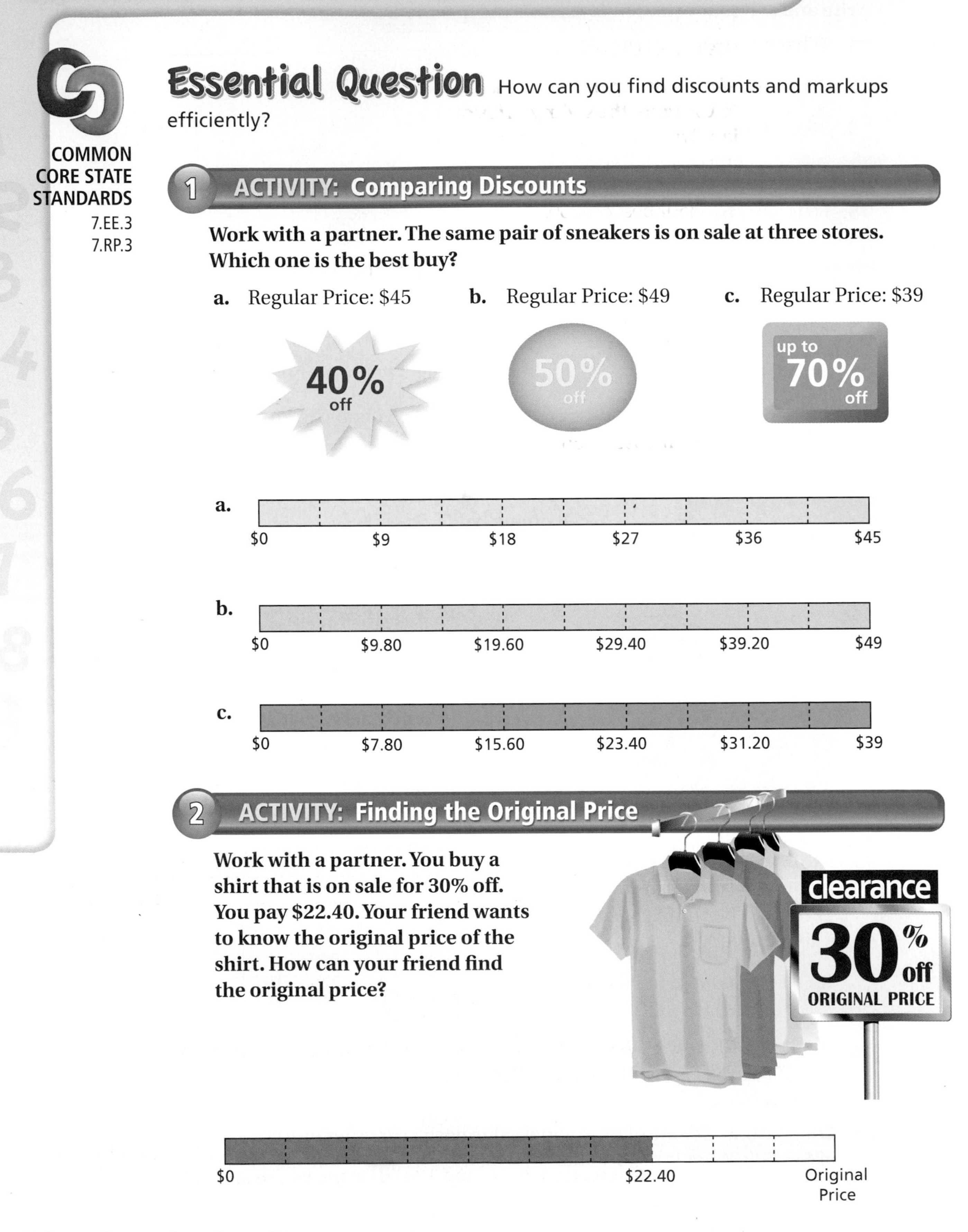

COMMON CORE STATE STANDARDS
7.EE.3
7.RP.3

Essential Question How can you find discounts and markups efficiently?

1 ACTIVITY: Comparing Discounts

Work with a partner. The same pair of sneakers is on sale at three stores. Which one is the best buy?

a. Regular Price: $45 **b.** Regular Price: $49 **c.** Regular Price: $39

a.

b.

c.

2 ACTIVITY: Finding the Original Price

Work with a partner. You buy a shirt that is on sale for 30% off. You pay $22.40. Your friend wants to know the original price of the shirt. How can your friend find the original price?

3 ACTIVITY: Calculating Markup

You own a small jewelry store. You increase the price of the jewelry by 125%.

Work with a partner. Use a model to estimate the selling price of the jewelry. Then use a calculator to find the selling price.

a. Your cost is $250.

b. Your cost is $50.

c. Your cost is $20.

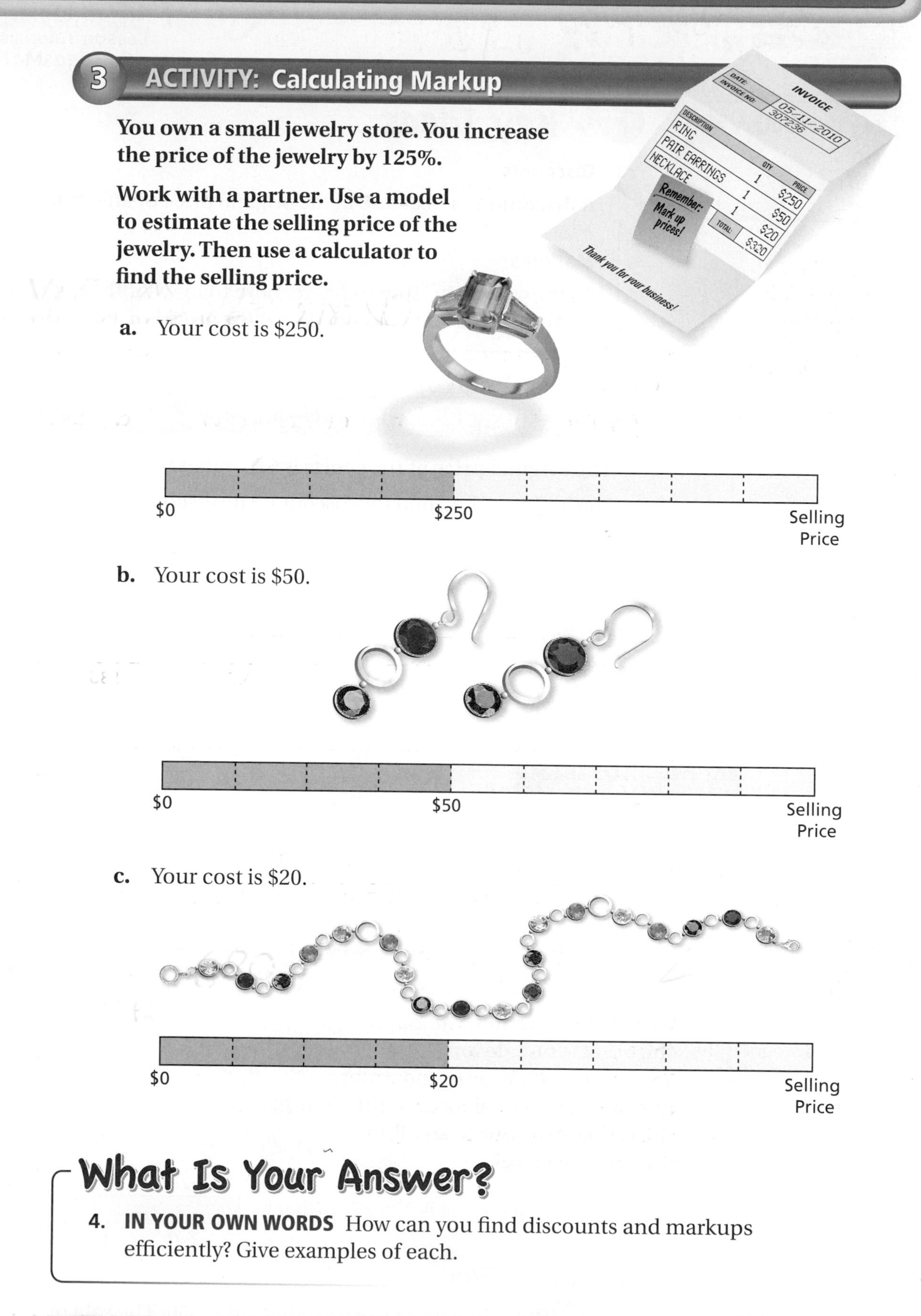

What Is Your Answer?

4. IN YOUR OWN WORDS How can you find discounts and markups efficiently? Give examples of each.

Practice Use what you learned about discounts and markups to complete Exercises 4, 9, 14, and 18–20 on pages 176 and 177.

Key Vocabulary
discount, *p. 174*
markup, *p. 174*

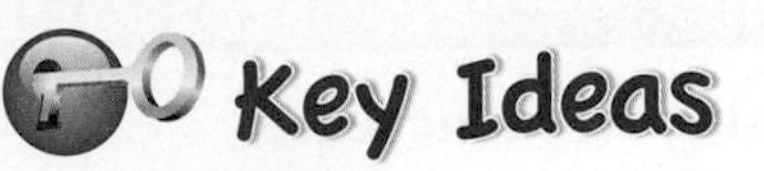

Discounts

A **discount** is a decrease in the original price of an item.

Markups

To make a profit, stores charge more than what they pay. The increase from what the store pays to the selling price is called a **markup**.

EXAMPLE 1 Finding a Sale Price

The original price of the shorts is $35. What is the sale price?

Method 1: First, find the discount. The discount is 25% of $35.

$a = p \cdot w$ Write percent equation.

$= 0.25 \cdot 35$ Subsitute 0.25 for *p* and 35 for *w*.

$= 8.75$ Multiply.

Next, find the sale price.

$$\text{sale price} = \text{original price} - \text{discount}$$
$$= 35 - 8.75$$
$$= 26.25$$

∴ The sale price is $26.25.

Method 2: First, find the percent of the original price.

$$100\% - 25\% = 75\%$$

Next, find the sale price.

$$\text{sale price} = 75\% \text{ of } \$35$$
$$= 0.75 \cdot 35$$
$$= 26.25$$

∴ The sale price is $26.25. **Check**

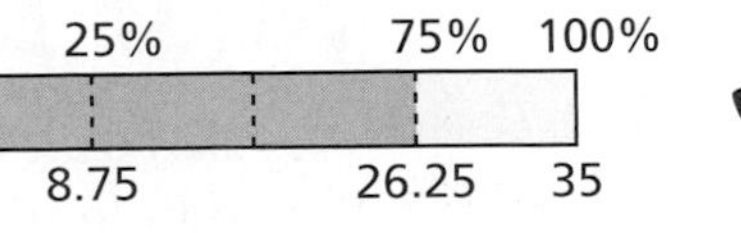

Study Tip

A 25% discount is the same as paying 75% of the original price.

On Your Own

Now You're Ready
Exercises 4–8

1. The original price of a skateboard is $50. The sale price includes a 20% discount. What is the sale price?

Multi-Language Glossary at BigIdeasMath.com.

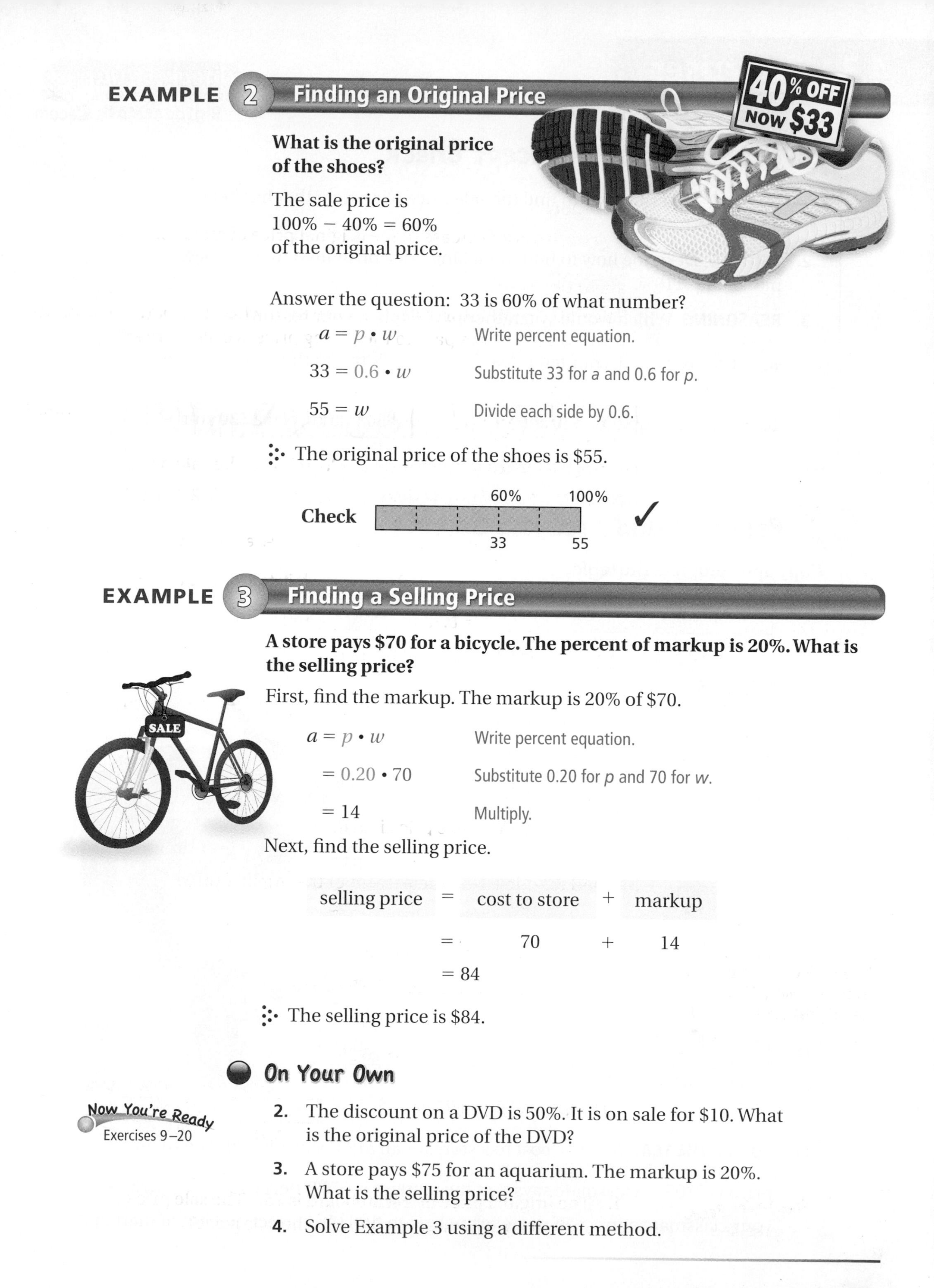

EXAMPLE 2 Finding an Original Price

What is the original price of the shoes?

The sale price is
$100\% - 40\% = 60\%$
of the original price.

Answer the question: 33 is 60% of what number?

$a = p \cdot w$	Write percent equation.
$33 = 0.6 \cdot w$	Substitute 33 for a and 0.6 for p.
$55 = w$	Divide each side by 0.6.

∴ The original price of the shoes is $55.

Check 60% 100% 33 55 ✓

EXAMPLE 3 Finding a Selling Price

A store pays $70 for a bicycle. The percent of markup is 20%. What is the selling price?

First, find the markup. The markup is 20% of $70.

$a = p \cdot w$	Write percent equation.
$= 0.20 \cdot 70$	Substitute 0.20 for p and 70 for w.
$= 14$	Multiply.

Next, find the selling price.

$$\text{selling price} = \text{cost to store} + \text{markup}$$
$$= 70 + 14$$
$$= 84$$

∴ The selling price is $84.

On Your Own

Now You're Ready
Exercises 9–20

2. The discount on a DVD is 50%. It is on sale for $10. What is the original price of the DVD?
3. A store pays $75 for an aquarium. The markup is 20%. What is the selling price?
4. Solve Example 3 using a different method.

4.3 Exercises

Vocabulary and Concept Check

1. **WRITING** Describe how to find the sale price of an item that has been discounted 25%.
2. **WRITING** Describe how to find the selling price of an item that has been marked up 110%.
3. **REASONING** Which would you rather pay? Explain your reasoning.

   a. 6% tax on a discounted price or 6% tax on the original price

   b. 30% markup on a $30 shirt or $30 markup on a $30 shirt

Practice and Problem Solving

Copy and complete the table.

	Original Price	Percent of Discount	Sale Price
① 4.	$80	20%	
5.	$42	15%	
6.	$120	80%	
7.	$112	32%	
8.	$69.80	60%	
② 9.		25%	$40
10.		5%	$57
11.		80%	$90
12.		64%	$72
13.		15%	$146.54
14.	$60		$45
15.	$82		$65.60
16.	$95		$61.75

17. **YOU BE THE TEACHER** The cost to a store for an MP3 player is $60. The selling price is $105. A classmate says that the markup is 175% because $\frac{\$105}{\$60} = 1.75$. Is your classmate correct? If not, explain how to find the correct percent of markup.

Find the cost to store, percent of markup, or selling price.

3 **18.** Cost to store: $70
Markup: 10%
Selling price: ▭

19. Cost to store: ▭
Markup: 75%
Selling price: $63

20. Cost to store: $75
Markup: ▭
Selling price: $180

21. SCOOTER The scooter is on sale for 90% off the original price. Which of the methods can you use to find the sale price? Which method do you prefer? Explain.

$45.85

Multiply $45.85 by 0.9.	Multiply $45.85 by 0.1.
Multiply $45.85 by 0.9, then add to $45.85.	Multiply $45.85 by 0.9, then subtract from $45.85.

22. GAMING You are shopping for a video game system.

a. At which store should you buy the system?

b. Store A has a weekend sale. How can this change your decision in part (a)?

Store	Cost to Store	Markup
A	$162	40%
B	$155	30%
C	$160	25%

23. STEREO A $129.50 stereo is discounted 40%. The next month, the sale price is discounted 60%. Is the stereo now "free"? If not, what is the sale price?

24. CLOTHING You buy a pair of jeans at a department store.

a. What is the percent of discount to the nearest percent?

b. What is the percent of sales tax to the nearest tenth of a percent?

c. The price of the jeans includes a 60% markup. After the discount, what is the percent of markup to the nearest percent?

Department Store

Jeans	39.99
Discount	-10.00
Subtotal	29.99
Sales Tax	1.95
Total	31.94

Thank You

25. Critical Thinking You buy a bicycle helmet for $22.26, which includes 6% sales tax. The helmet is discounted 30% off the selling price. What is the original price?

Fair Game Review What you learned in previous grades & lessons

Evaluate. *(Skills Review Handbook)*

26. 2000(0.085)

27. 1500(0.04)(3)

28. 3200(0.045)(8)

29. MULTIPLE CHOICE Which measurement is greater than 1 meter? *(Section 3.6)*

Ⓐ 38 inches Ⓑ 1 yard Ⓒ 3.4 feet Ⓓ 98 centimeters

4.4 Simple Interest

COMMON CORE STATE STANDARDS
7.EE.3
7.RP.3

Essential Question How can you find the amount of simple interest earned on a savings account? How can you find the amount of interest owed on a loan?

Simple interest is money earned on a savings account or an investment. It can also be money you pay for borrowing money.

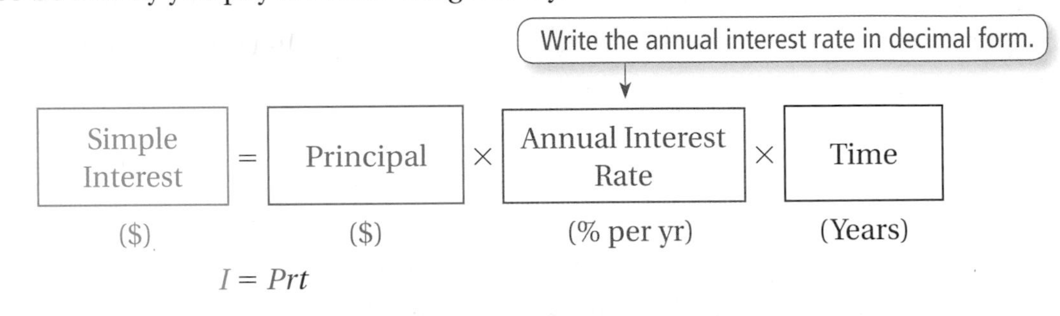

$I = Prt$

1 ACTIVITY: Finding Simple Interest

Work with a partner. You put $100 in a savings account. The account earns 6% simple interest per year. (a) Find the interest earned and the balance at the end of 6 months. (b) Copy and complete the table. Then make a bar graph that shows how the balance grows in 6 months.

a. $I = Prt$ — Write simple interest formula

$= 100(0.06)\left(\frac{6}{12}\right)$ — Substitute values.

$= 3$ — Multiply.

At the end of 6 months, you earn $3 in interest. So, your balance is $100 + $3 = $103.

b.

Time	Interest	Balance
0 month	$0	$100
1 month		
2 months		
3 months		
4 months		
5 months		
6 months	$3	$103

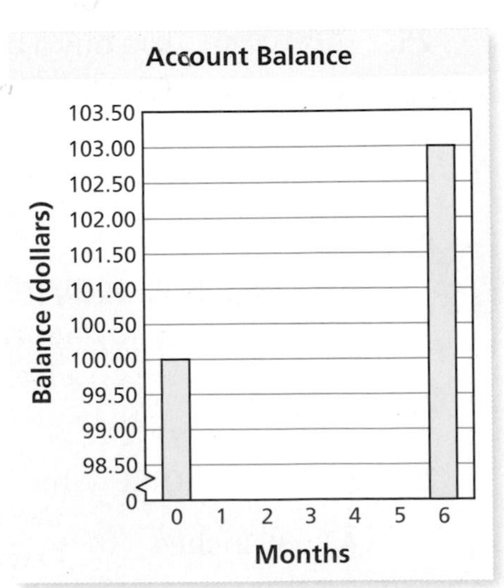

2 ACTIVITY: Financial Literacy

Work with a partner. Use the following information to write a report about credit cards. In the report, describe how a credit card works. Include examples that show the amount of interest paid each month on a credit card.

U.S. Credit Card Data

- A typical family in the United States owes about $5000 in credit card debt.
- A typical credit card interest rate is 18% to 20% per year. This is called the annual percentage rate.

3 ACTIVITY: The National Debt

Work with a partner. In 2010, the United States owed about $10 trillion in debt. The interest rate on the national debt is about 3% per year.

a. Write $10 trillion in decimal form. How many zeros does this number have?

b. How much interest does the United States pay each year on its national debt?

c. How much interest does the United States pay each day on its national debt?

d. The United States has a population of about 300 million people. Estimate the amount of interest that each person pays per year toward interest on the national debt.

What Is Your Answer?

4. **IN YOUR OWN WORDS** How can you find the amount of simple interest earned on a savings account? How can you find the amount of interest owed on a loan? Give examples with your answer.

Practice Use what you learned about simple interest to complete Exercises 4–7 on page 182.

4.4 Lesson

Key Vocabulary
interest, *p. 180*
principal, *p. 180*
simple interest, *p. 180*

Interest is money paid or earned for the use of money. The **principal** is the amount of money borrowed or deposited.

Key Idea

Simple Interest

Words **Simple interest** is money paid or earned only on the principal.

Algebra

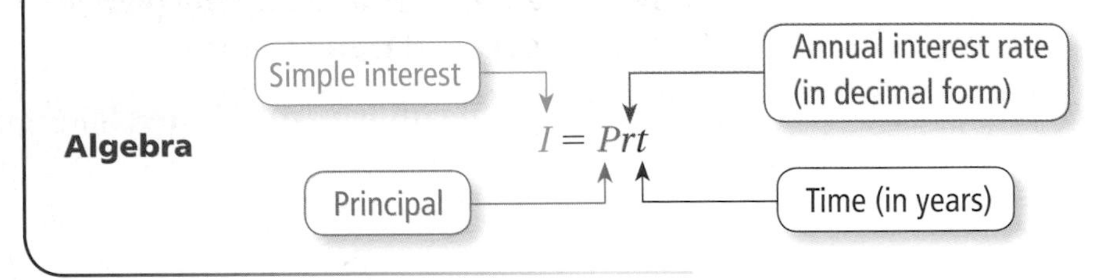

EXAMPLE 1 Finding Interest Earned

You put \$500 in a savings account. The account earns 3% simple interest per year. (a) What is the interest earned after 3 years? (b) What is the balance after 3 years?

a. $I = Prt$ — Write simple interest formula.

$= 500(0.03)(3)$ — Substitute 500 for P, 0.03 for r, and 3 for t.

$= 45$ — Multiply.

∴ The interest earned is \$45 after 3 years.

b. To find the balance, add the interest to the principal.

∴ So, the balance is $\$500 + \$45 = \$545$ after 3 years.

EXAMPLE 2 Finding an Annual Interest Rate

You put \$1000 in an account. The account earns \$100 simple interest in 4 years. What is the annual interest rate?

$I = Prt$ — Write simple interest formula.

$100 = 1000(r)(4)$ — Substitute 100 for I, 1000 for P, and 4 for t.

$100 = 4000r$ — Simplify.

$0.025 = r$ — Divide each side by 4000.

∴ The annual interest rate of the account is 0.025, or 2.5%.

On Your Own

Now You're Ready
Exercises 4–16

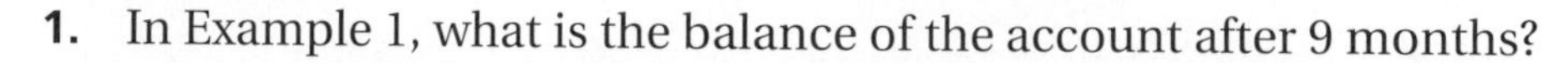

1. In Example 1, what is the balance of the account after 9 months?
2. You put \$350 in an account. The account earns \$17.50 simple interest in 2.5 years. What is the annual interest rate?

EXAMPLE 3 Finding an Amount of Time

A bank offers three savings accounts. The simple interest rate is determined by the principal. How long does it take an account with a principal of \$800 to earn \$100 interest?

1.5% Less than \$500
2.0% \$500-\$5000
3.0% More than \$5000

The pictogram shows that the interest rate for a principal of \$800 is 2%.

$I = Prt$	Write simple interest formula.
$100 = 800(0.02)(t)$	Substitute 100 for I, 800 for P, and 0.02 for r.
$100 = 16t$	Simplify.
$6.25 = t$	Divide each side by 16.

∴ The account earns \$100 in interest in 6.25 years.

EXAMPLE 4 Finding Amount Paid on a Loan

You borrow \$600 to buy a violin. The simple interest rate is 15%. You pay off the loan after 5 years. How much do you pay for the loan?

$I = Prt$	Write simple interest formula.
$= 600(0.15)(5)$	Substitute 600 for P, 0.15 for r, and 5 for t.
$= 450$	Multiply.

To find the amount you pay, add the interest to the loan amount.

∴ So, you pay \$600 + \$450 = \$1050 for the loan.

On Your Own

Now You're Ready
Exercises 17–27

3. In Example 3, how long does it take an account with a principal of \$10,000 to earn \$750 interest?
4. **WHAT IF?** In Example 4, you pay off the loan after 2 years. How much money do you save?

4.4 Exercises

Vocabulary and Concept Check

1. **VOCABULARY** Define each variable in $I = Prt$.
2. **WRITING** In each situation, tell whether you would want a *higher* or *lower* interest rate. Explain your reasoning.
   a. You borrow money
   b. You open a savings account
3. **REASONING** An account earns 6% simple interest. You want to find the interest earned on $200 after 8 months. What conversions do you need to make before you can use the formula $I = Prt$?

Practice and Problem Solving

An account earns simple interest. (a) Find the interest earned. (b) Find the balance of the account.

(1)
4. $600 at 5% for 2 years
5. $1500 at 4% for 5 years
6. $350 at 3% for 10 years
7. $1800 at 6.5% for 30 months
8. $700 at 8% for 6 years
9. $1675 at 4.6% for 4 years
10. $925 at 2% for 2.4 years
11. $5200 at 7.36% for 54 months

12. **ERROR ANALYSIS** Describe and correct the error in finding the simple interest earned on $500 at 6% for 18 months.

✗ $I = (500)(0.06)(18)$
$= \$540$

Find the annual simple interest rate.

(2)
13. $I = \$24, P = \$400, t = 2$ years
14. $I = \$562.50, P = \$1500, t = 5$ years
15. $I = \$54, P = \$900, t = 18$ months
16. $I = \$160.67, P = \$2000, t = 8$ months

Find the amount of time.

(3)
17. $I = \$30, P = \$500, r = 3\%$
18. $I = \$720, P = \$1000, r = 9\%$
19. $I = \$54, P = \$800, r = 4.5\%$
20. $I = \$450, P = \$2400, r = 7.5\%$

21. **BANKING** A savings account earns 5% annual simple interest. The principal is $1200. What is the balance after 4 years?
22. **SAVINGS** You put $400 in an account. The account earns $18 simple interest in 9 months. What is the annual interest rate?
23. **CD** You put $3000 in a CD (certificate of deposit) at the promotional rate. How long will it take to earn $336 in interest?

Find the amount paid for the loan.

④ **24.** $1500 at 9% for 2 years

25. $2000 at 12% for 3 years

26. $2400 at 10.5% for 5 years

27. $4800 at 9.9% for 4 years

Copy and complete the table.

	Principal	Interest Rate	Time	Simple Interest
28.	$12,000	4.25%	5 years	
29.		6.5%	18 months	$828.75
30.	$15,500	8.75%		$5425.00
31.	$18,000		54 months	$4252.50

32. ZOO A family charges a trip to the zoo on a credit card. The simple interest rate is 12%. The charges are paid after 3 months. What is the total amount paid for the trip?

Zoo Trip

Tickets	*67.70*
Food	*62.34*
Gas	*45.50*
Total Cost	*?*

33. MONEY MARKET You deposit $5000 in an account earning 7.5% simple interest. How long will it take for the balance of the account to be $6500?

34. LOANS A music company offers a loan to buy a drum set for $1500. What is the monthly payment?

35. REASONING How many years will it take for $2000 to double at a simple interest rate of 8%? Explain how you found your answer.

36. LOANS You have two loans, for 2 years each. The total interest for the two loans is $138. On the first loan, you pay 7.5% simple interest on a principal of $800. On the second loan, you pay 3% simple interest. What is the principal for the second loan?

37. **Critical Thinking** You put $500 in an account that earns 4% annual interest. The interest earned each year is added to the principal to create a new principal. Find the total amount in your account after each year for 3 years.

Fair Game Review What you learned in previous grades & lessons

Solve the proportion. *(Section 3.5)*

38. $\frac{4}{9} = \frac{12}{x}$

39. $\frac{15}{36} = \frac{n}{12}$

40. $\frac{m}{6.5} = \frac{14}{26}$

41. $\frac{2.4}{z} = \frac{3}{11.25}$

42. MULTIPLE CHOICE What is the solution of $4x + 5 = -11$? *(Section 2.6)*

Ⓐ -4 Ⓑ -1.5 Ⓒ 1.5 Ⓓ 4

4.3–4.4 Quiz

Find the price, discount, markup, or cost to store. *(Section 4.3)*

1. Original price: \$30
Discount: 10%
Sale price: ?

2. Original price: \$55
Discount: ?
Sale price: \$46.75

3. Original price: ?
Discount: 75%
Sale price: \$74.75

4. Cost to store: \$152
Markup: 50%
Selling price: ?

5. Cost to store: \$20
Markup: ?
Selling price: \$32

6. Cost to store: ?
Markup: 80%
Selling price: \$21.60

An account earns simple interest. Find the interest earned, principal, interest rate, or time. *(Section 4.4)*

7. Interest earned: ?
Principal: \$1200
Interest rate: 2%
Time: 5 years

8. Interest earned: \$25
Principal: \$500
Interest rate: 5%
Time: ?

9. Interest earned: \$76
Principal: \$800
Interest rate: ?
Time: 2 years

10. Interest earned: \$119.88
Principal: ?
Interest rate: 3.6%
Time: 3 years

11. DIGITAL CAMERA A digital camera costs \$229. The camera is on sale for 30% off and you have a coupon for an additional 15% off the original price. What is the final price? *(Section 4.3)*

12. WATER SKIS The original price of the water skis was \$200. What is the percent of discount? *(Section 4.3)*

SALE \$150

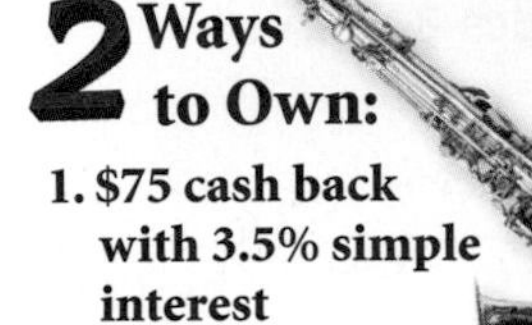

2 Ways to Own:
1. \$75 cash back with 3.5% simple interest
2. No interest for 2 years

13. SAXOPHONE A saxophone costs \$1200. A store offers two loan options. Which option saves more money if you pay the loan in 2 years? *(Section 4.4)*

14. LOAN You borrow \$200. The simple interest rate is 12%. You pay off the loan after 2 years. How much do you pay for the loan? *(Section 4.4)*

4 Chapter Review

Review Key Vocabulary

percent, *p. 160*
percent of change, *p. 166*
percent of increase, *p. 166*
percent of decrease, *p. 166*
discount, *p. 174*
markup, *p. 174*
interest, *p. 180*
principal, *p. 180*
simple interest, *p. 180*

Review Examples and Exercises

4.1 The Percent Equation *(pp. 158–163)*

What number is 72% of 25?

$a = p \cdot w$ — Write percent equation.

$= 0.72 \cdot 25$ — Substitute 0.72 for p and 25 for w.

$= 18$ — Multiply.

So, 72% of 25 is 18.

28 is what percent of 70?

$a = p \cdot w$ — Write percent equation.

$28 = p \cdot 70$ — Substitute 28 for a and 70 for w.

$0.4 = p$ — Divide each side by 70.

Because $0.4 = 40\%$, 28 is 40% of 70.

22.1 is 26% of what number?

$a = p \cdot w$ — Write percent equation.

$22.1 = 0.26 \cdot w$ — Substitute 22.1 for a and 0.26 for p.

$85 = w$ — Divide each side by 0.26.

So, 22.1 is 26% of 85.

Exercises

Write and solve an equation to answer the question.

1. What number is 24% of 25?
2. 9 is what percent of 20?
3. 85% of what number is 10.2?
4. 83% of 20 is what number?
5. **PARKING** 15% of the school parking spaces are handicap spaces. The school has 18 handicap spaces. How many parking spaces are there?

4.2 Percents of Increase and Decrease (pp. 164–169)

The table shows the number of skim boarders at a beach on Saturday and Sunday. What was the percent of change in boarders from Saturday to Sunday?

The number of skim boarders on Sunday is less than the number of skim boarders on Saturday. So, the percent of change is a percent of decrease.

Day	Number of Skim Boarders
Saturday	12
Sunday	9

$$\text{percent of decrease} = \frac{\text{original amount} - \text{new amount}}{\text{original amount}}$$

$= \frac{12 - 9}{12}$ Substitute.

$= \frac{3}{12}$ Subtract.

$= 0.25 = 25\%$ Write as a percent.

∴ The number of skim boarders decreased by 25% from Saturday to Sunday.

Exercises

Identify the percent of change as an *increase* or *decrease*. Then find the percent of change. Round to the nearest tenth of a percent, if necessary.

6. 6 yards to 36 yards

7. 6 hits to 3 hits

8. 120 meals to 52 meals

9. 35 words to 115 words

4.3 Discounts and Markups (pp. 172–177)

What is the original price of the tennis racquet?

The sale price is 100% − 30% = 70% of the original price.

Answer the question: 21 is 70% of what number?

$a = p \cdot w$ Write percent equation.

$21 = 0.7 \cdot w$ Substitute 21 for a and 0.7 for p.

$30 = w$ Divide each side by 0.7.

∴ The original price of the tennis racquet is $30.

Exercises

Find the price.

10. Original price: $50
Discount: 15%
Sale price: ?

11. Original price: ?
Discount: 20%
Sale price: $75

4.4 Simple Interest (pp. 178–183)

You put $200 in a savings account. The account earns 2% simple interest per year.

a. What is the interest after 4 years?

b. What is the balance after 4 years?

a. $I = Prt$ — Write simple interest formula.

$= 200(0.02)(4)$ — Substitute 200 for *P*, 0.02 for *r*, and 4 for *t*.

$= 16$ — Multiply.

∴ The interest earned is $16 after 4 years.

b. The balance is the principal plus the interest.

∴ So, the balance is $200 + $16 = $216 after 4 years.

You put $500 in an account. The account earns $55 simple interest in 5 years. What is the annual interest rate?

$I = Prt$ — Write simple interest formula.

$55 = 500(r)(5)$ — Substitute 55 for *I*, 500 for *P*, and 5 for *t*.

$55 = 2500r$ — Simplify.

$0.022 = r$ — Divide each side by 2500.

∴ The annual interest rate of the account is 0.022, or 2.2%.

Exercises

An account earns simple interest.

a. Find the interest earned.

b. Find the balance of the account.

12. $300 at 4% for 3 years

13. $2000 at 3.5% for 4 years

Find the annual simple interest rate.

14. $I = \$17, P = \$500, t = 2$ years

15. $I = \$426, P = \$1200, t = 5$ years

Find the amount of time.

16. $I = \$60, P = \$400, r = 5\%$

17. $I = \$237.90, P = \$1525, r = 2.6\%$

18. SAVINGS You put $100 in an account. The account earns $2 simple interest in 6 months. What is the annual interest rate?

4 Chapter Test

Write and solve an equation to answer the question.

1. 16% of 150 is what number?
2. 10 is 40% of what number?
3. 27 is what percent of 75?
4. What number is 35% of 56?

Identify the percent of change as an *increase* or *decrease*. Then find the percent of change. Round to the nearest tenth of a percent, if necessary.

5. 4 strikeouts to 10 strikeouts
6. \$24.00 to \$18.00

Find the price, discount, or markup.

7. Original price: \$15
   Discount: 5%
   Sale price: ?
8. Original price: \$189
   Discount: ?
   Sale price: \$75.60
9. Cost to store: \$15
   Markup: ?
   Selling price: \$24.75
10. Cost to store: \$5.50
    Markup: 75%
    Selling price: ?

An account earns simple interest. Find the interest earned, principal, interest rate, or time.

11. Interest earned: ?
    Principal: \$450
    Interest rate: 6%
    Time: 8 years
12. Interest earned: \$27
    Principal: ?
    Interest rate: 1.5%
    Time: 2 years
13. Interest earned: \$116.25
    Principal: \$1550
    Interest rate: ?
    Time: 9 months
14. Interest earned: \$45.60
    Principal: \$2400
    Interest rate: 3.8%
    Time: ?

15. **MOVIE PREVIEWS** There are eight previews before a movie. Seventy-five percent of the previews are for comedies. How many previews are for comedies?

16. **BOOK** What was the original price of the book?

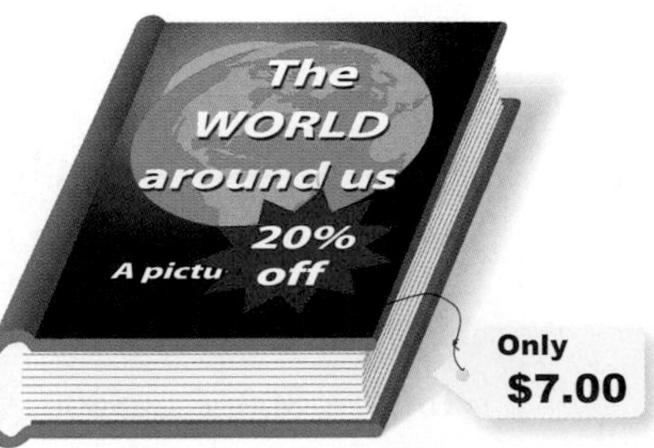

17. **TEXT MESSAGES** The cost of a text message increases from \$0.10 per message to \$0.25 per message. What is the percent increase in the cost of sending a text message?

18. **INVESTMENT** You put \$800 in an account that earns 4% simple interest. Find the total amount in your account after each year for 3 years.

4 Standardized Test Practice

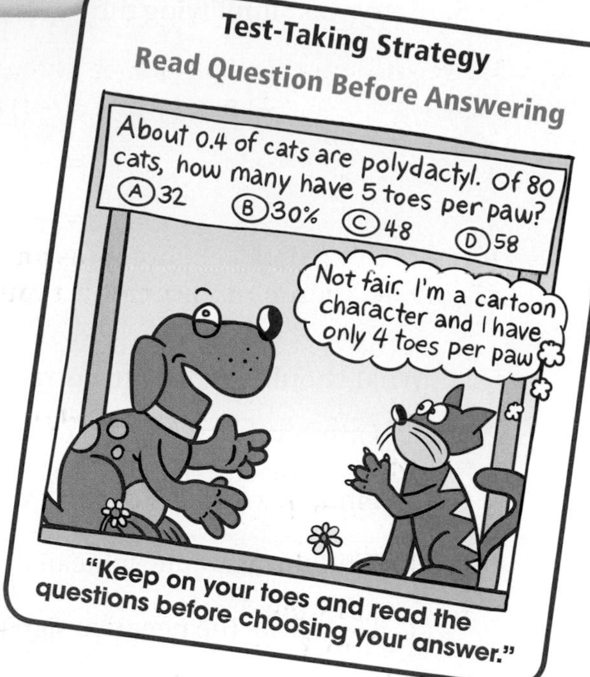

Test-Taking Strategy
Read Question Before Answering
About 0.4 of cats are polydactyl. Of 80 cats, how many have 5 toes per paw?
Ⓐ 32 Ⓑ 30% Ⓒ 48 Ⓓ 58
Not fair. I'm a cartoon character and I have only 4 toes per paw.
"Keep on your toes and read the questions before choosing your answer."

1. A movie theatre offers 30% off the price of a movie ticket to students from your school. The regular price of a movie ticket is \$8.50. What is the discounted price that you would pay for a ticket? *(7.RP.3)*

A. \$2.55 **C.** \$5.95

B. \$5.50 **D.** \$8.20

2. You are comparing the prices of four boxes of cereal. Two of the boxes contain free extra cereal.

- Box F costs \$3.59 and contains 16 ounces.
- Box G costs \$3.79 and contains 16 ounces, plus an additional 10% for free.
- Box H costs \$4.00 and contains 500 grams.
- Box I costs \$4.69 and contains 500 grams, plus an additional 20% for free.

Which box has the least unit cost? (1 ounce = 28.35 grams) *(7.RP.3)*

F. Box F **H.** Box H

G. Box G **I.** Box I

3. James is getting ready for wrestling season. As part of his preparation, he plans to lose 5% of his body weight. James currently weighs 160 pounds. How much will he weigh, in pounds, after he loses 5% of his weight? *(7.RP.3)*

4. Which proportion represents the problem below? *(7.RP.3)*

"17% of a number is 43. What is the number?"

A. $\frac{17}{43} = \frac{n}{100}$ **C.** $\frac{n}{43} = \frac{17}{100}$

B. $\frac{n}{17} = \frac{43}{100}$ **D.** $\frac{43}{n} = \frac{17}{100}$

5. Betty was simplifying the expression in the box below.

$$-|8 + (-13)| = -(|8| + |-13|)$$
$$= -|8 + 13|$$
$$= -|21|$$
$$= -21$$

What should Betty do to correct the error that she made? *(7.EE.3)*

F. Simplify $-|8 + (-13)|$ to get $-|-5|$.

G. Find the opposite of $|21|$, which is 21.

H. Find the absolute value of 8, which is -8.

I. Distribute the negative sign to get $|-8 + (-13)|$.

6. The students from the Math Club participated in a long-distance walk as a fundraiser. The number of hours each club member took to complete the walk is shown in the bar graph below.

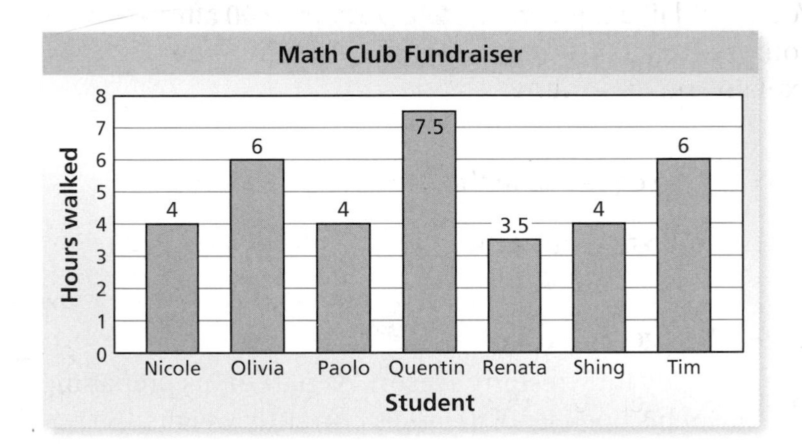

What is the mean number of hours the club members took to complete the walk? *(7.NS.3)*

A. 3.5 h

B. 4 h

C. 5 h

D. 7.5 h

7. A lighting store is holding a clearance sale. The store is offering discounts on all the lamps it sells. As the sale progresses, the store will increase the percent of discount it is offering.

You want to buy a lamp that has an original price of \$40. You will buy the lamp when its price is marked down to \$10. What percent discount will you have received? *(7.RP.3)*

8. A student scored 600 the first time she took the mathematics portion of her college entrance exam. The next time she took the exam, she scored 660. Her second score represents what percent increase over her first score? *(7.RP.3)*

F. 9.1%

G. 10%

H. 39.6%

I. 60%

9. Which inequality is represented by the graph on the number line below? *(7.EE.4b)*

−8 −6 −4 −2 0 2 4 6 8

A. $x < -4$

B. $x \leq -4$

C. $x > -4$

D. $x \geq -4$

10. You are planning to deposit \$4000 into an account that earns 5% simple interest per year. You will not make any other deposits or withdrawals. *(7.RP.3)*

Part A How long would it take for your account to contain \$4500? Show your work and explain your reasoning.

Part B You would like the account to contain \$5100 after 4 years. Would your initial \$4000 deposit be large enough? Show your work and explain your reasoning.

11. Brad was solving the equation in the box shown.

$$-3(2 - 5w) = -45$$
$$-6 + 15w = -45$$
$$9w = -45$$
$$\frac{9w}{9} = \frac{-45}{9}$$
$$w = -5$$

What should Brad do to correct the error that he made? *(7.EE.4a)*

F. Distribute −3 to get $6 - 15w$.

G. Distribute −3 to get $-6 - 15w$.

H. Add 6 to both sides to get $15w = -51$.

I. Add 6 to both sides to get $15w = -39$.

12. Which integer is closest to the value of the expression below? *(7.NS.2c)*

$$-7\frac{1}{4} \cdot \left(-9\frac{7}{8}\right)$$

A. −70

B. −63

C. 63

D. 70

5 Similarity and Transformations

5.1 Identifying Similar Figures

5.2 Perimeters and Areas of Similar Figures

5.3 Finding Unknown Measures in Similar Figures

5.4 Scale Drawings

5.5 Translations

5.6 Reflections

5.7 Rotations

"Just 2 more minutes. I'm almost done with my 'cat tessellation' painting."

"If you hold perfectly still..."

"...each frame becomes a horizontal..."

"...translation of the previous frame..."

What You Learned Before

Finding Perimeter (6.EE.2c)

Example 1 **Find the perimeter.**

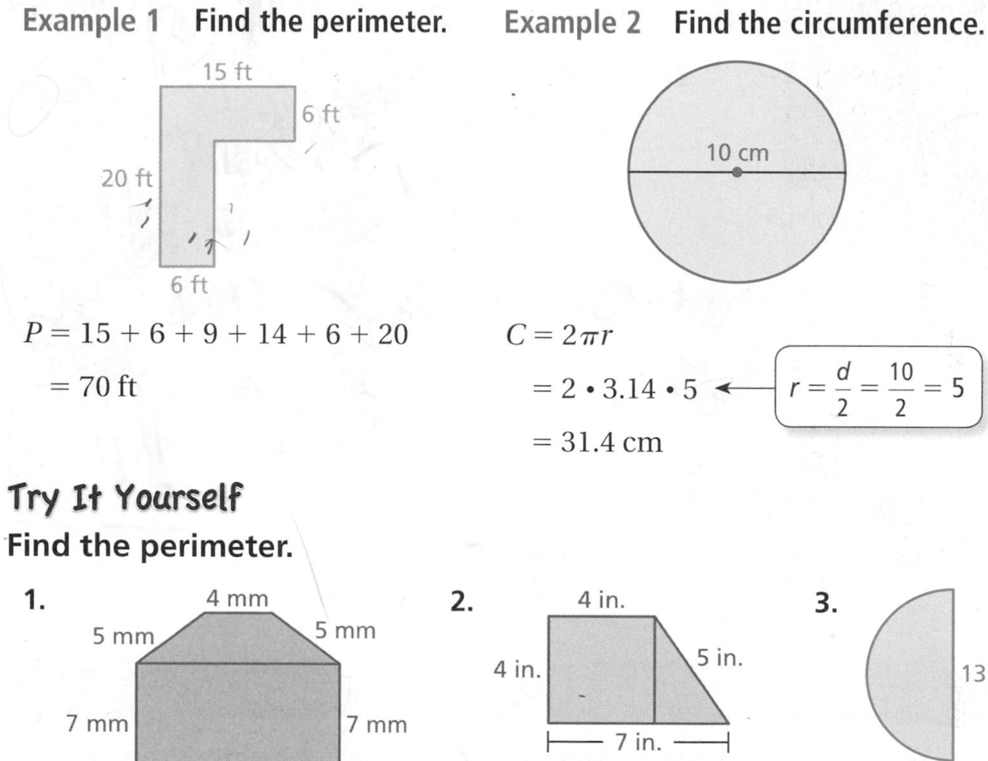

$P = 15 + 6 + 9 + 14 + 6 + 20$

$= 70$ ft

Example 2 **Find the circumference.**

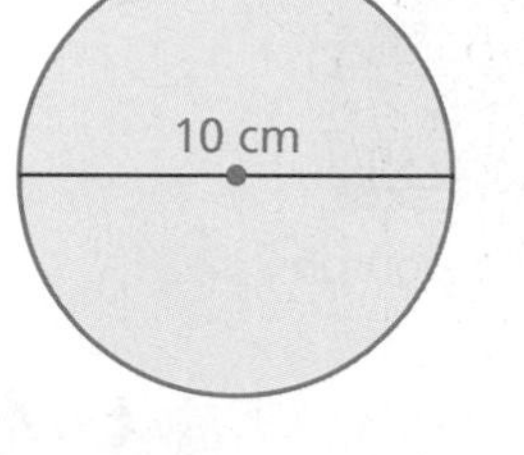

$C = 2\pi r$

$= 2 \cdot 3.14 \cdot 5$

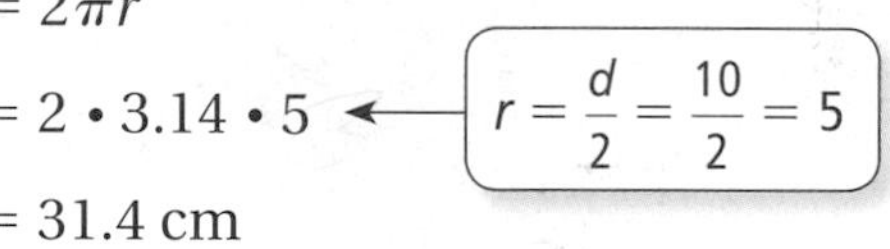

$= 31.4$ cm

Try It Yourself

Find the perimeter.

1.

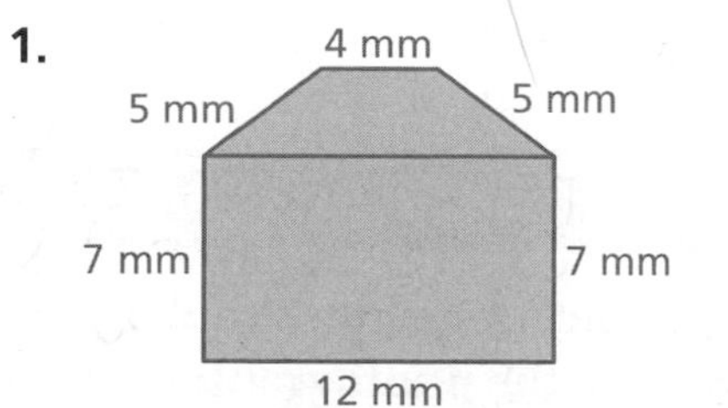

2.

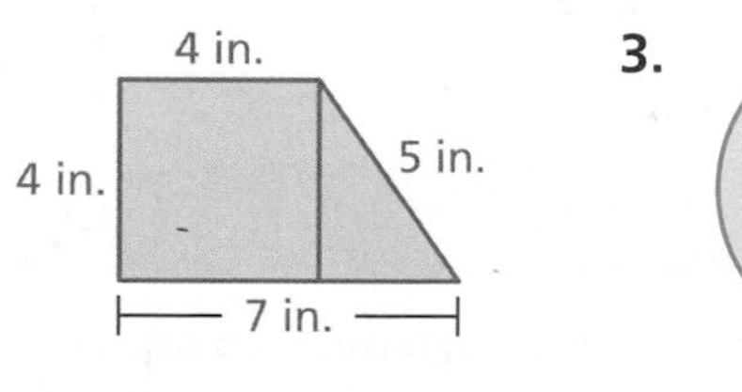

3. 13 cm

Solving Proportions (7.RP.2a)

Example 3 **Solve the proportion.**

a. $\frac{x}{32} = \frac{3}{4}$ **b.** $\frac{3x}{20} = \frac{3}{5}$

a.		b.
$\frac{x}{32} = \frac{3}{4}$	Write the proportion.	$\frac{3x}{20} = \frac{3}{5}$
$4x = 96$	Use the Cross Products Property.	$15x = 60$
$x = 24$	Solve for x.	$x = 4$

Try It Yourself

Solve the proportion.

4. $\frac{2}{7} = \frac{x}{21}$ **5.** $\frac{3}{4} = \frac{3y}{8}$ **6.** $\frac{3}{14} = \frac{9}{y}$ **7.** $\frac{8}{9x} = \frac{2}{9}$

5.1 Identifying Similar Figures

COMMON CORE STATE STANDARDS
7.G.1

Essential Question How can you use proportions to help make decisions in art, design, and magazine layouts?

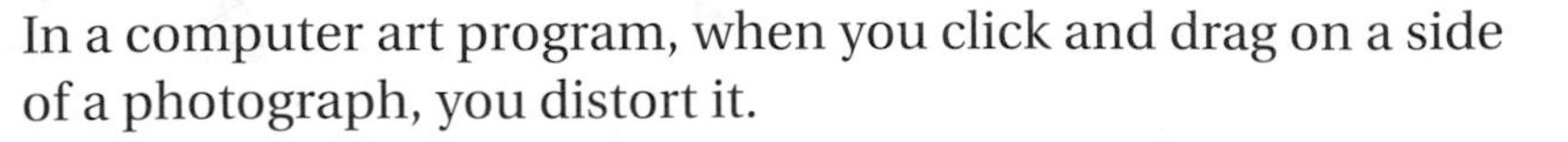

Original Photograph

In a computer art program, when you click and drag on a side of a photograph, you distort it.

But when you click and drag on a corner of the photograph, it remains proportional to the original.

Distorted

Distorted

Proportional

1 ACTIVITY: Reducing Photographs

Work with a partner. You are trying to reduce the photograph to the indicated size for a nature magazine. Can you reduce the photograph to the indicated size without distorting or cropping? Explain your reasoning.

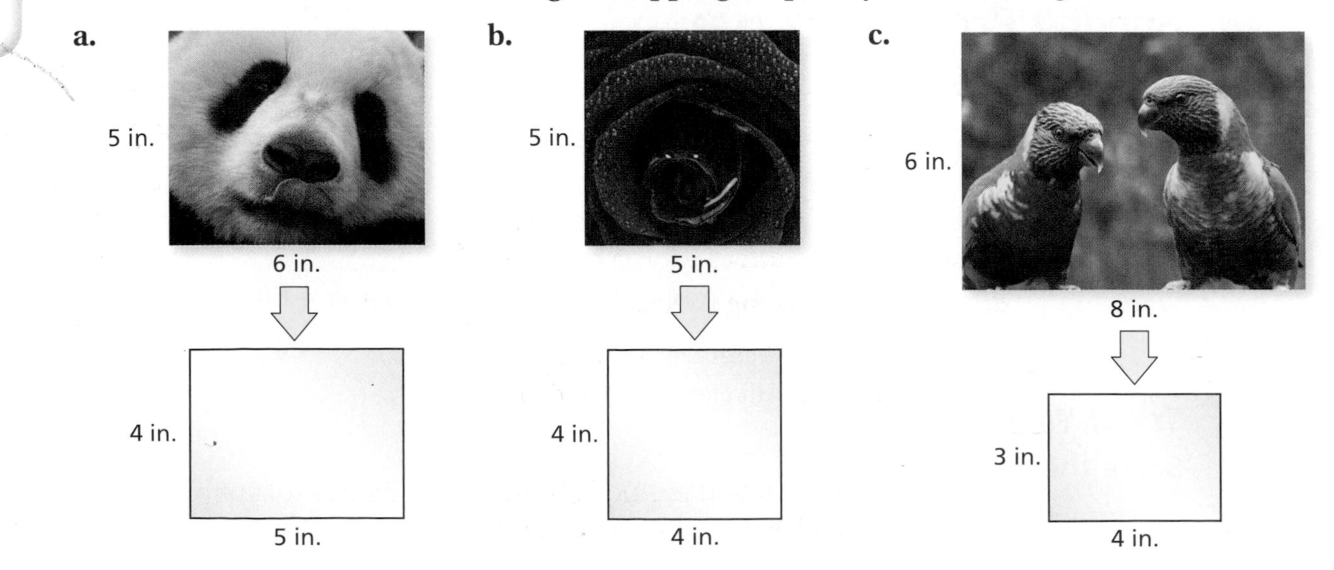

2 ACTIVITY: Proportional Designs

Work with a partner.

a. Tell whether the new designs are proportional to the original design. Explain your reasoning.

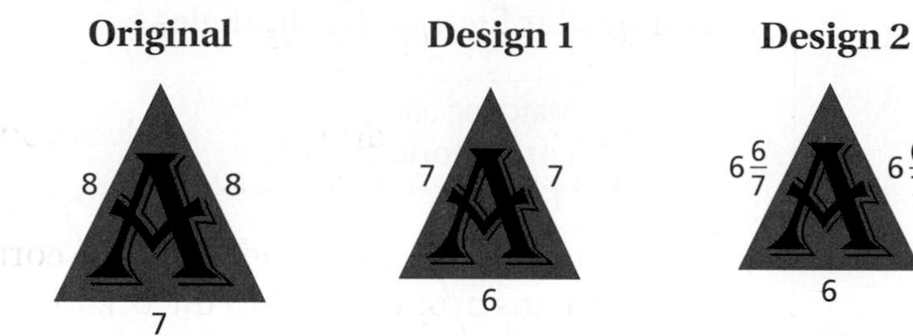

b. Draw two designs that are proportional to the given design. Make one bigger and one smaller. Label the sides of the designs with their lengths.

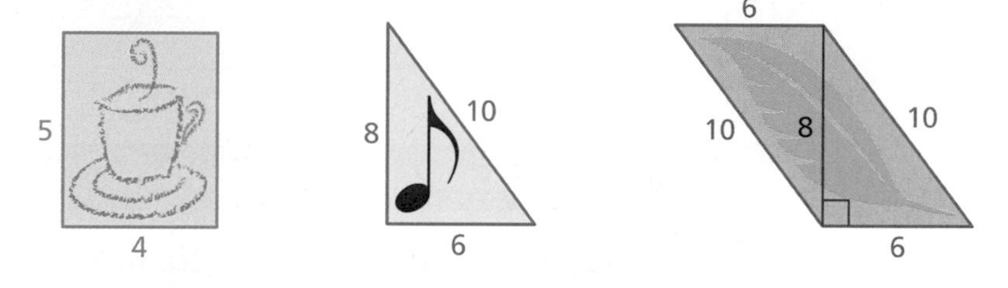

What Is Your Answer?

3. IN YOUR OWN WORDS How can you use proportions to help make decisions in art, design, and magazine layouts? Give two examples.

4. a. Use a computer art program to draw two rectangles that are proportional to each other.

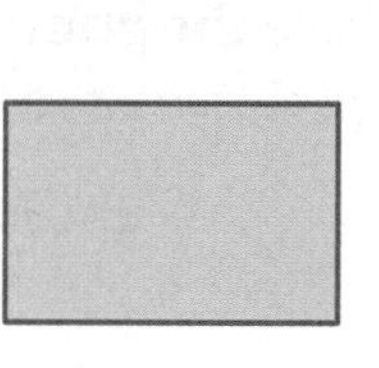

"I love this statue. It seems similar to a big statue I saw in New York."

b. Print the two rectangles on the same piece of paper.

c. Use a centimeter ruler to measure the length and width of each rectangle.

d. Find the following ratios. What can you conclude?

$$\frac{\text{Length of Larger}}{\text{Length of Smaller}} \qquad \frac{\text{Width of Larger}}{\text{Width of Smaller}}$$

Practice Use what you learned about similar figures to complete Exercises 9 and 10 on page 198.

5.1 Lesson

Check It Out
Lesson Tutorials
BigIdeasMath.com

Key Idea

Key Vocabulary
similar figures, *p. 196*
corresponding angles, *p. 196*
corresponding sides, *p. 196*

Similar Figures

Figures that have the same shape but not necessarily the same size are called **similar figures**. The triangles below are similar.

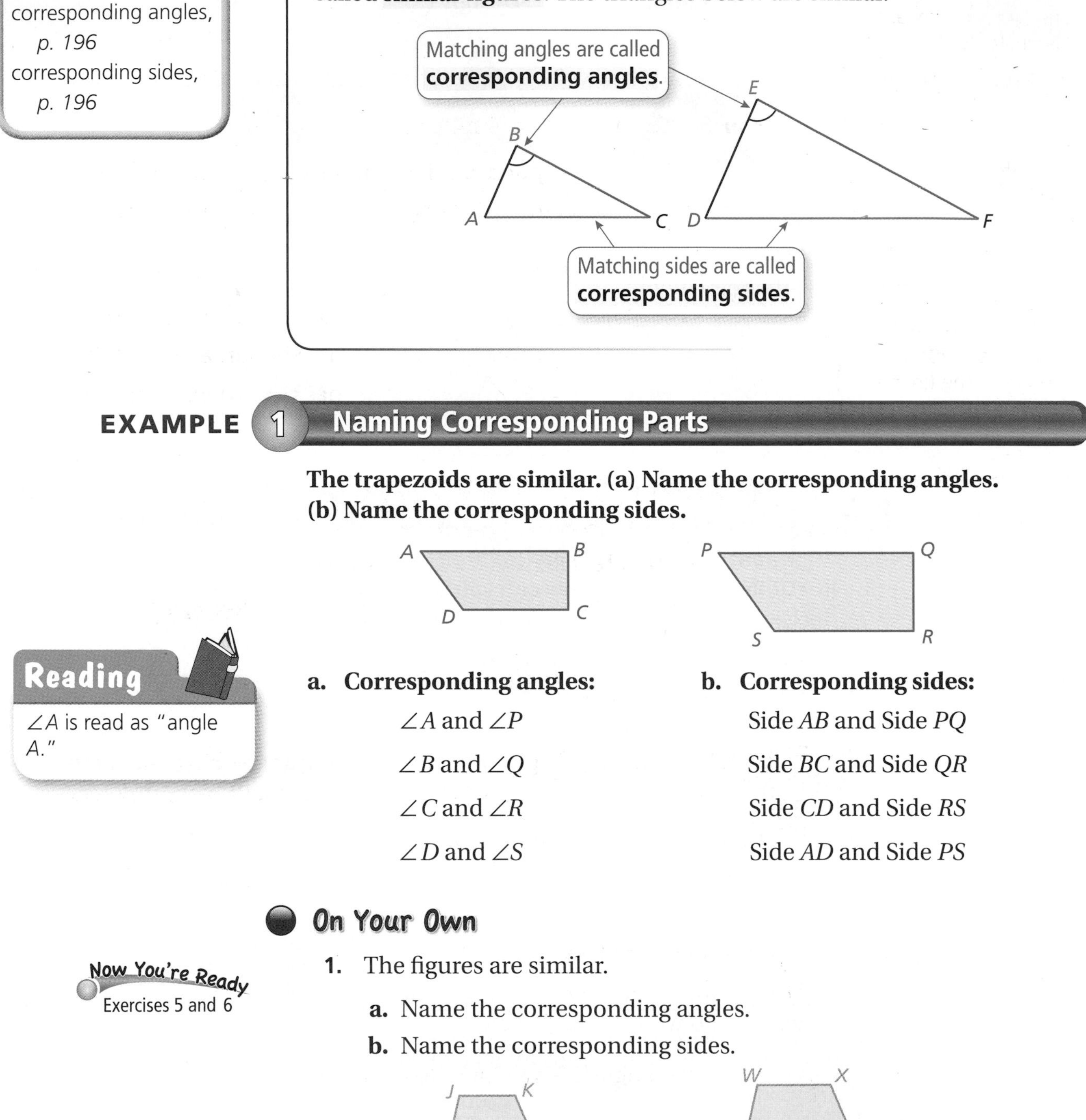

Matching angles are called **corresponding angles**.
A
B
C
D
E
F
Matching sides are called **corresponding sides**.

EXAMPLE 1 Naming Corresponding Parts

The trapezoids are similar. (a) Name the corresponding angles. (b) Name the corresponding sides.

Reading
$\angle A$ is read as "angle A."

a. Corresponding angles:

$\angle A$ and $\angle P$

$\angle B$ and $\angle Q$

$\angle C$ and $\angle R$

$\angle D$ and $\angle S$

b. Corresponding sides:

Side AB and Side PQ

Side BC and Side QR

Side CD and Side RS

Side AD and Side PS

On Your Own

Now You're Ready
Exercises 5 and 6

1. The figures are similar.
   - **a.** Name the corresponding angles.
   - **b.** Name the corresponding sides.

Reading

Red arcs are used to indicate angles that have the same measure.

The symbol $\sim$ means "is similar to."

Identifying Similar Figures

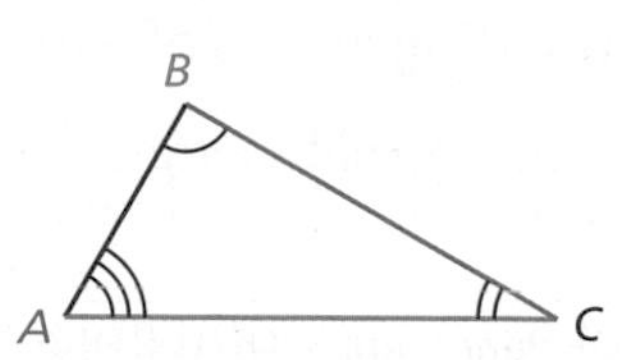

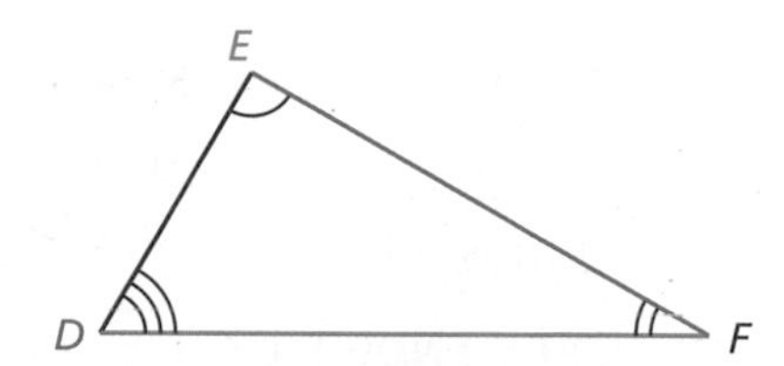

Triangle ABC is similar to triangle DEF: $\triangle ABC \sim \triangle DEF$

Words Two figures are similar if

- corresponding side lengths are proportional, and
- corresponding angles have the same measure.

Symbols ***Side Lengths***

$$\frac{AB}{DE} = \frac{BC}{EF} = \frac{AC}{DF}$$

Angles

$\angle A$ has the same measure as $\angle D$.

$\angle B$ has the same measure as $\angle E$.

$\angle C$ has the same measure as $\angle F$.

When writing a similarity statement, make sure to list the vertices of the figures in the same order.

EXAMPLE 2 Identifying Similar Figures

Which rectangle is similar to Rectangle A?

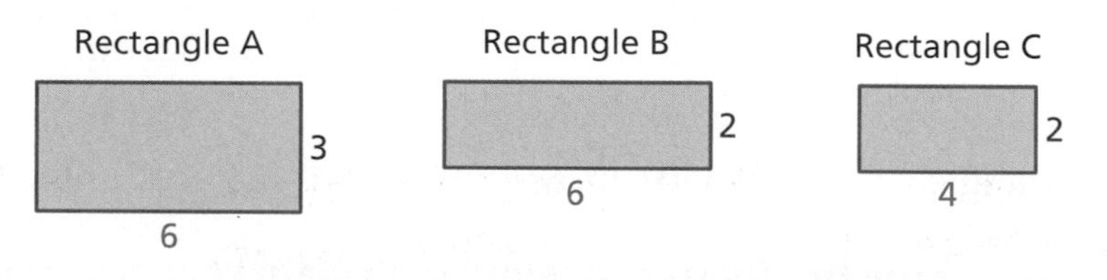

Each figure is a rectangle. So, corresponding angles have the same measure. Check to see if corresponding side lengths are proportional.

Rectangle A and Rectangle B

$$\frac{\text{Length of A}}{\text{Length of B}} = \frac{6}{6} = 1 \qquad \frac{\text{Width of A}}{\text{Width of B}} = \frac{3}{2}$$

Not proportional

Rectangle A and Rectangle C

$$\frac{\text{Length of A}}{\text{Length of C}} = \frac{6}{4} = \frac{3}{2} \qquad \frac{\text{Width of A}}{\text{Width of C}} = \frac{3}{2}$$

Proportional

So, Rectangle C is similar to Rectangle A.

On Your Own

Now You're Ready
Exercises 7–12

2. Rectangle D is 3 units long and 1 unit wide. Which rectangle in Example 2 is similar to Rectangle D?

5.1 Exercises

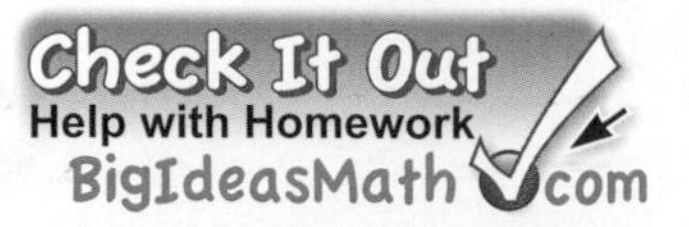

Vocabulary and Concept Check

1. **VOCABULARY** How are corresponding angles of two similar figures related?
2. **VOCABULARY** How are corresponding side lengths of two similar figures related?
3. **OPEN-ENDED** Give examples of two real-world objects whose shapes are similar.
4. **CRITICAL THINKING** Are two figures that have the same size and shape similar? Explain.

Practice and Problem Solving

Name the corresponding angles and the corresponding sides of the similar figures.

1 5.

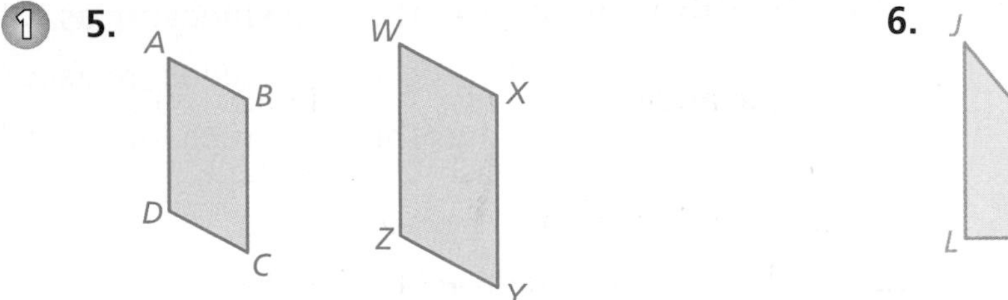

6.

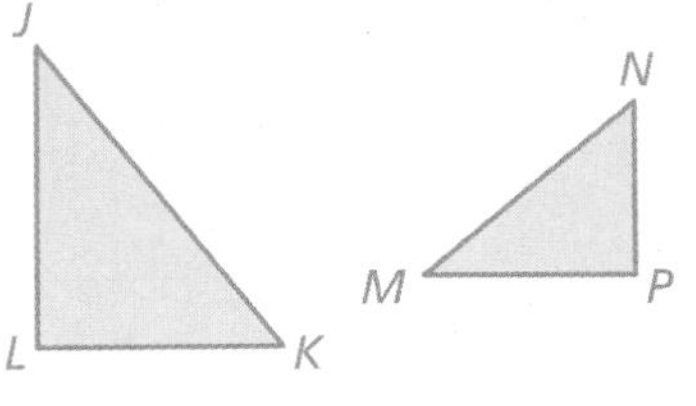

In a coordinate plane, draw the figures with the given vertices. Which figures are similar? Explain your reasoning.

2 7. Triangle A: (0, 0), (3, 0), (0, 3)
Triangle B: (0, 0), (5, 0), (0, 5)
Triangle C: (0, 0), (3, 0), (0, 6)

8. Rectangle A: (0, 0), (4, 0), (4, 2), (0, 2)
Rectangle B: (0, 0), (−6, 0), (−6, 3), (0, 3)
Rectangle C: (0, 0), (4, 0), (4, 2), (0, 2)

Tell whether the two figures are similar. Explain your reasoning.

9.

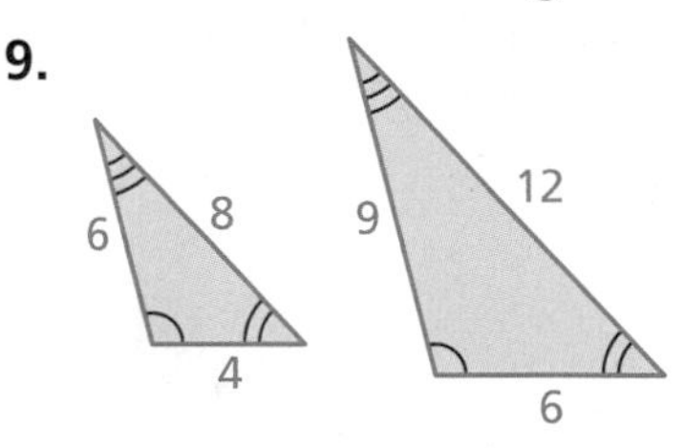

10.

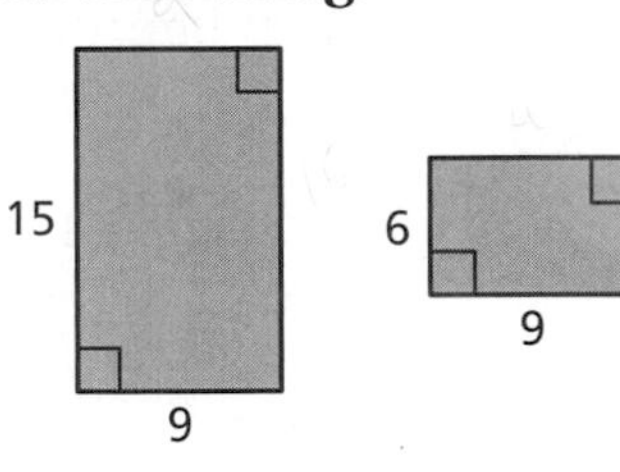

11. **MEXICO** A Mexican flag is 63 inches long and 36 inches high. Is the drawing at the right similar to the Mexican flag?

12. **DESKS** A student's rectangular desk is 30 inches long and 18 inches wide. The teacher's rectangular desk is 60 inches long and 36 inches wide. Are the desks similar?

The two triangles are similar. Find the measure of the angle.

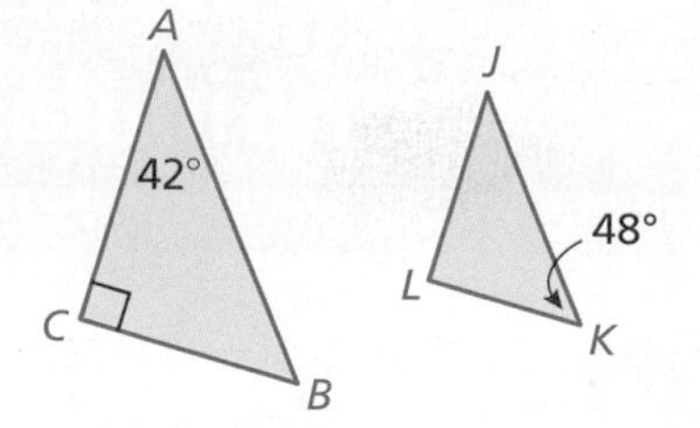

13. $\angle B$ **14.** $\angle L$ **15.** $\angle J$

16. REASONING Given $\triangle FGH \sim \triangle QRT$, name the corresponding angles and the corresponding sides.

17. PHOTOS You want to buy only photos that are similar rectangles. Which of the photo sizes should you buy?

Photo Size
4 in. × 5 in.
5 in. × 7 in.
8 in. × 12 in.
11 in. × 14 in.
18 in. × 27 in.

18. CRITICAL THINKING Are the following figures *always, sometimes,* or *never* similar? Explain.

a. Two triangles **b.** Two squares
c. Two rectangles **d.** A square and a triangle

19. LOGIC Can you draw two quadrilaterals each having two 130° angles and two 50° angles that are *not* similar? Justify your answer.

20. SIGN All of the angle measures in the sign are 90°.

a. Each side length is increased by 20%. Is the new sign similar to the original?

b. Each side length is increased by 6 inches. Is the new sign similar to the original?

21. GEOMETRY Use a ruler to draw two different isosceles triangles similar to the one shown. Measure the heights of each triangle to the nearest centimeter.

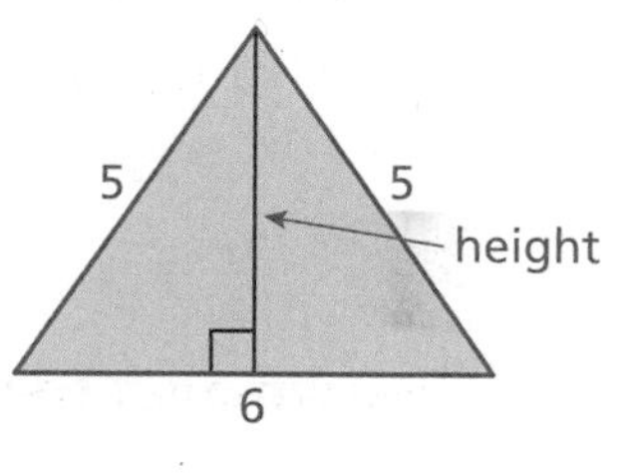

a. Is the ratio of the corresponding heights proportional to the ratio of the corresponding side lengths?

b. Do you think this is true for all similar triangles? Explain.

22. Critical Thinking Given $\triangle ABC \sim \triangle DEF$ and $\triangle DEF \sim \triangle JKL$, is $\triangle ABC \sim \triangle JKL$? Give an example or non-example.

Fair Game Review What you learned in previous grades & lessons

Simplify. *(Skills Review Handbook)*

23. $\left(\frac{4}{9}\right)^2$ **24.** $\left(\frac{3}{8}\right)^2$ **25.** $\left(\frac{7}{4}\right)^2$ **26.** $\left(\frac{6.5}{2}\right)^2$

27. MULTIPLE CHOICE Which equation shows inverse variation? *(Section 3.8)*

Ⓐ $3y = 8x$ Ⓑ $y = \frac{8}{3x}$ Ⓒ $\frac{y}{3} = \frac{x}{8}$ Ⓓ $y = 8x - 3$

5.2 Perimeters and Areas of Similar Figures

Essential Question How do changes in dimensions of similar geometric figures affect the perimeters and areas of the figures?

COMMON CORE STATE STANDARDS
7.G.1

1 ACTIVITY: Comparing Perimeters and Areas

Work with a partner. Use pattern blocks to make a figure whose dimensions are 2, 3, and 4 times greater than those of the original figure. Find the perimeter P and area A of each larger figure.

a. Sample: Square

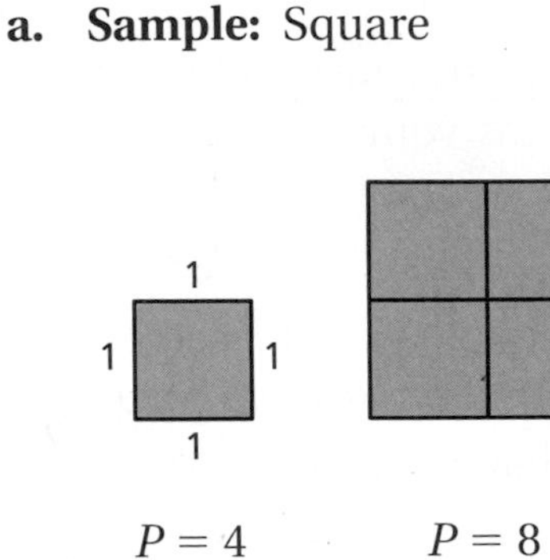

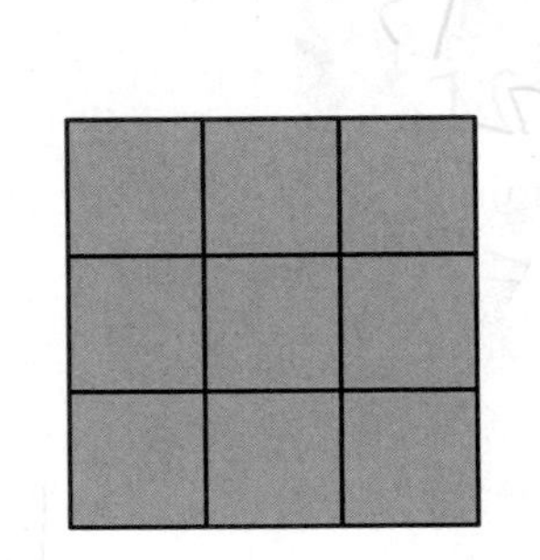

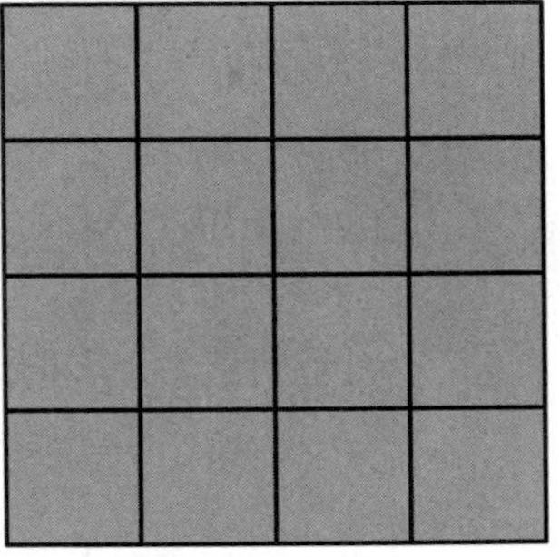

$P = 4$ $A = 1$

$P = 8$ $A = 4$

$P = 12$ $A = 9$

$P = 16$ $A = 16$

b. Triangle

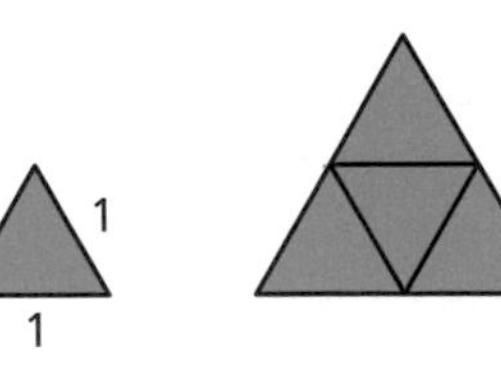

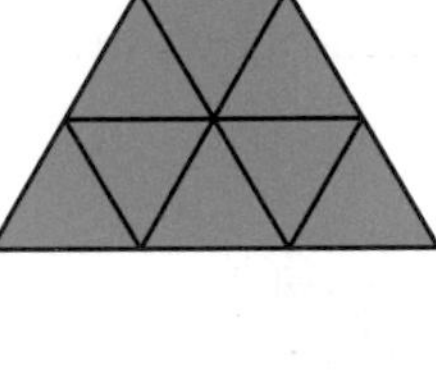

$P = 3$ $A = B$

$P = 6$ $A = 4B$

$P =$ ☐ $A =$ ☐

$P =$ ☐ $A =$ ☐

c. Rectangle

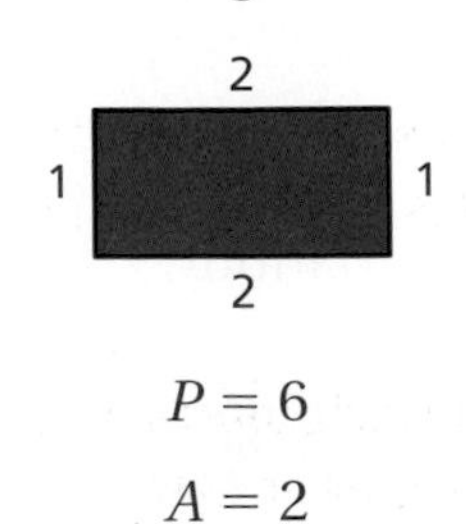

$P = 6$

$A = 2$

d. Parallelogram

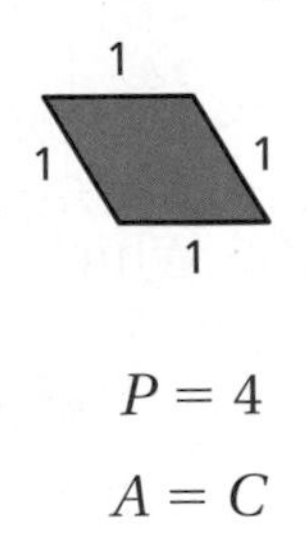

$P = 4$

$A = C$

2 ACTIVITY: Finding Patterns for Perimeters

Work with a partner. Copy and complete the table for the perimeters of the figures in Activity 1. Describe the pattern.

Perimeters	Figure	Original Side Lengths	Double Side Lengths	Triple Side Lengths	Quadruple Side Lengths
	square	$P = 4$	$P = 8$	$P = 12$	$P = 16$
	triangle	$P = 3$	$P = 6$		
	rectangle	$P = 6$			
	parallelogram	$P = 4$			

3 ACTIVITY: Finding Patterns for Areas

Work with a partner. Copy and complete the table for the areas of the figures in Activity 1. Describe the pattern.

Areas	Figure	Original Side Lengths	Double Side Lengths	Triple Side Lengths	Quadruple Side Lengths
	square	$A = 1$	$A = 4$	$A = 9$	$A = 16$
	triangle	$A = B$	$A = 4B$		
	rectangle	$A = 2$			
	parallelogram	$A = C$			

What Is Your Answer?

4. **IN YOUR OWN WORDS** How do changes in dimensions of similar geometric figures affect the perimeters and areas of the figures?

Use what you learned about perimeters and areas of similar figures to complete Exercises 8–11 on page 204.

Perimeters of Similar Figures

If two figures are similar, then the ratio of their perimeters is equal to the ratio of their corresponding side lengths.

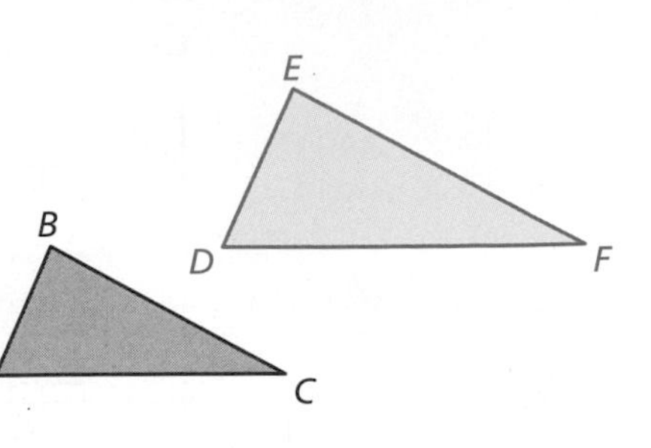

$$\frac{\text{Perimeter of } \triangle ABC}{\text{Perimeter of } \triangle DEF} = \frac{AB}{DE} = \frac{BC}{EF} = \frac{AC}{DF}$$

EXAMPLE 1 Finding Ratios of Perimeters

Find the ratio (red to blue) of the perimeters of the similar rectangles.

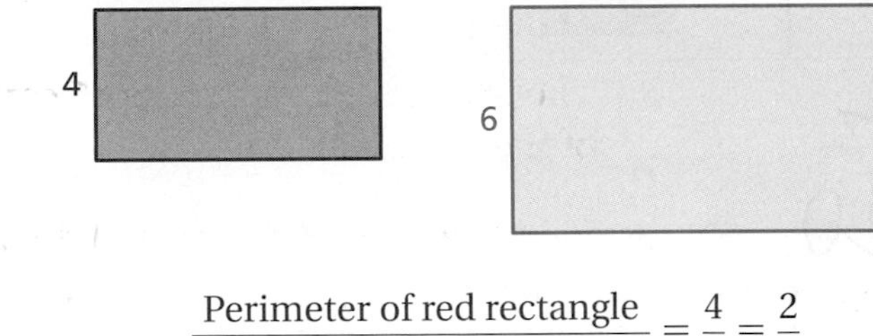

$$\frac{\text{Perimeter of red rectangle}}{\text{Perimeter of blue rectangle}} = \frac{4}{6} = \frac{2}{3}$$

∴ The ratio of the perimeters is $\frac{2}{3}$.

On Your Own

1. The height of Figure A is 9 feet. The height of a similar Figure B is 15 feet. What is the ratio of the perimeter of A to the perimeter of B?

Areas of Similar Figures

If two figures are similar, then the ratio of their areas is equal to the *square* of the ratio of their corresponding side lengths.

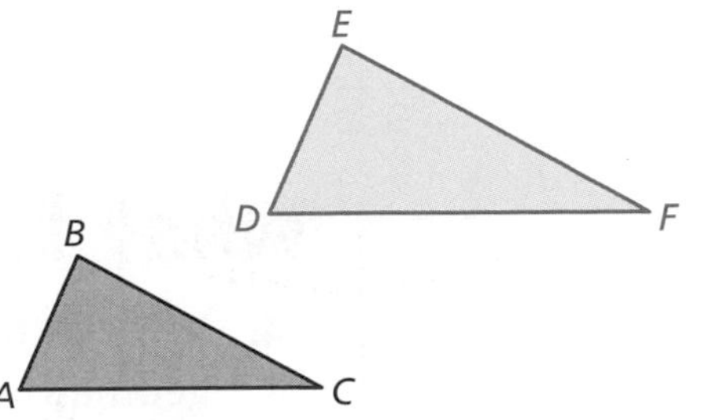

$$\frac{\text{Area of } \triangle ABC}{\text{Area of } \triangle DEF} = \left(\frac{AB}{DE}\right)^2 = \left(\frac{BC}{EF}\right)^2 = \left(\frac{AC}{DF}\right)^2$$

EXAMPLE 2 Finding Ratios of Areas

Find the ratio (red to blue) of the areas of the similar triangles.

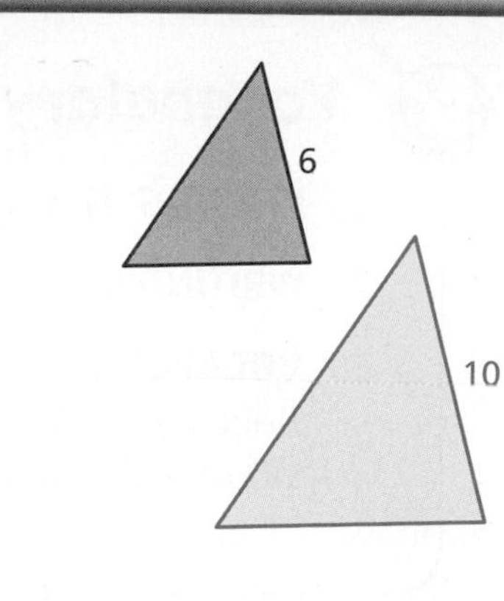

$$\frac{\text{Area of red triangle}}{\text{Area of blue triangle}} = \left(\frac{6}{10}\right)^2$$

$$= \left(\frac{3}{5}\right)^2 = \frac{9}{25}$$

∴ The ratio of the areas is $\frac{9}{25}$.

EXAMPLE 3 Real-Life Application

You place a picture on a page of a photo album. The page and the picture are similar rectangles.

a. How many times greater is the area of the page than the area of the picture?

b. The area of the picture is 45 square inches. What is the area of the page?

a. Find the ratio of the area of the page to the area of the picture.

$$\frac{\text{Area of page}}{\text{Area of picture}} = \left(\frac{\text{length of page}}{\text{length of picture}}\right)^2$$

$$= \left(\frac{8}{6}\right)^2 = \left(\frac{4}{3}\right)^2 = \frac{16}{9}$$

∴ The area of the page is $\frac{16}{9}$ times greater than the area of the picture.

b. Multiply the area of the picture by $\frac{16}{9}$.

$$45 \cdot \frac{16}{9} = 80$$

∴ The area of the page is 80 square inches.

On Your Own

Now You're Ready
Exercises 4–13

2. The base of Triangle P is 8 meters. The base of a similar Triangle Q is 7 meters. What is the ratio of the area of P to the area of Q?

3. In Example 3, the perimeter of the picture is 27 inches. What is the perimeter of the page?

5.2 Exercises

Vocabulary and Concept Check

1. **WRITING** How are the perimeters of two similar figures related?
2. **WRITING** How are the areas of two similar figures related?
3. **VOCABULARY** Rectangle $ABCD$ is similar to Rectangle $WXYZ$. The area of $ABCD$ is 30 square inches. What is the area of $WXYZ$? Explain.

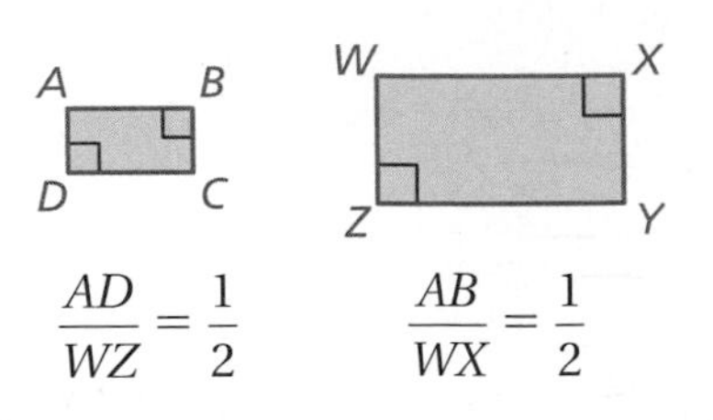

$\frac{AD}{WZ} = \frac{1}{2}$ $\frac{AB}{WX} = \frac{1}{2}$

Practice and Problem Solving

The two figures are similar. Find the ratios (red to blue) of the perimeters and of the areas.

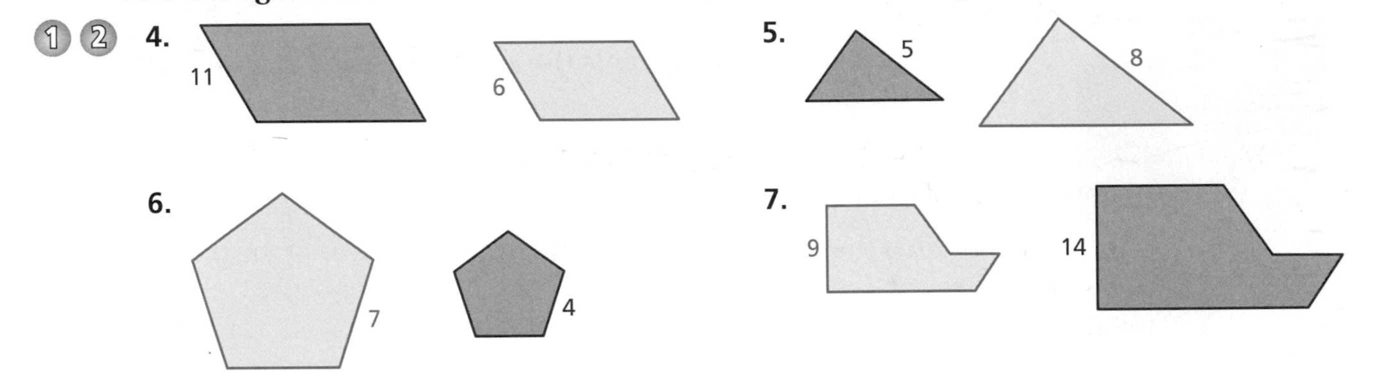

8. How does doubling the side lengths of a triangle affect its perimeter?
9. How does tripling the side lengths of a triangle affect its perimeter?
10. How does doubling the side lengths of a rectangle affect its area?
11. How does quadrupling the side lengths of a rectangle affect its area?

12. **FOOSBALL** The playing surfaces of two foosball tables are similar. The ratio of the corresponding side lengths is 10 : 7. What is the ratio of the areas?
13. **LAPTOP** The ratio of the corresponding side lengths of two similar computer screens is 13 : 15. The perimeter of the smaller screen is 39 inches. What is the perimeter of the larger screen?

Triangle ABC is similar to Triangle DEF. Tell whether the statement is *true* or *false*. Explain your reasoning.

14. $\frac{\text{Perimeter of } \triangle ABC}{\text{Perimeter of } \triangle DEF} = \frac{AB}{DE}$

15. $\frac{\text{Area of } \triangle ABC}{\text{Area of } \triangle DEF} = \frac{AB}{DE}$

16. **FABRIC** The cost of the fabric is $1.31. What would you expect to pay for a similar piece of fabric that is 18 inches by 42 inches?

17. **AMUSEMENT PARK** A model of a merry-go-round has a base area of about 450 square inches. What is the percent of increase of the base area from the model to the actual merry-go-round? Explain.

18. **CRITICAL THINKING** The circumference of Circle K is π. The circumference of Circle L is 4π.

    a. What is the ratio of their circumferences? of their radii? of their areas?

    b. What do you notice?

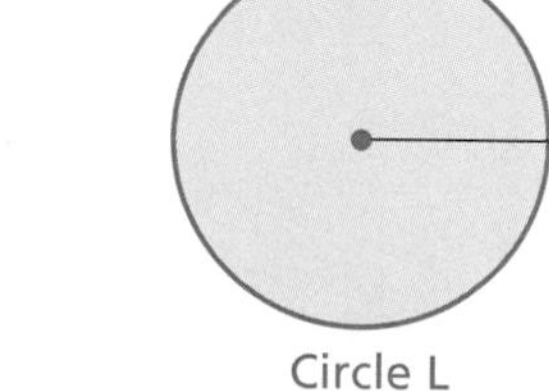

19. **GEOMETRY** Rhombus A is similar to Rhombus B. What is the ratio (A to B) of the corresponding side lengths?

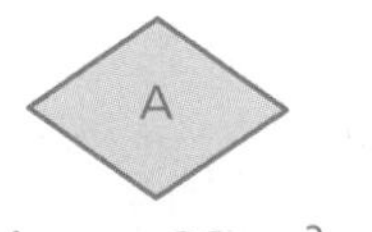

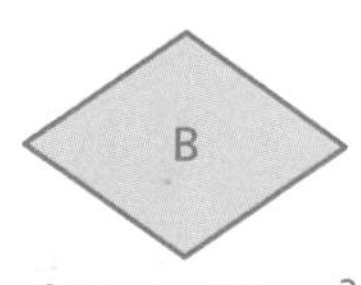

20. Geometry A triangle with an area of 10 square meters has a base of 4 meters. A similar triangle has an area of 90 square meters. What is the *height* of the larger triangle?

Fair Game Review What you learned in previous grades & lessons

Find the percent of change. Round to the nearest tenth of a percent, if necessary. *(Section 4.2)*

21. 24 feet to 30 feet

22. 90 miles to 63 miles

23. 150 liters to 86 liters

24. **MULTIPLE CHOICE** A runner completes an 800-meter race in 2 minutes 40 seconds. What is the runner's speed? *(Section 3.1)*

Ⓐ $\frac{3 \text{ sec}}{10 \text{ m}}$ Ⓑ $\frac{160 \text{ sec}}{1 \text{ m}}$ Ⓒ $\frac{5 \text{ m}}{1 \text{ sec}}$ Ⓓ $\frac{10 \text{ m}}{3 \text{ sec}}$

5.3 Finding Unknown Measures in Similar Figures

COMMON CORE STATE STANDARDS

7.G.1

Essential Question What information do you need to know to find the dimensions of a figure that is similar to another figure?

1 ACTIVITY: Drawing and Labeling Similar Figures

Work with a partner. You are given the red rectangle. Find a blue rectangle that is similar and has one side from (−1, −6) to (5, −6). Label the vertices.

a. Sample:

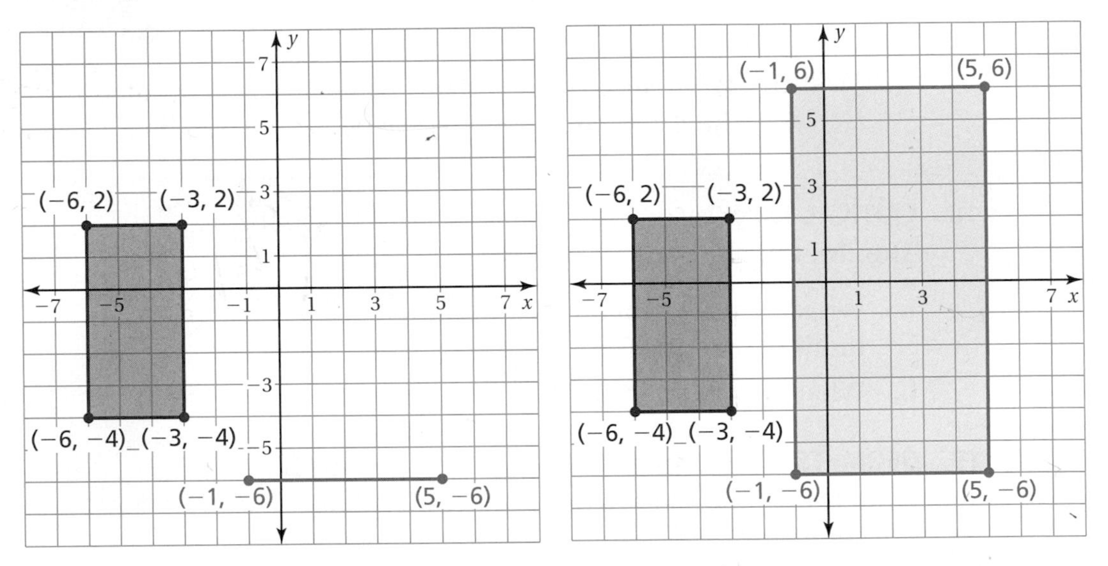

You can see that the two rectangles are similar by showing that ratios of corresponding sides are equal.

$$\frac{\text{Red Length}}{\text{Blue Length}} \stackrel{?}{=} \frac{\text{Red Width}}{\text{Blue Width}}$$

$$\frac{\text{change in } y}{\text{change in } y} \stackrel{?}{=} \frac{\text{change in } x}{\text{change in } x}$$

$$\frac{6}{12} \stackrel{?}{=} \frac{3}{6}$$

$$\frac{1}{2} = \frac{1}{2}$$

∴ The ratios are equal. So, the rectangles are similar.

b. There are three other blue rectangles that are similar to the red rectangle and have the given side.

- Draw each one. Label the vertices of each.
- Show that each is similar to the original red rectangle.

2 ACTIVITY: Reading a Map

Work with a partner.

a. The red and blue rectangles are similar. Find the length of the blue rectangle. Explain your reasoning.

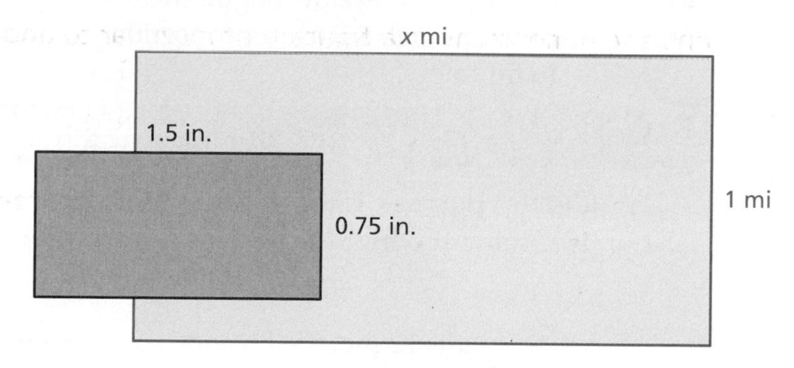

b. The distance marked by the vertical red line on the map is 1 mile. Find the distance marked by the horizontal red line. Explain your reasoning.

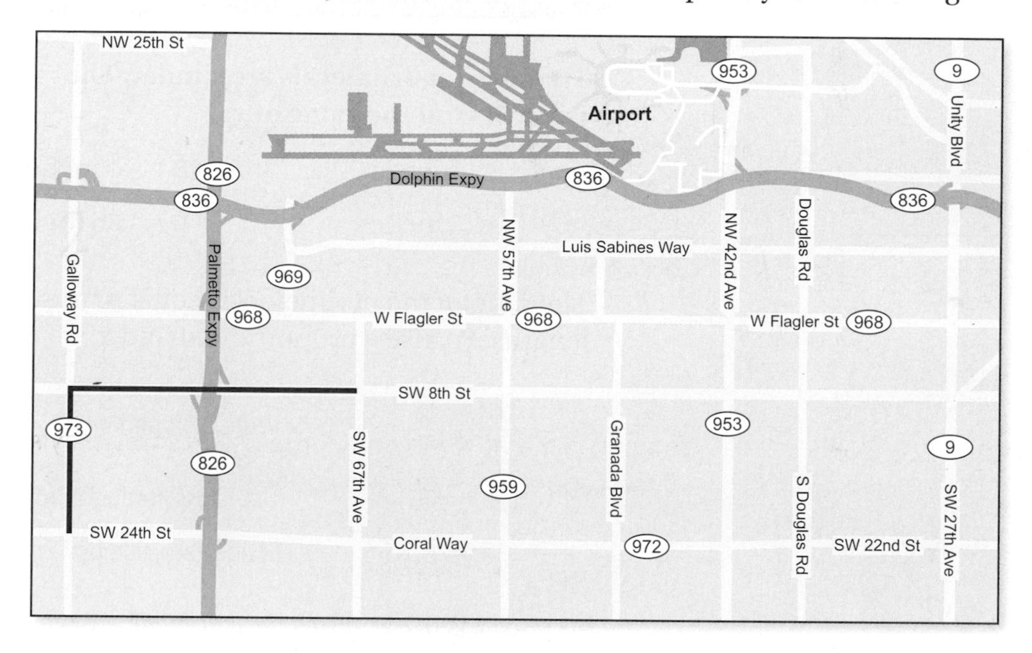

What Is Your Answer?

3. **IN YOUR OWN WORDS** What information do you need to know to find the dimensions of a figure that is similar to another figure? Give some examples using two rectangles.

4. When you know the length and width of one rectangle and the length of a similar rectangle, can you always find the missing width? Why or why not?

Practice Use what you learned about finding unknown measures in similar figures to complete Exercises 3 and 4 on page 210.

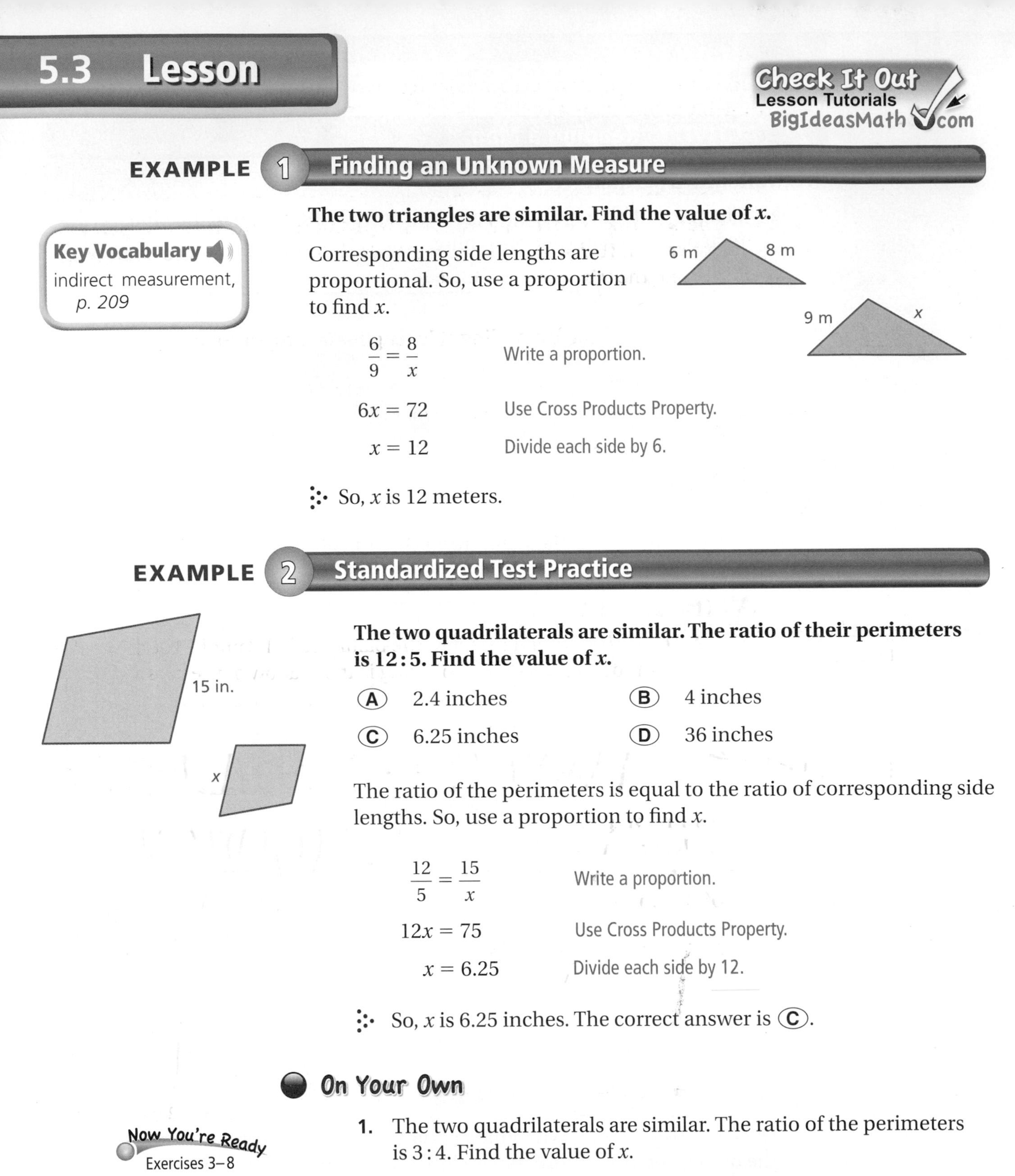

5.3 Lesson

Check It Out
Lesson Tutorials
BigIdeasMath.com

Key Vocabulary
indirect measurement, *p. 209*

EXAMPLE 1 Finding an Unknown Measure

The two triangles are similar. Find the value of x.

Corresponding side lengths are proportional. So, use a proportion to find x.

$\frac{6}{9} = \frac{8}{x}$ — Write a proportion.

$6x = 72$ — Use Cross Products Property.

$x = 12$ — Divide each side by 6.

∴ So, x is 12 meters.

EXAMPLE 2 Standardized Test Practice

The two quadrilaterals are similar. The ratio of their perimeters is 12 : 5. Find the value of x.

Ⓐ 2.4 inches Ⓑ 4 inches

Ⓒ 6.25 inches Ⓓ 36 inches

The ratio of the perimeters is equal to the ratio of corresponding side lengths. So, use a proportion to find x.

$\frac{12}{5} = \frac{15}{x}$ — Write a proportion.

$12x = 75$ — Use Cross Products Property.

$x = 6.25$ — Divide each side by 12.

∴ So, x is 6.25 inches. The correct answer is Ⓒ.

On Your Own

Now You're Ready
Exercises 3–8

1. The two quadrilaterals are similar. The ratio of the perimeters is 3 : 4. Find the value of x.

Indirect measurement uses similar figures to find a missing measure that is difficult to find directly.

EXAMPLE 3 Using Indirect Measurement

A person that is 6 feet tall casts a 3-foot-long shadow. A nearby palm tree casts a 15-foot-long shadow. What is the height h of the palm tree? Assume the triangles are similar.

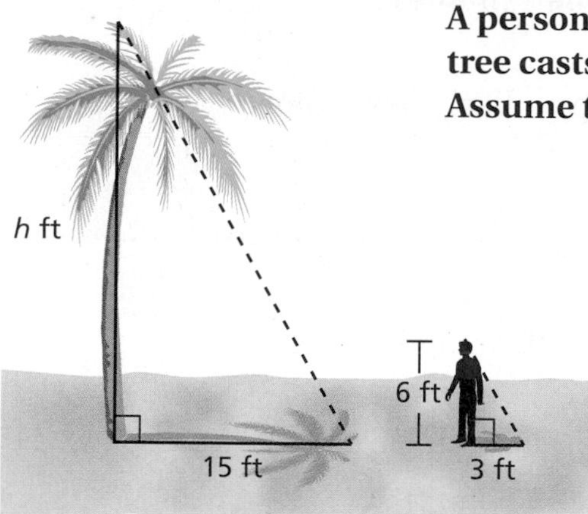

Corresponding side lengths are proportional.

$$\frac{h}{6} = \frac{15}{3}$$ Write a proportion.

$$6 \cdot \frac{h}{6} = \frac{15}{3} \cdot 6$$ Multiply each side by 6.

$$h = 30$$ Simplify.

The palm tree is 30 feet tall.

On Your Own

Now You're Ready
Exercise 9

2. **WHAT IF?** Later in the day, the palm tree in Example 3 casts a 25-foot-long shadow. How long is the shadow of the person?

EXAMPLE 4 Using Proportions to Find Area

A swimming pool is similar in shape to a volleyball court. What is the area A of the pool?

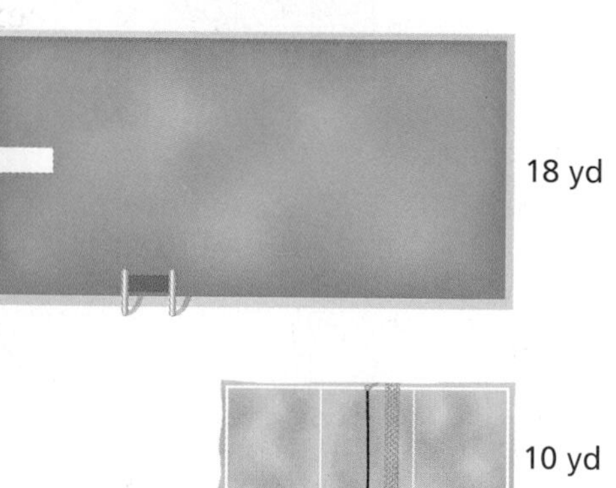

$$\frac{\text{Area of court}}{\text{Area of pool}} = \left(\frac{\text{width of court}}{\text{width of pool}}\right)^2$$

$$\frac{200}{A} = \left(\frac{10}{18}\right)^2$$ Substitute.

$$\frac{200}{A} = \frac{100}{324}$$ Simplify.

$$A = 648$$ Solve the proportion.

 The area of the pool is 648 square yards.

On Your Own

3. The length of the volleyball court in Example 4 is 20 yards. What is the perimeter of the pool?

5.3 Exercises

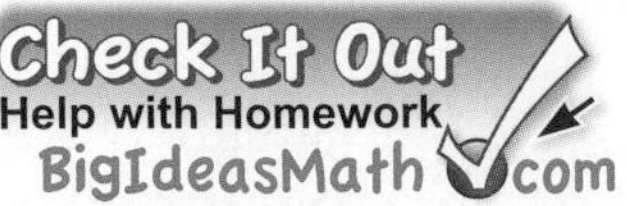

Vocabulary and Concept Check

1. **REASONING** How can you use corresponding side lengths to find unknown measures in similar figures?

2. **CRITICAL THINKING** In which of the situations would you likely use indirect measurement? Explain your reasoning.

Finding the height of a statue	Finding the width of a doorway
Finding the width of a river	Finding the length of a lake

Practice and Problem Solving

The polygons are similar. Find the value of *x*.

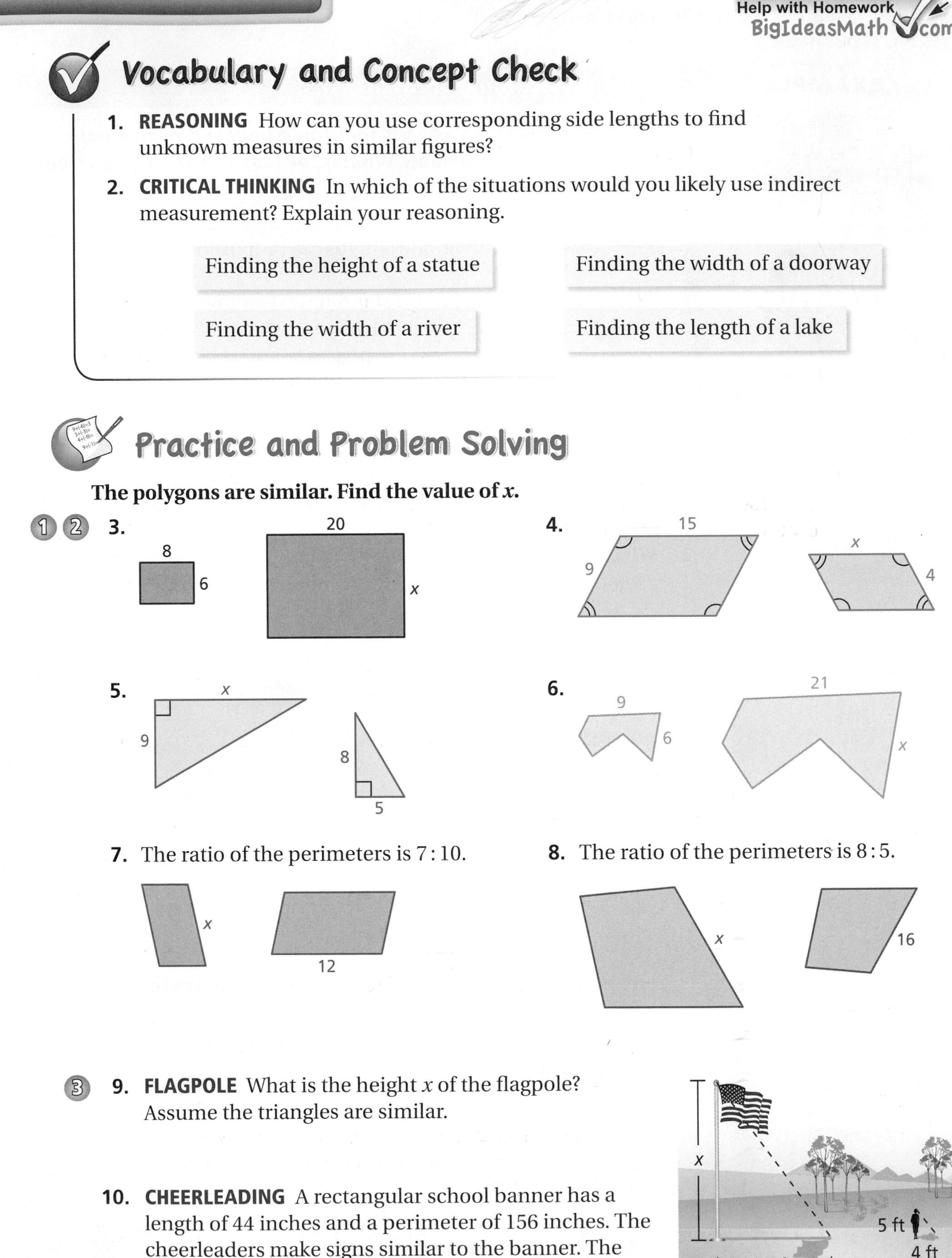

① ② 3.

4.

5.

6.

7. The ratio of the perimeters is 7 : 10.

8. The ratio of the perimeters is 8 : 5.

③ 9. **FLAGPOLE** What is the height *x* of the flagpole? Assume the triangles are similar.

10. **CHEERLEADING** A rectangular school banner has a length of 44 inches and a perimeter of 156 inches. The cheerleaders make signs similar to the banner. The length of a sign is 11 inches. What is its perimeter?

11. **SQUARE** The ratio of the side length of Square A to the side length of Square B is 4 : 9. The side length of Square A is 12 yards. What is the perimeter of Square B?

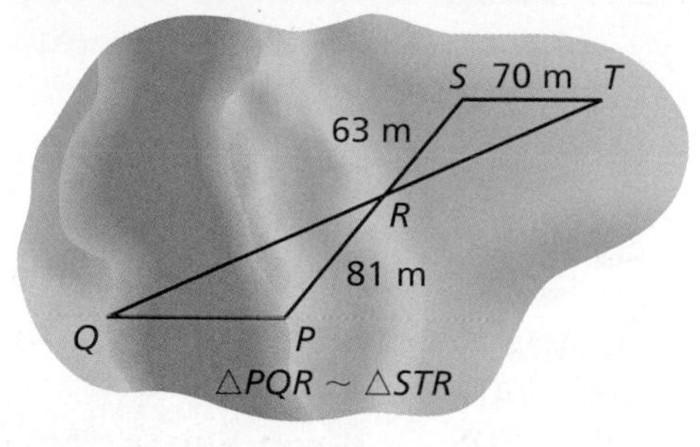

12. **RIVER** Is the distance QP across the river greater than 100 meters? Explain.

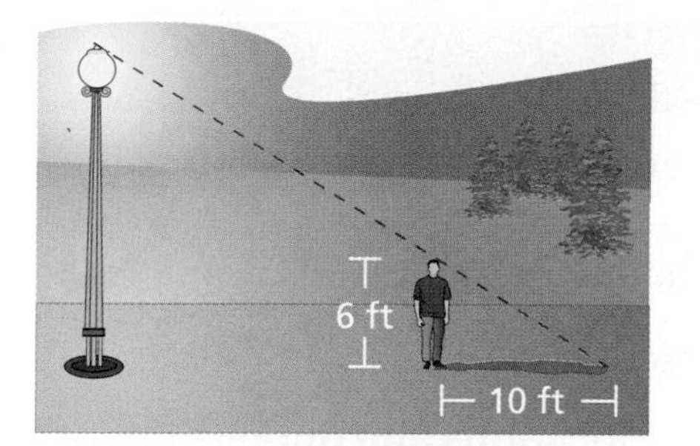

13. **STREET LIGHT** A person standing 20 feet from a street light casts a shadow as shown. How many times taller is the street light than the person? Assume the triangles are similar.

14. **AREA** A school playground is similar in shape to the community park. You can mow 250 square yards of grass in 15 minutes. How long would it take you to mow the grass on the playground?

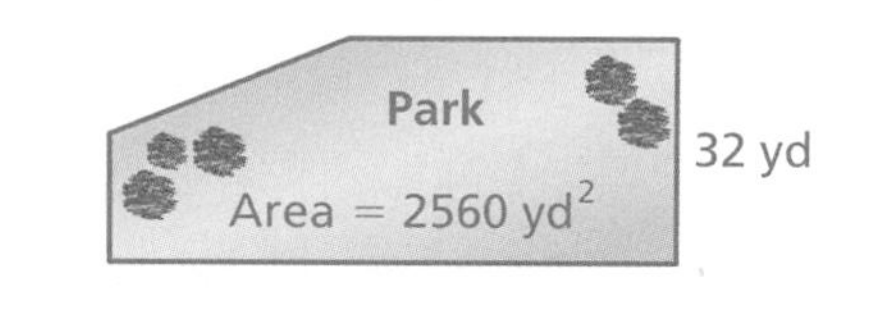

15. Critical Thinking Two bottles of fertilizer are needed to treat the flower garden shown. How many bottles are needed to treat a similar garden with a perimeter of 105 feet?

Fair Game Review What you learned in previous grades & lessons

Copy and complete the statement using a ratio. Round to the nearest hundredth, if necessary. *(Section 3.6)*

16. 4 mi ≈ ___ km
17. 12.5 in. ≈ ___ cm
18. 110 kg ≈ ___ lb
19. 6.2 km ≈ ___ mi
20. 10 cm ≈ ___ in.
21. 92.5 lb ≈ ___ kg

22. **MULTIPLE CHOICE** A recipe that makes 8 pints of salsa uses 22 tomatoes. Which proportion can you use to find the number n of tomatoes needed to make 12 pints of salsa? *(Section 3.4)*

Ⓐ $\frac{n}{8} = \frac{22}{12}$ Ⓑ $\frac{8}{22} = \frac{12}{n}$ Ⓒ $\frac{22}{n} = \frac{12}{8}$ Ⓓ $\frac{8}{22} = \frac{n}{12}$

5.4 Scale Drawings

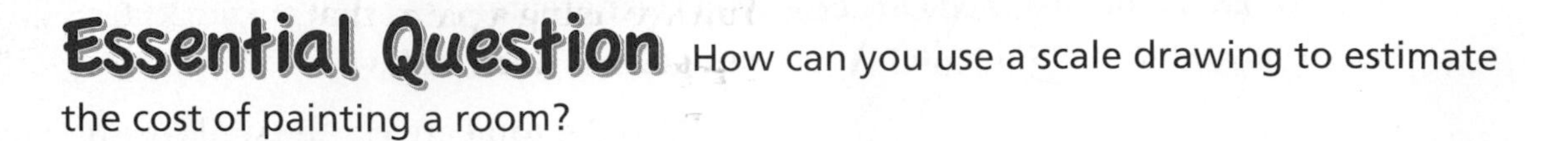

Essential Question How can you use a scale drawing to estimate the cost of painting a room?

COMMON CORE STATE STANDARDS
7.G.1

1 ACTIVITY: Making Scale Drawings

Work with a partner. You have decided that your classroom needs to be painted. Start by making a scale drawing of each of the four walls.

- **Measure each of the walls.**
- **Measure the locations and dimensions of parts that will *not* be painted.**
- **Decide on a scale for your drawings.**
- **Make a scale drawing of each of the walls.**

Sample: Wall #1

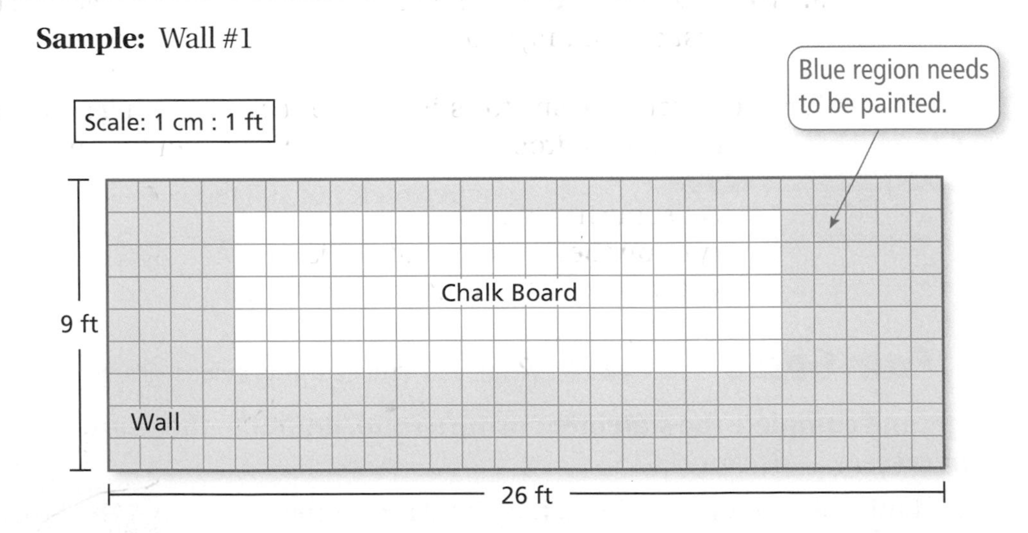

- **For each wall, find the area of the part that needs to be painted.**

	Dimensions	***Area***
Dimensions of the wall	9 ft by 26 ft	$9 \times 26 = 234$ sq ft
Dimensions of the part that will *not* be painted	5 ft by 17 ft	$5 \times 17 = 85$ sq ft
Area of painted part		149 sq ft

2 ACTIVITY: Using Scale Drawings

Work with a partner.

You are using a paint that covers 200 square feet per gallon. Each wall will need two coats of paint.

a. Find the total area of the walls from Activity 1 that needs to be painted.

b. Find the amount of paint you need to buy.

c. Estimate the total cost of painting your classroom.

Interior latex paint	$40 per gallon
Roller, pan, and brush set	$12

What Is Your Answer?

3. **IN YOUR OWN WORDS** How can you use a scale drawing to estimate the cost of painting a room?

4. Use a scale drawing to estimate the cost of painting another room, such as your bedroom or another room in your house.

5. Look at some maps in your school library or on the Internet. Make a list of the different scales used on the maps.

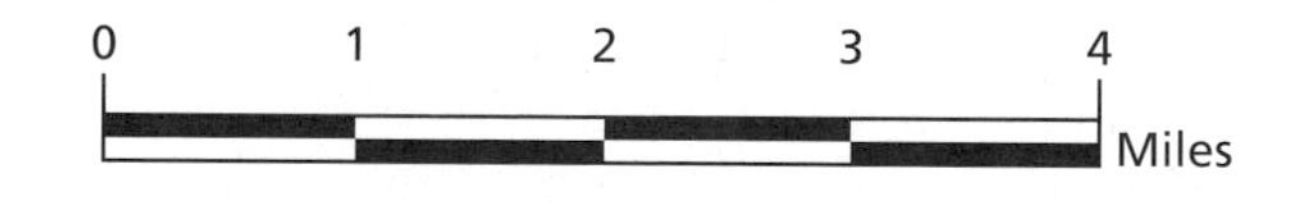

6. When you view a map on the Internet, how does the scale change when you zoom out? How does the scale change when you zoom in?

"I don't get it. According to this map, we only have to drive $8\frac{1}{2}$ inches."

Practice Use what you learned about scale drawings to complete Exercises 4–7 on page 216.

5.4 Lesson

Check It Out
Lesson Tutorials
BigIdeasMath.com

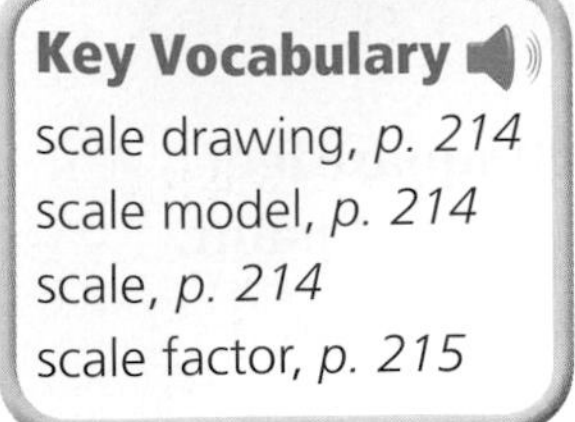

Scale Drawings and Models

A **scale drawing** is a proportional two-dimensional drawing of an object. A **scale model** is a proportional three-dimensional model of an object.

Scale

Measurements in scale drawings and models are proportional to the measurements of the actual object. The **scale** gives the ratio that compares the measurements of the drawing or model with the actual measurements.

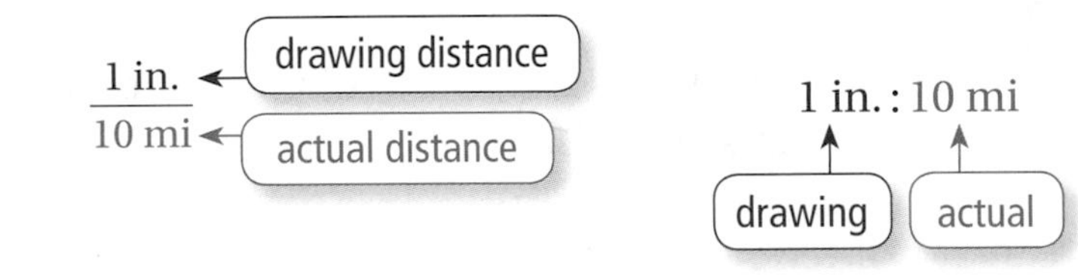

Scales are written so that the drawing distance comes first in the ratio.

EXAMPLE 1 Finding an Actual Distance

What is the actual distance d between Cadillac and Detroit?

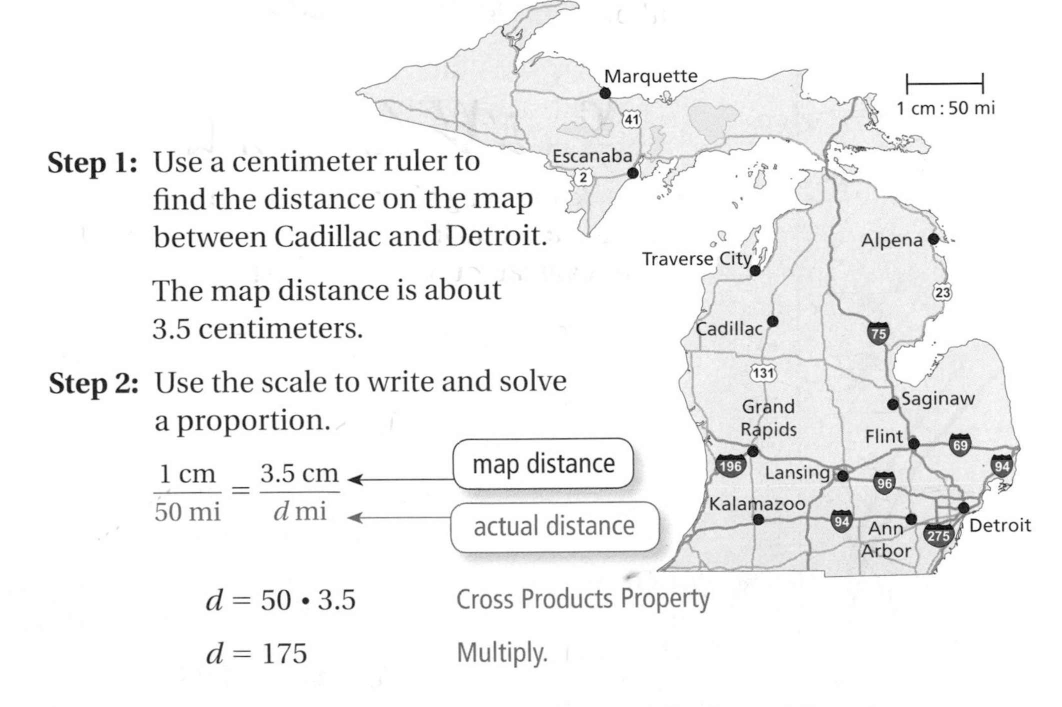

Step 1: Use a centimeter ruler to find the distance on the map between Cadillac and Detroit.

The map distance is about 3.5 centimeters.

Step 2: Use the scale to write and solve a proportion.

$$\frac{1 \text{ cm}}{50 \text{ mi}} = \frac{3.5 \text{ cm}}{d \text{ mi}}$$

$d = 50 \cdot 3.5$ Cross Products Property

$d = 175$ Multiply.

The distance between Cadillac and Detroit is about 175 miles.

On Your Own

Now You're Ready
Exercises 8–11

1. What is the actual distance between Traverse City and Marquette?

Multi-Language Glossary at BigIdeasMath.com.

EXAMPLE 2 Standardized Test Practice

The liquid outer core of Earth is 2300 kilometers thick. A scale model of the layers of Earth has a scale of 1 in. : 500 km. How thick is the liquid outer core of the model?

Ⓐ 0.2 in. Ⓑ 4.6 in. Ⓒ 0.2 km Ⓓ 4.6 km

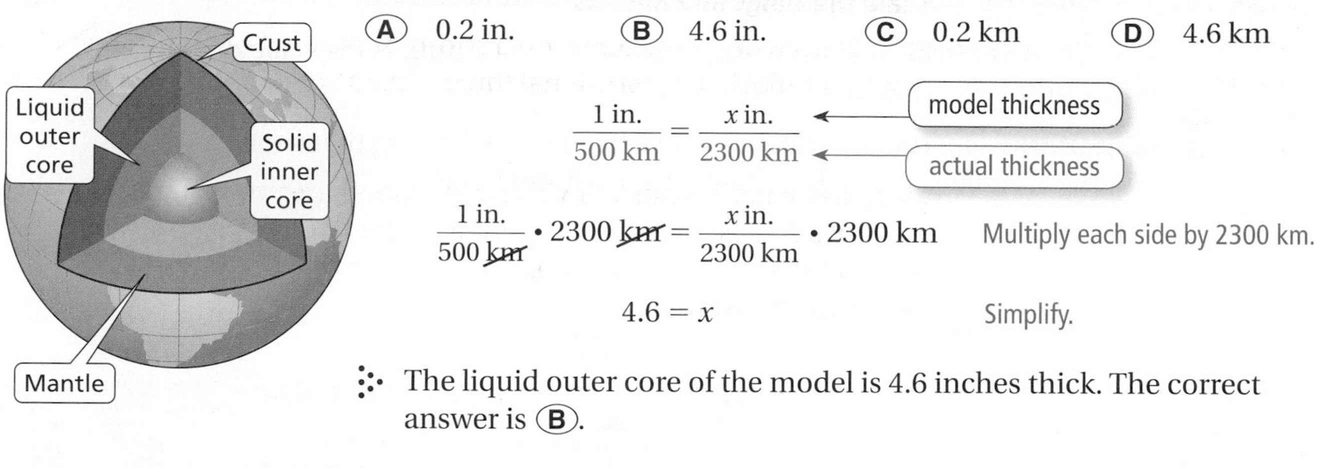

$$\frac{1 \text{ in.}}{500 \text{ km}} = \frac{x \text{ in.}}{2300 \text{ km}}$$ ← model thickness / ← actual thickness

$$\frac{1 \text{ in.}}{500 \cancel{\text{km}}} \cdot 2300 \cancel{\text{km}} = \frac{x \text{ in.}}{2300 \text{ km}} \cdot 2300 \text{ km}$$ Multiply each side by 2300 km.

$$4.6 = x$$ Simplify.

∴ The liquid outer core of the model is 4.6 inches thick. The correct answer is Ⓑ.

On Your Own

2. The mantle of Earth is 2900 kilometers thick. How thick is the mantle of the model?

A scale can be written without units when the units are the same. A scale without units is called a **scale factor**.

EXAMPLE 3 Finding a Scale Factor

A scale drawing of a spider is 5 centimeters long. The actual spider is 10 millimeters long. (a) What is the scale of the drawing? (b) What is the scale factor of the drawing?

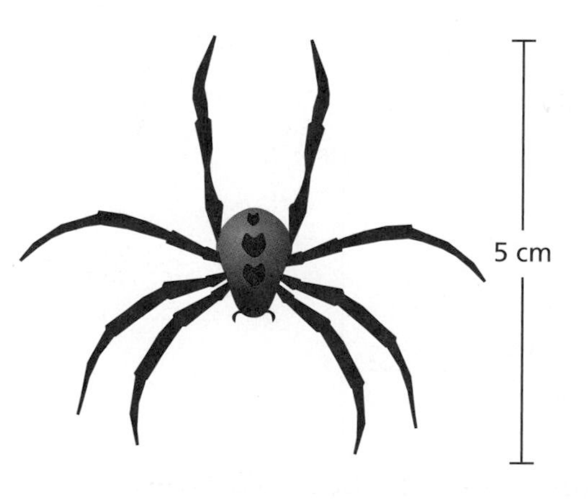

a. $$\frac{\text{drawing length}}{\text{actual length}} = \frac{5 \text{ cm}}{10 \text{ mm}} = \frac{1 \text{ cm}}{2 \text{ mm}}$$

∴ The scale is 1 cm : 2 mm.

b. Write the scale with the same units. Use the fact that 1 cm = 10 mm.

$$\text{scale factor} = \frac{1 \text{ cm}}{2 \text{ mm}} = \frac{10 \cancel{\text{mm}}}{2 \cancel{\text{mm}}} = \frac{5}{1}$$

∴ The scale factor is 5 : 1.

On Your Own

Now You're Ready
Exercises 12–16

3. A model has a scale of 1 mm : 20 cm. What is the scale factor of the model?

5.4 Exercises

Vocabulary and Concept Check

1. **VOCABULARY** Compare and contrast the terms *scale* and *scale factor*.
2. **CRITICAL THINKING** The scale of a drawing is 2 cm : 1 mm. Is the scale drawing *larger* or *smaller* than the actual object? Explain.
3. **REASONING** How would you find a scale factor of a drawing that shows a length of 4 inches when the actual object is 8 feet long?

Practice and Problem Solving

Use the drawing and a centimeter ruler.

4. What is the actual length of the flower garden?
5. What are the actual dimensions of the rose bed?
6. What are the actual perimeters of the perennial beds?
7. The area of the tulip bed is what percent of the area of the rose bed?

perennial bed | rose bed | tulip bed | perennial bed | 1 cm : 5 ft

Use the map in Example 1 to find the actual distance between the cities.

(1) 8. Kalamazoo and Ann Arbor
9. Lansing and Flint
10. Grand Rapids and Escanaba
11. Saginaw and Alpena

Find the missing dimension. Use the scale factor 1 : 12.

	Item	Model	Actual
(2) (3) 12.	Mattress	Length: 6.25 in.	Length: ____ in.
13.	Corvette	Length: ____ in.	Length: 15 ft
14.	Water Tower	Depth: 32 cm	Depth: ____ m
15.	Wingspan	Width: 5.4 ft	Width: ____ yd
16.	Football Helmet	Diameter: ____ mm	Diameter: 21 cm

17. **ERROR ANALYSIS** A scale is 1 cm : 20 m. Describe and correct the error in finding the actual distance that corresponds to 5 cm.

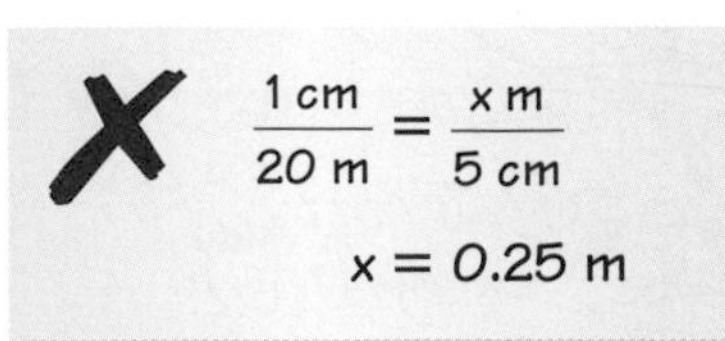

$\frac{1 \text{ cm}}{20 \text{ m}} = \frac{x \text{ m}}{5 \text{ cm}}$
$x = 0.25$ m

Use a centimeter ruler to measure the segment shown. Find the scale of the drawing.

18.

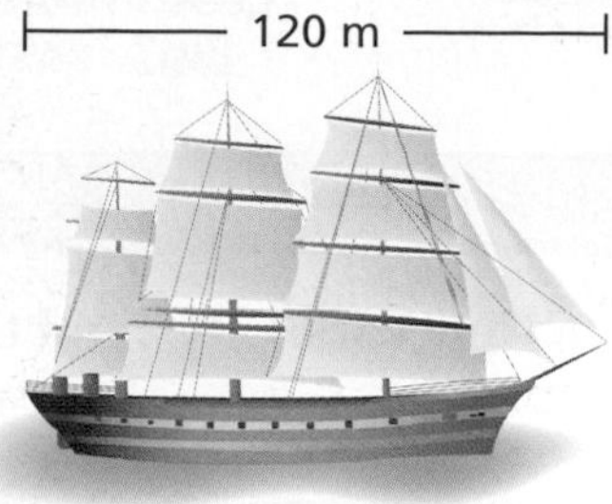

19.

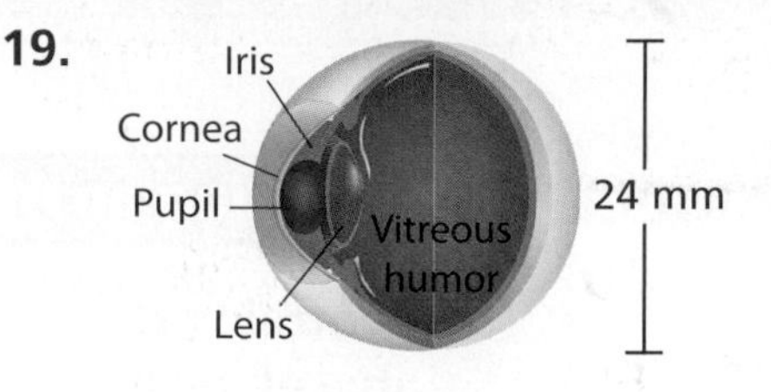

20. REASONING You know the length and width of a scale model. What additional information do you need to know to find the scale of the model?

21. OPEN-ENDED You are in charge of creating a billboard advertisement with the dimensions shown.

16 ft

8 ft

YOUR AD HERE

a. Choose a product. Then design the billboard using words and a picture.

b. What is the scale factor of your design?

Reduced drawing of blueprint

Bedroom

Living room

Bathroom

1 in. : 16 ft

22. BLUEPRINT In a blueprint, each square has a side length of $\frac{1}{4}$ inch.

a. Ceramic tile costs \$5 per square foot. How much would it cost to tile the bathroom?

b. Carpet costs \$18 per square yard. How much would it cost to carpet the bedroom and living room?

c. Which has a higher unit cost, the tile or the carpet? Explain.

23. REASONING You are making a scale model of the solar system. The radius of Earth is 6378 kilometers. The radius of the Sun is 695,500 kilometers. Is it reasonable to choose a baseball as a model of Earth? Explain your reasoning.

24. Critical Thinking A map on the Internet has a scale of 1 in. : 10 mi. You zoom out one level. The map has been reduced so that 2.5 inches on the old map appears as 1 inch on the new map. What is the scale of the new map?

Fair Game Review What you learned in previous grades & lessons

Plot and label the ordered pair in a coordinate plane. *(Section 1.6)*

25. $A(-4, 3)$ **26.** $B(2, -6)$ **27.** $C(5, 1)$ **28.** $D(-3, -7)$

29. MULTIPLE CHOICE A backpack is on sale for 15% off the original price. The original price is \$68. What is the sale price? *(Section 4.3)*

Ⓐ \$10.20 Ⓑ \$53 Ⓒ \$57.80 Ⓓ \$78.20

5.4b Scale Drawings

EXAMPLE 1 Finding an Actual Area

Central Park is a rectangular park in New York City. Find the actual area of the park.

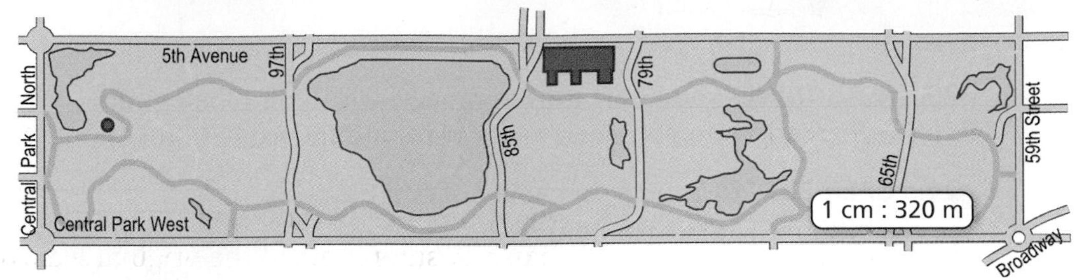

Step 1: Use a centimeter ruler to find the length and width of the park in the scale drawing.

The scale drawing of the park is 12.5 centimeters long and 2.5 centimeters wide.

Step 2: Use the scale to write and solve proportions to find the actual length and width of the park. Let ℓ be the actual length and let w be the actual width.

$$\frac{1 \text{ cm}}{320 \text{ m}} = \frac{12.5 \text{ cm}}{\ell \text{ m}} \qquad \text{drawing distance / actual distance} \qquad \frac{1 \text{ cm}}{320 \text{ m}} = \frac{2.5 \text{ cm}}{w \text{ m}}$$

$$\ell = 320 \cdot 12.5 \qquad\qquad w = 320 \cdot 2.5$$

$$\ell = 4000 \qquad\qquad w = 800$$

Step 3: Use a formula to find the area.

$A = \ell w$ — Write formula.

$= 4000(800)$ — Substitute 4000 for ℓ and 800 for w.

$= 3{,}200{,}000$ — Multiply.

The actual area of Central Park is 3,200,000 square meters.

Practice

The shuffleboard diagram has a scale of 1 cm : 1 ft. Find the actual area of the region.

1. red region
2. blue region
3. green region

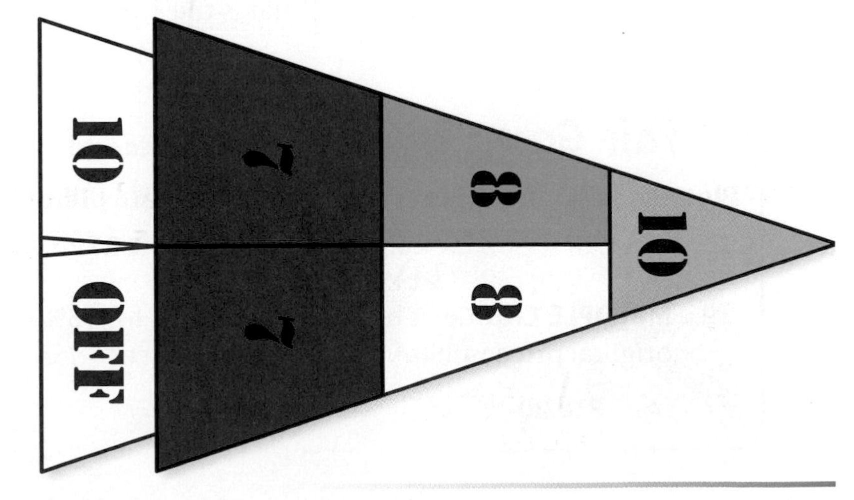

EXAMPLE 2 Recreating a Scale Drawing

1 in. : 8 ft

Recreate the scale drawing of a Romanian flag so that it has a scale of 1 in. : 4 ft.

Step 1: Compare measurements in the original scale drawing to measurements in the new scale drawing.

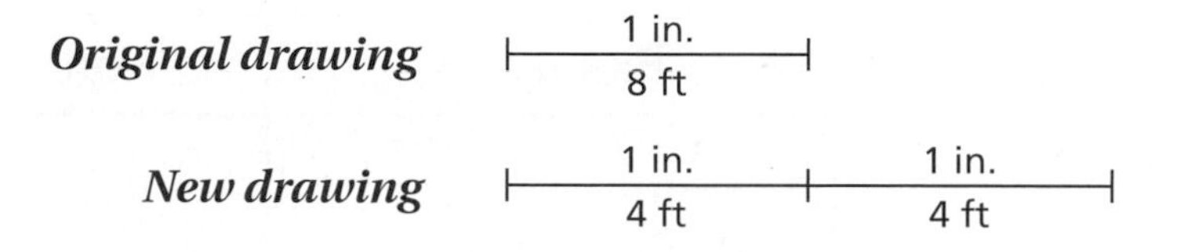

Measurements in the new scale drawing will be 2 times longer than measurements in the original scale drawing.

Step 2: Use an inch ruler to measure the original scale drawing. Multiply the measurements by 2 and create the new scale drawing.

Original scale drawing	*New scale drawing*
Length: 1.5 in.	Length: $1.5 \cdot 2 = 3$ in.
Width: 1 in.	Width: $1 \cdot 2 = 2$ in.
Blue bar width: 0.5 in.	Blue bar width: $0.5 \cdot 2 = 1$ in.
Yellow bar width: 0.5 in.	Yellow bar width: $0.5 \cdot 2 = 1$ in.
Red bar width: 0.5 in.	Red bar width: $0.5 \cdot 2 = 1$ in.

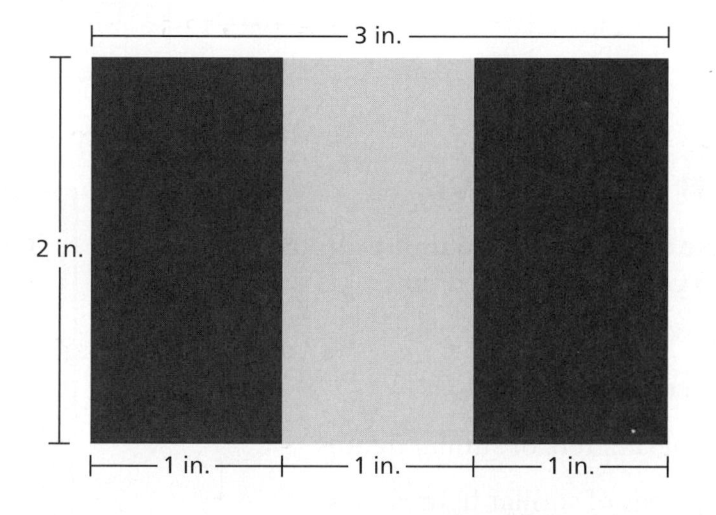

Practice

Recreate the scale drawing so that it has a scale of 1 cm : 4 m.

4.

1 cm : 8 m

5.

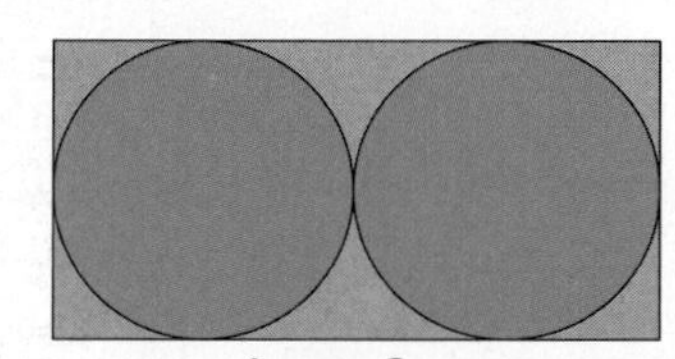

1 cm : 2 m

5 Study Help

You can use an **example and non-example chart** to list examples and non-examples of a vocabulary word or term. Here is an example and non-example chart for similar figures.

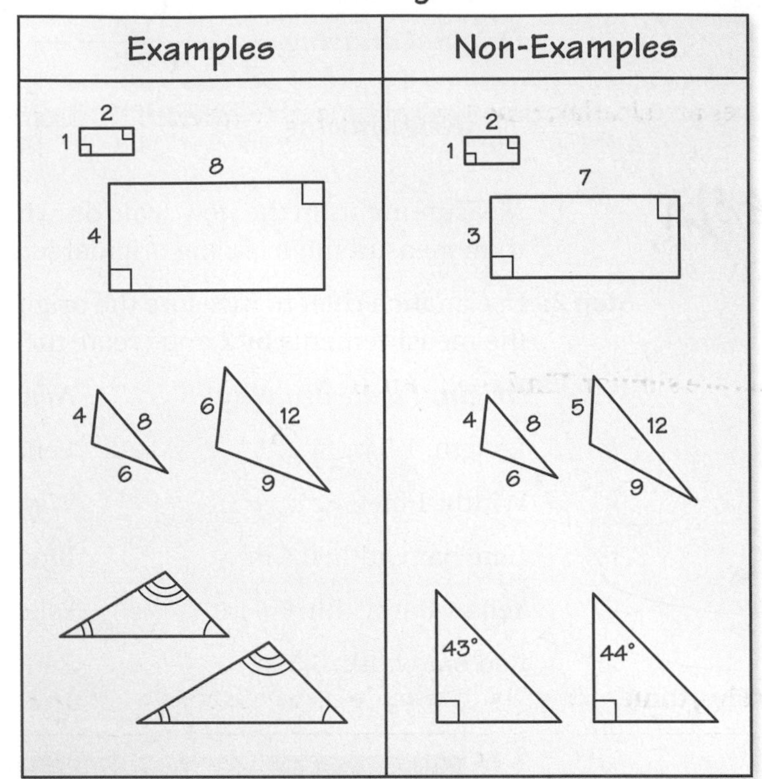

On Your Own

Make an example and non-example chart to help you study these topics.

1. corresponding angles
2. corresponding sides
3. perimeters of similar figures
4. areas of similar figures
5. indirect measurements
6. scale drawings

"I'm using an example and non-example chart for a talk on cat hygiene."

After you complete this chapter, make example and non-example charts for the following topics.

7. transformations

   a. translations **b.** reflections **c.** rotations

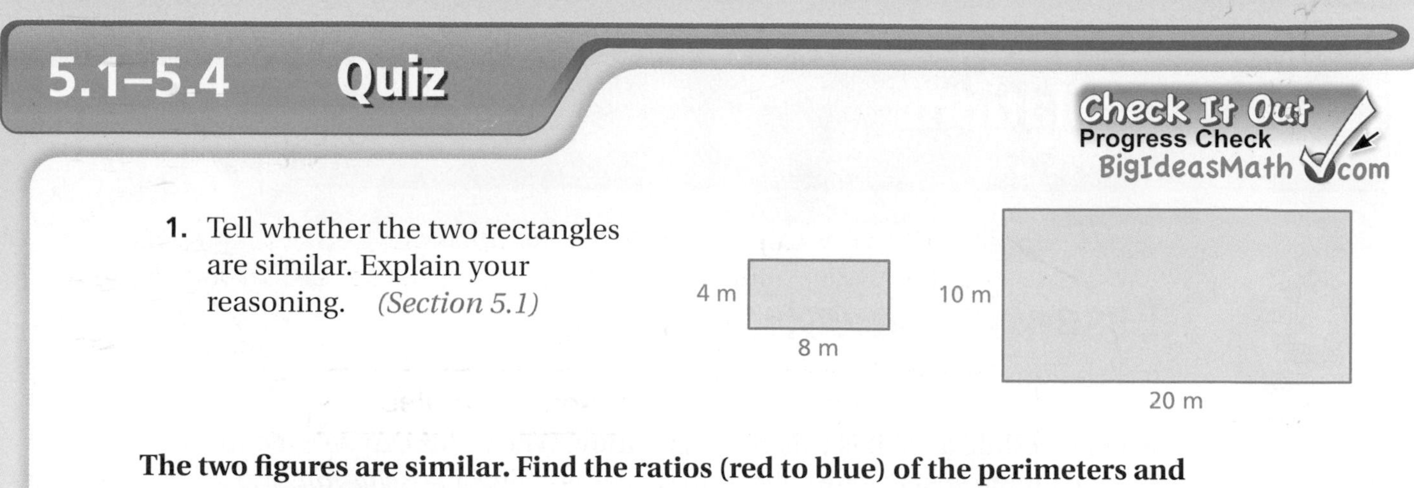

5.1–5.4 Quiz

1. Tell whether the two rectangles are similar. Explain your reasoning. *(Section 5.1)*

The two figures are similar. Find the ratios (red to blue) of the perimeters and of the areas. *(Section 5.2)*

2.

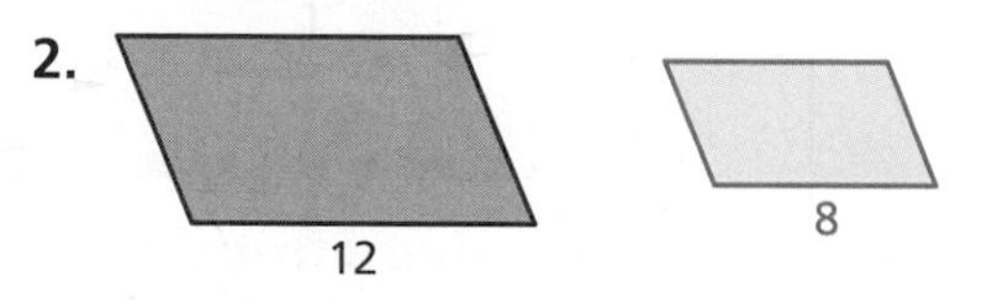

3.

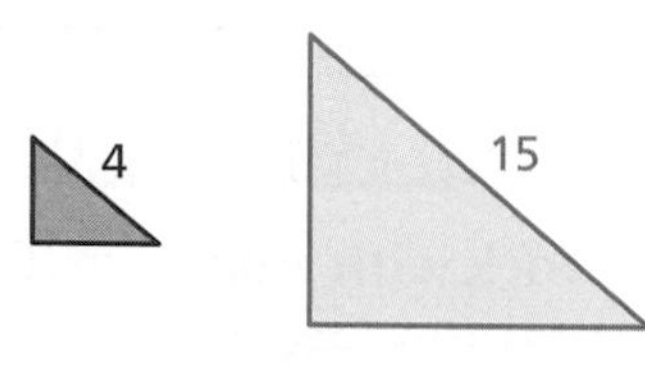

The polygons are similar. Find the value of x. *(Section 5.3)*

4.

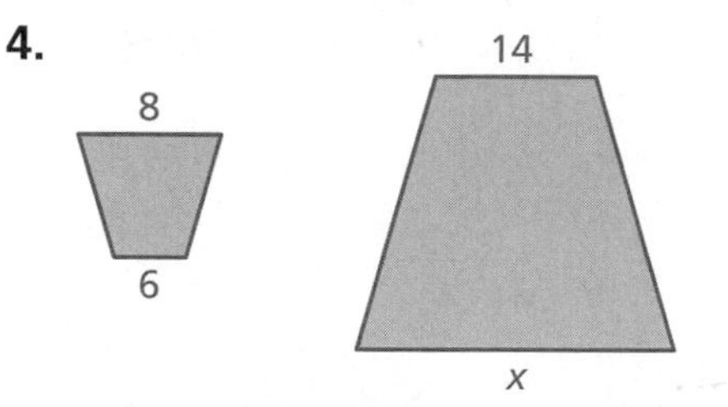

5.

Find the missing dimension. Use the scale factor 1 : 20. *(Section 5.4)*

	Item	Model	Actual
6.	Basketball Player	Height: ☐ in.	Height: 90 in.
7.	Dinosaur	Length: 3.75 ft	Length: ☐ ft

8. **POSTERS** The ratio of the corresponding side lengths of two similar posters is 7 in. : 20 in. What is the ratio of the perimeters? *(Section 5.2)*

9. **DOLPHIN** A dolphin in an aquarium is 12 feet long. A scale model of the dolphin is $3\frac{1}{2}$ inches long. What is the scale factor of the model? *(Section 5.4)*

10. **TENNIS COURT** The tennis courts for singles and doubles matches are different sizes. Are the courts similar? Explain. *(Section 5.1)*

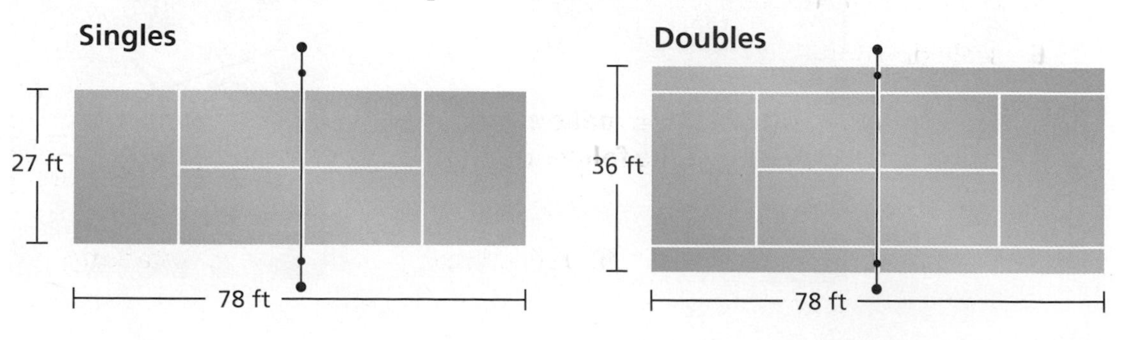

5.5 Translations

COMMON CORE STATE STANDARDS
7.G.2
8.G.1
8.G.3

Essential Question How can you use translations to make a tessellation?

Share Your
Work at...
My.BigIdeasMath.com

When you slide a tile it is called a **translation**. When tiles can be used to cover a floor with no empty spaces, the collection of tiles is called a *tessellation*.

1 ACTIVITY: Describing Tessellations

Work with a partner. Can you make the pattern by using a translation of single tiles that are all of the same shape and design? If so, show how.

a. Sample:

Tile Pattern

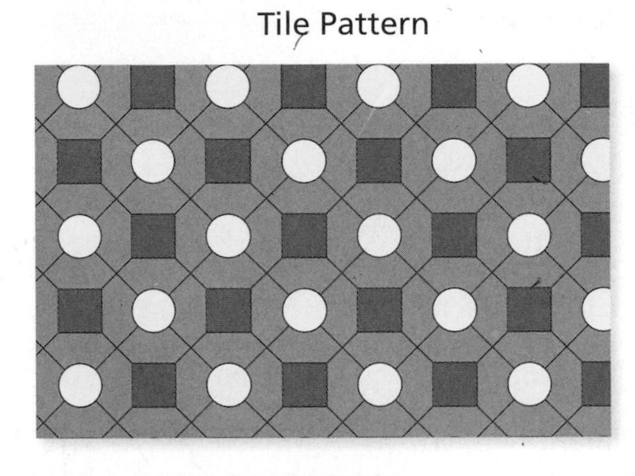

Single Tiles

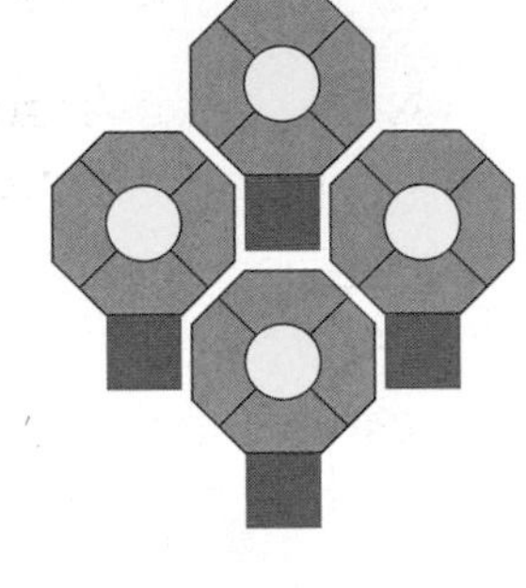

b.

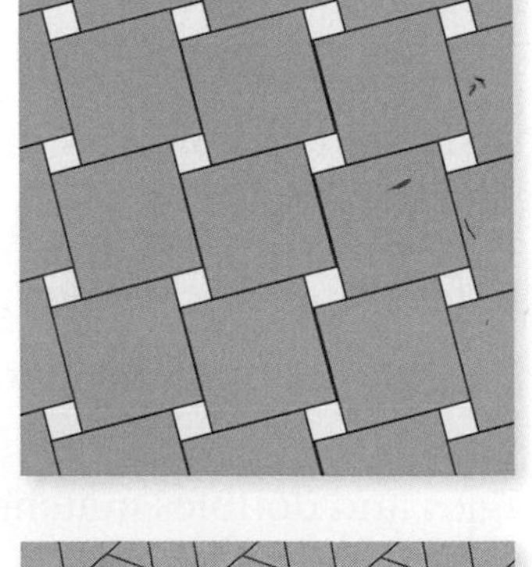

c.

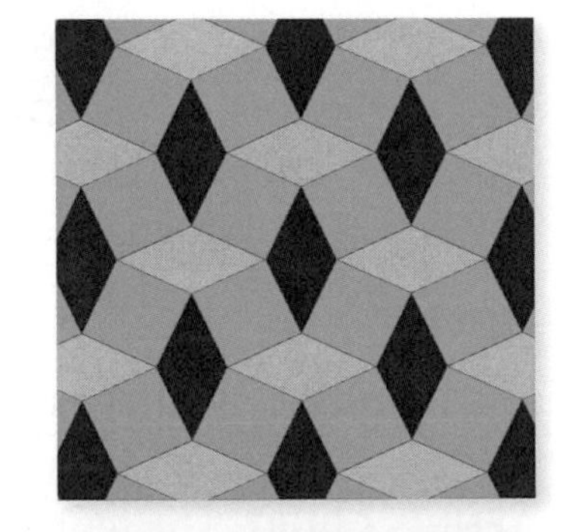

d.

e.

2 ACTIVITY: Tessellations and Basic Shapes

Work with a partner.

a. Which pattern blocks can you use to make a tessellation?

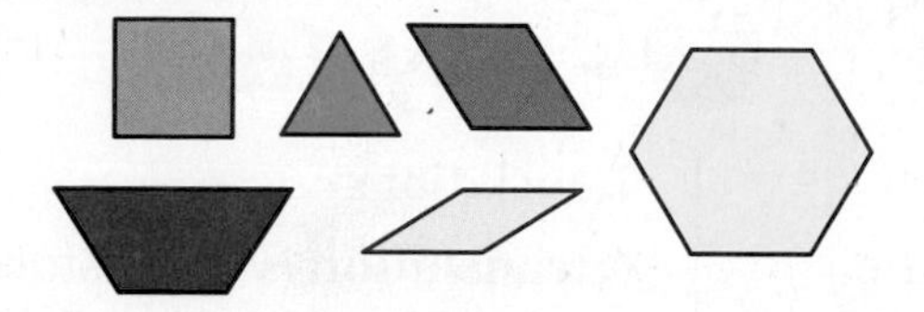

b. For each one that works, draw the tessellation.

c. Can you make the tessellation using only translation, or do you have to rotate or flip the pattern blocks?

3 ACTIVITY: Designing Tessellations

Work with a partner. Design your own tessellation. Use one of the basic shapes from Activity 2.

Sample:

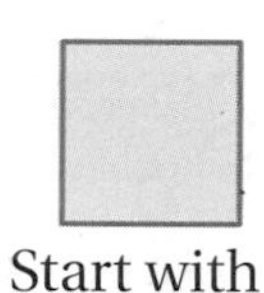

Start with a square.

Cut a design out of one side.

Tape it to the other side to make your pattern.

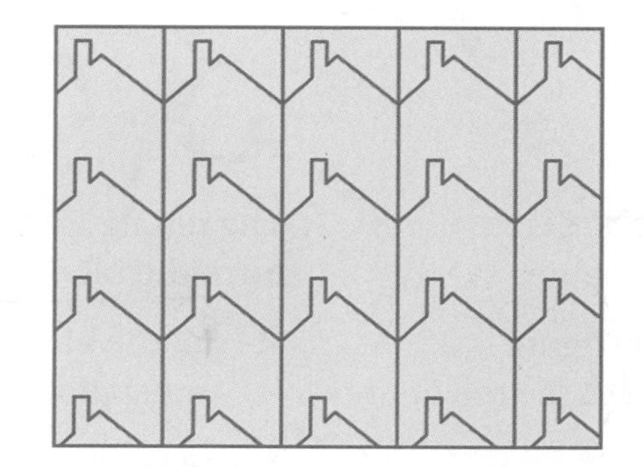

Use the pattern and translations to make your tessellation.

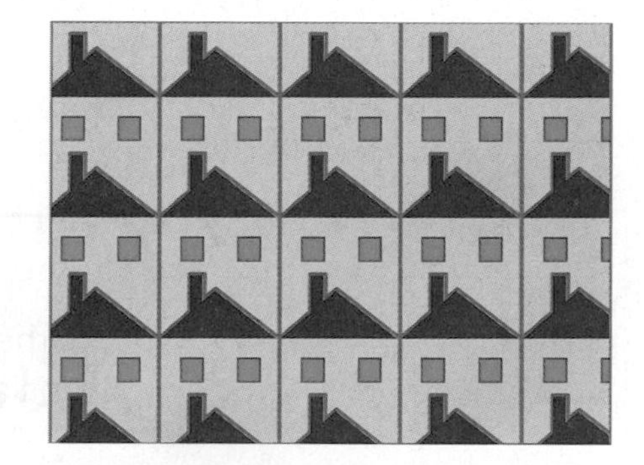

Color the tessellation.

What Is Your Answer?

4. **IN YOUR OWN WORDS** How can you use translations to make a tessellation? Give an example.

5. Draw any parallelogram. Does it tessellate? Is it true that any parallelogram can be translated to make a tessellation? Explain why.

Practice Use what you learned about translations to complete Exercises 4–6 on page 224.

5.5 Lesson

A **transformation** changes a figure into another figure. The new figure is called the **image**.

Key Vocabulary
transformation, *p. 222*
image, *p. 222*
translation, *p. 222*

Key Idea

Translations

A **translation** is a transformation in which a figure *slides* but does not turn. Every point of the figure moves the same distance and in the same direction.

The original figure and its image have the same size and shape.

EXAMPLE 1 Identifying a Translation

Tell whether the blue figure is a translation of the red figure.

a.

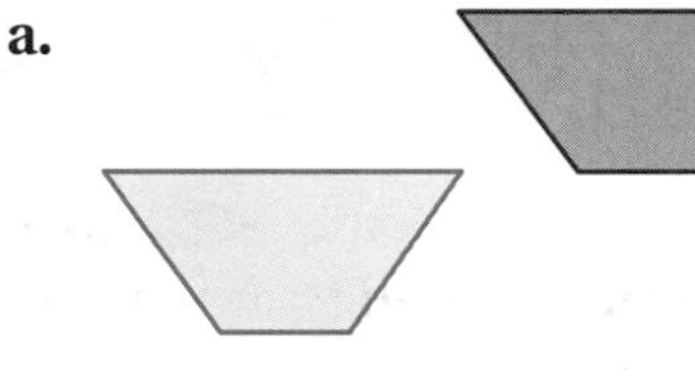

The red figure *slides* to form the blue figure.

So, the blue figure is a translation of the red figure.

b.

The red figure *turns* to form the blue figure.

So, the blue figure is *not* a translation of the red figure.

On Your Own

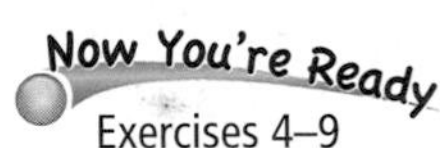

Tell whether the blue figure is a translation of the red figure. Explain.

1.

2.

3. 5 5

4. G G

EXAMPLE 2 Translating a Figure

Translate the red triangle 3 units right and 3 units down. What are the coordinates of the image?

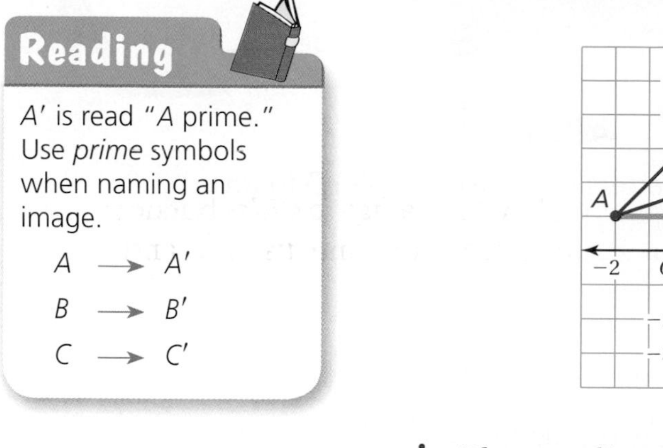

Reading
A′ is read "A prime."
Use prime symbols when naming an image.
A → A′
B → B′
C → C′

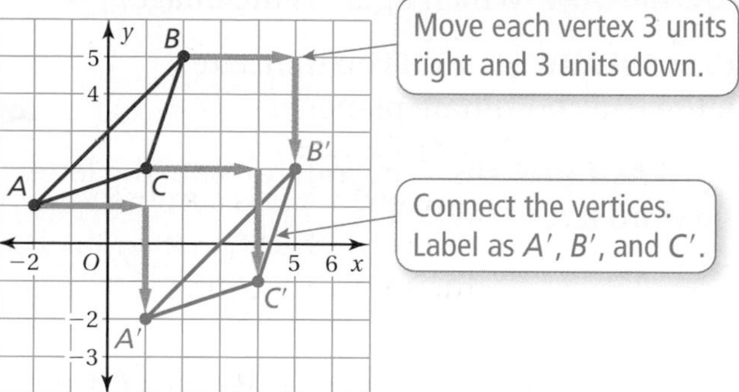

Connect the vertices. Label as A′, B′, and C′.

The coordinates of the image are $A'(1, -2)$, $B'(5, 2)$, and $C'(4, -1)$.

On Your Own

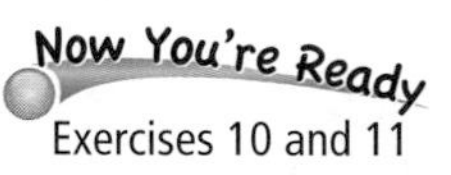

5. The red triangle is translated 4 units left and 2 units up. What are the coordinates of the image?

EXAMPLE 3 Translating a Figure

The vertices of a square are $A(1, -2)$, $B(3, -2)$, $C(3, -4)$, and $D(1, -4)$. Draw the figure and its image after a translation 4 units left and 6 units up.

Subtract 4 from each x-coordinate. Add 6 to each y-coordinate.

Vertices of $ABCD$	$(x - 4, y + 6)$	Vertices of $A'B'C'D'$
$A(1, -2)$	$(1 - 4, -2 + 6)$	$A'(-3, 4)$
$B(3, -2)$	$(3 - 4, -2 + 6)$	$B'(-1, 4)$
$C(3, -4)$	$(3 - 4, -4 + 6)$	$C'(-1, 2)$
$D(1, -4)$	$(1 - 4, -4 + 6)$	$D'(-3, 2)$

The figure and its image are shown at the right.

On Your Own

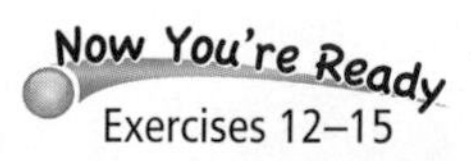

6. The vertices of a triangle are $A(-2, -2)$, $B(0, 2)$, and $C(3, 0)$. Draw the figure and its image after a translation 1 unit left and 2 units up.

5.5 Exercises

Vocabulary and Concept Check

1. **VOCABULARY** Which figure is the image?

2. **VOCABULARY** How do you translate a figure in a coordinate plane?

3. **CRITICAL THINKING** Can you translate the letters in the word TOKYO to form the word KYOTO? Explain.

Practice and Problem Solving

Tell whether the blue figure is a translation of the red figure.

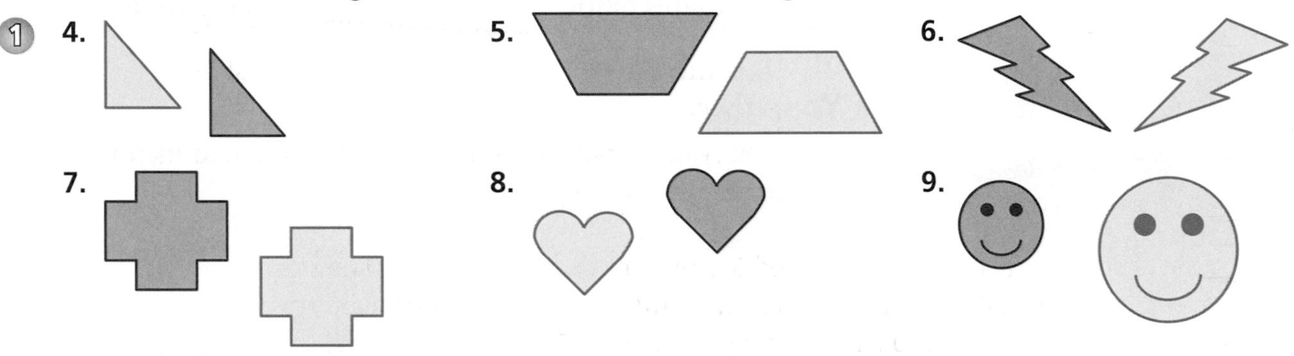

10. Translate the triangle 4 units right and 3 units down. What are the coordinates of the image?

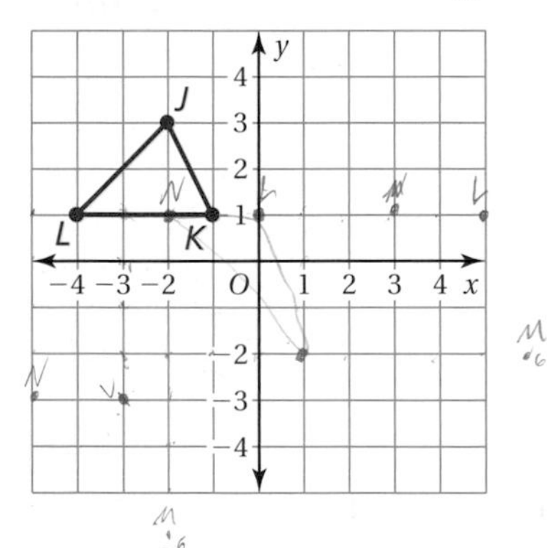

11. Translate the figure 2 units left and 4 units down. What are the coordinates of the image?

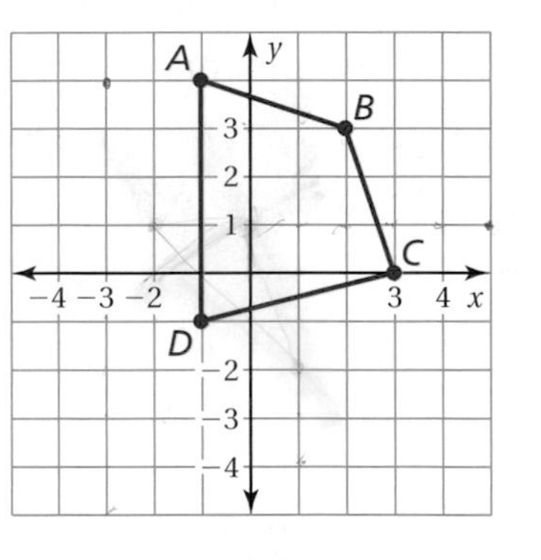

The vertices of a triangle are $L(0, 1)$, $M(1, -2)$, and $N(-2, 1)$. Draw the figure and its image after the translation.

12. 1 unit left and 6 units up

13. 5 units right

14. 2 units right and 3 units up

15. 3 units left and 4 units down

16. **ICONS** You can click and drag an icon on a computer screen. Is this an example of a translation? Explain.

Describe the translation of the point to its image.

17. $(3, -2) \rightarrow (1, 0)$

18. $(-8, -4) \rightarrow (-3, 5)$

Describe the translation from the red figure to the blue figure.

19.

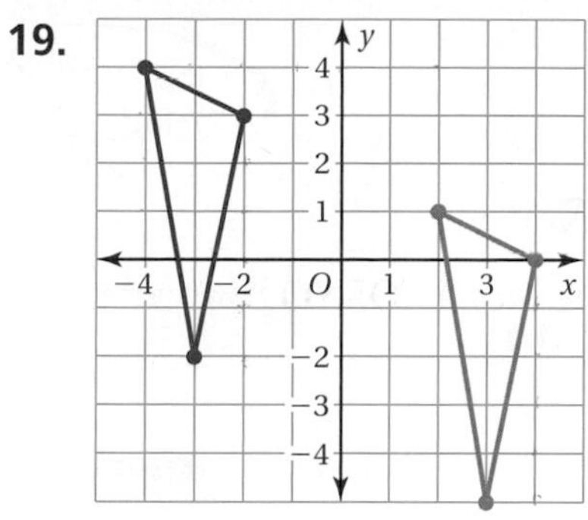

20.

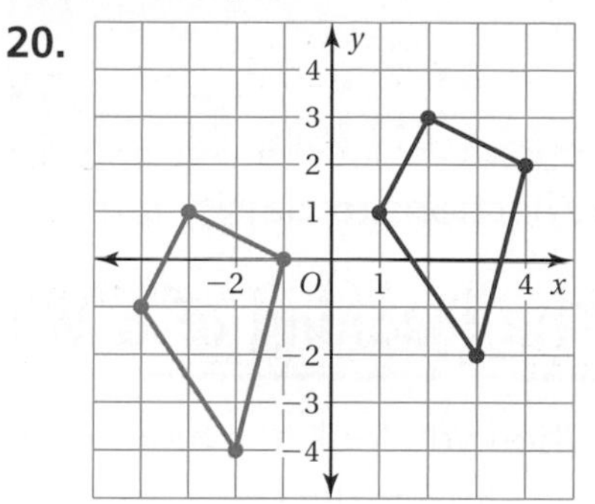

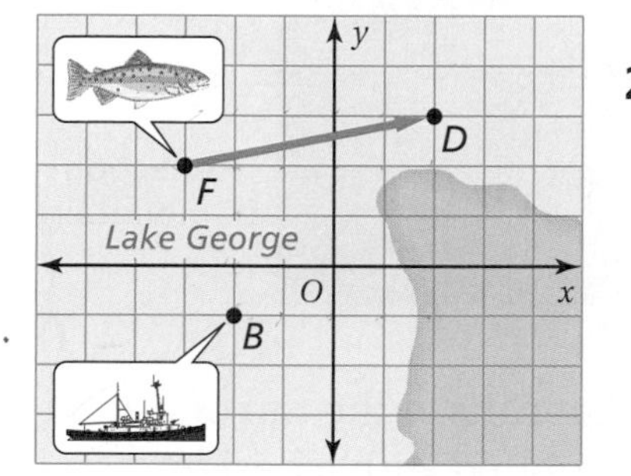

21. FISHING A school of fish translates from point F to point D.

a. Describe the translation of the school of fish.

b. Can the fishing boat make a similar translation? Explain.

c. Describe a translation the fishing boat could make to get to point D.

22. REASONING A triangle is translated 5 units right and 2 units up. Then the image is translated 3 units left and 8 units down. Write a translation of the original triangle to the ending position.

23. Critical Thinking In chess, a knight can move only in an L-shape pattern:

- *two* vertical squares then *one* horizontal square;
- *two* horizontal squares then *one* vertical square;
- *one* vertical square then *two* horizontal squares; or
- *one* horizontal square then *two* vertical squares.

Write a series of translations to move the knight from g8 to g5.

Fair Game Review What you learned in previous grades & lessons

Tell whether each figure can be folded in half so that one side matches the other. *(Skills Review Handbook)*

24.

25.

26.

27.

28. MULTIPLE CHOICE You put \$550 in an account that earns 4.4% simple interest per year. How much interest do you earn in 6 months? *(Section 4.4)*

Ⓐ \$1.21 Ⓑ \$12.10 Ⓒ \$121.00 Ⓓ \$145.20

5.6 Reflections

COMMON CORE STATE STANDARDS
7.G.2
8.G.1
8.G.3

Essential Question

How can you use reflections to classify a frieze pattern?

The Meaning of a Word ● Reflection

When you look at a mountain by a lake, you can see the **reflection**, or mirror image, of the mountain in the lake.

If you fold the photo on its axis, the mountain and its reflection will align.

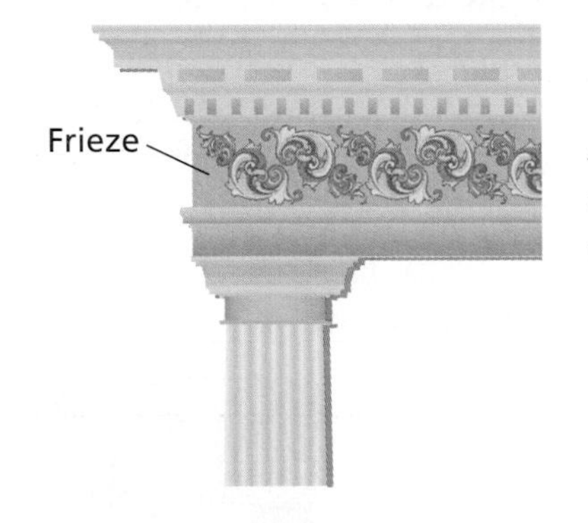

A *frieze* is a horizontal band that runs at the top of a building. A frieze is often decorated with a design that repeats.

- All frieze patterns are translations of themselves.
- Some frieze patterns are reflections of themselves.

1 EXAMPLE: Frieze Patterns

Is the frieze pattern a reflection of itself when folded horizontally? vertically?

- Fold (reflect) on horizontal axis. The pattern coincides.

- Fold (reflect) on vertical axis. The pattern coincides.

∴ This frieze pattern is a reflection of itself when folded horizontally *and* vertically.

2 ACTIVITY: Frieze Patterns and Reflections

Work with a partner. Is the frieze pattern a reflection of itself when folded *horizontally*, *vertically*, or *neither*?

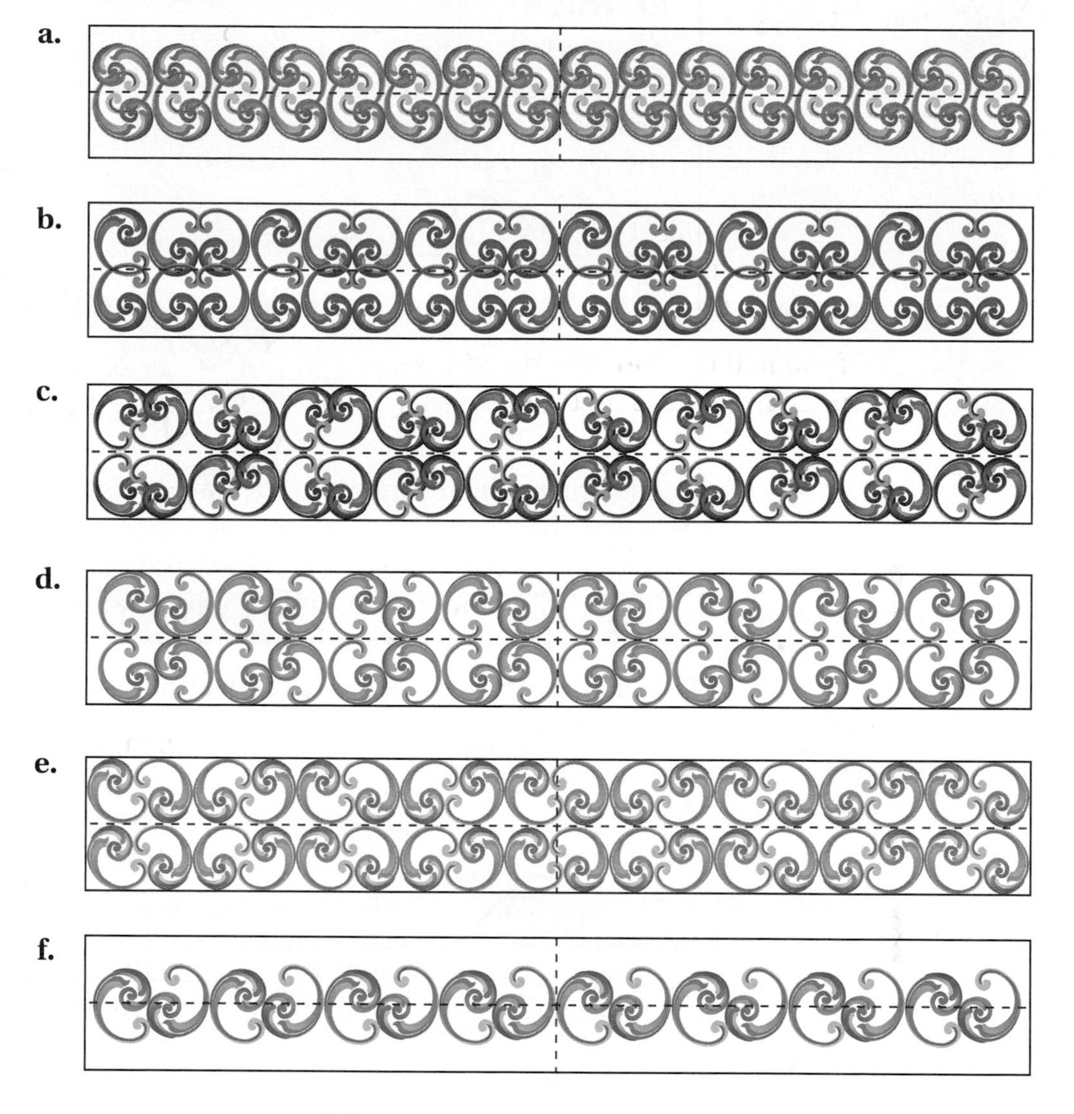

What Is Your Answer?

3. Draw a frieze pattern that is a reflection of itself when folded horizontally.
4. Draw a frieze pattern that is a reflection of itself when folded vertically.
5. Draw a frieze pattern that is not a reflection of itself when folded horizontally or vertically.
6. **IN YOUR OWN WORDS** How can you use reflections to classify a frieze pattern?

Practice

Use what you learned about reflections to complete Exercises 4–6 on page 230.

Check It Out
Lesson Tutorials
BigIdeasMath.com

Key Vocabulary
reflection, *p. 228*
line of reflection, *p. 228*

Reflections

A **reflection**, or flip, is a transformation in which a figure is reflected in a line called the **line of reflection**. A reflection creates a mirror image of the original figure.

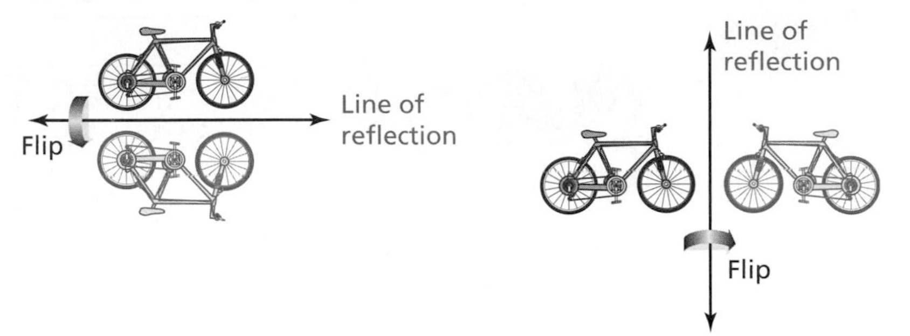

The original figure and its image have the same size and shape.

EXAMPLE 1 Identifying a Reflection

Tell whether the blue figure is a reflection of the red figure.

a.

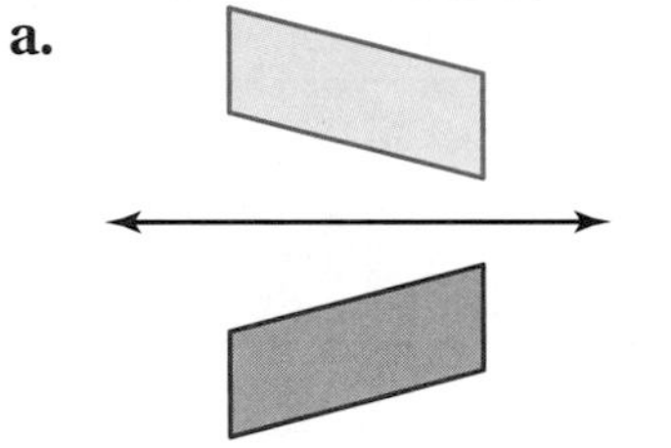

The red figure can be *flipped* to form the blue figure.

∴ So, the blue figure is a reflection of the red figure.

b.

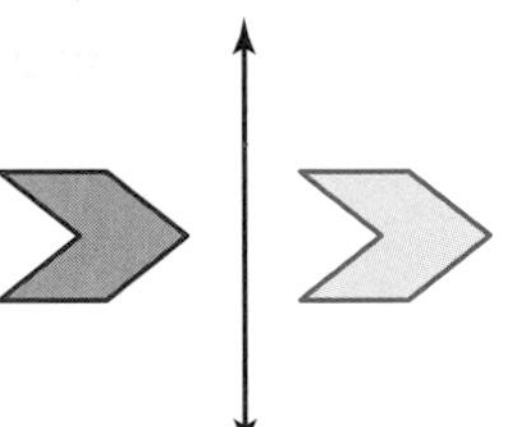

If the red figure were *flipped*, it would point to the left.

∴ So, the blue figure is *not* a reflection of the red figure.

On Your Own

Now You're Ready
Exercises 4–9

Tell whether the blue figure is a reflection of the red figure. Explain.

1.

2.

3.

EXAMPLE 2 Reflecting a Figure in the x-axis

The vertices of a triangle are $A(-1, 1)$, $B(2, 3)$, and $C(6, 3)$. Draw this triangle and its reflection in the x-axis. What are the coordinates of the image?

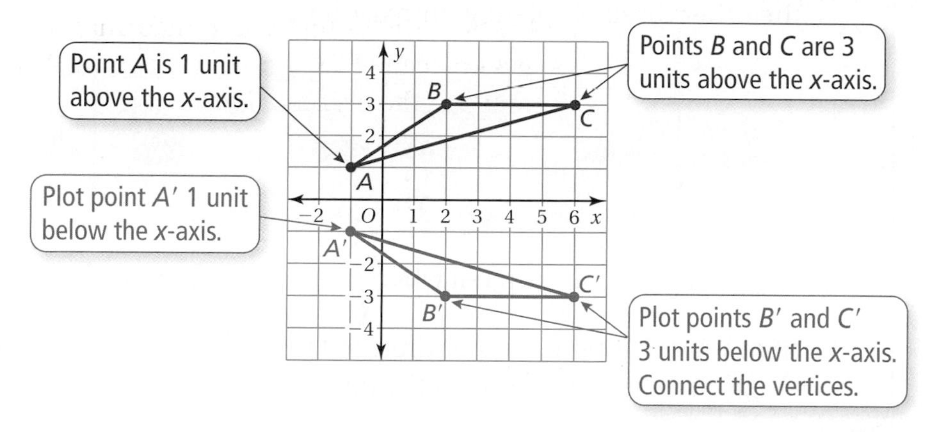

Point A is 1 unit above the x-axis.
Points B and C are 3 units above the x-axis.
Plot point A′ 1 unit below the x-axis.
Plot points B′ and C′ 3 units below the x-axis. Connect the vertices.

∴ The coordinates of the image are $A'(-1, -1)$, $B'(2, -3)$, and $C'(6, -3)$.

EXAMPLE 3 Reflecting a Figure in the y-axis

The vertices of a quadrilateral are $P(-2, 5)$, $Q(-1, -1)$, $R(-4, 2)$, and $S(-4, 4)$. Draw this quadrilateral and its reflection in the y-axis. What are the coordinates of the image?

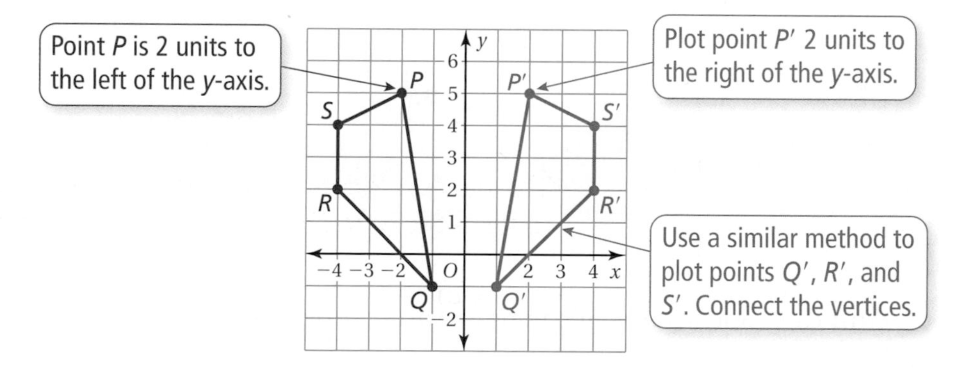

Point P is 2 units to the left of the y-axis.
Plot point P′ 2 units to the right of the y-axis.
Use a similar method to plot points Q′, R′, and S′. Connect the vertices.

∴ The coordinates of the image are $P'(2, 5)$, $Q'(1, -1)$, $R'(4, 2)$, and $S'(4, 4)$.

On Your Own

Now You're Ready
Exercises 10–17

4. The vertices of a rectangle are $A(-4, -3)$, $B(-4, -1)$, $C(-1, -1)$, and $D(-1, -3)$.

   a. Draw the rectangle and its reflection in the x-axis.

   b. Draw the rectangle and its reflection in the y-axis.

   c. Are the images in parts (a) and (b) the same size and shape? Explain.

Vocabulary and Concept Check

1. **WHICH ONE DOESN'T BELONG?** Which transformation does *not* belong with the other three? Explain your reasoning.

2. **WRITING** How can you tell when one figure is a reflection of another figure?

3. **REASONING** A figure lies entirely in Quadrant I. The figure is reflected in the x-axis. In which quadrant is the image?

Practice and Problem Solving

Tell whether the blue figure is a reflection of the red figure.

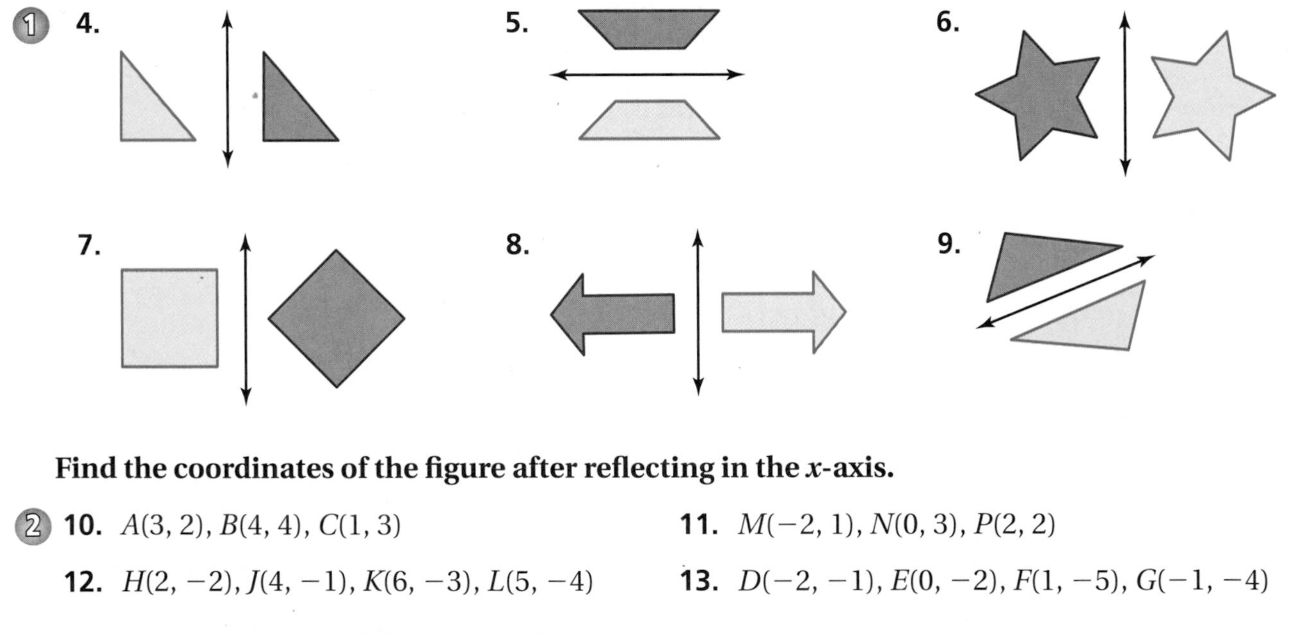

Find the coordinates of the figure after reflecting in the x-axis.

10. $A(3, 2)$, $B(4, 4)$, $C(1, 3)$
11. $M(-2, 1)$, $N(0, 3)$, $P(2, 2)$
12. $H(2, -2)$, $J(4, -1)$, $K(6, -3)$, $L(5, -4)$
13. $D(-2, -1)$, $E(0, -2)$, $F(1, -5)$, $G(-1, -4)$

Find the coordinates of the figure after reflecting in the y-axis.

14. $Q(-4, 2)$, $R(-2, 4)$, $S(-1, 1)$
15. $T(1, -1)$, $U(4, 2)$, $V(6, -2)$
16. $W(2, -1)$, $X(5, -2)$, $Y(5, -5)$, $Z(2, -4)$
17. $J(2, 2)$, $K(7, 4)$, $L(9, -2)$, $M(3, -1)$

18. **ALPHABET** Which letters look the same when reflected in the line?

A B C D E F G H I J K L M N O P Q R S T U V W X Y Z

The coordinates of a point and its image are given. Is the reflection in the *x-axis* or *y-axis*?

19. $(2, -2) \rightarrow (2, 2)$

20. $(-4, 1) \rightarrow (4, 1)$

21. $(-2, -5) \rightarrow (2, -5)$

22. $(-3, -4) \rightarrow (-3, 4)$

23. Translate the triangle 1 unit right and 5 units down. Then reflect the image in the y-axis.

24. CHOOSE TOOLS Use a computer drawing program to create photographs of people by copying one side of the person's face and reflecting it in a vertical line. Does the person look normal or very different?

25. MIRROR IMAGE One of the faces shown is an exact reflection of itself. Which one is it? How can you tell?

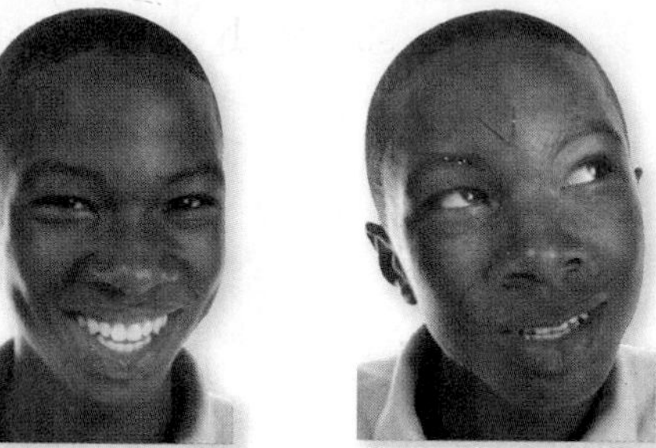

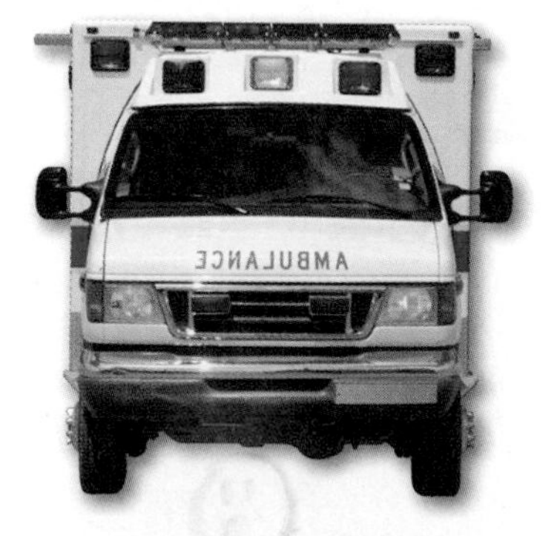

26. EMERGENCY VEHICLE Hold a mirror to the left side of the photo of the vehicle.

a. What word do you see in the mirror?

b. Why do you think it is written that way on the front of the vehicle?

27. Critical Thinking Reflect the triangle in the line $y = x$. How are the x- and y-coordinates of the image related to the x- and y-coordinates of the original triangle?

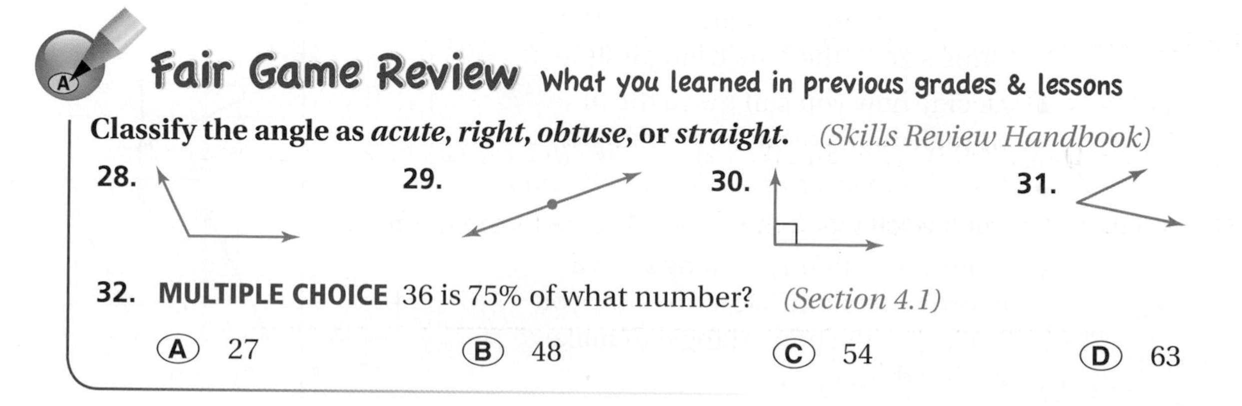

Fair Game Review What you learned in previous grades & lessons

Classify the angle as *acute, right, obtuse,* or *straight.* *(Skills Review Handbook)*

28.

29.

30.

31.

32. MULTIPLE CHOICE 36 is 75% of what number? *(Section 4.1)*

Ⓐ 27 Ⓑ 48 Ⓒ 54 Ⓓ 63

5.7 Rotations

COMMON CORE STATE STANDARDS
7.G.2
8.G.1
8.G.2
8.G.3

Essential Question What are the three basic ways to move an object in a plane?

The Meaning of a Word ● Rotate

A bicycle wheel can **rotate** clockwise or counterclockwise.

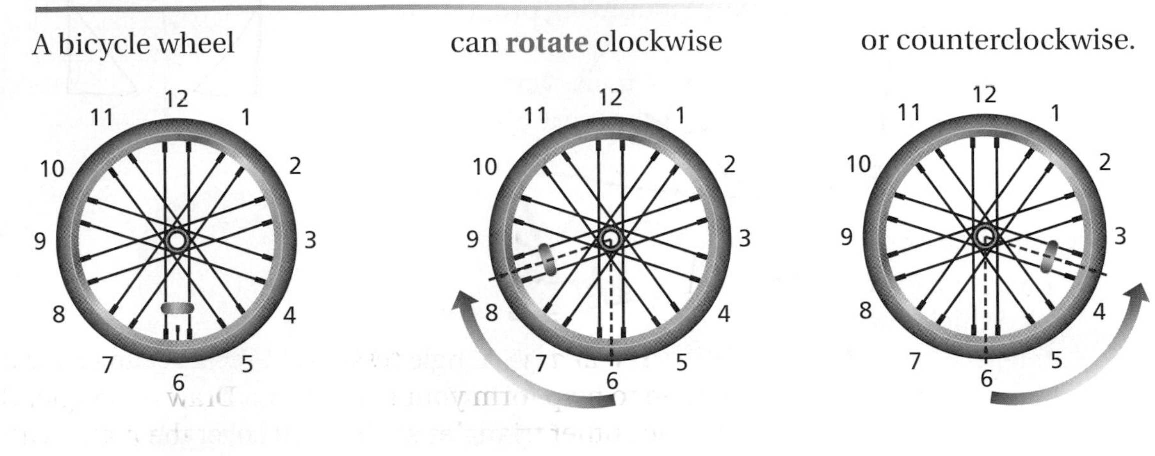

1 ACTIVITY: Three Basic Ways to Move Things

There are three basic ways to move objects on a flat surface.

1. Translate the object. **2.** Reflect the object. **3.** Rotate the object.

Work with a partner.

- Cut out a paper triangle that is the same size as the blue triangle shown.
- Decide how you can move the blue triangle to make each red triangle.
- Is each move a *translation*, a *reflection*, or a *rotation*?
- Draw four other red triangles in a coordinate plane. Describe how you can move the blue triangle to make each red triangle.

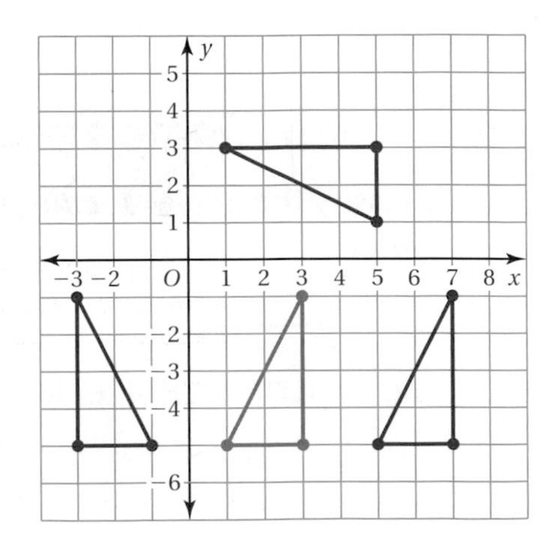

2 ACTIVITY: Tessellating a Plane

Work with a partner.

a. Describe how the figure labeled 1 in each diagram can be moved to make the other figures.

Triangles

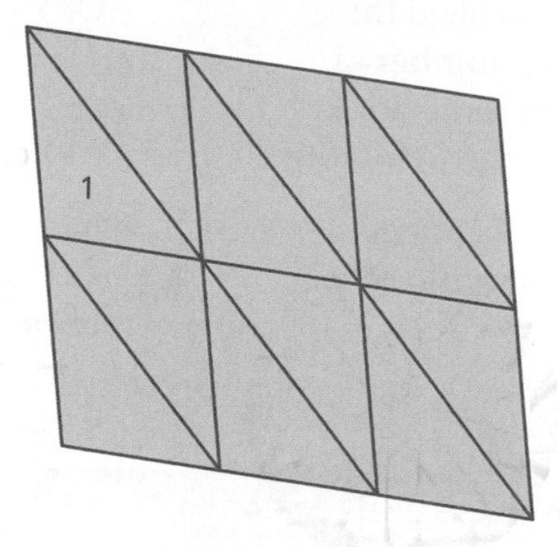

Quadrilaterals

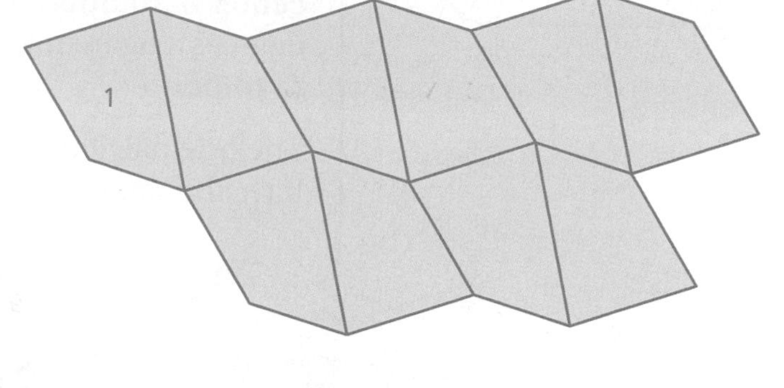

b. **EXPERIMENT** Will *any* triangle tessellate? Conduct an experiment to gather information to help form your conclusion. Draw a triangle. Cut it out. Then use it to trace other triangles so that you cover the plane with triangles that are all the same shape.

c. **EXPERIMENT** Will *any* quadrilateral tessellate? Conduct an experiment to gather information to help form your conclusion. Draw a quadrilateral. Cut it out. Then use it to trace other quadrilaterals so that you cover the plane with quadrilaterals that are all the same shape.

What Is Your Answer?

3. **IN YOUR OWN WORDS** What are the three basic ways to move an object in a plane? Draw an example of each.

"Dear Sub Shop: Why do you put the cheese on the subs so some parts have double coverage and some have none?"

"My suggestion is that you use the tessellation property of triangles for even cheese coverage."

Use what you learned about rotations to complete Exercises 7–9 on page 236.

5.7 Lesson

Key Vocabulary
rotation, *p. 234*
center of rotation, *p. 234*
angle of rotation, *p. 234*

Key Idea

Rotations

A **rotation**, or *turn*, is a transformation in which a figure is rotated about a point called the **center of rotation**. The number of degrees a figure rotates is the **angle of rotation**.

The original figure and its image have the same size and shape.

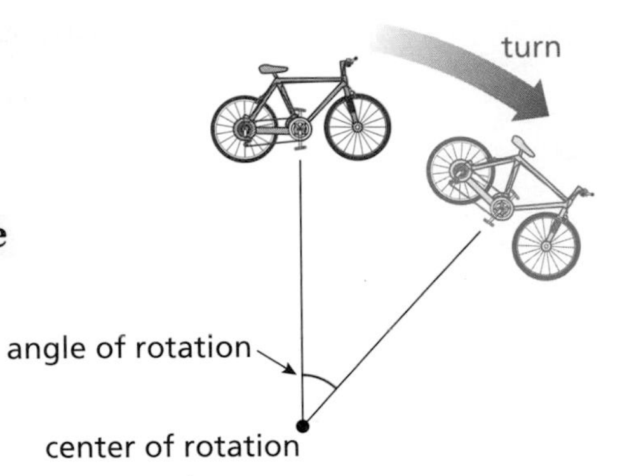

EXAMPLE 1 Standardized Test Practice

You must rotate the puzzle piece 270° clockwise about point P to fit it into a puzzle. Which piece fits in the puzzle as shown?

Ⓐ Ⓑ Ⓒ Ⓓ

Rotate the puzzle piece 270° clockwise about point P.

turn 270°

P

∴ The correct answer is Ⓒ.

Study Tip

When rotating figures, it may help to sketch the rotation in several steps, as shown in Example 1.

On Your Own

Now You're Ready
Exercises 7–12

1. Which piece is a 90° counterclockwise rotation about point P?
2. Is choice D a rotation of the original puzzle piece? If not, what kind of transformation does the image show?

EXAMPLE 2 Rotating a Figure

The vertices of a trapezoid are $W(-4, 2)$, $X(-3, 4)$, $Y(-1, 4)$, and $Z(-1, 2)$. Rotate the trapezoid 180° clockwise about the origin. What are the coordinates of the image?

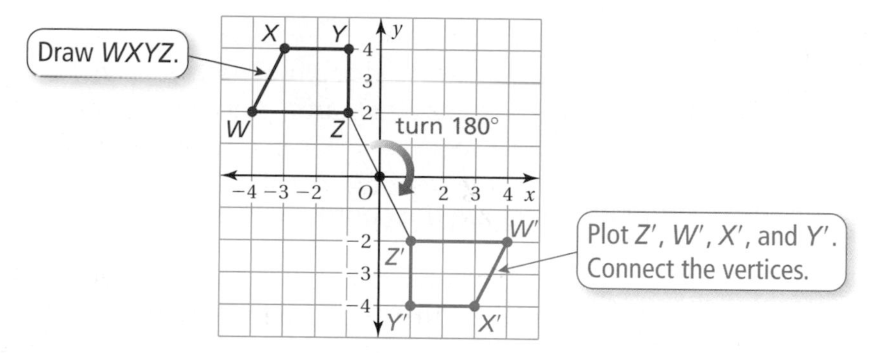
Draw WXYZ.
turn 180°
Plot Z′, W′, X′, and Y′. Connect the vertices.

∴ The coordinates of the image are $W'(4, -2)$, $X'(3, -4)$, $Y'(1, -4)$, and $Z'(1, -2)$.

EXAMPLE 3 Rotating a Figure

The vertices of a triangle are $J(1, 2)$, $K(4, 2)$, and $L(1, -3)$. Rotate the triangle 90° counterclockwise about vertex L. What are the coordinates of the image?

Common Error

Be sure to pay attention to whether a rotation is clockwise or counterclockwise.

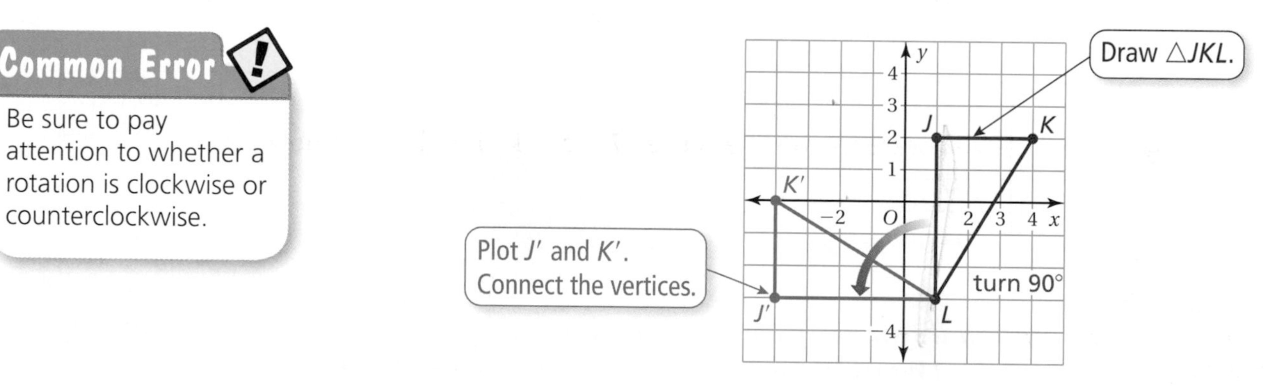
Draw △JKL.
turn 90°
Plot J′ and K′. Connect the vertices.

∴ The coordinates of the image are $J'(-4, -3)$, $K'(-4, 0)$, and $L'(1, -3)$.

On Your Own

Now You're Ready
Exercises 13–16

3. A triangle has vertices $Q(4, 5)$, $R(4, 0)$, and $S(1, 0)$.
   a. Rotate the triangle 90° counterclockwise about the origin.
   b. Rotate the triangle 180° about vertex S.
   c. Are the images in parts (a) and (b) the same size and shape? Explain.

5.7 Exercises

Check It Out
Help with Homework
BigIdeasMath.com

Vocabulary and Concept Check

1. **VOCABULARY** Identify the transformation shown.

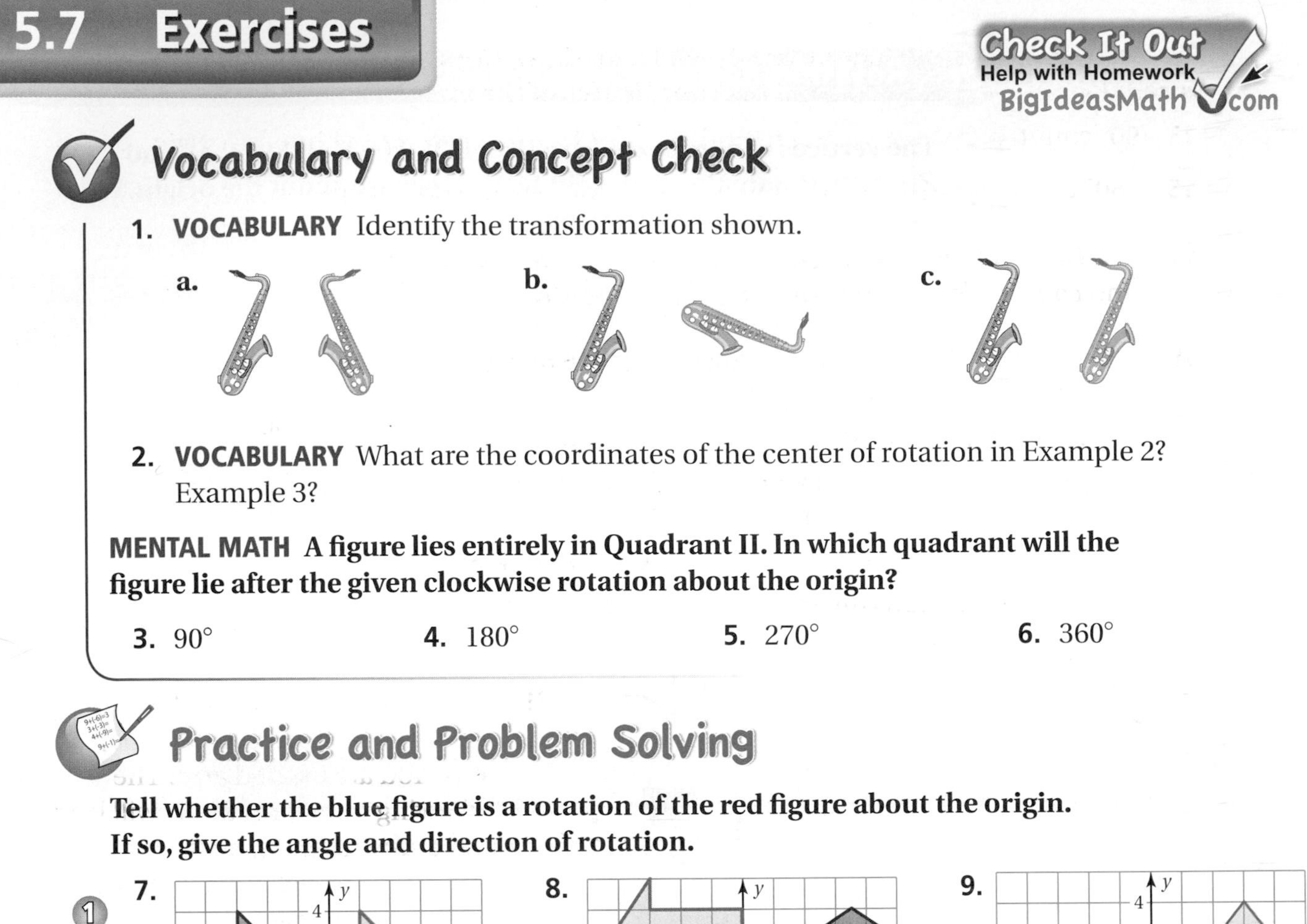

a. b. c.

2. **VOCABULARY** What are the coordinates of the center of rotation in Example 2? Example 3?

MENTAL MATH A figure lies entirely in Quadrant II. In which quadrant will the figure lie after the given clockwise rotation about the origin?

3. 90°
4. 180°
5. 270°
6. 360°

Practice and Problem Solving

Tell whether the blue figure is a rotation of the red figure about the origin. If so, give the angle and direction of rotation.

① 7.

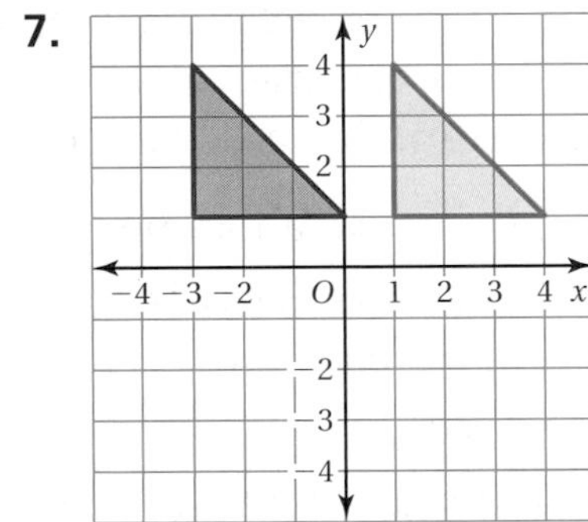

8.

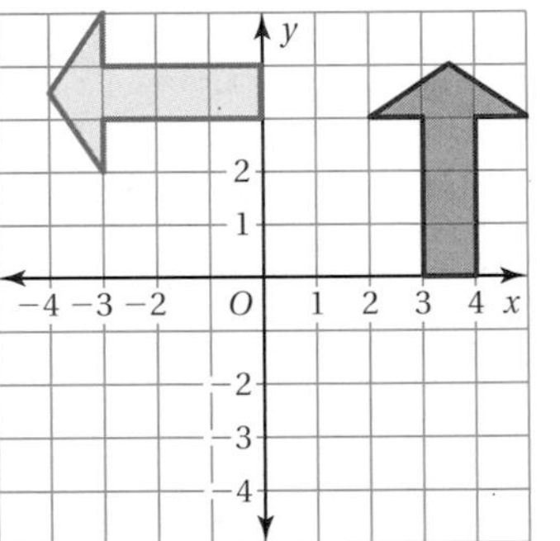

9.

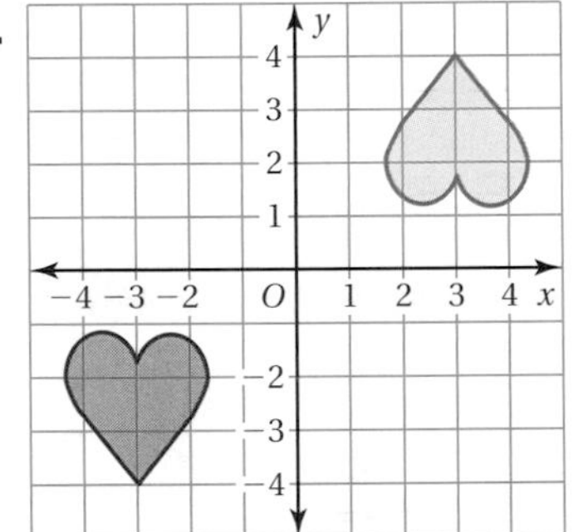

A figure has *rotational symmetry* if a rotation of 180° or less produces an image that fits exactly on the original figure. Explain why the figure has rotational symmetry.

10. 11. 12.

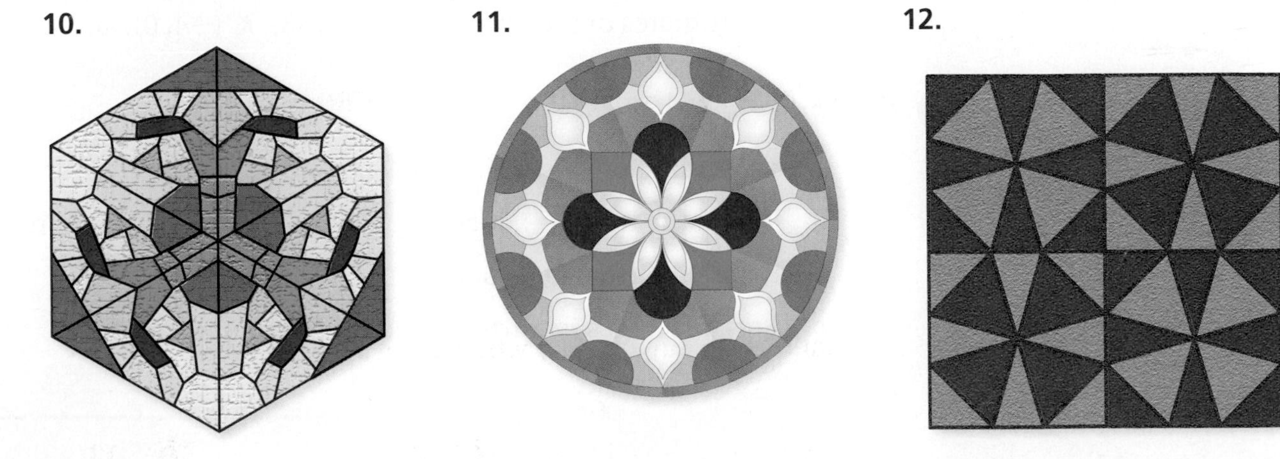

The vertices of a parallelogram are $A(-4, 1)$, $B(-3, 4)$, $C(-1, 4)$, and $D(-2, 1)$. Rotate the parallelogram as described. Find the coordinates of the image.

2 3 **13.** 90° counterclockwise about the origin

14. 270° clockwise about the origin

15. 180° clockwise about vertex D

16. 90° counterclockwise about vertex B

17. WRITING Why is it *not* necessary to use the words *clockwise* and *counterclockwise* when describing a rotation of 180°?

18. DILATIONS A *dilation* is a transformation in which a figure is enlarged or reduced.

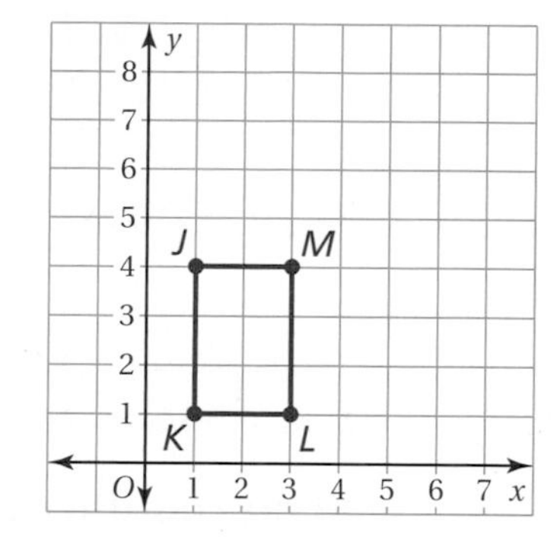

a. Dilate Rectangle $JKLM$ by multiplying the x- and y-coordinates of each vertex by 2. Compare the original figure and its image.

b. Are the rectangles identical? Are they similar? Explain.

c. How do dilations differ from translations, reflections, and rotations?

19. TREASURE MAP You want to find the treasure located on the map at ×. You are located at ●. The following transformations will lead you to the treasure, but they are not in the correct order. Find the correct order. Use each transformation exactly once.

- Rotate 180° about the origin.
- Reflect in the y-axis.
- Rotate 90° counterclockwise about the origin.
- Translate 1 unit right and 1 unit up.

20. **Reasoning** A triangle is rotated 90° counterclockwise about the origin. Its image is translated 1 unit left and 2 units down. The vertices of the final triangle are $(-5, 0)$, $(-2, 2)$, and $(-2, -1)$. What are the vertices of the original triangle?

Fair Game Review What you learned in previous grades & lessons

Identify the solid. *(Skills Review Handbook)*

21.

22.

23. MULTIPLE CHOICE What is the value of $x - y$ when $x = -5$ and $y = -8$? *(Section 1.3)*

Ⓐ -13 Ⓑ -3 Ⓒ 3 Ⓓ 13

5.5–5.7 Quiz

1. Translate the triangle 2 units right and 3 units down. What are the coordinates of the image? *(Section 5.5)*

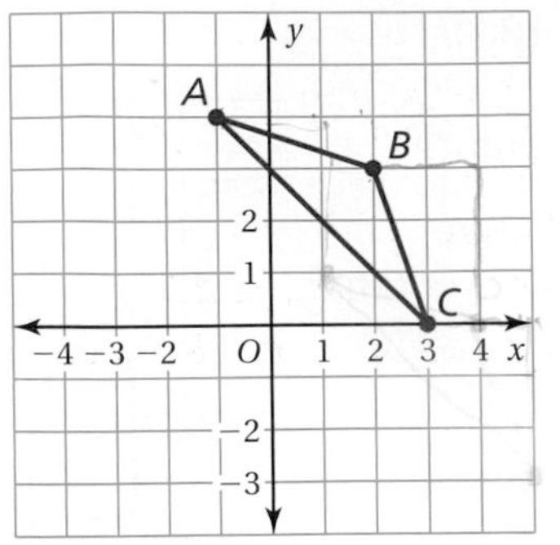

2. Translate the figure 2 units left and 4 units down. What are the coordinates of the image? *(Section 5.5)*

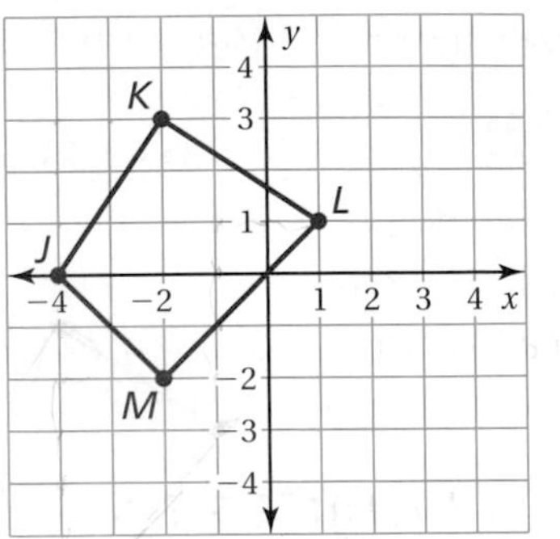

Find the coordinates of the figure after reflecting in the (a) x-axis and (b) y-axis. *(Section 5.6)*

3. $A(2, 0)$, $B(1, 5)$, $C(4, 3)$

4. $D(-2, -5)$, $E(-2, -2)$, $F(1, -2)$, $G(2, -5)$

Tell whether the blue figure is a rotation of the red figure about the origin. If so, give the angle and direction of rotation. *(Section 5.7)*

5.

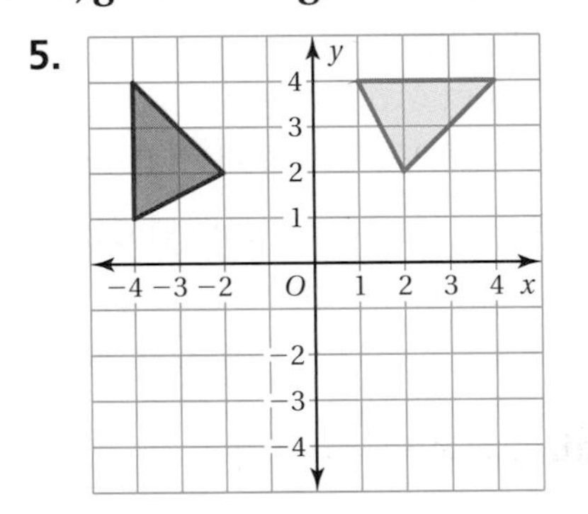

6.

7. AIRPLANE Describe a translation of the airplane from point A to point B. *(Section 5.5)*

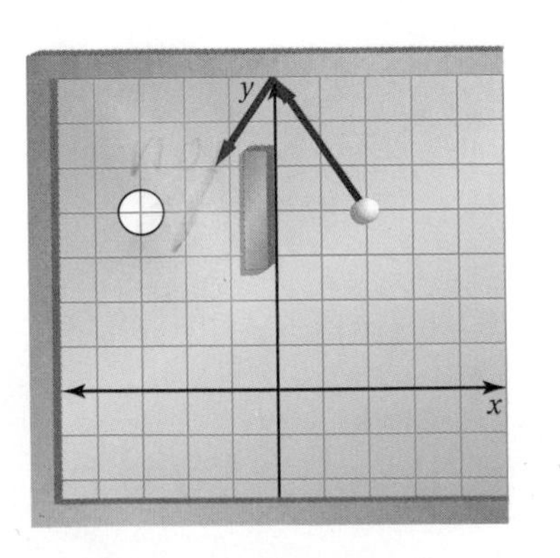

8. MINI GOLF You plan to hit the golf ball along the red path so that its image will be a reflection in the y-axis. Does the golf ball land in the hole? Explain. *(Section 5.6)*

9. GEOMETRY The pivot point of a compass is at the origin. A circle is drawn starting at (3, 6). What point is the compass pencil on when the compass has rotated 270° counterclockwise? *(Section 5.7)*

5 Chapter Review

Review Key Vocabulary

similar figures, *p. 196*
corresponding angles, *p. 196*
corresponding sides, *p. 196*
indirect measurement, *p. 209*
scale drawing, *p. 214*
scale model, *p. 214*
scale, *p. 214*
scale factor, *p. 215*
transformation, *p. 222*
image, *p. 222*
translation, *p. 222*
reflection, *p. 228*
line of reflection, *p. 228*
rotation, *p. 234*
center of rotation, *p. 234*
angle of rotation, *p. 234*

Review Examples and Exercises

5.1 Identifying Similar Figures *(pp. 194–199)*

Is Rectangle A similar to Rectangle B?

Each figure is a rectangle. So, corresponding angles have the same measure. Check to see if corresponding side lengths are proportional.

Rectangle A: 10 by 4. Rectangle B: 5 by 2.

$$\frac{\text{Length of A}}{\text{Length of B}} = \frac{10}{5} = 2 \qquad \frac{\text{Width of A}}{\text{Width of B}} = \frac{4}{2} = 2$$

Proportional

∴ So, Rectangle A is similar to Rectangle B.

Exercises

Tell whether the two figures are similar. Explain your reasoning.

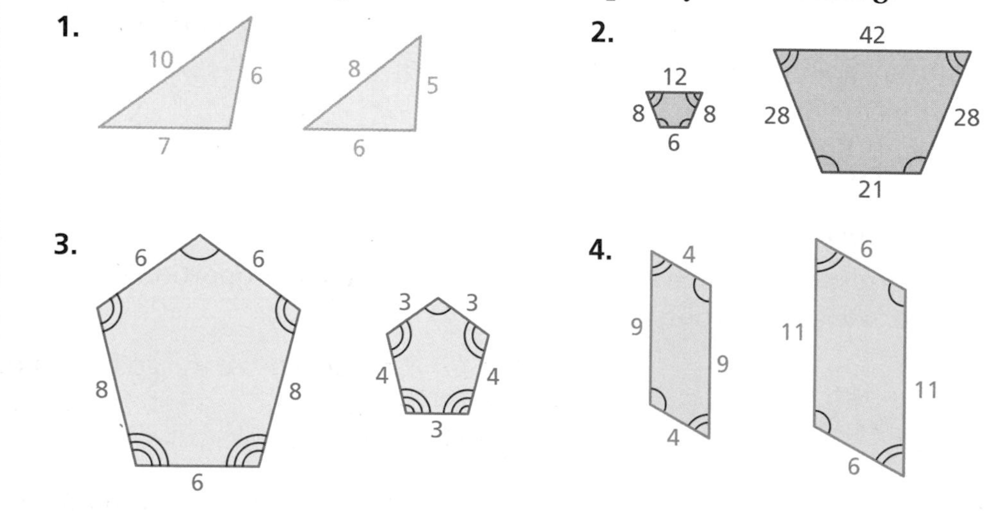

5.2 Perimeters and Areas of Similar Figures (pp. 200–205)

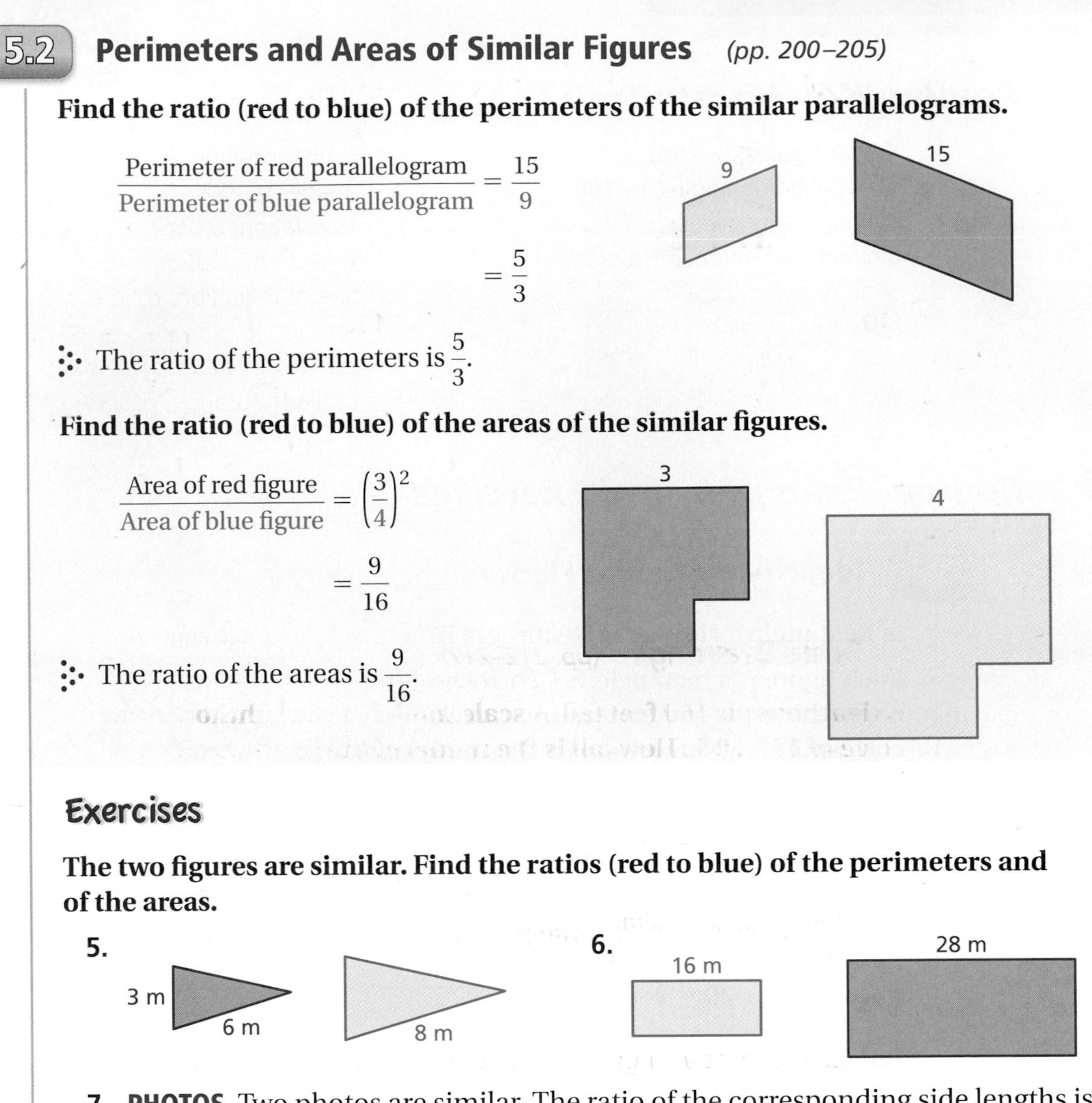

Find the ratio (red to blue) of the perimeters of the similar parallelograms.

$$\frac{\text{Perimeter of red parallelogram}}{\text{Perimeter of blue parallelogram}} = \frac{15}{9}$$

$$= \frac{5}{3}$$

∴ The ratio of the perimeters is $\frac{5}{3}$.

Find the ratio (red to blue) of the areas of the similar figures.

$$\frac{\text{Area of red figure}}{\text{Area of blue figure}} = \left(\frac{3}{4}\right)^2$$

$$= \frac{9}{16}$$

∴ The ratio of the areas is $\frac{9}{16}$.

Exercises

The two figures are similar. Find the ratios (red to blue) of the perimeters and of the areas.

5.

6.

7. PHOTOS Two photos are similar. The ratio of the corresponding side lengths is 3 : 4. What is the ratio of their areas?

5.3 Finding Unknown Measures in Similar Figures (pp. 206–211)

The two rectangles are similar. Find the value of x.

Corresponding side lengths of similar figures are proportional. So, use a proportion to find x.

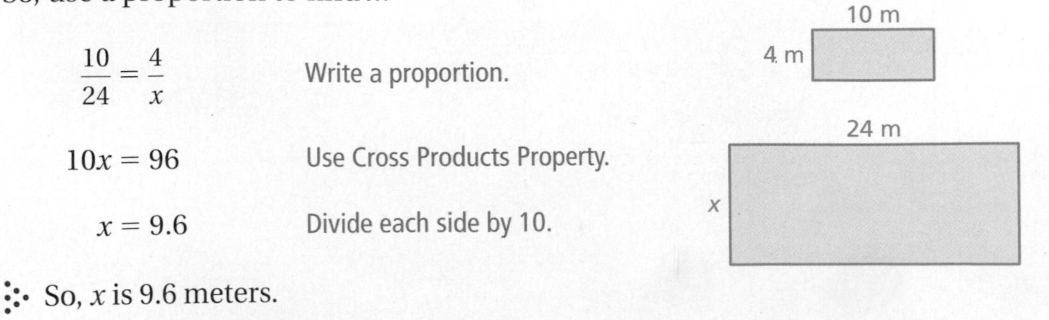

$\frac{10}{24} = \frac{4}{x}$ Write a proportion.

$10x = 96$ Use Cross Products Property.

$x = 9.6$ Divide each side by 10.

∴ So, x is 9.6 meters.

Exercises

The polygons are similar. Find the value of x.

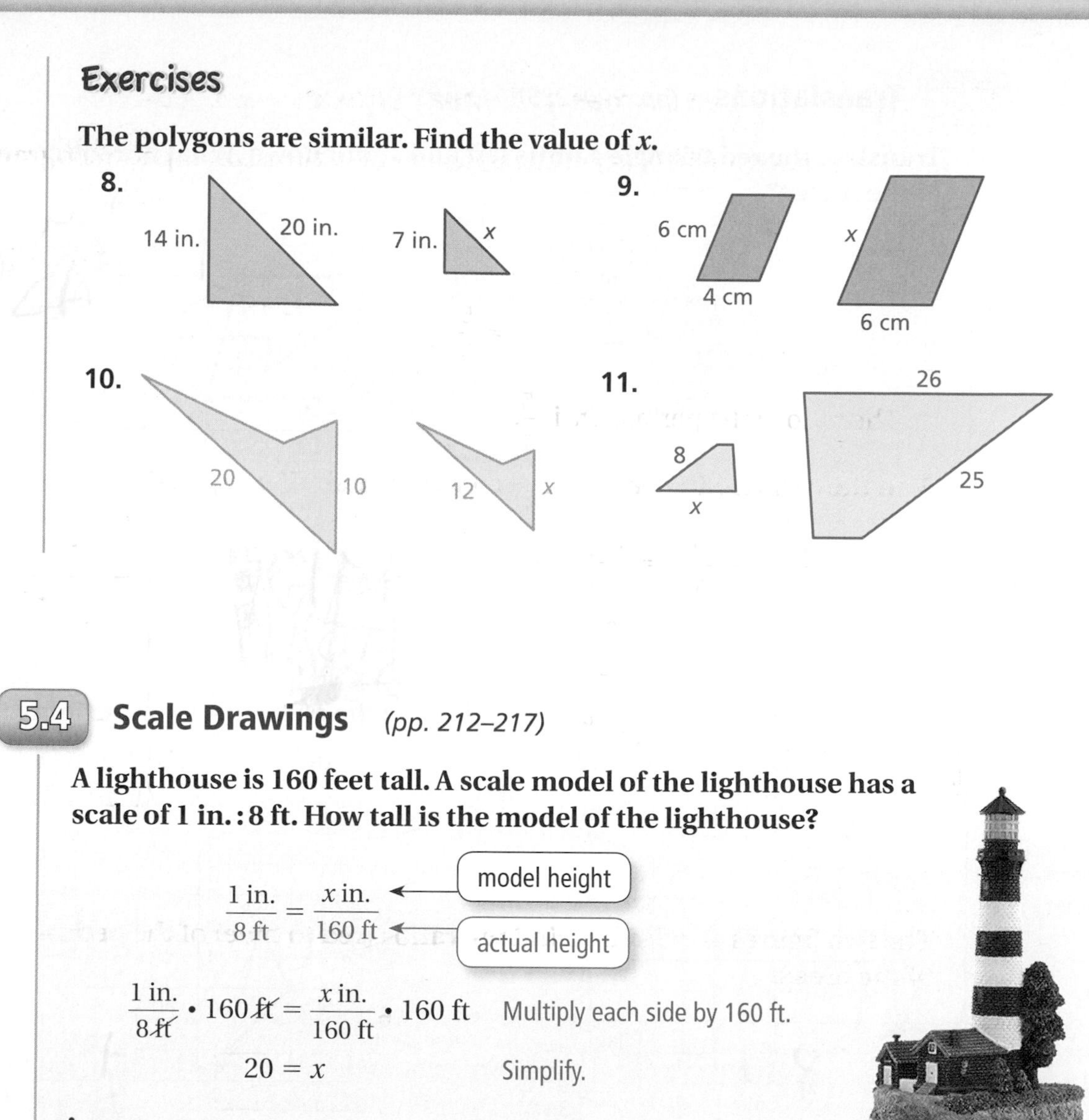

5.4 Scale Drawings *(pp. 212–217)*

A lighthouse is 160 feet tall. A scale model of the lighthouse has a scale of 1 in. : 8 ft. How tall is the model of the lighthouse?

$$\frac{1 \text{ in.}}{8 \text{ ft}} = \frac{x \text{ in.}}{160 \text{ ft}}$$

model height (points to x in.)

actual height (points to 160 ft)

$$\frac{1 \text{ in.}}{8 \cancel{\text{ft}}} \cdot 160 \cancel{\text{ft}} = \frac{x \text{ in.}}{160 \text{ ft}} \cdot 160 \text{ ft}$$ Multiply each side by 160 ft.

$$20 = x$$ Simplify.

∴ The model of the lighthouse is 20 inches tall.

Exercises

Use a centimeter ruler to measure the segment shown. Find the scale of the drawing.

12.

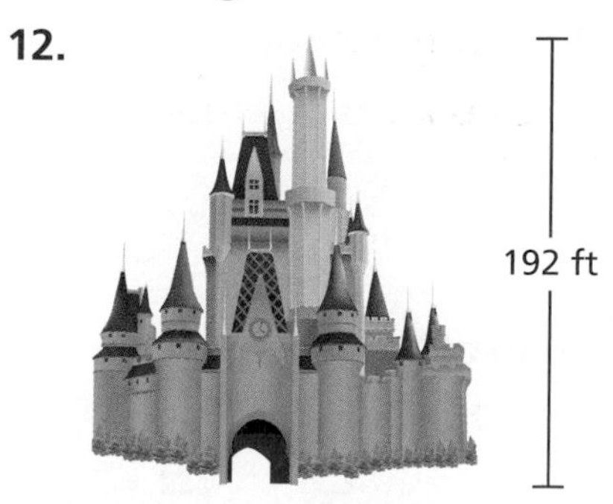

192 ft

13. 30 in.

5.5 Translations (pp. 220–225)

Translate the red triangle 4 units left and 1 unit down. What are the coordinates of the image?

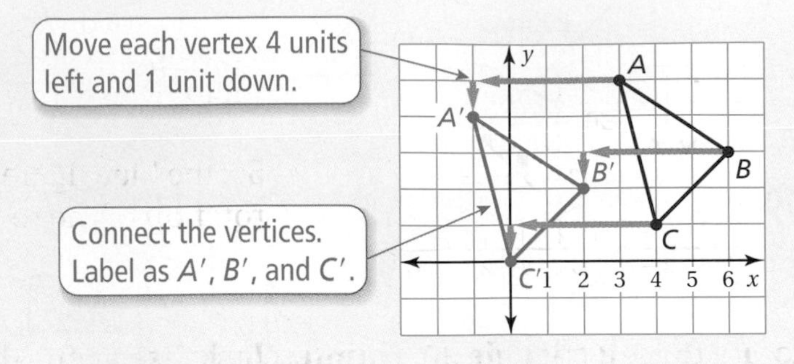

∴ The coordinates of the image are $A'(-1, 4)$, $B'(2, 2)$, and $C'(0, 0)$.

Exercises

Translate the figure as described. What are the coordinates of the image?

14. 3 units left and 2 units down

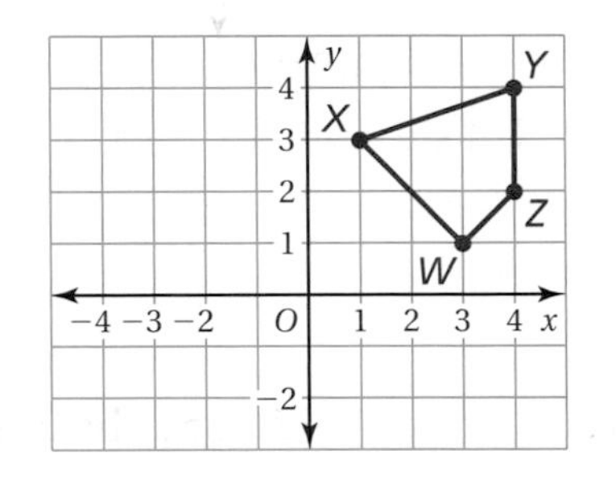

15. 5 units right and 4 units up

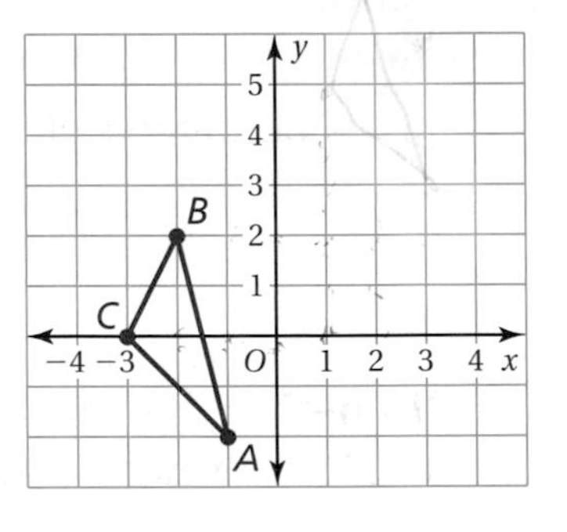

5.6 Reflections (pp. 226–231)

Tell whether the blue figure is a reflection of the red figure.

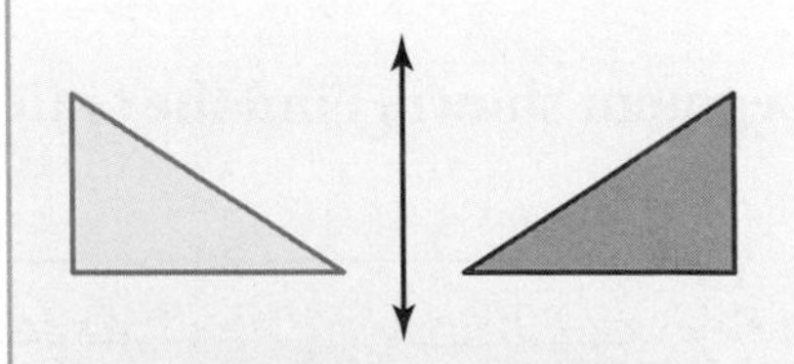

The red figure can be *flipped* to form the blue figure.

∴ So, the blue figure is a reflection of the red figure.

Exercises

Tell whether the blue figure is a reflection of the red figure.

16.

17.

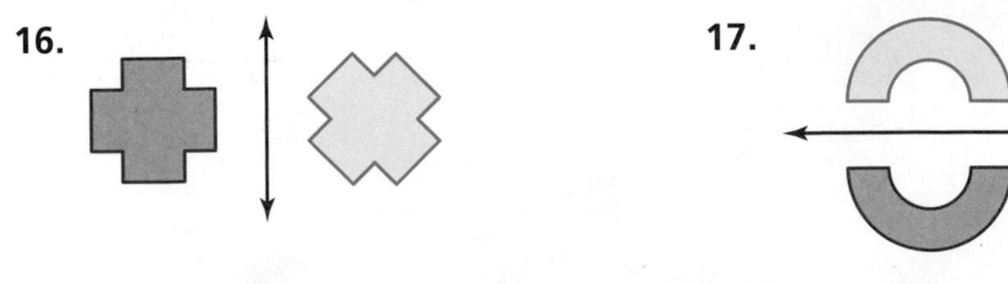

5.7 Rotations (pp. 232–237)

Tell whether the blue figure is a rotation of the red figure about a vertex. If so, give the angle and direction of rotation.

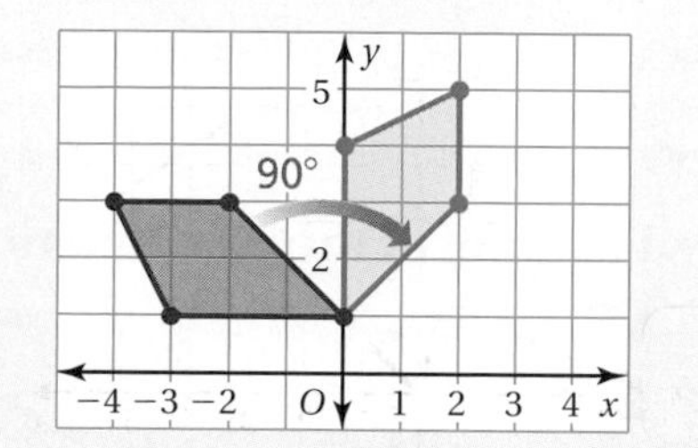

The red figure can be turned 90° clockwise about (0, 1) to form the blue figure.

∴ So, the blue figure is a 90° clockwise rotation of the red figure.

Rotate the red triangle 90° counterclockwise about the origin. What are the coordinates of the image?

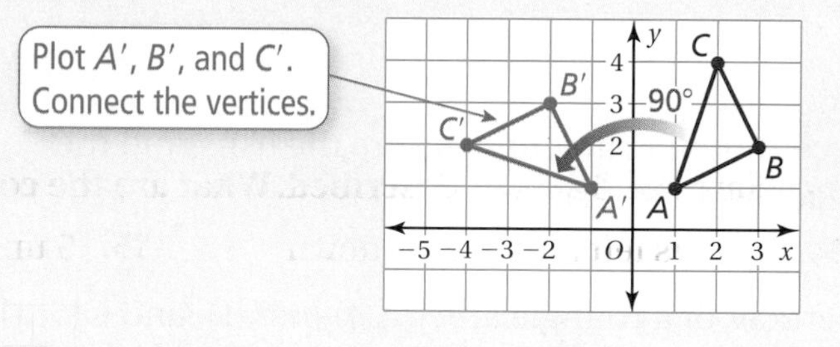

∴ The coordinates of the image are $A'(-1, 1)$, $B'(-2, 3)$, and $C'(-4, 2)$.

Exercises

Tell whether the blue figure is a rotation of the red figure about the origin. If so, give the angle and direction of rotation.

18.

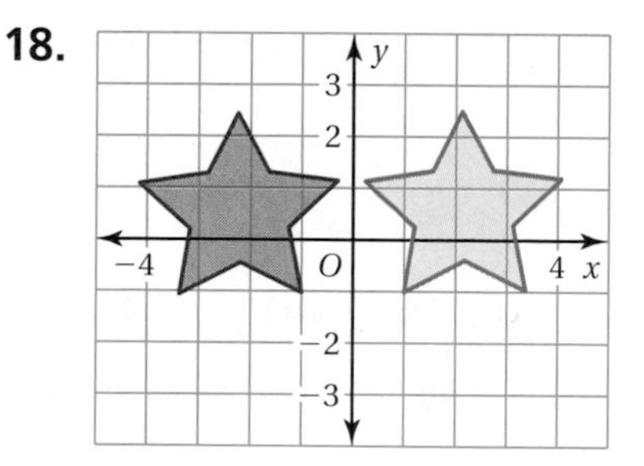

19.

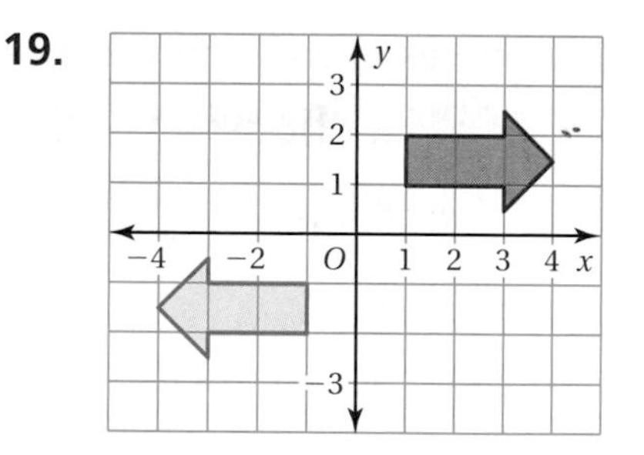

Rotate the figure as described. What are the coordinates of the image?

20. 270° counterclockwise about the origin

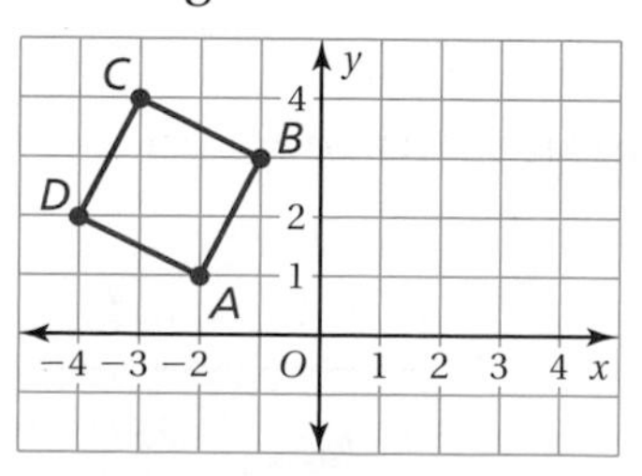

21. 180° clockwise about vertex M.

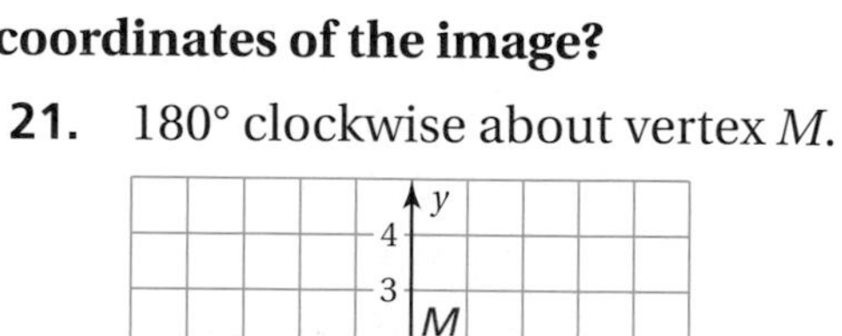

5 Chapter Test

Check It Out
Test Practice
BigIdeasMath.com

1. Tell whether the parallelograms are similar. Explain your reasoning.

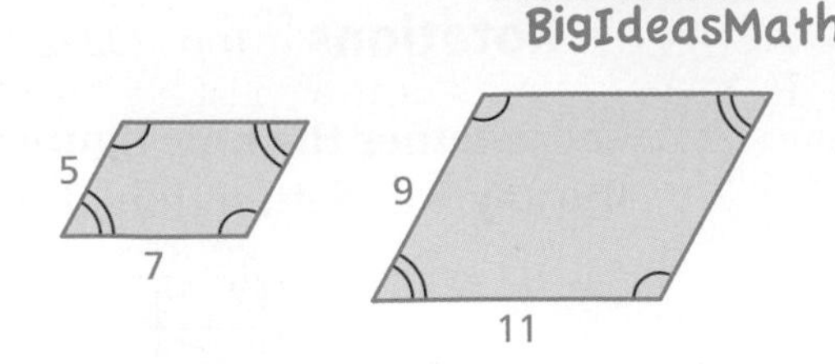

The two figures are similar. Find the ratios (red to blue) of the perimeters and of the areas.

2. 14, 8

3. 9, 12

4. Use a centimeter ruler to measure the fish. Find the scale factor of the drawing.

15 mm

5. The vertices of a triangle are $A(2, 4)$, $B(2, 1)$, and $C(5, 1)$. Draw the triangle and its image after a translation of 1 unit left and 3 units down.

6. Find the coordinates of the triangle whose vertices are $A(2, 5)$, $B(1, 2)$, and $C(3, 1)$ after reflecting in (a) the x-axis and (b) the y-axis.

The vertices of a triangle are $D(-2, -2)$, $E(-1, 1)$, and $F(1, -1)$. Rotate the triangle as described. Find the coordinates of the image.

7. 180° counterclockwise about the origin

8. 90° clockwise about the vertex D

9. **SCREENS** A wide screen television measures 36 inches by 54 inches. A movie theater screen measures 42 feet by 63 feet. Are the screens similar? Explain.

10. **HOCKEY** An air hockey table and an ice hockey rink are similar. The ratio of their corresponding side lengths is 1 inch : 2 feet. What is the ratio of their areas?

11. **HEIGHT** You are five feet tall and cast a seven-foot eight-inch shadow. At the same time, a basketball hoop casts a 19-foot shadow. How tall is the basketball hoop? Assume the triangles are similar.

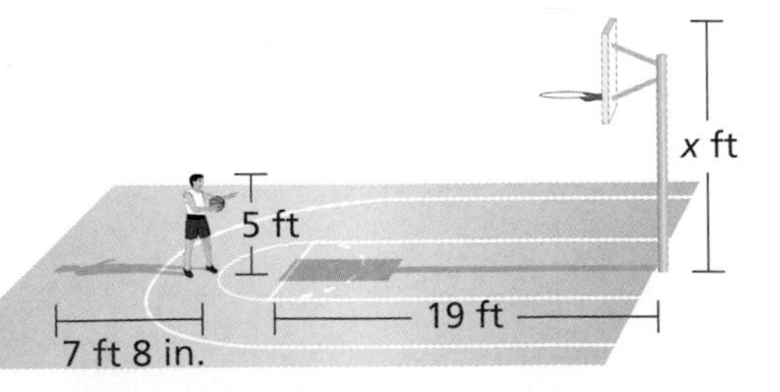

5 Standardized Test Practice

Test-Taking Strategy
Solve Problem Before Looking at Choices

Your paw has an area of 2 in.². A hyena's paw is twice as long. What is its area?

Ⓐ 4 in.² Ⓑ 6 in.² Ⓒ 8 in.² Ⓓ 10 in.²

4 times more area! Help!

CAT HYENA

"Solve the problem before looking at the choices. You know area increases as the square of the scale. So, it's 8 in.²."

1. A set of data is shown below. Two of the data are missing.

8, 2, 10, 4, 8, 4, 8, 8, ____, ____

The mean of the complete set of data is 6, and the median is 7. What are the two missing data? *(7.NS.3)*

A. 1 and 7 **C.** 4 and 4

B. 2 and 6 **D.** 6 and 6

2. What is the area of the shaded region in the figure below? (Use 3.14 for π.) *(7.NS.3)*

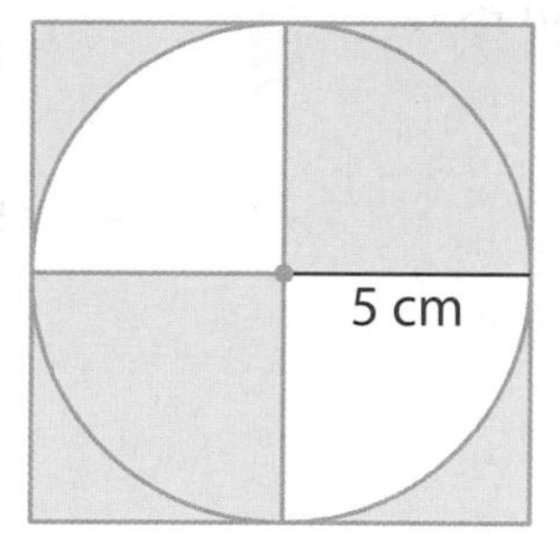

F. $21.5\ \text{cm}^2$ **H.** $80.375\ \text{cm}^2$

G. $60.75\ \text{cm}^2$ **I.** $84.3\ \text{cm}^2$

3. A clockwise rotation of 90° is equivalent to a counterclockwise rotation of how many degrees? *(7.G.2)*

4. You are building a scale model of a park that is planned for a city. The model uses the scale below.

1 centimeter = 2 meters

The park will have a rectangular reflecting pool with a length of 20 meters and a width of 12 meters. In your scale model, what will be the area of the reflecting pool? *(7.G.1)*

A. $60\ \text{cm}^2$ **C.** $480\ \text{cm}^2$

B. $120\ \text{cm}^2$ **D.** $960\ \text{cm}^2$

5. In the figure, $\triangle EFG \sim \triangle HIJ$.

Which proportion is *not* necessarily correct for $\triangle EFG$ and $\triangle HIJ$? *(7.G.1)*

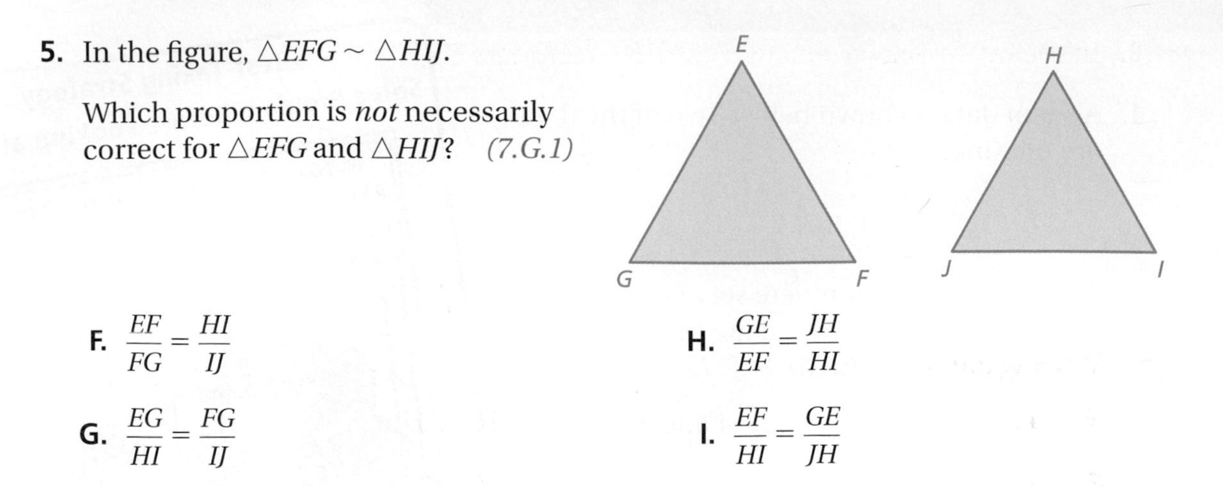

F. $\dfrac{EF}{FG} = \dfrac{HI}{IJ}$

G. $\dfrac{EG}{HI} = \dfrac{FG}{IJ}$

H. $\dfrac{GE}{EF} = \dfrac{JH}{HI}$

I. $\dfrac{EF}{HI} = \dfrac{GE}{JH}$

6. Brett was solving the equation in the box below.

$$\frac{c}{5} - (-15) = -35$$
$$\frac{c}{5} + 15 = -35$$
$$\frac{c}{5} + 15 - 15 = -35 - 15$$
$$\frac{c}{5} = -50$$
$$\frac{c}{5} = \frac{-50}{5}$$
$$c = -10$$

What should Brett do to correct the error that he made? *(7.EE.4a)*

A. Subtract 15 from -35 to get -20.

B. Rewrite $\frac{c}{5} - (-15)$ as $\frac{c}{5} - 15$.

C. Multiply both sides of the equation by 5 to get $c = -250$.

D. Multiply both sides of the equation by -5 to get $c = 250$.

7. In the figure below, $\triangle ABC \sim \triangle DEF$.

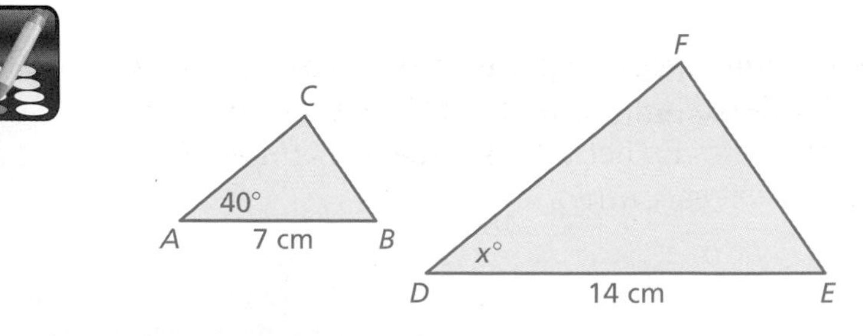

What is the value of x? *(7.G.1)*

8. In the figure below, rectangle $EFGH \sim$ rectangle $IJKL$.

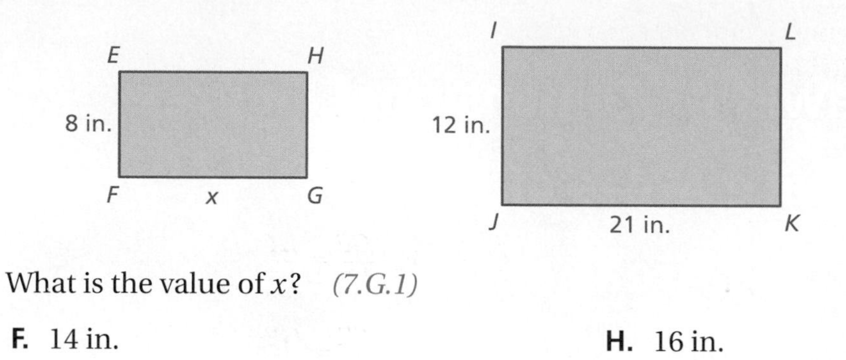

What is the value of x? *(7.G.1)*

F. 14 in.

G. 15 in.

H. 16 in.

I. 17 in.

9. Two cubes are shown below.

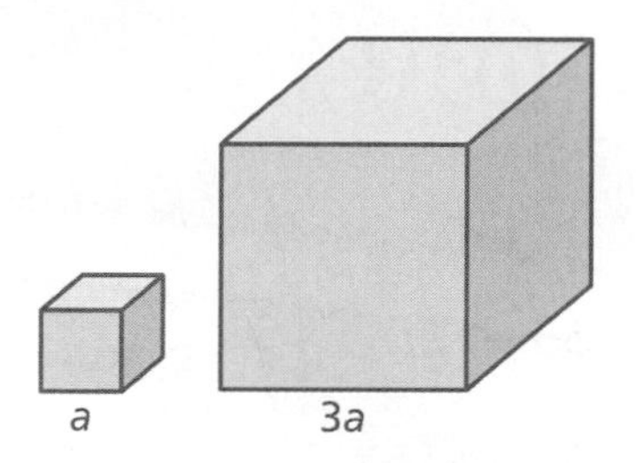

How many of the smaller cubes can be stacked to completely fill the larger cube? *(7.NS.3)*

A. 3

B. 9

C. 27

D. 54

10. A map of Donna's state has the following scale: *(7.G.1)*

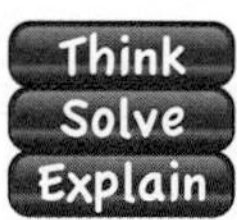

$$\frac{1}{2}\text{ inch} = 10\text{ miles}$$

Part A Donna measured the distances between her town and the state capitol on the map. Her measurement was $4\frac{1}{2}$ inches. Based on Donna's measurement, what is the actual distance, in miles, between her town and the state capitol? Show your work and explain your reasoning.

Part B Donna wants to mark her favorite campsite on the map. She knows that the campsite is 65 miles north of her town. What distance, in inches, on the map represents an actual distance of 65 miles? Show your work and explain your reasoning.

6 Surface Areas of Solids

"I want to paint my dog house. To make sure I buy the correct amount of paint, I want to calculate the lateral surface area."

"Then, because I want to paint the inside and the outside, I will multiply by 2. Does this seem right to you?"

"Dear Sir: Why do you sell dog food in tall cans and sell cat food in short cans?"

"Neither of these shapes is the optimal use of surface area when compared to volume."

What You Learned Before

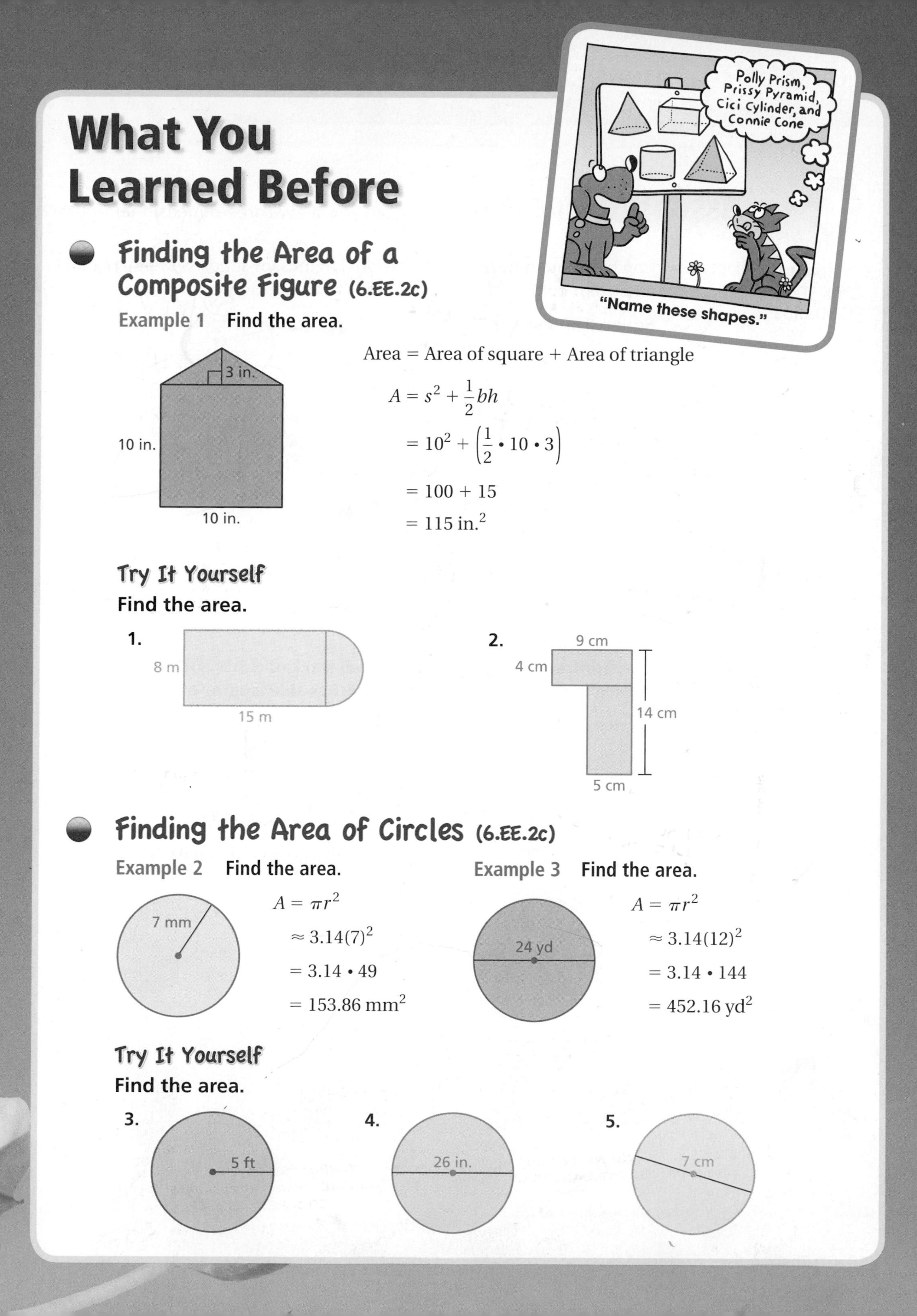

Polly Prism, Prissy Pyramid, Cici Cylinder, and Connie Cone
"Name these shapes."

Finding the Area of a Composite Figure (6.EE.2c)

Example 1 Find the area.

Area = Area of square + Area of triangle

$$A = s^2 + \frac{1}{2}bh$$
$$= 10^2 + \left(\frac{1}{2} \cdot 10 \cdot 3\right)$$
$$= 100 + 15$$
$$= 115 \text{ in.}^2$$

Try It Yourself

Find the area.

1. 8 m, 15 m
2. 9 cm, 4 cm, 14 cm, 5 cm

Finding the Area of Circles (6.EE.2c)

Example 2 Find the area.

$$A = \pi r^2$$
$$\approx 3.14(7)^2$$
$$= 3.14 \cdot 49$$
$$= 153.86 \text{ mm}^2$$

Example 3 Find the area.

$$A = \pi r^2$$
$$\approx 3.14(12)^2$$
$$= 3.14 \cdot 144$$
$$= 452.16 \text{ yd}^2$$

Try It Yourself

Find the area.

3. 5 ft
4. 26 in.
5. 7 cm

6.1 Drawing 3-Dimensional Figures

COMMON CORE STATE STANDARDS
7.G.3

Essential Question How can you draw three-dimensional figures?

Dot paper can help you draw three-dimensional figures, or solids. Shading parallel sides the same color helps create a three-dimensional illusion.

Square Dot Paper

Isometric Dot Paper

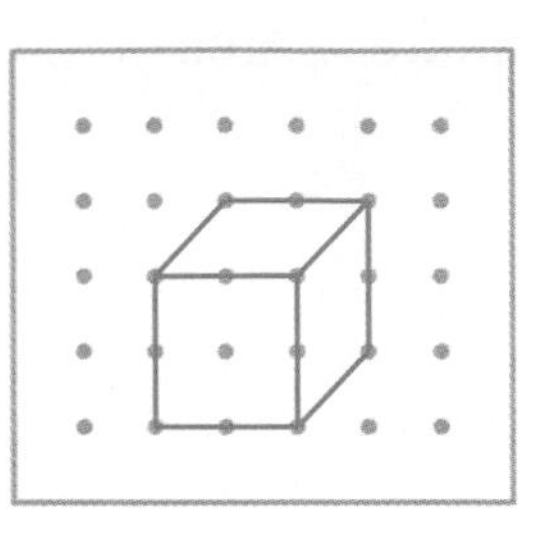

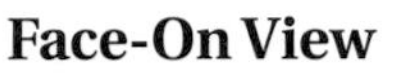
Face-On View

Corner View

1 ACTIVITY: Finding Surface Areas and Volumes

Work with a partner.

Draw the front, side, and top views of each stack of cubes. Then find the surface area and volume. Each small cube has side lengths of 1 unit.

a. Sample:

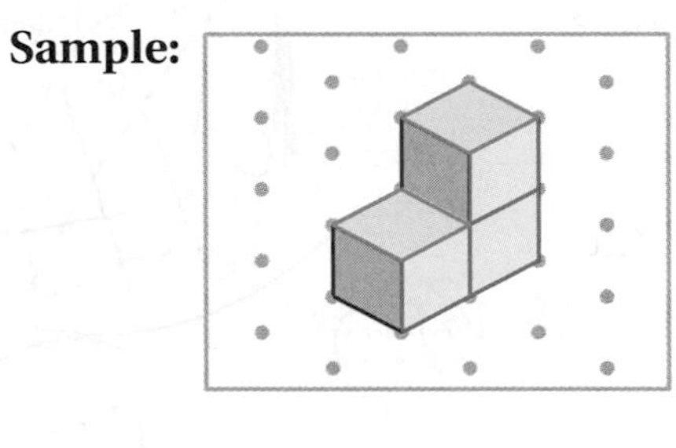

top front side

Volume: 3 cubic units

Surface Area: 14 square units

b.

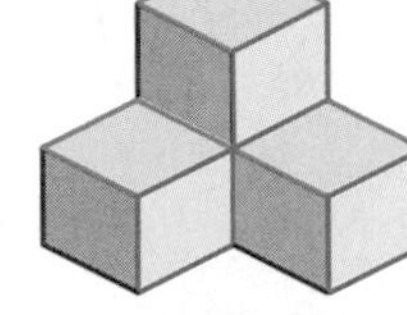

c.

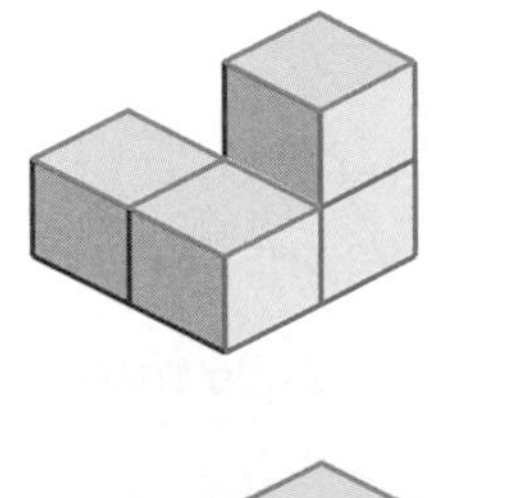

d.

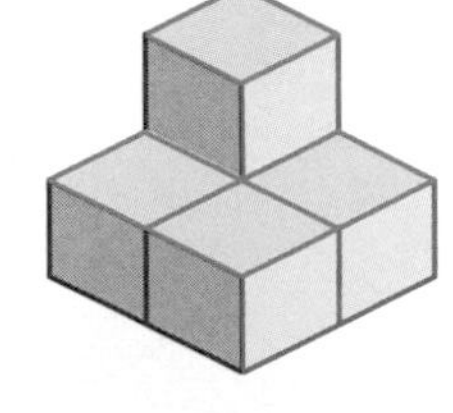

e.

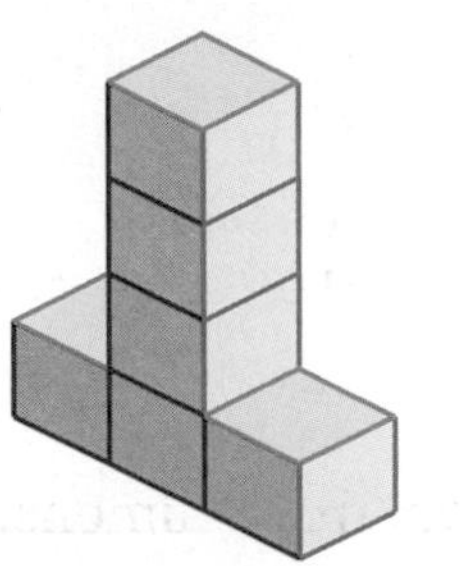

f.

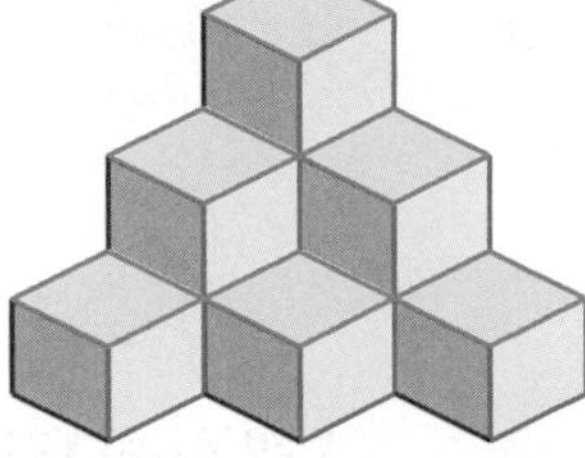

g.

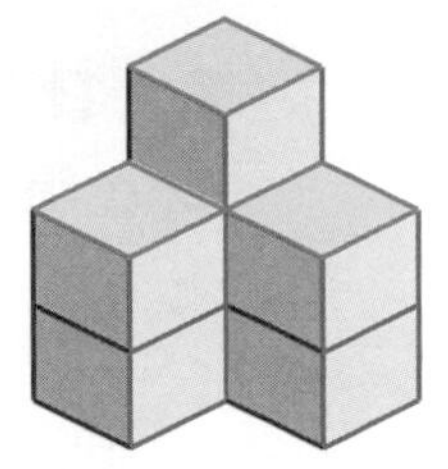

2 ACTIVITY: Drawing Solids

Work with a partner.

a. Draw all the different solids you can make by joining four cubes. (Two have been drawn.) Cubes must be joined on faces, not on edges only. Translations, reflections, and rotations do not count as different solids.

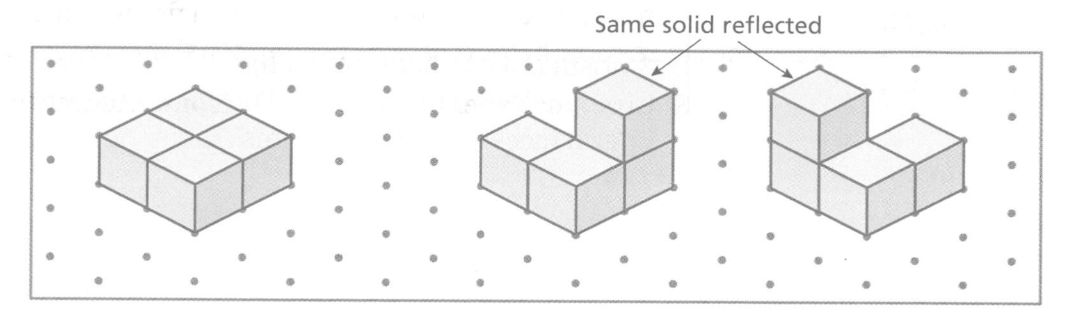

b. Do all the solids have the same surface area? Do all the solids have the same volume? Explain your reasoning.

What Is Your Answer?

3. **IN YOUR OWN WORDS** How can you draw three-dimensional figures? Draw and shade two prisms that have the same volume but different surface areas.

4. Maurits Escher (1898–1972) was a popular artist who drew optical illusions.

    a. What is the illusion in Escher's drawing?

    b. Why is the cartoon funny? What is the illusion in the cartoon?

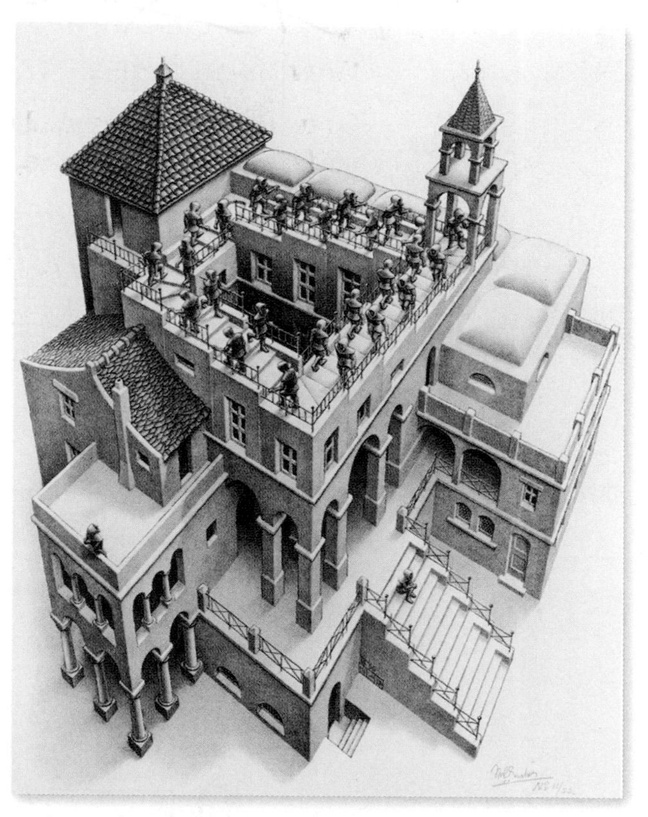

Use what you learned about three-dimensional figures to complete Exercises 7–9 on page 254.

6.1 Lesson

Key Vocabulary
three-dimensional figure, *p. 252*
polyhedron, *p. 252*
lateral face, *p. 252*

A **three-dimensional figure**, or *solid*, has length, width, and depth. A **polyhedron** is a three-dimensional figure whose faces are all polygons.

Key Ideas

Prisms

A prism is a polyhedron that has two parallel, identical bases. The **lateral faces** are parallelograms.

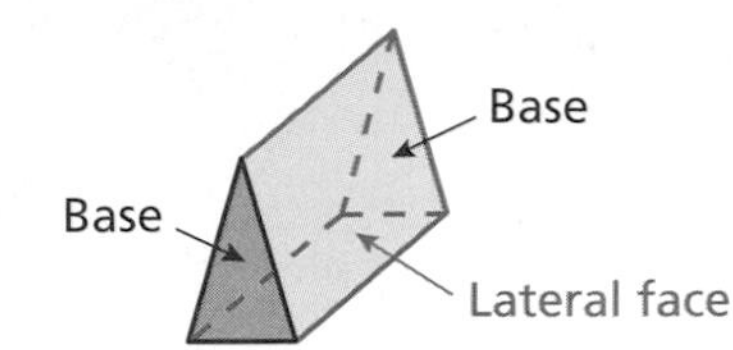

Triangular Prism

Pyramids

A pyramid is a polyhedron that has one base. The lateral faces are triangles.

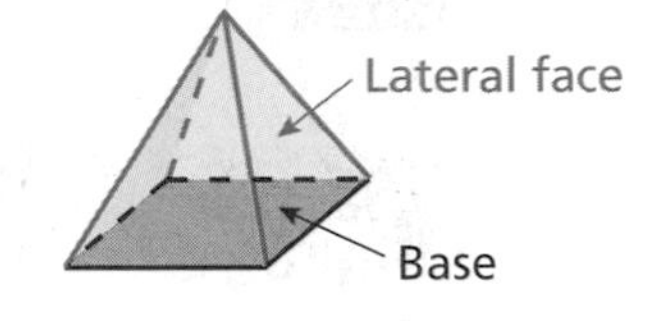

Rectangular Pyramid

The shape of the base tells the name of the prism or the pyramid.

Cylinders

A cylinder is a solid that has two parallel, identical circular bases.

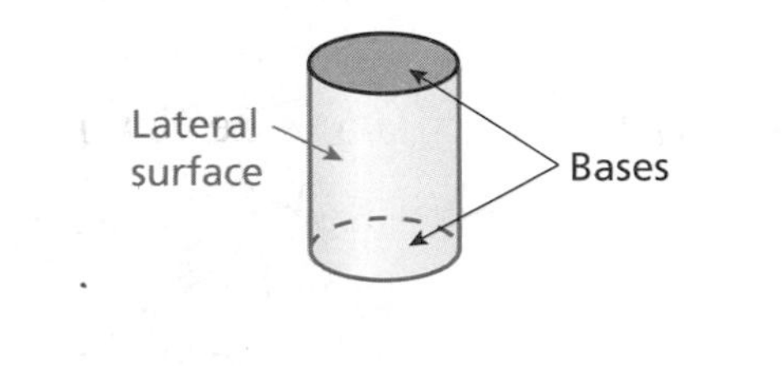

Cones

A cone is a solid that has one circular base and one vertex.

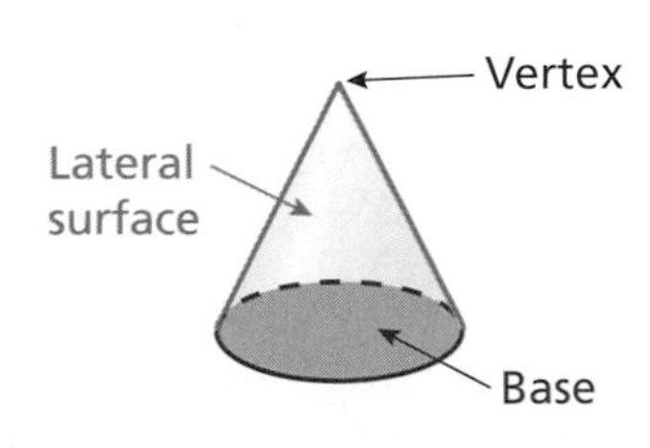

EXAMPLE 1 Drawing a Prism

Draw a rectangular prism.

Step 1

Draw identical rectangular bases.

Step 2

Connect corresponding vertices.

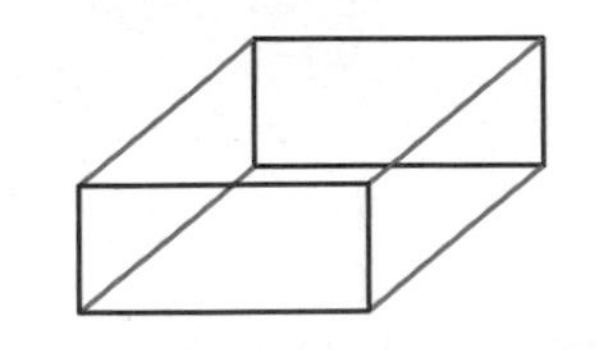

Step 3

Change any *hidden* lines to dashed lines.

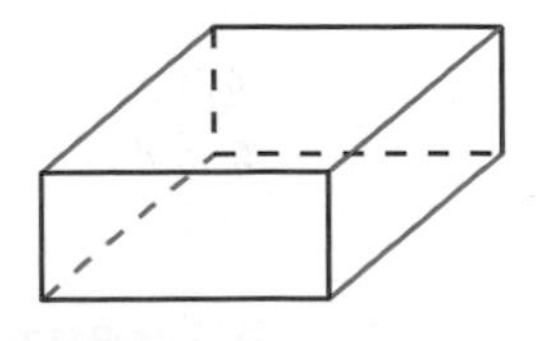

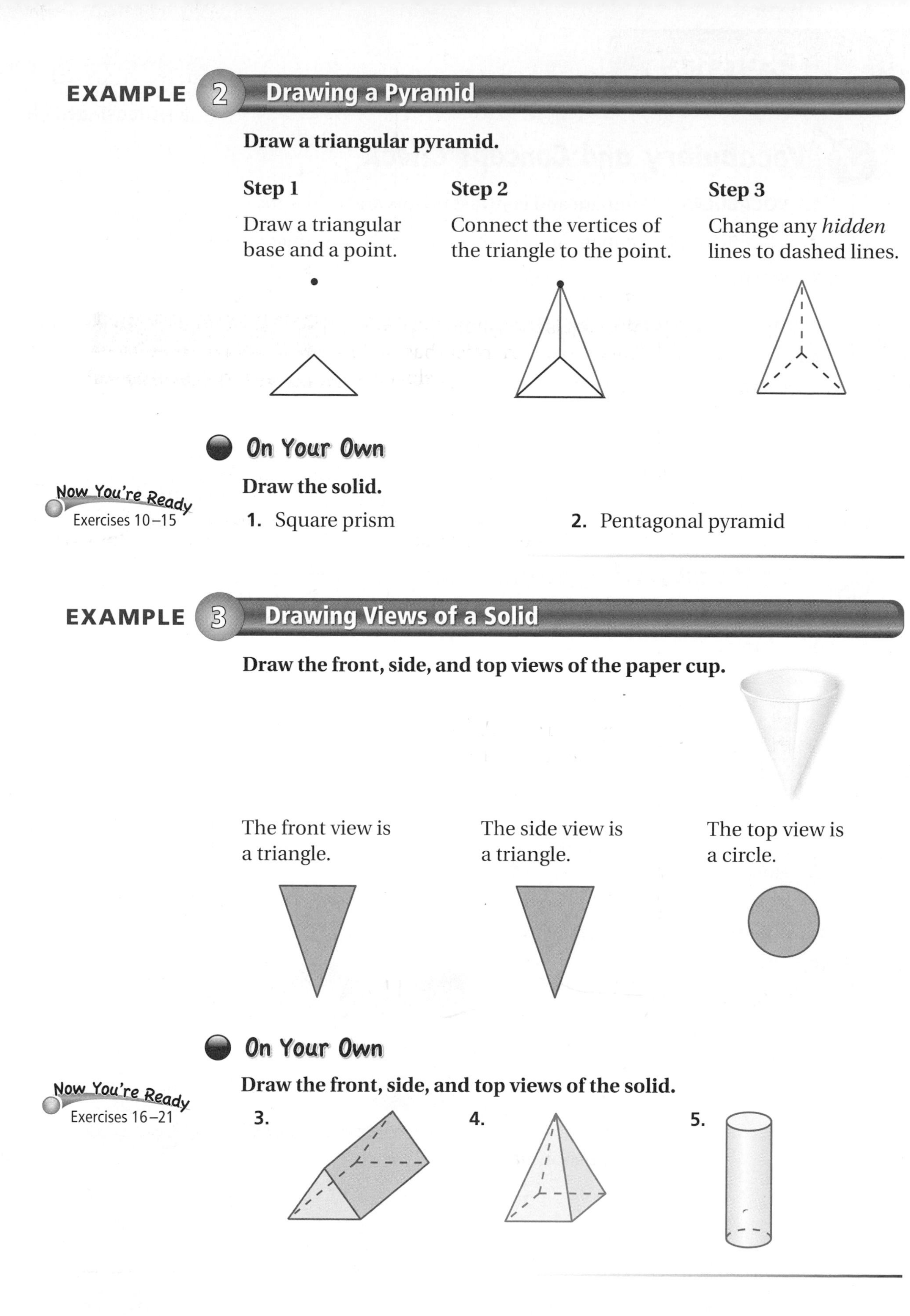

EXAMPLE 2 Drawing a Pyramid

Draw a triangular pyramid.

Step 1

Draw a triangular base and a point.

Step 2

Connect the vertices of the triangle to the point.

Step 3

Change any *hidden* lines to dashed lines.

On Your Own

Now You're Ready
Exercises 10–15

Draw the solid.

1. Square prism
2. Pentagonal pyramid

EXAMPLE 3 Drawing Views of a Solid

Draw the front, side, and top views of the paper cup.

The front view is a triangle.

The side view is a triangle.

The top view is a circle.

On Your Own

Now You're Ready
Exercises 16–21

Draw the front, side, and top views of the solid.

3.

4.

5.

6.1 Exercises

Vocabulary and Concept Check

1. **VOCABULARY** Compare and contrast prisms and cylinders.
2. **VOCABULARY** Compare and contrast pyramids and cones.
3. **WRITING** Give examples of prisms, pyramids, cylinders, and cones in real life.

Identify the shape of the base. Then name the solid.

4. 5. 6.

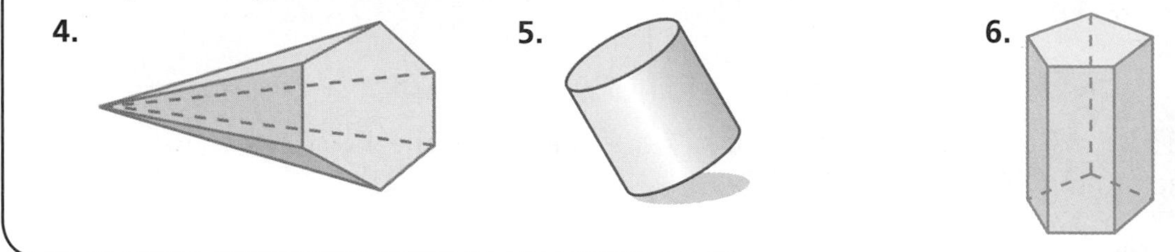

Practice and Problem Solving

Draw the front, side, and top views of the stack of cubes. Then find the surface area and volume.

7. 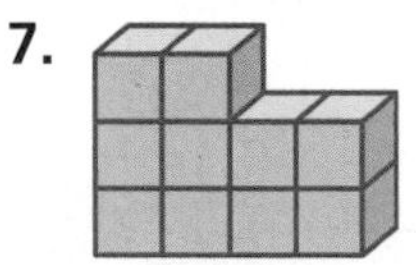

8.

9. 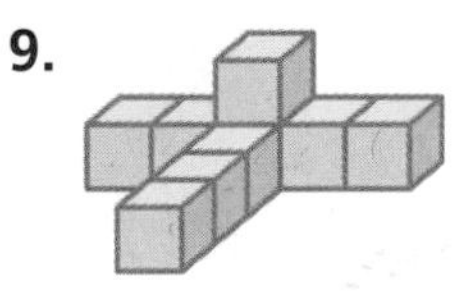

Draw the solid.

10. Triangular prism
11. Pentagonal prism
12. Rectangular pyramid
13. Hexagonal pyramid
14. Cone
15. Cylinder

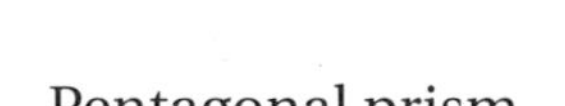

Draw the front, side, and top views of the solid.

16. 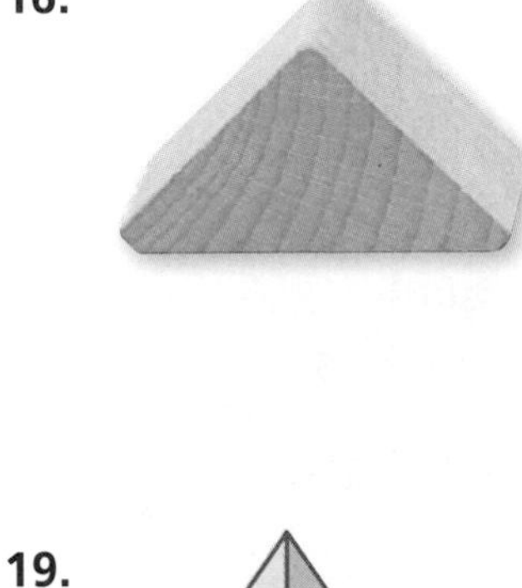

17.

18. 

19.

20. 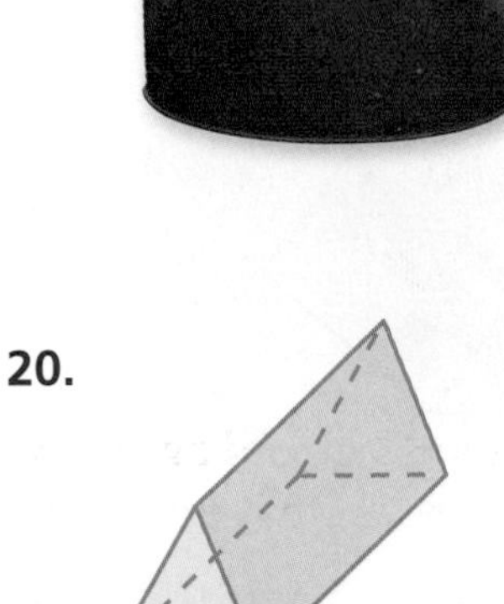

21. 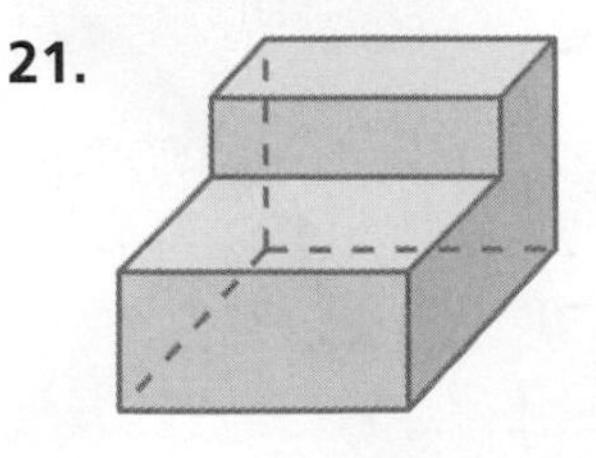

22. **PYRAMID ARENA** The Pyramid of Caius Cestius in Rome is in the shape of a square pyramid. Draw a sketch of the pyramid.

23. **RESEARCH** Use the Internet to find a picture of the Washington Monument. Describe its shape.

Draw a solid with the following front, side, and top views.

24.–25.

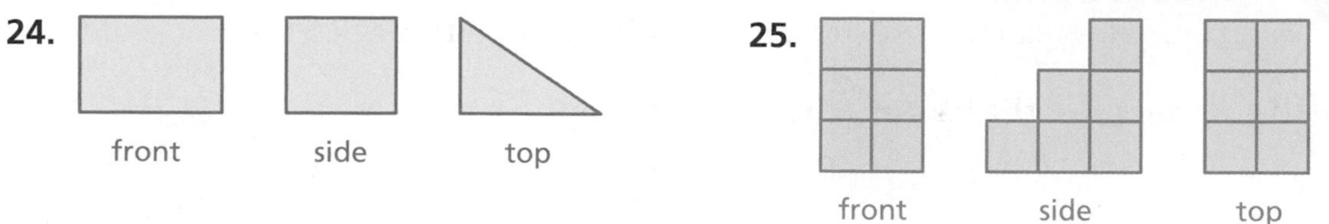

26. **PROJECT** Design and draw a house. Name the different solids that can be used to make a model of the house.

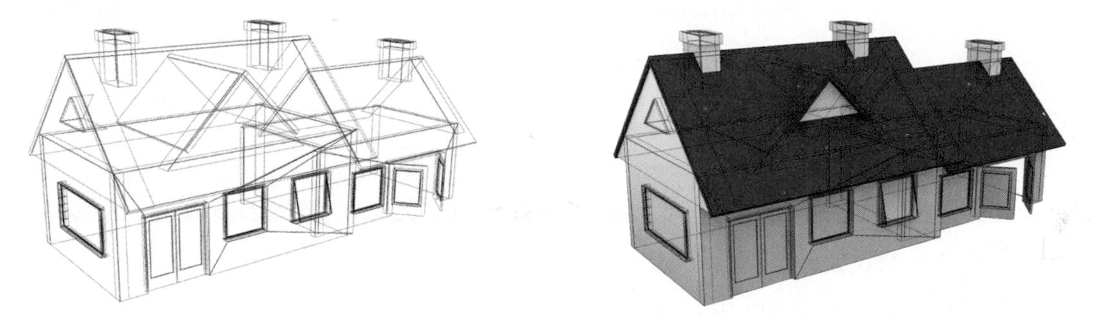

27. **REASONING** Two of the three views of a solid are shown.

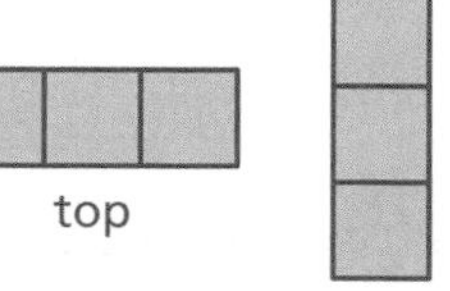

   a. What is the greatest number of unit cubes in the solid?
   b. What is the least number of unit cubes in the solid?
   c. Draw the front views of both solids in parts (a) and (b).

28. Reasoning Draw two different solids with five faces.
   a. Write the number of vertices and edges for each solid.
   b. Explain how knowing the numbers of edges and vertices helps you draw a three-dimensional figure.

Fair Game Review What you learned in previous grades & lessons

Find the area. *(Skills Review Handbook)*

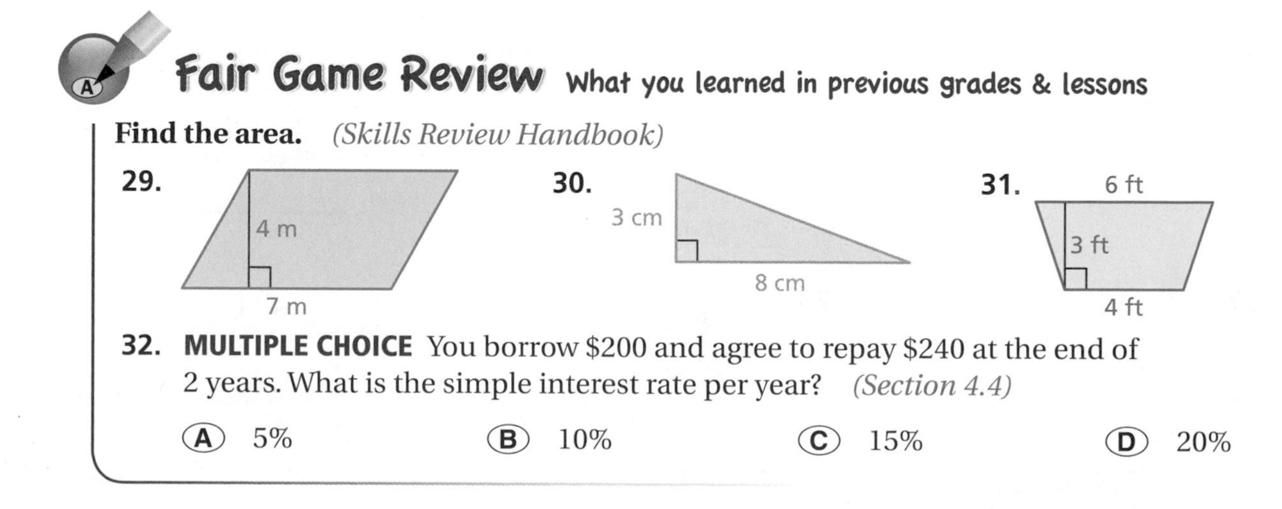

32. **MULTIPLE CHOICE** You borrow \$200 and agree to repay \$240 at the end of 2 years. What is the simple interest rate per year? *(Section 4.4)*

   Ⓐ 5% Ⓑ 10% Ⓒ 15% Ⓓ 20%

6.2 Surface Areas of Prisms

Essential Question How can you use a net to find the surface area of a prism?

COMMON CORE STATE STANDARDS
7.G.4
7.G.6

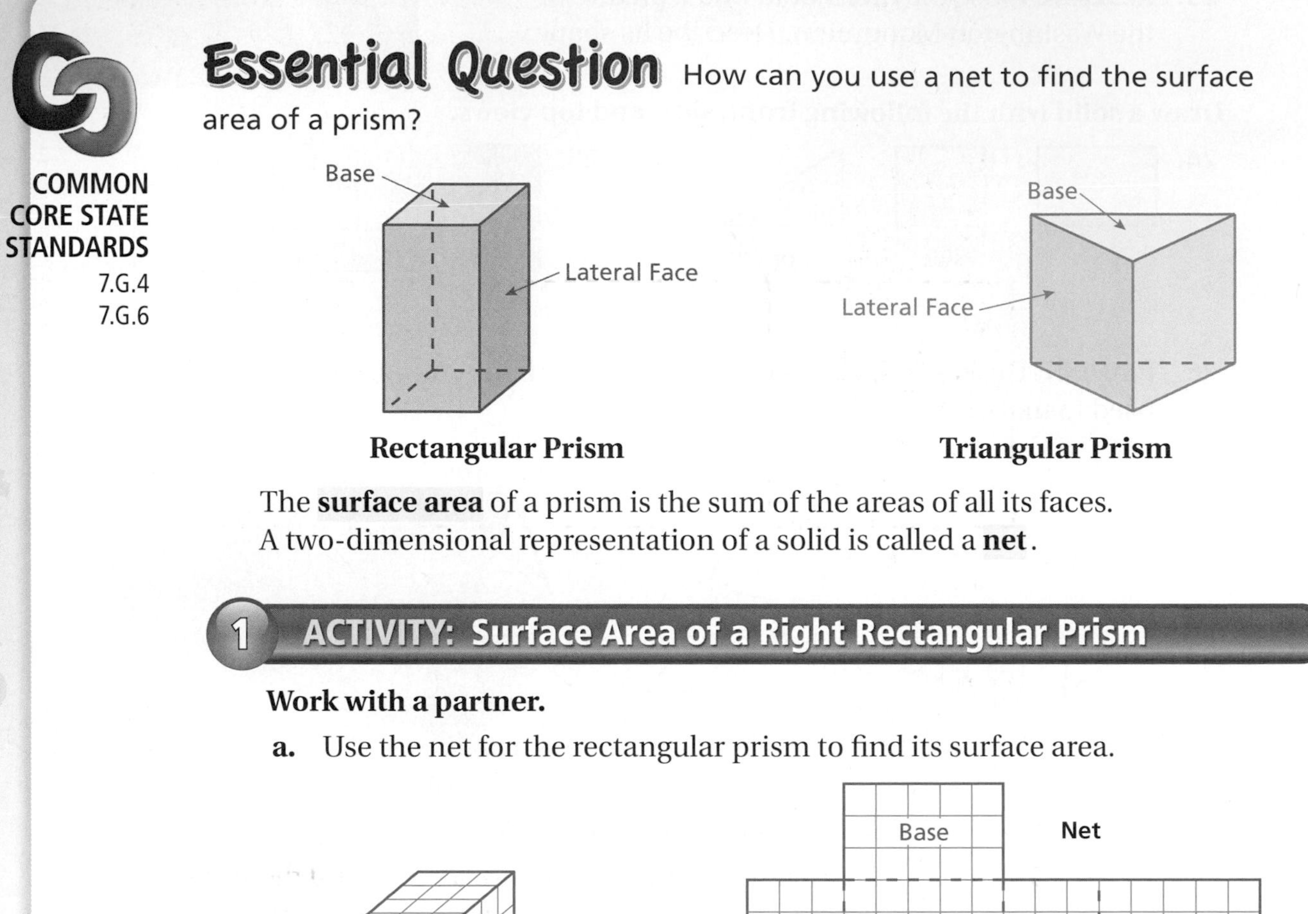

Rectangular Prism **Triangular Prism**

The **surface area** of a prism is the sum of the areas of all its faces. A two-dimensional representation of a solid is called a **net**.

1 ACTIVITY: Surface Area of a Right Rectangular Prism

Work with a partner.

a. Use the net for the rectangular prism to find its surface area.

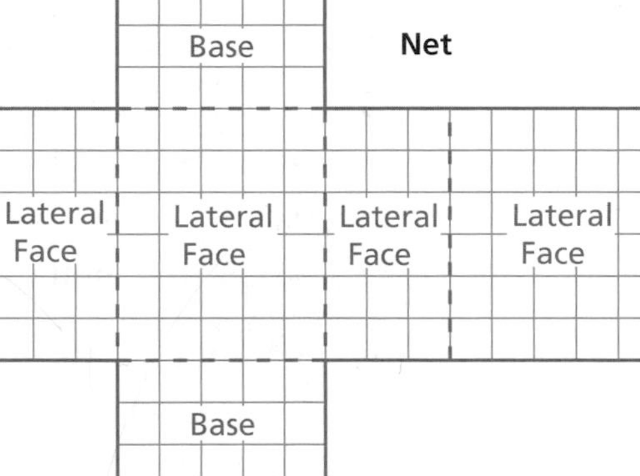

b. Copy the net for a rectangular prism. Label each side as h, w, or ℓ. Then use your drawing to write a formula for the surface area of a rectangular prism.

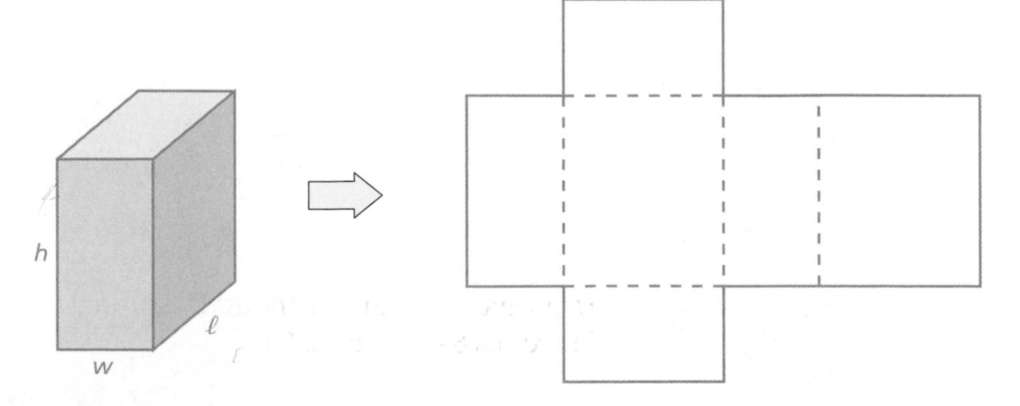

2 ACTIVITY: Finding Surface Area

Work with a partner. Find the surface area of the solid shown by the net. Copy the net, cut it out, and fold it to form a solid. Identify the solid.

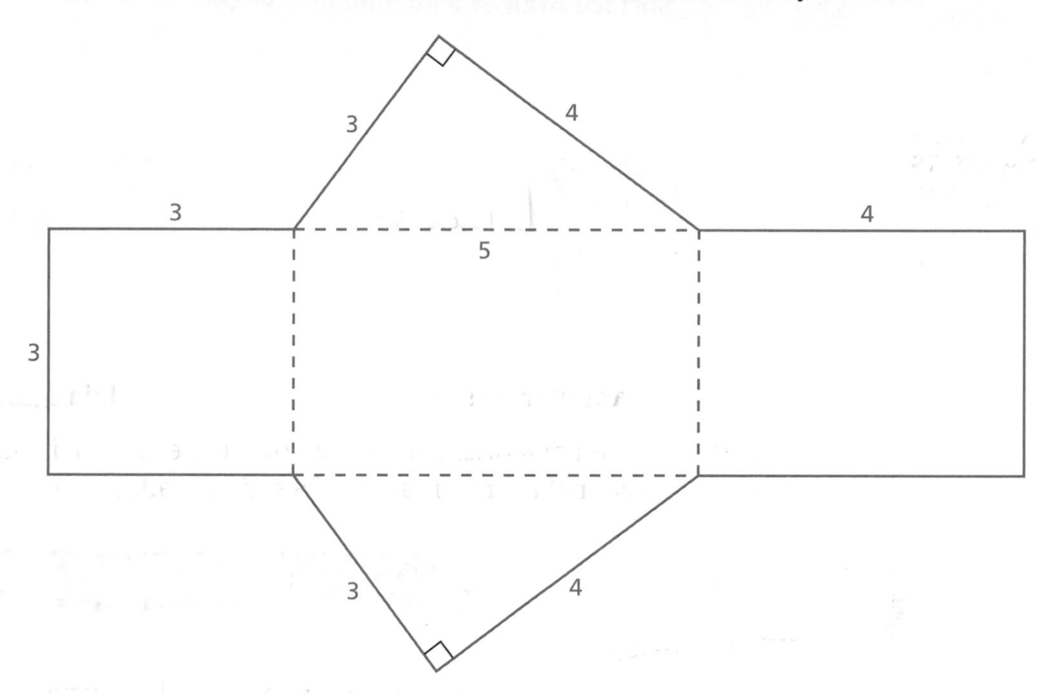

What Is Your Answer?

3. **IN YOUR OWN WORDS** How can you use a net to find the surface area of a prism? Draw a net, cut it out, and fold it to form a prism.

4. The greater the surface area of an ice block, the faster it will melt. Which will melt faster, the bigger block or the three smaller blocks? Explain your reasoning.

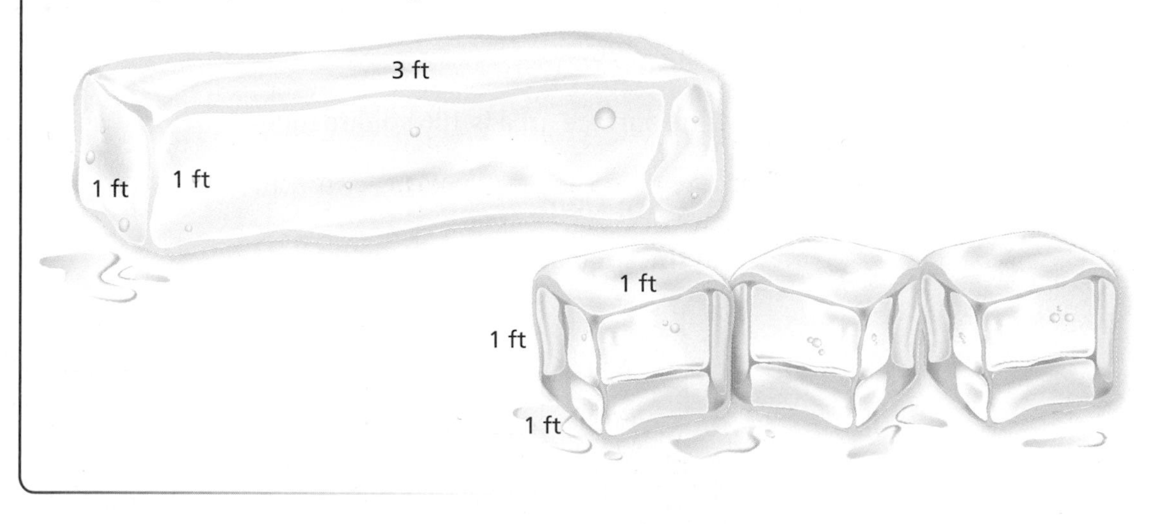

Practice Use what you learned about the surface area of a prism to complete Exercises 6–8 on page 260.

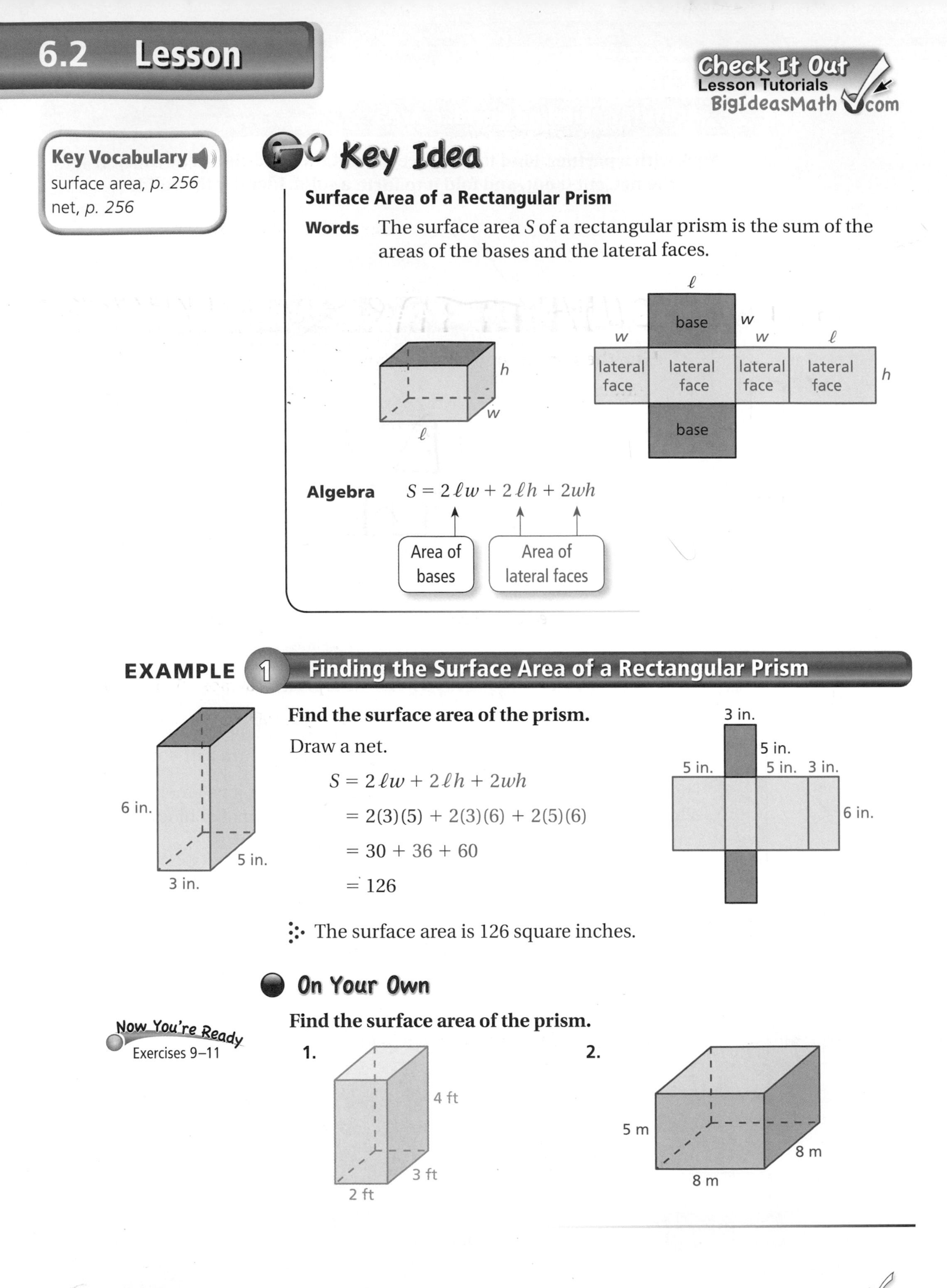

6.2 Lesson

Check It Out
Lesson Tutorials
BigIdeasMath.com

Key Vocabulary
surface area, *p. 256*
net, *p. 256*

Key Idea

Surface Area of a Rectangular Prism

Words The surface area S of a rectangular prism is the sum of the areas of the bases and the lateral faces.

Algebra $S = 2\ell w + 2\ell h + 2wh$

EXAMPLE 1 Finding the Surface Area of a Rectangular Prism

Find the surface area of the prism.

Draw a net.

$$S = 2\ell w + 2\ell h + 2wh$$
$$= 2(3)(5) + 2(3)(6) + 2(5)(6)$$
$$= 30 + 36 + 60$$
$$= 126$$

The surface area is 126 square inches.

On Your Own

Now You're Ready
Exercises 9–11

Find the surface area of the prism.

1.

2.

Multi-Language Glossary at BigIdeasMath.com.

Key Idea

Surface Area of a Prism

The surface area S of a prism is the sum of the areas of the bases and the lateral faces.

$$S = \text{areas of bases} + \text{areas of lateral faces}$$

EXAMPLE 2 Finding the Surface Area of a Triangular Prism

5 m, 6 m, 4 m, 3 m

Find the surface area of the prism.

Draw a net.

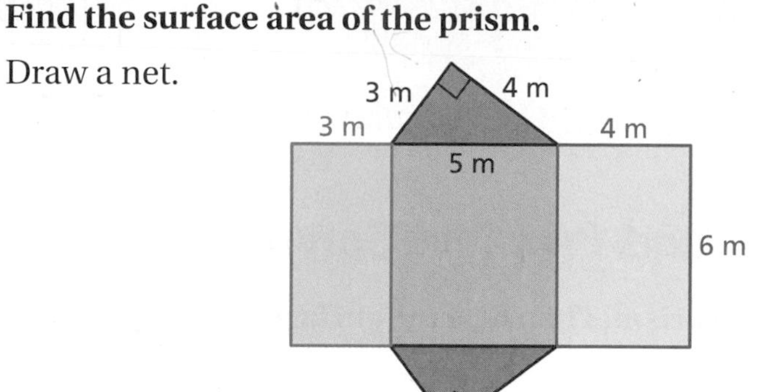

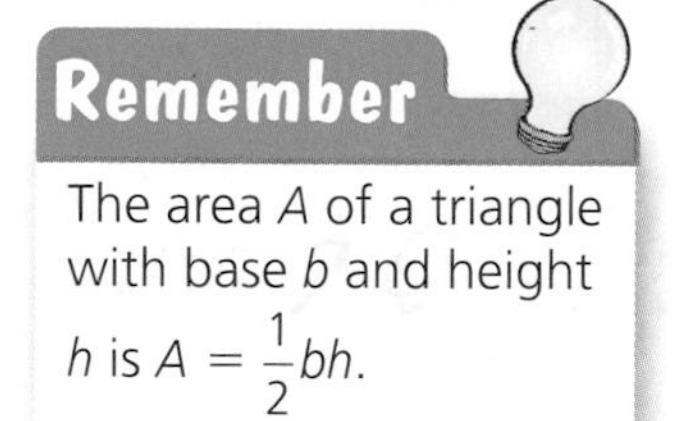

Remember
The area A of a triangle with base b and height h is A = 1/2 bh.

Area of a base

Red base: $\frac{1}{2} \cdot 3 \cdot 4 = 6$

Areas of lateral faces

Green lateral face: $3 \cdot 6 = 18$

Purple lateral face: $5 \cdot 6 = 30$

Blue lateral face: $4 \cdot 6 = 24$

Add the areas of the bases and the lateral faces.

$$S = \text{areas of bases} + \text{areas of lateral faces}$$

$$= 6 + 6 + 18 + 30 + 24$$

There are two identical bases. Count the area twice.

$$= 84$$

The surface area is 84 square meters.

On Your Own

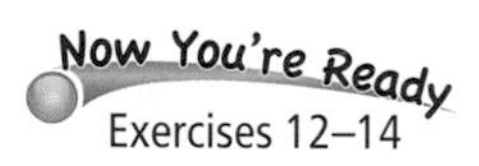

Find the surface area of the prism.

3. 5 m, 12 m, 3 m, 13 m

4. 4 cm, 5 cm, 4 cm, 3 cm

6.2 Exercises

Vocabulary and Concept Check

1. **OPEN-ENDED** Describe a real-world situation in which you would want to find the surface area of a prism.

Find the indicated area for the rectangular prism.

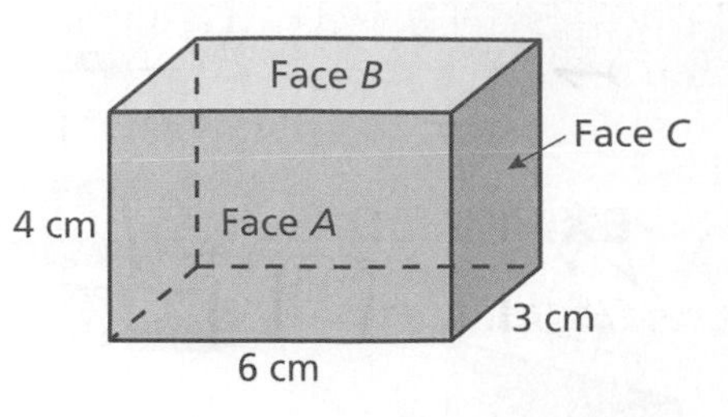

2. Area of Face *A*
3. Area of Face *B*
4. Area of Face *C*
5. Surface area of the prism

Practice and Problem Solving

Draw a net for the prism. Then find the surface area.

6. 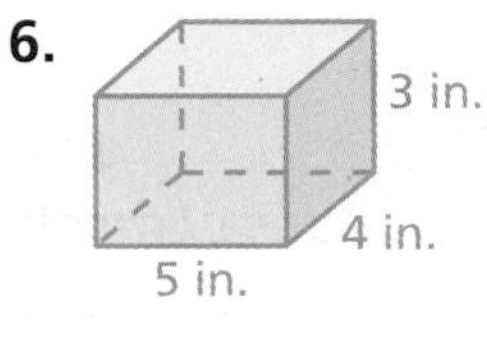

7. 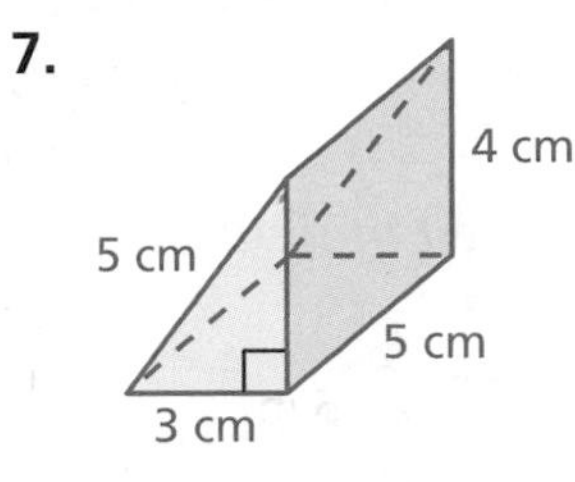

8. 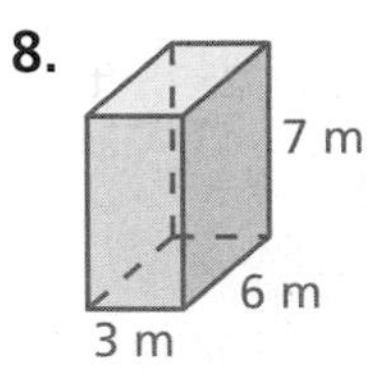

Find the surface area of the prism.

1 9. 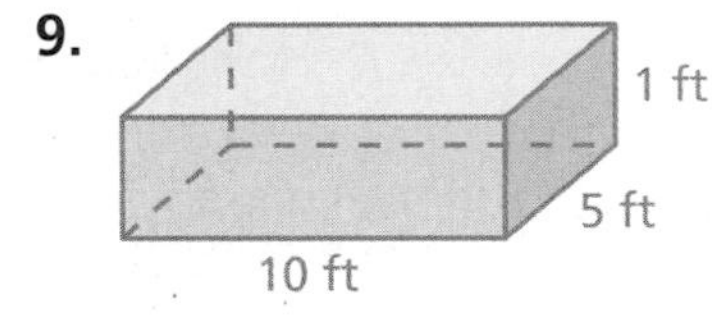

10. 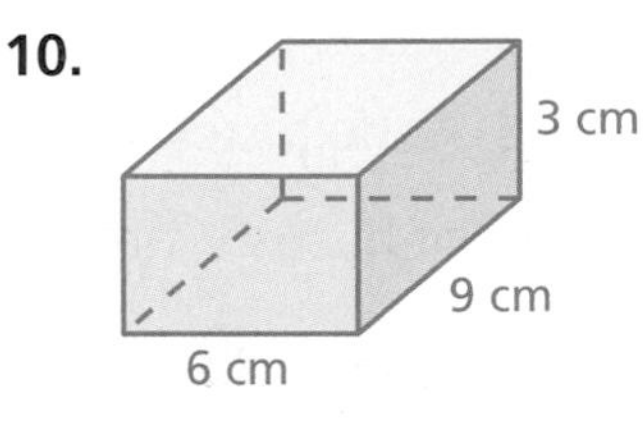

11. 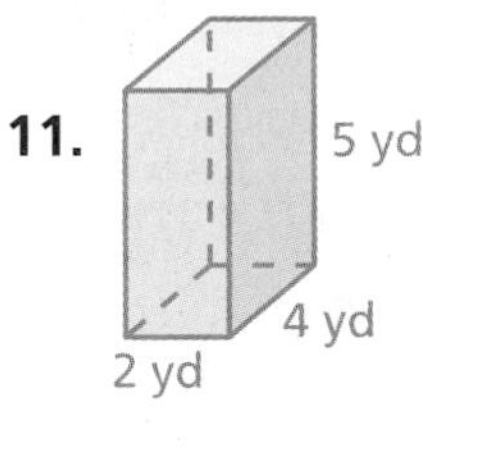

2 12. 

13. 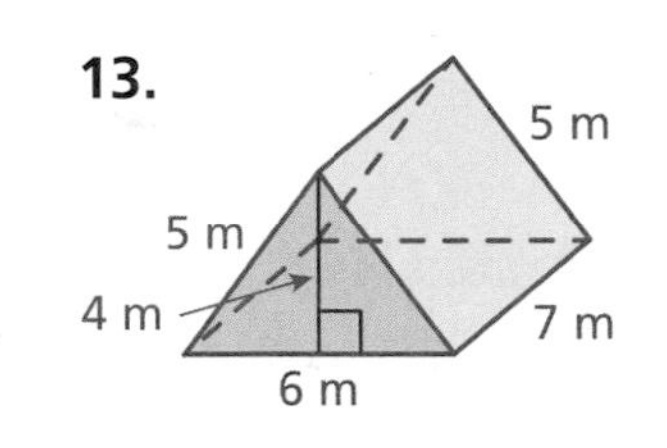

14. 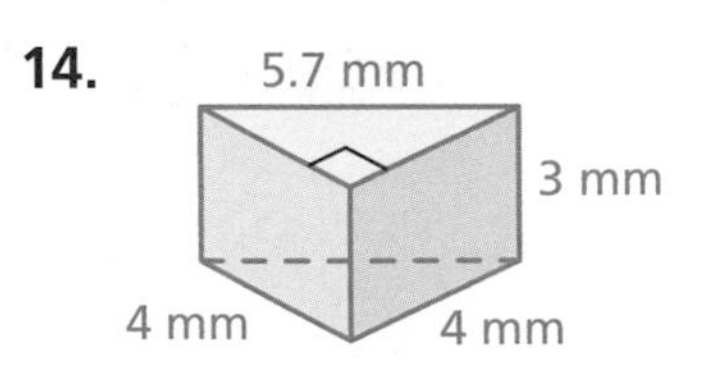

15. **GIFT BOX** What is the least amount of wrapping paper needed to wrap a gift box that measures 8 inches by 8 inches by 10 inches? Explain.

16. **TENT** What is the least amount of fabric needed to make the tent?

Find the surface area of the prism.

17.

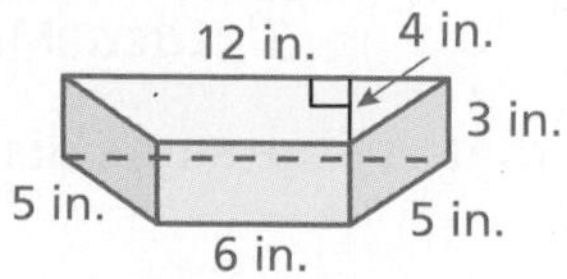

18.

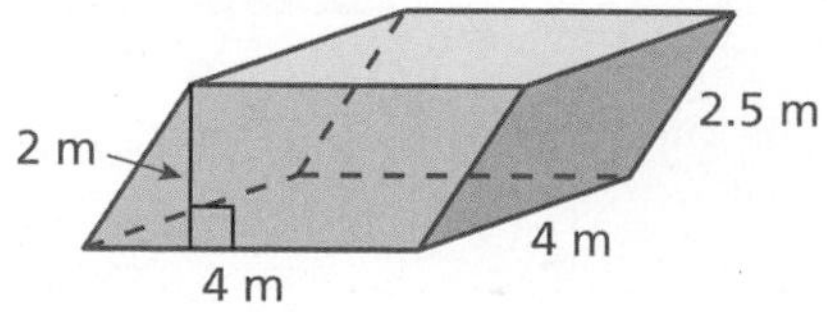

19. AQUARIUM A public library has an aquarium in the shape of a rectangular prism. The base is 6 feet by 2.5 feet. The height is 4 feet. How many square feet of glass were used to build the aquarium? (The top of the aquarium is open.)

20. STORAGE BOX The material used to make a storage box costs \$1.25 per square foot. The boxes have the same volume. How much does a company save by choosing to make 50 of Box 2 instead of 50 of Box 1?

	Length	Width	Height
Box 1	20 in.	6 in.	4 in.
Box 2	15 in.	4 in.	8 in.

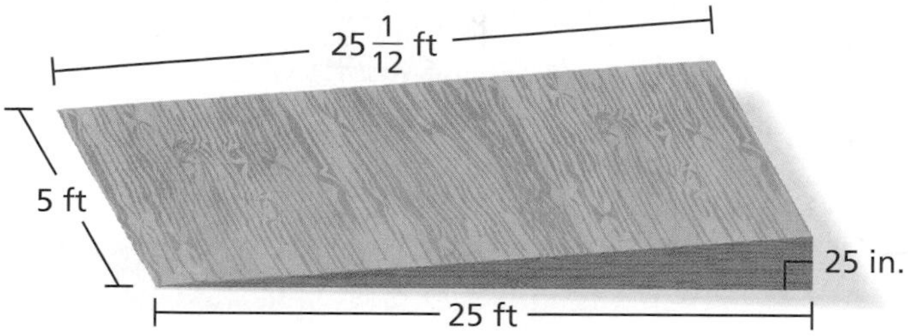

21. RAMP A quart of stain covers 100 square feet. How many quarts should you buy to stain the wheelchair ramp? (Assume you do not have to stain the bottom of the ramp.)

22. LABEL A label that wraps around a box of golf balls covers 75% of its lateral surface area. What is the value of x?

23. Critical Thinking Write a formula for the surface area of a rectangular prism using the height h, the perimeter P of a base, and the area B of a base.

Fair Game Review What you learned in previous grades & lessons

Find the perimeter. *(Skills Review Handbook)*

24.

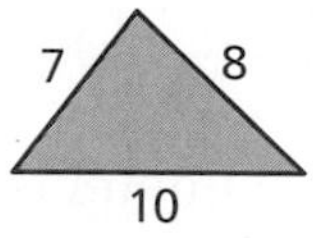

25.

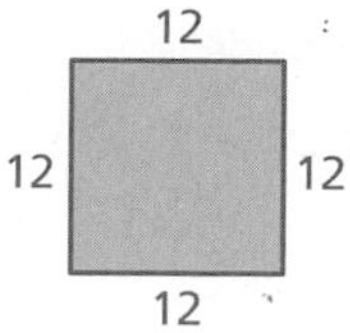

26.

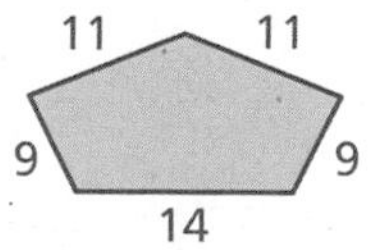

27. MULTIPLE CHOICE The class size increased 25% to 40 students. What was the original class size? *(Section 4.2)*

Ⓐ 10 Ⓑ 30 Ⓒ 32 Ⓓ 50

6.2b Circles

A **circle** is the set of all points in a plane that are the same distance from a point called the **center**.

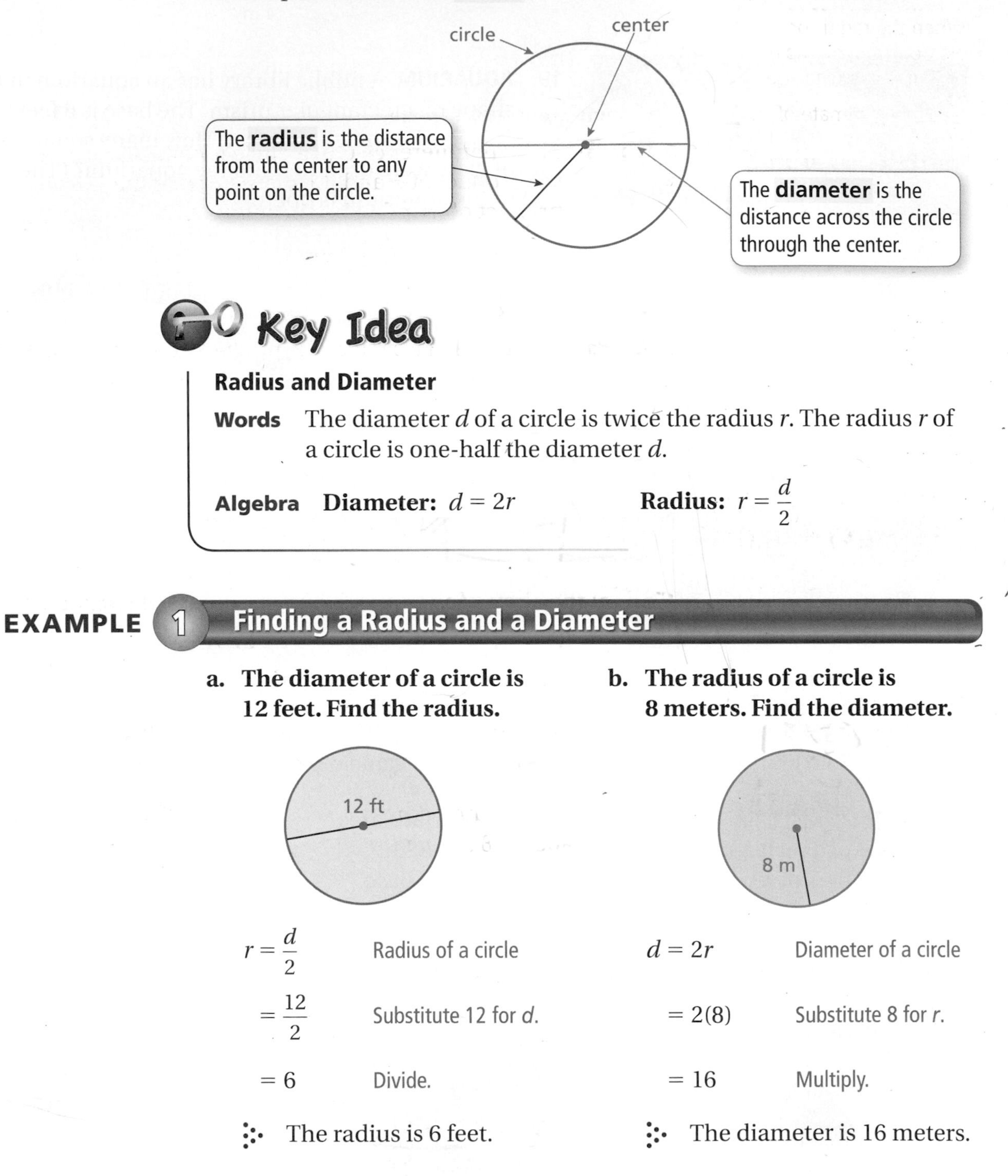

Key Idea

Radius and Diameter

Words The diameter d of a circle is twice the radius r. The radius r of a circle is one-half the diameter d.

Algebra **Diameter:** $d = 2r$ **Radius:** $r = \frac{d}{2}$

EXAMPLE 1 Finding a Radius and a Diameter

a. The diameter of a circle is 12 feet. Find the radius.

$r = \frac{d}{2}$ Radius of a circle

$= \frac{12}{2}$ Substitute 12 for d.

$= 6$ Divide.

The radius is 6 feet.

b. The radius of a circle is 8 meters. Find the diameter.

$d = 2r$ Diameter of a circle

$= 2(8)$ Substitute 8 for r.

$= 16$ Multiply.

The diameter is 16 meters.

Practice

1. **DIAMETER** The radius of a dartboard is 9 inches. Find the diameter.
2. **RADIUS** The diameter of a clock is 1 foot. Find the radius.

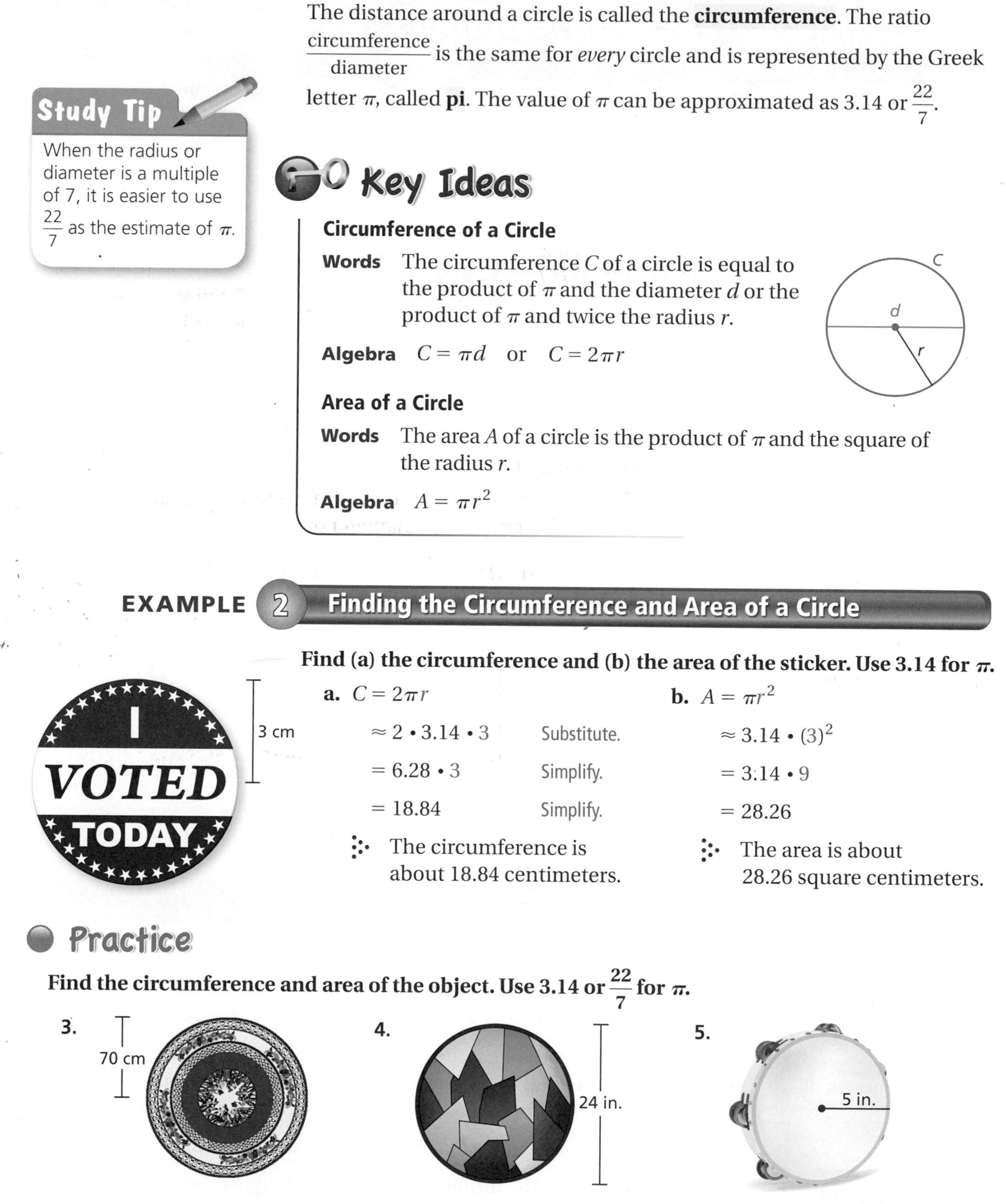

The distance around a circle is called the **circumference**. The ratio $\frac{\text{circumference}}{\text{diameter}}$ is the same for *every* circle and is represented by the Greek letter π, called **pi**. The value of π can be approximated as 3.14 or $\frac{22}{7}$.

Study Tip

When the radius or diameter is a multiple of 7, it is easier to use $\frac{22}{7}$ as the estimate of π.

Key Ideas

Circumference of a Circle

Words The circumference C of a circle is equal to the product of π and the diameter d or the product of π and twice the radius r.

Algebra $C = \pi d$ or $C = 2\pi r$

Area of a Circle

Words The area A of a circle is the product of π and the square of the radius r.

Algebra $A = \pi r^2$

EXAMPLE 2 Finding the Circumference and Area of a Circle

Find (a) the circumference and (b) the area of the sticker. Use 3.14 for π.

a. $C = 2\pi r$

$\approx 2 \cdot 3.14 \cdot 3$ Substitute.

$= 6.28 \cdot 3$ Simplify.

$= 18.84$ Simplify.

The circumference is about 18.84 centimeters.

b. $A = \pi r^2$

$\approx 3.14 \cdot (3)^2$

$= 3.14 \cdot 9$

$= 28.26$

The area is about 28.26 square centimeters.

Practice

Find the circumference and area of the object. Use 3.14 or $\frac{22}{7}$ for π.

3.

4.

5.

6. TIRE The diameter of a bicycle tire is 26 inches.

a. Find the circumference of the tire. Use 3.14 for π.

b. How many rotations does the tire make to travel 95 feet? Explain your reasoning.

6.3 Surface Areas of Cylinders

COMMON CORE STATE STANDARDS
7.G.4
7.G.6

Essential Question How can you find the surface area of a cylinder?

1 ACTIVITY: Finding Area

Work with a partner. Use a cardboard cylinder.

- **Talk about how you can find the area of the outside of the roll.**
- **Use a ruler to estimate the area of the outside of the roll.**
- **Cut the roll and press it out flat. Then find the area of the flattened cardboard. How close is your estimate to the actual area?**

Cut

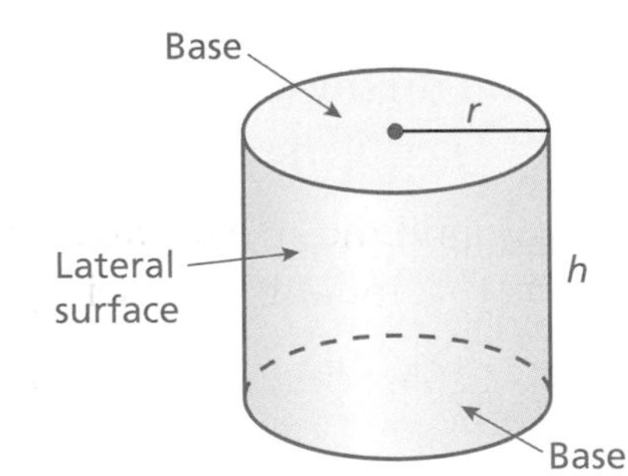

The surface area of a cylinder is the sum of the areas of the bases and the lateral surface.

2 ACTIVITY: Finding Surface Area

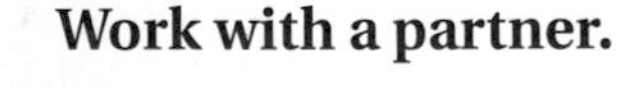

Work with a partner.

- **Trace the top and bottom of a can on paper. Cut out the two shapes.**
- **Cut out a long paper rectangle. Make the width the same as the height of the can. Wrap the rectangle around the can. Cut off the excess paper so the edges just meet.**
- **Make a net for the can. Name the shapes in the net.**
- **How are the dimensions of the rectangle related to the dimensions of the can?**
- **Explain how to use the net to find the surface area of the can.**

3 ACTIVITY: Estimation

Work with a partner. From memory, estimate the dimensions of the real-life item in inches. Then use the dimensions to estimate the surface area of the item in square inches.

What Is Your Answer?

4. **IN YOUR OWN WORDS** How can you find the surface area of a cylinder? Give an example with your description. Include a drawing of the cylinder.

5. To eight decimal places, $\pi \approx 3.14159265$. Which of the following is closest to π?

   a. 3.14 **b.** $\frac{22}{7}$ **c.** $\frac{355}{113}$

"To approximate the irrational number $\pi \approx 3.141593$, I simply remember 1, 1, 3, 3, 5, 5."

"Then I compute the rational number $\frac{355}{113} \approx 3.141593$."

Practice Use what you learned about the surface area of a cylinder to complete Exercises 5–7 on page 266.

6.3 Lesson

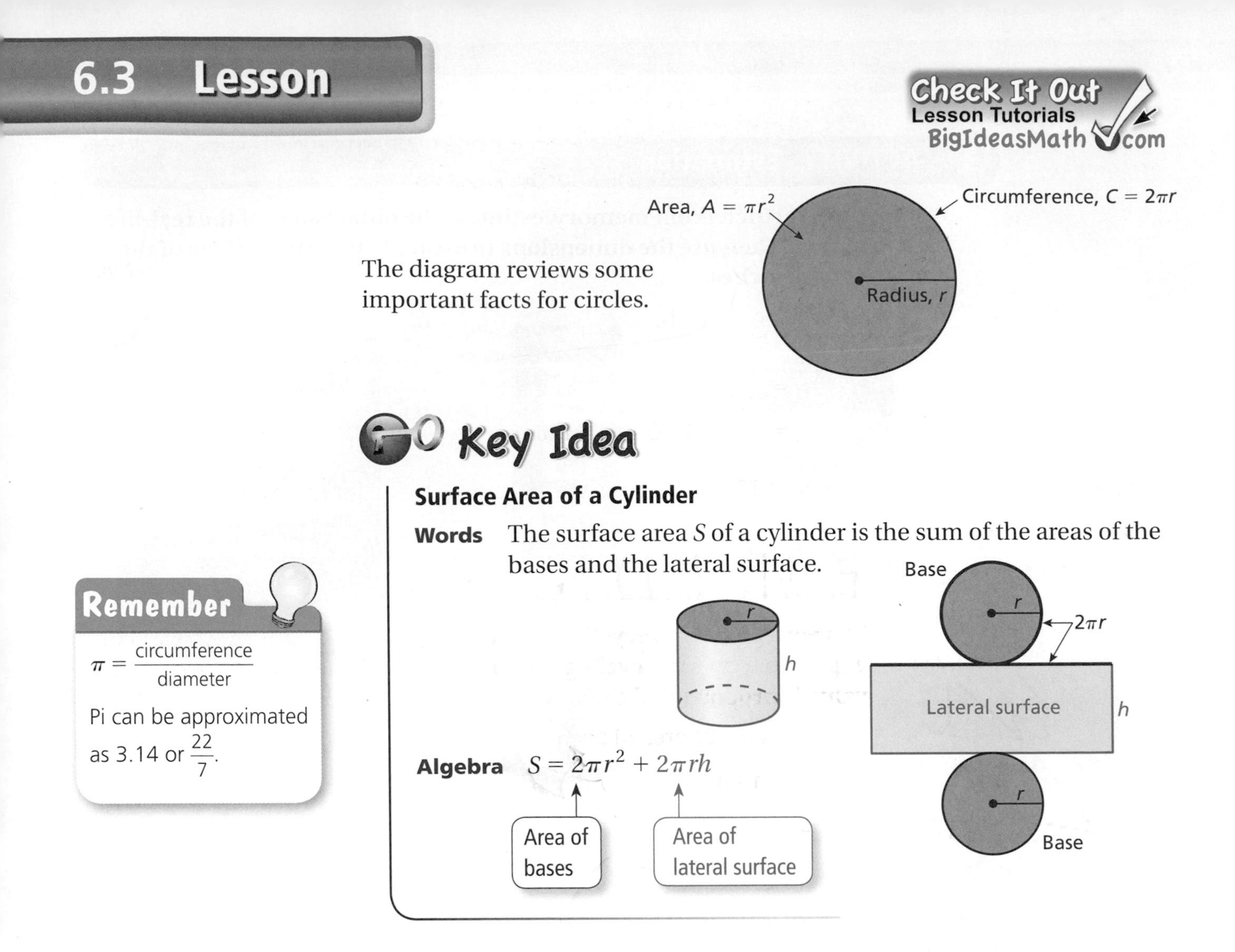

The diagram reviews some important facts for circles.

Key Idea

Surface Area of a Cylinder

Words The surface area S of a cylinder is the sum of the areas of the bases and the lateral surface.

Algebra $S = 2\pi r^2 + 2\pi rh$

Area of bases — Area of lateral surface

Remember

$\pi = \frac{\text{circumference}}{\text{diameter}}$

Pi can be approximated as 3.14 or $\frac{22}{7}$.

EXAMPLE 1 Finding the Surface Area of a Cylinder

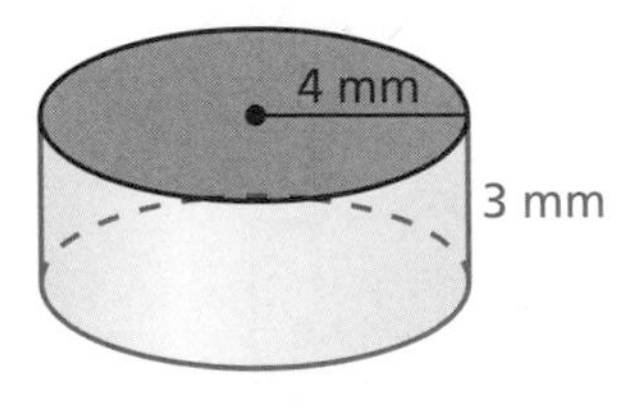

Find the surface area of the cylinder. Round your answer to the nearest tenth.

Draw a net.

$S = 2\pi r^2 + 2\pi rh$

$= 2\pi(4)^2 + 2\pi(4)(3)$

$= 32\pi + 24\pi$

$= 56\pi \approx 175.8$

The surface area is about 175.8 square millimeters.

4 mm
3 mm
4 mm

On Your Own

Now You're Ready Exercises 8–10

1. A cylinder has a radius of 2 meters and a height of 5 meters. Find the surface area of the cylinder. Round your answer to the nearest tenth.

EXAMPLE 2 Finding Surface Area

How much paper is used for the label on the can of peas?

1 in.

2 in.

Find the *lateral* surface area of the cylinder.

$$S = 2\pi rh$$ ← Do not include the area of the bases in the formula.

$= 2\pi(1)(2)$ Substitute.

$= 4\pi \approx 12.56$ Multiply.

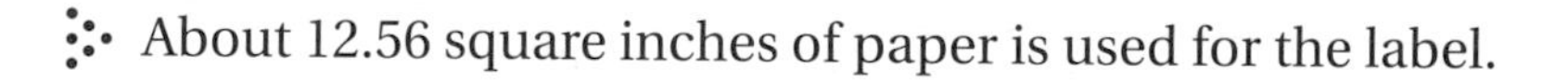

About 12.56 square inches of paper is used for the label.

EXAMPLE 3 Real-Life Application

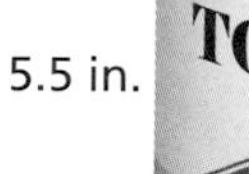

You earn \$0.01 for recycling the can in Example 2. How much can you expect to earn for recycling the tomato can? Assume that the recycle value is proportional to the surface area.

Find the surface area of each can.

Tomatoes

$S = 2\pi r^2 + 2\pi rh$

$= 2\pi(2)^2 + 2\pi(2)(5.5)$

$= 8\pi + 22\pi$

$= 30\pi$

Peas

$S = 2\pi r^2 + 2\pi rh$

$= 2\pi(1)^2 + 2\pi(1)(2)$

$= 2\pi + 4\pi$

$= 6\pi$

Use a proportion to find the recycle value x of the tomato can.

$$\frac{30\pi \text{ in.}^2}{x} = \frac{6\pi \text{ in.}^2}{\$0.01}$$ ← surface area ← recycle value

$30\pi \cdot 0.01 = x \cdot 6\pi$ Use Cross Products Property.

$5 \cdot 0.01 = x$ Divide each side by 6π.

$0.05 = x$ Simplify.

You can expect to earn \$0.05 for recycling the tomato can.

On Your Own

Now You're Ready
Exercises 11–13

2. **WHAT IF?** In Example 3, the height of the can of peas is doubled.
   a. Does the amount of paper used in the label double?
   b. Does the recycle value double? Explain.

6.3 Exercises

Vocabulary and Concept Check

1. **CRITICAL THINKING** Which part of the formula $S = 2\pi r^2 + 2\pi rh$ represents the lateral surface area of a cylinder?

2. **CRITICAL THINKING** Given the height and the circumference of the base of a cylinder, describe how to find the surface area of the entire cylinder.

Find the indicated area of the cylinder.

3. Area of a base

4. Surface area

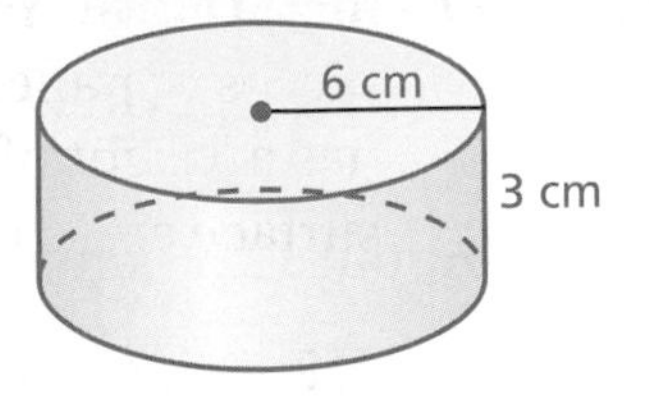

Practice and Problem Solving

Make a net for the cylinder. Then find the surface area of the cylinder. Round your answer to the nearest tenth.

5.

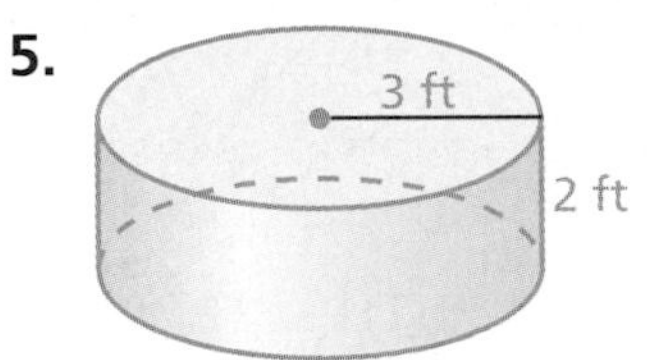

6.

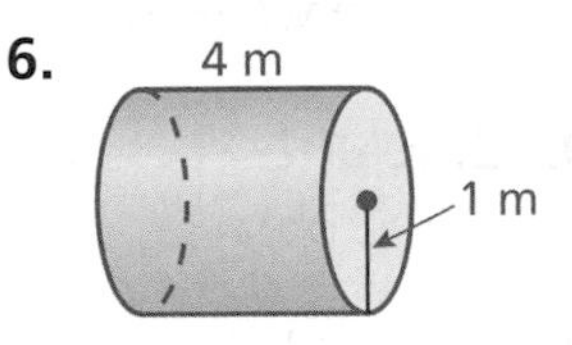

7.

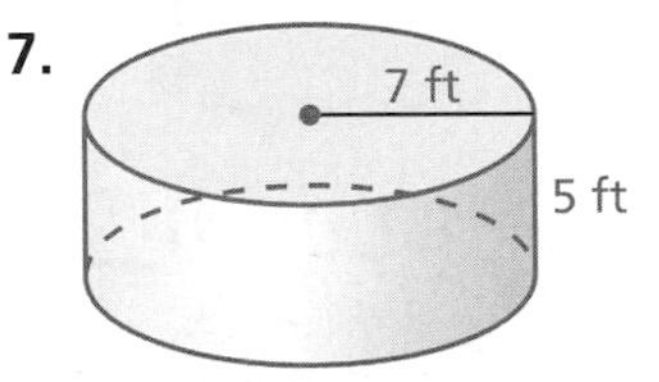

Find the surface area of the cylinder. Round your answer to the nearest tenth.

1 8.

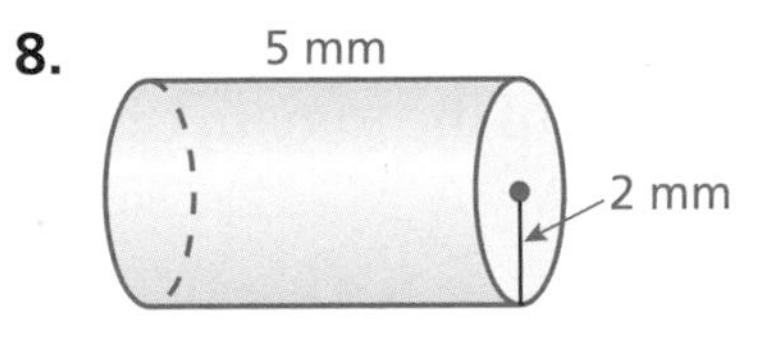

9.

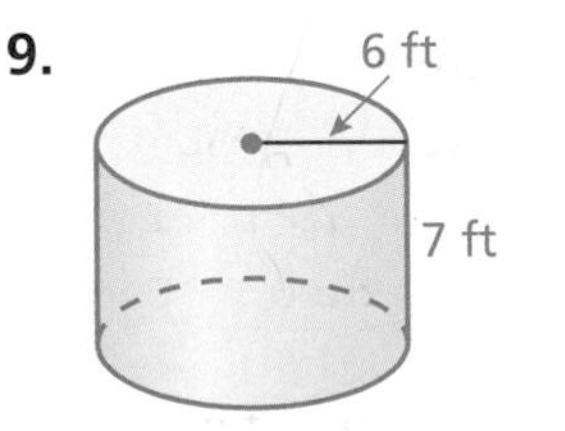

10.

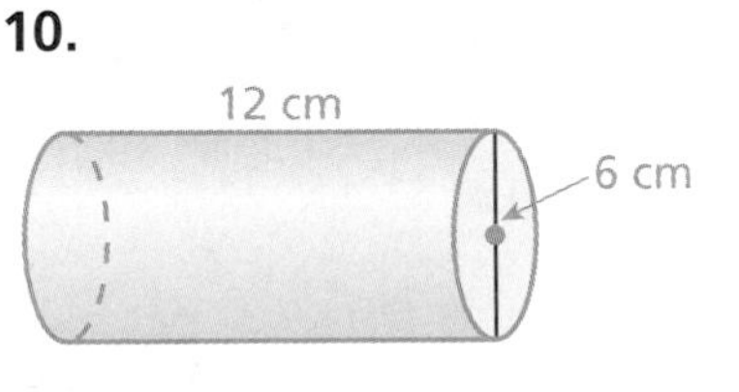

Find the lateral surface area of the cylinder. Round your answer to the nearest tenth.

2 11. 12. 13.

14. **TANKER** The truck's tank is a stainless steel cylinder. Find the surface area of the tank.

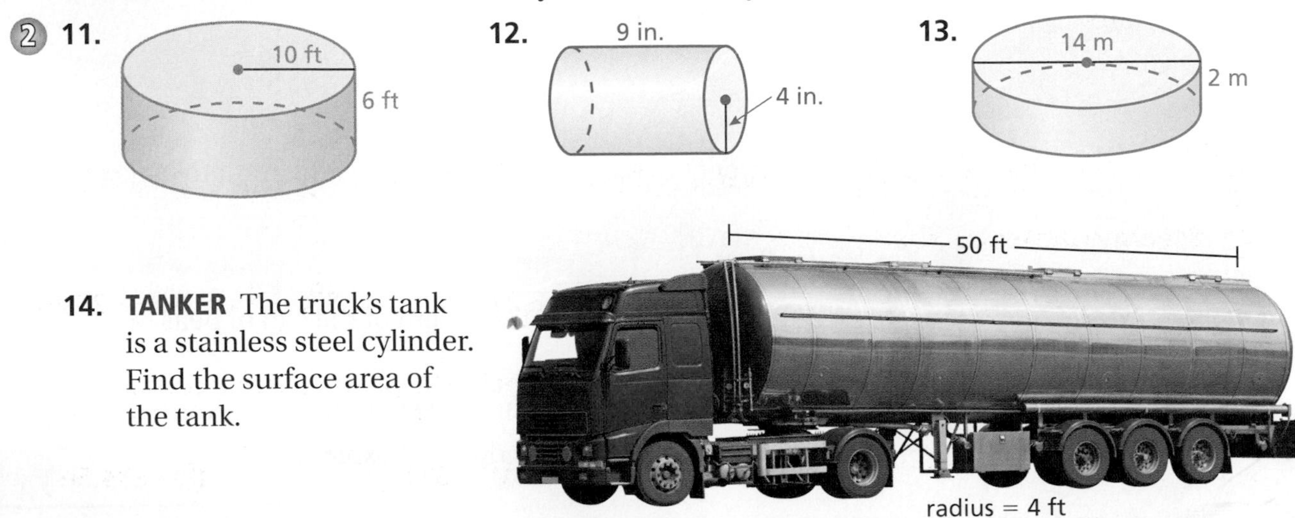

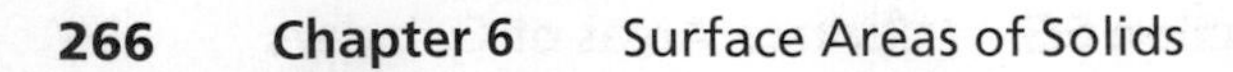

15. **ERROR ANALYSIS** Describe and correct the error in finding the surface area of the cylinder.

16 in.
6 in.
8 in.

16. **OTTOMAN** What percent of the surface area of the ottoman is green (not including the bottom)?

17. **REASONING** You make two cylinders using 8.5-inch by 11-inch pieces of paper. One has a height of 8.5 inches and the other has a height of 11 inches. Without calculating, compare the surface areas of the cylinders.

18. **INSTRUMENT** A ganza is a percussion instrument used in samba music.
   a. Find the surface area of each of the two labeled ganzas.
   b. The weight of the smaller ganza is 1.1 pounds. Assume that the surface area is proportional to the weight. What is the weight of the larger ganza?

19. **BRIE CHEESE** The cut wedge represents one-eighth of the cheese.
   a. Find the surface area of the cheese before it is cut.
   b. Find the surface area of the remaining cheese after the wedge is removed. Did the surface area increase, decrease, or remain the same?

3 in.
1 in.

20. Critical Thinking The lateral surface area of a cylinder is 184 square centimeters. The radius is 9 centimeters. What is the surface area of the cylinder? Explain how you found your answer.

Fair Game Review What you learned in previous grades & lessons

Evaluate the expression. *(Skills Review Handbook)*

21. $\frac{1}{2}(26)(9)$

22. $\frac{1}{2}(8.24)(3) + 8.24$

23. $\frac{1}{2}(18.84)(3) + 28.26$

24. **MULTIPLE CHOICE** A store pays $15 for a basketball. The percent of markup is 30%. What is the selling price? *(Section 4.3)*

(A) $10.50 (B) $19.50 (C) $30 (D) $34.50

6 Study Help

You can use a **four square** to organize information about a topic. Each of the four squares can be a category, such as *definition, vocabulary, example, non-example, words, algebra, table, numbers, visual, graph,* or *equation*. Here is an example of a four square for a three-dimensional figure.

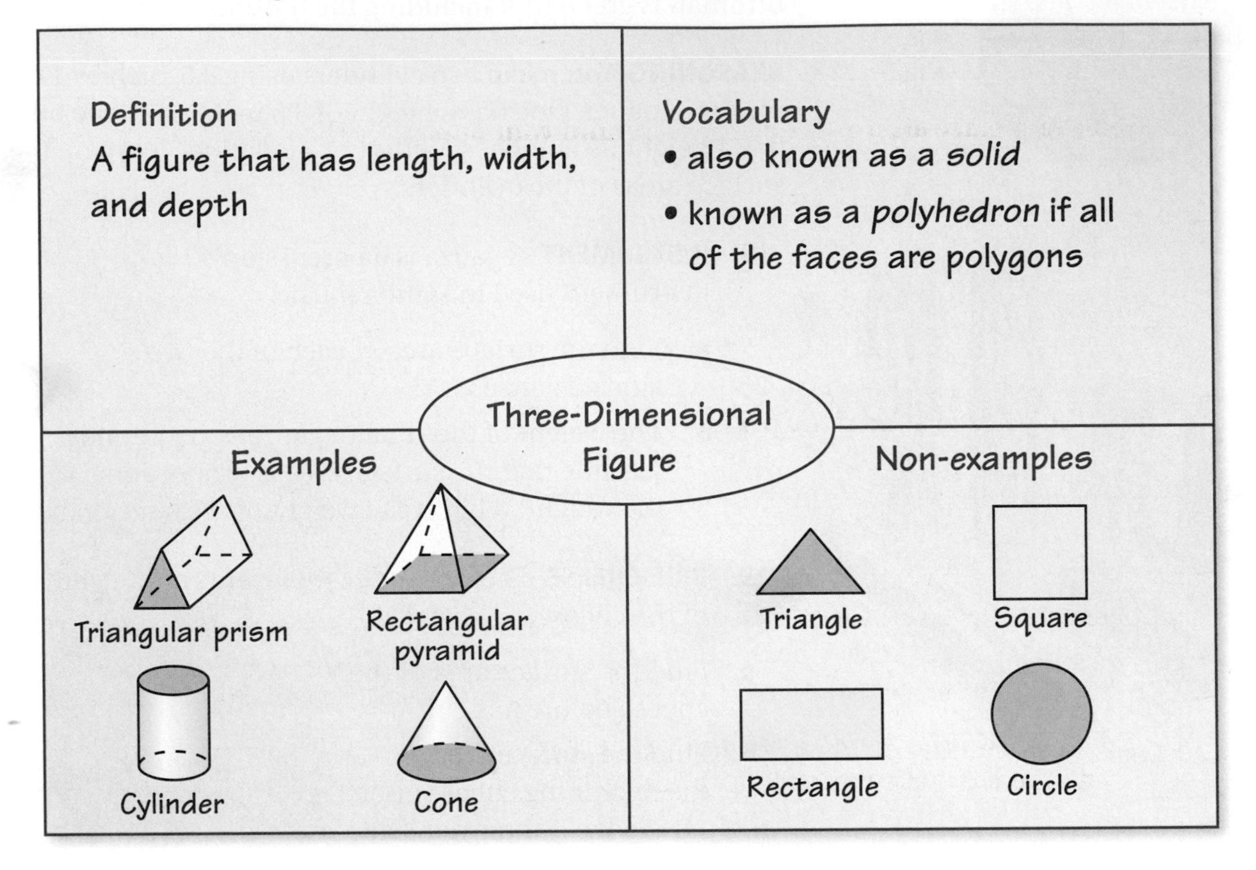

On Your Own

Make a four square to help you study these topics.

1. polyhedron
2. prism
3. pyramid
4. cylinder
5. cone
6. drawing a solid
7. surface area

   a. of a prism   **b.** of a cylinder

After you complete this chapter, make four squares for the following topics.

8. surface area

   a. of a pyramid   **b.** of a cone   **c.** of a composite solid

"My four square shows that my new red skateboard is faster than my old blue skateboard."

6.1–6.3 Quiz

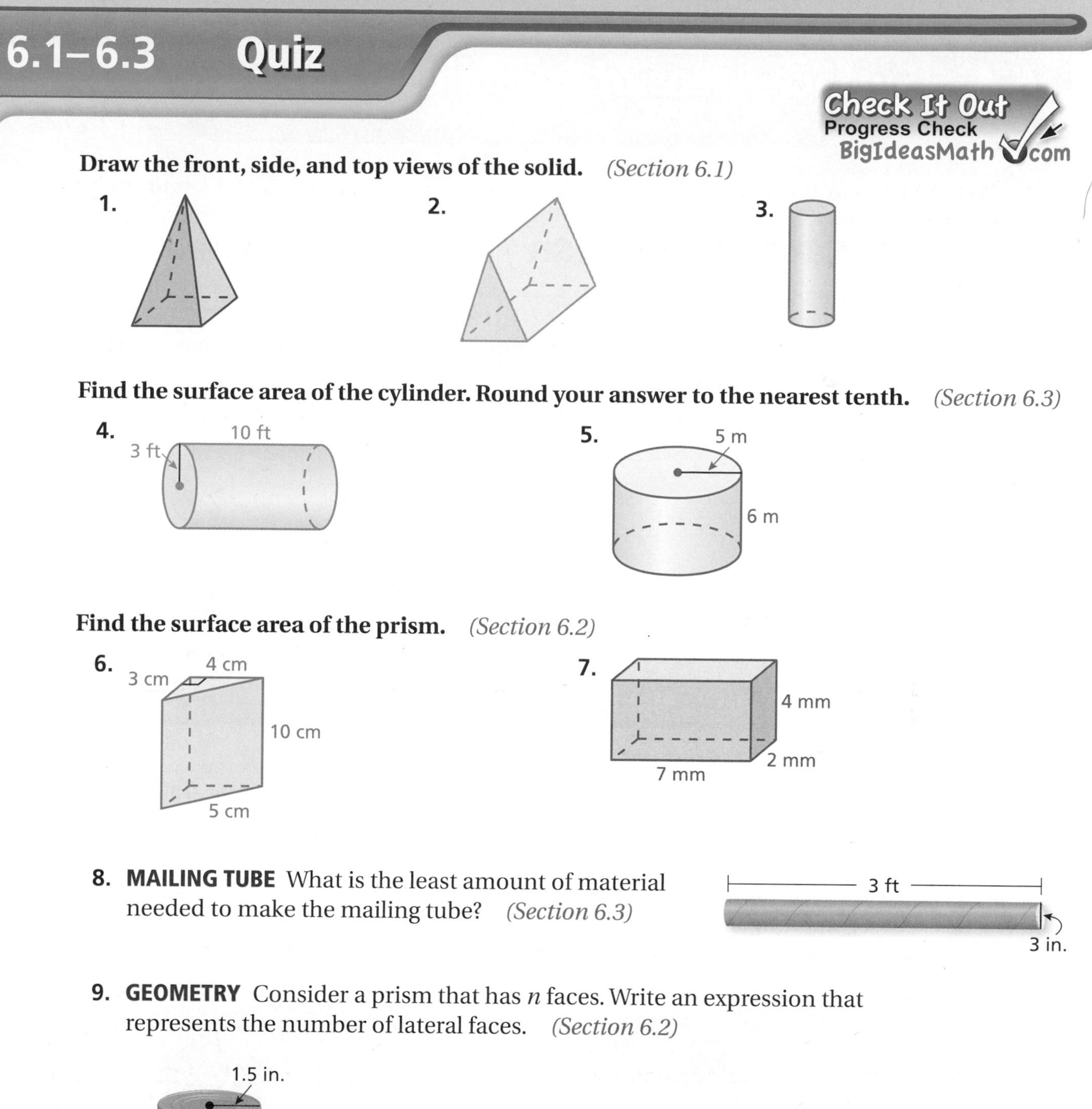

Draw the front, side, and top views of the solid. *(Section 6.1)*

1.

2.

3.

Find the surface area of the cylinder. Round your answer to the nearest tenth. *(Section 6.3)*

4.

5.

Find the surface area of the prism. *(Section 6.2)*

6.

7.

8. MAILING TUBE What is the least amount of material needed to make the mailing tube? *(Section 6.3)*

9. GEOMETRY Consider a prism that has n faces. Write an expression that represents the number of lateral faces. *(Section 6.2)*

10. TOMATO PASTE How much more paper is used for the label of the large can of tomato paste than for the label of the small can? *(Section 6.3)*

11. WOODEN CHEST All the faces of the wooden chest will be painted except for the bottom. Find the area to be painted, in *square inches*. *(Section 6.2)*

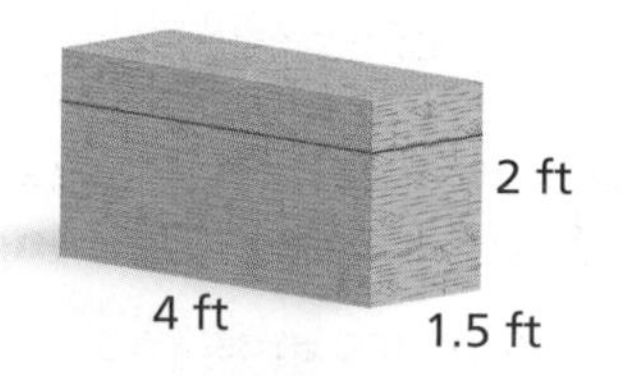

6.4 Surface Areas of Pyramids

COMMON CORE STATE STANDARDS
7.G.6

Essential Question How can you find the surface area of a pyramid?

Even though many well-known **pyramids** have square bases, the base of a pyramid can be any polygon.

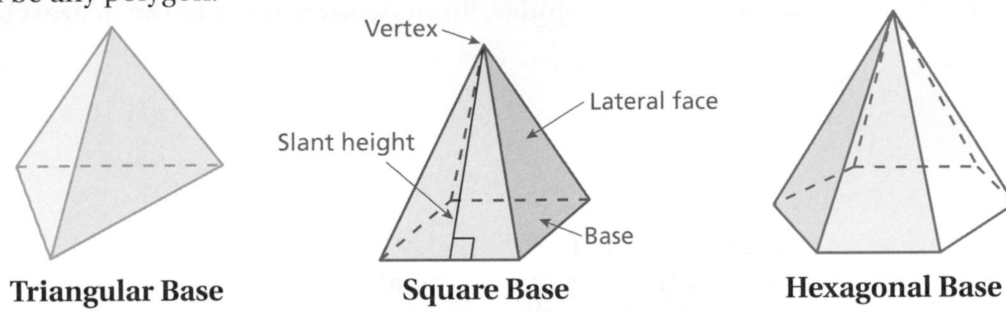

Triangular Base **Square Base** **Hexagonal Base**

1 ACTIVITY: Making a Scale Model

Work with a partner. Each pyramid has a square base.

- **Draw a net for a scale model of one of the pyramids. Describe your scale.**
- **Cut out the net and fold it to form a pyramid.**
- **Find the lateral surface area of the real-life pyramid.**

a. Cheops Pyramid in Egypt

Side = 230 m, Slant height ≈ 186 m

b. Muttart Conservatory in Edmonton

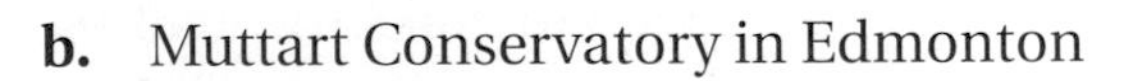

Side = 26 m, Slant height ≈ 27 m

c. Louvre Pyramid in Paris

Side = 35 m, Slant height ≈ 28 m

d. Pyramid of Caius Cestius in Rome

Side = 22 m, Slant height ≈ 29 m

2 ACTIVITY: Estimation

Work with a partner. There are many different types of gemstone cuts. Here is one called a brilliant cut.

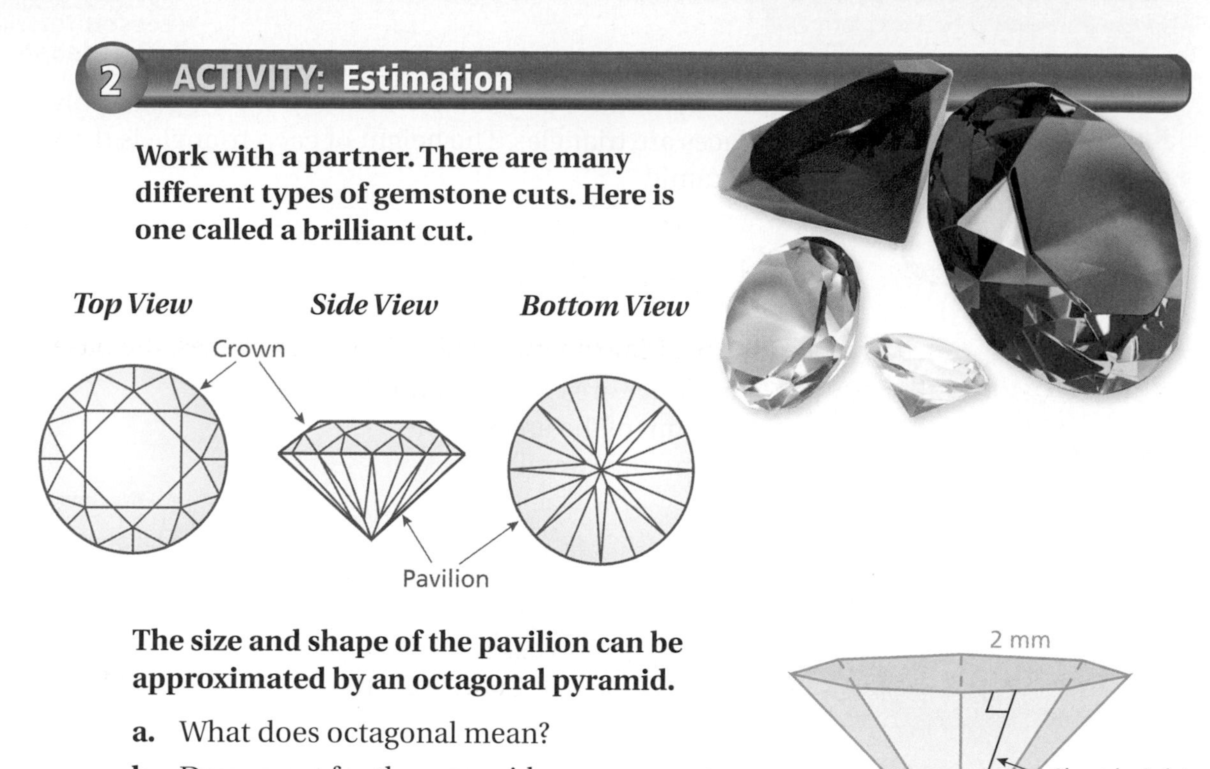

The size and shape of the pavilion can be approximated by an octagonal pyramid.

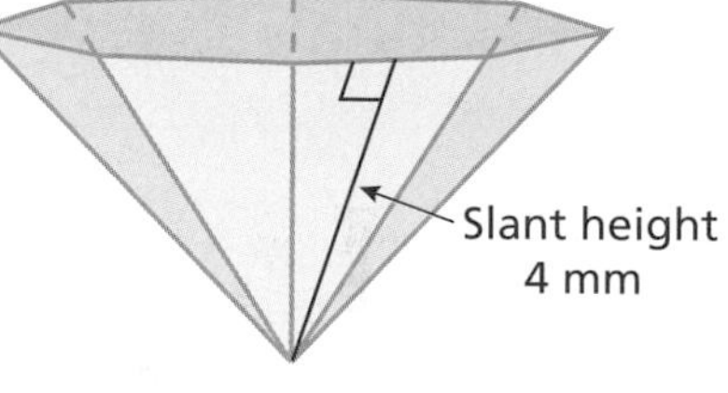

a. What does octagonal mean?

b. Draw a net for the pyramid.

c. Find the lateral surface area of the pyramid.

3 ACTIVITY: Building a Skylight

Work with a partner. The skylight has 12 triangular pieces of glass. Each piece has a base of 1 foot and a slant height of 3 feet.

a. How much glass will you need to make the skylight?

b. Can you cut the 12 glass triangles from a sheet of glass that is 4 feet by 8 feet? If so, draw a diagram showing how this can be done.

What Is Your Answer?

4. **IN YOUR OWN WORDS** How can you find the surface area of a pyramid? Draw a diagram with your explanation.

Use what you learned about the surface area of a pyramid to complete Exercises 4–6 on page 274.

6.4 Lesson

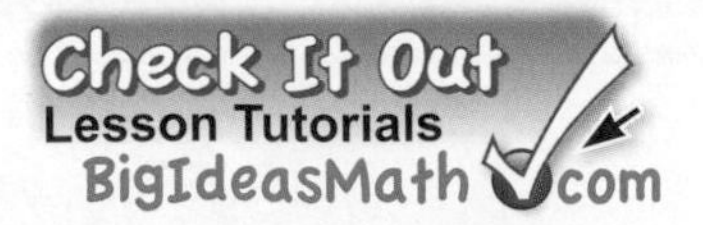

Key Vocabulary
regular pyramid, *p. 272*
slant height, *p. 272*

A **regular pyramid** is a pyramid whose base is a regular polygon. The lateral faces are triangles. The height of each triangle is the **slant height** of the pyramid.

Remember

In a regular polygon, all of the sides have the same length and all of the angles have the same measure.

Key Idea

Surface Area of a Pyramid

The surface area S of a pyramid is the sum of the areas of the base and the lateral faces.

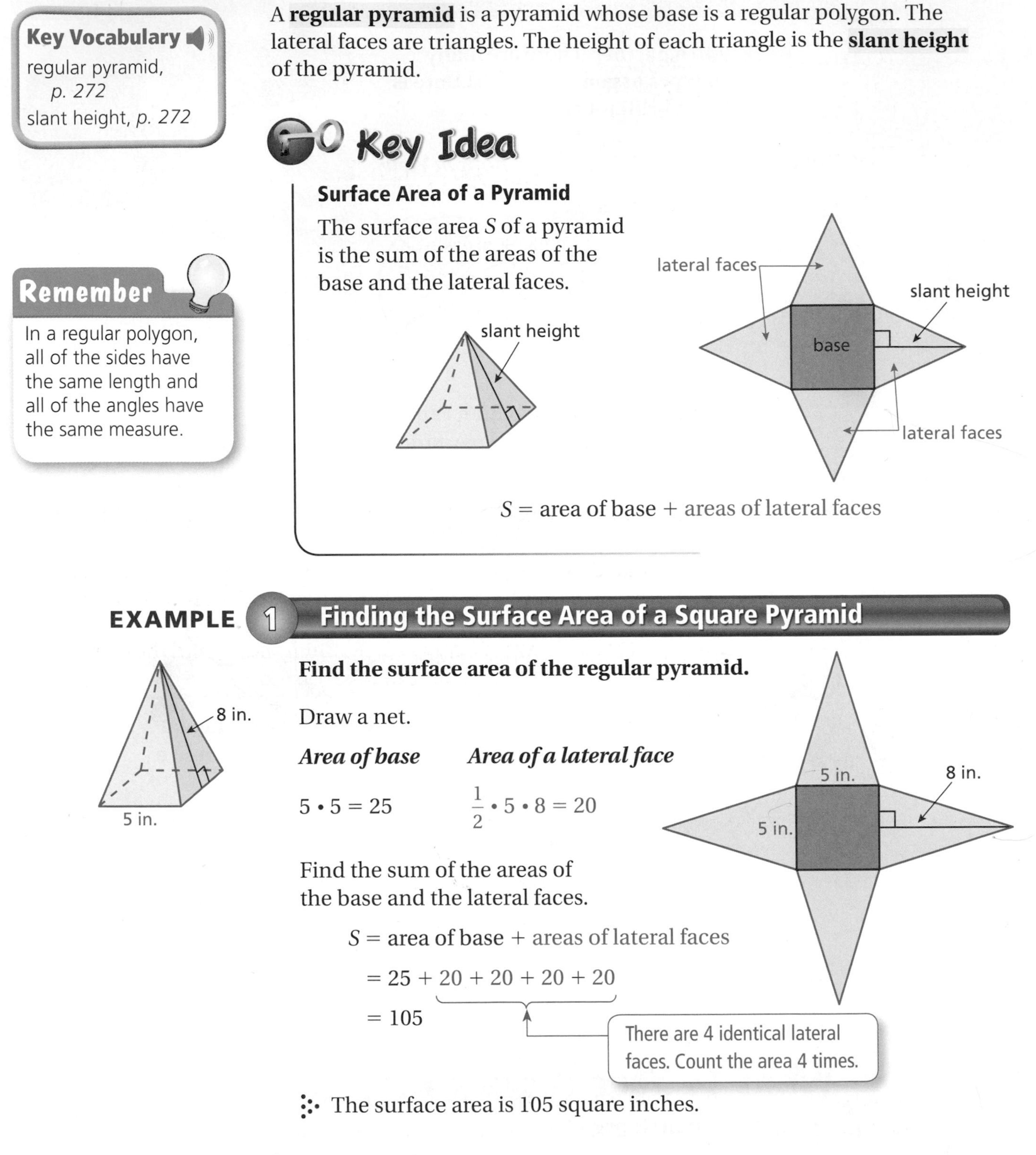

S = area of base + areas of lateral faces

EXAMPLE 1 Finding the Surface Area of a Square Pyramid

Find the surface area of the regular pyramid.

Draw a net.

Area of base	*Area of a lateral face*
$5 \cdot 5 = 25$	$\frac{1}{2} \cdot 5 \cdot 8 = 20$

Find the sum of the areas of the base and the lateral faces.

$$S = \text{area of base} + \text{areas of lateral faces}$$
$$= 25 + 20 + 20 + 20 + 20$$
$$= 105$$

There are 4 identical lateral faces. Count the area 4 times.

∴ The surface area is 105 square inches.

On Your Own

1. What is the surface area of a square pyramid with a base side length of 9 centimeters and a slant height of 7 centimeters?

EXAMPLE 2 Finding the Surface Area of a Triangular Pyramid

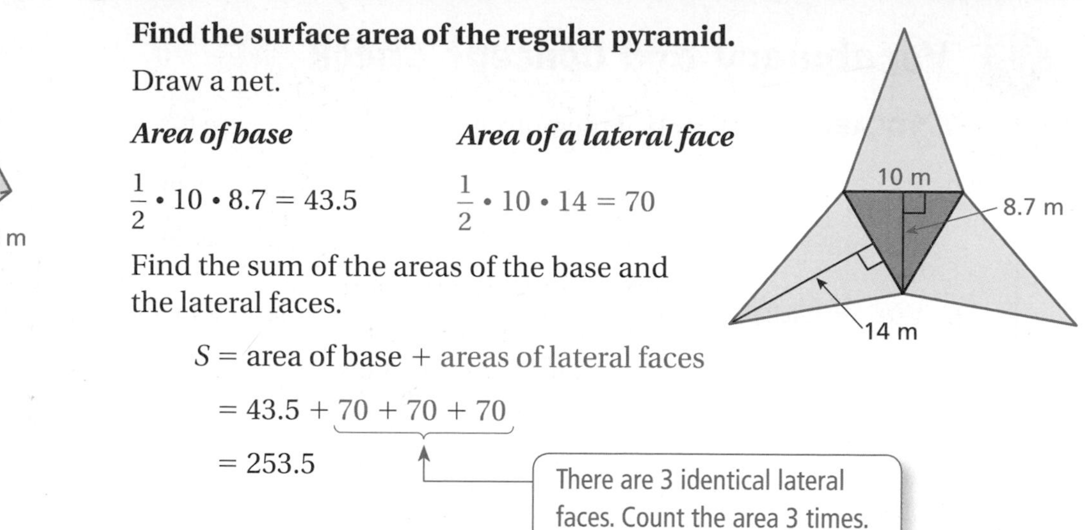

Find the surface area of the regular pyramid.

Draw a net.

Area of base

$\frac{1}{2} \cdot 10 \cdot 8.7 = 43.5$

Area of a lateral face

$\frac{1}{2} \cdot 10 \cdot 14 = 70$

Find the sum of the areas of the base and the lateral faces.

$$S = \text{area of base} + \text{areas of lateral faces}$$
$$= 43.5 + 70 + 70 + 70$$
$$= 253.5$$

There are 3 identical lateral faces. Count the area 3 times.

∴ The surface area is 253.5 square meters.

EXAMPLE 3 Real-Life Application

A roof is shaped like a square pyramid. One bundle of shingles covers 25 square feet. How many bundles should you buy to cover the roof?

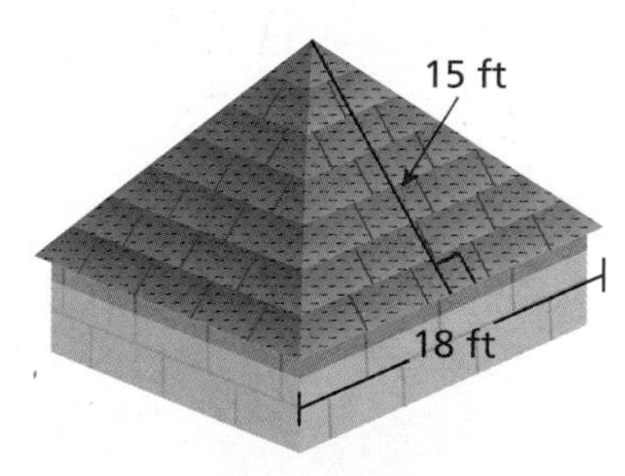

The base of the roof does not need shingles. So, find the sum of the areas of the lateral faces of the pyramid.

Area of a lateral face

$\frac{1}{2} \cdot 18 \cdot 15 = 135$

There are four identical lateral faces. So, the sum of the areas of the lateral faces is

$$135 + 135 + 135 + 135 = 540.$$

Because one bundle of shingles covers 25 square feet, it will take $540 \div 25 = 21.6$ bundles to cover the roof.

∴ So, you should buy 22 bundles of shingles.

On Your Own

Now You're Ready
Exercises 4–12

2. What is the surface area of the pyramid at the right?

3. **WHAT IF?** In Example 3, one bundle of shingles covers 32 square feet. How many bundles should you buy to cover the roof?

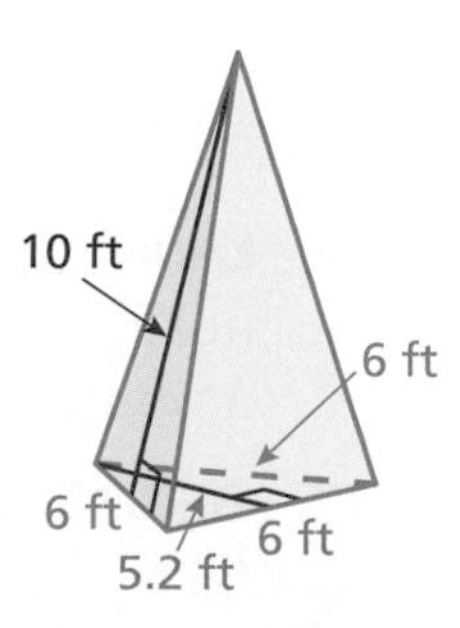

6.4 Exercises

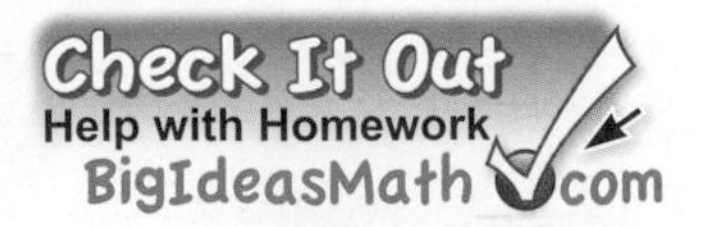

Vocabulary and Concept Check

1. **VOCABULARY** Which of the polygons could be the base for a regular pyramid?

 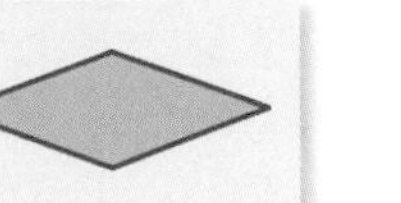  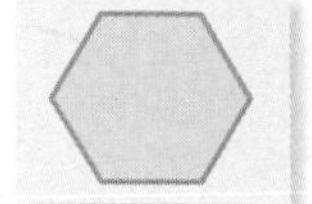

2. **VOCABULARY** Can a pyramid have rectangles as lateral faces? Explain.

3. **CRITICAL THINKING** Why is it helpful to know the slant height of a pyramid to find its surface area?

Practice and Problem Solving

Use the net to find the surface area of the regular pyramid.

4. 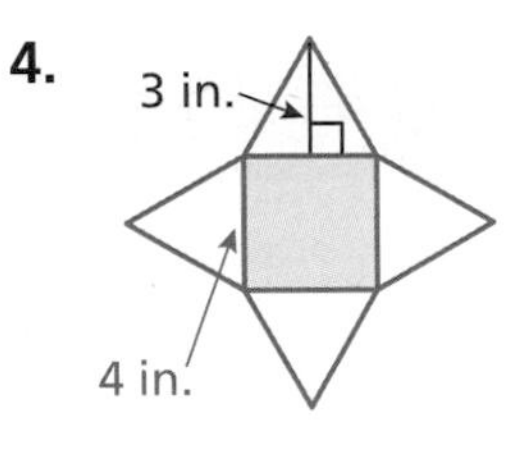

5. 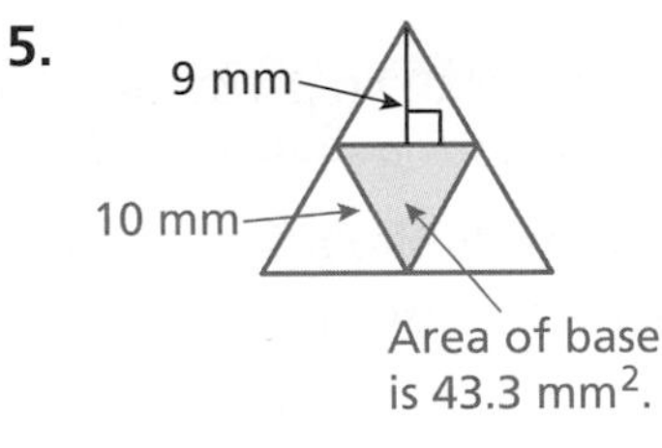

6. 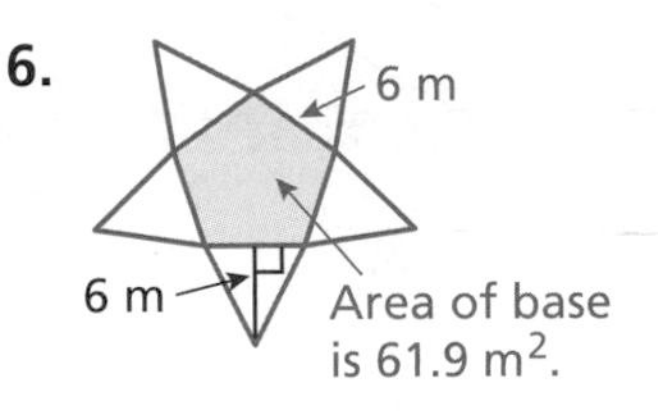

In Exercises 7–11, find the surface area of the regular pyramid.

① ② 7. 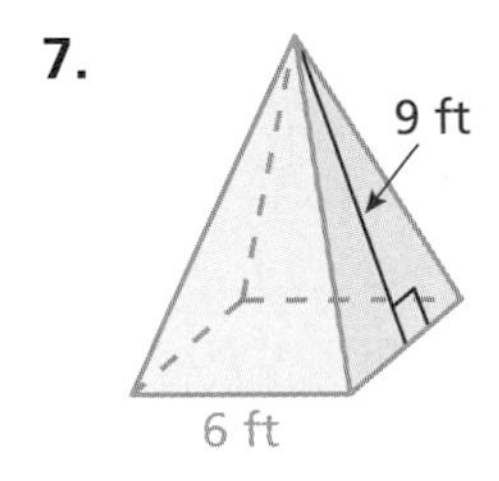

8. 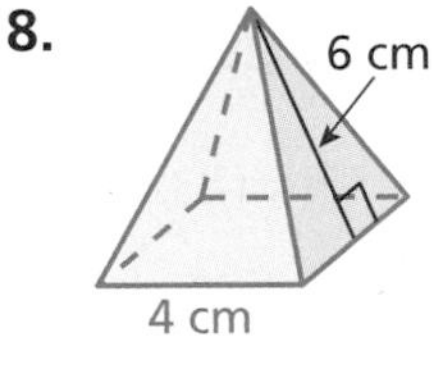

9. 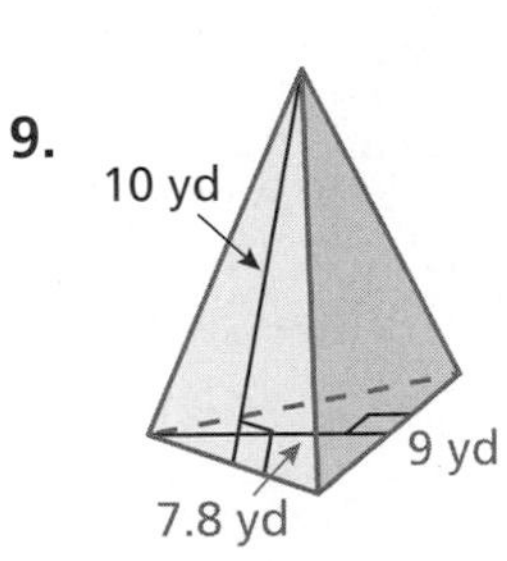

10. 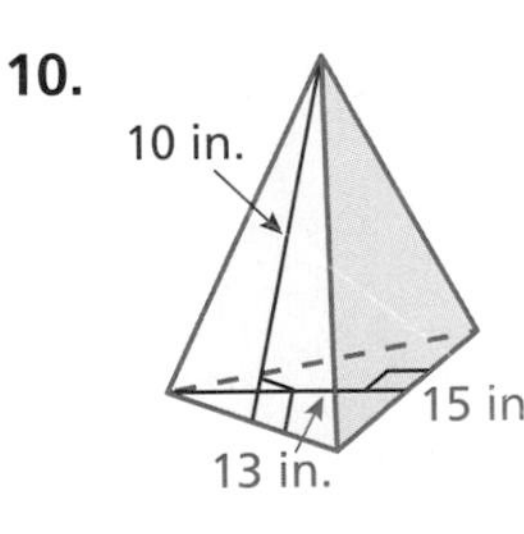

11. 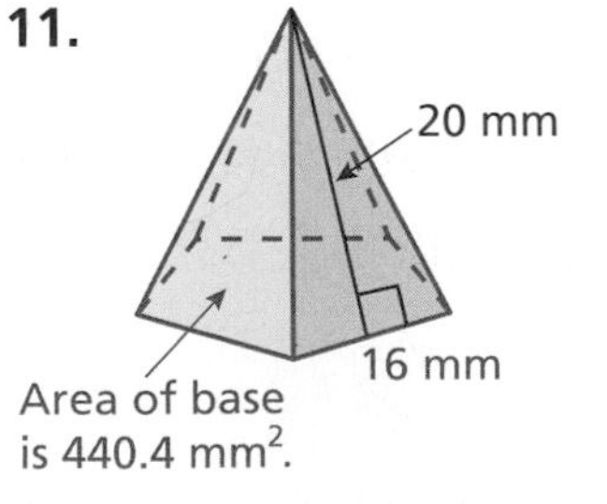

③ 12. **LAMPSHADE** The base of the lampshade is a regular hexagon with a side length of 8 inches. Estimate the amount of glass needed to make the lampshade.

13. **GEOMETRY** The surface area of a square pyramid is 85 square meters. The base length is 5 meters. What is the slant height?

14. **BMX** You are building a bike ramp that is shaped like a square pyramid. You use two 4-foot by 8-foot sheets of plywood. How much plywood do you have left over?

3 ft

5 ft

5 ft

15. **UMBRELLA** You are making an umbrella that is shaped like a regular octagonal pyramid.

5 ft

4 ft

   a. Estimate the amount of fabric that is needed to make the umbrella.

   b. The fabric comes in rolls that are 72 inches wide. You don't want to cut the fabric "on the bias". Find out what this means. Then, draw a diagram of how you can cut the fabric most efficiently.

   c. How much fabric is wasted?

16. **PRECISION** The *height* of a pyramid is the distance between the base and the top of the pyramid. Which is greater, the height of a pyramid or the slant height? Explain your reasoning.

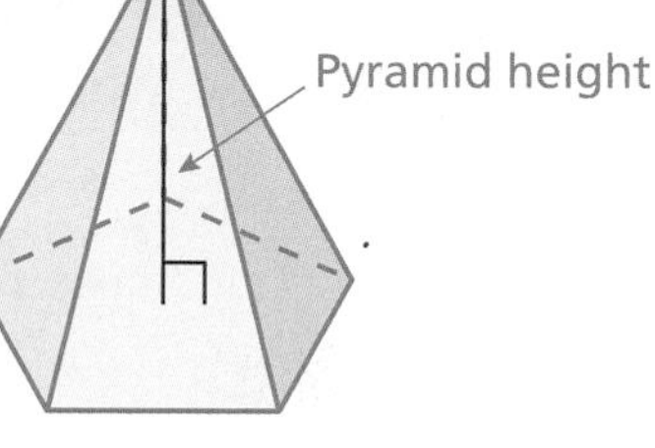

17. **TETRAHEDRON** A tetrahedron is a triangular pyramid whose four faces are identical equilateral triangles. The total lateral surface area is 93 square centimeters. Find the surface area of the tetrahedron.

18. **Reasoning** Is the total area of the lateral faces of a pyramid *greater than*, *less than*, or *equal* to the area of the base? Explain.

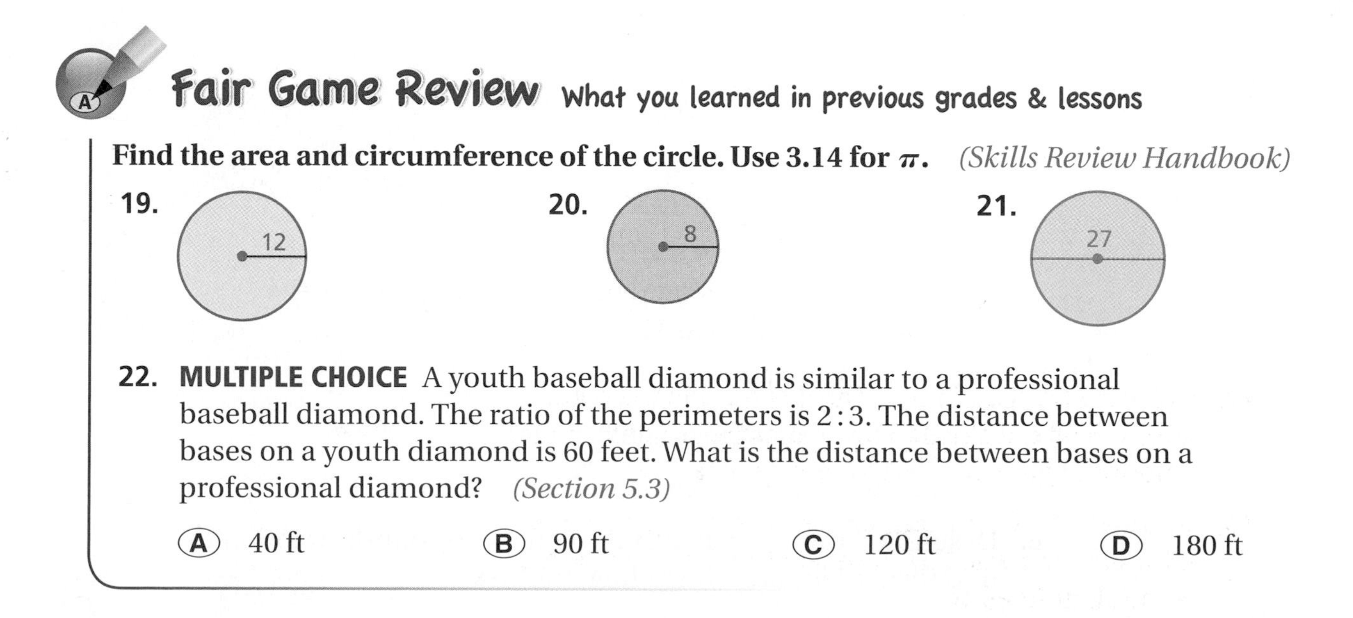

Fair Game Review What you learned in previous grades & lessons

Find the area and circumference of the circle. Use 3.14 for π. *(Skills Review Handbook)*

19. 12

20. 8

21. 27

22. **MULTIPLE CHOICE** A youth baseball diamond is similar to a professional baseball diamond. The ratio of the perimeters is 2 : 3. The distance between bases on a youth diamond is 60 feet. What is the distance between bases on a professional diamond? *(Section 5.3)*

   Ⓐ 40 ft   Ⓑ 90 ft   Ⓒ 120 ft   Ⓓ 180 ft

6.5 Surface Areas of Cones

COMMON CORE STATE STANDARDS
7.G.4
7.G.6

Essential Question How can you find the surface area of a cone?

A cone is a solid with one circular base and one vertex.

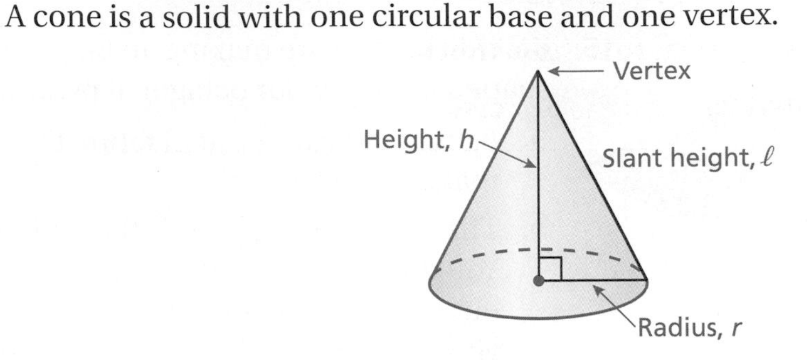

1 ACTIVITY: Finding the Surface Area of a Cone

Work with a partner.

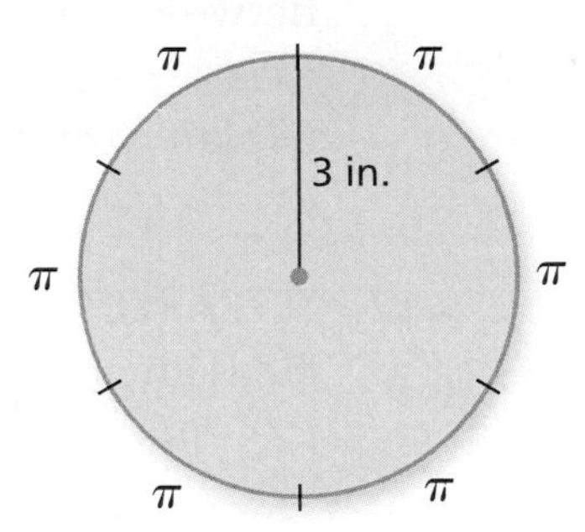

- **Draw a circle with a radius of 3 inches.**
- **Mark the circumference of the circle into six equal parts.**
- **The circumference of the circle is $2(\pi)(3) = 6\pi$. So each of the six parts on the circle has a length of π. Label each part.**
- **Cut out one part as shown. Then, make a cone.**

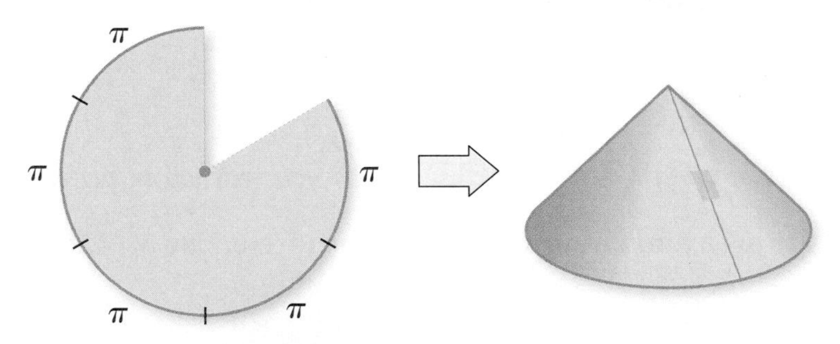

a. The base of the cone should be a circle. Explain why the circumference of the base is 5π.

b. Find the radius of the base.

c. What is the area of the original circle?

d. What is the area of the circle with one part missing?

e. Describe the surface area of the cone. Use your description to find the surface area, including the base.

2 ACTIVITY: Experimenting with Surface Area

Work with a partner.

- **Cut out another part from the circle in Activity 1 and make a cone.**
- **Find the radius of the base and the surface area of the cone.**
- **Record your results in the table.**
- **Repeat this three times.**
- **Describe the pattern.**

Shape					
Radius of Base					
Slant Height					
Surface Area					

3 ACTIVITY: Writing a Story

Write a story that uses real-life cones. Include a diagram and label the dimensions. In your story, explain why you would want to know the surface area of the cone. Then, estimate the surface area.

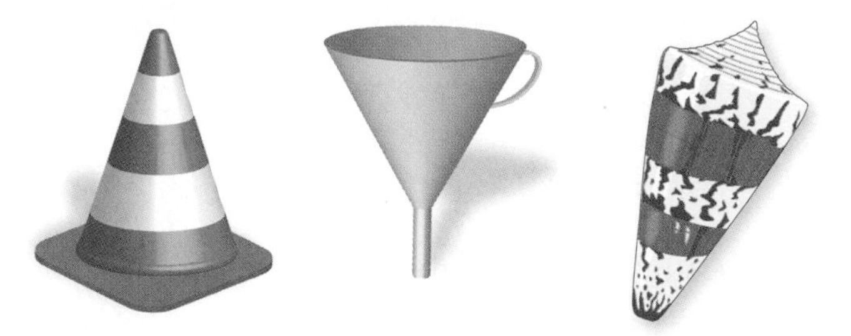

What Is Your Answer?

4. **IN YOUR OWN WORDS** How can you find the surface area of a cone? Draw a diagram with your explanation.

Practice Use what you learned about the surface area of a cone to complete Exercises 4–6 on page 280.

6.5 Lesson

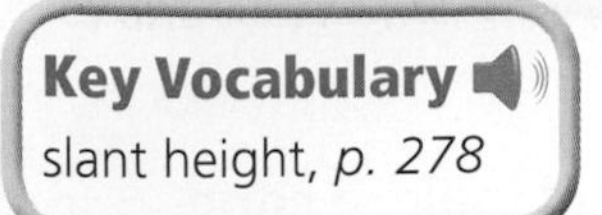

The distance from the vertex of a cone to any point on the edge of its base is called the **slant height** of the cone.

Surface Area of a Cone

Words The surface area S of a cone is the sum of the areas of the base and the lateral surface.

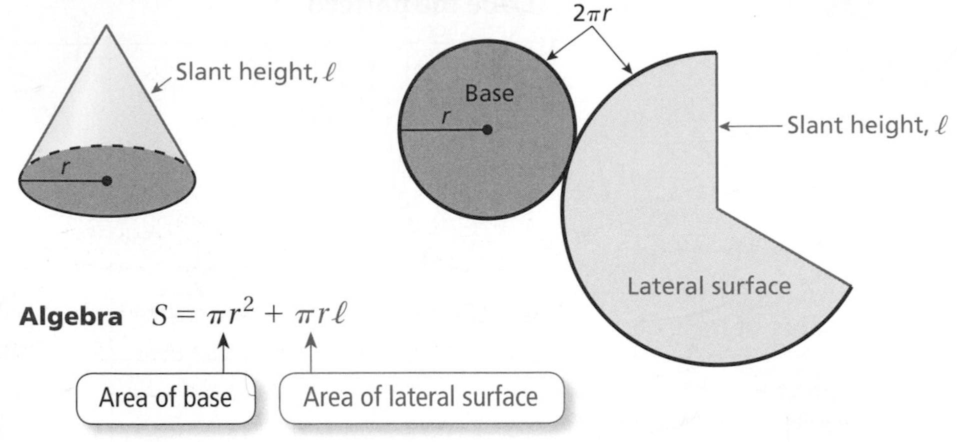

Algebra $S = \pi r^2 + \pi r \ell$

Area of base — Area of lateral surface

EXAMPLE 1 Finding the Surface Area of a Cone

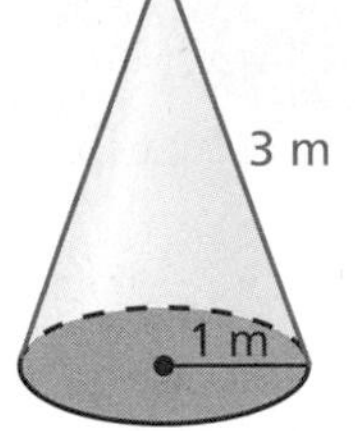

Find the surface area of the cone. Round your answer to the nearest tenth.

Draw a net.

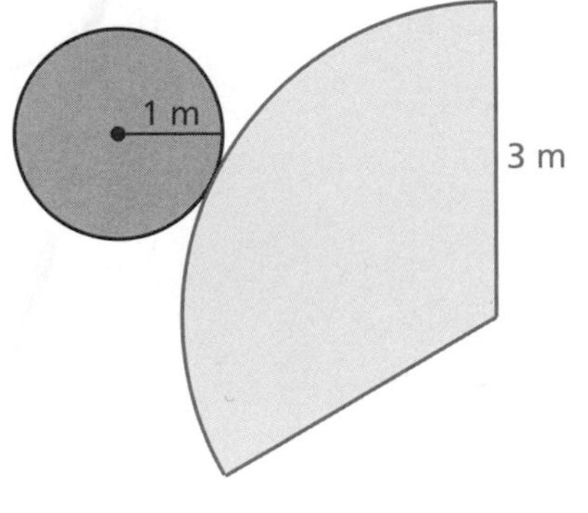

$$S = \pi r^2 + \pi r \ell$$
$$= \pi(1)^2 + \pi(1)(3)$$
$$= \pi + 3\pi$$
$$= 4\pi \approx 12.6$$

The surface area is about 12.6 square meters.

On Your Own

Now You're Ready
Exercises 4–9

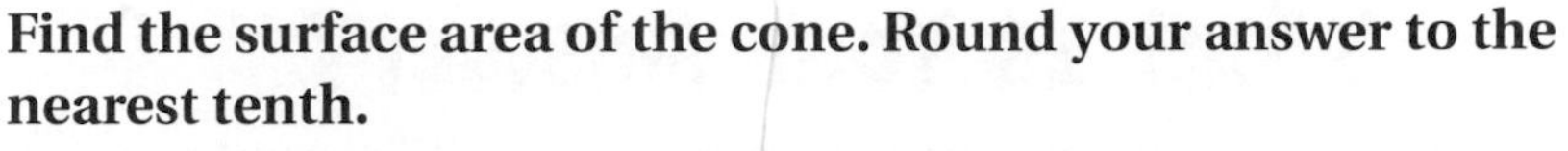

Find the surface area of the cone. Round your answer to the nearest tenth.

1.

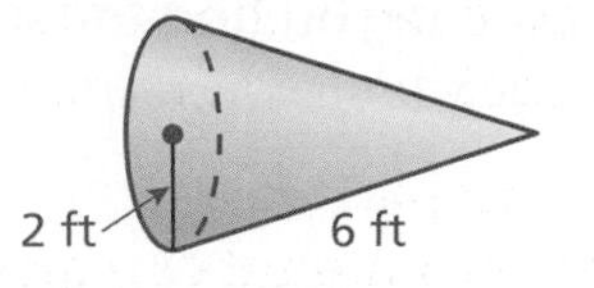

2.

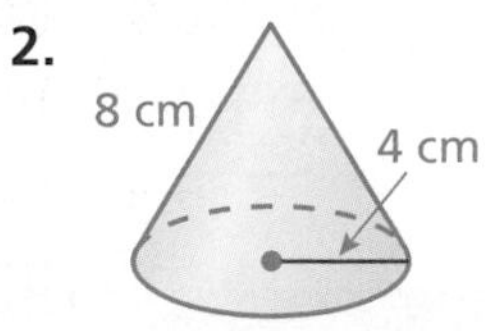

EXAMPLE 2 Finding the Slant Height of a Cone

The surface area of the cone is 100π square meters. What is the slant height ℓ of the cone?

$S = \pi r^2 + \pi r \ell$	Write formula.
$100\pi = \pi(5)^2 + \pi(5)(\ell)$	Substitute.
$100\pi = 25\pi + 5\pi\ell$	Simplify.
$75\pi = 5\pi\ell$	Subtract 25π from each side.
$15 = \ell$	Divide each side by 5π.

∴ The slant height is 15 meters.

EXAMPLE 3 Real-Life Application

You design a party hat. You attach a piece of elastic along a diameter. (a) How long is the elastic? (b) How much paper do you need to make the hat?

5 in.

C = 22 in.

a. To find the length of the elastic, find the diameter of the base.

$C = \pi d$	Write formula.
$22 \approx (3.14)d$	Substitute.
$7.0 \approx d$	Solve for d.

∴ The elastic is about 7 inches long.

b. To find how much paper you need, find the lateral surface area.

Remember

The diameter d of a circle is two times the radius r.

$d = 2r$

$S = \pi r \ell$ ← Do not include the area of the base in the formula.

$= \pi(3.5)(5)$	Substitute.
$= 17.5\pi \approx 55$	Multiply.

∴ You need about 55 square inches of paper to make the hat.

On Your Own

Now You're Ready
Exercises 10–14

3. **WHAT IF?** In Example 2, the surface area is 50π square meters. What is the slant height of the cone?

4. **WHAT IF?** In Example 3, the slant height of the party hat is doubled. Does the amount of paper used double? Explain.

6.5 Exercises

Vocabulary and Concept Check

1. **VOCABULARY** Is the base of a cone a polygon? Explain.
2. **CRITICAL THINKING** In the formula for the surface area of a cone, what does $\pi r \ell$ represent? What does πr^2 represent?
3. **REASONING** Write an inequality comparing the slant height ℓ and the radius r of a cone.

Practice and Problem Solving

Find the surface area of the cone. Round your answer to the nearest tenth.

① 4.

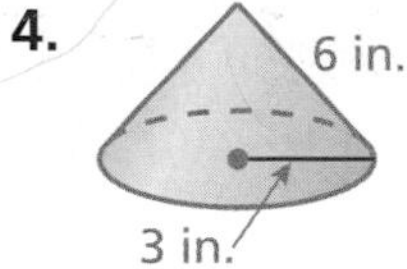

5. 5 m, 4 m

6.

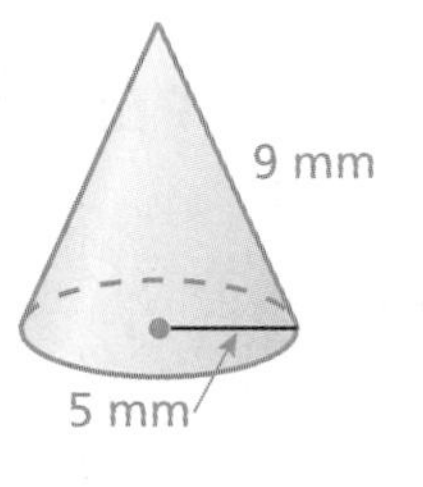

7.

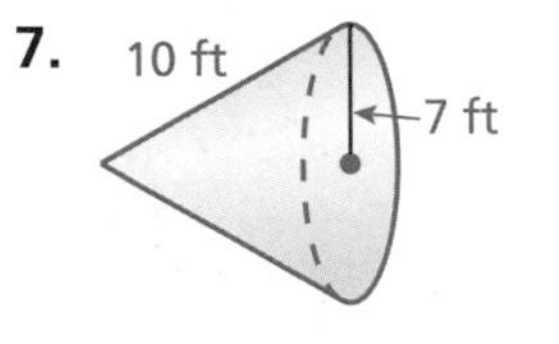

8.

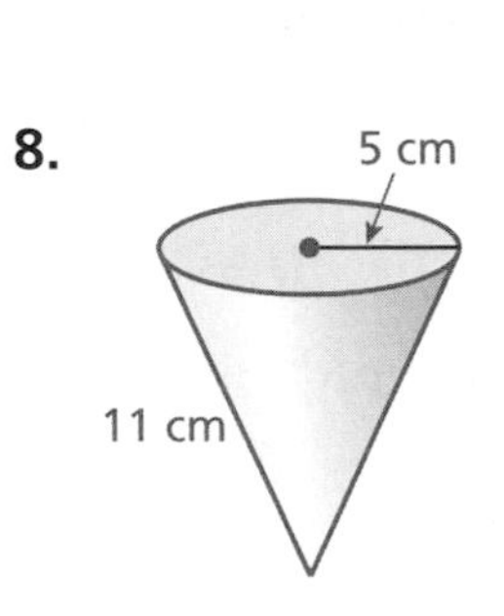

9.

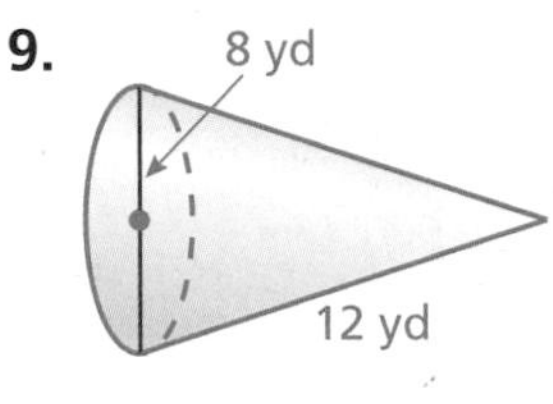

Find the slant height ℓ of the cone.

② 10. $S = 33\pi \text{ in.}^2$

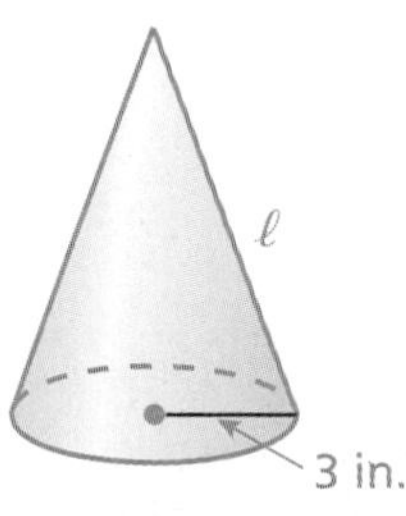

11. $S = 126\pi \text{ cm}^2$

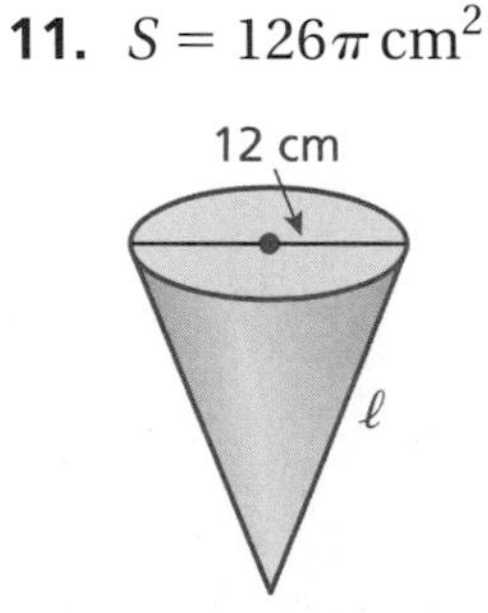

12. $S = 60\pi \text{ ft}^2$

ℓ, 5 ft

③ 13. **NÓN LÁ** How much material is needed to make the Nón Lá Vietnamese leaf hat?

14. **PAPER CUP** A paper cup shaped like a cone has a diameter of 6 centimeters and a slant height of 7.5 centimeters. How much paper is needed to make the cup?

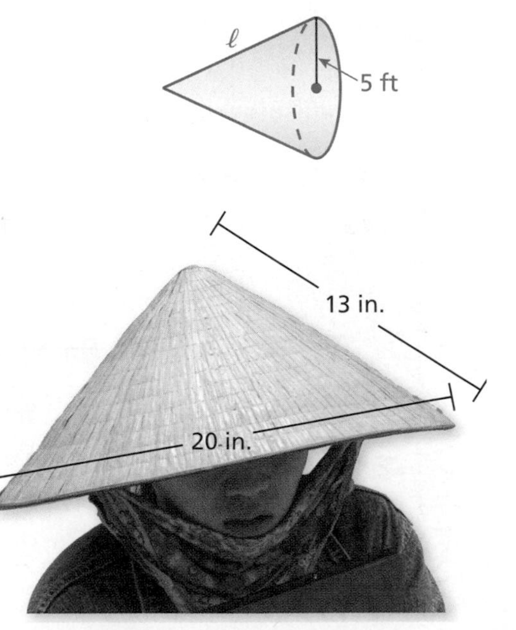

Find the surface area of the cone with diameter d and slant height ℓ.

15. $d = 2$ ft
$\ell = 18$ in.

16. $d = 12$ cm
$\ell = 85$ mm

17. $d = 4$ yd
$\ell = 10$ ft

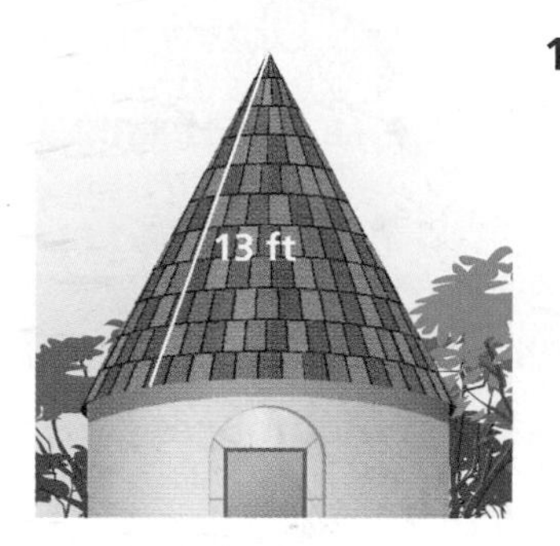

18. ROOF A roof is shaped like a cone with a diameter of 12 feet. One bundle of shingles covers 32 square feet. How many bundles should you buy to cover the roof?

19. MEGAPHONE Two stickers are placed on opposite sides of the megaphone. Estimate the percent of the surface area of the megaphone covered by the stickers. Round your answer to the nearest percent.

2.25 ft
6 in.
6 in.
1.2 ft

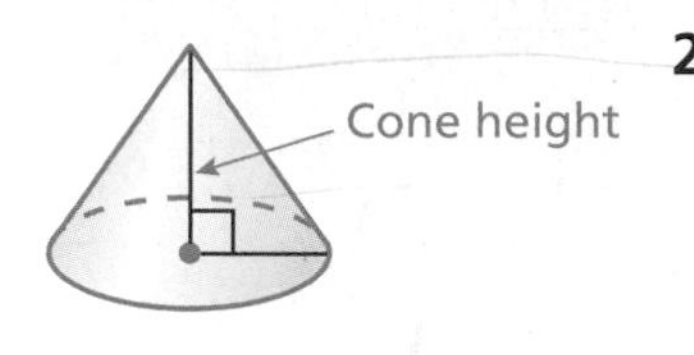

20. REASONING The height of a cone is the distance between the base and the vertex. Which is greater, the height of a cone or the slant height? Explain your reasoning.

21. GEOMETRY The surface area of a cone is also given as $S = \frac{1}{2}C\ell + B$, where C is the circumference and ℓ is the slant height. What does $\frac{1}{2}C\ell$ represent?

22. Critical Thinking A cone has a diameter of x millimeters and a slant height of y millimeters. A square pyramid has a base side length of x millimeters and a slant height of y millimeters. Which has the greater surface area? Explain.

Fair Game Review What you learned in previous grades & lessons

Find the area of the shaded region. Use 3.14 for π. *(Skills Review Handbook)*

23. **24.** **25.**

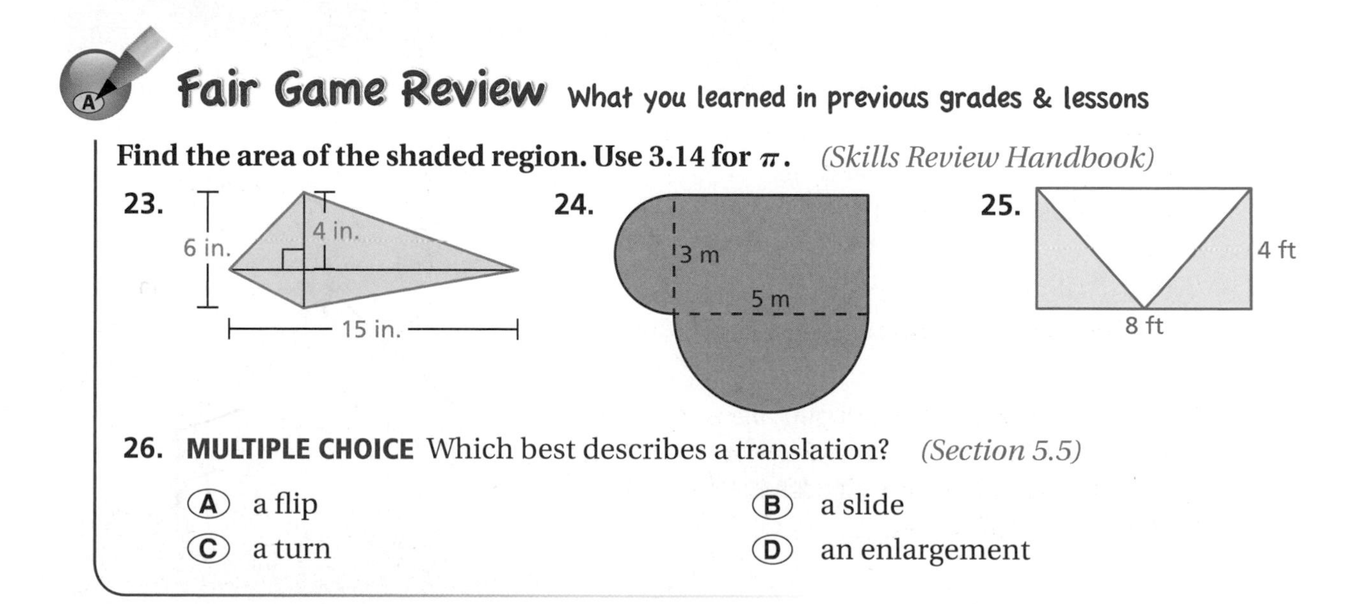

26. MULTIPLE CHOICE Which best describes a translation? *(Section 5.5)*

Ⓐ a flip
Ⓑ a slide
Ⓒ a turn
Ⓓ an enlargement

COMMON CORE STATE STANDARDS
7.G.4
7.G.6

Essential Question How can you find the surface area of a composite solid?

1 ACTIVITY: Finding a Surface Area

Work with a partner. You are manufacturing scale models of old houses.

a. Name the four basic solids of this composite figure.

b. Determine a strategy for finding the surface area of this model. Would you use a scale drawing? Would you use a net? Explain.

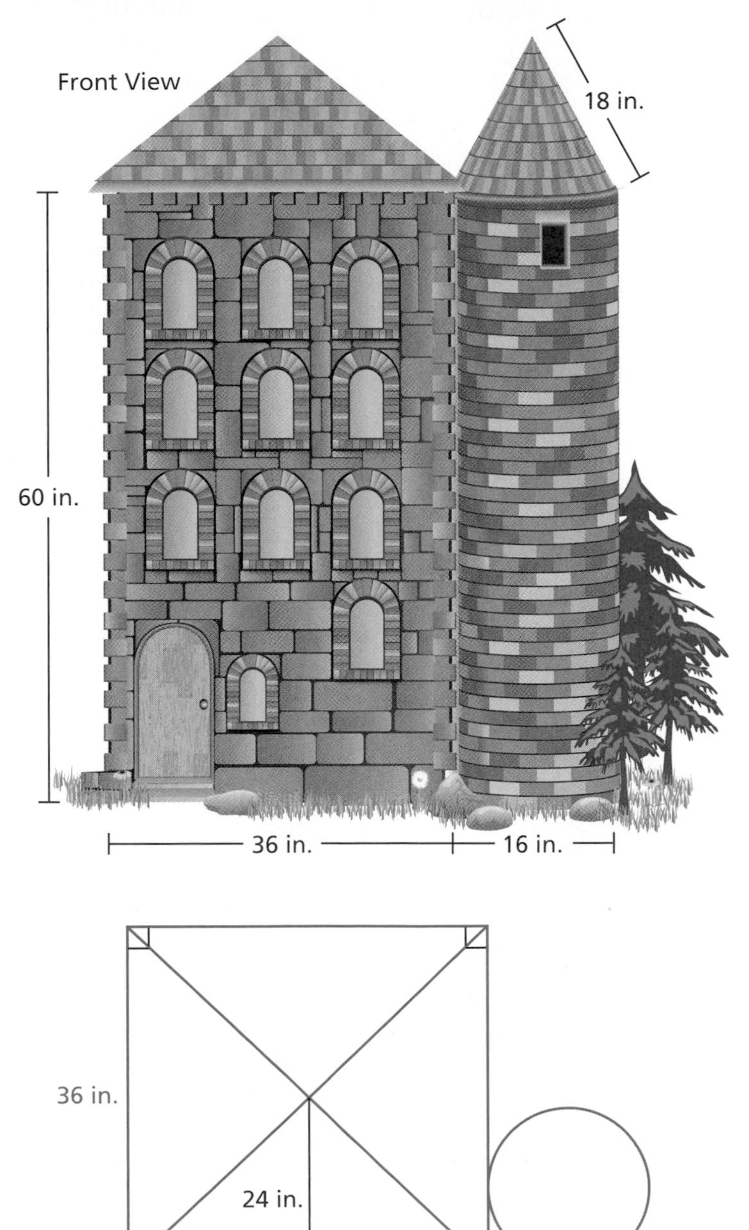

Many castles have cylindrical towers with conical roofs. These are called turrets.

2 ACTIVITY: Finding and Using a Pattern

Work with a partner.

- **Find the surface area of each figure.**
- **Use a table to organize your results.**
- **Describe the pattern in the table.**
- **Use the pattern to find the surface area of the figure that has a base of 10 blocks.**

3 ACTIVITY: Finding and Using a Pattern

Work with a partner. You own a roofing company. Each building has the same base area. Which roof would be cheapest? Which would be the most expensive? Explain your reasoning.

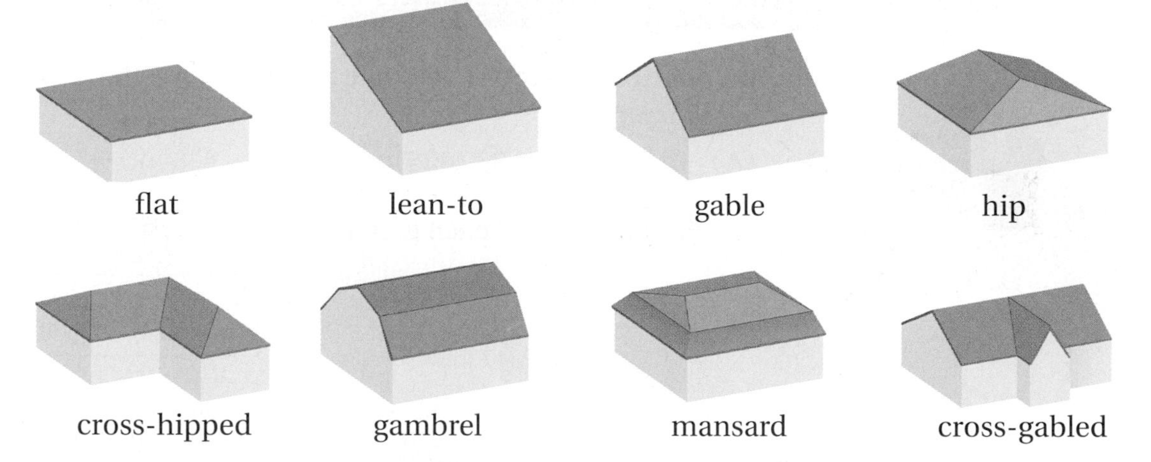

What Is Your Answer?

4. **IN YOUR OWN WORDS** How can you find the surface area of a composite solid?

5. Design a building that has a turret and also has a mansard roof. Find the surface area of the roof.

Use what you learned about the surface area of a composite solid to complete Exercises 6–8 on page 286.

6.6 Lesson

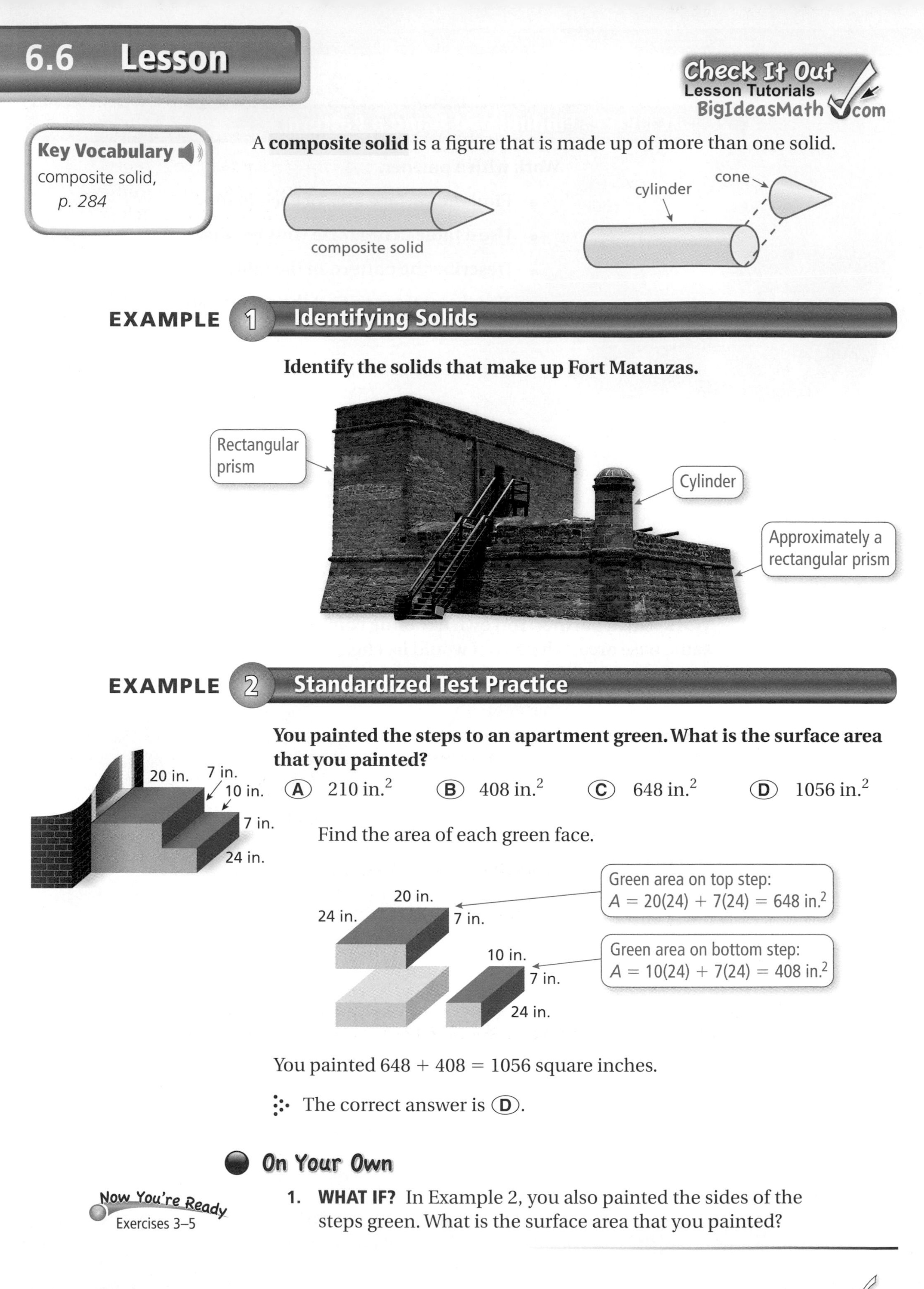

Check It Out
Lesson Tutorials
BigIdeasMath.com

Key Vocabulary
composite solid, *p. 284*

A **composite solid** is a figure that is made up of more than one solid.

EXAMPLE 1 Identifying Solids

Identify the solids that make up Fort Matanzas.

EXAMPLE 2 Standardized Test Practice

You painted the steps to an apartment green. What is the surface area that you painted?

Ⓐ 210 in.^{2} Ⓑ 408 in.^{2} Ⓒ 648 in.^{2} Ⓓ 1056 in.^{2}

Find the area of each green face.

Green area on top step:
$A = 20(24) + 7(24) = 648$ in.^{2}

Green area on bottom step:
$A = 10(24) + 7(24) = 408$ in.^{2}

You painted $648 + 408 = 1056$ square inches.

∴ The correct answer is Ⓓ.

On Your Own

Now You're Ready
Exercises 3–5

1. **WHAT IF?** In Example 2, you also painted the sides of the steps green. What is the surface area that you painted?

EXAMPLE 3 Finding the Surface Area of a Composite Solid

Find the surface area of the composite solid.

The solid is made up of a square prism and a square pyramid. Use the surface area formulas for a prism and a pyramid, but do not include the areas of the sides that overlap.

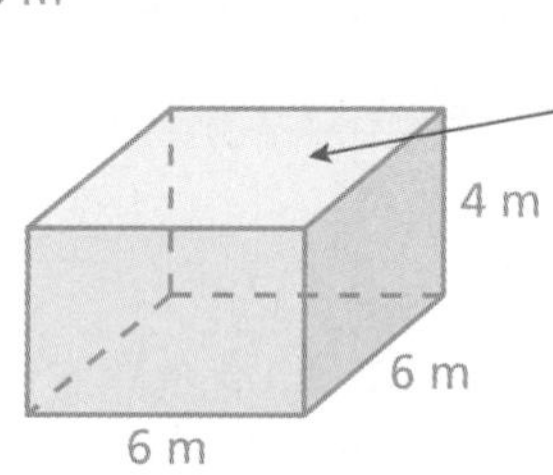

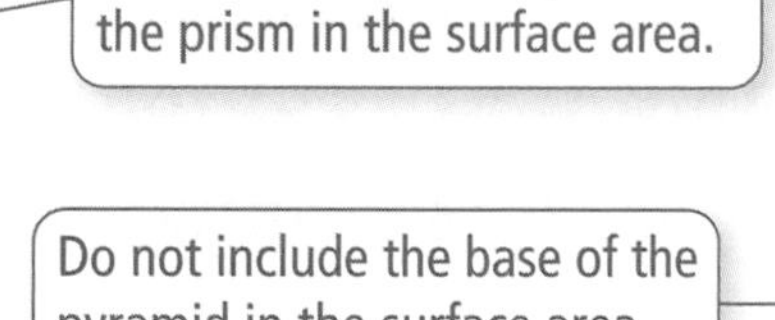

Do not include the top base of the prism in the surface area.

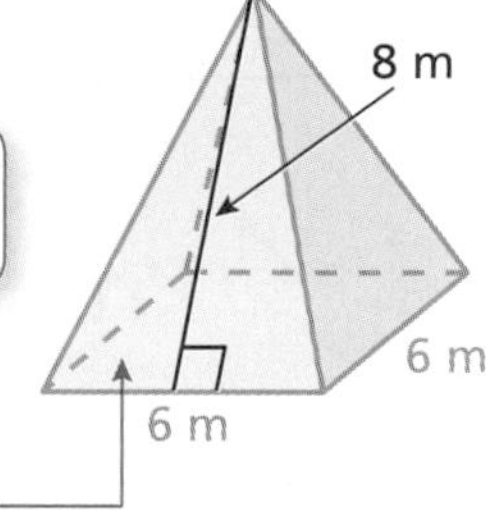

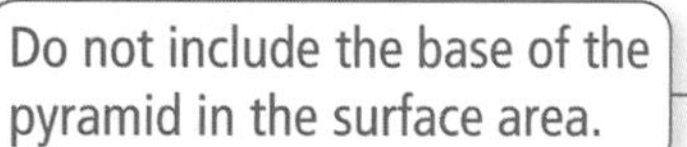

Do not include the base of the pyramid in the surface area.

Square prism

$S = \ell w + 2\ell h + 2wh$	Write formula.
$= 6(6) + 2(6)(4) + 2(6)(4)$	Substitute.
$= 36 + 48 + 48$	Multiply.
$= 132$	Add.

Square pyramid

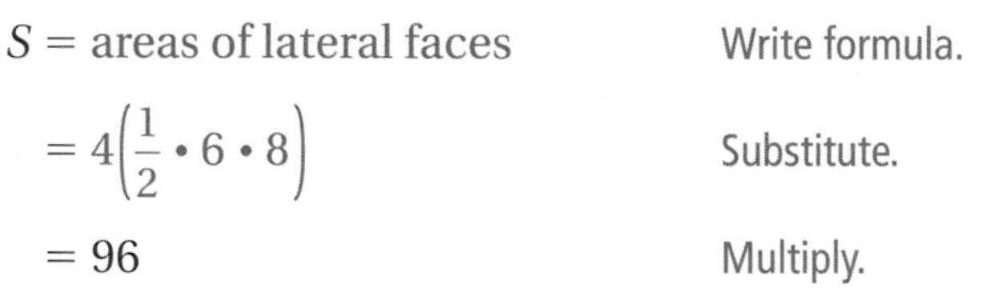

$S =$ areas of lateral faces	Write formula.
$= 4\left(\frac{1}{2} \cdot 6 \cdot 8\right)$	Substitute.
$= 96$	Multiply.

Find the sum of the surface areas: $132 + 96 = 228$.

∴ The surface area is 228 square meters.

On Your Own

Now You're Ready
Exercises 6–11

Identify the solids that make up the composite solid. Then find the surface area. Round your answer to the nearest tenth.

2. 3.

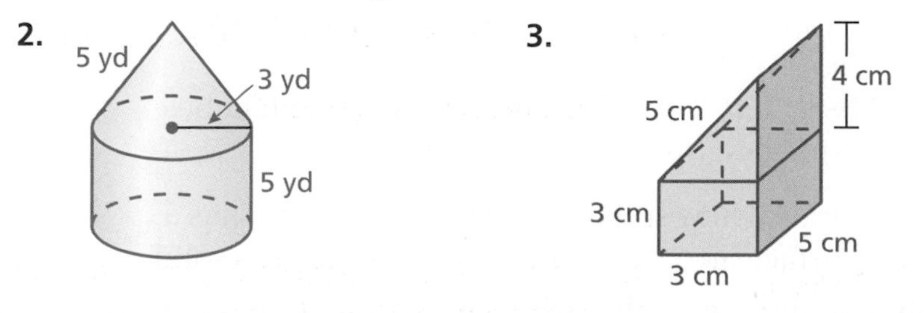

6.6 Exercises

Vocabulary and Concept Check

1. **OPEN-ENDED** Draw a composite solid formed by a triangular prism and a cone.
2. **REASONING** Explain how to find the surface area of the composite solid.

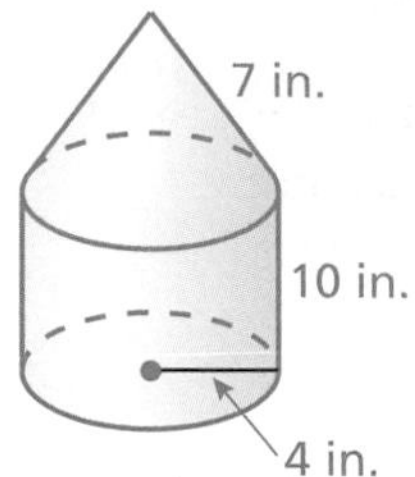

Practice and Problem Solving

Identify the solids that form the composite solid.

Identify the solids that form the composite solid. Then find the surface area. Round your answer to the nearest tenth.

6.

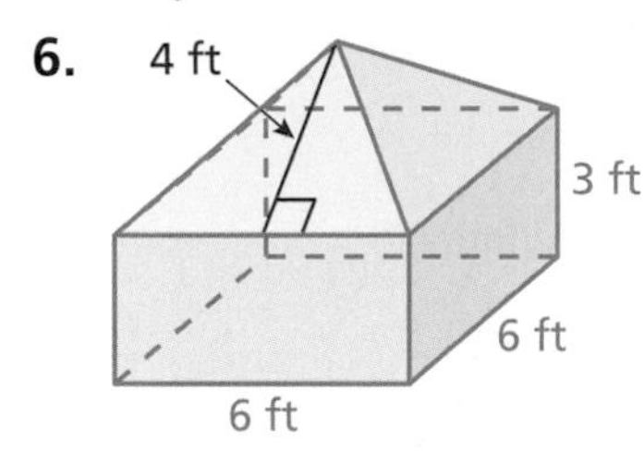

7.

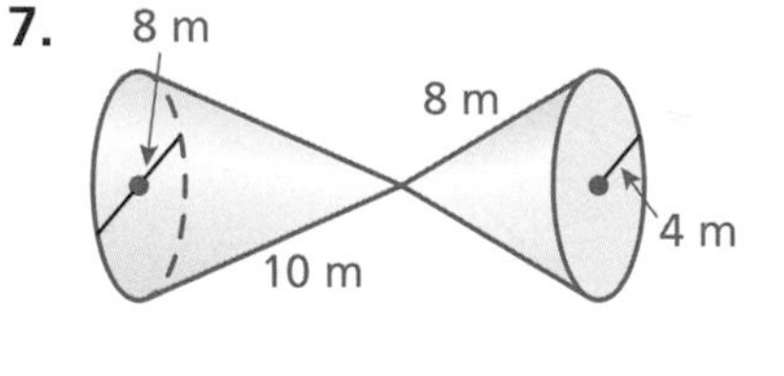

8.

4 in.
4 in.
5 in.
5 in.
5 in.

9.

2 cm
2.5 cm
2.5 cm
2 cm
3 cm
6 cm
5 cm

10.

7 in.
8 in.
8 in.
10 in.
6.9 in.
8 in.

11.

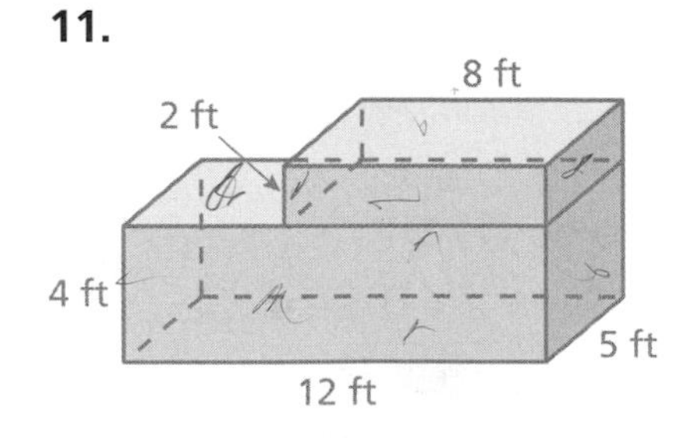

12. **OPEN-ENDED** The solid is made using eight cubes with side lengths of 1 centimeter.

    a. Draw a new solid using eight cubes that has a surface area less than that of the original solid.

    b. Draw a new solid using eight cubes that has a surface area greater than that of the original solid.

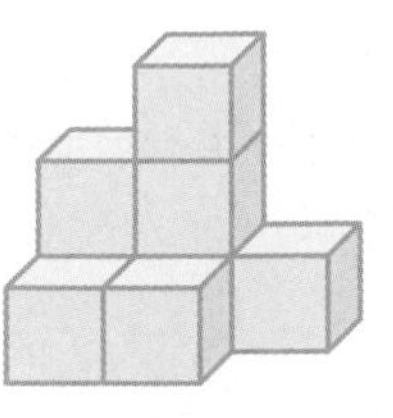

13. **BATTERIES** What is the percent increase in the surface area of the AAA battery to the AA battery? Round your answer to the nearest tenth of a percent.

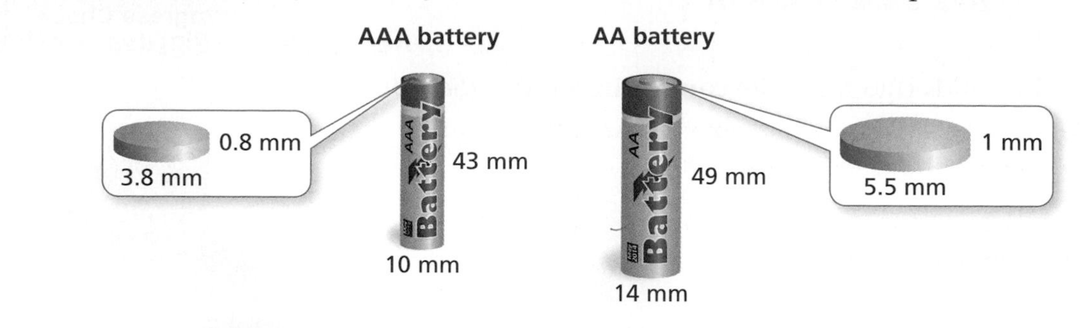

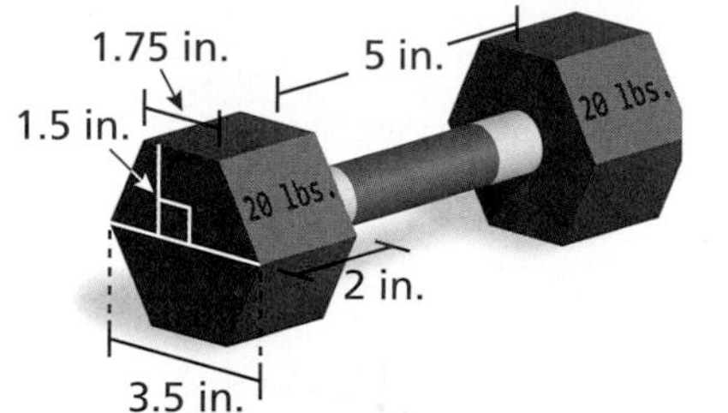

14. **BARBELL** The diameter of the handle of a barbell is 1 inch. The hexagonal weights are identical. What is the surface area of the barbell?

REASONING Find the surface area of the solid. Round your answer to the nearest tenth.

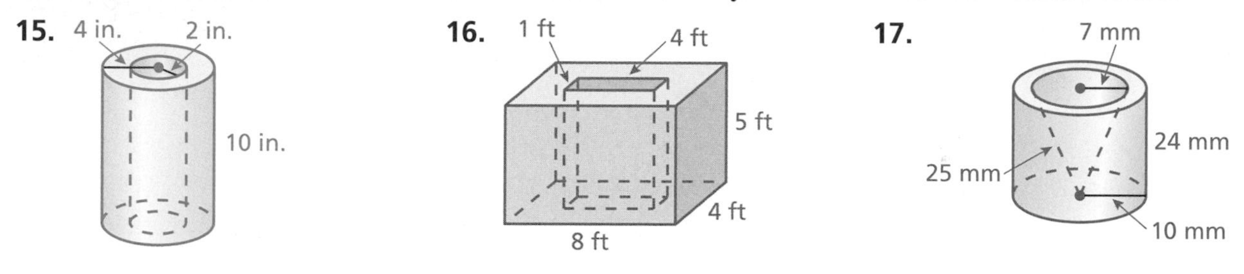

18. **Critical Thinking** The cube is made with 27 identical cubes. All cubes that cannot be seen are orange. Is the surface area of the solid formed without the purple cubes *greater than, less than,* or *equal to* the surface area of the solid formed without the green cubes? Explain your reasoning.

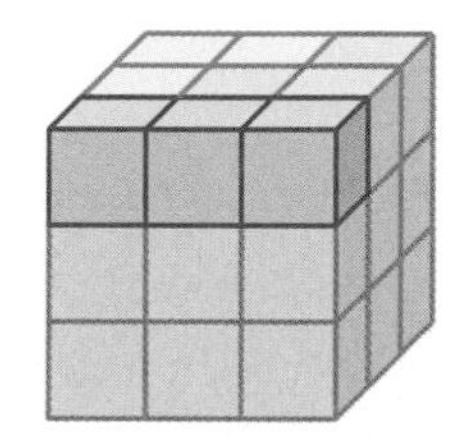

Fair Game Review What you learned in previous grades & lessons

Find the area. *(Skills Review Handbook)*

19. 2 ft, 5 ft

20. 4 cm, 8 cm

21. 7 in., 5 in., 12 in.

22. **MULTIPLE CHOICE** A cliff swallow nest is 86 meters above a canyon floor. The elevation of the nest is −56 meters. What is the elevation of the canyon floor? *(Section 2.4)*

Ⓐ −142   Ⓑ −30   Ⓒ 30   Ⓓ 142

6.4–6.6 Quiz

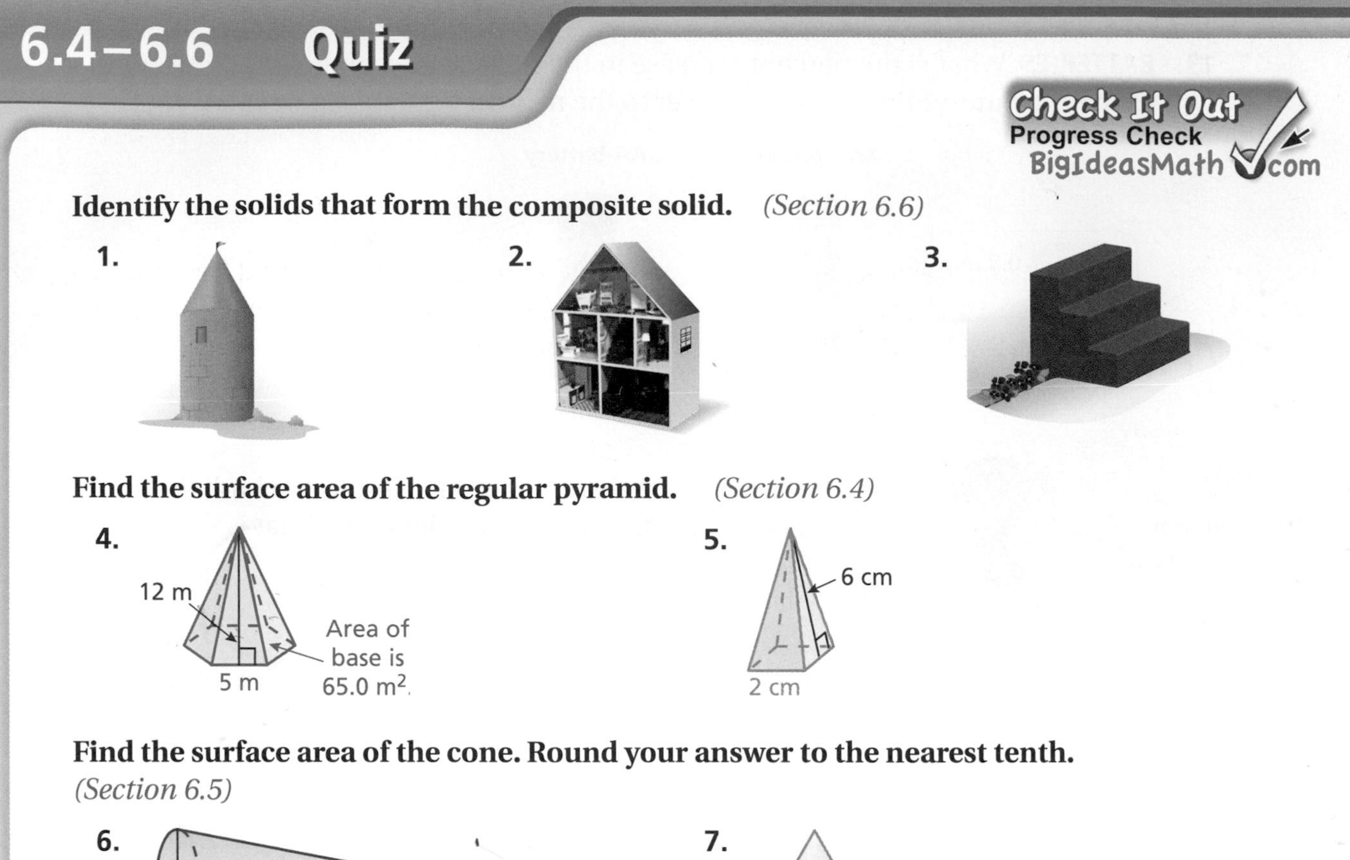

Identify the solids that form the composite solid. *(Section 6.6)*

1.

2.

3.

Find the surface area of the regular pyramid. *(Section 6.4)*

4. 12 m; 5 m; Area of base is 65.0 m².

5. 6 cm; 2 cm

Find the surface area of the cone. Round your answer to the nearest tenth. *(Section 6.5)*

6.

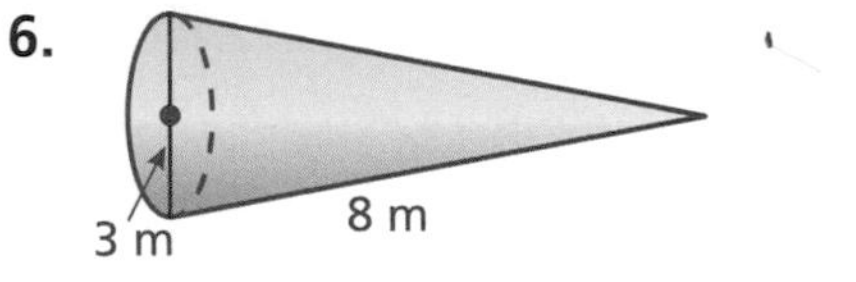

7.

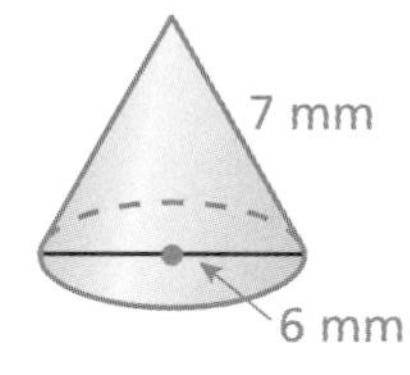

Find the surface area of the composite solid. Round your answer to the nearest tenth. *(Section 6.6)*

8.

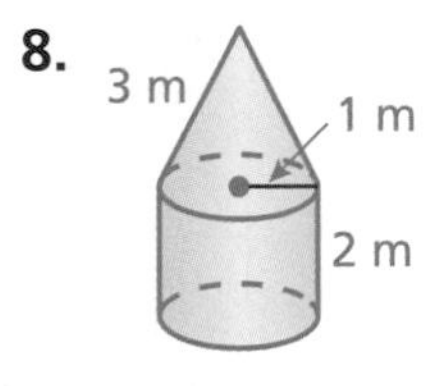

9.

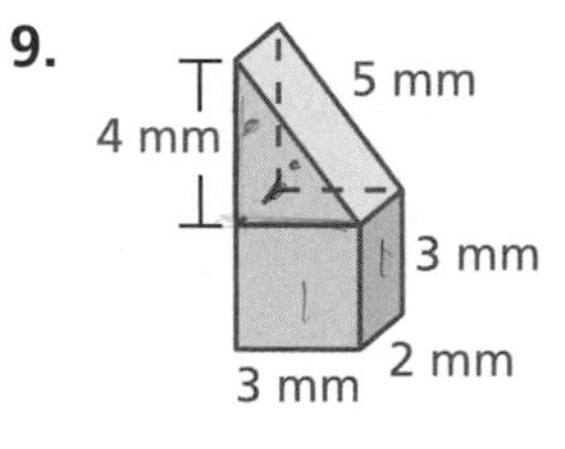

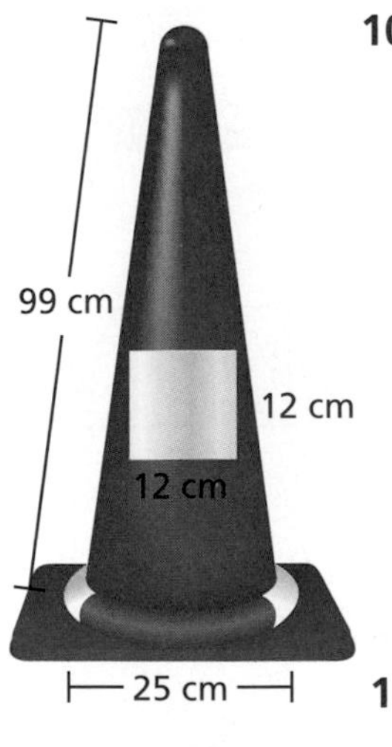

10. TRAFFIC CONE A square reflective sticker is placed on a traffic cone to make it more visible at night. Estimate the percent of the surface area of the traffic cone covered by the sticker to the nearest percent. *(Section 6.5)*

11. GEOMETRY The surface area of a cone is 90π square inches. The diameter of the base is 10 inches. What is the slant height? *(Section 6.5)*

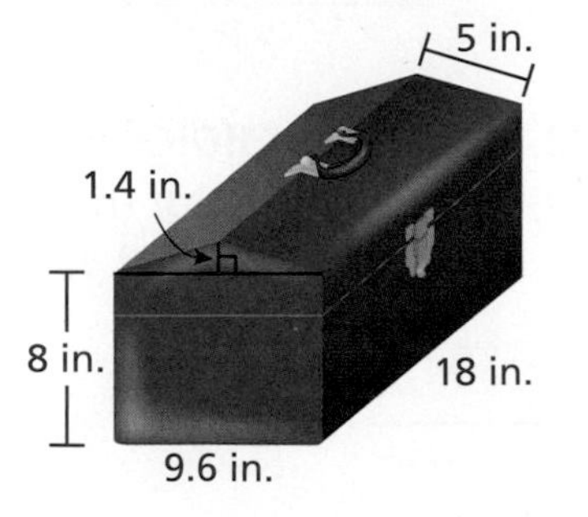

12. TOOLBOX Find the surface area of the toolbox. *(Section 6.6)*

6 Chapter Review

Review Key Vocabulary

Review Examples and Exercises

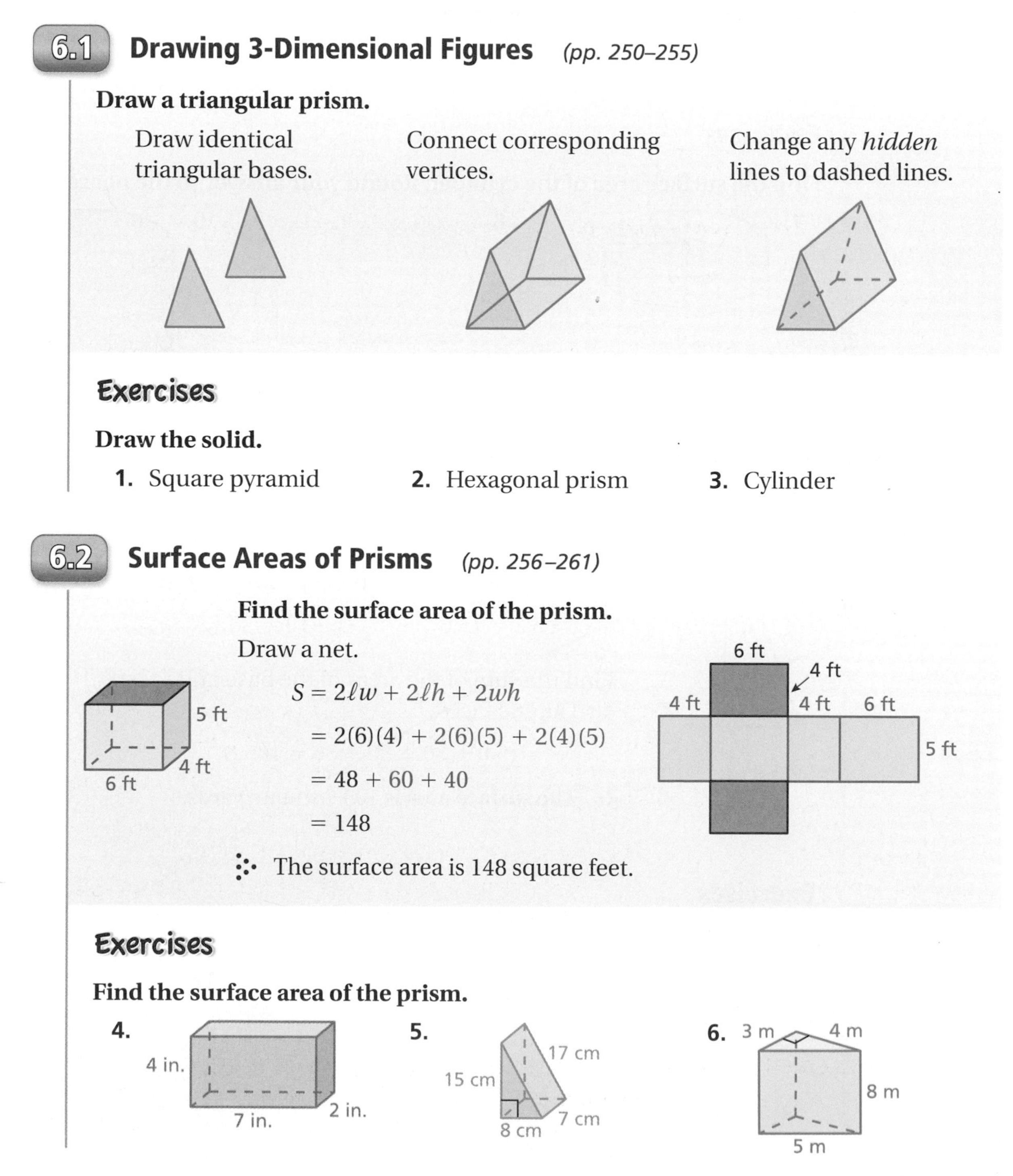

6.1 Drawing 3-Dimensional Figures *(pp. 250–255)*

Draw a triangular prism.

Draw identical triangular bases.

Connect corresponding vertices.

Change any *hidden* lines to dashed lines.

Exercises

Draw the solid.

1. Square pyramid
2. Hexagonal prism
3. Cylinder

6.2 Surface Areas of Prisms *(pp. 256–261)*

Find the surface area of the prism.

Draw a net.

$S = 2\ell w + 2\ell h + 2wh$

$= 2(6)(4) + 2(6)(5) + 2(4)(5)$

$= 48 + 60 + 40$

$= 148$

∴ The surface area is 148 square feet.

Exercises

Find the surface area of the prism.

4.
5.
6.

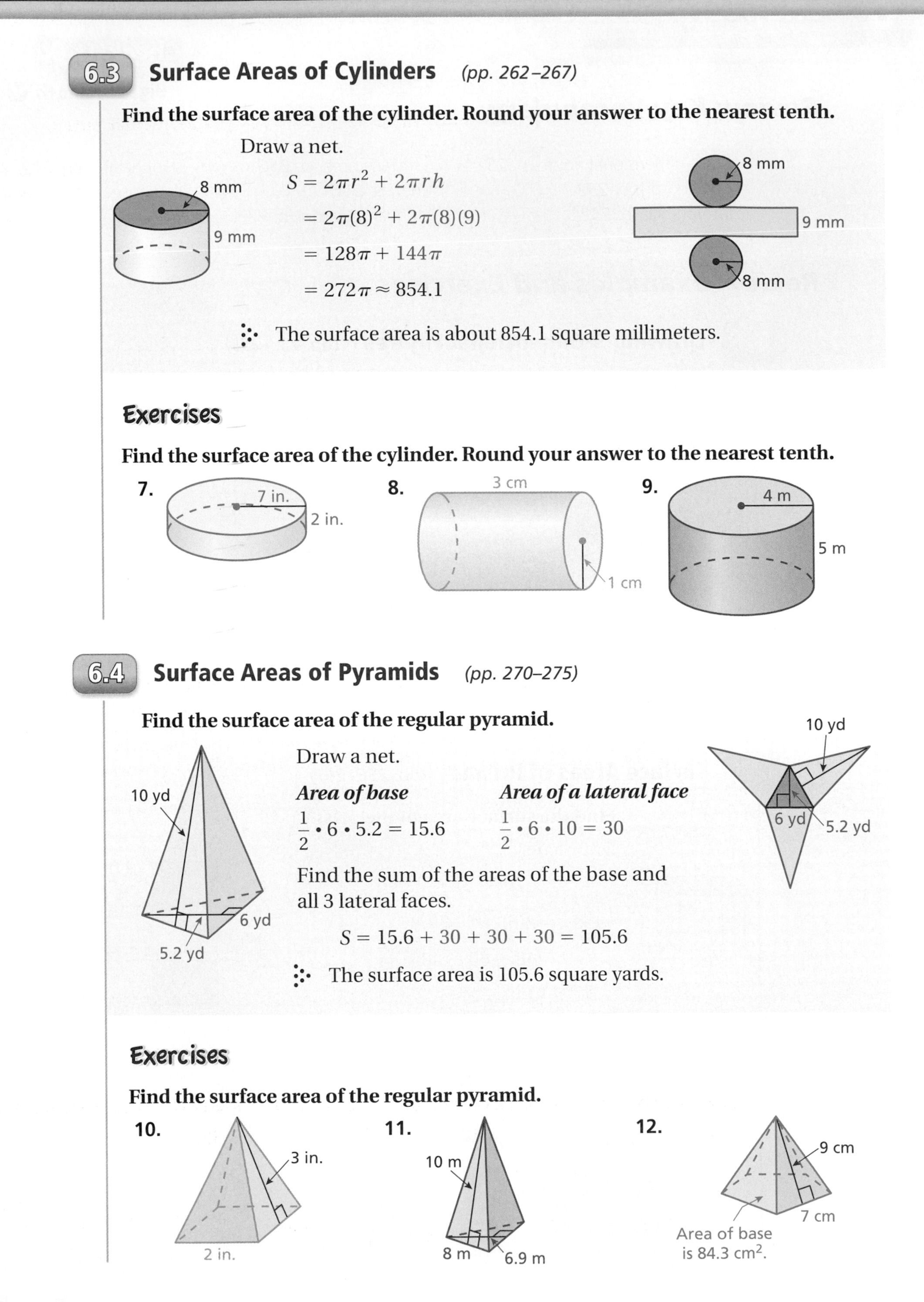

6.3 Surface Areas of Cylinders (pp. 262–267)

Find the surface area of the cylinder. Round your answer to the nearest tenth.

Draw a net.

$$S = 2\pi r^2 + 2\pi rh$$
$$= 2\pi(8)^2 + 2\pi(8)(9)$$
$$= 128\pi + 144\pi$$
$$= 272\pi \approx 854.1$$

∴ The surface area is about 854.1 square millimeters.

Exercises

Find the surface area of the cylinder. Round your answer to the nearest tenth.

7.

8.

9.

6.4 Surface Areas of Pyramids (pp. 270–275)

Find the surface area of the regular pyramid.

Draw a net.

Area of base

$\frac{1}{2} \cdot 6 \cdot 5.2 = 15.6$

Area of a lateral face

$\frac{1}{2} \cdot 6 \cdot 10 = 30$

Find the sum of the areas of the base and all 3 lateral faces.

$$S = 15.6 + 30 + 30 + 30 = 105.6$$

∴ The surface area is 105.6 square yards.

Exercises

Find the surface area of the regular pyramid.

10.

11.

12.

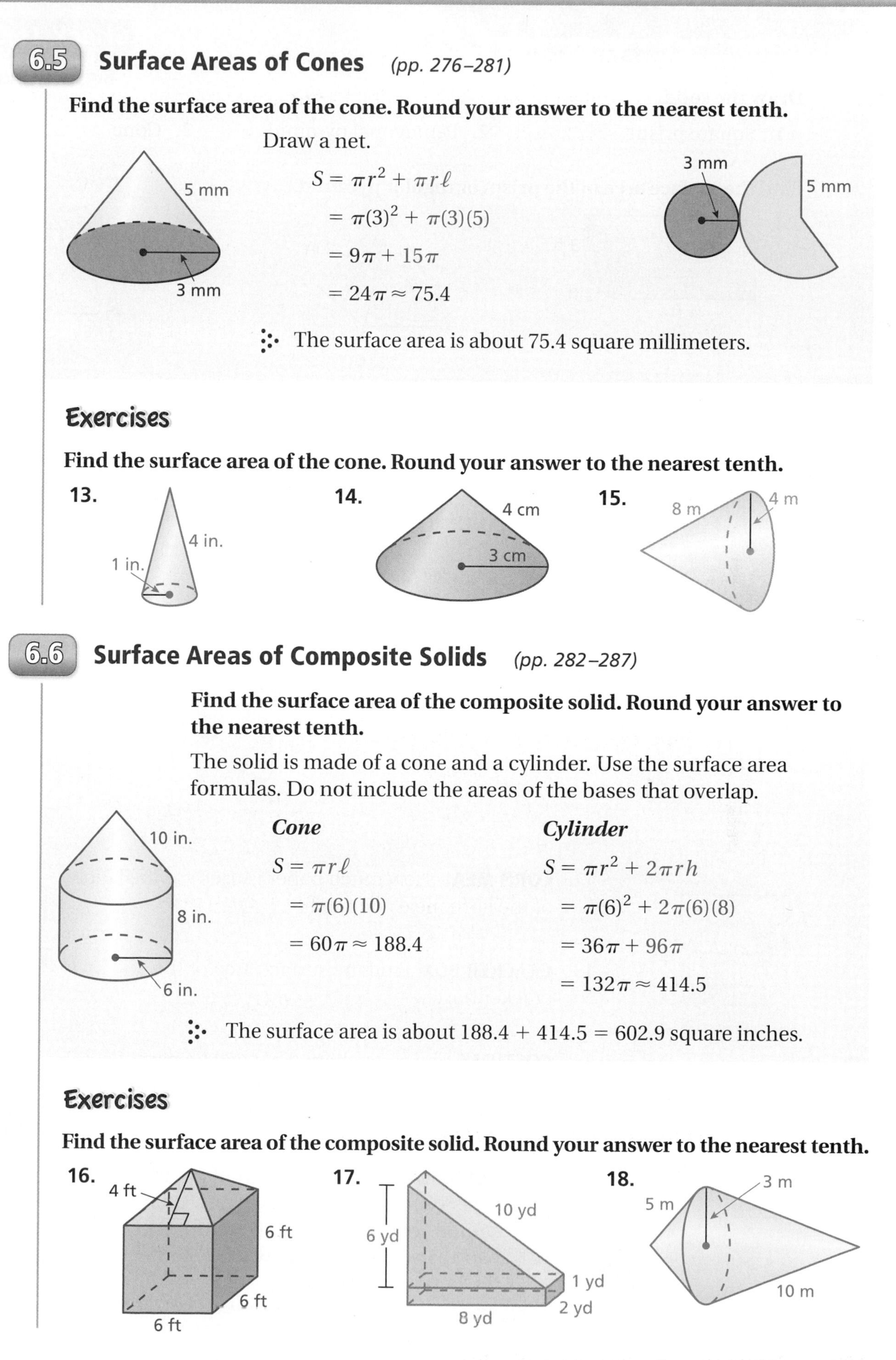

6.5 Surface Areas of Cones (pp. 276–281)

Find the surface area of the cone. Round your answer to the nearest tenth.

Draw a net.

$$S = \pi r^2 + \pi r \ell$$
$$= \pi(3)^2 + \pi(3)(5)$$
$$= 9\pi + 15\pi$$
$$= 24\pi \approx 75.4$$

∴ The surface area is about 75.4 square millimeters.

Exercises

Find the surface area of the cone. Round your answer to the nearest tenth.

13.

14.

15.

6.6 Surface Areas of Composite Solids (pp. 282–287)

Find the surface area of the composite solid. Round your answer to the nearest tenth.

The solid is made of a cone and a cylinder. Use the surface area formulas. Do not include the areas of the bases that overlap.

Cone

$$S = \pi r \ell$$
$$= \pi(6)(10)$$
$$= 60\pi \approx 188.4$$

Cylinder

$$S = \pi r^2 + 2\pi r h$$
$$= \pi(6)^2 + 2\pi(6)(8)$$
$$= 36\pi + 96\pi$$
$$= 132\pi \approx 414.5$$

∴ The surface area is about $188.4 + 414.5 = 602.9$ square inches.

Exercises

Find the surface area of the composite solid. Round your answer to the nearest tenth.

16.

17.

18.

6 Chapter Test

Check It Out
Test Practice
BigIdeasMath.com

Draw the solid.

1. Square prism
2. Pentagonal pyramid
3. Cone

Find the surface area of the prism or regular pyramid.

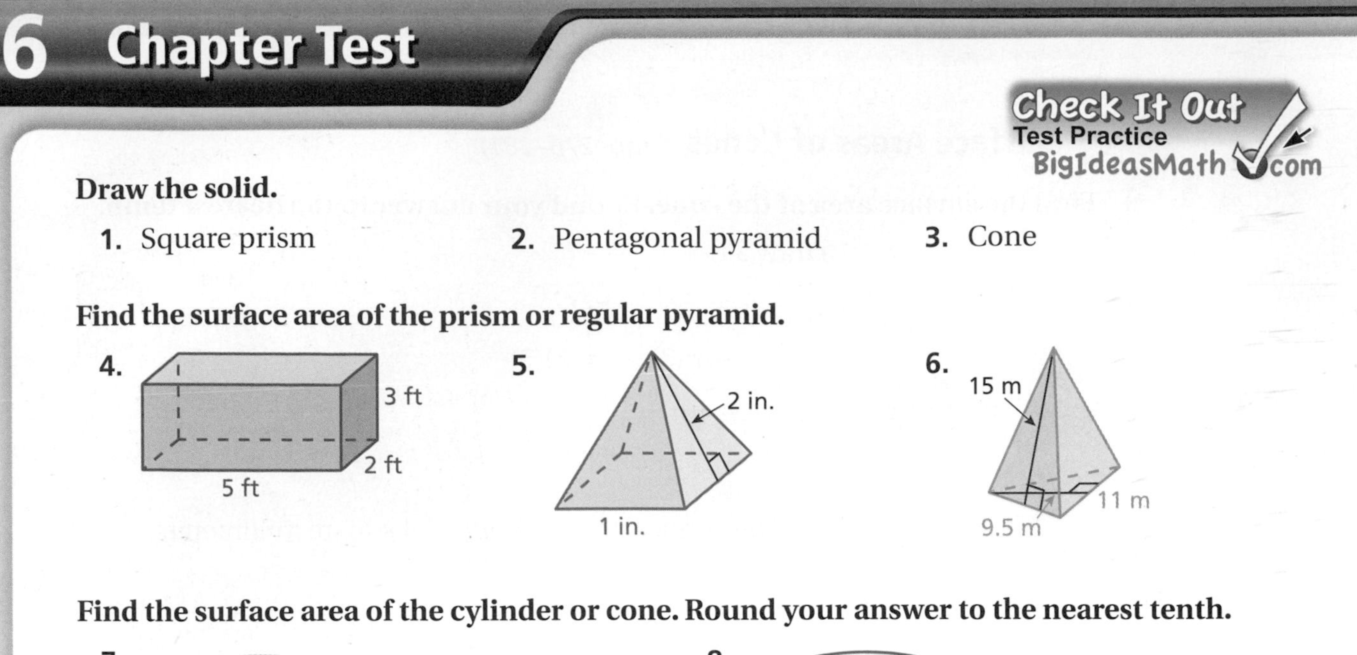

Find the surface area of the cylinder or cone. Round your answer to the nearest tenth.

9. Draw the front, side, and top views of the solid in Exercise 8.

Identify the solids that form the composite solid. Then find the surface area. Round your answer to the nearest tenth.

10. 9 ft, 7 ft, 1 ft

11. 13 cm, 5 cm, 6 cm, 12 cm, 1 cm

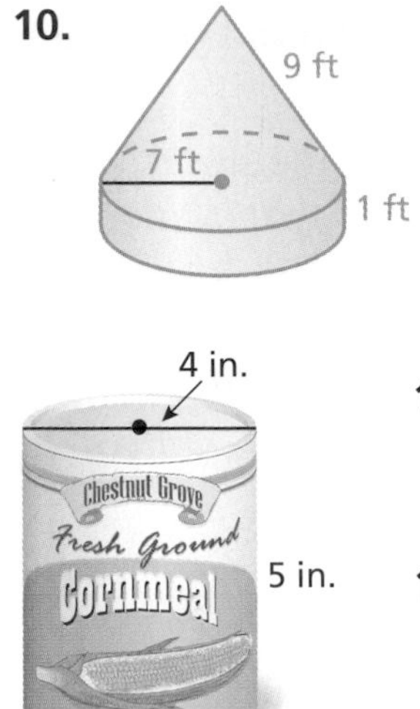

12. **CORN MEAL** How much paper is used for the label of the corn meal container?

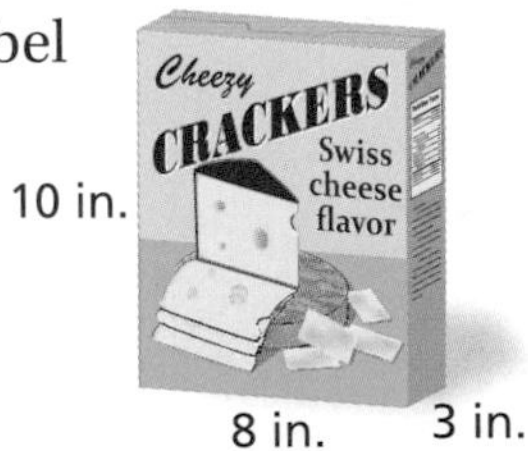

13. **CRACKER BOX** Find the surface area of the cracker box.

14. **COSTUME** The cone-shaped hat will be part of a costume for a school play. What is the least amount of material needed to make this hat?

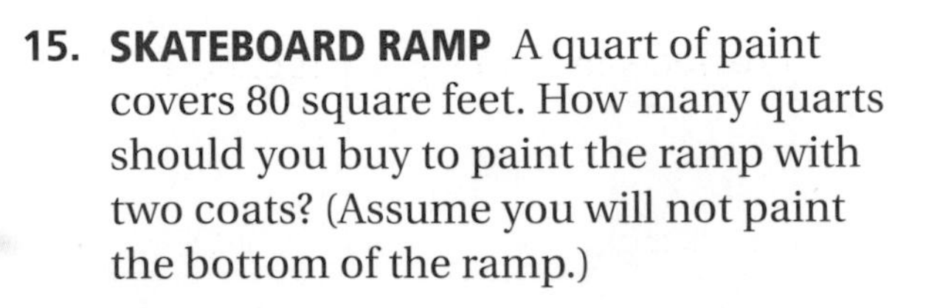

15. **SKATEBOARD RAMP** A quart of paint covers 80 square feet. How many quarts should you buy to paint the ramp with two coats? (Assume you will not paint the bottom of the ramp.)

6 Standardized Test Practice

1. In the figure below, $\triangle PQR \sim \triangle STU$.

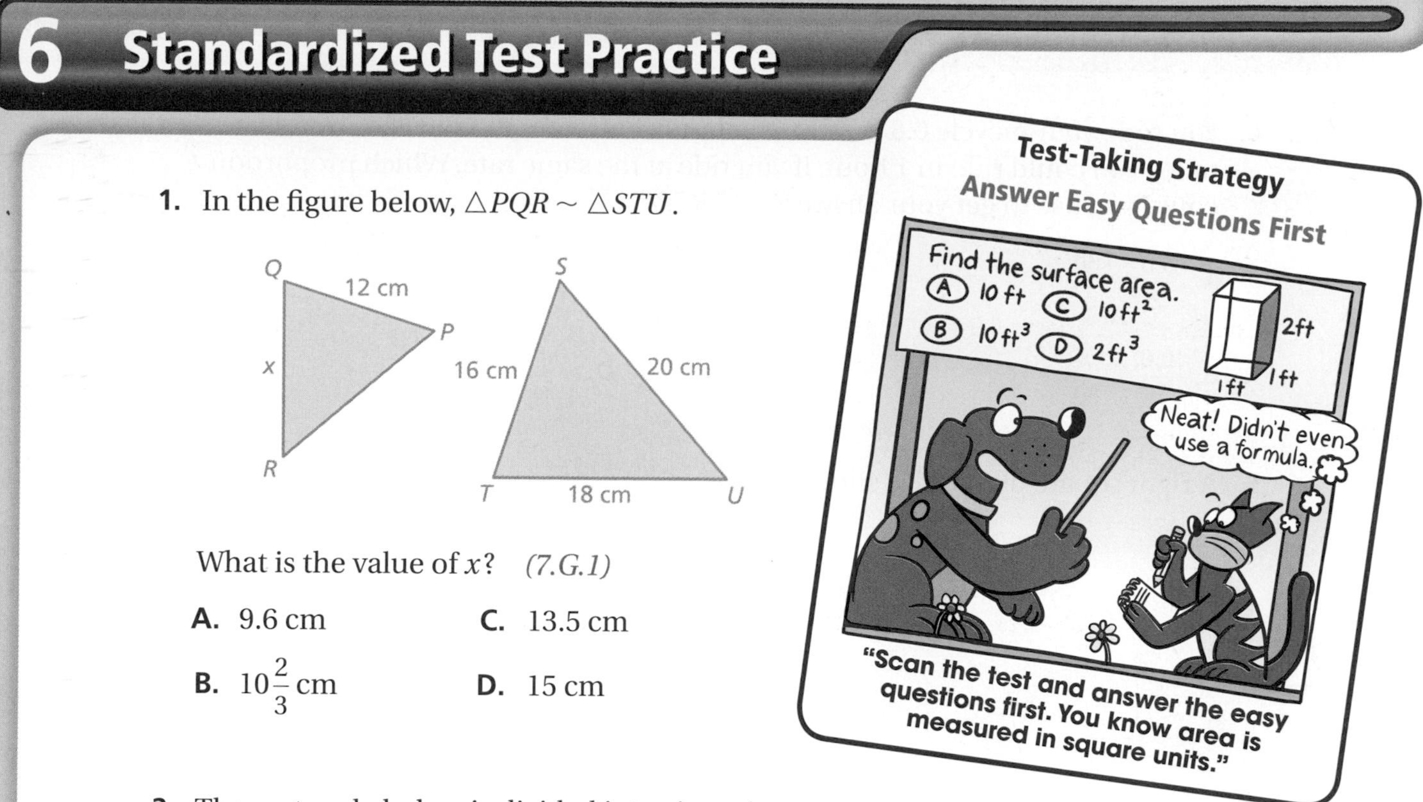

What is the value of x? *(7.G.1)*

A. 9.6 cm

B. $10\frac{2}{3}$ cm

C. 13.5 cm

D. 15 cm

2. The rectangle below is divided into six regions.

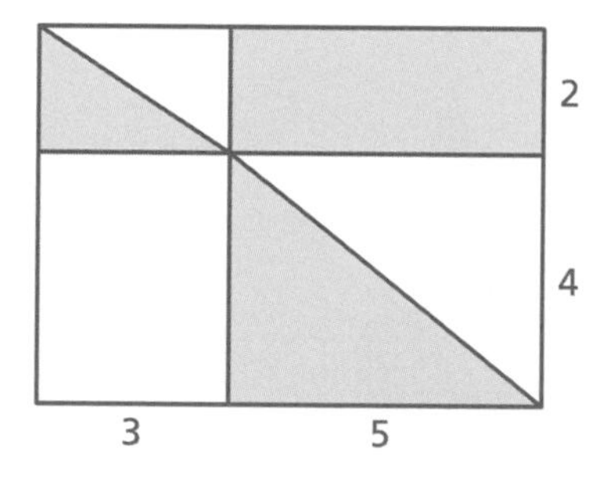

What is the area of the part of the figure that is shaded? *(7.NS.3)*

F. 23 units^2

G. 24 units^2

H. 25 units^2

I. 28 units^2

3. A right rectangular prism and its dimensions are shown below.

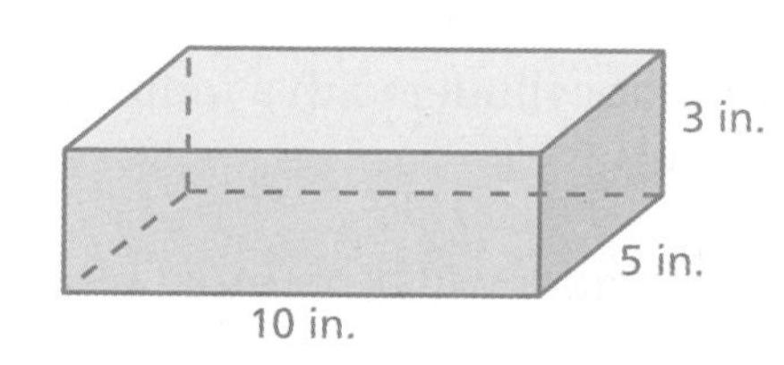

What is the total surface area, in square inches, of the right rectangular prism? *(7.G.6)*

4. You rode your bicycle 0.8 mile in 2 minutes. You want to know how many miles you could ride in 1 hour, if you ride at the same rate. Which proportion could you use to get your answer? *(7.RP.2d)*

A. $\frac{0.8}{2} = \frac{60}{x}$

B. $\frac{0.8}{2} = \frac{x}{60}$

C. $\frac{0.8}{2} = \frac{30}{x}$

D. $\frac{0.8}{2} = \frac{x}{30}$

5. A right square pyramid is shown below.

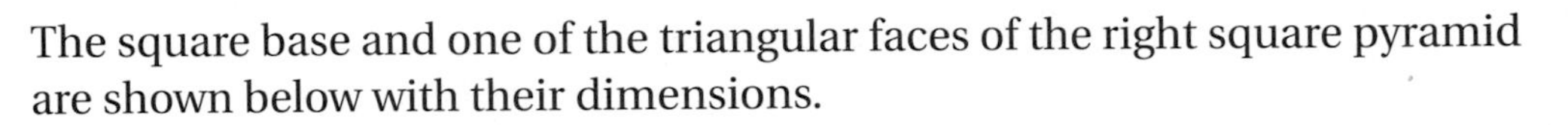

The square base and one of the triangular faces of the right square pyramid are shown below with their dimensions.

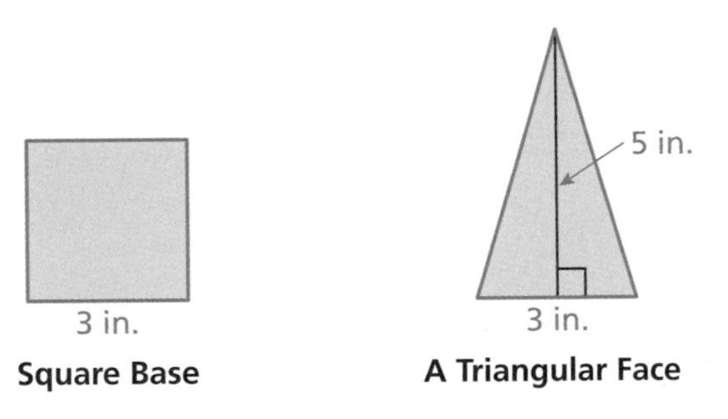

What is the total surface area of the right square pyramid? *(7.G.6)*

F. 16.5 in.^2

G. 31.5 in.^2

H. 39 in.^2

I. 69 in.^2

6. A right circular cylinder with a radius of 3 centimeters and a height of 7 centimeters will be carved out of wood. *(7.G.3, 7.G.6)*

Part A Draw and label a right circular cylinder with a radius of 3 centimeters and a height of 7 centimeters.

The two bases of the right circular cylinder will be painted blue. The rest of the cylinder will be painted red.

Part B What is the surface area, in square centimeters, that will be painted blue? Show your work and explain your reasoning. (Use 3.14 for π.)

Part C What is the surface area, in square centimeters, that will be painted red? Show your work and explain your reasoning. (Use 3.14 for π.)

7. Anna was simplifying the expression in the box below.

$$-\frac{3}{8} \cdot \left[\frac{2}{5} \div (-4)\right] = -\frac{3}{8} \cdot \left[\frac{2}{5} \cdot \left(-\frac{1}{4}\right)\right]$$
$$= -\frac{3}{8} \cdot \left(-\frac{1}{10}\right)$$
$$= -\frac{3}{80}$$

What should Anna do to correct the error that she made? *(7.NS.3)*

A. Make the product inside the brackets positive.

B. Multiply by -10 instead of $-\frac{1}{10}$.

C. Make the final product positive.

D. Multiply by 4 instead of $-\frac{1}{4}$.

8. Which equation has the greatest solution? *(7.EE.4a)*

F. $-3x + 9 = -15$

G. $12 = 2x + 28$

H. $\frac{x}{2} - 13 = -7$

I. $6 = \frac{x}{3} + 10$

9. A cube has a total surface area of 600 square inches. What is the length, in inches, of each edge of the cube? *(7.G.6)*

10. A line contains the two points plotted in the coordinate plane below.

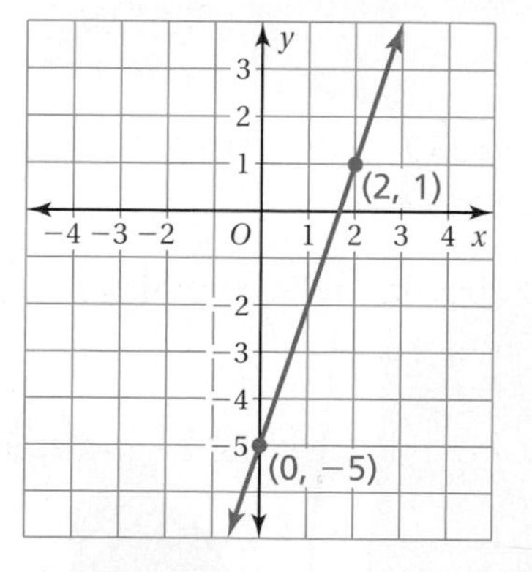

Another point on this line can be represented by the ordered pair $(-1, y)$. What is the value of y? *(7.NS.1b)*

A. -11

B. -8

C. -6

D. -2

7 Volumes of Solids

"I petitioned my owner for a dog house with greater volume."

"And this is what he built for me."

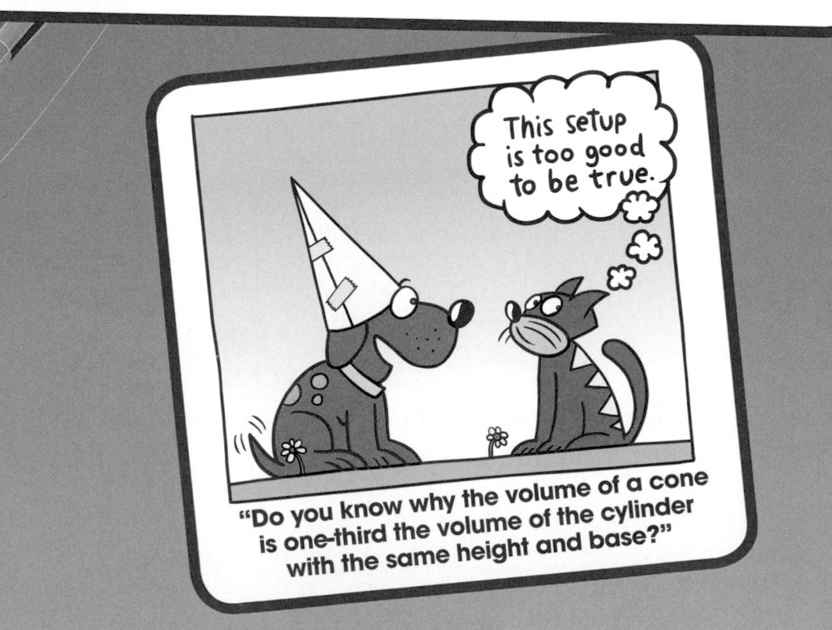

"Do you know why the volume of a cone is one-third the volume of the cylinder with the same height and base?"

What You Learned Before

Identifying Similar Figures (7.G.1)

Example 1 **Which rectangle is similar to Rectangle A?**

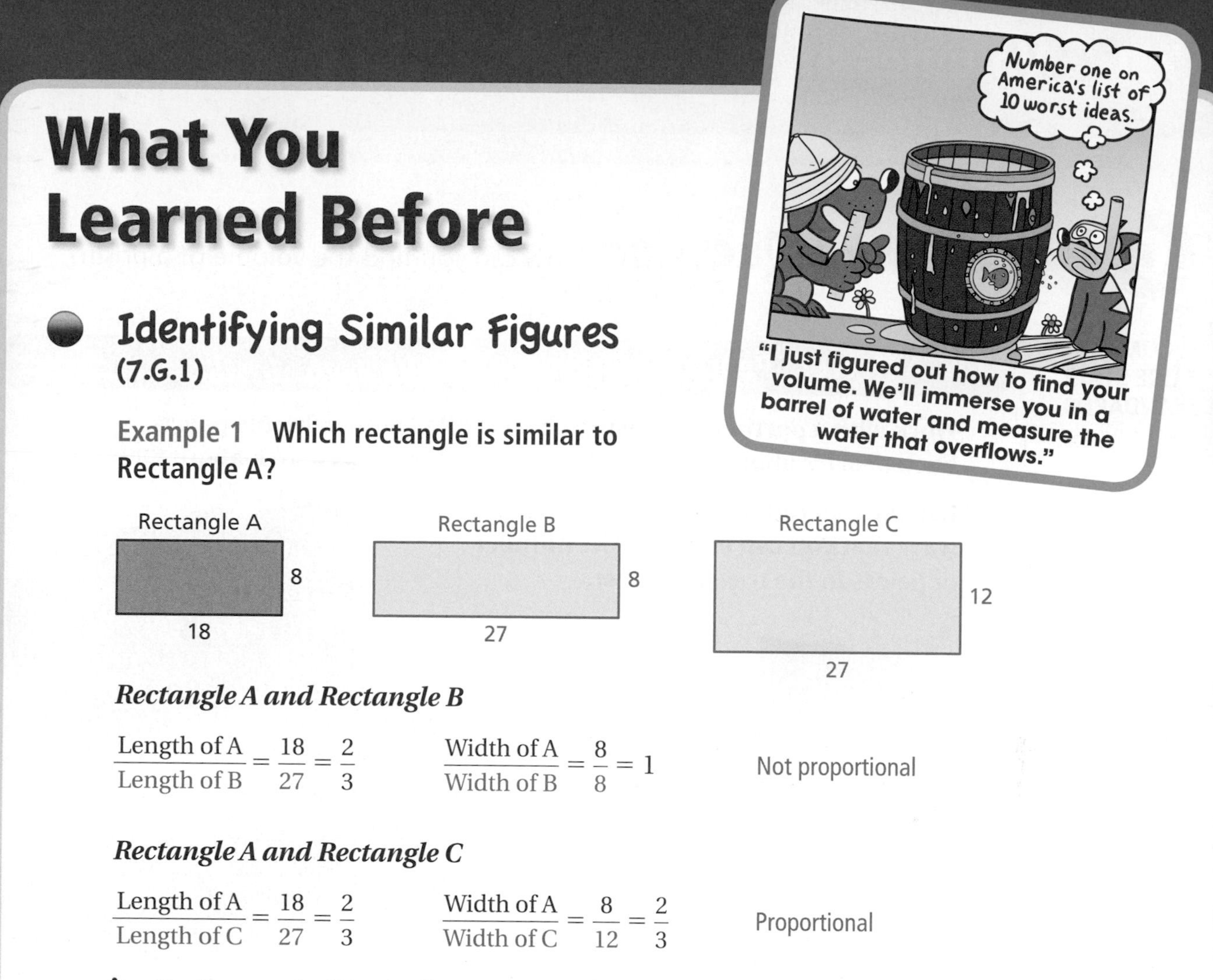

"I just figured out how to find your volume. We'll immerse you in a barrel of water and measure the water that overflows."

Rectangle A and Rectangle B

$$\frac{\text{Length of A}}{\text{Length of B}} = \frac{18}{27} = \frac{2}{3} \qquad \frac{\text{Width of A}}{\text{Width of B}} = \frac{8}{8} = 1$$

Not proportional

Rectangle A and Rectangle C

$$\frac{\text{Length of A}}{\text{Length of C}} = \frac{18}{27} = \frac{2}{3} \qquad \frac{\text{Width of A}}{\text{Width of C}} = \frac{8}{12} = \frac{2}{3}$$

Proportional

∴ So, Rectangle C is similar to Rectangle A.

Finding Measures in Similar Figures (7.G.1)

Example 2 **The two triangles are similar. Find the value of x.**

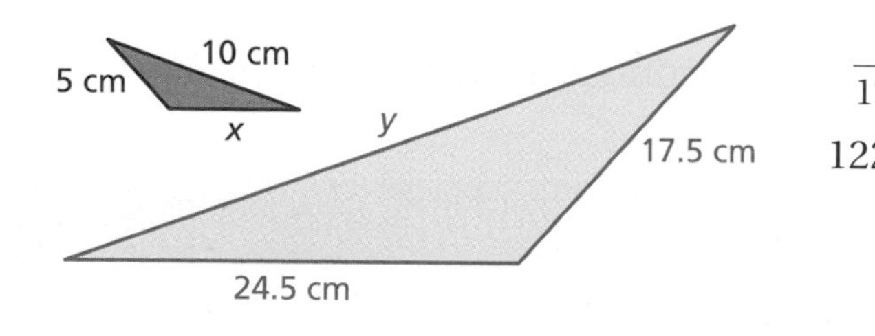

$\frac{5}{17.5} = \frac{x}{24.5}$ Write a proportion.

$122.5 = 17.5x$ Use Cross Products Property.

$7 = x$ Divide each side by 17.5.

∴ So, x is 7 centimeters.

Try It Yourself

1. Construct two more rectangles that are similar to Rectangle A in Example 1.
2. Find the value of y in Example 2.

7.1 Volumes of Prisms

Essential Question How can you find the volume of a prism?

COMMON CORE STATE STANDARDS
7.G.6

1 ACTIVITY: Pearls in a Treasure Chest

Work with a partner. A treasure chest is filled with valuable pearls. Each pearl is about 1 centimeter in diameter and is worth about $80.

Use the diagrams below to describe two ways that you can estimate the number of pearls in the treasure chest.

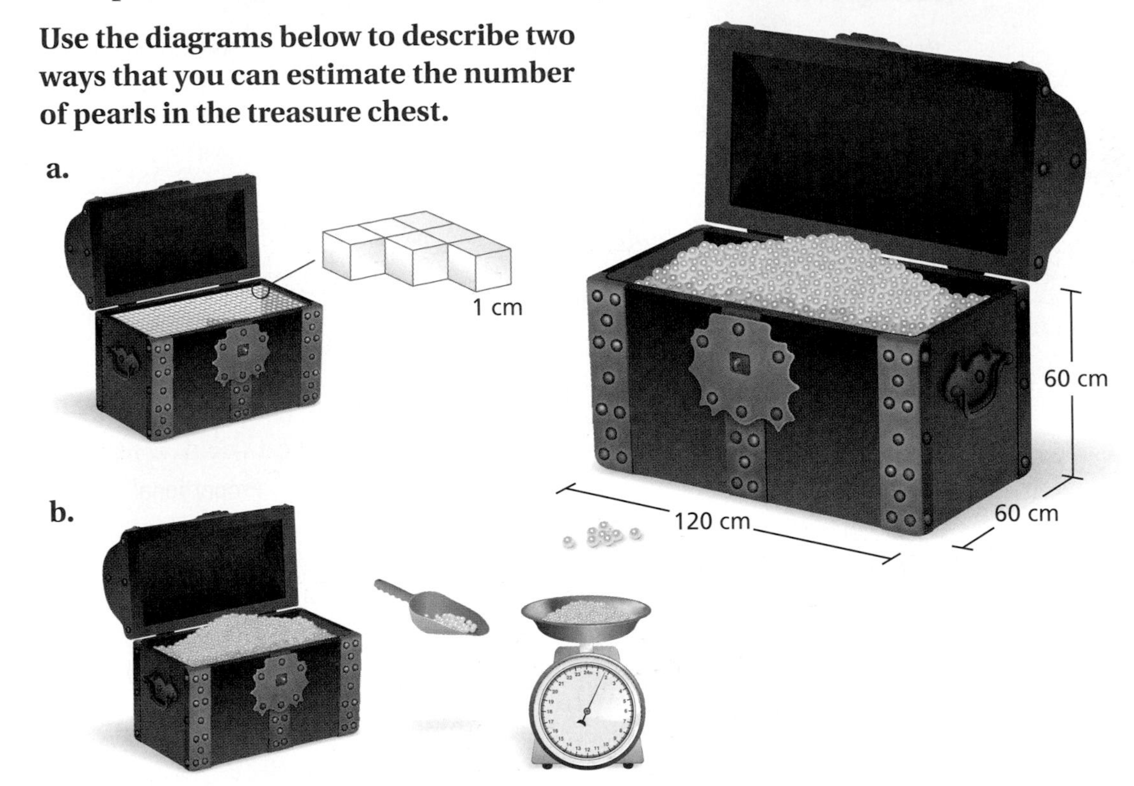

c. Use the method in part (a) to estimate the value of the pearls in the chest.

2 ACTIVITY: Finding a Formula for Volume

Work with a partner. You know that the formula for the volume of a rectangular prism is $V = \ell wh$.

a. Find a new formula that gives the volume in terms of the area of the base B and the height h.

b. Use both formulas to find the volume of each prism. Do both formulas give you the same volumes?

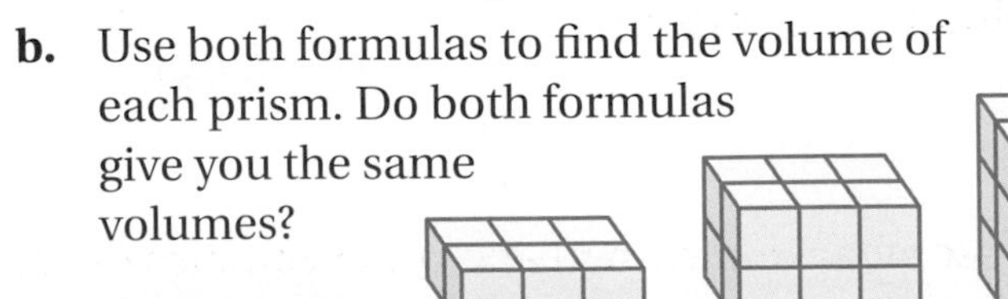

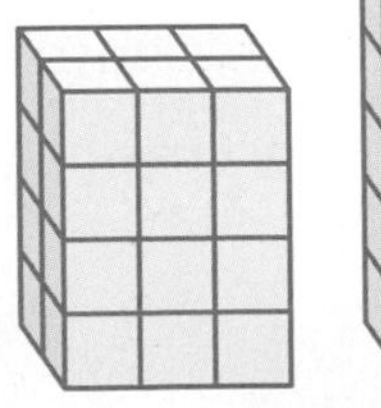

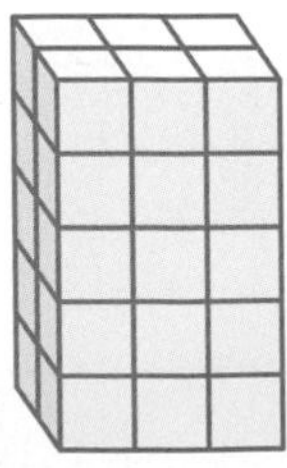

3 ACTIVITY: Finding a Formula for Volume

Work with a partner. Use the concept in Activity 2 to find a formula that gives the volume of any prism.

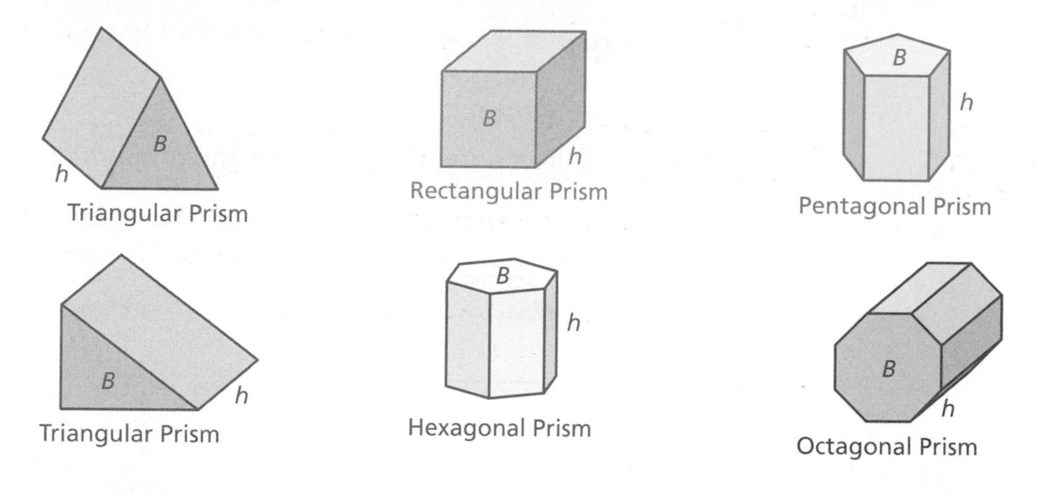

4 ACTIVITY: Using a Formula

Work with a partner. A ream of paper has 500 sheets.

a. Does a single sheet of paper have a volume? Why or why not?

b. If so, explain how you can find the volume of a single sheet of paper.

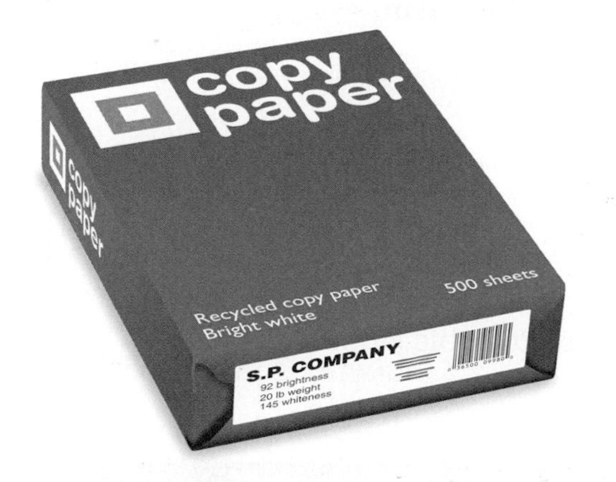

What Is Your Answer?

5. **IN YOUR OWN WORDS** How can you find the volume of a prism?

6. Draw a prism that has a trapezoid as its base. Use your formula to find the volume of the prism.

Use what you learned about the volumes of prisms to complete Exercises 4–6 on page 302.

7.1 Lesson

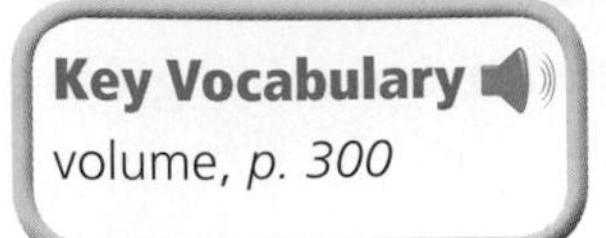

The **volume** of a three-dimensional figure is a measure of the amount of space that it occupies. Volume is measured in cubic units.

Key Idea

Volume of a Prism

Words The volume V of a prism is the product of the area of the base and the height of the prism.

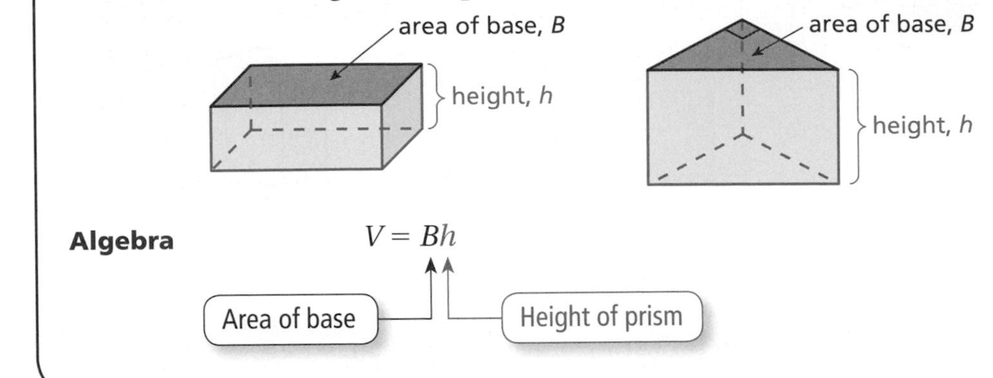

Algebra $V = Bh$

Area of base → B; Height of prism → h

EXAMPLE 1 Finding the Volume of a Prism

Study Tip

The area of the base of a rectangular prism is the product of the length ℓ and the width w.

You can use $V = \ell wh$ to find the volume of a rectangular prism.

Find the volume of the prism.

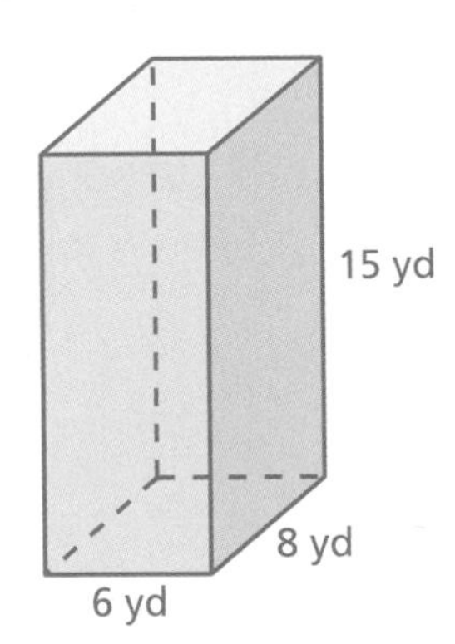

$V = Bh$	Write formula for volume.
$= 6(8) \cdot 15$	Substitute.
$= 48 \cdot 15$	Simplify.
$= 720$	Multiply.

∴ The volume is 720 cubic yards.

EXAMPLE 2 Finding the Volume of a Prism

Find the volume of the prism.

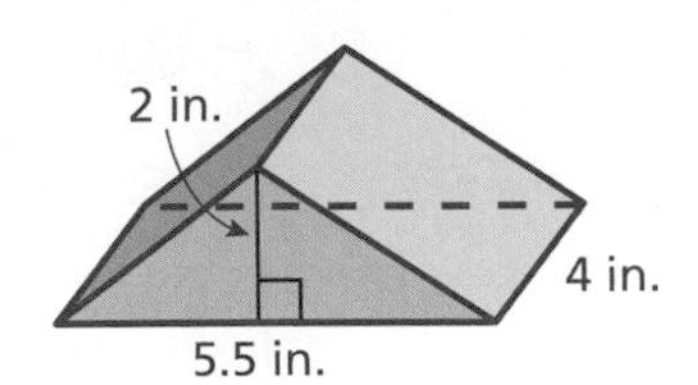

$V = Bh$	Write formula for volume.
$= \frac{1}{2}(5.5)(2) \cdot 4$	Substitute.
$= 5.5 \cdot 4$	Simplify.
$= 22$	Multiply.

∴ The volume is 22 cubic inches.

On Your Own

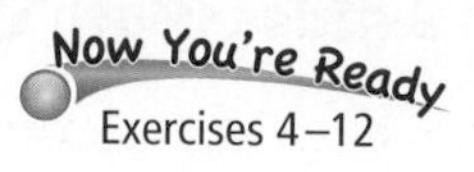

Find the volume of the prism.

1\.

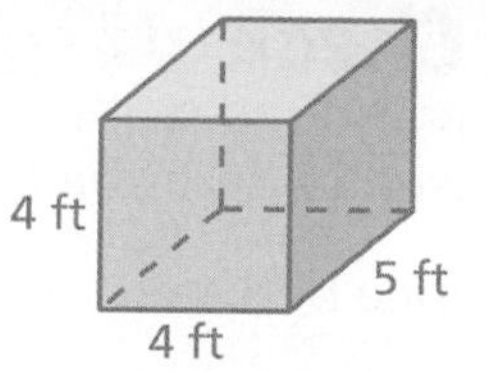

2\.

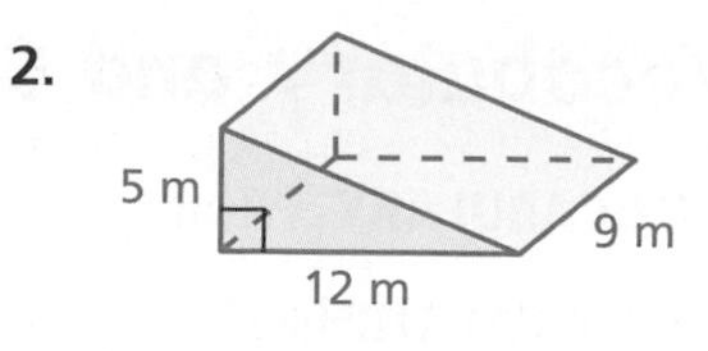

EXAMPLE 3 Real-Life Application

A movie theater designs two bags to hold 96 cubic inches of popcorn. (a) Find the height of each bag. (b) Which bag should the theater choose to reduce the amount of paper needed? Explain.

a. Find the height of each bag.

***Bag* A**

$V = Bh$

$96 = 4(3)(h)$

$96 = 12h$

$8 = h$

The height is 8 inches.

***Bag* B**

$V = Bh$

$96 = 4(4)(h)$

$96 = 16h$

$6 = h$

The height is 6 inches.

b. To determine the amount of paper needed, find the surface area of each bag. Do not include the top base.

***Bag* A**

$S = \ell w + 2\ell h + 2wh$

$= 4(3) + 2(4)(8) + 2(3)(8)$

$= 12 + 64 + 48$

$= 124 \text{ in.}^2$

***Bag* B**

$S = \ell w + 2\ell h + 2wh$

$= 4(4) + 2(4)(6) + 2(4)(6)$

$= 16 + 48 + 48$

$= 112 \text{ in.}^2$

The surface area of Bag B is less than the surface area of Bag A. So, the theater should choose Bag B.

On Your Own

3\. You design Bag C that has a volume of 96 cubic inches. Should the theater in Example 3 choose your bag? Explain.

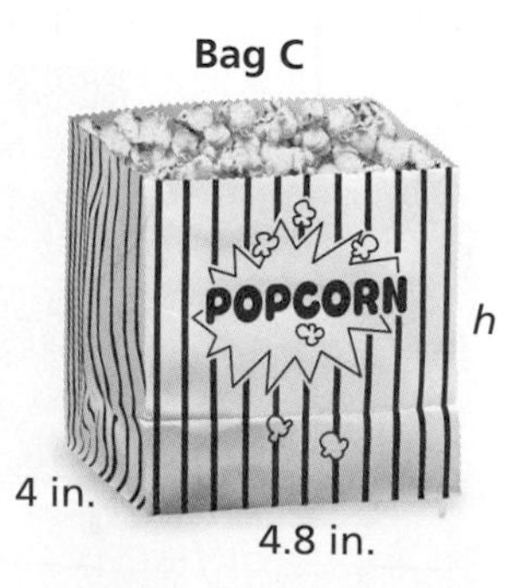

7.1 Exercises

Vocabulary and Concept Check

1. **VOCABULARY** What type of units are used to describe volume?
2. **CRITICAL THINKING** What is the difference between volume and surface area?
3. **CRITICAL THINKING** You are ordering packaging for a product. Should you be more concerned with volume or surface area? Explain.

Practice and Problem Solving

Find the volume of the prism.

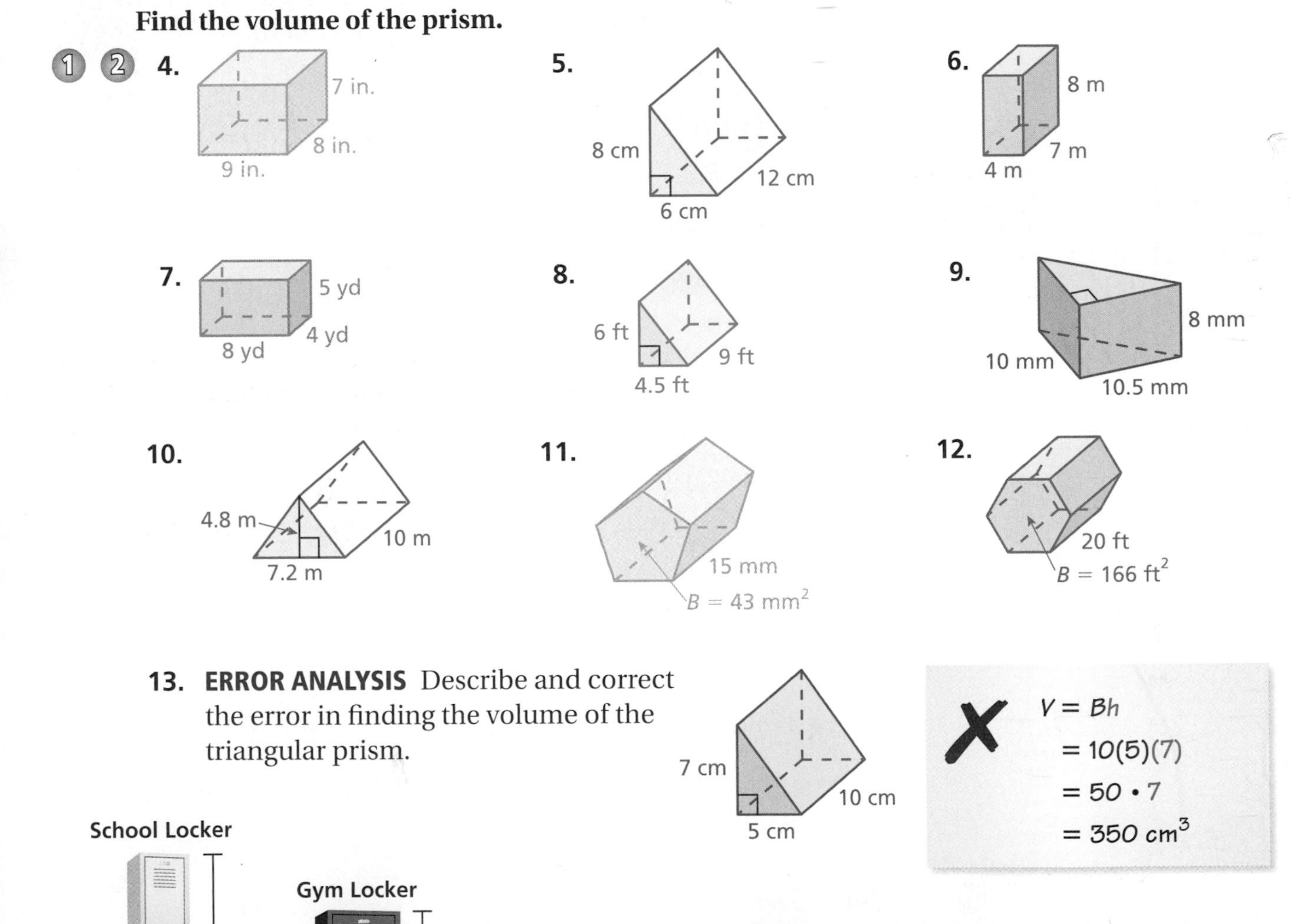

13. **ERROR ANALYSIS** Describe and correct the error in finding the volume of the triangular prism.

School Locker
60 in.
12 in.
10 in.

Gym Locker
48 in.
12 in.
15 in.

14. **LOCKER** Each locker is shaped like a rectangular prism. Which has more storage space? Explain.

15. **CEREAL BOX** A cereal box is 9 inches by 2.5 inches by 10 inches. What is the volume of the box?

Find the volume of the prism.

16. **17.**

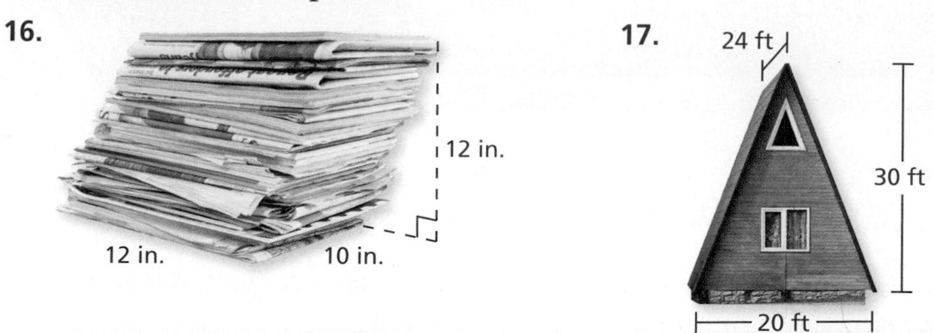

18. REASONING Two prisms have the same volume. Do they *always*, *sometimes*, or *never* have the same surface area? Explain.

19. CUBIC UNITS How many cubic inches are in a cubic foot? Use a sketch to explain your reasoning.

20. CAPACITY As a gift, you fill the calendar with packets of chocolate candy. Each packet has a volume of 2 cubic inches. Find the maximum number of packets you can fit inside the calendar.

21. HEIGHT Two liters of water are poured into an empty vase shaped like an octagonal prism. The base area is 100 square centimeters. What is the height of the water? ($1 \text{ L} = 1000 \text{ cm}^3$)

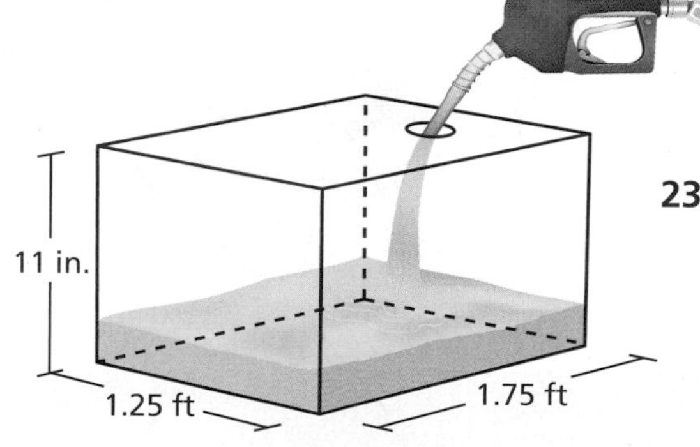

22. GAS TANK The gas tank is 20% full. Use the current price of gas in your community to find the cost to fill the tank. ($1 \text{ gal} = 231 \text{ in.}^3$)

23. OPEN-ENDED You visit an aquarium. One of the tanks at the aquarium holds 450 gallons of water. Draw a diagram to show one possible set of dimensions of the tank. ($1 \text{ gal} = 231 \text{ in.}^3$)

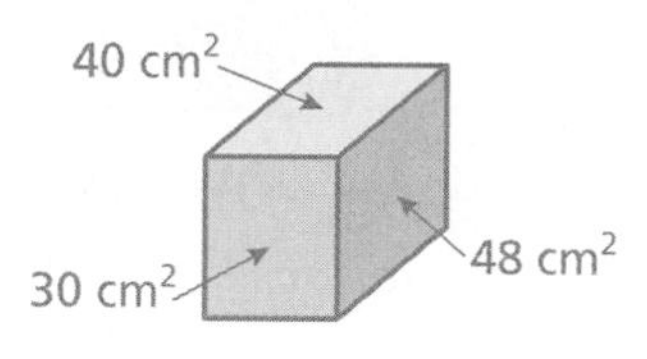

24. Critical Thinking What is the volume of the rectangular prism?

Fair Game Review What you learned in previous grades & lessons

Identify the transformation. *(Section 5.5, Section 5.6, and Section 5.7)*

25. **26.** **27.**

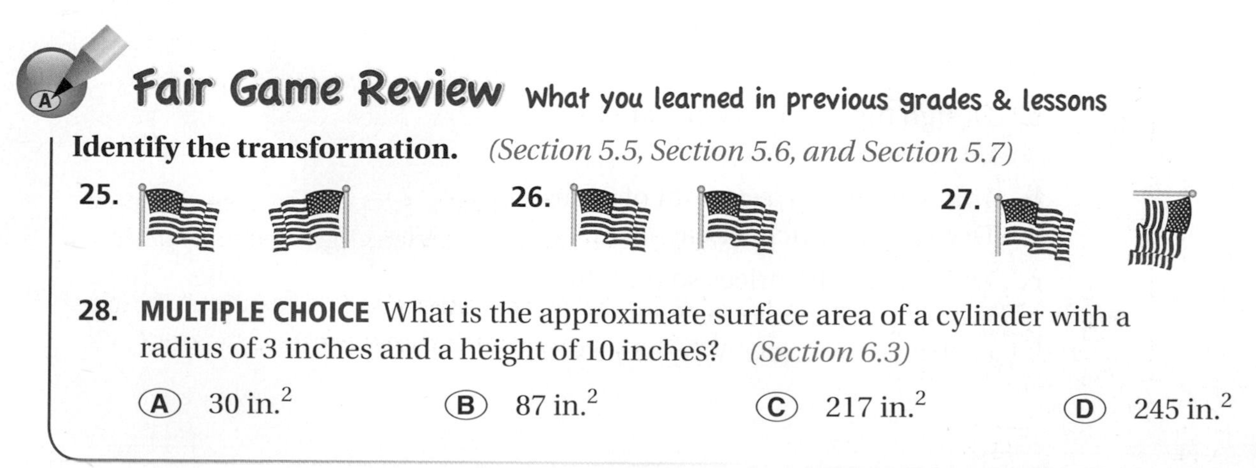

28. MULTIPLE CHOICE What is the approximate surface area of a cylinder with a radius of 3 inches and a height of 10 inches? *(Section 6.3)*

Ⓐ 30 in.^2 Ⓑ 87 in.^2 Ⓒ 217 in.^2 Ⓓ 245 in.^2

7.2 Volumes of Cylinders

COMMON CORE STATE STANDARDS

7.G.4
7.G.6
8.G.9

Essential Question How can you find the volume of a cylinder?

1 ACTIVITY: Finding a Formula Experimentally

Work with a partner.

a. Find the area of the face of a coin.

b. Find the volume of a stack of a dozen coins.

c. Generalize your results to find the volume of a cylinder.

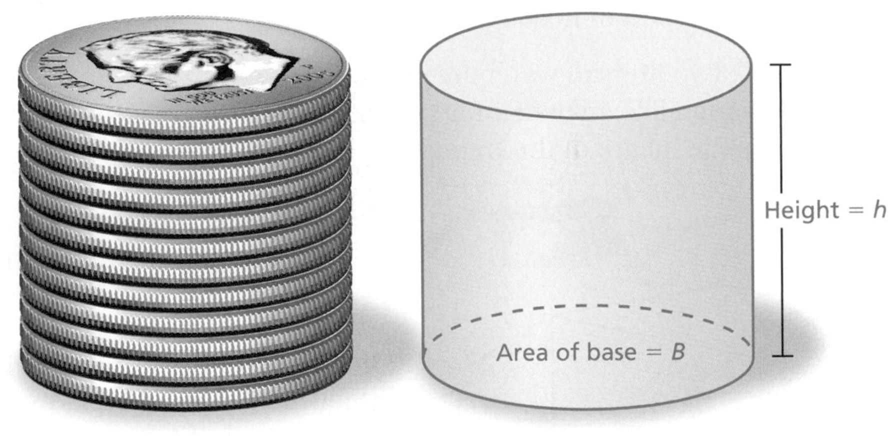

2 ACTIVITY: Making a Business Plan

Work with a partner. You are planning to make and sell 3 different sizes of cylindrical candles. You buy 1 cubic foot of candle wax for $20 to make 8 candles of each size.

a. Design the candles. What are the dimensions of each size?

b. You want to make a profit of $100. Decide on a price for each size.

c. Did you set the prices so that they are proportional to the volume of each size of candle? Why or why not?

3 ACTIVITY: Science Experiment

Work with a partner. Use the diagram to describe how you can find the volume of a small object.

ACTIVITY: Comparing Cylinders

Work with a partner.

a. Just by looking at the two cylinders, which one do you think has the greater volume? Explain your reasoning.

b. Find the volume of each cylinder. Was your prediction in part (a) correct? Explain your reasoning.

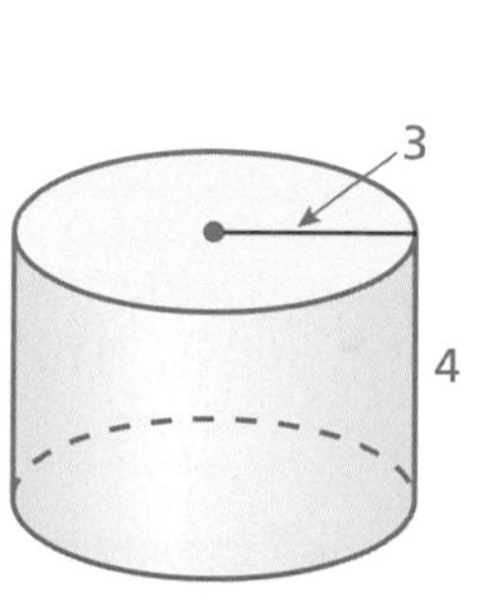

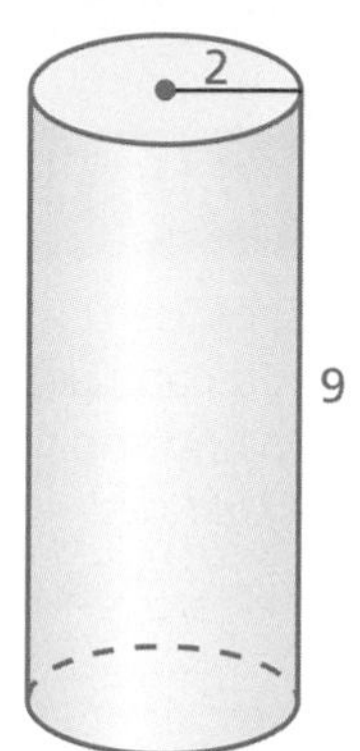

What Is Your Answer?

5. **IN YOUR OWN WORDS** How can you find the volume of a cylinder?

6. Compare your formula for the volume of a cylinder with the formula for the volume of a prism. How are they the same?

"Here's how I remember how to find the volume of any prism or cylinder."

"Base times tall, will fill 'em all."

Use what you learned about the volumes of cylinders to complete Exercises 3–5 on page 308.

7.2 Lesson

Check It Out
Lesson Tutorials
BigIdeasMath.com

Key Idea

Volume of a Cylinder

Words The volume V of a cylinder is the product of the area of the base and the height of the cylinder.

area of base, B

height, h

Algebra $V = Bh$

Area of base — Height of cylinder

EXAMPLE 1 Finding the Volume of a Cylinder

Find the volume of the cylinder. Round your answer to the nearest tenth.

$V = Bh$	Write formula for volume.
$= \pi(3)^2(6)$	Substitute.
$= 54\pi \approx 169.6$	Simplify.

3 m

6 m

The volume is about 169.6 cubic meters.

Study Tip

Because $B = \pi r^2$, you can use $V = \pi r^2 h$ to find the volume of a cylinder.

EXAMPLE 2 Finding the Height of a Cylinder

Find the height of the cylinder. Round your answer to the nearest whole number.

The diameter is 10 inches. So, the radius is 5 inches.

$V = Bh$	Write formula for volume.
$314 = \pi(5)^2(h)$	Substitute.
$314 = 25\pi h$	Simplify.
$4 \approx h$	Divide each side by 25π.

h

10 in.

Volume = 314 in.^{3}

The height is about 4 inches.

On Your Own

Now You're Ready
Exercises 3–11 and 13–15

Find the volume V or height h of the cylinder. Round your answer to the nearest tenth.

1. 15 ft

4 ft

$V \approx$

2. 8 cm

$h \approx$

Volume = 176 cm^3

EXAMPLE 3 Real-Life Application

How much salsa is missing from the jar?

The missing salsa fills a cylinder with a height of $10 - 4 = 6$ centimeters and a radius of 5 centimeters.

$V = Bh$	Write formula for volume.
$= \pi(5)^2(6)$	Substitute.
$= 150\pi \approx 471$	Simplify.

∴ About 471 cubic centimeters of salsa are missing from the jar.

EXAMPLE 4 Standardized Test Practice

About how many gallons of water does the water cooler bottle contain? ($1\ \text{ft}^3 \approx 7.5$ gal)

Ⓐ 5.3 gal Ⓑ 10 gal Ⓒ 17 gal Ⓓ 40 gal

Find the volume of the cylinder. The diameter is 1 foot. So, the radius is 0.5 foot.

$V = Bh$	Write formula for volume.
$= \pi(0.5)^2(1.7)$	Substitute.
$= 0.425\pi \approx 1.3345$	Simplify.

So, the cylinder contains about 1.3345 cubic feet of water. To find the number of gallons it contains, multiply by $\frac{7.5\ \text{gal}}{1\ \text{ft}^3}$.

$$1.3345\ \cancel{\text{ft}^3} \times \frac{7.5\ \text{gal}}{1\ \cancel{\text{ft}^3}} \approx 10\ \text{gal}$$

∴ The water cooler bottle contains about 10 gallons of water. The correct answer is Ⓑ.

On Your Own

3. **WHAT IF?** In Example 3, the height of the salsa in the jar is 5 centimeters. How much salsa is missing from the jar?
4. A cylindrical water tower has a diameter of 15 meters and a height of 5 meters. About how many gallons of water can the tower contain? ($1\ \text{m}^3 \approx 264$ gal)

Vocabulary and Concept Check

1. **DIFFERENT WORDS, SAME QUESTION** Which is different? Find "both" answers.

How much does it take to fill the cylinder?

What is the capacity of the cylinder?

How much does it take to cover the cylinder?

How much does the cylinder contain?

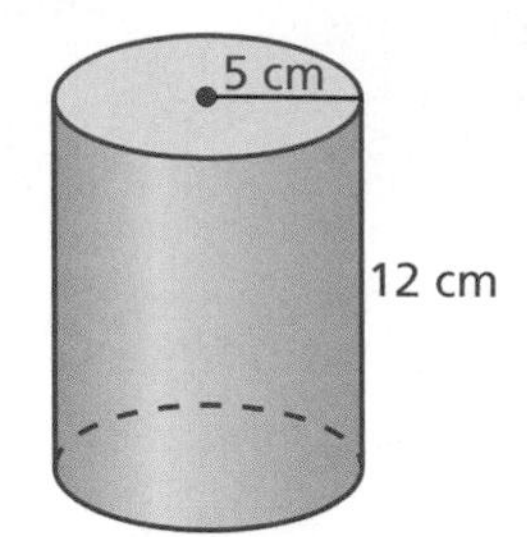

2. **REASONING** Without calculating, which of the solids has the greater volume? Explain.

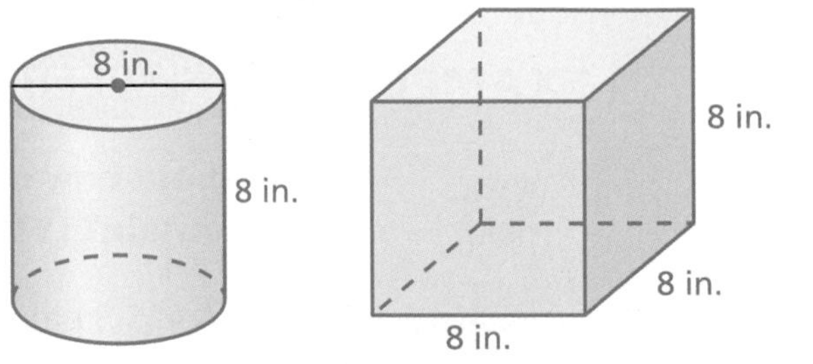

Practice and Problem Solving

Find the volume of the cylinder. Round your answer to the nearest tenth.

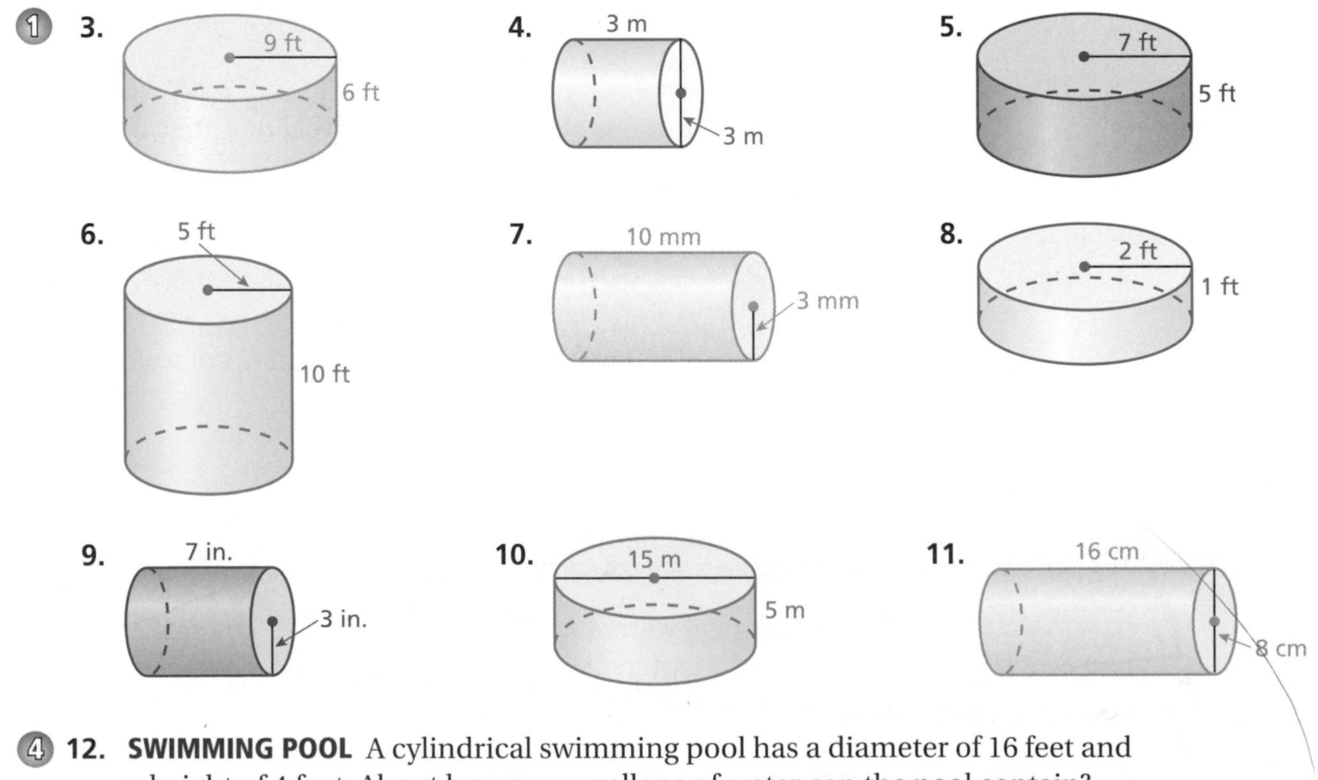

④ 12. **SWIMMING POOL** A cylindrical swimming pool has a diameter of 16 feet and a height of 4 feet. About how many gallons of water can the pool contain? Round your answer to the nearest whole number. ($1 \text{ ft}^3 \approx 7.5$ gal)

Find the height of the cylinder. Round your answer to the nearest whole number.

2 **13.** Volume $= 250 \text{ ft}^3$

14. Volume $= 32{,}000 \text{ in.}^3$

32 in.

h

15. Volume $= 600{,}000 \text{ cm}^3$

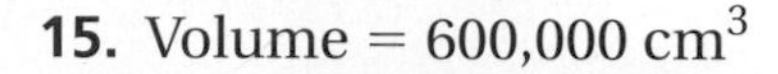

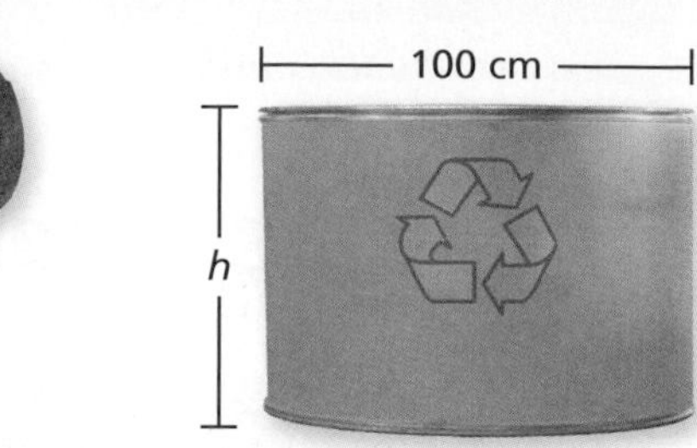

16. CRITICAL THINKING How does the volume of a cylinder change when its diameter is halved? Explain.

Round Hay Bale

17. HAY BALES A traditional "square" bale of hay is actually in the shape of a rectangular prism. Its dimensions are 2 feet by 2 feet by 4 feet. How many "square" bales contain the same amount of hay as one large "round" bale?

18. ROAD ROLLER A tank on the road roller is filled with water to make the roller heavy. The tank is a cylinder that has a height of 6 feet and a radius of 2 feet. One cubic foot of water weighs 62.5 pounds. Find the weight of the water in the tank.

19. VOLUME A cylinder has a surface area of 1850 square meters and a radius of 9 meters. Estimate the volume of the cylinder to the nearest whole number.

20. Critical Thinking Water flows at 2 feet per second through a pipe with a diameter of 8 inches. A cylindrical tank with a diameter of 15 feet and a height of 6 feet collects the water.

a. What is the volume, in cubic inches, of water flowing out of the pipe every second?

b. What is the height, in inches, of the water in the tank after 5 minutes?

c. How many minutes will it take to fill 75% of the tank?

Fair Game Review What you learned in previous grades & lessons

Write and solve an equation to answer the question. *(Section 4.1)*

21. 50% of 200 is what number?

22. 80% of 400 is what number?

23. MULTIPLE CHOICE The variables x and y vary directly. When x is 18, y is 24. Which equation relates x and y? *(Section 3.7)*

Ⓐ $y = \frac{3}{4}x$ Ⓑ $y = 2x - 12$ Ⓒ $y = 4x - 3$ Ⓓ $y = \frac{4}{3}x$

7.3 Volumes of Pyramids

COMMON CORE STATE STANDARDS
7.G.6

Essential Question How can you find the volume of a pyramid?

1 ACTIVITY: Finding a Formula Experimentally

Work with a partner.

- **Draw the two nets on cardboard and cut them out.**

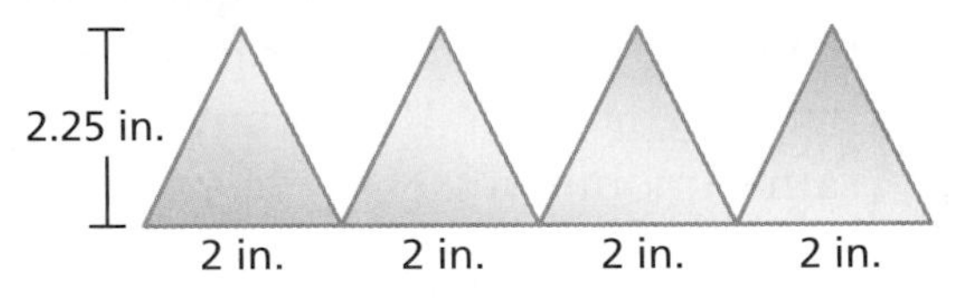

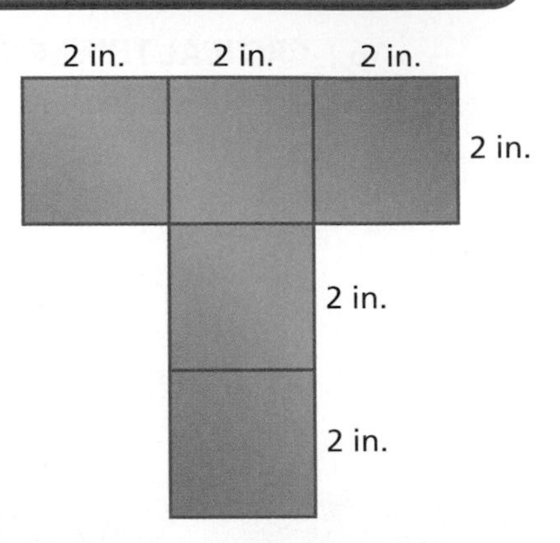

- **Fold and tape the nets to form an open square box and an open pyramid.**
- **Both figures should have the same size square base and the same height.**
- **Fill the pyramid with pebbles. Then pour the pebbles into the box. Repeat this until the box is full. How many pyramids does it take to fill the box?**
- **Use your result to find a formula for the volume of a pyramid.**

2 ACTIVITY: Comparing Volumes

Work with a partner. You are an archeologist studying two ancient pyramids. What factors would affect how long it took to build each pyramid? Given similar conditions, which pyramid took longer to build? Explain your reasoning.

Cholula Pyramid in Mexico
Height: about 217 ft
Base: about 1476 ft by 1476 ft

Cheops Pyramid in Egypt
Height: about 480 ft
Base: about 755 ft by 755 ft

3 ACTIVITY: Finding and Using a Pattern

Work with a partner.

- **Find the volumes of the pyramids.**
- **Organize your results in a table.**
- **Describe the pattern.**
- **Use your pattern to find the volume of a pyramid with a side length and height of 20.**

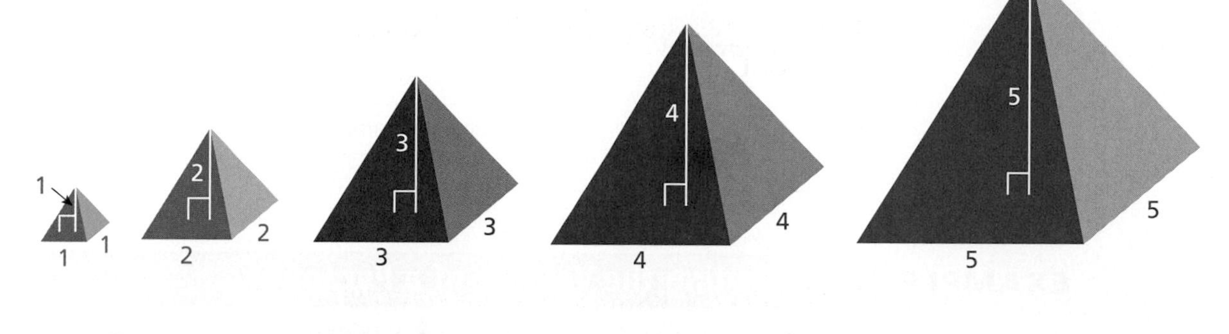

4 ACTIVITY: Breaking a Prism into Pyramids

Work with a partner. The rectangular prism can be cut to form three pyramids. Show that the sum of the volumes of the three pyramids is equal to the volume of the prism.

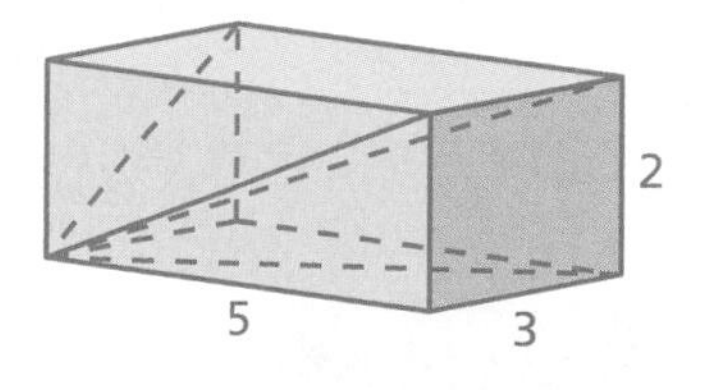

a. 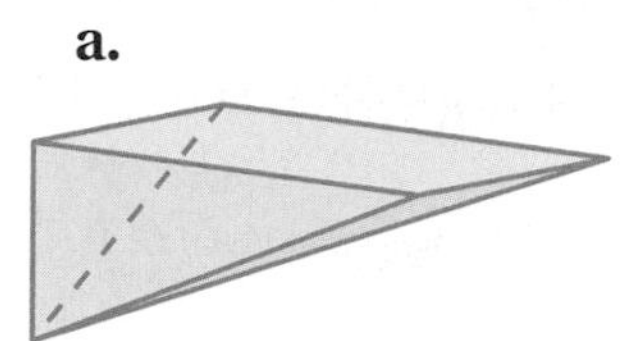

b. 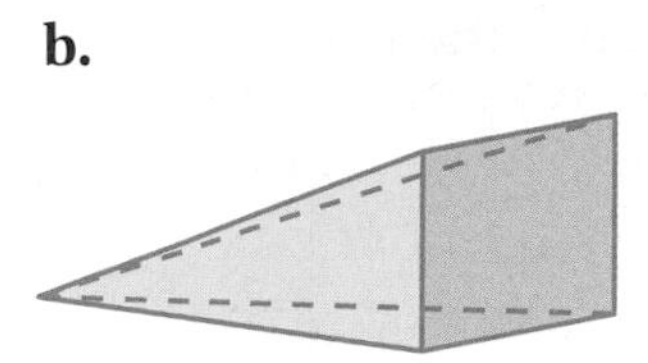

c. 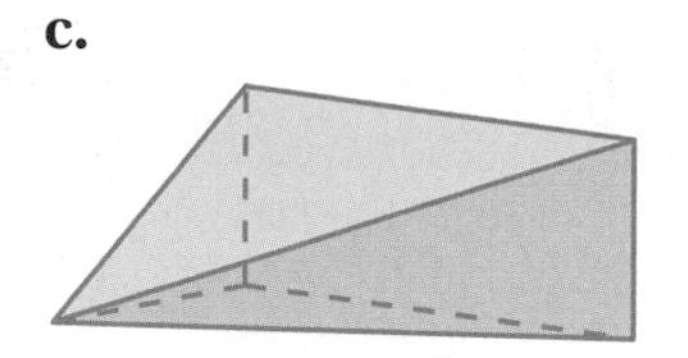

What Is Your Answer?

5. **IN YOUR OWN WORDS** How can you find the volume of a pyramid?
6. Write a general formula for the volume of a pyramid.

Practice Use what you learned about the volumes of pyramids to complete Exercises 4–6 on page 314.

7.3 Lesson

Key Idea

Volume of a Pyramid

Words The volume V of a pyramid is one-third the product of the area of the base and the height of the pyramid.

Algebra $V = \frac{1}{3}Bh$

Area of base (B)

Height of pyramid (h)

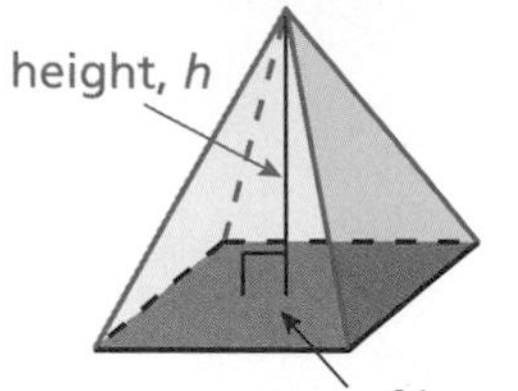

EXAMPLE 1 Finding the Volume of a Pyramid

Find the volume of the pyramid.

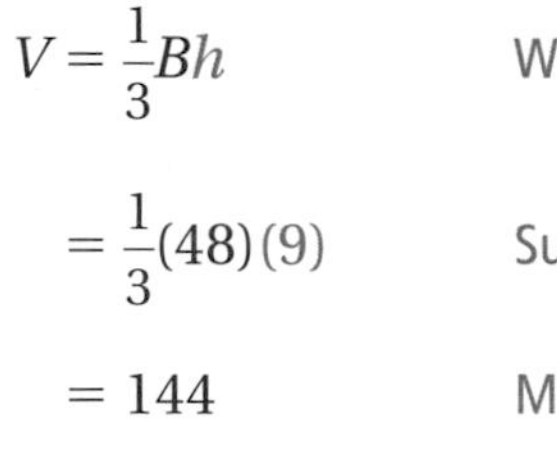

$V = \frac{1}{3}Bh$ Write formula for volume.

$= \frac{1}{3}(48)(9)$ Substitute.

$= 144$ Multiply.

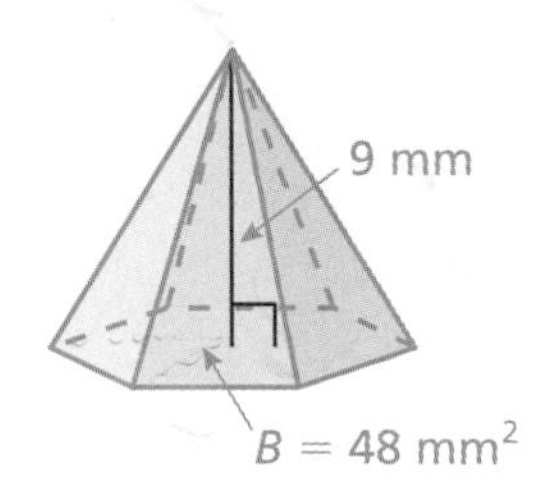

The volume is 144 cubic millimeters.

EXAMPLE 2 Finding the Volume of a Pyramid

Find the volume of the pyramid.

a.

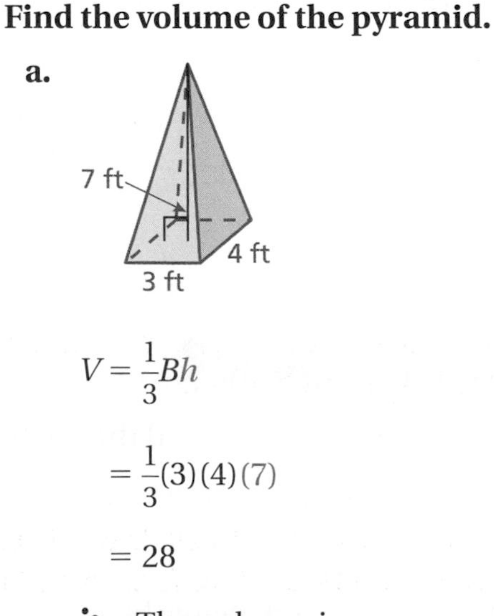

$V = \frac{1}{3}Bh$

$= \frac{1}{3}(3)(4)(7)$

$= 28$

The volume is 28 cubic feet.

b.

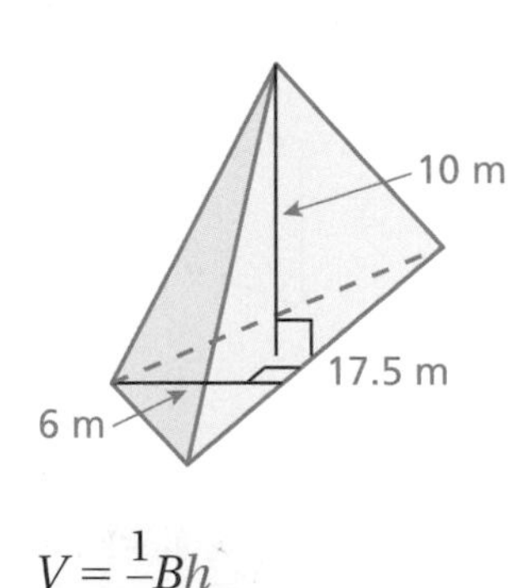

$V = \frac{1}{3}Bh$

$= \frac{1}{3}\left(\frac{1}{2}\right)(17.5)(6)(10)$

$= 175$

The volume is 175 cubic meters.

Study Tip

The area of the base of a rectangular pyramid is the product of the length ℓ and the width w.

You can use $V = \frac{1}{3}\ell wh$ to find the volume of a rectangular pyramid.

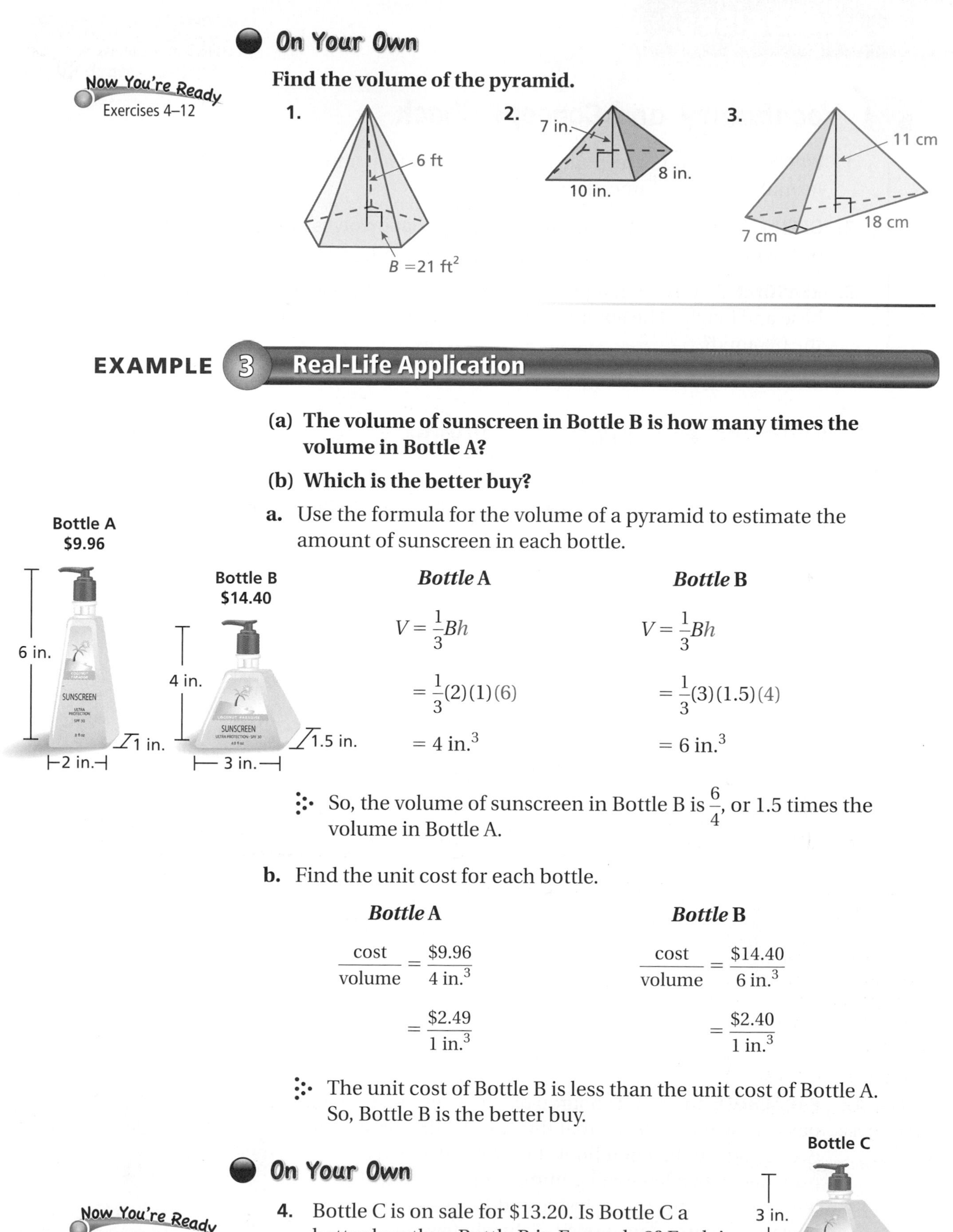

On Your Own

Now You're Ready
Exercises 4–12

Find the volume of the pyramid.

1. 6 ft; $B = 21 \text{ ft}^2$

2. 7 in.; 8 in.; 10 in.

3. 11 cm; 7 cm; 18 cm

EXAMPLE 3 Real-Life Application

(a) The volume of sunscreen in Bottle B is how many times the volume in Bottle A?

(b) Which is the better buy?

Bottle A $9.96: 6 in., 2 in., 1 in.

Bottle B $14.40: 4 in., 3 in., 1.5 in.

a. Use the formula for the volume of a pyramid to estimate the amount of sunscreen in each bottle.

***Bottle* A**	***Bottle* B**
$V = \frac{1}{3}Bh$	$V = \frac{1}{3}Bh$
$= \frac{1}{3}(2)(1)(6)$	$= \frac{1}{3}(3)(1.5)(4)$
$= 4 \text{ in.}^3$	$= 6 \text{ in.}^3$

So, the volume of sunscreen in Bottle B is $\frac{6}{4}$, or 1.5 times the volume in Bottle A.

b. Find the unit cost for each bottle.

***Bottle* A**	***Bottle* B**
$\frac{\text{cost}}{\text{volume}} = \frac{\$9.96}{4 \text{ in.}^3}$	$\frac{\text{cost}}{\text{volume}} = \frac{\$14.40}{6 \text{ in.}^3}$
$= \frac{\$2.49}{1 \text{ in.}^3}$	$= \frac{\$2.40}{1 \text{ in.}^3}$

The unit cost of Bottle B is less than the unit cost of Bottle A. So, Bottle B is the better buy.

On Your Own

Now You're Ready
Exercise 18

4. Bottle C is on sale for $13.20. Is Bottle C a better buy than Bottle B in Example 3? Explain.

Bottle C: 3 in., 3 in., 2 in.

7.3 Exercises

Vocabulary and Concept Check

1. **WRITING** How is the formula for the volume of a pyramid different from the formula for the volume of a prism?
2. **OPEN-ENDED** Describe a real-life situation that involves finding the volume of a pyramid.
3. **REASONING** A triangular pyramid and a triangular prism have the same base and height. The volume of the prism is how many times the volume of the pyramid?

Practice and Problem Solving

Find the volume of the pyramid.

1 2 **4.**

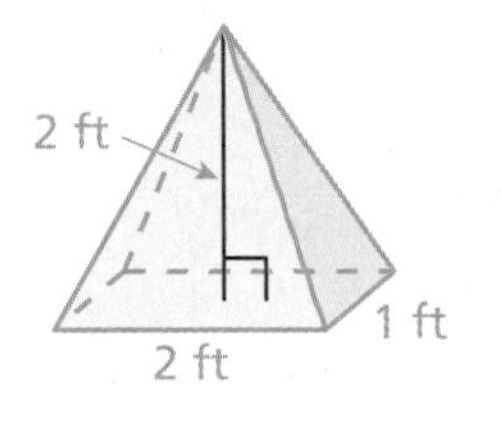

5.

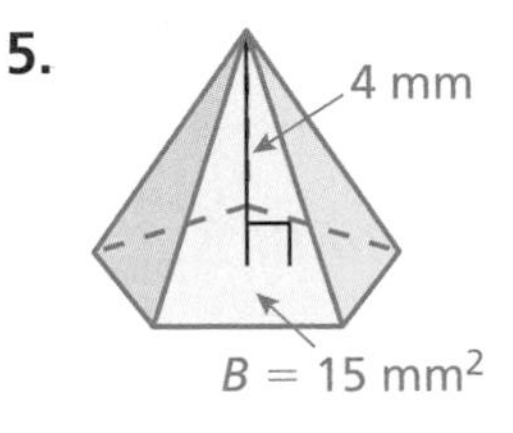

6.

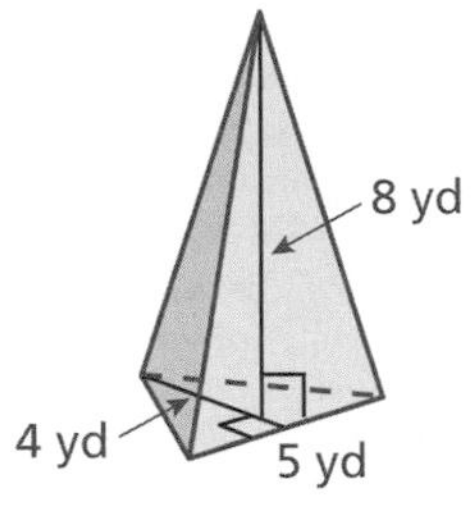

7.

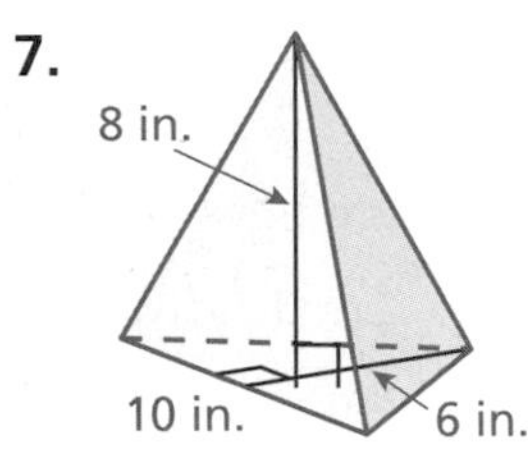

8.

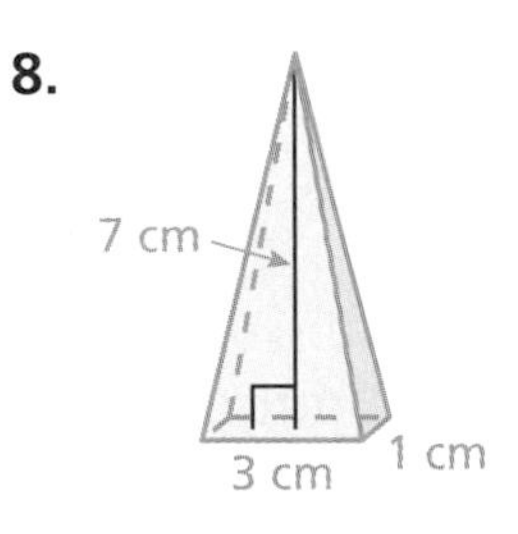

9.

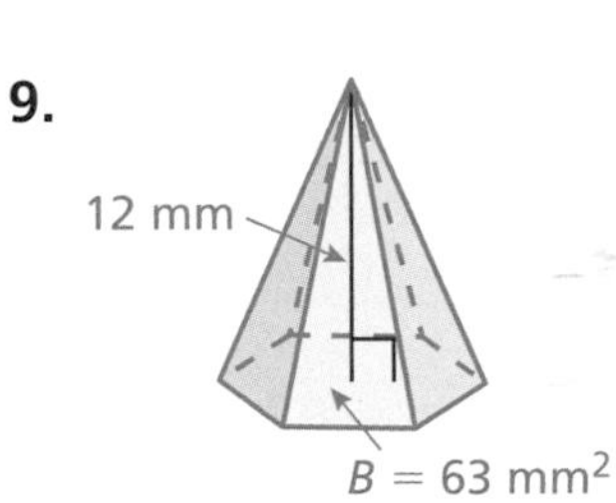

10.

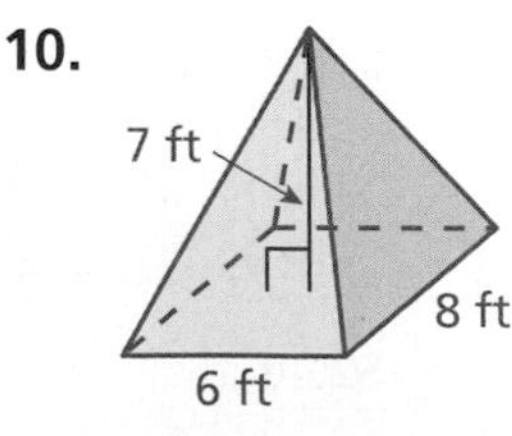

11.

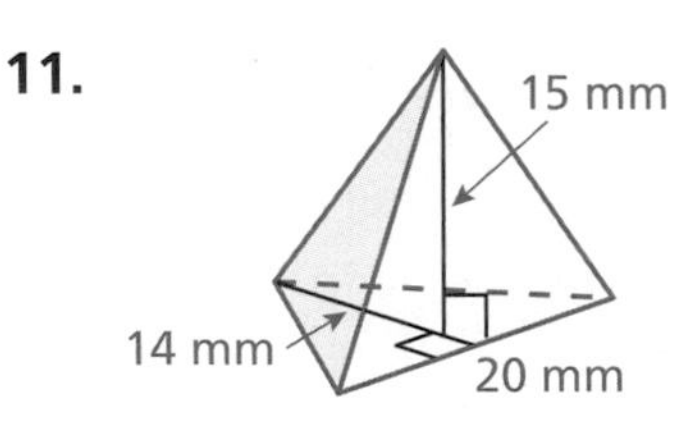

12. **PARACHUTE** In 1483, Leonardo da Vinci designed a parachute. It is believed that this was the first parachute ever designed. In a notebook, he wrote "If a man is provided with a length of gummed linen cloth with a length of 12 yards on each side and 12 yards high, he can jump from any great height whatsoever without injury." Find the volume of air inside Leonardo's parachute.

Not drawn to scale

Copy and complete the table to find the area of the base B or the height h of the pyramid.

	Volume, V	Area of Base, B	Height, h
13.	60 in.^{3}		6 in.
14.	144 cm^3	48 cm^2	
15.	135 ft^3	54 ft^2	

16. TEEPEE Use the photo of the teepee.

a. What is the shape of the base? How can you tell?

b. The teepee's height is about 10 feet. Estimate the volume of the teepee.

4 in.

3 in.

3 in.

Paperweight

17. PAPERWEIGHT How much glass is needed to manufacture 1000 paperweights? Explain your reasoning.

6 in.

8 in.

$B = 30$ in.^{2}

$B = 24$ in.^{2}

Spire A

Spire B

③ **18. SPIRE** Which sandcastle spire has a greater volume? How much more sand is required to make the spire with the greater volume?

19. OPEN-ENDED A pyramid has a volume of 40 cubic feet and a height of 6 feet. Find one possible set of dimensions of the rectangular base.

20. **Reasoning** Do the two solids have the same volume? Explain.

z y x

3z y x

Fair Game Review What you learned in previous grades & lessons

Simplify the expression. *(Skills Review Handbook)*

21. $\frac{1}{3} \times 12 \times 7$ **22.** $\frac{1}{3} \times 8 \times 27$ **23.** $\frac{1}{3} \times 6^2 \times 5$ **24.** $\frac{1}{3} \times 2^2 \times 15$

25. MULTIPLE CHOICE You spend 25% of your money on a shirt. Then you spend $\frac{1}{6}$ of the remainder on lunch. Lunch costs \$8. What percent of your money is spent on lunch? *(Section 4.1)*

Ⓐ 4.2% Ⓑ 12.5% Ⓒ 16.7% Ⓓ 32%

7.4 Volumes of Cones

COMMON CORE STATE STANDARDS
7.G.4
7.G.6
8.G.9

Essential Question How can you remember the formulas for surface area and volume?

You discovered that the volume of a pyramid is one-third the volume of a prism that has the same base and same height. You can use a similar activity to discover that the volume of a cone is one-third the volume of a cylinder that has the same base and height.

Volume of a Cone $= \frac{1}{3}$(Area of Base) $\times$ (Height)

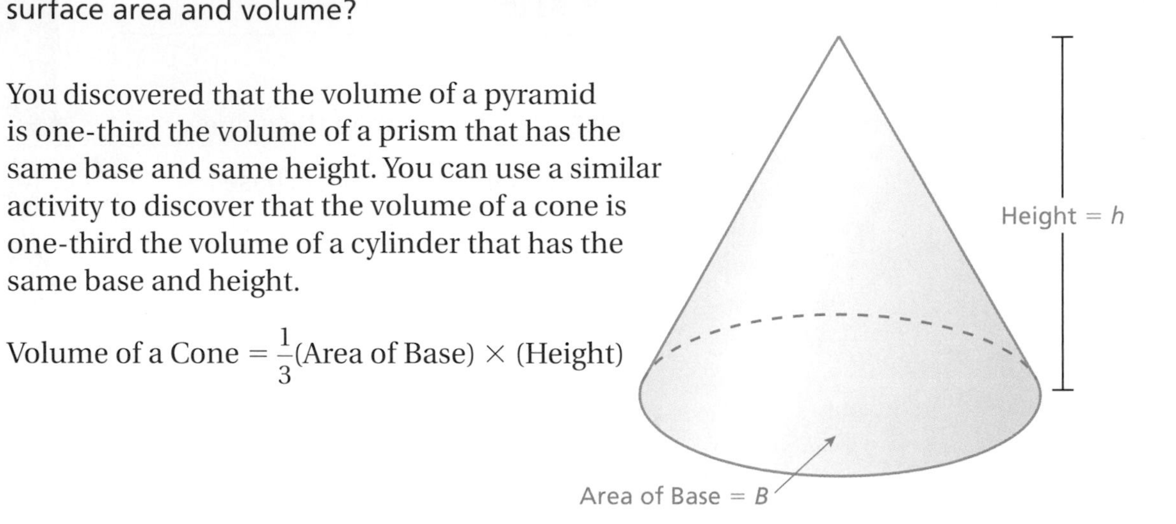

1 ACTIVITY: Summarizing Volume Formulas

Work with a partner. You can remember the volume formulas for all of the solids shown with just two concepts.

Volumes of Prisms and Cylinders

Volume = (Area of Base) $\times$ (Height)

Volumes of Pyramids and Cones

Volume = $\frac{1}{3}$ (Volume of Prism or Cylinder with same base and height)

Make a list of all the formulas you need to remember to find the area of a base. Talk about strategies for remembering these formulas.

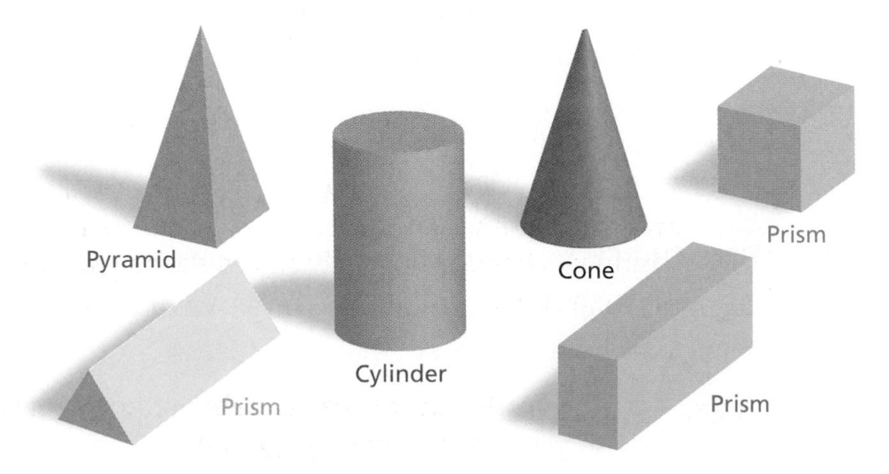

2 ACTIVITY: Volumes of Oblique Solids

Work with a partner. Think of a stack of paper. If you adjust the stack so that the sides are oblique (slanted), do you change the volume of the stack? If the volume of the stack does not change, then the formulas for volumes of right solids also apply to oblique solids.

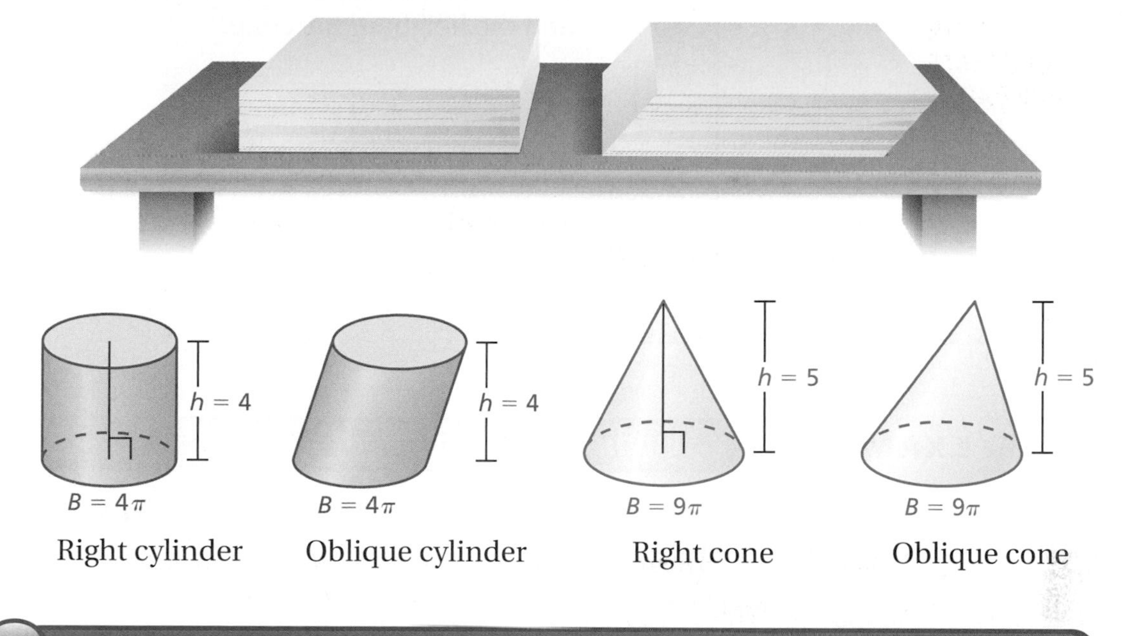

Right cylinder | Oblique cylinder | Right cone | Oblique cone

3 ACTIVITY: Summarizing Surface Area Formulas

Work with a partner. Make a list of the formulas for surface area that you studied in Chapter 6. Organize these formulas in a way similar to what you did in Activity 1.

Surface Area of a Right Prism =

Surface Area of a Right Pyramid =

Surface Area of a Right Cylinder =

Surface Area of a Right Cone =

What Is Your Answer?

4. **IN YOUR OWN WORDS** How can you remember the formulas for surface area and volume? Write all of the surface area and volume formulas on a summary sheet. Make the list short so that you do not have to memorize many formulas.

Use what you learned about the volumes of cones to complete Exercises 4–6 on page 320.

7.4 Lesson

Key Idea

Volume of a Cone

Words The volume V of a cone is one-third the product of the area of the base and the height of the cone.

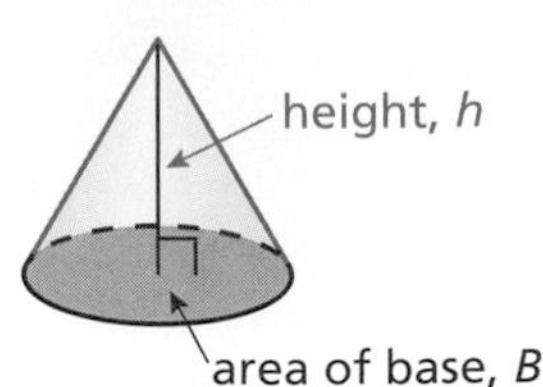

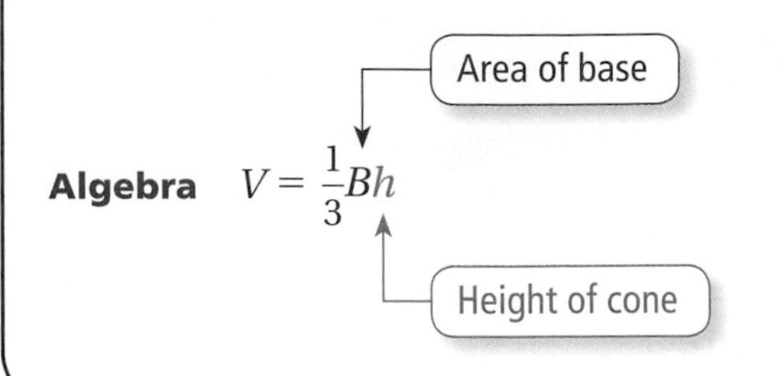

Algebra $V = \frac{1}{3}Bh$

EXAMPLE 1 Finding the Volume of a Cone

Study Tip

Because $B = \pi r^2$, you can use $V = \frac{1}{3}\pi r^2 h$ to find the volume of a cone.

Find the volume of the cone. Round your answer to the nearest tenth.

The diameter is 4 meters. So, the radius is 2 meters.

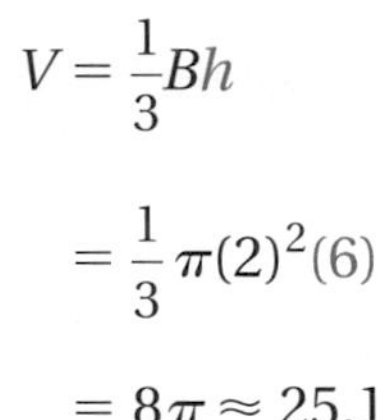

$V = \frac{1}{3}Bh$ Write formula.

$= \frac{1}{3}\pi(2)^2(6)$ Substitute.

$= 8\pi \approx 25.1$ Simplify.

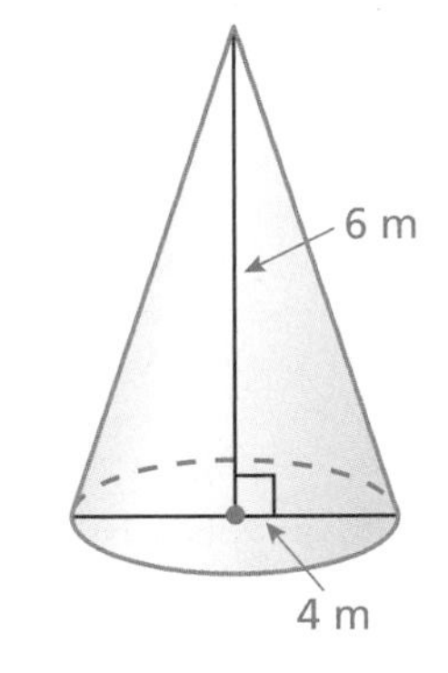

The volume is about 25.1 cubic meters.

EXAMPLE 2 Finding the Height of a Cone

Find the height of the cone. Round your answer to the nearest tenth.

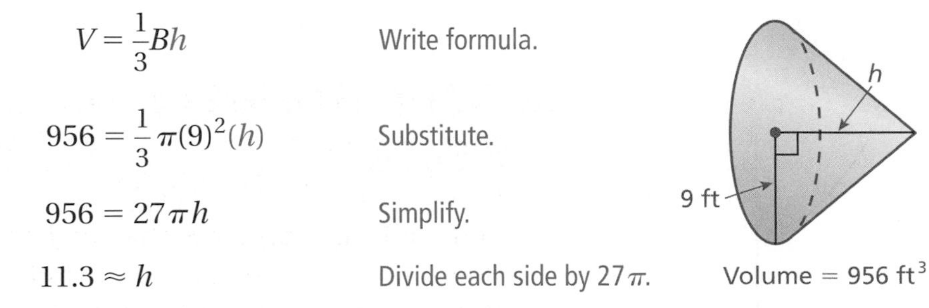

$V = \frac{1}{3}Bh$ Write formula.

$956 = \frac{1}{3}\pi(9)^2(h)$ Substitute.

$956 = 27\pi h$ Simplify.

$11.3 \approx h$ Divide each side by 27π.

The height is about 11.3 feet.

On Your Own

Now You're Ready
Exercises 4–17

Find the volume V or height h of the cone. Round your answer to the nearest tenth.

1. 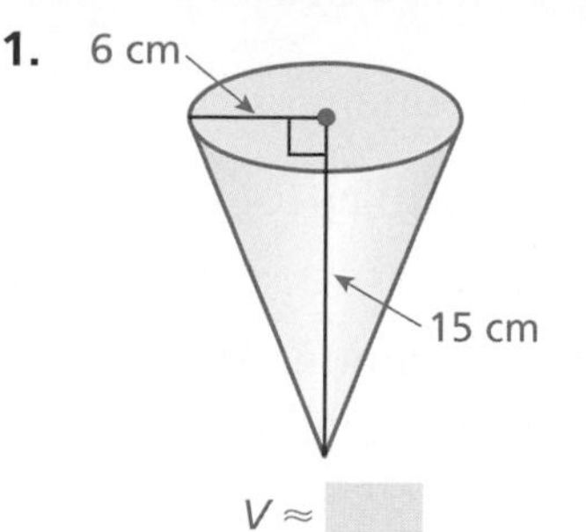

2. 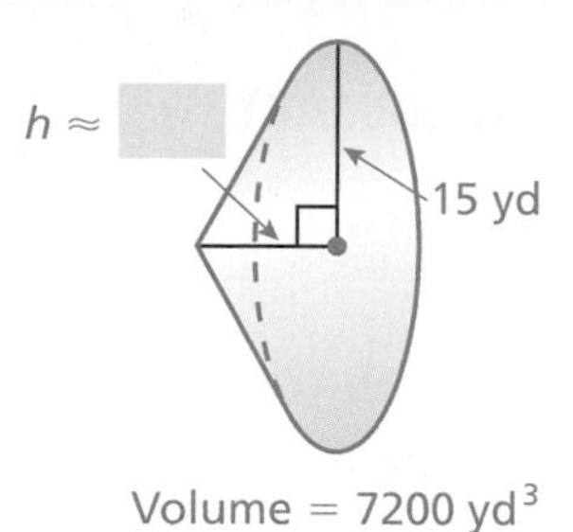

Volume = 7200 yd^3

EXAMPLE 3 Real-Life Application

30 mm

10 mm

24 mm

You must answer a trivia question before the sand in the timer falls to the bottom. The sand falls at a rate of 50 cubic millimeters per second. How much time do you have to answer the question?

Use the formula for the volume of a cone to find the volume of the sand in the timer.

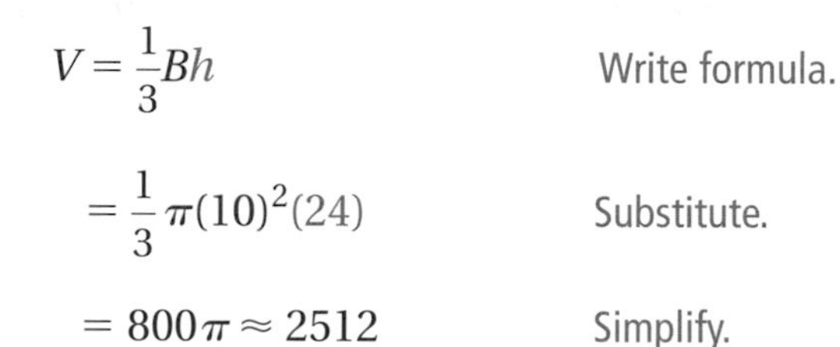

$$V = \frac{1}{3}Bh \qquad \text{Write formula.}$$

$$= \frac{1}{3}\pi(10)^2(24) \qquad \text{Substitute.}$$

$$= 800\pi \approx 2512 \qquad \text{Simplify.}$$

The volume of the sand is about 2512 cubic millimeters. To find the amount of time you have to answer the question, multiply the volume by the rate at which the sand falls.

$$2512 \text{ mm}^3 \times \frac{1 \text{ sec}}{50 \text{ mm}^3} = 50.24 \text{ sec}$$

∴ You have about 50 seconds to answer the question.

On Your Own

3. **WHAT IF?** In Example 3, the sand falls at a rate of 60 cubic millimeters per second. How much time do you have to answer the question?

4. **WHAT IF?** In Example 3, the height of the sand in the timer is 12 millimeters and the radius is 5 millimeters. How much time do you have to answer the question?

7.4 Exercises

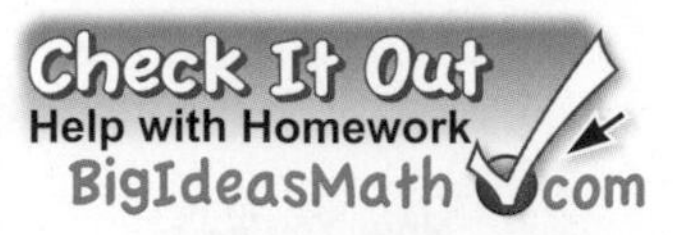

Vocabulary and Concept Check

1. **VOCABULARY** Describe the height of a cone.
2. **WRITING** Compare and contrast the formulas for the volume of a pyramid and the volume of a cone.
3. **REASONING** You know the volume of a cylinder. How can you find the volume of a cone with the same base and height?

Practice and Problem Solving

Find the volume of the cone. Round your answer to the nearest tenth.

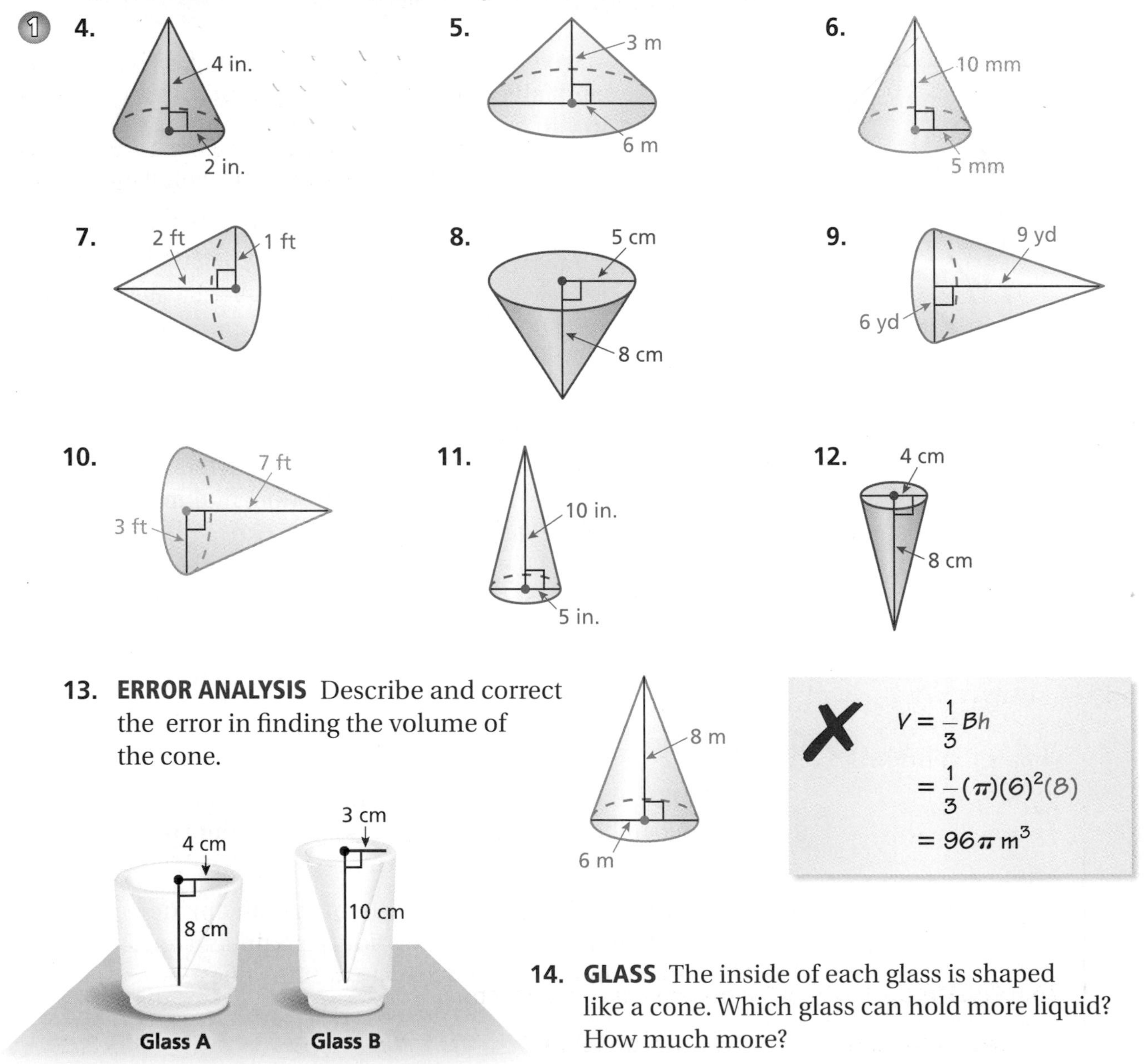

13. **ERROR ANALYSIS** Describe and correct the error in finding the volume of the cone.

14. **GLASS** The inside of each glass is shaped like a cone. Which glass can hold more liquid? How much more?

Find the height of the cone. Round your answer to the nearest tenth.

15. Volume $= \frac{1}{18}\pi$ ft^3

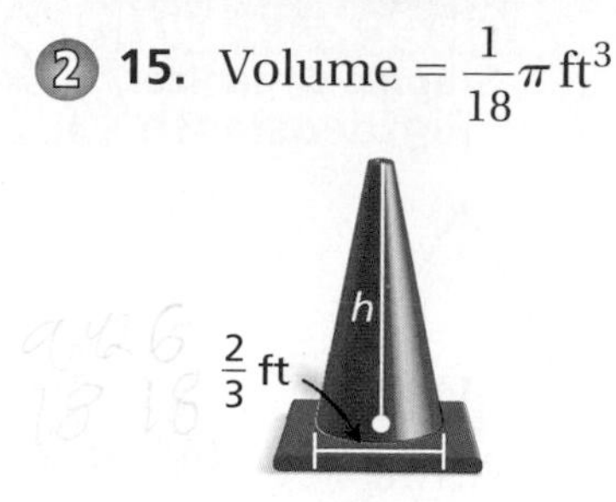

16. Volume $= 225$ cm^3

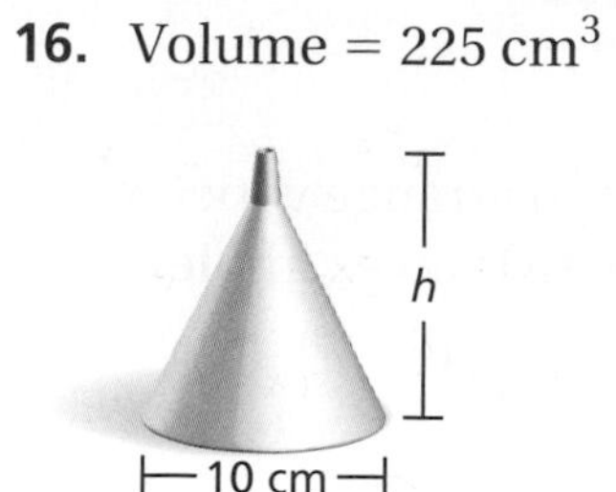

17. Volume $= 3.6$ in.^{3}

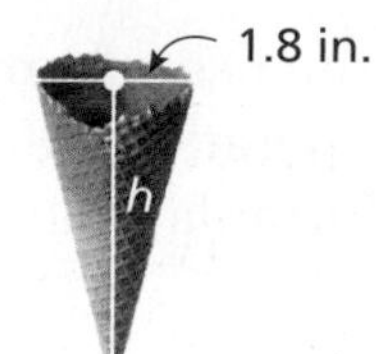

18. REASONING The volume of a cone is 20π cubic meters. What is the volume of a cylinder having the same base and same height?

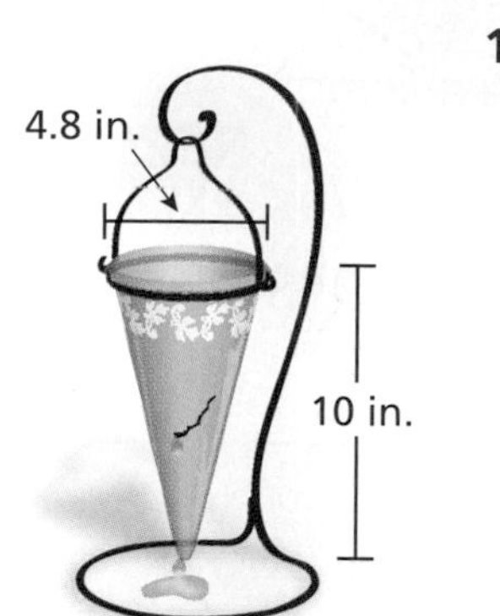

19. VASE Water leaks from a crack in a vase at a rate of 0.5 cubic inch per minute. How long does it take for 20% of the water to leak from a full vase?

20. PROBLEM SOLVING You have 10 gallons of lemonade to sell. (1 gal $\approx$ 3785 cm^3)

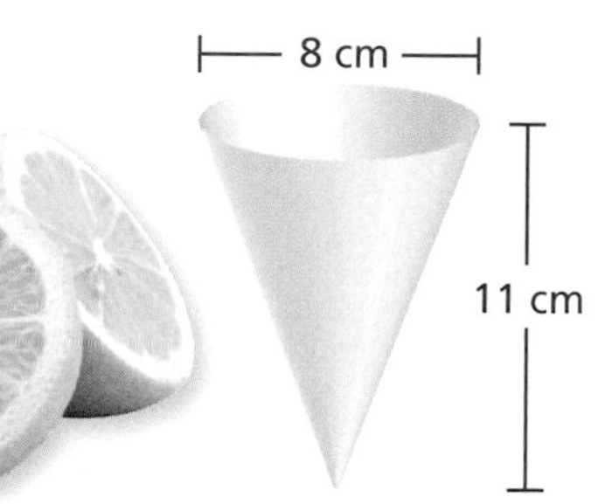

a. Each customer uses one paper cup. How many paper cups will you need?

b. The cups are sold in packages of 50. How many packages should you buy?

c. How many cups will be left over if you sell 80% of the lemonade?

21. REASONING The cylinder and the cone have the same volume. What is the height of the cone?

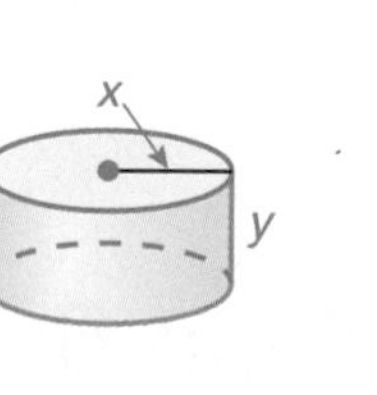

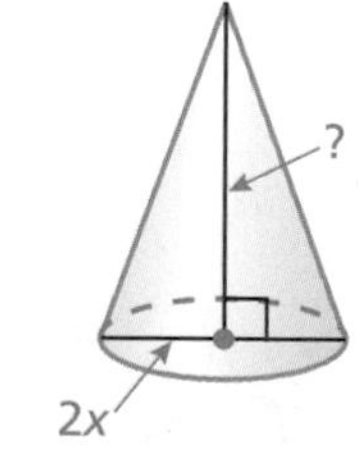

22. Critical Thinking Cone A has the same height but twice the radius of Cone B. What is the ratio of the volume of Cone A to the volume of Cone B?

Fair Game Review What you learned in previous grades & lessons

Find the volume of the solid. *(Section 7.1, Section 7.2, and Section 7.3)*

23. 9 m, 7 m, 5 m

24. 15 cm, 8 cm, 10 cm

25. 4 ft, 9.5 ft

26. MULTIPLE CHOICE Which scale has a scale factor of 3 : 1? *(Section 5.4)*

Ⓐ 1 in. : 2 ft Ⓑ 3 cm : 1 mm Ⓒ 5 ft : 15 yd Ⓓ 0.5 ft : 2 in.

7 Study Help

Check It Out
Graphic Organizer
BigIdeasMath.com

You can use a **formula triangle** to arrange variables and operations of a formula. Here is an example of a formula triangle for volume of a prism.

To find an unknown variable, use the other variables and the operation between them. For example, to find the area B of the base, cover up the B. Then you can see that you divide the volume V by the height h.

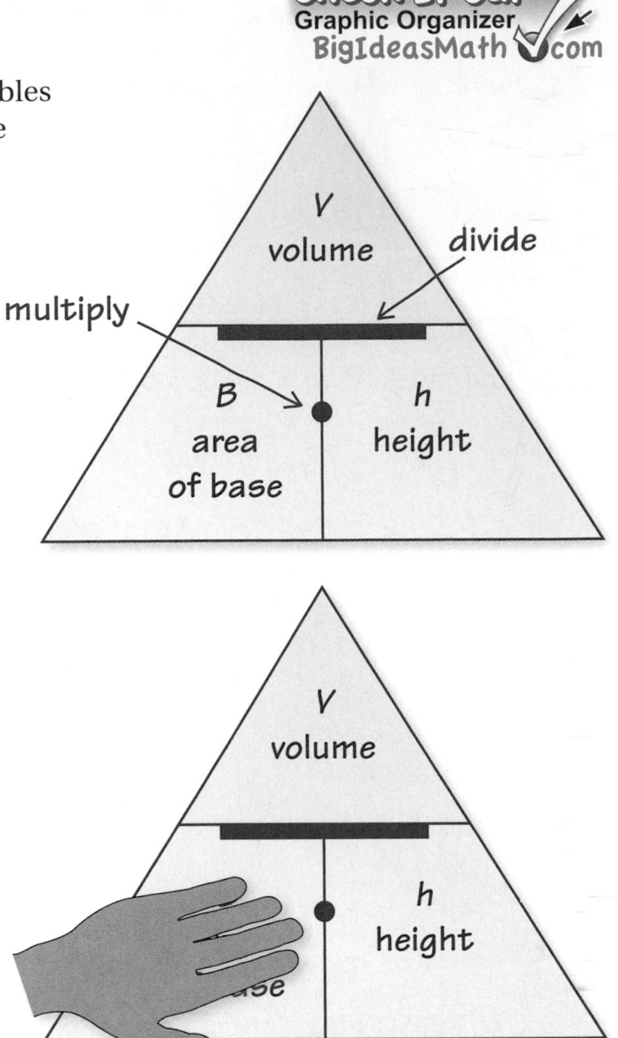

On Your Own

Make a formula triangle to help you study these topics. ***Hint:*** **Your formula triangles may have a different form than what is shown in the example.**

1. volume of a cylinder
2. volume of a pyramid
3. volume of a cone

After you complete this chapter, make formula triangles for the following topics.

4. volume of a composite solid
5. surface areas of similar solids
6. volumes of similar solids

"See how a formula triangle works? Cover any variable and you get its formula."

7.1–7.4 Quiz

Check It Out
Progress Check
BigIdeasMath.com

Find the volume of the prism. *(Section 7.1)*

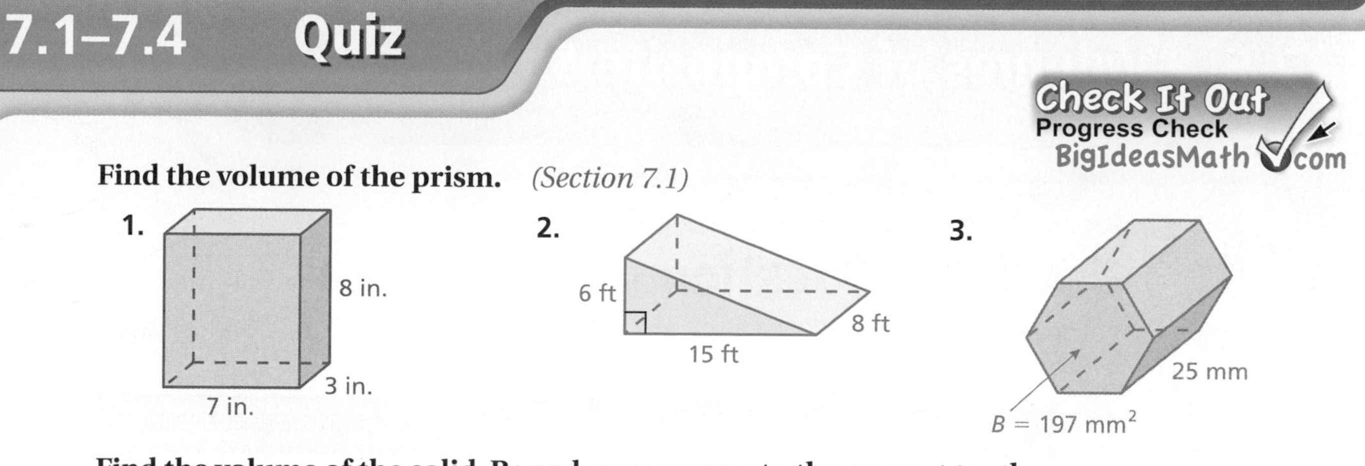

Find the volume of the solid. Round your answer to the nearest tenth. *(Section 7.2, Section 7.3, and Section 7.4)*

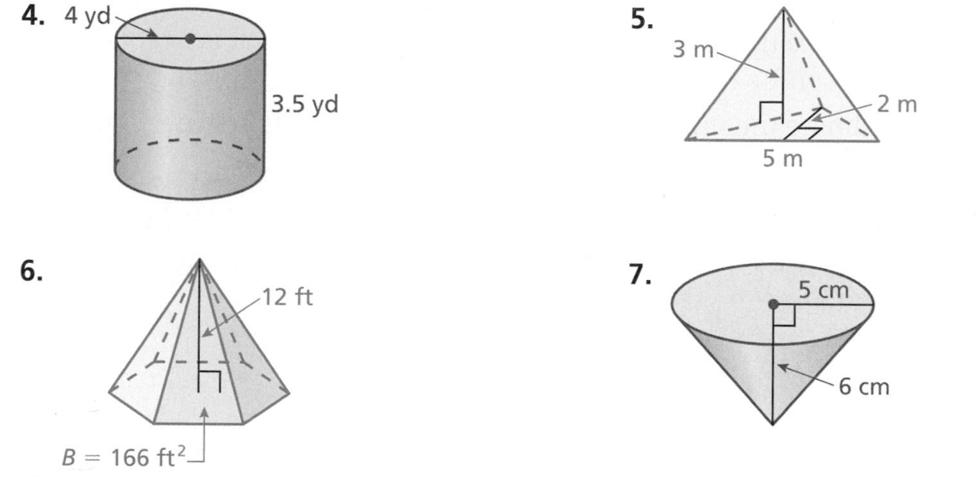

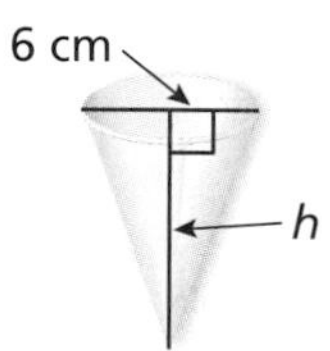

8. **PAPER CONE** The paper cone can hold 84.78 cubic centimeters of water. What is the height of the cone? *(Section 7.4)*

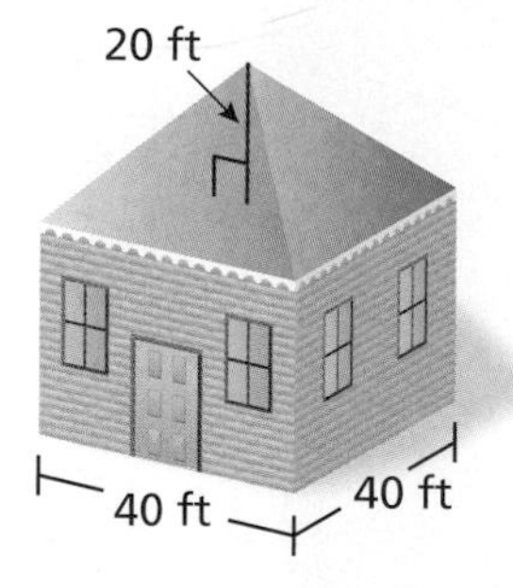

9. **ROOF** A pyramid hip roof is a good choice for a house in a hurricane area. What is the volume of the roof to the nearest tenth? *(Section 7.3)*

10. **CUBIC UNITS** How many cubic feet are there in a cubic yard? Use a sketch to explain your reasoning. *(Section 7.1)*

11. **JUICE CAN** You are buying two cylindrical cans of juice. Each can holds the same amount of juice. What is the height of Can B? *(Section 7.2)*

7.5 Volumes of Composite Solids

COMMON CORE STATE STANDARDS
7.G.4
7.G.6

Essential Question How can you estimate the volume of a composite solid?

1 ACTIVITY: Estimating Volume

Work with a partner. You work for a toy company and need to estimate the volume of a Minifigure that will be molded out of plastic.

a. Estimate the number of cubic inches of plastic that is needed to mold the Minifigure's head. Show your work.

b. Estimate the number of cubic inches of plastic that is needed to mold one of the Minifigure's legs. Show your work.

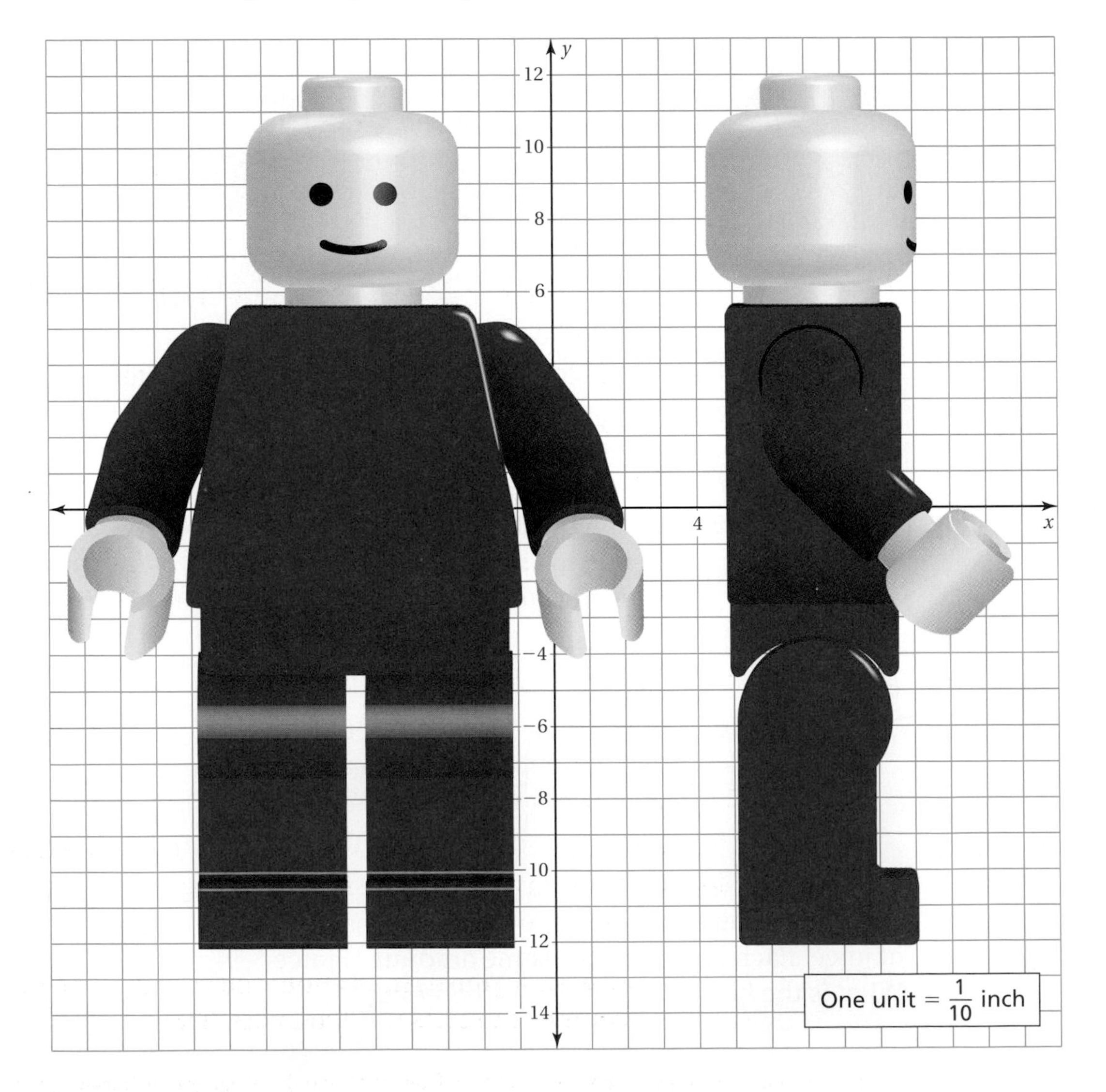

2 ACTIVITY: Finding the Volumes of Composite Solids

Work with a partner.

a. Make a plan for estimating the amount of plastic it takes to make a standard eight-stud LEGO® Brick.

© 2012 The LEGO Group, used with permission

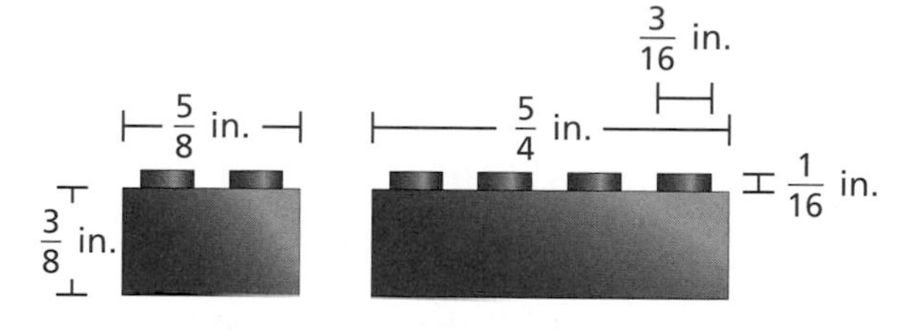

b. How much water, in cubic inches, would it take to make ten LEGO® Brick ice cubes?

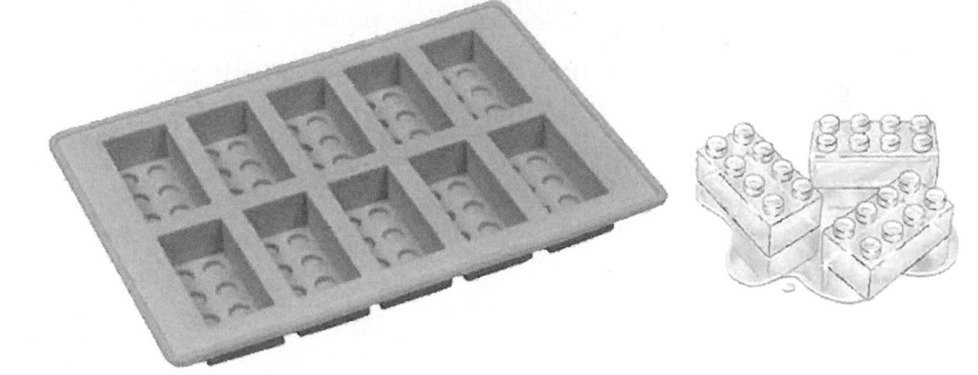

What Is Your Answer?

3. IN YOUR OWN WORDS How can you estimate the volume of a composite solid? Try thinking of some alternative strategies.

Use what you learned about the volumes of composite solids to complete Exercises 4–6 on page 328.

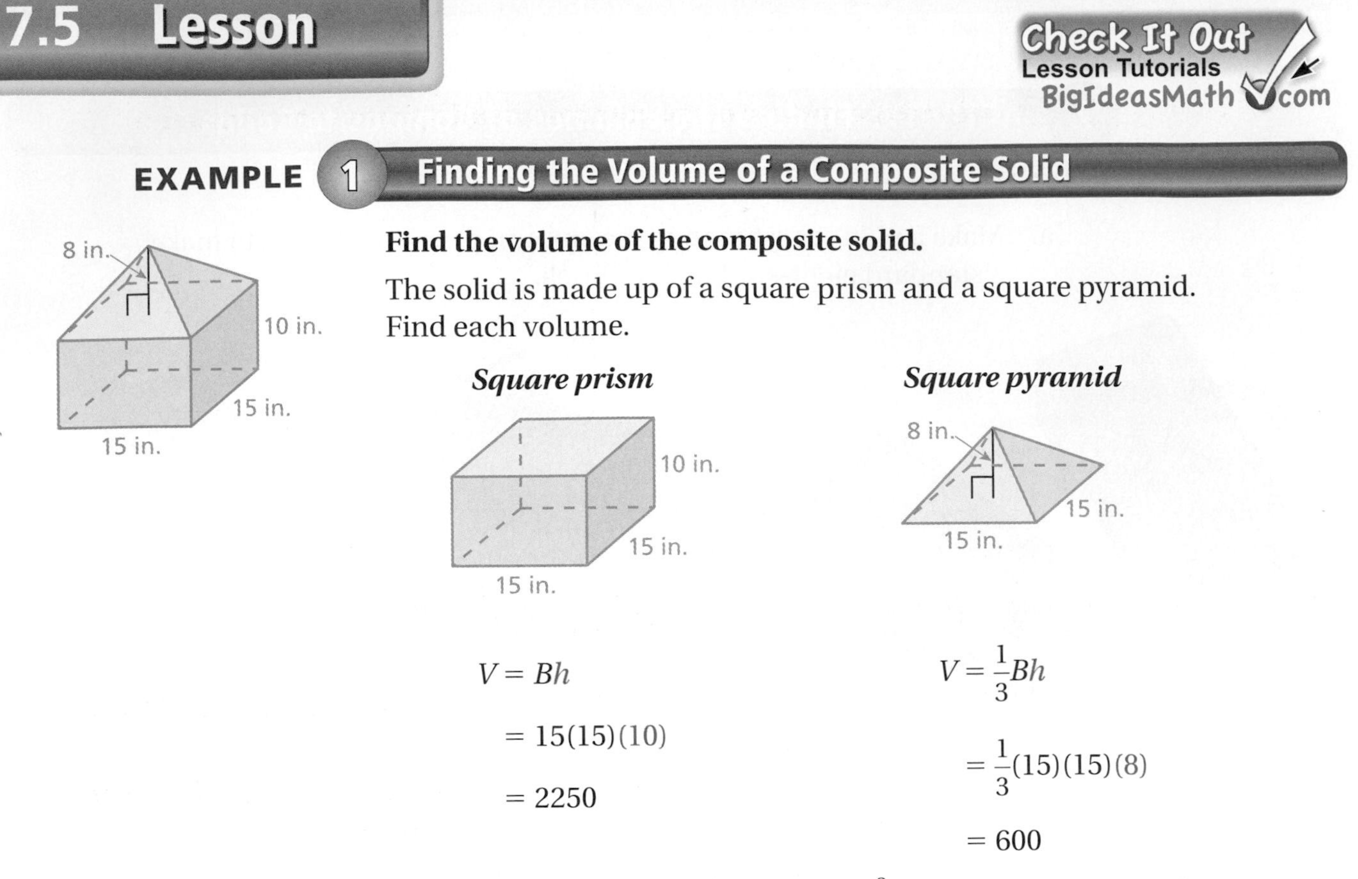

7.5 Lesson

Check It Out
Lesson Tutorials
BigIdeasMath.com

EXAMPLE 1 Finding the Volume of a Composite Solid

Find the volume of the composite solid.

The solid is made up of a square prism and a square pyramid. Find each volume.

Square prism

$$V = Bh$$
$$= 15(15)(10)$$
$$= 2250$$

Square pyramid

$$V = \frac{1}{3}Bh$$
$$= \frac{1}{3}(15)(15)(8)$$
$$= 600$$

Find the sum: $2250 + 600 = 2850 \text{ in.}^3$.

The volume of the composite solid is 2850 cubic inches.

EXAMPLE 2 Finding the Volume of a Composite Solid

Find the volume of the composite solid. Round your answer to the nearest tenth.

The solid is a cylinder with a cylinder-shaped hole. Find each volume.

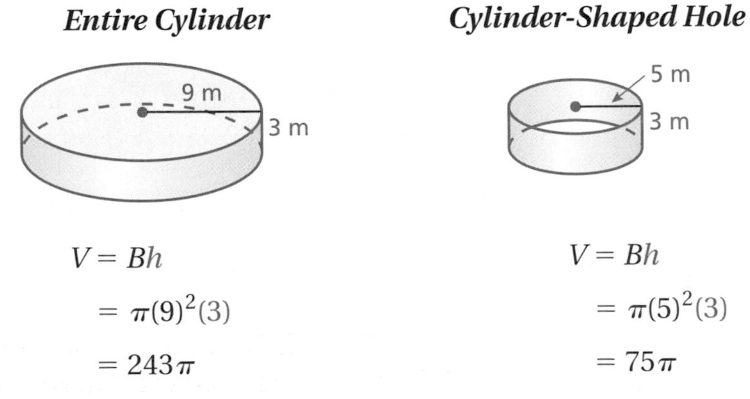

Entire Cylinder

$$V = Bh$$
$$= \pi(9)^2(3)$$
$$= 243\pi$$

Cylinder-Shaped Hole

$$V = Bh$$
$$= \pi(5)^2(3)$$
$$= 75\pi$$

Find the difference: $243\pi - 75\pi = 168\pi \approx 527.5 \text{ m}^3$.

The volume of the composite solid is about 527.5 cubic meters.

On Your Own

Now You're Ready
Exercises 4–11

Find the volume of the composite solid. Round your answer to the nearest tenth.

1. **2.**

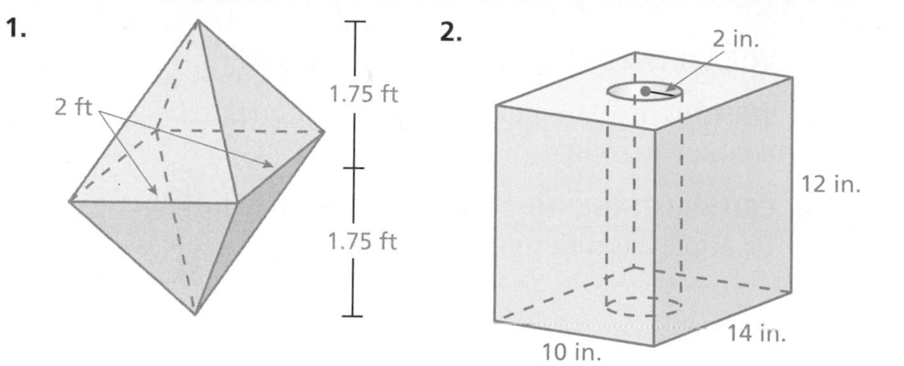

EXAMPLE 3 Real-Life Application

What is the volume of the silver ring in an Argentine peso? Round your answer to the nearest tenth.

The coin is a cylinder. The silver ring is the portion remaining when the inner cylinder is removed. Find the volume of each cylinder.

Entire cylinder

$$V = Bh$$
$$= \pi(11.5)^2(2.2)$$
$$= 290.95\pi$$

Inner cylinder

$$V = Bh$$
$$= \pi(8.5)^2(2.2)$$
$$= 158.95\pi$$

Subtract the volume of the inner cylinder from the volume of the entire cylinder: $290.95\pi - 158.95\pi = 132\pi \approx 414.5 \text{ mm}^3$.

∴ The volume of the silver ring is about 414.5 cubic millimeters.

On Your Own

3. **WHAT IF?** In Example 3, how would the volume of the silver ring change if the coin were only half as thick?

4. Find the volume of the composite solid. Round your answer to the nearest tenth.

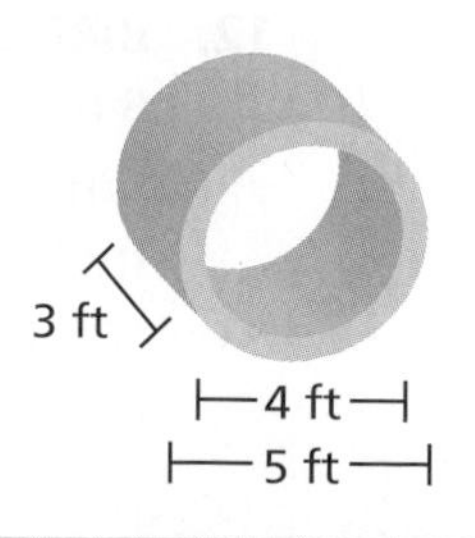

7.5 Exercises

Vocabulary and Concept Check

1. **VOCABULARY** What is a composite solid?
2. **WRITING** Explain how to find the volume of the composite solid.

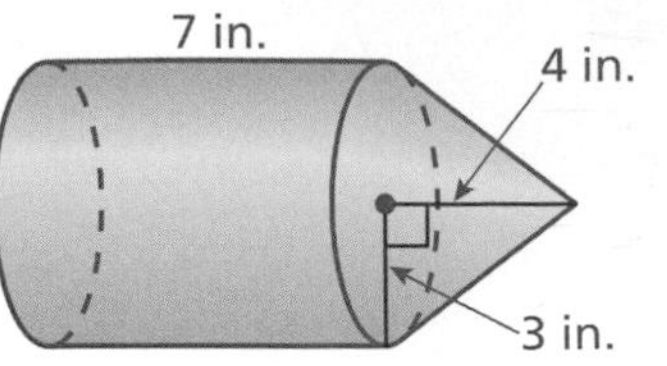

3. **CRITICAL THINKING** Explain how finding the volume in Example 2 is different from finding the volume in Example 1.

Practice and Problem Solving

Find the volume of the composite solid. Round your answer to the nearest tenth.

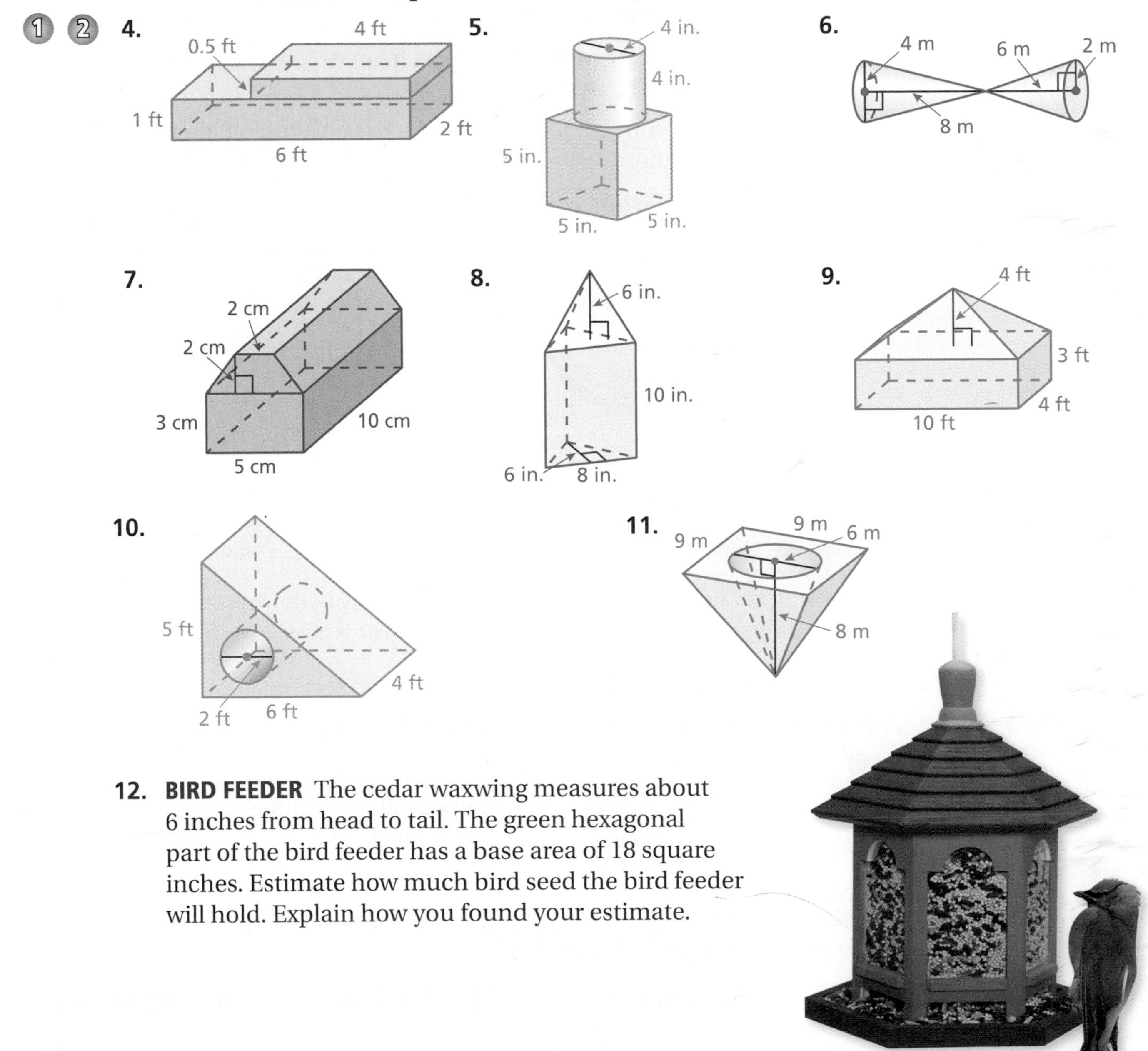

12. **BIRD FEEDER** The cedar waxwing measures about 6 inches from head to tail. The green hexagonal part of the bird feeder has a base area of 18 square inches. Estimate how much bird seed the bird feeder will hold. Explain how you found your estimate.

13. **CAKE** The raspberry layer cake has a diameter of 10 inches and a height of 5 inches.

   a. About what percent of the cake is remaining?

   b. Estimate the volume of the remaining cake.

14. **KICKBOARD** A foam kickboard used for swimming has two identical hand grips.

   a. Find the volume of the kickboard.

   b. One cubic inch of foam weighs about 0.007 pound. How much does the kickboard weigh?

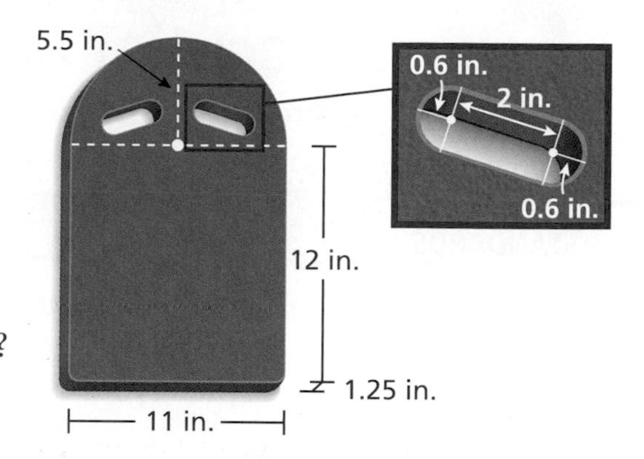

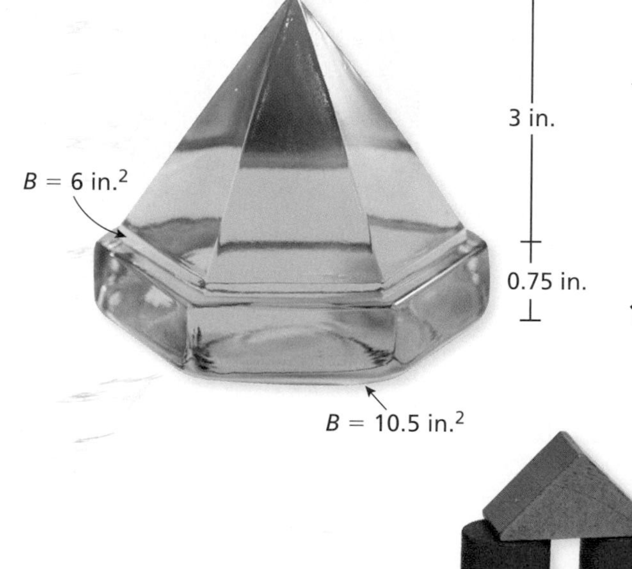

15. **PAPERWEIGHT** Estimate the amount of glass in the paperweight. Explain how you found your estimate.

16. **Puzzle** The volume of each group of solids is given. Find the volume of each of the four types of blocks.

$V = 8\pi + 8$

$V = 8\pi + 28$

$V = 8\pi + 20$

Fair Game Review What you learned in previous grades & lessons

The two figures are similar. Find the ratio (red to blue) of the areas. *(Section 5.2)*

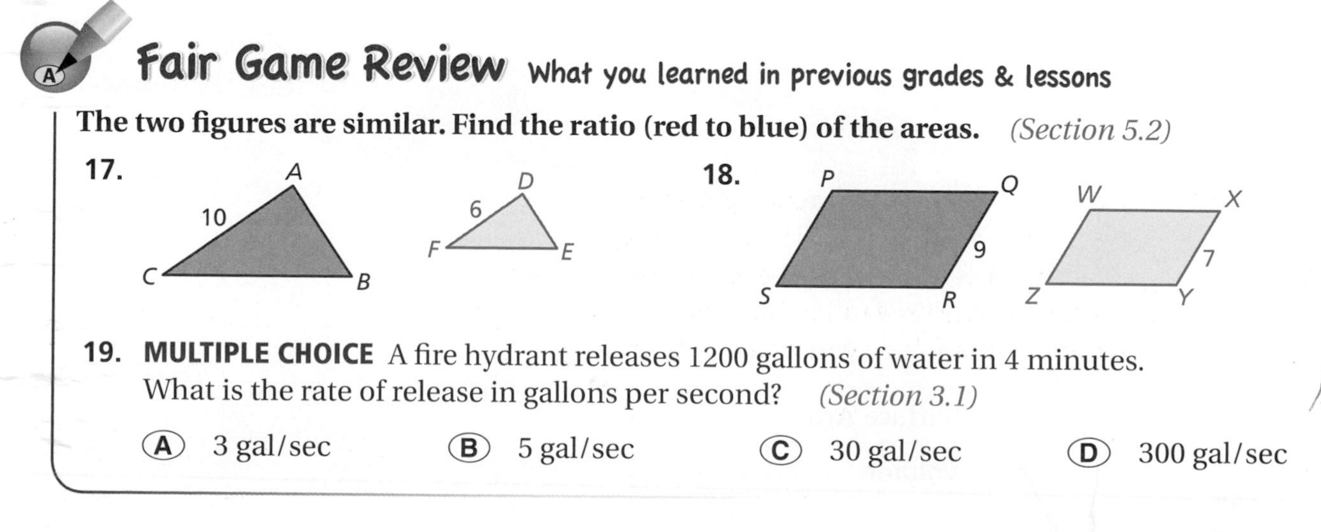

17.

18.

19. **MULTIPLE CHOICE** A fire hydrant releases 1200 gallons of water in 4 minutes. What is the rate of release in gallons per second? *(Section 3.1)*

   Ⓐ 3 gal/sec Ⓑ 5 gal/sec Ⓒ 30 gal/sec Ⓓ 300 gal/sec

7.6 Surface Areas and Volumes of Similar Solids

COMMON CORE STATE STANDARDS
7.G.4
7.G.6

Essential Question When the dimensions of a solid increase by a factor of k, how does the surface area change? How does the volume change?

1 ACTIVITY: Comparing Volumes and Surface Areas

Work with a partner. Copy and complete the table. Describe the pattern. Are the solids similar? Explain your reasoning.

a.

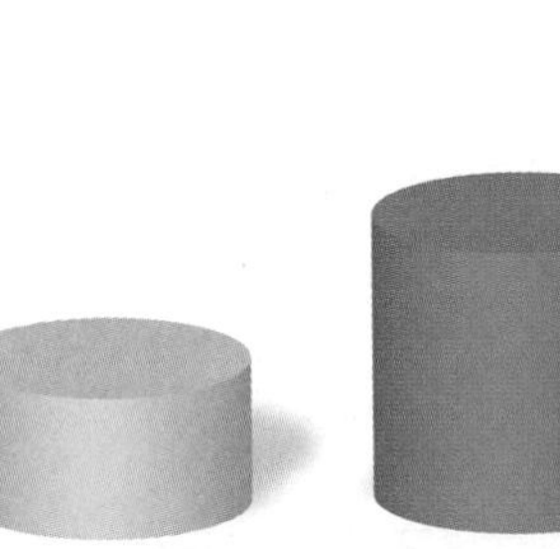

Radius	1	1	1	1	1
Height	1	2	3	4	5
Surface Area					
Volume					

b.

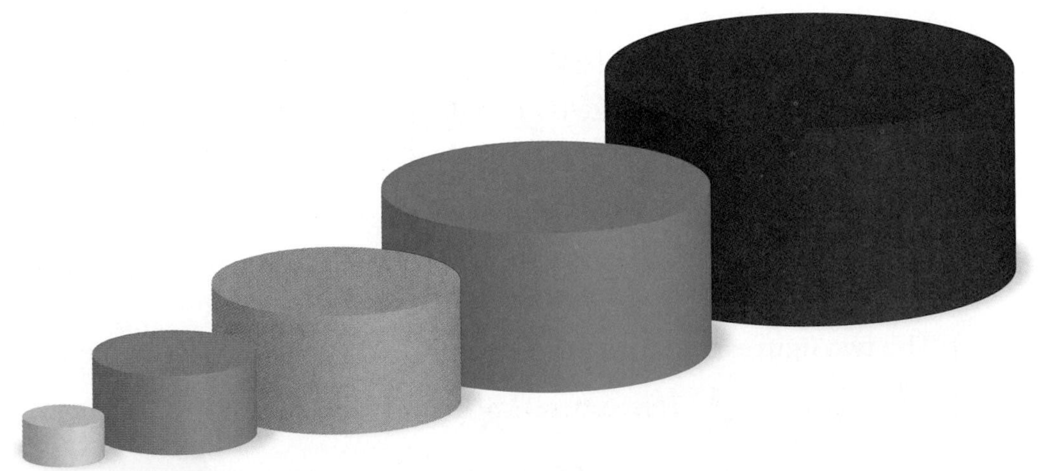

Radius	1	2	3	4	5
Height	1	2	3	4	5
Surface Area					
Volume					

2 ACTIVITY: Comparing Volumes and Surface Areas

Work with a partner. Copy and complete the table. Describe the pattern. Are the solids similar? Explain.

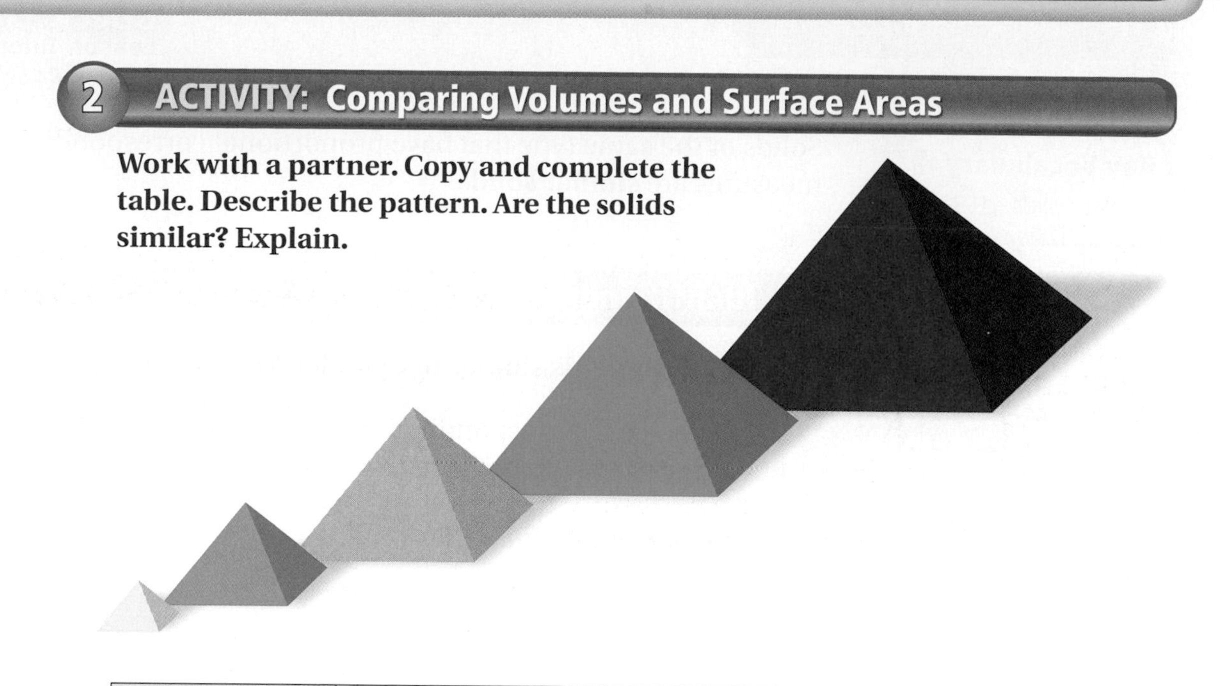

Base Side	6	12	18	24	30
Height	4	8	12	16	20
Slant Height	5	10	15	20	25
Surface Area					
Volume					

What Is Your Answer?

3. **IN YOUR OWN WORDS** When the dimensions of a solid increase by a factor of k, how does the surface area change?
4. **IN YOUR OWN WORDS** When the dimensions of a solid increase by a factor of k, how does the volume change?
5. All the dimensions of a cone increase by a factor of 5.

   a. How many times greater is the surface area? Explain.

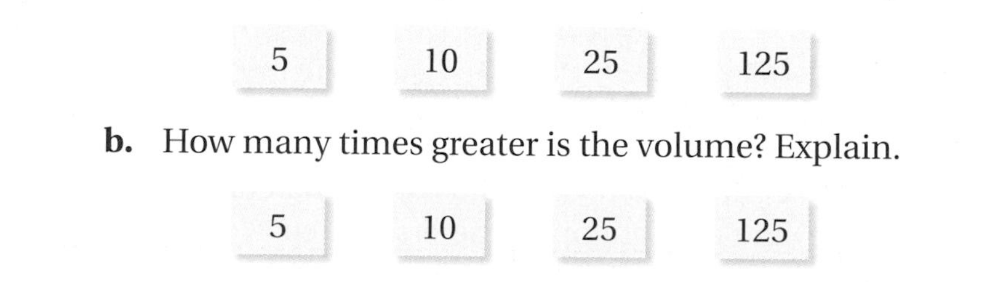

   5   10   25   125

   b. How many times greater is the volume? Explain.

   5   10   25   125

Use what you learned about the surface areas and volumes of similar solids to complete Exercises 4–6 on page 335.

7.6 Lesson

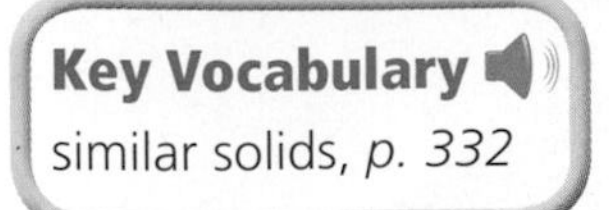

Solids of the same type that have proportional corresponding linear measures are **similar solids**.

EXAMPLE 1 Identifying Similar Solids

Which cylinder is similar to Cylinder A?

Cylinder A: radius 6 m, height 4 m

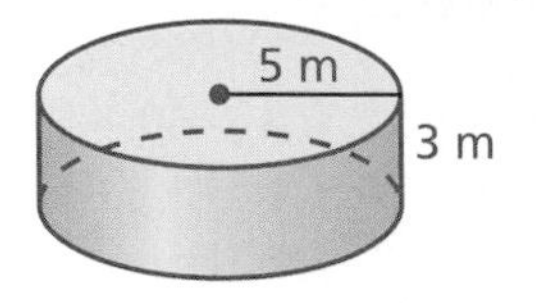

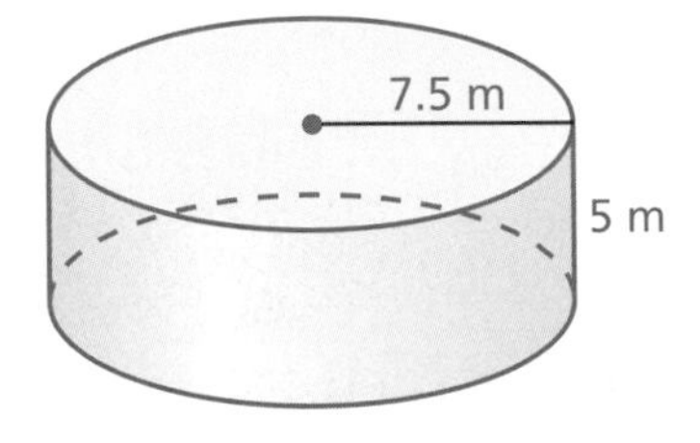

Check to see if corresponding linear measures are proportional.

***Cylinder* A *and Cylinder* B**

$$\frac{\text{Height of A}}{\text{Height of B}} = \frac{4}{3} \qquad \frac{\text{Radius of A}}{\text{Radius of B}} = \frac{6}{5}$$

***Cylinder* A *and Cylinder* C**

$$\frac{\text{Height of A}}{\text{Height of C}} = \frac{4}{5} \qquad \frac{\text{Radius of A}}{\text{Radius of C}} = \frac{6}{7.5} = \frac{4}{5}$$

Proportional

So, Cylinder C is similar to Cylinder A.

EXAMPLE 2 Finding Missing Measures in Similar Solids

The cones are similar. Find the missing slant height ℓ.

Cone X: radius 5 yd, slant height 13 yd. Cone Y: radius 7 yd, slant height ℓ.

$$\frac{\text{Radius of X}}{\text{Radius of Y}} = \frac{\text{Slant height of X}}{\text{Slant height of Y}}$$

$\frac{5}{7} = \frac{13}{\ell}$ Substitute.

$5\ell = 91$ Use Cross Products Property.

$\ell = 18.2$ Divide each side by 5.

The slant height is 18.2 yards.

On Your Own

Now You're Ready
Exercises 4–9

1. Cylinder D has a radius of 7.5 meters and a height of 4.5 meters. Which cylinder in Example 1 is similar to Cylinder D?
2. The prisms are similar. Find the missing width and length.

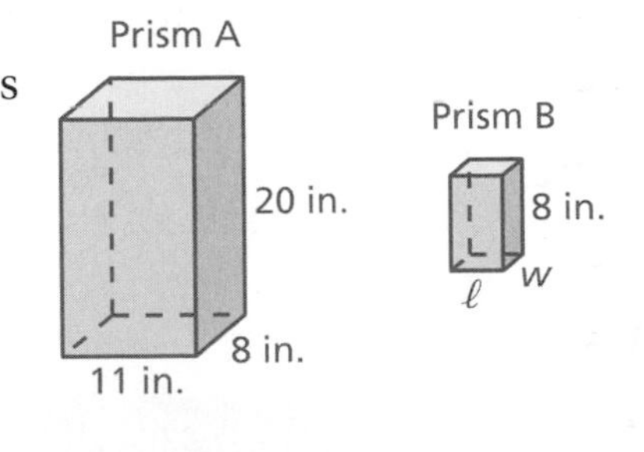

Multi-Language Glossary at BigIdeasMath.com.

Linear Measures

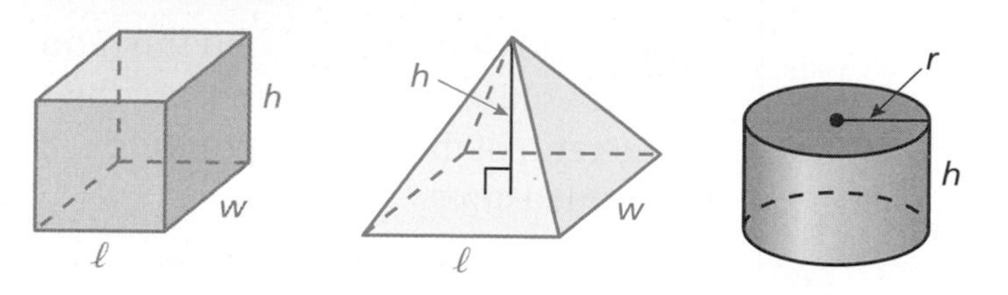

h

r

Surface Areas of Similar Solids

If two solids are similar, then the ratio of their surface areas is equal to the square of the ratio of their corresponding linear measures.

$$\frac{\text{Surface Area of A}}{\text{Surface Area of B}} = \left(\frac{a}{b}\right)^2$$

Solid A

a

Solid B

b

EXAMPLE 3 Finding Surface Area

Pyramid A

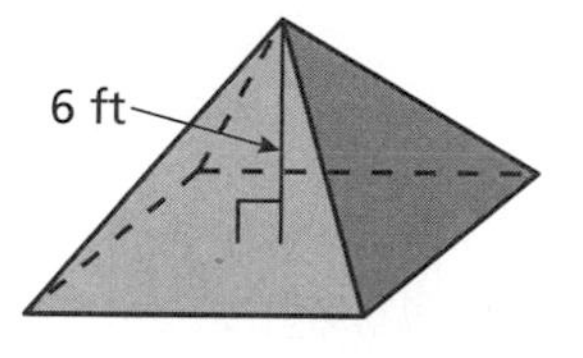

Pyramid B

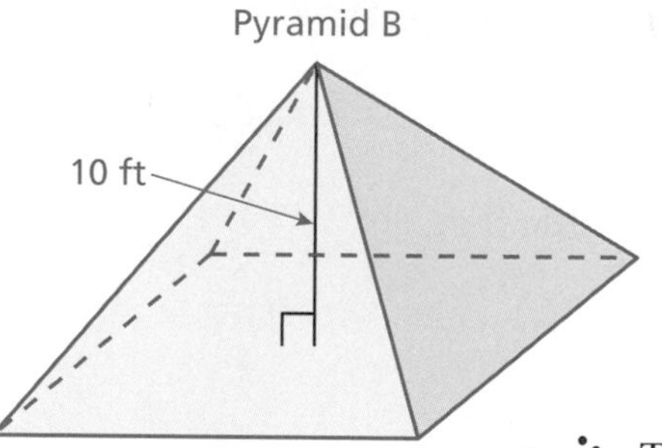

Surface Area = 600 ft²

The pyramids are similar. What is the surface area of Pyramid A?

$$\frac{\text{Surface Area of A}}{\text{Surface Area of B}} = \left(\frac{\text{Height of A}}{\text{Height of B}}\right)^2$$

$$\frac{S}{600} = \left(\frac{6}{10}\right)^2 \qquad \text{Substitute.}$$

$$\frac{S}{600} = \frac{36}{100} \qquad \text{Evaluate power.}$$

$$\frac{S}{600} \cdot 600 = \frac{36}{100} \cdot 600 \qquad \text{Multiply each side by 600.}$$

$$S = 216 \qquad \text{Simplify.}$$

∴ The surface area of Pyramid A is 216 square feet.

On Your Own

The solids are similar. Find the surface area of the red solid. Round your answer to the nearest tenth.

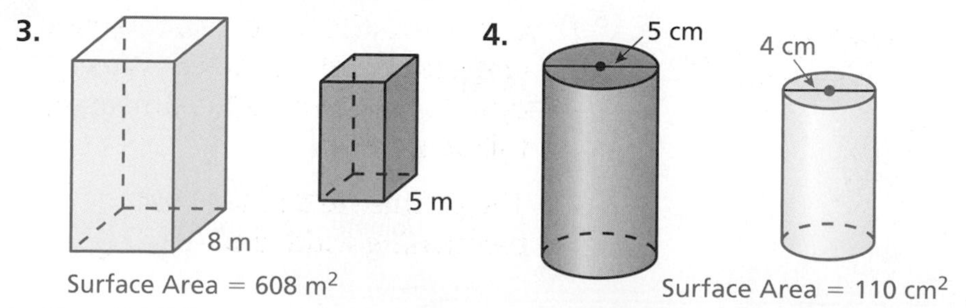

Volumes of Similar Solids

If two solids are similar, then the ratio of their volumes is equal to the cube of the ratio of their corresponding linear measures.

$$\frac{\text{Volume of A}}{\text{Volume of B}} = \left(\frac{a}{b}\right)^3$$

Solid A a

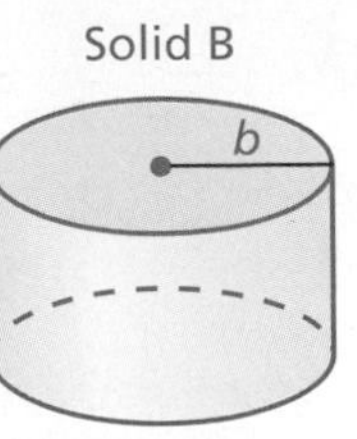

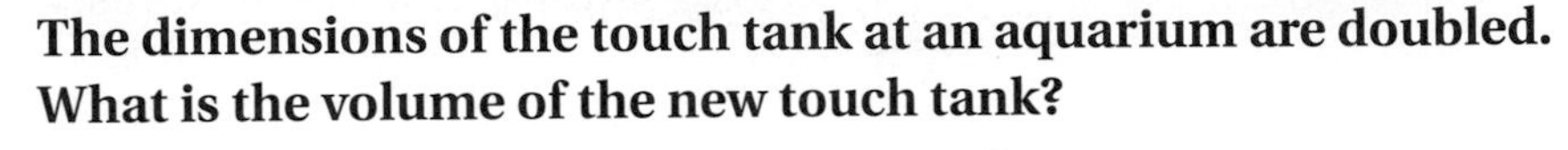

Original Tank

Volume = 2000 ft^3

The dimensions of the touch tank at an aquarium are doubled. What is the volume of the new touch tank?

Ⓐ 150 ft^3 Ⓑ 4000 ft^3 Ⓒ 8000 ft^3 Ⓓ 16,000 ft^3

The dimensions are doubled, so the ratio of the dimensions in the original tank to the dimensions in the new tank is 1 : 2.

$$\frac{\text{Original volume}}{\text{New volume}} = \left(\frac{\text{Original dimension}}{\text{New dimension}}\right)^3$$

$$\frac{2000}{V} = \left(\frac{1}{2}\right)^3 \qquad \text{Substitute.}$$

$$\frac{2000}{V} = \frac{1}{8} \qquad \text{Evaluate power.}$$

$$16{,}000 = V \qquad \text{Use Cross Products Property.}$$

∴ The volume of the new tank is 16,000 cubic feet. The correct answer is Ⓓ.

Study Tip

When the dimensions of a solid are multiplied by k, the surface area is multiplied by k^2 and the volume is multiplied by k^3.

On Your Own

The solids are similar. Find the volume of the red solid. Round your answer to the nearest tenth.

5.

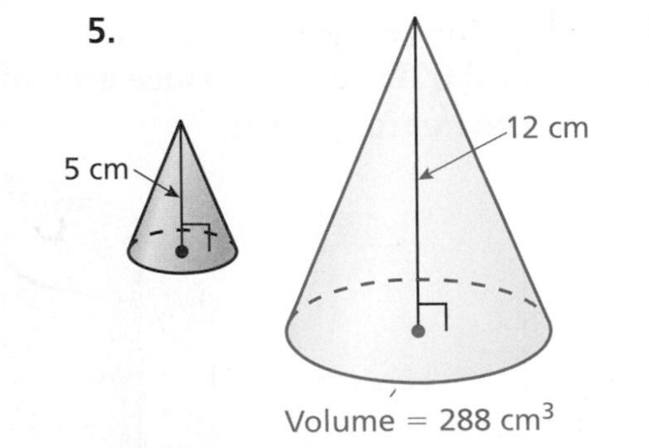

6.

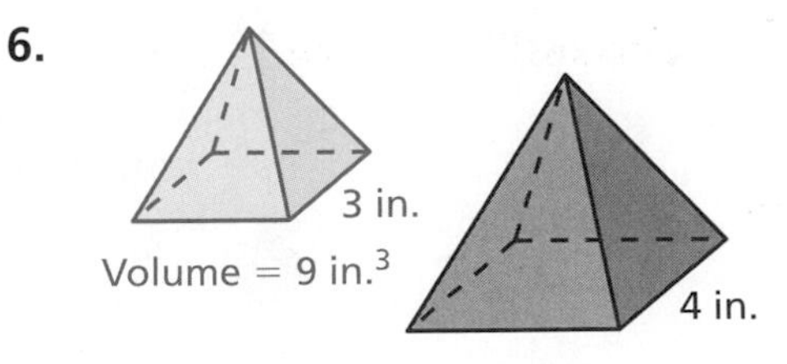

7.6 Exercises

Vocabulary and Concept Check

1. **VOCABULARY** What are similar solids?
2. **OPEN-ENDED** Draw two similar solids and label their corresponding linear measures.
3. **REASONING** The ratio of the corresponding linear measures of Cube A to Cube B is $\frac{2}{3}$.
   a. Find the ratio of the area of one face of Cube A to the area of one face of Cube B.
   b. Find the ratio of the volume of Cube A to the volume of Cube B.

Practice and Problem Solving

Determine whether the solids are similar.

4.

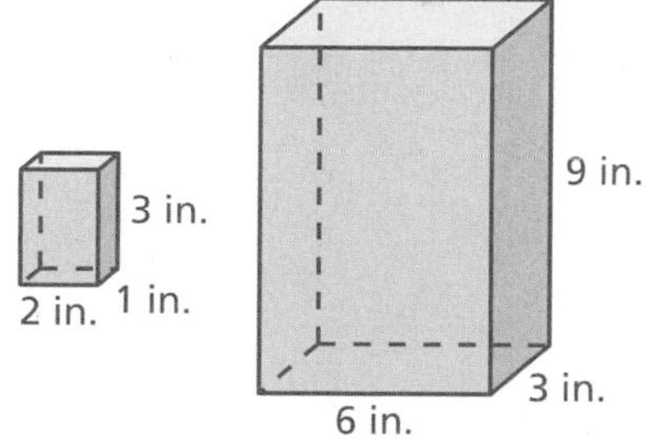

5.

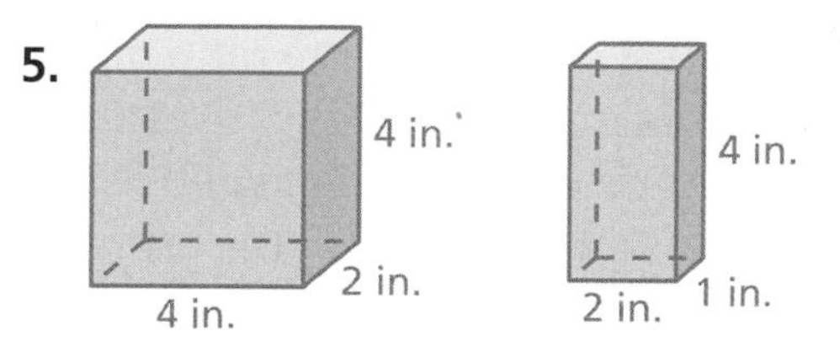

6.

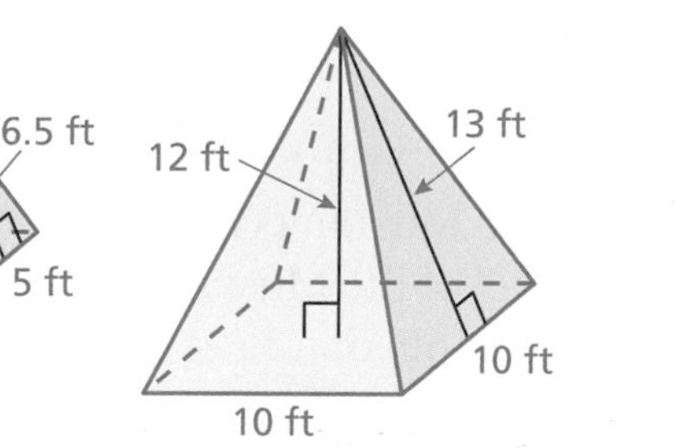

7.

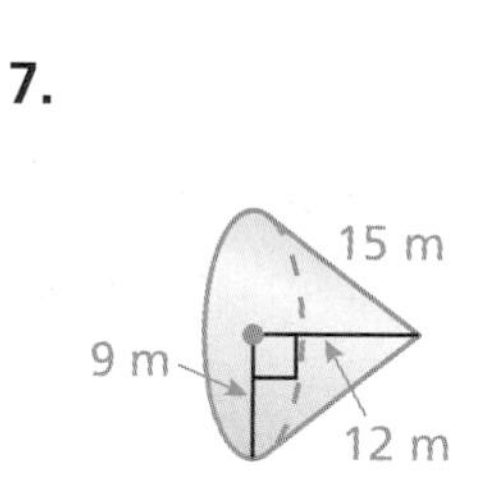

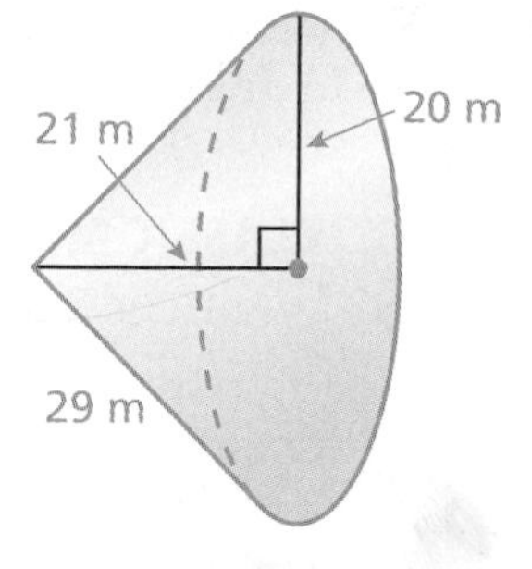

The solids are similar. Find the missing dimension(s).

8. 9.

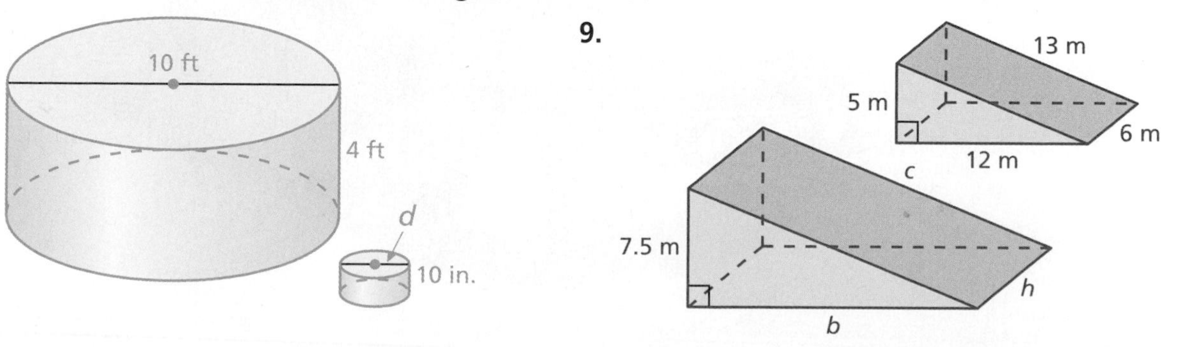

The solids are similar. Find the surface area S or volume V of the red solid. Round your answer to the nearest tenth.

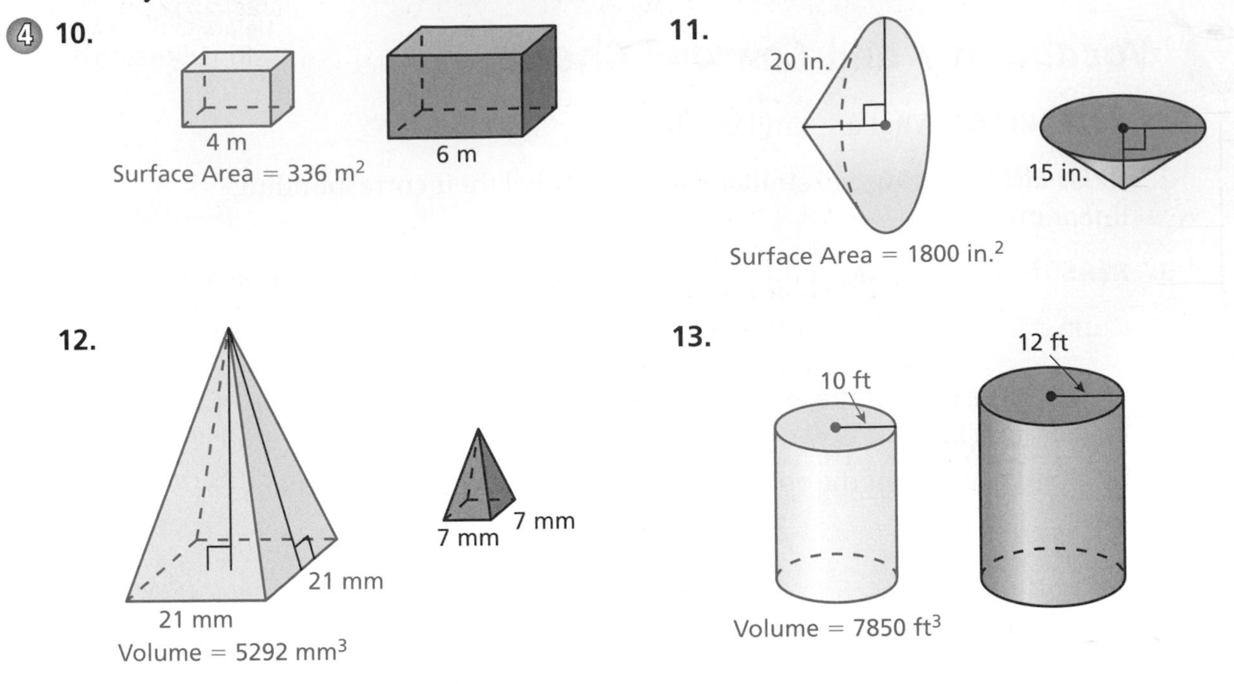

14. **ERROR ANALYSIS** The ratio of the corresponding linear measures of two similar solids is 3 : 5. The volume of the smaller solid is 108 cubic inches. Describe and correct the error in finding the volume of the larger solid.

> ✗
>
> $$\frac{108}{V} = \left(\frac{3}{5}\right)^2$$
>
> $$\frac{108}{V} = \frac{9}{25}$$
>
> $$300 = V$$
>
> The volume of the larger solid is 300 cubic inches.

15. **MIXED FRUIT** The ratio of the corresponding linear measures of two similar cans of fruit is 4 to 7. The smaller can has a surface area of 220 square centimeters. Find the surface area of the larger can.

16. **CLASSIC MUSTANG** The volume of a 1968 Ford Mustang GT engine is 390 cubic inches. Which scale model of the Mustang has the greater engine volume, a 1 : 18 scale model or a 1 : 24 scale model? How much greater?

17. **Critical Thinking** You and a friend make paper cones to collect beach glass. You cut out the largest possible three-fourths circle from each piece of paper.

a. Are the cones similar? Explain your reasoning.

b. Your friend says that because your sheet of paper is twice as large, your cone will hold exactly twice the volume of beach glass. Is this true? Explain your reasoning.

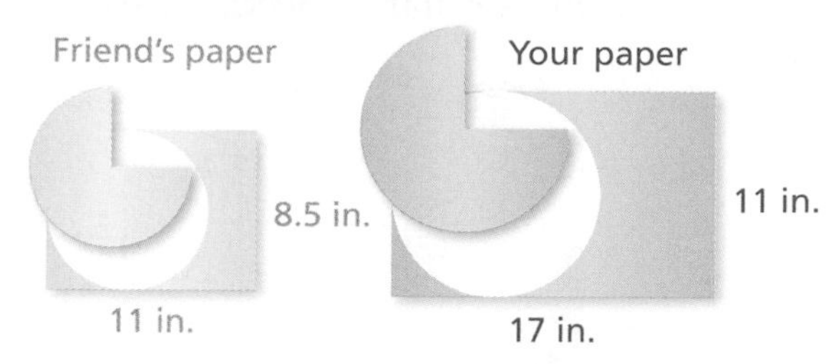

18. **MARBLE STATUE** You have a small marble statue of Wolfgang Mozart that is 10 inches tall and weighs 16 pounds. The original statue in Vienna is 7 feet tall.

a. Estimate the weight of the original statue. Explain your reasoning.

b. If the original statue were 20 feet tall, how much would it weigh?

Wolfgang Mozart

19. **REPEATED REASONING** The largest doll is 7 inches tall. Each of the other dolls is 1 inch shorter than the next larger doll. Make a table that compares the surface areas and volumes of the seven dolls.

Fair Game Review What you learned in previous grades & lessons

Add. *(Section 1.2)*

20. $69 + (-31) + 7 + (-6)$ **21.** $-2 + (-5) + (-12) + 20$ **22.** $10 + (-6) + (-5) + 1$

23. MULTIPLE CHOICE What is the mean of the numbers below? *(Skills Review Handbook)*

14, 6, 21, 8, 14, 19, 30

Ⓐ 6 Ⓑ 15 Ⓒ 16 Ⓓ 56

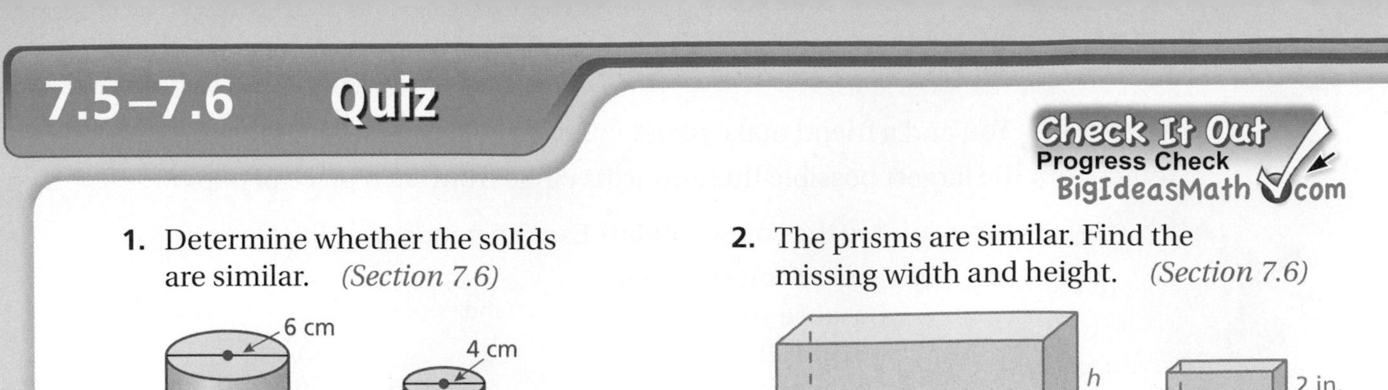

7.5–7.6 Quiz

1. Determine whether the solids are similar. *(Section 7.6)*

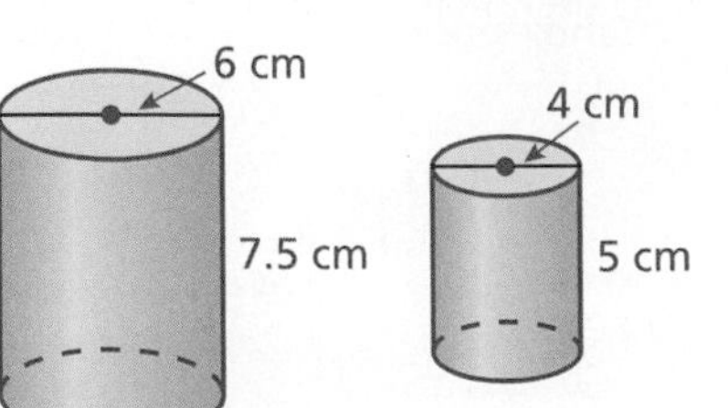

2. The prisms are similar. Find the missing width and height. *(Section 7.6)*

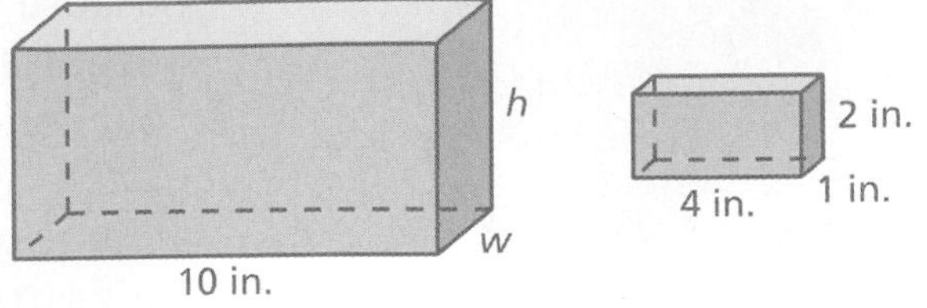

Find the volume of the composite solid. Round your answer to the nearest tenth. *(Section 7.5)*

3.

4.

5. The solids are similar. Find the surface area of the red solid. *(Section 7.6)*

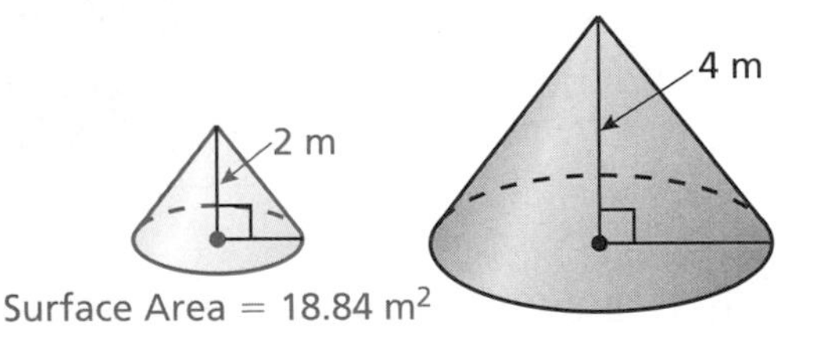

6. ARCADE You win a token after playing an arcade game. What is the volume of the gold ring? Round your answer to the nearest tenth. *(Section 7.5)*

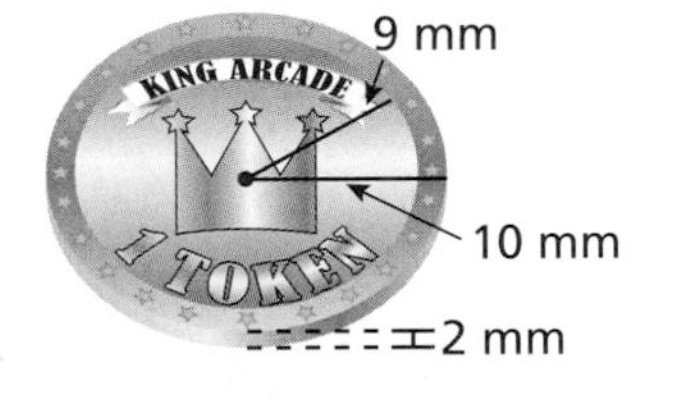

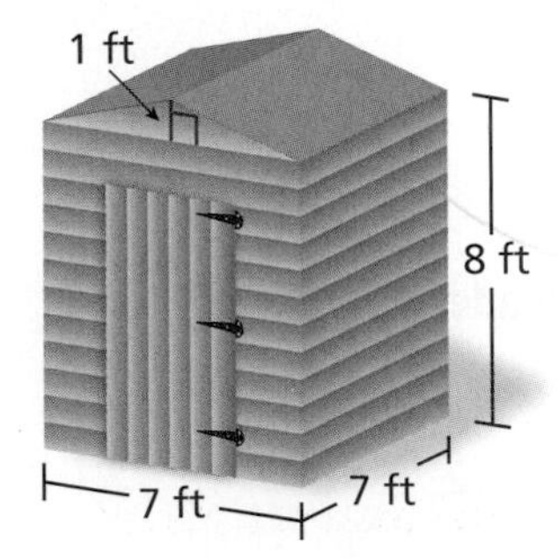

7. SHED What is the volume of the storage shed? *(Section 7.5)*

8. JEWELRY BOXES The ratio of the corresponding linear measures of two similar jewelry boxes is 2 to 3. The larger box has a volume of 162 cubic inches. Find the volume of the smaller jewelry box. *(Section 7.6)*

9. GELATIN You make a dessert with lemon gelatin and lime gelatin. What percent of the dessert is lime-flavored? Explain. *(Section 7.5)*

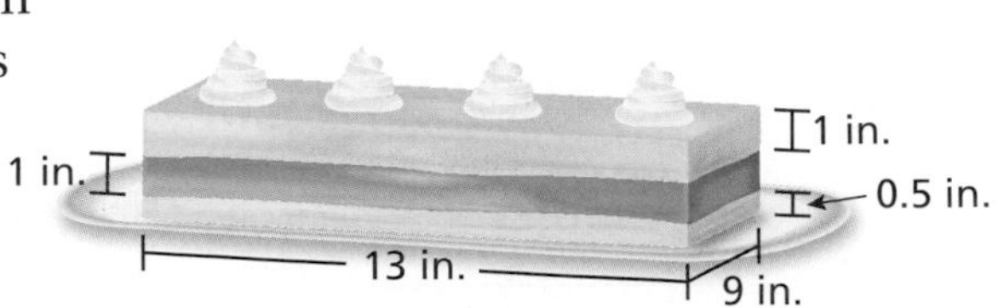

7 Chapter Review

Review Key Vocabulary

volume, *p. 300*

similar solids, *p. 332*

Review Examples and Exercises

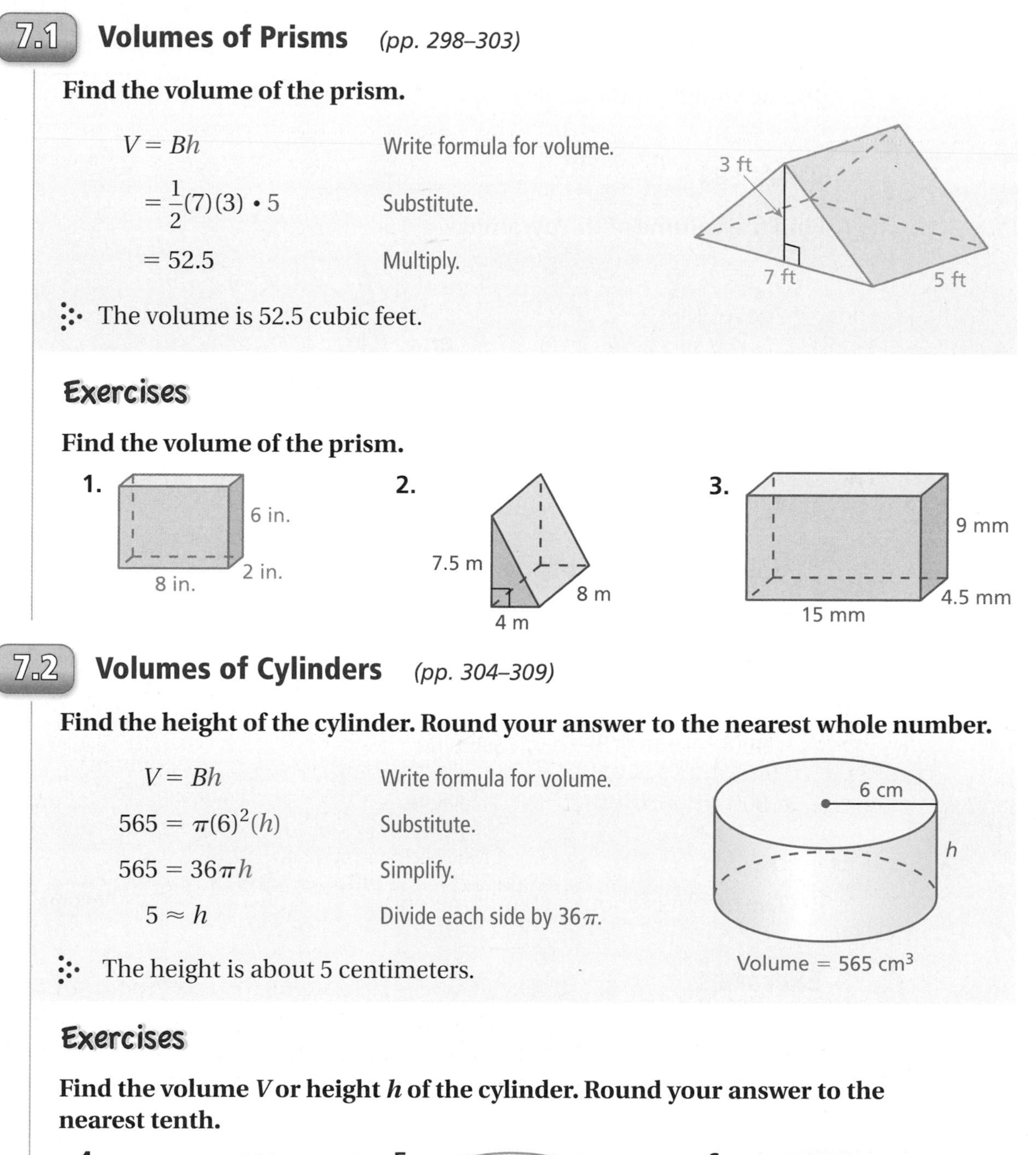

7.1 Volumes of Prisms *(pp. 298–303)*

Find the volume of the prism.

$V = Bh$ Write formula for volume.

$= \frac{1}{2}(7)(3) \cdot 5$ Substitute.

$= 52.5$ Multiply.

The volume is 52.5 cubic feet.

Exercises

Find the volume of the prism.

1.

2.

3.

7.2 Volumes of Cylinders *(pp. 304–309)*

Find the height of the cylinder. Round your answer to the nearest whole number.

$V = Bh$ Write formula for volume.

$565 = \pi(6)^2(h)$ Substitute.

$565 = 36\pi h$ Simplify.

$5 \approx h$ Divide each side by 36π.

The height is about 5 centimeters.

Exercises

Find the volume V or height h of the cylinder. Round your answer to the nearest tenth.

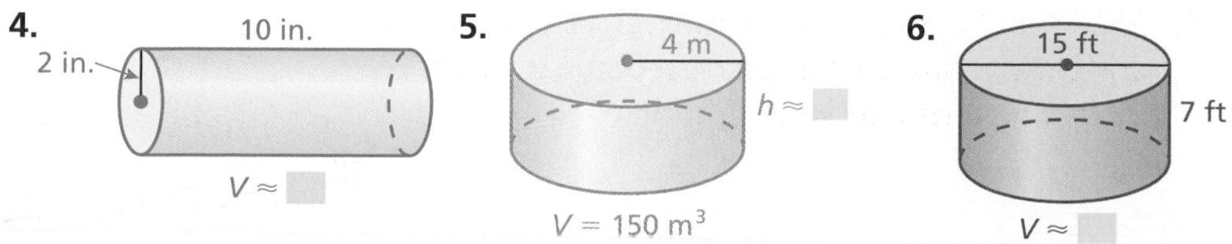

4.

5.

6.

7.3 Volumes of Pyramids (pp. 310–315)

Find the volume of the pyramid.

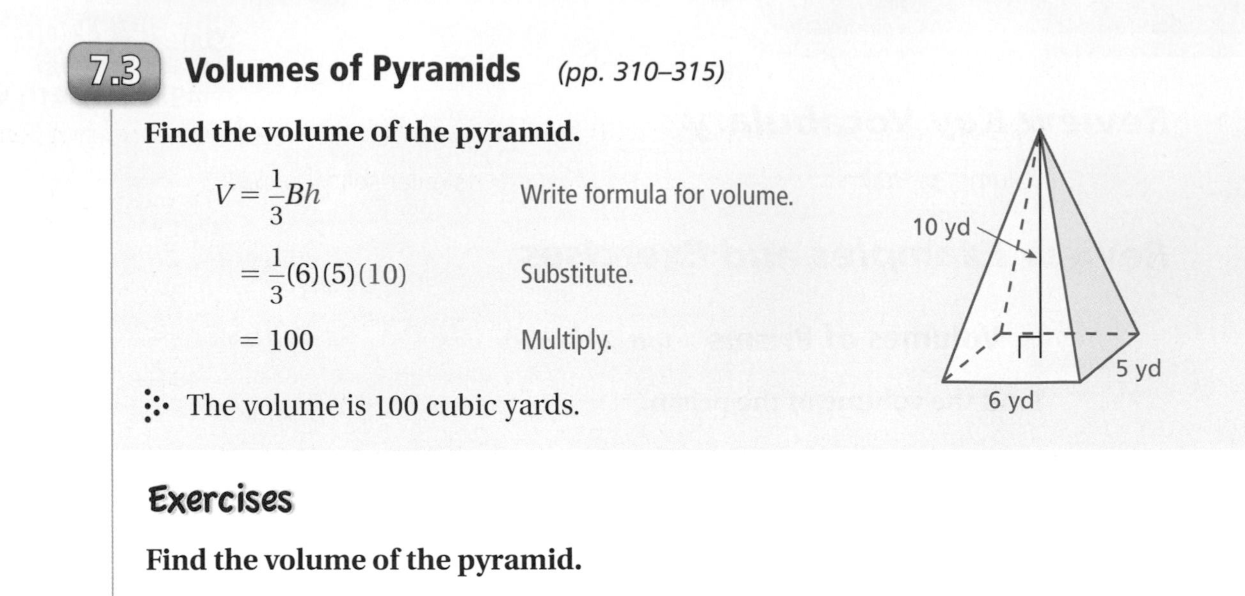

$V = \frac{1}{3}Bh$ — Write formula for volume.

$= \frac{1}{3}(6)(5)(10)$ — Substitute.

$= 100$ — Multiply.

∴ The volume is 100 cubic yards.

Exercises

Find the volume of the pyramid.

7. **8.** **9.**

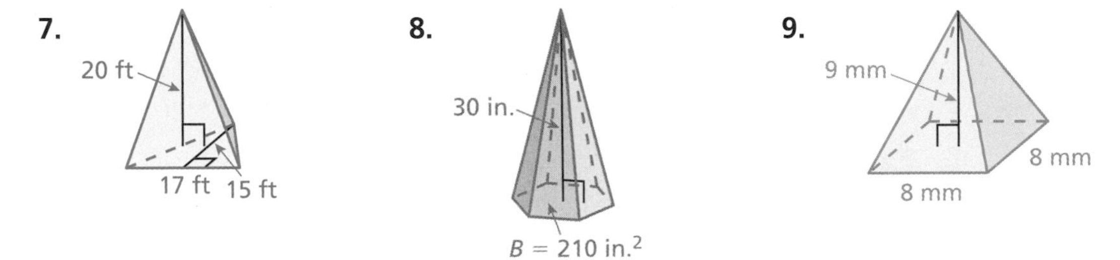

7.4 Volumes of Cones (pp. 316–321)

Find the height of the cone. Round your answer to the nearest tenth.

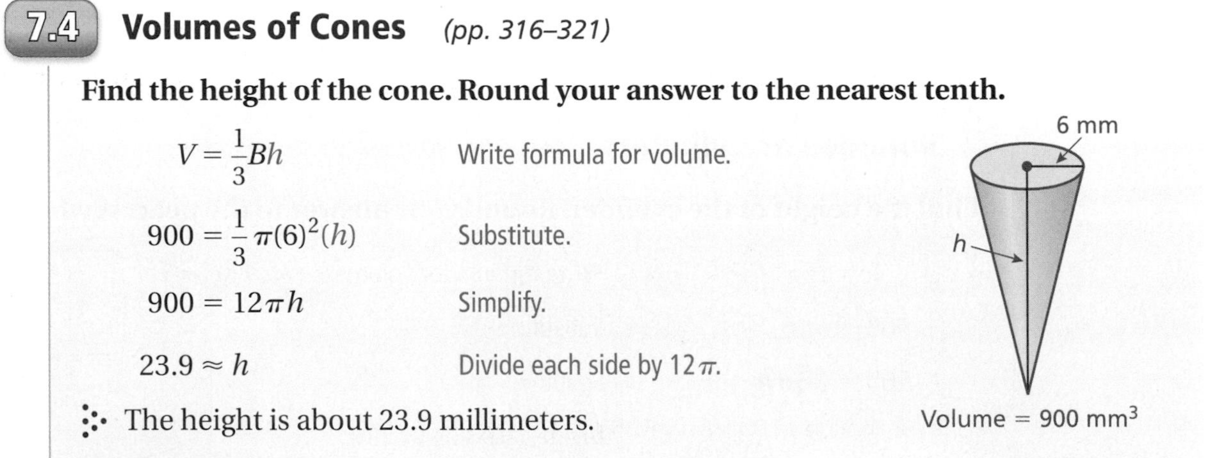

$V = \frac{1}{3}Bh$ — Write formula for volume.

$900 = \frac{1}{3}\pi(6)^2(h)$ — Substitute.

$900 = 12\pi h$ — Simplify.

$23.9 \approx h$ — Divide each side by 12π.

∴ The height is about 23.9 millimeters.

Exercises

Find the volume V or height h of the cone. Round your answer to the nearest tenth.

10. **11.** **12.**

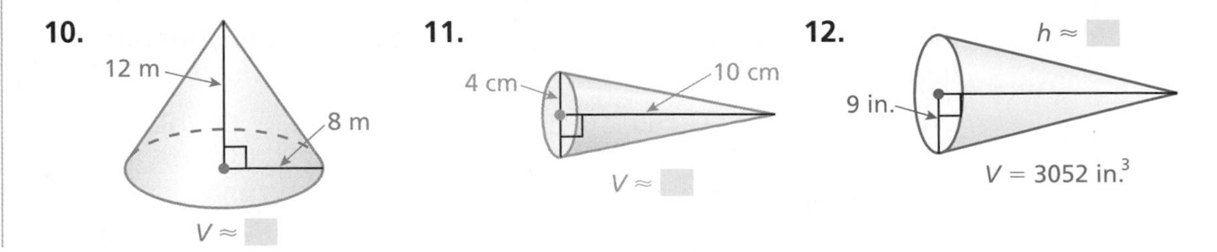

7.5 Volumes of Composite Solids (pp. 324–329)

Find the volume of the composite solid. Round your answer to the nearest tenth.

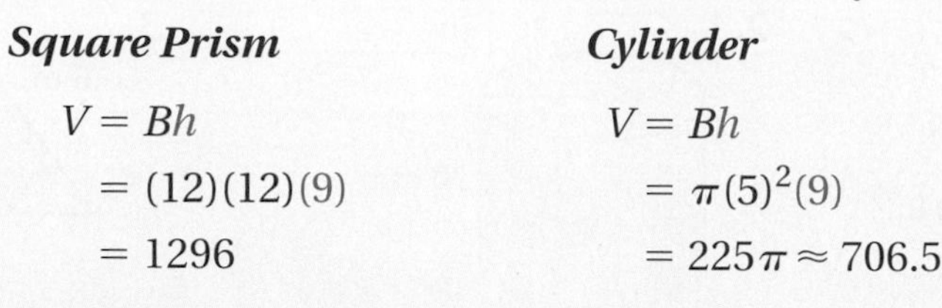

Square Prism

$V = Bh$

$= (12)(12)(9)$

$= 1296$

Cylinder

$V = Bh$

$= \pi(5)^2(9)$

$= 225\pi \approx 706.5$

Find the difference: $1296 - 706.5 = 589.5$.

∴ The volume of the composite solid is about 589.5 cubic feet.

Exercises

Find the volume of the composite solid. Round your answer to the nearest tenth.

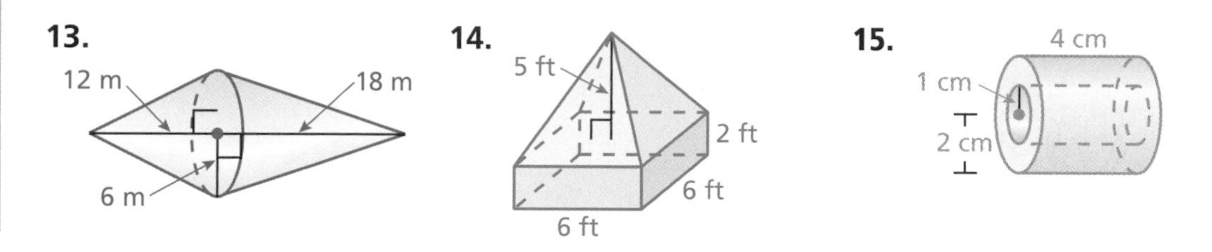

7.6 Surface Areas and Volumes of Similar Solids (pp. 330–337)

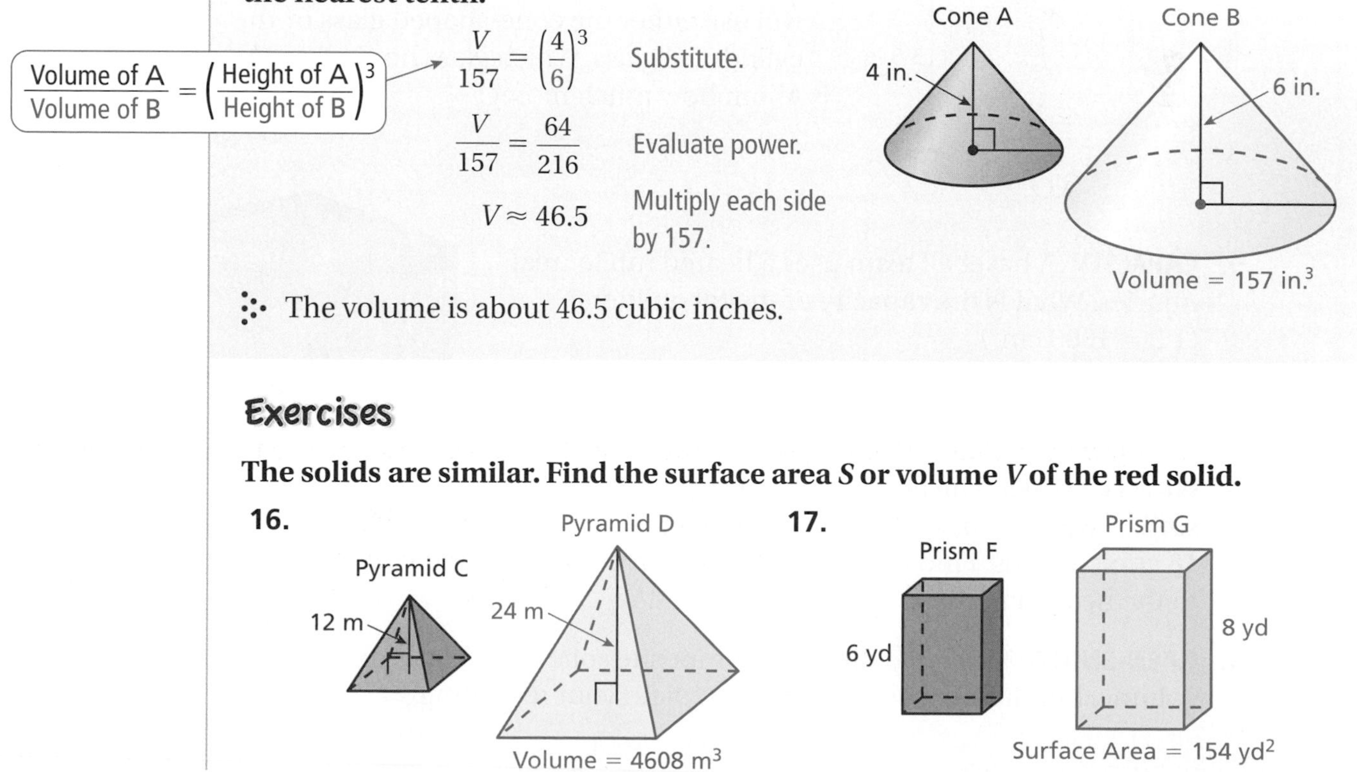

The cones are similar. What is the volume of the red cone? Round your answer to the nearest tenth.

$\dfrac{\text{Volume of A}}{\text{Volume of B}} = \left(\dfrac{\text{Height of A}}{\text{Height of B}}\right)^3$

$\dfrac{V}{157} = \left(\dfrac{4}{6}\right)^3$ Substitute.

$\dfrac{V}{157} = \dfrac{64}{216}$ Evaluate power.

$V \approx 46.5$ Multiply each side by 157.

∴ The volume is about 46.5 cubic inches.

Exercises

The solids are similar. Find the surface area S or volume V of the red solid.

Check It Out
Test Practice
BigIdeasMath.com

Find the volume of the solid. Round your answer to the nearest tenth.

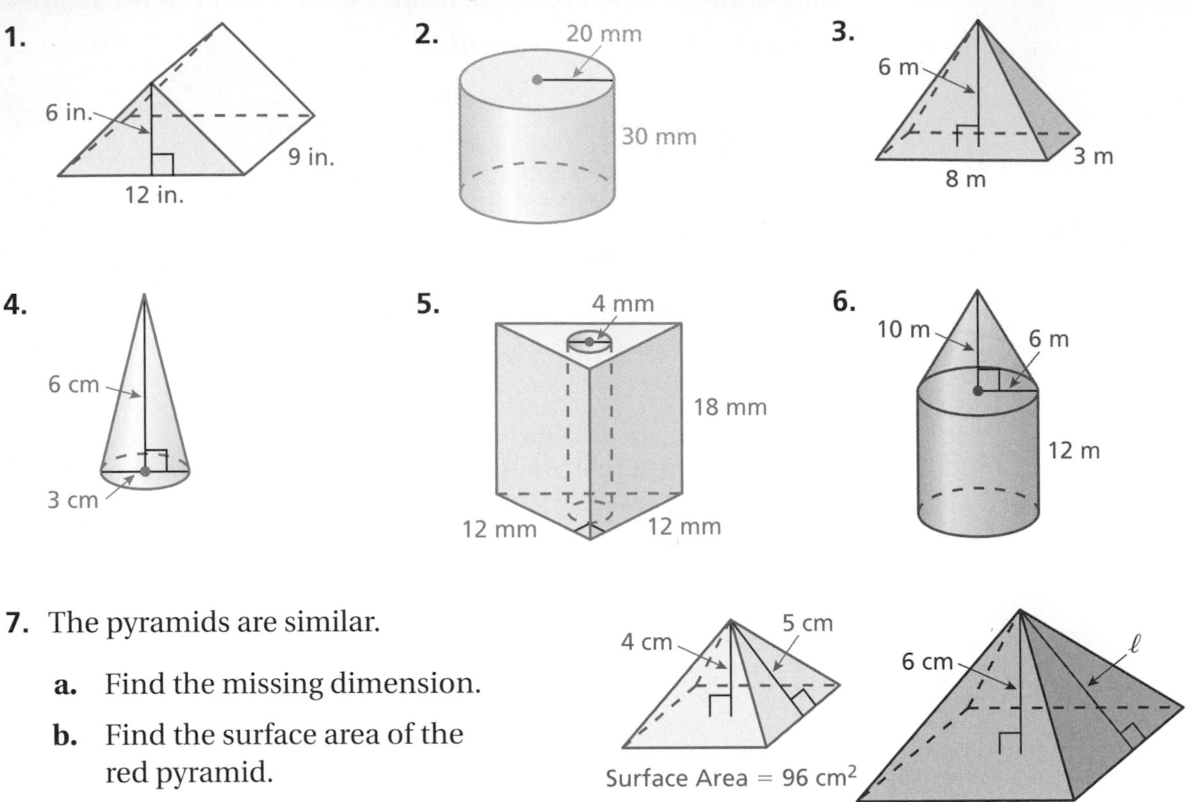

7. The pyramids are similar.

a. Find the missing dimension.

b. Find the surface area of the red pyramid.

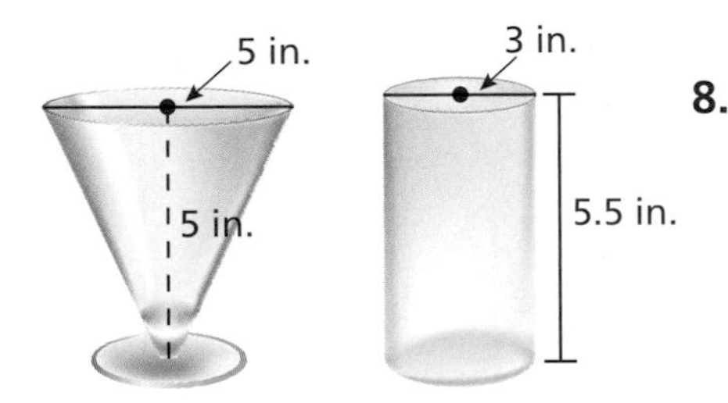

8. SMOOTHIES You are making smoothies. You will use either the cone-shaped glass or the cylindrical glass. Which glass holds more? About how much more?

9. CAPACITY A baseball team uses a heated tub to treat injuries. What is the capacity of the tub in liters? ($1 \text{ L} = 1000 \text{ cm}^3$)

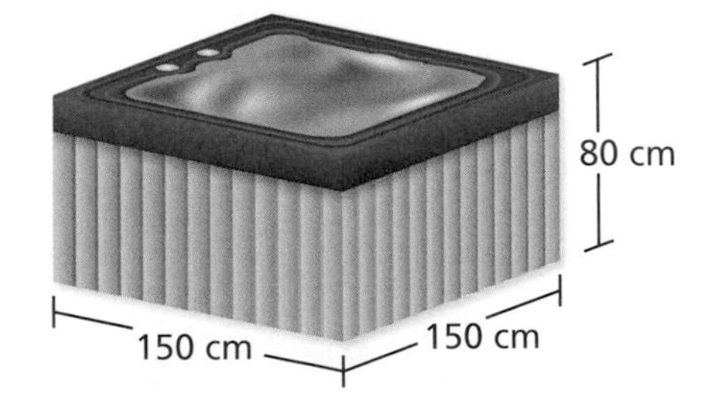

10. WAFFLE CONES The ratio of the corresponding linear measures of two similar waffle cones is 3 to 4. The smaller cone has a volume of about 18 cubic inches. Find the volume of the larger cone. Round your answer to the nearest tenth.

11. OPEN-ENDED Draw two different composite solids that have the same volume, but different surface areas. Explain your reasoning.

7 Standardized Test Practice

1. What is the value of the expression below when $h = \frac{1}{2}$ and $k = -\frac{1}{6}$? *(7.NS.2c)*

$$h \div 3 + 3 \div k$$

A. $-17\frac{5}{6}$

B. $-\frac{1}{3}$

C. 1

D. $17\frac{5}{6}$

2. A right circular cone and its dimensions are shown below.

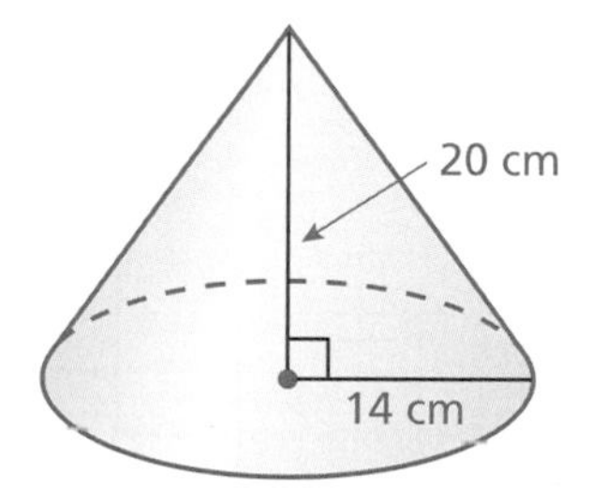

What is the volume of the right circular cone? $\left(\text{Use } \frac{22}{7} \text{ for } \pi.\right)$ *(7.G.6)*

F. $1{,}026\frac{2}{3}$ cm^3

G. 3,080 cm^3

H. $4{,}106\frac{2}{3}$ cm^3

I. 12,320 cm^3

3. A right triangular prism and its dimensions are shown below.

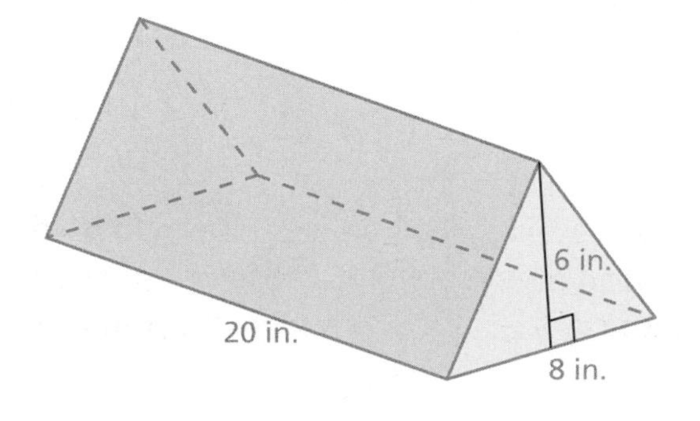

What is the volume of the right triangular prism? *(7.G.6)*

4. A tropical storm has maximum sustained surface winds of at least 39 miles per hour but less than 74 miles per hour. Which graph correctly represents the possible wind speeds of a tropical storm? *(7.EE.4b)*

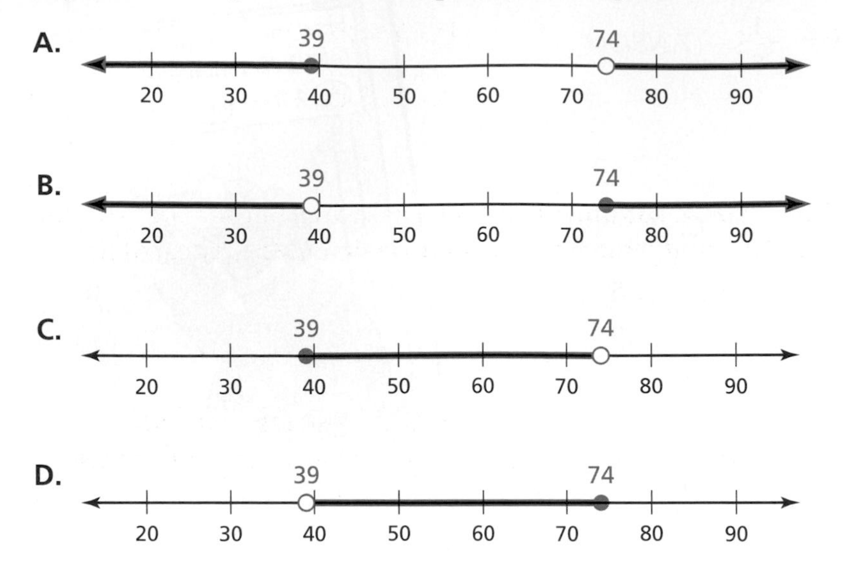

5. Use the coordinate plane to answer the question below.

Which point does *not* lie on the same line as the other three? *(7.NS.1b)*

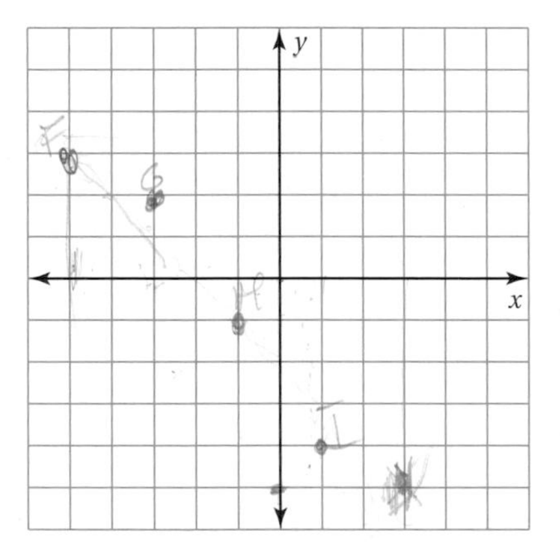

F. $(-5, 3)$ H. $(-1, -1)$

G. $(-3, 2)$ I. $(1, -4)$

6. Olga was solving an equation in the box shown.

What should Olga do to correct the error that she made? *(7.EE.4a)*

$$-\frac{2}{5}(10x - 15) = -30$$

$$10x - 15 = -30\left(-\frac{2}{5}\right)$$

$$10x - 15 = 12$$

$$10x - 15 + 15 = 12 + 15$$

$$10x = 27$$

$$\frac{10x}{10} = \frac{27}{10}$$

$$x = \frac{27}{10}$$

A. Multiply both sides by $-\frac{5}{2}$ instead of $-\frac{2}{5}$.

B. Multiply both sides by $\frac{2}{5}$ instead of $-\frac{2}{5}$.

C. Distribute $-\frac{2}{5}$ to get $-4x - 6$.

D. Add 15 to -30.

7. It has been raining at a rate of 0.08 inch per hour. At this rate, how much rain will fall in $2\frac{1}{2}$ hours? *(7.RP.2b)*

F. 0.032 in.

G. 0.2 in.

H. 0.3125 in.

I. 0.33 in.

8. A right circular cylinder has a volume of 1296 cubic inches. If the radius of the cylinder is divided by 12, what would be the volume, in cubic inches, of the smaller cylinder? *(7.G.6)*

9. If 9 friends share equally a large box of baseball cards, each friend gets 240 cards. If 6 friends share equally the same box of cards, how many cards does each friend get? *(7.RP.2a)*

A. 80

B. 160

C. 360

D. 400

10. All students in a class were surveyed to find out their preferences for writing instruments. The survey found that 12 students prefer to write with a pencil and 20 students prefer to write with a pen. What percent of students in the class prefer to write with a pencil? *(7.RP.3)*

F. 12%

G. 37.5%

H. 60%

I. 62.5%

11. The figure below is a diagram for making a tin lantern.

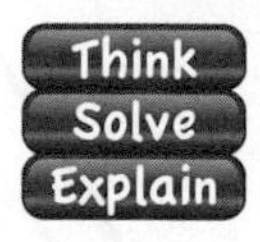

The figure consists of a right circular cylinder without its top base and a right circular cone without its base. What is the volume, in cubic inches, of the entire lantern? Show your work and explain your reasoning. (Use 3.14 for π.) *(7.G.6)*

8 Data Analysis and Samples

"I took a survey of pet owners on how many times per day you should treat your dog to a biscuit."

"What do you think?"

"I've completed a circle graph analyzing what you do each day."

What You Learned Before

"Mom, my owner, and Fluffy have agreed to participate in my random survey. Will you be my fourth participant?"

Finding Mean, Median, Mode, and Range (6.SP.5c)

The table shows the top ten Olympic pole vault heights for men and women.

Olympic Pole Vault Heights (meters)	
Men	**Women**
5.45	4.55
5.60	5.05
5.96	4.75
5.70	4.65
5.45	4.65
5.70	4.55
5.60	4.70
5.70	4.75
5.85	4.80
5.70	4.45

Example 1 What is the mean of the men's data?

$$\text{mean} = \frac{\text{sum of data}}{\text{number of data values}}$$

$$= \frac{56.71}{10} \approx 5.67$$

So, the mean height is about 5.67 meters.

Example 2 What is the median of the women's data?

$$4.45, 4.55, 4.55, 4.65, \underbrace{4.65, 4.70}, 4.75, 4.75, 4.80, 5.05$$

$$\frac{4.65 + 4.70}{2} \approx 4.68$$

So, the median is about 4.68 meters.

Example 3 What is the mode of the men's data?

Because it occurs most often, the mode is 5.70 meters.

Example 4 What is the range of the men's data?

$$\text{range} = \text{greatest data value} - \text{least data value}$$

$$= 5.96 - 5.45 = 0.51$$

So, the range is 0.51 meter.

Try It Yourself

Use the table to answer the question. Round your answer to the nearest hundredth.

1. What is the mean of the women's data?
2. What is the median of the men's data?
3. What is the mode(s) of the women's data?
4. What is the range of the women's data?

8.1 Stem-and-Leaf Plots

COMMON CORE STATE STANDARDS
7.SP.1
7.SP.2

Essential Question How can you use a stem-and-leaf plot to organize a set of numbers?

1 ACTIVITY: Decoding a Graph

Work with a partner. You intercept a secret message that contains two different types of plots. You suspect that each plot represents the same data. The graph with the dots indicates only ranges for the numbers.

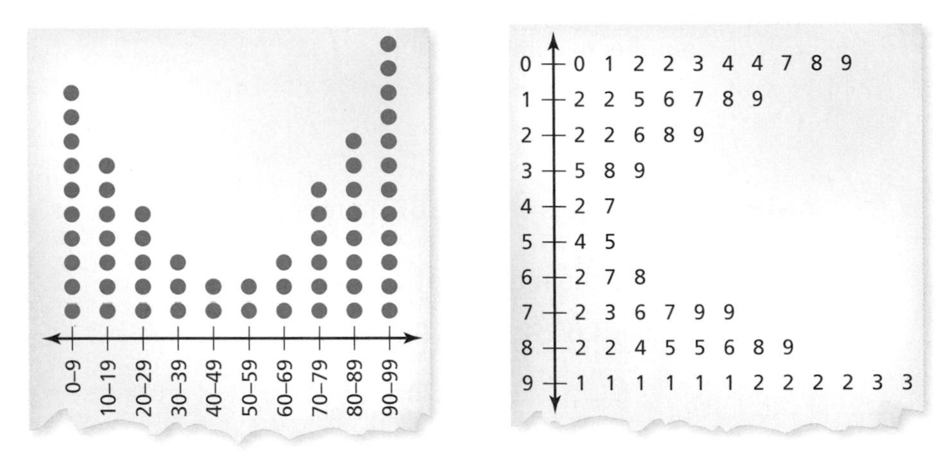

a. How many numbers are in the data set? How can you tell?

b. How many numbers are greater than or equal to 90? How can you tell?

c. Is 91 in the data set? If so, how many times is it in the set? How can you tell?

d. Make a list of all of the numbers in the data set.

e. You intercept a new secret message. Use the secret code shown below to decode the message.

Secret Code

A = 29	F = 31	K = 18	P = 4	U = 19
B = 33	G = 8	L = 26	Q = 10	V = 17
C = 7	H = 16	M = 22	R = 21	W = 12
D = 20	I = 5	N = 3	S = 2	X = 25
E = 15	J = 11	O = 9	T = 32	Y = 13
				Z = 1

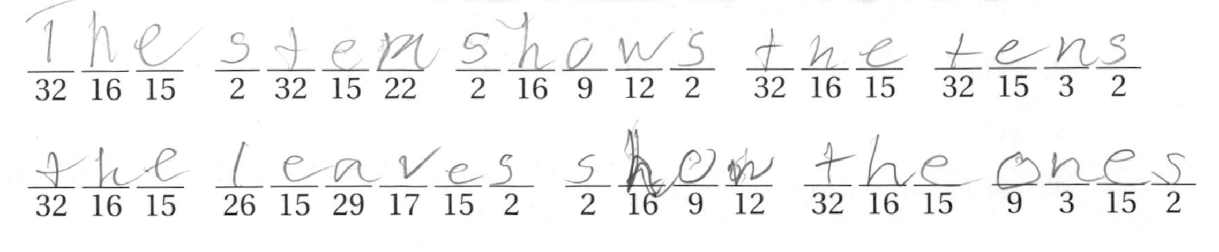

The stem shows the tens
32 16 15 2 32 15 22 2 16 9 12 2 32 16 15 32 15 3 2
the leaves show the ones
32 16 15 26 15 29 17 15 2 2 16 9 12 32 16 15 9 3 15 2

2 ACTIVITY: Organizing Data

Work with a partner. You are working on an archeological dig. You find several arrowheads.

18	61	62
42	42	42
23	41	40
45	45	45
37	28	50
35	39	34
37	32	26
63	24	54
58	58	60
52	53	72
17	73	

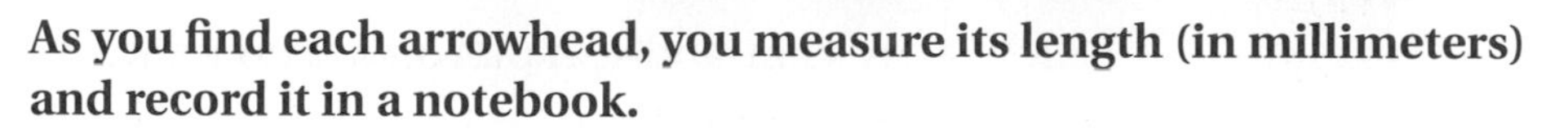

As you find each arrowhead, you measure its length (in millimeters) and record it in a notebook.

a. Use a stem-and-leaf plot to organize the lengths.

b. Find the mean length.

c. Find the median length.

d. Describe the distribution of the data.

3 ACTIVITY: Conducting an Experiment

Work with a partner. Use two number cubes to conduct the following experiment.

- **Toss the cubes four times and total the results.**

  Sample: $2 + 3 \quad + \quad 2 + 2 \quad + \quad 3 + 5 \quad + \quad 6 + 3 \quad = 26$

  1st toss, 2nd toss, 3rd toss, 4th toss

  So, 26 is the first number.

- **Repeat this process 29 more times.**
- **Use a stem-and-leaf plot to organize your results.**
- **Describe your results.**

What Is Your Answer?

4. IN YOUR OWN WORDS How can you use a stem-and-leaf plot to organize a set of numbers?

5. RESEARCH Find a career in which a person collects and organizes data. Describe how data are collected and organized in that career.

Use what you learned about stem-and-leaf plots to complete Exercises 4–7 on page 352.

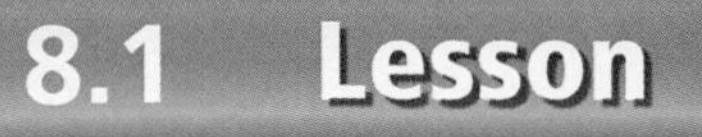

8.1 Lesson

Check It Out
Lesson Tutorials
BigIdeasMath.com

Key Vocabulary
stem-and-leaf plot, *p. 350*
stem, *p. 350*
leaf, *p. 350*

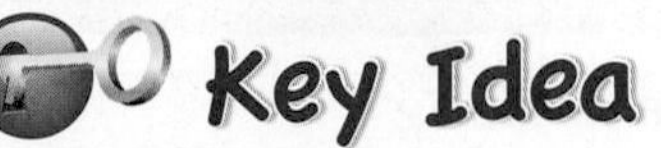

Key Idea

Stem-and-Leaf Plots

A **stem-and-leaf plot** uses the digits of data values to organize a data set. Each data value is broken into a **stem** (digit or digits on the left) and a **leaf** (digit or digits on the right).

A stem-and-leaf plot shows how data are distributed.

Stem	Leaf
2	0 0 1 2 5 7
3	1 4 8
4	2
5	8 9

Key: 2|0 = 20

The *key* explains what the stems and leaves represent.

EXAMPLE 1 Making a Stem-and-Leaf Plot

	A	B
1	DATE	MINUTES
2	JULY 9	55
3	JULY 9	3
4	JULY 9	6
5	JULY 10	14
6	JULY 10	18
7	JULY 10	5
8	JULY 10	23
9	JULY 11	30
10	JULY 11	23
11	JULY 11	10
12	JULY 11	2
13	JULY 11	36

Make a stem-and-leaf plot of the length of the 12 cell phone calls.

Step 1: Order the data.

2, 3, 5, 6, 10, 14, 18, 23, 23, 30, 36, 55

Step 2: Choose the stems and leaves. Because the data values range from 2 to 55, use the *tens* digits for the stems and the *ones* digits for the leaves.

Step 3: Write the stems to the *left* of the vertical line.

Step 4: Write the leaves for each stem to the *right* of the vertical line.

Phone call lengths

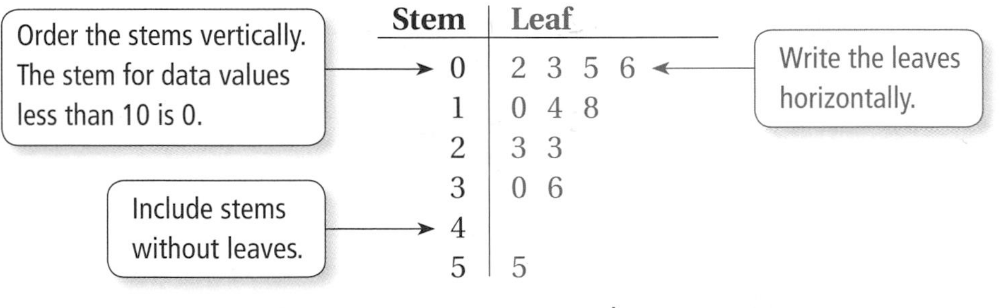
Order the stems vertically. The stem for data values less than 10 is 0.
Include stems without leaves.
Write the leaves horizontally.

Stem	Leaf
0	2 3 5 6
1	0 4 8
2	3 3
3	0 6
4	
5	5

Key: 1|4 = 14 minutes

On Your Own

Now You're Ready
Exercises 8–11

1. Make a stem-and-leaf plot of the hair lengths.

Hair Length (centimeters)									
5	1	20	12	27	2	30	5	7	38
40	47	1	2	1	32	4	44	33	23

EXAMPLE 2 Interpreting a Stem-and-Leaf Plot

Test Scores

Stem	Leaf
6	6
7	0 5 7 8
8	1 1 3 4 4 6 8 8 9
9	0 2 9
10	0

Key: 9|2 = 92 points

The stem-and-leaf plot shows student test scores. (a) How many students scored less than 80 points? (b) How many students scored at least 90 points? (c) How are the data distributed?

a. There are five scores less than 80 points: 66, 70, 75, 77, and 78.

∴ Five students scored less than 80 points.

b. There are four scores of at least 90 points: 90, 92, 99, and 100.

∴ Four students scored at least 90 points.

c. There are few low test scores and few high test scores. So, most of the scores are in the middle.

On Your Own

Now You're Ready
Exercises 16–19

2. Use the grading scale at the right.

a. How many students received a B on the test?

b. How many students received a C on the test?

A: 90–100
B: 80–89
C: 70–79
D: 60–69
F: 59 and below

EXAMPLE 3 Standardized Test Practice

Which statement is *not* true?

(A) Most of the plants are less than 20 inches tall.

(B) The median plant height is 11 inches.

(C) The range of the plant heights is 35 inches.

(D) The plant height that occurs most often is 11 inches.

Plant Heights

Stem	Leaf
0	1 2 4 5 6 8 9
1	0 1 1 5 7
2	2 5
3	6

Key: 1|5 = 15 inches

There are 15 plant heights. So, the median is the eighth data value, 10 inches.

∴ The correct answer is (B).

On Your Own

3. You are told that three plants are taller than 20 inches. Is the statement true? Explain.

8.1 Exercises

Vocabulary and Concept Check

1. **VOCABULARY** The key for a stem-and-leaf plot is $3|4 = 34$. Which number is the stem? the leaf?

2. **WRITING** Describe how to make a stem-and-leaf plot for the data values 14, 22, 9, 13, 30, 8, 25, and 29.

3. **WRITING** How does a stem-and-leaf plot show the distribution of data?

Practice and Problem Solving

Use the stem-and-leaf plot at the right.

4. How many data values are in the set?

5. What is the least value? greatest value?

6. What is the median? range?

7. Is the value 32 in the set? Explain.

Stem	Leaf
0	4 6 8
1	0
2	3 4
3	0 6 6 9
4	2

Key: $3|6 = 36$

Make a stem-and-leaf plot of the data.

1 8.

Books Read			
26	15	20	9
31	25	29	32
17	26	19	40

9.

Hours Online			
8	12	21	14
18	6	15	24
12	17	2	0

10.

Test Scores (%)				
87	82	95	91	69
88	68	87	65	81
97	85	80	90	62

11.

Points Scored				
58	50	42	71	75
45	51	43	38	71
42	70	56	58	43

12. **ERROR ANALYSIS** Describe and correct the error in making a stem-and-leaf plot of the data.

51, 25, 47, 42, 55, 26, 50, 44, 55

✗

Stem	Leaf
2	5 6
4	2 4 7
5	0 1 5 5

Key: $4|2 = 42$

13. **PUPPIES** The weights (in pounds) of eight puppies at a pet store are 12, 24, 17, 8, 18, 31, 24, and 15. Make a stem-and-leaf plot of the data. Describe the distribution of the data.

Make a stem-and-leaf plot of the data.

14.

Bikes Sold			
78	112	105	99
86	96	115	100
79	81	99	108

15.

Minutes in Line			
4.0	2.6	1.9	3.1
3.6	2.2	2.7	3.8
1.6	2.0	3.1	2.9

VOLLEYBALL The stem-and-leaf plot shows the number of digs for the top 15 players at a volleyball tournament.

Stem	Leaf
4	1 1 3 3 5
5	0 2 3 4
6	2 3 3 7
7	5
8	
9	7

Key: 5 | 0 = 50 digs

② **16.** How many players had more than 60 digs?

17. Find the mean, median, mode, and range of the data.

18. Describe the distribution of the data.

19. Which data value is the outlier? Describe how the outlier affects the mean.

20. **CHOOSE TOOLS** Use the Internet to find the heights of the players on your favorite professional sports team.

a. Make a stem-and-leaf plot of the data.

b. Analyze the stem-and-leaf plot and make two conclusions about the heights.

21. **OPEN-ENDED** Describe a real-life situation with eight data values that has a median of 33. Make a stem-and-leaf plot of the data.

Bowling Scores					
131	108	115	140	152	122
120	118	156	142	112	107
136	85	127	119	136	125

22. **Critical Thinking** Make a frequency table and a stem-and-leaf plot of the bowling scores in the table. Compare and contrast the two data displays. Which display is better for showing how the data are distributed? Explain.

Fair Game Review What you learned in previous grades & lessons

Draw the solid. *(Section 6.1)*

23. Square pyramid

24. Hexagonal prism

25. Cone

26. Cylinder

27. **MULTIPLE CHOICE** In a bar graph, what determines the length of each bar? *(Skills Review Handbook)*

Ⓐ Frequency Ⓑ Data value Ⓒ Leaf Ⓓ Change in data

8.2 Histograms

Essential Question How do histograms show the differences in distributions of data?

COMMON CORE STATE STANDARDS
7.SP.1
7.SP.2

1 ACTIVITY: Analyzing Distributions

Work with a partner. The graphs (histograms) show four different types of distributions.

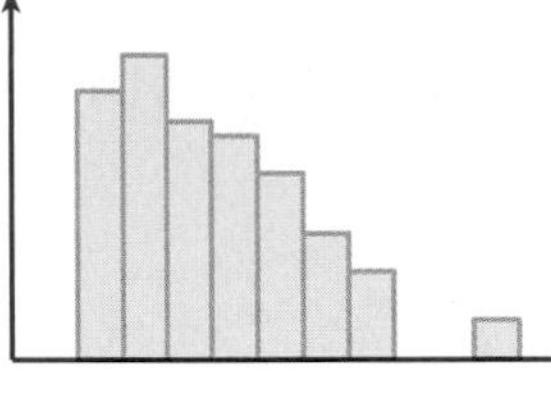
Skew Distribution

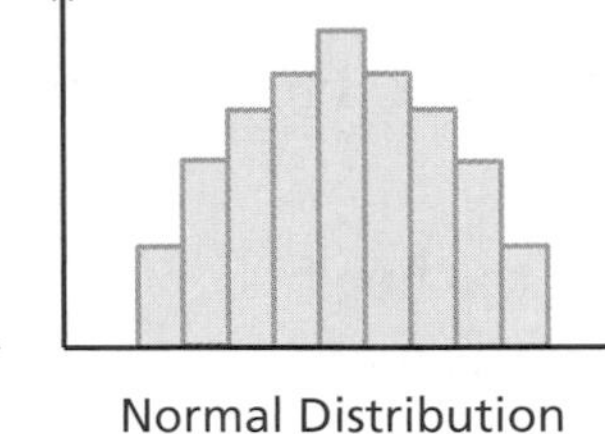
Normal Distribution

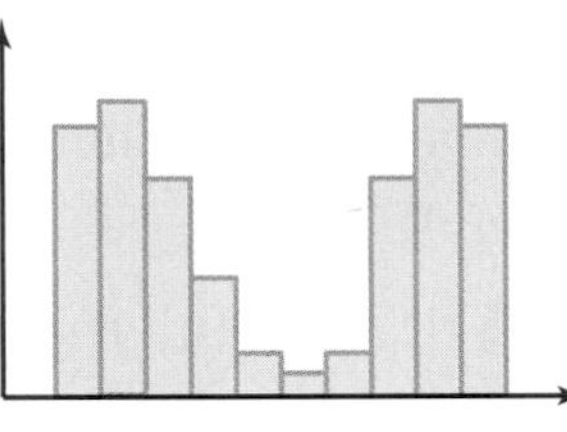
Bimodal Distribution

Flat Distribution

a. Describe a real-life example of each distribution.

b. Describe the mean, median, and mode of each distribution.

c. In which distributions are the mean and median about equal? Explain your reasoning.

d. How did each type of distribution get its name?

2 ACTIVITY: Analyzing Distributions

Work with a partner. A survey asked 100 adult men and 100 adult women to answer the following questions.

Question 1: What is your ideal weight?
Question 2: What is your ideal age?

Match the histogram to the question.

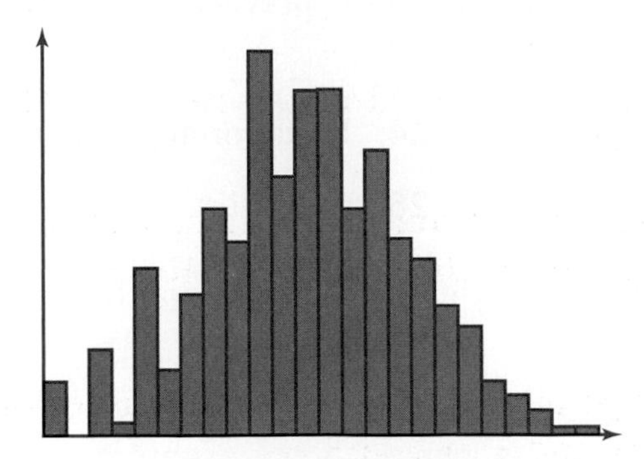

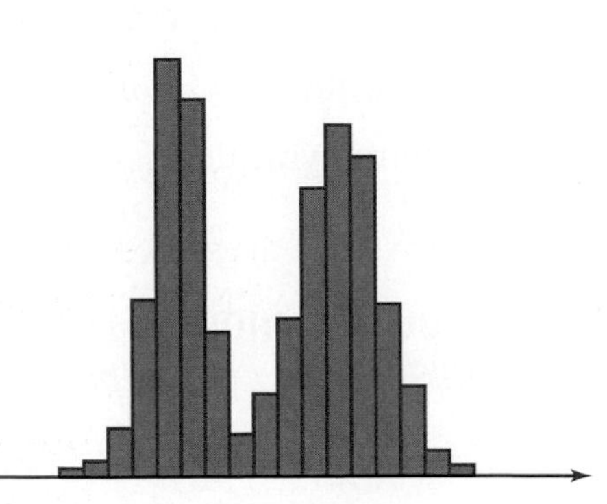

3 ACTIVITY: Conducting Experiments

Work with a partner. Conduct two experiments. Make a frequency table and a histogram for each experiment. Compare and contrast the results of the two experiments.

a. Toss one number cube 36 times. Record the numbers.

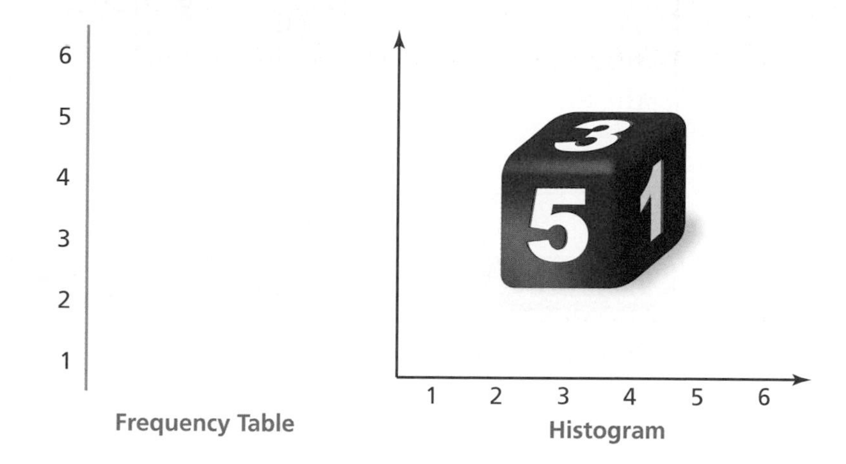

b. Toss two number cubes 36 times. Record the sums of the two numbers.

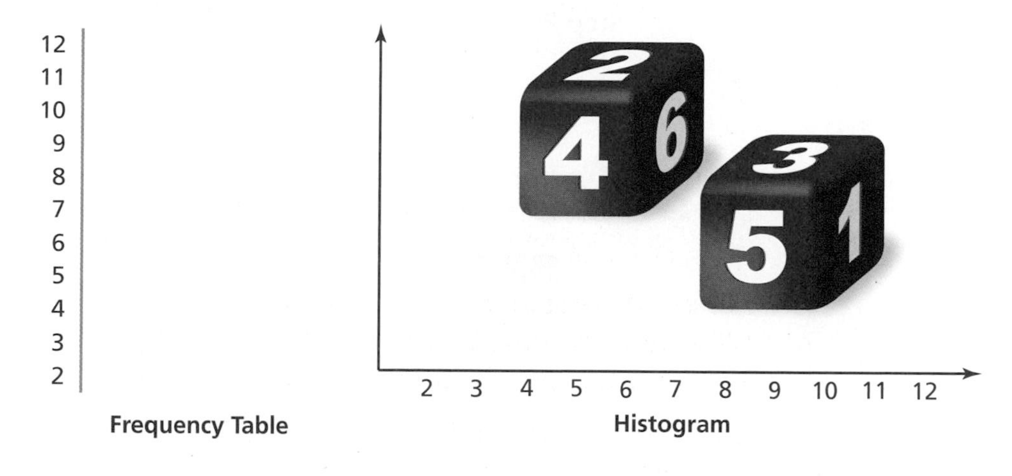

What Is Your Answer?

4. **IN YOUR OWN WORDS** How do histograms show the differences in distributions of data?
5. Describe an experiment that you can conduct to collect data. Predict the type of data distribution the results will create.

Practice Use what you learned about histograms to complete Exercises 4 and 5 on page 358.

8.2 Lesson

Key Vocabulary
histogram, *p. 356*

Key Idea

Histograms

A **histogram** is a bar graph that shows the frequency of data values in intervals of the same size.

The height of a bar represents the frequency of the values in the interval.

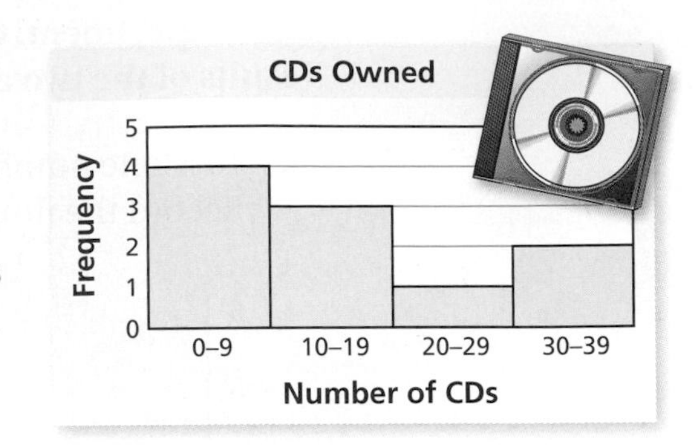

EXAMPLE 1 Making a Histogram

Remember

A *frequency table* groups data values into intervals. The *frequency* is the number of data values in an interval.

The frequency table shows the number of pairs of shoes that each person in a class owns. Display the data in a histogram.

Pairs of Shoes	Frequency
1–3	11
4–6	4
7–9	0
10–12	3
13–15	6

Step 1: Draw and label the axes.

Step 2: Draw a bar to represent the frequency of each interval.

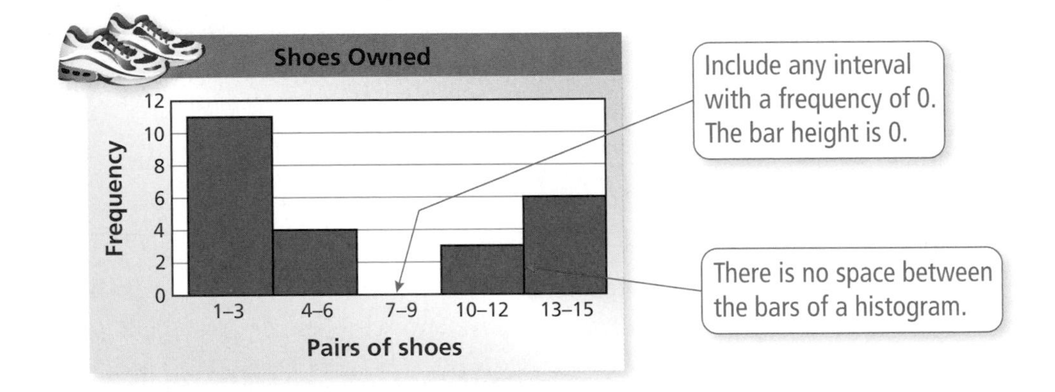

On Your Own

Now You're Ready
Exercises 6–8

1. The frequency table shows the ages of people riding a roller coaster. Display the data in a histogram.

Age	10–19	20–29	30–39	40–49	50–59
Frequency	16	11	5	2	4

EXAMPLE 2 Using a Histogram

The histogram shows the winning speeds at the Daytona 500. (a) Which interval contains the most data values? (b) How many of the winning speeds are less than 140 miles per hour? (c) How many of the winning speeds are at least 160 miles per hour?

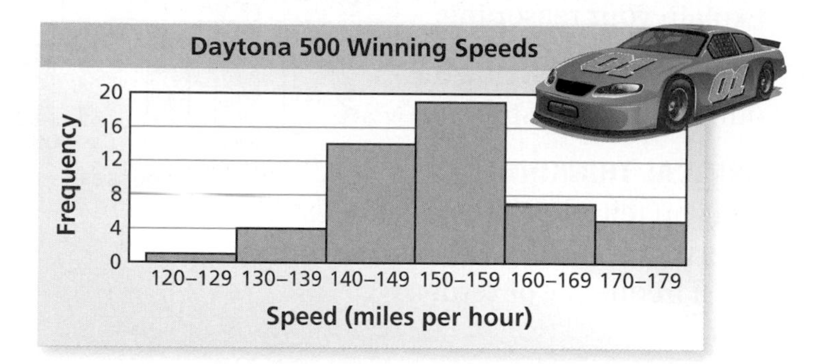

a. The interval with the tallest bar contains the most data values.

So, the 150–159 miles per hour interval contains the most data values.

b. One winning speed is in the 120–129 miles per hour interval and four winning speeds are in the 130–139 miles per hour interval.

So, $1 + 4 = 5$ winning speeds are less than 140 miles per hour.

c. Seven winning speeds are in the 160–169 miles per hour interval and five winning speeds are in the 170–179 miles per hour interval.

So, $7 + 5 = 12$ winning speeds are at least 160 miles per hour.

On Your Own

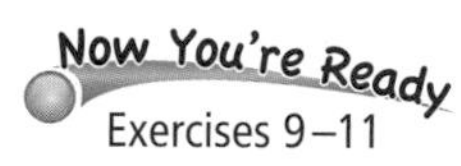

Exercises 9–11

2. The histogram shows the number of hours that students in a class slept last night.

a. How many students slept at least 8 hours?

b. How many students slept less than 12 hours?

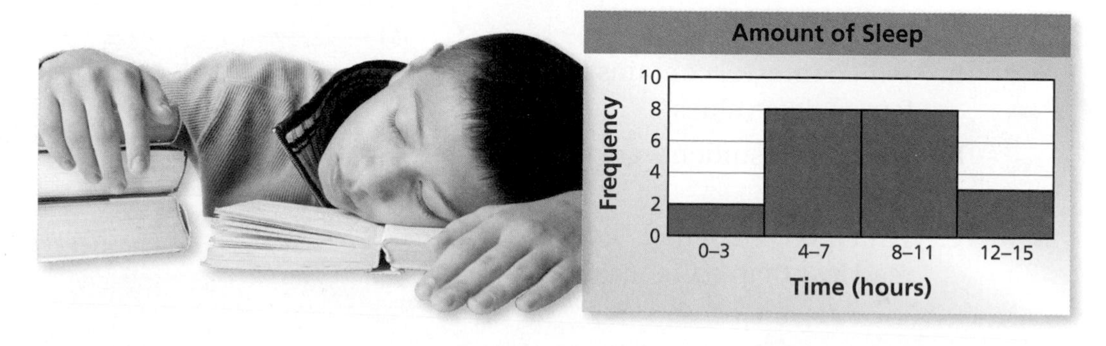

8.2 Exercises

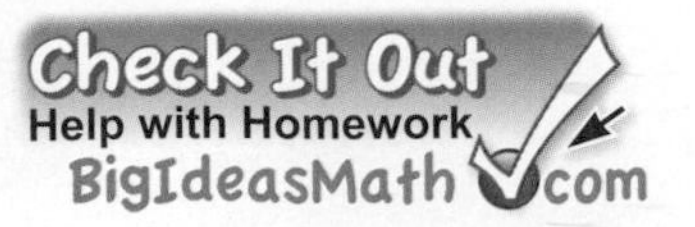

Vocabulary and Concept Check

1. **VOCABULARY** Which graph is a histogram? Explain your reasoning.

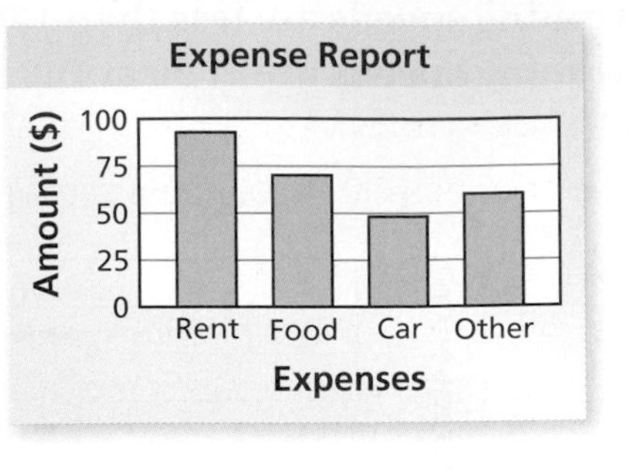

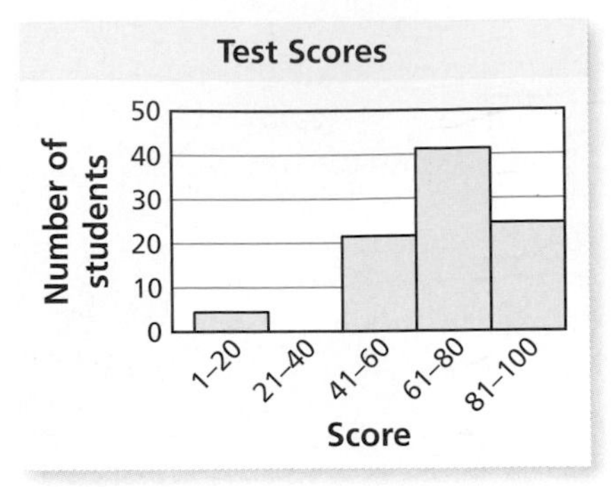

2. **REASONING** Describe the outliers in the histogram.
3. **CRITICAL THINKING** How can you tell when an interval of a histogram has a frequency of zero?

Practice and Problem Solving

Determine the type of distribution shown by the histogram.

4. 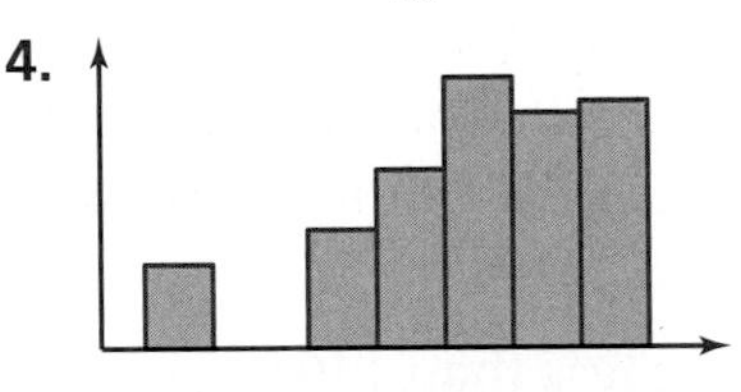

5. 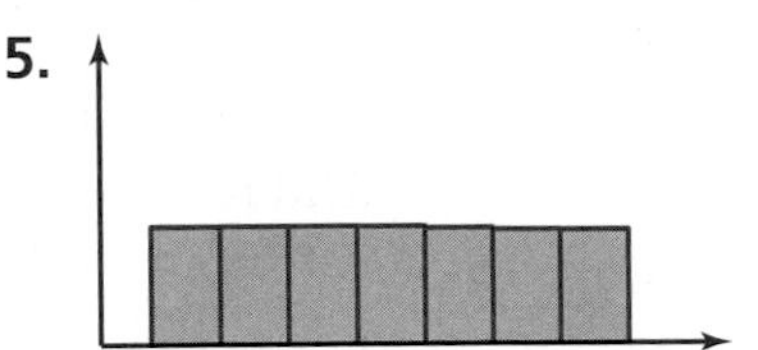

Display the data in a histogram.

1 6.

States Visited	
States	**Frequency**
1–5	12
6–10	14
11–15	6
16–20	3

7.

Chess Team	
Wins	**Frequency**
10–13	3
14–17	4
18–21	4
22–25	2

8.

Movies Watched	
Movies	**Frequency**
0–1	5
2–3	11
4–5	8
6–7	1

2 9. **MAGAZINES** The histogram shows the number of magazines read last month by students in a class.

a. Which interval contains the fewest data values?

b. How many students are in the class?

c. What percent of the students read less than six magazines?

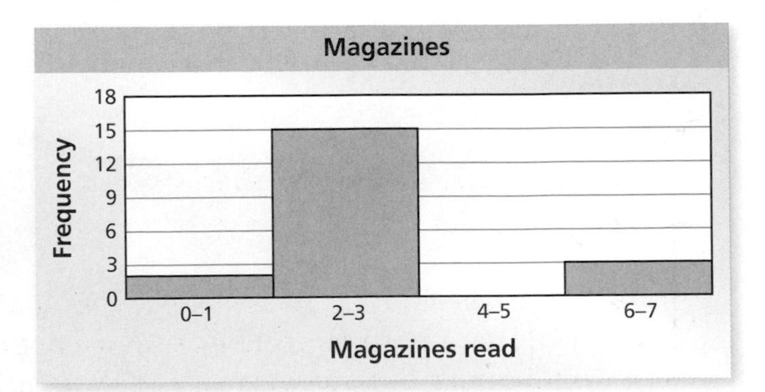

10. VOTING The histogram shows the percent of the voting age population that voted in a recent presidential election. Explain whether each statement is supported by the graph.

a. Only 40% of one state voted.

b. Most states had between 50% and 64.9% that voted.

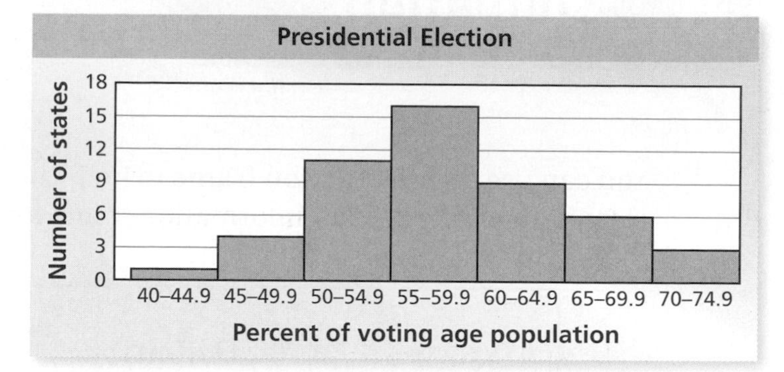

11. AREA The histograms show the areas of counties in Pennsylvania and Indiana. Which state do you think has the greater area? Explain.

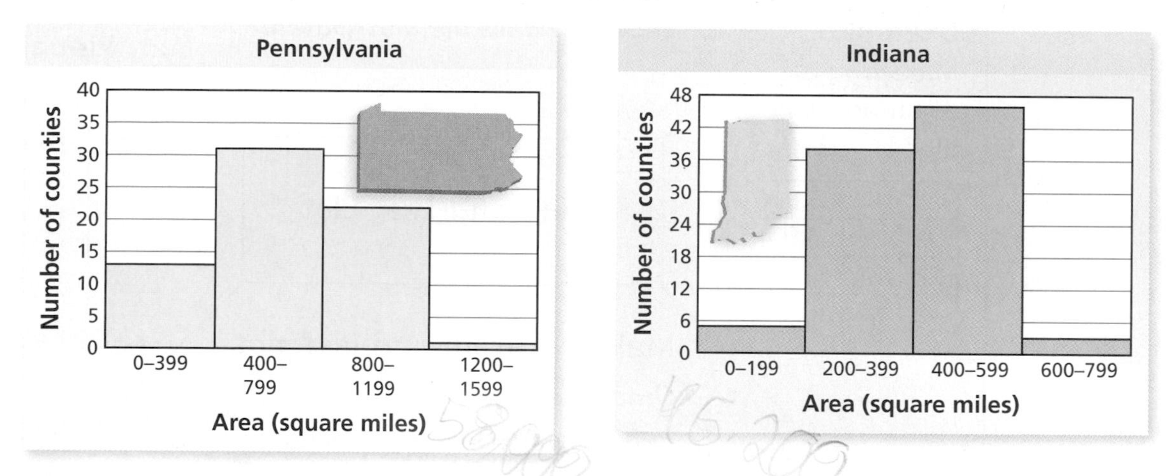

12. REASONING Can you find the mean, median, mode, and range of the data in Exercise 7? If so, find them. If not, explain why.

13. Critical Thinking The table shows the weights of guide dogs enrolled in a training program.

a. Make a histogram of the data starting with the interval 51–55.

b. Make another histogram of the data using a different sized interval.

c. Compare and contrast the two histograms.

Weight (lb)					
81	88	57	82	70	85
71	51	82	77	79	77
83	80	54	80	81	73
59	84	75	76	68	78
83	78	55	67	85	79

Fair Game Review What you learned in previous grades & lessons

Find the percent of the number. *(Section 4.1)*

14. 25% of 180 **15.** 30% of 90 **16.** 16% of 140 **17.** 64% of 80

18. MULTIPLE CHOICE Two rectangles are similar. The smaller rectangle has a length of 8 feet. The larger rectangle has a length of 14 feet. What is the ratio of the area of the smaller rectangle to the area of the larger rectangle? *(Section 5.3)*

Ⓐ 7 : 4 Ⓑ 4 : 7 Ⓒ 9 : 16 Ⓓ 16 : 49

8 Study Help

Check It Out
Graphic Organizer
BigIdeasMath.com

You can use an **information frame** to help you organize and remember concepts. Here is an example of an information frame for a stem-and-leaf plot.

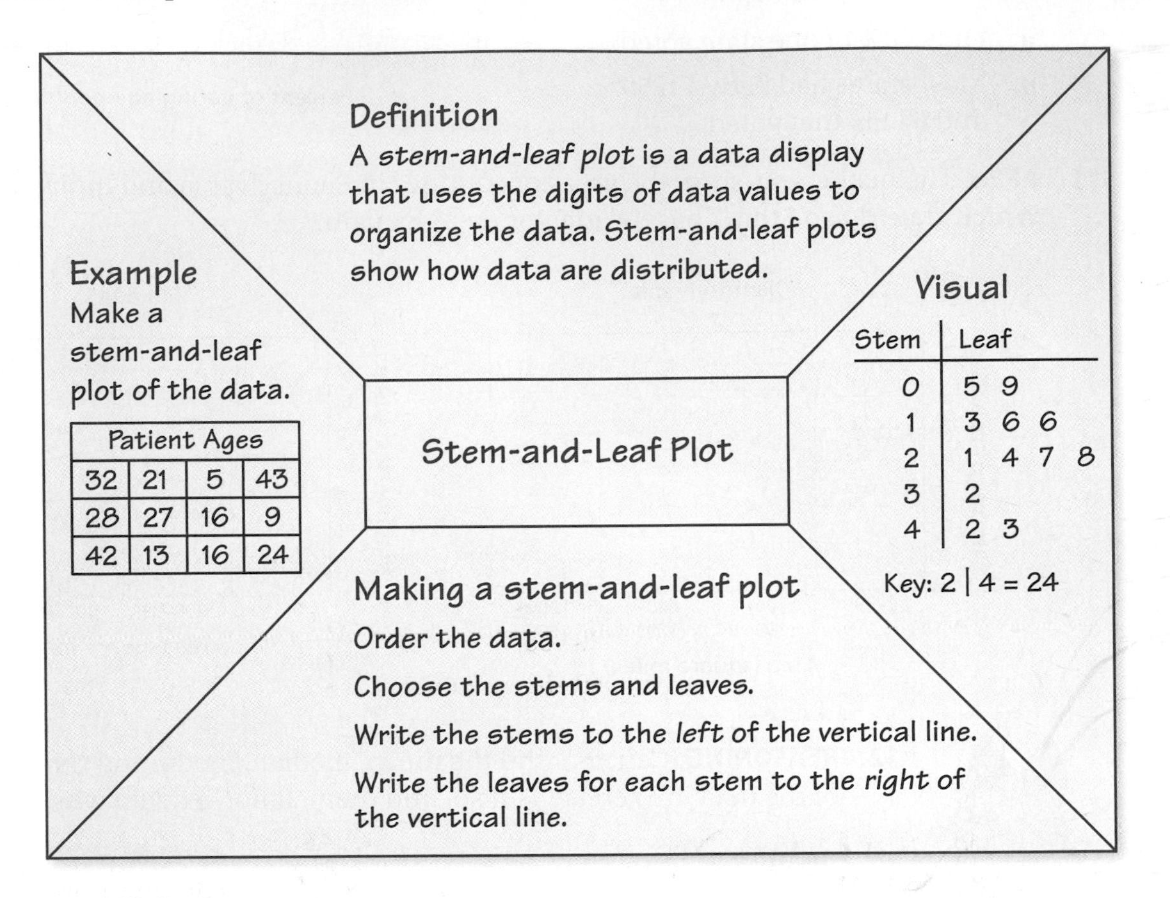

Definition
A *stem-and-leaf plot* is a data display that uses the digits of data values to organize the data. Stem-and-leaf plots show how data are distributed.
Example
Make a stem-and-leaf plot of the data.
Patient Ages
32 21 5 43
28 27 16 9
42 13 16 24
Stem-and-Leaf Plot
Visual
Stem | Leaf
0 | 5 9
1 | 3 6 6
2 | 1 4 7 8
3 | 2
4 | 2 3
Key: 2 | 4 = 24
Making a stem-and-leaf plot
Order the data.
Choose the stems and leaves.
Write the stems to the *left* of the vertical line.
Write the leaves for each stem to the *right* of the vertical line.

On Your Own

Make an information frame to help you study the topic.

1. histogram

After you complete this chapter, make information frames for the following topics.

2. circle graph
3. making a prediction about a population
4. Pick three other topics that you studied earlier in this course. Make an information frame for each topic.

"I'm having trouble thinking of a good title for my information frame."

8.1–8.2 Quiz

Check It Out
Progress Check
BigIdeasMath.com

Make a stem-and-leaf plot of the data. *(Section 8.1)*

1.

Cans Collected Each Month			
80	90	84	92
76	83	79	59
68	55	58	61

2.

Miles Driven Each Day				
21	18	12	16	10
16	9	15	20	28
35	50	37	20	11

3.

Ages of Tortoises			
86	99	100	124
92	85	110	130
115	129	83	104

4.

Kilometers Run Each Day				
6.0	5.6	6.2	3.0	2.5
3.5	2.0	5.0	3.9	3.1
6.2	3.1	4.5	3.8	6.1

Display the data in a histogram. *(Section 8.2)*

5.

Soccer Team Goals	
Goals per Game	**Frequency**
0–1	5
2–3	4
4–5	0
6–7	1

6.

Minutes Practiced	
Minutes	**Frequency**
0–19	8
20–39	10
40–59	11
60–79	2

7.

Poems Written for Class	
Poems	**Frequency**
0–4	6
5–9	16
10–14	4
15–19	2
20–24	2

8. WEIGHTS The weights (in ounces) of nine packages are 7, 22, 16, 12, 6, 18, 15, 13, and 25. Make a stem-and-leaf plot of the data. Describe the distribution of the data. *(Section 8.1)*

9. REBOUNDS The histogram shows the number of rebounds per game for a middle school basketball player this season. *(Section 8.2)*

a. Which interval contains the most data values?

b. How many games were played by the player this season?

c. What percent of the games did the player have 4 or more rebounds?

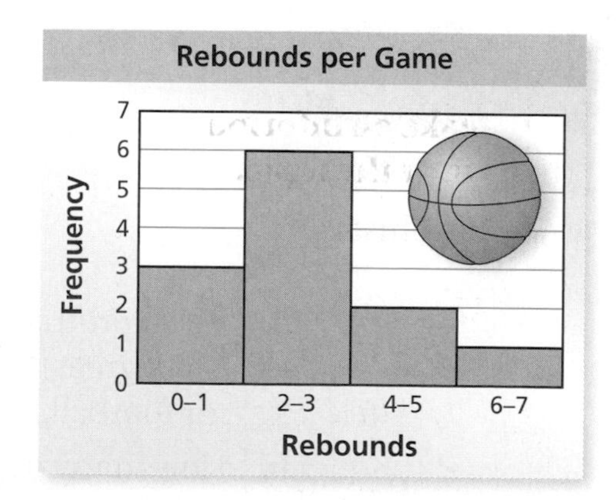

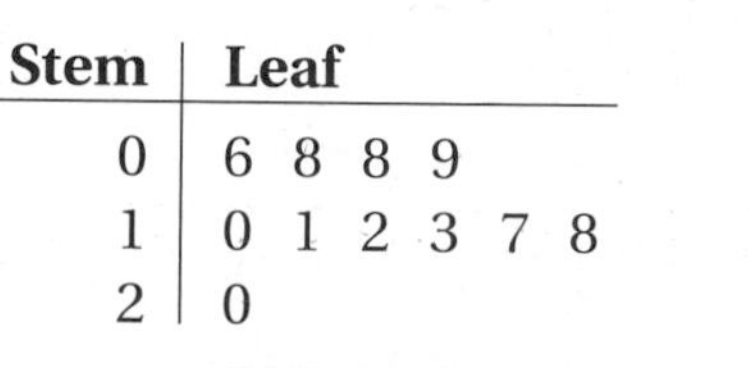

Stem	Leaf
0	6 8 8 9
1	0 1 2 3 7 8
2	0

Key: 0 | 9 = 9 hours

10. STAGE CREW The stem-and-leaf plot shows the number of hours 11 stage crew members spent building sets. Find the mean, median, mode, and range of the data. *(Section 8.1)*

8.3 Circle Graphs

Essential Question How can you use a circle graph to show the results of a survey?

Share Your Work at... My.BigIdeasMath.com

COMMON CORE STATE STANDARDS
7.SP.1
7.SP.2

1 ACTIVITY: Reading a Circle Graph

Work with a partner. Six hundred middle school students were asked "What is your favorite sport?" The circle graph shows the results of the survey.

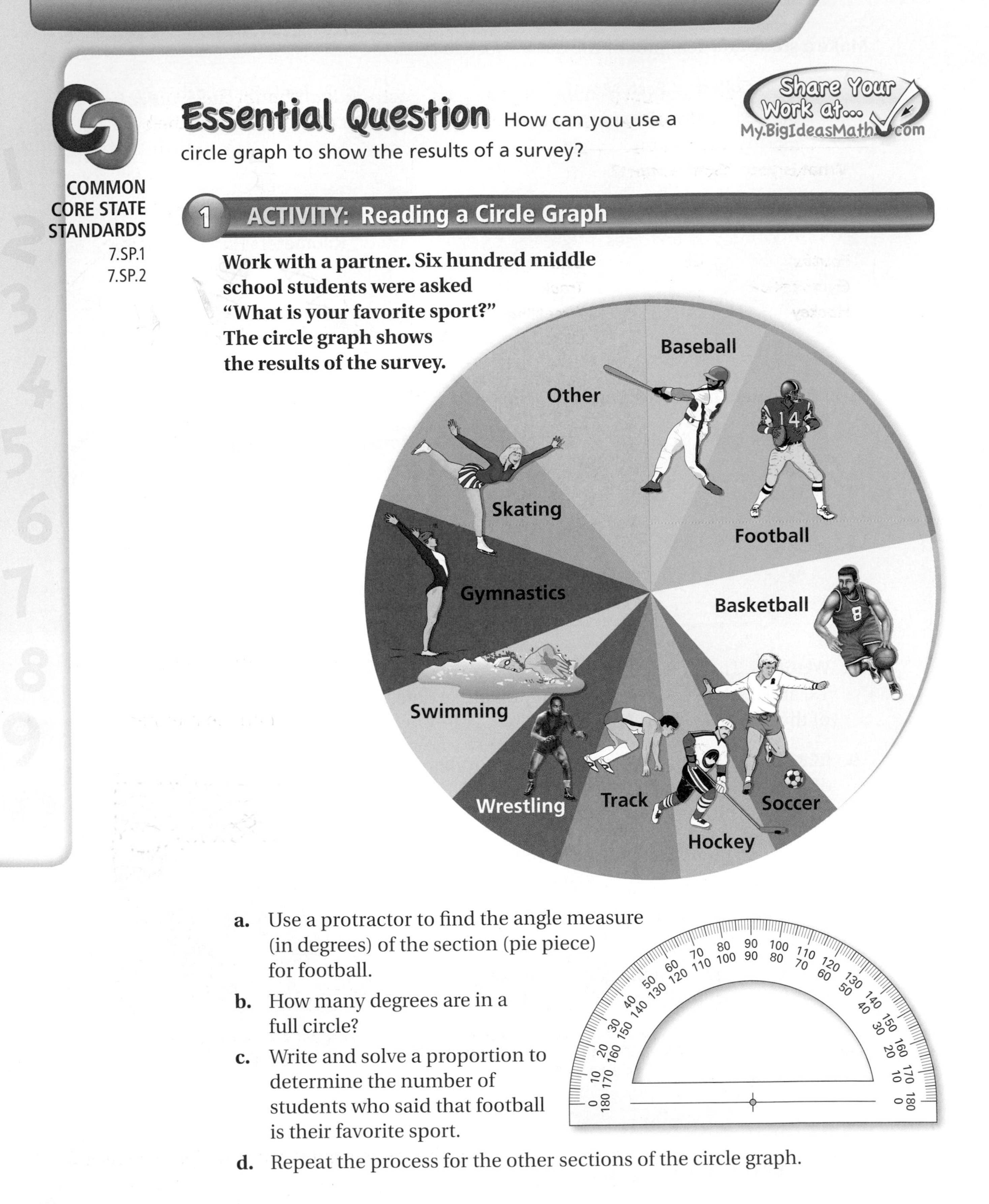

a. Use a protractor to find the angle measure (in degrees) of the section (pie piece) for football.

b. How many degrees are in a full circle?

c. Write and solve a proportion to determine the number of students who said that football is their favorite sport.

d. Repeat the process for the other sections of the circle graph.

2 ACTIVITY: Making a Circle Graph

Work with a partner.

a. Conduct a survey in your class. Each student should check his or her favorite sport on a piece of paper similar to the one shown below.

What is your favorite sport?

Baseball	❑	Skating	❑
Basketball	❑	Soccer	❑
Football	❑	Swimming	❑
Gymnastics	❑	Track	❑
Hockey	❑	Wrestling	❑
		Other	❑

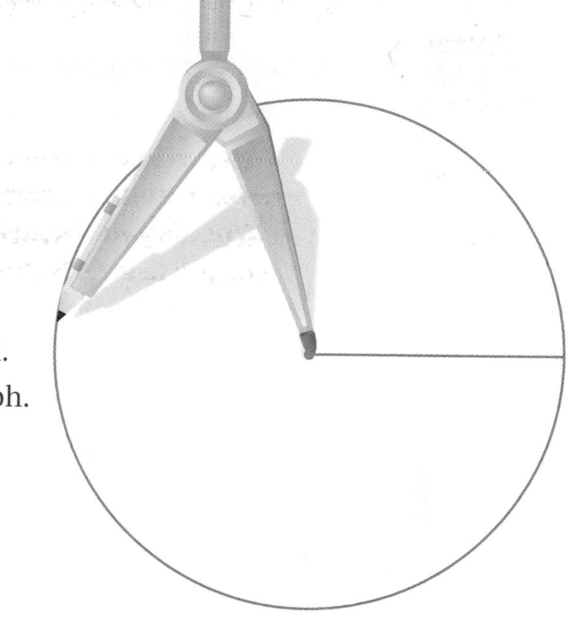

b. Organize the results on the board.

c. Display the results in a circle graph.

d. Compare and contrast your class survey with the survey in Activity 1.

What Is Your Answer?

3. **IN YOUR OWN WORDS** How can you use a circle graph to show the results of a survey?

4. Find a circle graph in a newspaper, in a magazine, or on the Internet. Copy it and describe the results that it shows.

"I conducted a survey and asked 30 people if they would like a million dollars."

"I organized the results in a circle graph."

Use what you learned about circle graphs to complete Exercises 5–7 on page 366.

8.3 Lesson

Key Vocabulary
circle graph, *p. 364*

Key Idea

Circle Graphs

A **circle graph** displays data as sections of a circle. The sum of the angle measures in a circle graph is 360°.

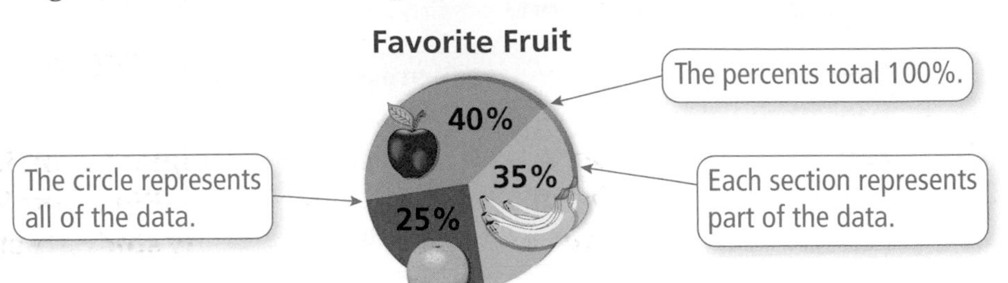

When the data are given in percents, multiply the decimal form of each percent by 360° to find the angle measure for each section.

EXAMPLE 1 Making a Circle Graph

Favorite Amusement Park	People
Disney World	25
Busch Gardens	15
Universal Studios	12
Marineland	8

The table shows the results of a survey. Display the data in a circle graph.

Step 1: Find the total number of people.

$$25 + 15 + 12 + 8 = 60$$

Step 2: Find the angle measure for each section of the graph. Multiply the fraction of people that chose each park by 360°.

Disney World

$$\frac{25}{60} \cdot 360° = 150°$$

Busch Gardens

$$\frac{15}{60} \cdot 360° = 90°$$

Universal Studios

$$\frac{12}{60} \cdot 360° = 72°$$

Marineland

$$\frac{8}{60} \cdot 360° = 48°$$

Check

$$\frac{25}{60} + \frac{15}{60} + \frac{12}{60} + \frac{8}{60} = 1$$

$$150° + 90° + 72° + 48° = 360° \checkmark$$

Step 3: Use a protractor to draw the angle measures found in Step 2 on a circle. Then label the sections.

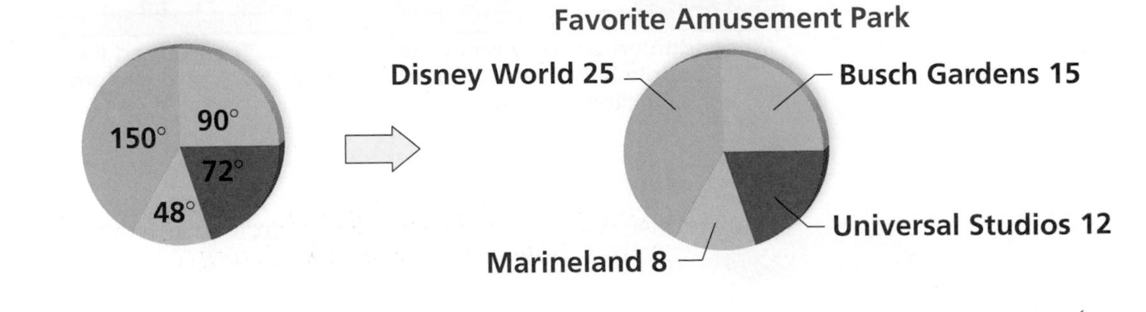

Multi-Language Glossary at BigIdeasMath.com.

On Your Own

1. The table shows the dog and cat ownership among teachers in a school. Display the data in a circle graph.

Kind of pet	Dogs only	Cats only	Both	Neither
Percent	30%	30%	15%	25%

EXAMPLE 2 Using a Circle Graph

Students chose one of four topics for their science projects. (a) What fraction of the students chose *Biotechnology*? (b) How many students are in the class? (c) How many students chose *Plate Tectonics*?

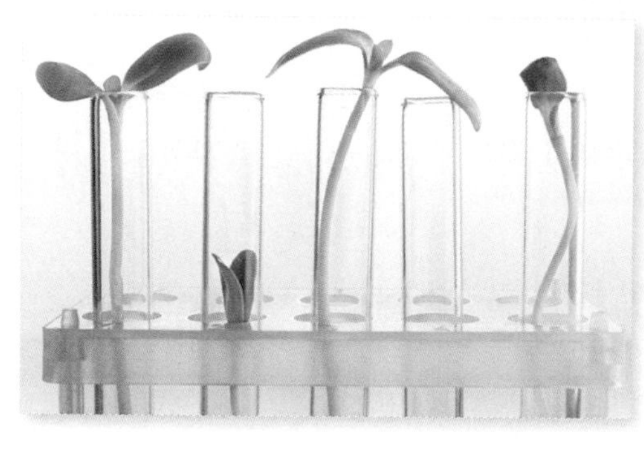

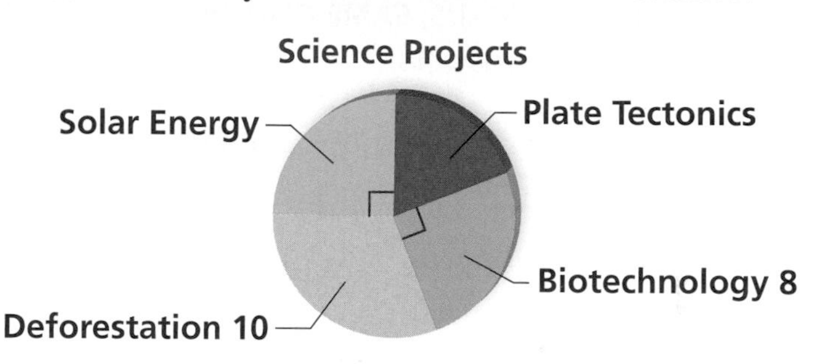

a. Because the *Biotechnology* section has a right angle, this section represents $\frac{90^\circ}{360^\circ} = \frac{1}{4}$ of the data.

∴ One-fourth of the students chose *Biotechnology*.

b. Let x be the number of students.

$\frac{1}{4}x = 8$ Use the circle graph and the results of part (a) to write an equation.

$x = 32$ Multiply each side by 4.

∴ There are 32 students in the class.

c. Because the *Solar Energy* and *Biotechnology* sections have the same angle measure, the same number of students chose each project. So, 8 students chose *Solar Energy*. Subtract to find the number of students who chose *Plate Tectonics*.

$32 - 8 - 8 - 10 = 6$

∴ Six students chose *Plate Tectonics*.

On Your Own

2. What percent chose *Deforestation*? What percent chose either *Biotechnology* or *Solar Energy*?

8.3 Exercises

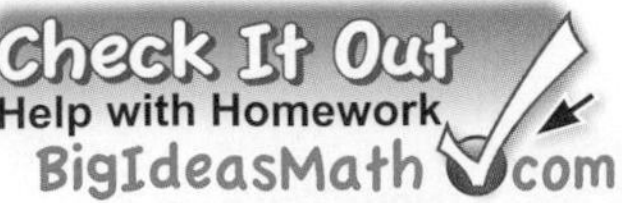

Vocabulary and Concept Check

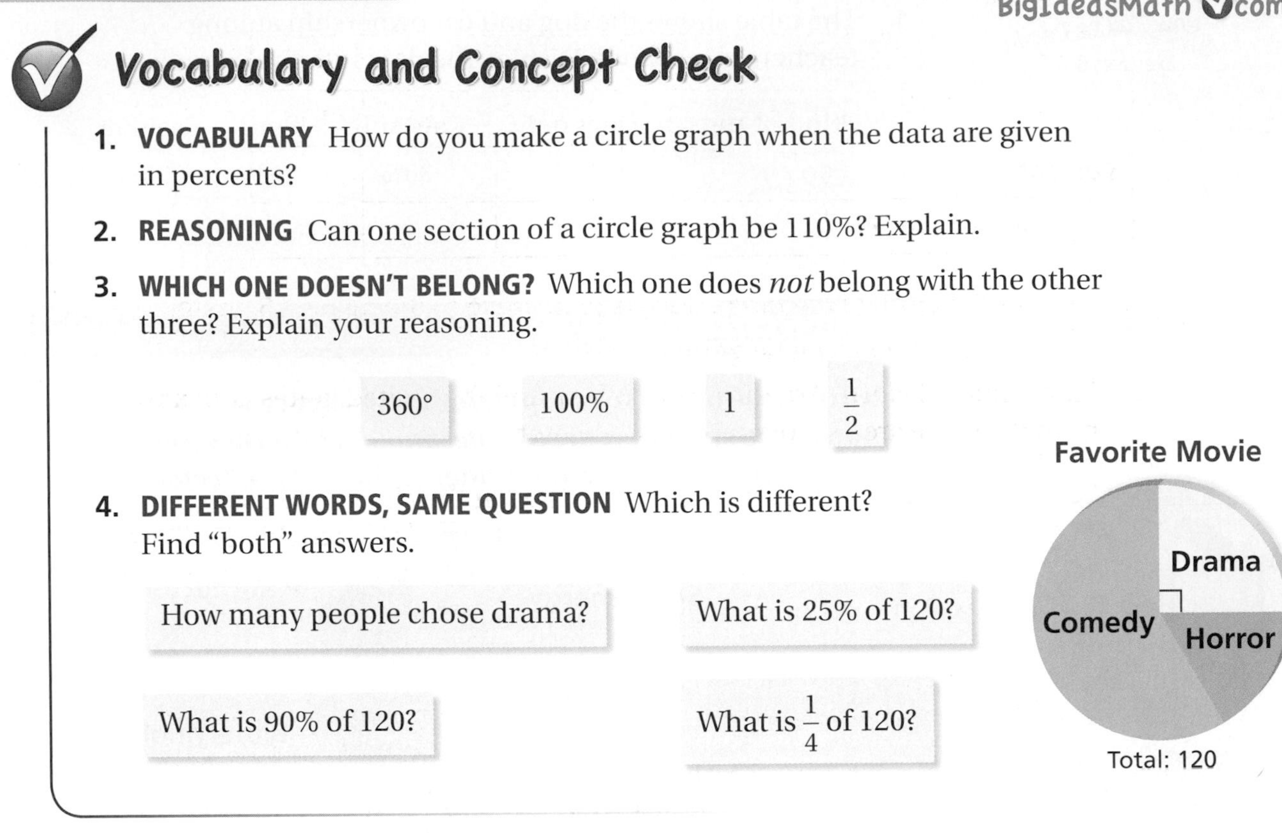

1. **VOCABULARY** How do you make a circle graph when the data are given in percents?
2. **REASONING** Can one section of a circle graph be 110%? Explain.
3. **WHICH ONE DOESN'T BELONG?** Which one does *not* belong with the other three? Explain your reasoning.

   360°   100%   1 $\frac{1}{2}$

4. **DIFFERENT WORDS, SAME QUESTION** Which is different? Find "both" answers.

   How many people chose drama?

   What is 25% of 120?

   What is 90% of 120?

   What is $\frac{1}{4}$ of 120?

Practice and Problem Solving

The circle graph shows the results of a survey on favorite fruit.

5. Which fruit is the most popular?
6. Compare the number of students who chose oranges with the number of students who chose apples.
7. The survey included 80 students. How many students chose bananas?

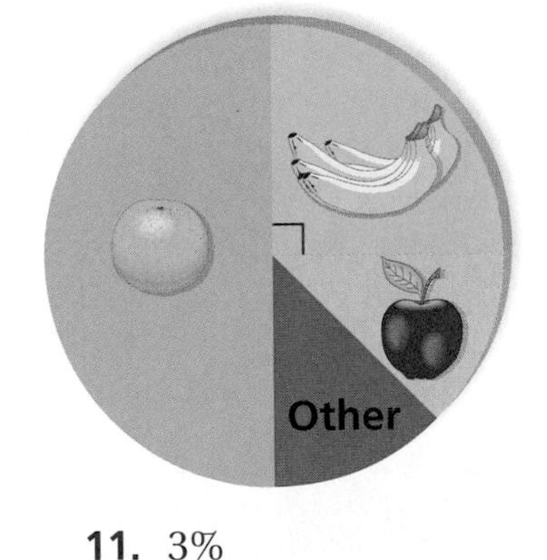

Find the angle measure that corresponds to the percent of a circle.

① 8. 20%   9. 15%   10. 70%   11. 3%

Display the data in a circle graph.

12.

Season	Rainfall (inches)
Spring	9
Summer	18
Fall	6
Winter	3

13.

Expense	Cost (dollars)
Play rights	400
Costume rental	650
Programs/tickets	300
Advertising	250
Other	400

14. LAND AREA The table shows the land areas, in millions of square miles, of all the countries in North America.

Country	United States	Mexico	Canada	Other
Land area	3.72	0.76	3.85	1.12
Percent				
Angle in circle graph				

a. Copy and complete the table. Round each angle to the nearest 5 degrees.

b. Display the data in a circle graph.

c. Find a map of North America. Do Canada and the United States appear to have the same area? Explain why or why not.

15. REASONING A survey asks a group of students what they like to do during summer vacation. The results show that 68% like to go to the beach, 45% like to go camping, 72% like to go to amusement parks, and 29% like to go to the mall. Can a circle graph be used to display these data? Explain your reasoning.

Department Store Sales

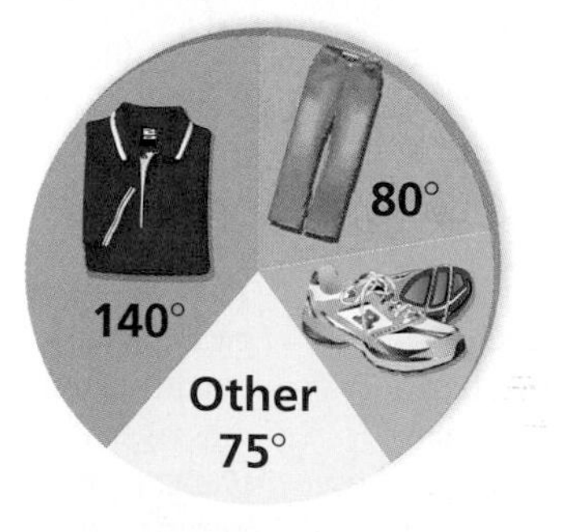

16. RETAIL A department store had $7200 in sales.

a. Find the amount collected for each category.

b. Long sleeve shirts were $\frac{1}{4}$ of the shirt sales. Find the angle measure of the section that would represent long sleeve shirts on the circle graph.

17. WRITING What math skills are needed to interpret data in a circle graph?

18. **Critical Thinking** Make a circle graph and a bar graph of the data in the table. Compare and contrast the two data displays. Which of the two better represents the data? Explain your reasoning.

Favorite Subject	Students
Art	12
English	56
Math	82
Music	28
Science	22

Fair Game Review What you learned in previous grades & lessons

Solve the proportion. *(Section 3.5)*

19. $\frac{5}{8} = \frac{x}{64}$

20. $\frac{6}{n} = \frac{51}{127.5}$

21. $\frac{138}{23} = \frac{9}{w}$

22. MULTIPLE CHOICE Which formula gives the surface area of a prism? *(Section 6.2)*

Ⓐ $S = 2\ell w + 2\ell h + 2wh$

Ⓑ $S = Ch + 2B$

Ⓒ $S = 2\pi r^2 + 2\pi rh$

Ⓓ $S = \frac{1}{2}C\ell + B$

8.4 Samples and Populations

Essential Question How can you use a survey to make conclusions about the general population?

COMMON CORE STATE STANDARDS
7.SP.1
7.SP.2
7.SP.3

1 ACTIVITY: Interpreting a Survey

Work with a partner. Read the newspaper article. Analyze the survey by answering the following questions.

a. The article does not say how many "teens and young adults" were surveyed. How many do you think need to be surveyed so that the results can represent all teens and young adults in your state? in the United States? Explain your reasoning.

b. Outline the newspaper article. List all of the important points.

c. Write a questionnaire that could have been used for the survey. Do not include leading questions. For example, "Do you think your cell phone plan is restrictive?" is a leading question.

The Daily Ti

VOL 01 No. 279 WEDNESDAY, OCTOBER 6, 2010

TEXT MESSAGING SURVEY RESULTS

A survey reports that almost one-third of teens and young adults believe that their text messaging plans are restrictive.

About 40% say their plans lead to higher cell phone bills. According to those participating in the survey, the average number of text messages sent per day is between 6 and 7.

The majority of survey participants say they would send more text messages if their cell phone plans were not as restrictive.

2 ACTIVITY: Conducting a Survey

Work with a partner. The newspaper article in Activity 1 states that the average number of text messages sent per day is between 6 and 7.

a. Does this statement seem correct to you? Explain your reasoning.

b. Plan a survey to check this statement. How will you conduct the survey?

c. Survey your classmates. Organize your data using one of the types of graphs you have studied in this chapter.

d. Write a newspaper article summarizing the results of your survey.

3 ACTIVITY: Conducting and Summarizing a Survey

Work with a partner.

- **Plan a survey to determine how many of the following texting shortcuts people know.**
- **Write a questionnaire to use in your survey.**
- **In the survey, try to determine whether *teenagers* or *people over 30* know more of the shortcuts.**
- **Conduct your survey. What can you conclude from the results? Do the results confirm your prediction?**

Lol Thx Call u l8r 2nite?

Texting Shortcuts

R	Are	U	You
4	For	L8R	Later
SUP	What's up	TTYL	Talk to you later
PLZ	Please	BRB	Be right back
C	See	LOL	Laugh out loud
IDK	I don't know	BFF	Best friends forever
JK	Just kidding	THX	Thanks
2NITE	Tonight	GR8	Great
QPSA?	Que Pasa?	4COL	For crying out loud

What Is Your Answer?

4. **IN YOUR OWN WORDS** How can you use a survey to make conclusions about the general population?

5. Find a survey in a newspaper, in a magazine, or on the Internet. Decide whether you think the conclusion of the survey is correct. Explain your reasoning.

"I'm sending my Mom a text message for Mother's Day."

"2 GR8 2 ME 2 EVR B 4GOT10. XXOO"

Use what you learned about samples and populations to complete Exercises 3–5 on page 372.

Check It Out
Lesson Tutorials
BigIdeasMath.com

Key Vocabulary
population, *p. 370*
sample, *p. 370*

Key Idea

Samples and Population

A **population** is an entire group of people or objects. A **sample** is a part of the population.

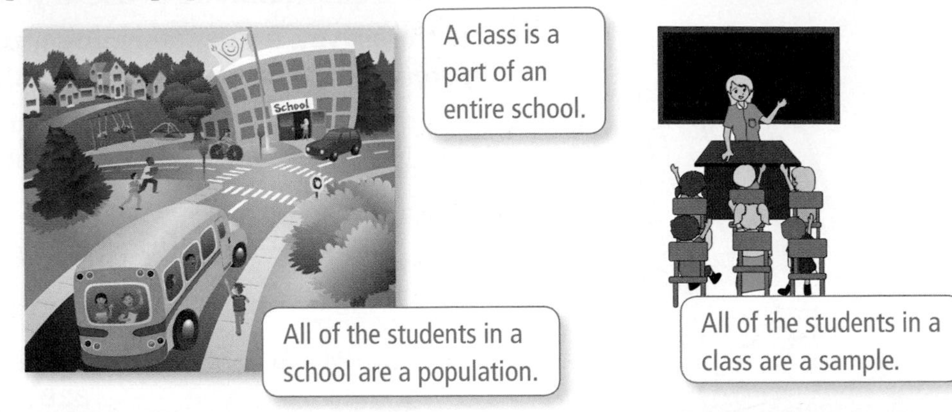

A class is a part of an entire school.
School
All of the students in a school are a population.
All of the students in a class are a sample.

EXAMPLE 1 Identifying a Population and a Sample

Response	Residents
Favor road	533
Oppose road	267

An agency wants to know the opinions of county residents on the construction of a new road. The agency surveys 800 residents. Identify the population and the sample.

The population is all county residents. The sample consists of the 800 residents surveyed by the agency.

On Your Own

Now You're Ready
Exercises 6–9

1. You want to know how many students in your school are going to the volleyball game. You survey 50 students. Ten are going to the game. The rest are not going to the game. Identify the population and the sample.

Key Idea

Reasonable Samples

A reasonable sample is

- selected at random,
- representative of the population, and
- large enough to provide accurate data.

The results of a reasonable sample are proportional to the results of the population. So, reasonable samples can be used to make predictions about the population.

Multi-Language Glossary at BigIdeasMath.com.

EXAMPLE 2 Standardized Test Practice

You want to estimate the number of students in a high school who ride the school bus. Which sample is best?

Ⓐ 4 students in the hallway

Ⓑ All students in the marching band

Ⓒ 50 seniors at random

Ⓓ 100 students at random during lunch

Choice A is not large enough to provide accurate data.

Choice B is not selected at random.

Choice C is not representative of the population because seniors are more likely to drive to school than other students.

∴ Choice Ⓓ is best. It is large and random.

EXAMPLE 3 Making Predictions

Movies per Week

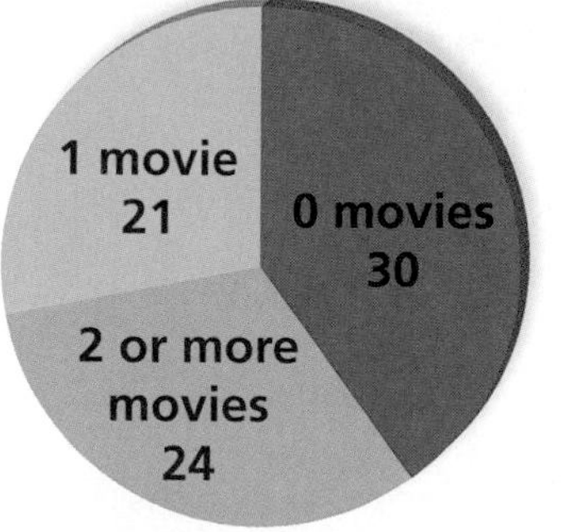

You ask 75 randomly chosen students how many movies they watch each week. There are 1200 students in the school. (a) Predict the number n of students in the school who watch one movie each week. (b) Is the prediction appropriate? Explain.

a. Find the fraction of students in the sample who watch one movie.

$$\frac{\text{Students who watch one movie}}{\text{Number of students in sample}} = \frac{21}{75}$$

Multiply to find n.

$$n = \frac{21}{75}(1200) = 336$$

∴ About 336 students watch one movie each week.

b. The sample is selected at random, representative of the population, and large enough to provide accurate data.

∴ The sample is reasonable, so the prediction is appropriate.

On Your Own

Now You're Ready
Exercises 10–13

2. **WHAT IF?** In Example 2, you want to estimate the number of seniors in a high school who ride the school bus. Which sample should you use to make a prediction?

3. In Example 3, predict the number n of students in the school who watch two or more movies each week.

8.4 Exercises

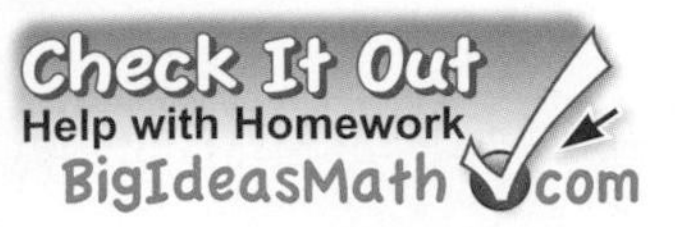

Vocabulary and Concept Check

1. **VOCABULARY** Why would you survey a sample instead of a population?
2. **CRITICAL THINKING** What should you consider when conducting a survey?

Practice and Problem Solving

The circle graph shows the results of a survey of 960 adults randomly chosen from different parts of the United States. In the survey, each adult was asked to name his or her favorite nut.

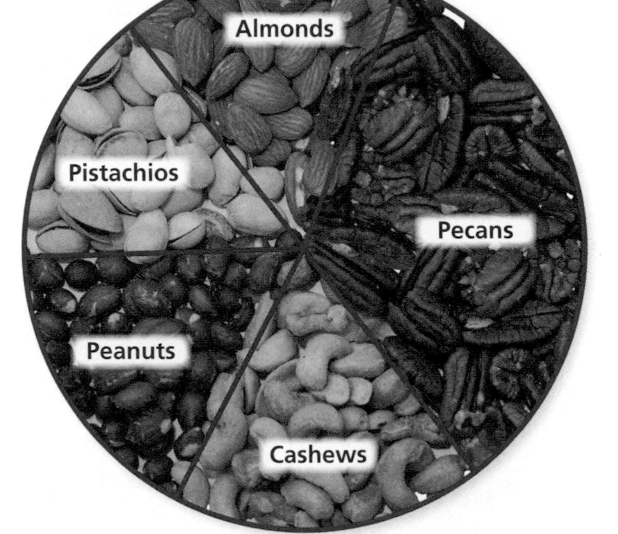

3. Do you think the results would be similar if the survey were conducted using middle school students? children in first grade? Explain your reasoning.
4. What other type of data display could be used to show the data?
5. Plan a survey to check the results of the survey. How could you conduct the survey so that the people surveyed would be chosen at random?

Identify the population and the sample.

(1) 6.

7.

150 Quarters

All quarters in circulation

8.

5 Ants

Colony of ants

9.

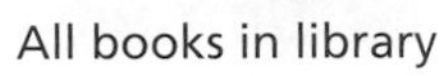

All books in library

10 library books

(3) 10. **ERROR ANALYSIS** Consider the information given in Example 3. Describe and correct the error in predicting the number n of students in the school who watch zero movies each week.

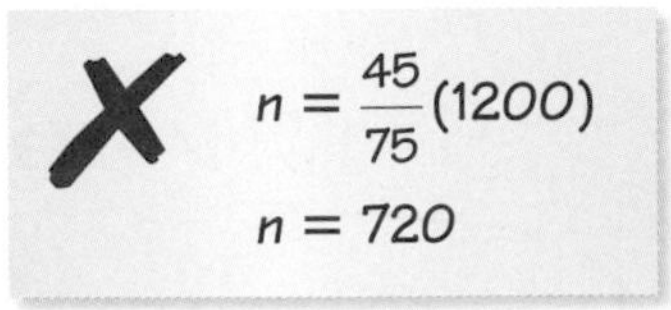

② **11. INSTRUMENT** You want to know the number of students in your school who play a musical instrument. You survey the first 15 students who arrive at a band class.

a. What is the population of your survey? the sample?

b. Is the sample reasonable? Explain.

Which sample is better for making a prediction? Explain.

12.

Predict the number of students in a school who like gym class.	
Sample A	A random sample of 8 students from the yearbook
Sample B	A random sample of 80 students from the yearbook

13.

Predict the number of defective pencils produced per day.	
Sample A	A random sample of 500 pencils from 20 machines
Sample B	A random sample of 500 pencils from 1 machine

Determine whether you would survey the population or a sample. Explain.

14. You want to know the average height of seventh-graders in the United States.

15. You want to know the favorite types of music of students in your homeroom.

16. You want to know the number of students in your state who have summer jobs.

Ticket Sales	
Adults	Students
522	210

17. THEATER A survey asked 72 randomly chosen students if they were going to attend the school play. Twelve said yes. Predict the number of students who attend the school.

18. CRITICAL THINKING Explain why 200 people with email addresses may not be a random sample.

19. Reasoning A guidance counselor surveys a random sample of 60 out of 900 high school students. Using the survey results, the counselor predicts that approximately 675 students plan to attend college. Do you agree with her prediction? Explain.

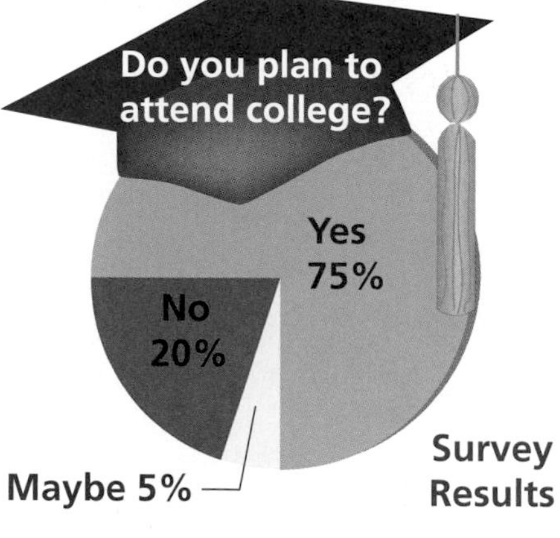

Fair Game Review What you learned in previous grades & lessons

Write the fraction as a percent. *(Skills Review Handbook)*

20. $\frac{5}{8}$ **21.** $\frac{5}{16}$ **22.** $\frac{21}{27}$ **23.** $\frac{36}{44}$

24. MULTIPLE CHOICE What is the volume of the cone? *(Section 7.4)*

Ⓐ $16\pi\text{ cm}^3$ Ⓑ $108\pi\text{ cm}^3$

Ⓒ $48\pi\text{ cm}^3$ Ⓓ $144\pi\text{ cm}^3$

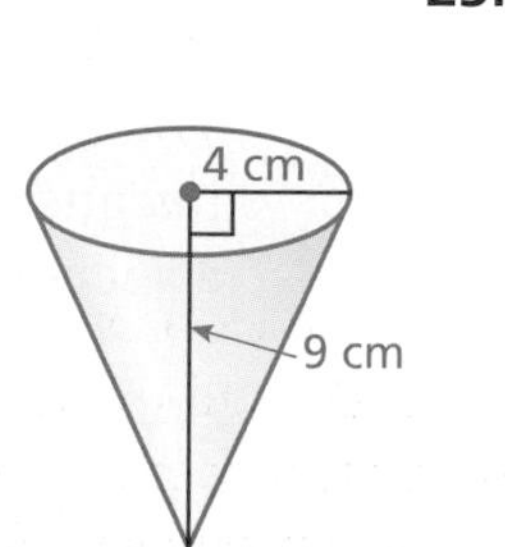

8.4b Comparing Populations

EXAMPLE 1 Making Predictions from Samples

You want to know if students at a school are in favor of building a new soccer field. You conduct two surveys. For Survey 1, you randomly ask 50 students in the lunch room. For Survey 2, you randomly ask 50 student athletes. There are 1200 students in the school.

Response	Survey 1	Survey 2
Yes	17	32
No	25	11
Not Sure	8	7

a. Use the results of each survey to predict the number n of students in the school that are in favor of building a new soccer field.

Find the fraction of students in each sample that responded "yes."

Survey 1		*Survey 2*
$\frac{17}{50}$	$\frac{\text{Responded "yes"}}{\text{Number in sample}}$	$\frac{32}{50}$

Multiply each fraction by the total number of students in the school to find n for each survey.

Survey 1

$$n = \frac{17}{50}(1200) = 408$$

Survey 2

$$n = \frac{32}{50}(1200) = 768$$

∴ Using Survey 1, you can predict that 408 students are in favor of building a new soccer field. Using Survey 2, you can predict that 768 students are in favor of building a new soccer field.

b. Which prediction is more reliable? Explain.

The sample in Survey 1 is selected at random, representative of the population, and large enough to provide accurate data. This sample is reasonable.

The sample in Survey 2 is not representative of the population because student athletes are more likely to be in favor of building a soccer field. This sample is not reasonable.

∴ The prediction from Survey 1 is more reliable because it uses a reasonable sample.

EXAMPLE 2 Using Samples to Compare Populations

The double box-and-whisker plot shows the gas mileages of random samples of cars and trucks. Compare the gas mileages of cars to the gas mileages of trucks.

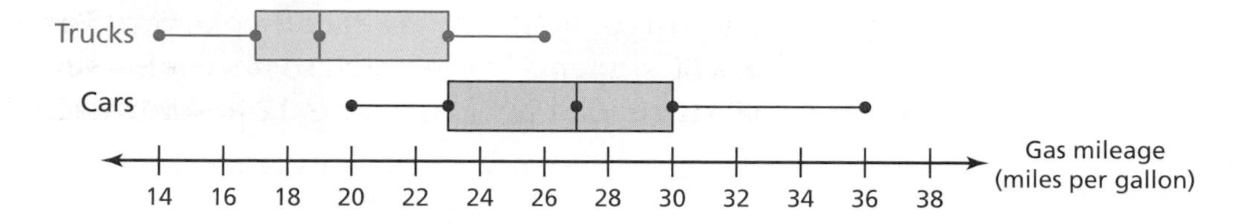

Trucks: The third quartile is 23 miles per gallon. So, 25% of the trucks get 23 miles per gallon or *more*.

Cars: The first quartile is 23 miles per gallon. So, 75% of the cars get 23 miles per gallon or *more*.

∴ In general, cars get better gas mileage than trucks.

Practice

1. **SPORTS** You want to survey students in your grade about their favorite sport. Describe samples that are (a) reasonable and (b) not reasonable.
2. **PROJECT** Conduct the surveys in Exercise 1 at your school.
   a. Use the results of each survey to predict the favorite sport of the students in your grade.
   b. Do you think your predictions are reasonable? Explain.
3. **HEIGHT** The double box-and-whisker plot shows the heights of random samples of boys and girls in a school. Compare the heights of boys to the heights of girls.

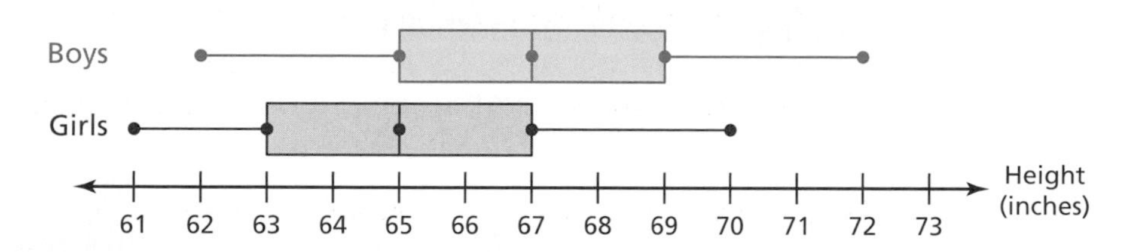

4. **GRADES** Each table shows a random sample of test grades in a class. Create a double box-and-whisker plot of the data. Compare the two data sets.

Grades in Mr. Smith's Class			
72	68	84	87
76	75	52	73
88	84	69	71
76	78	86	82

Grades in Mrs. Higsbee's Class			
85	88	93	78
76	65	71	86
90	96	85	88
79	90	82	94

8.3–8.4 Quiz

Check It Out
Progress Check
BigIdeasMath.com

Identify the population and the sample. *(Section 8.4)*

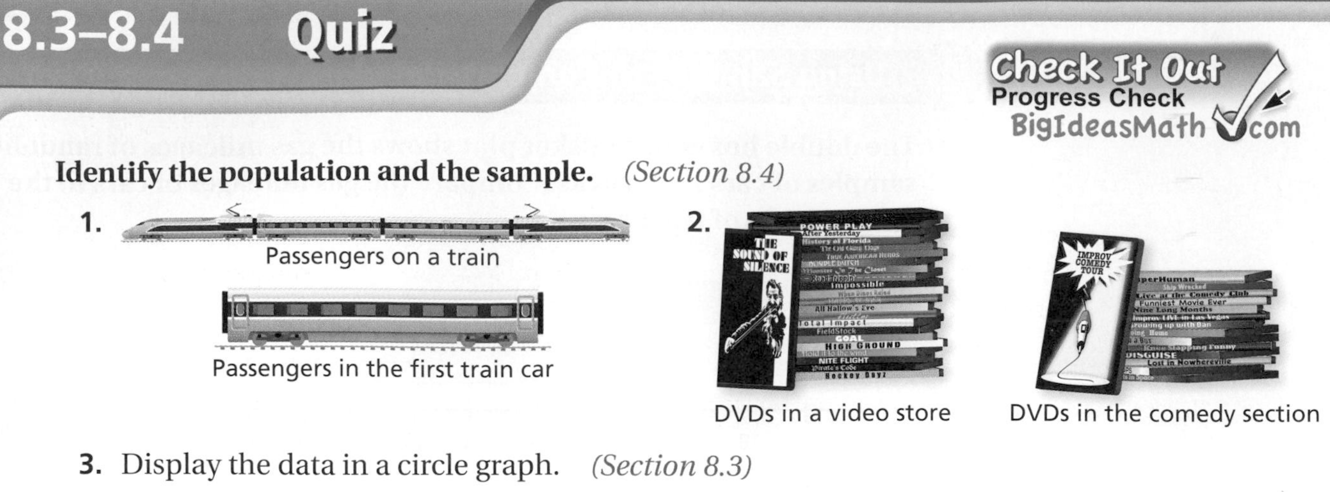

3. Display the data in a circle graph. *(Section 8.3)*

Favorite Book Genre	Students
Fantasy	20
Historical Fiction	10
Mystery	15
Nonfiction	5

Which sample is better for making a prediction? Explain. *(Section 8.4)*

4.

Predict the number of people who plan to vote in this year's election.	
Sample A	A random sample of 5000 registered voters
Sample B	A random sample of 50,000 United States citizens

5.

Predict the number of defective light bulbs produced per day.	
Sample A	A sample of the last 1000 light bulbs produced during a day
Sample B	A random sample of 1000 light bulbs produced throughout a day

6. EXERCISE EQUIPMENT You want to know how many students in your school support the purchase of new exercise equipment for the gym this year. You survey the first 20 students who arrive for football team tryouts. *(Section 8.4)*

a. What is the population for your survey? the sample?

b. Is the sample reasonable? Explain.

7. WORLD LANGUAGES Every student in a middle school takes one world language class. *(Section 8.3)*

a. What fraction of the students take French?

b. How many students are in the school?

c. How many students take Spanish?

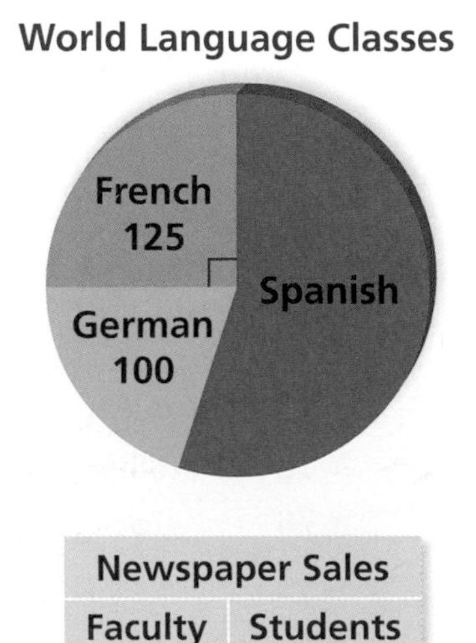

8. NEWSPAPERS A survey asks 48 randomly chosen students if they plan to buy a school newspaper this week. Of the 48 surveyed, 32 plan to buy a school newspaper. Predict the number of students enrolled at the school. *(Section 8.4)*

Newspaper Sales	
Faculty	Students
18	360

8 Chapter Review

Check It Out
Vocabulary Help
BigIdeasMath.com

Review Key Vocabulary

stem-and-leaf plot, *p. 350*
stem, *p. 350*
leaf, *p. 350*
histogram, *p. 356*
circle graph, *p. 364*
population, *p. 370*
sample, *p. 370*

Review Examples and Exercises

8.1 Stem-and-Leaf Plots (pp. 348–353)

Day	DVDs Rented
Sun.	50
Mon.	19
Tue.	25
Wed.	28
Thu.	39
Fri.	53
Sat.	50

Make a stem-and-leaf plot of the number of DVDs rented each day at a store.

Step 1: Order the data. 19, 25, 28, 39, 50, 50, 53

Step 2: Choose the stems and leaves. Because the data range from 19 to 53, use the *tens* digits for the stems and the *ones* digits for the leaves.

Step 3: Write the stems to the *left* of the vertical line.

Step 4: Write the leaves for each stem to the *right* of the vertical line.

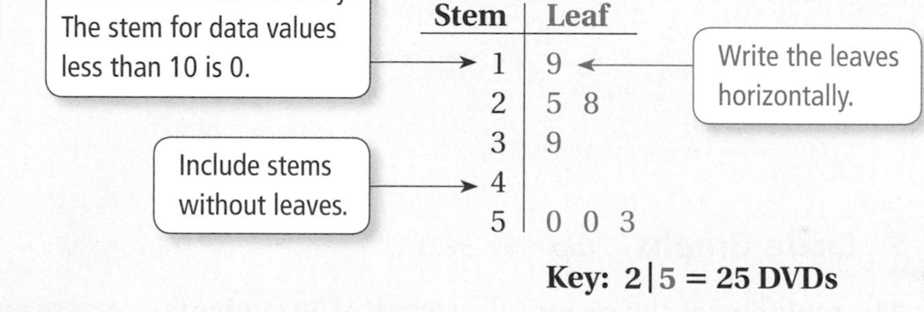
Order the stems vertically. The stem for data values less than 10 is 0.
Include stems without leaves.
Write the leaves horizontally.

DVDs Rented

Stem	Leaf
1	9
2	5 8
3	9
4	
5	0 0 3

Key: 2|5 = 25 DVDs

Exercises

Make a stem-and-leaf plot of the data.

1.

Hats Sold Each Day			
5	18	12	15
21	30	8	12
13	9	14	25

2.

Ages of Park Volunteers			
13	17	40	15
48	21	19	52
13	55	60	20

The stem-and-leaf plot shows the weights (in pounds) of yellowfin tuna caught during a fishing contest.

3. How many tuna weigh less than 90 pounds?
4. What is the median weight of the tuna?

Weights of Tuna

Stem	Leaf
7	6
8	0 2 5 7 9
9	5 6
10	2

Key: 8|5 = 85 pounds

8.2 Histograms (pp. 354–359)

The frequency table shows the number of crafts each member of the Craft Club made for a fundraiser. Display the data in a histogram.

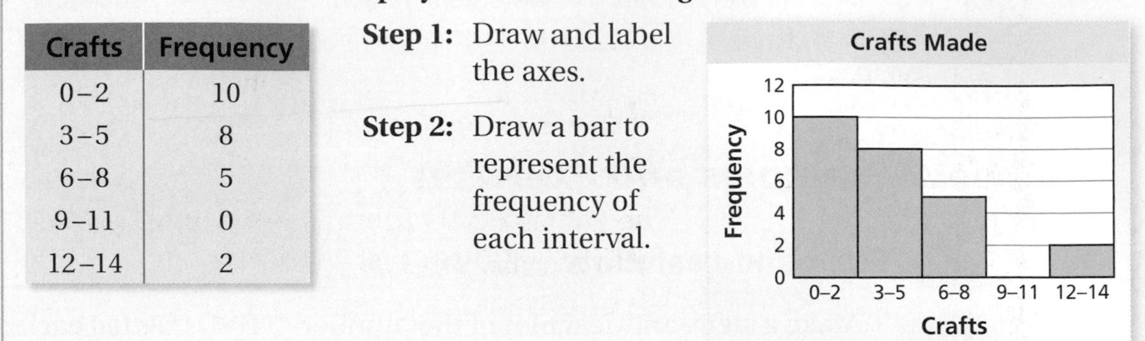

Crafts	Frequency
0–2	10
3–5	8
6–8	5
9–11	0
12–14	2

Step 1: Draw and label the axes.

Step 2: Draw a bar to represent the frequency of each interval.

Exercises

Display the data in a histogram.

5.

Heights of Gymnasts	
Heights (in.)	**Frequency**
50–54	1
55–59	8
60–64	5
65–69	2

6.

Minutes Studied	
Minutes	**Frequency**
0–19	5
20–39	9
40–59	12
60–79	3

8.3 Circle Graphs (pp. 362–367)

The table shows the results of a survey of 50 students. Display the data in a circle graph.

Favorite P.E. Activity	Students
Badminton	15
Volleyball	10
Kickball	25

Step 1: Find the angle measure for each section of the graph.

Multiply the fraction of students who chose each activity by 360°.

Badminton

$\frac{15}{50} \cdot 360^\circ = 108^\circ$

Volleyball

$\frac{10}{50} \cdot 360^\circ = 72^\circ$

Kickball

$\frac{25}{50} \cdot 360^\circ = 180^\circ$

Step 2: Use a protractor to draw the angle measures on a circle. Label the sections.

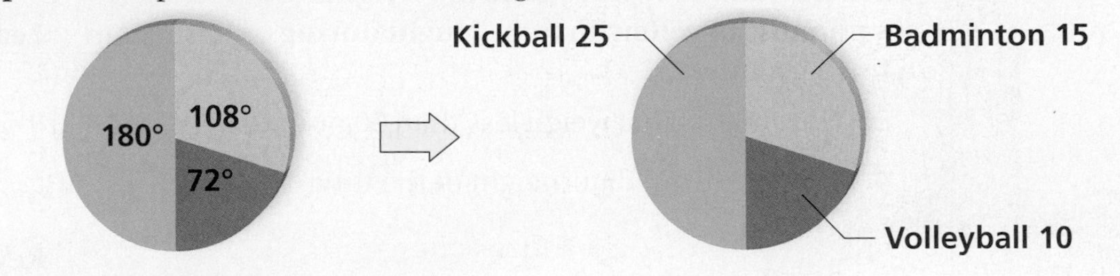

Exercises

Display the data in a circle graph.

7.

Singing Part	Students
Soprano	12
Alto	20
Tenor	18
Bass	10

8.

Candidate	Votes
Jon	60
Isabelle	35
Carmen	50
Ernesto	55

8.4 Samples and Populations *(pp. 368–373)*

You ask 80 randomly chosen students how many pets they have. There are 600 students in the school. (a) Predict the number n of students in the school who have exactly one pet. (b) Is the prediction appropriate? Explain.

a. Find the fraction of students in the sample who have exactly one pet.

$$\frac{\text{Students who have exactly one pet}}{\text{Number of students in sample}} = \frac{42}{80}$$

Multiply to find n.

$$n = \frac{42}{80}(600) = 315$$

∴ About 315 students in the school have exactly one pet.

b. The sample is selected at random, is representative of the population, and is large enough to provide accurate data.

∴ The sample is reasonable, so the prediction is appropriate.

Pets

2 or more pets 24

0 pets 14

1 pet 42

Exercises

9. Use the information in the Example above. Predict the number x of students in the school who have two or more pets.

10. Your principal wants to know how many parents plan to attend Back-to-School Night. The principal surveys 50 parents and finds that 40 plan to attend. Identify the population and the sample.

11. Which sample is better for making a prediction? Explain.

Predict the number of people in your town who support building a new library.	
Sample A	A random sample of 500 people in your town
Sample B	A random sample of 5000 people in your state

8 Chapter Test

Make a stem-and-leaf plot of the data.

1.

Quiz Scores (%)			
96	88	80	72
80	94	92	100
76	80	68	90

2.

CDs Sold Each Day				
45	31	29	38	38
67	40	62	45	60
40	39	60	43	48

3. Display the data in a histogram.

Television Watched Per Week	
Hours	**Frequency**
0–9	14
10–19	16
20–29	10
30–39	8

4. Display the data in a circle graph.

Category	Amount Spent ($)
Clothing	30
Entertainment	10
Food	5
Savings	15

5. Which sample is better for making a prediction? Explain.

Predict the number of students in your school who play at least one sport.	
Sample A	A random sample of 10 students from the school student roster
Sample B	A random sample of 80 students from the school student roster

6. WATER The histogram shows the number of glasses of water that the students in a class drink in one day.

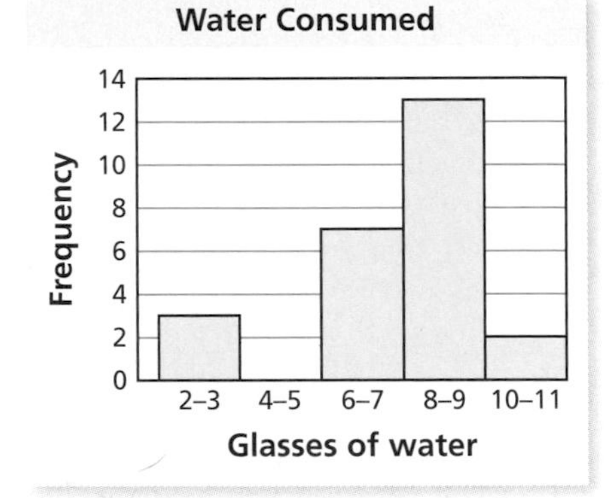

a. Which interval contains the fewest data values?

b. How many students are in the class?

c. Health experts recommend drinking at least 8 glasses of water per day. What percent of the students drink the recommended amount?

7. FIELD TRIP Of 60 randomly chosen students surveyed, 16 chose the aquarium as their favorite field trip. There are 720 students in the school. Predict the number of students in the school who would choose the aquarium as their favorite field trip.

8. MALL There are 240 stores in a mall.

a. Find the number of stores in each category.

b. Electronics stores make up $\frac{1}{5}$ of the "Other" category. Find the angle measure of the section that would represent electronics.

8 Standardized Test Practice

1. Which deposit will earn the most simple interest for the given terms and annual interest rates? *(7.RP.3)*

   A. \$3300 for 1 year at 4%

   B. \$2000 for 1 year at 5.5%

   C. \$2500 for 6 months at 8%

   D. \$3000 for 18 months at 3%

2. The band instructor made the circle graph below to show the percent of students in each section of the Jazz Band.

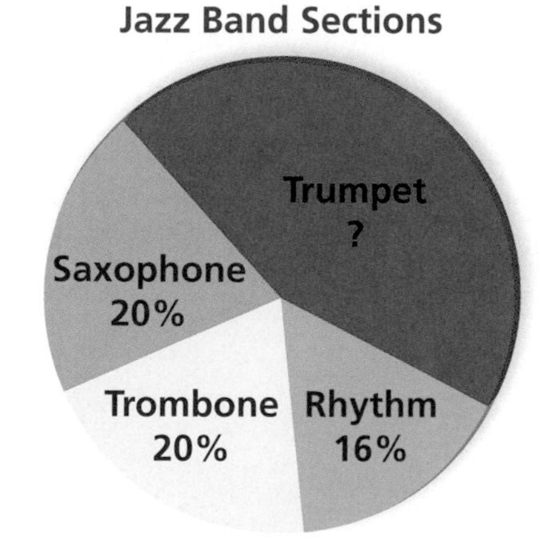

   There are 4 students in the rhythm section. What is the number of students in the trumpet section? *(7.RP.3)*

   F. 11

   G. 12

   H. 28

   I. 44

3. One of the leaves is missing in the stem-and-leaf plot.

   The median of the data set represented by the stem-and-leaf plot is 38. What is the value of the missing leaf? *(7.SP.1)*

Stem	Leaf
1	3 4
2	
3	4 5 7 7 7 ? 9
4	0 1 1 4
5	0 2 3

Key: 1 | 4 = 14

4. Mario was solving the equation in the box below.

$$3y + 4 + y = 5 - 11$$
$$3y + 4 = -6$$
$$3y = -10$$
$$y = -\frac{10}{3}$$

What should Mario do to correct the error that he made? *(7.EE.4a)*

A. Divide -6 by 3 to get 2.

B. Add $3y$ and y to get $4y$.

C. Subtract 4 from -6 to get -2.

D. Simplify $3y + 4 + y$ to get $7y$.

5. The president of a service organization made a large bowl of fruit punch for a party. She needs to decide whether to serve the punch in 6-ounce servings or 4-ounce servings. She determined that she could serve 96 guests with 6-ounce servings. How many more guests could she serve with 4-ounce servings than with 6-ounce servings? *(7.RP.2a)*

F. 144

G. 64

H. 48

I. 32

6. At the end of the school year, your teacher counted up the number of absences for each student. The results are shown in the histogram below.

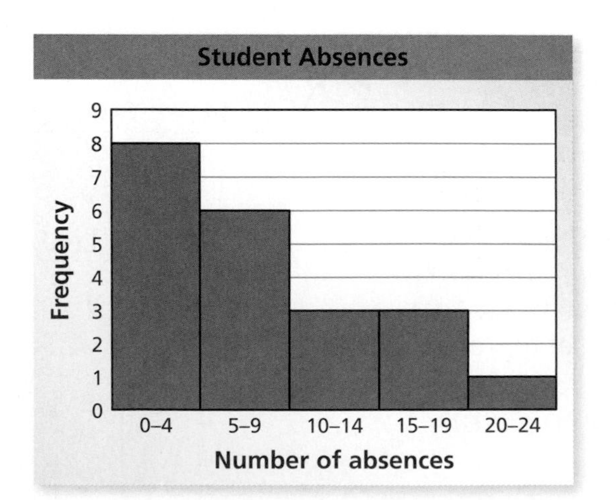

Based on the histogram, how many students had fewer than 10 absences? *(7.SP.2)*

7. What is the value of the expression below when $a = -6$ and $b = 4$? *(7.EE.3)*

$$\frac{|2 - a|}{-|b|}$$

A. -2

B. -1

C. 1

D. 2

8. A right square pyramid and its dimensions are shown below.

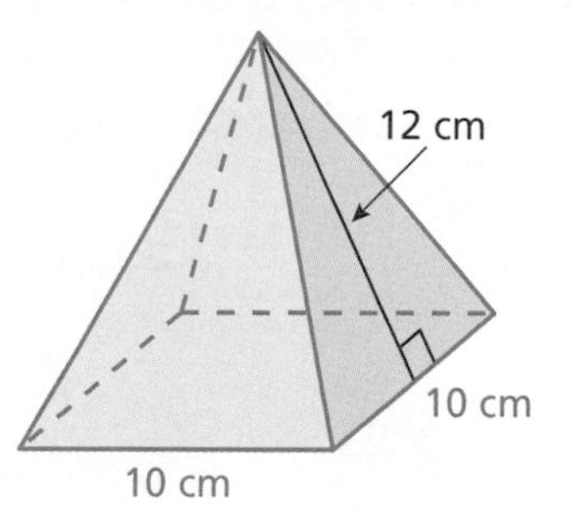

What is the surface area of the pyramid? *(7.G.6)*

F. 160 cm^2

G. 240 cm^2

H. 340 cm^2

I. 400 cm^2

9. Suppose that the volume of a right rectangular prism is V. The lengths of all the edges of the prism are then doubled. What is the volume of the new prism that is created? *(7.G.6)*

A. $2V$

B. $4V$

C. $6V$

D. $8V$

10. A stem-and-leaf plot is shown.

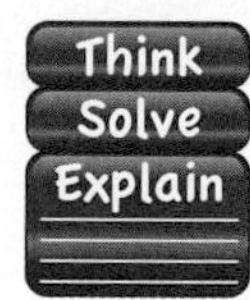

For Parts A–D, explain how the stem-and-leaf plot can help to find each measure. Then find each measure using the stem-and-leaf plot. Show your work and explain your reasoning for each answer. *(7.SP.2)*

Part A range

Part B mode

Part C median

Part D mean

Stem	Leaf
4	5 8 8
5	2 7 7 7 8
6	0 1 6 8
7	
8	3 3 4 8 9 9
9	0 8

Key: 5|2 = 52

9 Probability

9.1 Introduction to Probability

9.2 Theoretical Probability

9.3 Experimental Probability

9.4 Independent and Dependent Events

RAWR!

"If there are 7 cats in a sack and I draw one at random,..."

It's ZERO! Because I'm not getting in a sack with 6 other cats!

"... what is the probability that I will draw you?"

CUBE CONSTRUCTION PLANS

"I'm just about finished making my two number cubes."

"Now, here's how the game works. You toss the two cubes."

I've got a better idea. Let's just toss one number cube.

"If the sum is even I win. If it's odd, you win."

What You Learned Before

"Let's spin to decide what we will have for lunch."

Simplifying Fractions (6.NS.4)

Example 1 Simplify $\frac{12}{36}$.

$$\frac{12 \div 12}{36 \div 12} = \frac{1}{3}$$

Simplify fractions by using the Greatest Common Factor.

Example 2 Simplify $\frac{33}{60}$.

$$\frac{33 \div 3}{60 \div 3} = \frac{11}{20}$$

Writing Ratios (7.RP.1)

Example 3

a. Write the ratio of girls to boys in Classroom A.

	Boys	Girls
Classroom A	14	11
Classroom B	12	8

$$\frac{\text{Girls in Classroom A}}{\text{Boys in Classroom A}} = \frac{11}{14}$$

So, the ratio of girls to boys in Classroom A is $\frac{11}{14}$.

b. Write the ratio of boys in Classroom B to the total number of students in both classes.

$$\frac{\text{Boys in Classroom B}}{\text{Total number of students}} = \frac{12}{14 + 11 + 12 + 8} = \frac{12}{45} = \frac{4}{15}$$

Write in simplest form.

So, the ratio of boys in Classroom B to the total students is $\frac{4}{15}$.

Try It Yourself

Write the ratio in simplest form.

1. Baseballs to footballs
2. Footballs to total pieces of equipment
3. Sneakers to ballet slippers
4. Sneakers to total number of shoes

9.1 Introduction to Probability

COMMON CORE STATE STANDARDS
7.SP.5

Essential Question

How can you predict the results of spinning a spinner?

1 ACTIVITY: Helicopter Flight

Play with a partner.

- **You begin flying the helicopter at (0, 0) on the coordinate plane. Your goal is to reach the cabin at (20, 14).**
- **Spin any one of the spinners. Move one unit in the indicated direction.**
- **If the helicopter encounters any obstacles, you must start over.**
- **Record the number of moves it takes to land exactly on (20, 14).**
- **After you have played once, it is your partner's turn to play.**
- **The player who finishes in the fewest moves wins.**

Up
Forward
Reverse
Down

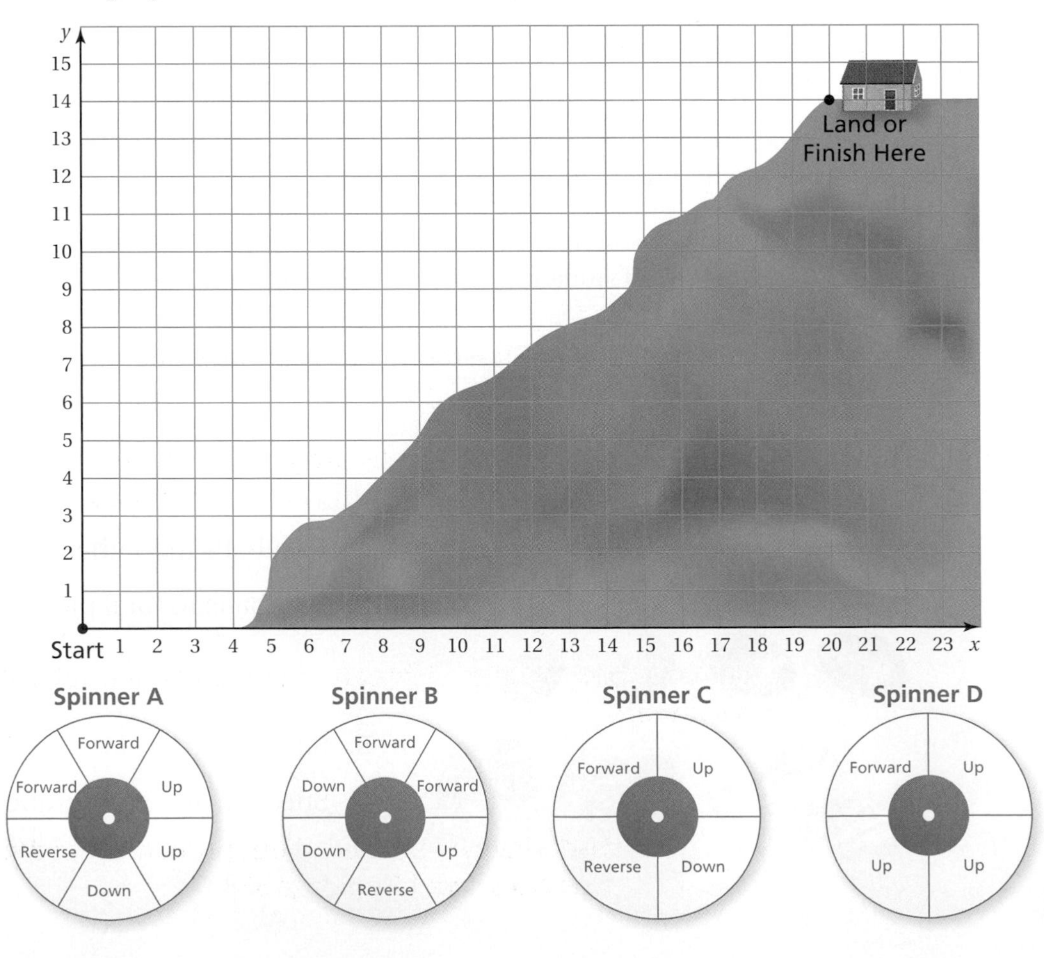

2 ACTIVITY: Analyzing the Spinners

Work with a partner.

a. How are the spinners in Activity 1 alike? How are they different?

b. Which spinner will advance the helicopter to the finish fastest? Why?

c. If you want to move up, which spinner should you spin? Why?

d. Spin each spinner 50 times and record the results.

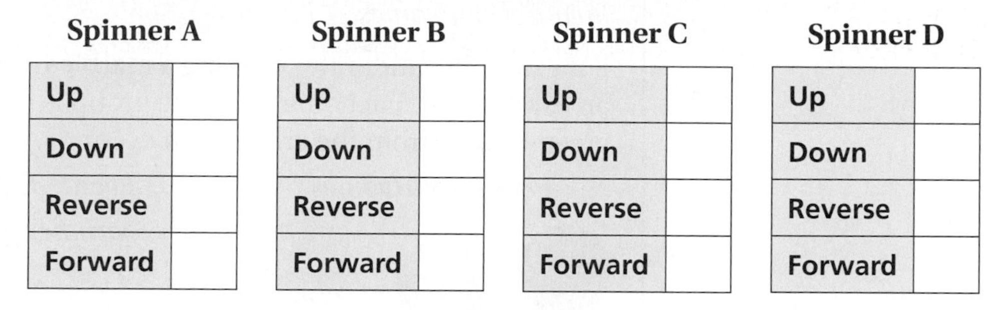

Spinner A

Up	
Down	
Reverse	
Forward	

Spinner B

Up	
Down	
Reverse	
Forward	

Spinner C

Up	
Down	
Reverse	
Forward	

Spinner D

Up	
Down	
Reverse	
Forward	

e. Organize the results from part (d) in a bar graph for each spinner.

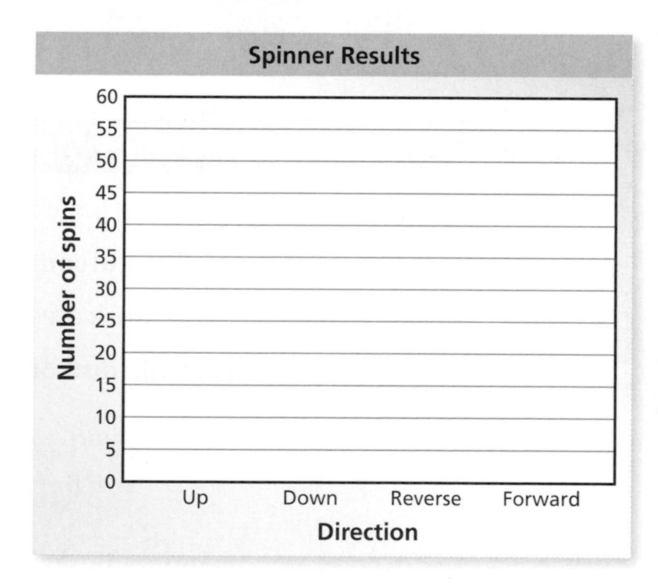

f. After analyzing the results, would you change your strategy in the helicopter flight game? Explain why or why not.

What Is Your Answer?

3. **IN YOUR OWN WORDS** How can you predict the results of spinning a spinner?

Practice Use what you learned about probability and spinners to complete Exercises 4 and 5 on page 388.

Key Vocabulary
experiment, *p. 386*
outcomes, *p. 386*
event, *p. 386*
probability, *p. 387*

Key Ideas

Outcomes and Events

An **experiment** is an activity with varying results. The possible results of an experiment are called **outcomes**. A collection of one or more outcomes is an **event**. The outcomes of a specific event are called *favorable outcomes*.

For example, randomly selecting a marble from a group of marbles is an experiment. Each marble in the group is an outcome. Selecting a green marble from the group is an event.

Possible outcomes

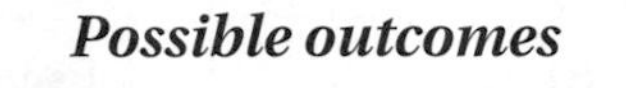

Event: Choosing a green marble

Number of favorable outcomes: 2

EXAMPLE 1 Identifying Outcomes

You roll the number cube.

a. What are the possible outcomes?

The six possible outcomes are rolling a 1, 2, 3, 4, 5, and 6.

b. What are the favorable outcomes of rolling an even number?

even	*not* even
2, 4, 6	1, 3, 5

∴ The favorable outcomes of the event are rolling a 2, 4, and 6.

c. What are the favorable outcomes of rolling a number greater than 5?

greater than 5	*not* greater than 5
6	1, 2, 3, 4, 5

∴ The favorable outcome of the event is rolling a 6.

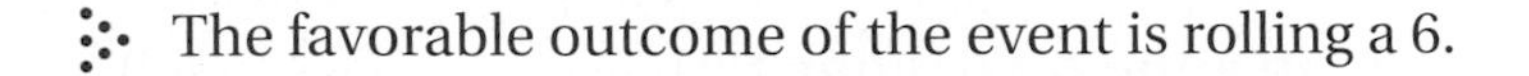

On Your Own

Now You're Ready
Exercises 6–12

1. You randomly choose a letter from a hat that contains the letters A through K. (a) What are the possible outcomes? (b) What are the favorable outcomes of choosing a vowel?

Multi-Language Glossary at BigIdeasMath.com.

2 Counting Outcomes

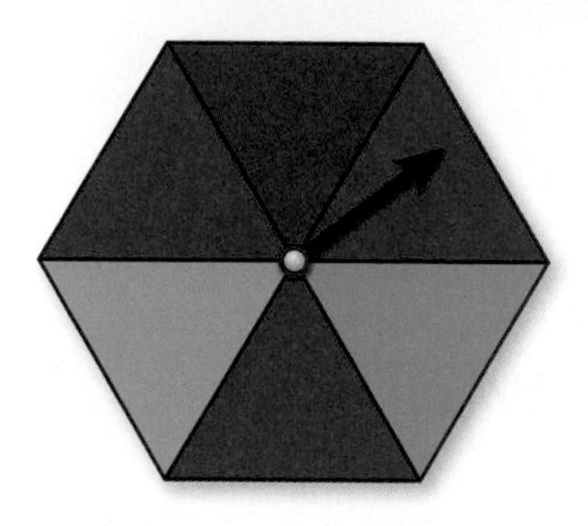

You spin the spinner.

a. How many possible outcomes are there?

The spinner has 6 sections. So, there are 6 possible outcomes.

b. In how many ways can spinning red occur?

The spinner has 3 red sections. So, spinning red can occur in 3 ways.

c. In how many ways can spinning *not* purple occur? What are the favorable outcomes of spinning *not* purple?

The spinner has 5 sections that are *not* purple. So, spinning *not* purple can occur in 5 ways.

purple	*not* purple
purple	red, red, red, green, blue

The favorable outcomes of the event are red, red, red, green, and blue.

On Your Own

2. You randomly choose a marble.
   a. How many possible outcomes are there?
   b. In how many ways can choosing blue occur?
   c. In how many ways can choosing *not* yellow occur? What are the favorable outcomes of choosing *not* yellow?

Key Idea

Probability

The **probability** of an event is a number that measures the likelihood that the event will occur. Probabilities are between 0 and 1, including 0 and 1. The diagram relates likelihoods (above the diagram) and probabilities (below the diagram).

Impossible	Unlikely	Equally likely to happen or not happen	Likely	Certain
0	$\frac{1}{4}$	$\frac{1}{2}$	$\frac{3}{4}$	1
0	0.25	0.5	0.75	1
0%	25%	50%	75%	100%

Study Tip

Probabilities can be written as fractions, decimals, or percents.

9.1 Exercises

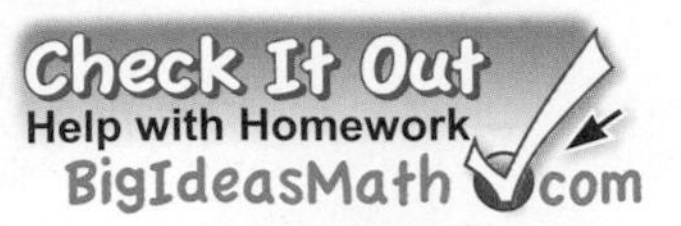

Vocabulary and Concept Check

1. **VOCABULARY** Is rolling an even number on a number cube an *outcome* or an *event*? Explain.
2. **REASONING** Can the probability of an event be 1.5? Explain.
3. **OPEN-ENDED** Give a real-life example of an event that is impossible. Give a real-life example of an event that is certain.

Practice and Problem Solving

Use the spinners shown.

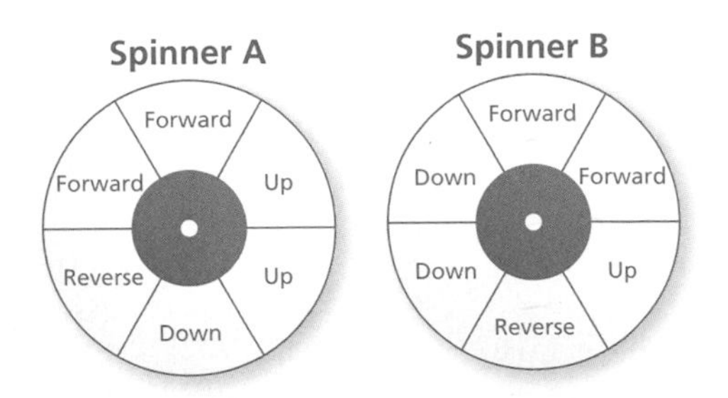

4. You want to move down. Which spinner should you spin? Explain.
5. You want to move forward. Does it matter which spinner you spin? Explain.

① 6. What are the possible outcomes of randomly choosing one of the tiles shown below?

You randomly choose one of the tiles shown above. Find the favorable outcomes of the event.

7. Choosing a 6
8. Choosing an odd number
9. Choosing a number greater than 5
10. Choosing an odd number less than 5
11. Choosing a number less than 3
12. Choosing a number divisible by 3

You randomly choose one marble from the bag. (a) Find the number of ways the event can occur. (b) Find the favorable outcomes of the event.

② 13. Choosing blue
14. Choosing green
15. Choosing purple
16. Choosing yellow
17. Choosing *not* red
18. Choosing *not* blue

19. **ERROR ANALYSIS** Describe and correct the error in finding the number of ways that choosing *not* purple can occur.

✗

purple	*not* purple
purple	red, blue, green, yellow

Choosing not purple can occur in 4 ways.

Color Distinguishment Help at BigIdeasMath.com.

20. COINS You have 10 coins in your pocket. Five are Susan B. Anthony Dollars, two are Golden Dollars featuring Sacagawea, and three are Presidential Dollars. You randomly choose a coin. In how many ways can choosing *not* a Presidential Dollar occur?

Susan B. Anthony Dollar

Golden Dollar Featuring Sacagawea*

Presidential Dollar

Spinner A

Tell whether the statement is *true* or *false*. If it is false, change the italicized word to make the statement true.

21. There are *three* possible outcomes of spinning Spinner A.

22. Spinning *red* can occur in four ways on Spinner B.

23. Spinning blue and spinning *green* are equally likely on Spinner A.

24. It is *impossible* to spin purple on Spinner B, so it is certain to spin not purple on spinner B.

25. LIKELIHOOD There are more red sections on Spinner B than on Spinner A. Does this mean that you are more likely to spin red on Spinner B? Explain.

Spinner B

26. MUSIC A bargain bin contains classical and rock CDs. There are 60 CDs in the bin. You are equally likely to randomly choose a classical CD or a rock CD from the bin. How many of the CDs are classical CDs?

27. Reasoning You randomly choose one of the cards. Then, you randomly choose a second card. Describe how the number of possible outcomes changes after the first card is chosen.

Fair Game Review What you learned in previous grades & lessons

Multiply. *(Section 2.3)*

28. $\frac{1}{2} \times 2$ **29.** $\frac{5}{6} \times 36$ **30.** $-\frac{4}{5} \times 25$ **31.** $\frac{1}{8} \times (-28)$

32. MULTIPLE CHOICE You are making half of a recipe that requires $\frac{3}{4}$ cup of sugar. How much sugar should you use? *(Section 2.3)*

Ⓐ $\frac{3}{8}$ cup Ⓑ $\frac{5}{8}$ cup Ⓒ $\frac{5}{4}$ cups Ⓓ $\frac{3}{2}$ cups

9.2 Theoretical Probability

Essential Question How can you find a theoretical probability?

COMMON CORE STATE STANDARDS
7.SP.7a
7.SP.7b

1 ACTIVITY: Black and White Spinner Game

Work with a partner. You work for a game company. You need to create a game that uses the spinner below.

a. Write rules for a game that uses the spinner. Then play it.

b. After playing the game, do you want to revise the rules? Explain.

c. Each pie-shaped section of the spinner is the same size. What is the measure of the central angle of each section?

d. What is the probability that the spinner will land on 1? Explain.

2 ACTIVITY: Changing the Spinner

Work with a partner. For each spinner, find the probability of landing on each number. Do your rules from Activity 1 make sense for these spinners? Explain.

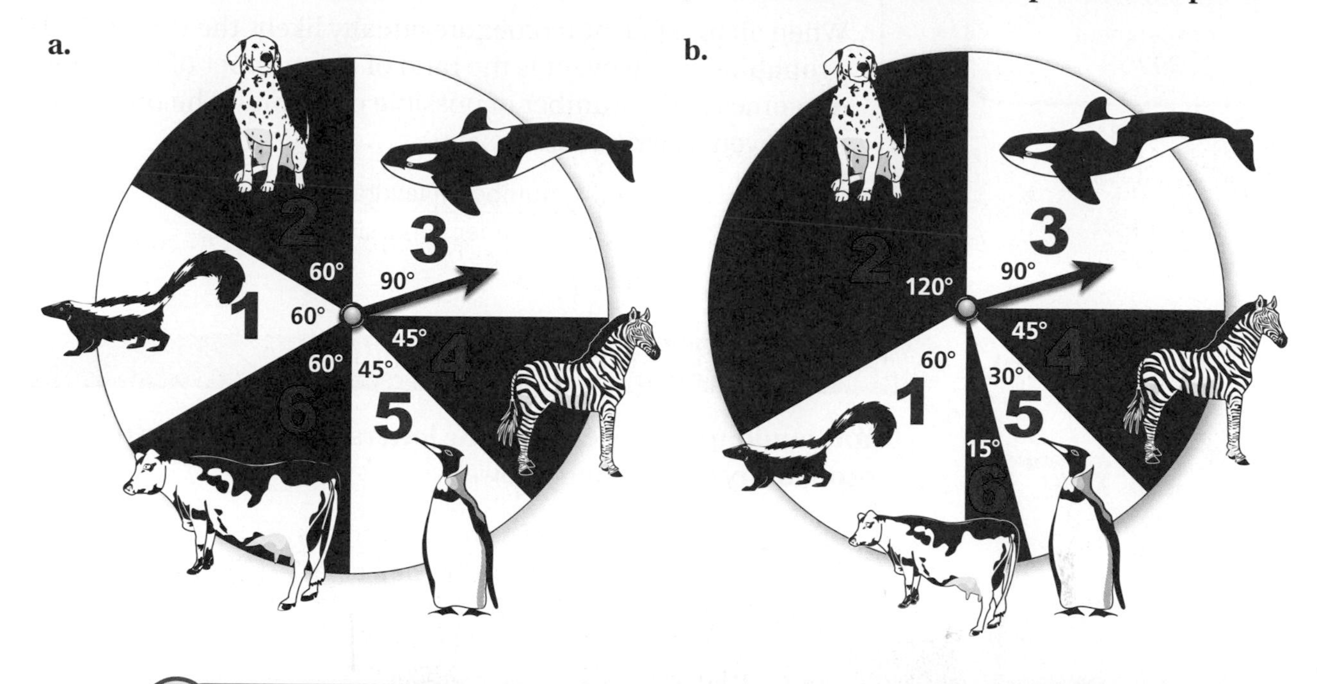

3 ACTIVITY: Is This Game Fair?

Work with a partner. Apply the following rules to each spinner in Activities 1 and 2. Is the game fair? If not, who has the better chance of winning?

- **Take turns spinning the spinner.**
- **If the spinner lands on an odd number, Player 1 wins.**
- **If the spinner lands on an even number, Player 2 wins.**

What Is Your Answer?

4. **IN YOUR OWN WORDS** How can you find a theoretical probability?
5. Find and describe a career in which probability is used. Explain why probability is used in that career.
6. Two people play the following game.

   Each player has 6 cards numbered 1, 2, 3, 4, 5, and 6. At the same time, each player holds up one card. If the product of the two numbers is odd, Player 1 wins. If the product is even, Player 2 wins. Continue until both players are out of cards. Which player is more likely to win? Why?

Practice Use what you learned about theoretical probability to complete Exercises 4–7 on page 394.

Check It Out
Lesson Tutorials
BigIdeasMath.com

Key Vocabulary
theoretical probability, *p. 392*
fair experiment, *p. 393*

Key Idea

Theoretical Probability

When all possible outcomes are equally likely, the **theoretical probability** of an event is the ratio of the number of favorable outcomes to the number of possible outcomes. The probability of an event is written as $P(\text{event})$.

$$P(\text{event}) = \frac{\text{number of favorable outcomes}}{\text{number of possible outcomes}}$$

EXAMPLE 1 Finding a Theoretical Probability

You randomly choose one of the letters shown. What is the theoretical probability of choosing a vowel?

$$P(\text{event}) = \frac{\text{number of favorable outcomes}}{\text{number of possible outcomes}}$$

$$P(\text{vowel}) = \frac{3}{7}$$

There are 3 vowels.
There is a total of 7 letters.

∴ The probability of choosing a vowel is $\frac{3}{7}$ or about 43%.

EXAMPLE 2 Using a Theoretical Probability

The theoretical probability that you randomly choose a green marble from a bag is $\frac{3}{8}$. There are 40 marbles in the bag. How many are green?

$$P(\text{green}) = \frac{\text{number of green marbles}}{\text{total number of marbles}}$$

$$\frac{3}{8} = \frac{n}{40}$$ Substitute. Let n be the number of green marbles.

$$15 = n$$ Multiply each side by 40.

∴ There are 15 green marbles in the bag.

On Your Own

Now You're Ready
Exercises 4–11

1. In Example 1, what is the theoretical probability of choosing an X?
2. The theoretical probability that you spin an odd number on a spinner is 0.6. The spinner has 10 sections. How many sections have odd numbers?

Multi-Language Glossary at BigIdeasMath.com.

A game is fair if every player has the same probability of winning.

An experiment is **fair** if all of its possible outcomes are equally likely.

The spinner is equally likely to land on 1 or 2. The spinner is fair.

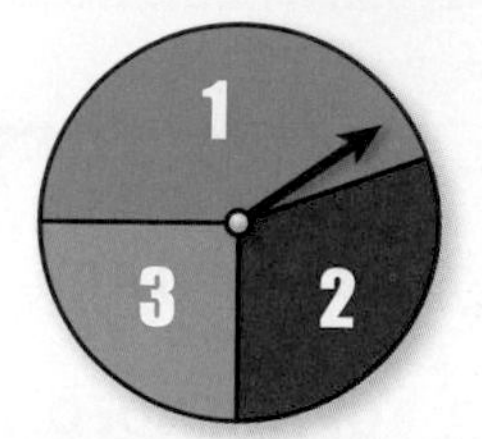

The spinner is more likely to land on 1 than on either 2 or 3. The spinner is *not* fair.

EXAMPLE 3 Making a Prediction

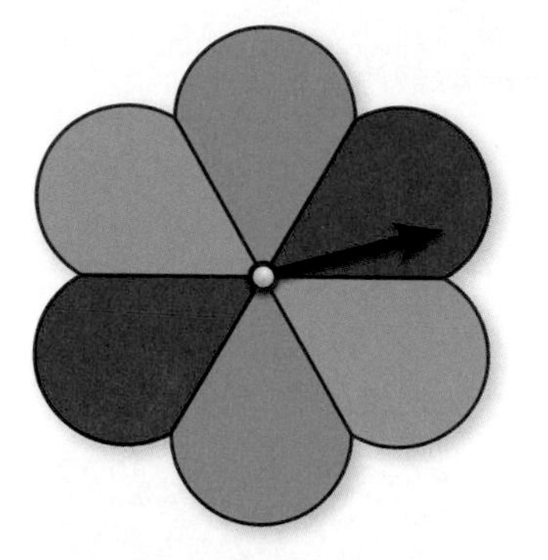

Scoring Rules:

- You get one point when the spinner lands on blue or green.
- Your friend gets one point when the spinner lands on red.
- The first person to get 10 points wins.

You and your friend play the game. (a) Is the spinner fair? (b) Is the game fair? (c) Predict the number of turns it will take you to win.

a. Yes, the spinner is fair because it is equally likely to land on red, blue, or green.

b. Find and compare the theoretical probabilities of the events.

$$\textbf{You: } P(\text{blue or green}) = \frac{\text{number of blue or green sections}}{\text{total number of sections}}$$

$$= \frac{4}{6} = \frac{2}{3}$$

$$\textbf{Your friend: } P(\text{red}) = \frac{\text{number of red sections}}{\text{total number of sections}}$$

$$= \frac{2}{6} = \frac{1}{3}$$

It is more likely that the spinner will land on blue or green than on red. Because your probability is greater, the game is *not* fair.

c. Write and solve an equation using P(blue or green) found in part (b). Let x be the number of turns it will take you to win.

$\frac{2}{3}x = 10$ Write equation.

$x = 15$ Multiply each side by $\frac{3}{2}$.

So, you can predict that it will take you 15 turns to win.

On Your Own

Now You're Ready
Exercises 12–14

3. WHAT IF? In Example 3, you get one point when the spinner lands on blue or green. Your friend gets one point when the spinner lands on red or blue. The first person to get 5 points wins. Is the game fair? Explain.

9.2 Exercises

Vocabulary and Concept Check

1. **VOCABULARY** An event has a theoretical probability of 0.5. What does this mean?
2. **OPEN-ENDED** Describe an event that has a theoretical probability of $\frac{1}{4}$.
3. **WHICH ONE DOESN'T BELONG?** Which spinner does *not* belong with the other three? Explain your reasoning.

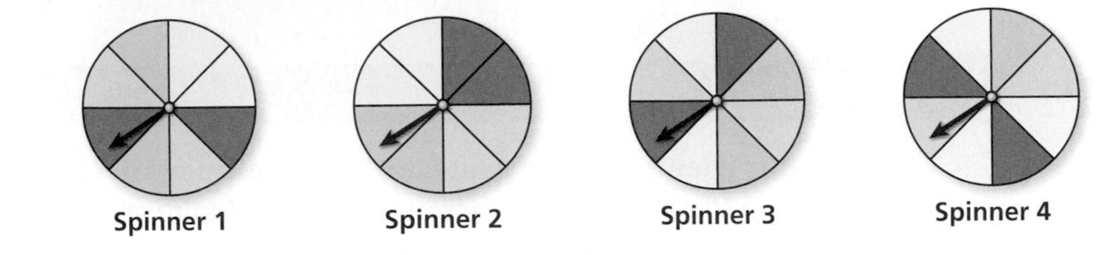

Practice and Problem Solving

Use the spinner to determine the theoretical probability of the event.

① 4. Spinning red

5. Spinning a 1
6. Spinning an odd number
7. Spinning a multiple of 2
8. Spinning a number less than 7
9. Spinning a 7
10. **LETTERS** Each letter of the alphabet is printed on an index card. What is the theoretical probability of randomly choosing any letter except Z?

② 11. **GAME SHOW** On a game show, a contestant randomly chooses a chip from a bag that contains numbers and strikes. The theoretical probability of choosing a strike is $\frac{3}{10}$. There are 30 chips in the bag. How many are strikes?

A number cube is rolled. Determine if the game is fair. If it is *not* fair, who has the greater probability of winning?

③ 12. You win if the number is odd. Your friend wins if the number is even.

13. You win if the number is less than 3. If it is not less than 3, your friend wins.
14. **SCORING POINTS** You get one point if a 1 or a 2 is rolled on the number cube. Your friend gets one point if a 5 or a 6 is rolled. The first person to 5 points wins.
    a. Is the number cube fair? Is the game fair? Explain.
    b. Predict the number of turns it will take you to win.

15. **HISTORY** You write a report about your favorite president. Your friend writes a report on a randomly chosen president. What is the theoretical probability that you write reports on the same president?

16. **BIRTHDAYS** The bar graph shows the birthday months of all 200 employees at a local business.

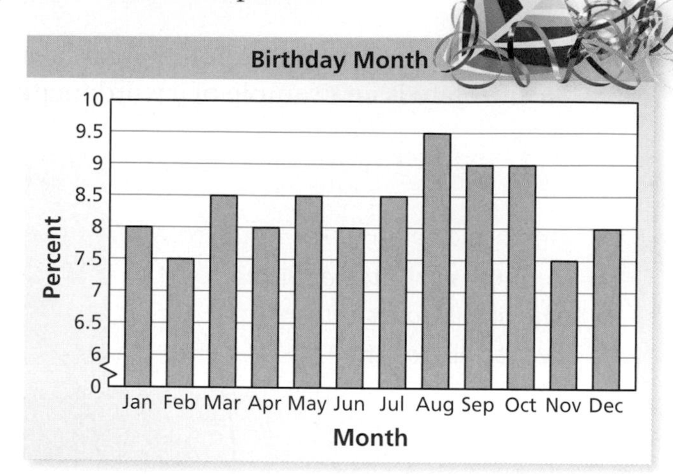

   a. What is the theoretical probability of randomly choosing a person at the business who was born in a month with an R in its name?

   b. What is the theoretical probability of randomly choosing a person at the business who has a birthday in the first half of the year?

17. **SCHEDULING** There are 16 females and 20 males in a class.

   a. What is the theoretical probability of randomly choosing a female from the class?

   b. One week later, there are 45 students in the class. The theoretical probability of randomly selecting a female is the same as last week. How many males joined the class?

Mother's Genes

Father's Genes	*X*	*X*
X	*XX*	
Y		

A Punnett square is a grid used to show possible gene combinations for the offspring of two parents. In the Punnett square shown, a boy is represented by *XY*. A girl is represented by *XX*.

18. Complete the Punnett square.

19. **PRECISION** Explain why the probability of two parents having a boy or having a girl is equally likely.

20. Critical Thinking Two parents each have the gene combination *Cs*. The gene *C* is for curly hair. The gene *s* is for straight hair.

   a. Make a Punnett square for the two parents. If all outcomes are equally likely, what is the probability of a child having the gene combination *CC*?

   b. Any gene combination that includes a *C* results in curly hair. If all outcomes are equally likely, what is the probability of a child having curly hair?

Fair Game Review What you learned in previous grades & lessons

Multiply. *(Section 2.3)*

21. $\frac{1}{2} \times \frac{1}{2}$

22. $-\frac{1}{6} \times \frac{2}{3}$

23. $-\frac{3}{5} \times \frac{7}{8}$

24. $\frac{4}{5} \times \frac{1}{36}$

25. **MULTIPLE CHOICE** What is the mean of the numbers 11, 6, 12, 22, 7, 8, and 4? *(Skills Review Handbook)*

   (A) 4 (B) 8 (C) 10 (D) 70

9 Study Help

Check It Out
Graphic Organizer
BigIdeasMath.com

You can use a **word magnet** to organize information associated with a vocabulary word or term. Here is an example of a word magnet for probability.

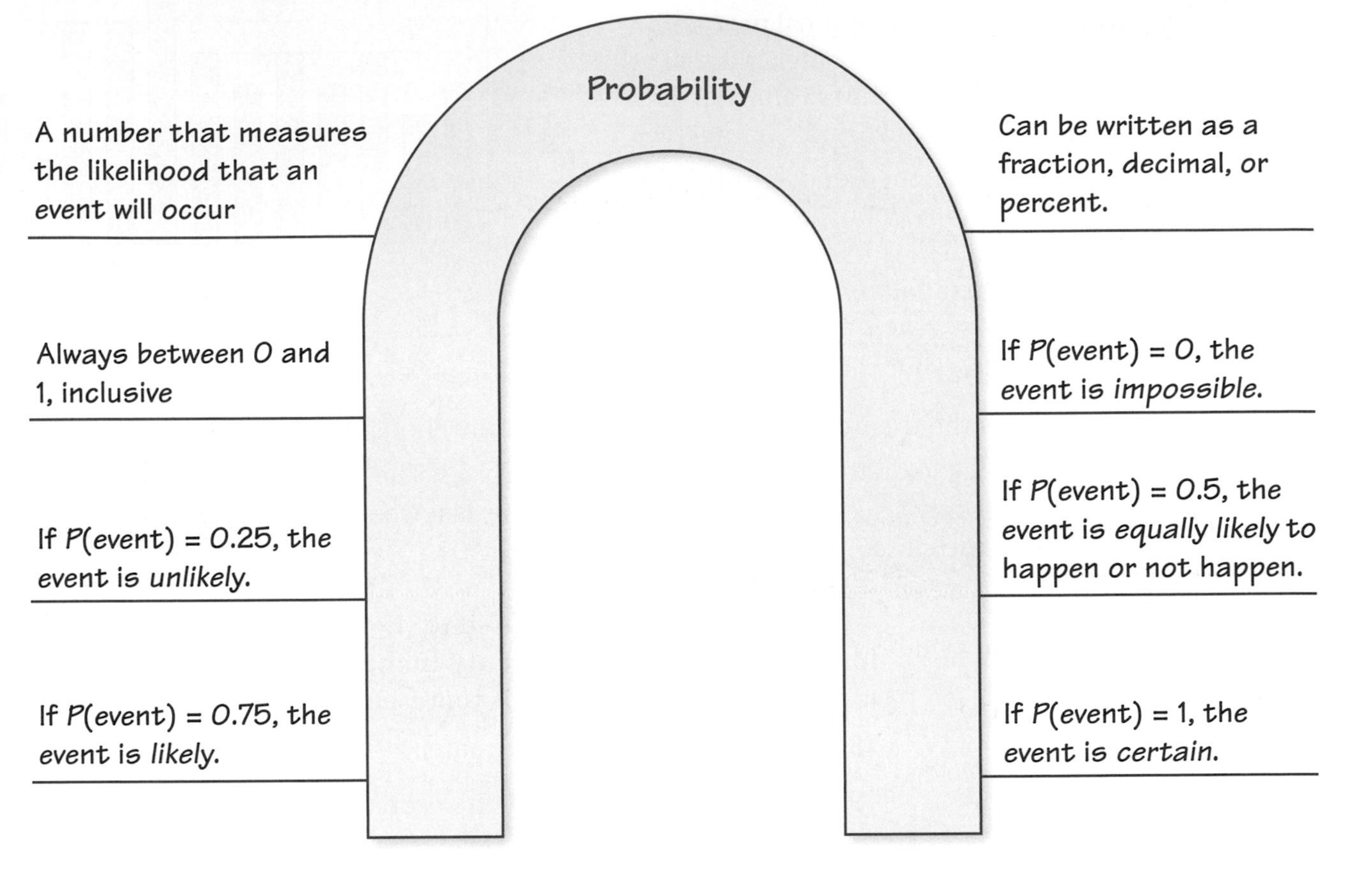

Probability
A number that measures the likelihood that an event will occur
Can be written as a fraction, decimal, or percent.
Always between 0 and 1, inclusive
If P(event) = 0, the event is impossible.
If P(event) = 0.25, the event is unlikely.
If P(event) = 0.5, the event is equally likely to happen or not happen.
If P(event) = 0.75, the event is likely.
If P(event) = 1, the event is certain.

On Your Own

Make a word magnet to help you study these topics.

1. event
2. outcome
3. theoretical probability

After you complete this chapter, make word magnets for the following topics.

4. experimental probability
5. independent events
6. dependent events
7. Choose three other topics that you studied earlier in this course. Make a word magnet for each topic.

HYENA
Beast
Wild
Teeth
Sneaks up
Eats cats
Predator
I'm going to warn Fluffy not to attend.

"I'm going to sell my word magnet poster at the Fraidy Cat Festival."

9.1–9.2 Quiz

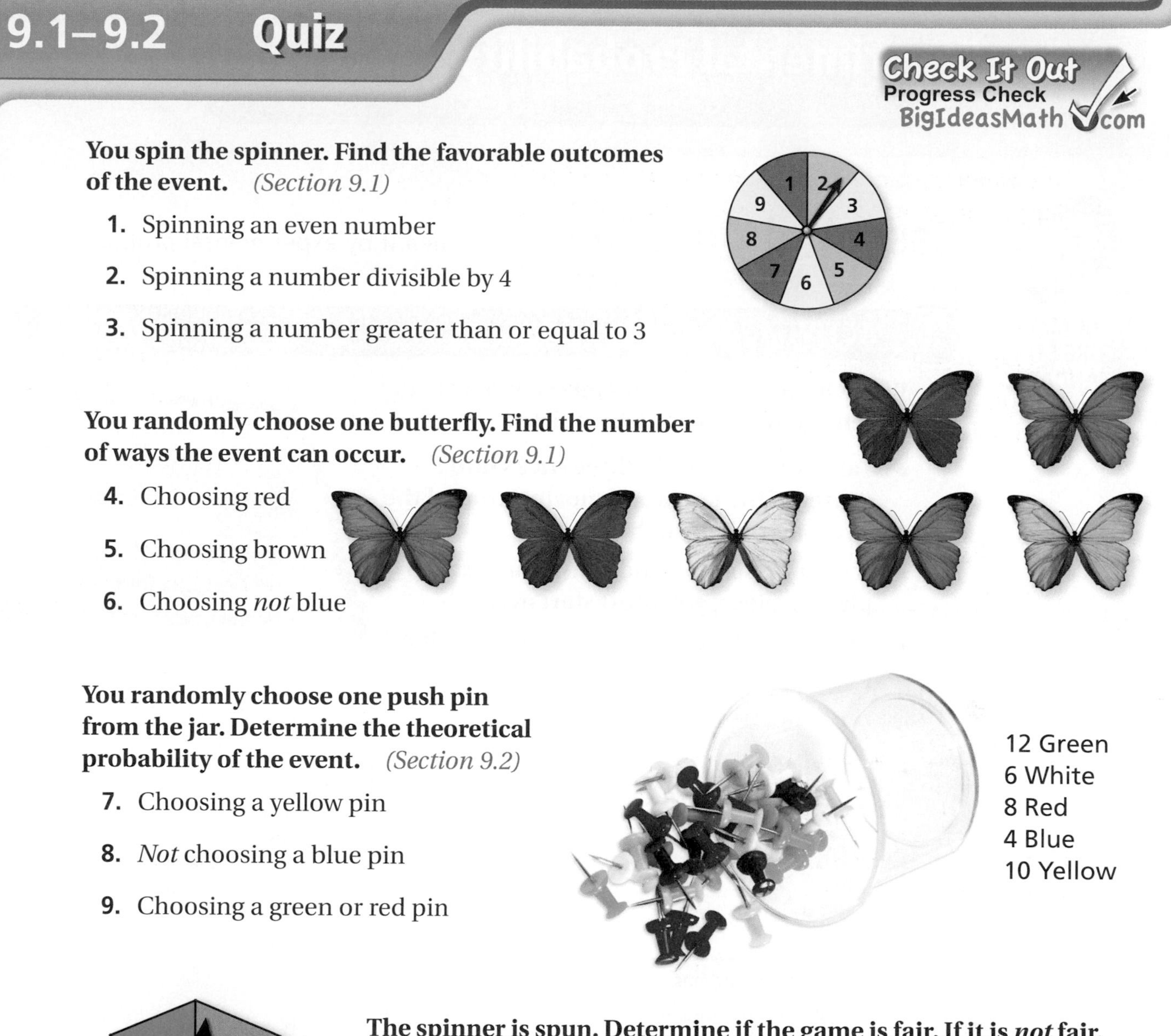

You spin the spinner. Find the favorable outcomes of the event. *(Section 9.1)*

1. Spinning an even number
2. Spinning a number divisible by 4
3. Spinning a number greater than or equal to 3

You randomly choose one butterfly. Find the number of ways the event can occur. *(Section 9.1)*

4. Choosing red
5. Choosing brown
6. Choosing *not* blue

You randomly choose one push pin from the jar. Determine the theoretical probability of the event. *(Section 9.2)*

12 Green
6 White
8 Red
4 Blue
10 Yellow

7. Choosing a yellow pin
8. *Not* choosing a blue pin
9. Choosing a green or red pin

The spinner is spun. Determine if the game is fair. If it is *not* fair, who has the greater probability of winning? *(Section 9.2)*

10. You win if the number is even. Your friend wins if the number is odd.
11. You win if the number is less than 4. Your friend wins if the number is 4 or greater.

12. **TICKETS** The theoretical probability that your ticket will be drawn from a bucket to win a bicycle is $\frac{1}{35}$. There are 665 tickets in the bucket. How many tickets are yours? *(Section 9.2)*

13. **APPLES** There are 104 apples in a bushel. The probability of randomly choosing a Granny Smith apple from the bushel is 25%. How many of the apples are *not* Granny Smith apples? *(Section 9.2)*

9.3 Experimental Probability

Essential Question What is meant by experimental probability?

COMMON CORE STATE STANDARDS
7.SP.6

1 ACTIVITY: Throwing Sticks

Play with a partner. This game is based on an Apache game called "Throw Sticks."

- **Take turns throwing three sticks into the center of the circle and moving around the circle according to the chart.**
- **If your opponent lands on or passes your playing piece, you must start over.**
- **The first player to pass his or her starting point wins.**

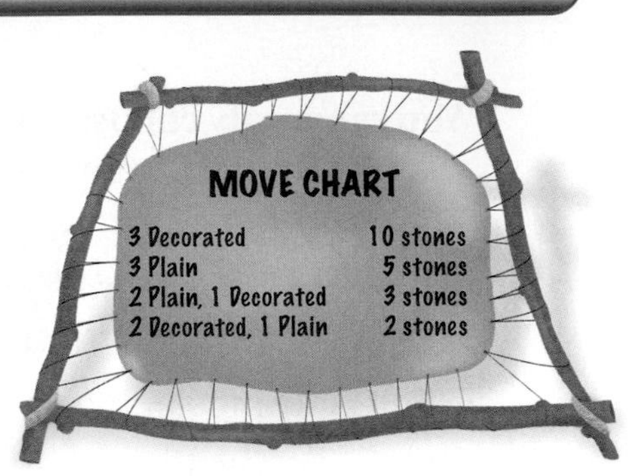

MOVE CHART

3 Decorated	10 stones
3 Plain	5 stones
2 Plain, 1 Decorated	3 stones
2 Decorated, 1 Plain	2 stones

Each stick has one plain side and one decorated side.

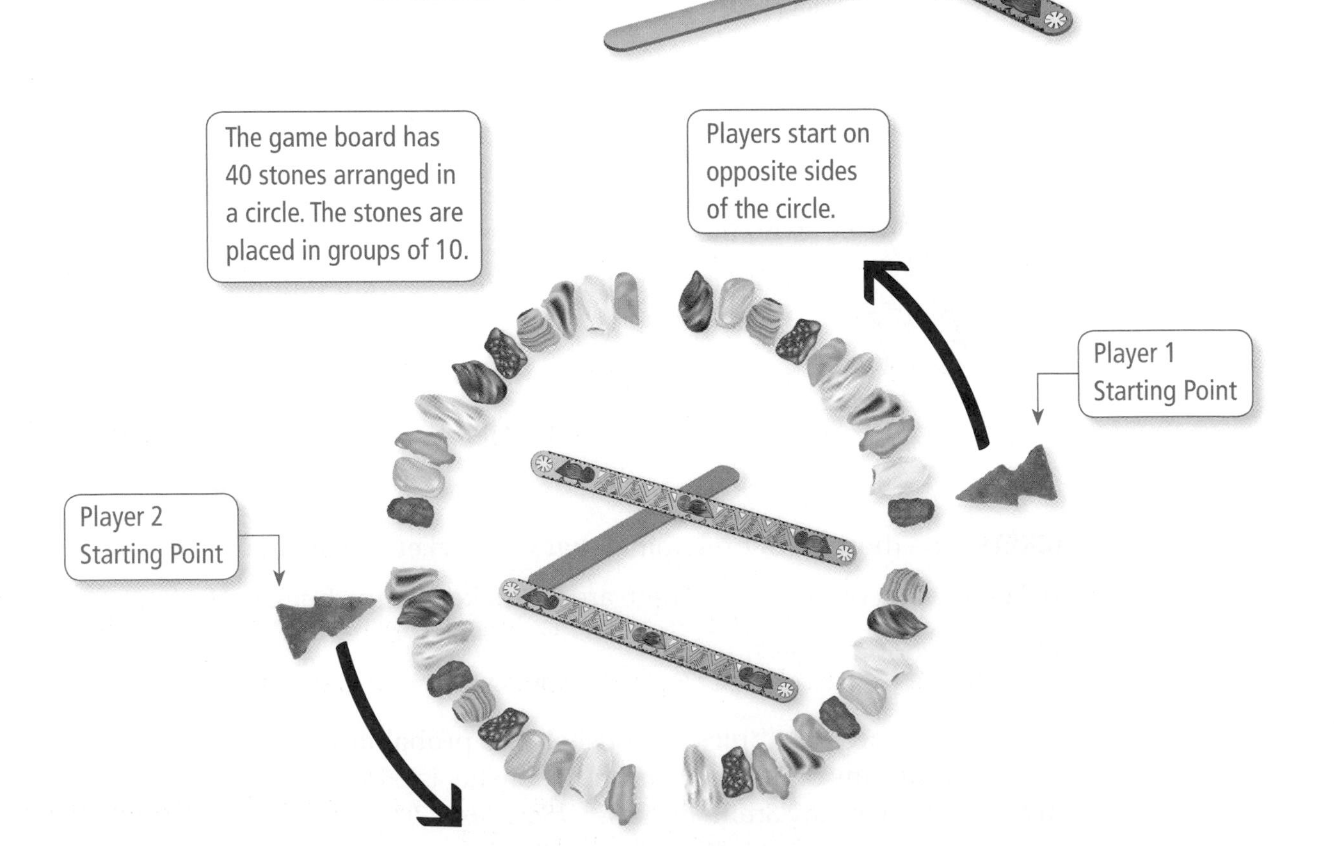

The game board has 40 stones arranged in a circle. The stones are placed in groups of 10.

Players start on opposite sides of the circle.

Player 1 Starting Point

Player 2 Starting Point

2 ACTIVITY: Conducting an Experiment

Work with a partner. Throw the 3 sticks 32 times. Tally the results using the outcomes listed below. Organize the results in a bar graph. Use the bar graph to estimate the probability of each outcome. These are called experimental probabilities.

a. PPP

b. DPP

c. DDP

d. DDD

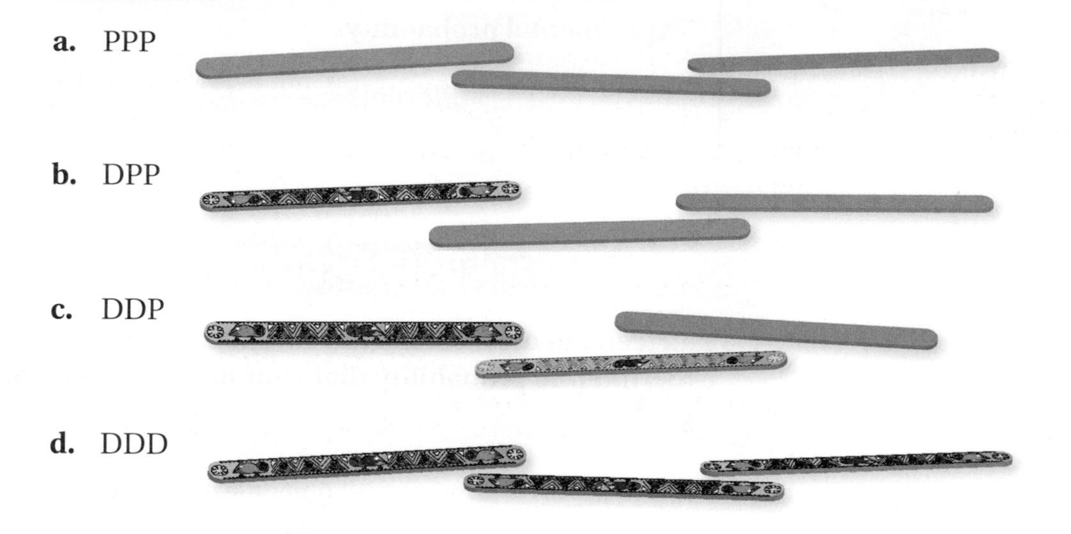

3 ACTIVITY: Analyzing the Possibilities

Work with a partner. A tree diagram helps you see different ways that the same outcome can occur.

a. Find the number of ways that each outcome can occur.

- Three Ps
- One D and two Ps
- Two Ds and one P
- Three Ds

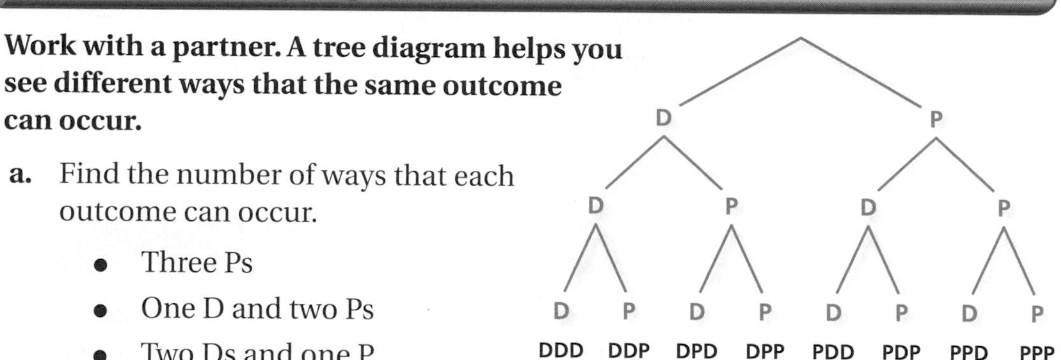

b. Find the theoretical probability of each outcome.

c. Compare and contrast your experimental and theoretical probabilities.

What Is Your Answer?

4. IN YOUR OWN WORDS What is meant by experimental probability?

5. Give a real-life example of experimental probability.

Practice Use what you learned about experimental probability to complete Exercises 3–6 on page 402.

9.3 Lesson

Check It Out
Lesson Tutorials
BigIdeasMath.com

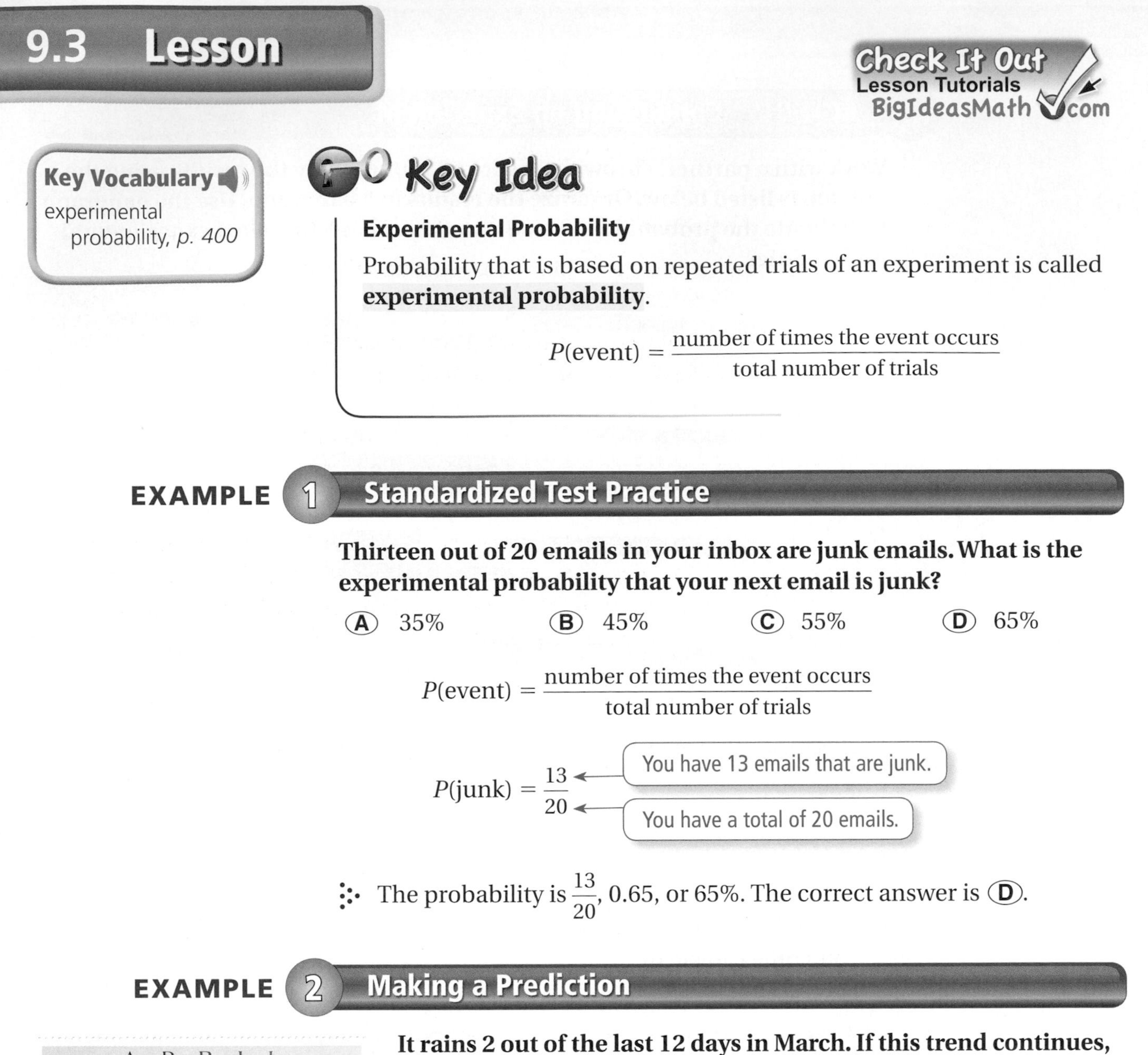

Key Vocabulary
experimental probability, *p. 400*

Key Idea

Experimental Probability

Probability that is based on repeated trials of an experiment is called **experimental probability**.

$$P(\text{event}) = \frac{\text{number of times the event occurs}}{\text{total number of trials}}$$

EXAMPLE 1 Standardized Test Practice

Thirteen out of 20 emails in your inbox are junk emails. What is the experimental probability that your next email is junk?

(A) 35% (B) 45% (C) 55% (D) 65%

$$P(\text{event}) = \frac{\text{number of times the event occurs}}{\text{total number of trials}}$$

$$P(\text{junk}) = \frac{13}{20}$$

You have 13 emails that are junk.
You have a total of 20 emails.

∴ The probability is $\frac{13}{20}$, 0.65, or 65%. The correct answer is (D).

EXAMPLE 2 Making a Prediction

"April showers bring May flowers." Old Proverb, 1557

It rains 2 out of the last 12 days in March. If this trend continues, how many rainy days would you expect in April?

Find the experimental probability of a rainy day.

$$P(\text{event}) = \frac{\text{number of times the event occurs}}{\text{total number of trials}}$$

$$P(\text{rain}) = \frac{2}{12} = \frac{1}{6}$$

It rains 2 days.
There is a total of 12 days.

To make a prediction, multiply the probability of a rainy day by the number of days in April.

$$\frac{1}{6} \cdot 30 = 5$$

∴ You can predict that there will be 5 rainy days in April.

Multi-Language Glossary at BigIdeasMath.com.

On Your Own

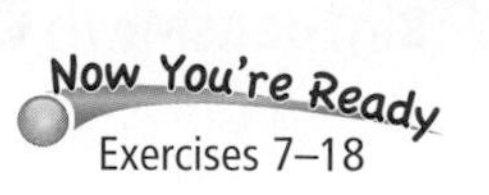

1. In Example 1, what is the experimental probability that your next email is *not* junk?
2. At a clothing company, an inspector finds 5 defective pairs in a shipment of 200 jeans.
   a. What is the experimental probability of a pair of jeans being defective?
   b. About how many would you expect to be defective in a shipment of 5000 pairs of jeans?

EXAMPLE 3 Comparing Experimental and Theoretical Probabilities

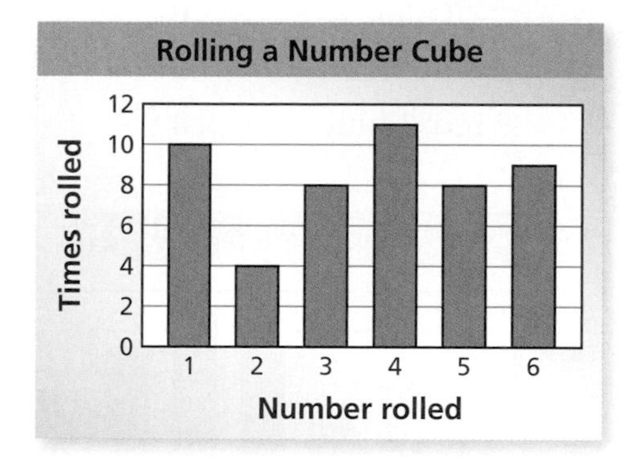

The bar graph shows the results of rolling a number cube 50 times. What is the experimental probability of rolling an odd number? How does this compare with the theoretical probability of rolling an odd number?

Find the experimental probability of rolling a 1, 3, or 5.

The bar graph shows 10 ones, 8 threes, and 8 fives. So, an odd number was rolled $10 + 8 + 8 = 26$ times in a total of 50 rolls.

Experimental Probability

$$P(\text{event}) = \frac{\text{number of times the event occurs}}{\text{total number of trials}}$$

$$P(\text{odd}) = \frac{26}{50}$$

An odd number was rolled 26 times.

There was a total of 50 rolls.

$$= \frac{13}{25}$$

Theoretical Probability

$$P(\text{event}) = \frac{\text{number of favorable outcomes}}{\text{number of possible outcomes}}$$

$$P(\text{odd}) = \frac{3}{6}$$

There are 3 odd numbers.

There is a total of 6 numbers.

$$= \frac{1}{2}$$

∴ The experimental probability is $\frac{13}{25} = 0.52 = 52\%$. The theoretical probability is $\frac{1}{2} = 0.5 = 50\%$. The experimental and theoretical probabilities are similar.

On Your Own

Now You're Ready
Exercise 19

3. In Example 3, what is the experimental probability of rolling a number greater than 1? How does this compare with the theoretical probability of rolling a number greater than 1?

9.3 Exercises

Vocabulary and Concept Check

1. **VOCABULARY** Describe how to find the experimental probability of an event.
2. **REASONING** You flip a coin 10 times and find the experimental probability of flipping tails to be 0.7. Does this seem reasonable? Explain.

Practice and Problem Solving

You have three sticks. Each stick has one red side and one blue side. You throw the sticks 10 times and record the results. Use the table to find the experimental probability of the event.

Outcome	Frequency
3 red	4
3 blue	0
2 blue, 1 red	2
2 red, 1 blue	4

3. Tossing 3 red
4. Tossing 2 blue, 1 red
5. Tossing 2 red, 1 blue
6. *Not* tossing all red

Use the bar graph to find the experimental probability of the event.

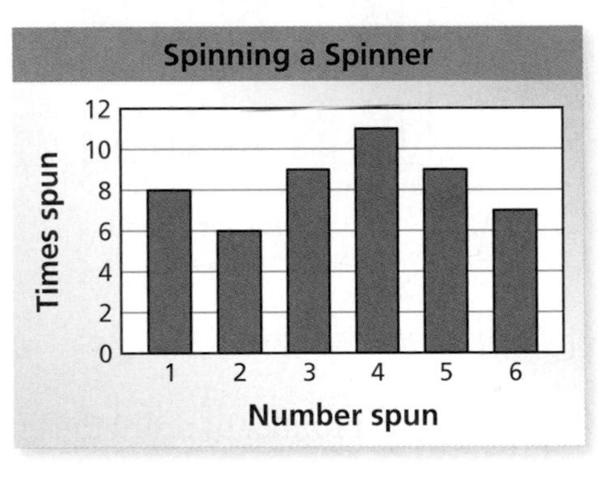

1 7. Spinning a 6
8. Spinning an even number
9. *Not* spinning a 1
10. Spinning a number less than 3
11. Spinning a 1 or a 3
12. Spinning a 7

13. **ERROR ANALYSIS** Describe and correct the error in finding $P(4)$ using the bar graph.

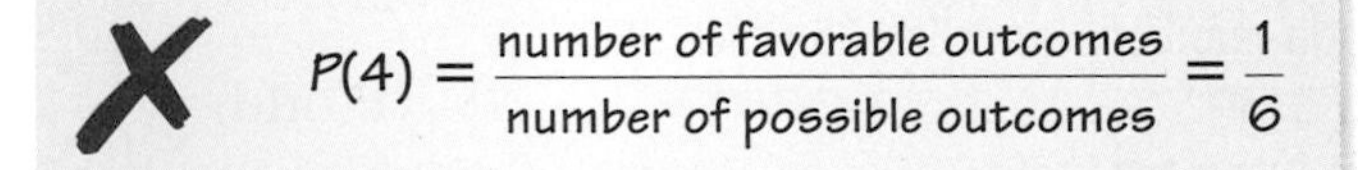

14. **EGGS** You check 20 cartons of eggs. Three of the cartons have at least one cracked egg. What is the experimental probability that a carton of eggs has at least one cracked egg?

2 15. **BOARD GAME** There are 105 lettered tiles in a board game. You choose the tiles shown. How many of the 105 tiles would you expect to be vowels?

16. **CARDS** You have a package of 20 assorted thank-you cards. You pick the four cards shown. How many of the 20 cards would you expect to have flowers on them?

17. **QUALITY CONTROL** An inspector estimates that $\frac{1}{2}$% of MP3 players are defective. In a shipment of 5000 MP3 players, predict the number that are defective.

18. **MUSIC** During a 24-hour period, the ratio of pop songs played to rap songs played on a radio station is 60 : 75.

    a. What is the experimental probability that the next song played is rap?

    b. Out of the next 90 songs, how many would you expect to be pop?

③ 19. **FLIPPING A COIN** You flip a coin 20 times. You flip heads 12 times. Compare your experimental probability of flipping heads with the theoretical probability of flipping heads.

You roll a pair of number cubes 60 times. You record your results in the bar graph shown.

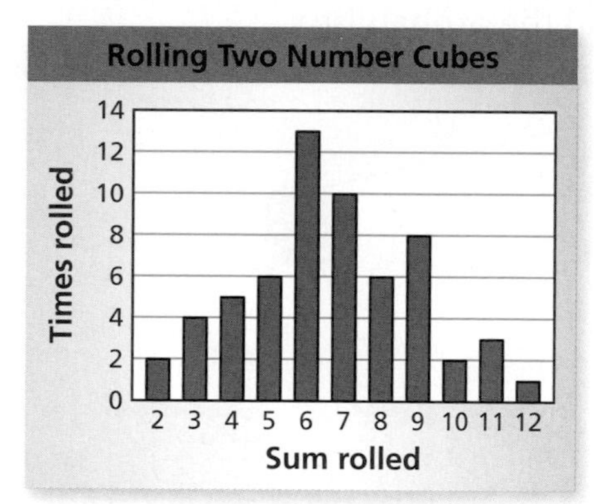

20. Use the bar graph to find the experimental probability of rolling each sum. Which sum is most likely?

21. Use the table to find the theoretical probability of rolling each sum. Which sum is most likely?

22. Compare the probabilities you found in Exercises 20 and 21.

23. Critical Thinking You roll two number cubes. Describe and perform an experiment to find the probability that the product of the two numbers rolled is at least 12. How many times did you roll the number cubes?

Fair Game Review What you learned in previous grades & lessons

Solve the equation. *(Section 2.5)*

24. $5x = 100$ 25. $75 = 15x$ 26. $2x = -26$ 27. $-4x = -96$

28. **MULTIPLE CHOICE** What is the least common denominator of the fractions $\frac{1}{16}$, $\frac{2}{19}$, and $\frac{3}{76}$? *(Skills Review Handbook)*

Ⓐ 16   Ⓑ 76   Ⓒ 304   Ⓓ 1216

9.4 Independent and Dependent Events

COMMON CORE STATE STANDARDS
7.SP.8a
7.SP.8b
7.SP.8c

Essential Question What is the difference between dependent and independent events?

1 ACTIVITY: Dependent Events

Work with a partner. You have three marbles in a bag. There are two green marbles and one purple marble. You randomly draw two marbles from the bag.

a. Use the tree diagram to find the probability that both marbles are green.

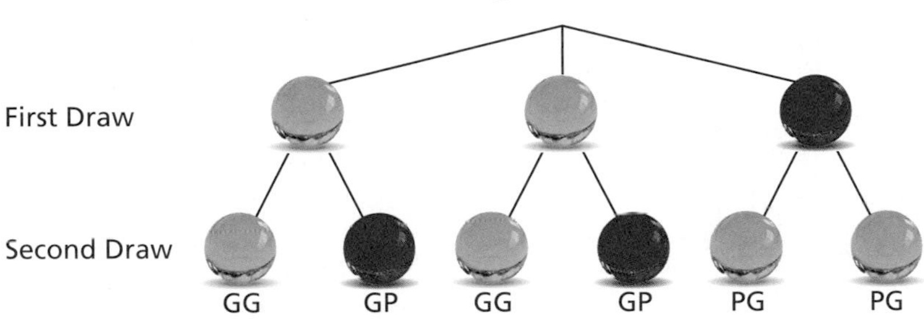

b. In the tree diagram, does the probability of getting a green marble on the second draw *depend* on the color of the first marble? Explain.

2 ACTIVITY: Independent Events

Work with a partner. Using the same marbles from Activity 1, randomly draw a marble from the bag. Then put the marble back in the bag and draw a second marble.

a. Use the tree diagram to find the probability that both marbles are green.

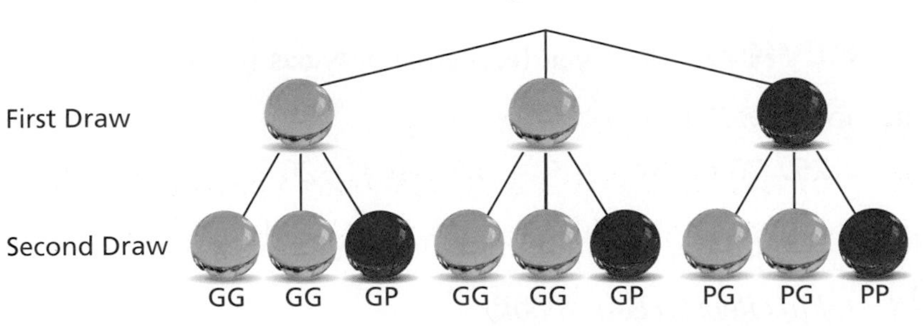

b. In the tree diagram, does the probability of getting a green marble on the second draw *depend* on the color of the first marble? Explain.

3 ACTIVITY: Conducting an Experiment

Work with a partner. Conduct two experiments.

a. In the first experiment, randomly draw two marbles from the bag 36 times. Record each result as GG or GP. Make a bar graph of your results.

b. What is the experimental probability of drawing two green marbles? Does this answer seem reasonable? Explain.

c. In the second experiment, randomly draw one marble from the bag. Put it back. Draw a second marble. Repeat this 36 times. Record each result as GG, GP, or PP. Make a bar graph of your results.

d. What is the experimental probability of drawing two green marbles? Does this answer seem reasonable? Explain.

1st Experiment

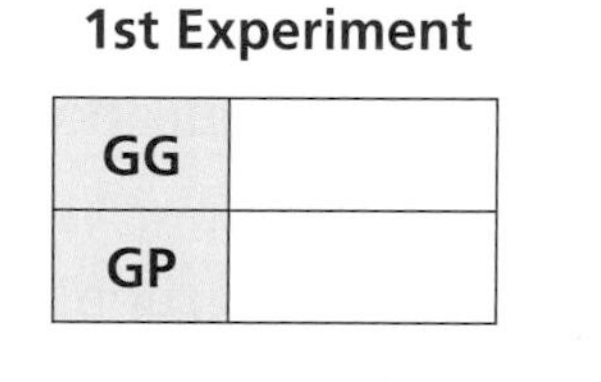

GG	
GP	

2nd Experiment

GG	
GP	
PP	

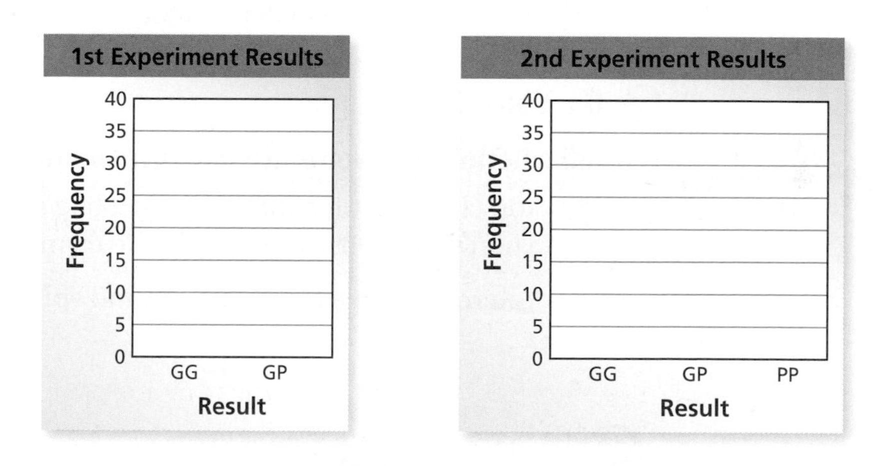

What Is Your Answer?

4. **IN YOUR OWN WORDS** What is the difference between dependent and independent events? Describe a real-life example of each.

Practice Use what you learned about independent and dependent events to complete Exercises 5 and 6 on page 409.

9.4 Lesson

Check It Out
Lesson Tutorials
BigIdeasMath.com

Key Vocabulary
independent events, p. 406
dependent events, p. 406

Two events are **independent events** if the occurrence of one event *does not* affect the likelihood that the other event will occur.

Two events are **dependent events** if the occurrence of one event *does* affect the likelihood that the other event will occur.

EXAMPLE 1 Identifying Independent and Dependent Events

Tell whether the events are *independent* or *dependent*. Explain.

a. You flip heads on one coin and tails on another coin.

The outcome of flipping one coin does not affect the outcome of flipping the other coin.

∴ So, the events are independent.

b. Your teacher chooses one student to lead a group, and then chooses another student to lead another group.

The teacher cannot pick the same student to lead both groups. So, there are fewer students to choose from when the leader of the second group is chosen.

∴ So, the events are dependent.

On Your Own

Tell whether the events are *independent* or *dependent*. Explain.

1. You choose a blue marble from a bag and set it aside. Then you choose a green marble from the bag.
2. You roll a 5 on a number cube and spin blue on a spinner.

Key Idea

Probability of Independent Events

Words The probability of two independent events A and B is the probability of A times the probability of B.

Symbols $P(A \text{ and } B) = P(A) \cdot P(B)$

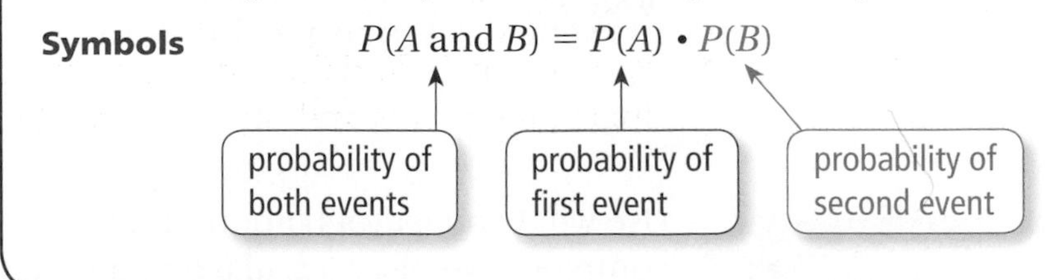

Multi-Language Glossary at BigIdeasMath.com.

EXAMPLE 2 Finding the Probability of Independent Events

You flip two quarters. What is the probability that you flip two heads?

Method 1: Use a tree diagram to find the probability.

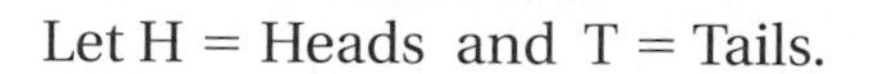

Let H = Heads and T = Tails.

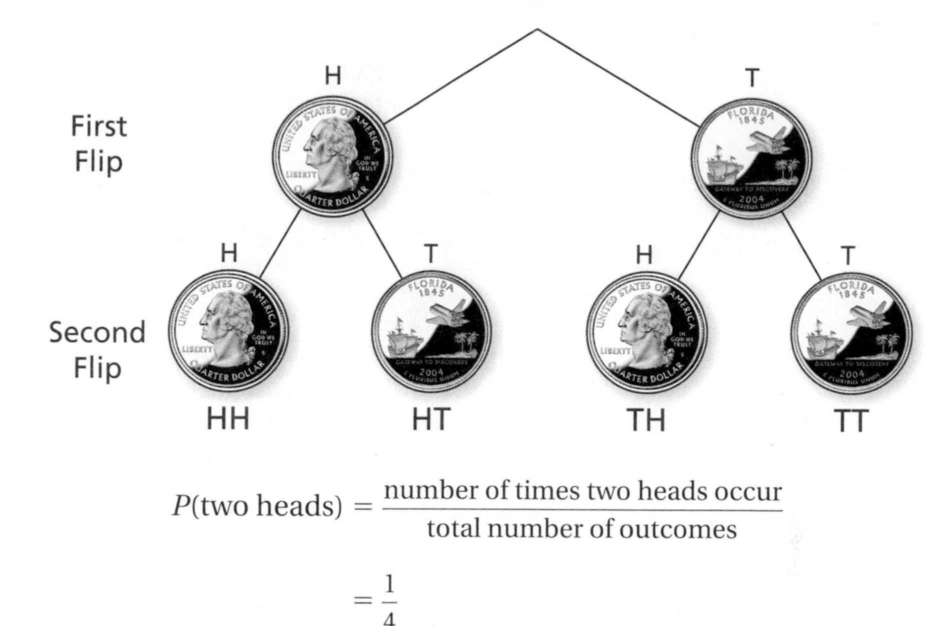

$$P(\text{two heads}) = \frac{\text{number of times two heads occur}}{\text{total number of outcomes}}$$

$$= \frac{1}{4}$$

∴ The probability that you flip two heads is $\frac{1}{4}$.

Method 2: Use the formula for independent events.

$$P(A \text{ and } B) = P(A) \cdot P(B)$$

$$P(\text{heads and heads}) = P(\text{heads}) \cdot P(\text{heads})$$

$$= \frac{1}{2} \cdot \frac{1}{2} \quad \text{Substitute.}$$

$$= \frac{1}{4} \quad \text{Multiply.}$$

∴ The probability that you flip two heads is $\frac{1}{4}$.

On Your Own

Now You're Ready
Exercises 10–18

3. You flip two coins. What is the probability that you flip one heads and one tails?
4. You flip a coin and roll a number cube. What is the probability that you flip tails and roll a number less than 5?

Key Idea

Probability of Dependent Events

Words The probability of two dependent events A and B is the probability of A times the probability of B after A occurs.

Symbols

$$P(A \text{ and } B) = P(A) \cdot P(B \text{ after } A)$$

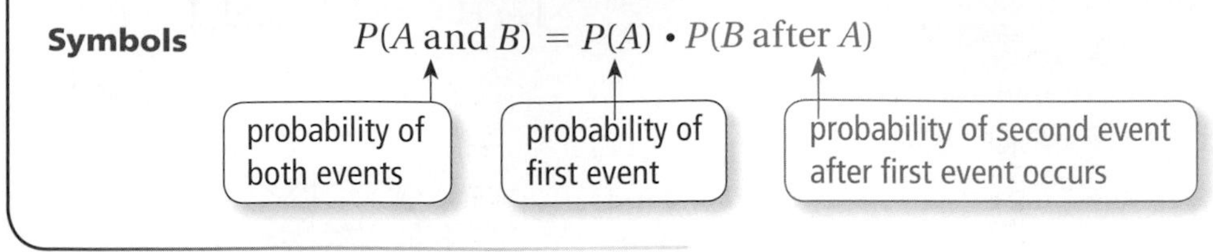

probability of both events
probability of first event
probability of second event after first event occurs

EXAMPLE 3 Finding the Probability of Dependent Events

You randomly choose a flower from the vase to take home. Your friend randomly chooses another flower from the vase to take home. What is the probability that you choose a purple flower and your friend chooses a yellow flower?

Choosing a flower changes the number of flowers left in the vase. So, the events are dependent.

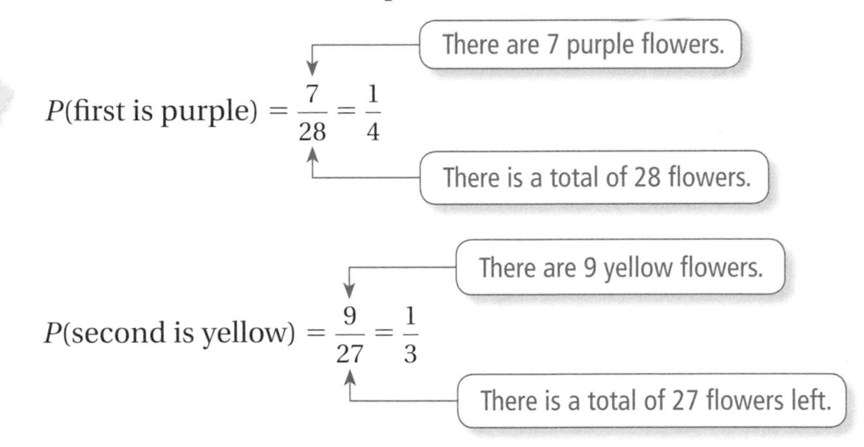

There are 7 purple flowers.
There is a total of 28 flowers.
There are 9 yellow flowers.
There is a total of 27 flowers left.

$$P(\text{first is purple}) = \frac{7}{28} = \frac{1}{4}$$

$$P(\text{second is yellow}) = \frac{9}{27} = \frac{1}{3}$$

Purple: 7
Yellow: 9
Pink: 12

Use the formula to find the probability.

$$P(A \text{ and } B) = P(A) \cdot P(B \text{ after } A)$$

$$P(\text{purple and yellow}) = P(\text{purple}) \cdot P(\text{yellow after purple})$$

$$= \frac{1}{4} \cdot \frac{1}{3} \qquad \text{Substitute.}$$

$$= \frac{1}{12} \qquad \text{Simplify.}$$

The probability of choosing a purple flower and then a yellow flower is $\frac{1}{12}$, or about 8%.

On Your Own

Now You're Ready Exercises 19–25

5. **WHAT IF?** In Example 3, what is the probability that both flowers are purple?

9.4 Exercises

Vocabulary and Concept Check

1. **VOCABULARY** Events A and B are independent. Describe two ways to find $P(A \text{ and } B)$.

2. **FILL IN THE BLANKS** Copy and complete the tree diagram to find the possible outcomes for flipping a coin three times.

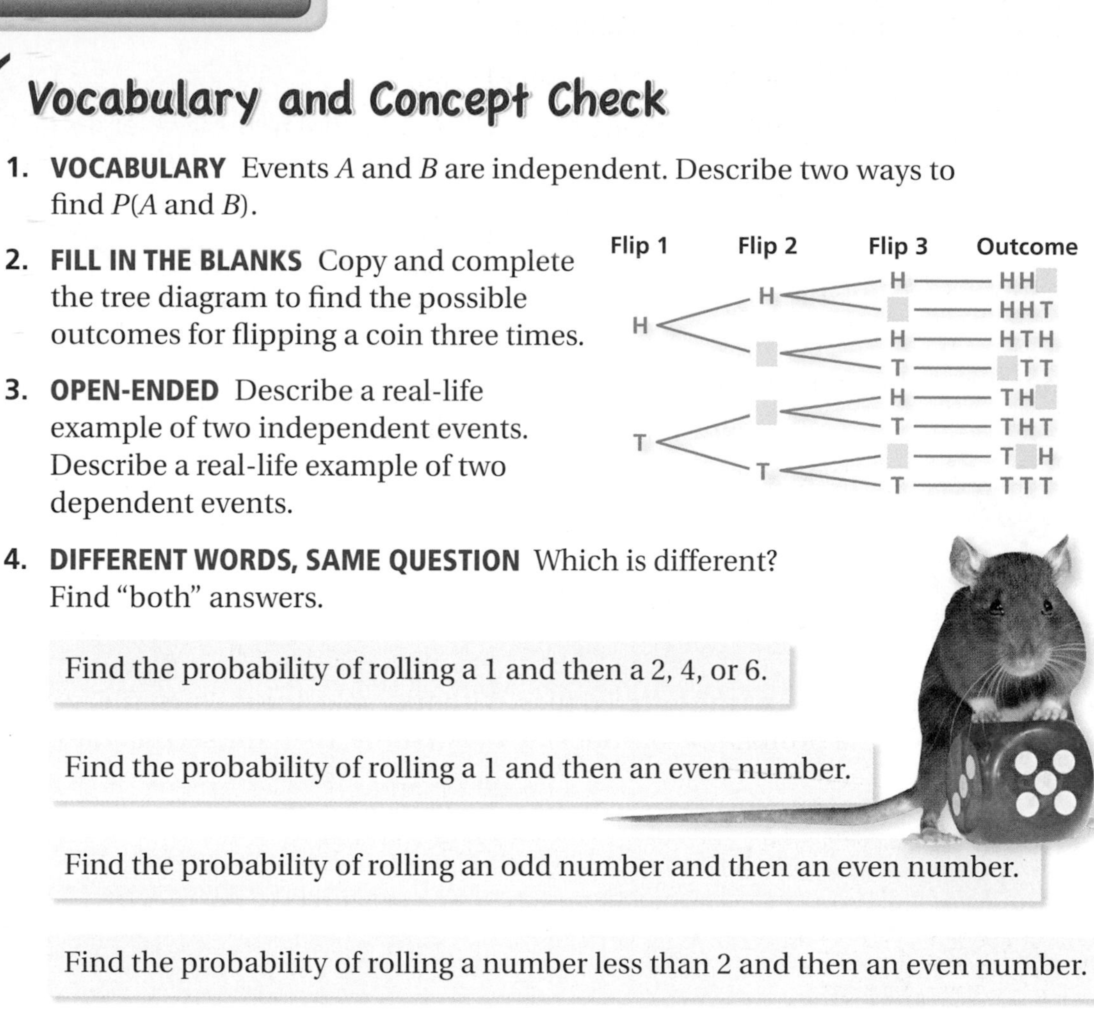

3. **OPEN-ENDED** Describe a real-life example of two independent events. Describe a real-life example of two dependent events.

4. **DIFFERENT WORDS, SAME QUESTION** Which is different? Find "both" answers.

   Find the probability of rolling a 1 and then a 2, 4, or 6.

   Find the probability of rolling a 1 and then an even number.

   Find the probability of rolling an odd number and then an even number.

   Find the probability of rolling a number less than 2 and then an even number.

Practice and Problem Solving

Tell whether the events are *independent* or *dependent*. Explain.

1 5. You roll a number cube twice.

   First Roll: You roll a 4.
   Second Roll: You roll an even number.

6. You flip a coin twice.

   First Flip: Heads
   Second Flip: Heads

7. You randomly draw a marble from a bag containing 2 red marbles and 5 green marbles. You put the marble back and then draw a second marble.

   First Draw: Green Second Draw: Red

8. You randomly draw a marble from a bag containing 2 red marbles and 5 green marbles. You keep the marble and then draw a second marble.

   First Draw: Green Second Draw: Red

9. You and your friend are in a drawing for two door prizes. You can win only one prize.

   First Draw: Your name is drawn. Second Draw: Your friend's name is drawn.

A spinner has three equal sections numbered 1, 2, and 3. You spin it twice. Use the tree diagram to find the probability of the events.

First Spin: 1, 2, 3
Second Spin: 1 2 3, 1 2 3, 1 2 3

② **10.** Spinning a 1 and then a 3

11. Spinning an odd number and then a 2

12. Spinning a 3 and then an even number

13. Spinning an even number and then an odd number

14. Spinning an odd number on each spin

You spin the spinner and flip a coin. Find the probability of the events.

15. Spinning a 4 and flipping heads

16. Spinning an even number and flipping tails

17. Spinning a multiple of 3 and flipping heads

18. Spinning white and *not* flipping tails

You randomly choose one of the lettered tiles. Without replacing the first tile, you choose a second tile. Find the probability of choosing the first tile, then the second tile.

③ **19.** R and N

20. A and L

21. D and O

22. N and yellow

23. O and *not* yellow

24. *Not* O and O

25. If you randomly choose all seven tiles in order, what is the probability that you will spell the name of a popular vacation destination in Florida?

26. EARRINGS A jewelry box contains two gold hoop earrings and two silver hoop earrings. You randomly choose two earrings. What is the probability that both are silver hoop earrings?

27. PASSWORD You forgot the last two digits of your password for a website.

a. You choose a two-digit number at random. What is the probability that your choice is correct?

b. Suppose you remember that both digits are even numbers. How does this change the probability that your choice is correct?

28. **FISH** You randomly choose two fish from the bowl. What is the probability that the first is red and the second is gold?

29. **TAKING A TEST** You are guessing at two questions on a multiple choice test. Each question has three choices: A, B, and C.

    a. What is the probability that you guess the correct answers to both questions?

    b. Suppose you can eliminate one of the choices for each question. How does this change the probability that your guesses are correct?

30. **REASONING** The probability of winning a spelling bee *and* winning a checkers game is 10%. The probability of winning a checkers game is $\frac{1}{2}$. (a) What is the probability of winning a spelling bee? (b) You enter 10 spelling bees. How many do you expect to win?

31. **SHOES** Twenty percent of the shoes manufactured by a company are black. One shoe is chosen and replaced. Then a second shoe is chosen. What is the probability that *neither* shoe is black?

32. **Critical Thinking** You randomly choose a pair of sunglasses from the shelf below. Then you randomly choose a second pair of sunglasses without replacing the first pair. List all of the possible outcomes.

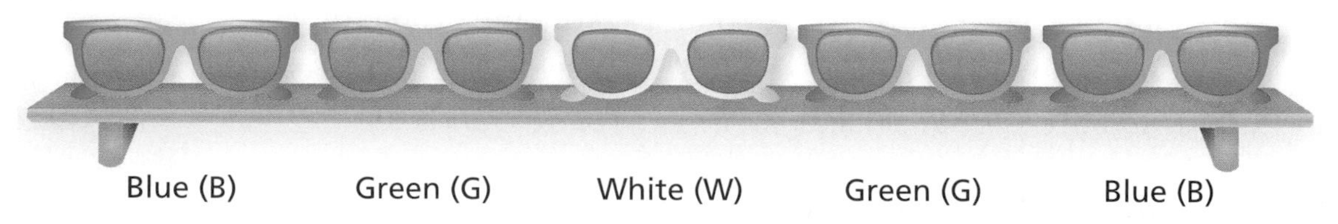

ODDS The *odds in favor of* an event is the ratio of the number of favorable outcomes to the number of unfavorable outcomes. The *odds against* an event is the ratio of the number of unfavorable outcomes to the number of favorable outcomes. Find the *odds in favor of* and the *odds against* the event when rolling a number cube.

33. Rolling a 6

34. Rolling a number less than 5

35. Rolling a 6, then rolling a 3

Fair Game Review What you learned in previous grades & lessons

Solve the equation. *(Section 2.4, Section 2.5, and Section 2.6)*

36. $6 = 9.3 + x$

37. $\frac{n}{2} = -5.4$

38. $-4p + 6 = -10$

39. **MULTIPLE CHOICE** Which intervals can be used to make a histogram? *(Section 8.2)*

    Ⓐ 16–18, 19–21, 22–26, 27–32

    Ⓑ 91–110, 111–130, 131–150

    Ⓒ 11–20, 21–40, 41–50, 51–70

    Ⓓ 50–60, 60–70, 70–80, 80–90

9.3–9.4 Quiz

Use the bar graph to find the experimental probability of the event. *(Section 9.3)*

1. Rolling a 4
2. Rolling a multiple of 3
3. Rolling a 2 or a 3
4. Rolling a number less than 7

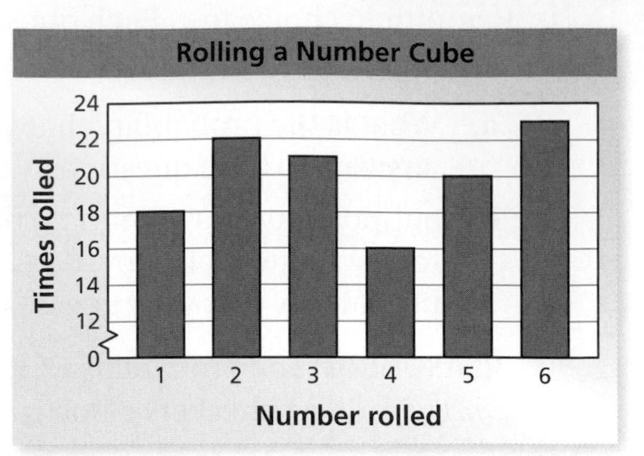

You randomly choose a playing piece and flip a coin. Find the probability of the events. *(Section 9.4)*

5. Choosing red and flipping tails
6. Choosing black and flipping heads
7. *Not* choosing red and *not* flipping heads

You randomly choose one of the letter blocks. Without replacing the first block, you choose a second block. Find the probability of choosing the first block, then the second block. *(Section 9.4)*

8. E and M
9. M and *not* E
10. Red and blue

11. **PENS** There are 30 pens in a box. You choose the five pens shown. How many of the 30 pens would you expect to have red ink? *(Section 9.3)*

12. **SWEATERS** A drawer contains three tan sweaters and two black sweaters. You randomly choose two sweaters. What is the probability that both sweaters are black? *(Section 9.4)*

13. **SPINNER** You spin the spinner 40 times. It lands on red 32 times. Compare the experimental probability of the spinner landing on red with the theoretical probability of the spinner landing on red. *(Section 9.3)*

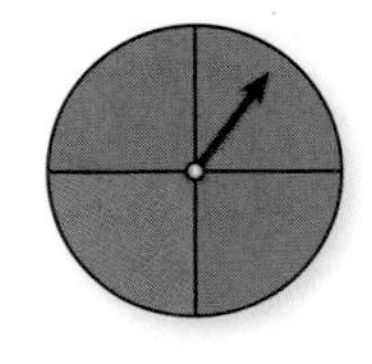

14. **FISH** You randomly choose one fish from each bowl for your aquarium. What is the probability of choosing two gold fish? *(Section 9.4)*

9 Chapter Review

Review Key Vocabulary

experiment, *p. 386*
outcomes, *p. 386*
event, *p. 386*
probability, *p. 387*
theoretical probability, *p. 392*
fair experiment, *p. 393*
experimental probability, *p. 400*
independent events, *p. 406*
dependent events, *p. 406*

Review Examples and Exercises

9.1 Introduction to Probability *(pp. 384–389)*

You randomly choose one toy racecar.

a. In how many ways can choosing a green car occur?

b. In how many ways can choosing a car that is *not* green occur? What are the favorable outcomes of choosing a car that is *not* green?

a. There are 5 green cars. So, choosing a green car can occur in 5 ways.

b. There are 2 cars that are *not* green. So, choosing a car that is *not* green can occur in 2 ways.

green	*not* green
green, green, green, green, green	blue, red

∴ The favorable outcomes of the event are blue and red.

Exercises

You spin the spinner. Find the number of ways the event can occur.

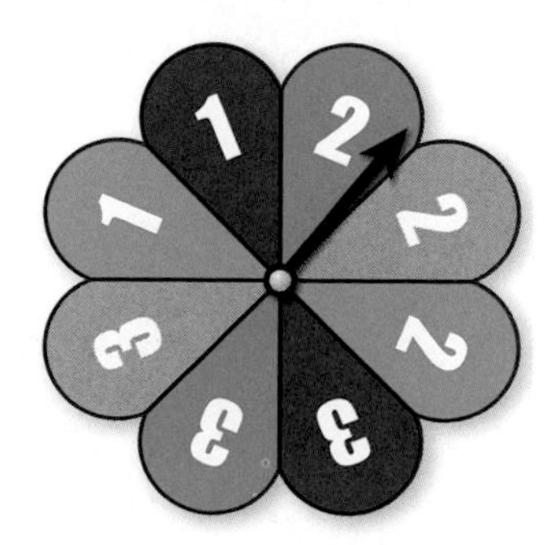

1. Spinning a 1
2. Spinning a 3
3. Spinning an odd number
4. Spinning a number greater than 0
5. On the spinner, what are the favorable outcomes of spinning a number less than 3?

9.2 Theoretical Probability (pp. 390–395)

The theoretical probability that you choose a purple grape from a bag is $\frac{2}{9}$. There are 36 grapes in the bag. How many are purple?

$$P(\text{purple}) = \frac{\text{number of purple grapes}}{\text{total number of grapes}}$$

$\frac{2}{9} = \frac{n}{36}$ Substitute. Let n be the number of purple grapes.

$8 = n$ Multiply each side by 36.

∴ There are 8 purple grapes in the bag.

Exercises

6. You get one point when the spinner at the right lands on an odd number. Your friend gets one point when it lands on an even number. The first person to get 5 points wins. Is the game fair? If it is not fair, who has the greater probability of winning?

7. The probability that you spin an even number on a spinner is $\frac{2}{3}$. The spinner has 12 sections. How many sections have even numbers?

9.3 Experimental Probability (pp. 398–403)

The bar graph shows the results of spinning the spinner 70 times. What is the experimental probability of spinning a 2?

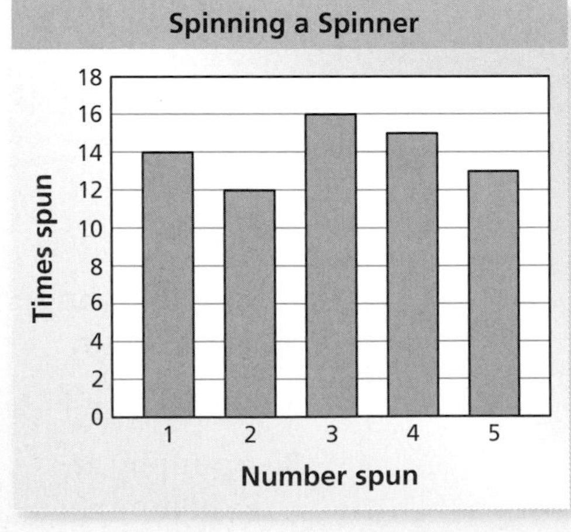

The bar graph shows 14 ones, 12 twos, 16 threes, 15 fours, and 13 fives. So, a two was spun 12 times in 70 spins.

$$P(\text{event}) = \frac{\text{number of times the event occurs}}{\text{total number of trials}}$$

$$P(2) = \frac{12}{70}$$

A 2 was spun 12 times.

There was a total of 70 spins.

$$= \frac{6}{35}$$

∴ The experimental probability is $\frac{6}{35}$, or about 17%.

Exercises

Use the bar graph on page 414 to find the experimental probability of the event.

8. Spinning a 3

9. Spinning an odd number

10. *Not* spinning a 5

11. Spinning a number greater than 3

9.4 Independent and Dependent Events *(pp. 404–411)*

You randomly choose a marble without replacing it. Your friend then chooses another marble. What is the probability that you choose a red marble and your friend chooses a blue marble?

Choosing a marble changes the number of marbles left. So, the events are dependent.

$P(\text{first is red}) = \frac{5}{12}$ — There are 5 red marbles. There is a total of 12 marbles.

$P(\text{second is blue}) = \frac{3}{11}$ — There are 3 blue marbles. There is a total of 11 marbles left.

Use the formula to find the probability.

$P(\text{red and blue}) = P(\text{red}) \cdot P(\text{blue after red})$

$= \frac{5}{12} \cdot \frac{3}{11}$ Substitute.

$= \frac{5}{44}$ Simplify.

∴ The probability of choosing a red marble followed by a blue marble is $\frac{5}{44}$, or about 11%.

Exercises

You randomly choose one of the lettered tiles. Without replacing the first tile, you choose a second tile. Find the probability of choosing the first tile, then the second tile.

12. R and A

13. A and A

14. R and *not* D

15. You choose one of the lettered tiles and flip a coin. What is the probability of choosing an A and flipping heads?

9 Chapter Test

Check It Out
Test Practice
BigIdeasMath.com

You randomly choose one game piece. (a) Find the number of ways the event can occur. (b) Find the favorable outcomes of the event.

1. Choosing green
2. Choosing *not* yellow

The spinner is spun. Determine if the game is fair. If it is *not* fair, who has the greater probability of winning?

3. You win if the number is odd. Your friend wins if the number is even.
4. You win if the number is less than 5. Your friend wins if the number is greater than 5. If the number is 5, nobody wins.

Use the bar graph to find the experimental probability of the event.

5. Rolling a 1 or a 2
6. Rolling an odd number
7. *Not* rolling a 5

Rolling a Number Cube

Times rolled: 0, 10, 12, 14, 16, 18, 20

Number rolled: 1, 2, 3, 4, 5, 6

You randomly choose one chess piece. Without replacing the first piece, you choose a second piece. Find the probability of choosing the first piece, then the second piece.

8. Bishop and bishop
9. King and queen
10. King and pawn
11. King and *not* pawn

12. **MINTS** You have a bag of 60 assorted mints. You randomly choose six mints. Two of the mints you choose are peppermints. How many of the 60 mints would you expect to be peppermints?

13. **NAMES** The names of 49 middle school students are placed in a hat. The probability of randomly drawing the name of a seventh-grade student is $\frac{3}{7}$. How many seventh-grade students' names are in the hat?

14. **BEADS** Thirty percent of the beads in a bag are blue. One bead is randomly chosen and replaced. Then a second bead is chosen. What is the probability that *neither* bead is blue?

Test-Taking Strategy
Use Intelligent Guessing
What's the probability of drawing 1 hyena out of a bag with 2 hyenas and 3 mice?
A -10% B 40% C 60% D 500%
40% < 60% I'm hoping 40%.
"You know it can't be -10% or 500%. So, you can intelligently guess between 40% and 60%."

1. A school athletic director asked each athletic team member to name his or her favorite professional sports team. The results are below:

- Columbus Crew: 3
- Detroit Shock: 4
- Florida Marlins: 20
- Florida Panthers: 8
- Jacksonville Jaguars: 26
- Miami Dolphins: 22
- Miami Heat: 15
- New York Buzz: 5
- Orlando Magic: 18
- Pensacola Power: 7
- Tampa Bay Buccaneers: 17
- Tampa Bay Lightning: 12
- Tampa Bay Rays: 28
- Other: 6

One athletic team member is picked at random. What is the likelihood that this team member's favorite professional sports team is *not* located in Florida? *(7.SP.7b)*

A. certain

B. likely, but not certain

C. unlikely, but not impossible

D. impossible

2. Trapezoid $KLMN$ is graphed in the coordinate plane shown.

Rotate trapezoid $KLMN$ 90° clockwise about the origin. What are the coordinates of point M', the image of point M after the rotation? *(7.G.2)*

F. $(-3, -2)$

G. $(-2, -3)$

H. $(-2, 3)$

I. $(3, 2)$

3. Each student in your class voted for his or her favorite day of the week. Their votes are shown below:

- Sunday: 6
- Friday: 8
- Saturday: 10
- Other day: 6

A student from your class is picked at random. What is the probability that this student's favorite day of the week is Sunday? *(7.SP.7b)*

4. A formula for converting a temperature in degrees Celsius C to a temperature in degrees Fahrenheit F is shown below.

$$F = \frac{9}{5}C + 32$$

When the temperature in degrees Celsius is $-9°$, what is the temperature, to the nearest degree, in degrees Fahrenheit? *(7.EE.4a)*

A. $-27°$

B. $-16°$

C. $16°$

D. $27°$

5. A nutritionist calculated the calories in a meal that came from three sources, as shown below.

- Calories from carbohydrates: 240
- Calories from fat: 180
- Calories from protein: 180

The nutritionist wants to make a circle graph to display this data. What should be the angle measure for the section labeled "carbohydrates"? *(7.SP.1)*

F. $120°$

G. $144°$

H. $180°$

I. $240°$

6. A right circular cone has a diameter of 10 centimeters and a slant height of 13 centimeters. Stan was computing its surface area in the box below.

$$\begin{aligned} \pi r^2 + \pi r\ell &= 3.14 \cdot 5^2 + 3.14 \cdot 5 \cdot 13 \\ &= 15.7^2 + 15.7 \cdot 13 \\ &= 246.49 + 204.1 \\ &= 450.59 \text{ cm}^2 \end{aligned}$$

What should Stan do to correct the error that he made? *(7.G.6)*

A. Use the formula $\frac{1}{3}\pi r^2 h$.

B. Label the answer with the unit cm^3.

C. Square the 5 before multiplying by 3.14.

D. Distribute the 3.14 to get $3.14 \cdot 5 + 3.14 \cdot 13$.

7. Which expression is *not* equal to the other three? *(7.EE.3)*

F. 6

G. -6

H. $|6|$

I. $|-6|$

8. A spinner is divided into 8 congruent sections, as shown below.

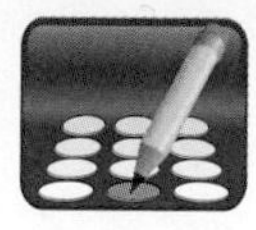

You spin the spinner twice. What is the probability that the arrow will stop in a yellow section both times? *(7.SP.7b)*

9. For a vacation trip, 4 cousins decided to rent a minivan and share the cost equally. They determined that each cousin would have to pay $300.

The minivan had 2 unused seats, so 2 additional cousins joined them. If the larger group of cousins now share the cost equally, what would each cousin have to pay? *(7.RP.2a)*

A. $150 **C.** $300

B. $200 **D.** $450

10. A travel store has 16 umbrellas in stock. Ten of the umbrellas are black. What percent of the umbrellas are black? *(7.RP.3)*

F. 1.6% **H.** 58%

G. 10% **I.** 62.5%

11. A professional golfer hit 1125 tee shots, of which 675 landed in the fairway. A statistic called "driving accuracy" is found by using the formula below. *(7.SP.2)*

Think
Solve
Explain

$$\text{driving accuracy} = \frac{\text{number of fairways hit}}{\text{number of tee shots}}$$

Part A What was the golfer's driving accuracy, expressed as a percent? Show your work and explain your reasoning.

Part B Based only on your answer to Part A, how many of the golfer's next 500 tee shots would you expect to land in the fairway? Explain your reasoning.

10 Linear Equations and Functions

"Let *x* be the number of my dog biscuits that I am sharing with you."

"Now, double that number and add 4 and you get 4 times the sum of 1 and the number."

"How many biscuits do you get?"

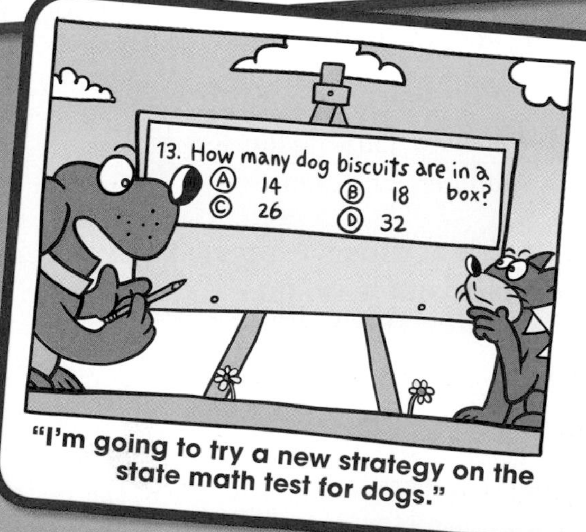

"I'm going to try a new strategy on the state math test for dogs."

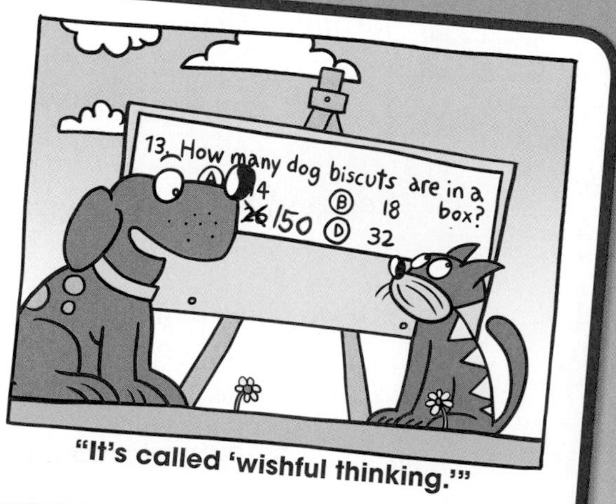

"It's called 'wishful thinking.'"

What You Learned Before

"I'm analyzing our new business. When the income line intersects the expense line, we will start making a profit. How many months will it take?"

Solving One-Step Equations (6.EE.7)

Example 1 **Solve $x - 5 = 7$.**

$x - 5 = 7$	Write the equation.
$\underline{+5} = \underline{+5}$	Add 5 to each side.
$x = 12$	Simplify.

∴ The solution is $x = 12$.

Example 2 **Solve $6x = 48$.**

$6x = 48$	Write the equation.
$\frac{6x}{6} = \frac{48}{6}$	Divide each side by 6.
$x = 8$	Simplify.

∴ The solution is $x = 8$.

Try It Yourself

Solve the equation.

1. $x - 8 = 2$ **2.** $x + 4 = 17$ **3.** $4x = 20$ **4.** $\frac{x}{3} = 7$

Plotting Points (6.NS.6c)

Example 3 **Write the ordered pair that corresponds to Point *D*.**

Point D is 3 units to the left of the origin and 1 unit up. So, the x-coordinate is -3 and the y-coordinate is 1.

∴ The ordered pair $(-3, 1)$ corresponds to Point D.

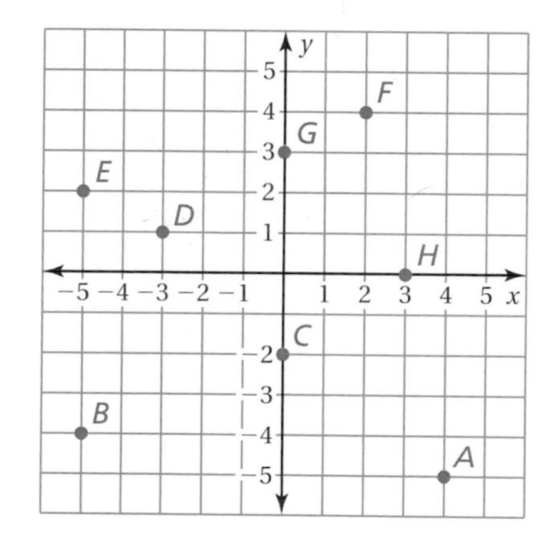

Example 4 **Which point is located at (2, 4)?**

Start at the origin. Move 2 units right and 4 units up.

∴ Point F is located at (2, 4).

Try It Yourself

Use the graph to answer the question.

5. Write the ordered pair that corresponds to Point A.

6. Write the ordered pair that corresponds to Point C.

7. Which point is located at (0, 3)?

8. Which point is located in Quadrant III.

10.1 Solving Multi-Step Equations

COMMON CORE STATE STANDARDS
7.EE.4a
8.EE.7a
8.EE.7b

Essential Question How can you convert temperatures between the Fahrenheit and Celsius scales?

1 ACTIVITY: Comparing Fahrenheit and Celsius

Work with a partner. The temperature scales show the relationship between the Fahrenheit and Celsius scales. Use the two scales to complete the table.

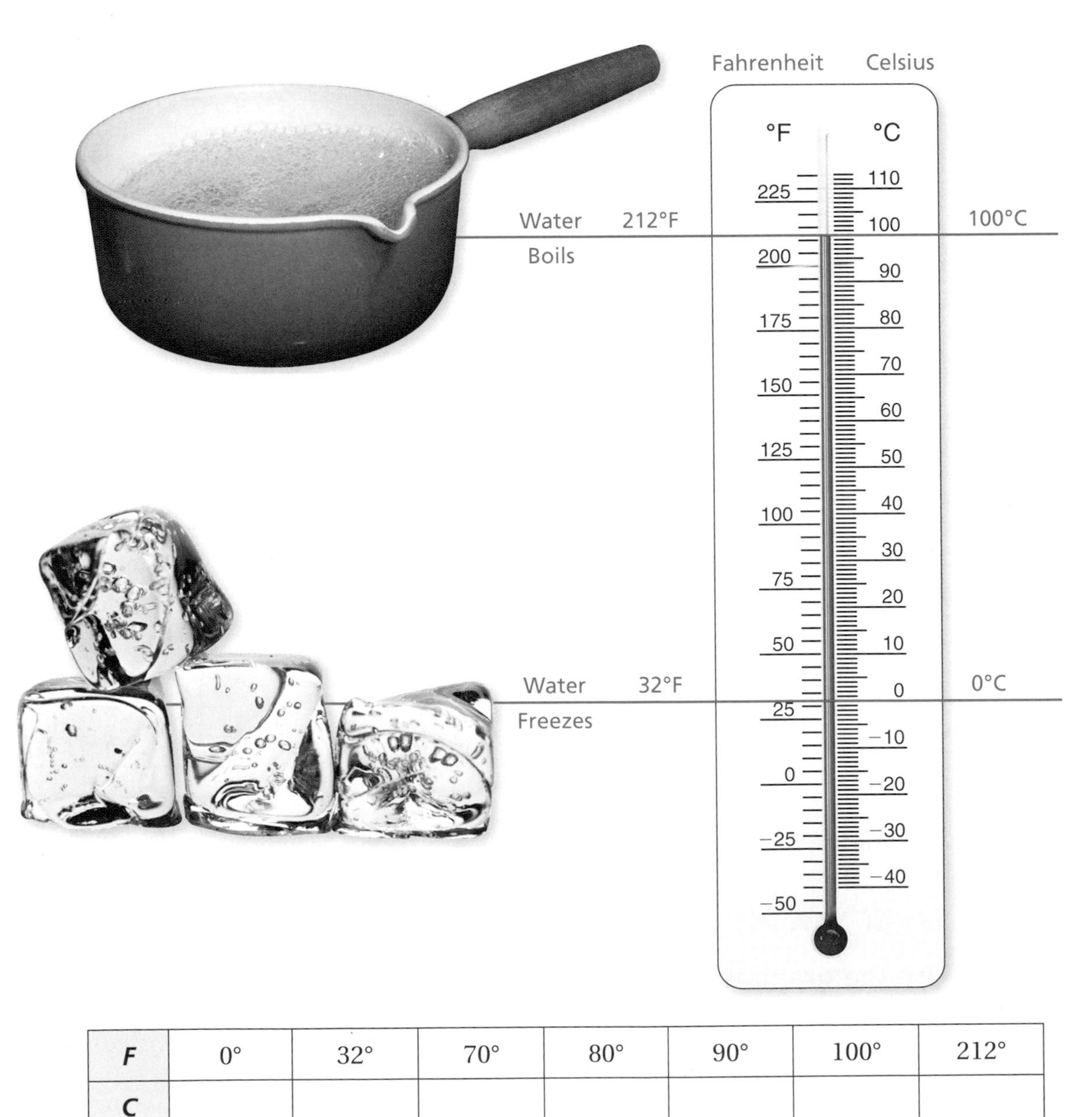

F	0°	32°	70°	80°	90°	100°	212°
C							

2 ACTIVITY: Comparing Fahrenheit and Celsius

Work with a partner.

a. Plot the points from the table in Activity 1.

b. Draw a line through the points.

c. Find the slope of the line. Write the slope as a fraction in simplest form.

d. Which of the following shows the relationship between C and F?

$$C = \frac{5}{9}(F + 32)$$

$$C = \frac{5}{9}(F - 32)$$

$$C = \frac{9}{5}(F + 32)$$

$$C = \frac{9}{5}(F - 32)$$

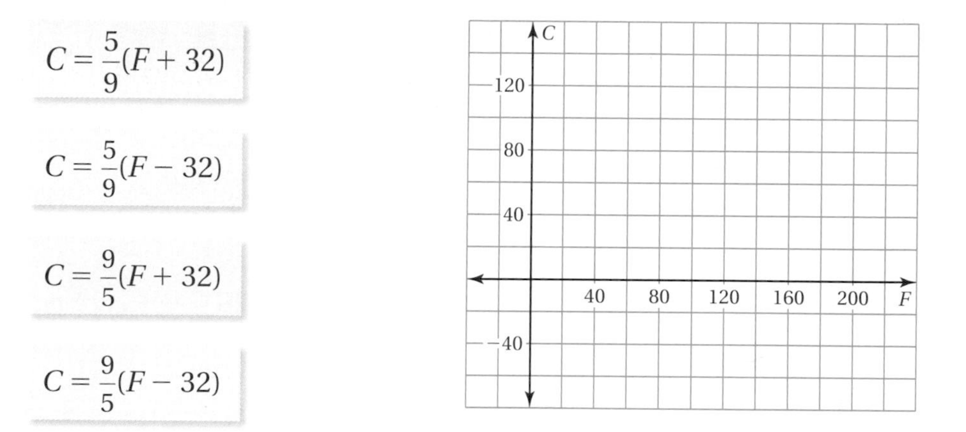

3 ACTIVITY: Converting Temperatures

Work with a partner. You have email pals in four countries that use the Celsius scale. Write each temperature in degrees Fahrenheit. Then use the scale in Activity 1 to check that your answer is reasonable.

a. Canada: 19°C

b. Mexico: 35°C

c. Japan: 28°C

d. Russia: 6°C

What Is Your Answer?

4. IN YOUR OWN WORDS How can you convert temperatures between the Fahrenheit and Celsius scales? Give two examples.

Use what you learned about multi-step equations to complete Exercises 3–5 on page 426.

10.1 Lesson

Check It Out
Lesson Tutorials
BigIdeasMath.com

EXAMPLE 1 Combining Like Terms to Solve an Equation

Solve $-2x + 4x - 12 = 40$. Check your solution.

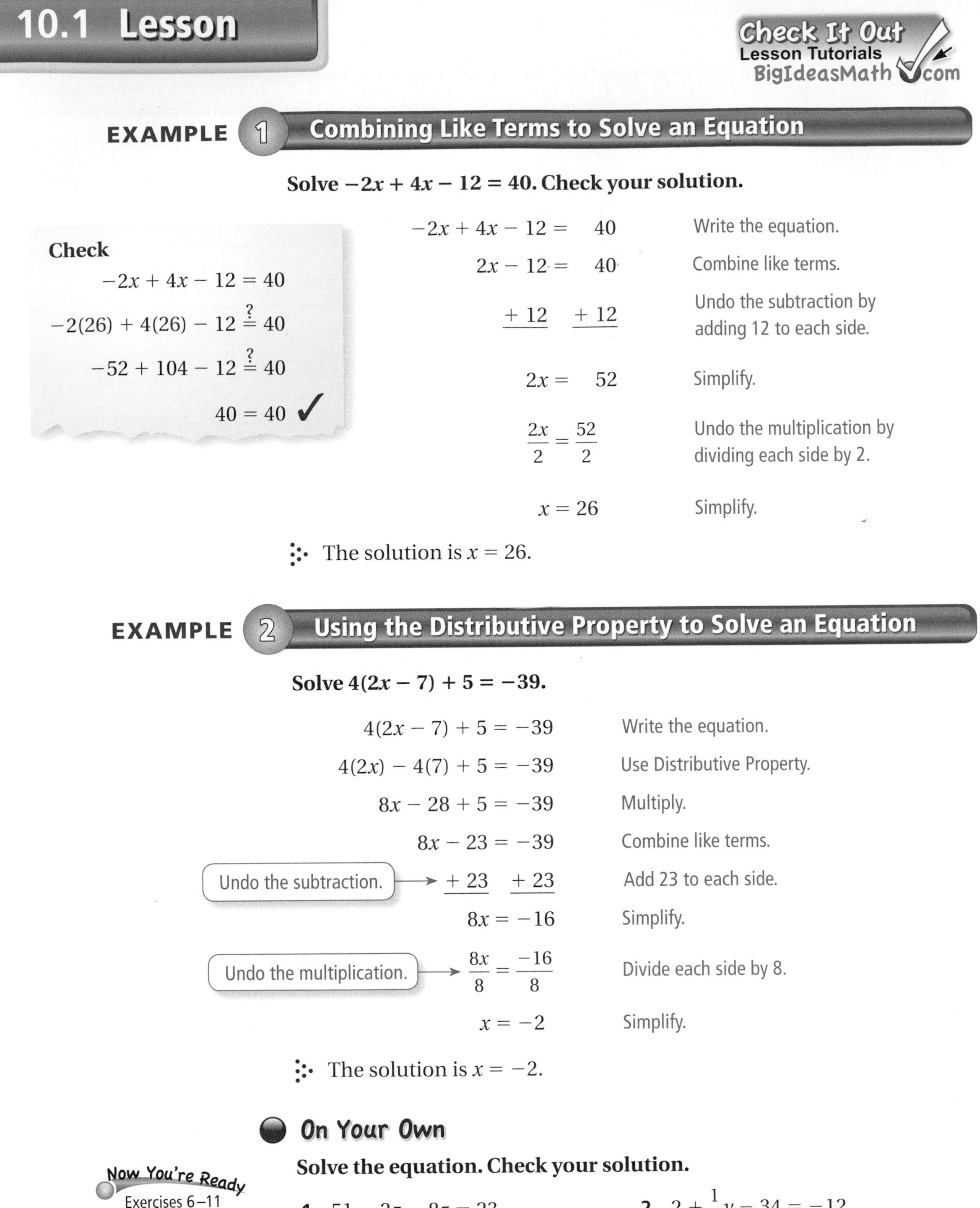

$-2x + 4x - 12 = 40$	Write the equation.
$2x - 12 = 40$	Combine like terms.
$\underline{+\,12} \quad \underline{+\,12}$	Undo the subtraction by adding 12 to each side.
$2x = 52$	Simplify.
$\frac{2x}{2} = \frac{52}{2}$	Undo the multiplication by dividing each side by 2.
$x = 26$	Simplify.

∴ The solution is $x = 26$.

Check

$-2x + 4x - 12 = 40$

$-2(26) + 4(26) - 12 \stackrel{?}{=} 40$

$-52 + 104 - 12 \stackrel{?}{=} 40$

$40 = 40$ ✓

EXAMPLE 2 Using the Distributive Property to Solve an Equation

Solve $4(2x - 7) + 5 = -39$.

$4(2x - 7) + 5 = -39$	Write the equation.
$4(2x) - 4(7) + 5 = -39$	Use Distributive Property.
$8x - 28 + 5 = -39$	Multiply.
$8x - 23 = -39$	Combine like terms.
Undo the subtraction. → $\underline{+\,23} \quad \underline{+\,23}$	Add 23 to each side.
$8x = -16$	Simplify.
Undo the multiplication. → $\frac{8x}{8} = \frac{-16}{8}$	Divide each side by 8.
$x = -2$	Simplify.

∴ The solution is $x = -2$.

On Your Own

Now You're Ready
Exercises 6–11

Solve the equation. Check your solution.

1. $51 - 2z - 8z = 23$
2. $2 + \frac{1}{2}y - 34 = -12$
3. $-2(x + 5) - 4x = 11$
4. $3 - 2.5(3 - 2d) = 5.5$

EXAMPLE 3 Using a Formula

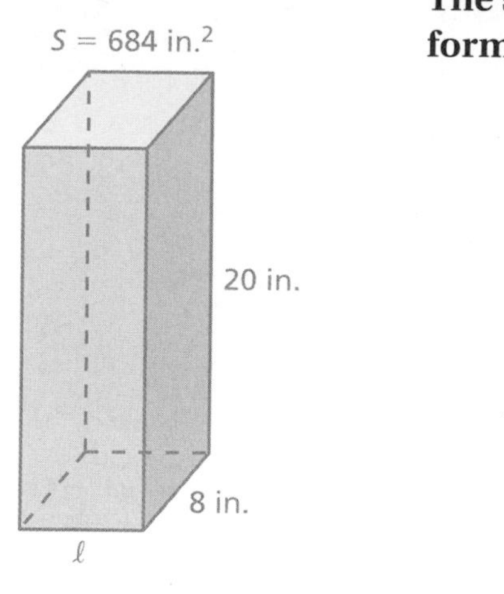

S = 684 in.²
20 in.
8 in.
ℓ

The surface area of a rectangular prism can be found using the formula $S = 2\ell w + 2\ell h + 2wh$. What is the length of the prism?

$S = 2\ell w + 2\ell h + 2wh$	Write the formula.
$684 = 2\ell(8) + 2\ell(20) + 2(8)(20)$	Substitute.
$684 = 16\ell + 40\ell + 320$	Simplify.
$684 = 56\ell + 320$	Combine like terms.
$-320 \qquad -320$	Subtract 320 from each side.
$364 = 56\ell$	Simplify.
$\frac{364}{56} = \frac{56\ell}{56}$	Divide each side by 56.
$6.5 = \ell$	Simplify.

∴ The length is 6.5 inches.

EXAMPLE 4 Real-Life Application

Your order for concert tickets is shown. You receive one free song download per ticket. How many song downloads do you receive?

Let x be the number of tickets.

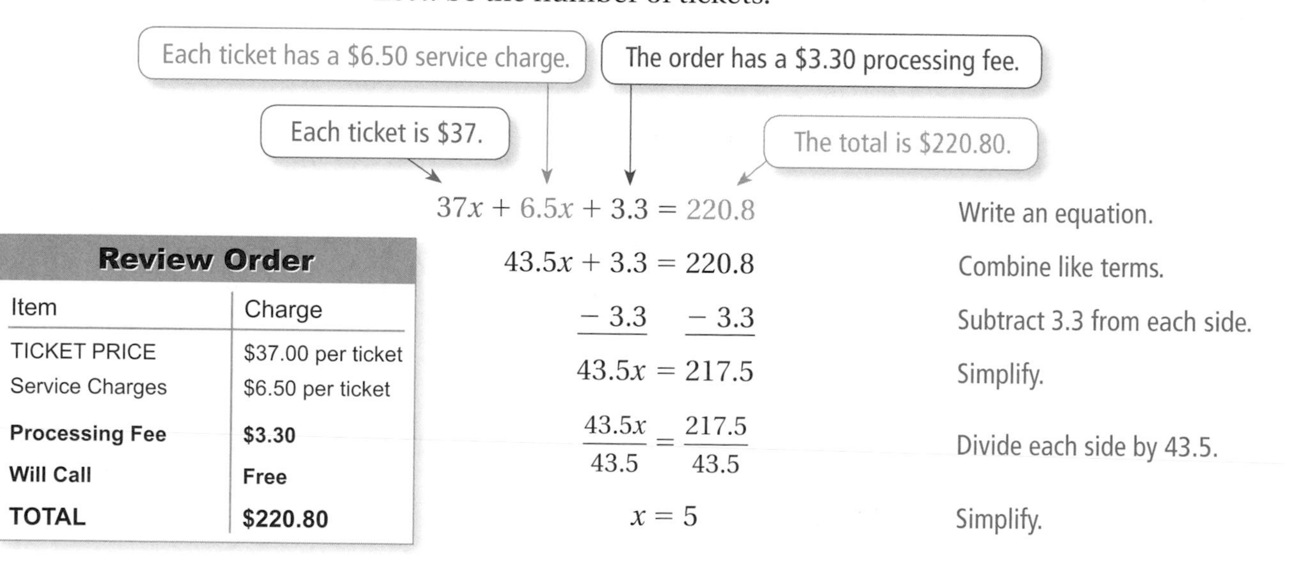

Each ticket has a $6.50 service charge.
The order has a $3.30 processing fee.
Each ticket is $37.
The total is $220.80.

$37x + 6.5x + 3.3 = 220.8$	Write an equation.
$43.5x + 3.3 = 220.8$	Combine like terms.
$-3.3 \qquad -3.3$	Subtract 3.3 from each side.
$43.5x = 217.5$	Simplify.
$\frac{43.5x}{43.5} = \frac{217.5}{43.5}$	Divide each side by 43.5.
$x = 5$	Simplify.

Review Order

Item	Charge
TICKET PRICE	$37.00 per ticket
Service Charges	$6.50 per ticket
Processing Fee	**$3.30**
Will Call	**Free**
TOTAL	**$220.80**

∴ You purchase 5 tickets. So, you receive 5 song downloads.

On Your Own

5. **WHAT IF?** In Example 3, the surface area of the prism is 656 square inches. What is the length of the prism?
6. **WHAT IF?** In Example 4, you change your order. The total is $133.80. How many song downloads do you receive?

10.1 Exercises

Vocabulary and Concept Check

1. **REASONING** Describe the steps you would use to solve the equation $6x + 3x + 2 = 8$.
2. **REASONING** Describe the steps you would use to solve the equation $2(4x - 5) + 11 = -63$.

Practice and Problem Solving

Use the temperature scales to convert between Celsius and Kelvin.

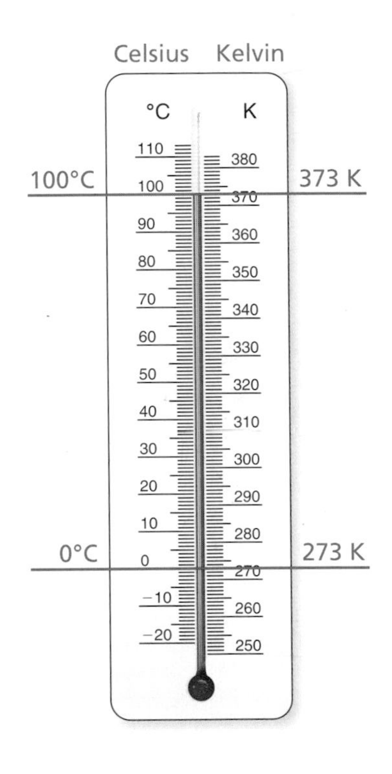

3. Kelvin is a temperature scale used by scientists. Convert 40°C to Kelvin.
4. Water boils at 100°C. At what temperature does water boil on the Kelvin scale?
5. Write an equation that describes the relationship between Kelvin and Celsius.

Solve the equation. Check your solution.

①② 6. $-3x + 7x = 17 + 11$

7. $3x + 12x - 20 = 25$

8. $-5n + 6n + 15 - 3n = -3$

9. $3(2m + 9) - 1 = 2$

10. $4(x - 3) + 7 - 10x = 5$

11. $\frac{2}{3}(y - 12) = -6$

12. **ERROR ANALYSIS** Describe and correct the error in solving the equation.

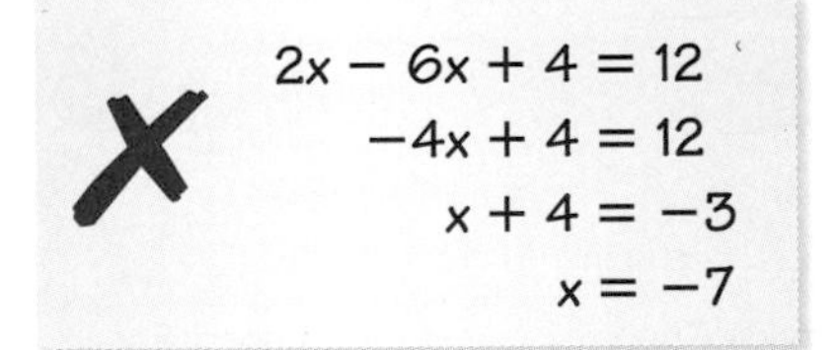

13. **PUZZLE** Twenty-five more than 3 times the quantity of a number decreased by 9 is 43. Find the number.

③ 14. **CYLINDER** What is the diameter of a soup can if the lateral surface area is 188.4 square centimeters and the height is 10 centimeters?

15. **MOVIES** You and three friends go to the movies.
    a. Your group purchases 2 specials, 2 single tickets, and one additional box of popcorn. The total is $36.50. What is the price of a single ticket?
    b. The price of 2 boxes of popcorn and 3 drinks is fifty cents more than purchasing one special. What is the price of a drink?

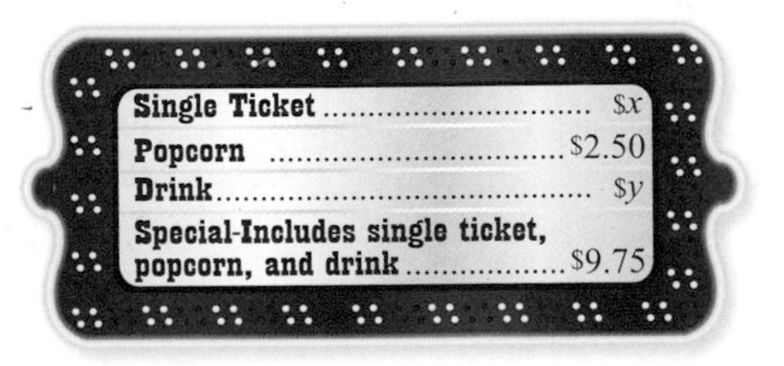

16. PUZZLE During a recent season, there were 18 girls on the high school softball team. There were equal numbers of freshmen, sophomores, and seniors on the team. The number of juniors was two less than the number of seniors. How many juniors were on the team?

17. SALE A store is having a clearance sale. The total for a skirt, a pair of shoes, and a pair of socks is $25.60 before tax. The pair of socks costs $3.50. The original prices of the skirt and the shoes were the same. What is the sale price of the skirt?

Back To School Sale
10% OFF
All Shoes In stock
30% OFF
All Pants & T-Shirts In stock
60% OFF
All Skirts In stock

18. RUNNING Use the clues given by the table.

a. You ran a total of 12 miles this week. Make a table showing the number of miles you ran each day.

b. Did you run every day? Explain.

Day	Mon.	Tues.	Wed.	Thurs.	Fri.
Distance	x	$\frac{1}{6}x$	$1 - \frac{1}{4}x$	$x - \frac{2}{3}$	x

19. Critical Thinking Use the graphic organizer to solve the equation in two different ways. What are the advantages and disadvantages of each way?

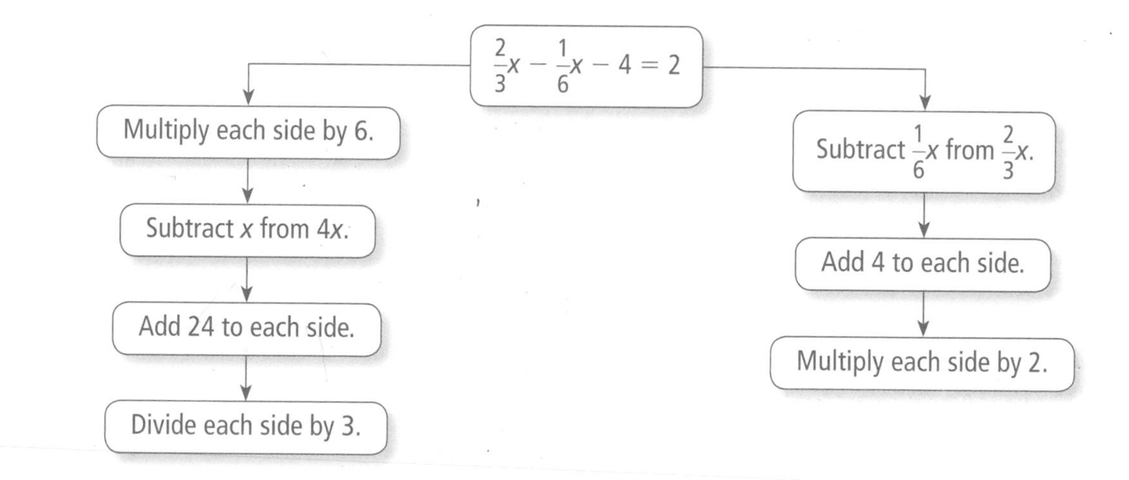

$\frac{2}{3}x - \frac{1}{6}x - 4 = 2$
Multiply each side by 6.
Subtract x from $4x$.
Add 24 to each side.
Divide each side by 3.
Subtract $\frac{1}{6}x$ from $\frac{2}{3}x$.
Add 4 to each side.
Multiply each side by 2.

Fair Game Review What you learned in previous grades & lessons

Simplify. *(Skills Review Handbook)*

20. $12[10 - (3 + 6)] + 14$

21. $2[4 + (3 + 6)] - 12 - 2$

22. MULTIPLE CHOICE What is the value of the expression below? *(Section 1.4)*

$$-4^2 - (-3)^2 + 1$$

Ⓐ -24 Ⓑ -6 Ⓒ 3 Ⓓ 6

10.2 Solving Equations with Variables on Both Sides

COMMON CORE STATE STANDARDS
7.EE.4a
8.EE.7a
8.EE.7b

Essential Question How can you solve an equation that has variables on both sides?

1 ACTIVITY: Using a Table, Graph, and Algebra

Work with a partner. You have an email pal in Antarctica. Your email pal tells you the temperature in McMurdo. You ask whether he gave the temperature in Celsius or Fahrenheit. He says "It's the same on both scales." What is the temperature?

a. TABLE Use "Guess, Check, and Revise" with a table to find the only temperature that is the same on both scales.

F							
C							

b. GRAPH Draw the line given by $C = F$ in the coordinate plane. Locate the point at which the graph of $C = F$ intersects the graph of

$$C = \frac{5}{9}(F - 32).$$

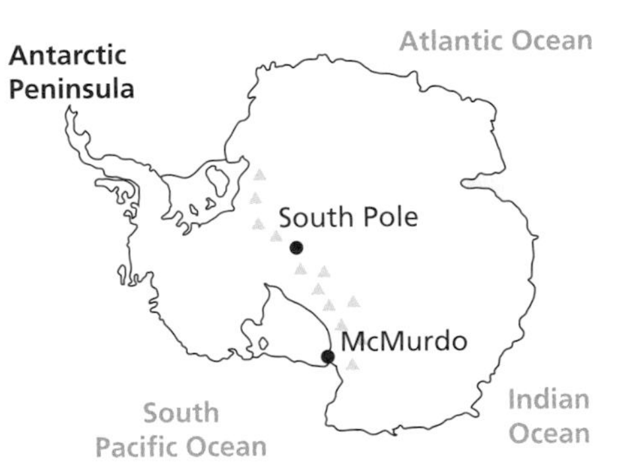

c. ALGEBRA Let x be the temperature that is the same on both scales. Substitute x for C and F in the equation

$$C = \frac{5}{9}(F - 32).$$

Then solve for x.

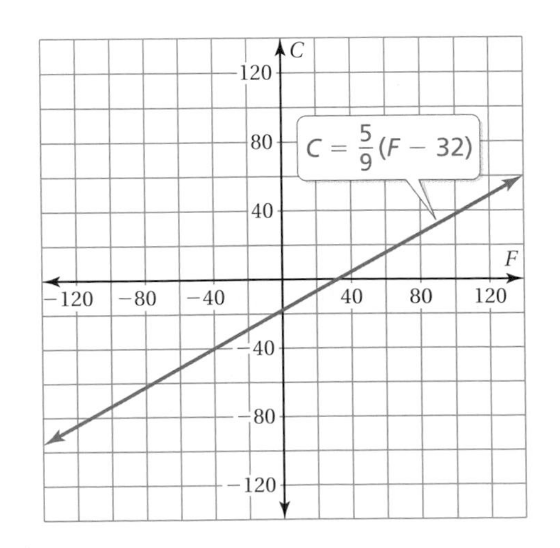

d. Compare your solutions from parts (a)–(c). Did you get the same solution with each method? Which method do you prefer? Why?

2 GAME: Race to the South Pole

Play with a partner.

- Write each expression on a scrap of brown or blue paper. Place the brown pieces of paper in one bag and the blue pieces of paper in another bag.
- Draw an expression from each bag and set them equal to each other.
- If you can solve the equation, you move one space. If you cannot solve the equation, your partner gets a chance to solve it and move one space.
- Put the expressions back into their bags.
- Take turns. The first person to reach the South Pole wins.

Brown Papers	Blue Papers
x	$2x$
$x + 1$	$2x + 4$
$x - 1$	$-2x$
$x + 2$	$-2x + 4$
$x - 2$	$3x$
$x + 3$	$3x + 6$
$x - 3$	$-3x$
	$-3x + 6$

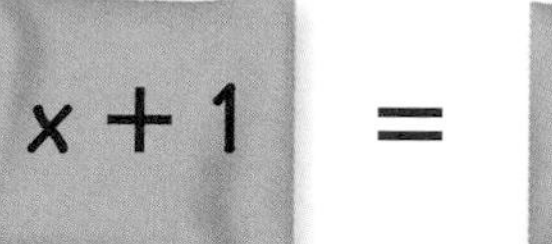

=

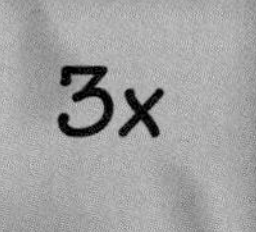

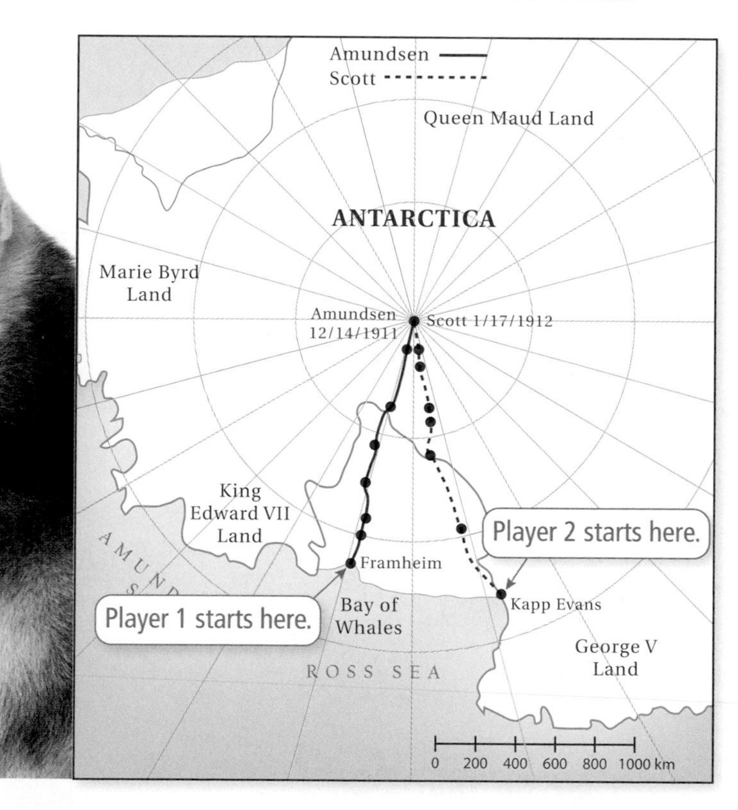

What Is Your Answer?

3. **IN YOUR OWN WORDS** How can you solve an equation that has variables on both sides? Give an example and solve it.

Use what you learned about solving equations with variables on both sides to complete Exercises 4–9 on page 432.

10.2 Lesson

To solve an equation with variables on both sides, collect the variable terms on one side and the constant terms on the other side.

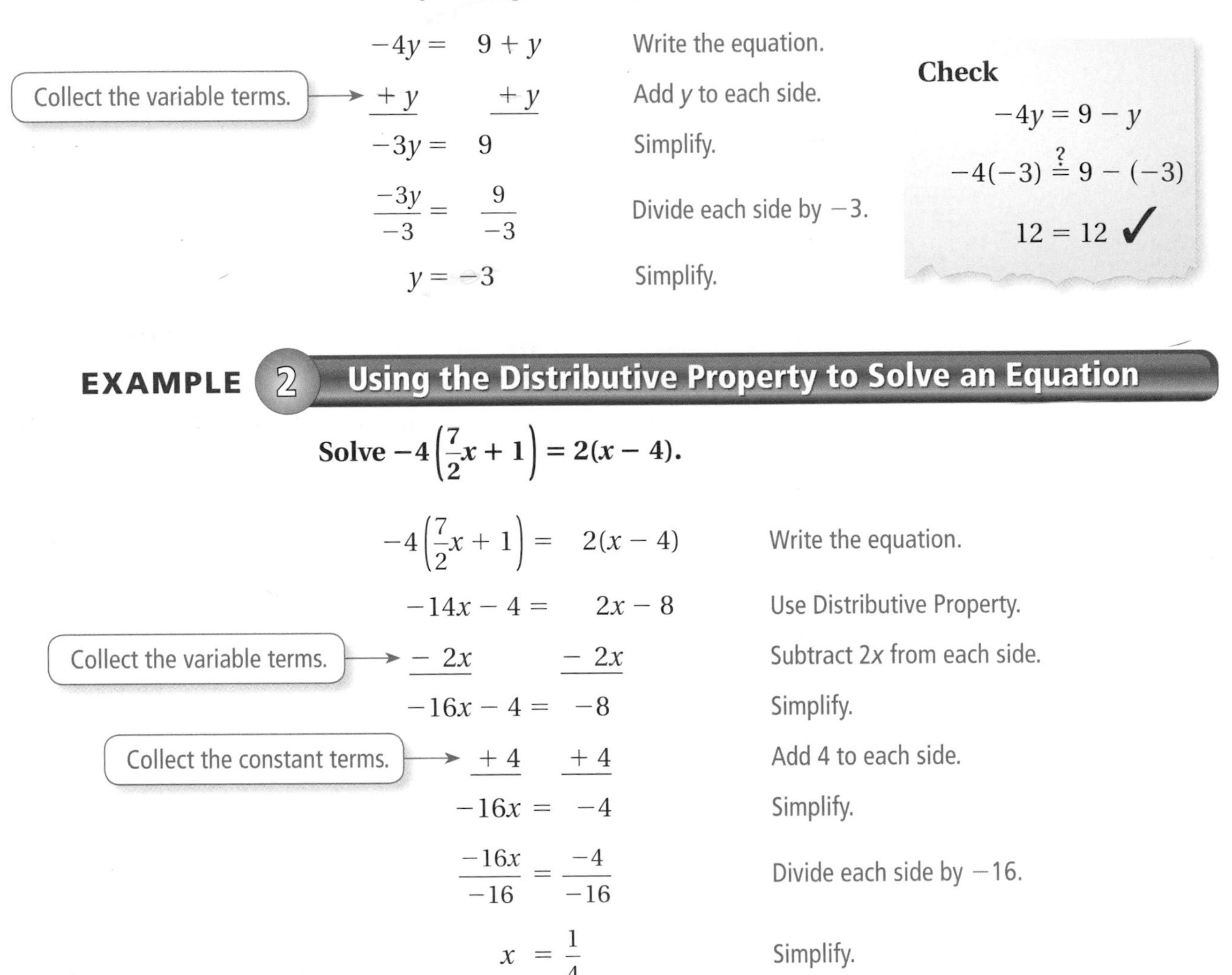

EXAMPLE 1 Solving an Equation with Variables on Both Sides

Solve $-4y = 9 - y$. Check your solution.

$$-4y = 9 + y \quad \text{Write the equation.}$$

Collect the variable terms. → $+y \qquad +y$ Add y to each side.

$$-3y = 9 \quad \text{Simplify.}$$

$$\frac{-3y}{-3} = \frac{9}{-3} \quad \text{Divide each side by } -3.$$

$$y = -3 \quad \text{Simplify.}$$

Check

$$-4y = 9 - y$$

$$-4(-3) \stackrel{?}{=} 9 - (-3)$$

$$12 = 12 \checkmark$$

EXAMPLE 2 Using the Distributive Property to Solve an Equation

Solve $-4\left(\frac{7}{2}x + 1\right) = 2(x - 4)$.

$$-4\left(\frac{7}{2}x + 1\right) = 2(x - 4) \quad \text{Write the equation.}$$

$$-14x - 4 = 2x - 8 \quad \text{Use Distributive Property.}$$

Collect the variable terms. → $-\,2x \qquad -\,2x$ Subtract $2x$ from each side.

$$-16x - 4 = -8 \quad \text{Simplify.}$$

Collect the constant terms. → $+\,4 \qquad +\,4$ Add 4 to each side.

$$-16x = -4 \quad \text{Simplify.}$$

$$\frac{-16x}{-16} = \frac{-4}{-16} \quad \text{Divide each side by } -16.$$

$$x = \frac{1}{4} \quad \text{Simplify.}$$

On Your Own

Now You're Ready
Exercises 4–22

Solve the equation. Check your solution.

1. $-5x = x + 12$
2. $15 - 4z = z - 8$
3. $6x = 2(7x + 3)$
4. $-3(1 - x) = \frac{1}{2}(2x + 1)$

EXAMPLE 3 Standardized Test Practice

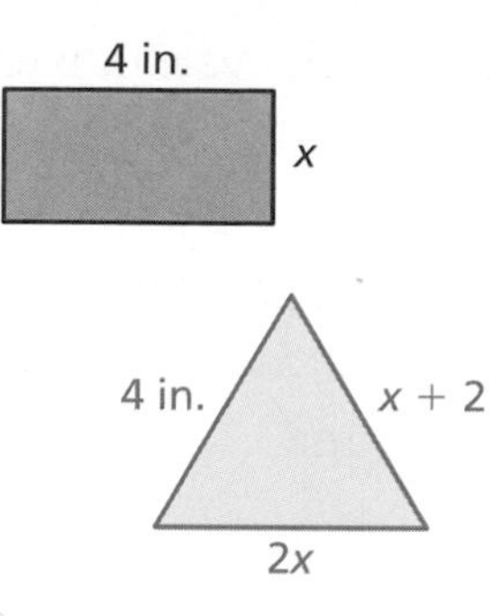

The polygons have the same perimeter. What is the area of the rectangle?

Ⓐ 12 in. Ⓑ 8 in.^{2} Ⓒ 12 in.^{2} Ⓓ 24 in.^{2}

The perimeter of the rectangle is $x + x + 4 + 4 = 2x + 8$.

The perimeter of the triangle is $4 + 2x + x + 2 = 3x + 6$.

Perimeter of rectangle — Perimeter of triangle

$2x + 8 = 3x + 6$ Write an equation.

$8 = x + 6$ Subtract $2x$ from each side.

$2 = x$ Subtract 6 from each side.

So, the area of the rectangle is $2(4) = 8$ square inches. The correct answer is Ⓑ.

EXAMPLE 4 Real-Life Application

A boat travels x miles per hour from Toledo to Buffalo. The boat travels 3 miles per hour slower on the return trip. What is the distance from Toledo to Buffalo?

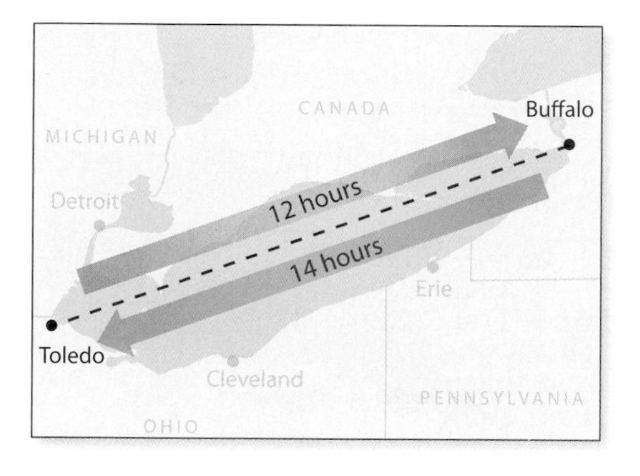

The speed, or rate, of the boat on the return trip is $(x - 3)$ miles per hour.

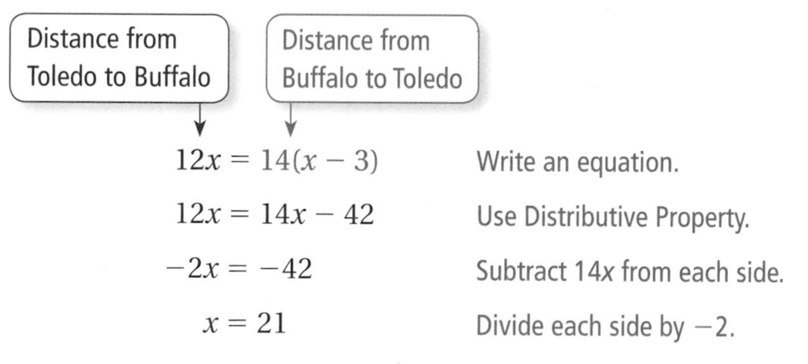

$12x = 14(x - 3)$ Write an equation.

$12x = 14x - 42$ Use Distributive Property.

$-2x = -42$ Subtract $14x$ from each side.

$x = 21$ Divide each side by -2.

The boat travels 21 miles per hour from Toledo to Buffalo in 12 hours. So, the boat travels $12(21) = 252$ miles.

The distance from Toledo to Buffalo is 252 miles.

On Your Own

5. In Example 3, what is the perimeter of the triangle?
6. In Example 4, what is the speed of the boat on the return trip?

10.2 Exercises

Vocabulary and Concept Check

1. **REASONING** Describe the steps you would use to solve $5x = 3x + 4$.
2. **OPEN-ENDED** Write an equation with variables on both sides that has a solution of -1.
3. **WRITING** To solve $-3x + 4 = -2x$, is it easier to add $3x$ to each side or add $2x$ to each side? Explain.

Practice and Problem Solving

Solve the equation. Check your solution.

(1)

4. $2x = -x - 6$
5. $-9 + c = 4c$
6. $-3y = 7y$
7. $-5a = 2 - a$
8. $6w - 5 = 8w$
9. $3z = 14 + 10z$
10. $4f + 8 = 9f - 12$
11. $3p - 11 = 5p + 6$
12. $-7k + 10 = 9k + 18$
13. $-2 + 2d = 6d + 6$
14. $8b - 7 = -11 + 3b$
15. $12h - 7 = 6h + 8$

16. **ERROR ANALYSIS** Describe and correct the error in solving the equation.

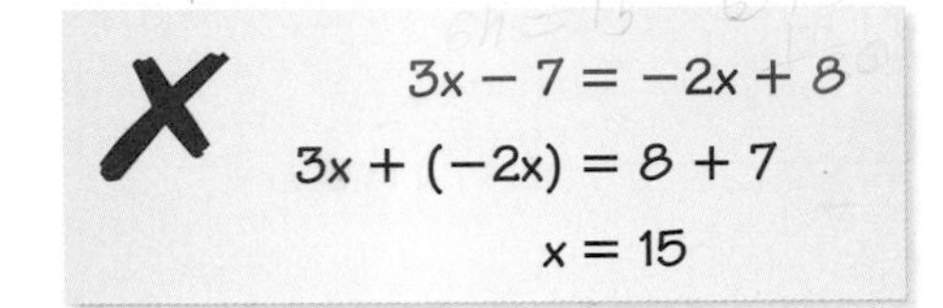

Solve the equation. Check your solution.

(2)

17. $2k - 8 = 4(k + 1)$
18. $3(g - 3) = 2(6 - 2g)$
19. $-5(f + 7) = 3(3f - 1)$
20. $9.2 - 4w = -2(3w + 5)$
21. $2.5(3b - 4) = 3.5b - 6b$
22. $6(1.5h - 1) = 5(2.2h + 3)$

23. **ERROR ANALYSIS** Describe and correct the error in solving the equation.

✗

$$2(v - 5) = -(3v + 5)$$
$$2v - 10 = -3v + 5$$
$$5v = 15$$
$$v = 3$$

24. **MUSIC LESSONS** It costs \$50 to be a member of a music club. A member of the club pays \$10 per music lesson. A nonmember pays \$20 per music lesson. How many music lessons must a member and a nonmember take so that the cost for each is the same?

In a regular polygon each side has the same length. Find the perimeter of the regular polygon.

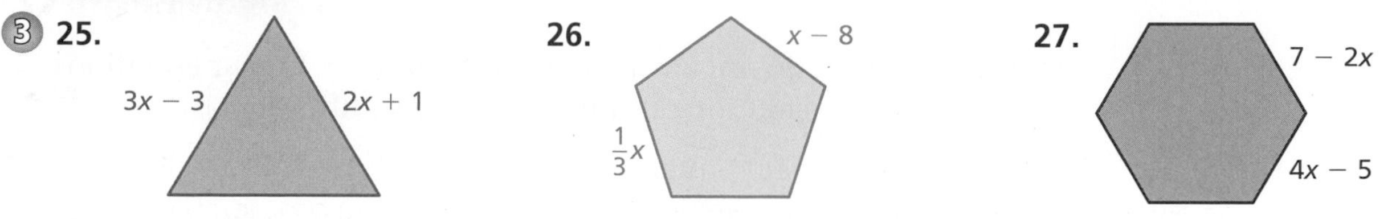

Write an equation that describes the situation. Then solve the equation.

28. Two more than the product of a number n and 3 is equal to the product of the number n and 4.

29. Three less than 5 times a number n is equal to 8 more than negative 6 times the number n.

30. SAVINGS The table shows the beginning balance in each savings account. You save \$10 per week and your friend saves \$17 per week. How many weeks will it take for you and your friend to have the same balance?

Savings Account Balance	
You	**Your Friend**
\$126	\$84

31. FLIGHT A nonstop flight from Billings to Minneapolis takes x hours. The return flight takes 30 minutes longer. What is the distance from Billings to Minneapolis?

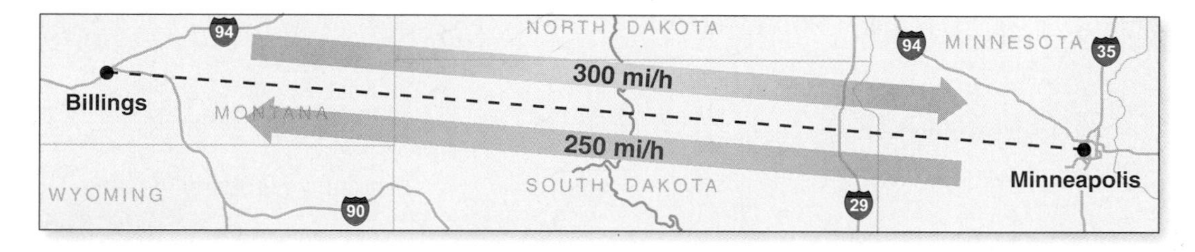

32. PLAYGROUND There are 87 children at a playground. There are 23 more boys than girls. How many of each are at the playground?

33. GEOMETRY $\triangle ABC$ is similar to $\triangle DEF$. The ratio of AC to DF is $3:5$. What is the perimeter of each triangle?

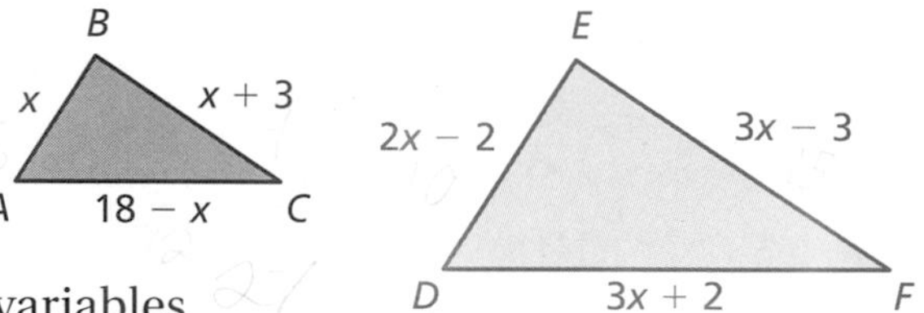

34. Critical Thinking Is it possible for an equation with variables on both sides to have no solution? If so, give an example.

Fair Game Review What you learned in previous grades & lessons

Graph the function. *(Skills Review Handbook)*

35. $y = 2x - 1$ **36.** $y = 4 - x$ **37.** $y = 0.5x + 2$

38. MULTIPLE CHOICE What is the theoretical probability of rolling an even number on a number cube? *(Section 9.2)*

Ⓐ $\frac{1}{6}$ Ⓑ $\frac{1}{3}$ Ⓒ $\frac{1}{2}$ Ⓓ $\frac{2}{3}$

10.2b Solutions of Linear Equations

Linear equations do not always have one solution. Linear equations can also have no solution or infinitely many solutions.

When solving a linear equation that has no solution, you will obtain an equivalent equation that is not true for any value of x, such as $0 = 2$.

EXAMPLE 1 Solving Equations with No Solution

a. Solve $3 - 4x = -7 - 4x$.

	$3 - 4x = -7 - 4x$	Write the equation.
Undo the subtraction. →	$+ 4x \qquad + 4x$	Add $4x$ to each side.
	$3 = -7$ ✗	Simplify.

The equation $3 = -7$ is never true. So, the equation has no solution.

b. Solve $\frac{1}{2}(10x + 7) = 5x$.

	$\frac{1}{2}(10x + 7) = 5x$	Write the equation.
	$5x + \frac{7}{2} = 5x$	Distributive Property
Undo the addition. →	$- 5x \qquad - 5x$	Subtract $5x$ from each side.
	$\frac{7}{2} = 0$ ✗	Simplify.

The equation $\frac{7}{2} = 0$ is never true. So, the equation has no solution.

Practice

Solve the equation.

1. $x + 6 = x$
2. $2x + 1 = 2x - 1$
3. $3x - 1 = 1 - 3x$
4. $4x - 9 = 3.5x - 9$
5. $\frac{1}{3}(2x + 9) = \frac{2}{3}x$
6. $6(5 - 2x) = -4(3x + 1)$
7. **GEOMETRY** Are there any values of x for which the areas of the figures are the same? Explain.

When solving a linear equation that has infinitely many solutions, you will obtain an equivalent equation that is true for all values of x, such as $-5 = -5$.

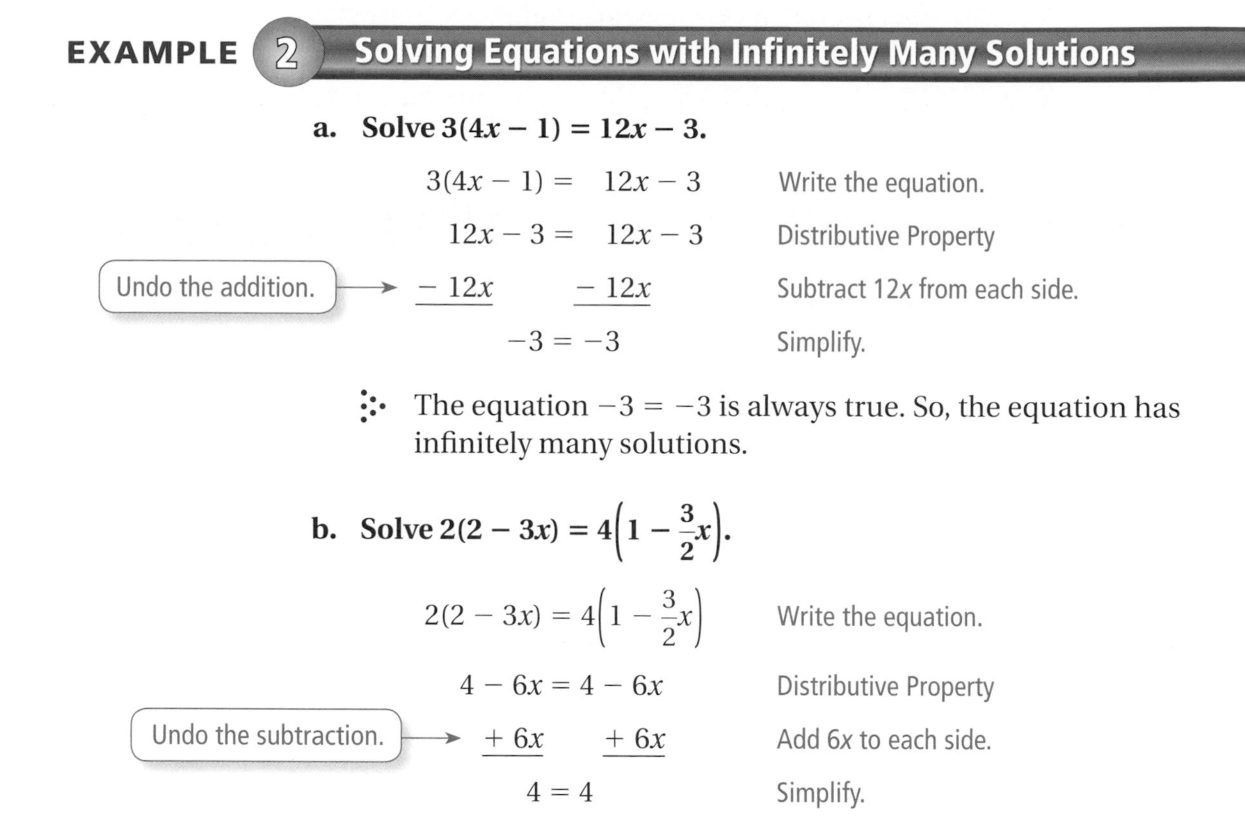

EXAMPLE 2 Solving Equations with Infinitely Many Solutions

a. Solve $3(4x - 1) = 12x - 3$.

$3(4x - 1) = 12x - 3$	Write the equation.
$12x - 3 = 12x - 3$	Distributive Property
$-12x \qquad -12x$	Subtract $12x$ from each side.
$-3 = -3$	Simplify.

Undo the addition.

The equation $-3 = -3$ is always true. So, the equation has infinitely many solutions.

b. Solve $2(2 - 3x) = 4\left(1 - \frac{3}{2}x\right)$.

$2(2 - 3x) = 4\left(1 - \frac{3}{2}x\right)$	Write the equation.
$4 - 6x = 4 - 6x$	Distributive Property
$+6x \qquad +6x$	Add $6x$ to each side.
$4 = 4$	Simplify.

Undo the subtraction.

The equation $4 = 4$ is always true. So, the equation has infinitely many solutions.

Practice

Solve the equation.

8. $x + 8 - x = 9$

9. $\frac{1}{2}x + \frac{1}{2}x = x + 1$

10. $3x + 15 = 3(x + 5)$

11. $\frac{1}{2}(6x - 4) = 3x - 2$

12. $5x - 7 = 4x - 1$

13. $2x + 4 = -(-7x + 6)$

14. $5.5 - x = -4.5 - x$

15. $10x - \frac{8}{3} - 4x = 6x$

16. $-3(2x - 3) = -6x + 9$

17. $6(7x + 7) = 7(6x + 6)$

18. $\frac{3}{4}(4x - 8) = -10$

19. $-\frac{1}{8} = 2(x - 1)$

10.3 Solving Equations Using Tables and Graphs

COMMON CORE STATE STANDARDS
7.EE.4a
8.EE.7a
8.EE.7b

Essential Question How can you use tables and graphs to solve equations?

1 ACTIVITY: Using a Table, Graph, and Algebra

Work with a partner. You start a website design company. How many sites must you design before you start making a profit?

- **You pay $4000 for a new computer and software.**
- **It costs you $100 to design each website.**
- **You charge $500 to design each website.**

Let x represent the number of sites you design.

$C = 4000 + 100x$ Cost of designing x sites

$R = 500x$ Income for designing x sites

You will start making a profit when $C = R$. That is, when you have designed enough websites to cover your start-up cost of $4000 and $100 for each site.

a. TABLE Use "Guess, Check, and Revise" with a table to find the value of x for which $C = R$.

x							
C							
R							

b. GRAPH Graph $C = 4000 + 100x$ and $R = 500x$ in the same coordinate plane. Find the value of x for which the two lines intersect.

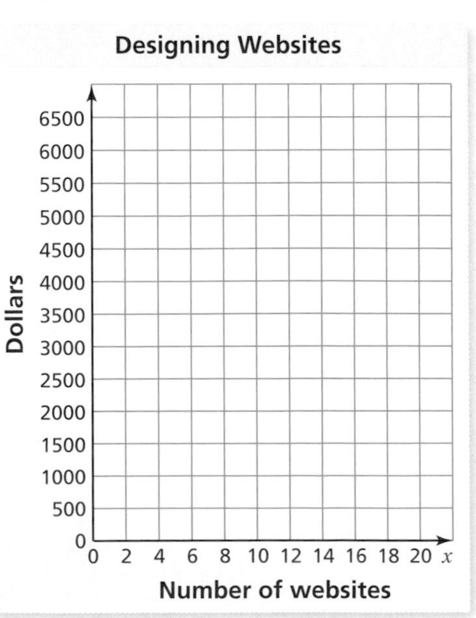

c. ALGEBRA Set C equal to R.

$$C = R$$

$$4000 + 100x = 500x$$

Solve for x.

d. The point at which the two lines intersect is called the "break-even" point. Why is it called this?

2 ACTIVITY: Planning Your Own Business

Make a plan to start your own business.

- Describe your business.
- Are you providing a product or a service?
- Make a list of the things you need to start the business. Find the cost of each item or service.
- Write an equation that represents the cost of making x items. Write an equation that represents the income for selling x items.
- Use a table to compare the cost and income for several values of x.

x							
C							
R							

- Draw a graph that shows when your company will reach the break-even point.

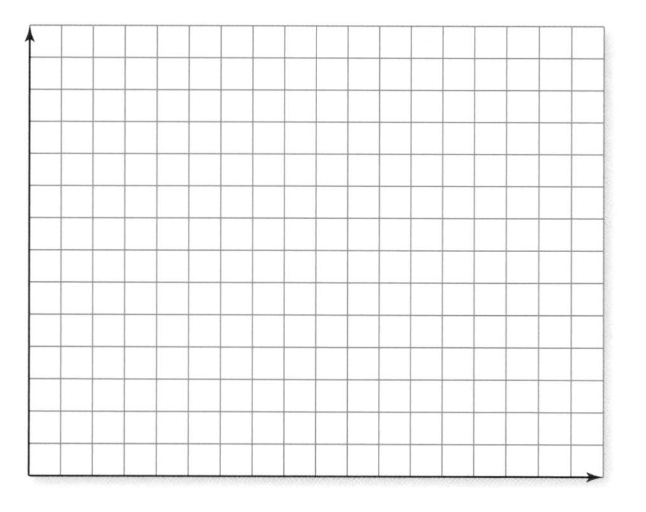

- Organize all of your planning in a folder. Include your company name, logo, and a plan for advertising and selling your product or service.

What Is Your Answer?

3. **IN YOUR OWN WORDS** How can you use tables and graphs to solve equations? Describe a real-life example.

Practice Use what you learned about solving equations using tables and graphs to complete Exercises 4–6 and 13–15 on page 438.

10.3 Lesson

Solving Equations

Method 1 **Use algebra.** *(Section 10.1 and Section 10.2)*

Method 2 **Use a table.** Find the value of the variable that makes the value of each side of the equation the same. *(Section 10.3)*

Method 3 **Use a graph.** Graph each side of the equation. The x-coordinate of the point of intersection is the solution of the equation. *(Section 10.3)*

EXAMPLE 1 Solving an Equation Using a Table

Use a table to solve $4x + 2 = 5x$. Check your solution.

Find the x-value that makes $4x + 2$ equal to $5x$.

Try different values of x.

x	$4x + 2$	$5x$
-1	$4(-1) + 2 = -2$	$5(-1) = -5$
0	$4(0) + 2 = 2$	$5(0) = 0$
1	$4(1) + 2 = 6$	$5(1) = 5$
2	$4(2) + 2 = 10$	$5(2) = 10$

Each side of the equation equals 10 when $x = 2$.

∴ The solution is $x = 2$.

Check

$4x + 2 = 5x$

$-4x \quad -4x$

$2 = x$ ✓

EXAMPLE 2 Solving an Equation Using Graphs

Use a graph to solve $x + 1 = 5 - x$. Check your solution.

Write equations for each side of the original equation.

$$x + 1 = 5 - x$$

$y = x + 1$ $\quad$ $y = 5 - x$

Use a graphing calculator to graph $y = x + 1$ and $y = 5 - x$.

X=2 Y=3

∴ The lines intersect at (2, 3). So, the solution is $x = 2$.

Check

$x + 1 = 5 - x$

$2 + 1 \stackrel{?}{=} 5 - 2$

$3 = 3$ ✓

On Your Own

Now You're Ready
Exercises 4–15

Use a table or a graph to solve the equation. Check your solution. Then explain your choice of method.

1. $1 - 2z = 2z - 3$

2. $-3x + 2 = x + 18$

3. $-x = 2x - 3$

4. $x + 2 = 2x - 3$

EXAMPLE 3 Real-Life Application

Hourly Rentals	
Canoe	Kayak
\$15	\$10

You pay \$30 for a workshop on environmental conservation. Then you rent a canoe for x hours. Your friend pays \$45 for a kayak paddling lesson, then rents a kayak for x hours. At the end of the day, you both spent the same amount. How many hours did you rent the canoe?

Write equations for your total cost and your friend's total cost.

Your total cost

$y = 15x + 30$

(canoe cost: $15x$; workshop cost: 30)

Your friend's total cost

$y = 10x + 45$

(kayak cost: $10x$; lesson cost: 45)

Use a graphing calculator to graph each equation.

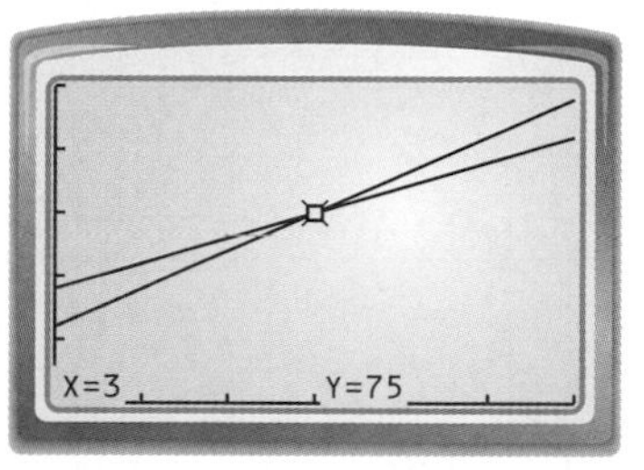

The lines intersect at (3, 75). So, you rented the canoe for 3 hours and paid a total of \$75.

Check Find the total costs for 3 hours.

Your total cost

$y = 15x + 30$

$= 15(3) + 30$

$= 75$

Your friend's total cost

$y = 10x + 45$

$= 10(3) + 45$

$= 75$ ✓

On Your Own

5. WHAT IF? In Example 3, your friend receives a \$5 discount on the paddling lesson. At the end of the day, you both spent the same amount. How many hours did you rent the canoe?

10.3 Exercises

Vocabulary and Concept Check

1. **REASONING** What equation is being solved by the table? What is the solution?

x	$3x - 2$	$x + 2$
0	−2	2
1	1	3
2	4	4

2. **VOCABULARY** You graph each side of an equation. The point of intersection is $(4, -2)$. What is the solution of the equation?

3. **WRITING** Would you rather use a table or a graph to solve an equation? Explain.

Practice and Problem Solving

Use a table to solve the equation. Check your solution.

① 4. $4x = 3x + 1$ 5. $7h - 6 = 4h$ 6. $-3k - 4 = k$

7. $9 = 3g - 6$ 8. $6z = 8z + 2$ 9. $3a + 5 = -11 - 5a$

10. $3 - 4q = 2q - 21$ 11. $-5p - 3 = 2p + 18$ 12. $-7w + 15 = 3w - 5$

Use the graph to solve the equation. Check your solution.

② 13. $-2x - 6 = x + 3$ 14. $2x - 4 = -3x + 11$ 15. $x - 1 = 4x + 5$

$y = x + 3$, $(-3, 0)$, $y = -2x - 6$

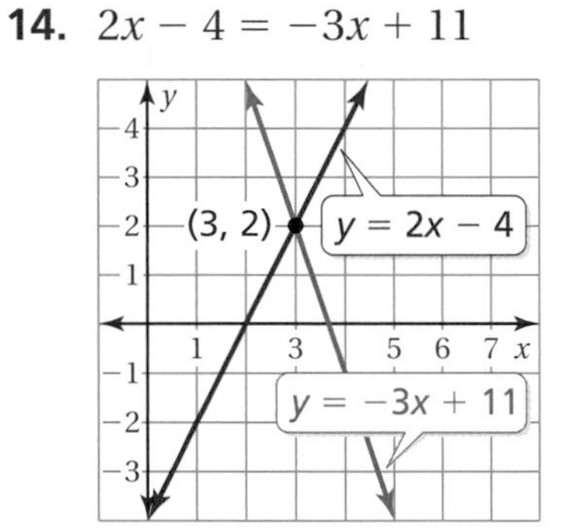

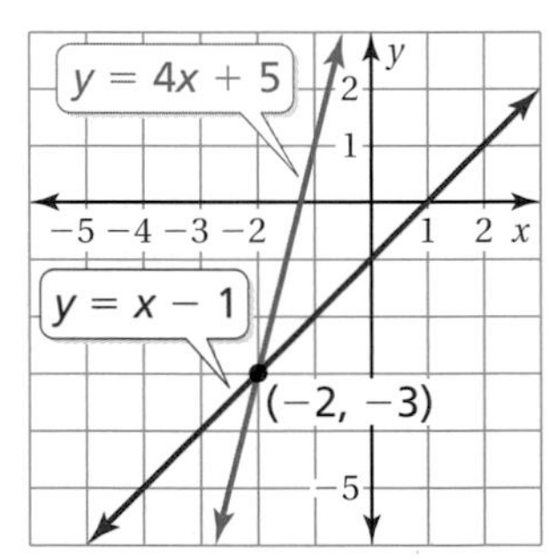

16. **SHOPPING** You buy 5 shirts and your friend buys 8 shirts. You spend \$48 on shoes. You both spend the same amount of money. All the shirts cost the same amount. Use a table to find the cost of each shirt.

17. **CELL PHONE** A wireless carrier offers a free cell phone with Plan A or a \$120 cell phone with Plan B. Use a graph to find the number of months it takes for the cost of Plan A to equal the cost of Plan B.

Find x. Then find the perimeter of the square.

18. **19.** **20.**

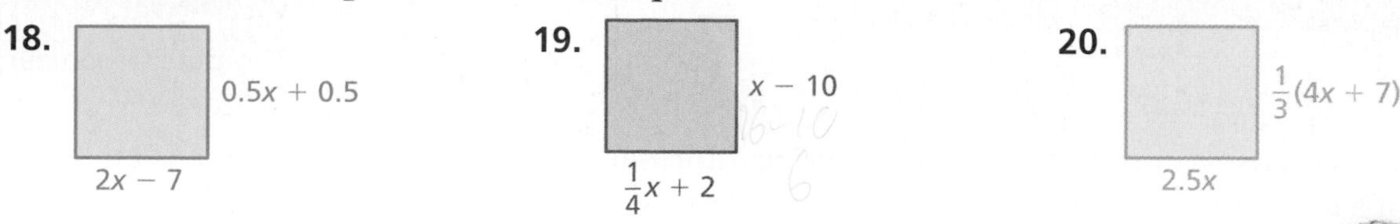

21. CRITICAL THINKING Explain how to solve the equation $5x + 1 = 10x - 10$ using a table and "Guess, Check, and Revise." What is the solution?

22. RIBBON You cut a piece of ribbon into two pieces so that one piece is 24 inches longer than the other. The longer piece is four times as long as the shorter piece.

a. How long is each piece of ribbon?

b. How long was the original ribbon?

23. BICYCLING You and a friend are participating in a bicycling fundraiser. Both of you start at the same location. You leave 30 minutes before your friend. How many hours will it take for your friend to catch you?

Cycling Rate (mi/h)	
You	Your Friend
15	18

24. REASONING Use the graph to determine whether the equation $x + 3 = x - 1$ has a solution. Explain your reasoning.

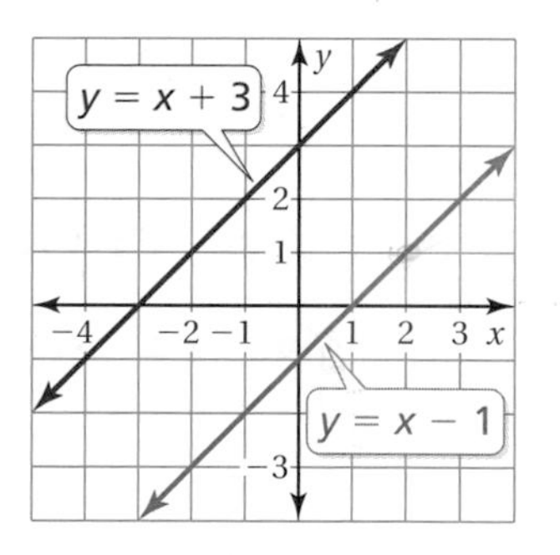

25. Geometry The area of the rectangle is twice the area of the triangle. Find the area of each figure.

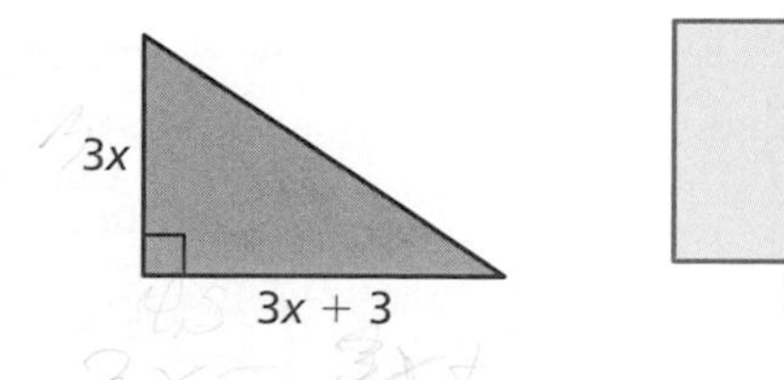

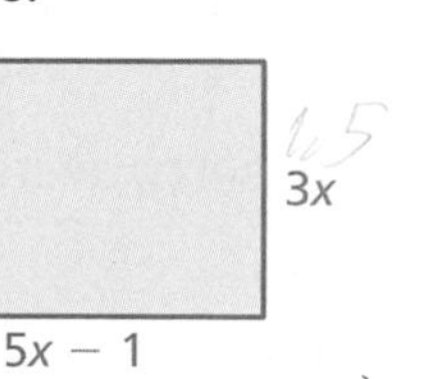

Fair Game Review What you learned in previous grades & lessons

Find the slope of the line that passes through the two points. *(Section 3.2)*

26. (0, 0), (4, 6)

27. (2, 1), (6, 3)

28. (−3, −1), (6, 2)

29. MULTIPLE CHOICE Your backpack contains 3 blue pens and 2 black pens. You randomly choose one pen to give to a friend and then you randomly choose another pen to use yourself. What is the probability that both pens are blue? *(Section 9.4)*

Ⓐ $\frac{6}{25}$ Ⓑ $\frac{3}{10}$ Ⓒ $\frac{9}{25}$ Ⓓ $\frac{3}{5}$

10 Study Help

Check It Out
Graphic Organizer
BigIdeasMath.com

You can use a **concept circle** to organize information about a concept. Here is an example of a concept circle for combining like terms to solve an equation.

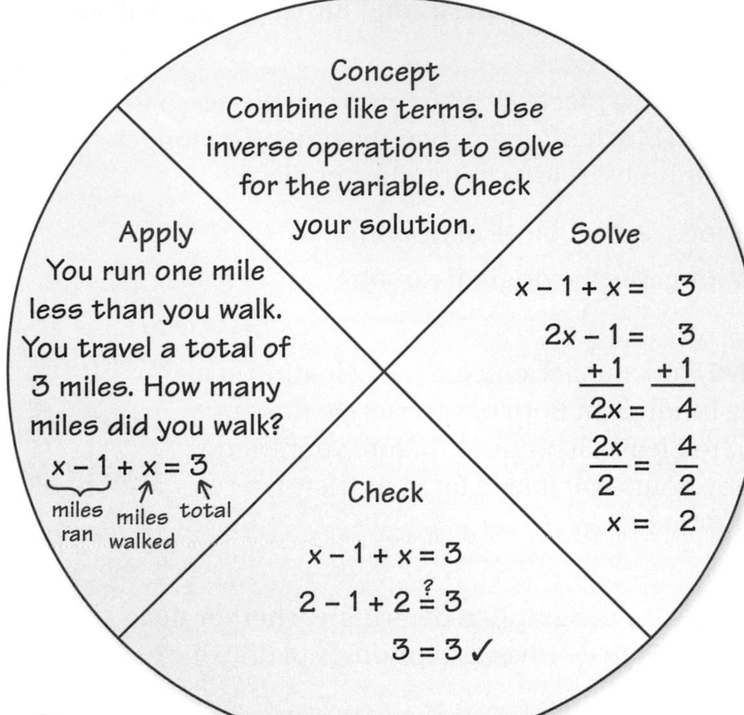

Combining Like Terms to Solve an Equation
Concept
Combine like terms. Use inverse operations to solve for the variable. Check your solution.
Apply
You run one mile less than you walk. You travel a total of 3 miles. How many miles did you walk?
x − 1 + x = 3
miles ran
miles walked
total
Solve
x − 1 + x = 3
2x − 1 = 3
+ 1 + 1
2x = 4
2x/2 = 4/2
x = 2
Check
x − 1 + x = 3
2 − 1 + 2 ≟ 3
3 = 3 ✓

On Your Own

Make a concept circle to help you study these topics.

1. using the distributive property to solve an equation
2. solving an equation with variables on both sides
3. solving an equation using a table
4. solving an equation using a graph

After you complete this chapter, make concept circles for the following topics.

5. finding the slope of a line
6. graphing a line using a point and a slope
7. identifying slope and y-intercept
8. graphing a line using slope-intercept form
9. graphing a linear function

"Do you think this concept circle will help my owner understand that 'Speak' and 'Sit' need motivation?"

10.1–10.3 Quiz

Solve the equation. Check your solution. *(Section 10.1 and Section 10.2)*

1. $-9x + 5x - 18 = 2$
2. $5(2t - 5) - 6 = 9$
3. $-5y = 8 - y$
4. $3a - 8 = 2(6 - a)$

Use a table to solve the equation. Check your solution. *(Section 10.3)*

5. $2x + 1 = 3x$
6. $-7w - 4 = 2 - 4w$

Use the graph to solve the equation. Check your solution. *(Section 10.3)*

7. $-2x + 7 = 3x - 3$

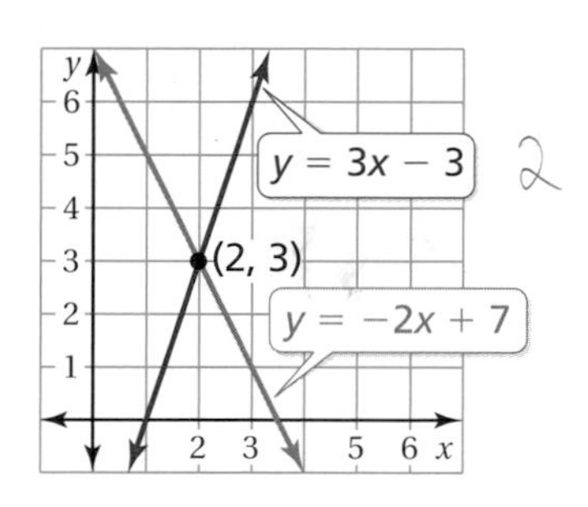

8. $\frac{1}{2}x + 3 = -2x - 2$

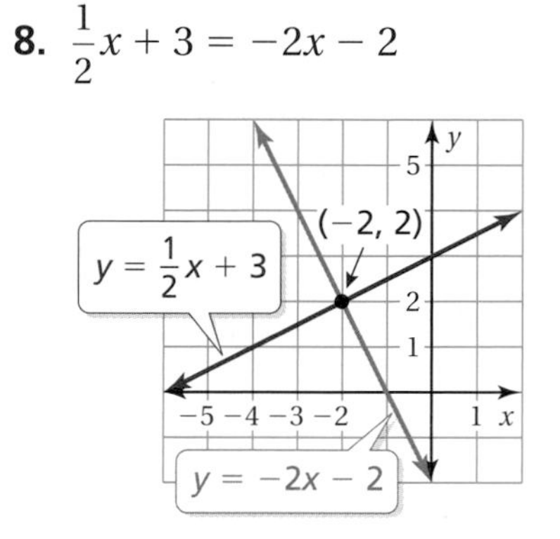

9. The rectangles have the same perimeter. What is the area of the red rectangle? *(Section 10.2)*

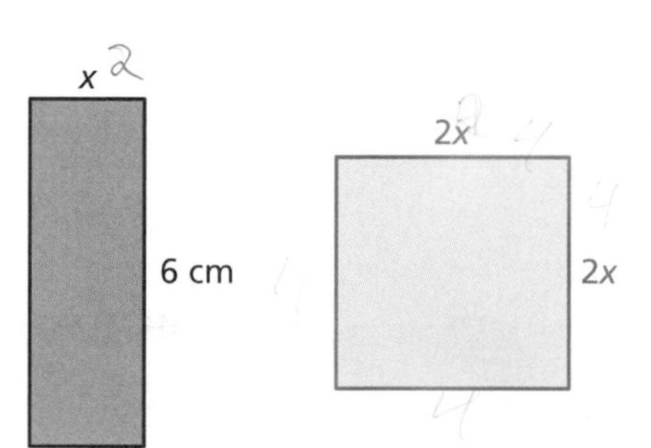

10. **PUZZLE** Sixteen more than 5 times the quantity of a number minus 6 is 20. Find the number. *(Section 10.1)*

11. **WOOD** You cut a piece of wood into two smaller pieces so that one piece is 40 centimeters longer than the other piece. The longer piece is three times as long as the shorter piece. *(Section 10.3)*

    a. What are the lengths of the pieces of wood?

    b. How long was the original piece of wood?

12. **BETTAS** An aquarium store sells red, blue, and orange bettas. There are 14 bettas in the store. Two bettas are orange. The number of blue bettas is 3 times the number of red bettas. Find the number of blue bettas in the store. *(Section 10.1)*

10.4 Slope of a Line

COMMON CORE STATE STANDARDS
7.RP.2b

Essential Question How can the slope of a line be used to describe the line?

You studied the following definition of the slope of a line.

Slope is the rate of change between any two points on a line. It is a measure of the *steepness* of a line. To find the slope of a line, find the ratio of the change in y (vertical change) to the change in x (horizontal change).

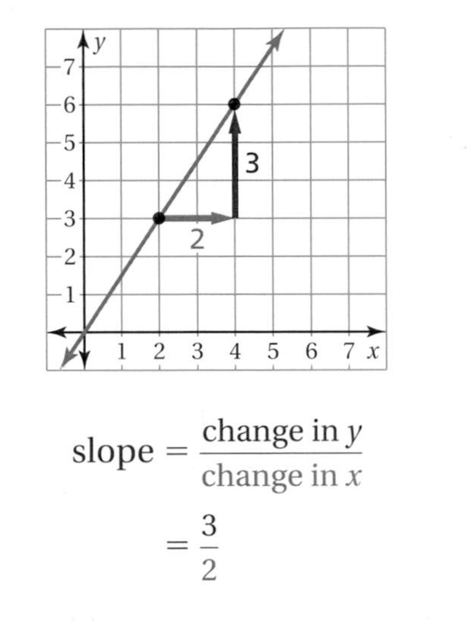

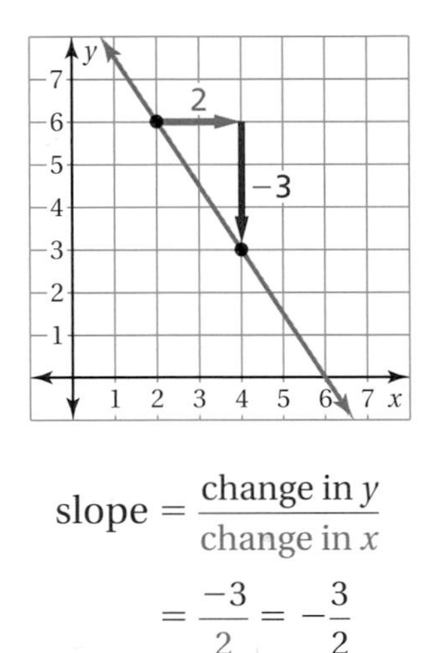

$$\text{slope} = \frac{\text{change in } y}{\text{change in } x} = \frac{3}{2}$$

$$\text{slope} = \frac{\text{change in } y}{\text{change in } x} = \frac{-3}{2} = -\frac{3}{2}$$

1 ACTIVITY: Extending the Concept of a Slope

Work with a partner. Find the slope of each line.

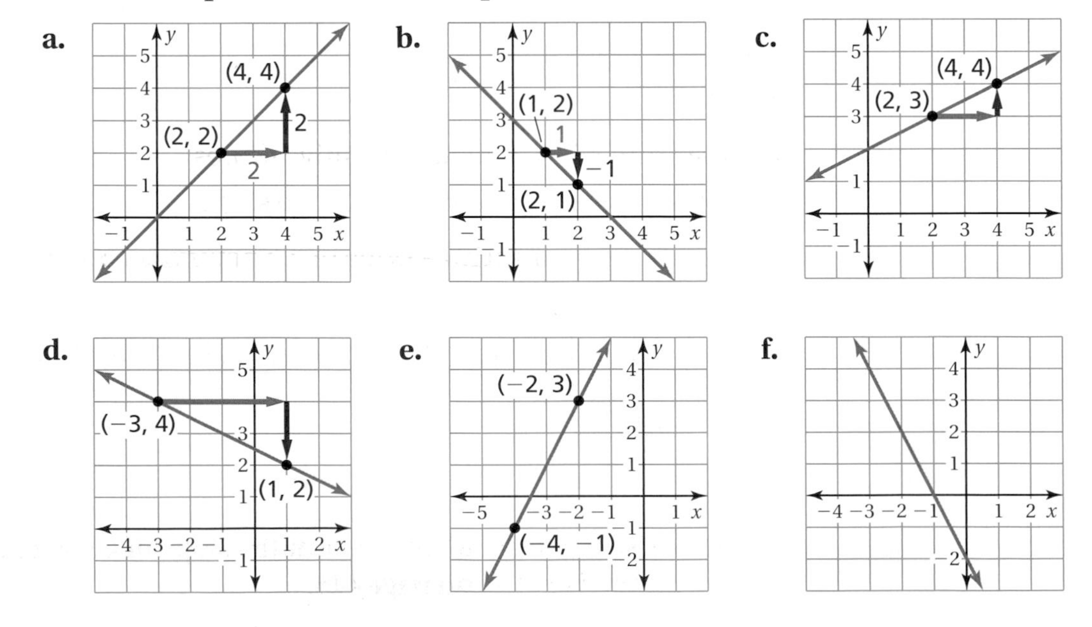

Inductive Reasoning

Work with a partner. Copy and complete the table.

	Two Points	Change in y	Change in x	Slope of Line
1a	**2.** (2, 2), (4, 4)	2	2	
1b	**3.** (1, 2), (2, 1)	-1	1	
1c	**4.** (2, 3), (4, 4)			
1d	**5.** $(-3, 4)$, (1, 2)			
1e	**6.** $(-4, -1)$, $(-2, 3)$			
1f	**7.**			
	8. $(-4, 0)$, (0, 1)			
	9. $(-3, 4)$, $(6, -2)$			
	10. $(-4, 2)$, $(8, -1)$			
	11. $(-6, -1)$, (3, 5)			
	12. $(-5, 7)$, $(10, -5)$			
	13. (0, 1), (4, 1)			
	14. $(-4, -2)$, $(-3, -6)$			

What Is Your Answer?

15. **IN YOUR OWN WORDS** How can the slope of a line be used to describe the line?

a. Draw three lines that have positive slopes.

b. Draw three lines that have negative slopes.

16. Compare a slope of 1 with a slope of 2. Show your comparison on a graph.

17. Compare a slope of -1 with a slope of -2. Show your comparison on a graph.

Practice Use what you learned about the slope of a line to complete Exercises 4–7 on page 446.

10.4 Lesson

Check It Out
Lesson Tutorials
BigIdeasMath.com

Key Vocabulary
slope, *p. 444*
rise, *p. 444*
run, *p. 444*

Key Idea

Slope

The **slope** of a line is a ratio of the change in y (the **rise**) to the change in x (the **run**) between any two points on the line.

$$\text{slope} = \frac{\text{change in } y}{\text{change in } x} = \frac{\text{rise}}{\text{run}}$$

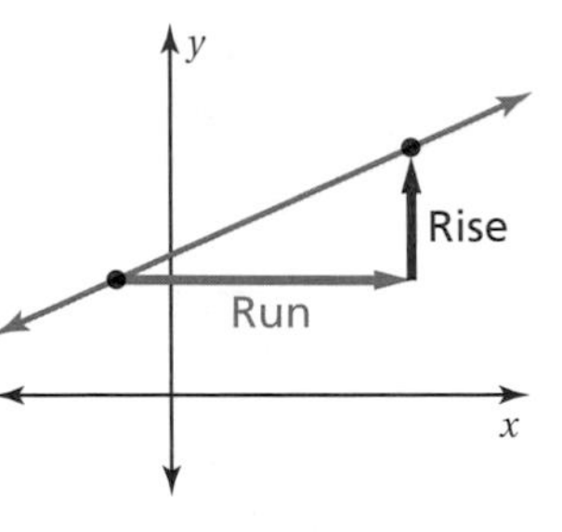

EXAMPLE 1 Real-Life Application

What is the slope of the stairs when walking up to the apartment?

$\text{slope} = \frac{\text{rise}}{\text{run}}$ Write formula for slope.

$= \frac{8 \text{ ft}}{12 \text{ ft}}$ Substitute.

$= \frac{2}{3}$ Simplify.

The slope of the stairs is $\frac{2}{3}$.

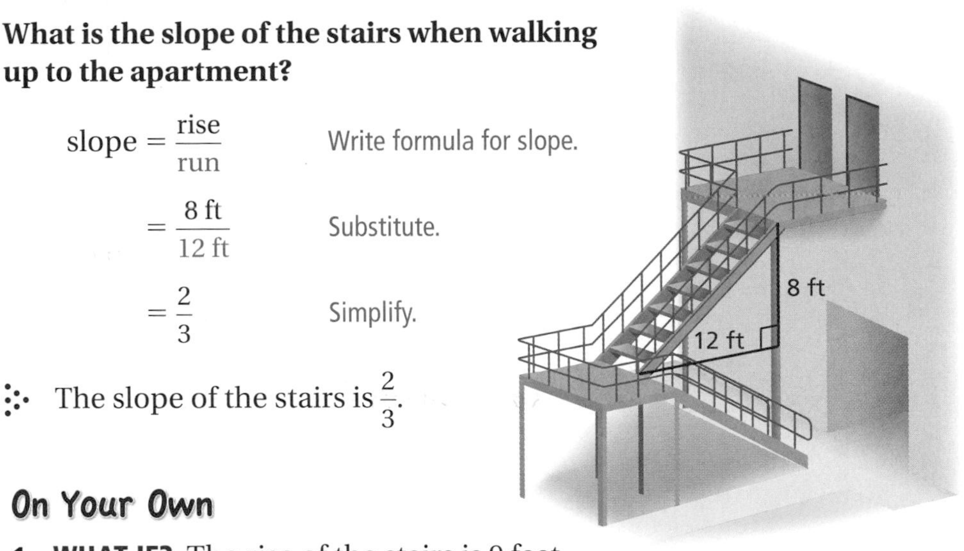

On Your Own

1. **WHAT IF?** The rise of the stairs is 9 feet. What is the slope of the stairs?

EXAMPLE 2 Finding the Slope of a Line

Study Tip

A line that rises from left to right has a *positive* slope.

A line that falls from left to right has a *negative* slope.

Find the slope of the line.

$\text{slope} = \frac{\text{rise}}{\text{run}}$ Write formula for slope.

$= \frac{-2}{4}$ Substitute.

$= -\frac{1}{2}$ Simplify.

The slope is $-\frac{1}{2}$.

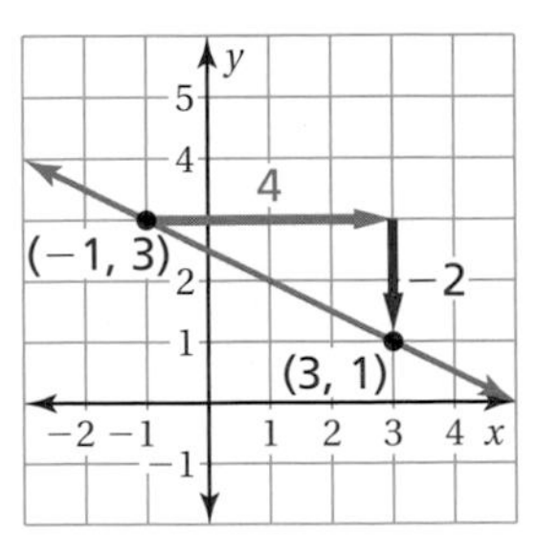

Multi-Language Glossary at BigIdeasMath.com.

On Your Own

Now You're Ready
Exercises 8–10

Find the slope of the line.

2.

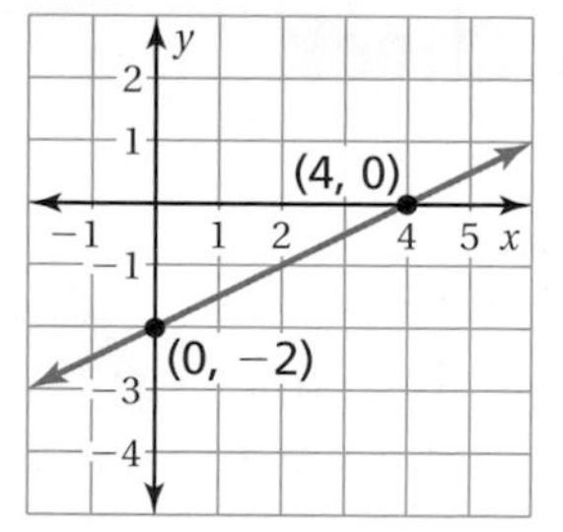

3.

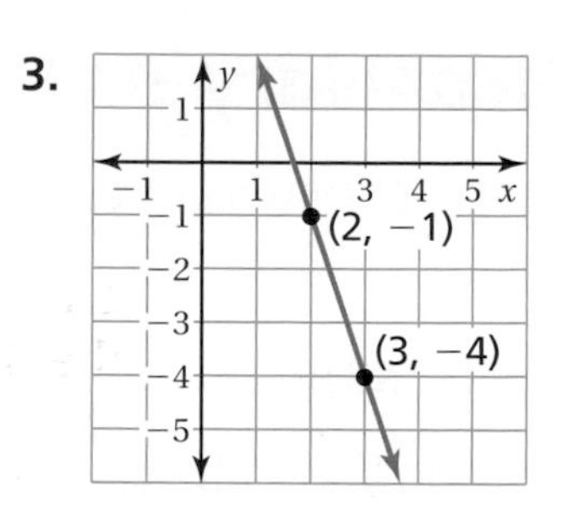

EXAMPLE 3 Graphing Lines Using a Point and a Slope

a. Graph the line with a slope of 3 that passes through (−3, −4).

Step 1: Plot (−3, −4).

Step 2: Find the rise and the run.

$$\text{slope} = \frac{\text{rise}}{\text{run}} = \frac{3}{1}$$

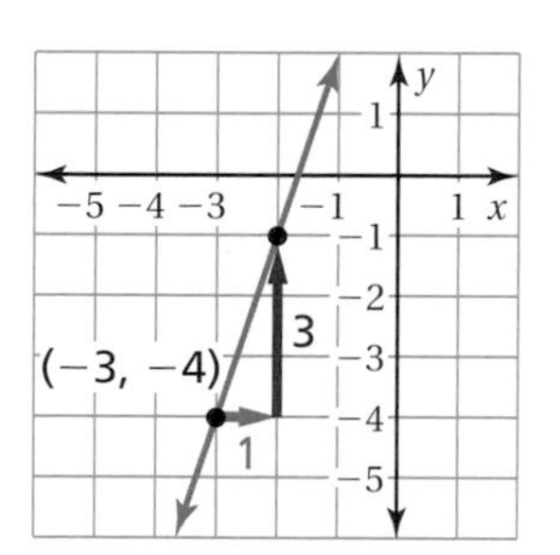

Step 3: Plot the point that is 1 unit right and 3 units up from (−3, −4).

Step 4: Draw a line through the two points.

∴ The graph is shown at the left.

b. Graph the line with a slope of $-\frac{4}{3}$ that passes through (2, 3).

Step 1: Plot (2, 3).

Step 2: Find the rise and the run.

$$\text{slope} = \frac{\text{rise}}{\text{run}} = \frac{-4}{3}$$

Step 3: Plot the point that is 3 units right and 4 units down from (2, 3).

Step 4: Draw a line through the two points.

∴ The graph is shown at the left.

On Your Own

Now You're Ready
Exercises 11–14

Graph the line with the given slope that passes through the given point.

4. slope = 1; (1, 3)

5. slope = $\frac{2}{5}$; (2, −1)

6. slope = −2; (0, −2)

7. slope = $-\frac{3}{2}$; (−4, −4)

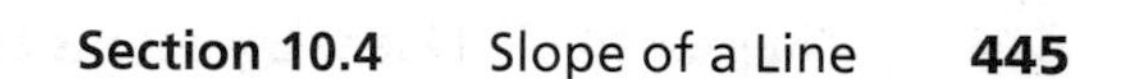

10.4 Exercises

Vocabulary and Concept Check

1. **OPEN-ENDED** How could you measure the slope of the roof of a house?

2. **REASONING** The slope of a line is $\frac{1}{2}$. What do you know about the graph of the line?

3. **OPEN-ENDED** A line with slope $-\frac{2}{3}$ passes through the point $(2, -6)$. Find two additional points on the line.

Practice and Problem Solving

Copy and complete the table.

	Two Points	Change in y	Change in x	Slope of Line
4.	$(0, 5), (2, 7)$			
5.	$(-10, 4), (5, -20)$			
6.	$(-8, -4), (-6, 5)$			
7.	$(1, 9), (7, 6)$			

Find the slope of the line.

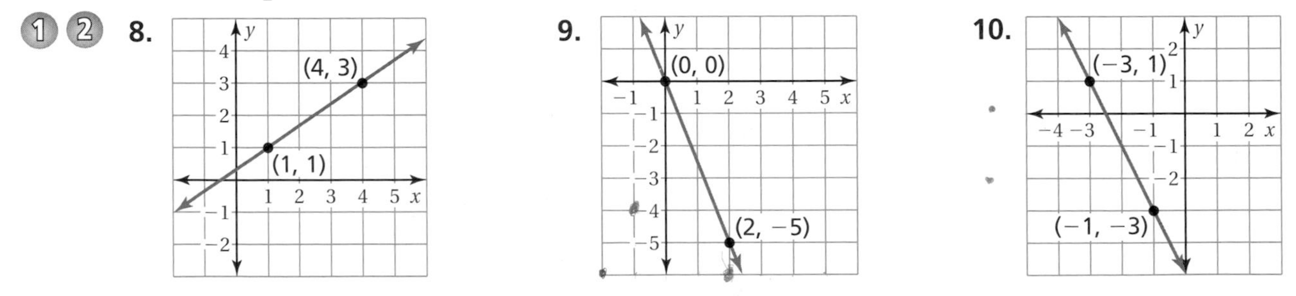

Graph the line with the given slope that passes through the given point.

3 **11.** slope $= 3$; $(1, 1)$

12. slope $= \frac{1}{5}$; $(-2, 2)$

13. slope $= -2$; $(4, 7)$

14. slope $= -\frac{3}{7}$; $(-6, 1)$

15. POINTS What point is 3 units right and 5 units down from $(-6, -2)$?

16. ERROR ANALYSIS Describe and correct the error made in finding the slope of the line.

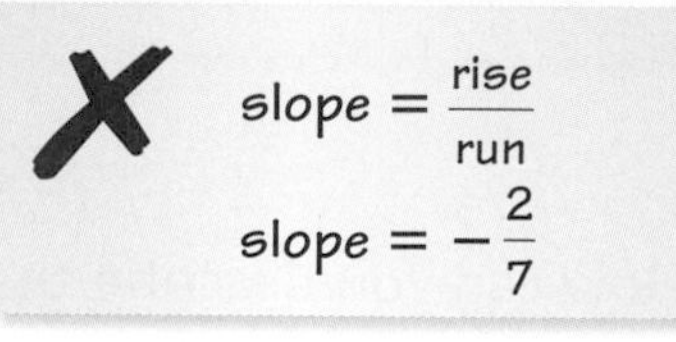

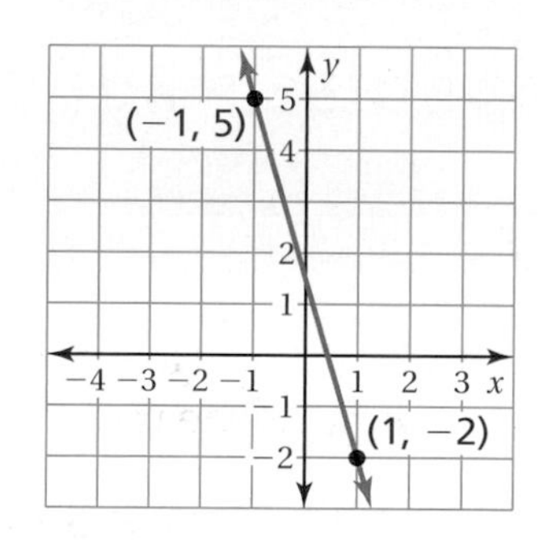

17. RACING The bankings of two race tracks are shown.

a. What is the slope of each track?

b. Which track has the steeper slope?

c. Why are the tracks sloped and not flat?

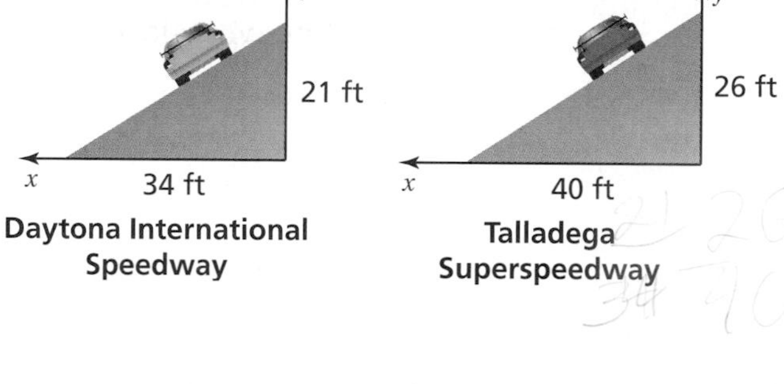

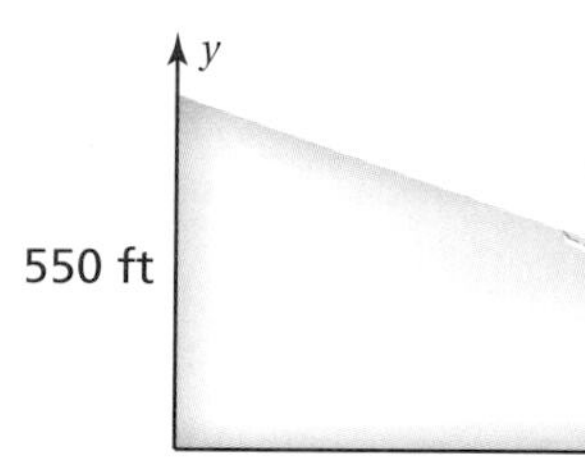

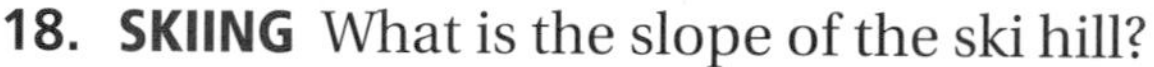

18. SKIING What is the slope of the ski hill?

19. ROADWAY A construction company is hired to build the roadway shown at the right.

a. The gradient of a hill can be calculated by converting its slope to a percent. What is the gradient of the roadway?

b. The gradient of the new roadway cannot exceed 18%. What changes could you make to accommodate this restriction?

70 m

200 m

20. **Reasoning** A line has a slope of 0.5 and passes through the point $(-4, 5)$. Does the line also pass through the point $(2, 7)$? Explain.

Fair Game Review What you learned in previous grades & lessons

Solve the equation. Check your solution. *(Section 2.6)*

21. $4x - 8 = -24$

22. $2x - 19 = 5$

23. $15 - 3x = -6$

24. MULTIPLE CHOICE What is the solution of the equation? *(Section 10.1)*

$$6[4(3 + x) - 14] + 2^2 = 0$$

Ⓐ $\frac{1}{2}$ Ⓑ $\frac{1}{3}$ Ⓒ $\frac{4}{3}$ Ⓓ $\frac{5}{3}$

10.5 Linear Functions

COMMON CORE STATE STANDARDS

8.EE.6

Essential Question

How can you describe the graph of an equation of the form $y = mx + b$?

The Meaning of a Word ● Function

Your score on a test is usually a **function** of how much you study.
If you study a lot, your score is usually high.
If you study a little, your score is usually low.

1 ACTIVITY: Using an Input-Output Table

Work with a partner.

a. Copy and complete the input-output table for the equation $y = -\frac{1}{2}x + 2$.

Input, x	-3	-2	-1	0	1	2	3
Output, y							

b. Plot the points from the table.

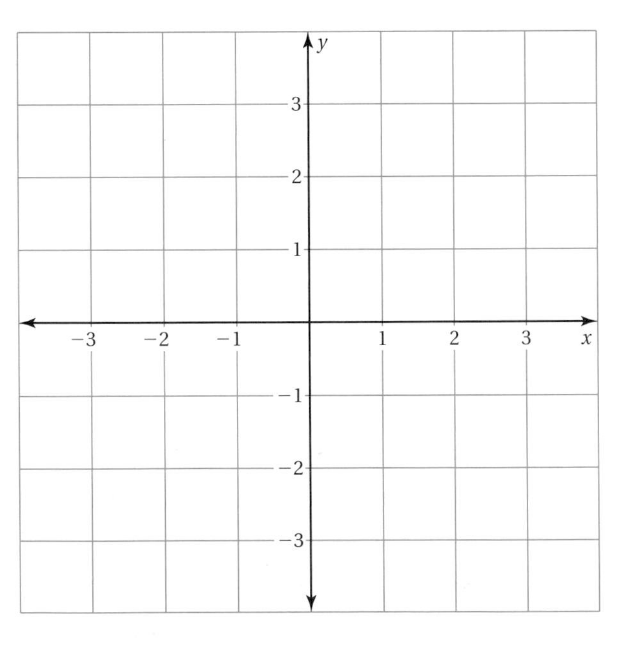

c. Describe the pattern of the points. Draw a graph that represents the pattern.

d. Choose three values of x that are not in the table. Find their corresponding y-values and plot the points. Do the points lie on the graph you made in part (c)?

Inductive Reasoning

Work with a partner. Sketch the graph of each equation. Then copy and complete the table.

Equation	Description of Graph	Point of Intersection with *y*-axis	Slope of Graph
1 **2.** $y = -\frac{1}{2}x + 2$	Line	(0, 2)	$-\frac{1}{2}$
3. $y = -x + 2$			
4. $y = -x + 1$			
5. $y = -\frac{1}{2}x + 1$			
6. $y = x + 1$			
7. $y = x - 1$			
8. $y = \frac{1}{2}x - 1$			
9. $y = \frac{1}{2}x + 1$			
10. $y = 2x + 1$			
11. $y = 2x - 2$			
12. $y = -2x + 3$			

What Is Your Answer?

13. IN YOUR OWN WORDS How can you describe the graph of an equation of the form $y = mx + b$?

a. How does the value of m affect the graph?

b. How does the value of b affect the graph?

c. Test your answers to parts (a) and (b) with three equations that are not in the table.

14. Why is an equation of the form $y = mx + b$ called a linear function? What does the word *linear* mean? What does the word *function* mean?

Use what you learned about linear functions to complete Exercises 12–17 on page 452.

10.5 Lesson

Key Vocabulary
linear function, *p. 450*
y-intercept, *p. 450*
slope-intercept form, *p. 450*

A **linear function** is a function whose graph is a line. The ***y*-intercept** is the *y*-coordinate of the point where the line crosses the *y*-axis.

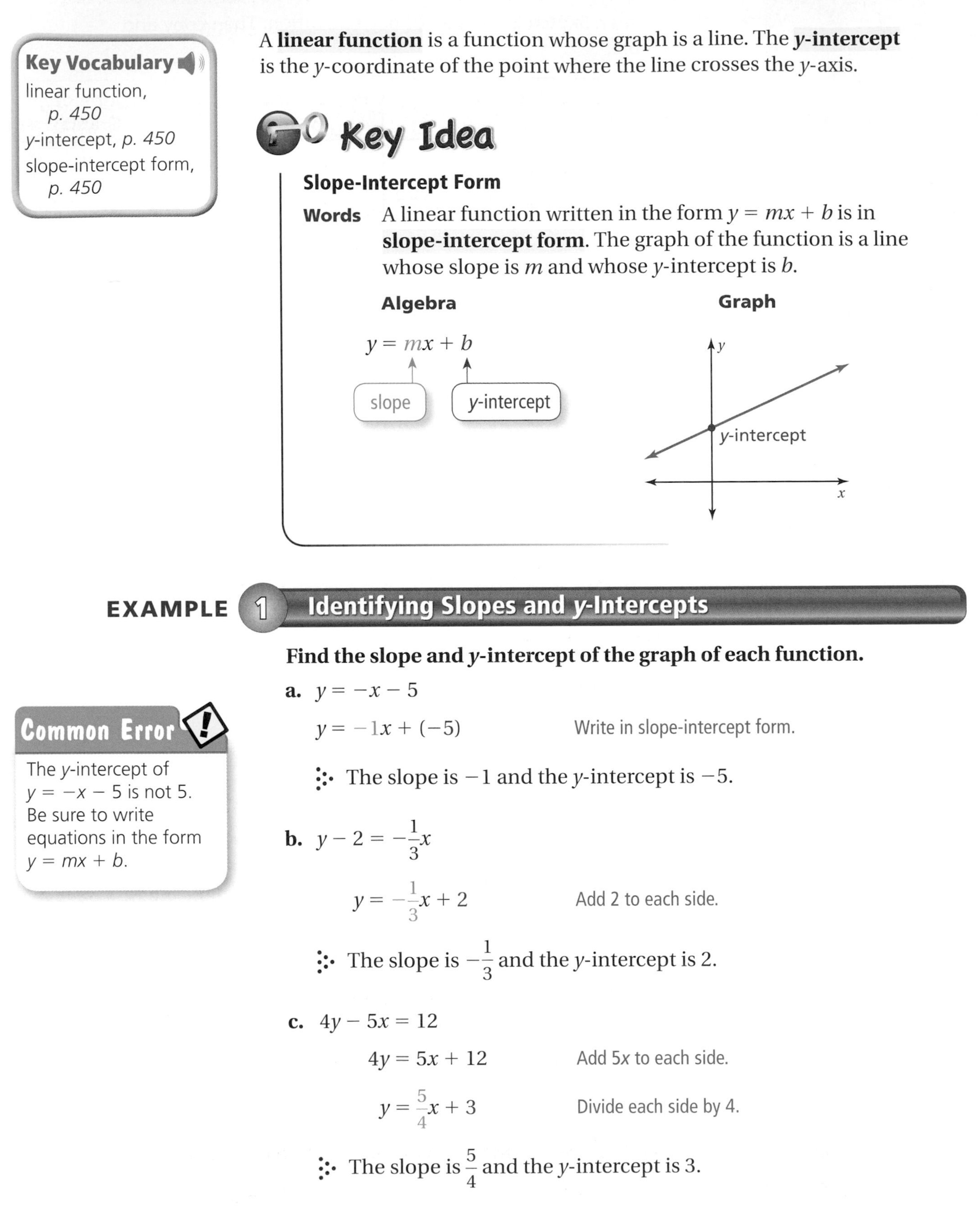

Key Idea

Slope-Intercept Form

Words A linear function written in the form $y = mx + b$ is in **slope-intercept form**. The graph of the function is a line whose slope is m and whose y-intercept is b.

Algebra

$y = mx + b$

slope — m; *y*-intercept — b

Graph

EXAMPLE 1 Identifying Slopes and *y*-Intercepts

Find the slope and *y*-intercept of the graph of each function.

a. $y = -x - 5$

$y = -1x + (-5)$ Write in slope-intercept form.

∴ The slope is -1 and the y-intercept is -5.

b. $y - 2 = -\frac{1}{3}x$

$y = -\frac{1}{3}x + 2$ Add 2 to each side.

∴ The slope is $-\frac{1}{3}$ and the y-intercept is 2.

c. $4y - 5x = 12$

$4y = 5x + 12$ Add $5x$ to each side.

$y = \frac{5}{4}x + 3$ Divide each side by 4.

∴ The slope is $\frac{5}{4}$ and the y-intercept is 3.

Common Error

The *y*-intercept of $y = -x - 5$ is not 5. Be sure to write equations in the form $y = mx + b$.

On Your Own

Now You're Ready
Exercises 6–11

Find the slope and y-intercept of the graph of the linear function.

1. $y = -3x + 10$
2. $-5y + x = 25$

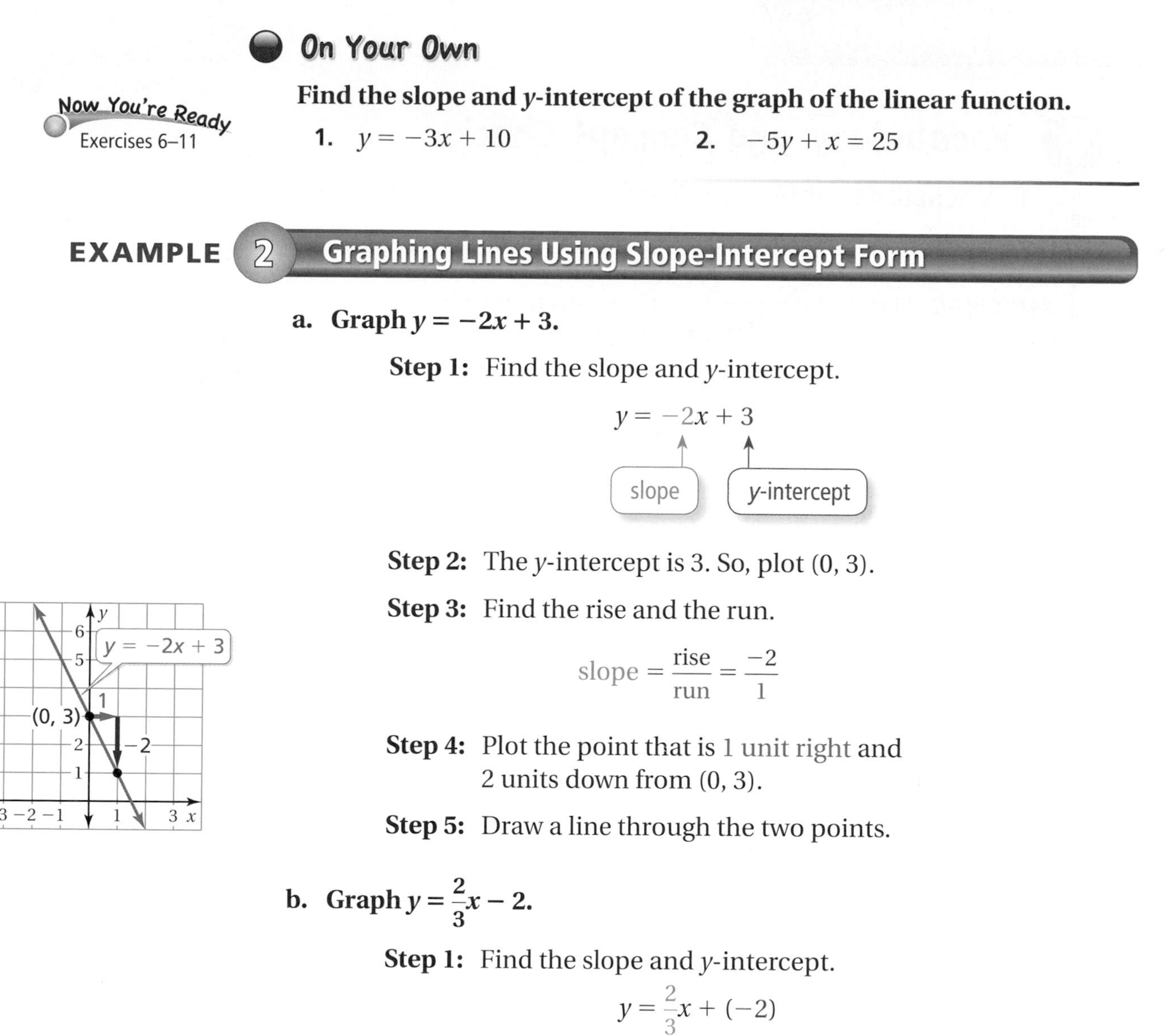

EXAMPLE 2 Graphing Lines Using Slope-Intercept Form

a. Graph $y = -2x + 3$.

Step 1: Find the slope and y-intercept.

$$y = -2x + 3$$

slope, y-intercept

Step 2: The y-intercept is 3. So, plot (0, 3).

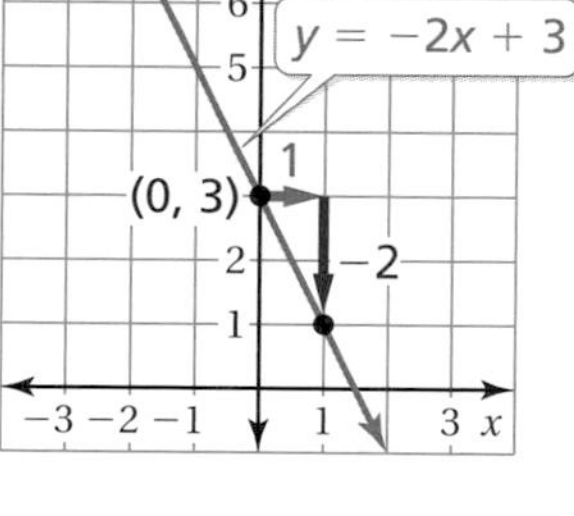

Step 3: Find the rise and the run.

$$\text{slope} = \frac{\text{rise}}{\text{run}} = \frac{-2}{1}$$

Step 4: Plot the point that is 1 unit right and 2 units down from (0, 3).

Step 5: Draw a line through the two points.

b. Graph $y = \frac{2}{3}x - 2$.

Step 1: Find the slope and y-intercept.

$$y = \frac{2}{3}x + (-2)$$

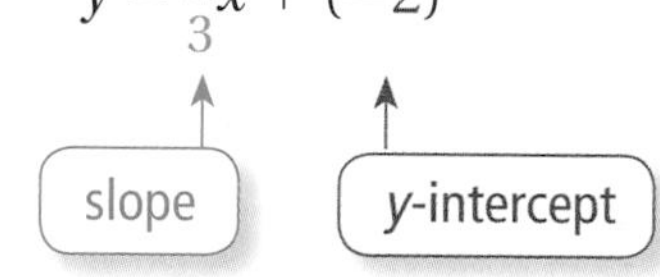

Step 2: The y-intercept is -2. So, plot $(0, -2)$.

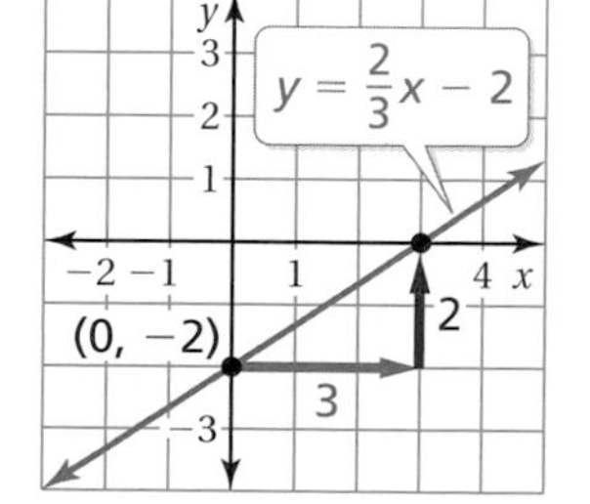

Step 3: Find the rise and the run.

$$\text{slope} = \frac{\text{rise}}{\text{run}} = \frac{2}{3}$$

Step 4: Plot the point that is 3 units right and 2 units up from $(0, -2)$.

Step 5: Draw a line through the two points.

On Your Own

Now You're Ready
Exercises 12–23

Graph the linear function using slope-intercept form.

3. $y = -x - 1$
4. $y = \frac{3}{2}x - 5$

10.5 Exercises

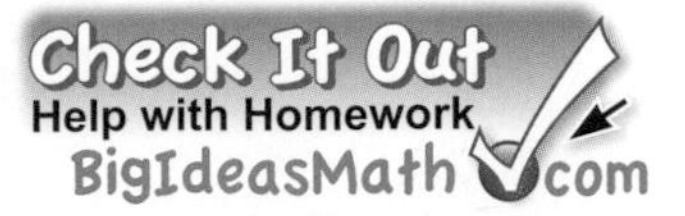

Vocabulary and Concept Check

1. **VOCABULARY** What is the y-intercept of a line?

2. **WRITING** Why is $y = mx + b$ called the slope-intercept form of a line?

MATCHING Match the linear function with its graph.

3. $y = 2x$

4. $y = 2x - 1$

5. $y = -2x + 1$

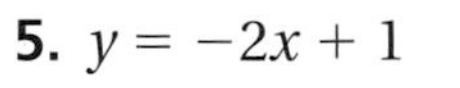

A.

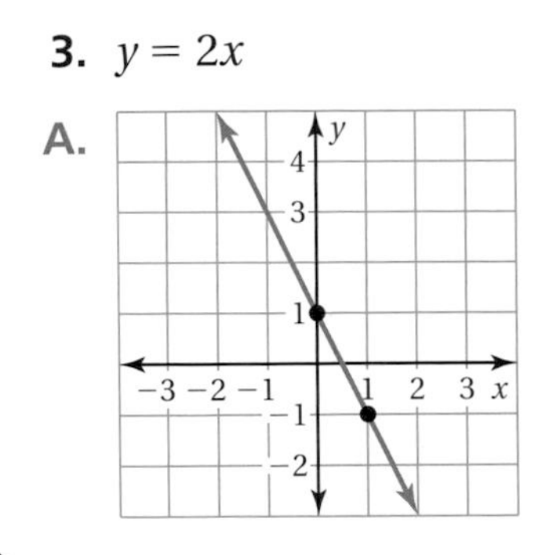

B.

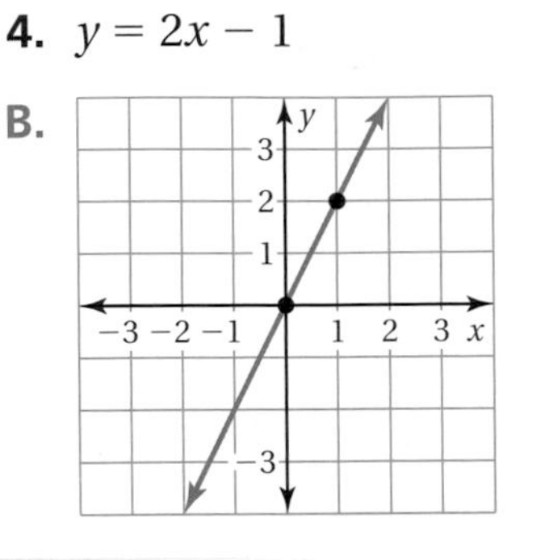

C.

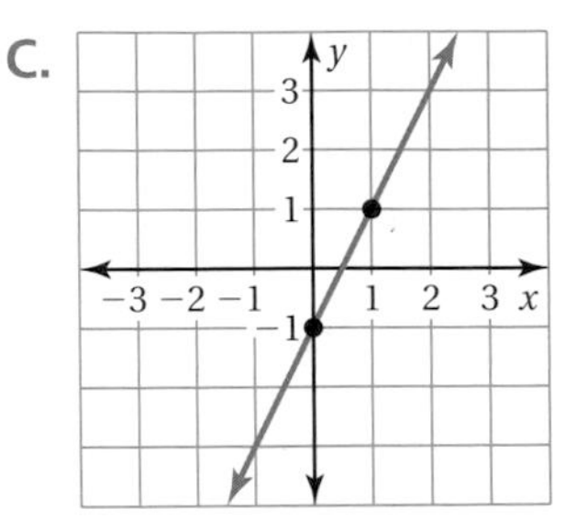

Practice and Problem Solving

Find the slope and y-intercept of the graph of the linear function.

① 6. $y = 4x + 1$

7. $y = -2x + 6$

8. $y = \frac{2}{3}x - 3$

9. $5x + y = 3$

10. $9x - 3y = 24$

11. $-4y + 10x = 36$

Graph the linear function using slope-intercept form.

② 12. $y = 3x - 3$

13. $y = 2x + 5$

14. $y = -x + 4$

15. $y = -\frac{3}{2}x - 1$

16. $y = -\frac{1}{5}x + 2$

17. $y = \frac{1}{4}x - 4$

18. $y = -4x + 1$

19. $y = 6x - 5$

20. $y = -3x - 2$

21. $5y - 4x = -15$

22. $5x + 3y = -6$

23. $3x + 4y = 12$

24. **ERROR ANALYSIS** Describe and correct the error in graphing the linear function $y = \frac{1}{4}x - 1$.

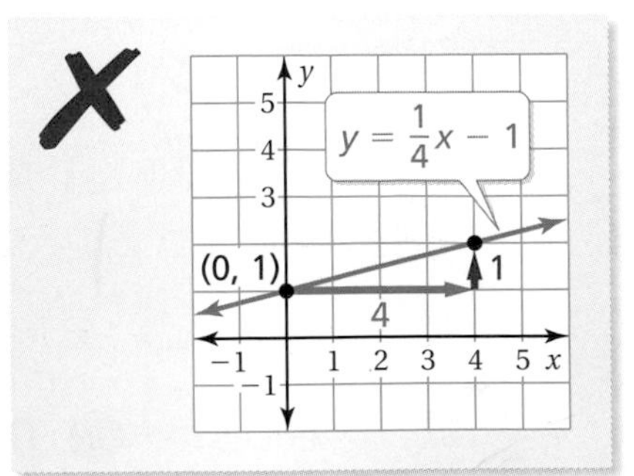

25. **CARTOONIST** The number c of cartoons a cartoonist plans to complete by the nth day of the month is given by $c = 24 + 4n$. What does the y-intercept represent?

Write an equation of the linear function in slope-intercept form.

26.

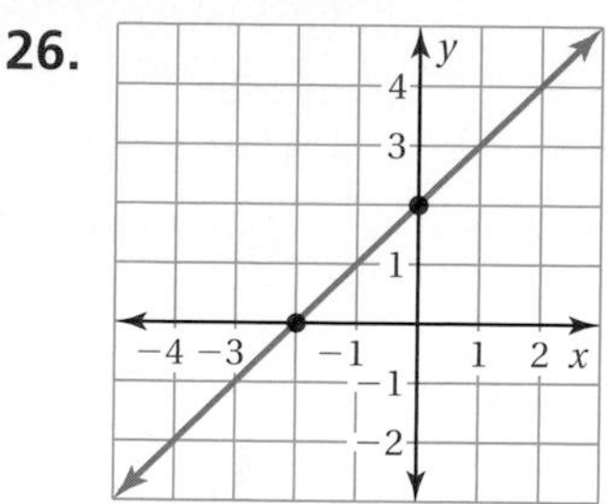

27.

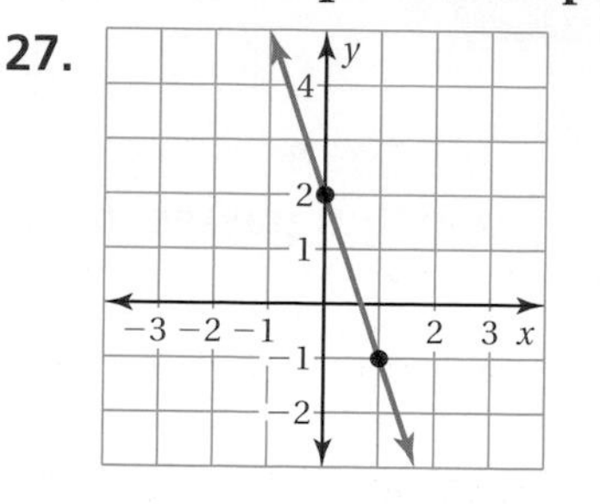

28.

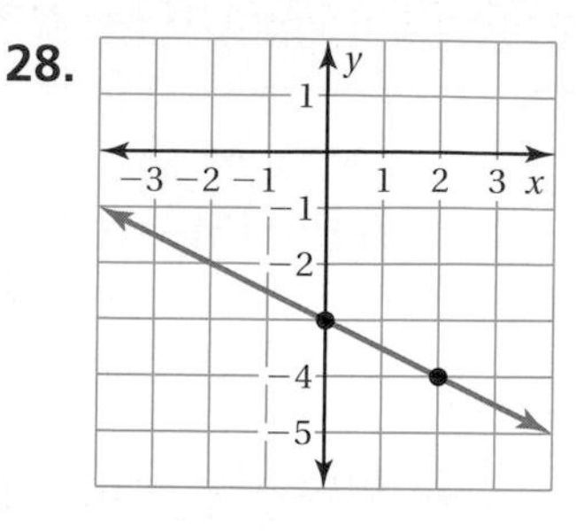

29. PERIMETER The perimeter of the rectangle can be modeled by the linear function $y = 2x + 7$.

3.5

x

a. Find the slope and y-intercept of the graph of the linear function.

b. Graph the linear function.

c. Is it possible for the rectangle to have a perimeter of 5 units? Examine the graph and explain.

30. ESCALATOR To get from the second floor to the first floor in a mall, you can either ride the escalator or take the stairs. The graph shows the vertical distance y (in feet) you have left to travel on the escalator after x seconds.

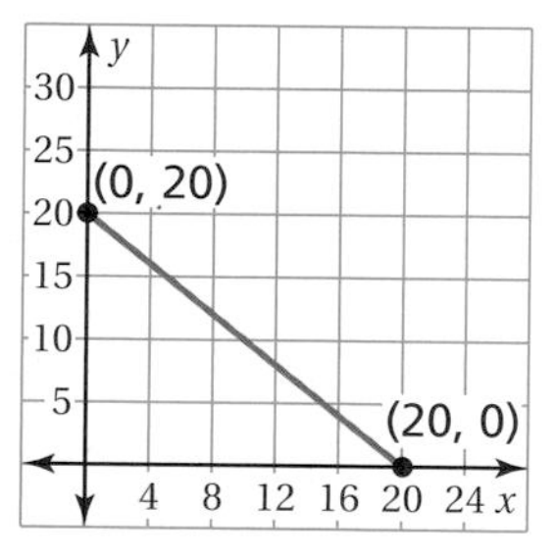

a. Write an equation in slope-intercept form of the linear function representing the amount of time you have left on the escalator.

b. How long does it take to ride the escalator from the second floor to the first floor?

c. The equation $y = -1.6x + 20$ represents the vertical distance y (in feet) you have left to travel on the stairs after x seconds. How much time do you save by taking the stairs?

31. Reasoning Find the slope and y-intercept of the graph of $Ax + By = C$ in terms of A, B, and C.

Fair Game Review What you learned in previous grades & lessons

Solve the equation. *(Section 10.2)*

32. $3x - 6 = 9 - 2x$

33. $5 - 8v = 3v - \frac{1}{2}$

34. $-5w - 4 = 4(w - 7)$

35. MULTIPLE CHOICE Which does *not* describe the sum of the sections of a circle graph? *(Section 8.3)*

Ⓐ $\frac{1}{2}$ Ⓑ 1 Ⓒ 100% Ⓓ 360°

10.4–10.5 Quiz

Check It Out
Progress Check
BigIdeasMath.com

Graph the line with the given slope that passes through the given point. *(Section 10.4)*

1. slope $= -3$; $(2, 4)$

2. slope $= \frac{2}{3}$; $(-1, 0)$

3. Find the slope of the line. *(Section 10.4)*

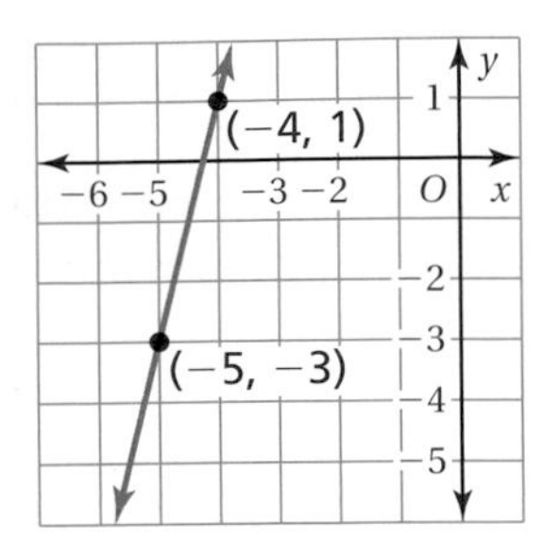

Find the slope and y-intercept of the graph of the linear function. *(Section 10.5)*

4. $y = 2x + 9$

5. $y = -5x + 4$

6. $y = \frac{3}{5}x - 6$

7. $3x - y = 8$

8. $2x + 8y = -16$

9. $4x - 3y = 15$

Graph the linear function using slope-intercept form. *(Section 10.5)*

10. $y = -x + 2$

11. $y = \frac{3}{2}x - 4$

12. $-x + 3y = 3$

13. PERIMETER The perimeter of the rectangle can be modeled by the linear function $y = 2x + 11$. *(Section 10.5)*

a. Find the slope and y-intercept of the graph of the linear function.

b. Graph the linear function.

c. Is it possible for the rectangle to have a perimeter of 9 units? Explain.

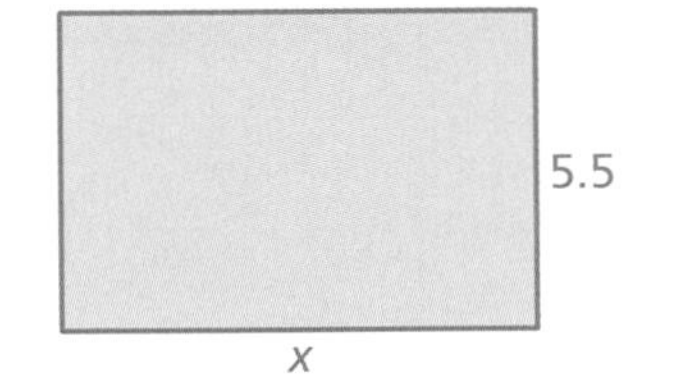

14. RAMP What is the slope of the skateboard ramp? What is the slope of the skateboard ramp if the length is doubled? *(Section 10.4)*

2 ft

4 ft

10 Chapter Review

Review Key Vocabulary

slope, *p. 444*
rise, *p. 444*
run, *p. 444*
linear function, *p. 450*
y-intercept, *p. 450*
slope-intercept form, *p. 450*

Review Examples and Exercises

10.1 Solving Multi-Step Equations *(pp. 422–427)*

a. Solve $-5x + 9x + 30 = 14$. Check your solution.

$-5x + 9x + 30 = 14$	Write the equation.
$4x + 30 = 14$	Combine like terms.
$-30 \quad -30$	Subtract 30 from each side.
$4x = -16$	Simplify.
$\frac{4x}{4} = \frac{-16}{4}$	Divide each side by 4.
$x = -4$	Simplify.

The solution is $x = -4$.

Check

$$-5x + 9x + 30 = 14$$
$$-5(-4) + 9(-4) + 30 \stackrel{?}{=} 14$$
$$20 - 36 + 30 \stackrel{?}{=} 14$$
$$14 = 14 \checkmark$$

b. Solve $2(n + 5) - 3 = 9$. Check your solution.

$2(n + 5) - 3 = 9$	Write the equation.
$2(n) + 2(5) - 3 = 9$	Use Distributive Property.
$2n + 10 - 3 = 9$	Multiply.
$2n + 7 = 9$	Subtract.
$-7 \quad -7$	Subtract 7 from each side.
$2n = 2$	Simplify.
$\frac{2n}{2} = \frac{2}{2}$	Divide each side by 2.
$n = 1$	Simplify.

The solution is $n = 1$.

Check

$$2(n + 5) - 3 = 9$$
$$2(1 + 5) - 3 \stackrel{?}{=} 9$$
$$2(1) + 2(5) - 3 \stackrel{?}{=} 9$$
$$2 + 10 - 3 \stackrel{?}{=} 9$$
$$9 = 9 \checkmark$$

Exercises

Solve the equation. Check your solution.

1. $-4x + 6x - 15 = -25$
2. $11k - 8k - 3 = 9$
3. $3(2n - 7) - 3 = 36$
4. $7 - 1.5(4 - 10d) = 31$

10.2 Solving Equations with Variables on Both Sides (pp. 428–433)

Solve $z - 42 = -6z$. Check your solution.

$z - 42 = -6z$	Write the equation.
$\underline{-z} \qquad \underline{-z}$	Subtract z from each side.
$-42 = -7z$	Simplify.
$\dfrac{-42}{-7} = \dfrac{-7z}{-7}$	Divide each side by -7.
$6 = z$	Simplify.

Check

$z - 42 = -6z$

$6 - 42 \stackrel{?}{=} -6(6)$

$-36 = -36$ ✓

The solution is $z = 6$.

Exercises

Solve the equation. Check your solution.

5. $3x = x - 18$
6. $-5n + 4 = 24 - n$
7. $8s = 3(s + 5)$
8. $-2(1 - 2a) = 4\left(\frac{5}{4}a - 2\right)$
9. $5w + 6 = -4 - 5w$
10. $-6(m + 2) = m - 2$

10.3 Solving Equations Using Tables and Graphs (pp. 434–439)

Use a table to solve $7x = 6x + 1$. Check your solution.

Find the x-value that makes $7x$ equal to $6x + 1$.

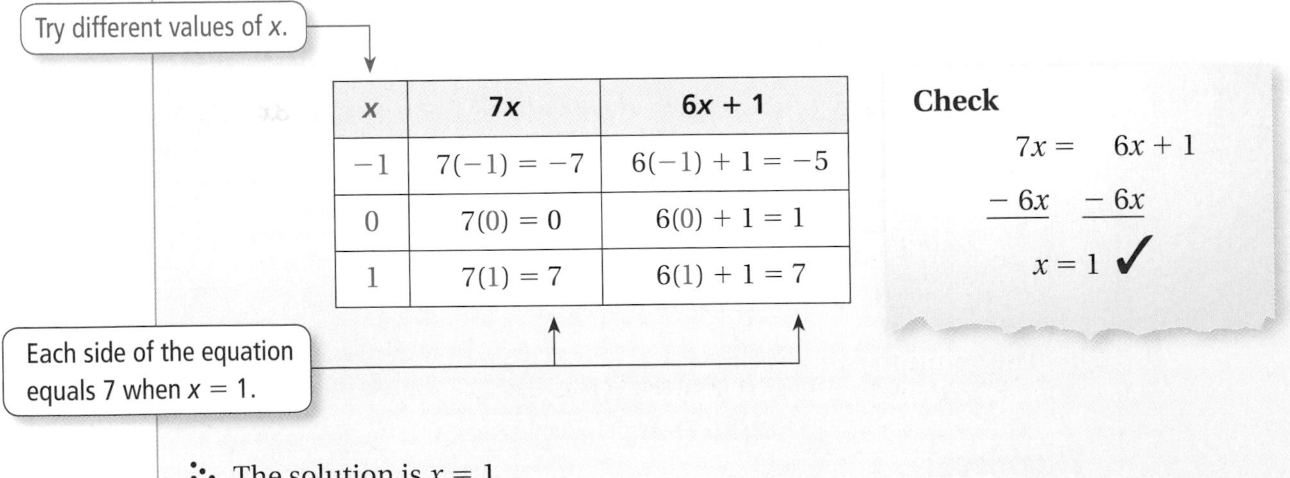
Try different values of x.
Each side of the equation equals 7 when x = 1.

x	$7x$	$6x + 1$
-1	$7(-1) = -7$	$6(-1) + 1 = -5$
0	$7(0) = 0$	$6(0) + 1 = 1$
1	$7(1) = 7$	$6(1) + 1 = 7$

Check

$7x = 6x + 1$

$\underline{-6x} \quad \underline{-6x}$

$x = 1$ ✓

The solution is $x = 1$.

Exercises

Use a table to solve the equation. Check your solution.

11. $6x - 4 = 8x$

12. $x + 7 = 3x + 1$

13. $-13 - 2b = b - 7$

14. $4 + 4m = m + 13$

10.4 Slope of a Line *(pp. 442–447)*

Find the slope of the line.

$$\text{slope} = \frac{\text{rise}}{\text{run}} \qquad \text{Write formula for slope.}$$

$$= \frac{1}{3} \qquad \text{Substitute.}$$

∴ The slope is $\frac{1}{3}$.

Exercises

Find the slope of the line.

15.

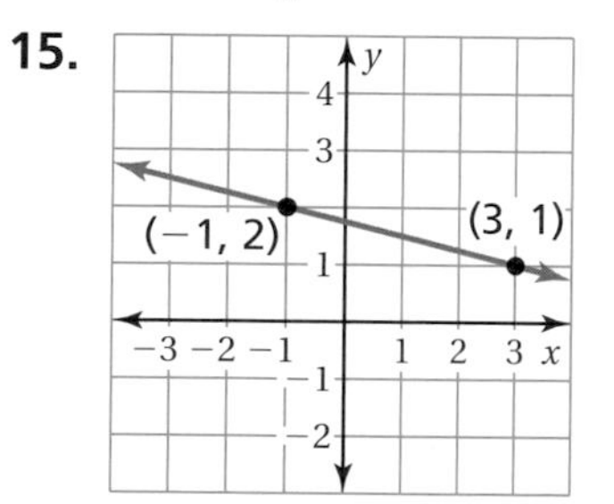

16.

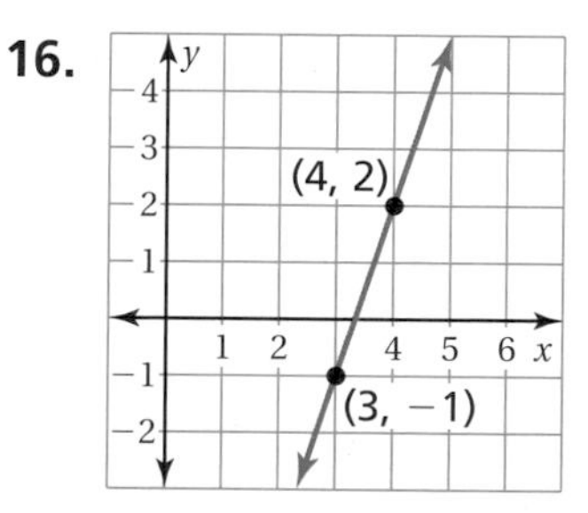

10.5 Linear Functions *(pp. 448–453)*

Find the slope and y-intercept of the graph of the function $3x + 4y = 8$.

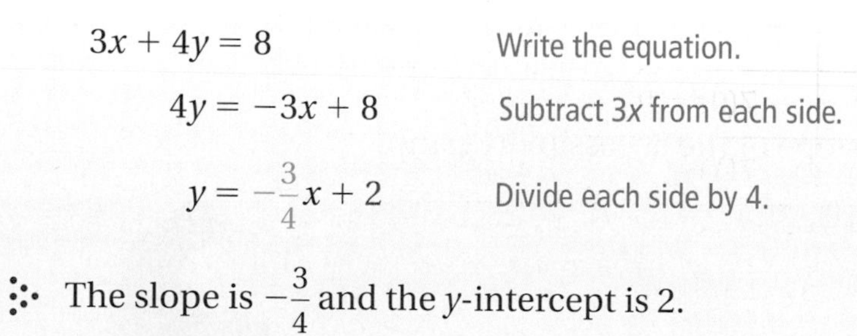

$3x + 4y = 8$ — Write the equation.

$4y = -3x + 8$ — Subtract $3x$ from each side.

$y = -\frac{3}{4}x + 2$ — Divide each side by 4.

∴ The slope is $-\frac{3}{4}$ and the y-intercept is 2.

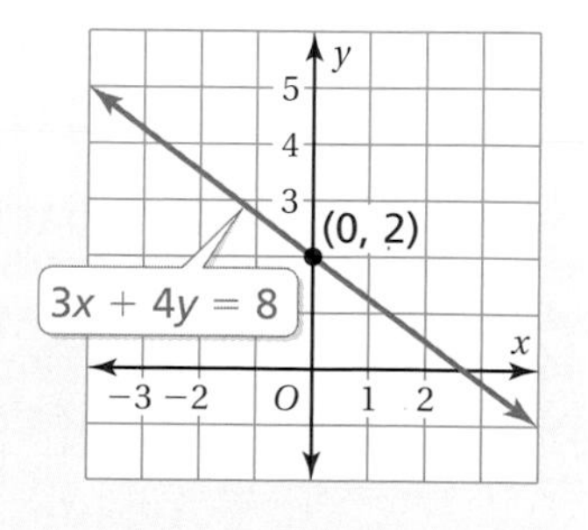

Exercises

17. Find the slope and y-intercept of the graph of the function $5x + 5y = 10$.

18. Graph the linear function $y = 4x - 1$ using slope-intercept form.

10 Chapter Test

Solve the equation. Check your solution.

1. $-3x + 6x - 17 = 10$
2. $-7x - 5x + 2 = -22$
3. $2(k + 4) - 6 = -12$
4. $-3y = 8 - y$

5. Use a table to solve $5x + 2 = 4x$. Check your solution.

6. Use the graph to solve the equation $x + 6 = -2x - 9$. Check your solution.

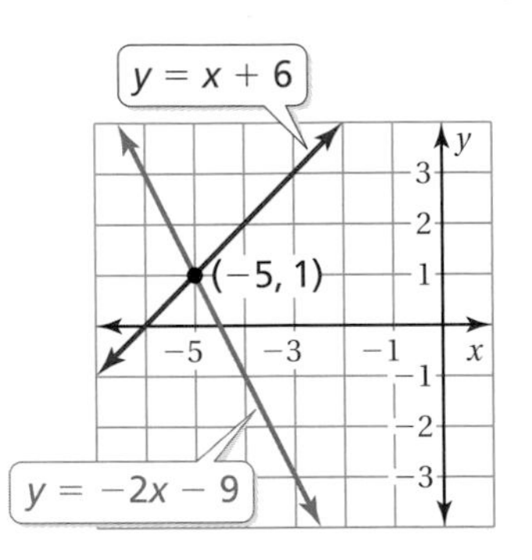

7. Find the slope of the line.

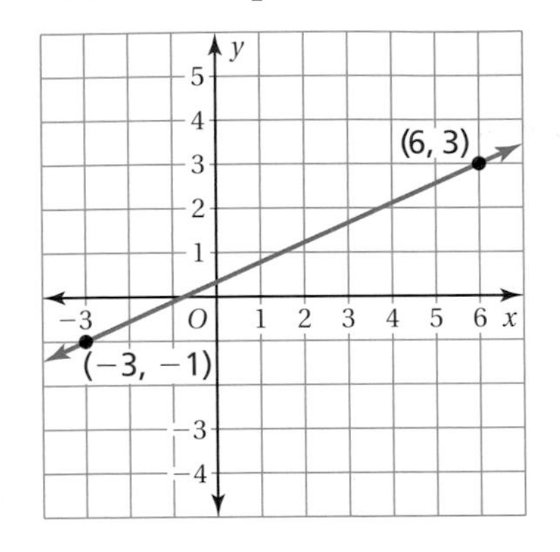

Graph the line with the given slope that passes through the given point.

8. slope = 2; $(-3, 0)$
9. slope = $-\frac{3}{4}$; $(1, 2)$

Find the slope and y-intercept of the graph of the linear function.

10. $y = 3x - 6$
11. $y = -2x + 3$

Graph the linear function using slope-intercept form.

12. $y = -x + 3$
13. $2x + 6y = 12$

14. **DRIVING** Two cars are driving on the same highway. Car A is 5 miles ahead of Car B and is traveling at a speed of 45 miles per hour. Car B is traveling at a speed of 55 miles per hour. How many minutes will it take for Car B to catch up with Car A?

15. **RAMP** What is the slope of the accessibility ramp?

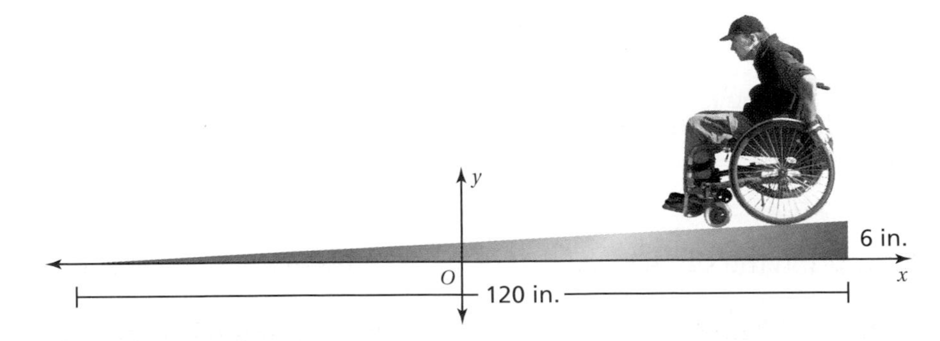

10 Standardized Test Practice

1. In 2000, the population of a small town was 1600 people. In 2010, the population of the town was 1200 people. What is the percent of decrease? *(7.RP.3)*

   A. 25% C. 75%

   B. 33.3% D. 133.3%

2. What is the surface area of the composite solid? *(7.G.6)*

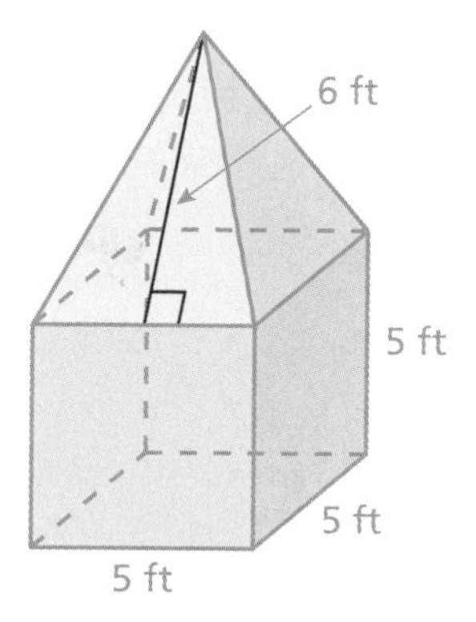

   F. 160 ft^2 H. 210 ft^2

   G. 185 ft^2 I. 270 ft^2

3. The vertices of a triangle are $A(-1, 2)$, $B(1, 4)$, and $C(6, 2)$. The triangle is reflected in the x-axis. Which ordered pair is not a vertex of its image? *(8.G.3)*

   A. $(-1, 2)$ C. $(-1, -2)$

   B. $(1, -4)$ D. $(6, -2)$

4. What is the y-intercept of the graph of the linear function below? *(8.EE.6)*

$$4x + 6y = 6$$

5. What could be the first step to solve the equation below? *(8.EE.7b)*

$$5x + 2 = 3(x + 4)$$

   F. Multiply x by 3 and 4 by 3. H. Subtract 2 from 4.

   G. Subtract x from $5x$. I. Combine $5x$ and 2.

6. The pyramids are similar. What is the volume of Pyramid B? *(7.G.6)*

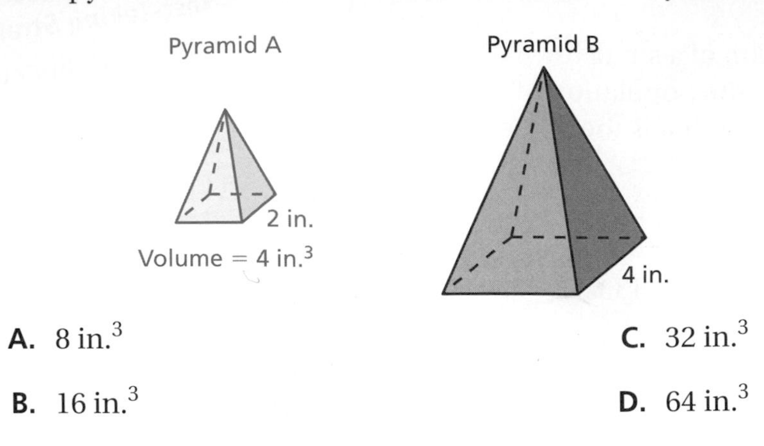

A. 8 in.^{3}

B. 16 in.^{3}

C. 32 in.^{3}

D. 64 in.^{3}

7. What is the slope of the line shown below? *(7.RP.2b)*

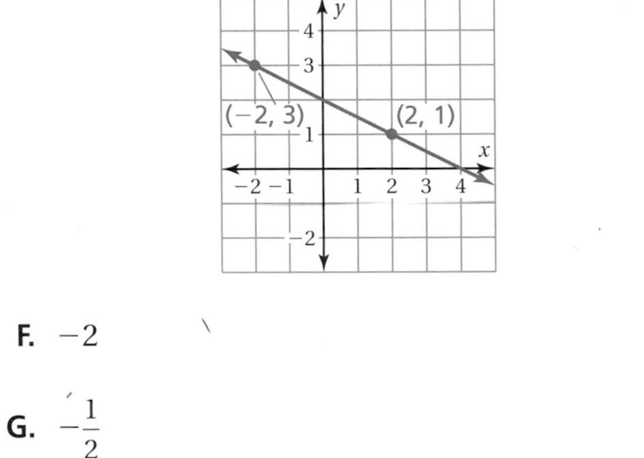

F. -2

G. $-\frac{1}{2}$

H. $\frac{1}{2}$

I. 2

8. Which figure is a rotation of the figure shown below? *(8.G.2)*

A.

B.

C.

D.

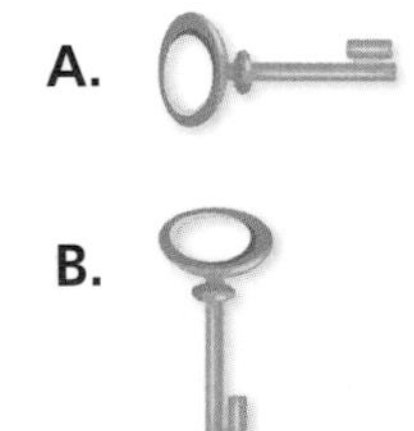

9. You randomly choose one of the six marbles shown without replacing it. Your friend then randomly chooses a marble. What is the probability that you choose a green marble and your friend chooses a yellow marble? *(7.SP.8a)*

10. What is the perimeter of the square shown below? *(8.EE.7b)*

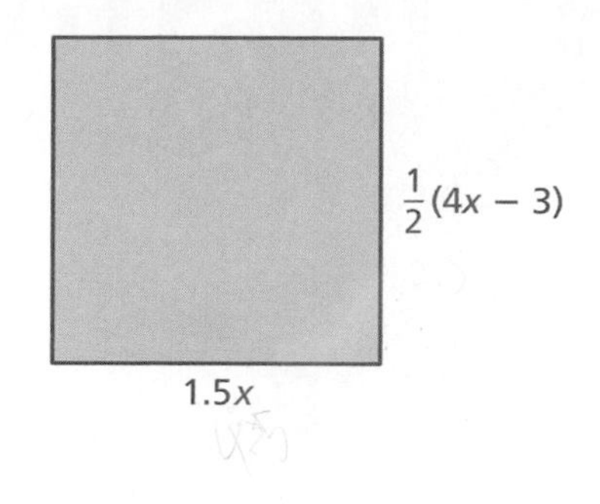

F. 3 units

G. 4.5 units

H. 12 units

I. 18 units

11. The stem-and-leaf plot shows the heights of 24 randomly selected students at a high school. Which statement is *not* true? *(7.SP.2)*

Stem	Leaf
5	4
5	9
6	2 4 4 4
6	5 6 6 6 8 8 8 8 9
7	0 0 0 3 3 4
7	5 7 9

Key: 5 | 4 = 54 inches

A. The median is less than 70 inches.

B. The mean is greater than 60 inches.

C. The range is greater than 20 inches.

D. The mode is less than the median.

12. Ernesto was solving the equation in the box shown at the right.

What should Ernesto do to correct the error that he made? *(8.EE.7b)*

$$3(q - 2) = -\frac{1}{2}(2q - 4)$$
$$3q - 6 = -q - 2$$
$$4q = 4$$
$$q = 1$$

F. Rewrite $3(q - 2)$ as $3q - 2$.

G. Rewrite $-\frac{1}{2}(2q - 4)$ as $-q - 4$.

H. Rewrite $-\frac{1}{2}(2q - 4)$ as $-q + 2$.

I. Rewrite $-\frac{1}{2}(2q - 4)$ as $-q + 4$.

13. A boat travels x miles per hour from Town M to Town N. The boat travels 5 miles per hour faster on the return trip. *(8.EE.7b)*

Think Solve Explain

Part A Write an equation that could be used to find x.

Part B Using your equation from Part A, find x. Show all steps clearly.

Part C What is the distance between Town M and Town N? Explain your reasoning.

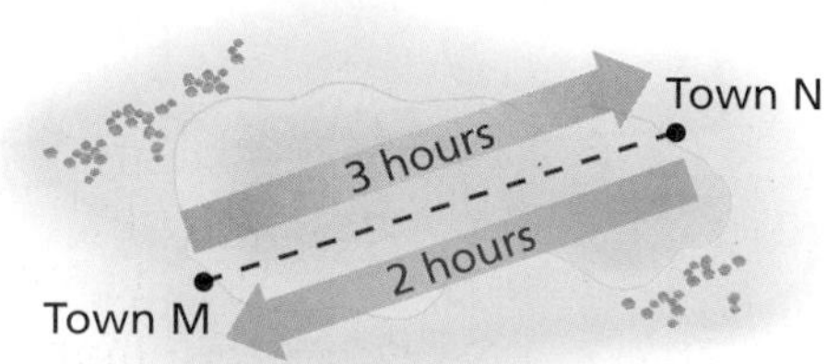

11 Angles and Similarity

11.1 Classifying Angles

11.2 Angles and Sides of Triangles

11.3 Angles of Polygons

11.4 Using Similar Triangles

11.5 Parallel Lines and Transversals

"Let's use shadows and similar triangles to indirectly measure the height of the giant hyena standing right behind you."

"Start with any triangle."

"Tear off the angles. You can always rearrange the angles so that they form a straight line."

"What does that prove?"

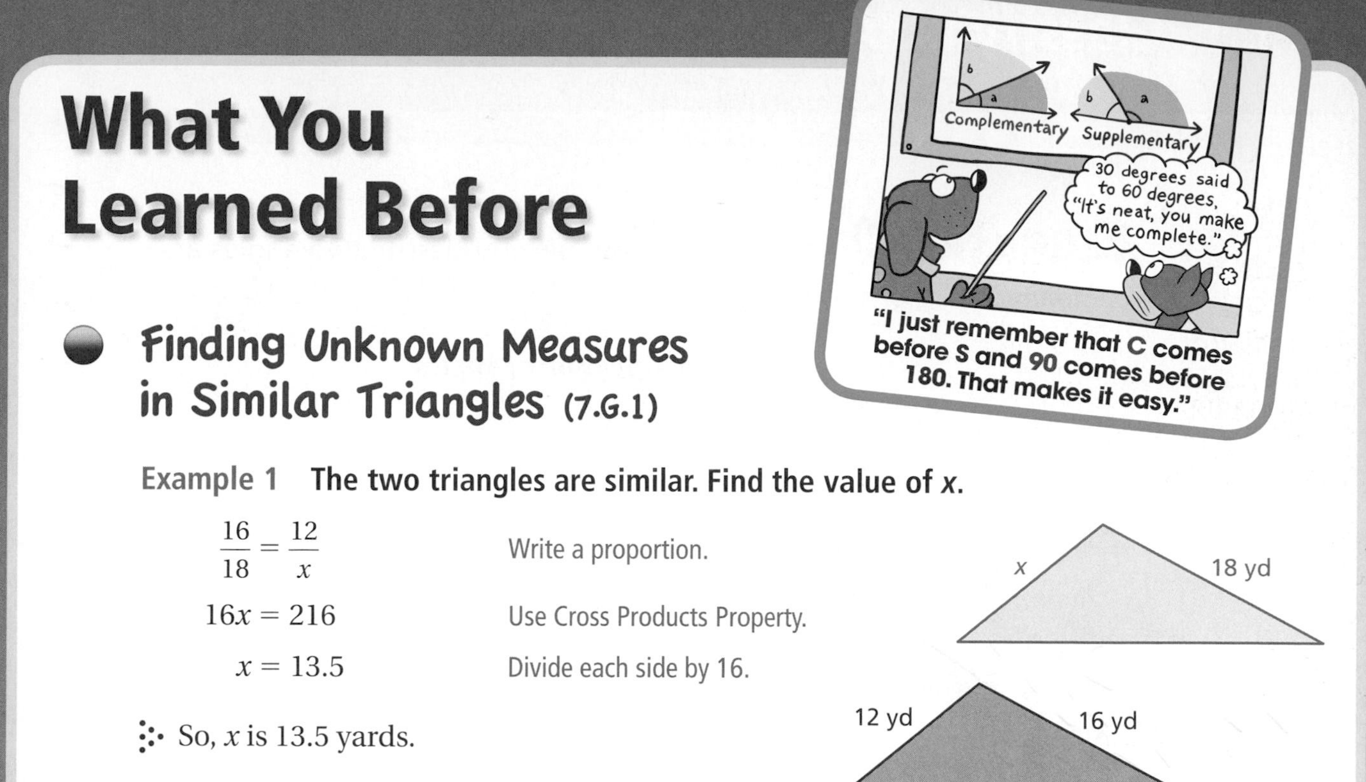

What You Learned Before

Finding Unknown Measures in Similar Triangles (7.G.1)

Example 1 **The two triangles are similar. Find the value of *x*.**

$\frac{16}{18} = \frac{12}{x}$ Write a proportion.

$16x = 216$ Use Cross Products Property.

$x = 13.5$ Divide each side by 16.

∴ So, x is 13.5 yards.

Example 2 **The two quadrilaterals are similar. The ratio of their perimeters is 4 : 5. Find the value of *x*.**

$\frac{4}{5} = \frac{x}{25}$ Write a proportion.

$100 = 5x$ Use Cross Products Property.

$20 = x$ Divide each side by 5.

∴ So, x is 20 centimeters.

Try It Yourself

The polygons are similar. Find the value of *x*.

1.

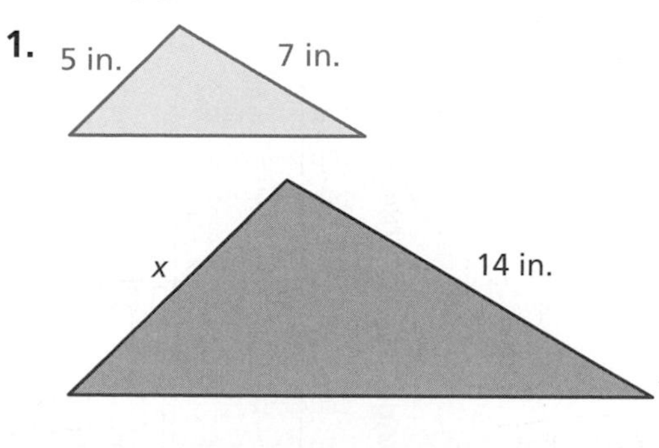

2. The ratio of the perimeters is 2 : 1.

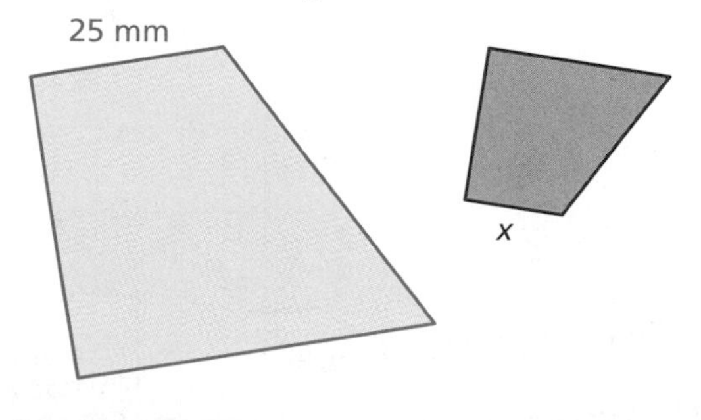

11.1 Classifying Angles

COMMON CORE STATE STANDARDS
7.G.5
8.G.5

Essential Question How can you classify two angles as complementary or supplementary?

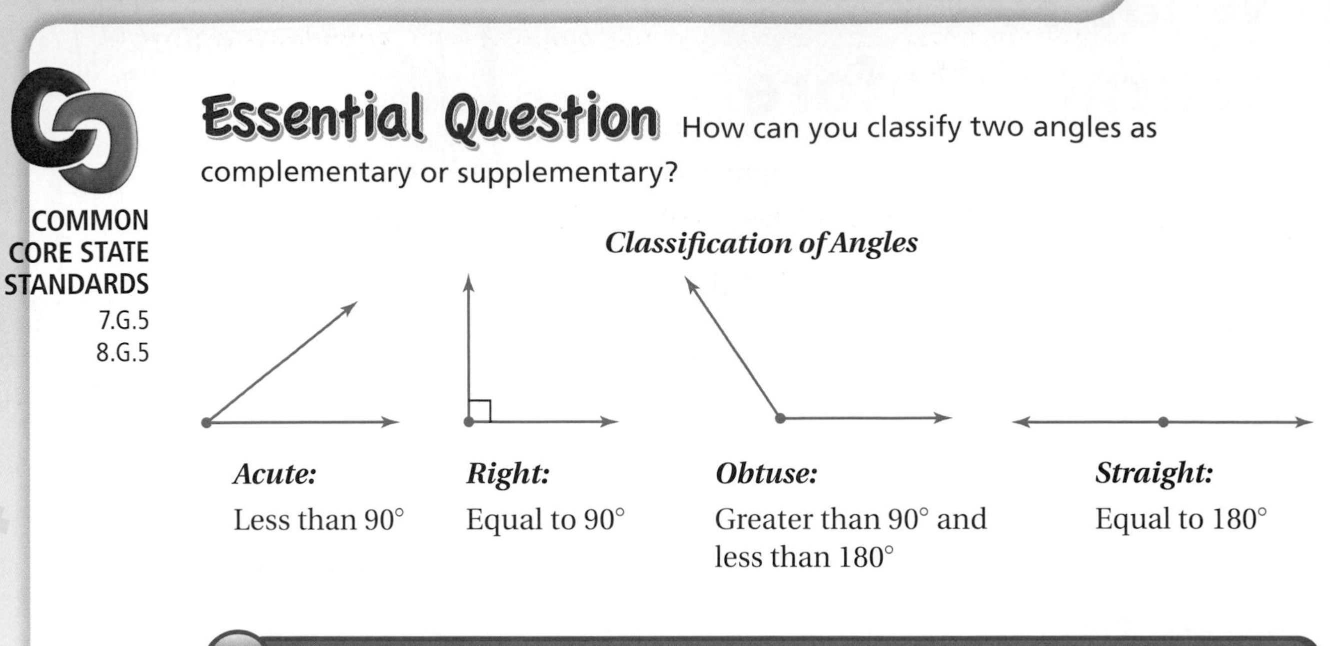

1 ACTIVITY: Complementary and Supplementary Angles

Work with a partner.

- **Copy and complete each table.**
- **Graph each function. Is the function linear?**
- **Write an equation for y as a function of x.**
- **Describe the values of x that make sense for each function.**

a. Two angles are **complementary** if the sum of their measures is 90°. In the table, x and y are complementary.

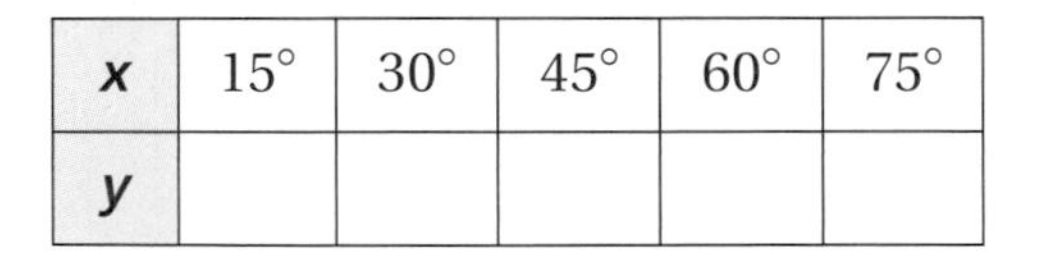

x	15°	30°	45°	60°	75°
y					

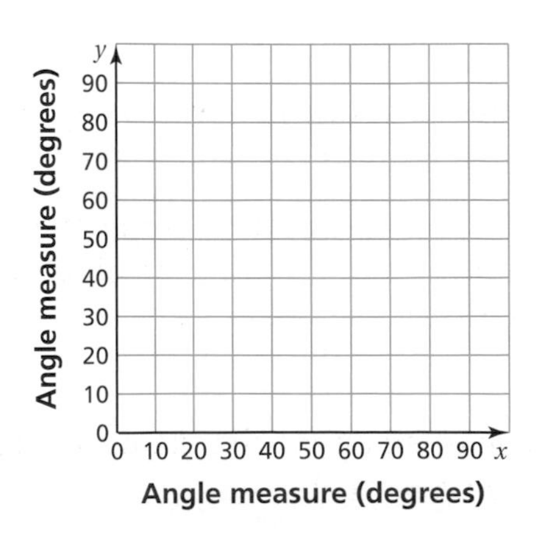

b. Two angles are **supplementary** if the sum of their measures is 180°. In the table, x and y are supplementary.

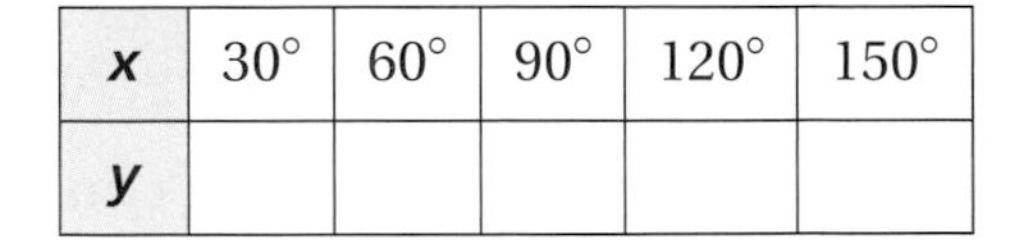

x	30°	60°	90°	120°	150°
y					

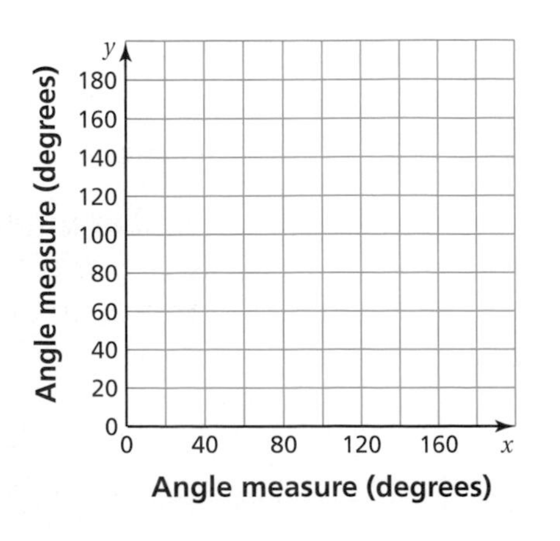

2 ACTIVITY: Exploring Rules About Angles

Work with a partner. Copy and complete each sentence with *always*, *sometimes*, or *never*.

a. If x and y are complementary angles, then both x and y are __________ acute.

b. If x and y are supplementary angles, then x is __________ acute.

c. If x is a right angle, then x is __________ acute.

3 ACTIVITY: Naming Angles

Some angles, such as $\angle A$, can be named by a single letter. When this does not clearly identify an angle, you should use three letters, as follows.

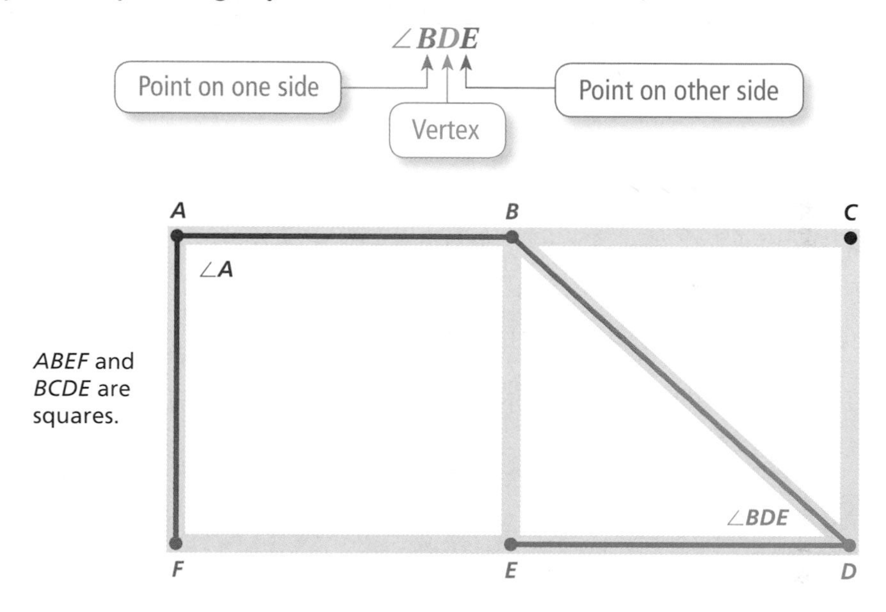

Work with a partner.

a. Name all pairs of complementary angles in the diagram above.

b. Name all pairs of supplementary angles in the diagram above.

What Is Your Answer?

4. IN YOUR OWN WORDS How can you classify two angles as complementary or supplementary? Give examples of each type.

5. Find examples of real-life objects that use complementary and supplementary angles. Make a drawing of each object and approximate the degree measure of each angle.

Practice Use what you learned about classifying angles to complete Exercises 3–5 on page 468.

11.1 Lesson

Check It Out
Lesson Tutorials
BigIdeasMath.com

Key Vocabulary
complementary angles, *p. 466*
supplementary angles, *p. 466*
congruent angles, *p. 467*
vertical angles, *p. 467*

Key Ideas

Complementary Angles

Words Two angles are **complementary angles** if the sum of their measures is 90°.

Examples

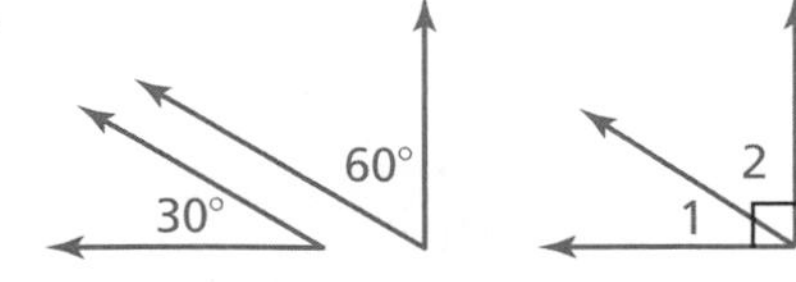

∠1 and ∠2 are complementary angles.

Supplementary Angles

Words Two angles are **supplementary angles** if the sum of their measures is 180°.

Examples

135° 45° 3 4

∠3 and ∠4 are supplementary angles.

EXAMPLE 1 Classifying Pairs of Angles

Tell whether the angles are *complementary*, *supplementary*, or *neither*.

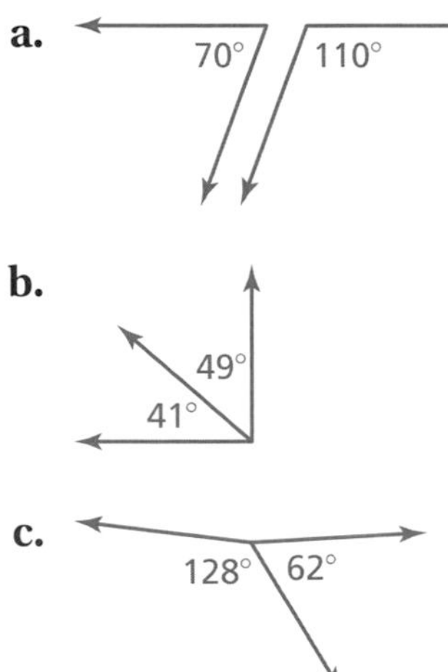

a. $70° + 110° = 180°$

So, the angles are supplementary.

b. $41° + 49° = 90°$

So, the angles are complementary.

c. $128° + 62° = 190°$

So, the angles are *neither* complementary nor supplementary.

On Your Own

Now You're Ready
Exercises 6–11

Tell whether the angles are *complementary*, *supplementary*, or *neither*.

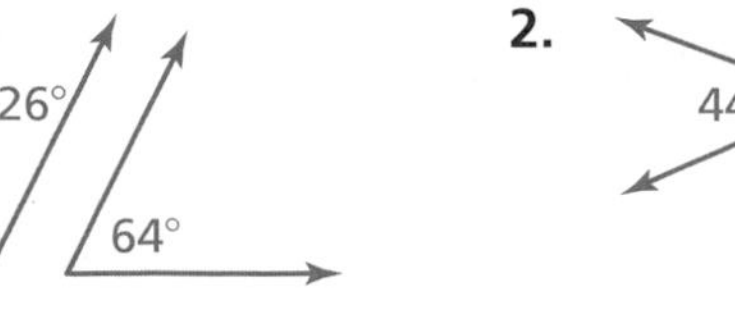

1.

2.

3. 70° 19°

Multi-Language Glossary at BigIdeasMath.com.

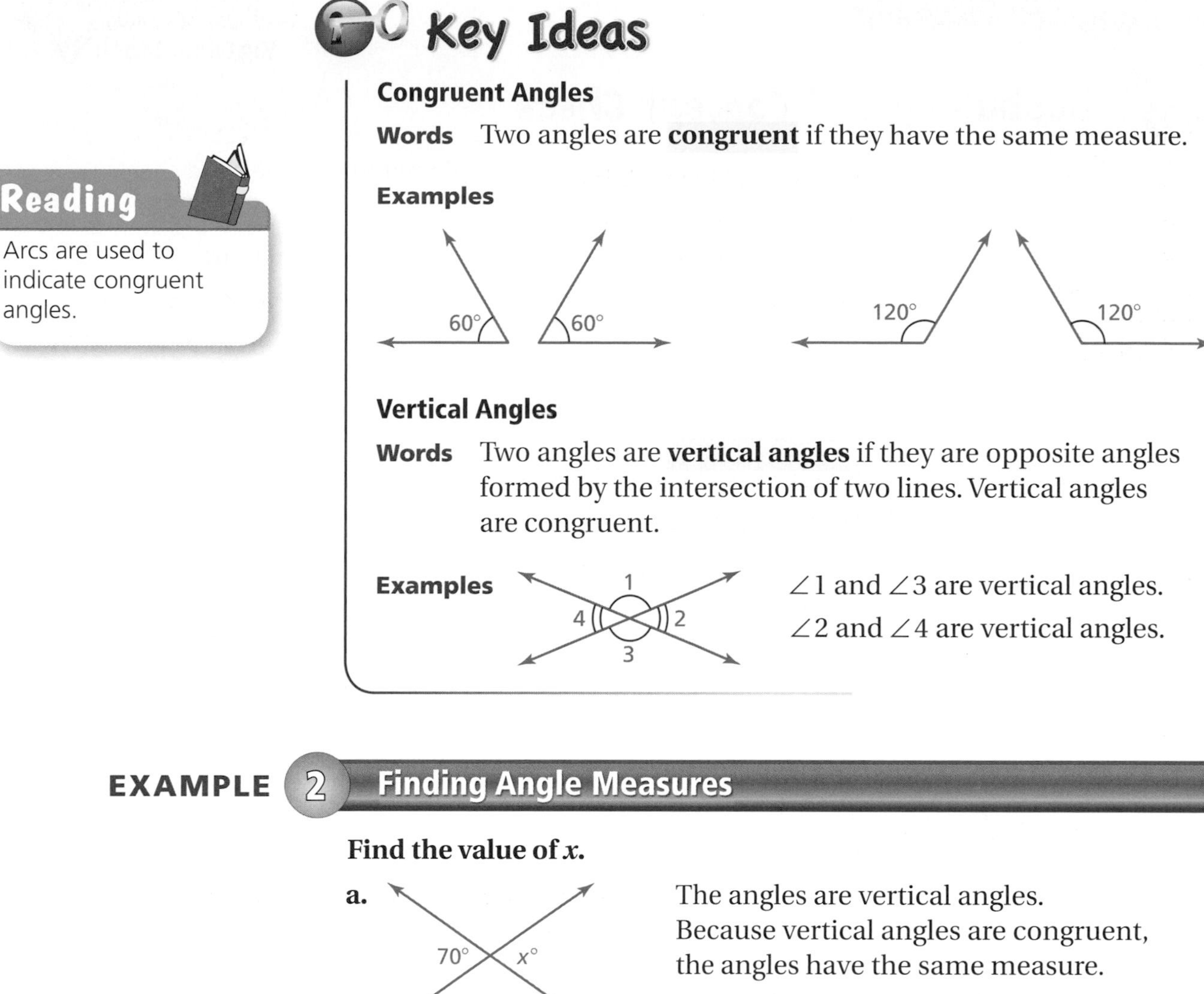

Key Ideas

Reading

Arcs are used to indicate congruent angles.

Congruent Angles

Words Two angles are **congruent** if they have the same measure.

Examples

Vertical Angles

Words Two angles are **vertical angles** if they are opposite angles formed by the intersection of two lines. Vertical angles are congruent.

Examples

$\angle 1$ and $\angle 3$ are vertical angles.

$\angle 2$ and $\angle 4$ are vertical angles.

EXAMPLE 2 Finding Angle Measures

Find the value of x.

a.

The angles are vertical angles. Because vertical angles are congruent, the angles have the same measure.

So, x is 70.

b.

The angles are complementary. So, the sum of their measures is 90°.

$$x + 50 = 90$$

$$x = 40$$

So, x is 40.

On Your Own

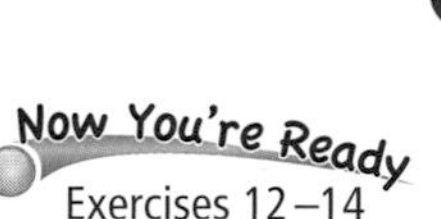

Find the value of x.

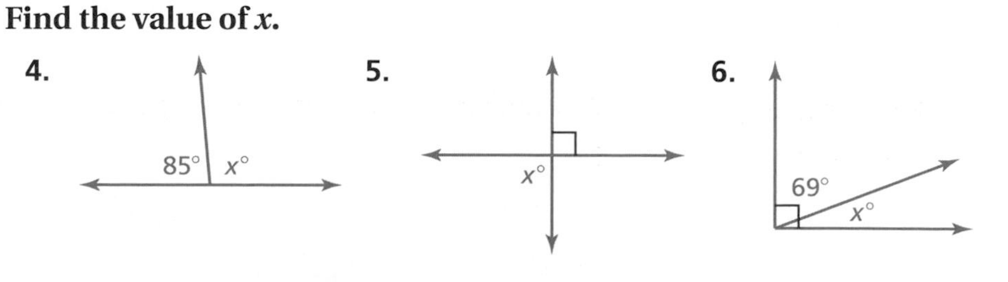

11.1 Exercises

Vocabulary and Concept Check

1. **VOCABULARY** Explain the difference between complementary angles and supplementary angles.
2. **WRITING** When two lines intersect, how many pairs of vertical angles are formed? Explain.

Practice and Problem Solving

Tell whether the statement is *always*, *sometimes*, or *never* true. Explain.

3. If x and y are supplementary angles, then x is obtuse.
4. If x and y are right angles, then x and y are supplementary angles.
5. If x and y are complementary angles, then y is a right angle.

Tell whether the angles are *complementary*, *supplementary*, or *neither*.

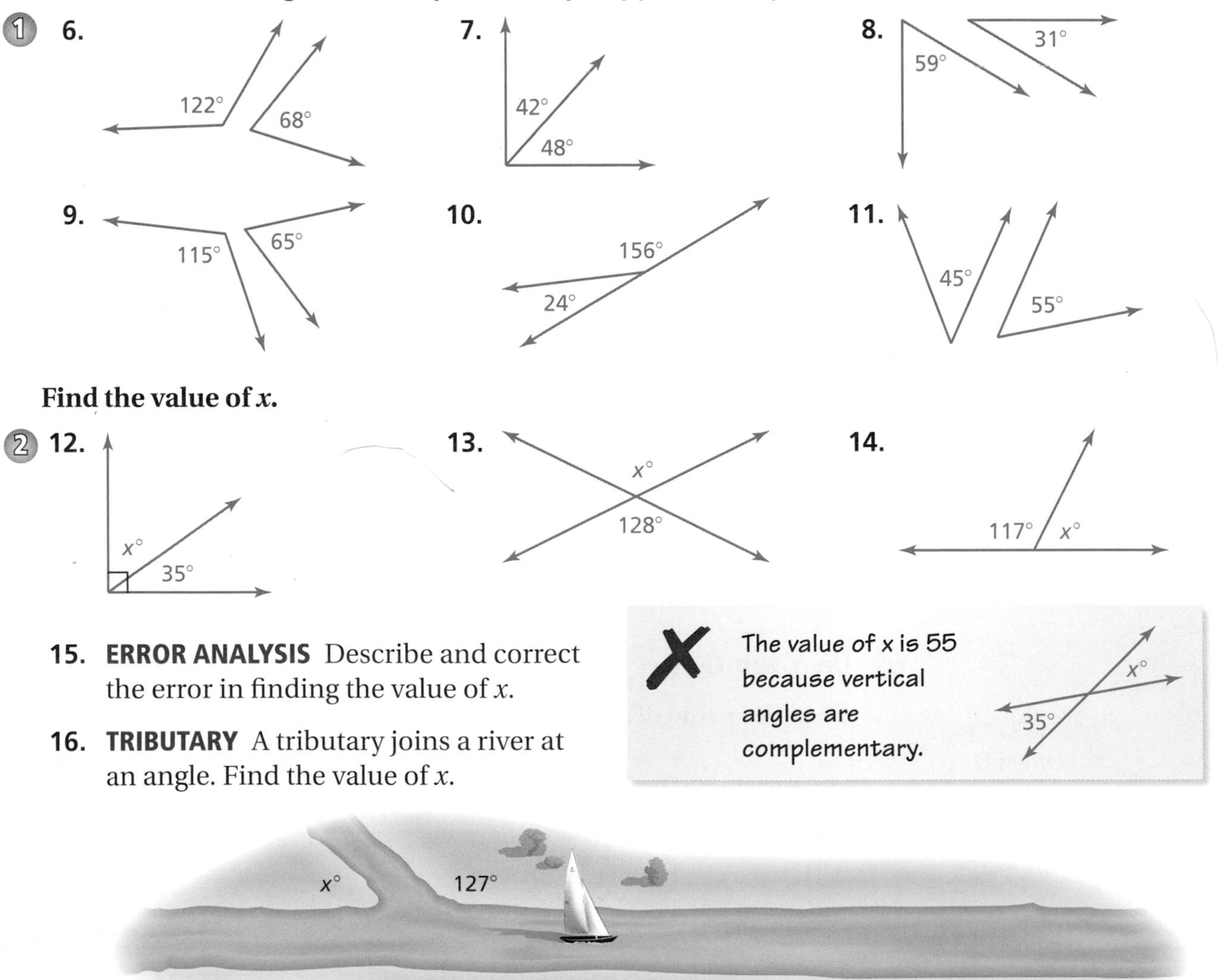

Find the value of x.

17.

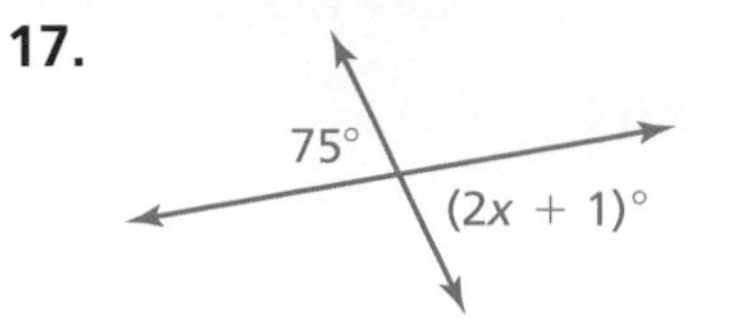

18. $4x°$, $2x°$

19. $7x°$, $(x + 20)°$

20. OPEN-ENDED Give an example of an angle that can be a supplementary angle but cannot be a complementary angle. Explain.

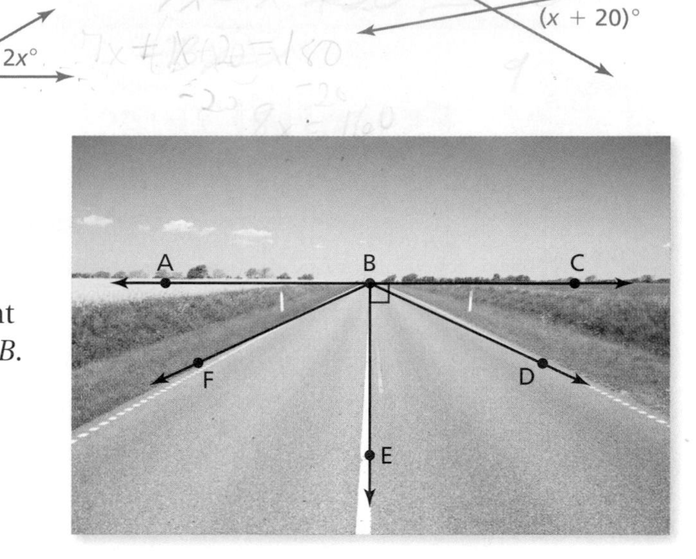

21. VANISHING POINT The vanishing point of the picture is represented by point B.

a. Name two pairs of complementary angles.

b. Name three pairs of supplementary angles.

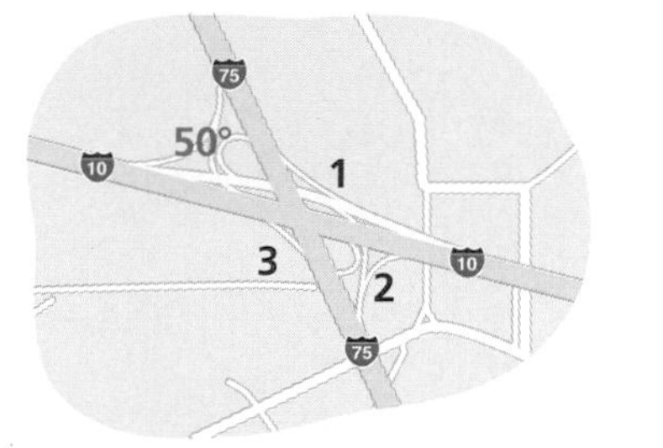

22. INTERSECTION What are the measures of the other three angles formed by the intersection?

23. RATIO The measures of two complementary angles have a ratio of 3 : 2. What is the measure of the larger angle?

24. REASONING Two angles are vertical angles. What are their measures if they are also complementary angles? supplementary angles?

25. Critical Thinking Find the values of x, y, and z.

$4x°$, $x°$, $2x°$, $9z°$, $2y°$

Fair Game Review What you learned in previous grades & lessons

Solve the equation. Check your solution. *(Section 2.4 and Section 2.6)*

26. $x + 60 + 45 = 180$ **27.** $x + 58.5 + 92.2 = 180$ **28.** $x + x + 110 = 180$

29. MULTIPLE CHOICE The graph of which equation has a slope of $-\frac{1}{2}$ and a y-intercept of 4? *(Section 10.5)*

Ⓐ $y = \frac{1}{2}x + 4$ Ⓑ $y = -\frac{1}{2}x - 4$ Ⓒ $y = -\frac{1}{2}x + 4$ Ⓓ $y = \frac{1}{2}x - 4$

11.2 Angles and Sides of Triangles

COMMON CORE STATE STANDARDS
8.G.5

Essential Question How can you classify triangles by their angles?

1 ACTIVITY: Exploring the Angles of a Triangle

Work with a partner.

a. Draw a triangle that has an obtuse angle. Label the angles *A*, *B*, and *C*.

b. Carefully cut out the triangle. Tear off the three corners of the triangle.

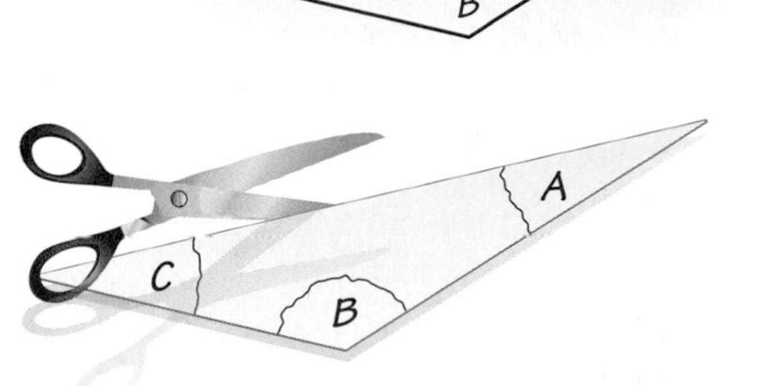

c. Draw a straight line on a piece of paper. Arrange angles *A* and *B* as shown.

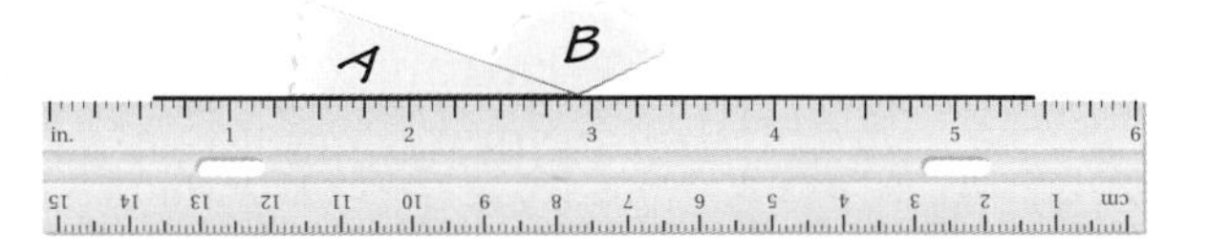

d. Place the third angle as shown. What does this tell you about the sum of the measures of the angles?

e. Draw three other triangles that have different shapes. Repeat parts (b)–(d) for each one. Do you get the same result as in part (d)? Explain.

f. Write a rule about the sum of the measures of the angles of a triangle.

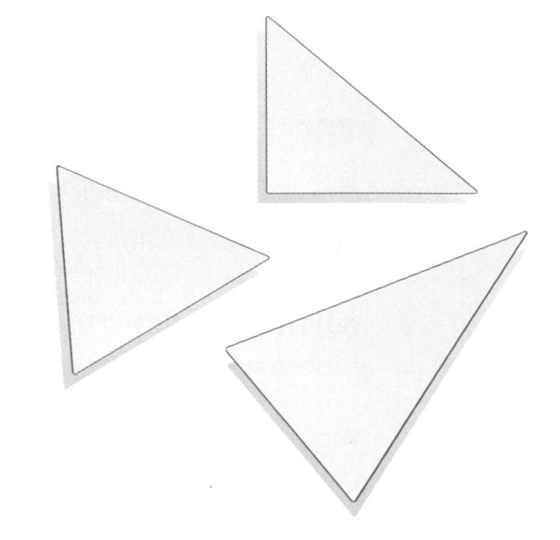

2 ACTIVITY: Thinking About Vocabulary

Work with a partner. Talk about the meaning of each name. Use reasoning to define each name. Then match each name with a triangle.

Note: Each triangle has at least one name, but some have more than one name.

a. Right triangle

b. Acute triangle

c. Obtuse triangle

d. Equiangular triangle

e. Equilateral triangle

f. Isosceles triangle

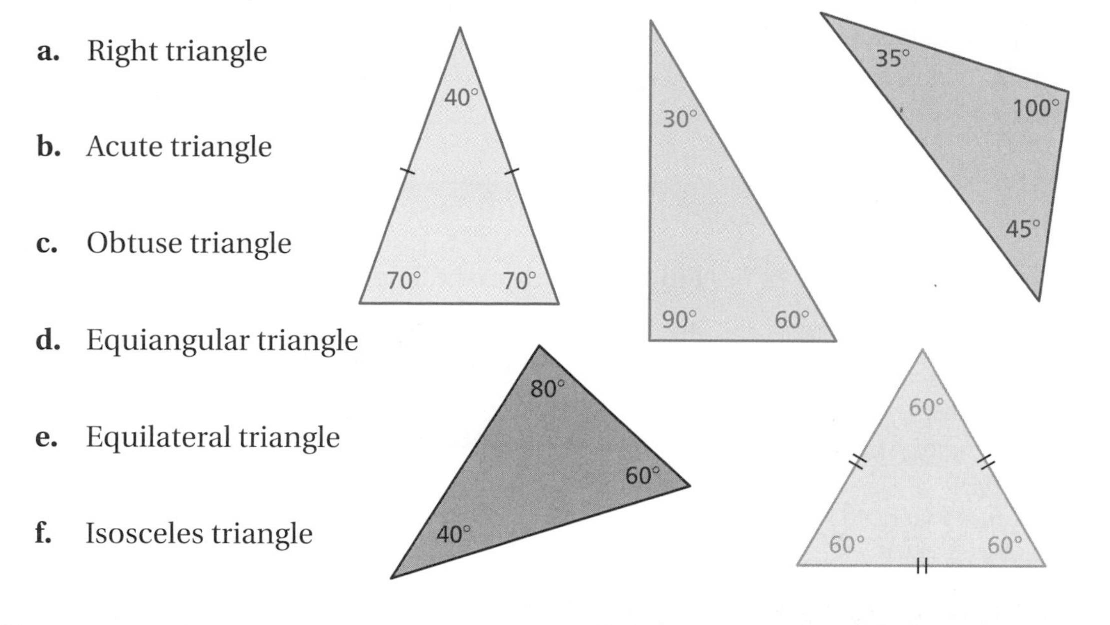

3 ACTIVITY: Triangles in Art

Work with a partner.

a. Trace four triangles in the painting. Classify each triangle using the names in Activity 2.

b. Design your own abstract art painting. How many different types of triangles did you use in your painting?

Abstract II by Linda Bahner
www.spiritartist.com

What Is Your Answer?

4. IN YOUR OWN WORDS How can you classify triangles by their angles?

5. Find examples of real-life triangles in architecture. Name each type of triangle that you find.

Practice Use what you learned about angles of triangles to complete Exercises 3–5 on page 474.

11.2 Lesson

Key Vocabulary
isosceles triangle, *p. 472*
congruent sides, *p. 472*
equilateral triangle, *p. 472*
equiangular triangle, *p. 472*

Key Idea

Angle Measures of a Triangle

Words The sum of the angle measures of a triangle is 180°.

Algebra $x + y + z = 180$

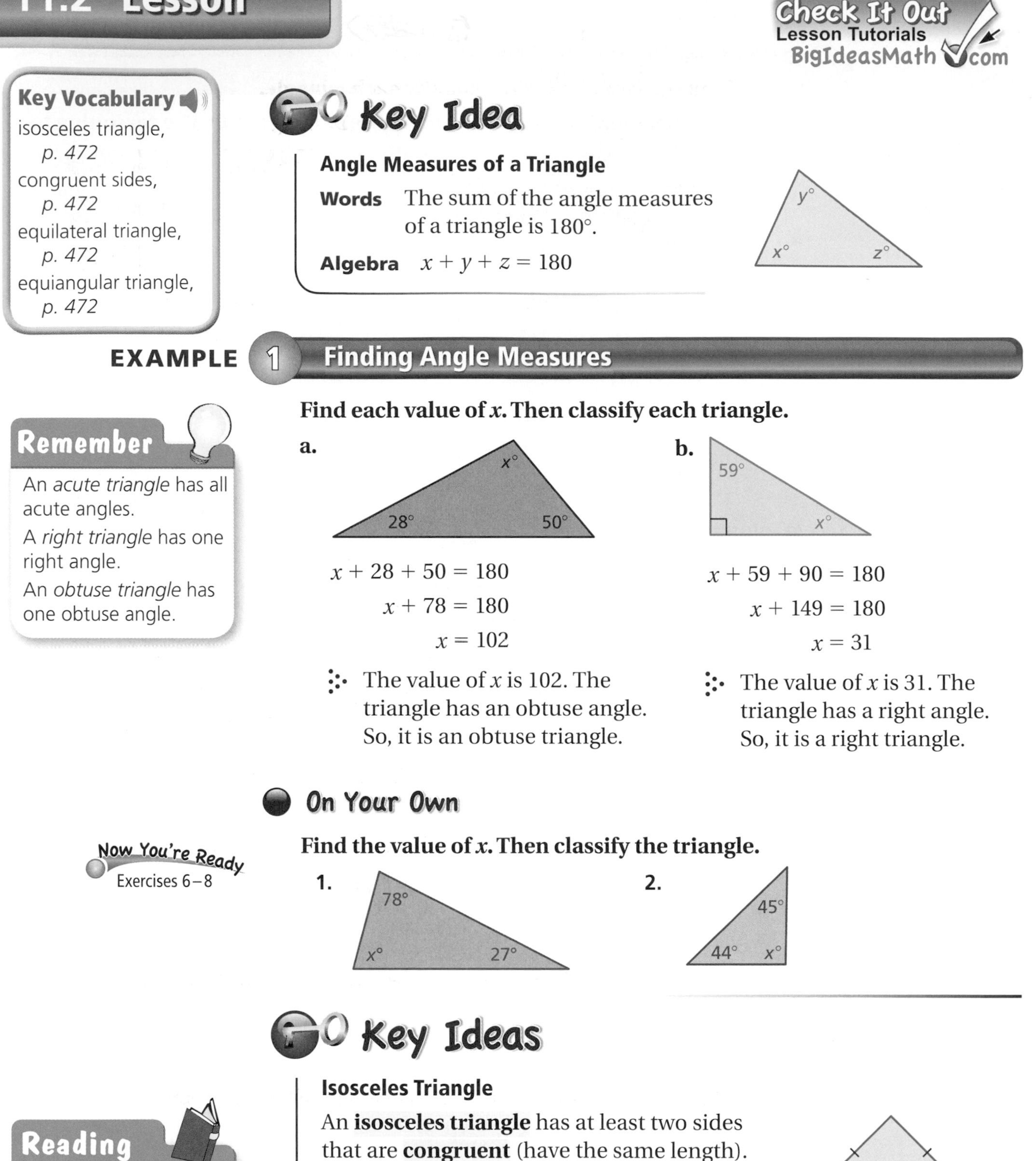

EXAMPLE 1 Finding Angle Measures

Remember

An *acute triangle* has all acute angles.

A *right triangle* has one right angle.

An *obtuse triangle* has one obtuse angle.

Find each value of *x*. Then classify each triangle.

a.

$$x + 28 + 50 = 180$$
$$x + 78 = 180$$
$$x = 102$$

The value of x is 102. The triangle has an obtuse angle. So, it is an obtuse triangle.

b.

$$x + 59 + 90 = 180$$
$$x + 149 = 180$$
$$x = 31$$

The value of x is 31. The triangle has a right angle. So, it is a right triangle.

On Your Own

Now You're Ready
Exercises 6–8

Find the value of *x*. Then classify the triangle.

1.

2.

Key Ideas

Isosceles Triangle

An **isosceles triangle** has at least two sides that are **congruent** (have the same length).

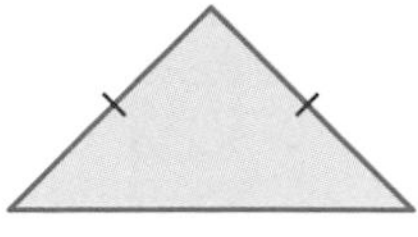

Equilateral Triangle

An **equilateral triangle** has three congruent sides.

An equilateral triangle is also **equiangular** (three congruent angles).

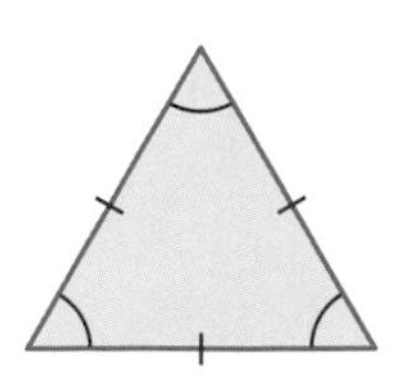

Reading

Small line segments are used to indicate congruent sides.

EXAMPLE 2 Finding Angle Measures

Find the value of x. Then classify each triangle.

a. Flag of Jamaica

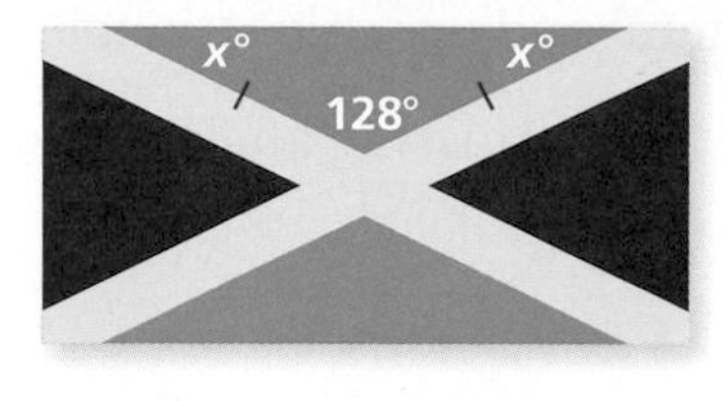

$$x + x + 128 = 180$$
$$2x + 128 = 180$$
$$2x = 52$$
$$x = 26$$

The value of x is 26. Two of the sides are congruent. So, it is an isosceles triangle.

b. Flag of Cuba

$$x + x + 60 = 180$$
$$2x + 60 = 180$$
$$2x = 120$$
$$x = 60$$

The value of x is 60. All three angles are congruent. So, it is an equilateral and equiangular triangle.

EXAMPLE 3 Standardized Test Practice

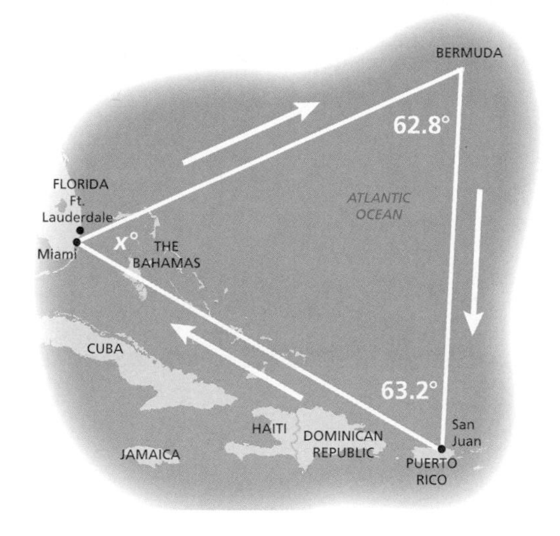

An airplane leaves from Miami and travels around the Bermuda Triangle. What is the value of x?

Ⓐ 26.8 Ⓑ 27.2 Ⓒ 54 Ⓓ 64

Use what you know about the angle measures of a triangle to write an equation.

$x + 62.8 + 63.2 = 180$	Write equation.
$x + 126 = 180$	Add.
$x = 54$	Subtract 126 from each side.

The value of x is 54. The correct answer is Ⓒ.

On Your Own

Now You're Ready
Exercises 9–11

Find the value of x. Then classify the triangle in as many ways as possible.

3. **4.**

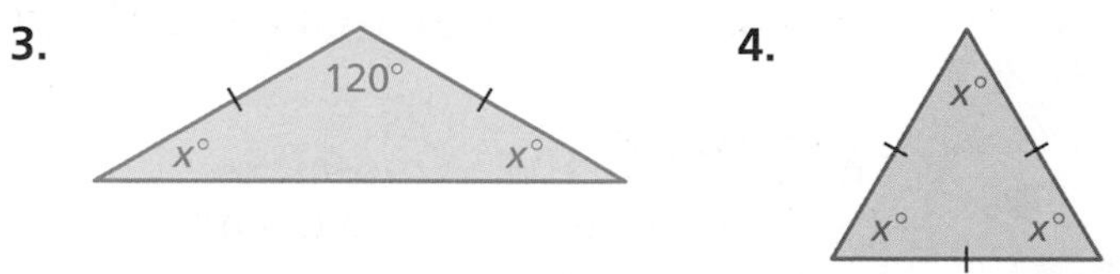

5. In Example 3, the airplane leaves from Fort Lauderdale. The angle measure at Bermuda is 63.9° and the angle measure at San Juan is 61.8°. Find the value of x.

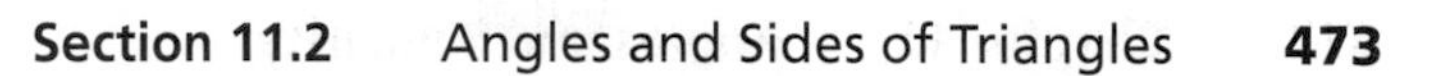

11.2 Exercises

Check It Out
Help with Homework
BigIdeasMath.com

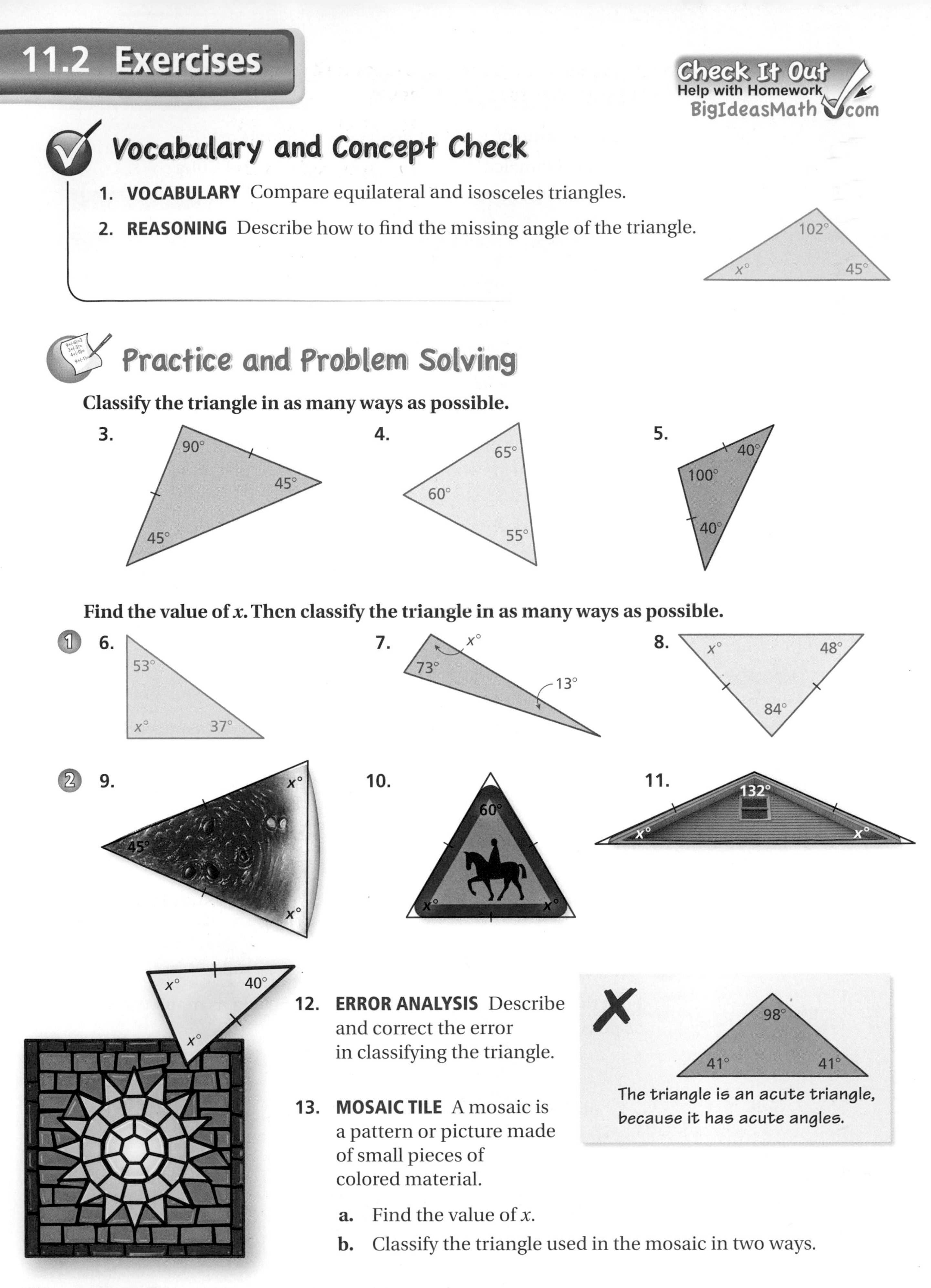

Vocabulary and Concept Check

1. **VOCABULARY** Compare equilateral and isosceles triangles.
2. **REASONING** Describe how to find the missing angle of the triangle.

Practice and Problem Solving

Classify the triangle in as many ways as possible.

3.

4.

5.

Find the value of *x*. Then classify the triangle in as many ways as possible.

1 6.

7.

8.

2 9.

10.

11.

12. **ERROR ANALYSIS** Describe and correct the error in classifying the triangle.

98°
41°
41°
The triangle is an acute triangle, because it has acute angles.

13. **MOSAIC TILE** A mosaic is a pattern or picture made of small pieces of colored material.

   a. Find the value of x.

   b. Classify the triangle used in the mosaic in two ways.

Tell whether a triangle can have the given angle measures. If not, change the first angle measure so that the angle measures form a triangle.

14. $76.2°, 81.7°, 22.1°$

15. $115.1°, 47.5°, 93°$

16. $5\frac{2}{3}°, 64\frac{1}{3}°, 87°$

17. $31\frac{3}{4}°, 53\frac{1}{2}°, 94\frac{3}{4}°$

18. CRITICAL THINKING Consider the three isosceles triangles.

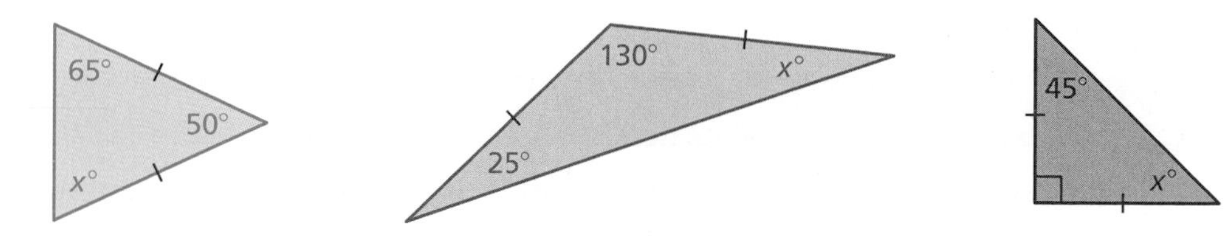

a. Find the value of x for each triangle.

b. What do you notice about the angle measures of each triangle?

c. Write a rule about the angle measures of an isosceles triangle.

19. REASONING Explain why all triangles have at least two acute angles.

20. CARDS One method of stacking cards is shown.

a. Find the value of x.

b. Critical Thinking Describe how to stack the cards with different angles. Is the value of x limited? If so, what are the limitations? Explain your reasoning.

Fair Game Review What you learned in previous grades & lessons

Write and solve an equation to find x. Use 3.14 for π. *(Skills Review Handbook)*

21. $P = 48$ cm

22. $P = 28$ in.

23. $P = 25.42$ m

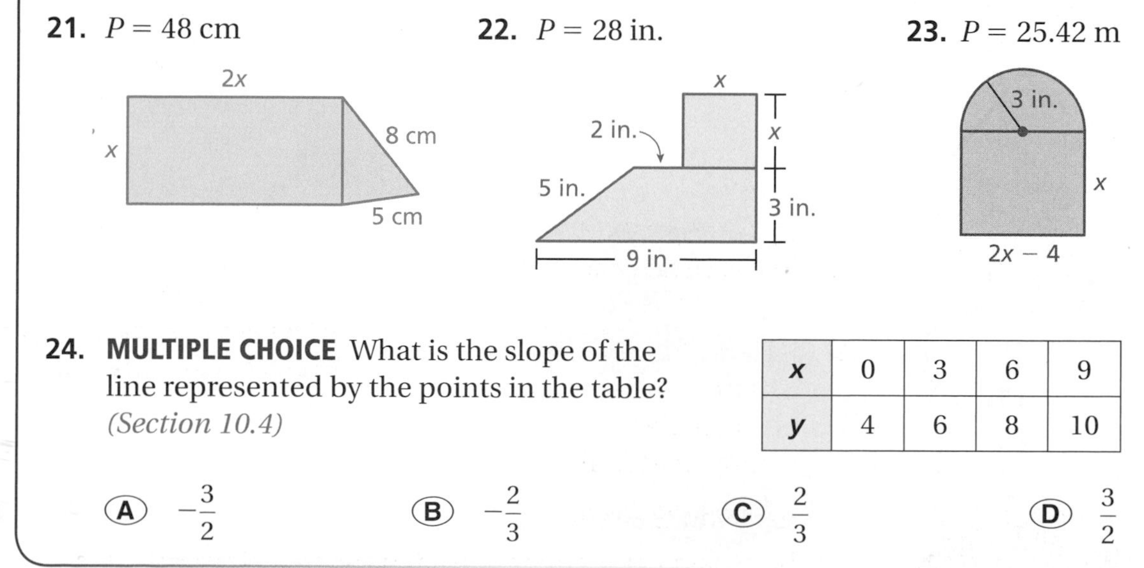

24. MULTIPLE CHOICE What is the slope of the line represented by the points in the table? *(Section 10.4)*

x	0	3	6	9
y	4	6	8	10

Ⓐ $-\frac{3}{2}$ Ⓑ $-\frac{2}{3}$ Ⓒ $\frac{2}{3}$ Ⓓ $\frac{3}{2}$

11.3 Angles of Polygons

COMMON CORE STATE STANDARDS
8.G.5

Essential Question How can you find a formula for the sum of the angle measures of any polygon?

1 ACTIVITY: The Sum of the Angle Measures of a Polygon

Work with a partner. Find the sum of the angle measures of each polygon with *n* sides.

a. Sample: Quadrilateral: $n = 4$

Draw a line that divides the quadrilateral into two triangles.

Because the sum of the angle measures of each triangle is 180°, the sum of the angle measures of the quadrilateral is 360°.

$$(A + B + C) + (D + E + F) = 180^\circ + 180^\circ$$
$$= 360^\circ$$

b. Pentagon: $n = 5$

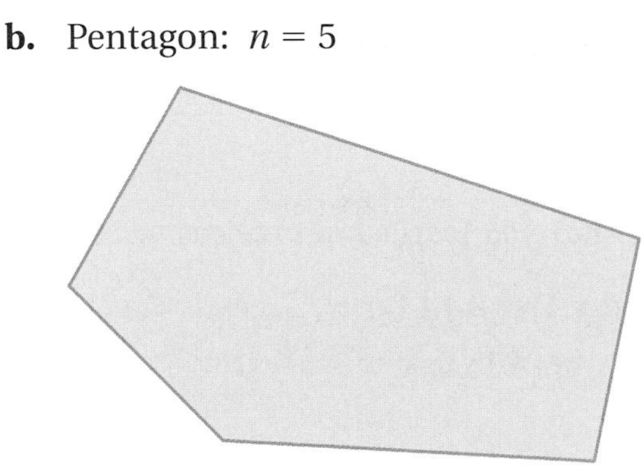

c. Hexagon: $n = 6$

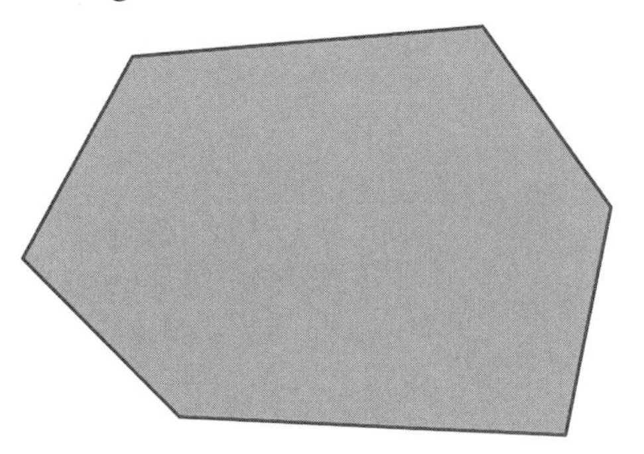

d. Heptagon: $n = 7$

e. Octagon: $n = 8$

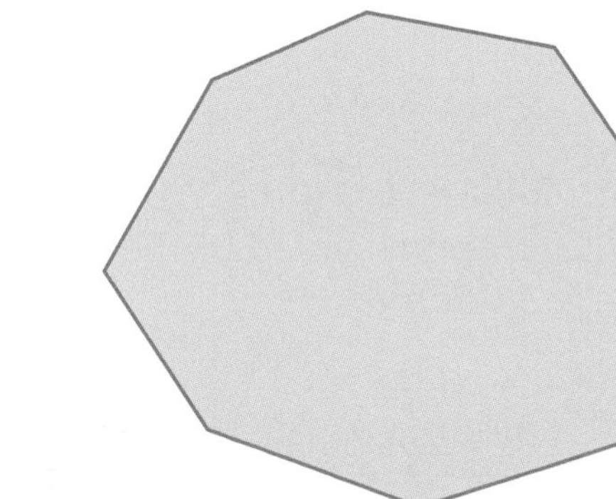

2 ACTIVITY: The Sum of the Angle Measures of a Polygon

Work with a partner.

a. Use the table to organize your results from Activity 1.

Sides, n	3	4	5	6	7	8
Angle Sum, S						

b. Plot the points in the table in a coordinate plane.

c. Write a linear equation that relates S to n.

d. What values of n make sense for the function? Explain your reasoning.

e. Use the function to find the sum of the angle measures of a polygon with 10 sides.

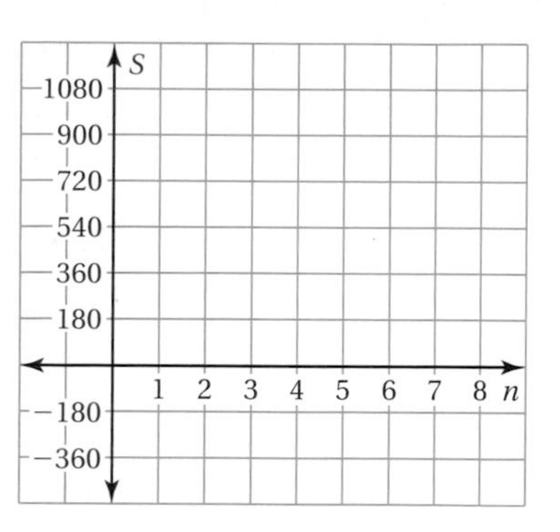

3 ACTIVITY: The Sum of the Angle Measures of a Polygon

Work with a partner.

A polygon is convex if the line segment connecting any two vertices lies entirely inside the polygon. A polygon that is not convex is called concave.

Does the equation you found in Activity 2 apply to concave polygons? Explain.

How can you define the measure of an angle so that your equation applies to *any* polygon?

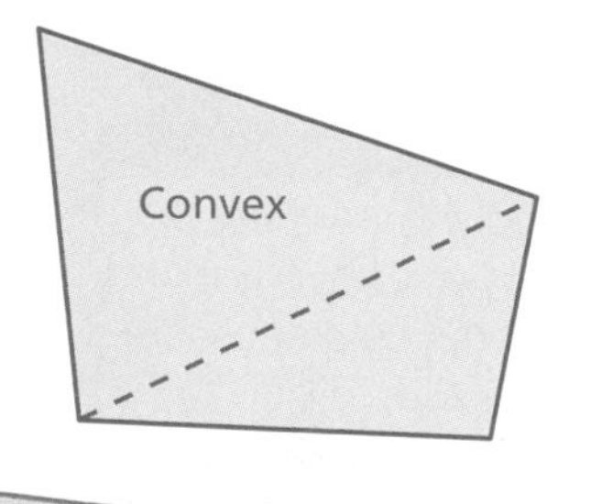

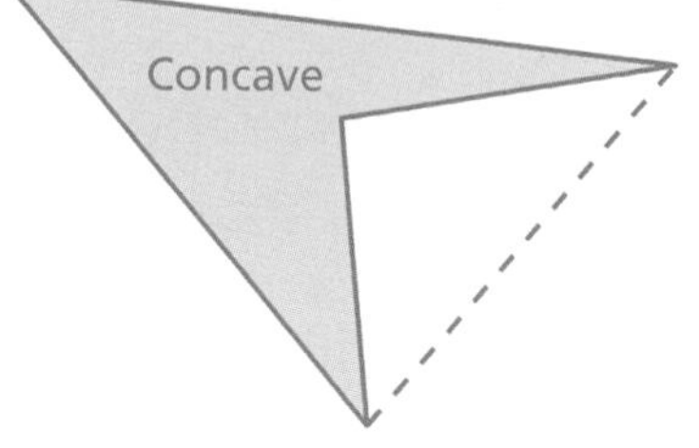

What Is Your Answer?

4. **IN YOUR OWN WORDS** How can you find a formula for the sum of the angle measures of any polygon?

Use what you learned about angles of polygons to complete Exercises 4–6 on page 481.

11.3 Lesson

Key Vocabulary

A **polygon** is a closed plane figure made up of three or more line segments that intersect only at their endpoints.

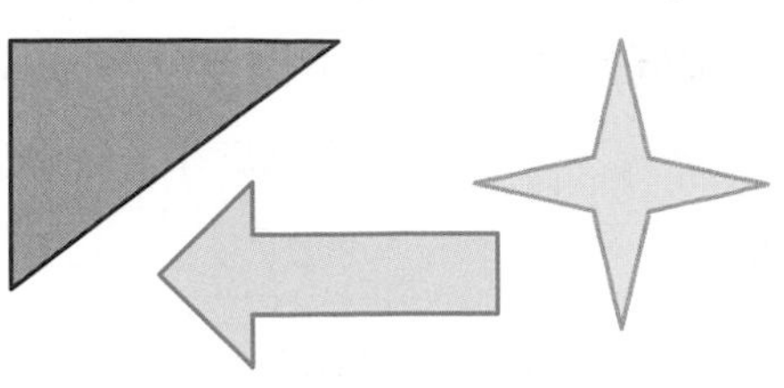

Polygons **Not polygons**

Key Idea

Angle Measures of a Polygon

The sum S of the angle measures of a polygon with n sides is

$$S = (n - 2) \cdot 180^\circ.$$

EXAMPLE 1 Finding the Sum of the Angle Measures of a Polygon

Find the sum of the angle measures of the school crossing sign.

The sign is in the shape of a pentagon. It has 5 sides.

$S = (n - 2) \cdot 180^\circ$	Write the formula.
$= (5 - 2) \cdot 180^\circ$	Substitute 5 for n.
$= 3 \cdot 180^\circ$	Subtract.
$= 540^\circ$	Multiply.

The sum of the angle measures is 540°.

Reading

For polygons whose names you have not learned, you can use the phrase "*n*-gon," where *n* is the number of sides. For example, a 15-gon is a polygon with 15 sides.

On Your Own

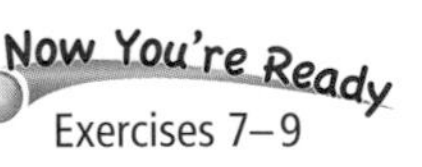

Exercises 7–9

Find the sum of the angle measures of the green polygon.

1. 

2. 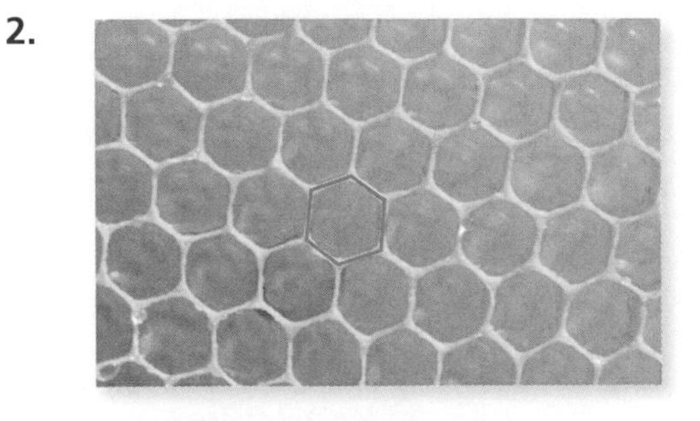

EXAMPLE 2 Finding an Angle Measure of a Polygon

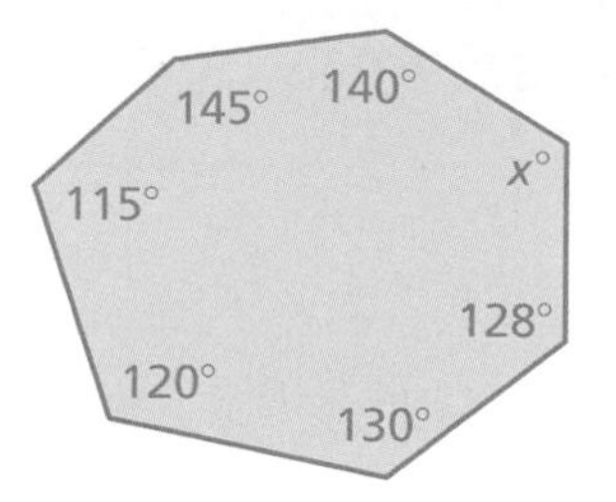

Find the value of x.

Step 1: The polygon has 7 sides. Find the sum of the angle measures.

$S = (n - 2) \cdot 180°$ Write the formula.

$= (7 - 2) \cdot 180°$ Substitute 7 for n.

$= 900°$ Simplify. The sum of the angle measures is 900°.

Step 2: Write and solve an equation.

$$140 + 145 + 115 + 120 + 130 + 128 + x = 900$$

$$778 + x = 900$$

$$x = 122$$

∴ The value of x is 122.

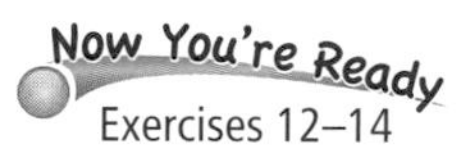

Find the value of x.

3. 4. 5.

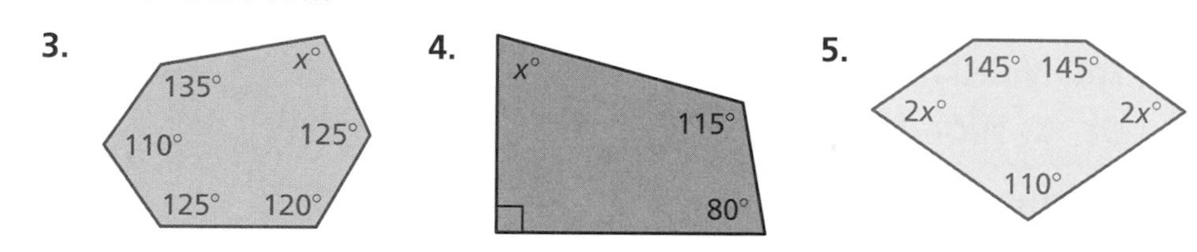

In a **regular polygon**, all of the sides are congruent and all of the angles are congruent.

EXAMPLE 3 Real-Life Application

The hexagon is about 15,000 miles across. Approximately four Earths could fit inside it.

A cloud system discovered on Saturn is in the approximate shape of a regular hexagon. Find the measure of each angle of the hexagon.

Step 1: A hexagon has 6 sides. Find the sum of the angle measures.

$S = (n - 2) \cdot 180°$ Write the formula.

$= (6 - 2) \cdot 180°$ Substitute 6 for n.

$= 720°$ Simplify. The sum of the angle measures is 720°.

Step 2: Divide the sum by the number of angles, 6.

$$720° \div 6 = 120°$$

∴ The measure of each angle is 120°.

On Your Own

Now You're Ready
Exercises 16–18

Find the measure of each angle of the regular polygon.

6. octagon **7.** decagon **8.** 18-gon

Key Idea

Convex and Concave Polygons

A polygon is **convex** if every line segment connecting any two vertices lies entirely inside the polygon.

A polygon is **concave** if at least one line segment connecting any two vertices lies outside the polygon.

EXAMPLE 4 Identifying Convex and Concave Polygons

Tell whether the polygon is *convex* or *concave*. Explain.

a. **b.**

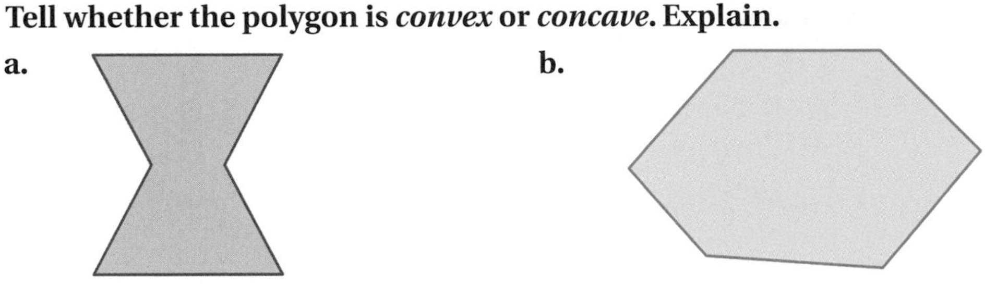

A line segment connecting two vertices lies outside the polygon. So, the polygon is concave.

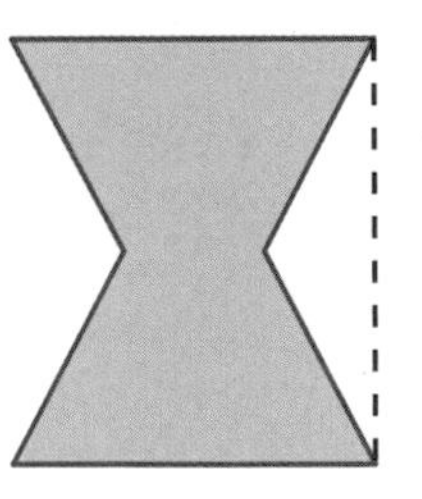

No line segment connecting two vertices lies outside the polygon. So, the polygon is convex.

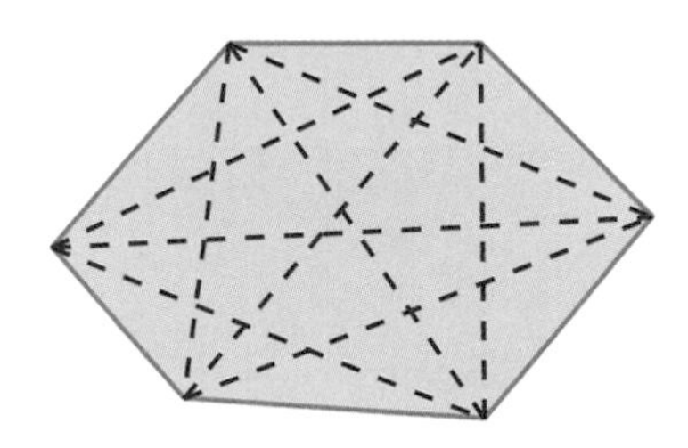

The Meaning of a Word

Concave

To remember the term con**cave**, think of a polygon that is "**cave**d in."

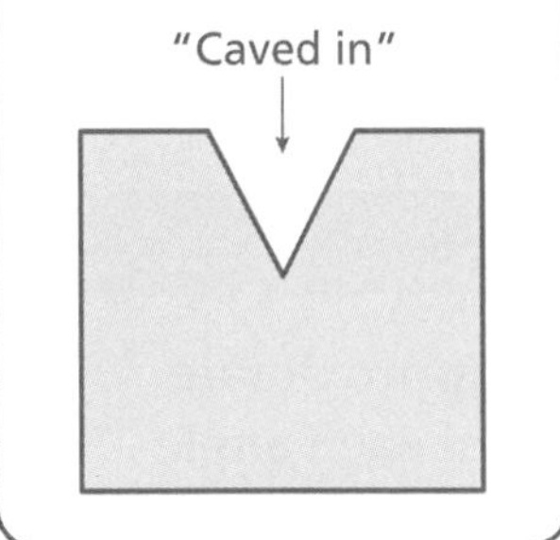

On Your Own

Now You're Ready
Exercises 22–24

Tell whether the polygon is *convex* or *concave*. Explain.

9. **10.** **11.**

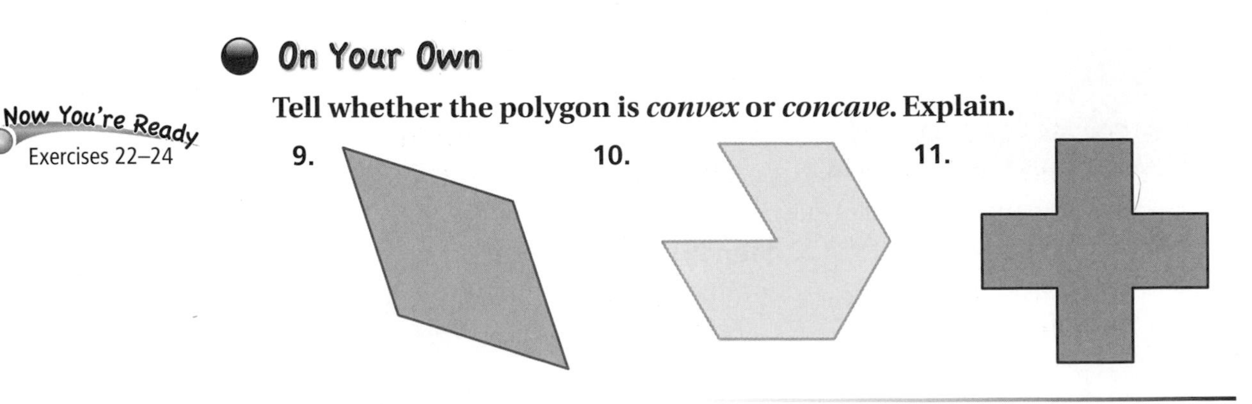

11.3 Exercises

Check It Out
Help with Homework
BigIdeasMath.com

Vocabulary and Concept Check

1. **VOCABULARY** Draw a regular polygon that has three sides.
2. **WHICH ONE DOESN'T BELONG?** Which figure does *not* belong with the other three? Explain your reasoning.

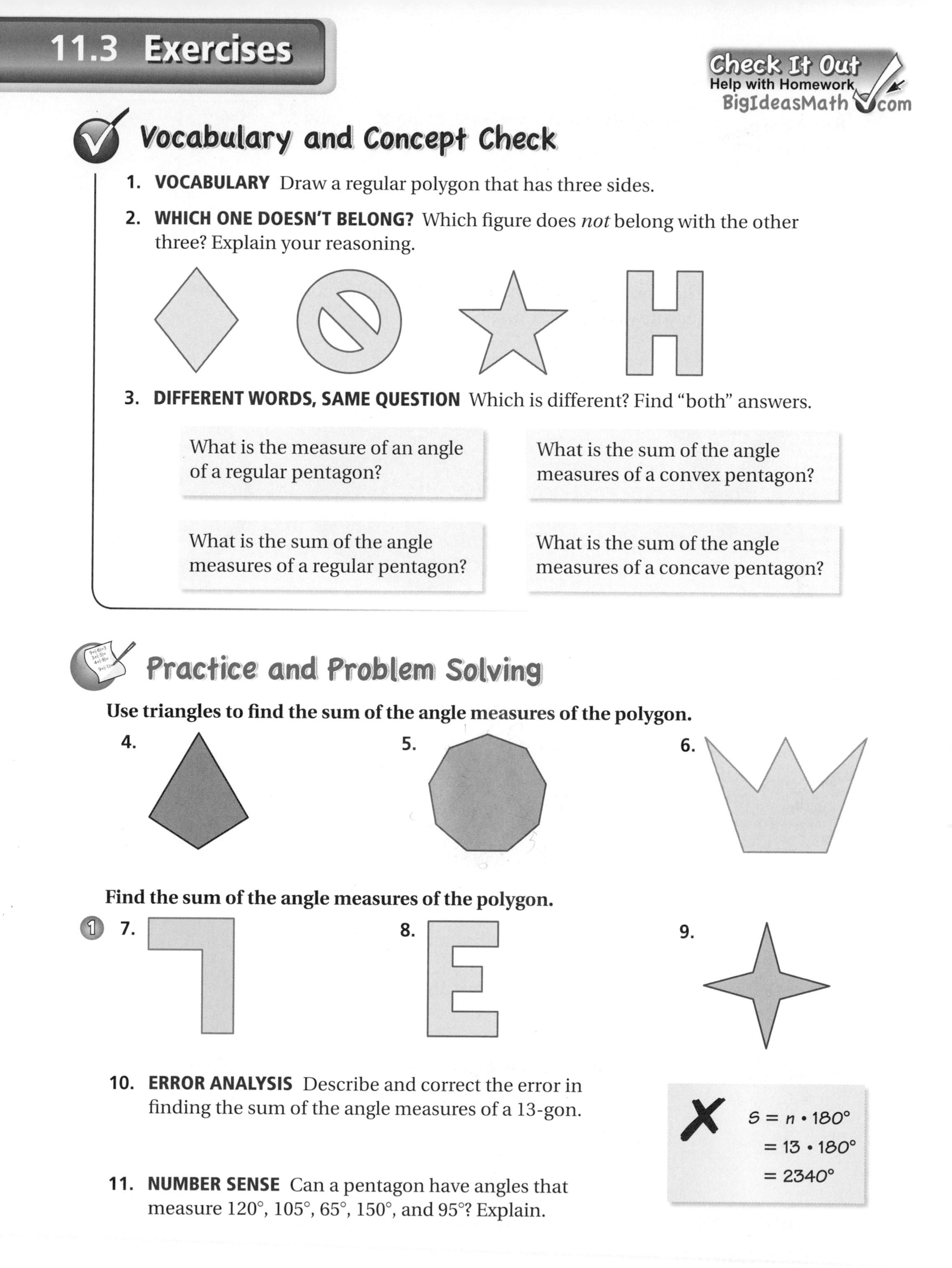

3. **DIFFERENT WORDS, SAME QUESTION** Which is different? Find "both" answers.

What is the measure of an angle of a regular pentagon?	What is the sum of the angle measures of a convex pentagon?
What is the sum of the angle measures of a regular pentagon?	What is the sum of the angle measures of a concave pentagon?

Practice and Problem Solving

Use triangles to find the sum of the angle measures of the polygon.

4.
5.
6.

Find the sum of the angle measures of the polygon.

1 7.
8.
9.

10. **ERROR ANALYSIS** Describe and correct the error in finding the sum of the angle measures of a 13-gon.

$$S = n \cdot 180^\circ$$
$$= 13 \cdot 180^\circ$$
$$= 2340^\circ$$

11. **NUMBER SENSE** Can a pentagon have angles that measure 120°, 105°, 65°, 150°, and 95°? Explain.

Find the value of x.

(2) **12.** **13.** **14.**

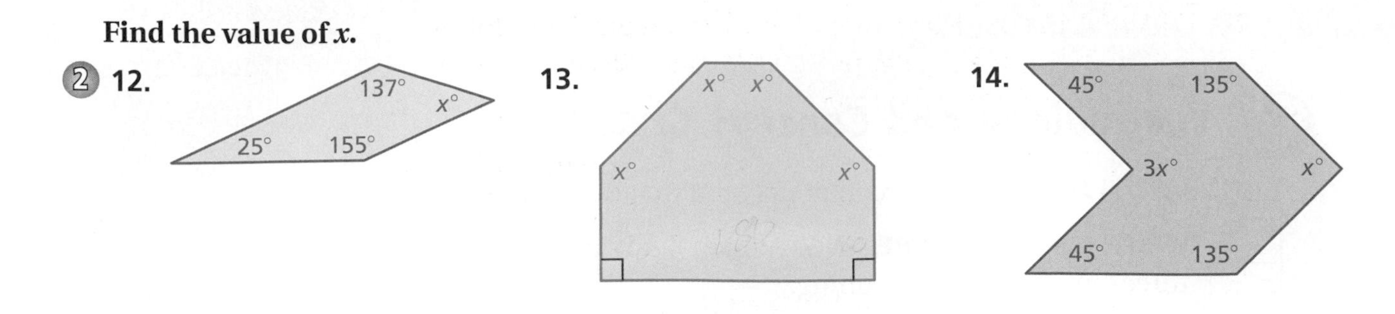

15. REASONING The sum of the angle measures in a regular polygon is 1260°. What is the measure of one of the angles of the polygon?

Find the measure of each angle of the regular polygon.

(3) **16.** **17.** **18.**

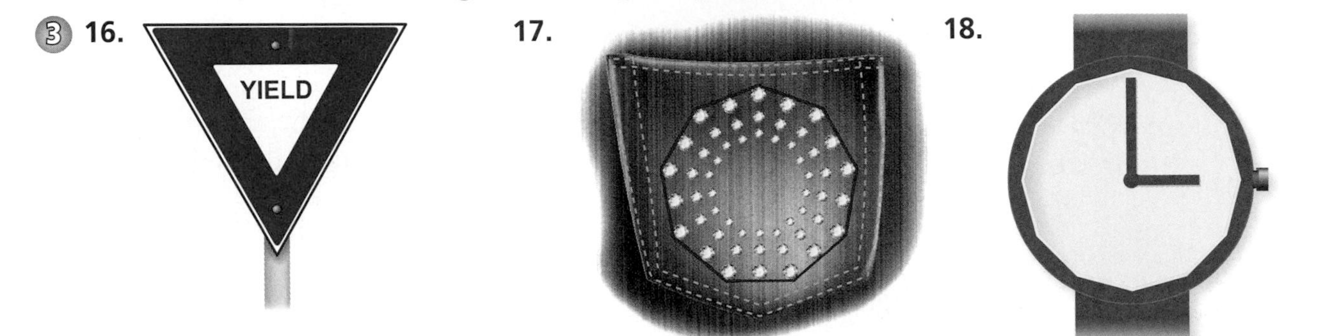

19. ERROR ANALYSIS Describe and correct the error in finding the measure of each angle of a regular 20-gon.

✗
$S = (n - 2) \cdot 180°$
$= (20 - 2) \cdot 180°$
$= 18 \cdot 180°$
$= 3240°$
$3240° \div 18 = 180$
The measure of each angle is 180°.

20. FIRE HYDRANT A fire hydrant bolt is in the shape of a regular pentagon.

a. What is the measure of each angle?

b. Why are fire hydrants made this way?

21. PUZZLE The angles of a regular polygon each measure 165°. How many sides does the polygon have?

Tell whether the polygon is *convex* or *concave*. Explain.

(4) **22.** **23.** **24.**

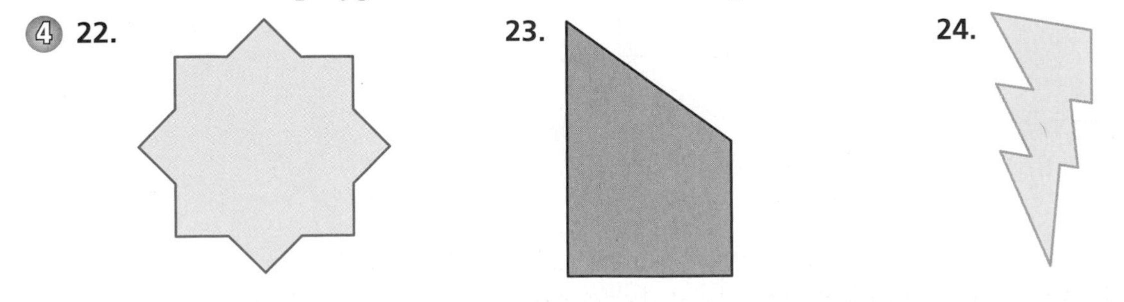

25. CRITICAL THINKING Can a concave polygon be regular? Explain.

26. OPEN-ENDED Draw a polygon that has congruent sides but is not regular.

27. **STAINED GLASS** The center of the stained glass window is in the shape of a regular polygon. What is the measure of each angle of the polygon?

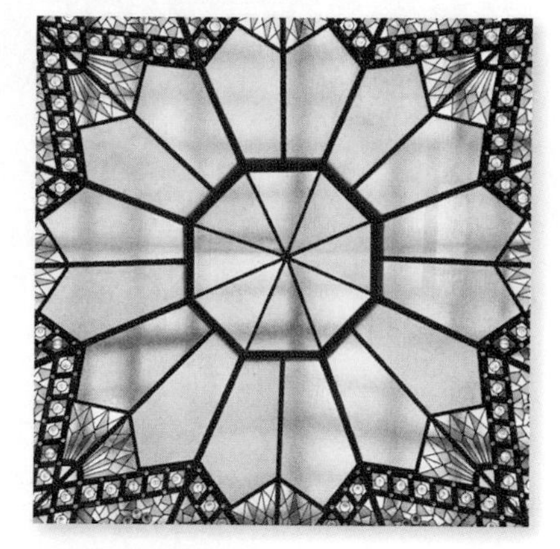

28. **PENTAGON** Draw a pentagon that has two right angles, two 45° angles, and one 270° angle.

29. **GAZEBO** The floor of a gazebo is in the shape of a heptagon. Four of the angles measure 135°. The other angles have equal measures. Find the measure of each of the remaining angles.

30. **MONEY** The border of a Susan B. Anthony dollar is in the shape of a regular polygon.

   a. How many sides does the polygon have?

   b. What is the measure of each angle of the border? Round your answer to the nearest degree.

31. **REASONING** Copy and complete the table. Does the table represent a linear function? Explain.

Sides of a Regular Polygon, n	3	4	5	6	7	8	9	10
Measure of One Angle, a								

32. **Geometry** When tiles can be used to cover a floor with no empty spaces, the collection of tiles is called a *tessellation*.

   a. Create a tessellation using equilateral triangles.

   b. Find two more regular polygons that form tessellations.

   c. Create a tessellation that uses two different regular polygons.

Fair Game Review What you learned in previous grades & lessons

Solve the proportion. *(Section 3.3 and Section 3.5)*

33. $\frac{x}{12} = \frac{3}{4}$

34. $\frac{14}{21} = \frac{x}{3}$

35. $\frac{x}{9} = \frac{2}{6}$

36. $\frac{4}{10} = \frac{x}{15}$

37. **MULTIPLE CHOICE** The ratio of tulips to daisies is 3 : 5. Which of the following could be the total number of tulips and daisies? *(Skills Review Handbook)*

   Ⓐ 6 Ⓑ 10 Ⓒ 15 Ⓓ 16

11 Study Help

You can use an **example and non-example chart** to list examples and non-examples of a vocabulary word or item. Here is an example and non-example chart for complementary angles.

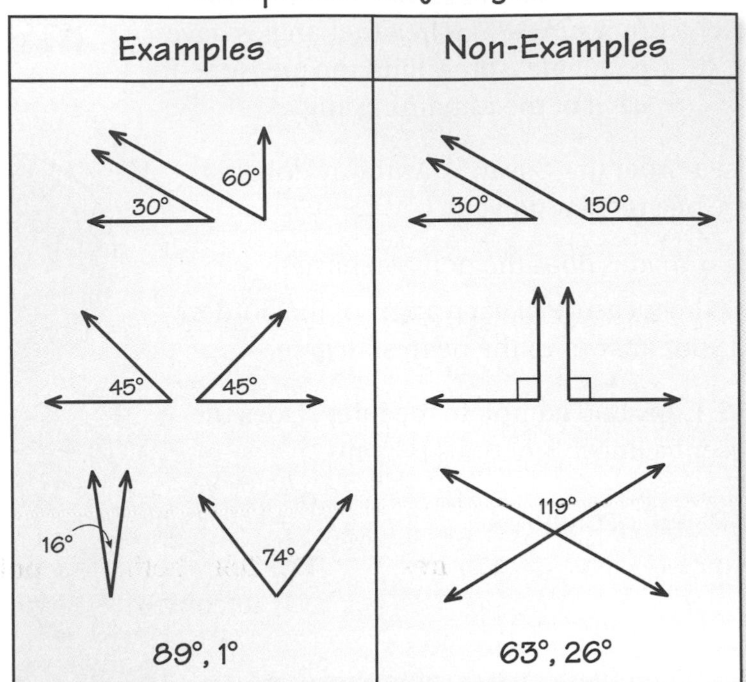

On Your Own

Make an example and non-example chart to help you study these topics.

1. isosceles triangles
2. equilateral triangles
3. regular polygons
4. convex polygons
5. concave polygons

After you complete this chapter, make example and non-example charts for the following topics.

6. similar triangles
7. transversals
8. interior angles

"What do you think of my example & non-example chart for popular cat toys?"

11.1–11.3 Quiz

Check It Out
Progress Check
BigIdeasMath.com

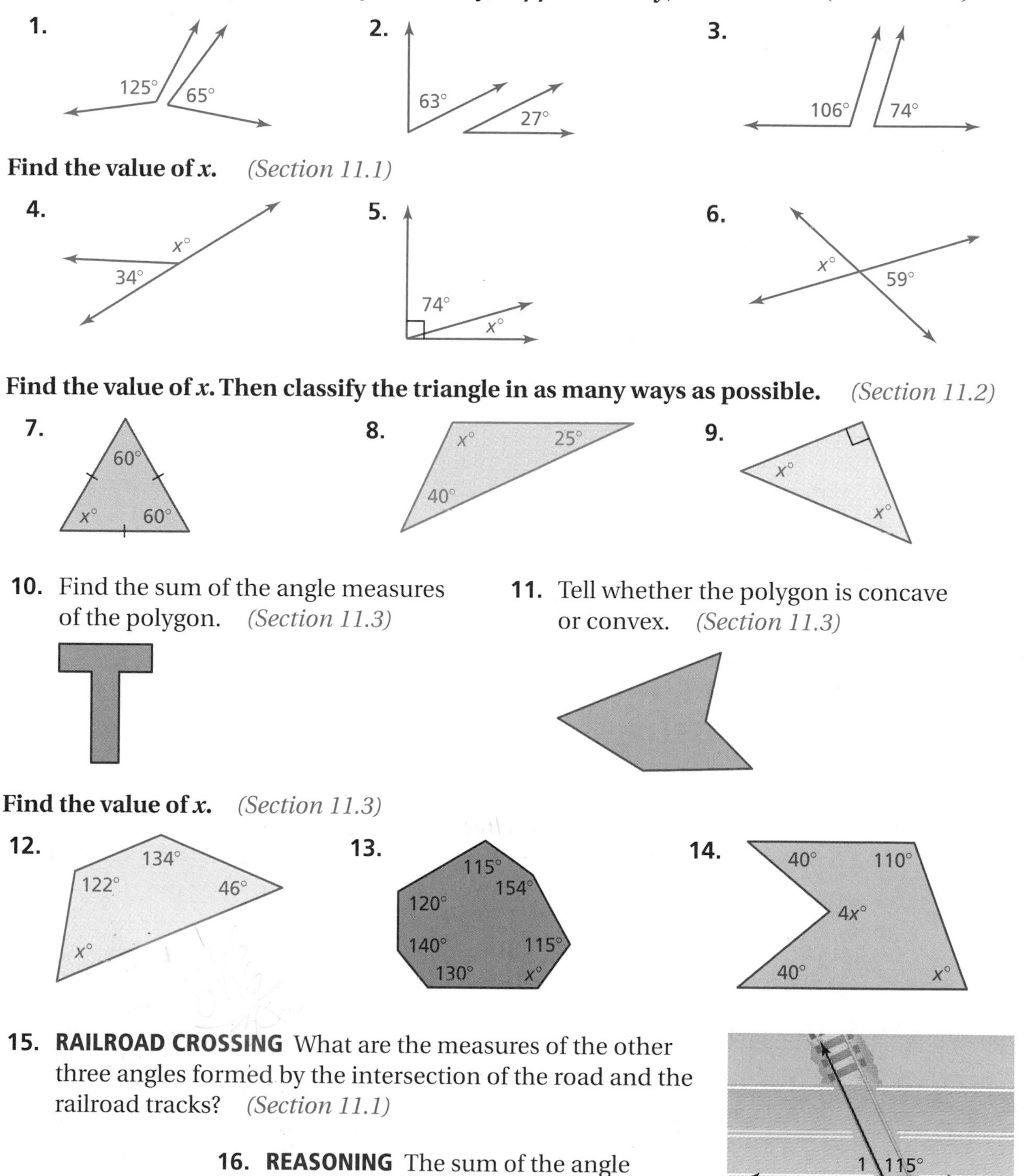

Tell whether the angles are *complementary*, *supplementary*, or *neither*. *(Section 11.1)*

1. 125°, 65°

2. 63°, 27°

3. 106°, 74°

Find the value of *x*. *(Section 11.1)*

4. $x°$, 34°

5. 74°, $x°$

6. $x°$, 59°

Find the value of *x*. Then classify the triangle in as many ways as possible. *(Section 11.2)*

7. 60°, $x°$, 60°

8. $x°$, 25°, 40°

9. $x°$, $x°$

10. Find the sum of the angle measures of the polygon. *(Section 11.3)*

11. Tell whether the polygon is concave or convex. *(Section 11.3)*

Find the value of *x*. *(Section 11.3)*

12. 134°, 122°, 46°, $x°$

13. 115°, 154°, 120°, 140°, 115°, 130°, $x°$

14. 40°, 110°, $4x°$, 40°, $x°$

15. RAILROAD CROSSING What are the measures of the other three angles formed by the intersection of the road and the railroad tracks? *(Section 11.1)*

16. REASONING The sum of the angle measures of a polygon is 4140°. How many sides does the polygon have? *(Section 11.3)*

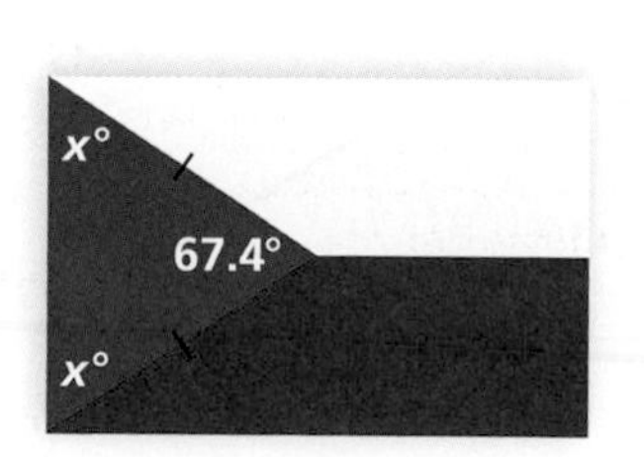

17. FLAG Classify the triangle on the flag of the Czech Republic in as many ways as possible. *(Section 11.2)*

11.4 Using Similar Triangles

COMMON CORE STATE STANDARDS
8.G.5

Essential Question Which properties of triangles make them special among all other types of polygons?

You already know that two triangles are **similar** if and only if the ratios of their corresponding side lengths are equal.

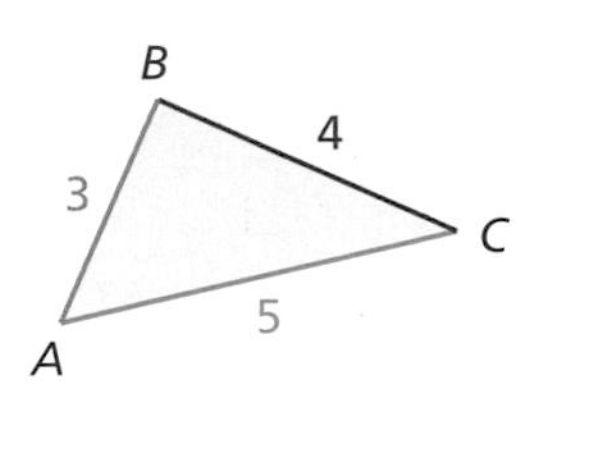

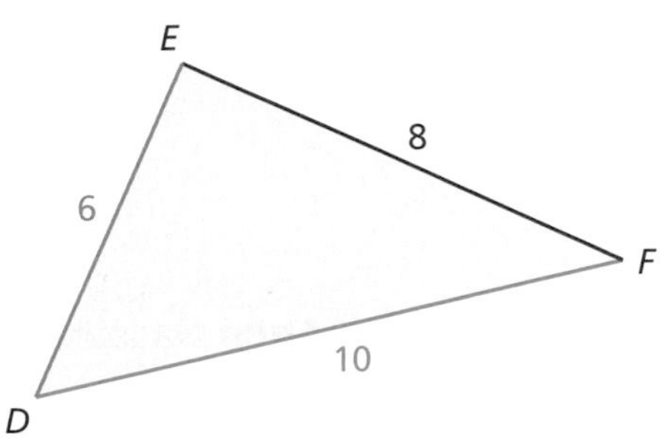

For example, $\triangle ABC$ is similar to $\triangle DEF$ because the ratios of their corresponding side lengths are equal.

$$\frac{6}{3} = \frac{10}{5} = \frac{8}{4}$$

1 ACTIVITY: Angles of Similar Triangles

Work with a partner.

- **Discuss how to make a triangle that is larger than $\triangle XYZ$ and has the *same* angle measures as $\triangle XYZ$.**
- **Measure the lengths of the sides of the two triangles.**
- **Find the ratios of the corresponding side lengths. Are they all the same? What can you conclude?**

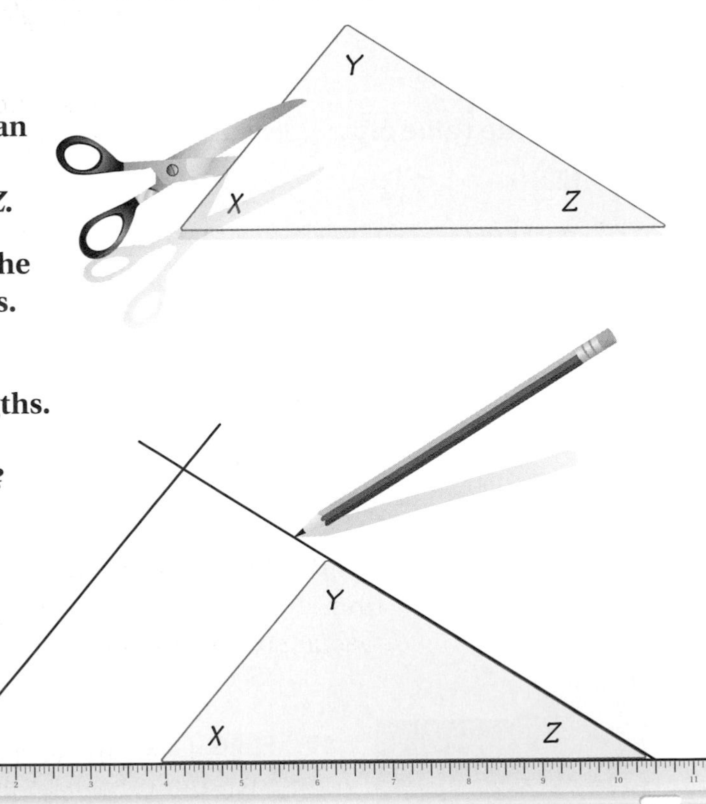

2 ACTIVITY: Amazing Triangles

Work with a partner. Use what you know about polygons to decide whether each statement is true. In each case, explain your reasoning.

a. If two triangles are similar, then the ratios of their corresponding side lengths are equal.

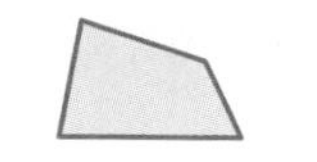

If two quadrilaterals are similar, then the ratios of their corresponding side lengths are equal.

b. 

If the ratios of the corresponding sides of two triangles are equal, then the triangles are similar.

If the ratios of the corresponding sides of two quadrilaterals are equal, then the quadrilaterals are similar.

c. If two triangles are similar, then their corresponding angles are congruent.

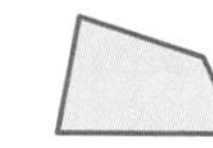

If two quadrilaterals are similar, then their corresponding angles are congruent.

d. 

If the corresponding angles in two triangles are congruent, then the triangles are similar.

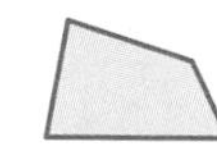

If the corresponding angles in two quadrilaterals are congruent, then the quadrilaterals are similar.

e. 

If the corresponding sides of two triangles are congruent, then the two triangles have identical shapes.

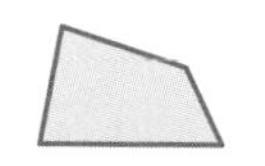

If the corresponding sides of two quadrilaterals are congruent, then the two quadrilaterals have identical shapes.

What Is Your Answer?

3. **IN YOUR OWN WORDS** Which properties of triangles make them special among all other types of polygons? Describe two careers in which the special properties of triangles are used.

Use what you learned about similar triangles to complete Exercises 3 and 4 on page 490.

11.4 Lesson

Key Vocabulary
similar triangles, *p. 488*
indirect measurement, *p. 489*

Triangles that have the same shape but not necessarily the same size are **similar triangles**.

Study Tip
If two angles in one triangle are congruent to two angles in another triangle, then the third angles are also congruent.

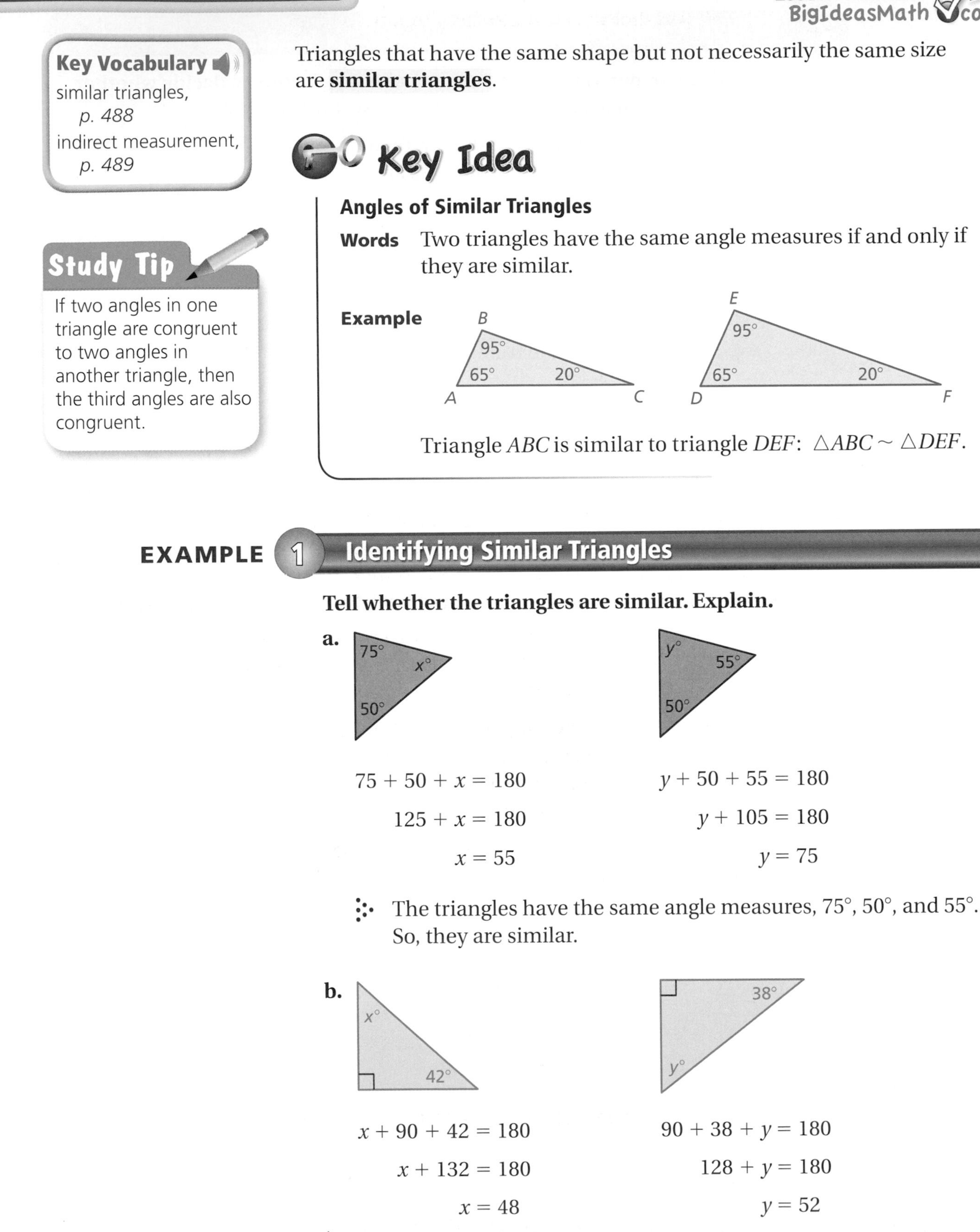

Key Idea

Angles of Similar Triangles

Words Two triangles have the same angle measures if and only if they are similar.

Example

Triangle ABC is similar to triangle DEF: $\triangle ABC \sim \triangle DEF$.

EXAMPLE 1 Identifying Similar Triangles

Tell whether the triangles are similar. Explain.

a.

$75 + 50 + x = 180$
$125 + x = 180$
$x = 55$

$y + 50 + 55 = 180$
$y + 105 = 180$
$y = 75$

The triangles have the same angle measures, 75°, 50°, and 55°. So, they are similar.

b.

$x + 90 + 42 = 180$
$x + 132 = 180$
$x = 48$

$90 + 38 + y = 180$
$128 + y = 180$
$y = 52$

The triangles do not have the same angle measures. So, they are not similar.

Multi-Language Glossary at BigIdeasMath.com.

On Your Own

Now You're Ready
Exercises 5–8

Tell whether the triangles are similar. Explain.

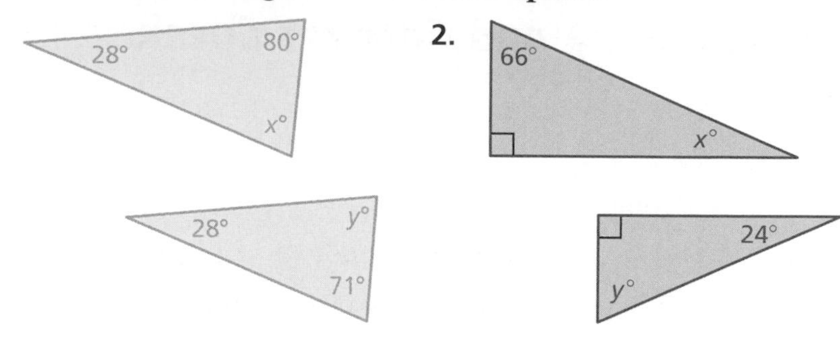

Indirect measurement uses similar figures to find a missing measure when it is difficult to find directly.

EXAMPLE 2 Using Indirect Measurement

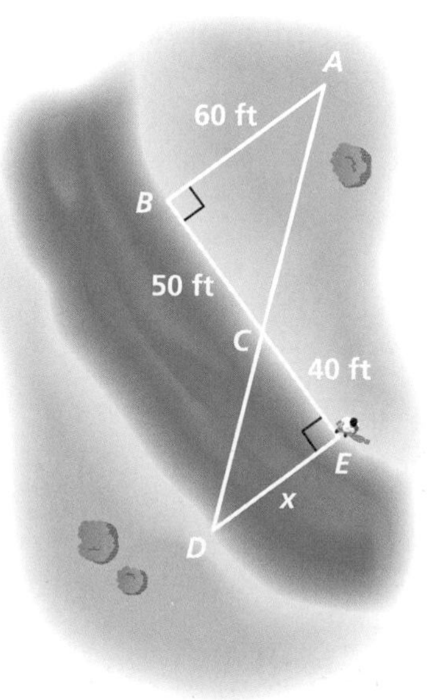

You plan to cross a river and want to know how far it is to the other side. You take measurements on your side of the river and make the drawing shown. (a) Explain why $\triangle ABC$ and $\triangle DEC$ are similar. (b) What is the distance x across the river?

a. $\angle B$ and $\angle E$ are right angles, so they are congruent. $\angle ACB$ and $\angle DCE$ are vertical angles, so they are congruent.

Because two angles in $\triangle ABC$ are congruent to two angles in $\triangle DEC$, the third angles are also congruent. The triangles have the same angle measures, so they are similar.

b. The ratios of the corresponding side lengths in similar triangles are equal. Write and solve a proportion to find x.

$$\frac{x}{60} = \frac{40}{50}$$ Write a proportion.

$$60 \cdot \frac{x}{60} = 60 \cdot \frac{40}{50}$$ Multiply each side by 60.

$$x = 48$$ Simplify.

∴ The distance across the river is 48 feet.

On Your Own

Now You're Ready
Exercises 10 and 11

3. **WHAT IF?** In Example 2, the distance from vertex A to vertex B is 55 feet. What is the distance across the river?

11.4 Exercises

Vocabulary and Concept Check

1. **REASONING** How can you use similar triangles to find a missing measurement?

2. **WHICH ONE DOESN'T BELONG?** Which triangle does *not* belong with the other three? Explain your reasoning.

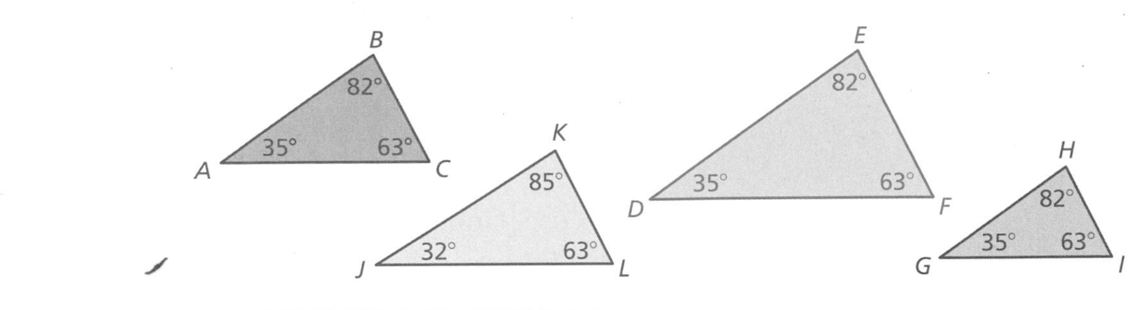

Practice and Problem Solving

Make a triangle that is larger than the one given and has the same angle measures. Find the ratios of the corresponding side lengths.

3. 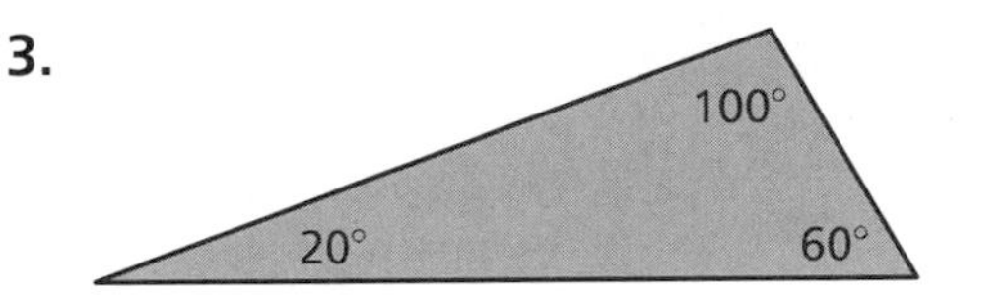

4. 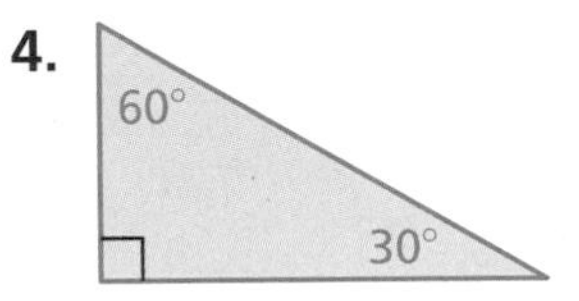

Tell whether the triangles are similar. Explain.

1 5. 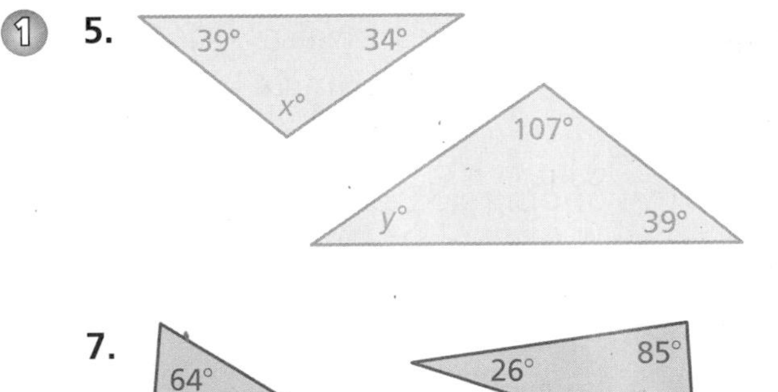

6. 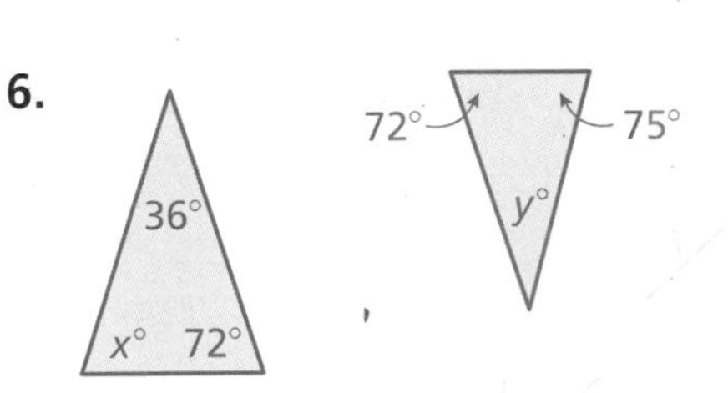

7. 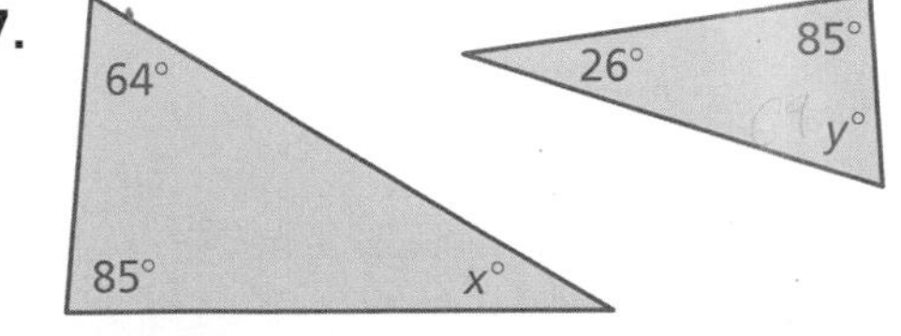

8. 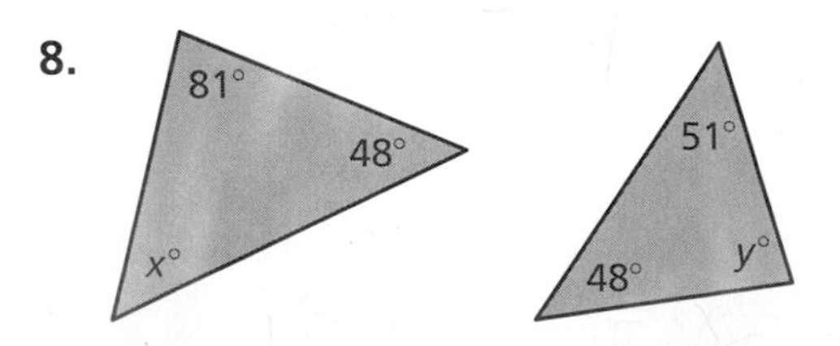

9. **ERROR ANALYSIS** Describe and correct the error in using indirect measurement.

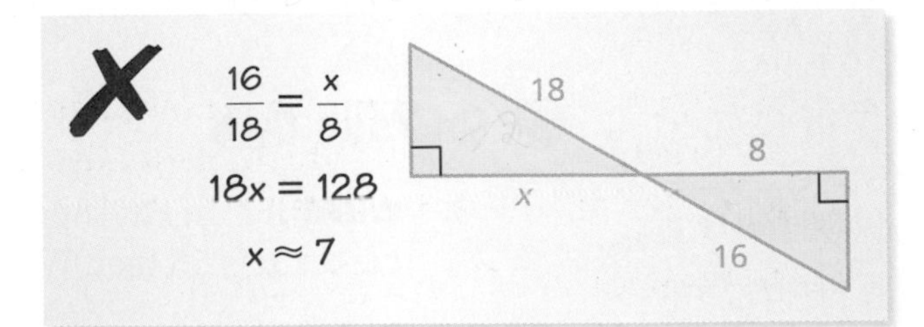

The triangles are similar. Find the value of x.

10.

11.

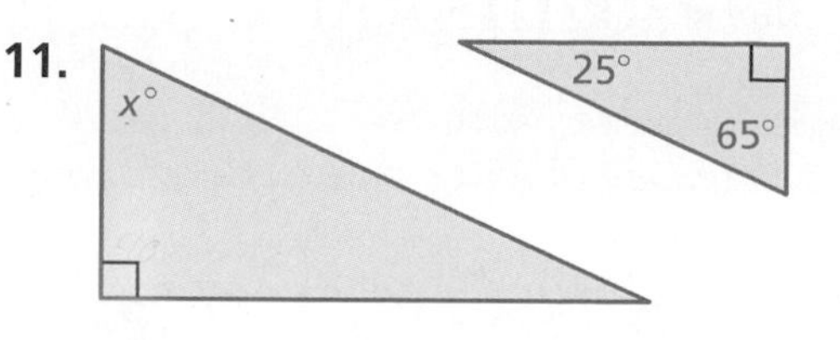

12. TREASURE The map shows the number of steps you must take to get to the treasure. However, the map is old and the last dimension is unreadable. How many steps do you take from the pyramids to the treasure?

13. CRITICAL THINKING The side lengths of a triangle are increased by 50% to make a similar triangle. Does the area increase by 50% as well? Explain.

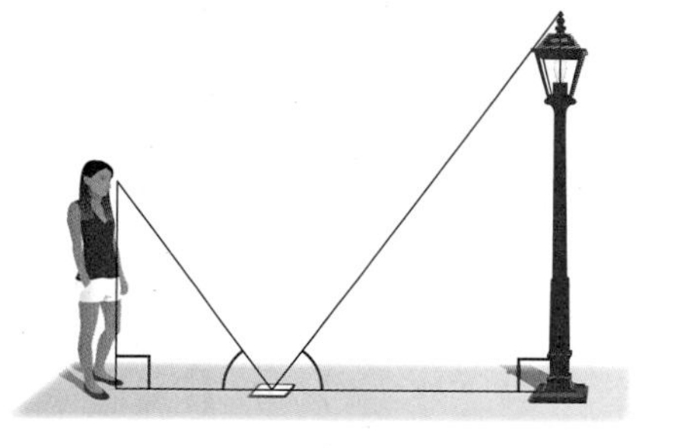

14. PROJECT Using a mirror, a tape measure, and indirect measurement, you can find the height of a lamppost. Place the mirror flat on the ground 6 feet from the lamppost. Move away from the mirror and the lamppost until you can see the top of the lamppost in the mirror. Measure the distance between yourself and the mirror. Then use similar triangles to find the height of the lamppost.

15. Geometry The drawing shows the scoring zone of a standard shuffleboard court. $\triangle DAE \sim \triangle BAG \sim \triangle CAF$. The lengths of segments AG, GF, and FE are equal.

a. Find x.

b. Find CF.

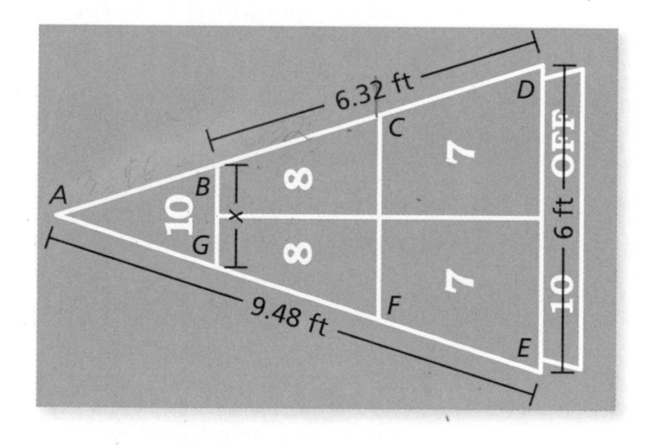

Fair Game Review What you learned in previous grades & lessons

Write and solve an equation to answer the question. *(Section 4.1)*

16. 30% of 20 is what number?

17. 24 is what percent of 96?

18. 5% of what number is 16?

19. 18% of 250 is what number?

20. MULTIPLE CHOICE The minimum number of people needed for a group rate at a baseball game is 20. Which inequality represents the number of people needed to get the group rate? *(Skills Review Handbook)*

Ⓐ $x < 20$ Ⓑ $x > 20$ Ⓒ $x \leq 20$ Ⓓ $x \geq 20$

11.5 Parallel Lines and Transversals

Essential Question How can you use properties of parallel lines to solve real-life problems?

COMMON CORE STATE STANDARDS
8.G.5

1 ACTIVITY: A Property of Parallel Lines

Work with a partner.

- Talk about what it means for two lines to be parallel. Decide on a strategy for drawing two parallel lines.
- Use your strategy to carefully draw two lines that are parallel.
- Now, draw a third line that intersects the two parallel lines. This line is called a **transversal**.
- The two parallel lines and the transversal form eight angles. Which of these angles have equal measures? Explain your reasoning.

parallel lines

1 2 3 4 5 6 7 8

transversal

2 ACTIVITY: Creating Parallel Lines

Work with a partner.

a. If you were building the house in the photograph, how could you make sure that the studs are parallel to each other?

b. Identify sets of parallel lines and transversals in the photograph.

3 ACTIVITY: Indirect Measurement

Work with a partner.

a. Use the fact that two rays from the Sun are parallel to explain why $\triangle ABC$ and $\triangle DEF$ are similar.

b. Explain how to use similar triangles to find the height of the flagpole.

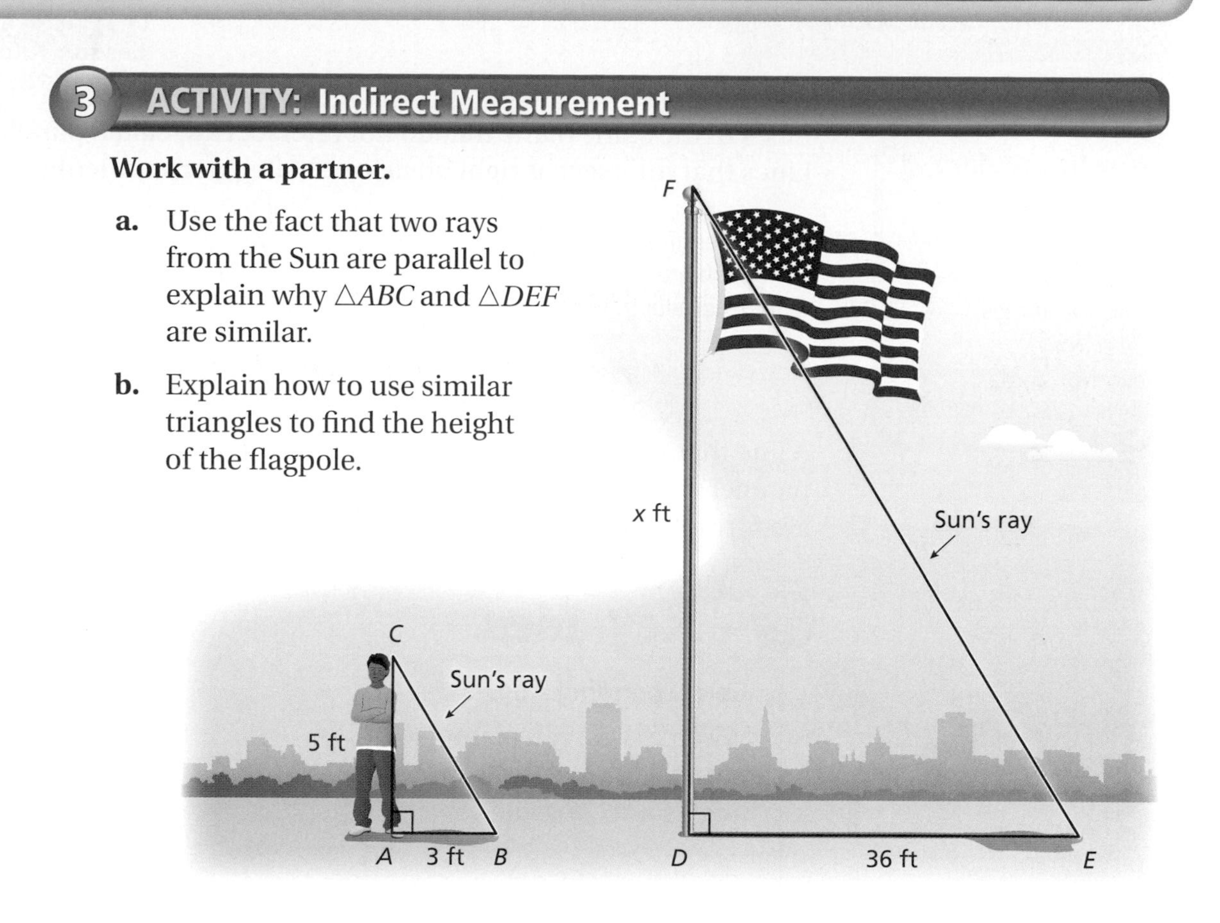

What Is Your Answer?

4. **IN YOUR OWN WORDS** How can you use properties of parallel lines to solve real-life problems? Describe some examples.

5. **INDIRECT MEASUREMENT PROJECT** Work with a partner or in a small group.

   a. Explain why the process in Activity 3 is called "indirect" measurement.

   b. Use indirect measurement to measure the height of something outside your school (a tree, a building, a flagpole). Before going outside, decide what you need to take with you to do the measurement.

   c. Draw a diagram of the indirect measurement process you used. In the diagram, label the lengths that you actually measured and also the lengths that you calculated.

Practice Use what you learned about parallel lines and transversals to complete Exercises 3–6 on page 497.

Key Vocabulary
perpendicular lines, p. 494
transversal, p. 494
interior angles, p. 495
exterior angles, p. 495

Lines in the same plane that do not intersect are called parallel lines. Lines that intersect at right angles are called **perpendicular lines**.

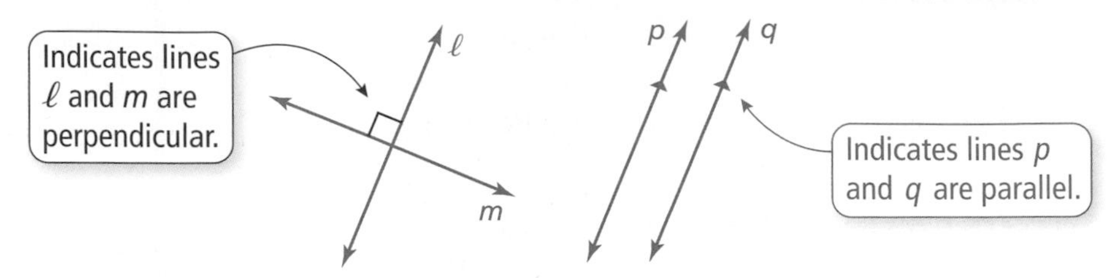

A line that intersects two or more lines is called a **transversal**. When parallel lines are cut by a transversal, several pairs of congruent angles are formed.

Study Tip
Corresponding angles lie on the same side of the transversal in corresponding positions.

Corresponding Angles

When a transversal intersects parallel lines, corresponding angles are congruent.

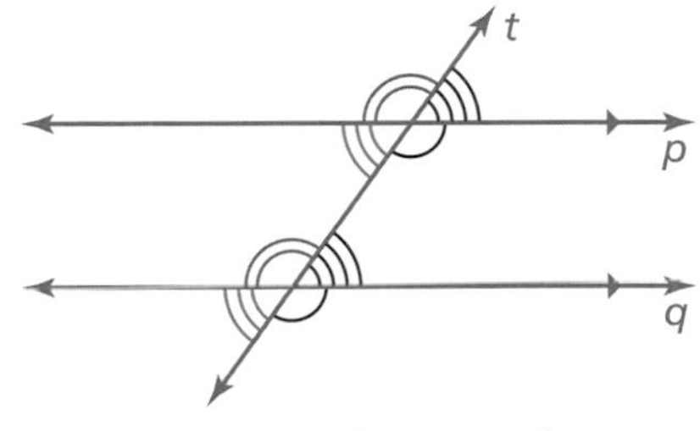

Corresponding angles

EXAMPLE 1 Finding Angle Measures

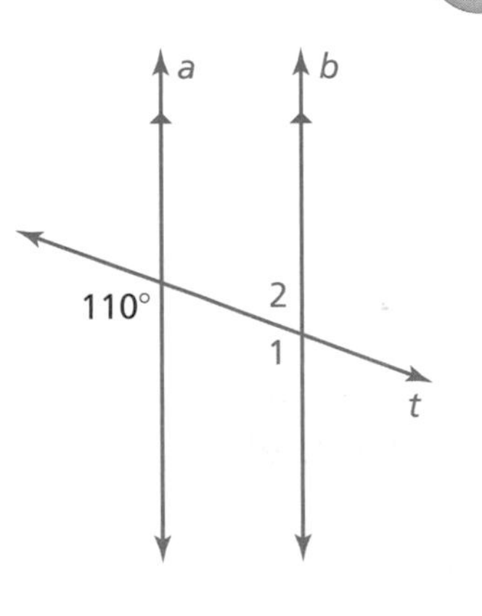

Use the figure to find the measures of (a) $\angle 1$ and (b) $\angle 2$.

a. $\angle 1$ and the 110° angle are corresponding angles. They are congruent.

∴ So, the measure of $\angle 1$ is 110°.

b. $\angle 1$ and $\angle 2$ are supplementary.

$\angle 1 + \angle 2 = 180^\circ$	Definition of supplementary angles
$110^\circ + \angle 2 = 180^\circ$	Substitute 110° for $\angle 1$.
$\angle 2 = 70^\circ$	Subtract 110° from each side.

∴ So, the measure of $\angle 2$ is 70°.

On Your Own

Now You're Ready
Exercises 7–9

Use the figure to find the measure of the angle. Explain your reasoning.

t, ℓ, m, 63°, 1, 2

1. $\angle 1$
2. $\angle 2$

EXAMPLE 2 Using Corresponding Angles

Use the figure to find the measures of the numbered angles.

∠1: ∠1 and the 75° angle are vertical angles. They are congruent.

∴ So, the measure of ∠1 is 75°.

∠2 and ∠3: The 75° angle is supplementary to both ∠2 and ∠3.

$75° + \angle 2 = 180°$ Definition of supplementary angles

$\angle 2 = 105°$ Subtract 75° from each side.

∴ So, the measures of ∠2 and ∠3 are 105°.

∠4, ∠5, ∠6, and ∠7: Using corresponding angles, the measures of ∠4 and ∠6 are 75°, and the measures of ∠5 and ∠7 are 105°.

On Your Own

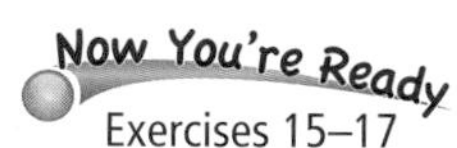

3. Use the figure to find the measures of the numbered angles.

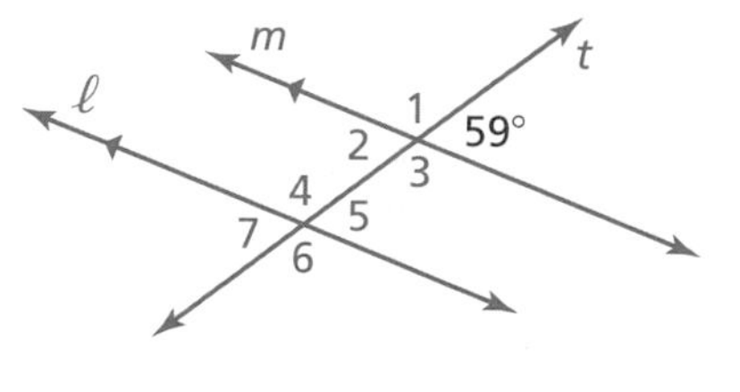

When two parallel lines are cut by a transversal, four **interior angles** are formed on the inside of the parallel lines and four **exterior angles** are formed on the outside of the parallel lines.

∠3, ∠4, ∠5, and ∠6 are interior angles.
∠1, ∠2, ∠7, and ∠8 are exterior angles.

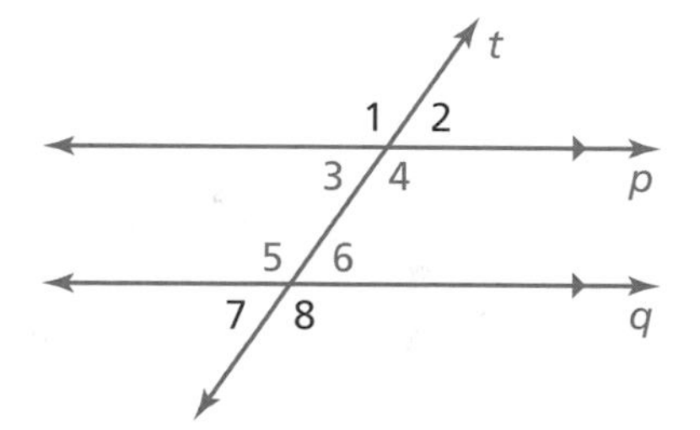

EXAMPLE 3 Standardized Test Practice

A store owner uses pieces of tape to paint a window advertisement. The letters are slanted at an 80° angle. What is the measure of ∠1?

Ⓐ 80° Ⓑ 100° Ⓒ 110° Ⓓ 120°

Because all of the letters are slanted at an 80° angle, the dashed lines are parallel. The piece of tape is the transversal.

Using the corresponding angles, the 80° angle is congruent to the angle that is supplementary to ∠1, as shown.

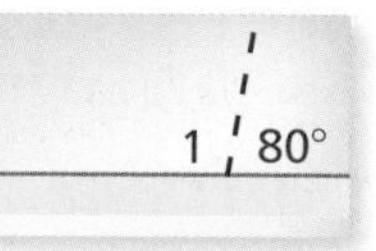

∴ The measure of ∠1 is 180° − 80° = 100°. The correct answer is Ⓑ.

On Your Own

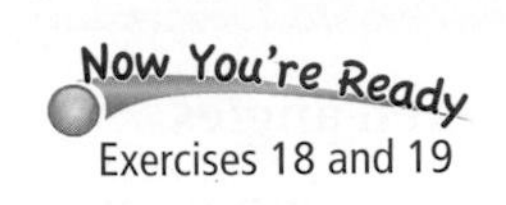

Exercises 18 and 19

4. **WHAT IF?** In Example 3, the letters are slanted at a 65° angle. What is the measure of $\angle 1$?

Key Idea

Alternate interior angles and alternate exterior angles lie on opposite sides of the transversal.

Alternate Interior Angles and Alternate Exterior Angles

When a transversal intersects parallel lines, alternate interior angles are congruent and alternate exterior angles are congruent.

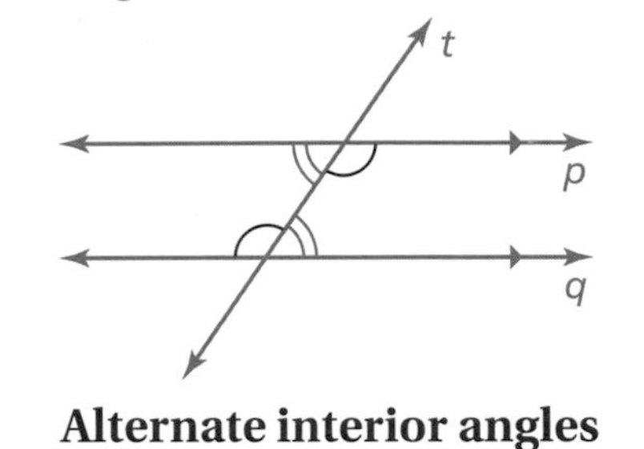

Alternate interior angles

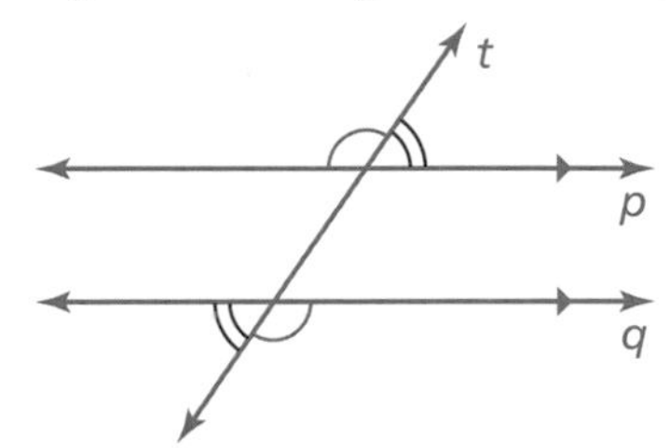

Alternate exterior angles

EXAMPLE 4 Identifying Alternate Interior and Alternate Exterior Angles

The photo shows a portion of an airport. Describe the relationship between each pair of angles.

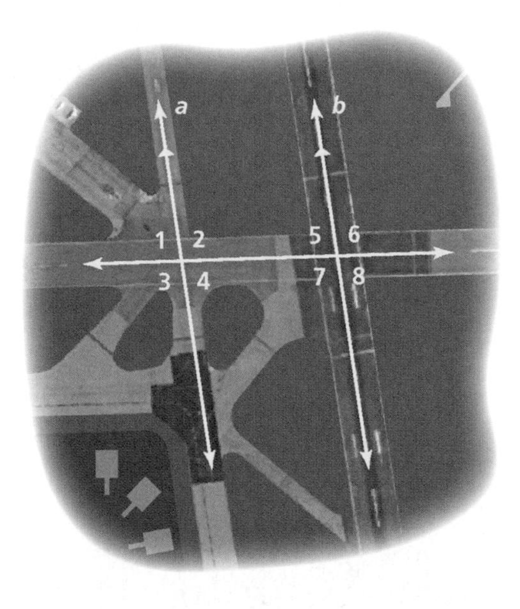

a. $\angle 3$ **and** $\angle 6$

$\angle 3$ and $\angle 6$ are alternate exterior angles.

So, $\angle 3$ is congruent to $\angle 6$.

b. $\angle 2$ **and** $\angle 7$

$\angle 2$ and $\angle 7$ are alternate interior angles.

So, $\angle 2$ is congruent to $\angle 7$.

On Your Own

Now You're Ready
Exercises 20 and 21

In Example 4, the measure of $\angle 4$ is 84°. Find the measure of the angle. Explain your reasoning.

5. $\angle 3$
6. $\angle 5$
7. $\angle 6$

11.5 Exercises

Vocabulary and Concept Check

1. **VOCABULARY** Draw two parallel lines and a transversal. Label a pair of corresponding angles.

2. **WHICH ONE DOES NOT BELONG?** Which statement does *not* belong with the other three? Explain your reasoning. Refer to the figure for Exercises 3–6.

The measure of $\angle 2$	The measure of $\angle 5$
The measure of $\angle 6$	The measure of $\angle 8$

Practice and Problem Solving

In Exercises 3–6, use the figure.

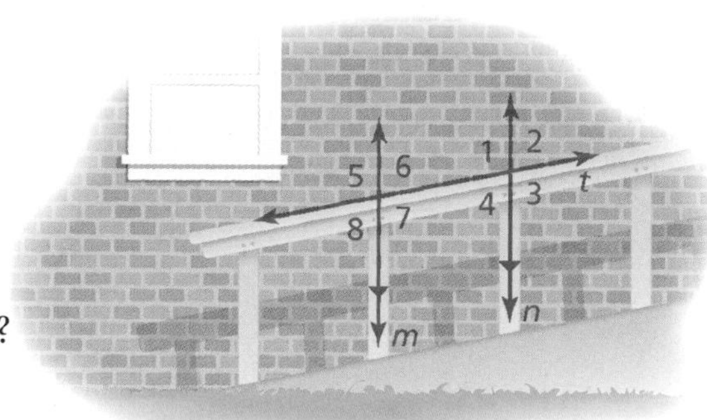

3. Identify the parallel lines.
4. Identify the transversal.
5. How many angles are formed by the transversal?
6. Which of the angles are congruent?

Use the figure to find the measures of the numbered angles.

7.

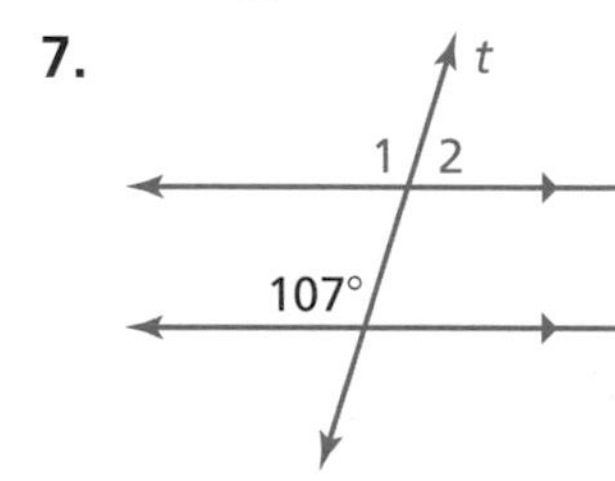

8. a, b, t, 95°, 3, 4

9. a, b, t, 49°, 5, 6

10. **ERROR ANALYSIS** Describe and correct the error in describing the relationship between the angles.

✗ $\angle 5$ is congruent to $\angle 6$.

11. **PARKING** The painted lines that separate parking spaces are parallel. The measure of $\angle 1$ is 60°. What is the measure of $\angle 2$? Explain.

12. **OPEN-ENDED** Describe two real-life situations that use parallel lines.

13. PROJECT Draw two horizontal lines and a transversal on a piece of notebook paper. Label the angles as shown. Use a pair of scissors to cut out the angles. Compare the angles to determine which angles are congruent.

14. REASONING Refer to the figure for Exercise 13. What is the least number of angle measures you need to know in order to find the measure of every angle? Explain your reasoning.

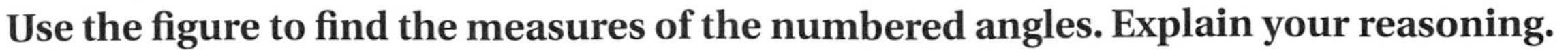

Use the figure to find the measures of the numbered angles. Explain your reasoning.

15. **16.**

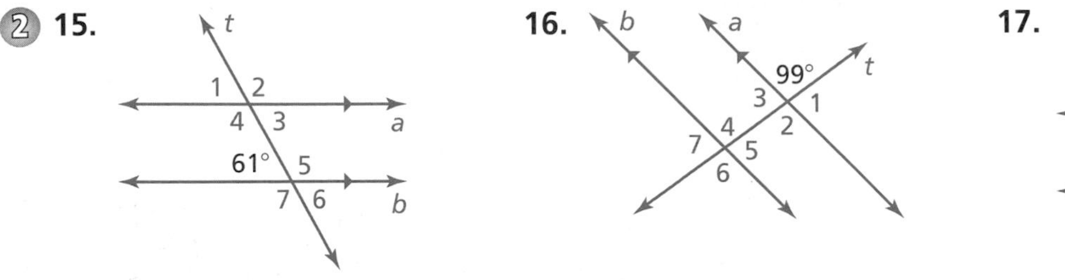

17.

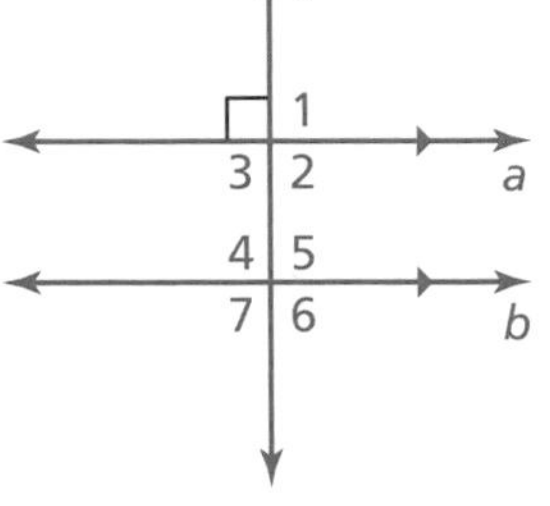

Complete the statement. Explain your reasoning.

18. If the measure of $\angle 1 = 124°$, then the measure of $\angle 4 =$ ___.

19. If the measure of $\angle 2 = 48°$, then the measure of $\angle 3 =$ ___.

20. If the measure of $\angle 4 = 55°$, then the measure of $\angle 2 =$ ___.

21. If the measure of $\angle 6 = 120°$, then the measure of $\angle 8 =$ ___.

22. If the measure of $\angle 7 = 50.5°$, then the measure of $\angle 6 =$ ___.

23. If the measure of $\angle 3 = 118.7°$, then the measure of $\angle 2 =$ ___.

24. RAINBOW A rainbow is formed when sunlight reflects off raindrops at different angles. For blue light, the measure of $\angle 2$ is 40°. What is the measure of $\angle 1$?

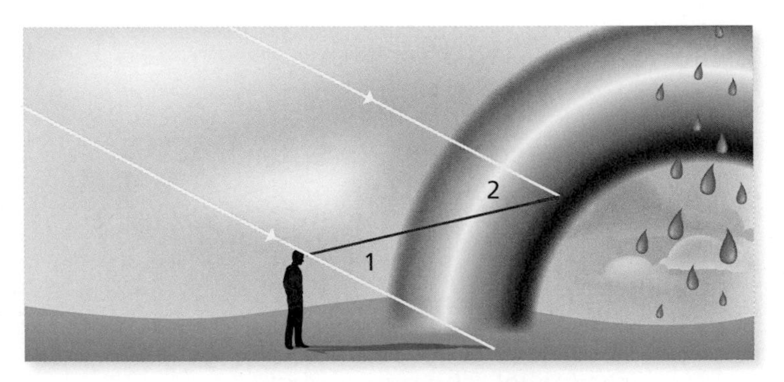

25. REASONING If a transversal is perpendicular to two parallel lines, what can you conclude about the angles formed? Explain.

26. WRITING Describe two ways you can show that $\angle 1$ is congruent to $\angle 7$.

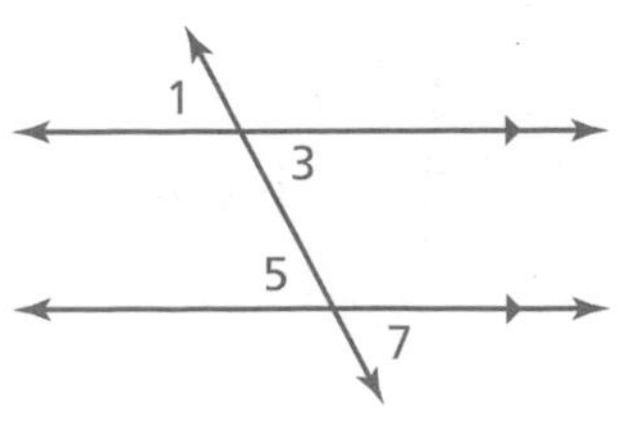

CRITICAL THINKING Find the value of x.

27.

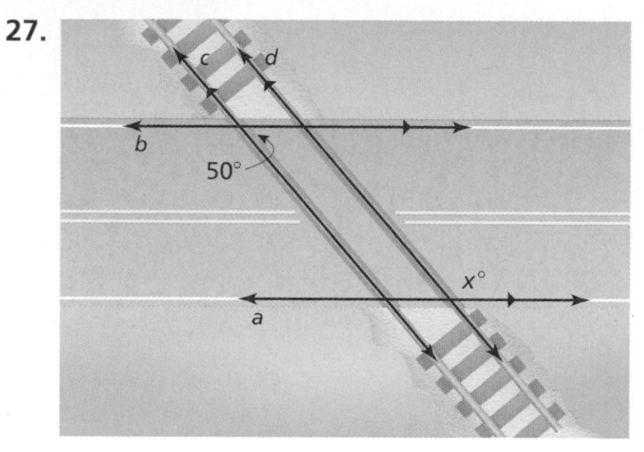

28.

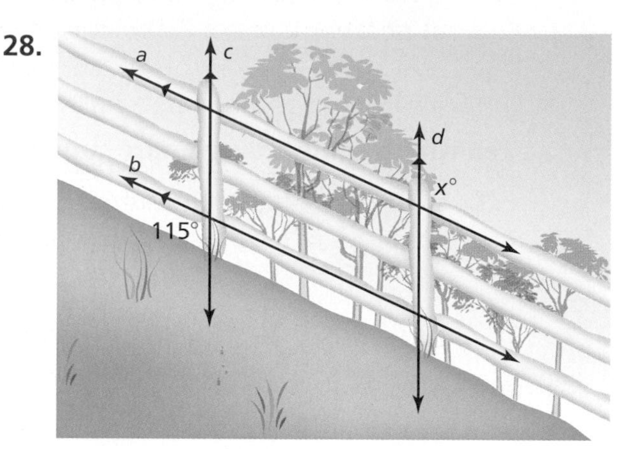

29. **OPTICAL ILLUSION** Refer to the figure.

a. Do the horizontal lines appear to be parallel? Explain.

b. Draw your own optical illusion using parallel lines.

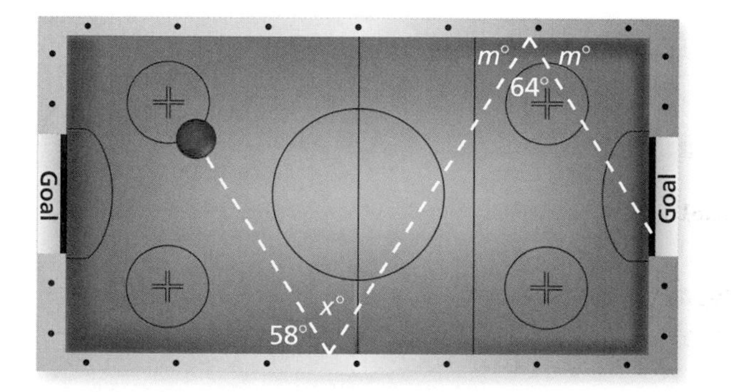

30. **Geometry** The figure shows the angles used to make a double bank shot in an air hockey game.

a. Find the value of x.

b. Can you still get the red puck in the goal if x is increased by a little? by a lot? Explain.

Fair Game Review What you learned in previous grades & lessons

Evaluate the expression. *(Section 1.4)*

31. $4 + 3^2$ 32. $5(2)^2 - 6$ 33. $11 + (-7)^2 - 9$ 34. $8 \div 2^2 + 1$

35. **MULTIPLE CHOICE** The volume of the cylinder is 20π cubic inches. What is the radius of the base? *(Section 7.2)*

Ⓐ 1 inch Ⓑ 2 inches

Ⓒ 3 inches Ⓓ 4 inches

5 in.

11.4–11.5 Quiz

Tell whether the triangles are similar. Explain. *(Section 11.4)*

1. 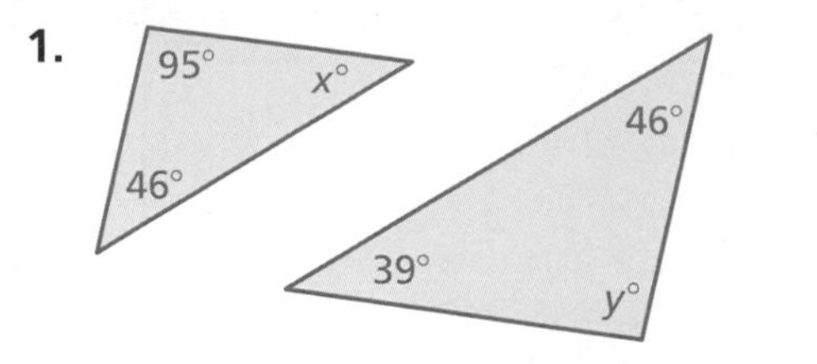

2. 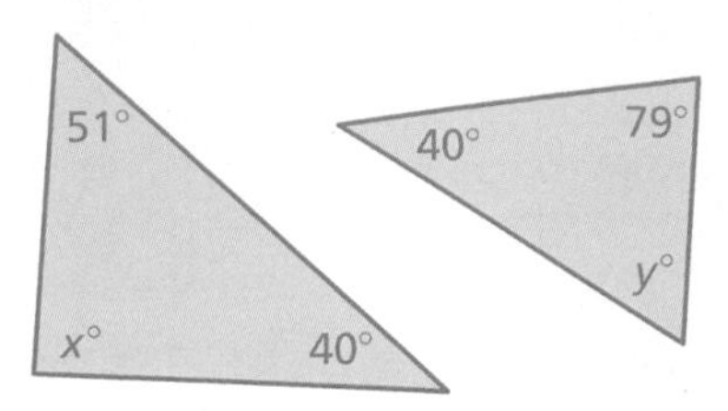

The triangles are similar. Find the value of *x*. *(Section 11.4)*

3. 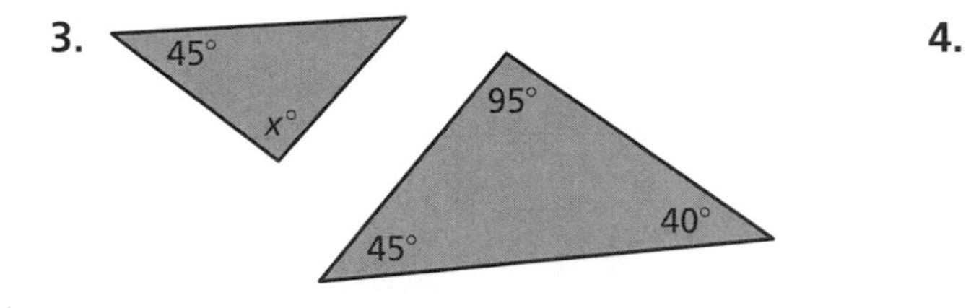

4. 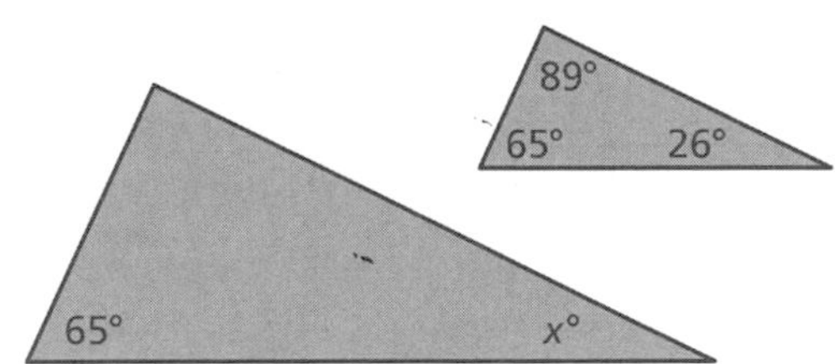

Use the figure to find the measure of the angle. Explain your reasoning. *(Section 11.5)*

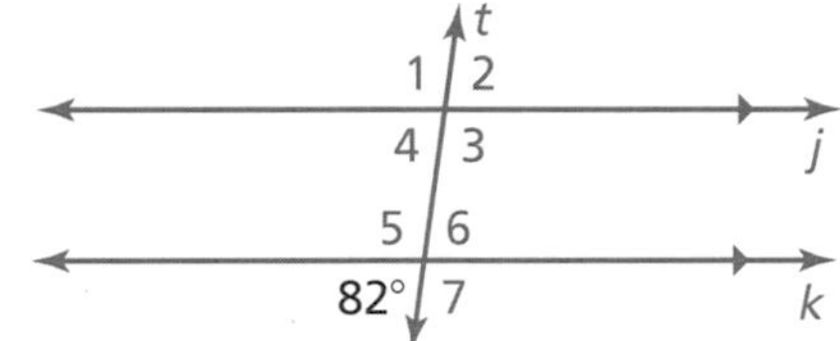

5. $\angle 2$

6. $\angle 6$

7. $\angle 4$

8. $\angle 1$

Complete the statement. Explain your reasoning. *(Section 11.5)*

9. If the measure of $\angle 1 = 123°$, then the measure of $\angle 7 =$ ▭.

10. If the measure of $\angle 2 = 58°$, then the measure of $\angle 5 =$ ▭.

11. If the measure of $\angle 5 = 119°$, then the measure of $\angle 3 =$ ▭.

12. If the measure of $\angle 4 = 60°$, then the measure of $\angle 6 =$ ▭.

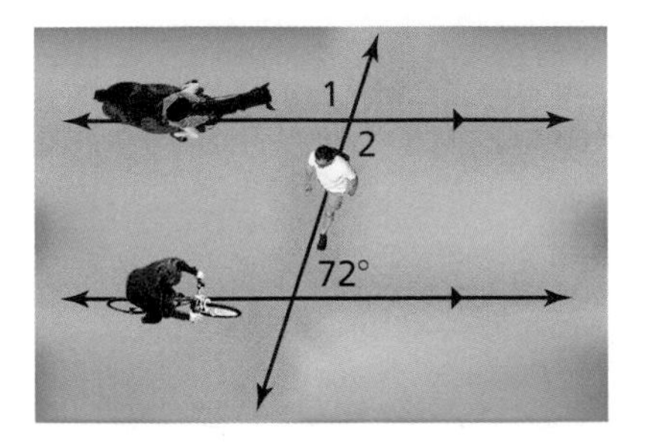

13. PARK In a park, a bike path and a horse riding path are parallel. In one part of the park, a hiking trail intersects the two paths. Find the measures of $\angle 1$ and $\angle 2$. Explain your reasoning. *(Section 11.5)*

14. PERIMETER The side lengths of a right triangle are doubled to make a similar triangle. Does the perimeter double as well? Explain. *(Section 11.4)*

11 Chapter Review

Review Key Vocabulary

complementary angles, *p. 466*
supplementary angles, *p. 466*
congruent angles, *p. 467*
vertical angles, *p. 467*
isosceles triangle, *p. 472*
congruent sides, *p. 472*
equilateral triangle, *p. 472*
equiangular triangle, *p. 472*
polygon, *p. 478*
regular polygon, *p. 479*
convex polygon, *p. 480*
concave polygon, *p. 480*
similar triangles, *p. 488*
indirect measurement, *p. 489*
perpendicular lines, *p. 494*
transversal, *p. 494*
interior angles, *p. 495*
exterior angles, *p. 495*

Review Examples and Exercises

11.1 Classifying Angles *(pp. 464–469)*

Find the value of x.

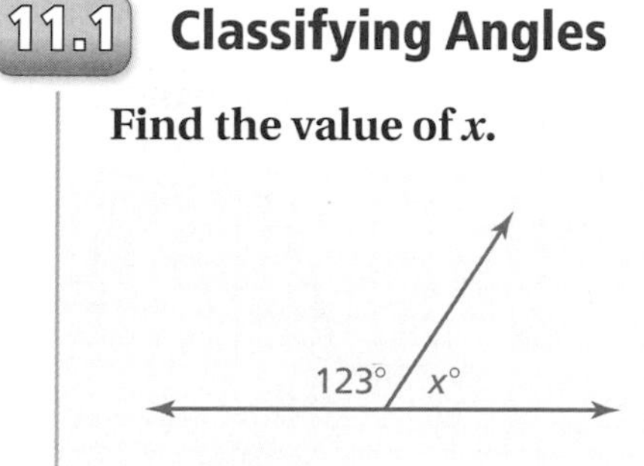

The angles are supplementary angles. So, the sum of their measures is 180°.

$$x + 123 = 180$$

$$x = 57$$

So, x is 57.

Exercises

Find the value of x.

1.

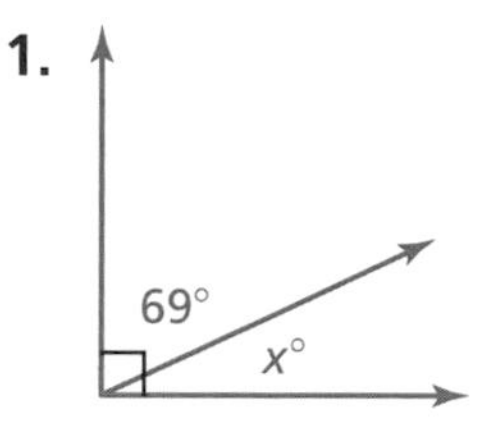

2.

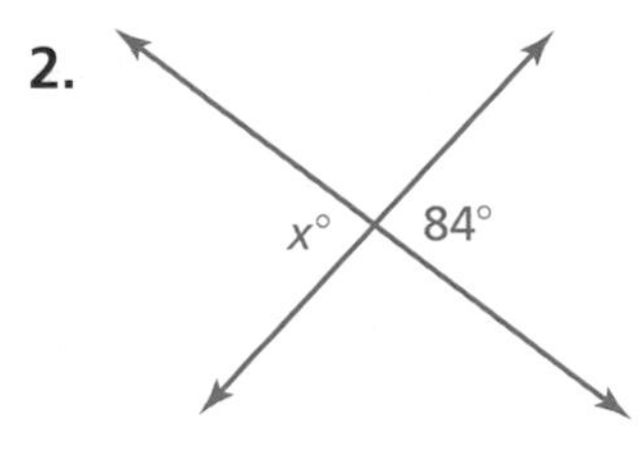

11.2 Angles and Sides of Triangles *(pp. 470–475)*

Find the value of x. Then classify the triangle.

$$x + 50 + 55 = 180$$

$$x + 105 = 180$$

$$x = 75$$

The value of x is 75. The triangle has three acute angle measures, 50°, 55°, and 75°. So, it is an acute triangle.

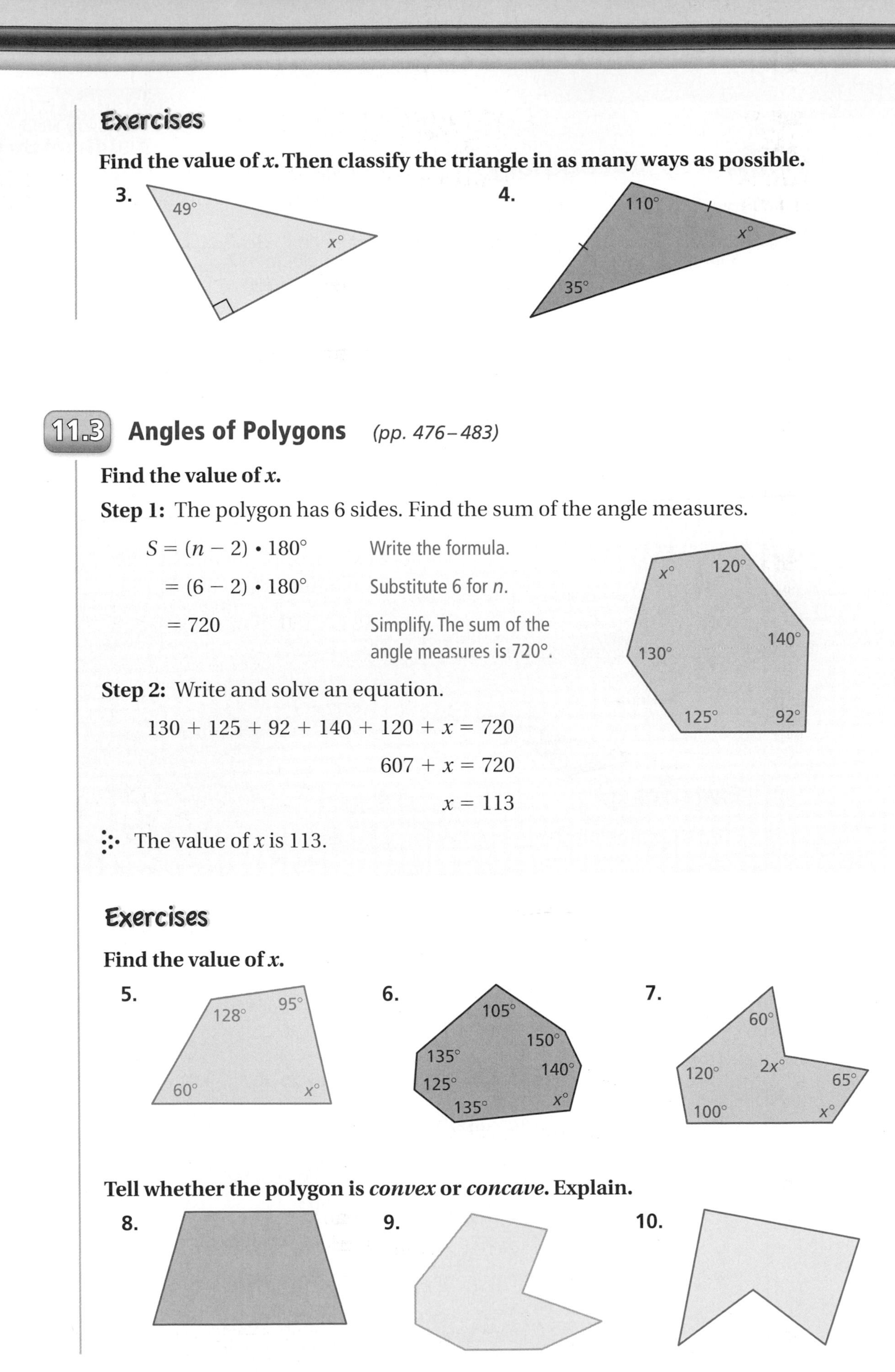

Exercises

Find the value of x. Then classify the triangle in as many ways as possible.

3. (Triangle with angles 49°, x°, and a right angle)

4. (Triangle with angles 110°, 35°, x°; two sides marked congruent)

11.3 Angles of Polygons *(pp. 476–483)*

Find the value of x.

Step 1: The polygon has 6 sides. Find the sum of the angle measures.

$S = (n - 2) \cdot 180°$	Write the formula.
$= (6 - 2) \cdot 180°$	Substitute 6 for *n*.
$= 720$	Simplify. The sum of the angle measures is 720°.

(Hexagon with angles x°, 120°, 140°, 92°, 125°, 130°)

Step 2: Write and solve an equation.

$$130 + 125 + 92 + 140 + 120 + x = 720$$

$$607 + x = 720$$

$$x = 113$$

∴ The value of x is 113.

Exercises

Find the value of x.

5. (Quadrilateral with angles 128°, 95°, x°, 60°)

6. (Heptagon with angles 105°, 150°, 140°, x°, 135°, 125°, 135°)

7. (Hexagon with angles 60°, $2x$°, 65°, x°, 100°, 120°)

Tell whether the polygon is *convex* or *concave*. Explain.

8.

9.

10.

11.4 Using Similar Triangles (pp. 486–491)

Tell whether the triangles are similar. Explain.

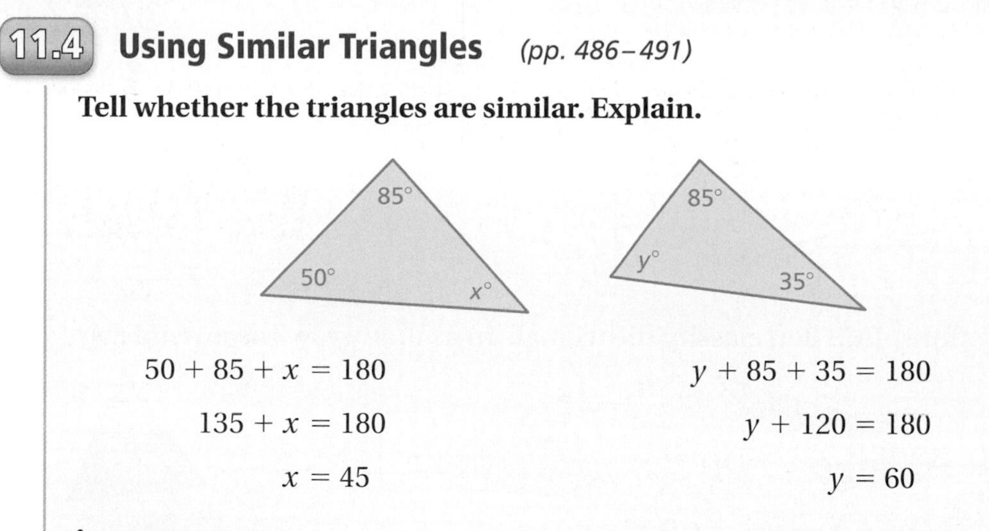

$$50 + 85 + x = 180 \qquad y + 85 + 35 = 180$$

$$135 + x = 180 \qquad y + 120 = 180$$

$$x = 45 \qquad y = 60$$

∴ The triangles do not have the same angle measures. So, they are not similar.

Exercises

11. Tell whether the triangles are similar. Explain.

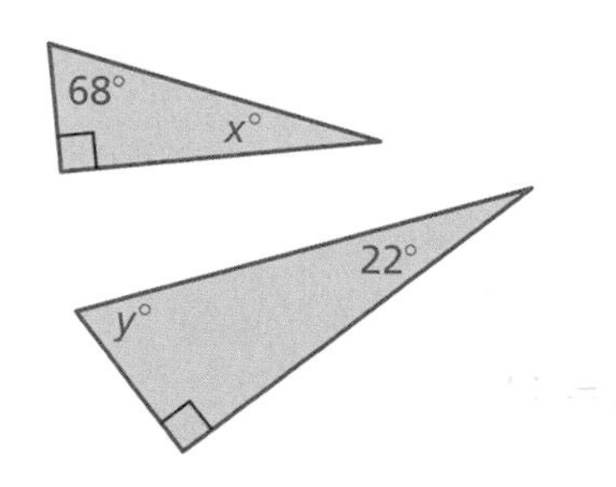

12. The triangles are similar. Find the value of x.

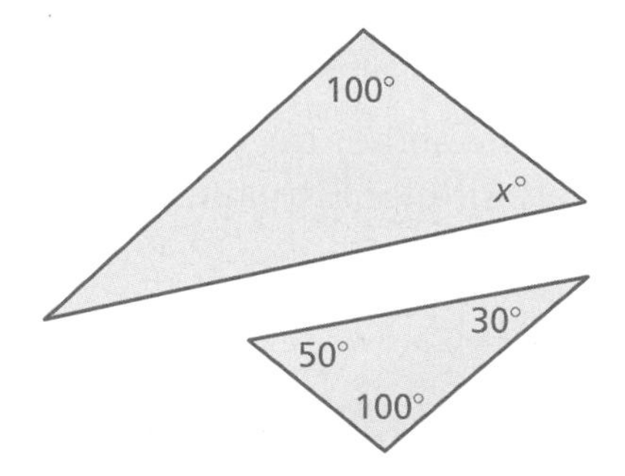

11.5 Parallel Lines and Transversals (pp. 492–499)

Use the figure to find the measure of ∠6.

∠2 and the 55° angle are supplementary. So, the measure of ∠2 is 180° − 55° = 125°.

∠2 and ∠6 are corresponding angles. They are congruent.

∴ So, the measure of ∠6 is 125°.

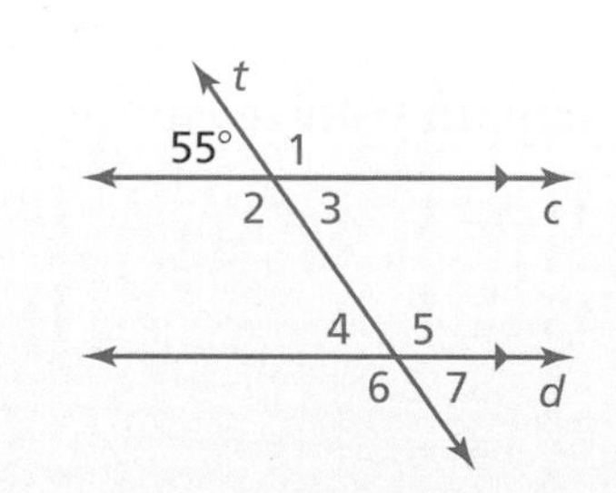

Exercises

Use the figure to find the measure of the angle. Explain your reasoning.

13. ∠8 **14.** ∠5

15. ∠7 **16.** ∠2

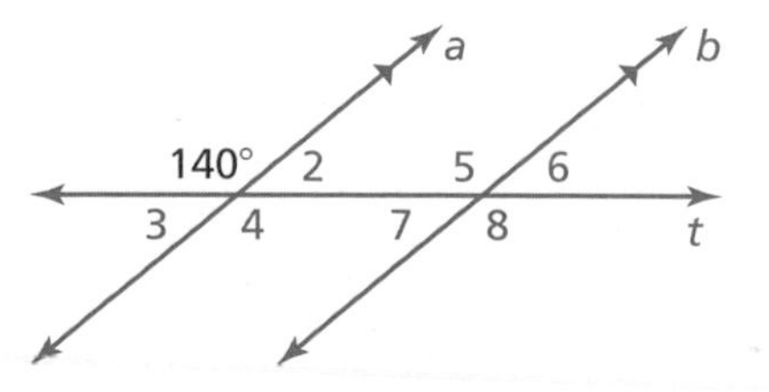

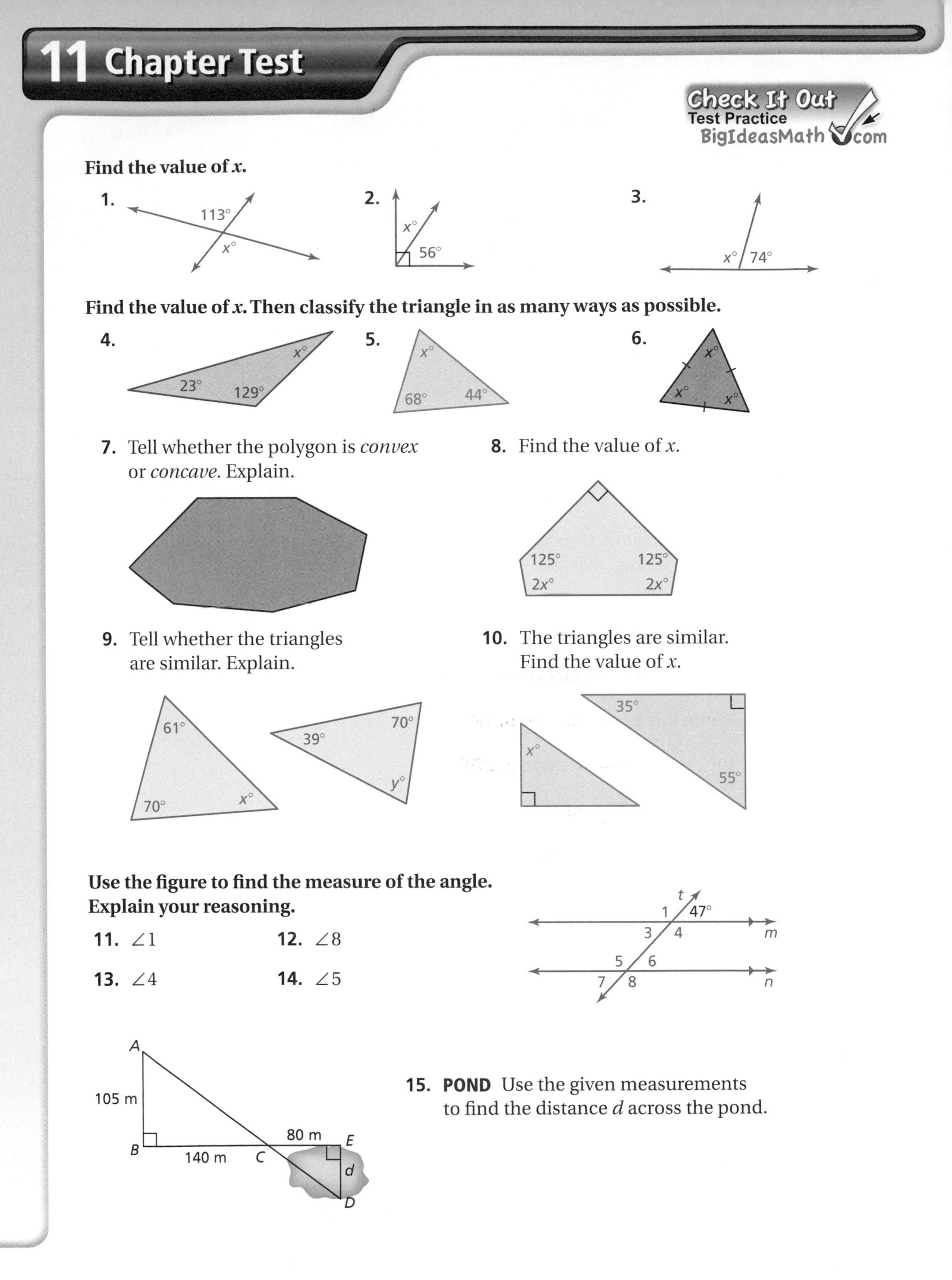

11 Chapter Test

Check It Out
Test Practice
BigIdeasMath.com

Find the value of x.

1.

2.

3.

Find the value of x. Then classify the triangle in as many ways as possible.

4.

5.

6.

7. Tell whether the polygon is *convex* or *concave*. Explain.

8. Find the value of x.

9. Tell whether the triangles are similar. Explain.

10. The triangles are similar. Find the value of x.

Use the figure to find the measure of the angle. Explain your reasoning.

11. $\angle 1$

12. $\angle 8$

13. $\angle 4$

14. $\angle 5$

15. **POND** Use the given measurements to find the distance d across the pond.

11 Standardized Test Practice

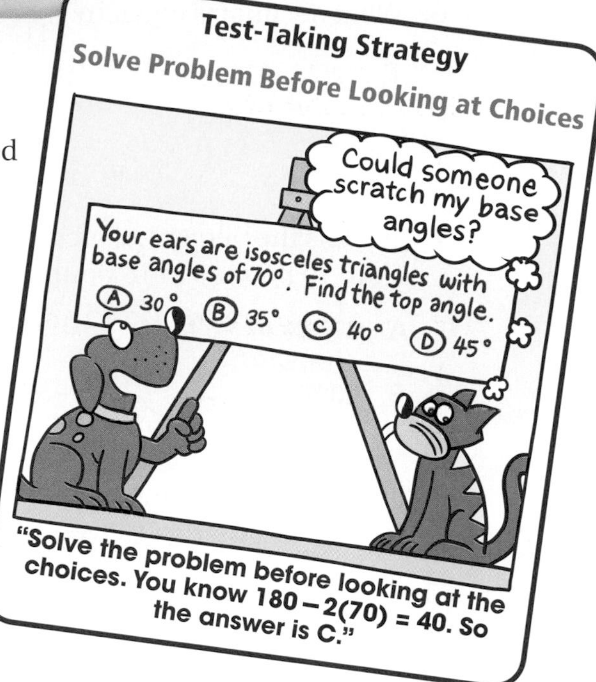

Test-Taking Strategy
Solve Problem Before Looking at Choices
Could someone scratch my base angles?
Your ears are isosceles triangles with base angles of 70°. Find the top angle.
Ⓐ 30° Ⓑ 35° Ⓒ 40° Ⓓ 45°

"Solve the problem before looking at the choices. You know 180 − 2(70) = 40. So the answer is C."

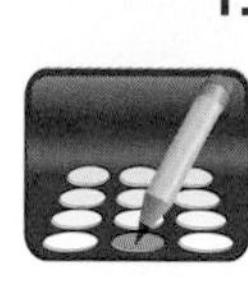

1. The border of a Canadian one-dollar coin is in the shape of an 11-sided regular polygon. The shape was chosen to help visually-impaired people identify the coin. Find the measure of each angle of the border, in degrees. Round your answer to the nearest degree. *(8.G.5)*

2. A city mayor wants to estimate the number of residents in the city who are in favor of adding fluoride to the public water supply. Which sample is best? *(7.SP.1)*

A. 120 dental employees in the city

B. 5 people chosen randomly from the city hall cafeteria

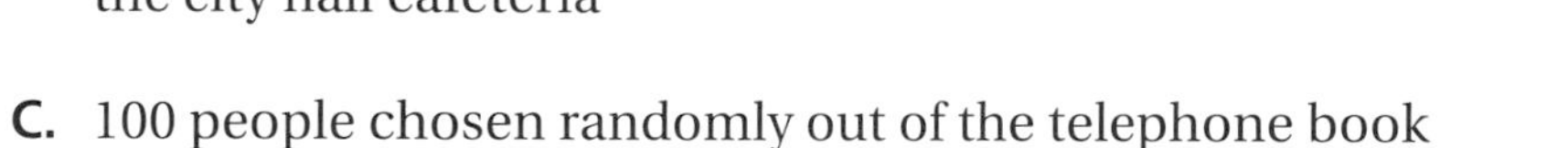

C. 100 people chosen randomly out of the telephone book

D. 140 women chosen randomly from the voter registration list

3. Which is equivalent to a clockwise rotation of 180°? *(8.G.2)*

F. A counterclockwise rotation of 180°

G. A counterclockwise rotation of 270°

H. A clockwise rotation of 360°

I. A counterclockwise rotation of 360°

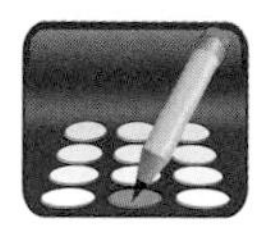

4. $\angle 1$ has a measure of 28°. Find the measure of $\angle 2$, in degrees. *(7.G.5)*

2 1

5. Which value of x makes the equation below true? *(8.EE.7b)*

$$5x - 3 = 11$$

A. $x = 1.6$

B. $x = 2.8$

C. $x = 40$

D. $x = 70$

6. What decimal is equal to $7\frac{3}{11}$? *(7.NS.2d)*

F. 7.27

G. $7.\overline{27}$

H. 7.3

I. $7.\overline{3}$

7. What is the percent of increase when a baby's body weight changes from 10 pounds to 16 pounds? *(7.RP.3)*

A. 37.5%

B. 60%

C. 137.5%

D. 160%

8. The sum S of the angle measures of a polygon with n sides can be found using a formula. *(8.G.5)*

Think
Solve
Explain

Part A Write the formula.

Part B A quadrilateral has angles measuring 100, 90, and 90 degrees. Find the measure of its fourth angle. Show your work and explain your reasoning.

Part C The sum of the measures of the angles of the pentagon shown is 540 degrees. Divide the pentagon into triangles to show why this must be true. Show your work and explain your reasoning.

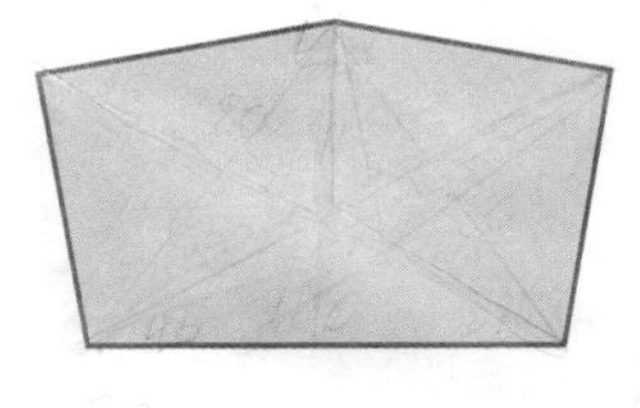

9. The line shown in the graph has a slope of $\frac{2}{5}$.

What is the equation of the line? *(8.EE.6)*

F. $x = \frac{2}{5}y + 5$

G. $y = \frac{2}{5}x + 5$

H. $x = \frac{2}{5}y + 1$

I. $y = \frac{2}{5}x + 1$

10. Megan flips three quarters. What is the probability that she flips exactly two heads? *(7.SP.8a)*

A. $\frac{1}{4}$

B. $\frac{3}{8}$

C. $\frac{1}{2}$

D. $\frac{2}{3}$

11. Jamie makes a candle in the shape of the cylinder shown. Jamie was computing the volume of the candle in the box below.

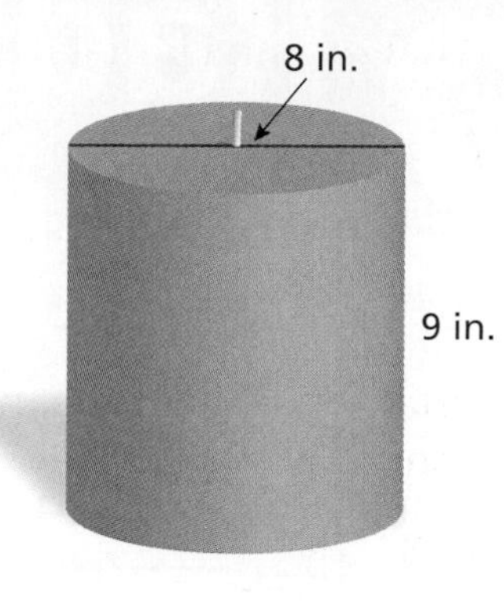

$$V = Bh$$
$$= \pi(8)^2(9)$$
$$= 576\pi$$
$$\approx 1809 \text{ in.}^3$$

What should Jamie do to correct the error that she made? *(8.G.9)*

F. Use the formula $2\pi r^2 + 2\pi rh$ for the volume.

G. Label the answer with the units in.^{2}

H. Multiply π by 9 before multiplying by 8^2.

I. Use $r = \frac{d}{2} = \frac{8}{2} = 4$ in the formula $B = \pi r^2$ for the area of the base.

12. In the diagram below, $\triangle ABC \sim \triangle DEF$. What is the value of x? *(8.G.5)*

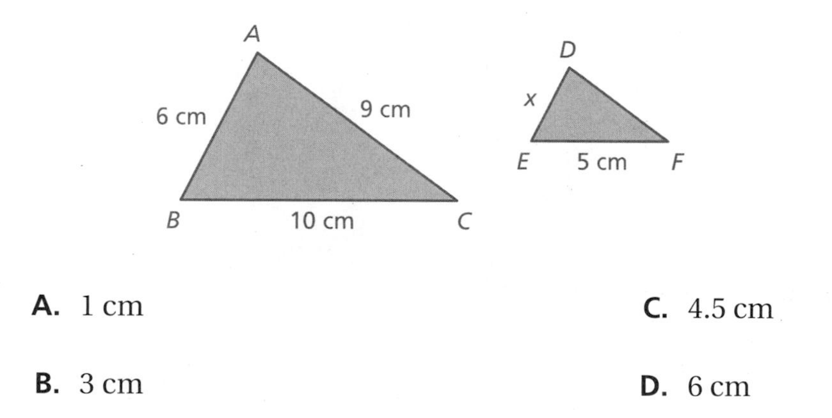

A. 1 cm

B. 3 cm

C. 4.5 cm

D. 6 cm

13. The cost of the gold chain for a necklace is proportional to the length of the chain. How much does a 17-inch length gold chain cost? *(7.RP.2b)*

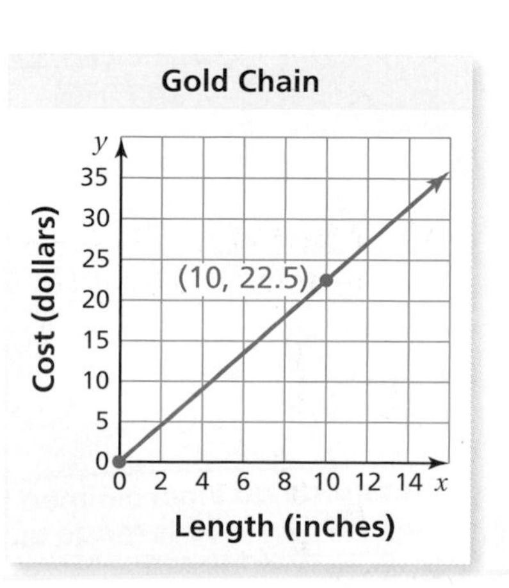

F. $38.25

G. $40.00

H. $42.25

I. $42.50

12 Radicals and the Pythagorean Theorem

- 12.1 Finding Square Roots
- 12.2 The Pythagorean Theorem
- 12.3 Approximating Square Roots
- 12.4 Simplifying Square Roots
- 12.5 Using the Pythagorean Theorem

"I'm pretty sure that Pythagoras was a Greek."

"I said 'Greek', not 'Geek'."

"Leonardo da Vinci claimed that the human face is made up of golden ratios."

"Let's see if the same is true of a cat's face."

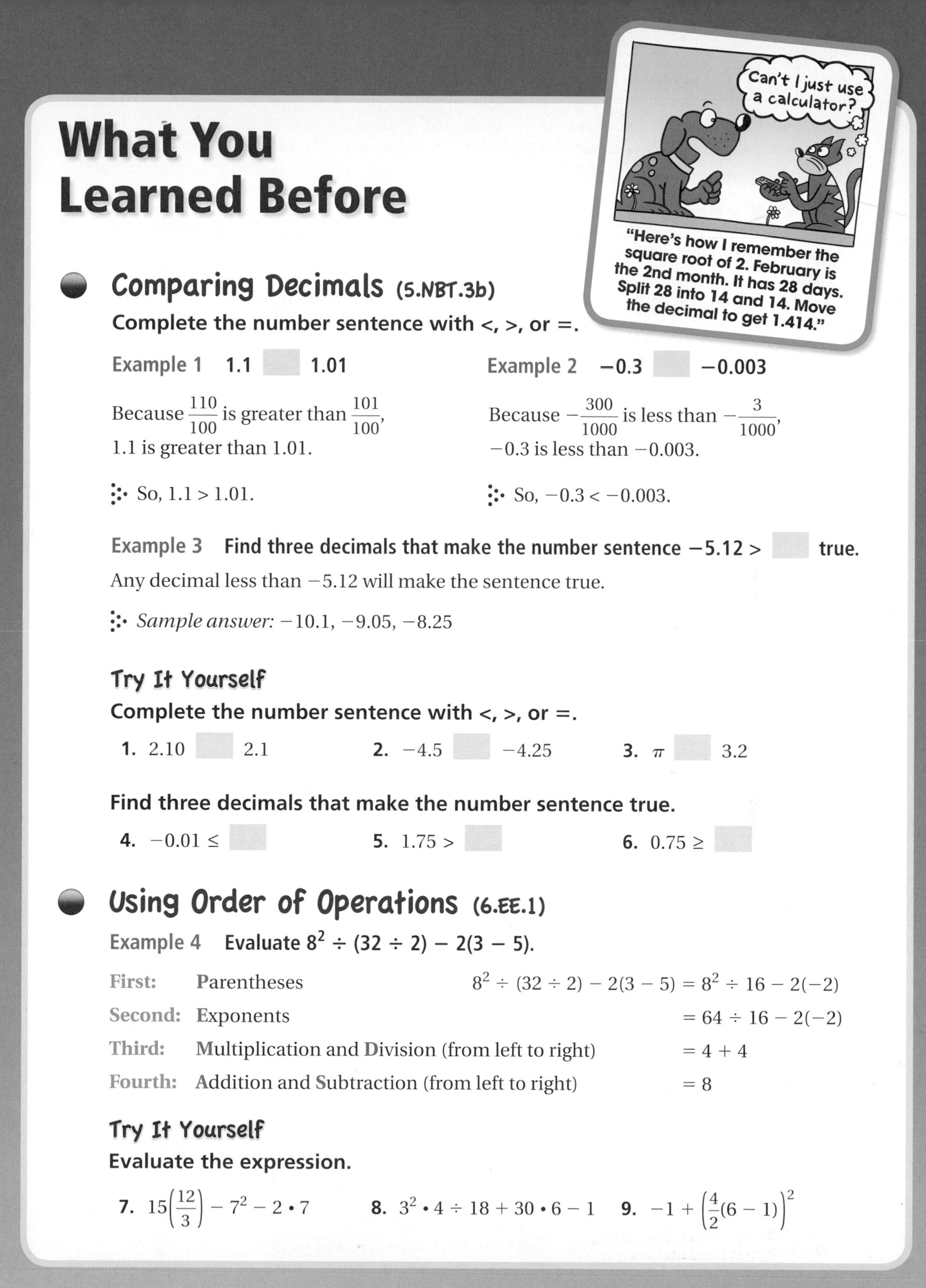

What You Learned Before

Can't I just use a calculator?
"Here's how I remember the square root of 2. February is the 2nd month. It has 28 days. Split 28 into 14 and 14. Move the decimal to get 1.414."

Comparing Decimals (5.NBT.3b)

Complete the number sentence with <, >, or =.

Example 1 $\quad 1.1 \ \square \ 1.01$

Because $\frac{110}{100}$ is greater than $\frac{101}{100}$, 1.1 is greater than 1.01.

∴ So, $1.1 > 1.01$.

Example 2 $\quad -0.3 \ \square \ -0.003$

Because $-\frac{300}{1000}$ is less than $-\frac{3}{1000}$, -0.3 is less than -0.003.

∴ So, $-0.3 < -0.003$.

Example 3 **Find three decimals that make the number sentence $-5.12 > \square$ true.**

Any decimal less than -5.12 will make the sentence true.

∴ *Sample answer:* $-10.1, -9.05, -8.25$

Try It Yourself

Complete the number sentence with <, >, or =.

1. $2.10 \ \square \ 2.1$ **2.** $-4.5 \ \square \ -4.25$ **3.** $\pi \ \square \ 3.2$

Find three decimals that make the number sentence true.

4. $-0.01 \le \square$ **5.** $1.75 > \square$ **6.** $0.75 \ge \square$

Using Order of Operations (6.EE.1)

Example 4 **Evaluate $8^2 \div (32 \div 2) - 2(3 - 5)$.**

First:	Parentheses	$8^2 \div (32 \div 2) - 2(3 - 5) = 8^2 \div 16 - 2(-2)$
Second:	Exponents	$= 64 \div 16 - 2(-2)$
Third:	Multiplication and Division (from left to right)	$= 4 + 4$
Fourth:	Addition and Subtraction (from left to right)	$= 8$

Try It Yourself

Evaluate the expression.

7. $15\left(\frac{12}{3}\right) - 7^2 - 2 \cdot 7$ **8.** $3^2 \cdot 4 \div 18 + 30 \cdot 6 - 1$ **9.** $-1 + \left(\frac{4}{2}(6 - 1)\right)^2$

12.1 Finding Square Roots

COMMON CORE STATE STANDARDS
8.EE.2

Essential Question How can you find the side length of a square when you are given the area of the square?

When you multiply a number by itself, you square the number.

Symbol for squaring is 2nd power. → $4^2 = 4 \cdot 4$

$= 16$ 4 squared is 16.

To "undo" this, take the **square root** of the number.

Symbol for square root is a radical sign. → $\sqrt{16} = \sqrt{4^2} = 4$ The square root of 16 is 4.

1 ACTIVITY: Finding Square Roots

Work with a partner. Use a square root symbol to write the side length of the square. Then find the square root. Check your answer by multiplying.

a. Sample: $s = \sqrt{121} = 11$ ft

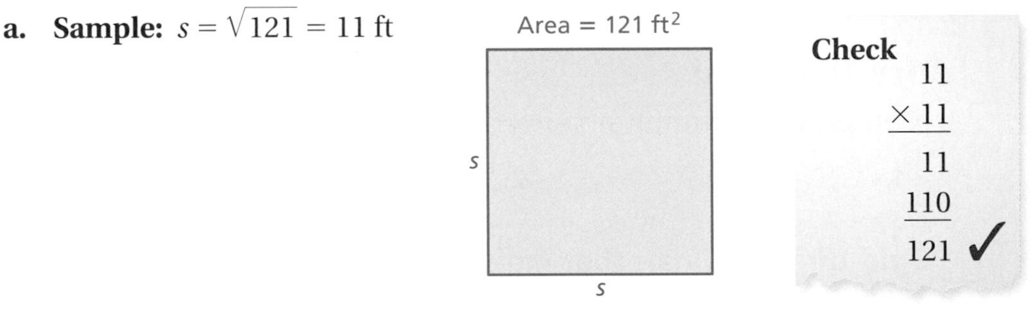

The side length of the square is 11 feet.

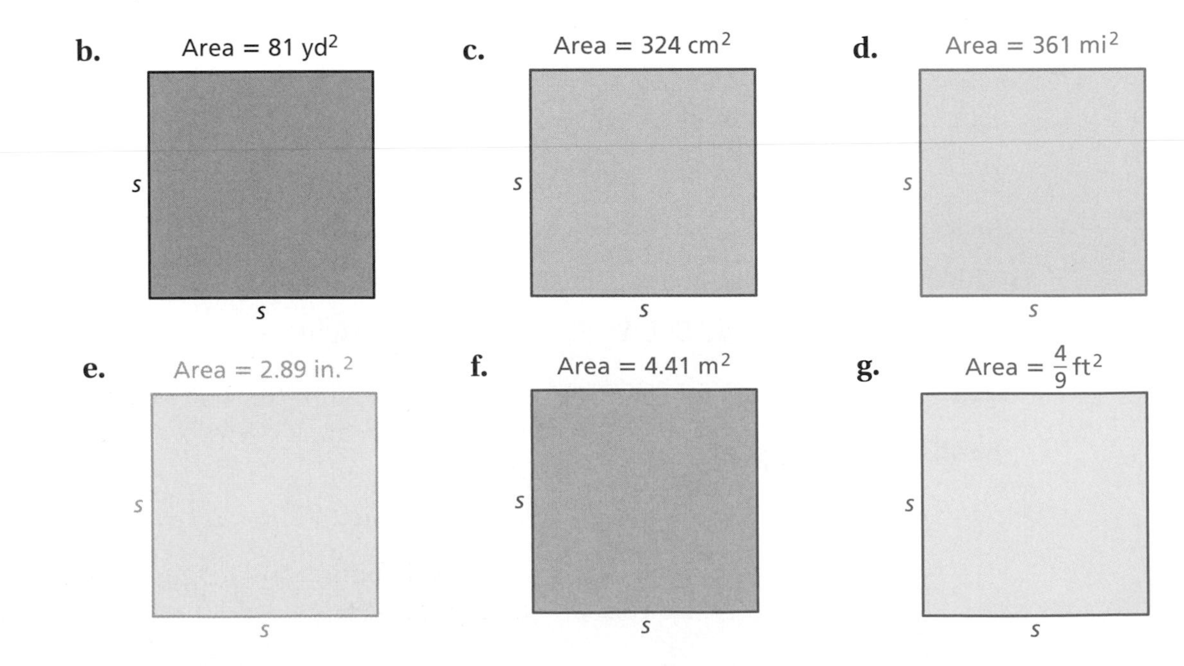

2 ACTIVITY: The Period of a Pendulum

Work with a partner.

The period of a pendulum is the time (in seconds) it takes the pendulum to swing back *and* forth.

The period T is represented by $T = 1.1\sqrt{L}$, where L is the length of the pendulum (in feet).

Copy and complete the table. Then graph the function. Is the function linear?

L	1.00	1.96	3.24	4.00	4.84	6.25	7.29	7.84	9.00
T									

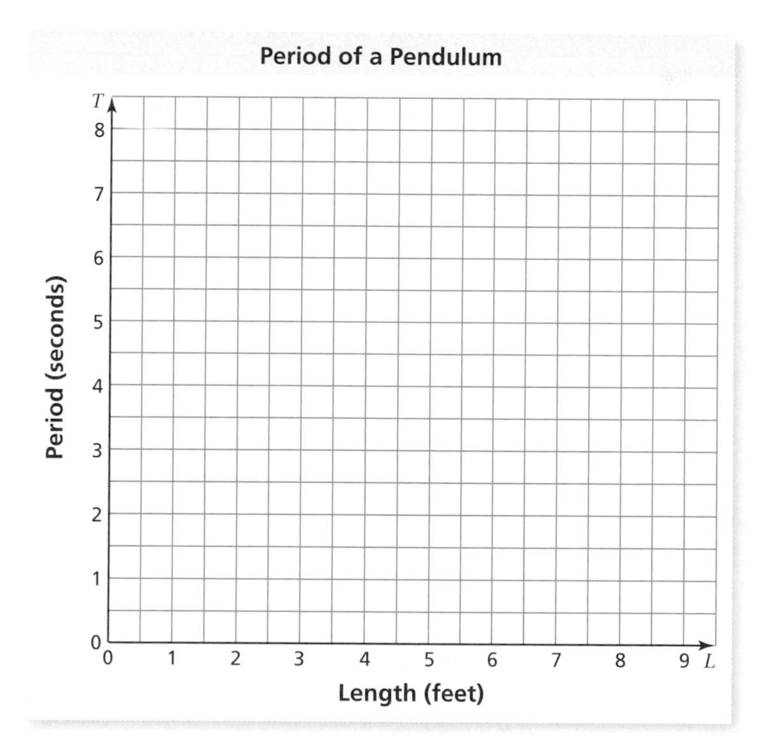

What Is Your Answer?

3. **IN YOUR OWN WORDS** How can you find the side length of a square when you are given the area of the square? Give an example. How can you check your answer?

Practice Use what you learned about finding square roots to complete Exercises 4–6 on page 514.

12.1 Lesson

Check It Out
Lesson Tutorials
BigIdeasMath.com

Key Vocabulary
square root, *p. 512*
perfect square, *p. 512*
radical sign, *p. 512*
radicand, *p. 512*

A **square root** of a number is a number that when multiplied by itself, equals the given number. Every positive number has a positive *and* a negative square root. A **perfect square** is a number with integers as its square roots.

EXAMPLE 1 Finding Square Roots of a Perfect Square

Find the two square roots of 49.

$$7 \cdot 7 = 49 \text{ and } (-7) \cdot (-7) = 49$$

So, the square roots of 49 are 7 and −7.

Study Tip
Zero has one square root, which is 0.

The symbol $\sqrt{}$ is called a **radical sign**. It is used to represent a square root. The number under the radical sign is called the **radicand**.

Positive Square Root $\sqrt{}$	Negative Square Root $-\sqrt{}$	Both Square Roots $\pm\sqrt{}$
$\sqrt{16} = 4$	$-\sqrt{16} = -4$	$\pm\sqrt{16} = \pm 4$

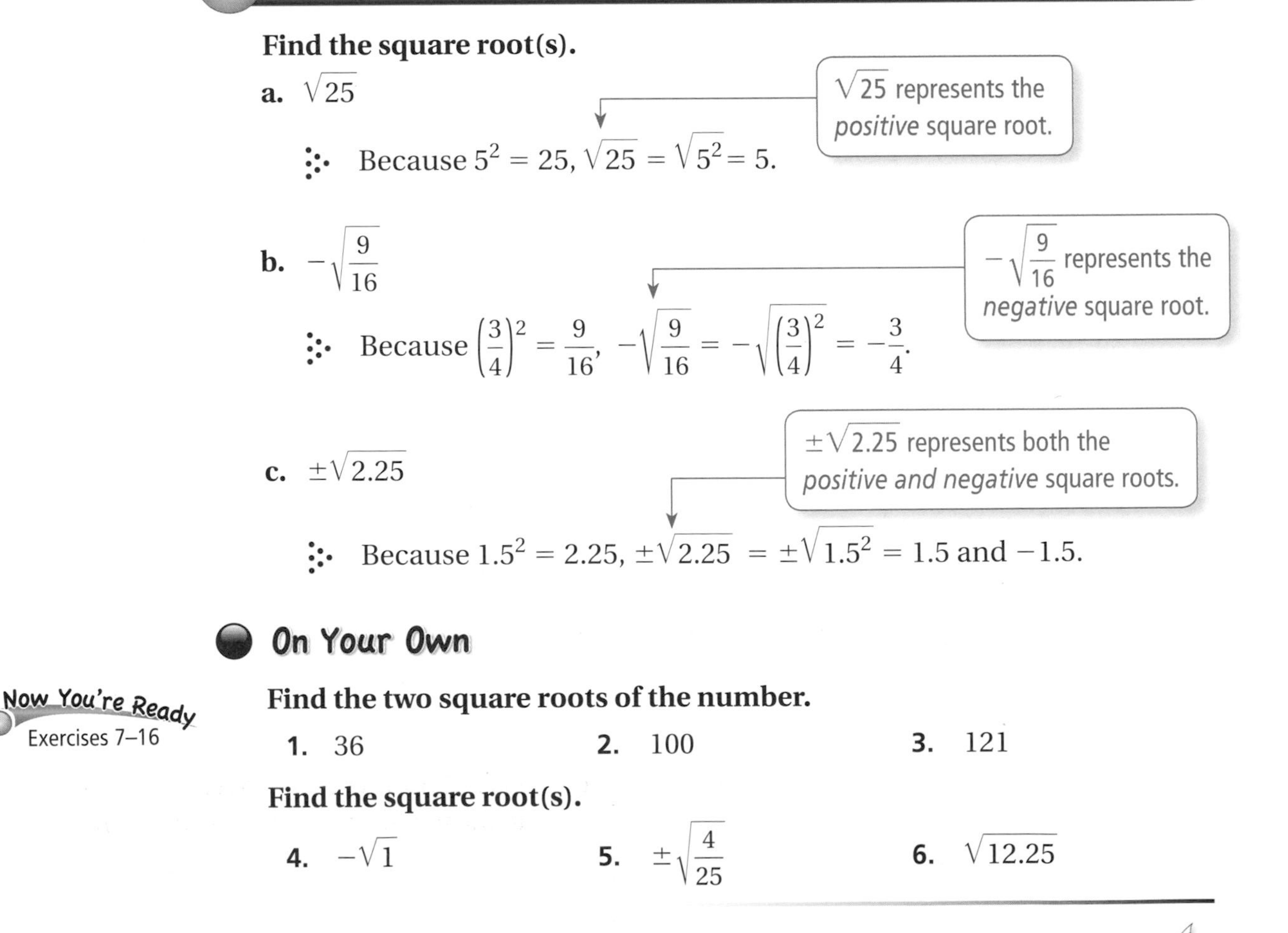

EXAMPLE 2 Finding Square Roots

Find the square root(s).

a. $\sqrt{25}$

Because $5^2 = 25$, $\sqrt{25} = \sqrt{5^2} = 5$.

$\sqrt{25}$ represents the *positive* square root.

b. $-\sqrt{\frac{9}{16}}$

Because $\left(\frac{3}{4}\right)^2 = \frac{9}{16}$, $-\sqrt{\frac{9}{16}} = -\sqrt{\left(\frac{3}{4}\right)^2} = -\frac{3}{4}$.

$-\sqrt{\frac{9}{16}}$ represents the *negative* square root.

c. $\pm\sqrt{2.25}$

Because $1.5^2 = 2.25$, $\pm\sqrt{2.25} = \pm\sqrt{1.5^2} = 1.5$ and -1.5.

$\pm\sqrt{2.25}$ represents both the *positive and negative* square roots.

On Your Own

Now You're Ready
Exercises 7–16

Find the two square roots of the number.

1. 36
2. 100
3. 121

Find the square root(s).

4. $-\sqrt{1}$
5. $\pm\sqrt{\frac{4}{25}}$
6. $\sqrt{12.25}$

Multi-Language Glossary at BigIdeasMath.com.

EXAMPLE 3 Evaluating Expressions Involving Square Roots

Evaluate the expression.

a. $5\sqrt{36} + 7$

$$\begin{aligned} 5\sqrt{36} + 7 &= 5(6) + 7 && \text{Evaluate the square root.} \\ &= 30 + 7 && \text{Multiply.} \\ &= 37 && \text{Add.} \end{aligned}$$

b. $\frac{1}{4} + \sqrt{\frac{18}{2}}$

$$\begin{aligned} \frac{1}{4} + \sqrt{\frac{18}{2}} &= \frac{1}{4} + \sqrt{9} && \text{Simplify.} \\ &= \frac{1}{4} + 3 && \text{Evaluate the square root.} \\ &= 3\frac{1}{4} && \text{Add.} \end{aligned}$$

Squaring a positive number and finding a square root are inverse operations. Use this relationship to solve equations involving squares.

EXAMPLE 4 Real-Life Application

The area of a crop circle is 45,216 square feet. What is the radius of the crop circle? Use 3.14 for π.

$$\begin{aligned} A &= \pi r^2 && \text{Write the formula for the area of a circle.} \\ 45{,}216 &\approx 3.14r^2 && \text{Substitute 45,216 for } A \text{ and 3.14 for } \pi. \\ 14{,}400 &= r^2 && \text{Divide each side by 3.14.} \\ \sqrt{14{,}400} &= \sqrt{r^2} && \text{Take positive square root of each side.} \\ 120 &= r && \text{Simplify.} \end{aligned}$$

∴ The radius of the crop circle is about 120 feet.

On Your Own

Now You're Ready
Exercises 18–23

Evaluate the expression.

7. $12 - 3\sqrt{25}$ **8.** $\sqrt{\frac{28}{7}} + 2.4$ **9.** $5(\sqrt{49} - 10)$

10. The area of a circle is 2826 square feet. Write and solve an equation to find the radius of the circle. Use 3.14 for π.

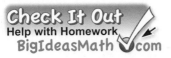

12.1 Exercises

Check It Out
Help with Homework
BigIdeasMath.com

Vocabulary and Concept Check

1. **VOCABULARY** Is 26 a perfect square? Explain.
2. **REASONING** Can the square of an integer be a negative number? Explain.
3. **NUMBER SENSE** Does $\sqrt{256}$ represent the positive square root of 256, the negative square root of 256, or both? Explain.

Practice and Problem Solving

Find the side length of the square. Check your answer by multiplying.

4. Area = 441 cm^2

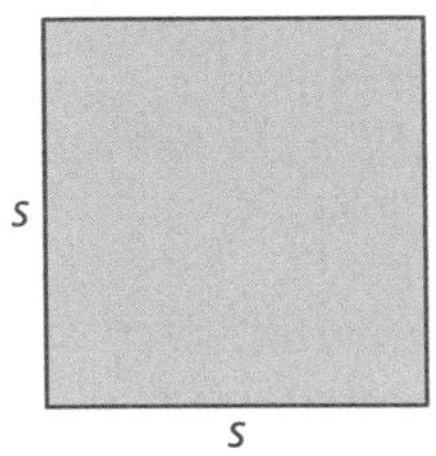

5. Area = 1.69 km^2

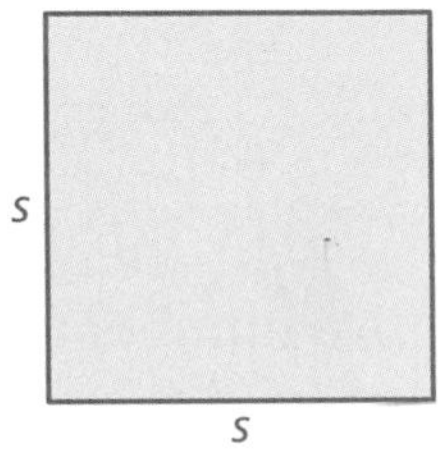

6. Area = $\frac{25}{36}$ yd^2

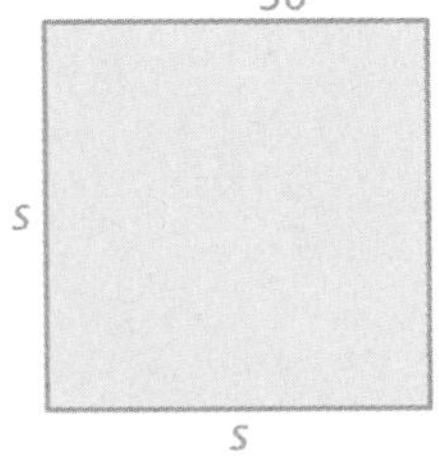

Find the two square roots of the number.

(1) 7. 9 8. 64 9. 4 10. 144

Find the square root(s).

(2) 11. $\sqrt{625}$ 12. $-\sqrt{\frac{9}{100}}$ 13. $\pm\sqrt{\frac{1}{961}}$

14. $\sqrt{7.29}$ 15. $\pm\sqrt{4.84}$ 16. $-\sqrt{361}$

17. **ERROR ANALYSIS** Describe and correct the error in finding the square roots.

✗ $\pm\sqrt{\frac{1}{4}} = \frac{1}{2}$

Evaluate the expression.

(3) 18. $3\sqrt{16} - 5$ 19. $10 - 4\sqrt{\frac{1}{16}}$ 20. $\sqrt{6.76} + 5.4$

21. $8\sqrt{8.41} + 1.8$ 22. $2\left(\sqrt{\frac{80}{5}} - 5\right)$ 23. $4\left(\sqrt{\frac{147}{3}} + 3\right)$

24. **NOTEPAD** The area of the base of a square notepad is 9 square inches. What is the length of one side of the base of the notepad?

25. **CRITICAL THINKING** There are two square roots of 25. Why is there only one answer for the radius of the button?

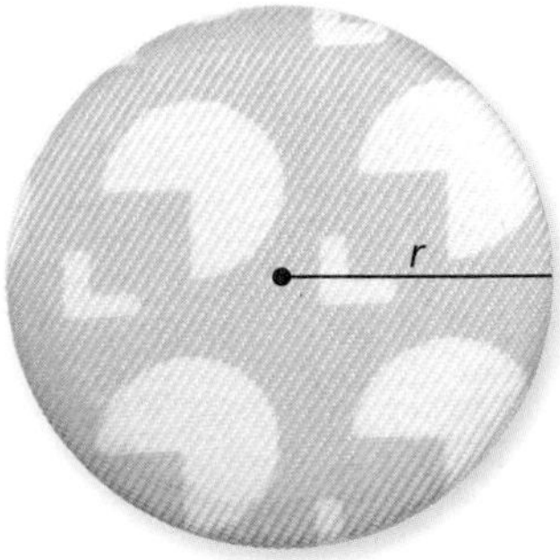

$A = 25\pi\ \text{mm}^2$

Copy and complete the statement with <, >, or =.

26. $\sqrt{81}$ $\square$ 8

27. 0.5 $\square$ $\sqrt{0.25}$

28. $\frac{3}{2}$ $\square$ $\sqrt{\frac{25}{4}}$

29. SAILBOAT The area of a sail is $40\frac{1}{2}$ square feet. The base and the height of the sail are equal. What is the height of the sail (in feet)?

30. REASONING Is the product of two perfect squares always a perfect square? Explain your reasoning.

31. ENERGY The kinetic energy K (in joules) of a falling apple is represented by $K = \frac{v^2}{2}$, where v is the speed of the apple (in meters per second). How fast is the apple traveling when the kinetic energy is 32 joules?

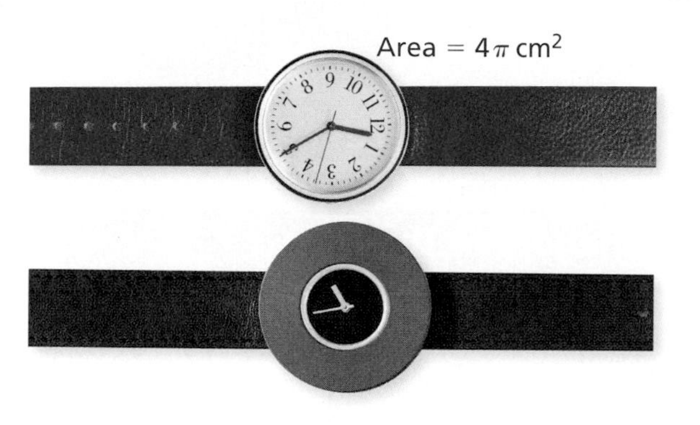

32. WATCHES The areas of the two watch faces have a ratio of 16 : 25.

a. What is the ratio of the radius of the smaller watch face to the radius of the larger watch face?

b. What is the radius of the larger watch face?

33. WINDOW The cost C (in dollars) of making a square window with a side length of n inches is represented by $C = \frac{n^2}{5} + 175$. A window costs \$355. What is the length (in feet) of the window?

34. Geometry The area of the triangle is represented by the formula $A = \sqrt{s(s-21)(s-17)(s-10)}$, where s is equal to half the perimeter. What is the height of the triangle?

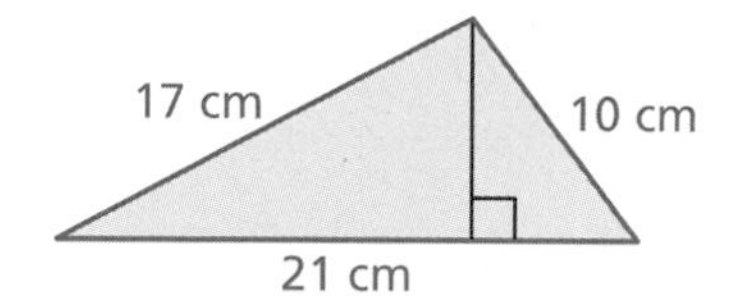

Fair Game Review What you learned in previous grades & lessons

Evaluate the expression. *(Skills Review Handbook)*

35. $3^2 + 4^2$

36. $8^2 + 15^2$

37. $13^2 - 5^2$

38. $25^2 - 24^2$

39. MULTIPLE CHOICE Which of the following describes the triangle? *(Section 11.2)*

Ⓐ Acute Ⓑ Right

Ⓒ Obtuse Ⓓ Equiangular

12.2 The Pythagorean Theorem

COMMON CORE STATE STANDARDS
8.G.6
8.G.7

Essential Question How are the lengths of the sides of a right triangle related?

Pythagoras was a Greek mathematician and philosopher who discovered one of the most famous rules in mathematics. In mathematics, a rule is called a **theorem**. So, the rule that Pythagoras discovered is called the Pythagorean Theorem.

Pythagoras
(c. 570 B.C.–c. 490 B.C.)

1 ACTIVITY: Discovering the Pythagorean Theorem

Work with a partner.

a. On grid paper, draw any right triangle. Label the lengths of the two shorter sides (the **legs**) a and b.

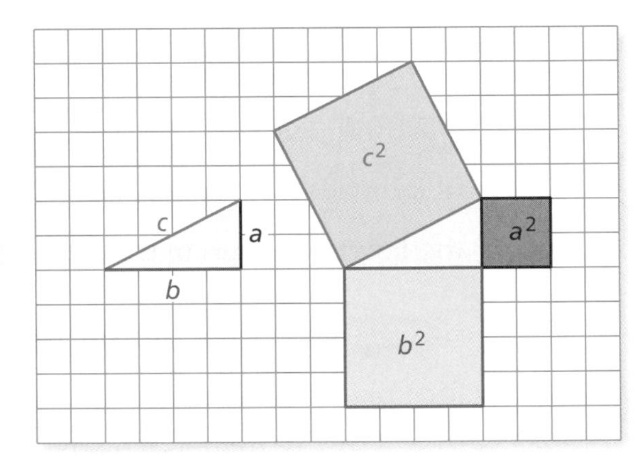

b. Label the length of the longest side (the **hypotenuse**) c.

c. Draw squares along each of the three sides. Label the areas of the three squares a^2, b^2, and c^2.

d. Cut out the three squares. Make eight copies of the right triangle and cut them out. Arrange the figures to form two identical larger squares.

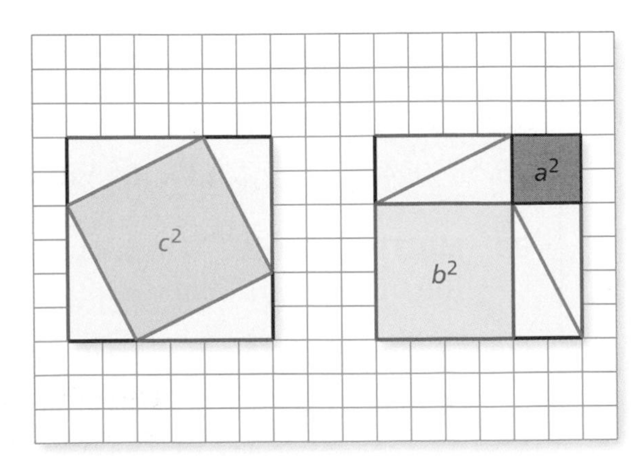

e. What does this tell you about the relationship among a^2, b^2, and c^2?

2 ACTIVITY: Finding the Length of the Hypotenuse

Work with a partner. Use the result of Activity 1 to find the length of the hypotenuse of each right triangle.

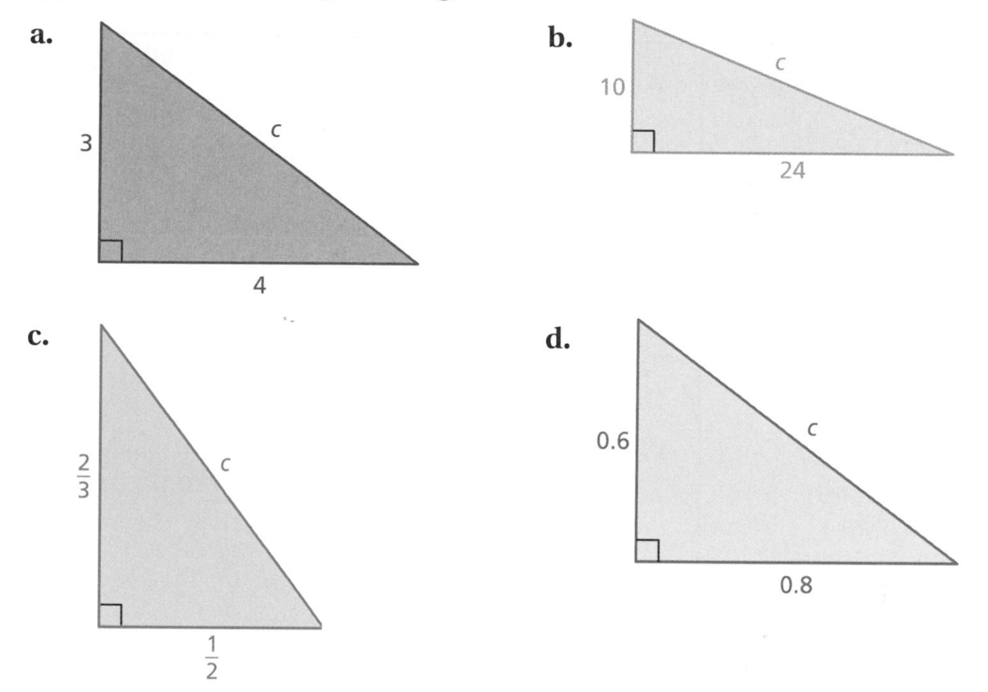

3 ACTIVITY: Finding the Length of a Leg

Work with a partner. Use the result of Activity 1 to find the length of the leg of each right triangle.

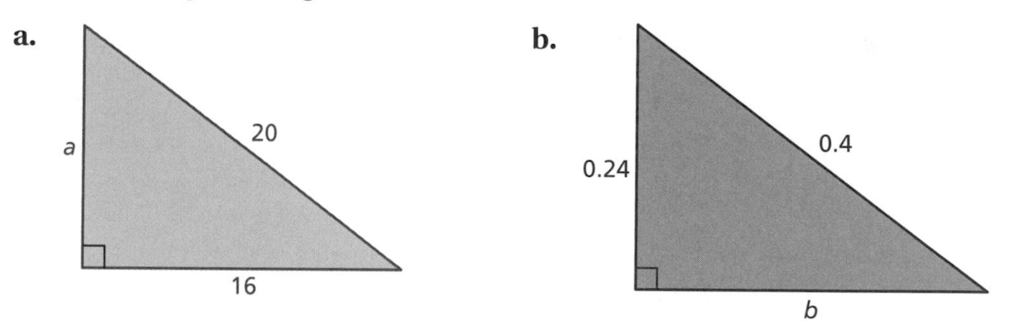

What Is Your Answer?

4. **IN YOUR OWN WORDS** How are the lengths of the sides of a right triangle related? Give an example using whole numbers.

Practice Use what you learned about the Pythagorean Theorem to complete Exercises 3–5 on page 520.

12.2 Lesson

Key Vocabulary
theorem, *p. 516*
legs, *p. 518*
hypotenuse, *p. 518*
Pythagorean Theorem, *p. 518*

Sides of a Right Triangle

The sides of a right triangle have special names.

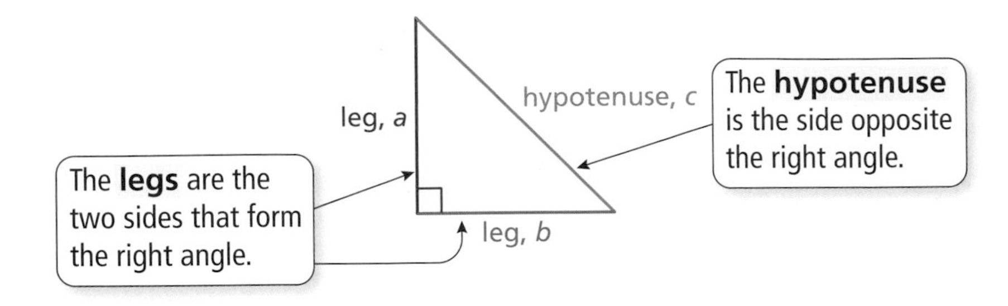

The Pythagorean Theorem

Words In any right triangle, the sum of the squares of the lengths of the legs is equal to the square of the length of the hypotenuse.

Algebra $a^2 + b^2 = c^2$

In a right triangle, the legs are the shorter sides and the hypotenuse is always the longest side.

EXAMPLE 1 Finding the Length of a Hypotenuse

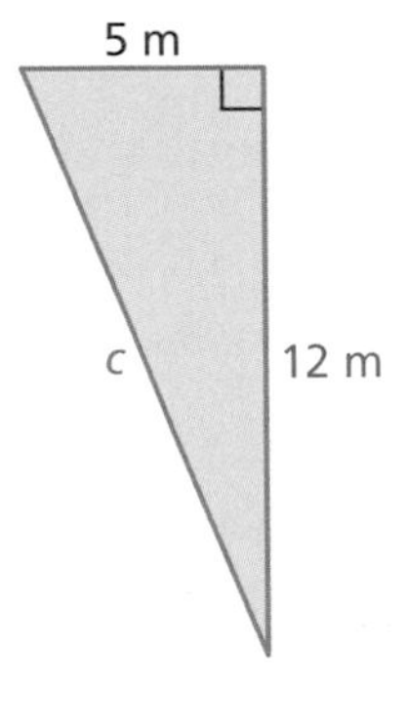

Find the length of the hypotenuse of the triangle.

$a^2 + b^2 = c^2$	Write the Pythagorean Theorem.
$5^2 + 12^2 = c^2$	Substitute 5 for *a* and 12 for *b*.
$25 + 144 = c^2$	Evaluate powers.
$169 = c^2$	Add.
$\sqrt{169} = \sqrt{c^2}$	Take positive square root of each side.
$13 = c$	Simplify.

∴ The length of the hypotenuse is 13 meters.

On Your Own

Find the length of the hypotenuse of the triangle.

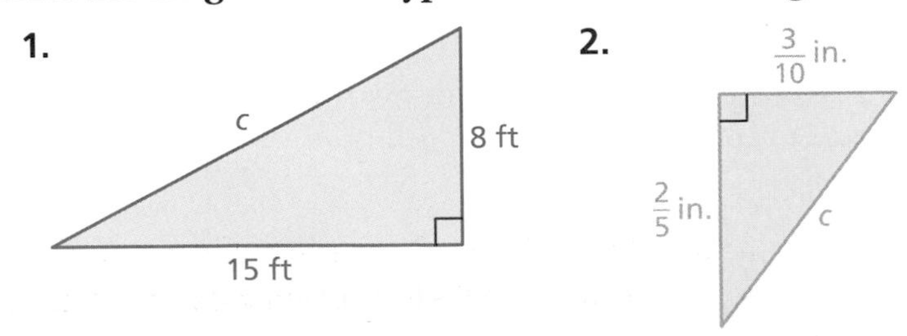

EXAMPLE 2 Finding the Length of a Leg

Find the missing length of the triangle.

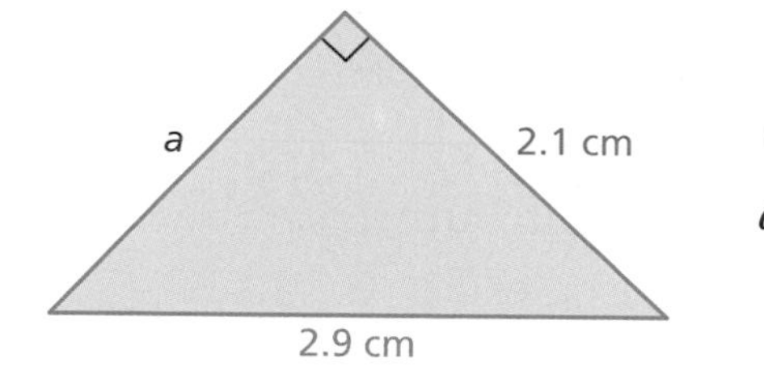

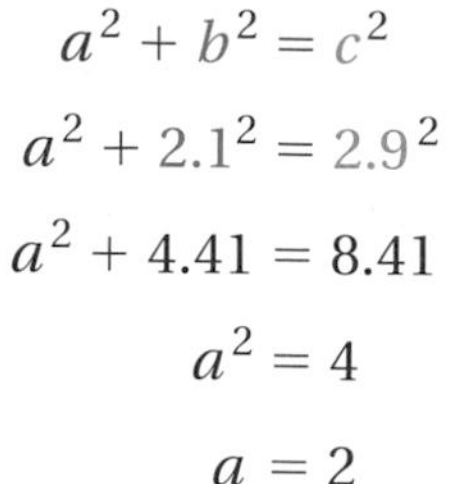

$a^2 + b^2 = c^2$	Write the Pythagorean Theorem.
$a^2 + 2.1^2 = 2.9^2$	Substitute 2.1 for b and 2.9 for c.
$a^2 + 4.41 = 8.41$	Evaluate powers.
$a^2 = 4$	Subtract 4.41 from each side.
$a = 2$	Take positive square root of each side.

∴ The length of the leg is 2 centimeters.

EXAMPLE 3 Standardized Test Practice

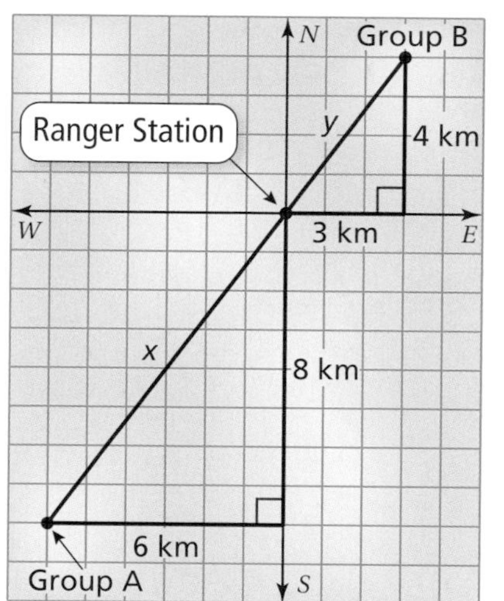

Hiking Group A leaves a ranger station and hikes 8 kilometers south then 6 kilometers west. Group B leaves the station and hikes 3 kilometers east then 4 kilometers north. Using the figure, how far apart are the two groups of hikers?

Ⓐ 5 km Ⓑ 10 km Ⓒ 15 km Ⓓ 21 km

The distance between the groups is the sum of the hypotenuses, x and y. Use the Pythagorean Theorem to find x and y.

$a^2 + b^2 = c^2$	Write the Pythagorean Theorem.	$a^2 + b^2 = c^2$
$6^2 + 8^2 = x^2$	Substitute.	$3^2 + 4^2 = y^2$
$36 + 64 = x^2$	Evaluate powers.	$9 + 16 = y^2$
$100 = x^2$	Add.	$25 = y^2$
$10 = x$	Take positive square root of each side.	$5 = y$

∴ The distance between the groups of hikers is $10 + 5 = 15$ kilometers. So, the correct answer is Ⓒ.

On Your Own

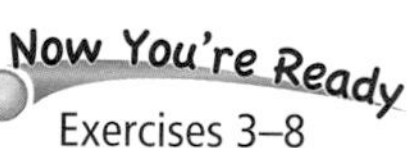

Find the missing length of the triangle.

3. 34 yd, 16 yd, b

4. a, 9.6 m, 10.4 m

5. **WHAT IF?** In Example 3, Group A hikes 12 kilometers south and 9 kilometers west. How far apart are the hikers?

12.2 Exercises

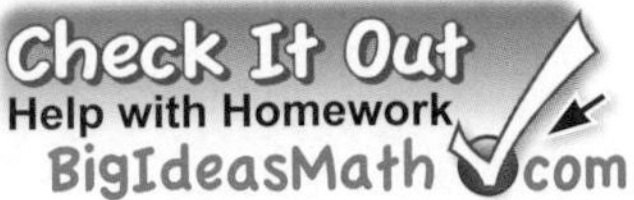

Vocabulary and Concept Check

1. **VOCABULARY** In a right triangle, how can you tell which sides are the legs and which side is the hypotenuse?
2. **DIFFERENT WORDS, SAME QUESTION** Which is different? Find "both" answers.

Which side is the hypotenuse?

Which side is the longest?

Which side is a leg?

Which side is opposite the right angle?

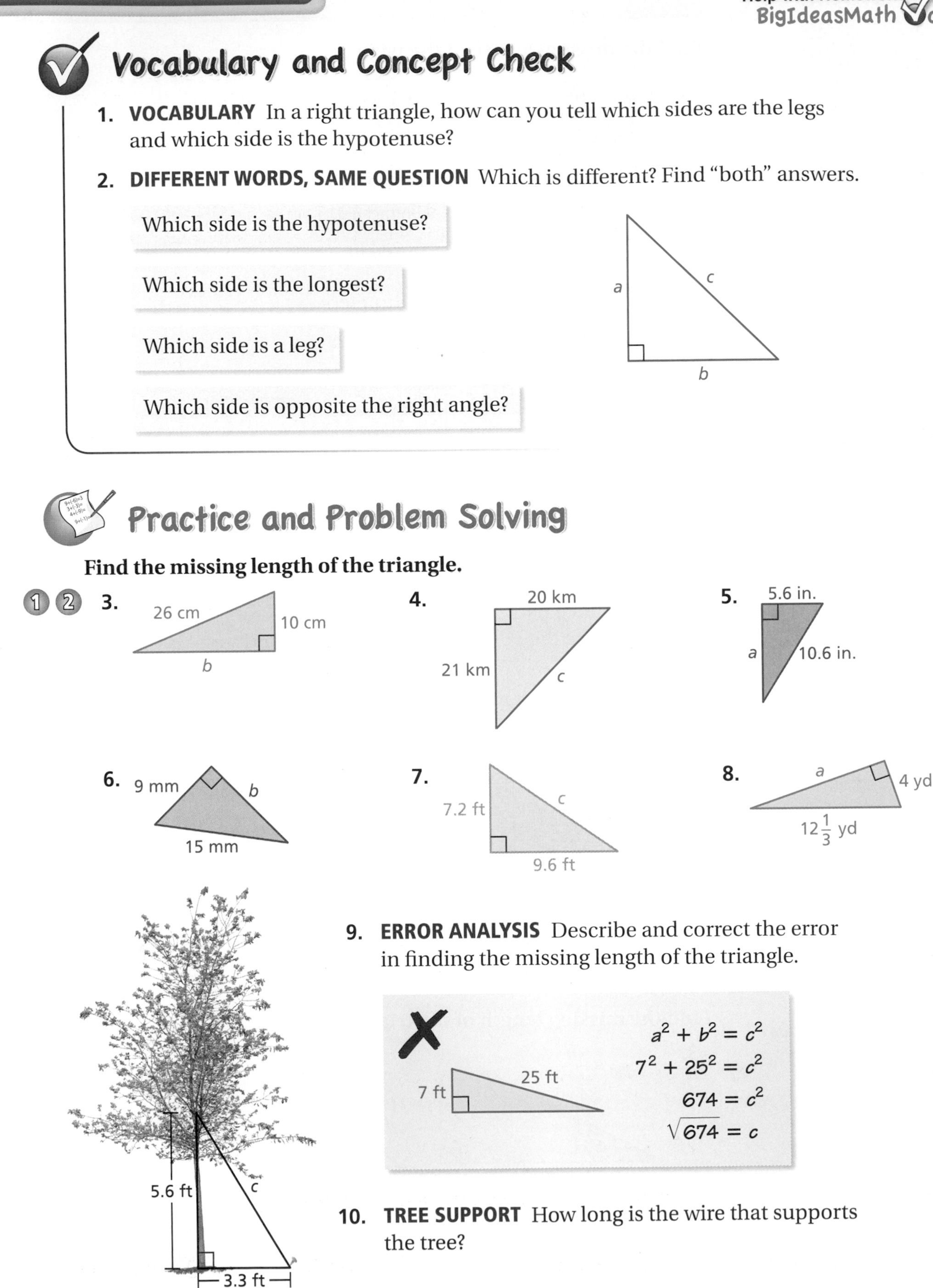

Practice and Problem Solving

Find the missing length of the triangle.

1 2 **3.**

4.

5.

6.

7.

8.

9. **ERROR ANALYSIS** Describe and correct the error in finding the missing length of the triangle.

10. **TREE SUPPORT** How long is the wire that supports the tree?

Find the value of x.

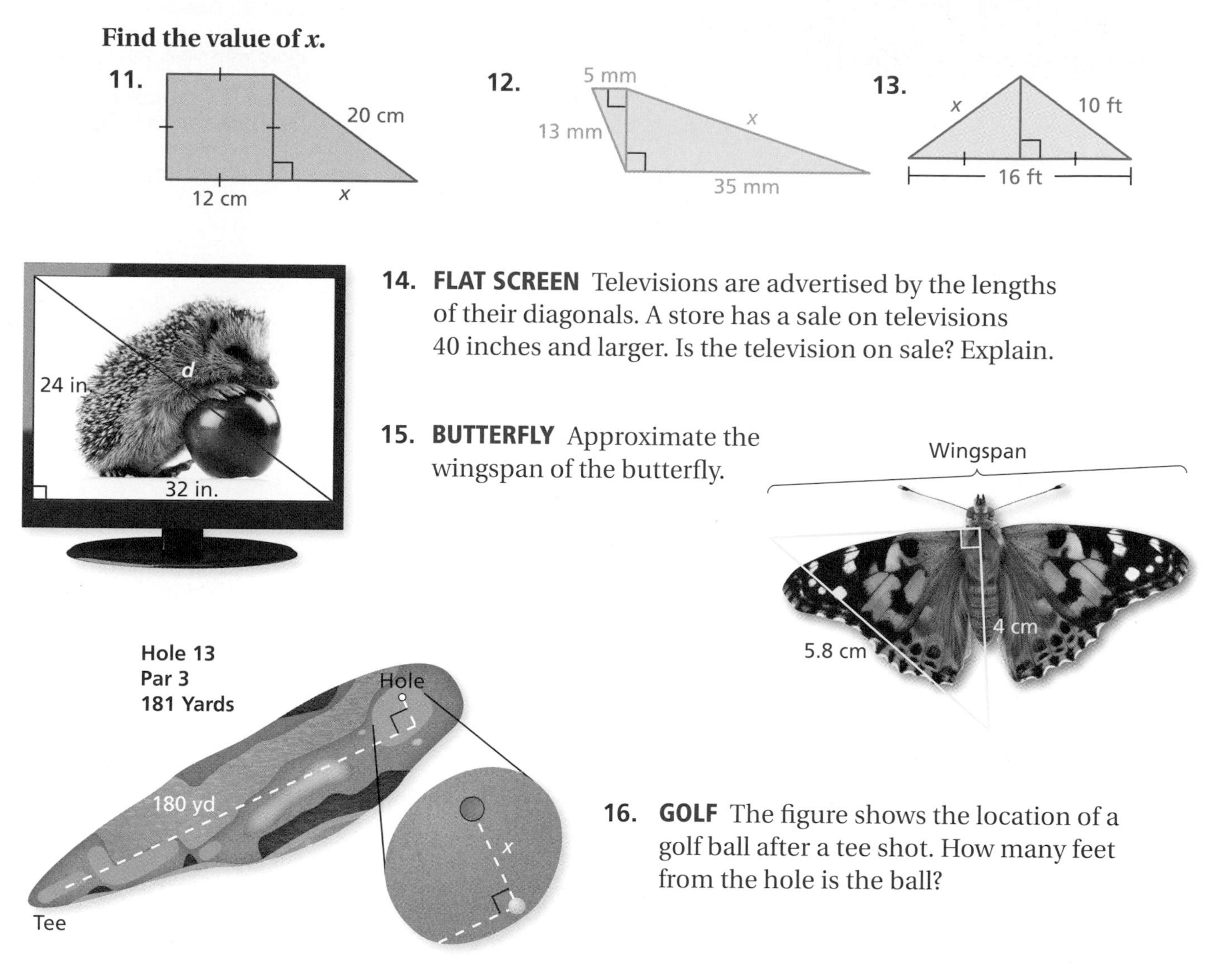

14. FLAT SCREEN Televisions are advertised by the lengths of their diagonals. A store has a sale on televisions 40 inches and larger. Is the television on sale? Explain.

15. BUTTERFLY Approximate the wingspan of the butterfly.

16. GOLF The figure shows the location of a golf ball after a tee shot. How many feet from the hole is the ball?

17. SNOWBALLS You and a friend stand back-to-back. You run 20 feet forward then 15 feet to your right. At the same time, your friend runs 16 feet forward then 12 feet to her right. She stops and hits you with a snowball.

a. Draw the situation in a coordinate plane.

b. How far does your friend throw the snowball?

18. **Algebra** The legs of a right triangle have lengths of 28 meters and 21 meters. The hypotenuse has a length of $5x$ meters. What is the value of x?

Fair Game Review What you learned in previous grades & lessons

Find the square root(s). *(Section 12.1)*

19. $\pm\sqrt{36}$ **20.** $-\sqrt{121}$ **21.** $\sqrt{169}$ **22.** $-\sqrt{225}$

23. MULTIPLE CHOICE Which type of triangle can have an obtuse angle? *(Section 11.2)*

Ⓐ equiangular Ⓑ right Ⓒ isosceles Ⓓ equilateral

12 Study Help

You can use a **summary triangle** to explain a topic. Here is an example of a summary triangle for finding the length of the hypotenuse of a triangle.

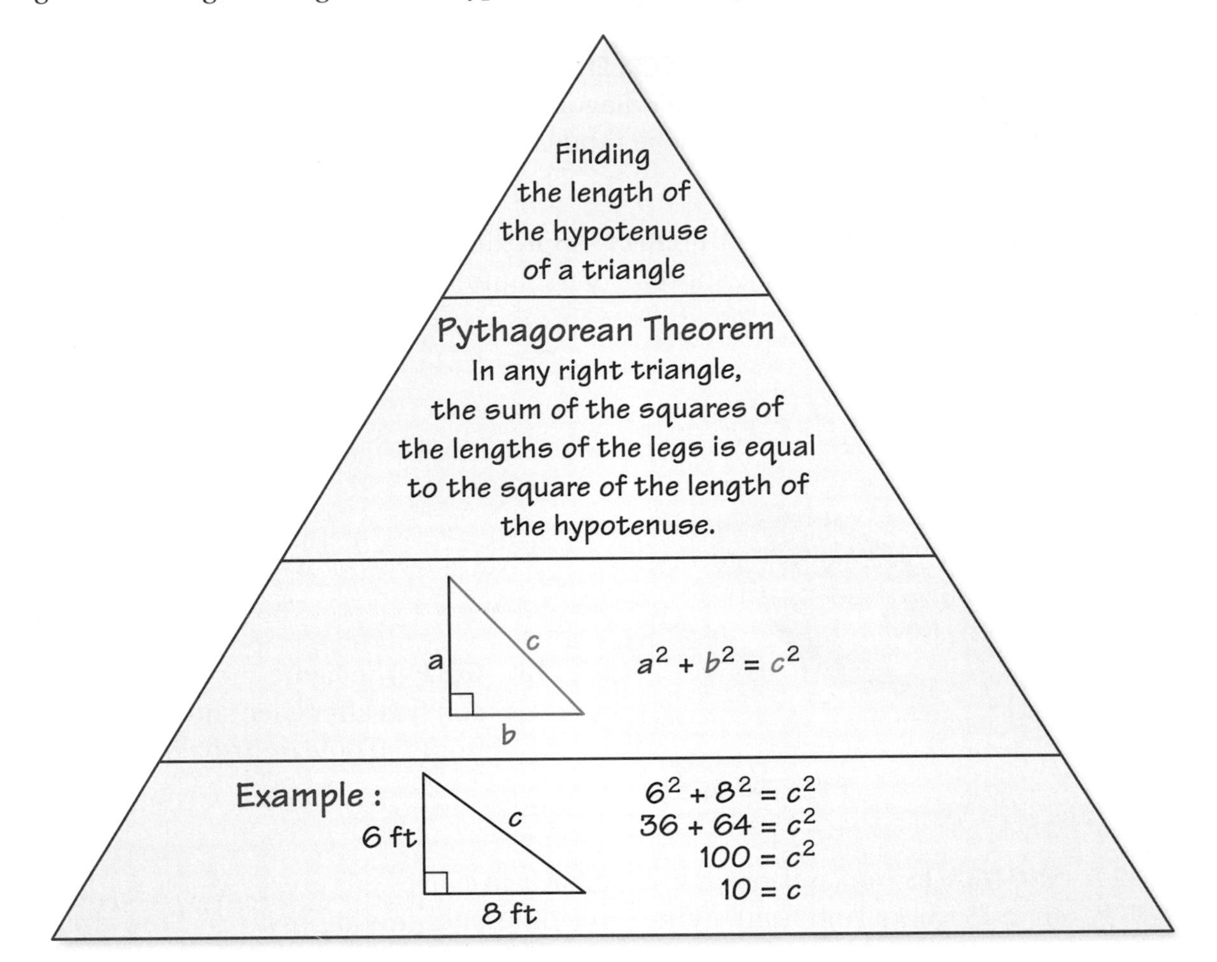

Finding the length of the hypotenuse of a triangle
Pythagorean Theorem
In any right triangle, the sum of the squares of the lengths of the legs is equal to the square of the length of the hypotenuse.
a
b
c
$a^2 + b^2 = c^2$
Example :
6 ft
8 ft
c
$6^2 + 8^2 = c^2$
$36 + 64 = c^2$
$100 = c^2$
$10 = c$

On Your Own

Make a summary triangle to help you study these topics.

1. finding square roots
2. evaluating expressions involving square roots
3. finding the length of a leg of a right triangle

After you complete this chapter, make summary triangles for the following topics.

4. approximating square roots
5. simplifying square roots

"What do you call a cheese summary triangle that isn't yours?"

12.1–12.2 Quiz

Check It Out
Progress Check
BigIdeasMath.com

Find the two square roots of the number. *(Section 12.1)*

1. 196 **2.** 49 **3.** 400

Find the square root(s). *(Section 12.1)*

4. $-\sqrt{4}$ **5.** $\sqrt{\frac{16}{25}}$ **6.** $\pm\sqrt{6.25}$

Evaluate the expression. *(Section 12.1)*

7. $3\sqrt{49} + 5$ **8.** $10 - 4\sqrt{16}$ **9.** $\frac{1}{4} + \sqrt{\frac{100}{4}}$

Find the missing length of the triangle. *(Section 12.2)*

10. **12.**

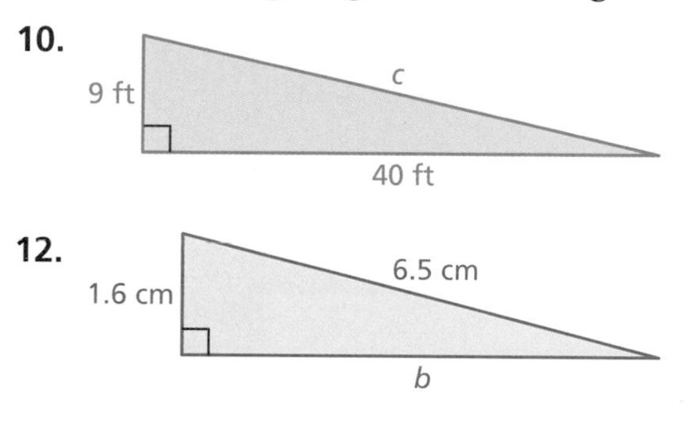

11. **13.**

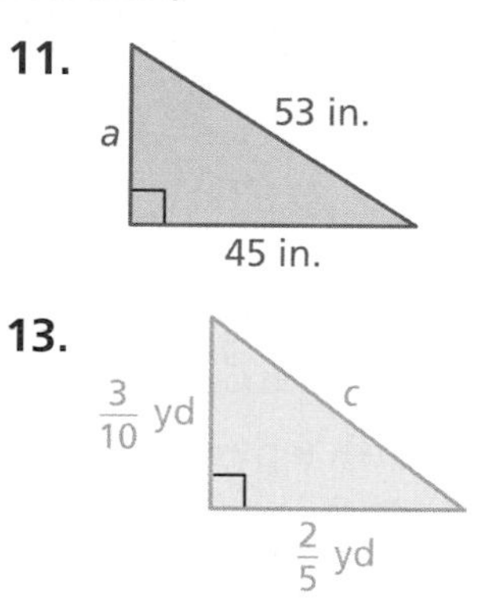

14. POOL The area of a circular pool cover is 314 square feet. Write and solve an equation to find the diameter of the pool cover. Use 3.14 for π. *(Section 12.1)*

15. LAND A square parcel of land has an area of 1 million square feet. What is the length of one side of the parcel? *(Section 12.1)*

16. FABRIC You are cutting a rectangular piece of fabric in half along the diagonal. The fabric measures 28 inches wide and $1\frac{1}{4}$ yards long. What is the length (in inches) of the diagonal? *(Section 12.2)*

12.3 Approximating Square Roots

COMMON CORE STATE STANDARDS
8.EE.2
8.NS.1
8.NS.2

Essential Question How can you find decimal approximations of square roots that are irrational?

You already know that a rational number is a number that can be written as the ratio of two integers. Numbers that cannot be written as the ratio of two integers are called **irrational**.

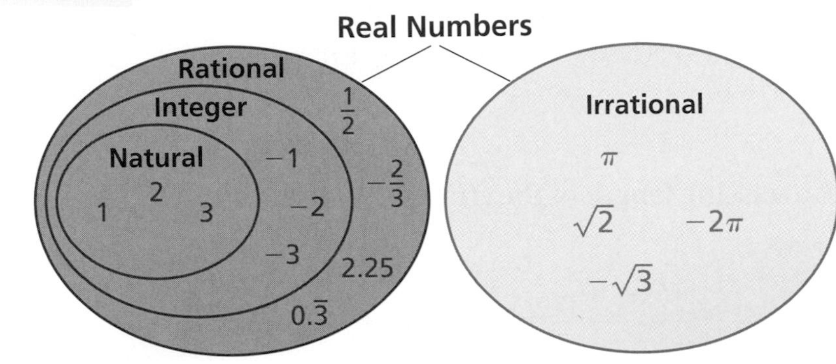

1 ACTIVITY: Approximating Square Roots

Work with a partner.

Archimedes was a Greek mathematician, physicist, engineer, inventor, and astronomer.

Archimedes
(c. 287 B.C.–c. 212 B.C.)

a. Archimedes tried to find a rational number whose square is 3. Here are two that he tried.

$$\frac{265}{153} \text{ and } \frac{1351}{780}$$

Are either of these numbers equal to $\sqrt{3}$? How can you tell?

b. Use a calculator with a square root key to approximate $\sqrt{3}$.

Write the number on a piece of paper. Then enter it into the calculator and square it. Then subtract 3. Do you get 0? Explain.

c. Calculators did not exist in the time of Archimedes. How do you think he might have approximated $\sqrt{3}$?

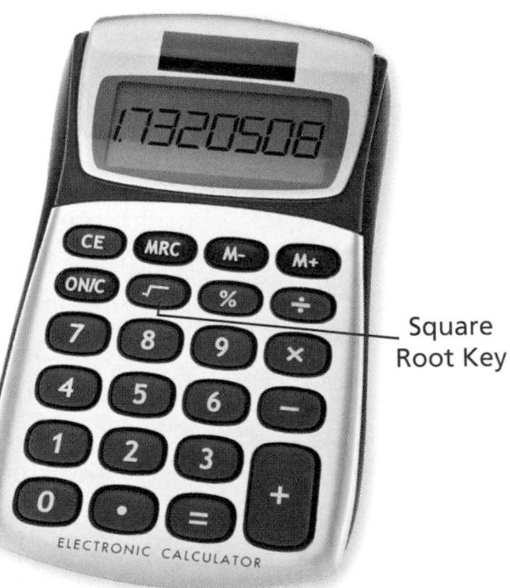

2 ACTIVITY: Approximating Square Roots Geometrically

Work with a partner.

a. Use grid paper and the given scale to draw a horizontal line segment 1 unit in length. Label this segment *AC*.

b. Draw a vertical line segment 2 units in length. Label this segment *DC*.

c. Set the point of a compass on *A*. Set the compass to 2 units. Swing the compass to intersect segment *DC*. Label this intersection as *B*.

d. Use the Pythagorean Theorem to show that the length of segment *BC* is $\sqrt{3}$ units.

e. Use the grid paper to approximate $\sqrt{3}$.

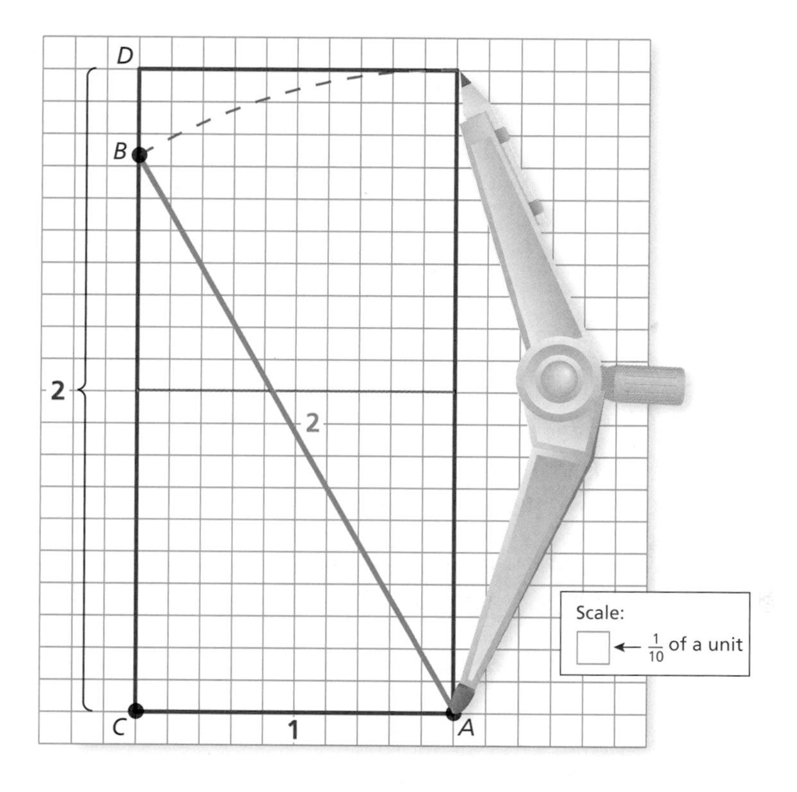

What Is Your Answer?

3. Repeat Activity 2 for a triangle in which segment *CA* is 2 units and segment *BA* is 3 units. Use the Pythagorean Theorem to show that segment *BC* is $\sqrt{5}$ units. Use the grid paper to approximate $\sqrt{5}$.

4. **IN YOUR OWN WORDS** How can you find decimal approximations of square roots that are irrational?

Use what you learned about approximating square roots to complete Exercises 5–8 on page 529.

12.3 Lesson

Key Vocabulary
irrational number, *p. 526*
real numbers, *p. 526*

A rational number is a number that can be written as the ratio of two integers. An **irrational number** cannot be written as the ratio of two integers.

- The square root of any whole number that is not a perfect square is irrational.
- The decimal form of an irrational number neither terminates nor repeats.

Key Idea

Real Numbers

Rational numbers and irrational numbers together form the set of **real numbers**.

Remember
Decimals that *terminate* or *repeat* are rational.

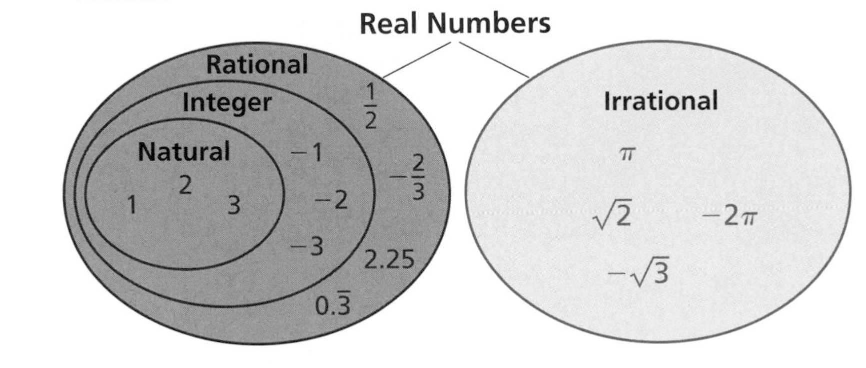

EXAMPLE 1 Classifying Real Numbers

Tell whether the number is *rational* or *irrational*. Explain.

	Number	Rational or Irrational	Reasoning
a.	$\sqrt{12}$	Irrational	12 is not a perfect square.
b.	$-0.36\overline{4}$	Rational	$-0.36\overline{4}$ is a repeating decimal.
c.	$-1\frac{3}{7}$	Rational	$-1\frac{3}{7}$ can be written as $\frac{-10}{7}$.
d.	0.85	Rational	0.85 can be written as $\frac{17}{20}$.

On Your Own

Now You're Ready
Exercises 9–14

Tell whether the number is *rational* or *irrational*. Explain.

1. 0.121221222. . . **2.** $-\sqrt{196}$ **3.** $\sqrt{2}$

EXAMPLE 2 Approximating Square Roots

Estimate $\sqrt{52}$ to the nearest integer.

Use a number line and the square roots of the perfect squares nearest to the radicand. The nearest perfect square less than 52 is 49. The nearest perfect square greater than 52 is 64.

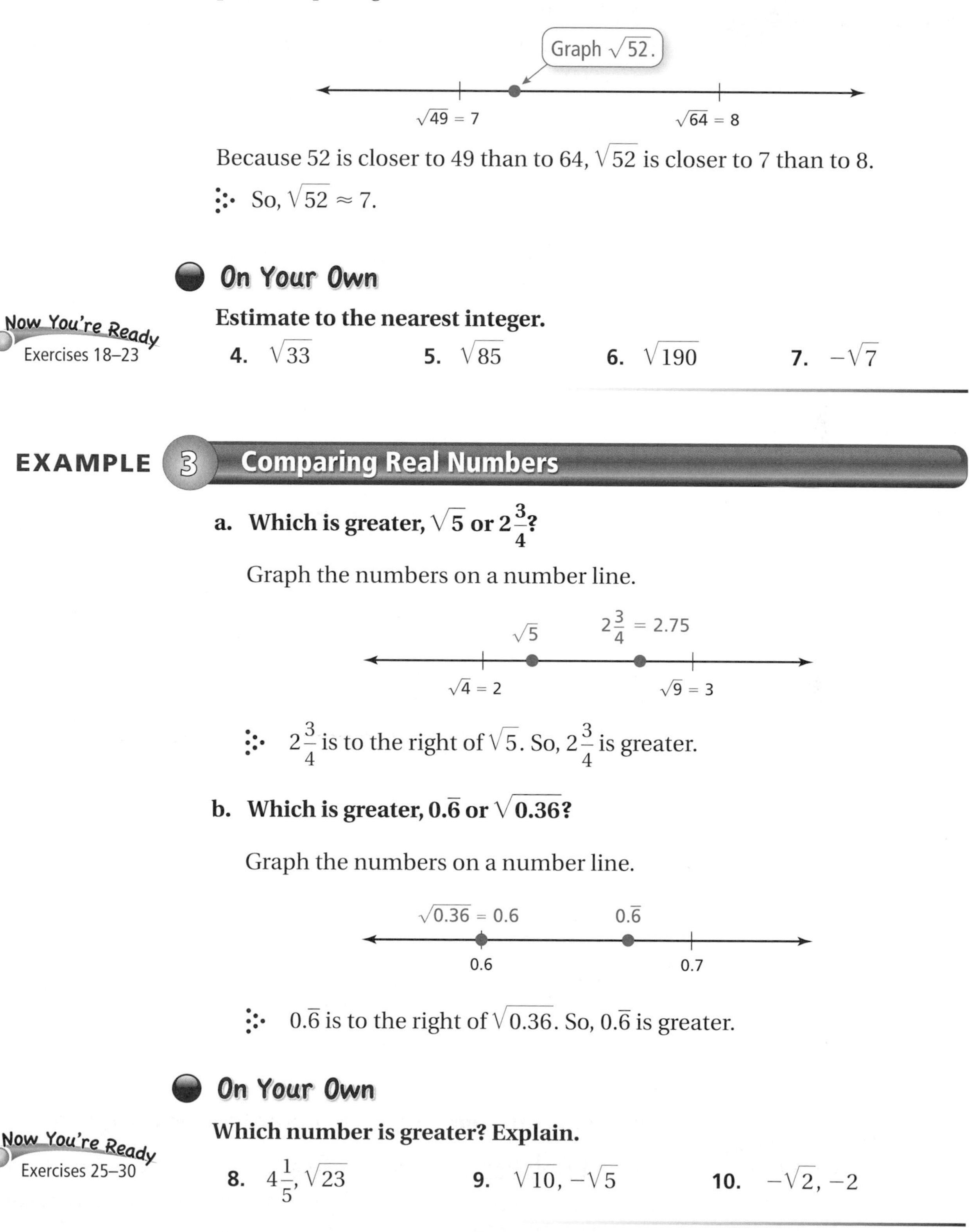

Because 52 is closer to 49 than to 64, $\sqrt{52}$ is closer to 7 than to 8.

∴ So, $\sqrt{52} \approx 7$.

On Your Own

Now You're Ready
Exercises 18–23

Estimate to the nearest integer.

4. $\sqrt{33}$ **5.** $\sqrt{85}$ **6.** $\sqrt{190}$ **7.** $-\sqrt{7}$

EXAMPLE 3 Comparing Real Numbers

a. Which is greater, $\sqrt{5}$ or $2\frac{3}{4}$?

Graph the numbers on a number line.

$\sqrt{5}$ $\quad 2\frac{3}{4} = 2.75$

$\sqrt{4} = 2$ $\quad \sqrt{9} = 3$

∴ $2\frac{3}{4}$ is to the right of $\sqrt{5}$. So, $2\frac{3}{4}$ is greater.

b. Which is greater, $0.\overline{6}$ or $\sqrt{0.36}$?

Graph the numbers on a number line.

$\sqrt{0.36} = 0.6$ $\quad 0.\overline{6}$

0.6 $\quad$ 0.7

∴ $0.\overline{6}$ is to the right of $\sqrt{0.36}$. So, $0.\overline{6}$ is greater.

On Your Own

Now You're Ready
Exercises 25–30

Which number is greater? Explain.

8. $4\frac{1}{5}$, $\sqrt{23}$ **9.** $\sqrt{10}$, $-\sqrt{5}$ **10.** $-\sqrt{2}$, -2

EXAMPLE 4 Approximating an Expression

The radius of a circle with area A is approximately $\sqrt{\frac{A}{3}}$. The area of a circular mouse pad is 51 square inches. Estimate its radius.

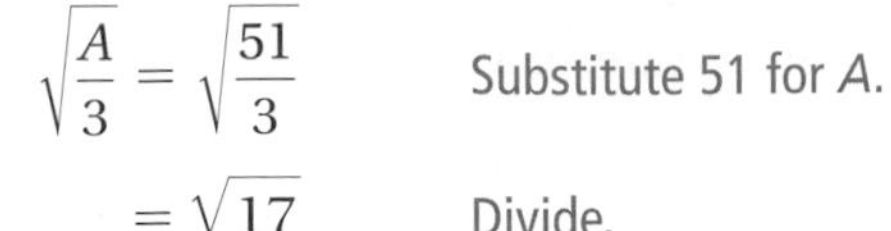

$$\sqrt{\frac{A}{3}} = \sqrt{\frac{51}{3}} \qquad \text{Substitute 51 for } A.$$

$$= \sqrt{17} \qquad \text{Divide.}$$

The nearest perfect square less than 17 is 16. The nearest perfect square greater than 17 is 25.

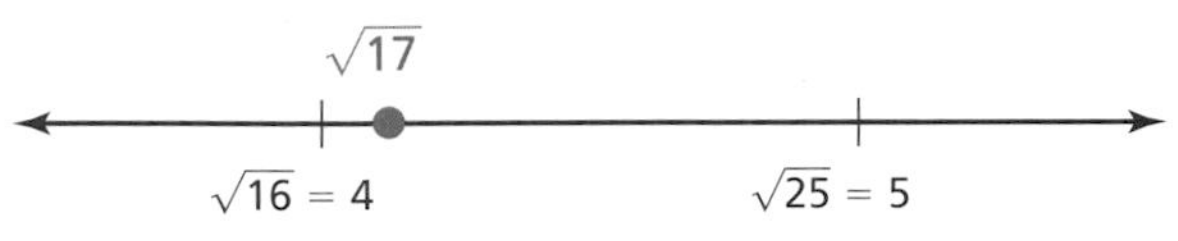

Because 17 is closer to 16 than to 25, $\sqrt{17}$ is closer to 4 than to 5.

∴ The radius is about 4 inches.

On Your Own

11. **WHAT IF?** The area of a circular mouse pad is 64 square inches. Estimate its radius.

EXAMPLE 5 Real-Life Application

The distance (in nautical miles) you can see with a periscope is $1.17\sqrt{h}$, where h is the height of the periscope above the water. Can a periscope that is 6 feet above the water see twice as far as a periscope that is 3 feet above the water? Explain.

Use a calculator to find the distances.

3 feet above water		*6 feet above water*
$1.17\sqrt{h} = 1.17\sqrt{3}$	Substitute for h.	$1.17\sqrt{h} = 1.17\sqrt{6}$
≈ 2.03	Use a calculator.	≈ 2.87

You can see $\frac{2.87}{2.03} \approx 1.41$ times farther with the periscope that is 6 feet above the water than with the periscope that is 3 feet above the water.

∴ No, the periscope that is 6 feet above the water cannot see twice as far.

On Your Own

12. You use a periscope that is 10 feet above the water. Can you see farther than 4 nautical miles? Explain.

12.3 Exercises

Check It Out
Help with Homework
BigIdeasMath.com

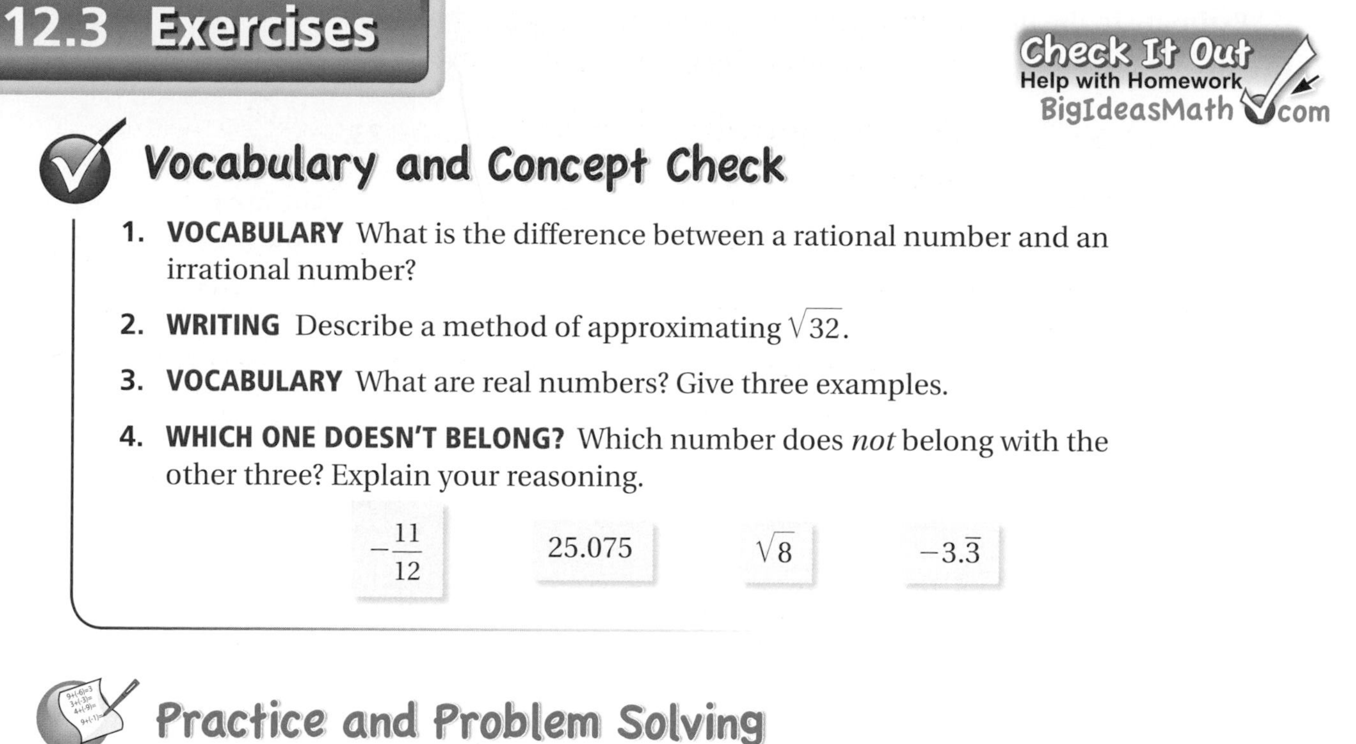

Vocabulary and Concept Check

1. **VOCABULARY** What is the difference between a rational number and an irrational number?
2. **WRITING** Describe a method of approximating $\sqrt{32}$.
3. **VOCABULARY** What are real numbers? Give three examples.
4. **WHICH ONE DOESN'T BELONG?** Which number does *not* belong with the other three? Explain your reasoning.

$-\dfrac{11}{12}$ $\quad$ 25.075 $\quad$ $\sqrt{8}$ $\quad$ $-3.\overline{3}$

Practice and Problem Solving

Tell whether the rational number is a reasonable approximation of the square root.

5. $\dfrac{559}{250}, \sqrt{5}$
6. $\dfrac{3021}{250}, \sqrt{11}$
7. $\dfrac{678}{250}, \sqrt{28}$
8. $\dfrac{1677}{250}, \sqrt{45}$

Tell whether the number is *rational* or *irrational*. Explain.

1 9. $3.66666\overline{6}$
10. $\dfrac{\pi}{6}$
11. $-\sqrt{7}$
12. -1.125
13. $-3\dfrac{8}{9}$
14. $\sqrt{15}$

15. **ERROR ANALYSIS** Describe and correct the error in classifying the number.

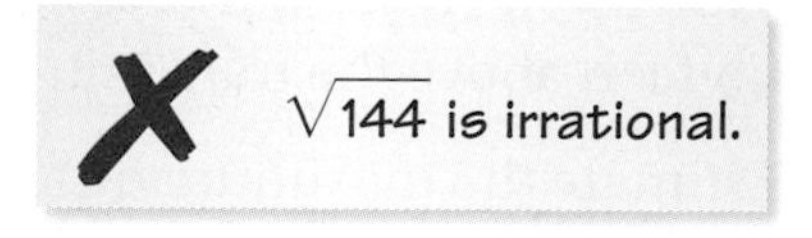

16. **SCRAPBOOKING** You cut a picture into a right triangle for your scrapbook. The lengths of the legs of the triangle are 4 inches and 6 inches. Is the length of the hypotenuse a rational number? Explain.

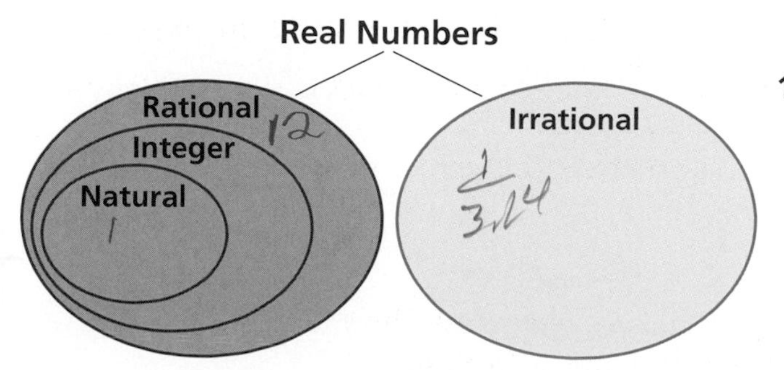

17. **VENN DIAGRAM** Place each number in the correct area of the Venn Diagram.
    a. Your age
    b. The square root of any prime number
    c. The ratio of the circumference of a circle to its diameter

Estimate to the nearest integer.

18. $\sqrt{24}$
19. $\sqrt{685}$
20. $-\sqrt{61}$
21. $-\sqrt{105}$
22. $\sqrt{\frac{27}{4}}$
23. $-\sqrt{\frac{335}{2}}$
24. **CHECKERS** A checkerboard is 8 squares long and 8 squares wide. The area of each square is 14 square centimeters. Estimate the perimeter of the checkerboard.

Which number is greater? Explain.

25. $\sqrt{20}$, 10
26. $\sqrt{15}$, -3.5
27. $\sqrt{133}$, $10\frac{3}{4}$
28. $\frac{2}{3}$, $\sqrt{\frac{16}{81}}$
29. $-\sqrt{0.25}$, -0.25
30. $-\sqrt{182}$, $-\sqrt{192}$

31. **FOUR SQUARE** The area of a four square court is 66 square feet. Estimate the length s of one of the sides of the court.

32. **RADIO SIGNAL** The maximum distance (in nautical miles) that a radio transmitter signal can be sent is represented by the expression $1.23\sqrt{h}$, where h is the height (in feet) above the transmitter.

Estimate the maximum distance x (in nautical miles) between the plane that is receiving the signal and the transmitter. Round your answer to the nearest tenth.

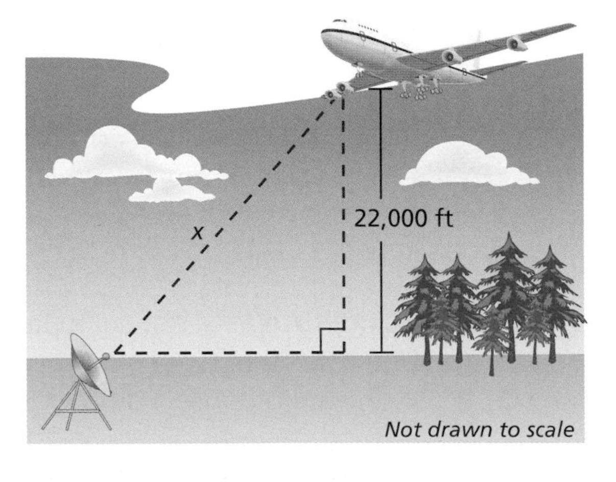

33. **OPEN-ENDED** Find two numbers a and b that satisfy the diagram.

9 $\sqrt{a}$ $\sqrt{b}$ 10

Estimate to the nearest tenth.

34. $\sqrt{0.39}$ **35.** $\sqrt{1.19}$ **36.** $\sqrt{1.52}$

37. ROLLER COASTER The velocity v (in meters per second) of a roller coaster is represented by the equation $v = 3\sqrt{6r}$, where r is the radius of the loop. Estimate the velocity of a car going around the loop. Round your answer to the nearest tenth.

38. Is $\sqrt{\frac{1}{4}}$ a rational number? Is $\sqrt{\frac{3}{16}}$ a rational number? Explain.

39. WATER BALLOON The time t (in seconds) it takes a water balloon to fall d meters is represented by the equation $t = \sqrt{\frac{d}{4.9}}$. Estimate the time it takes the balloon to fall to the ground from a window that is 14 meters above the ground. Round your answer to the nearest tenth.

40. Number Sense Determine if the statement is *sometimes, always,* or *never* true. Explain your reasoning and give an example of each.

a. A rational number multiplied by a rational number is rational.

b. A rational number multiplied by an irrational number is rational.

c. An irrational number multiplied by an irrational number is rational.

Fair Game Review What you learned in previous grades & lessons

Simplify the expression. *(Skills Review Handbook)*

41. $2x + 3y - 5x$ **42.** $3\pi + 8(t - \pi) - 4t$ **43.** $17k - 9 + 23k$

44. MULTIPLE CHOICE What is the ratio (red to blue) of the corresponding side lengths of the similar triangles? *(Section 5.1)*

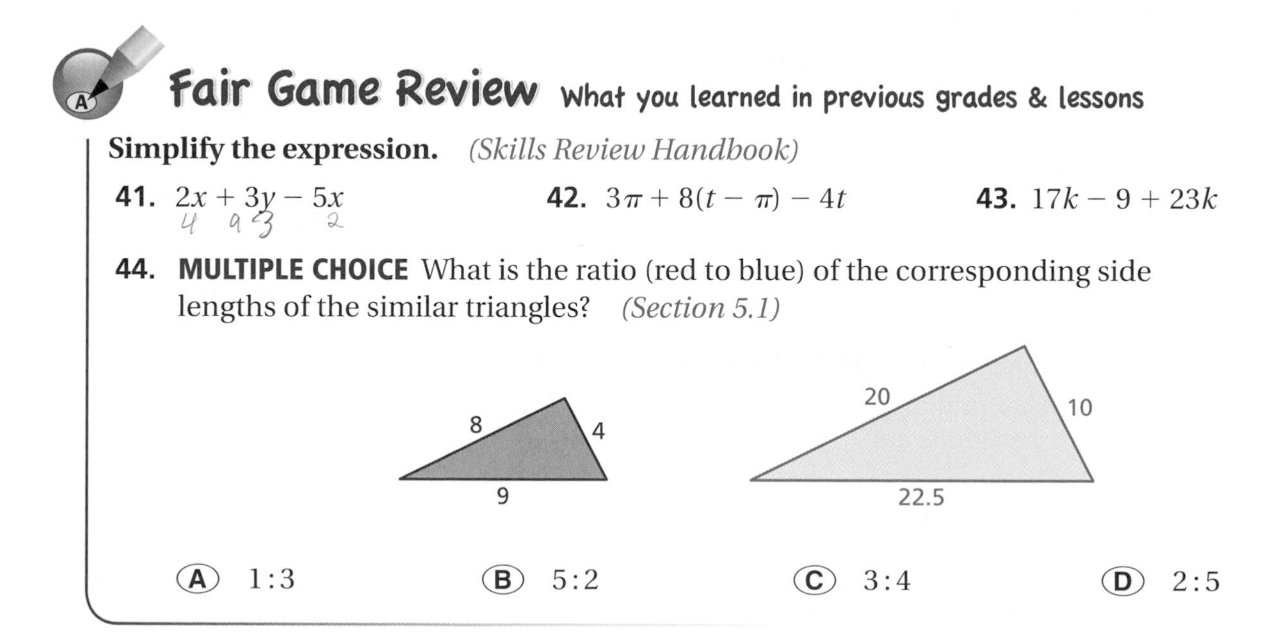

Ⓐ 1 : 3 Ⓑ 5 : 2 Ⓒ 3 : 4 Ⓓ 2 : 5

12.4 Simplifying Square Roots

COMMON CORE STATE STANDARDS
8.NS.2

Essential Question How can you use a square root to describe the golden ratio?

Two quantities are in the *golden ratio* if the ratio between the sum of the quantities and the greater quantity is the same as the ratio between the greater quantity and the lesser quantity.

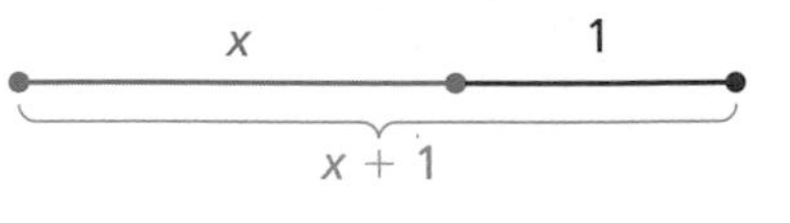

$$\frac{x+1}{x} = \frac{x}{1}$$

In a future algebra course, you will be able to prove that the golden ratio is

$$\frac{1+\sqrt{5}}{2}$$ Golden ratio.

1 ACTIVITY: Constructing a Golden Ratio

Work with a partner.

a. Use grid paper and the given scale to draw a square that is 1 unit by 1 unit (blue).

b. Draw a line from midpoint C of one side of the square to the opposite corner D, as shown.

c. Use the Pythagorean Theorem to find the length of segment CD.

d. Set the point of a compass on C. Set the compass radius to the length of segment CD. Swing the compass to intersect line BC at point E.

e. The rectangle $ABEF$ is called a *golden rectangle* because the ratio of its side lengths is the golden ratio.

f. Use a calculator to find a decimal approximation of the golden ratio. Round your answer to two decimal places.

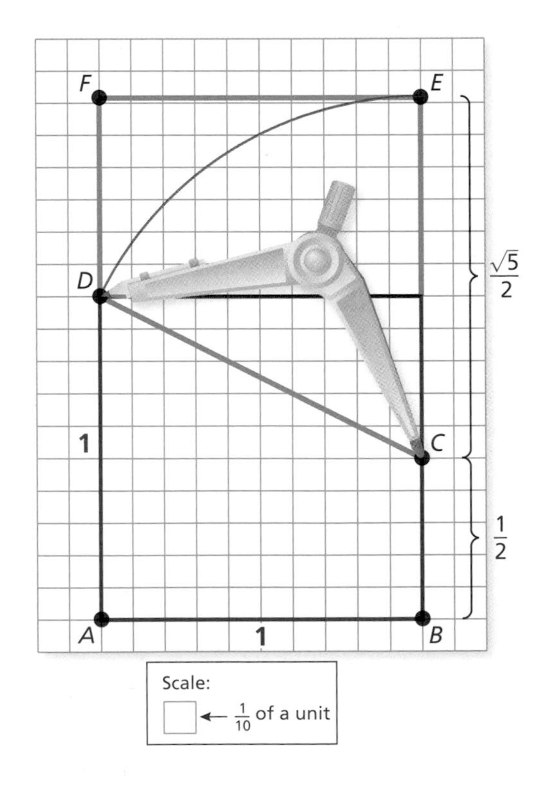

2 ACTIVITY: The Golden Ratio and the Human Body

Work with a partner.

Leonardo da Vinci was one of the first to notice that there are several ratios in the human body that approximate the golden ratio.

a. Use a tape measure or two yardsticks to measure the lengths shown in the diagram for both you and your partner. (Take your shoes off before measuring.)

b. Copy the tables below. Record your results in the first two columns.

c. Calculate the ratios shown in the tables.

d. Leonardo da Vinci stated that for many people, the ratios are close to the golden ratio. How close are your ratios?

f
e
g
h
b
c
Navel
a
d

You		
$a =$	$b =$	$\frac{a}{b} =$
$c =$	$d =$	$\frac{c}{d} =$
$e =$	$f =$	$\frac{e}{f} =$
$g =$	$h =$	$\frac{g}{h} =$

Partner		
$a =$	$b =$	$\frac{a}{b} =$
$c =$	$d =$	$\frac{c}{d} =$
$e =$	$f =$	$\frac{e}{f} =$
$g =$	$h =$	$\frac{g}{h} =$

What Is Your Answer?

3. **IN YOUR OWN WORDS** How can you use a square root to describe the golden ratio? Use the Internet or some other reference to find examples of the golden ratio in art and architecture.

Practice Use what you learned about square roots to complete Exercises 3–5 on page 536.

12.4 Lesson

You can add or subtract radical expressions the same way you combine like terms, such as $5x + 4x = 9x$.

EXAMPLE 1 Adding and Subtracting Square Roots

Reading

Do not assume that radicals that have different radicands cannot be simplified.

An expression such as $2\sqrt{4} + \sqrt{1}$ can easily be simplified.

a. Simplify $5\sqrt{2} + 4\sqrt{2}$.

$5\sqrt{2} + 4\sqrt{2} = (5 + 4)\sqrt{2}$ Use the Distributive Property.

$= 9\sqrt{2}$ Simplify.

b. Simplify $2\sqrt{3} - 7\sqrt{3}$.

$2\sqrt{3} - 7\sqrt{3} = (2 - 7)\sqrt{3}$ Use the Distributive Property.

$= -5\sqrt{3}$ Simplify.

On Your Own

Now You're Ready
Exercises 6–14

Simplify the expression.

1. $\sqrt{5} + \sqrt{5}$
2. $6\sqrt{10} + 4\sqrt{10}$
3. $2\sqrt{7} - \sqrt{7}$

To simplify square roots that are not perfect squares, use the following property.

Key Idea

Product Property of Square Roots

Algebra $\sqrt{xy} = \sqrt{x} \cdot \sqrt{y}$, where $x, y \geq 0$

Numbers $\sqrt{4 \cdot 3} = \sqrt{4} \cdot \sqrt{3} = 2\sqrt{3}$

EXAMPLE 2 Simplifying Square Roots

Study Tip

A square root is simplified when the radicand has no perfect square factors other than 1.

Simplify $\sqrt{50}$.

$\sqrt{50} = \sqrt{25 \cdot 2}$ Factor using the greatest perfect square factor.

$= \sqrt{25} \cdot \sqrt{2}$ Use the Product Property of Square Roots.

$= 5\sqrt{2}$ Simplify.

On Your Own

Now You're Ready
Exercises 16–20

Simplify the expression.

4. $\sqrt{24}$
5. $\sqrt{45}$
6. $\sqrt{98}$

Key Idea

Quotient Property of Square Roots

Algebra $\sqrt{\frac{x}{y}} = \frac{\sqrt{x}}{\sqrt{y}}$, where $x \geq 0$ and $y > 0$

Numbers $\sqrt{\frac{7}{9}} = \frac{\sqrt{7}}{\sqrt{9}} = \frac{\sqrt{7}}{3}$

EXAMPLE 3 Simplifying Square Roots

Simplify $\sqrt{\frac{11}{16}}$.

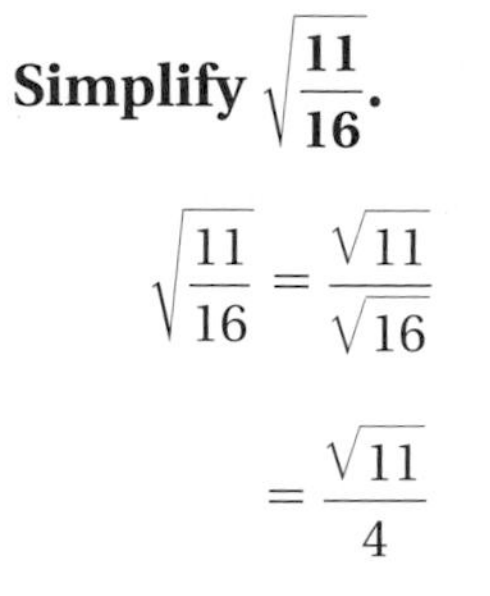

$$\sqrt{\frac{11}{16}} = \frac{\sqrt{11}}{\sqrt{16}}$$ Use the Quotient Property of Square Roots.

$$= \frac{\sqrt{11}}{4}$$ Simplify.

EXAMPLE 4 Finding a Volume

Remember

The volume V of a rectangular prism is the product of the area of its base B and its height h.

$V = Bh$

Find the volume of the rectangular prism.

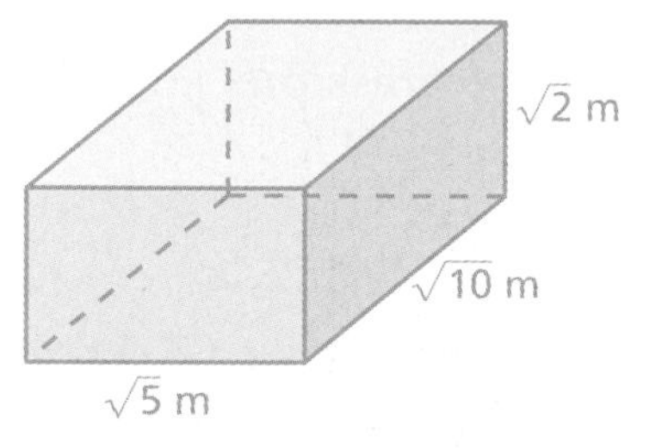

$V = Bh$ Write formula for volume.

$= (\sqrt{5})(\sqrt{10})(\sqrt{2})$ Substitute.

$= \sqrt{5 \cdot 10 \cdot 2}$ Use the Product Property of Square Roots.

$= \sqrt{100}$ Multiply.

$= 10$ Simplify.

∴ The volume is 10 cubic meters.

On Your Own

Simplify the expression.

7. $\sqrt{\frac{35}{36}}$
8. $\sqrt{\frac{13}{4}}$
9. $\sqrt{\frac{5}{b^2}}$

10. **WHAT IF?** In Example 4, the height of the rectangular prism is $\sqrt{8}$ meters. Find the volume of the prism.

12.4 Exercises

Vocabulary and Concept Check

1. **WRITING** Describe how combining like terms is similar to adding and subtracting square roots.
2. **WRITING** How are the Product Property of Square Roots and the Quotient Property of Square Roots similar?

Practice and Problem Solving

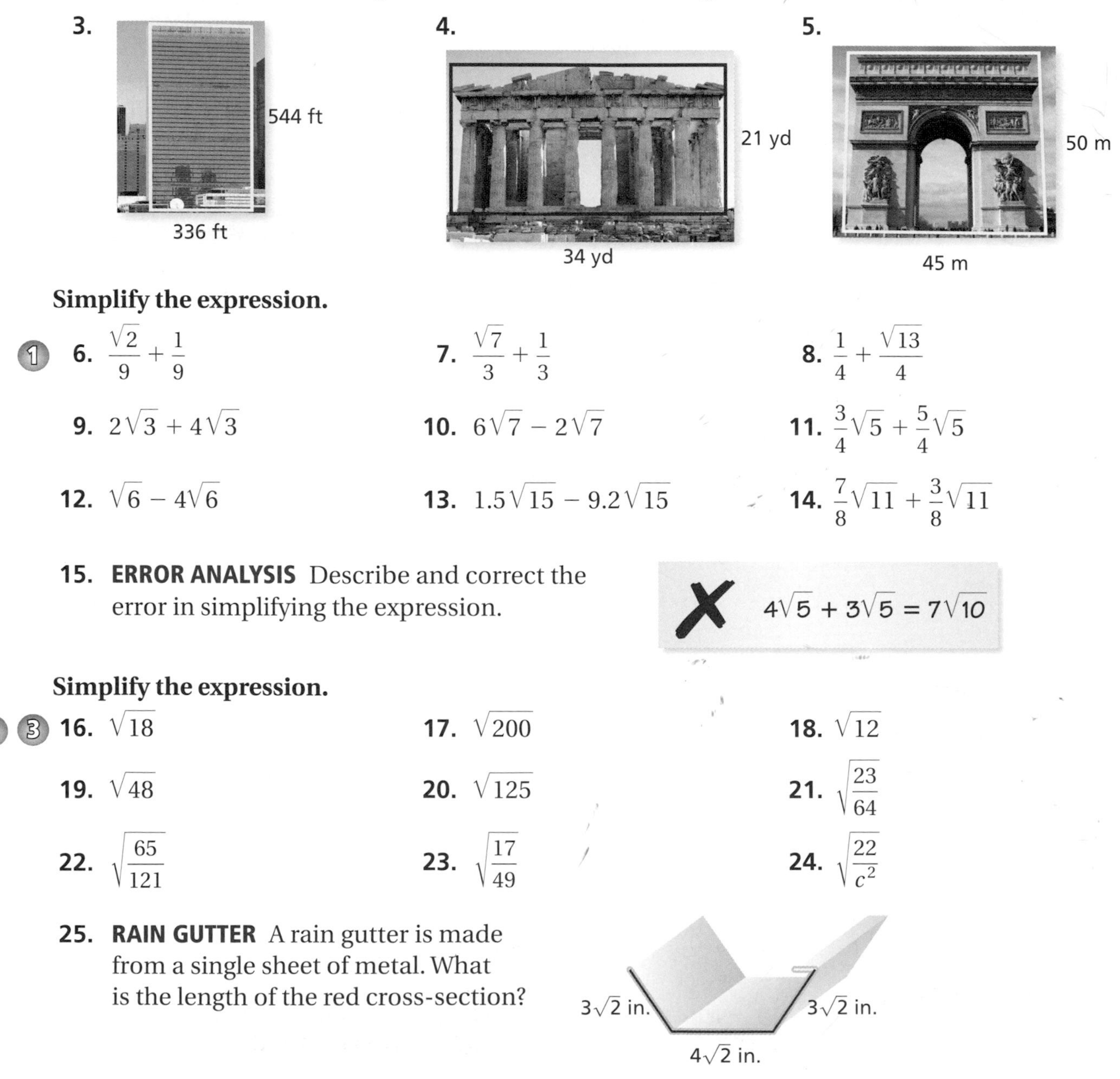

Find the ratio of the side lengths. Is the ratio close to the golden ratio?

3. 544 ft; 336 ft

4. 21 yd; 34 yd

5. 50 m; 45 m

Simplify the expression.

(1) 6. $\frac{\sqrt{2}}{9} + \frac{1}{9}$

7. $\frac{\sqrt{7}}{3} + \frac{1}{3}$

8. $\frac{1}{4} + \frac{\sqrt{13}}{4}$

9. $2\sqrt{3} + 4\sqrt{3}$

10. $6\sqrt{7} - 2\sqrt{7}$

11. $\frac{3}{4}\sqrt{5} + \frac{5}{4}\sqrt{5}$

12. $\sqrt{6} - 4\sqrt{6}$

13. $1.5\sqrt{15} - 9.2\sqrt{15}$

14. $\frac{7}{8}\sqrt{11} + \frac{3}{8}\sqrt{11}$

15. **ERROR ANALYSIS** Describe and correct the error in simplifying the expression.

✗ $4\sqrt{5} + 3\sqrt{5} = 7\sqrt{10}$

Simplify the expression.

(2) (3) 16. $\sqrt{18}$

17. $\sqrt{200}$

18. $\sqrt{12}$

19. $\sqrt{48}$

20. $\sqrt{125}$

21. $\sqrt{\frac{23}{64}}$

22. $\sqrt{\frac{65}{121}}$

23. $\sqrt{\frac{17}{49}}$

24. $\sqrt{\frac{22}{c^2}}$

25. **RAIN GUTTER** A rain gutter is made from a single sheet of metal. What is the length of the red cross-section?

$3\sqrt{2}$ in.; $3\sqrt{2}$ in.; $4\sqrt{2}$ in.

Simplify the expression.

26. $3\sqrt{5} - \sqrt{45}$

27. $\sqrt{24} + 4\sqrt{6}$

28. $\frac{4}{3}\sqrt{7} + \sqrt{28}$

29. VOLUME What is the volume of the aquarium (in cubic feet)?

$\sqrt{42}$ ft

$\sqrt{30}$ ft

$\sqrt{35}$ ft

30. RATIO The ratio $3 : x$ is equivalent to the ratio $x : 5$. What are the possible values of x?

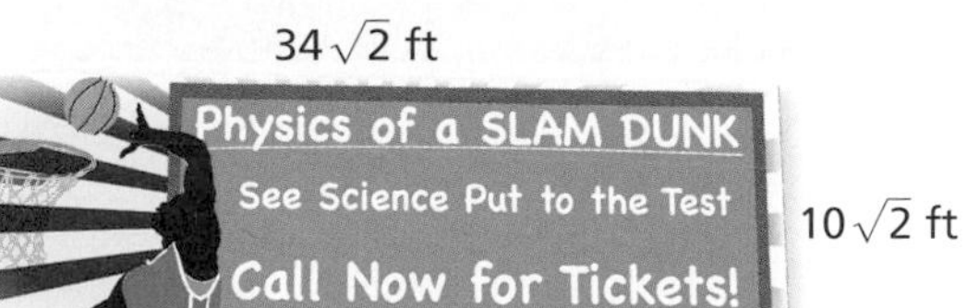

31. BILLBOARD The billboard has the shape of a rectangle.

a. What is the perimeter of the billboard?

b. What is the area of the billboard?

32. MT. FUJI Mt. Fuji is in the shape of a cone with a volume of about 475π cubic kilometers. What is the radius of the base of Mt. Fuji?

The height of Mt. Fuji is 3.8 kilometers.

33. Geometry A block of ice is in the shape of a square prism. You want to put the block of ice in a cylindrical cooler. The equation $s^2 = 2r^2$ represents the minimum radius r needed for the block of ice with side length s to fit in the cooler.

a. Solve the equation for r.

b. Use the equation in part (a) to find the minimum radius needed when the side length of the block of ice is $\sqrt{98}$ inches.

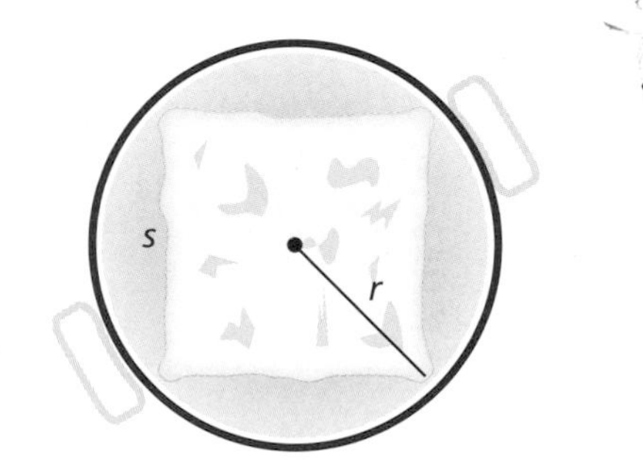

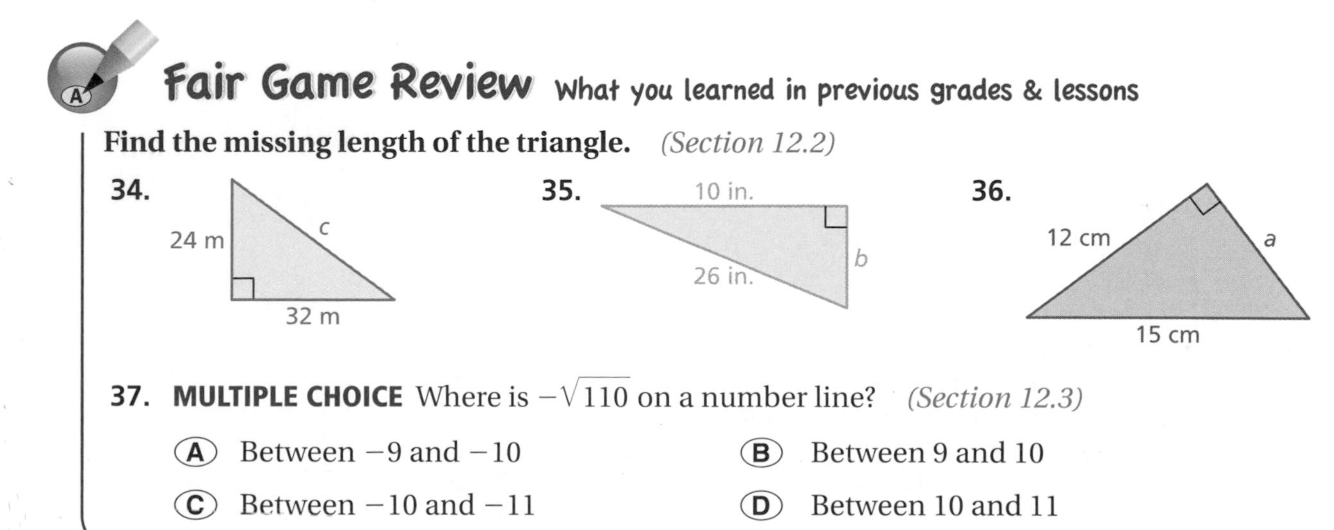

Fair Game Review What you learned in previous grades & lessons

Find the missing length of the triangle. *(Section 12.2)*

34.

35.

36.

37. MULTIPLE CHOICE Where is $-\sqrt{110}$ on a number line? *(Section 12.3)*

Ⓐ Between -9 and -10

Ⓑ Between 9 and 10

Ⓒ Between -10 and -11

Ⓓ Between 10 and 11

12.4b Cube Roots

A **cube root** of a number is a number that when multiplied by itself, and then multiplied by itself again, equals the given number. A **perfect cube** is a number that can be written as the cube of an integer. The symbol $\sqrt[3]{}$ is used to represent a cube root.

EXAMPLE 1 Finding Cube Roots

The symbol for cubing is the third power. So, $2^3 = 2 \cdot 2 \cdot 2$.

Find the cube root.

a. $\sqrt[3]{8}$

$2 \cdot 2 \cdot 2 = 8$

∴ Because $2^3 = 8$, $\sqrt[3]{8} = \sqrt[3]{2^3} = 2$.

b. $\sqrt[3]{-27}$

$-3 \cdot (-3) \cdot (-3) = -27$

∴ Because $(-3)^3 = -27$, $\sqrt[3]{-27} = \sqrt[3]{(-3)^3} = -3$.

EXAMPLE 2 Simplifying Expressions Involving Cube Roots

Simplify the expression.

a. $9\sqrt[3]{2} - 5\sqrt[3]{2}$

$9\sqrt[3]{2} - 5\sqrt[3]{2} = (9 - 5)\sqrt[3]{2}$ — Use the Distributive Property.

$= 4\sqrt[3]{2}$ — Simplify.

b. $8\sqrt[3]{1000} + 3$

$8\sqrt[3]{1000} + 3 = 8(10) + 3$ — Evaluate the cube root.

$= 80 + 3$ — Multiply.

$= 83$ — Add.

Practice

Find the cube root.

1. $\sqrt[3]{-1}$ **2.** $\sqrt[3]{64}$ **3.** $\sqrt[3]{-216}$ **4.** $\sqrt[3]{125}$

5. $\sqrt[3]{0}$ **6.** $\sqrt[3]{-343}$ **7.** $\sqrt[3]{729}$ **8.** $\sqrt[3]{0.008}$

Simplify the expression.

9. $8\sqrt[3]{5} + 3\sqrt[3]{5}$ **10.** $2\sqrt[3]{3} - 12\sqrt[3]{3}$ **11.** $5\sqrt[3]{10} + 9\sqrt[3]{10}$

12. $5\sqrt[3]{27} + 1$ **13.** $7\sqrt[3]{-512} + 6$ **14.** $5 + 8\sqrt[3]{-\frac{27}{64}}$

EXAMPLE 3 Comparing Real Numbers

Which is greater, $\sqrt[3]{50}$ or 3.1?

Graph the numbers on a number line. To graph $\sqrt[3]{50}$, use the cube roots of the perfect cubes nearest to the radicand. The nearest perfect cube less than 50 is 27. The nearest perfect cube greater than 50 is 64.

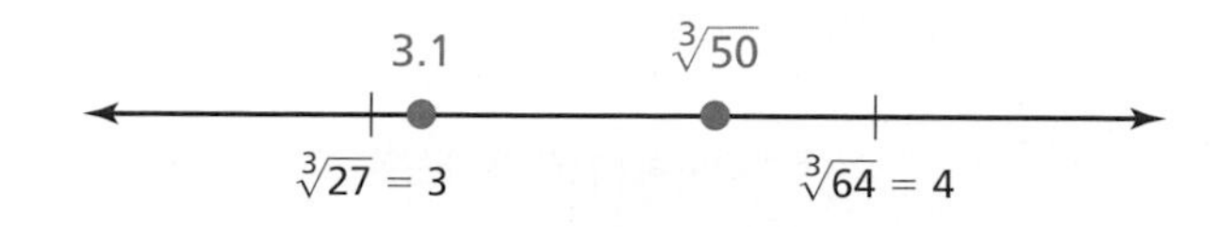

∴ $\sqrt[3]{50}$ is to the right of 3.1. So, $\sqrt[3]{50}$ is greater.

Cubing a number and finding a cube root are inverse operations. Use this relationship to solve equations involving cubes.

EXAMPLE 4 Real-Life Application

Find the surface area of the baseball display case.

Use the formula for the volume of a cube to find the side length s.

> **Remember**
>
> The volume V of a cube with side length s is given by $V = s^3$. The surface area S is given by $S = 6s^2$.

$V = s^3$	Write formula for volume.
$125 = s^3$	Substitute 125 for V.
$\sqrt[3]{125} = \sqrt[3]{s^3}$	Take the cube root of each side.
$5 = s$	Simplify.

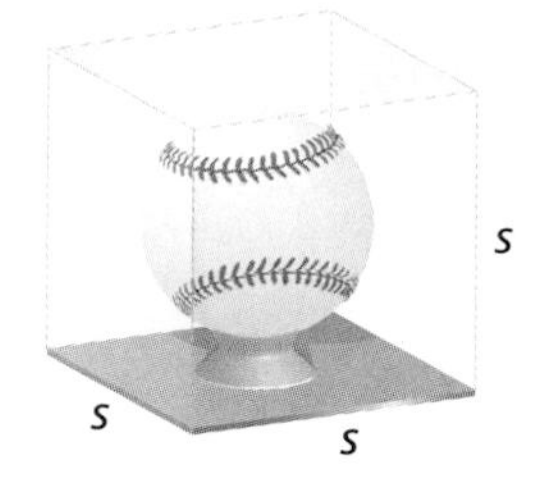

Volume = 125 in.^{3}

The side length is 5 inches. Use a formula to find the surface area of the cube.

$S = 6s^2$	Write formula for surface area.
$= 6(5)^2$	Substitute 5 for s.
$= 6(25)$	Evaluate 5^2.
$= 150$	Simplify.

∴ The surface area of the baseball display is 150 square inches.

Practice

Which number is greater? Explain.

15. $1.9, \sqrt[3]{5}$ **16.** $2.5, \sqrt[3]{21}$ **17.** $\sqrt[3]{100}, 4.25$

18. $\sqrt[3]{35}, 4$ **19.** $-2, \sqrt[3]{-7}$ **20.** $-3.5, \sqrt[3]{-60}$

21. The volume of a cube is 512 cubic centimeters. Find the surface area of the cube.

12.5 Using the Pythagorean Theorem

Essential Question How can you use the Pythagorean Theorem to solve real-life problems?

COMMON CORE STATE STANDARDS
8.G.6
8.G.7
8.G.8

1 ACTIVITY: Using the Pythagorean Theorem

Work with a partner.

a. A baseball player throws a ball from second base to home plate. How far does the player throw the ball? Include a diagram showing how you got your answer. Decide how many decimal points of accuracy are reasonable. Explain your reasoning.

b. The distance from the pitcher's mound to home plate is 60.5 feet. Does this form a right triangle with first base? Explain your reasoning.

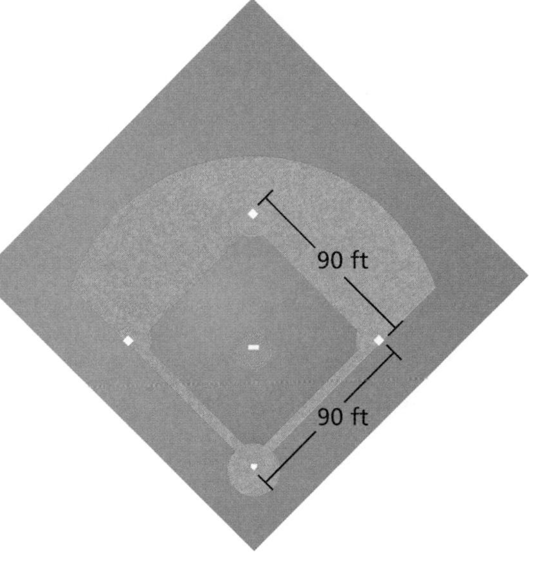

2 ACTIVITY: Firefighting and Ladders

Work with a partner.

The recommended angle for a firefighting ladder is 75°.

When a 110-foot ladder is put up against a building at this angle, the base of the ladder is about 28 feet from the building.

The base of the ladder is 8 feet above the ground.

How high on the building will the ladder reach? Round your answer to the nearest tenth.

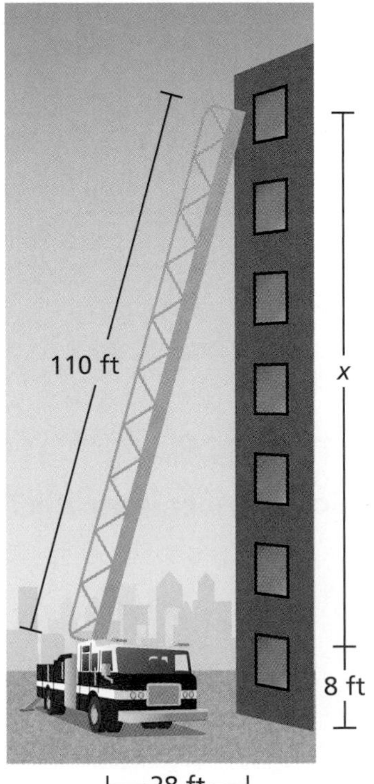

3 ACTIVITY: Finding Perimeters

Work with a partner.

Find the perimeter of each figure. Round your answer to the nearest tenth. Did you use the Pythagorean Theorem? If so, explain.

a. Right triangle **b.** Trapezoid **c.** Parallelogram

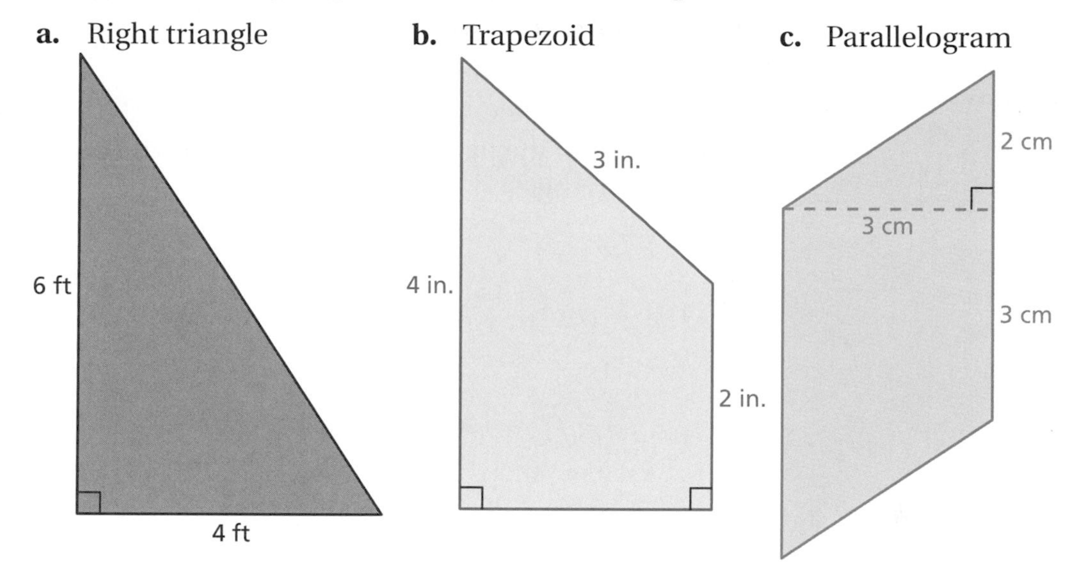

4 ACTIVITY: Writing a Formula

Work with a partner.

a. Write a formula for the area of an equilateral triangle with side length *s*.

b. Use your formula to find the area of an equilateral triangle with a side length of 10 inches.

What Is Your Answer?

5. **IN YOUR OWN WORDS** How can you use the Pythagorean Theorem to solve real-life problems?
6. Describe a situation in which you could use the Pythagorean Theorem to help make decisions. Give an example of a real-life problem.

Use what you learned about using the Pythagorean Theorem to complete Exercises 3–5 on page 542.

12.5 Lesson

EXAMPLE 1 Finding a Distance in a Coordinate Plane

Key Vocabulary
Pythagorean triple, *p. 541*

The park is 5 miles east of your home. The library is 4 miles north of the park. How far is your home from the library? Round your answer to the nearest tenth.

Plot a point for your home at the origin in a coordinate plane. Then plot points for the locations of the park and the library to form a right triangle.

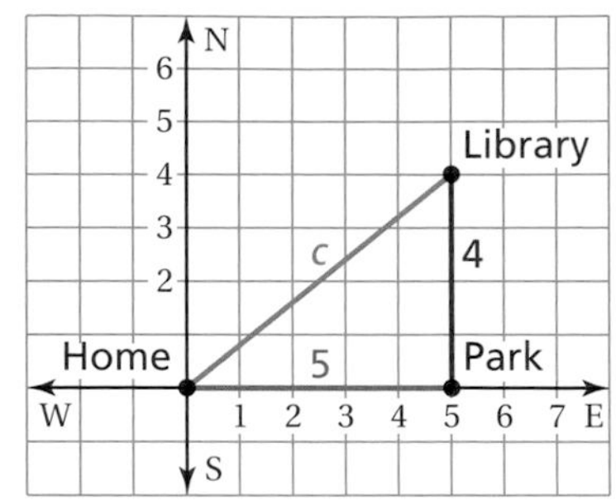

$a^2 + b^2 = c^2$	Write the Pythagorean Theorem.
$4^2 + 5^2 = c^2$	Substitute 4 for *a* and 5 for *b*.
$16 + 25 = c^2$	Evaluate powers.
$41 = c^2$	Add.
$\sqrt{41} = \sqrt{c^2}$	Take positive square root of each side.
$6.4 \approx c$	Use a calculator.

Your home is about 6.4 miles from the library.

On Your Own

1. The post office is 3 miles west of your home. Your school is 2 miles north of the post office. How far is your home from your school? Round your answer to the nearest tenth.

EXAMPLE 2 Real-Life Application

Find the height of the firework. Round your answer to the nearest tenth.

$a^2 + b^2 = c^2$	Write the Pythagorean Theorem.
$x^2 + 300^2 = 335^2$	Substitute.
$x^2 + 90{,}000 = 112{,}225$	Evaluate powers.
$x^2 = 22{,}225$	Subtract 90,000 from each side.
$\sqrt{x^2} = \sqrt{22{,}225}$	Take positive square root of each side.
$x \approx 149.1$	Use a calculator.

The height of the firework is about $149.1 + 1.5 = 150.6$ meters.

On Your Own

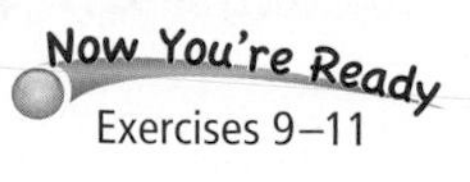

2. **WHAT IF?** In Example 2, the distance between you and the firework is 350 meters. Find the height of the firework. Round your answer to the nearest tenth.

A **Pythagorean triple** is a set of three positive integers a, b, and c where $a^2 + b^2 = c^2$.

Key Idea

Converse of the Pythagorean Theorem

If the equation $a^2 + b^2 = c^2$ is true for the side lengths of a triangle, then the triangle is a right triangle.

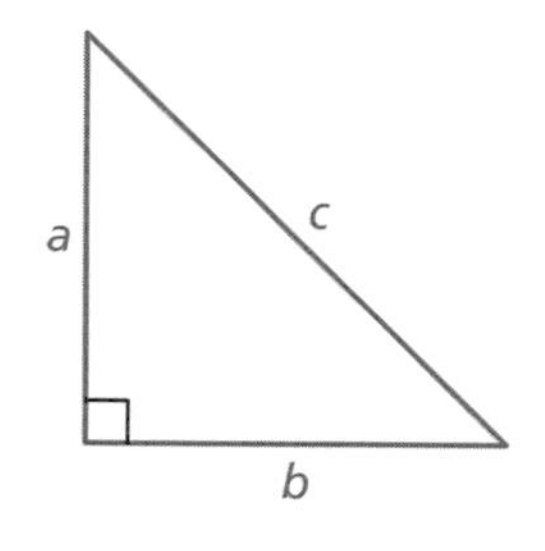

When using the converse of the Pythagorean Theorem, always substitute the length of the longest side for c.

EXAMPLE 3 Identifying a Right Triangle

Tell whether the given triangle is a right triangle.

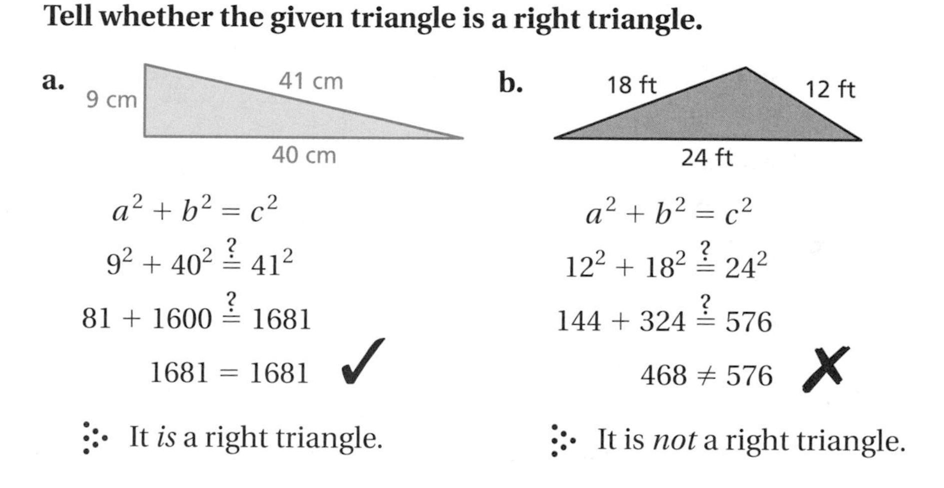

a.

$$a^2 + b^2 = c^2$$
$$9^2 + 40^2 \stackrel{?}{=} 41^2$$
$$81 + 1600 \stackrel{?}{=} 1681$$
$$1681 = 1681 \checkmark$$

∴ It *is* a right triangle.

b.

$$a^2 + b^2 = c^2$$
$$12^2 + 18^2 \stackrel{?}{=} 24^2$$
$$144 + 324 \stackrel{?}{=} 576$$
$$468 \neq 576 \times$$

∴ It is *not* a right triangle.

On Your Own

Tell whether the triangle with the given side lengths is a right triangle.

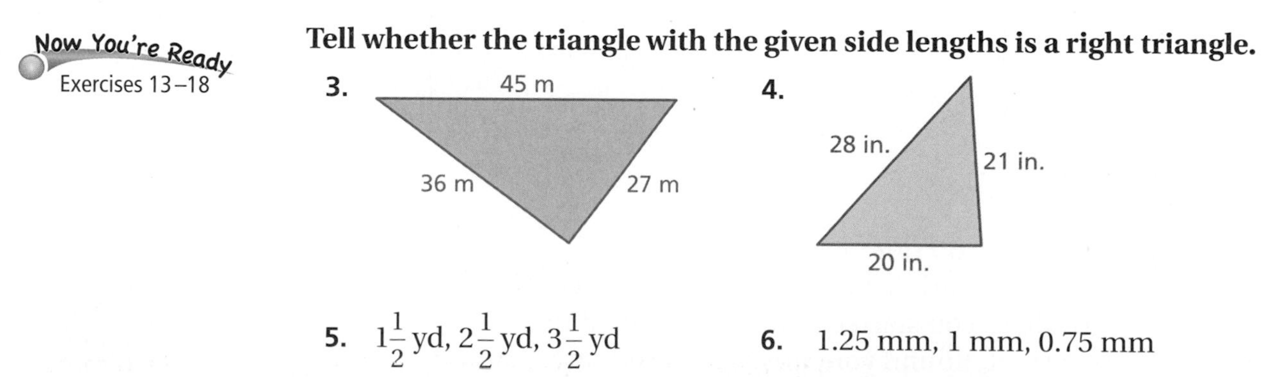

5. $1\frac{1}{2}$ yd, $2\frac{1}{2}$ yd, $3\frac{1}{2}$ yd

6. 1.25 mm, 1 mm, 0.75 mm

12.5 Exercises

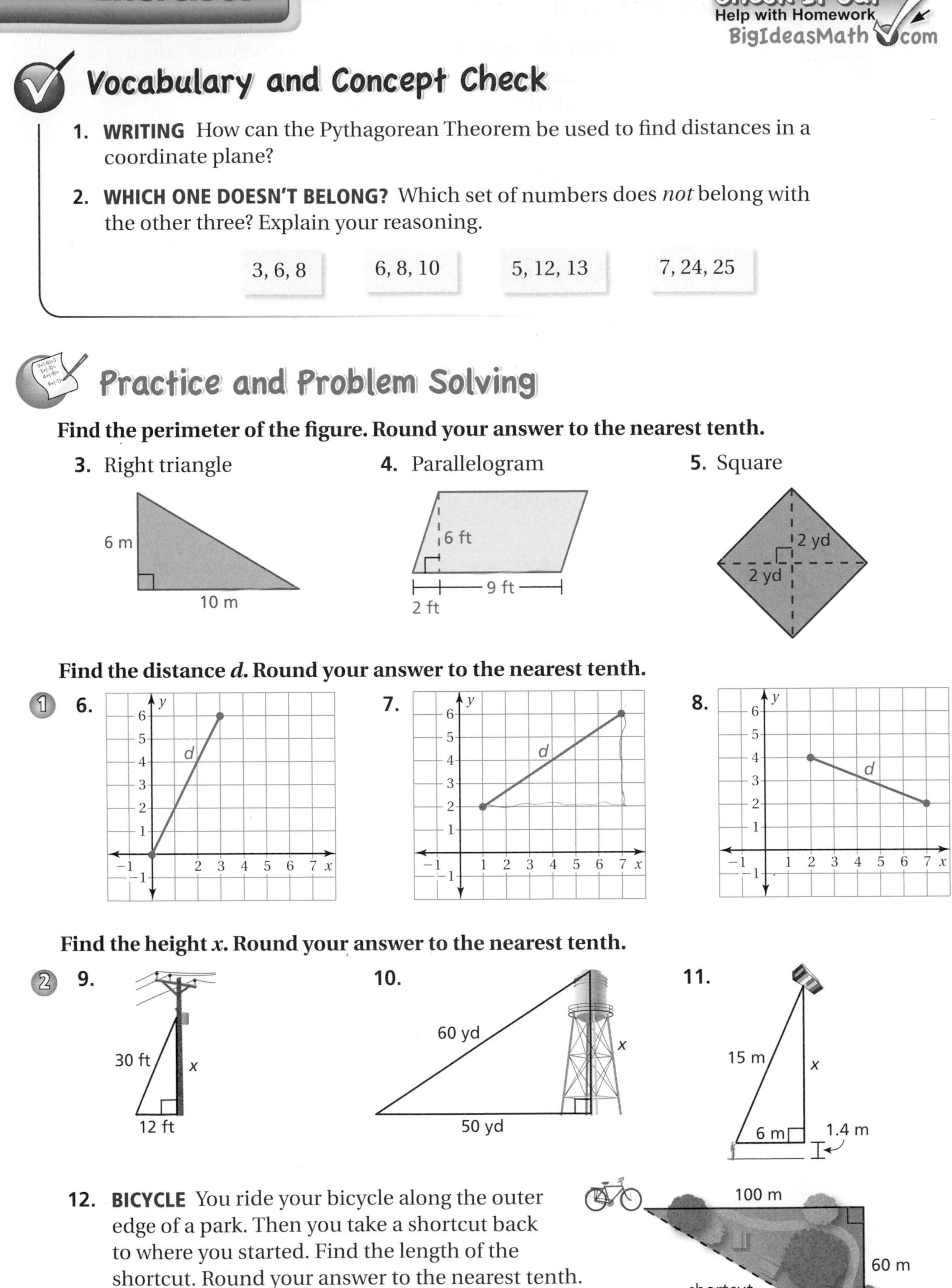

Vocabulary and Concept Check

1. **WRITING** How can the Pythagorean Theorem be used to find distances in a coordinate plane?

2. **WHICH ONE DOESN'T BELONG?** Which set of numbers does *not* belong with the other three? Explain your reasoning.

3, 6, 8	6, 8, 10	5, 12, 13	7, 24, 25

Practice and Problem Solving

Find the perimeter of the figure. Round your answer to the nearest tenth.

3. Right triangle

4. Parallelogram

5. Square

Find the distance d. Round your answer to the nearest tenth.

① 6.

7.

8.

Find the height x. Round your answer to the nearest tenth.

② 9.

10.

11.

12. **BICYCLE** You ride your bicycle along the outer edge of a park. Then you take a shortcut back to where you started. Find the length of the shortcut. Round your answer to the nearest tenth.

Tell whether the triangle with the given side lengths is a right triangle.

3 **13.**

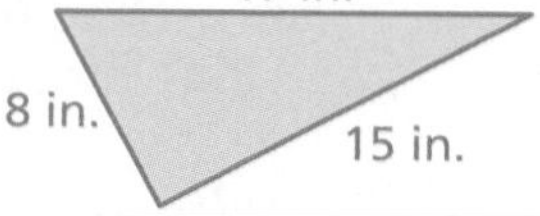

14.

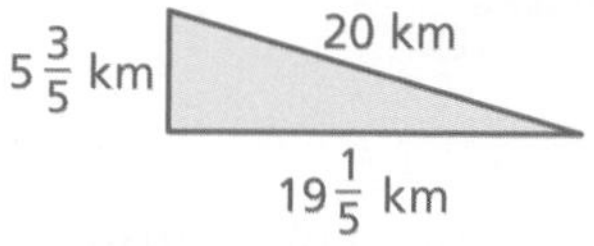

15.

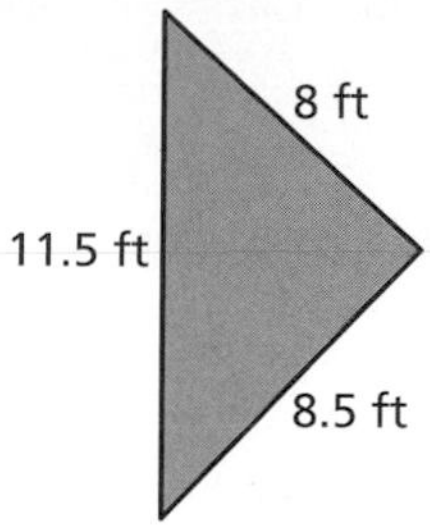

16. 14 mm, 19 mm, 23 mm

17. $\frac{9}{10}$ mi, $1\frac{1}{5}$ mi, $1\frac{1}{2}$ mi

18. 1.4 m, 4.8 m, 5 m

19. **STAIRS** There are 12 steps in the staircase. Find the distance from point *A* to point *B* (in feet). Round your answer to the nearest tenth.

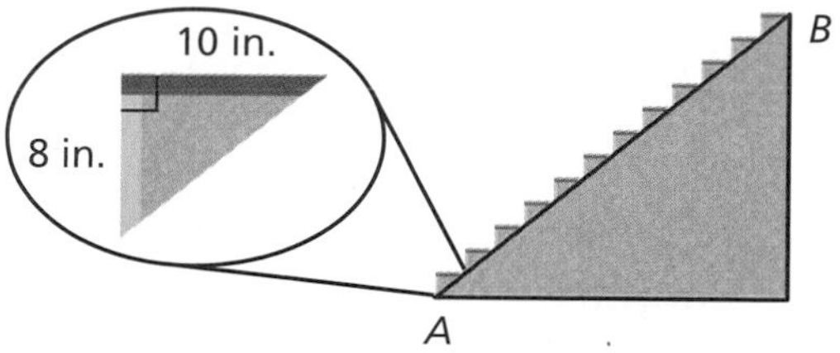

20. **AIRPORT** Which plane is closer to the tower? Explain.

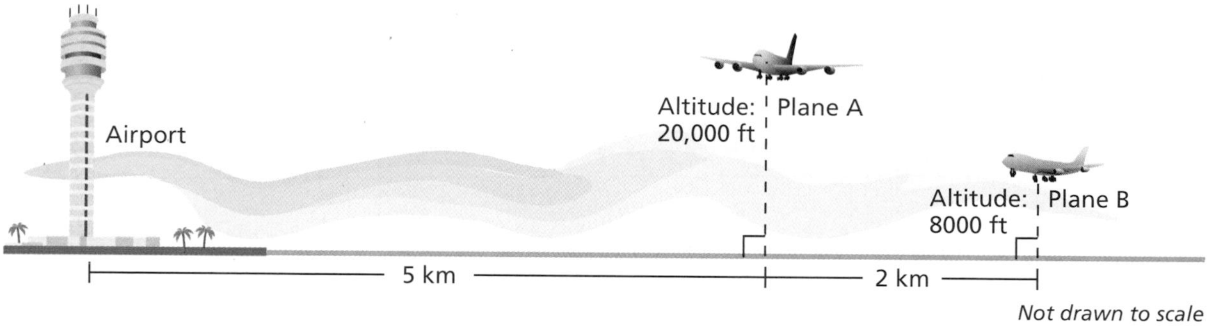

21. **PROJECT** Find a shoebox or some other small box.

a. Measure the dimensions of the box.

b. Without measuring, find length *BC* and length *AB*.

c. Use a piece of string and a ruler to check the lengths you found in part (b).

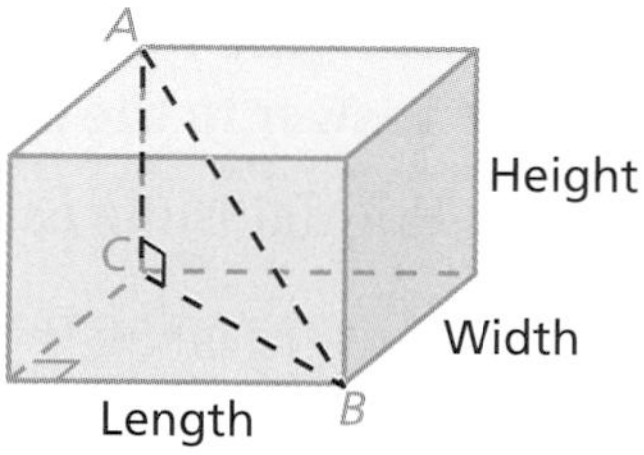

22. Critical Thinking Plot the points (−1, −2), (2, 1), and (−3, 6) in a coordinate plane. Are the points the vertices of a right triangle? Explain.

Fair Game Review What you learned in previous grades & lessons

Find the mean, median, and mode of the data. *(Skills Review Handbook)*

23. 12, 9, 17, 15, 12, 13

24. 21, 32, 16, 27, 22, 19, 10

25. 67, 59, 34, 71, 59

26. **MULTIPLE CHOICE** What is the sum of the angle measures of an octagon? *(Section 11.3)*

Ⓐ 720° Ⓑ 1080° Ⓒ 1440° Ⓓ 1800°

12.3–12.5 Quiz

Check It Out
Progress Check
BigIdeasMath.com

Tell whether the number is *rational* or *irrational*. Explain. *(Section 12.3)*

1. $-\sqrt{225}$
2. $-1\frac{1}{9}$
3. $\sqrt{41}$

Estimate to the nearest integer. *(Section 12.3)*

4. $\sqrt{38}$
5. $-\sqrt{99}$
6. $\sqrt{172}$

Which number is greater? Explain. *(Section 12.3)*

7. $\sqrt{11}, 3\frac{3}{5}$
8. $\sqrt{1.44}, 1.1\overline{8}$

Simplify the expression. *(Section 12.4)*

9. $\sqrt{2} + 2\sqrt{2}$
10. $3\sqrt{15} - 7\sqrt{15}$
11. $\sqrt{\frac{6}{25}}$

Find the volume of the rectangular prism. *(Section 12.4)*

12.
13.

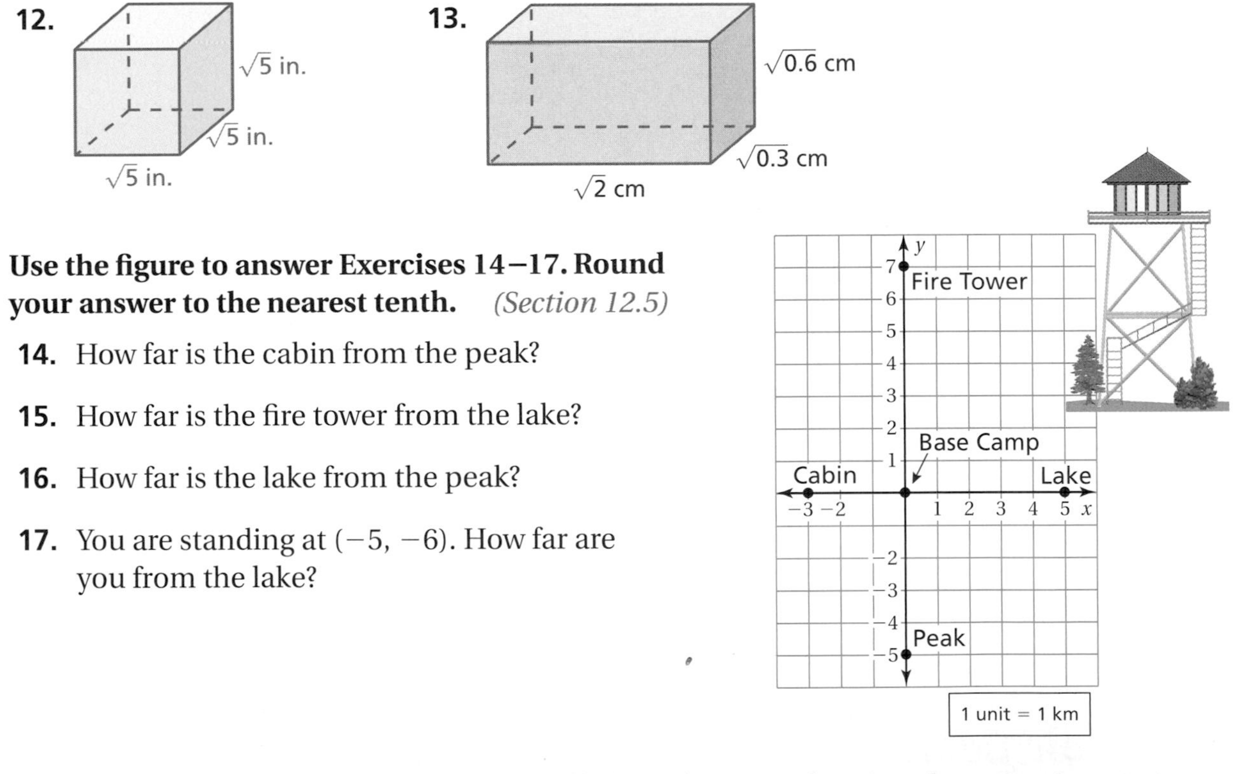

Use the figure to answer Exercises 14–17. Round your answer to the nearest tenth. *(Section 12.5)*

14. How far is the cabin from the peak?
15. How far is the fire tower from the lake?
16. How far is the lake from the peak?
17. You are standing at $(-5, -6)$. How far are you from the lake?

Tell whether the triangle with the given side lengths is a right triangle. *(Section 12.5)*

18.
19.

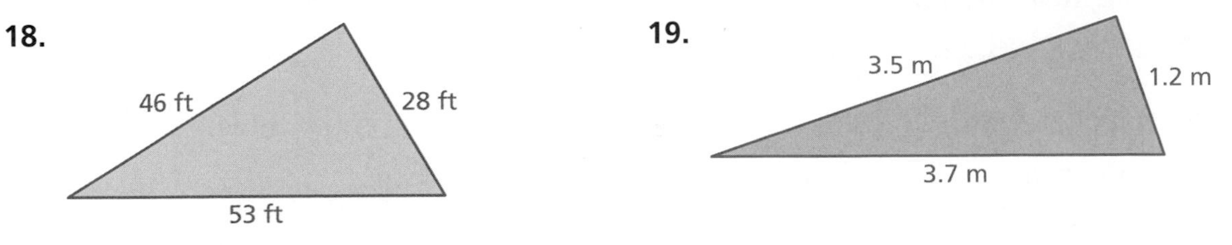

12 Chapter Review

Review Key Vocabulary

square root, *p. 512*
perfect square, *p. 512*
radical sign, *p. 512*
radicand, *p. 512*
theorem, *p. 516*
legs, *p. 518*
hypotenuse, *p. 518*
Pythagorean Theorem, *p. 518*
irrational number, *p. 526*
real numbers, *p. 526*
Pythagorean triple, *p. 541*

Review Examples and Exercises

12.1 Finding Square Roots *(pp. 510–515)*

Find the square root(s).

a. $-\sqrt{36}$

$-\sqrt{36}$ represents the *negative* square root.

Because $6^2 = 36$, $-\sqrt{36} = -\sqrt{6^2} = -6$.

b. $\sqrt{1.96}$

$\sqrt{1.96}$ represents the *positive* square root.

Because $1.4^2 = 1.96$, $\sqrt{1.96} = \sqrt{1.4^2} = 1.4$.

c. $\pm\sqrt{\frac{16}{81}}$

$\pm\sqrt{\frac{16}{81}}$ represents both the *positive and negative* square roots.

Because $\left(\frac{4}{9}\right)^2 = \frac{16}{81}$, $\pm\sqrt{\frac{16}{81}} = \pm\sqrt{\left(\frac{4}{9}\right)^2} = \frac{4}{9}$ and $-\frac{4}{9}$.

Exercises

Find the two square roots of the number.

1. 16 **2.** 900 **3.** 2500

Find the square root(s).

4. $\sqrt{1}$ **5.** $-\sqrt{\frac{9}{25}}$ **6.** $\pm\sqrt{1.96}$

Evaluate the expression.

7. $15 - 4\sqrt{16}$ **8.** $\sqrt{\frac{54}{6}} + \frac{2}{3}$ **9.** $10(\sqrt{81} - 12)$

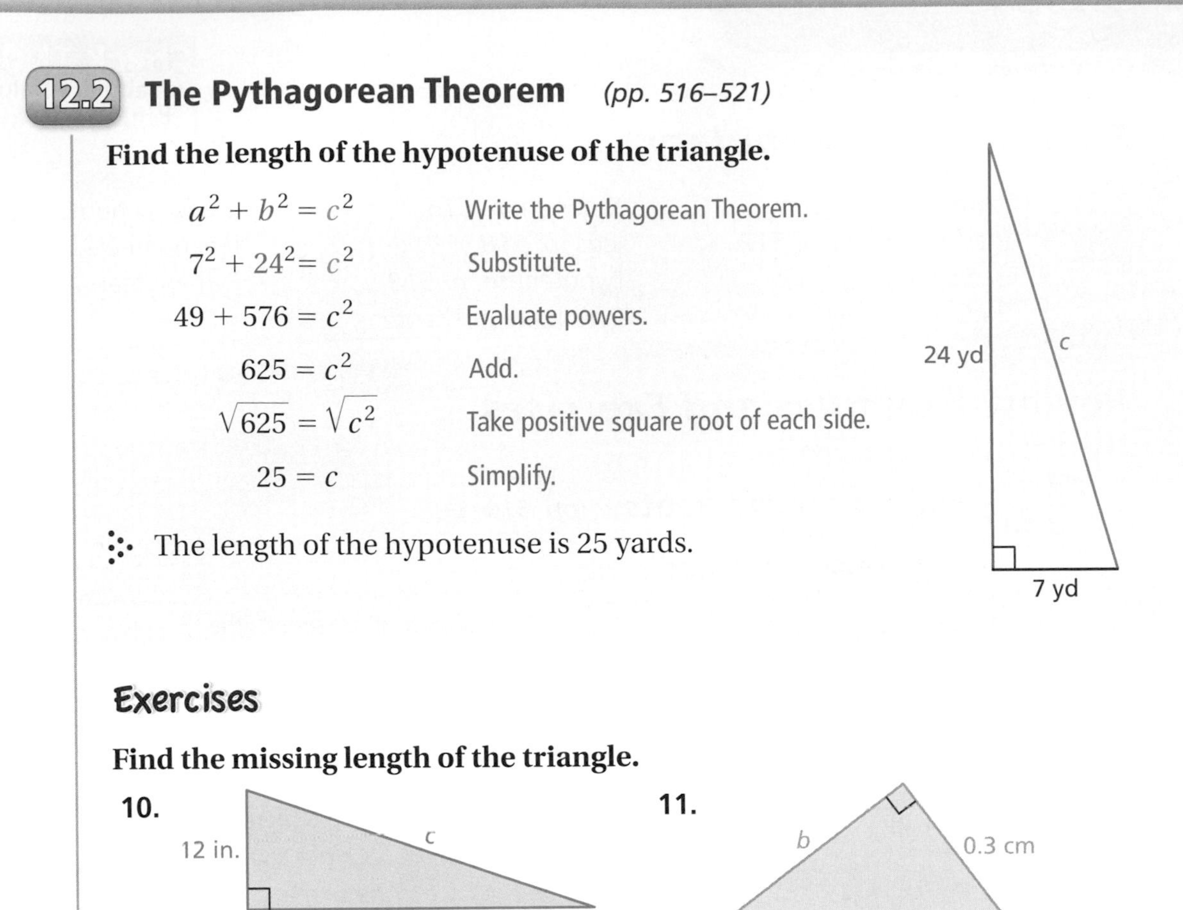

12.2 The Pythagorean Theorem (pp. 516–521)

Find the length of the hypotenuse of the triangle.

$a^2 + b^2 = c^2$	Write the Pythagorean Theorem.
$7^2 + 24^2 = c^2$	Substitute.
$49 + 576 = c^2$	Evaluate powers.
$625 = c^2$	Add.
$\sqrt{625} = \sqrt{c^2}$	Take positive square root of each side.
$25 = c$	Simplify.

∴ The length of the hypotenuse is 25 yards.

Exercises

Find the missing length of the triangle.

10.

11.

12.3 Approximating Square Roots (pp. 524–531)

Estimate $\sqrt{34}$ to the nearest integer.

Use a number line and the square roots of the perfect squares nearest to the radicand. The nearest perfect square less than 34 is 25. The nearest perfect square greater than 34 is 36.

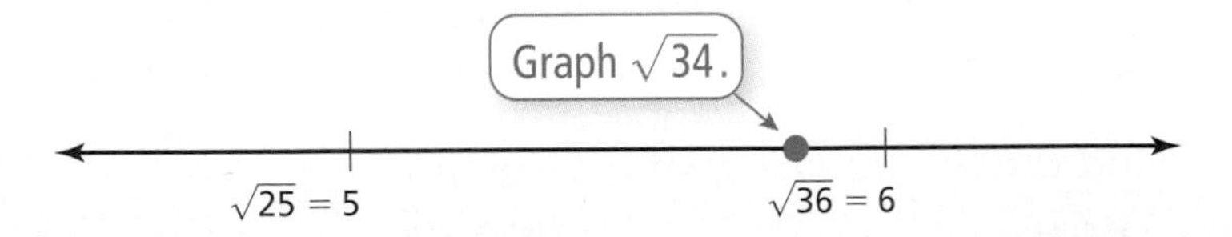

Because 34 is closer to 36 than to 25, $\sqrt{34}$ is closer to 6 than to 5.

∴ So, $\sqrt{34} \approx 6$.

Exercises

Estimate to the nearest integer.

12. $\sqrt{14}$ **13.** $\sqrt{90}$ **14.** $\sqrt{175}$

12.4 Simplifying Square Roots (pp. 532–537)

Simplify $\sqrt{28}$.

$\sqrt{28} = \sqrt{4 \cdot 7}$ Factor using the greatest perfect square factor.

$= \sqrt{4} \cdot \sqrt{7}$ Use the Product Property of Square Roots.

$= 2\sqrt{7}$ Simplify.

Simplify $\sqrt{\dfrac{13}{64}}$.

$\sqrt{\dfrac{13}{64}} = \dfrac{\sqrt{13}}{\sqrt{64}}$ Use the Quotient Property of Square Roots.

$= \dfrac{\sqrt{13}}{8}$ Simplify.

Exercises

Simplify the expression.

15. $\sqrt{\dfrac{99}{100}}$ **16.** $\sqrt{96}$ **17.** $\sqrt{75}$

12.5 Using the Pythagorean Theorem (pp. 538–543)

Find the height of the stilt walker. Round your answer to the nearest tenth.

$a^2 + b^2 = c^2$ Write the Pythagorean Theorem.

$6^2 + x^2 = 13^2$ Substitute.

$36 + x^2 = 169$ Evaluate powers.

$x^2 = 133$ Subtract 36 from each side.

$\sqrt{x^2} = \sqrt{133}$ Take positive square root of each side.

$x \approx 11.5$ Use a calculator.

∴ The height of the stilt walker is about 11.5 feet.

13 ft

x

6 ft

Exercises

Find the height x. Round your answer to the nearest tenth, if necessary.

18.

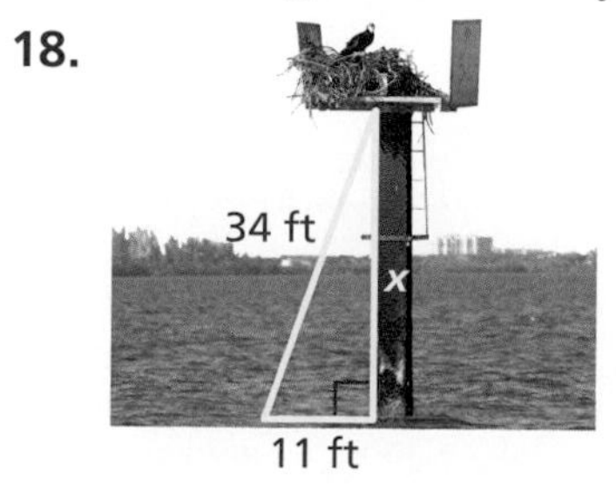

19.

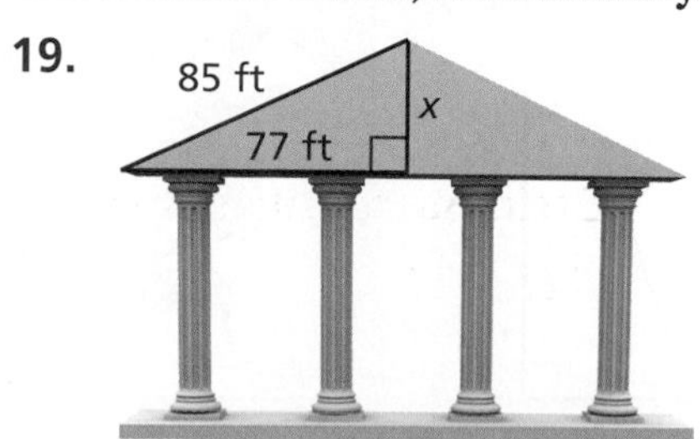

12 Chapter Test

Check It Out
Test Practice
BigIdeasMath.com

Find the square root(s).

1. $-\sqrt{1600}$
2. $\sqrt{\dfrac{25}{49}}$
3. $\pm\sqrt{\dfrac{100}{9}}$

Evaluate the expression.

4. $12 + 8\sqrt{16}$
5. $\dfrac{1}{2} + \sqrt{\dfrac{72}{2}}$

6. Find the missing length of the triangle.

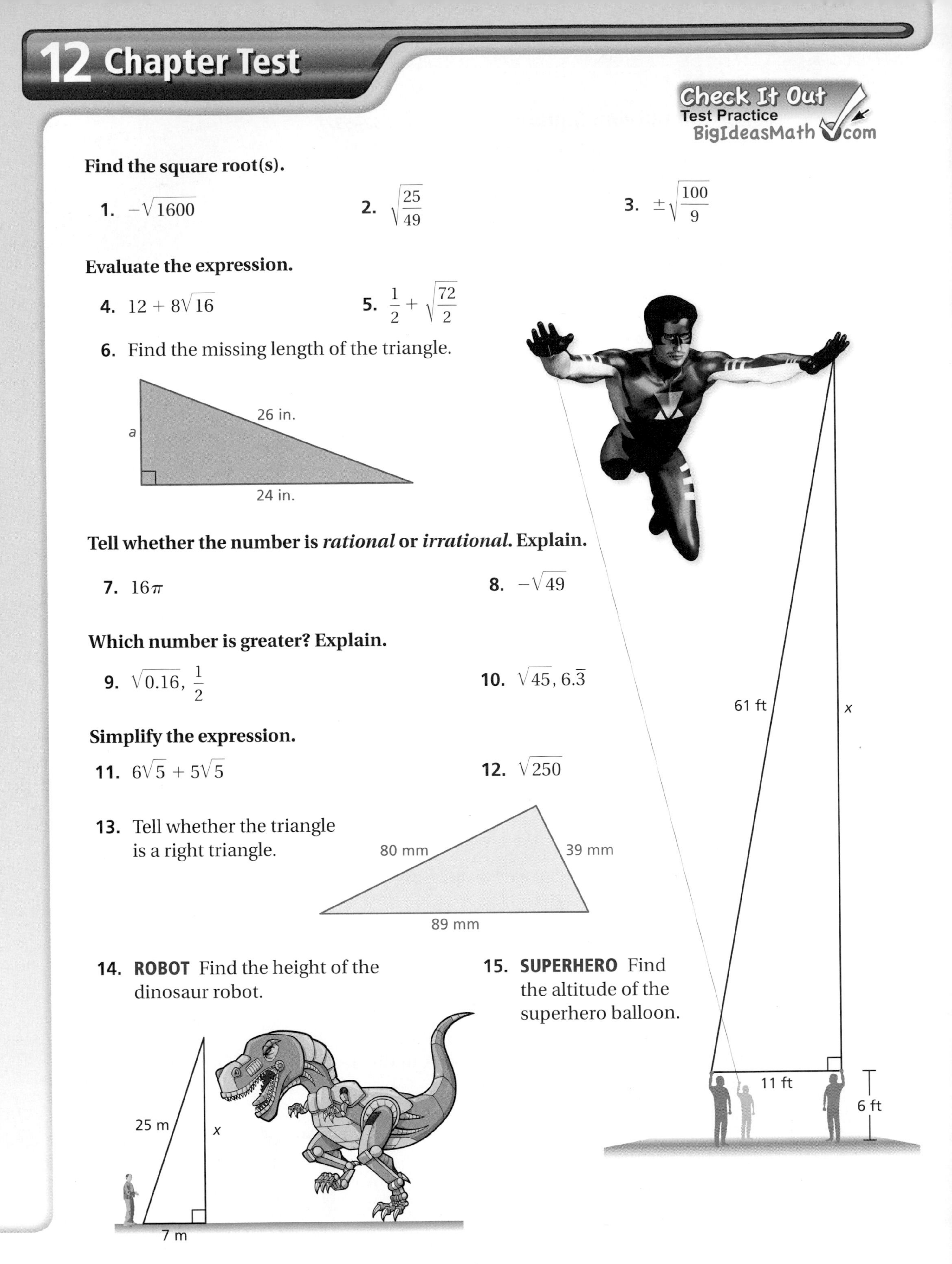

Tell whether the number is *rational* or *irrational*. Explain.

7. 16π
8. $-\sqrt{49}$

Which number is greater? Explain.

9. $\sqrt{0.16}, \dfrac{1}{2}$
10. $\sqrt{45}, 6.\overline{3}$

Simplify the expression.

11. $6\sqrt{5} + 5\sqrt{5}$
12. $\sqrt{250}$

13. Tell whether the triangle is a right triangle.

14. **ROBOT** Find the height of the dinosaur robot.

15. **SUPERHERO** Find the altitude of the superhero balloon.

12 Standardized Test Practice

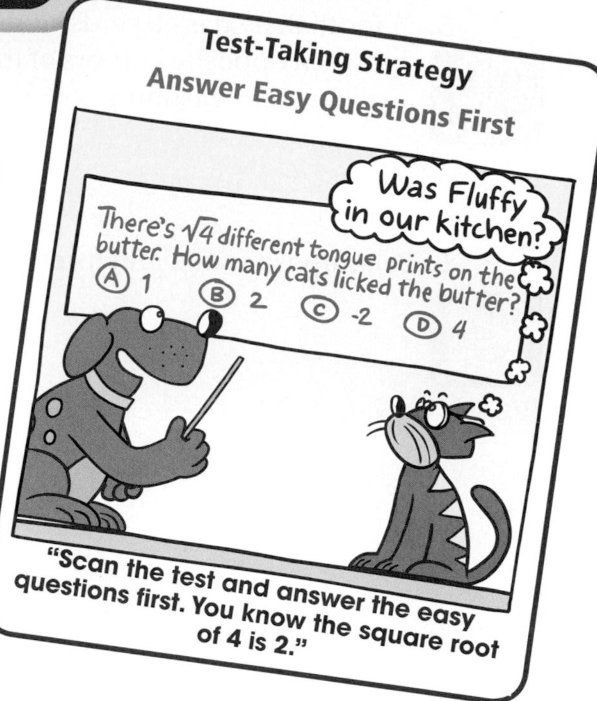

1. The period T of a pendulum is the time, in seconds, it takes the pendulum to swing back and forth. The period can be found using the formula $T = 1.1\sqrt{L}$, where L is the length, in feet, of the pendulum. A pendulum has a length of 4 feet. Find its period. *(8.EE.2)*

A. 5.1 sec

B. 4.4 sec

C. 3.1 sec

D. 2.2 sec

2. The steps Pat took to write the equation in slope-intercept form are shown below. What should Pat change in order to correctly rewrite the equation in slope-intercept form? *(8.EE.6)*

$$3x - 6y = 1$$
$$3x = 6y + 1$$
$$x = 2y + \frac{1}{3}$$

F. Use the formula $m = \frac{\text{rise}}{\text{run}}$.

G. Use the formula $m = \frac{\text{run}}{\text{rise}}$.

H. Subtract $3x$ from both sides of the equation and divide every term by -6.

I. Subtract 1 from both sides of the equation and divide every term by 3.

3. You deposit \$1500 in a savings account that earns 6% simple interest per year. Assuming you do not make any other deposits or withdrawals, how much interest will your account have earned after 4 years? *(7.RP.3)*

A. \$360

B. \$625

C. \$3600

D. \$6250

4. What is the mean of the data in the box below? *(7.NS.3)*

$-8.75, -4.43, -2.25, 4.46, 5.97$

F. -5

G. -1

H. 1

I. 5

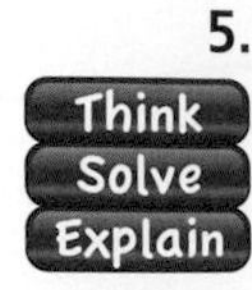

5. A football field is 40 yards wide and 120 yards long. Find the distance between opposite corners of the football field. Show your work and explain your reasoning. *(8.G.7)*

6. A composite solid and its dimensions are shown below.

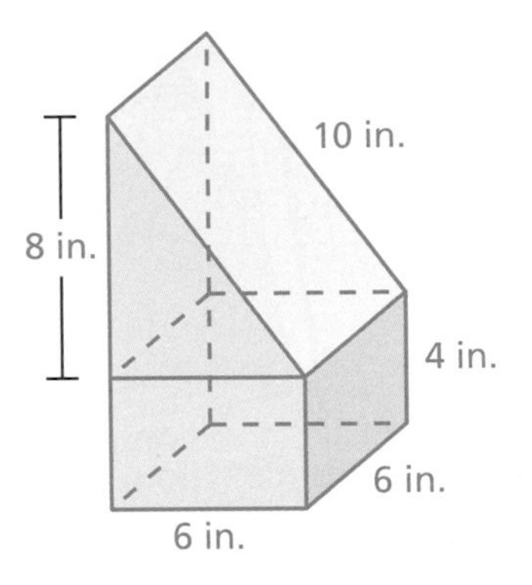

What is the surface area, in square inches, of the composite solid? *(7.G.6)*

7. What is the solution to the equation shown below? *(8.EE.7b)*

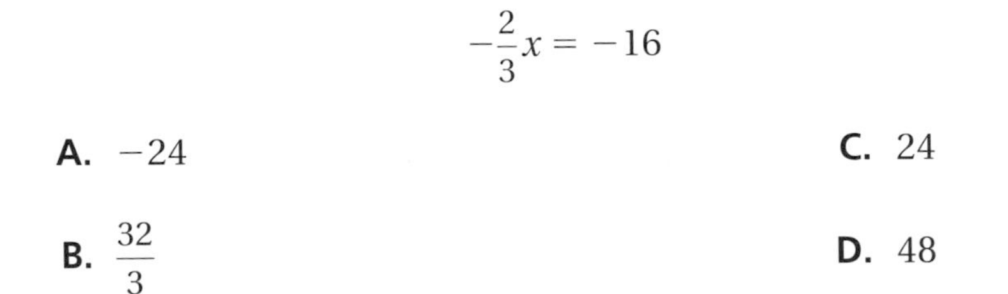

$$-\frac{2}{3}x = -16$$

A. -24

B. $\frac{32}{3}$

C. 24

D. 48

8. What is the value of x in the right triangle shown? *(8.G.7)*

F. 16 cm

G. 18 cm

H. 24 cm

I. $\sqrt{674}$ cm

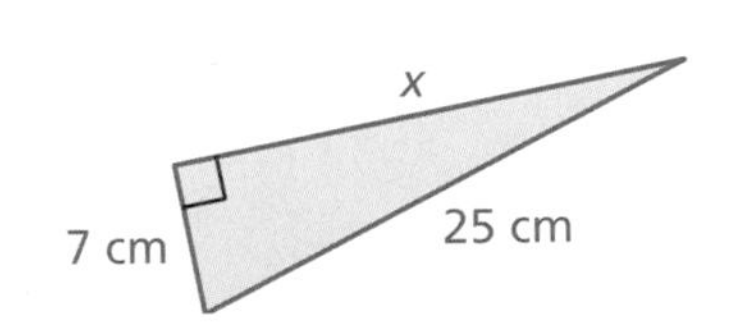

9. Find the height of the tree in the diagram. *(7.G.1)*

A. 22.5 ft

B. 31.5 ft

C. 35 ft

D. 40 ft

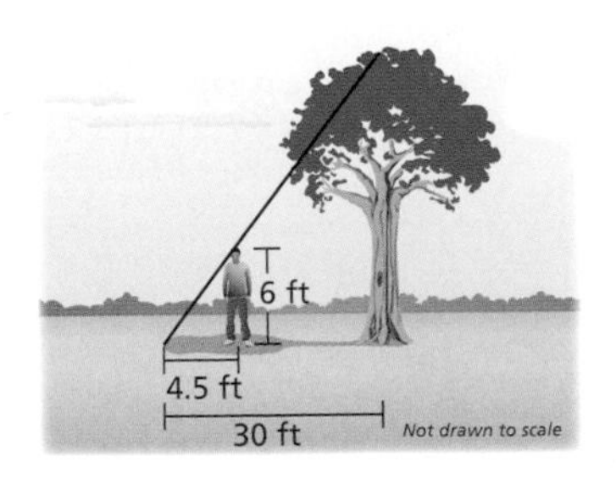

10. Which expression is equivalent to $12\sqrt{24}$? *(8.NS.2)*

F. $48\sqrt{6}$

G. $24\sqrt{12}$

H. $24\sqrt{6}$

I. 6

11. The measure of an angle is x degrees. What is the measure of its complement? *(7.G.5)*

A. $(90 - x)°$

B. $(180 - x)°$

C. $(x - 90)°$

D. $(x - 180)°$

12. A spinner is divided into 6 congruent sections, as shown at the right. You spin the spinner. What is the probability that the arrow will land on a red section? *(7.SP.7a)*

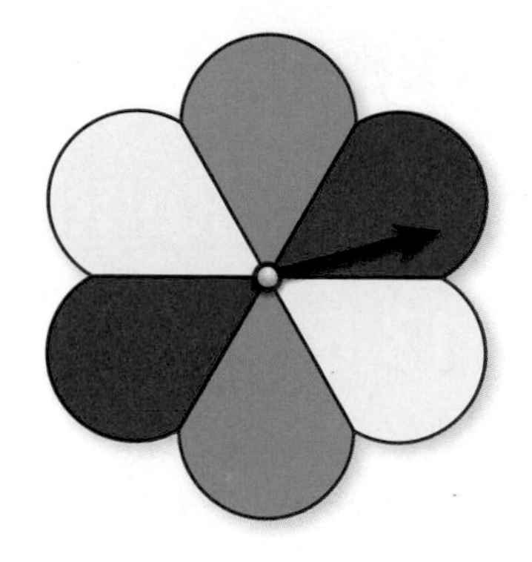

F. $\frac{1}{6}$

G. $\frac{1}{4}$

H. $\frac{1}{3}$

I. $\frac{1}{2}$

13. An airplane flies 56 miles due north and then 33 miles due east. How many miles is the plane from its starting point? *(8.G.8)*

14. Which graph represents the linear equation $y = -2x - 2$? *(8.EE.6)*

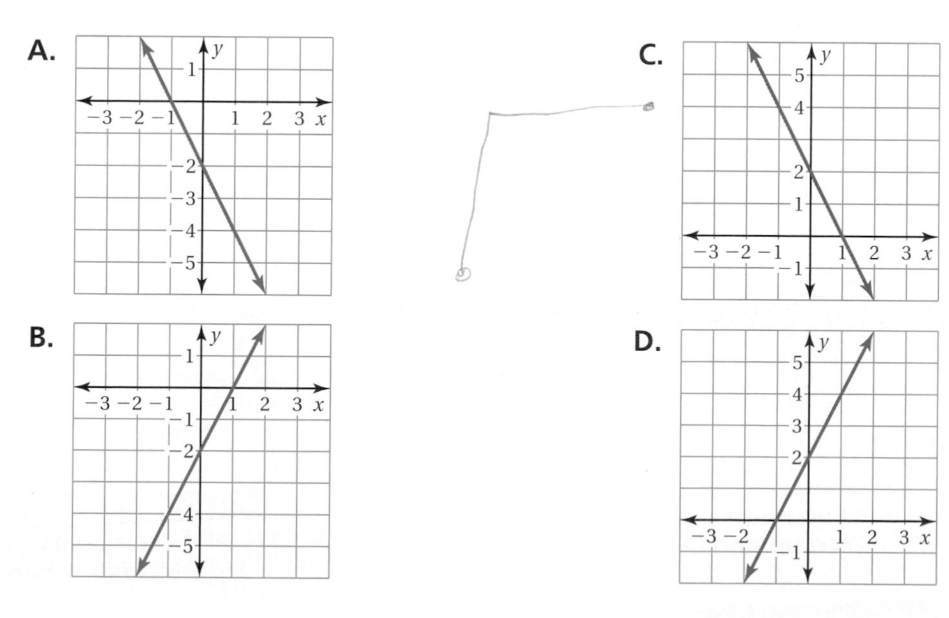

13 Exponents and Scientific Notation

"Here's how it goes, Descartes."

"The friends of my friends are my friends. The friends of my enemies are my enemies."

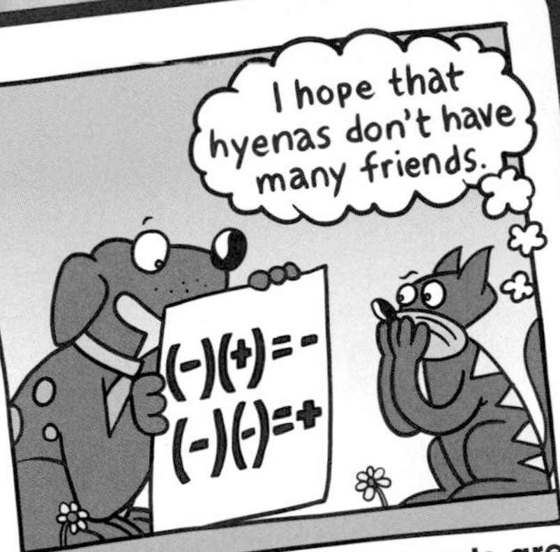

"The enemies of my friends are my enemies. The enemies of my enemies are my friends."

"If one flea had 100 babies, and each baby grew up and had 100 babies, ..."

"... and each of those babies grew up and had 100 babies, you would have 1,010,101 fleas."

What You Learned Before

"It's called the *Power of Negative One*, Descartes!"

Adding and Subtracting Decimals (6.NS.3)

Example 1 Find 2.65 + 5.012.

$$\begin{array}{r} 2.650 \\ +\ 5.012 \\ \hline 7.662 \end{array}$$

Example 2 Find 3.7 − 0.48.

$$\begin{array}{r} \overset{6}{\cancel{7}}\ \overset{10}{\cancel{0}} \\ 3.7\,0 \\ -\ 0.4\,8 \\ \hline 3.2\,2 \end{array}$$

Try It Yourself

Find the sum or difference.

1. 2.73 + 1.007 **2.** 3.4 − 1.27 **3.** 0.35 + 0.749 **4.** 1.019 + 0.09

5. 6.03 − 1.008 **6.** 4.21 − 0.007 **7.** 0.228 + 1.205 **8.** 3.003 − 1.9

Multiplying and Dividing Decimals (6.NS.3)

Example 3 Find 2.1 • 0.35.

$$\begin{array}{r} 2.1 \\ \times\ 0.3\,5 \\ \hline 1\,0\,5 \\ 6\,3\ \ \\ \hline 0.7\,3\,5 \end{array}$$

2.1 ← 1 decimal place
× 0.35 ← + 2 decimal places
0.735 ← 3 decimal places

Example 4 Find 1.08 ÷ 0.9.

$0.9\overline{)1.08}$ Multiply each number by 10.

$$\begin{array}{r} 1.2 \\ 9\overline{)10.8} \\ -\ 9\ \ \ \\ \hline 1\,8 \\ -\ 1\,8 \\ \hline 0 \end{array}$$

Place the decimal point above the decimal point in the dividend 10.8.

Try It Yourself

Find the product or quotient.

9. 1.75 • 0.2 **10.** 1.4 • 0.6 **11.** $\begin{array}{r} 7.03 \\ \times\ 4.3 \\ \hline \end{array}$ **12.** $\begin{array}{r} 0.894 \\ \times\ 0.2 \\ \hline \end{array}$

13. 5.40 ÷ 0.09 **14.** 4.17 ÷ 0.3 **15.** $0.15\overline{)3.6}$ **16.** $0.004\overline{)7.2}$

13.1 Exponents

COMMON CORE STATE STANDARDS
8.EE.1

Essential Question
How can you use exponents to write numbers?

The expression 3^5 is called a **power**. The **base** is 3. The **exponent** is 5.

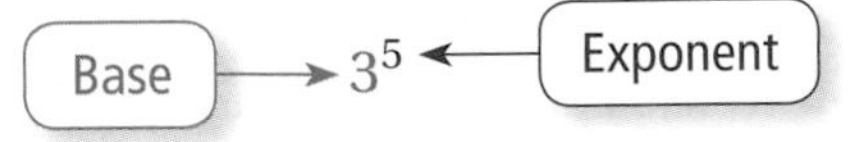

1 ACTIVITY: Using Exponent Notation

Work with a partner.

a. Copy and complete the table.

Power	Repeated Multiplication Form	Value
$(-3)^1$	-3	-3
$(-3)^2$	$(-3) \cdot (-3)$	9
$(-3)^3$		
$(-3)^4$		
$(-3)^5$		
$(-3)^6$		
$(-3)^7$		

b. Describe what is meant by the expression $(-3)^n$. How can you find the value of $(-3)^n$?

2 ACTIVITY: Using Exponent Notation

Work with a partner.

a. The cube at the right has \$3 in each of its small cubes. Write a single power that represents the total amount of money in the large cube.

b. Evaluate the power to find the total amount of money in the large cube.

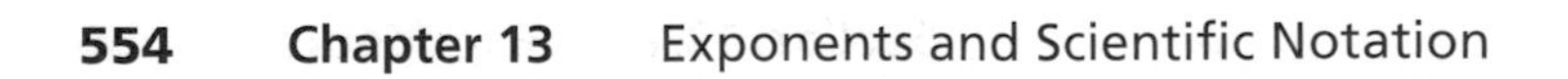

3 ACTIVITY: Writing Powers as Whole Numbers

Work with a partner. Write each distance as a whole number. Which numbers do you know how to write in words? For instance, in words, 10^3 is equal to *one thousand*.

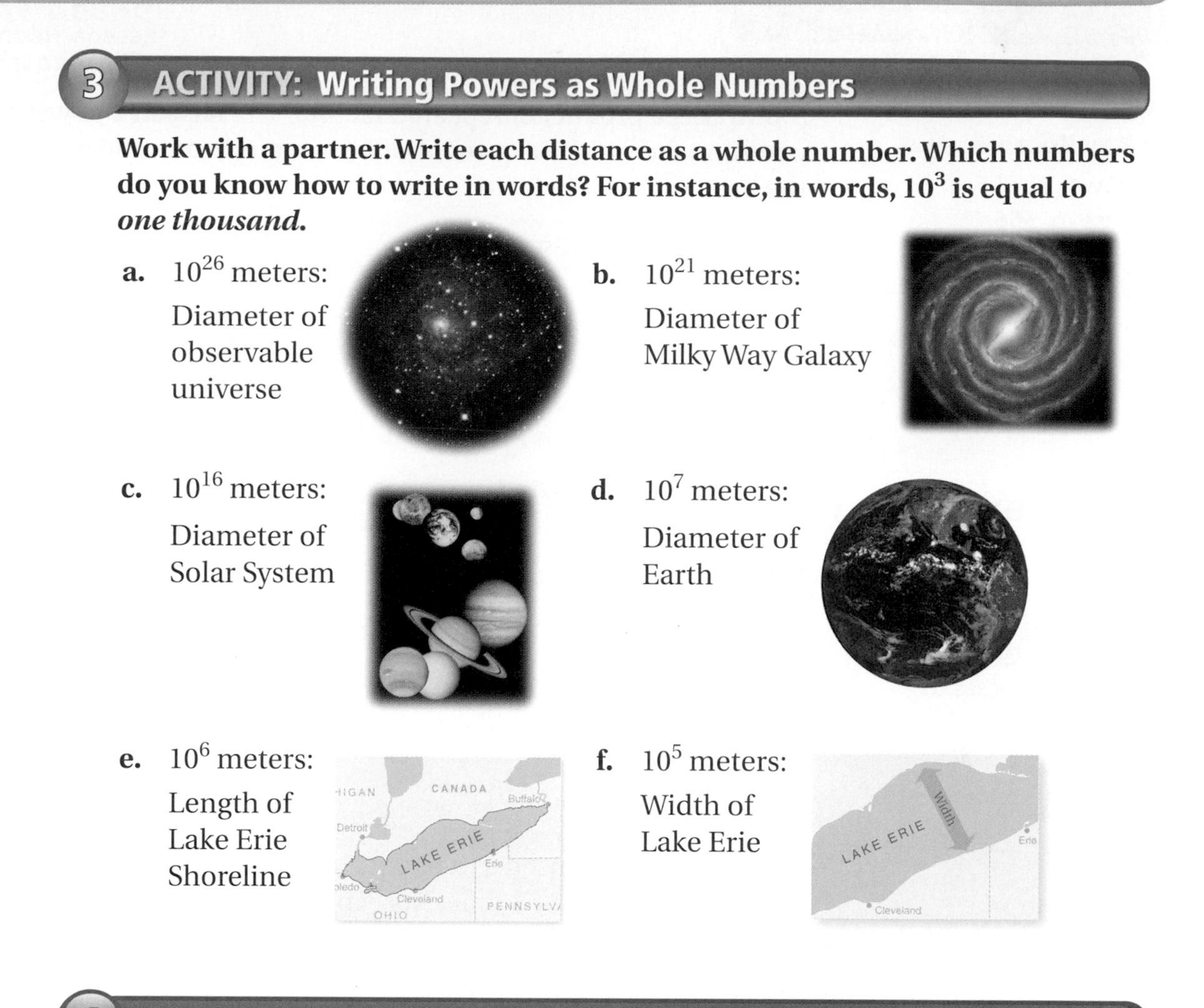

a. 10^{26} meters: Diameter of observable universe

b. 10^{21} meters: Diameter of Milky Way Galaxy

c. 10^{16} meters: Diameter of Solar System

d. 10^{7} meters: Diameter of Earth

e. 10^{6} meters: Length of Lake Erie Shoreline

f. 10^{5} meters: Width of Lake Erie

4 ACTIVITY: Writing a Power

Work with a partner. Write the number of kits, cats, sacks, and wives as a power.

As I was going to St. Ives
I met a man with seven wives
And every wife had seven sacks
And every sack had seven cats
And every cat had seven kits
Kits, cats, sacks, wives
How many were going to St. Ives?

Nursery Rhyme, 1730

What Is Your Answer?

5. **IN YOUR OWN WORDS** How can you use exponents to write numbers? Give some examples of how exponents are used in real life.

Practice Use what you learned about exponents to complete Exercises 3–5 on page 558.

13.1 Lesson

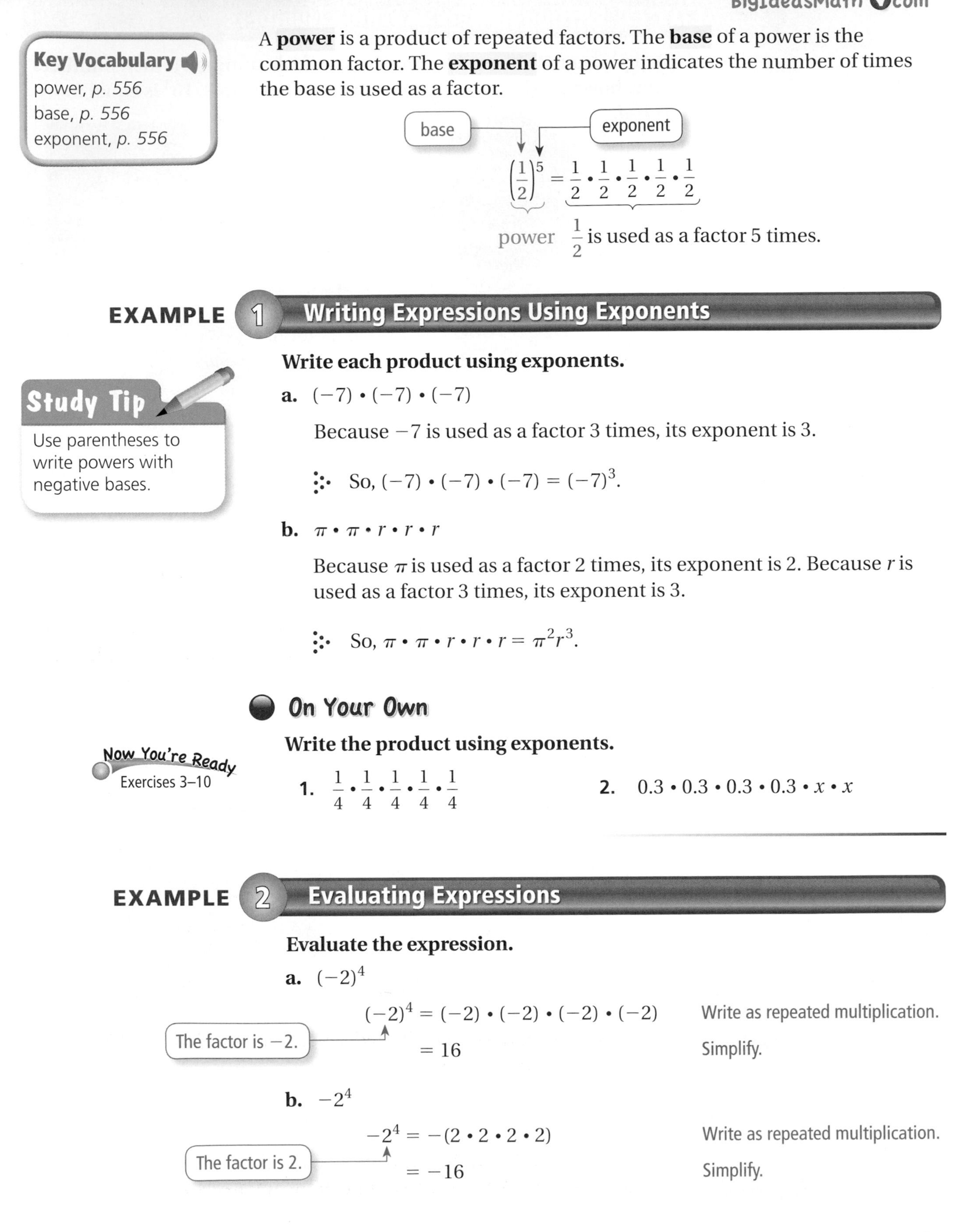

Key Vocabulary
power, *p. 556*
base, *p. 556*
exponent, *p. 556*

A **power** is a product of repeated factors. The **base** of a power is the common factor. The **exponent** of a power indicates the number of times the base is used as a factor.

$$\left(\frac{1}{2}\right)^5 = \frac{1}{2} \cdot \frac{1}{2} \cdot \frac{1}{2} \cdot \frac{1}{2} \cdot \frac{1}{2}$$

$\frac{1}{2}$ is used as a factor 5 times.

EXAMPLE 1 Writing Expressions Using Exponents

Write each product using exponents.

a. $(-7) \cdot (-7) \cdot (-7)$

Because -7 is used as a factor 3 times, its exponent is 3.

So, $(-7) \cdot (-7) \cdot (-7) = (-7)^3$.

Study Tip

Use parentheses to write powers with negative bases.

b. $\pi \cdot \pi \cdot r \cdot r \cdot r$

Because π is used as a factor 2 times, its exponent is 2. Because r is used as a factor 3 times, its exponent is 3.

So, $\pi \cdot \pi \cdot r \cdot r \cdot r = \pi^2 r^3$.

On Your Own

Now You're Ready
Exercises 3–10

Write the product using exponents.

1. $\frac{1}{4} \cdot \frac{1}{4} \cdot \frac{1}{4} \cdot \frac{1}{4} \cdot \frac{1}{4}$

2. $0.3 \cdot 0.3 \cdot 0.3 \cdot 0.3 \cdot x \cdot x$

EXAMPLE 2 Evaluating Expressions

Evaluate the expression.

a. $(-2)^4$

$(-2)^4 = (-2) \cdot (-2) \cdot (-2) \cdot (-2)$ Write as repeated multiplication.

$= 16$ Simplify.

The factor is -2.

b. -2^4

$-2^4 = -(2 \cdot 2 \cdot 2 \cdot 2)$ Write as repeated multiplication.

$= -16$ Simplify.

The factor is 2.

Multi-Language Glossary at BigIdeasMath.com.

EXAMPLE 3 Using Order of Operations

Evaluate the expression.

a. $3 + 2 \cdot 3^4$

$$\begin{aligned} 3 + 2 \cdot 3^4 &= 3 + 2 \cdot 81 && \text{Evaluate the power.} \\ &= 3 + 162 && \text{Multiply.} \\ &= 165 && \text{Add.} \end{aligned}$$

b. $3^3 - 8^2 \div 2$

$$\begin{aligned} 3^3 - 8^2 \div 2 &= 27 - 64 \div 2 && \text{Evaluate the powers.} \\ &= 27 - 32 && \text{Divide.} \\ &= -5 && \text{Subtract.} \end{aligned}$$

On Your Own

Now You're Ready
Exercises 11–16 and 21–26

Evaluate the expression.

3. -5^4 **4.** $\left(-\frac{1}{6}\right)^3$ **5.** $\left| -3^3 \div 27 \right|$ **6.** $9 - 2^5 \cdot 0.5$

EXAMPLE 4 Real-Life Application

In sphering, a person is secured inside a small, hollow sphere that is surrounded by a larger sphere. The space between the spheres is inflated with air. What is the volume of the inflated space? (The volume V of a sphere is $V = \frac{4}{3}\pi r^3$. Use 3.14 for π.)

Outer sphere		*Inner sphere*
$V = \frac{4}{3}\pi r^3$	Write formula.	$V = \frac{4}{3}\pi r^3$
$= \frac{4}{3}\pi (1.5)^3$	Substitute.	$= \frac{4}{3}\pi (1)^3$
$= \frac{4}{3}\pi (3.375)$	Evaluate the power.	$= \frac{4}{3}\pi (1)$
≈ 14.13	Multiply.	≈ 4.19

So, the volume of the inflated space is about $14.13 - 4.19$, or 9.94 cubic meters.

On Your Own

7. WHAT IF? In Example 4, the diameter of the inner sphere is 1.8 meters. What is the volume of the inflated space?

13.1 Exercises

Vocabulary and Concept Check

1. **VOCABULARY** Describe the difference between an exponent and a power. Can the two words be used interchangeably?

2. **WHICH ONE DOESN'T BELONG?** Which one does *not* belong with the other three? Explain your reasoning.

5^3 The exponent is 3.	5^3 The power is 5.	5^3 The base is 5.	5^3 Five is used as a factor 3 times.

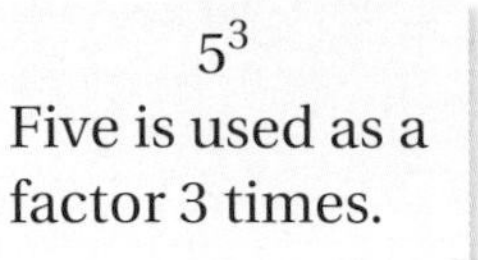

Practice and Problem Solving

Write the product using exponents.

(1) 3. $3 \cdot 3 \cdot 3 \cdot 3$

4. $(-6) \cdot (-6)$

5. $\left(-\frac{1}{2}\right) \cdot \left(-\frac{1}{2}\right) \cdot \left(-\frac{1}{2}\right)$

6. $\frac{1}{3} \cdot \frac{1}{3} \cdot \frac{1}{3}$

7. $\pi \cdot \pi \cdot \pi \cdot x \cdot x \cdot x \cdot x$

8. $(-4) \cdot (-4) \cdot (-4) \cdot y \cdot y$

9. $8 \cdot 8 \cdot 8 \cdot 8 \cdot b \cdot b \cdot b$

10. $(-t) \cdot (-t) \cdot (-t) \cdot (-t) \cdot (-t)$

Evaluate the expression.

(2) 11. 5^2

12. -11^3

13. $(-1)^6$

14. $\left(\frac{1}{2}\right)^6$

15. $\left(-\frac{1}{12}\right)^2$

16. $-\left(\frac{1}{9}\right)^3$

17. **ERROR ANALYSIS** Describe and correct the error in evaluating the expression.

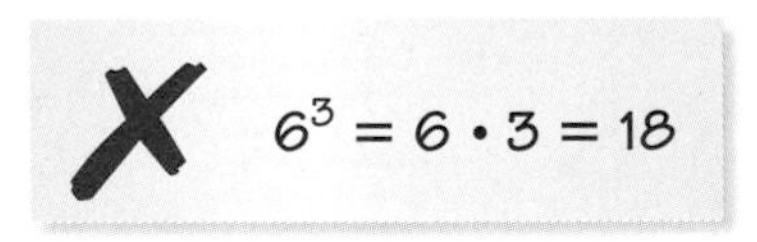

18. **PRIME FACTORIZATION** Write the prime factorization of 675 using exponents.

19. **NUMBER SENSE** Write $-\left(\frac{1}{4} \cdot \frac{1}{4} \cdot \frac{1}{4} \cdot \frac{1}{4}\right)$ using exponents.

20. **RUSSIAN DOLLS** The largest doll is 12 inches tall. The height of each of the other dolls is $\frac{7}{10}$ the height of the next larger doll. Write an expression for the height of the smallest doll. What is the height of the smallest doll?

Evaluate the expression.

3 **21.** $5 + 3 \cdot 2^3$

22. $2 + 7 \cdot (-3)^2$

23. $(13^2 - 12^2) \div 5$

24. $\frac{1}{2}(4^3 - 6 \cdot 3^2)$

25. $\left| \frac{1}{2}(7 + 5^3) \right|$

26. $\left| \left(-\frac{1}{2}\right)^3 \div \left(\frac{1}{4}\right)^2 \right|$

27. MONEY You have a part-time job. One day your boss offers to pay you either $2^h - 1$ or 2^{h-1} dollars for each hour h you work that day. Copy and complete the table. Which option should you choose? Explain.

h	1	2	3	4	5
$2^h - 1$					
2^{h-1}					

28. CARBON-14 DATING Carbon-14 dating is used by scientists to determine the age of a sample.

a. The amount C (in grams) of a 100-gram sample of carbon-14 remaining after t years is represented by the equation $C = 100(0.99988)^t$. Use a calculator to find the amount of carbon-14 remaining after 4 years.

b. What percent of the carbon-14 remains after 4 years?

29. Critical Thinking The frequency (in vibrations per second) of a note on a piano is represented by the equation $F = 440(1.0595)^n$, where n is the number of notes above A-440. Each black or white key represents one note.

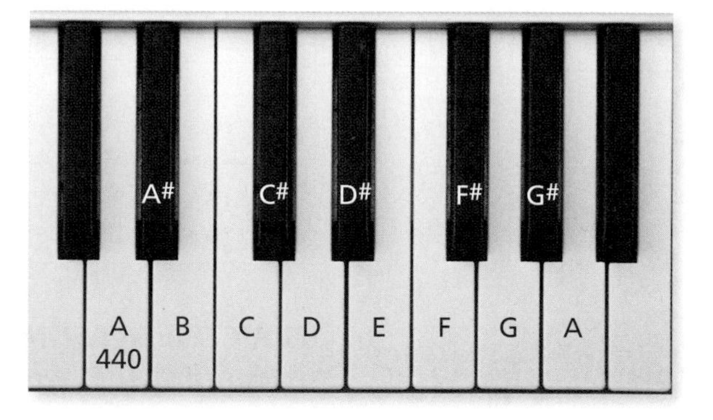

a. How many notes do you take to travel from A-440 to A?

b. What is the frequency of A?

c. Describe the relationship between the number of notes between A-440 and A and the frequency of the notes.

Fair Game Review What you learned in previous grades & lessons

Tell which property is illustrated by the statement. *(Skills Review Handbook)*

30. $8 \cdot x = x \cdot 8$

31. $(2 \cdot 10)x = 2(10 \cdot x)$

32. $3(x \cdot 1) = 3x$

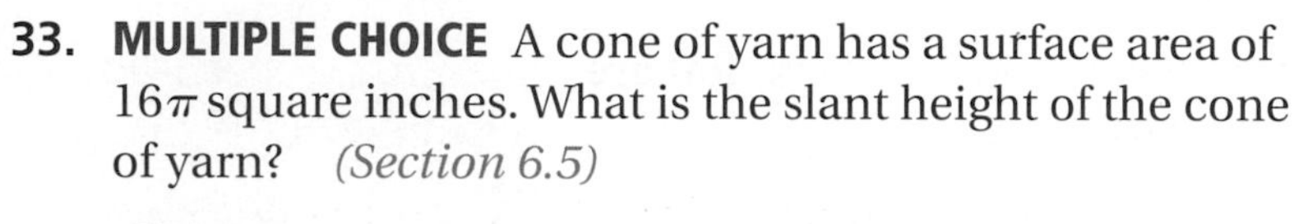

33. MULTIPLE CHOICE A cone of yarn has a surface area of 16π square inches. What is the slant height of the cone of yarn? *(Section 6.5)*

Ⓐ 4 in. Ⓑ 6 in.

Ⓒ 8 in. Ⓓ 10 in.

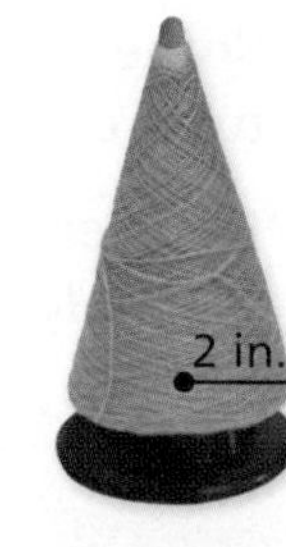

13.2 Product of Powers Property

COMMON CORE STATE STANDARDS
8.EE.1

Essential Question How can you multiply two powers that have the same base?

1 ACTIVITY: Finding Products of Powers

Work with a partner.

a. Copy and complete the table.

Product	Repeated Multiplication Form	Power
$2^2 \cdot 2^4$	$2 \cdot 2 \cdot 2 \cdot 2 \cdot 2 \cdot 2$	2^6
$(-3)^2 \cdot (-3)^4$	$(-3) \cdot (-3) \cdot (-3) \cdot (-3) \cdot (-3) \cdot (-3)$	$(-3)^6$
$7^3 \cdot 7^2$		
$5.1^1 \cdot 5.1^6$		
$(-4)^2 \cdot (-4)^2$		
$10^3 \cdot 10^5$		
$\left(\frac{1}{2}\right)^5 \cdot \left(\frac{1}{2}\right)^5$		

b. **INDUCTIVE REASONING** Describe the pattern in the table. Then write a rule for multiplying two powers that have the same base.

$$a^m \cdot a^n = a^{\square}$$

c. Use your rule to simplify the products in the first column of the table above. Does your rule give the results in the third column?

2 ACTIVITY: Using a Calculator

Work with a partner.

Some calculators have *exponent keys* that are used to evaluate powers.

Use a calculator with an exponent key to evaluate the products in Activity 1.

Exponent Key

3 ACTIVITY: The Penny Puzzle

Work with a partner.

- The rows y and columns x of a chess board are numbered as shown.
- Each position on the chess board has a stack of pennies. (Only the first row is shown.)
- The number of pennies in each stack is

  $2^x \cdot 2^y$.

a. How many pennies are in the stack in location (3, 5)?

b. Which locations have 32 pennies in their stacks?

c. How much money (in dollars) is in the location with the tallest stack?

d. A penny is about 0.06 inch thick. About how tall (in inches) is the tallest stack?

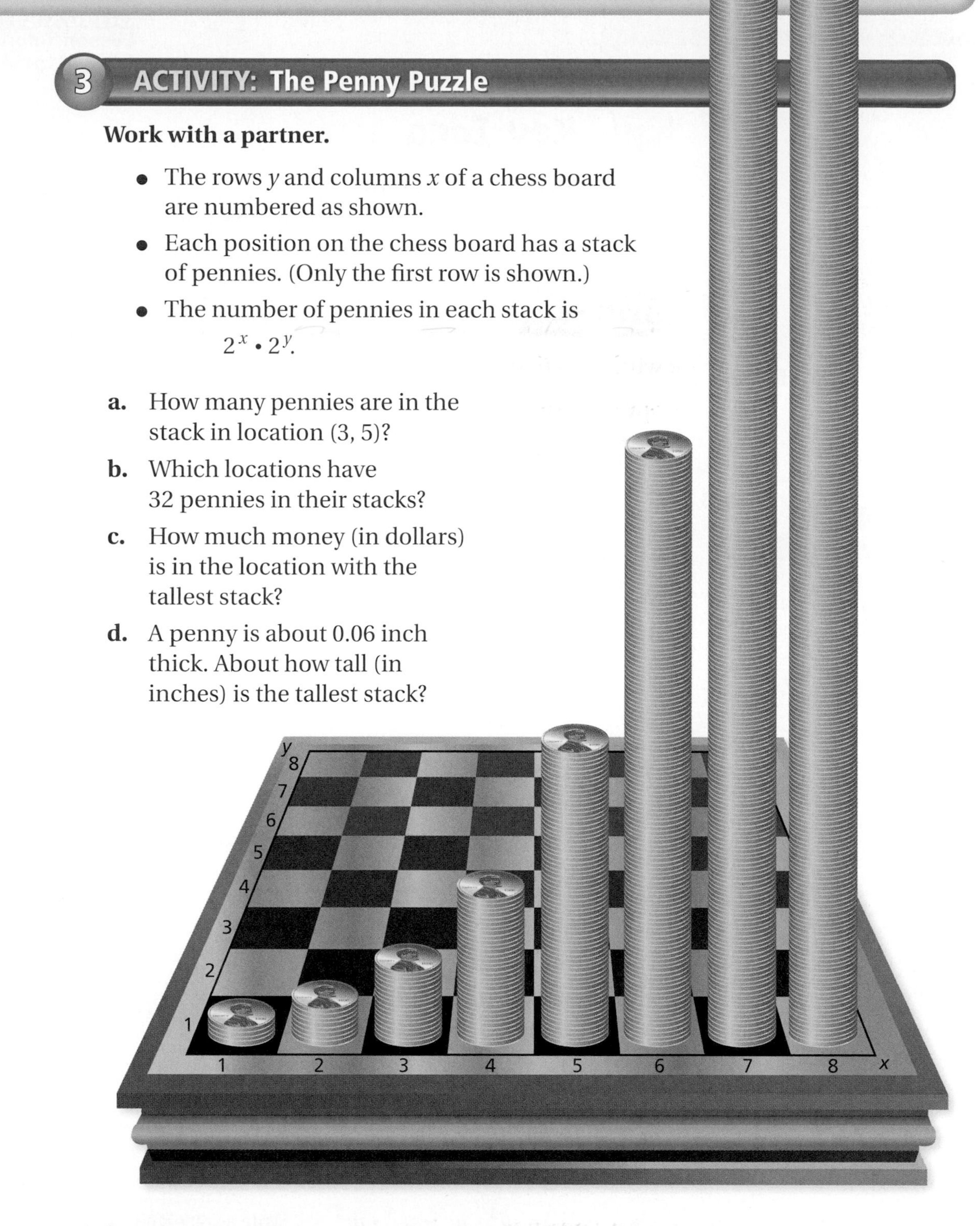

What Is Your Answer?

4. IN YOUR OWN WORDS How can you multiply two powers that have the same base? Give two examples of your rule.

Use what you learned about the Product of Powers Property to complete Exercises 3–5 on page 564.

13.2 Lesson

Check It Out
Lesson Tutorials
BigIdeasMath.com

Key Idea

Product of Powers Property

Words To multiply powers with the same base, add their exponents.

Numbers $4^2 \cdot 4^3 = 4^{2+3} = 4^5$ **Algebra** $a^m \cdot a^n = a^{m+n}$

EXAMPLE 1 Multiplying Powers with the Same Base

a. $2^4 \cdot 2^5 = 2^{4+5}$ The base is 2. Add the exponents.

$= 2^9$ Simplify.

Study Tip

When a number is written without an exponent, its exponent is 1.

b. $-5 \cdot (-5)^6 = (-5)^1 \cdot (-5)^6$ Rewrite -5 as $(-5)^1$.

$= (-5)^{1+6}$ The base is -5. Add the exponents.

$= (-5)^7$ Simplify.

c. $x^3 \cdot x^7 = x^{3+7}$ The base is x. Add the exponents.

$= x^{10}$ Simplify.

On Your Own

Simplify the expression. Write your answer as a power.

1. $6^2 \cdot 6^4$
2. $\left(-\frac{1}{2}\right)^3 \cdot \left(-\frac{1}{2}\right)^6$
3. $z \cdot z^{12}$

EXAMPLE 2 Raising a Power to a Power

a. $(3^4)^3 = 3^4 \cdot 3^4 \cdot 3^4$ Write as repeated multiplication.

$= 3^{4+4+4}$ The base is 3. Add the exponents.

$= 3^{12}$ Simplify.

b. $(w^5)^4 = w^5 \cdot w^5 \cdot w^5 \cdot w^5$ Write as repeated multiplication.

$= w^{5+5+5+5}$ The base is w. Add the exponents.

$= w^{20}$ Simplify.

On Your Own

Now You're Ready
Exercises 3–14

Simplify the expression. Write your answer as a power.

4. $(4^4)^3$
5. $(y^2)^4$
6. $(\pi^3)^3$
7. $((-4)^3)^2$

EXAMPLE 3 Raising a Product to a Power

a. $(2x)^3 = 2x \cdot 2x \cdot 2x$ — Write as repeated multiplication.

$= (2 \cdot 2 \cdot 2) \cdot (x \cdot x \cdot x)$ — Group like bases using properties of multiplication.

$= 2^{1+1+1} \cdot x^{1+1+1}$ — The bases are 2 and x. Add the exponents.

$= 2^3 \cdot x^3 = 8x^3$ — Simplify.

b. $(xy)^2 = xy \cdot xy$ — Write as repeated multiplication.

$= (x \cdot x) \cdot (y \cdot y)$ — Group like bases using properties of multiplication.

$= x^{1+1} \cdot y^{1+1}$ — The bases are x and y. Add the exponents.

$= x^2y^2$ — Simplify.

On Your Own

Now You're Ready
Exercises 17–22

Simplify the expression.

8. $(5y)^4$ **9.** $(0.5n)^2$ **10.** $(ab)^5$

EXAMPLE 4 Standardized Test Practice

Details

Local Disk (C:)
Local Disk

Free Space: 16GB

Total Space: 64GB

A gigabyte (GB) of computer storage space is 2^{30} bytes. The details of a computer are shown. How many bytes of total storage space does the computer have?

Ⓐ 2^{34} Ⓑ 2^{36} Ⓒ 2^{180} Ⓓ 128^{30}

The computer has 64 gigabytes of total storage space. Notice that 64 can be written as a power, 2^6. Use a model to solve the problem.

Total number of bytes = Number of bytes in a gigabyte • Number of gigabytes

$= 2^{30} \cdot 2^6$ — Substitute.

$= 2^{30+6}$ — Add exponents.

$= 2^{36}$ — Simplify.

The computer has 2^{36} bytes of total storage space. The correct answer is Ⓑ.

On Your Own

11. How many bytes of free storage space does the computer have?

13.2 Exercises

Vocabulary and Concept Check

1. **REASONING** When should you use the Product of Powers Property?
2. **CRITICAL THINKING** Can you use the Product of Powers Property to multiply powers with different bases? Explain.

Practice and Problem Solving

Simplify the expression. Write your answer as a power.

① ② 3. $3^2 \cdot 3^2$

4. $8^{10} \cdot 8^4$

5. $(-4)^5 \cdot (-4)^7$

6. $a^3 \cdot a^3$

7. $h^6 \cdot h$

8. $\left(\frac{2}{3}\right)^2 \cdot \left(\frac{2}{3}\right)^6$

9. $\left(-\frac{5}{7}\right)^8 \cdot \left(-\frac{5}{7}\right)^9$

10. $(-2.9) \cdot (-2.9)^7$

11. $(5^4)^3$

12. $(b^{12})^3$

13. $(3.8^3)^4$

14. $\left(\left(-\frac{3}{4}\right)^5\right)^2$

ERROR ANALYSIS **Describe and correct the error in simplifying the expression.**

15. ✗ $5^2 \cdot 5^9 = (5 \cdot 5)^{2+9}$
$= 25^{11}$

16. ✗ $(r^6)^4 = r^{6+4}$
$= r^{10}$

Simplify the expression.

③ 17. $(6g)^3$

18. $(-3v)^5$

19. $\left(\frac{1}{5}k\right)^2$

20. $(1.2m)^4$

21. $(rt)^{12}$

22. $\left(-\frac{3}{4}p\right)^3$

23. **CRITICAL THINKING** Is $3^2 + 3^3$ equal to 3^5? Explain.

24. **ARTIFACT** A display case for the artifact is in the shape of a cube. Each side of the display case is three times longer than the width of the artifact.

    a. Write an expression for the volume of the case. Write your answer as a power.

    b. Simplify the expression.

w in.

w in.

Simplify the expression.

25. $2^4 \cdot 2^5 - (2^2)^2$

26. $16\left(\frac{1}{2}x\right)^4$

27. $5^2(5^3 \cdot 5^2)$

28. CLOUDS The lowest altitude of an altocumulus cloud is about 3^8 feet. The highest altitude of an altocumulus cloud is about 3 times the lowest altitude. What is the highest altitude of an altocumulus cloud? Write your answer as a power.

29. PYTHON EGG The volume V of a python egg is given by the formula $V = \frac{4}{3}\pi abc$. For the python egg shown, $a = 2$ inches, $b = 2$ inches, and $c = 3$ inches.

a. Find the volume of the python egg.

b. Square the dimensions of the python egg. Then evaluate the formula. How does this volume compare to your answer in part (a)?

30. PYRAMID The volume of a square pyramid is $V = \frac{1}{3}b^2h$, where b is the length of one side of the base and h is the height of the pyramid. The length of each side of the base increases by 50%. Write a formula for the volume of the new pyramid.

31. MAIL The United States Postal Service delivers about $2^6 \cdot 5^3$ pieces of mail each second. There are $2^8 \cdot 3^4 \cdot 5^2$ seconds in 6 days. How many pieces of mail does the United States Postal Service deliver in 6 days? Write your answer as a power.

32. Critical Thinking Find the value of x in the equation without evaluating the power.

a. $2^5 \cdot 2^x = 256$

b. $\left(\frac{1}{3}\right)^2 \cdot \left(\frac{1}{3}\right)^x = \frac{1}{729}$

Fair Game Review What you learned in previous grades & lessons

Simplify. *(Skills Review Handbook)*

33. $\frac{4 \cdot 4}{4}$

34. $\frac{5 \cdot 5 \cdot 5}{5}$

35. $\frac{2 \cdot 3}{2}$

36. $\frac{8 \cdot 6 \cdot 6}{6 \cdot 8}$

37. MULTIPLE CHOICE What is the measure of each angle of the regular polygon? *(Section 11.3)*

Ⓐ 45° Ⓑ 135°

Ⓒ 1080° Ⓓ 1440°

13.3 Quotient of Powers Property

Essential Question How can you divide two powers that have the same base?

COMMON CORE STATE STANDARDS
8.EE.1

1 ACTIVITY: Finding Quotients of Powers

Work with a partner.

a. Copy and complete the table.

Quotient	Repeated Multiplication Form	Power
$\frac{2^4}{2^2}$	$\frac{\overset{1}{\cancel{2}} \cdot \overset{1}{\cancel{2}} \cdot 2 \cdot 2}{\underset{1}{\cancel{2}} \cdot \underset{1}{\cancel{2}}}$	2^2
$\frac{(-4)^5}{(-4)^2}$	$\frac{\overset{1}{\cancel{(-4)}} \cdot \overset{1}{\cancel{(-4)}} \cdot (-4) \cdot (-4) \cdot (-4)}{\underset{1}{\cancel{(-4)}} \cdot \underset{1}{\cancel{(-4)}}}$	$(-4)^3$
$\frac{7^7}{7^3}$		
$\frac{8.5^9}{8.5^6}$		
$\frac{10^8}{10^5}$		
$\frac{3^{12}}{3^4}$		
$\frac{(-5)^7}{(-5)^5}$		
$\frac{11^4}{11^1}$		

b. **INDUCTIVE REASONING** Describe the pattern in the table. Then write a rule for dividing two powers that have the same base.

$$\frac{a^m}{a^n} = a^{\square}$$

c. Use your rule to simplify the quotients in the first column of the table above. Does your rule give the results in the third column?

2 ACTIVITY: Comparing Volumes

Work with a partner.

How many of the smaller cubes will fit inside the larger cube? Record your results in the table. Describe the pattern in the table.

a. Sample:

$s = 4$ $s = 4^2$

b.

$s = 3$ $s = 3^2$

c.

$s = 6$ $s = 6^2$

d.

$s = 10$ $s = 10^2$

	Volume of Smaller Cube	Volume of Larger Cube	$\frac{\text{Larger Volume}}{\text{Smaller Volume}}$	Answer
a.	4^3	$(4^2)^3 = 4^6$	$\frac{4^6}{4^3}$	4^3
b.				
c.				
d.				

What Is Your Answer?

3. **IN YOUR OWN WORDS** How can you divide two powers that have the same base? Give two examples of your rule.

Use what you learned about the Quotient of Powers Property to complete Exercises 3–6 on page 570.

13.3 Lesson

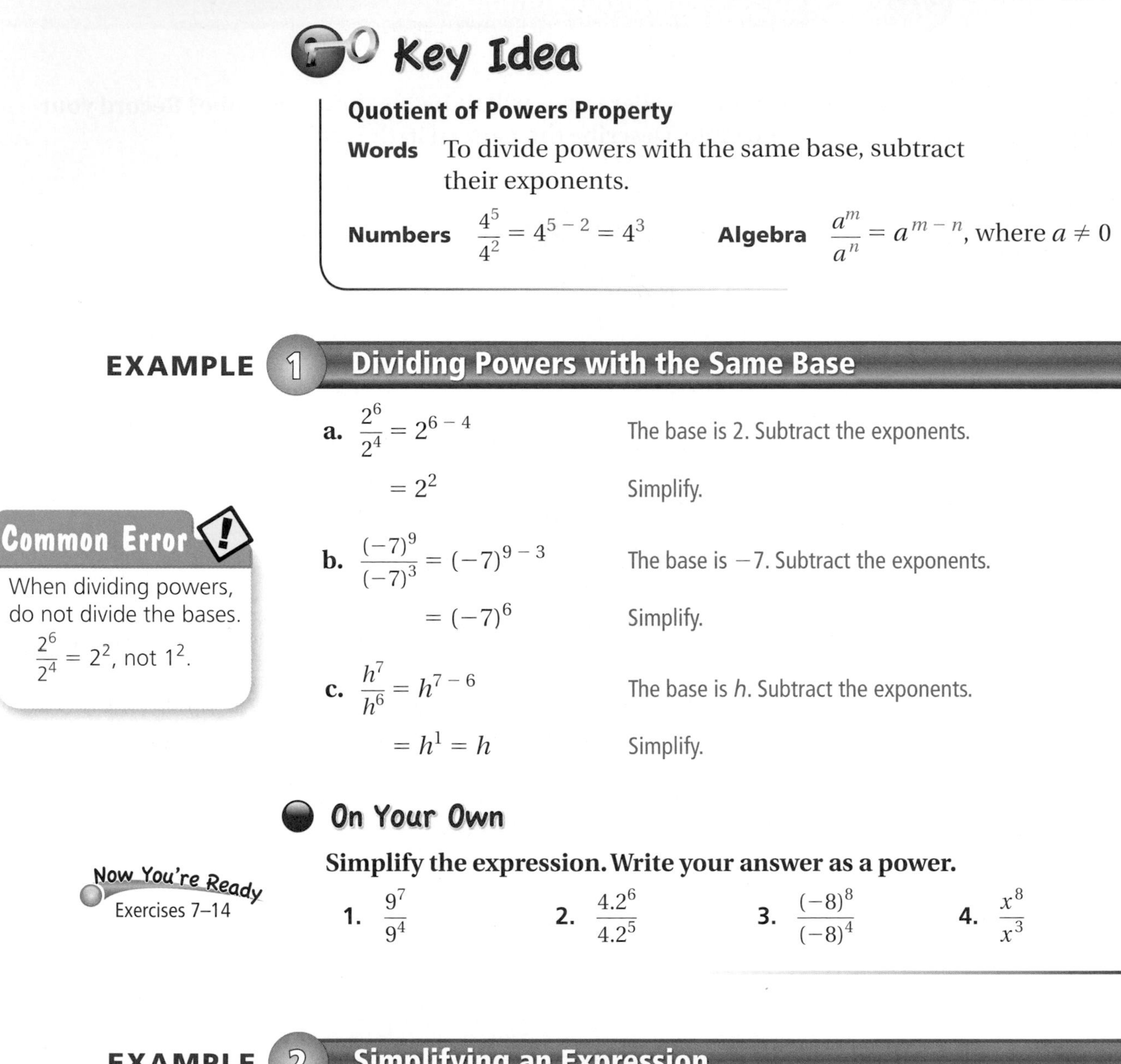

Key Idea

Quotient of Powers Property

Words To divide powers with the same base, subtract their exponents.

Numbers $\dfrac{4^5}{4^2} = 4^{5-2} = 4^3$ **Algebra** $\dfrac{a^m}{a^n} = a^{m-n}$, where $a \neq 0$

EXAMPLE 1 Dividing Powers with the Same Base

a. $\dfrac{2^6}{2^4} = 2^{6-4}$ The base is 2. Subtract the exponents.

$= 2^2$ Simplify.

b. $\dfrac{(-7)^9}{(-7)^3} = (-7)^{9-3}$ The base is -7. Subtract the exponents.

$= (-7)^6$ Simplify.

c. $\dfrac{h^7}{h^6} = h^{7-6}$ The base is h. Subtract the exponents.

$= h^1 = h$ Simplify.

Common Error

When dividing powers, do not divide the bases.

$\dfrac{2^6}{2^4} = 2^2$, not 1^2.

On Your Own

Simplify the expression. Write your answer as a power.

1. $\dfrac{9^7}{9^4}$ **2.** $\dfrac{4.2^6}{4.2^5}$ **3.** $\dfrac{(-8)^8}{(-8)^4}$ **4.** $\dfrac{x^8}{x^3}$

Now You're Ready
Exercises 7–14

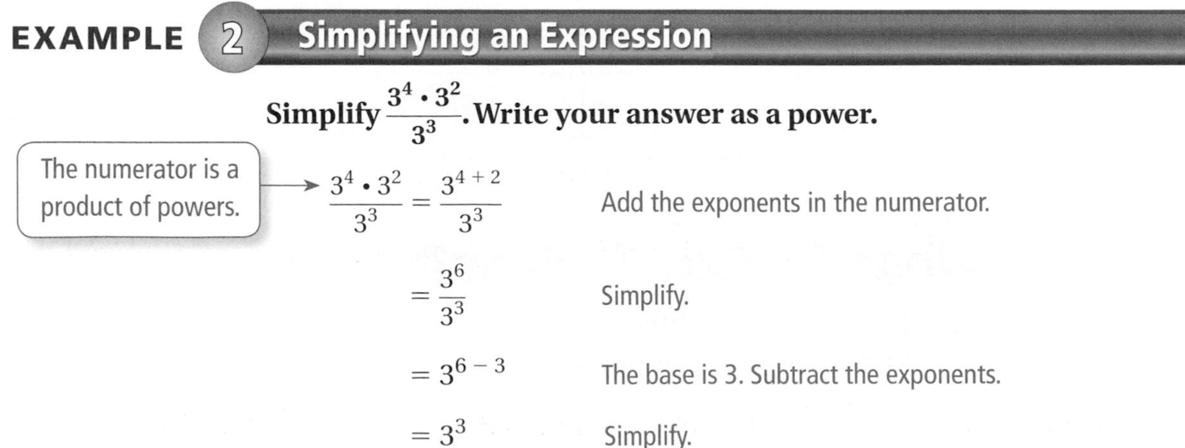

EXAMPLE 2 Simplifying an Expression

Simplify $\dfrac{3^4 \cdot 3^2}{3^3}$. Write your answer as a power.

The numerator is a product of powers.

$\dfrac{3^4 \cdot 3^2}{3^3} = \dfrac{3^{4+2}}{3^3}$ Add the exponents in the numerator.

$= \dfrac{3^6}{3^3}$ Simplify.

$= 3^{6-3}$ The base is 3. Subtract the exponents.

$= 3^3$ Simplify.

EXAMPLE 3 Simplifying an Expression

Study Tip

You can also simplify the expression in Example 3 as follows.

$$\frac{a^{10}}{a^6} \cdot \frac{a^7}{a^4} = \frac{a^{10} \cdot a^7}{a^6 \cdot a^4}$$
$$= \frac{a^{17}}{a^{10}}$$
$$= a^{17-10}$$
$$= a^7$$

Simplify $\frac{a^{10}}{a^6} \cdot \frac{a^7}{a^4}$. Write your answer as a power.

$\frac{a^{10}}{a^6} \cdot \frac{a^7}{a^4} = a^{10-6} \cdot a^{7-4}$ Subtract the exponents.

$= a^4 \cdot a^3$ Simplify.

$= a^{4+3}$ Add the exponents.

$= a^7$ Simplify.

On Your Own

Now You're Ready Exercises 16–21

Simplify the expression. Write your answer as a power.

5. $\frac{2^{15}}{2^3 \cdot 2^5}$

6. $\frac{d^5}{d} \cdot \frac{d^9}{d^8}$

EXAMPLE 4 Real-Life Application

The projected population of Tennessee in 2030 is about $5 \cdot 5.9^8$. Predict the average number of people per square mile in 2030.

Use a model to solve the problem.

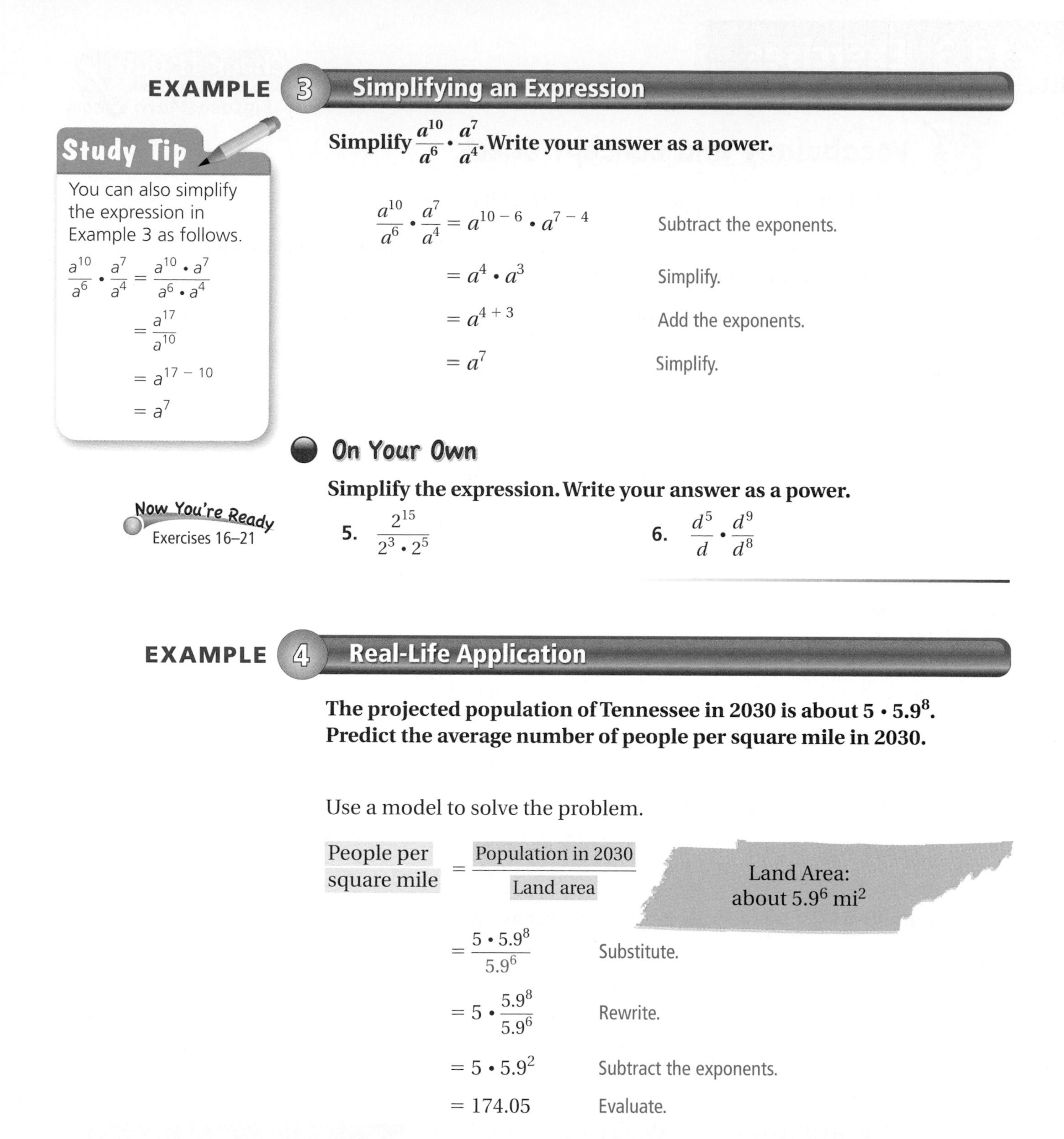

People per square mile $=$ $\frac{\text{Population in 2030}}{\text{Land area}}$

$= \frac{5 \cdot 5.9^8}{5.9^6}$ Substitute.

$= 5 \cdot \frac{5.9^8}{5.9^6}$ Rewrite.

$= 5 \cdot 5.9^2$ Subtract the exponents.

$= 174.05$ Evaluate.

There will be about 174 people per square mile in Tennessee in 2030.

On Your Own

Now You're Ready Exercises 23–28

7. The projected population of Alabama in 2020 is about $2.25 \cdot 2^{21}$. The land area of Alabama is about 2^{17} square kilometers. Predict the average number of people per square kilometer in 2020.

13.3 Exercises

Vocabulary and Concept Check

1. **WRITING** Explain in your own words what it means to divide powers.

2. **WHICH ONE DOESN'T BELONG?** Which quotient does *not* belong with the other three? Explain your reasoning.

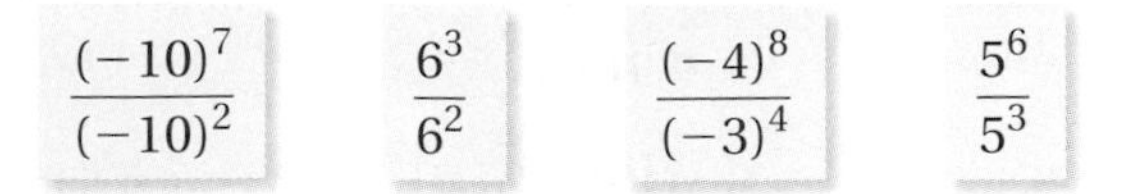

$\dfrac{(-10)^7}{(-10)^2}$ $\quad$ $\dfrac{6^3}{6^2}$ $\quad$ $\dfrac{(-4)^8}{(-3)^4}$ $\quad$ $\dfrac{5^6}{5^3}$

Practice and Problem Solving

Simplify the expression. Write your answer as a power.

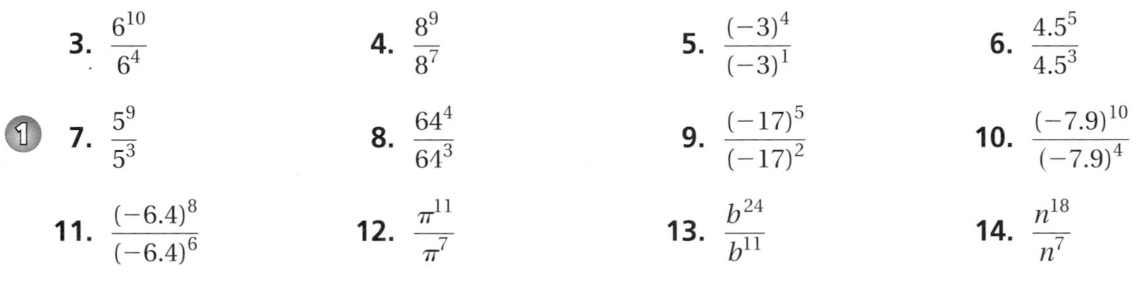

3. $\dfrac{6^{10}}{6^4}$
4. $\dfrac{8^9}{8^7}$
5. $\dfrac{(-3)^4}{(-3)^1}$
6. $\dfrac{4.5^5}{4.5^3}$
7. (1) $\dfrac{5^9}{5^3}$
8. $\dfrac{64^4}{64^3}$
9. $\dfrac{(-17)^5}{(-17)^2}$
10. $\dfrac{(-7.9)^{10}}{(-7.9)^4}$
11. $\dfrac{(-6.4)^8}{(-6.4)^6}$
12. $\dfrac{\pi^{11}}{\pi^7}$
13. $\dfrac{b^{24}}{b^{11}}$
14. $\dfrac{n^{18}}{n^7}$

15. **ERROR ANALYSIS** Describe and correct the error in simplifying the quotient.

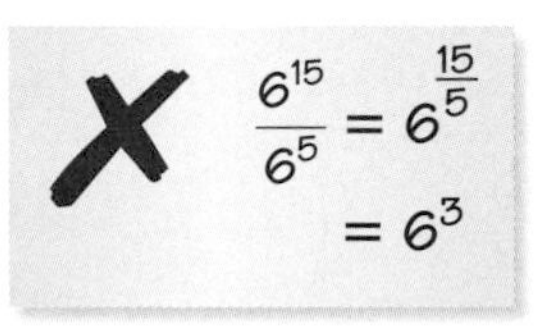

Simplify the expression. Write your answer as a power.

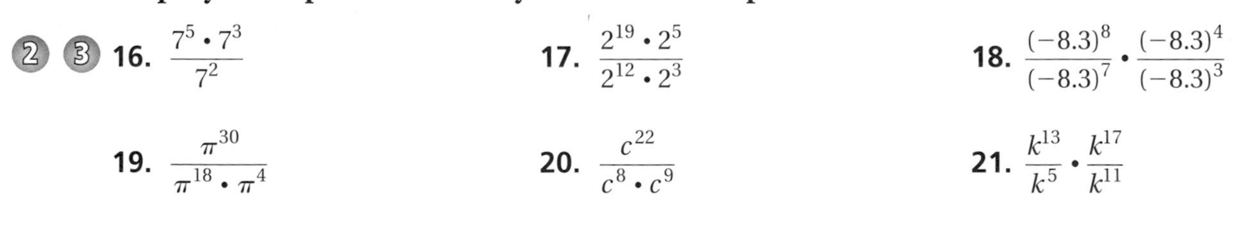

16. (2) (3) $\dfrac{7^5 \cdot 7^3}{7^2}$
17. $\dfrac{2^{19} \cdot 2^5}{2^{12} \cdot 2^3}$
18. $\dfrac{(-8.3)^8}{(-8.3)^7} \cdot \dfrac{(-8.3)^4}{(-8.3)^3}$
19. $\dfrac{\pi^{30}}{\pi^{18} \cdot \pi^4}$
20. $\dfrac{c^{22}}{c^8 \cdot c^9}$
21. $\dfrac{k^{13}}{k^5} \cdot \dfrac{k^{17}}{k^{11}}$

22. **SOUND INTENSITY** The sound intensity of a normal conversation is 10^6 times greater than the quietest noise a person can hear. The sound intensity of a jet at takeoff is 10^{14} times greater than the quietest noise a person can hear. How many times more intense is the sound of a jet at takeoff than the sound of a normal conversation?

Simplify the expression.

4 **23.** $\dfrac{x \cdot 4^8}{4^5}$

24. $\dfrac{6^3 \cdot w}{6^2}$

25. $\dfrac{a^3 \cdot b^4 \cdot 5^4}{b^2 \cdot 5}$

26. $\dfrac{5^{12} \cdot c^{10} \cdot d^2}{5^9 \cdot c^9}$

27. $\dfrac{x^{15}y^9}{x^8y^3}$

28. $\dfrac{m^{10}n^7}{m^1n^6}$

29. MEMORY The memory capacities and prices of five MP3 players are shown in the table.

MP3 Player	Memory (GB)	Price
A	2^1	\$70
B	2^2	\$120
C	2^3	\$170
D	2^4	\$220
E	2^5	\$270

a. How many times more memory does MP3 Player D have than MP3 Player B?

b. Do the differences in price between consecutive sizes reflect a constant rate of change?

30. CRITICAL THINKING Consider the equation $\dfrac{9^m}{9^n} = 9^2$.

a. Find two numbers m and n that satisfy the equation.

b. Are there any other pairs of numbers that satisfy the equation? Explain.

Milky Way Galaxy
$10 \cdot 10^{10}$ stars

31. STARS There are about 10^{24} stars in the Universe. Each galaxy has approximately the same number of stars as the Milky Way Galaxy. About how many galaxies are in the Universe?

32. Number Sense Find the value of x that makes $\dfrac{8^{3x}}{8^{2x+1}} = 8^9$ true. Explain how you found your answer.

Fair Game Review What you learned in previous grades & lessons

Subtract. *(Section 1.2 and Section 1.3)*

33. $-4 - 5$ **34.** $-23 - (-15)$ **35.** $33 - (-28)$ **36.** $18 - 22$

37. MULTIPLE CHOICE What is the value of x? *(Section 11.1)*

(A) 20 (B) 30

(C) 45 (D) 60

13 Study Help

Check It Out
Graphic Organizer
BigIdeasMath.com

You can use an **information wheel** to organize information about a topic. Here is an example of an information wheel for exponents.

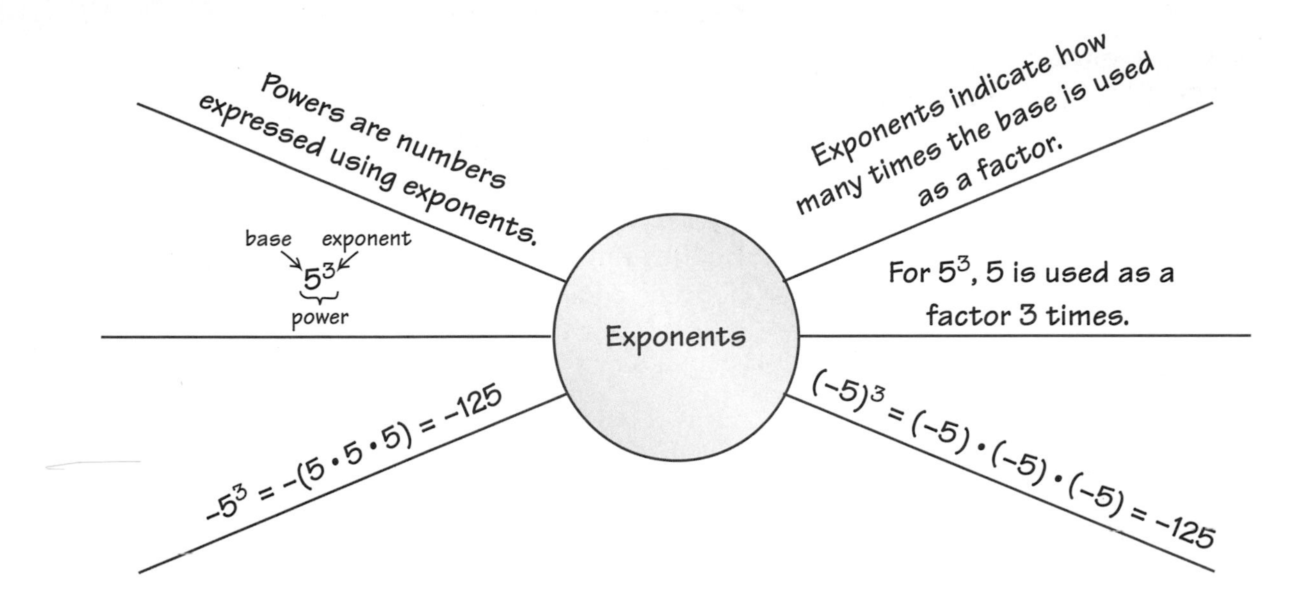

On Your Own

Make an information wheel to help you study these topics.

1. order of operations
2. Product of Powers Property
3. Quotient of Powers Property

After you complete this chapter, make information wheels for the following topics.

4. zero exponents
5. negative exponents
6. writing numbers in scientific notation
7. writing numbers in standard form
8. Choose three other topics you studied earlier in this course. Make an information wheel for each topic to summarize what you know about them.

"My information wheel for Fluffy has matching adjectives and nouns."

13.1–13.3 Quiz

Check It Out
Progress Check
BigIdeasMath.com

Write the product using exponents. *(Section 13.1)*

1. $(-5) \cdot (-5) \cdot (-5) \cdot (-5)$

2. $\frac{1}{6} \cdot \frac{1}{6} \cdot \frac{1}{6} \cdot \frac{1}{6} \cdot \frac{1}{6}$

3. $(-x) \cdot (-x) \cdot (-x) \cdot (-x) \cdot (-x) \cdot (-x)$

4. $7 \cdot 7 \cdot m \cdot m \cdot m$

Evaluate the expression. *(Section 13.1)*

5. 5^4

6. $(-2)^6$

Simplify the expression. Write your answer as a power. *(Section 13.2)*

7. $3^8 \cdot 3$

8. $(a^5)^3$

Simplify the expression. *(Section 13.2)*

9. $(3c)^4$

10. $\left(-\frac{2}{7}p\right)^2$

Simplify the expression. Write your answer as a power. *(Section 13.3)*

11. $\frac{8^7}{8^4}$

12. $\frac{6^3 \cdot 6^7}{6^2}$

13. $\frac{\pi^{15}}{\pi^3 \cdot \pi^9}$

14. $\frac{t^{13}}{t^5} \cdot \frac{t^8}{t^6}$

15. SEQUENCE The nth term of a sequence can be found by evaluating $10^n - 1$. Copy and complete the table to find the first four terms of the sequence. *(Section 13.1)*

n	$10^n - 1$
1	
2	
3	
4	

16. CRITICAL THINKING Is $(ab)^2$ equivalent to ab^2? Explain. *(Section 13.2)*

17. EARTHQUAKES An earthquake of magnitude 3.0 is 10^2 times stronger than an earthquake of magnitude 1.0. An earthquake of magnitude 8.0 is 10^7 times stronger than an earthquake of magnitude 1.0. How many times stronger is an earthquake of magnitude 8.0 than an earthquake of magnitude 3.0? *(Section 13.3)*

13.4 Zero and Negative Exponents

COMMON CORE STATE STANDARDS
8.EE.1

Essential Question How can you define zero and negative exponents?

1 ACTIVITY: Finding Patterns and Writing Definitions

Work with a partner.

a. Talk about the following notation.

Thousands Hundreds Tens Ones

$$4327 = 4 \cdot 10^3 + 3 \cdot 10^2 + 2 \cdot 10^1 + 7 \cdot 10^{\square}$$

What patterns do you see in the first three exponents?
Continue the pattern to find the fourth exponent.
How would you define 10^0? Explain.

b. Copy and complete the table.

n	5	4	3	2	1	0
2^n						

What patterns do you see in the first six values of 2^n?
How would you define 2^0? Explain.

c. Use the Quotient of Powers Property to complete the table.

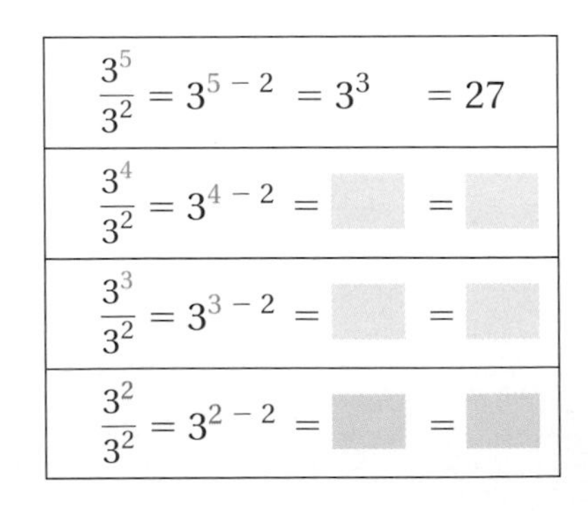

$\frac{3^5}{3^2} = 3^{5-2} = 3^3 = 27$
$\frac{3^4}{3^2} = 3^{4-2} = \square = \square$
$\frac{3^3}{3^2} = 3^{3-2} = \square = \square$
$\frac{3^2}{3^2} = 3^{2-2} = \square = \square$

What patterns do you see in the first four rows of the table?
How would you define 3^0? Explain.

2 ACTIVITY: Comparing Volumes

Work with a partner.

The quotients show three ratios of the volumes of the solids. Identify each ratio, find its value, and describe what it means.

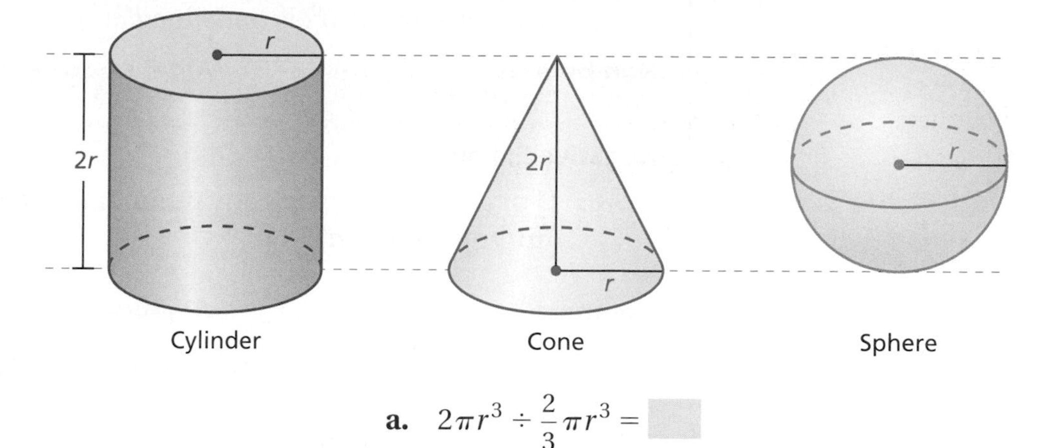

a. $2\pi r^3 \div \frac{2}{3}\pi r^3 = $ ▢

b. $\frac{4}{3}\pi r^3 \div \frac{2}{3}\pi r^3 = $ ▢

c. $2\pi r^3 \div \frac{4}{3}\pi r^3 = $ ▢

3 ACTIVITY: Writing a Definition

Work with a partner.

Compare the two methods used to simplify $\frac{3^2}{3^5}$. Then describe how you can rewrite a power with a negative exponent as a fraction.

Method 1

$$\frac{3^2}{3^5} = \frac{\overset{1}{\cancel{3}} \cdot \overset{1}{\cancel{3}}}{\underset{1}{\cancel{3}} \cdot \underset{1}{\cancel{3}} \cdot 3 \cdot 3 \cdot 3}$$

$$= \frac{1}{3^3}$$

Method 2

$$\frac{3^2}{3^5} = 3^{2-5}$$

$$= 3^{-3}$$

What Is Your Answer?

4. **IN YOUR OWN WORDS** How can you define zero and negative exponents? Give two examples of each.

Practice Use what you learned about zero and negative exponents to complete Exercises 5–8 on page 578.

13.4 Lesson

Key Ideas

Zero Exponents

Words Any nonzero number to the zero power is equal to 1. Zero to the zero power, 0^0, is *undefined*.

Numbers $4^0 = 1$ **Algebra** $a^0 = 1$, where $a \neq 0$

Negative Exponents

Words For any integer n and any number a not equal to 0, a^{-n} is equal to 1 divided by a^n.

Numbers $4^{-2} = \frac{1}{4^2}$ **Algebra** $a^{-n} = \frac{1}{a^n}$, where $a \neq 0$

EXAMPLE 1 Evaluating Expressions

a. $3^{-4} = \frac{1}{3^4}$ Definition of negative exponent

$= \frac{1}{81}$ Evaluate power.

b. $(-8.5)^{-4} \cdot (-8.5)^4 = (-8.5)^{-4+4}$ Add the exponents.

$= (-8.5)^0$ Simplify.

$= 1$ Definition of zero exponent

c. $\frac{2^6}{2^8} = 2^{6-8}$ Subtract the exponents.

$= 2^{-2}$ Simplify.

$= \frac{1}{2^2}$ Definition of negative exponent

$= \frac{1}{4}$ Evaluate power.

On Your Own

Now You're Ready
Exercises 9–16

Evaluate the expression.

1. 4^{-2}
2. $(-2)^{-5}$
3. $6^{-8} \cdot 6^8$
4. $\frac{(-3)^5}{(-3)^6}$
5. $\frac{1}{5^7} \cdot \frac{1}{5^{-4}}$
6. $\frac{4^5 \cdot 4^{-3}}{4^2}$

EXAMPLE 2 Simplifying Expressions

a. $-5x^0 = -5(1)$ Definition of zero exponent

$= -5$ Multiply.

b. $\dfrac{9y^{-3}}{y^5} = 9y^{-3-5}$ Subtract the exponents.

$= 9y^{-8}$ Simplify.

$= \dfrac{9}{y^8}$ Definition of negative exponent

On Your Own

Now You're Ready
Exercises 20–27

Simplify. Write the expression using only positive exponents.

7. $8x^{-2}$

8. $b^0 \cdot b^{-10}$

9. $\dfrac{z^6}{15z^9}$

EXAMPLE 3 Real-Life Application

A drop of water leaks from a faucet every second. How many liters of water leak from the faucet in 1 hour?

Convert 1 hour to seconds.

$$1 \cancel{h} \times \frac{60 \cancel{min}}{1 \cancel{h}} \times \frac{60 \text{ sec}}{1 \cancel{min}} = 3600 \text{ sec}$$

Drop of water: 50^{-2} L

Water leaks from the faucet at a rate of 50^{-2} liter per second. Multiply the time by the rate.

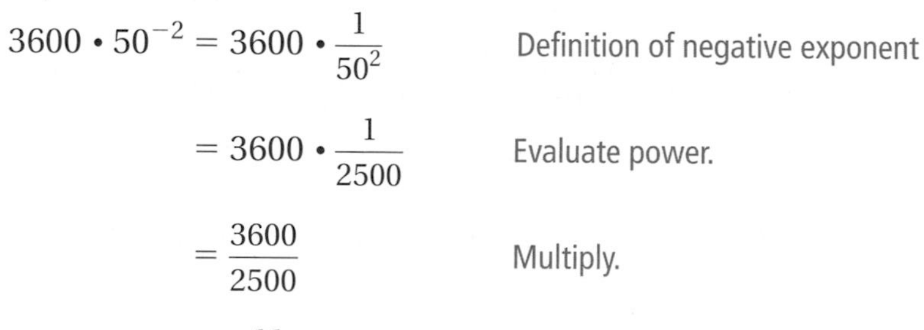

$3600 \cdot 50^{-2} = 3600 \cdot \dfrac{1}{50^2}$ Definition of negative exponent

$= 3600 \cdot \dfrac{1}{2500}$ Evaluate power.

$= \dfrac{3600}{2500}$ Multiply.

$= 1\dfrac{11}{25} = 1.44$ Simplify.

So, 1.44 liters of water leak from the faucet in 1 hour.

On Your Own

10. **WHAT IF?** In Example 3, the faucet leaks water at a rate of 5^{-5} liter per second. How many liters of water leak from the faucet in 1 hour?

13.4 Exercises

Vocabulary and Concept Check

1. **VOCABULARY** If a is a nonzero number, does the value of a^0 depend on the value of a? Explain.
2. **WRITING** Explain how to evaluate 10^{-3}.
3. **NUMBER SENSE** Without evaluating, order 5^0, 5^4, and 5^{-5} from least to greatest.
4. **DIFFERENT WORDS, SAME QUESTION** Which is different? Find "both" answers.

Rewrite $\frac{1}{3 \cdot 3 \cdot 3}$ using a negative exponent.

Write 3 to the negative third power.

Write $\frac{1}{3}$ cubed as a power.

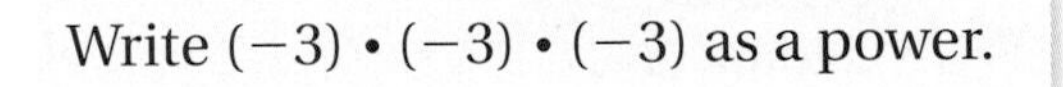
Write $(-3) \cdot (-3) \cdot (-3)$ as a power.

Practice and Problem Solving

5. Use the Quotient of Powers Property to copy and complete the table.

n	4	3	2	1
$\frac{5^n}{5^2}$				

6. What patterns do you see?
7. How would you define 5^0? Why?
8. How can you rewrite 5^{-1} as a fraction?

Evaluate the expression.

1 9. 6^{-2}

10. 158^0

11. $\frac{4^3}{4^5}$

12. $\frac{-3}{(-3)^2}$

13. $(-2)^{-8} \cdot (-2)^8$

14. $3^{-3} \cdot 3^{-2}$

15. $\frac{1}{5^{-3}} \cdot \frac{1}{5^6}$

16. $\frac{(1.5)^2}{(1.5)^{-2} \cdot (1.5)^4}$

17. **ERROR ANALYSIS** Describe and correct the error in evaluating the expression.

$$(4)^{-3} = (-4)(-4)(-4)$$
$$= -64$$

18. **SAND** The mass of a grain of sand is about 10^{-3} gram. About how many grains of sand are in the bag of sand?

19. **CRITICAL THINKING** How can you write the number 1 as 2 to a power? 10 to a power?

Simplify. Write the expression using only positive exponents.

② **20.** $6y^{-4}$ **21.** $8^{-2} \cdot a^7$ **22.** $\frac{9c^3}{c^{-4}}$ **23.** $\frac{5b^{-2}}{b^{-3}}$

24. $\frac{8x^3}{2x^9}$ **25.** $3d^{-4} \cdot 4d^4$ **26.** $m^{-2} \cdot n^3$ **27.** $\frac{3^{-2} \cdot k^0 \cdot w^0}{w^{-6}}$

METRIC UNITS In Exercises 28–31, use the table.

28. How many millimeters are in a decimeter?

29. How many micrometers are in a centimeter?

30. How many nanometers are in a millimeter?

31. How many micrometers are in a meter?

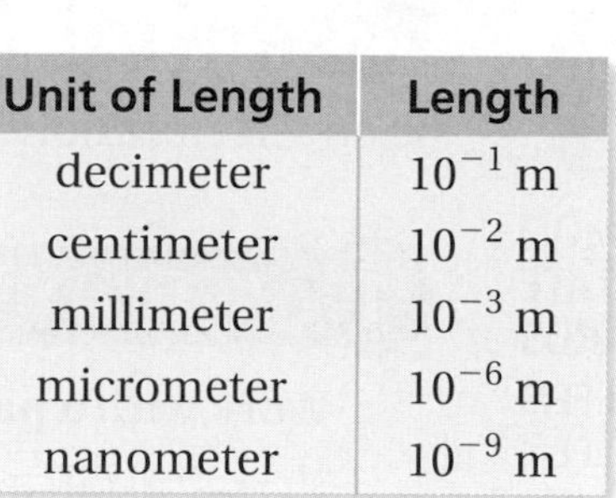

Unit of Length	Length
decimeter	10^{-1} m
centimeter	10^{-2} m
millimeter	10^{-3} m
micrometer	10^{-6} m
nanometer	10^{-9} m

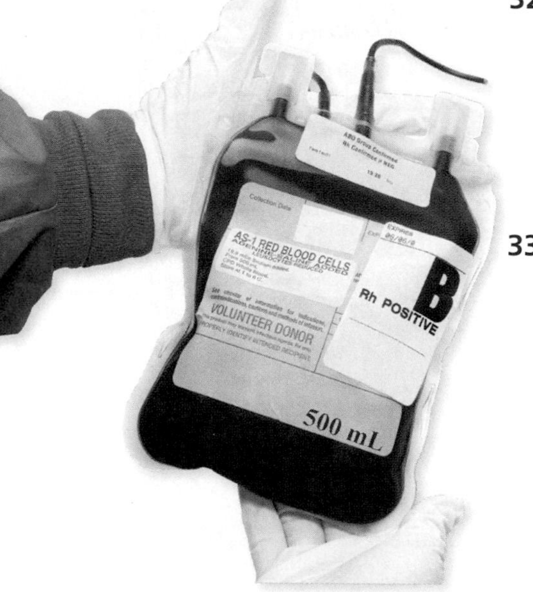

32. MICROBES A species of bacteria is 10 micrometers long. A virus is 10,000 times smaller than the bacteria.

a. Using the table above, find the length of the virus in meters.

b. Is the answer to part (a) *less than*, *greater than*, or *equal to* one nanometer?

33. BLOOD DONATION Every 2 seconds, someone in the United States needs blood. A sample blood donation is shown. ($1 \text{ mm}^3 = 10^{-3}$ mL)

a. One cubic millimeter of blood contains about 10^4 white blood cells. How many white blood cells are in the donation? Write your answer in words.

b. One cubic millimeter of blood contains about 5×10^6 red blood cells. How many red blood cells are in the donation? Write your answer in words.

c. Compare your answers for parts (a) and (b).

34. OPEN-ENDED Write two different powers with negative exponents that have the same value.

35. **Reasoning** The rule for negative exponents states that $a^{-n} = \frac{1}{a^n}$. Explain why this rule does not apply when $a = 0$.

Fair Game Review What you learned in previous grades & lessons

Simplify the expression. *(Section 13.2 and Section 13.3)*

36. $10^3 \cdot 10^6$ **37.** $10^2 \cdot 10$ **38.** $\frac{10^8}{10^4}$

39. MULTIPLE CHOICE What value of x makes the equation $5(x - 6) = 3x + 8$ true? *(Section 10.2)*

Ⓐ -19 Ⓑ -2.75 Ⓒ 7 Ⓓ 19

13.5 Reading Scientific Notation

COMMON CORE STATE STANDARDS
8.EE.3
8.EE.4

Essential Question How can you read numbers that are written in scientific notation?

1 ACTIVITY: Very Large Numbers

Work with a partner.

- Use a calculator. Experiment with multiplying large numbers until your calculator gives an answer that is *not* in standard form.
- When the calculator at the right was used to multiply 2 billion by 3 billion, it listed the result as

  6.0E+18.

- Multiply 2 billion by 3 billion by hand. Use the result to explain what 6.0E+18 means.
- Check your explanation using products of other large numbers.
- Why didn't the calculator show the answer in standard form?
- Experiment to find the maximum number of digits your calculator displays. For instance, if you multiply 1000 by 1000 and your calculator shows 1,000,000, then it can display 7 digits.

2 ACTIVITY: Very Small Numbers

Work with a partner.

- Use a calculator. Experiment with multiplying very small numbers until your calculator gives an answer that is *not* in standard form.
- When the calculator at the right was used to multiply 2 billionths by 3 billionths, it listed the result as

  6.0E–18.

- Multiply 2 billionths by 3 billionths by hand. Use the result to explain what 6.0E–18 means.
- Check your explanation using products of other very small numbers.

3 ACTIVITY: Reading Scientific Notation

Work with a partner.

Each description gives an example of a number written in scientific notation. Answer the question in the description. Write your answer in standard form.

a. Nearly 1.0×10^5 dust mites can live in 1 square yard of carpet.

How many dust mites can live in 100 square yards of carpet?

b. A micron is about 4.0×10^{-5} inch. The length of a dust mite is 250 microns.

How long is a dust mite in inches?

c. About 1.0×10^{15} bacteria live in a human body.

How many bacteria are living in the humans in your classroom?

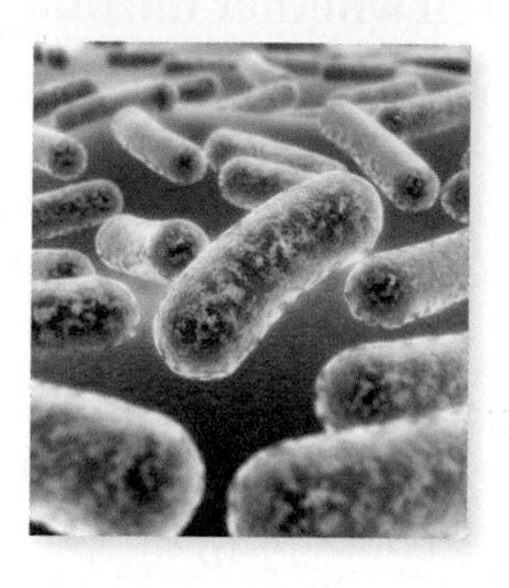

d. A micron is about 4.0×10^{-5} inch. The length of a bacterium is about 0.5 micron.

How many bacteria could lie end-to-end on your finger?

e. Earth has only about 1.5×10^8 kilograms of gold. Earth has a mass of 6.0×10^{24} kilograms.

What percent of Earth's mass is gold?

f. A gram is about 0.035 ounce. An atom of gold weighs about 3.3×10^{-22} gram.

How many atoms are in an ounce of gold?

What Is Your Answer?

4. **IN YOUR OWN WORDS** How can you read numbers that are written in scientific notation? Why do you think this type of notation is called "scientific notation?" Why is scientific notation important?

Use what you learned about reading scientific notation to complete Exercises 3–5 on page 584.

Check It Out
Lesson Tutorials
BigIdeasMath.com

Key Vocabulary
scientific notation, *p. 582*

Study Tip
Scientific notation is used to write very small and very large numbers.

Key Idea

Scientific Notation

A number is written in **scientific notation** when it is represented as the product of a factor and a power of 10. The factor must be at least 1 and less than 10.

The factor is at least 1 and less than 10.
8.3 × 10^-7
The power of 10 has an integer exponent.

EXAMPLE 1 Identifying Numbers Written in Scientific Notation

Tell whether the number is written in scientific notation. Explain.

a. 5.9×10^{-6}

∴ The factor is at least 1 and less than 10. The power of 10 has an integer exponent. So, the number is written in scientific notation.

b. 0.9×10^{8}

∴ The factor is less than 1. So, the number is not written in scientific notation.

Key Idea

Writing Numbers in Standard Form

When writing a number from scientific notation to standard form, the absolute value of the exponent tells you how many places to move the decimal point.

- If the exponent is negative, move the decimal point to the left.
- If the exponent is positive, move the decimal point to the right.

EXAMPLE 2 Writing Numbers in Standard Form

a. Write 3.22×10^{-4} in standard form.

$3.22 \times 10^{-4} = 0.000322$ (4) — Move decimal point $|-4| = 4$ places to the left.

b. Write 7.9×10^{5} in standard form.

$7.9 \times 10^{5} = 790{,}000$ (5) — Move decimal point $|5| = 5$ places to the right.

On Your Own

Now You're Ready
Exercises 6–23

1. Is 12×10^4 written in scientific notation? Explain.

Write the number in standard form.

2. 6×10^7
3. 9.9×10^{-5}
4. 1.285×10^4

EXAMPLE 3 Comparing Numbers in Scientific Notation

An object with a lesser density than water will float. An object with a greater density than water will sink. Use each given density (in kilograms per cubic meter) to explain what happens when you place a brick and an apple in water.

Water: 1.0×10^3 **Brick:** 1.84×10^3 **Apple:** 6.41×10^2

Write each density in standard form.

Water	Brick	Apple
$1.0 \times 10^3 = 1000$	$1.84 \times 10^3 = 1840$	$6.41 \times 10^2 = 641$

∴ The apple is less dense than water, so it will float. The brick is denser than water, so it will sink.

EXAMPLE 4 Real-Life Application

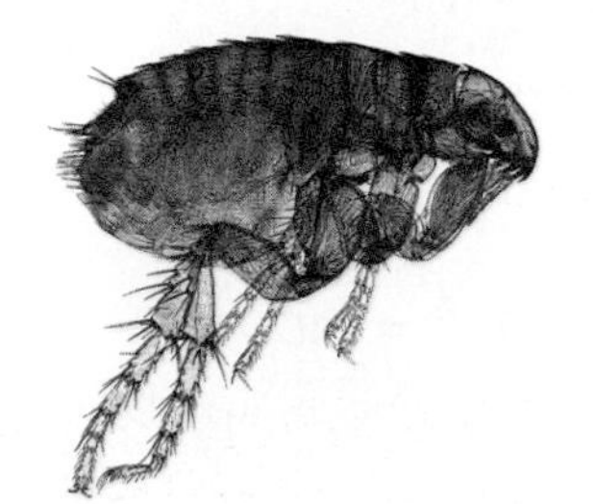

A female flea consumes about 1.4×10^{-5} liter of blood per day.

A dog has 100 female fleas. How many milliliters of blood do the fleas consume per day?

$$1.4 \times 10^{-5} \cdot 100 = 0.000014 \cdot 100 \quad \text{Write in standard form.}$$

$$= 0.0014 \quad \text{Multiply.}$$

∴ The fleas consume about 0.0014 liter, or 1.4 milliliters of blood per day.

On Your Own

5. **WHAT IF?** In Example 3, the density of lead is 1.14×10^4 kilograms per cubic meter. What happens when lead is placed in water?
6. **WHAT IF?** In Example 4, a dog has 75 female fleas. How many milliliters of blood do the fleas consume per day?

13.5 Exercises

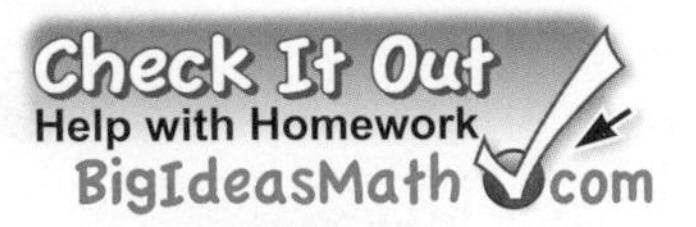

Vocabulary and Concept Check

1. **WRITING** Describe the difference between scientific notation and standard form.
2. **WHICH ONE DOESN'T BELONG?** Which number does *not* belong with the other three? Explain.

2.8×10^{15} | 4.3×10^{-30} | 1.05×10^{28} | 10×9.2^{-13}

Practice and Problem Solving

Write your answer in standard form.

3. A micrometer is 1.0×10^{-6} meter. How long is 150 micrometers in meters?
4. An acre is about 4.05×10^{7} square centimeters. How many square centimeters are in 4 acres?
5. A cubic millimeter is about 6.1×10^{-5} cubic inches. How many cubic millimeters are in 1.22 cubic inches?

Tell whether the number is written in scientific notation. Explain.

(1)
6. 1.8×10^{9}
7. 3.45×10^{14}
8. 0.26×10^{-25}
9. 10.5×10^{12}
10. 46×10^{-17}
11. 5×10^{-19}
12. 7.814×10^{-36}
13. 0.999×10^{42}
14. 6.022×10^{23}

Write the number in standard form.

(2)
15. 7×10^{7}
16. 8×10^{-3}
17. 5×10^{2}
18. 2.7×10^{-4}
19. 4.4×10^{-5}
20. 2.1×10^{3}
21. 1.66×10^{9}
22. 3.85×10^{-8}
23. 9.725×10^{6}

24. **ERROR ANALYSIS** Describe and correct the error in writing the number in standard form.

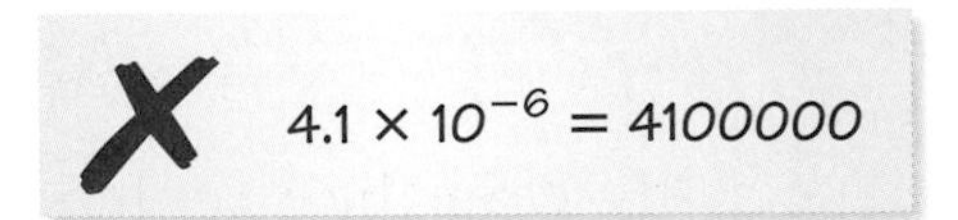

2.7×10^{8} platelets per milliliter

25. **PLATELETS** Platelets are cell-like particles in the blood that help form blood clots.
    a. How many platelets are in 3 milliliters of blood? Write your answer in standard form.
    b. An adult body contains about 5 liters of blood. How many platelets are in an adult body?

26. **REASONING** A googol is 1.0×10^{100}. How many zeros are in a googol?

3 27. **STARS** The table shows the surface temperatures of five stars.

a. Which star has the highest surface temperature?

b. Which star has the lowest surface temperature?

Star	Betelgeuse	Bellatrix	Sun	Aldebaran	Rigel
Surface Temperature (°F)	6.2×10^3	3.8×10^4	1.1×10^4	7.2×10^3	2.2×10^4

28. **CORAL REEF** The area of the Florida Keys National Marine Sanctuary is about 9.6×10^3 square kilometers. The area of the Florida Reef Tract is about 16.2% of the area of the sanctuary. What is the area of the Florida Reef Tract in square kilometers?

29. **REASONING** A gigameter is 1.0×10^6 kilometers. How many square kilometers are in 5 square gigameters?

30. **WATER** There are about 1.4×10^9 cubic kilometers of water on Earth. About 2.5% of the water is fresh water. How much fresh water is on Earth?

31. **Critical Thinking** The table shows the speed of light through five media.

a. In which medium does light travel the fastest?

b. In which medium does light travel the slowest?

Medium	Speed
Air	6.7×10^8 mi/h
Glass	6.6×10^8 ft/sec
Ice	2.3×10^5 km/sec
Vacuum	3.0×10^8 m/sec
Water	2.3×10^{10} cm/sec

Fair Game Review What you learned in previous grades & lessons

Simplify. Write the expression using only positive exponents. *(Section 13.2 and Section 13.4)*

32. $6^3 \cdot 6^5$

33. $10^2 \cdot 10^5$

34. $8^{-1} \cdot 8^{-11}$

35. $10^{-6} \cdot 10^{-10}$

36. **MULTIPLE CHOICE** What is the length of the hypotenuse of the right triangle? *(Section 12.5)*

Ⓐ $\sqrt{18}$ in. Ⓑ $\sqrt{41}$ in.

Ⓒ 18 in. Ⓓ 41 in.

4 in.

5 in.

13.6 Writing Scientific Notation

Essential Question How can you write a number in scientific notation?

COMMON CORE STATE STANDARDS
8.EE.3
8.EE.4

1 ACTIVITY: Finding pH Levels

Work with a partner. In chemistry, pH is a measure of the activity of dissolved hydrogen ions (H^+). Liquids with low pH values are called acids. Liquids with high pH values are called bases.

Find the pH of each liquid. Is the liquid a base, neutral, or an acid?

a. Lime juice:
$[H^+] = 0.01$

b. Egg:
$[H^+] = 0.00000001$

c. Distilled water:
$[H^+] = 0.0000001$

d. Ammonia water:
$[H^+] = 0.00000000001$

e. Tomato juice:
$[H^+] = 0.0001$

f. Hydrochloric acid:
$[H^+] = 1$

pH	$[H^+]$	
14	1×10^{-14}	Bases
13	1×10^{-13}	
12	1×10^{-12}	
11	1×10^{-11}	
10	1×10^{-10}	
9	1×10^{-9}	
8	1×10^{-8}	
7	1×10^{-7}	Neutral
6	1×10^{-6}	Acids
5	1×10^{-5}	
4	1×10^{-4}	
3	1×10^{-3}	
2	1×10^{-2}	
1	1×10^{-1}	
0	1×10^{0}	

2 ACTIVITY: Writing Scientific Notation

Work with a partner. Match each planet with its description. Then write each of the following in scientific notation.

- **Distance from the Sun (in miles)**
- **Distance from the Sun (in feet)**
- **Mass (in kilograms)**

a. Distance: 1,800,000,000 miles
Mass: 87,000,000,000,000,000,000,000,000 kg

b. Distance: 67,000,000 miles
Mass: 4,900,000,000,000,000,000,000,000 kg

c. Distance: 890,000,000 miles
Mass: 570,000,000,000,000,000,000,000,000 kg

d. Distance: 93,000,000 miles
Mass: 6,000,000,000,000,000,000,000,000 kg

e. Distance: 140,000,000 miles
Mass: 640,000,000,000,000,000,000,000 kg

f. Distance: 2,800,000,000 miles
Mass: 100,000,000,000,000,000,000,000,000 kg

g. Distance: 480,000,000 miles
Mass: 1,900,000,000,000,000,000,000,000,000 kg

h. Distance: 36,000,000 miles
Mass: 330,000,000,000,000,000,000,000 kg

3 ACTIVITY: Making a Scale Drawing

Work with a partner. The illustration in Activity 2 is not drawn to scale. Make a scale drawing of the distances in our solar system.

- **Cut a sheet of paper into three strips of equal width. Tape the strips together.**
- **Draw a long number line. Label the number line in hundreds of millions of miles.**
- **Locate each planet's position on the number line.**

What Is Your Answer?

4. **IN YOUR OWN WORDS** How can you write a number in scientific notation?

Practice Use what you learned about writing scientific notation to complete Exercises 3–5 on page 590.

13.6 Lesson

Key Idea

Writing Numbers in Scientific Notation

Step 1: Move the decimal point to the right of the first nonzero digit.

Step 2: Count the number of places you moved the decimal point. This determines the exponent of the power of 10.

Number greater than or equal to 10

Use a positive exponent when you move the decimal point to the left.

$8600 = 8.6 \times 10^3$ (3)

Number between 0 and 1

Use a negative exponent when you move the decimal point to the right.

$0.0024 = 2.4 \times 10^{-3}$ (3)

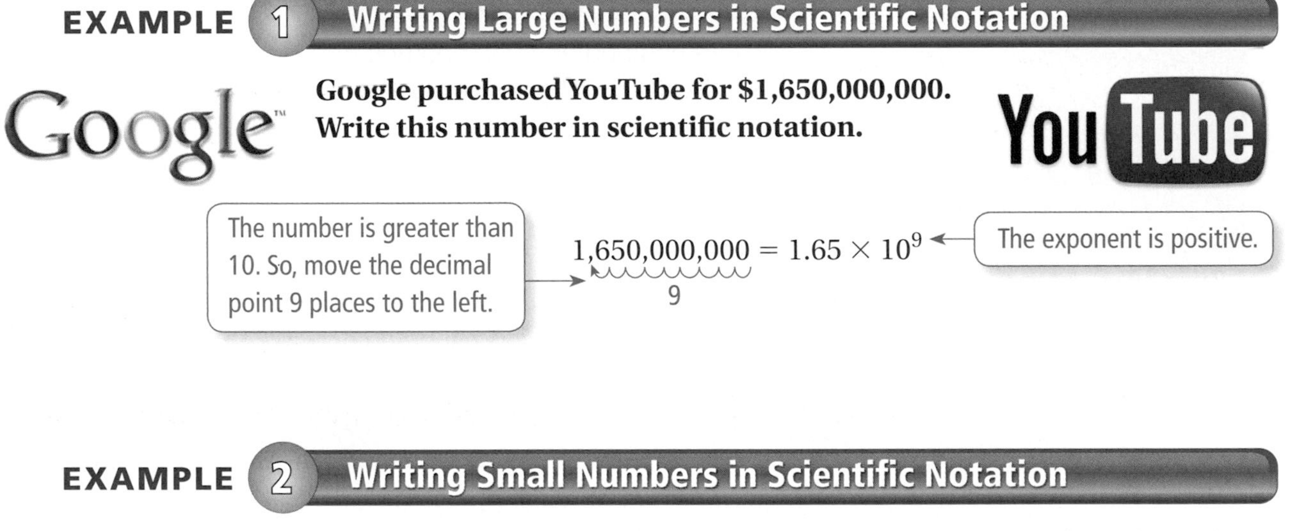

EXAMPLE 1 Writing Large Numbers in Scientific Notation

Google purchased YouTube for \$1,650,000,000. Write this number in scientific notation.

The number is greater than 10. So, move the decimal point 9 places to the left.

$1,650,000,000 = 1.65 \times 10^9$ (9)

The exponent is positive.

EXAMPLE 2 Writing Small Numbers in Scientific Notation

The 2004 Indonesian earthquake slowed the rotation of Earth, making the length of a day 0.00000268 second shorter. Write this number in scientific notation.

The number is between 0 and 1. So, move the decimal point 6 places to the right.

$0.00000268 = 2.68 \times 10^{-6}$ (6)

The exponent is negative.

On Your Own

Now You're Ready
Exercises 3–11

Write the number in scientific notation.

1. 50,000
2. 25,000,000
3. 683
4. 0.005
5. 0.00000033
6. 0.000506

EXAMPLE 3 Standardized Test Practice

An album has sold 8,780,000 copies. How many more copies does it need to sell to receive the award?

Ⓐ 1.22×10^{-7} Ⓑ 1.22×10^{-6}

Ⓒ 1.22×10^{6} Ⓓ 1.22×10^{7}

Use a model to solve the problem.

Remaining sales needed for award = Sales required for award − Current sales total

$= 10{,}000{,}000 - 8{,}780{,}000$

$= 1{,}220{,}000$

$= 1.22 \times 10^{6}$

∴ The album must sell 1.22×10^{6} more copies to receive the award. The correct answer is Ⓒ.

An album receives an award when it sells 10,000,000 copies.

On Your Own

7. An album has sold 955,000 copies. How many more copies does it need to sell to receive the award? Write your answer in scientific notation.

EXAMPLE 4 Multiplying Numbers in Scientific Notation

Find $(3 \times 10^{-5}) \times (5 \times 10^{-2})$. Write your answer in scientific notation.

$(3 \times 10^{-5}) \times (5 \times 10^{-2})$

$= 3 \times 5 \times 10^{-5} \times 10^{-2}$ Commutative Property of Multiplication

$= (3 \times 5) \times (10^{-5} \times 10^{-2})$ Associative Property of Multiplication

$= 15 \times 10^{-7}$ Simplify.

$= 1.5 \times 10^{1} \times 10^{-7}$ Write factor in scientific notation.

$= 1.5 \times 10^{-6}$ Simplify.

Study Tip

You can check your answer using standard form.

$(3 \times 10^{-5}) \times (5 \times 10^{-2})$

$= 0.00003 \times 0.05$

$= 0.0000015$

$= 1.5 \times 10^{-6}$

On Your Own

Now You're Ready
Exercises 14–19

Multiply. Write your answer in scientific notation.

8. $(2.5 \times 10^{8}) \times (2 \times 10^{3})$
9. $(2 \times 10^{-4}) \times (1 \times 10^{-4})$
10. $(5 \times 10^{-4}) \times (5.4 \times 10^{-9})$
11. $(7 \times 10^{2}) \times (3 \times 10^{5})$

13.6 Exercises

Vocabulary and Concept Check

1. **REASONING** How do you know whether a number written in standard form will have a positive or negative exponent when written in scientific notation?
2. **WRITING** Describe how to write a number in scientific notation.

Practice and Problem Solving

Write the number in scientific notation.

1 2
3. 0.0021
4. 5,430,000
5. 321,000,000
6. 0.00000625
7. 0.00004
8. 10,700,000
9. 45,600,000,000
10. 0.000000000009256
11. 840,000

ERROR ANALYSIS Describe and correct the error in writing the number in scientific notation.

12.

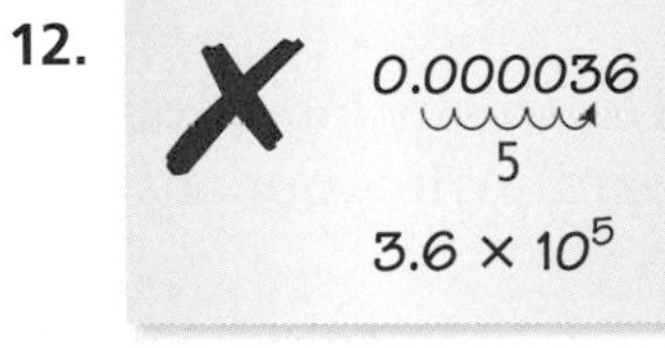

13.

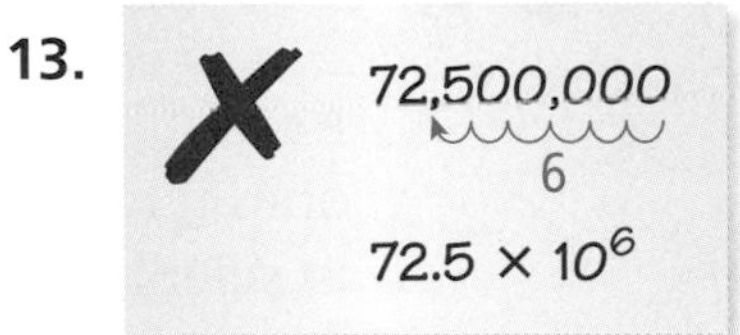

Multiply. Write your answer in scientific notation.

4
14. $(4 \times 10^4) \times (2 \times 10^6)$
15. $(3 \times 10^{-8}) \times (3 \times 10^{-2})$
16. $(5 \times 10^{-7}) \times (3 \times 10^6)$
17. $(8 \times 10^3) \times (2 \times 10^4)$
18. $(6 \times 10^8) \times (1.4 \times 10^{-5})$
19. $(7.2 \times 10^{-1}) \times (4 \times 10^{-7})$

20. **HAIR** What is the diameter of a human hair in scientific notation?

Diameter: 0.000099 meter

21. **EARTH** What is the circumference of Earth in scientific notation?

Circumference at the equator: about 40,100,000 meters

22. **WATERFALLS** During high flow, more than 44,380,000 gallons of water go over Niagara Falls every minute. Write this number in scientific notation.

Find the area of the figure. Write your answer in scientific notation.

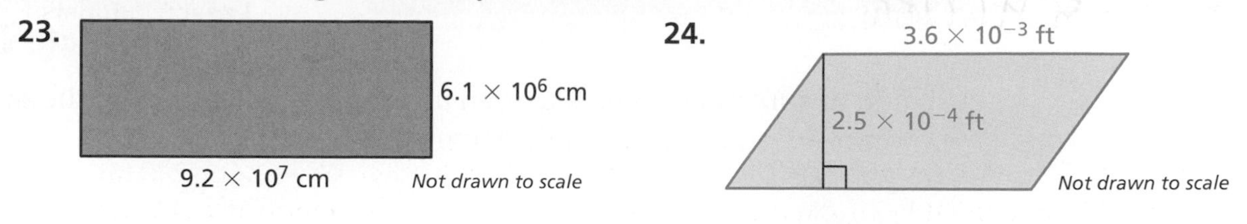

25. **SPACE SHUTTLE** The power of a space shuttle during launch is the force of the solid rocket boosters multiplied by the velocity. The velocity is 3.75×10^2 meters per second. What is the power (in newton-meters per second) of the shuttle shown during launch?

Force = 2.6×10^7 N

26. **NUMBER SENSE** Write 670 million in three ways.

27. **PROJECT** Use the Internet or some other reference to find the populations of India, China, Argentina, the United States, and Egypt. Round each population to the nearest million.
    - **a.** Write each population in scientific notation.
    - **b.** Use the Internet or some other reference to find the population density for each country.
    - **c.** Use the results of parts (a) and (b) to find the area of each country.

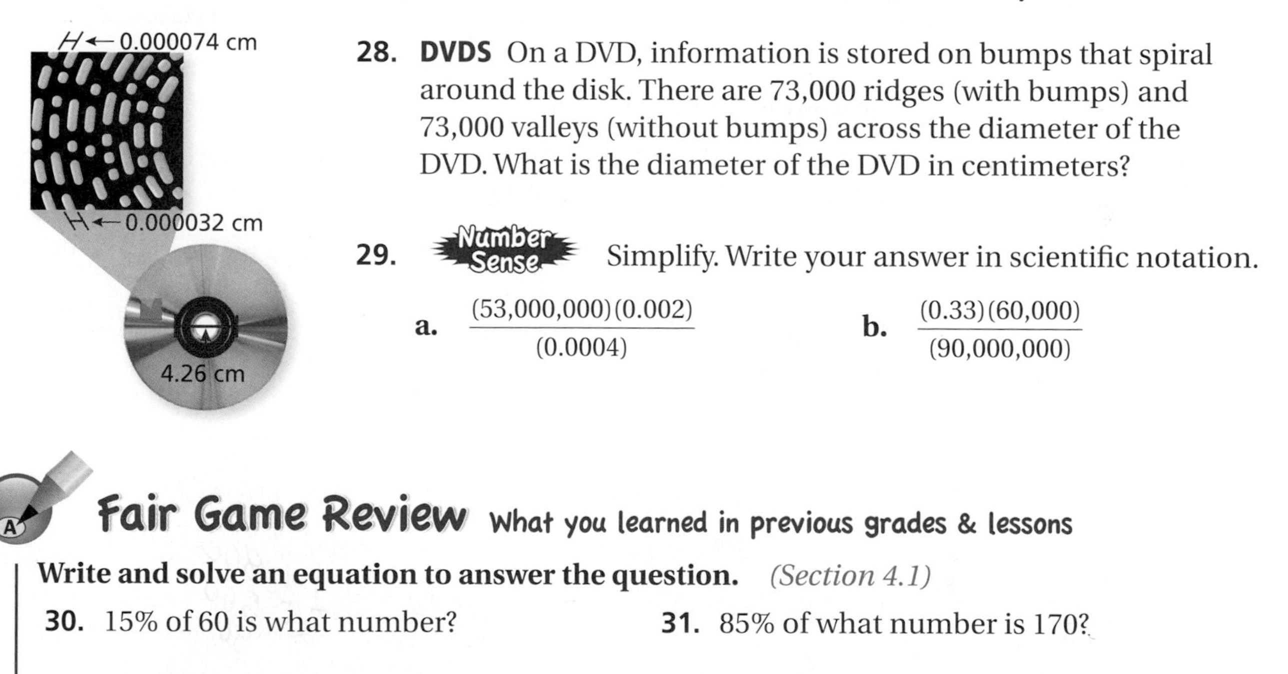

28. **DVDS** On a DVD, information is stored on bumps that spiral around the disk. There are 73,000 ridges (with bumps) and 73,000 valleys (without bumps) across the diameter of the DVD. What is the diameter of the DVD in centimeters?

29. **Number Sense** Simplify. Write your answer in scientific notation.
    - **a.** $\dfrac{(53{,}000{,}000)(0.002)}{(0.0004)}$
    - **b.** $\dfrac{(0.33)(60{,}000)}{(90{,}000{,}000)}$

Fair Game Review What you learned in previous grades & lessons

Write and solve an equation to answer the question. *(Section 4.1)*

30. 15% of 60 is what number?

31. 85% of what number is 170?

32. **MULTIPLE CHOICE** Which of the following is the slope of the line? *(Section 10.4)*

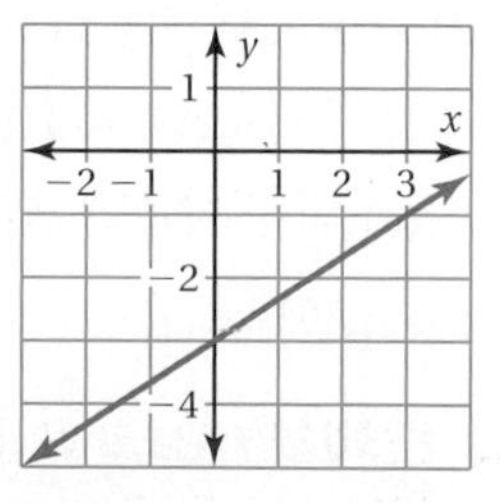

Ⓐ $-\dfrac{3}{2}$ Ⓑ $-\dfrac{2}{3}$

Ⓒ $\dfrac{2}{3}$ Ⓓ $\dfrac{3}{2}$

13.6b Scientific Notation

To add or subtract numbers written in scientific notation with the same power of 10, add or subtract the factors.

EXAMPLE 1 Adding Numbers Written in Scientific Notation

Find $(4.6 \times 10^3) + (8.72 \times 10^3)$. Write your answer in scientific notation.

$(4.6 \times 10^3) + (8.72 \times 10^3)$	
$= (4.6 + 8.72) \times 10^3$	Distributive Property
$= 13.32 \times 10^3$	Add.
$= (1.332 \times 10^1) \times 10^3$	Write 13.32 in scientific notation.
$= 1.332 \times 10^4$	Product of Powers Property

To add or subtract numbers written in scientific notation with different powers of 10, first rewrite the numbers so they have the same power of 10.

EXAMPLE 2 Subtracting Numbers Written in Scientific Notation

Find $(3.5 \times 10^{-2}) - (6.6 \times 10^{-3})$. Write your answer in scientific notation.

The numbers do not have the same power of 10. Rewrite 6.6×10^{-3} so that it has the same power of 10 as 3.5×10^{-2}.

$6.6 \times 10^{-3} = 6.6 \times 10^{-1} \times 10^{-2}$	Rewrite 10^{-3} as $10^{-1} \times 10^{-2}$.
$= 0.66 \times 10^{-2}$	Rewrite 6.6×10^{-1} as 0.66.

Subtract the factors.

$(3.5 \times 10^{-2}) - (0.66 \times 10^{-2})$	
$= (3.5 - 0.66) \times 10^{-2}$	Distributive Property
$= 2.84 \times 10^{-2}$	Subtract.

Practice

Add or subtract. Write your answer in scientific notation.

1. $(3 \times 10^7) + (2.4 \times 10^7)$
2. $(7.2 \times 10^{-6}) + (5.44 \times 10^{-6})$
3. $(9.2 \times 10^8) - (4 \times 10^8)$
4. $(7.8 \times 10^{-5}) - (4.5 \times 10^{-5})$
5. $(9.7 \times 10^6) + (6.7 \times 10^5)$
6. $(8.2 \times 10^2) + (3.41 \times 10^{-1})$
7. $(1.1 \times 10^5) - (4.3 \times 10^4)$
8. $(2.4 \times 10^{-1}) - (5.5 \times 10^{-2})$

To divide numbers written in scientific notation, divide the factors and powers of 10 separately.

EXAMPLE 3 Dividing Numbers Written in Scientific Notation

Find $\dfrac{1.5 \times 10^{-8}}{6 \times 10^{7}}$. Write your answer in scientific notation.

$$\frac{1.5 \times 10^{-8}}{6 \times 10^{7}} = \frac{1.5}{6} \times \frac{10^{-8}}{10^{7}}$$ Rewrite as a product of fractions.

$$= 0.25 \times \frac{10^{-8}}{10^{7}}$$ Divide 1.5 by 6.

$$= 0.25 \times 10^{-15}$$ Quotient of Powers Property

$$= 2.5 \times 10^{-1} \times 10^{-15}$$ Write 0.25 in scientific notation.

$$= 2.5 \times 10^{-16}$$ Product of Powers Property

EXAMPLE 4 Real-Life Application

How many times greater is the diameter of the Sun than the diameter of Earth?

Diameter $= 1.28 \times 10^4$ km

Diameter $= 1.4 \times 10^6$ km

Divide the diameter of the Sun by the diameter of Earth.

$$\frac{1.4 \times 10^{6}}{1.28 \times 10^{4}} = \frac{1.4}{1.28} \times \frac{10^{6}}{10^{4}}$$ Rewrite as a product of fractions.

$$= 1.09375 \times 10^{2}$$ Divide and use Quotient of Powers Property.

$$= 109.375$$ Write in standard form.

The diameter of the Sun is about 109 times greater than the diameter of Earth.

Practice

Divide. Write your answer in scientific notation.

9. $(6 \times 10^4) \div (3 \times 10^4)$
10. $(2.3 \times 10^7) \div (9.2 \times 10^7)$
11. $(1.5 \times 10^{-3}) \div (7.5 \times 10^2)$
12. $(5.8 \times 10^{-6}) \div (2 \times 10^{-3})$

13. **MONEY** How many times greater is the thickness of a dime than the thickness of a dollar bill?

Thickness $= 1.35 \times 10^{-1}$ cm

Thickness $= 1.0922 \times 10^{-2}$ cm

13.4–13.6 Quiz

Evaluate the expression. *(Section 13.4)*

1. $(-4.8)^{-9} \cdot (-4.8)^{9}$

2. $\dfrac{5^4}{5^7}$

Simplify. Write the expression using only positive exponents. *(Section 13.4)*

3. $8d^{-6}$

4. $\dfrac{12x^5}{4x^7}$

Tell whether the number is written in scientific notation. Explain. *(Section 13.5)*

5. 23×10^9

6. 0.6×10^{-7}

Write the number in standard form. *(Section 13.5)*

7. 8×10^6

8. 1.6×10^{-2}

Write the number in scientific notation. *(Section 13.6)*

9. 0.00524

10. 892,000,000

Multiply. Write your answer in scientific notation. *(Section 13.6)*

11. $(9 \times 10^3) \times (4 \times 10^4)$

12. $(2 \times 10^{-5}) \times (3.1 \times 10^{-2})$

13. PLANETS The table shows the equatorial radii of the eight planets in our solar system. *(Section 13.5)*

a. Which planet has the second smallest equatorial radius?

b. Which planet has the second greatest equatorial radius?

Planet	Equatorial Radius (km)
Mercury	2.44×10^3
Venus	6.05×10^3
Earth	6.38×10^3
Mars	3.4×10^3
Jupiter	7.15×10^4
Saturn	6.03×10^4
Uranus	2.56×10^4
Neptune	2.48×10^4

14. OORT CLOUD The Oort cloud is a spherical cloud that surrounds our solar system. It is about 2×10^5 astronomical units from the Sun. An astronomical unit is about 1.5×10^8 kilometers. How far is the Oort cloud from the Sun in kilometers? *(Section 13.6)*

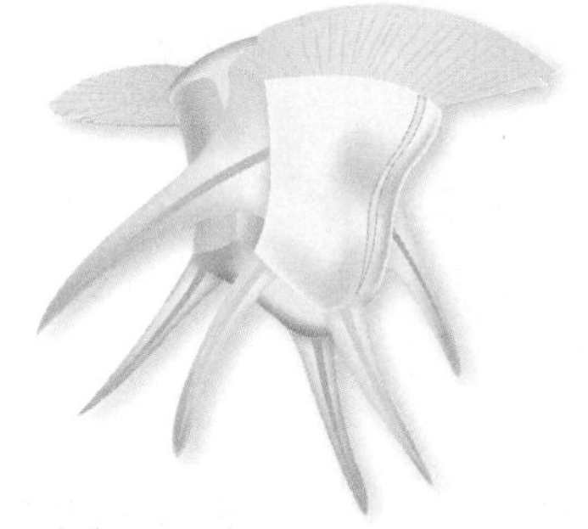

15. ORGANISM A one-celled, aquatic organism called a dinoflagellate is 1000 micrometers long. *(Section 13.4)*

a. One micrometer is 10^{-6} meter. What is the length of the dinoflagellate in meters?

b. Is the length of the dinoflagellate equal to 1 millimeter or 1 kilometer? Explain.

13 Chapter Review

Review Key Vocabulary

power, *p. 556*
base, *p. 556*
exponent, *p. 556*
scientific notation, *p. 582*

Review Examples and Exercises

13.1 Exponents *(pp. 554–559)*

Write $(-4) \cdot (-4) \cdot (-4) \cdot y \cdot y$ using exponents.

Because -4 is used as a factor 3 times, its exponent is 3. Because y is used as a factor 2 times, its exponent is 2.

∴ So, $(-4) \cdot (-4) \cdot (-4) \cdot y \cdot y = (-4)^3y^2$.

Exercises

Write the product using exponents.

1. $(-9) \cdot (-9) \cdot (-9) \cdot (-9) \cdot (-9)$
2. $2 \cdot 2 \cdot 2 \cdot n \cdot n$

Evaluate the expression.

3. 6^3
4. $-\left(\frac{1}{2}\right)^4$
5. $\left|\frac{1}{2}(16 - 6^3)\right|$

13.2 Product of Powers Property *(pp. 560–565)*

a. $\left(-\frac{1}{8}\right)^7 \cdot \left(-\frac{1}{8}\right)^4 = \left(-\frac{1}{8}\right)^{7+4}$ The base is $-\frac{1}{8}$. Add the exponents.

$= \left(-\frac{1}{8}\right)^{11}$ Simplify.

b. $(3m)^2 = 3m \cdot 3m$ Write as repeated multiplication.

$= (3 \cdot 3) \cdot (m \cdot m)$ Use properties of multiplication.

$= 3^{1+1} \cdot m^{1+1}$ The bases are 3 and m. Add the exponents.

$= 3^2 \cdot m^2 = 9m^2$ Simplify.

Exercises

Simplify the expression.

6. $p^5 \cdot p^2$
7. $(n^{11})^2$
8. $(5y)^3$
9. $(-2k)^4$

13.3 Quotient of Powers Property (pp. 566–571)

a. $\dfrac{(-4)^9}{(-4)^6} = (-4)^{9-6}$ — The base is −4. Subtract the exponents.

$= (-4)^3$ — Simplify.

b. $\dfrac{x^4}{x^3} = x^{4-3}$ — The base is *x*. Subtract the exponents.

$= x^1$

$= x$ — Simplify.

Exercises

Simplify the expression. Write your answer as a power.

10. $\dfrac{8^8}{8^3}$

11. $\dfrac{5^2 \cdot 5^9}{5}$

12. $\dfrac{w^8}{w^7} \cdot \dfrac{w^5}{w^2}$

Simplify the expression.

13. $\dfrac{2^2 \cdot 2^5}{2^3}$

14. $\dfrac{(6c)^3}{c}$

15. $\dfrac{m^8}{m^6} \cdot \dfrac{m^{10}}{m^9}$

13.4 Zero and Negative Exponents (pp. 574–579)

a. $10^{-3} = \dfrac{1}{10^3}$ — Definition of negative exponent

$= \dfrac{1}{1000}$ — Evaluate power.

b. $(-0.5)^{-5} \cdot (-0.5)^5 = (-0.5)^{-5+5}$ — Add the exponents.

$= (-0.5)^0$ — Simplify.

$= 1$ — Definition of zero exponent

Exercises

Evaluate the expression.

16. 2^{-4}

17. 95^0

18. $\dfrac{8^2}{8^4}$

19. $(-12)^{-7} \cdot (-12)^7$

20. $\dfrac{1}{7^9} \cdot \dfrac{1}{7^{-6}}$

21. $\dfrac{9^4 \cdot 9^{-2}}{9^2}$

13.5 Reading Scientific Notation *(pp. 580–585)*

a. Write 5.9×10^4 in standard form.

$5.9 \times 10^4 = 59{,}000$ (4) — Move decimal point 4 places to the right.

b. Write 7.31×10^{-6} in standard form.

$7.31 \times 10^{-6} = 0.00000731$ (6) — Move decimal point 6 places to the left.

Exercises

Tell whether the number is written in scientific notation. Explain.

22. 0.9×10^9 **23.** 3.04×10^{-11} **24.** 15×10^{26}

Write the number in standard form.

25. 2×10^7 **26.** 4.8×10^{-3} **27.** 6.25×10^5

13.6 Writing Scientific Notation *(pp. 586–591)*

a. In 2010, the population of the United States was about 309,000,000. Write this number in scientific notation.

The number is greater than 10. So, move the decimal point 8 places to the left.

$309{,}000{,}000 = 3.09 \times 10^8$ (8)

The exponent is positive.

b. The cornea of an eye is 0.00056 meter thick. Write this number in scientific notation.

Cornea

The number is between 0 and 1. So, move the decimal point 4 places to the right.

$0.00056 = 5.6 \times 10^{-4}$ (4)

The exponent is negative.

Exercises

Write the number in scientific notation.

28. 0.00036 **29.** 800,000 **30.** 79,200,000

Multiply. Write your answer in scientific notation.

31. $(4 \times 10^3) \times (2 \times 10^2)$ **32.** $(1.5 \times 10^{-9}) \times (8 \times 10^{-3})$

13 Chapter Test

Write the product using exponents.

1. $(-15) \cdot (-15) \cdot (-15)$

2. $\left(\frac{1}{12}\right) \cdot \left(\frac{1}{12}\right) \cdot \left(\frac{1}{12}\right) \cdot \left(\frac{1}{12}\right) \cdot \left(\frac{1}{12}\right)$

Evaluate the expression.

3. -2^3

4. $10 + 3^3 \div 9$

Simplify the expression. Write your answer as a power.

5. $9^{10} \cdot 9$

6. $\frac{(-3.5)^{13}}{(-3.5)^9}$

Evaluate the expression.

7. $5^{-2} \cdot 5^2$

8. $\frac{-8}{(-8)^3}$

Write the number in standard form.

9. 3×10^7

10. 9.05×10^{-3}

Multiply. Write your answer in scientific notation.

11. $(7 \times 10^3) \times (5 \times 10^2)$

12. $(3 \times 10^{-5}) \times (2 \times 10^{-3})$

13. **HAMSTER** A hamster toy is in the shape of a sphere. The volume V of a sphere is represented by $V = \frac{4}{3}\pi r^3$, where r is the radius of the sphere. What is the volume of the toy? Round your answer to the nearest cubic centimeter. Use 3.14 for π.

14. **CRITICAL THINKING** Is $(xy^2)^3$ the same as $(xy^3)^2$? Explain.

15. **RICE** A grain of rice weighs about 3^3 milligrams. About how many grains of rice are in one scoop?

16. **TASTE BUDS** There are about 10,000 taste buds on a human tongue. Write this number in scientific notation.

17. **LEAD** From 1978 to 2008, the amount of lead allowed in the air in the United States was 1.5×10^{-6} gram per cubic meter. In 2008, the amount allowed was reduced by 90%. What is the new amount of lead allowed in the air?

One scoop of rice weighs about 3^9 milligrams.

13 Standarized Test Practice

Test-Taking Strategy
Use Intelligent Guessing
Cats were first tamed 3·2^10 years ago in Egypt. How long ago was that?
Ⓐ 3000 Ⓑ 3072 Ⓒ 5000 Ⓓ 40
Who says I am tame? Growl. Hiss.
"It can't be 40 or 5000 because they aren't divisible by 3. So, you can intelligently guess between 3000 and 3072."

1. Mercury's distance to the Sun is approximately 5.79×10^7 kilometers. Write this distance in standard form. *(8.EE.4)*

   A. 5,790,000,000 km　　**C.** 57,900,000 km

   B. 579,000,000 km　　**D.** 5,790,000 km

2. The steps Jim took to answer the question are shown below. What should Jim change to correctly answer the question? *(8.G.5)*

   How many degrees are in the largest angle in the triangle below?

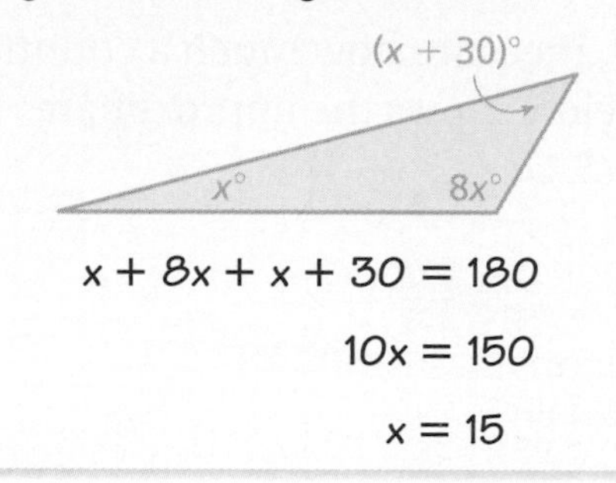

$$x + 8x + x + 30 = 180$$
$$10x = 150$$
$$x = 15$$

   F. The left side of the equation should equal 360° instead of 180°.

   G. The sum of the acute angles should equal 90°.

   H. Evaluate the smallest angle when $x = 15$.

   I. Evaluate the largest angle when $x = 15$.

3. Which expression is equivalent to the expression below? *(8.EE.1)*

$$2^4 2^3$$

   A. 2^{12}　　**C.** 48

   B. 4^7　　**D.** 128

4. A rectangular prism and its dimensions are shown below.

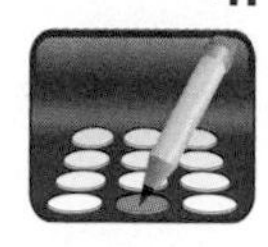

$\sqrt{6}$ ft, $\sqrt{3}$ ft, $\sqrt{8}$ ft

   What is the volume, in cubic feet, of the rectangular prism? *(8.NS.2)*

5. A bank account pays interest so that the amount in the account doubles every 10 years. The account started with $5000 in 1940. How much would be in the account in the year 2010? *(8.EE.1)*

F. $40,000 **H.** $640,000

G. $320,000 **I.** $1,280,000

6. Which expression is equivalent to $5\sqrt{5} + 2\sqrt{5}$? *(8.NS.2)*

A. $7\sqrt{5}$ **C.** $7\sqrt{10}$

B. $10\sqrt{5}$ **D.** $10\sqrt{10}$

Think Solve Explain

7. The gross domestic product (GDP) is a way to measure how much a country produces economically in a year. The table below shows the approximate population and GDP for the United States. *(8.EE.4)*

United States 2008	
Population	300 million (300,000,000)
GDP	14.4 trillion dollars ($14,400,000,000,000)

Part A Find the GDP per person for the United States. Show your work and explain your reasoning.

Part B Write the population and GDP using scientific notation.

Part C Find the GDP per person for the United States using your answers from Part B. Write your answer in scientific notation. Show your work and explain your reasoning.

8. What is the slope of the line shown in the graph? *(7.RP.2b)*

F. $-\frac{1}{3}$ **H.** -3

G. $\frac{1}{3}$ **I.** 3

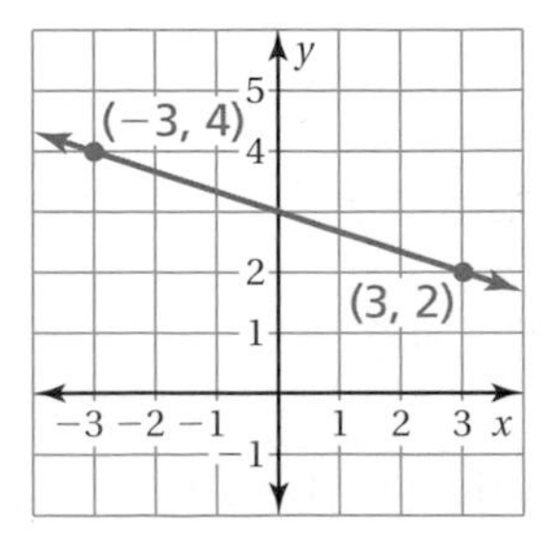

9. The two quadrilaterals shown below are similar.

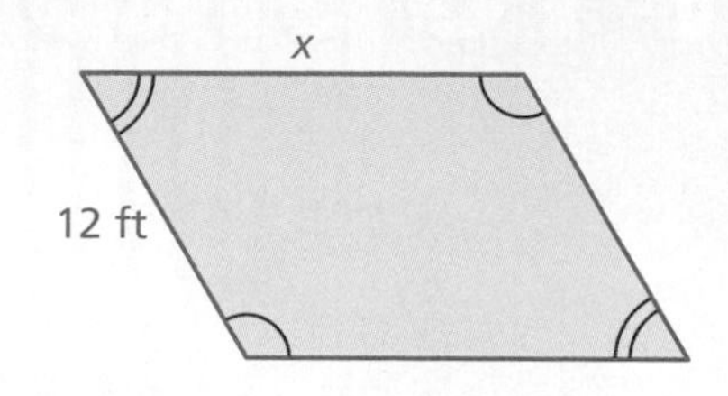

What is the value of x? *(7.G.1)*

A. 4 feet

B. 8 feet

C. 9 feet

D. 16 feet

10. Find $(-2.5)^{-2}$. *(8.EE.1)*

11. Triangle *LMN* is shown at the right.

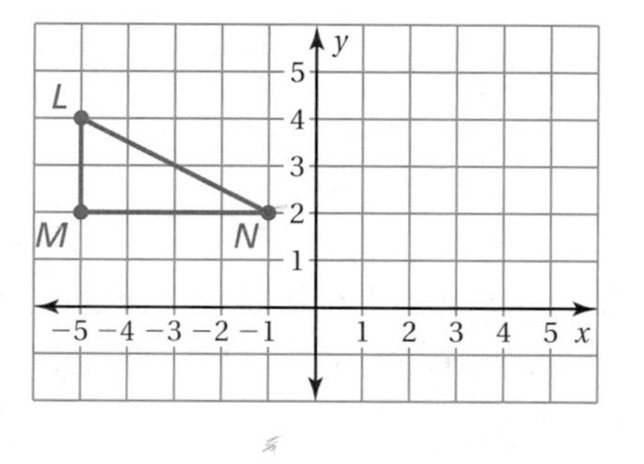

Reflect triangle *LMN* in the y-axis. What are the coordinates of point N', the image of point N? *(8.G.3)*

F. $(-1, -2)$

G. $(1, 2)$

H. $(1, -2)$

I. $(2, -1)$

12. The dimensions of Prism B are double the dimensions of Prism A. What is the surface area of Prism B? *(7.G.6)*

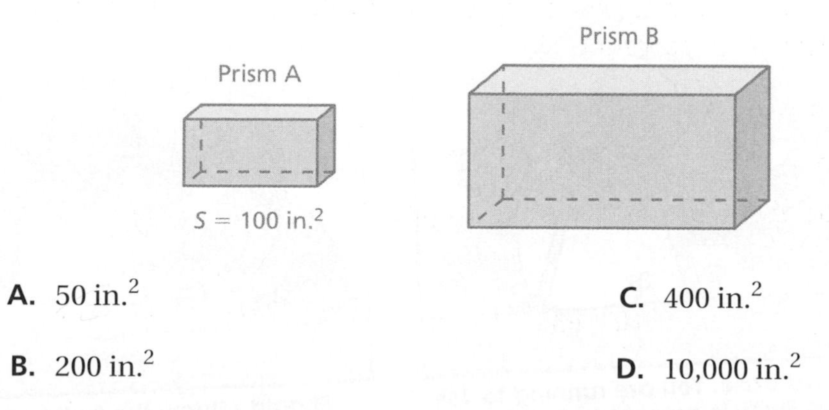

A. 50 in.^{2}

B. 200 in.^{2}

C. 400 in.^{2}

D. 10,000 in.^{2}

Additional Topics

- Topic 1 Angles
- Topic 2 Geometry
- Topic 3 Converting Units of Measure
- Topic 4 Comparing Rates
- Topic 5 Transformations
- Topic 6 Volumes of Spheres
- Topic 7 Triangles and Slope

"Move 4 of the lines to make 3 equilateral triangles."

"Well done, Descartes!"

"I'm at 3rd base. You are running to 1st base and Fluffy is running to 2nd base."

"Should I throw the ball to 2nd to get Fluffy out or throw it to 1st to get you out?"

What You Learned Before

Measuring Angles (4.MD.6)

Example 1 **Use a protractor to find the measure of each angle. Then classify the angle as *acute, obtuse, right,* or *straight*.**

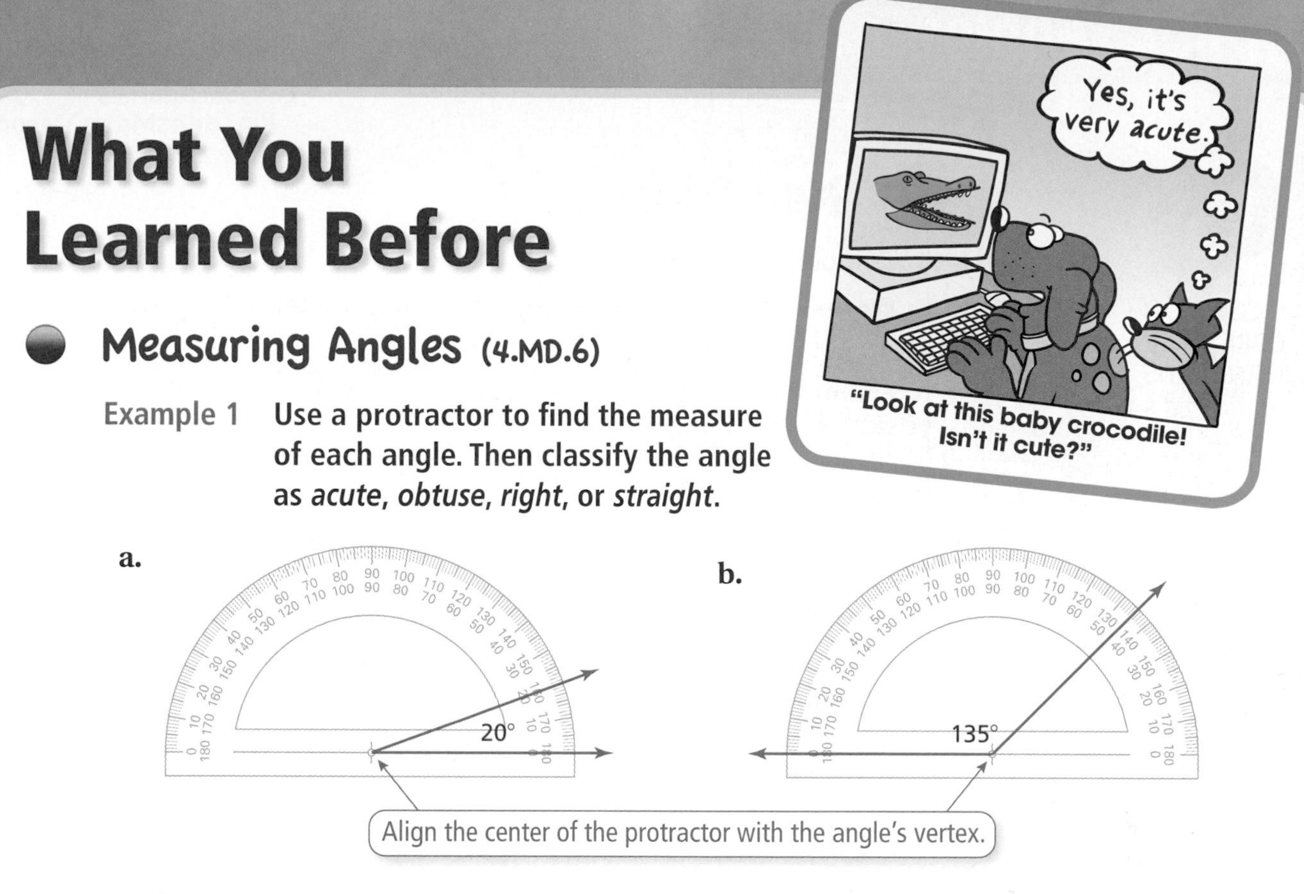

Yes, it's very acute.
"Look at this baby crocodile! Isn't it cute?"
a.
20°
b.
135°
Align the center of the protractor with the angle's vertex.

∴ The angle measure is 20°. So, the angle is acute.

∴ The angle measure is 135°. So, the angle is obtuse.

Drawing Angles (4.G.1)

Example 2 **Use a protractor to draw a 45° angle.**

Draw a ray. Place the center of the protractor on the endpoint of the ray and align the protractor so the ray passes through the 0° mark. Make a mark at 45°. Then draw a ray from the endpoint at the center of the protractor through the mark at 45°.

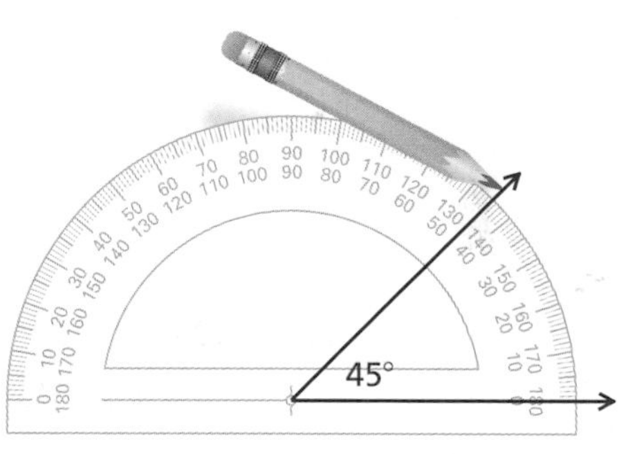

45°

Try It Yourself

Use a protractor to find the measure of the angle. Then classify the angle as *acute, obtuse, right,* or *straight*.

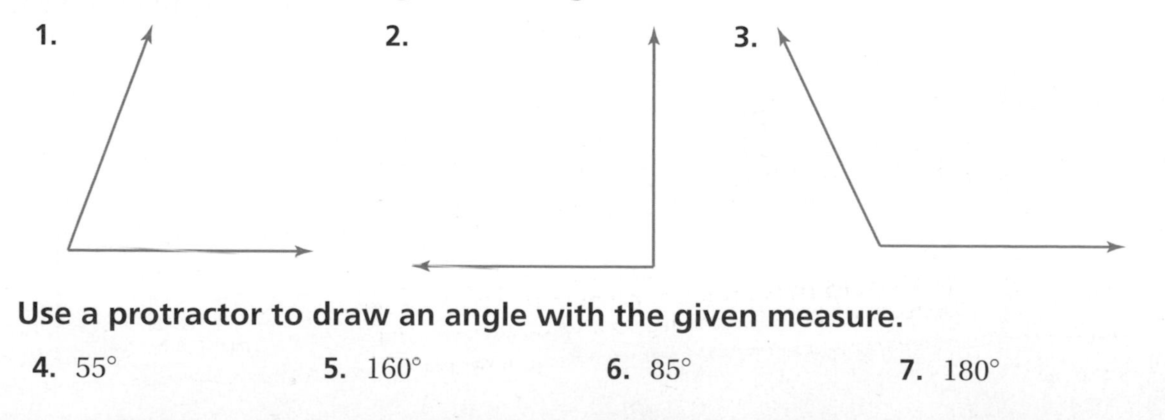

1.
2.
3.

Use a protractor to draw an angle with the given measure.

4. 55° **5.** 160° **6.** 85° **7.** 180°

Topic 1 Angles

COMMON CORE STATE STANDARDS
7.G.5

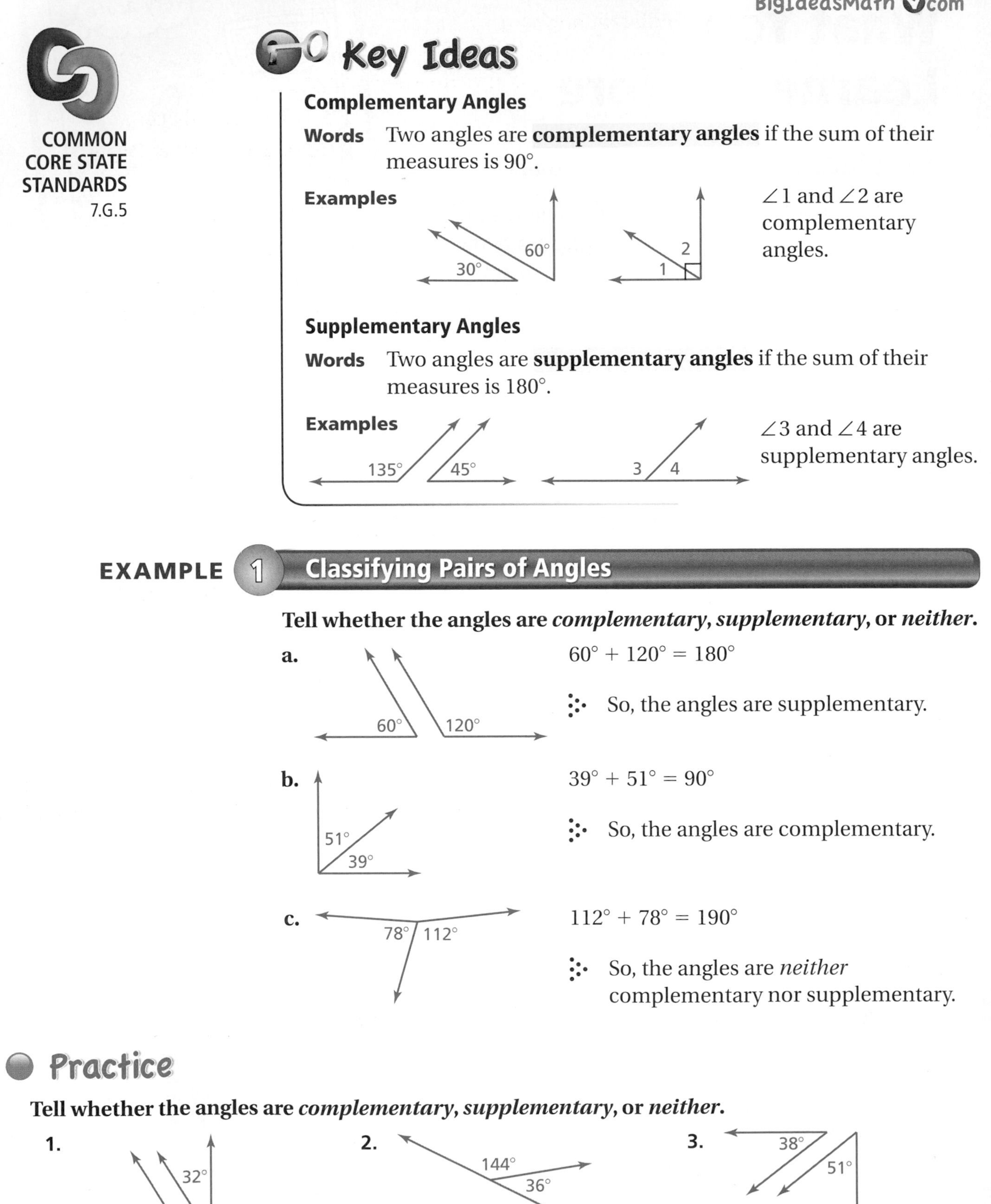

Key Ideas

Complementary Angles

Words Two angles are **complementary angles** if the sum of their measures is 90°.

Examples

∠1 and ∠2 are complementary angles.

Supplementary Angles

Words Two angles are **supplementary angles** if the sum of their measures is 180°.

Examples

∠3 and ∠4 are supplementary angles.

EXAMPLE 1 Classifying Pairs of Angles

Tell whether the angles are *complementary*, *supplementary*, or *neither*.

a. $60^\circ + 120^\circ = 180^\circ$

So, the angles are supplementary.

b. $39^\circ + 51^\circ = 90^\circ$

So, the angles are complementary.

c. $112^\circ + 78^\circ = 190^\circ$

So, the angles are *neither* complementary nor supplementary.

Practice

Tell whether the angles are *complementary*, *supplementary*, or *neither*.

1. 32°, 58°

2. 144°, 36°

3. 38°, 51°

Key Ideas

Adjacent Angles

Words Two angles are **adjacent angles** if they share a common side and have the same vertex.

Examples

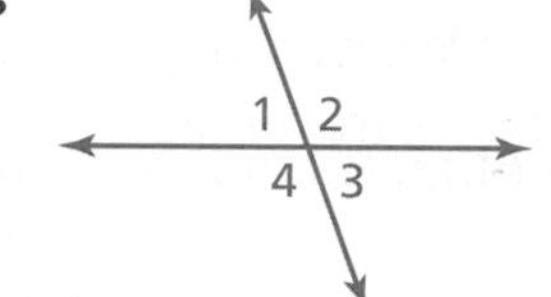

$\angle 1$ and $\angle 2$ are adjacent.

$\angle 2$ and $\angle 4$ are not adjacent.

Vertical Angles

Words Two angles are **vertical angles** if they are opposite angles formed by the intersection of two lines. Vertical angles have the same measure.

Examples

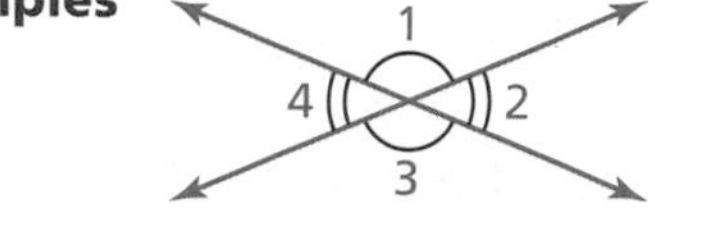

$\angle 1$ and $\angle 3$ are vertical angles.

$\angle 2$ and $\angle 4$ are vertical angles.

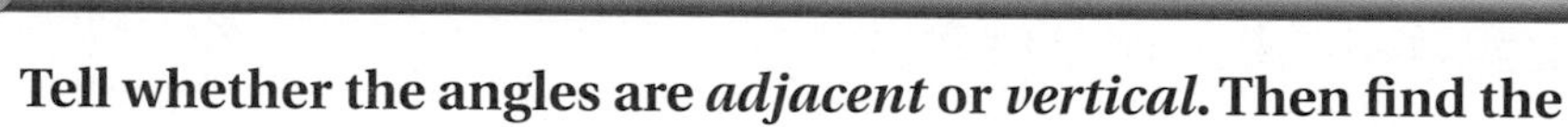

EXAMPLE 2 Finding Angle Measures

Tell whether the angles are *adjacent* or *vertical*. Then find the value of x.

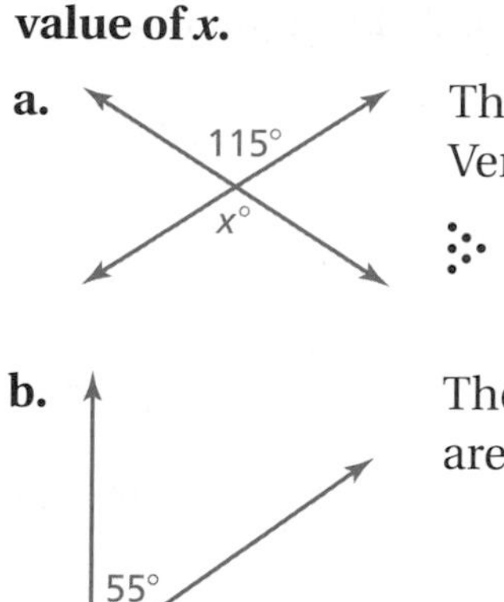

a. The angles are vertical angles.
Vertical angles have the same measure.

So, x is 115.

b. The angles are adjacent angles. Because the angles are complementary, the sum of their measures is 90°.

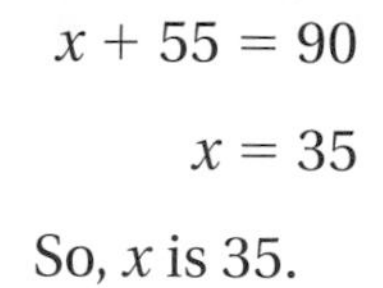

$$x + 55 = 90$$

$$x = 35$$

So, x is 35.

Practice

Tell whether the angles are *adjacent* or *vertical*. Then find the value of x.

4. **5.** **6.**

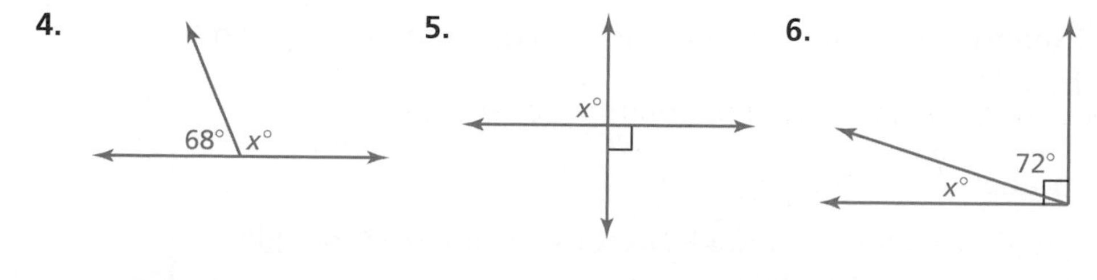

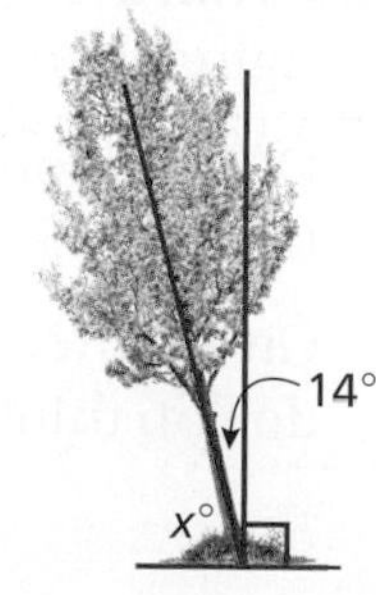

7. LANDSCAPING The tree is tilted 14°. Find the value of x.

EXAMPLE 1 Constructing a Triangle with Given Side Lengths

COMMON CORE STATE STANDARDS
7.G.2
7.G.3

Construct a triangle with side lengths of 3 centimeters, 4 centimeters, and 5 centimeters.

Step 1: Cut different colored straws to the appropriate lengths.

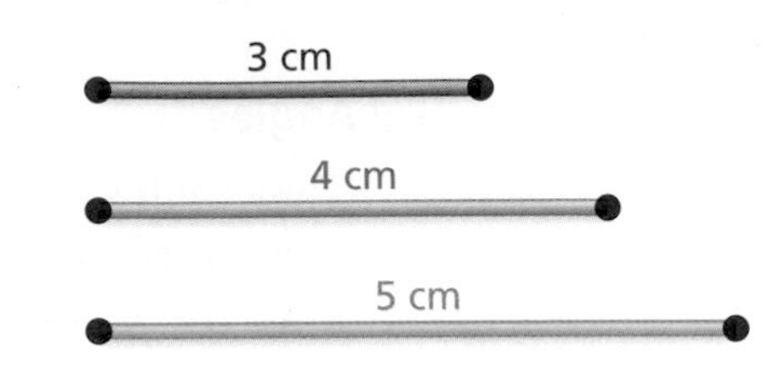

Step 2: Arrange the straws side by side.

Step 3: Form the triangle by connecting the ends of the red and green straws.

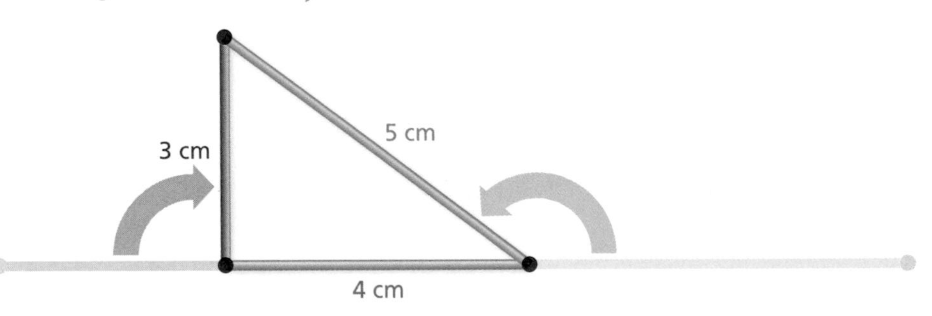

Practice

Construct a triangle with the given side lengths.

1. 6 cm, 5 cm, 3 cm

2. 5 cm, 12 cm, 13 cm

3. 2 in., 3 in., 2 in.

4. **REASONING** Repeat Exercises 1–3, but place the sides in a different order when arranging them side by side.

a. Did you construct any triangles that look different from the original triangles? Do the triangles have the same shape and size?

b. Given three side lengths that form a triangle, how many different triangles do you think you can construct? Explain.

EXAMPLE 2 Constructing a Triangle with Given Side Lengths

Construct a triangle with side lengths of 1 inch, 2.5 inches, and 1 inch, if possible.

Step 1: Cut different colored straws to the appropriate lengths.

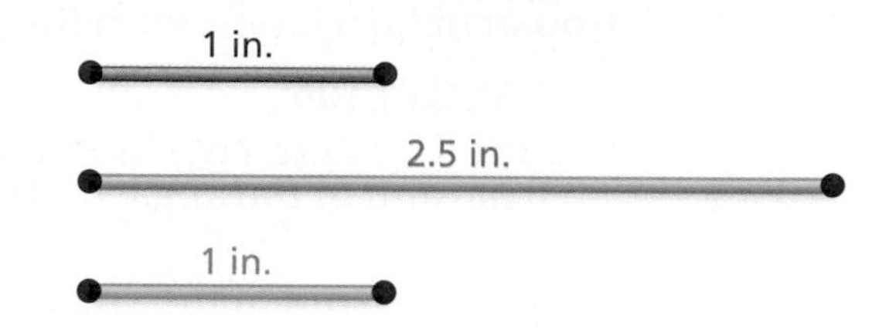

Step 2: Arrange the straws side by side.

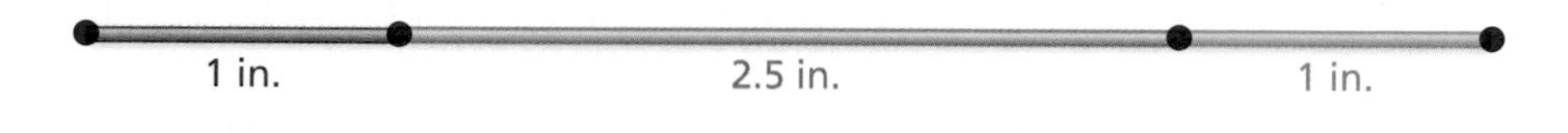

Step 3: The red and green straws are not long enough for their ends to connect. So, a triangle cannot be constructed.

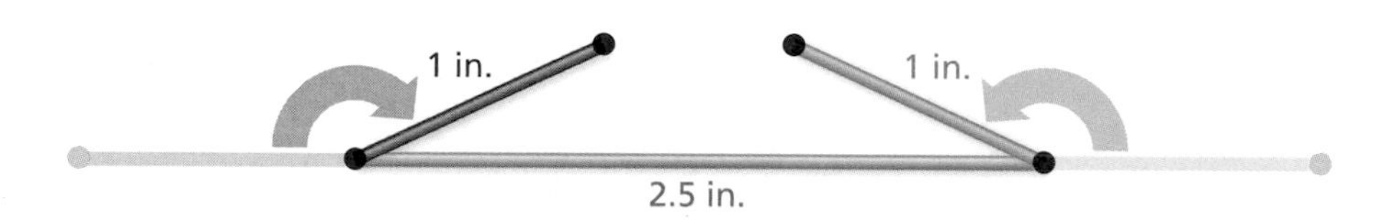

Practice

Construct a triangle with the given side lengths, if possible.

5. 2 cm, 4 cm, 1 cm

6. 6 cm, 8 cm, 10 cm

7. 1 in., 2 in., 1 in.

8. 5 cm, 7 cm, 4 cm

9. 2 in., 2 in., 2 in.

10. 1 in., 5 in., 3 in.

11. **REASONING** Complete the table below for each set of side lengths in Exercises 5–10. Write a rule that compares the sum of any two side lengths to the third side length.

Side Length			
Sum of Other Two Side Lengths			

The sum of the angle measures of any triangle is 180°. You can use a protractor to construct a triangle given three angle measures.

EXAMPLE 3 Constructing a Triangle with Given Angle Measures

Construct a triangle with the given angle measures, if possible.

a. 35°, 25°, 100°

The sum of the angle measures is 35° + 25° + 100° = 160°.

∴ The sum of the angle measures is not 180°. So, you cannot construct a triangle with these angle measures.

b. 30°, 60°, 90°

The sum of the angle measures is 30° + 60° + 90° = 180°. Use a protractor to construct a triangle with these angle measures.

Step 1: Use a protractor to draw the 30° angle.

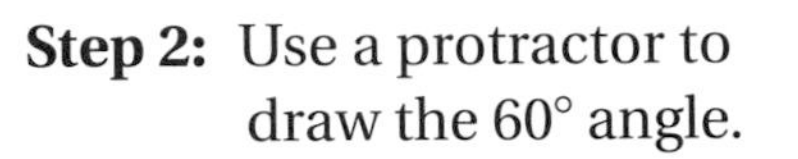

Step 2: Use a protractor to draw the 60° angle.

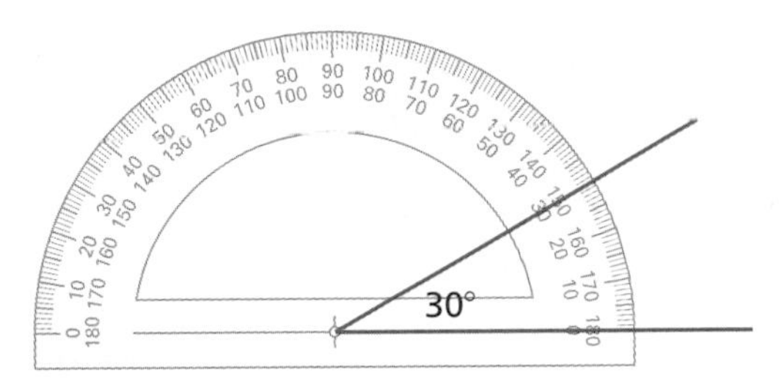

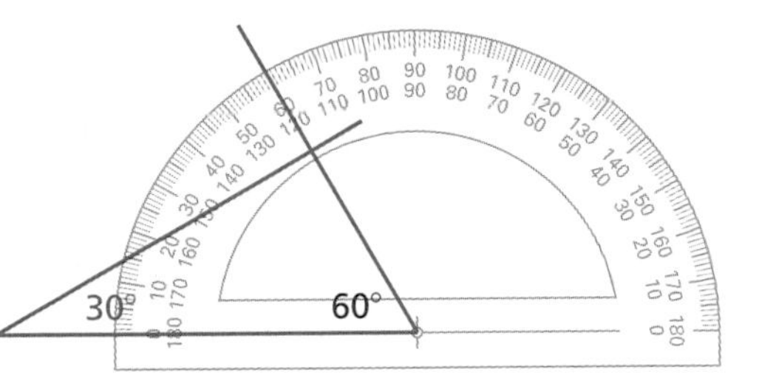

Step 3: The protractor shows that the measure of the remaining angle is 90°.

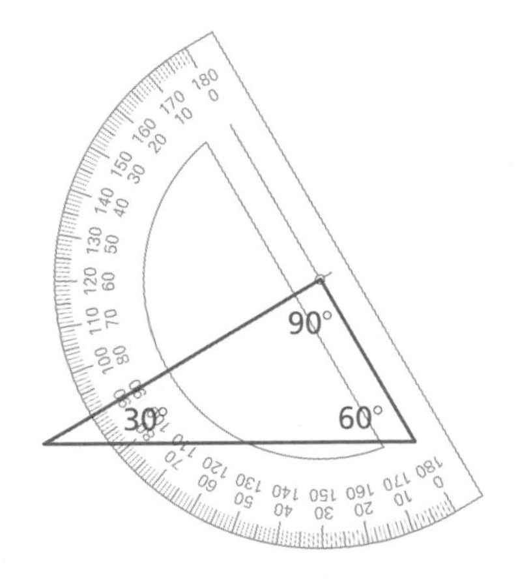

Practice

Construct a triangle with the given angle measures, if possible.

12. 50°, 30°, 110°

13. 60°, 60°, 60°

14. 60°, 40°, 80°

15. 45°, 75°, 100°

16. 20°, 20°, 120°

17. 70°, 70°, 40°

18. REASONING Construct a triangle with angle measures 90°, 45°, and 45°. Could you have drawn the sides longer? How many triangles can you construct given three angle measures with a sum of 180°?

Consider a plane "slicing" through a solid. The intersection of the plane and the solid is a two-dimensional shape. For example, the diagram shows that the intersection of the plane and the rectangular prism is a rectangle.

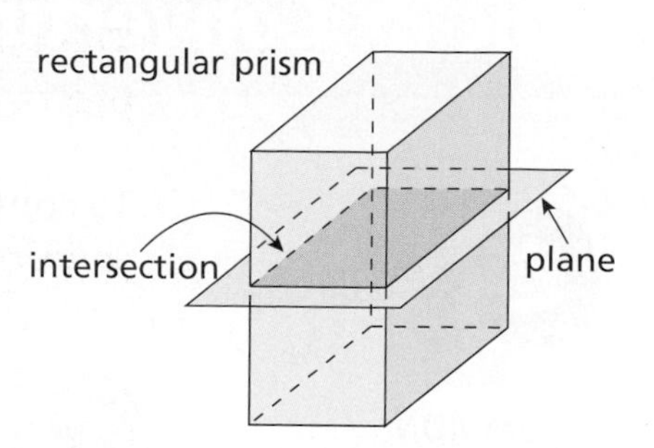

EXAMPLE 4 Describing the Intersection of a Plane and a Solid

Describe the intersection of the plane and the solid.

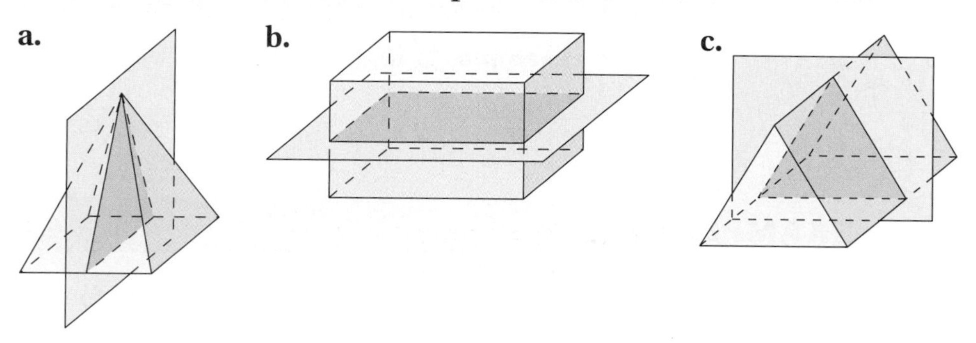

a. The intersection is a triangle.

b. The intersection is a rectangle.

c. The intersection is a triangle.

Practice

Describe the intersection of the plane and the solid.

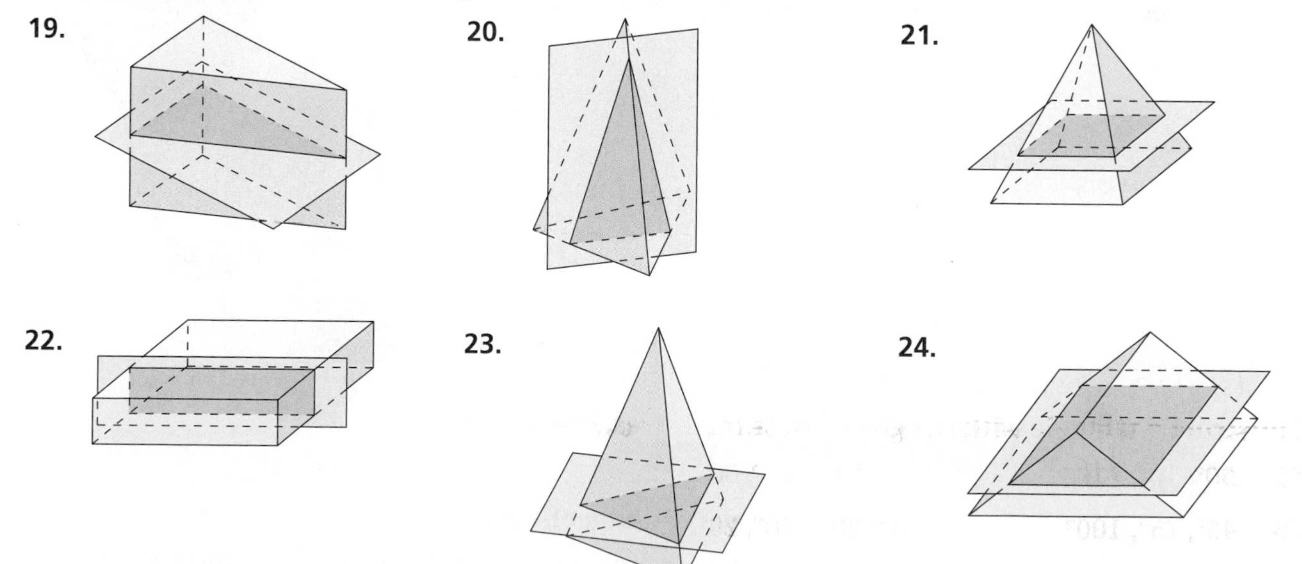

25. REASONING A plane that intersects a prism is parallel to the bases of the prism. Describe the intersection of the plane and the prism.

Topic 3 Converting Units of Measure

COMMON CORE STATE STANDARDS
7.RP.3

To convert between customary and metric units, multiply by one or more *conversion factors.*

Key Idea

Conversion Factor

A **conversion factor** is a rate that equals 1.

	Relationship	*Conversion factors*
Example	$1 \text{ m} \approx 3.28 \text{ ft}$	$\frac{1 \text{ m}}{3.28 \text{ ft}}$ and $\frac{3.28 \text{ ft}}{1 \text{ m}}$

EXAMPLE 1 Converting a Rate

Convert the speed of the zip liner to feet per second.

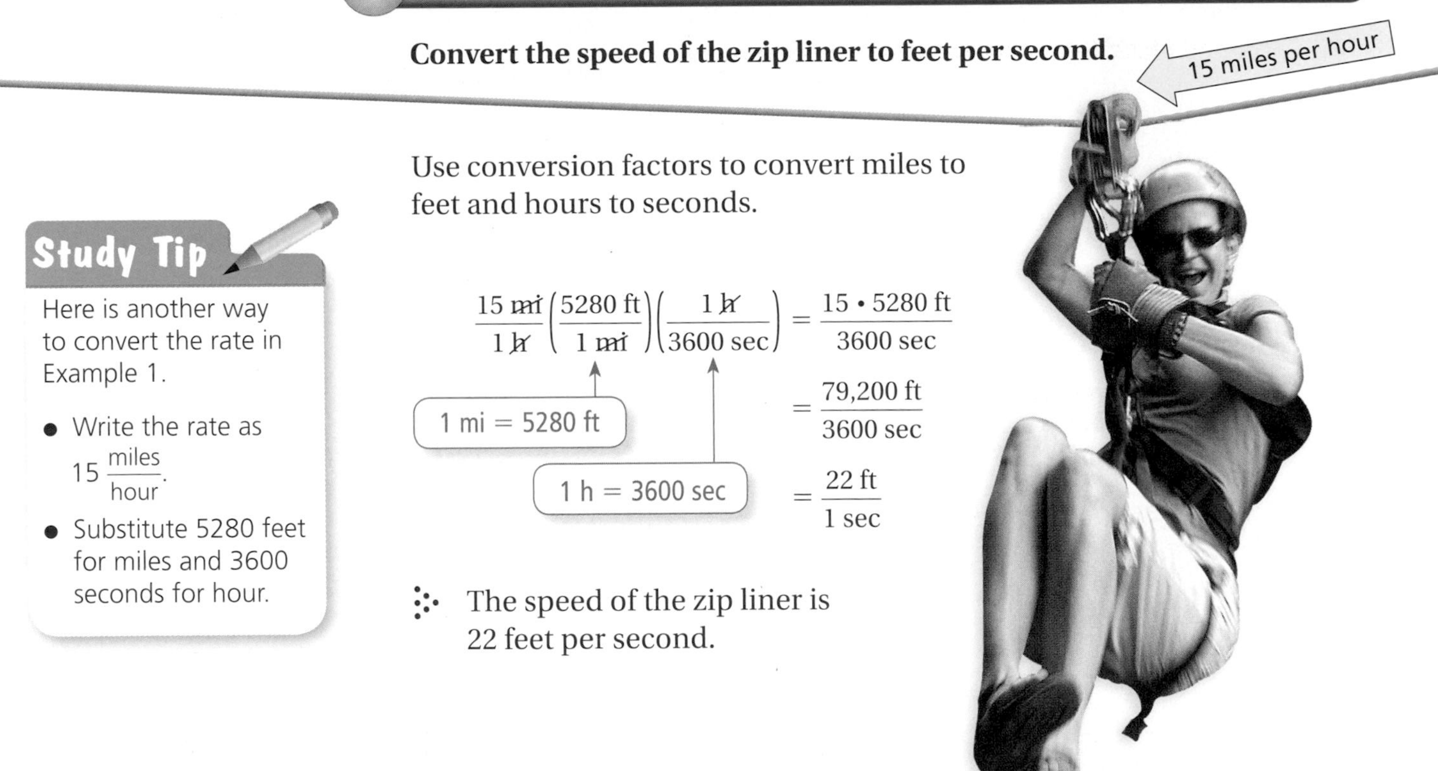

Use conversion factors to convert miles to feet and hours to seconds.

$$\frac{15 \cancel{\text{mi}}}{1 \cancel{\text{h}}}\left(\frac{5280 \text{ ft}}{1 \cancel{\text{mi}}}\right)\left(\frac{1 \cancel{\text{h}}}{3600 \text{ sec}}\right) = \frac{15 \cdot 5280 \text{ ft}}{3600 \text{ sec}}$$

1 mi = 5280 ft

1 h = 3600 sec

$$= \frac{79{,}200 \text{ ft}}{3600 \text{ sec}}$$

$$= \frac{22 \text{ ft}}{1 \text{ sec}}$$

∴ The speed of the zip liner is 22 feet per second.

Study Tip

Here is another way to convert the rate in Example 1.

- Write the rate as $15 \frac{\text{miles}}{\text{hour}}$.
- Substitute 5280 feet for miles and 3600 seconds for hour.

Key Idea

Converting Units for Area or Volume

To convert units for area, multiply the area by the *square* of the conversion factor.

To convert units for volume, multiply the volume by the *cube* of the conversion factor.

EXAMPLE 2 Converting Units for Area

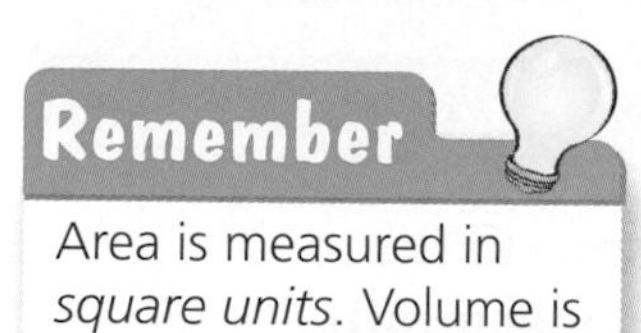

Remember

Area is measured in *square units*. Volume is measured in *cubic units*.

The painting *Fracture* by Benedict Gibson has an area of 2880 square inches. What is the area of the painting in square feet?

1 ft = 12 in.

$$2880 \text{ in.}^2 = 2880 \text{ in.}^2 \cdot \left(\frac{1 \text{ ft}}{12 \text{ in.}}\right)^2$$

$$= 2880 \cancel{\text{in.}^2} \cdot \frac{1 \text{ ft}^2}{144 \cancel{\text{in.}^2}}$$

$$= \frac{2880}{144} \text{ ft}^2$$

$$= 20 \text{ ft}^2$$

∴ The area of the painting is 20 square feet.

EXAMPLE 3 Converting Units for Volume

What is the volume of the rectangular prism in cubic centimeters?

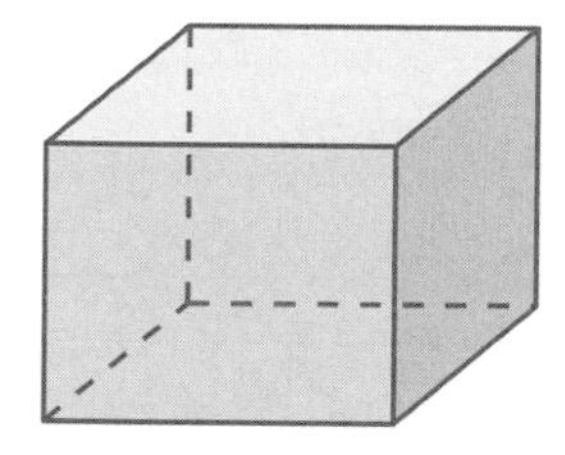

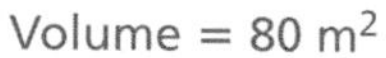

Volume = 80 m²

1 m = 100 cm

$$80 \text{ m}^3 = 80 \text{ m}^3 \cdot \left(\frac{100 \text{ cm}}{1 \text{ m}}\right)^3$$

$$= 80 \cancel{\text{m}^3} \cdot \frac{1{,}000{,}000 \text{ cm}^3}{1 \cancel{\text{m}^3}}$$

$$= 80{,}000{,}000 \text{ cm}^3$$

∴ The volume is 80,000,000 cubic centimeters.

Practice

Copy and complete the statement.

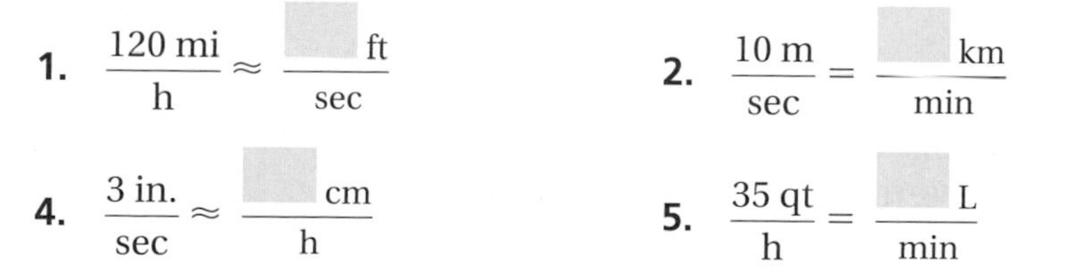

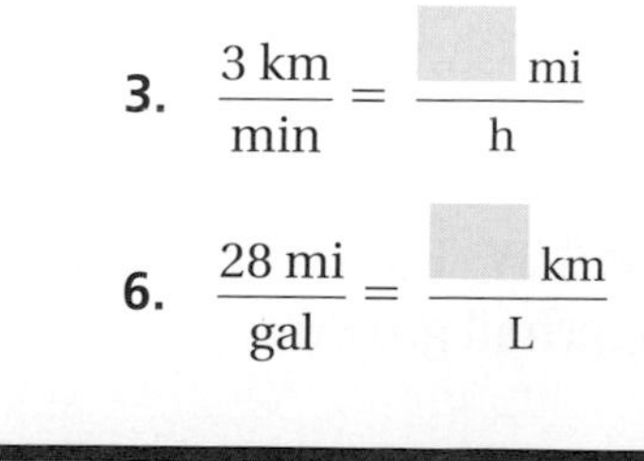

1. $\frac{120 \text{ mi}}{\text{h}} \approx \frac{\square \text{ ft}}{\text{sec}}$

2. $\frac{10 \text{ m}}{\text{sec}} = \frac{\square \text{ km}}{\text{min}}$

3. $\frac{3 \text{ km}}{\text{min}} = \frac{\square \text{ mi}}{\text{h}}$

4. $\frac{3 \text{ in.}}{\text{sec}} \approx \frac{\square \text{ cm}}{\text{h}}$

5. $\frac{35 \text{ qt}}{\text{h}} = \frac{\square \text{ L}}{\text{min}}$

6. $\frac{28 \text{ mi}}{\text{gal}} = \frac{\square \text{ km}}{\text{L}}$

7. The painting *Buzy Market* by Haitian painter Frantz Petion has an area of 6 square feet. What is the area of the painting in square inches?

8. The volume of a rectangular prism is 50 cubic centimeters. What is the volume of the rectangular prism in cubic millimeters?

Topic 4 Comparing Rates

EXAMPLE 1 Comparing Proportional Relationships

COMMON CORE STATE STANDARDS
8.EE.5

The distance y (in meters) traveled by a four-person ski lift in x seconds is represented by the equation $y = 2.5x$. The graph shows the distance traveled by a two-person ski lift.

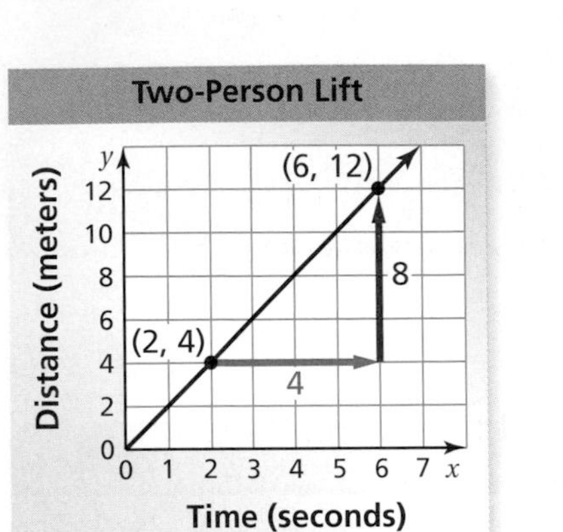

a. Which ski lift is faster?

Four-Person Lift

The equation is written in slope-intercept form.

$$y = 2.5x$$

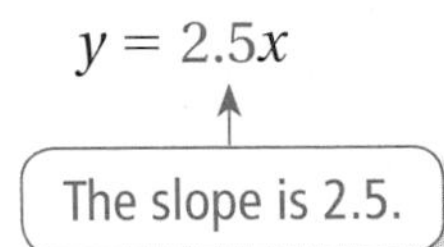

The four-person lift travels 2.5 meters per second.

Two-Person Lift

$$\text{slope} = \frac{\text{rise}}{\text{run}} = \frac{8}{4} = 2$$

The two-person lift travels 2 meters per second.

So, the four-person lift is faster than the two-person lift.

b. Graph the equation that represents the four-person lift in the same coordinate plane as the two-person lift. Compare the steepness of the graphs. What does this mean in the context of the problem?

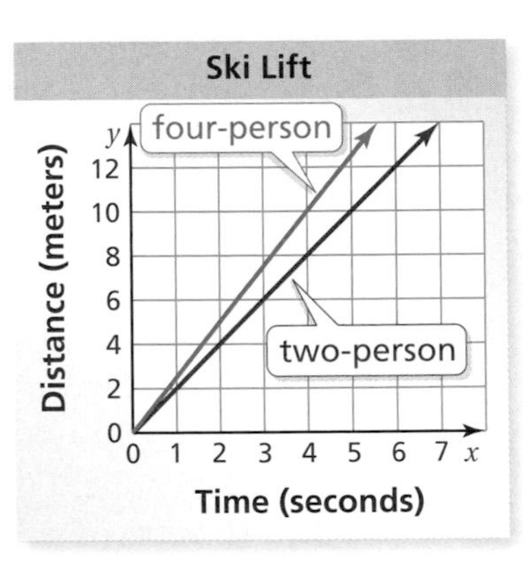

The graph that represents the four-person lift is steeper than the graph that represents the two-person lift. So, the four-person lift is faster.

Practice

1. **BIOLOGY** Toenails grow about 13 millimeters per year. The table shows fingernail growth.

Weeks	1	2	3	4
Fingernail Growth (millimeters)	0.7	1.4	2.1	2.8

   a. Do fingernails or toenails grow faster?

   b. Graph equations that represent the growth rates of toenails and fingernails in the same coordinate plane. Compare the steepness of the graphs. What does this mean in the context of the problem?

EXAMPLE 2 Comparing Equations

The earnings y (in dollars) of a nighttime employee working x hours is represented by the equation $y = 7.5x + 30$. The table shows the earnings of a daytime employee.

Time (hours)	1	2	3	4
Earnings (dollars)	12.50	25.00	37.50	50.00

+1 +1 +1

+12.50 +12.50 +12.50

a. Which employee has a higher hourly wage?

Nighttime Employee

$y = 7.5x + 30$

The slope is 7.5.

The nighttime employee earns \$7.50 per hour.

Daytime Employee

$\dfrac{\text{change in earnings}}{\text{change in time}} = \dfrac{\$12.50}{1 \text{ hour}}$

The daytime employee earns \$12.50 per hour.

So, the daytime employee has a higher hourly wage.

b. Write an equation that relates the daytime employee's earnings to the number of hours worked. Graph the equations that represent the earnings of the two employees in the same coordinate plane. Interpret the graphs.

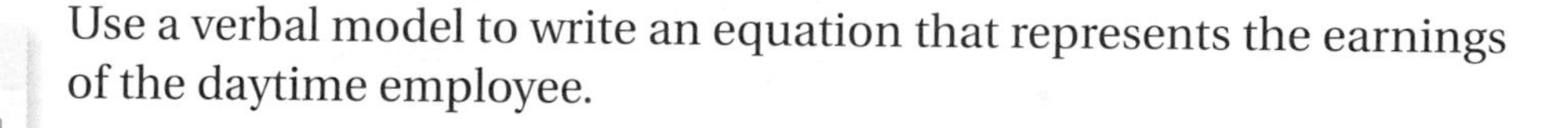

Employee Earnings

Use a verbal model to write an equation that represents the earnings of the daytime employee.

$$\text{Earnings} = \text{Hourly wage} \cdot \text{Hours worked}$$

$$y = 12.5x$$

The graph shows that the daytime employee has a higher hourly wage, but does not earn more money than the nighttime employee until each person has worked more than 6 hours.

Practice

2. **EMPLOYMENT** Manager A earns \$15 per hour and receives a \$50 bonus. The graph shows the earnings of Manager B.
   - **a.** Which manager has a higher hourly wage?
   - **b.** After how many hours does Manager B earn more money than Manager A?

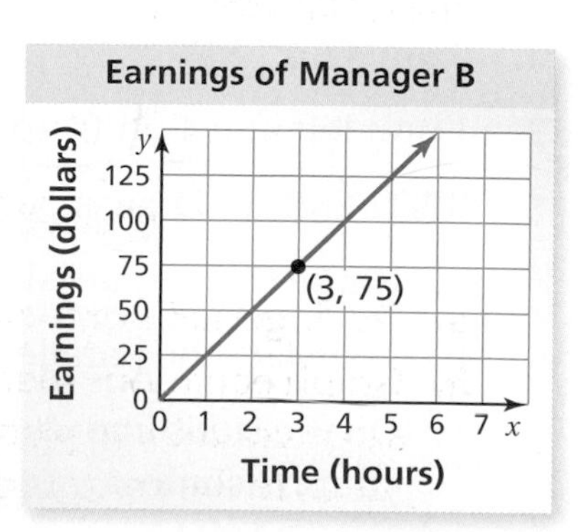

COMMON CORE STATE STANDARDS
8.G.1
8.G.2
8.G.3
8.G.4

Key Idea

Translations

A **translation**, or *slide*, is a transformation in which a figure moves but does not turn. Every point of the figure moves the same distance and in the same direction.

For translations, the original figure and its image have the same size and shape. Figures with the same size and shape are called **congruent figures**.

EXAMPLE 1 Translating a Figure

The vertices of a parallelogram are $A(-4, -3)$, $B(-2, -2)$, $C(3, -4)$, and $D(1, -5)$. Translate the parallelogram 2 units left and 4 units up. What are the coordinates of the image?

A' is read "A prime." Use *prime* symbols when naming an image.

$A \longrightarrow A'$
$B \longrightarrow B'$
$C \longrightarrow C'$
$D \longrightarrow D'$

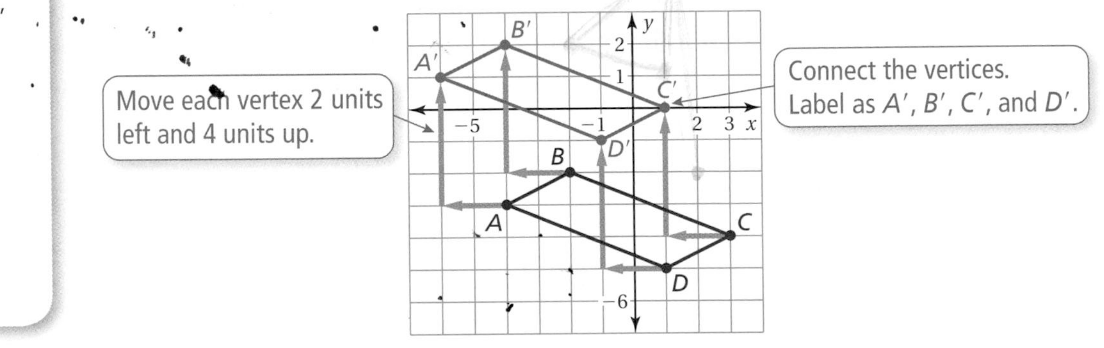

The coordinates of the image are $A'(-6, 1)$, $B'(-4, 2)$, $C'(1, 0)$, and $D'(-1, -1)$.

Practice

The vertices of a triangle are $P(-2, 2)$, $Q(1, 4)$, and $R(1, 1)$. Draw the triangle and its image after the translation. Find the coordinates of the image.

1. 6 units up
2. 2 units right
3. 1 unit left and 4 units up
4. 3 units right and 5 units down
5. **OPEN-ENDED** Draw a parallelogram $ABCD$ in a coordinate plane.
   a. Name the parallel sides.
   b. Translate the parallelogram to a different location in the coordinate plane.
   c. Do the sides in part (a) remain parallel after the translation? Explain your reasoning.

Reflections

A **reflection**, or *flip*, is a transformation in which a figure is reflected in a line called the *line of reflection*. A reflection creates a mirror image of the original figure.

For reflections, the original figure and its image are congruent.

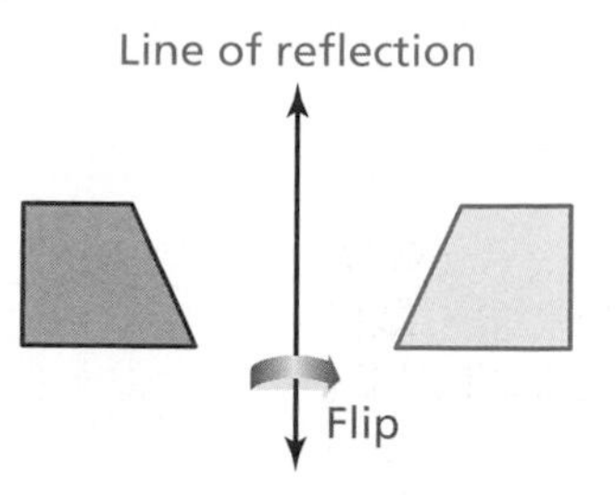

EXAMPLE 2 Reflecting a Figure

The vertices of a pentagon are $V(-4, -5)$, $W(-4, -1)$, $X(-2, -1)$, $Y(-1, -3)$, and $Z(-2, -5)$. Reflect the pentagon in the y-axis. What are the coordinates of the image?

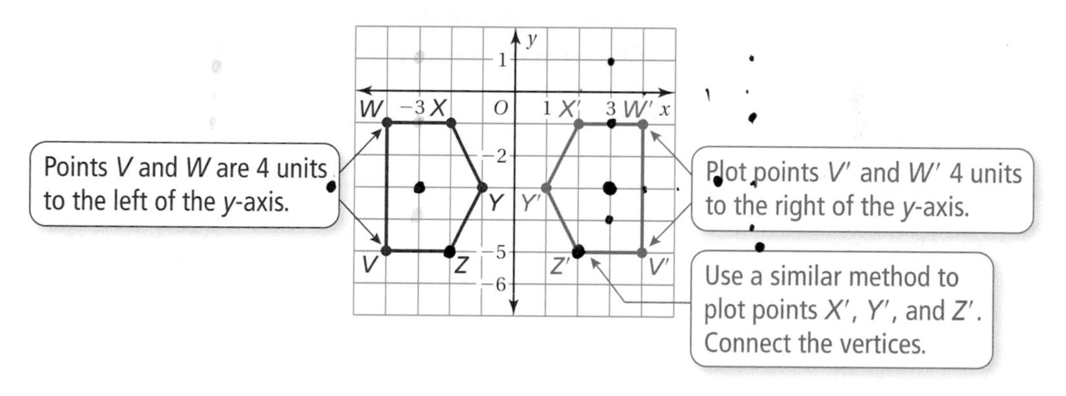

Points V and W are 4 units to the left of the y-axis.
Plot points V′ and W′ 4 units to the right of the y-axis.
Use a similar method to plot points X′, Y′, and Z′. Connect the vertices.

∴ The coordinates of the image are $V'(4, -5)$, $W'(4, -1)$, $X'(2, -1)$, $Y'(1, -3)$, and $Z'(2, -5)$.

Practice

Find the coordinates of the figure after reflecting in the x-axis.

6. $A(-8, 1)$, $B(-3, 4)$, $C(-3, 1)$

7. $L(3, 1)$, $M(3, 4)$, $N(7, 4)$, $P(7, 1)$

Find the coordinates of the figure after reflecting in the y-axis.

8. $W(2, -5)$, $X(3, -3)$, $Y(6, -3)$, $Z(7, -5)$

9. $H(-6, -7)$, $I(-6, -2)$, $J(-3, -3)$, $K(-3, -8)$

10. **REASONING** The coordinates of a figure and its image are given. Is the reflection in the *x-axis* or the *y-axis*?

    $W(2, -3), X(2, -1), Y(4, -1), Z(4, -3) \longrightarrow W'(2, 3), X'(2, 1), Y'(4, 1), Z'(4, 3)$

11. **OPEN-ENDED** Draw a rectangle $ABCD$ in a coordinate plane. Reflect rectangle $ABCD$ in the x-axis or y-axis.

    a. Is angle B congruent to angle B'? Explain your reasoning.

    b. Is side CD congruent to side $C'D'$? Explain your reasoning.

Key Idea

Rotations

A **rotation**, or *turn*, is a transformation in which a figure is rotated about a point called the *center of rotation*. The number of degrees a figure rotates is the *angle of rotation*.

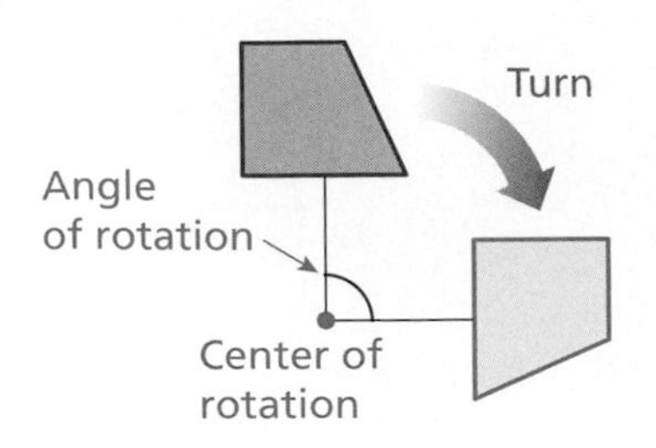

For rotations, the original figure and its image are congruent.

EXAMPLE 3 Rotating a Figure

The vertices of a trapezoid are $P(2, -2)$, $Q(4, -2)$, $R(5, -5)$, and $S(4, -5)$. Rotate the trapezoid 90° clockwise about the origin. What are the coordinates of the image?

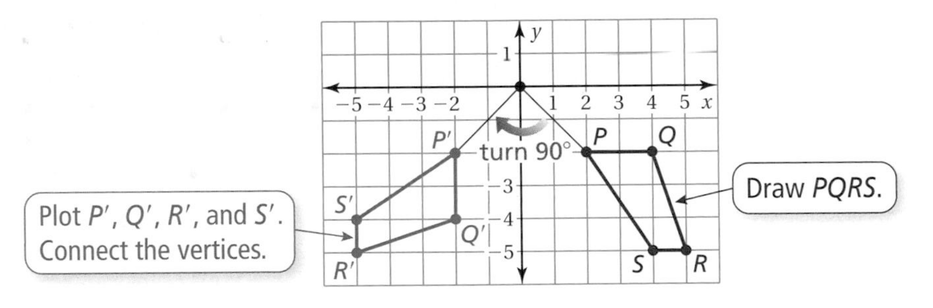

The coordinates of the image are $P'(-2, -2)$, $Q'(-2, -4)$, $R'(-5, -5)$, and $S'(-5, -4)$.

Practice

The vertices of a trapezoid are $L(1, 1)$, $M(2, 4)$, $N(4, 4)$, and $P(5, 1)$. Rotate the trapezoid as described. Find the coordinates of the image.

12. 90° clockwise about the origin

13. 180° counterclockwise about the origin

14. **REASONING** A figure is congruent to another figure if you can create the second figure from the first by a sequence of translations, reflections, and rotations.

    a. Is triangle ABC congruent to triangle DEF? Explain your reasoning.

    b. Is triangle ABC congruent to triangle GHJ? Explain your reasoning.

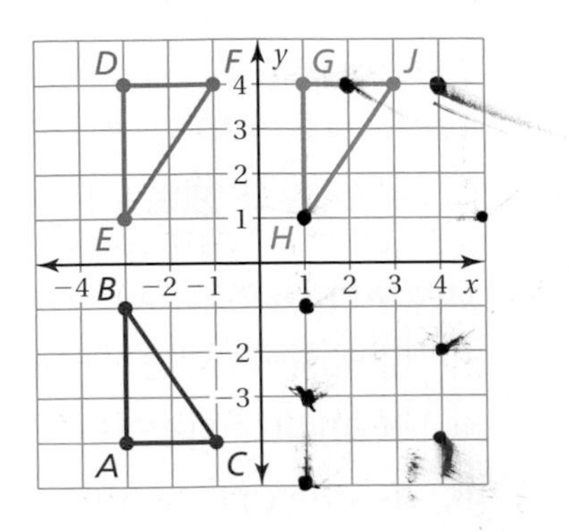

Dilations

A **dilation** is a transformation in which a figure is made larger or smaller with respect to a fixed point called the *center of dilation.*

For dilations, the original figure and its image are similar.

DILATE

Center of dilation

You can check your answer by drawing a line from the origin through each vertex of the original figure. The vertices of the image should lie on these lines.

The ratio of the side lengths of the image to the corresponding side lengths of the original figure is the *scale factor* of the dilation. To dilate a figure in the coordinate plane with respect to the origin, multiply the coordinates of each vertex by the scale factor k.

- When $k > 1$, the dilation is called an *enlargement.*
- When $k > 0$ and $k < 1$, the dilation is called a *reduction.*

EXAMPLE 4 Dilating a Figure

Draw the image of quadrilateral *FGHJ* after a dilation with a scale factor of 2. Identify the type of dilation.

Multiply each x- and y-coordinate by the scale factor 2.

Vertices of $FGHJ$	$(x \cdot 2, y \cdot 2)$	Vertices of $F'G'H'J'$
$F(1, 3)$	$(1 \cdot 2, 3 \cdot 2)$	$F'(2, 6)$
$G(2, 4)$	$(2 \cdot 2, 4 \cdot 2)$	$G'(4, 8)$
$H(3, 3)$	$(3 \cdot 2, 3 \cdot 2)$	$H'(6, 6)$
$J(2, 1)$	$(2 \cdot 2, 1 \cdot 2)$	$J'(4, 2)$

∴ The dilation is an *enlargement* because the scale factor is greater than 1.

Practice

The vertices of a rectangle are $E(2, -4)$, $F(2, -1)$, $G(6, -1)$, and $H(6, -4)$. Dilate the rectangle using the given scale factor. Find the coordinates of the image. Identify the type of dilation.

15. scale factor $= \frac{1}{2}$

16. scale factor $= 3$

17. REASONING A figure is similar to another figure if you can create the second figure from the first by a sequence of translations, reflections, rotations, and dilations.

a. Is triangle XYZ congruent to triangle JKL? Explain your reasoning.

b. Is triangle XYZ similar to triangle PQR? Explain your reasoning.

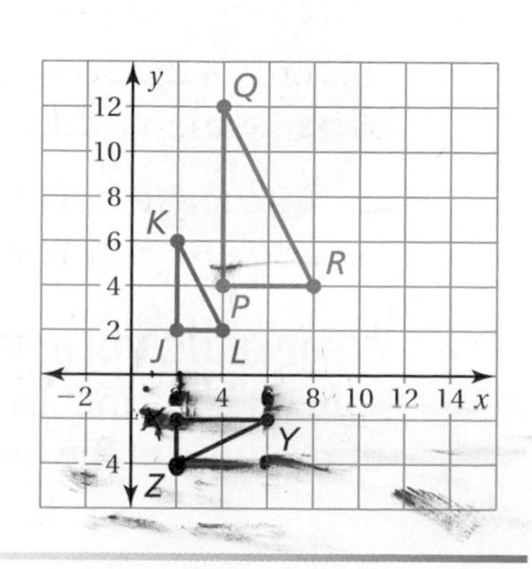

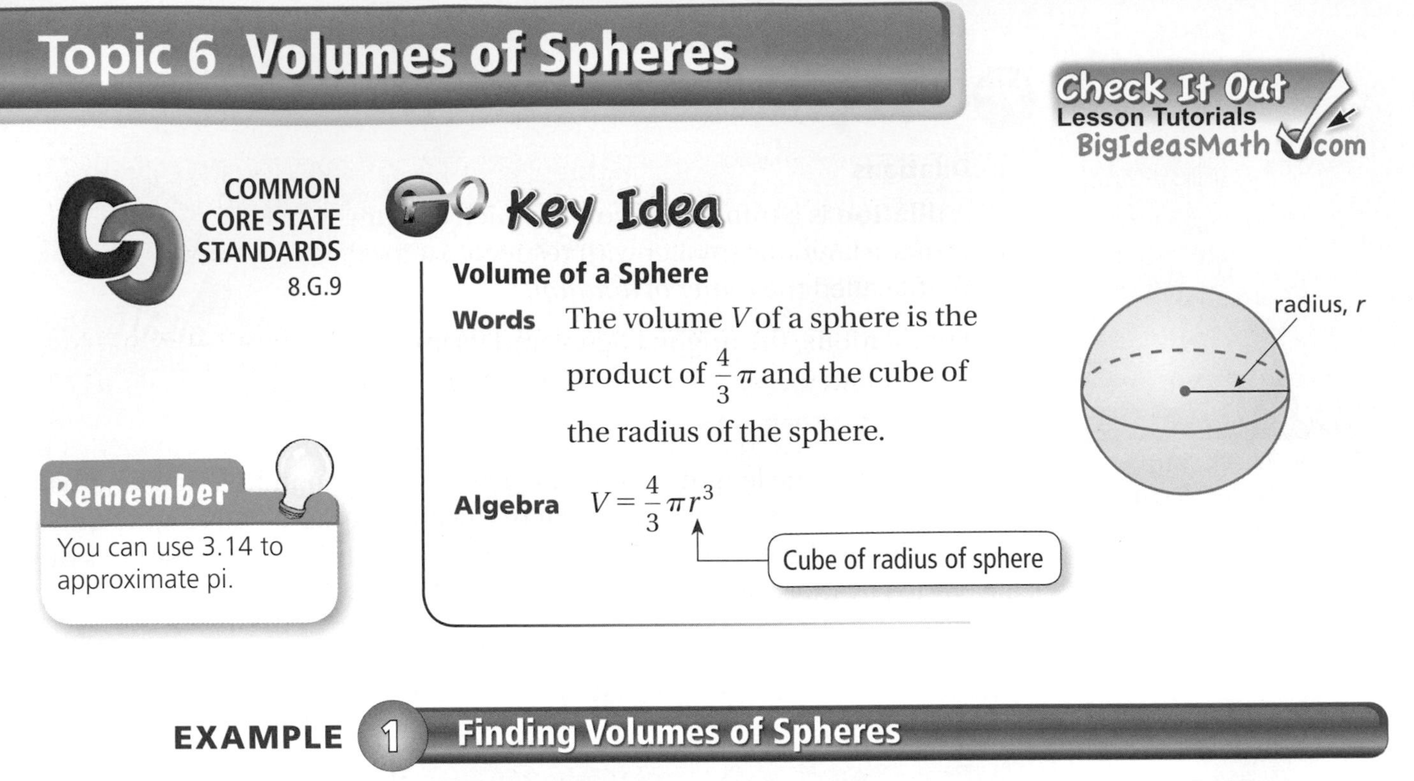

Topic 6 Volumes of Spheres

Check It Out
Lesson Tutorials
BigIdeasMath.com

COMMON CORE STATE STANDARDS
8.G.9

Key Idea

Volume of a Sphere

Words The volume V of a sphere is the product of $\frac{4}{3}\pi$ and the cube of the radius of the sphere.

Algebra $V = \frac{4}{3}\pi r^3$ ← Cube of radius of sphere

Remember

You can use 3.14 to approximate pi.

EXAMPLE 1 Finding Volumes of Spheres

Find the volume of the sphere. Round your answer to the nearest tenth.

15 mm

The diameter is 15 millimeters.
So, the radius is 7.5 millimeters.

$V = \frac{4}{3}\pi r^3$ Write the formula.

$= \frac{4}{3}\pi (7.5)^3$ Substitute.

$= 562.5\pi \approx 1766.3$ Simplify.

∴ The volume is about 1766.3 cubic millimeters.

Practice

Find the volume of the sphere. Round your answer to the nearest tenth.

4. **COMPARING VOLUMES** The radius of a baseball is 1.45 inches. The radius of a softball is 1.9 inches. How many times greater is the volume of the softball than the volume of the baseball?

A **hemisphere** is one-half of a sphere.

EXAMPLE 2 Real-Life Application

Find the volume of the silo. Round your answer to the nearest whole number.

The silo is made up of a cylinder and a hemisphere. Find each volume.

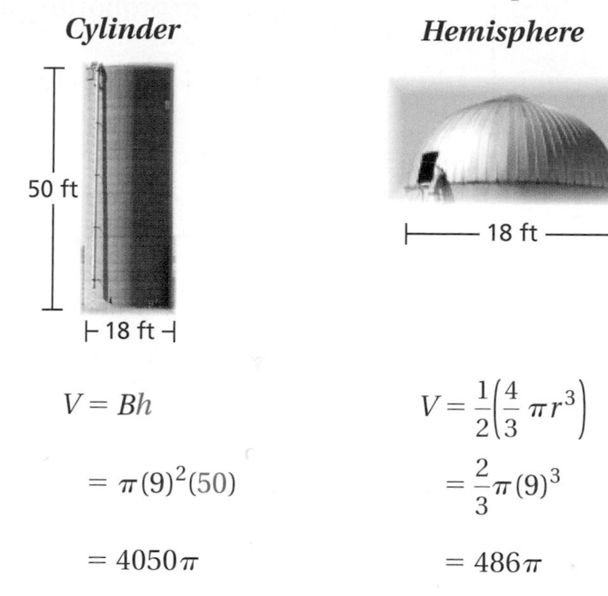

Cylinder	*Hemisphere*
$V = Bh$	$V = \frac{1}{2}\left(\frac{4}{3}\pi r^3\right)$
$= \pi(9)^2(50)$	$= \frac{2}{3}\pi(9)^3$
$= 4050\pi$	$= 486\pi$

Study Tip

Because the diameter of the cylinder and hemisphere is 18 feet, the radius is 9 feet.

Find the sum: $4050\pi + 486\pi = 4536\pi \approx 14{,}243 \text{ ft}^3$.

∴ The volume of the silo is about 14,243 cubic feet.

Practice

5. **WHAT IF?** In Example 2, how would the volume of the silo change if the diameter were 16 feet?

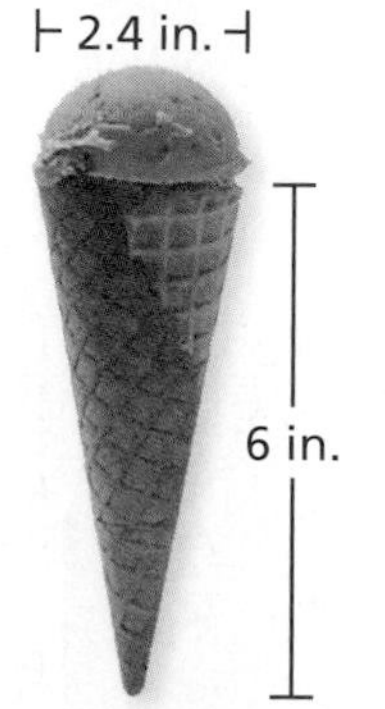

6. **ICE CREAM CONE** Ice cream is packed into a cone along with a hemisphere of ice cream on top of the cone. Find the volume of ice cream used for the ice cream cone. Round your answer to the nearest whole number.

7. **PACKAGING** A cylindrical container of three rubber balls has a height of 18 centimeters and a diameter of 6 centimeters. Each ball in the container has a radius of 3 centimeters. Find the amount of space in the container that is not occupied by rubber balls. Round your answer to the nearest whole number.

COMMON CORE STATE STANDARDS
8.EE.6

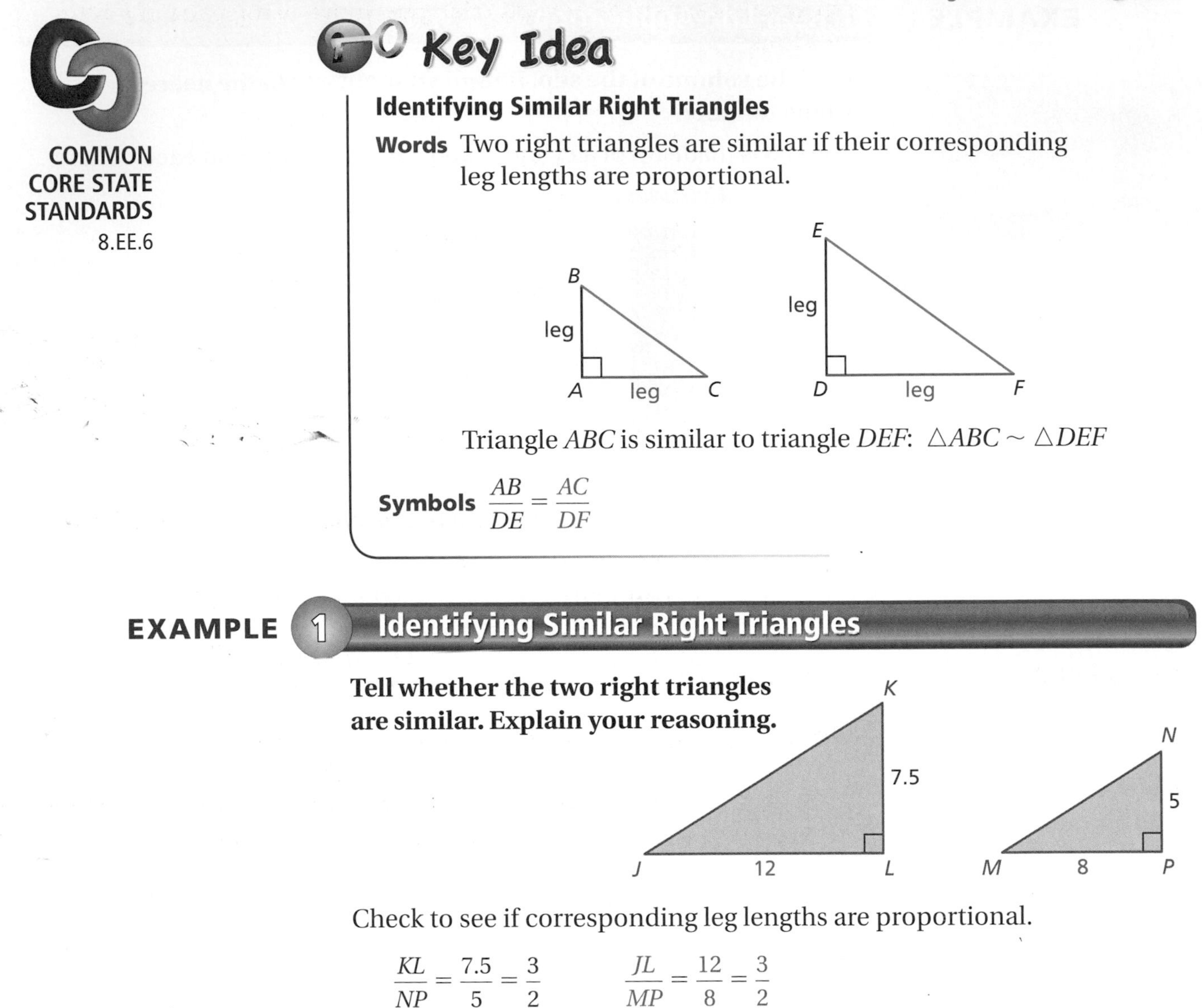

Key Idea

Identifying Similar Right Triangles

Words Two right triangles are similar if their corresponding leg lengths are proportional.

Triangle ABC is similar to triangle DEF: $\triangle ABC \sim \triangle DEF$

Symbols $\dfrac{AB}{DE} = \dfrac{AC}{DF}$

EXAMPLE 1 Identifying Similar Right Triangles

Tell whether the two right triangles are similar. Explain your reasoning.

Check to see if corresponding leg lengths are proportional.

$$\frac{KL}{NP} = \frac{7.5}{5} = \frac{3}{2} \qquad \frac{JL}{MP} = \frac{12}{8} = \frac{3}{2}$$

∴ Corresponding leg lengths are proportional. So, $\triangle JKL \sim \triangle MNP$.

Practice

Tell whether the two right triangles are similar. Explain your reasoning.

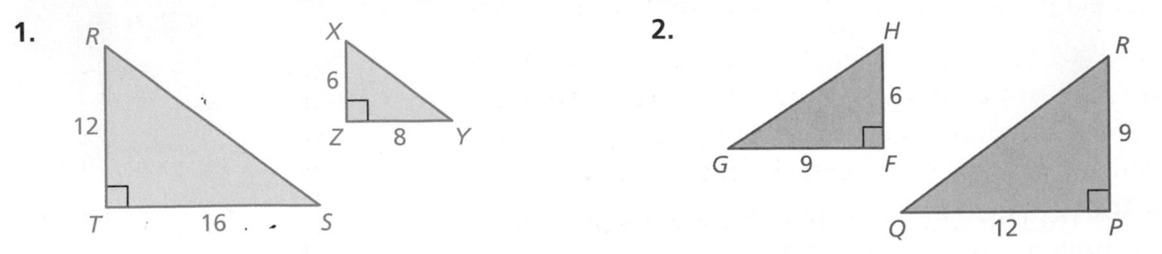

3. **REASONING** How does the ratio of the leg lengths of a right triangle compare to the ratio of the corresponding leg lengths of a similar right triangle? Explain.

EXAMPLE 2 Using Similar Triangles to Find Slope

The graph shows similar right triangles drawn using pairs of points on a line.

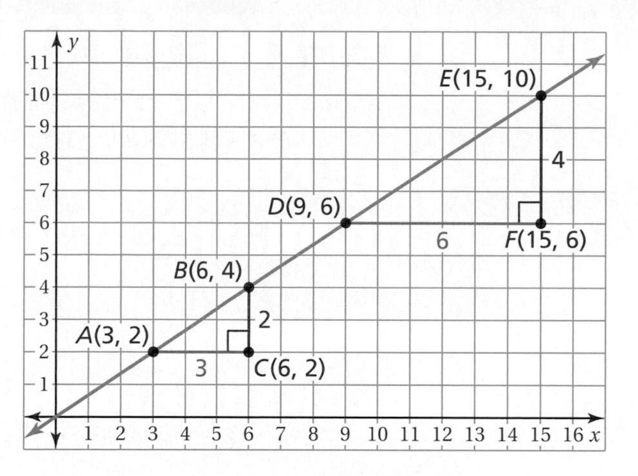

a. For each triangle, find the ratio of the length of the vertical leg to the length of the horizontal leg.

Triangle *ABC*	**Triangle *DEF***
$\dfrac{\text{vertical leg}}{\text{horizontal leg}} = \dfrac{BC}{AC} = \dfrac{2}{3}$	$\dfrac{\text{vertical leg}}{\text{horizontal leg}} = \dfrac{EF}{DF} = \dfrac{4}{6} = \dfrac{2}{3}$

b. Relate the ratios in part (a) to the slope of the line.

The ratios in part (a) represent rise over run, or the slope of the line between points A and B, and between points D and E.

∴ So, the slope of the line is $\frac{2}{3}$.

Practice

4. **SLOPE** Consider the line shown in the graph.
   a. Draw two triangles that show the rise and the run of the line using points A and B and points M and N.
   b. Use the triangles to find the slope of the line.
   c. Repeat parts (a) and (b) using different pairs of points.

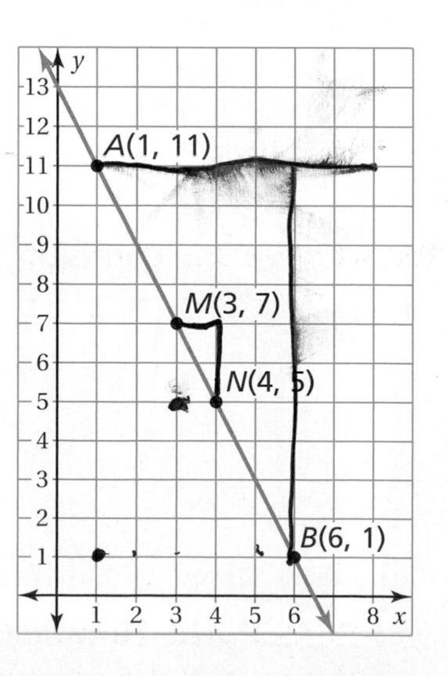

5. **REASONING** You draw a triangle that shows the slope of a line using two points. Then you draw another triangle that shows the slope using a different pair of points on the same line. Are the triangles similar? Explain.

6. **WRITING** Explain why you can find the slope of a line using any two points on the line.

Appendix A
My Big Ideas Projects

A.1 Literature Project
The Mathematics of Jules Verne

A.2 History Project
Mathematics in Ancient Greece

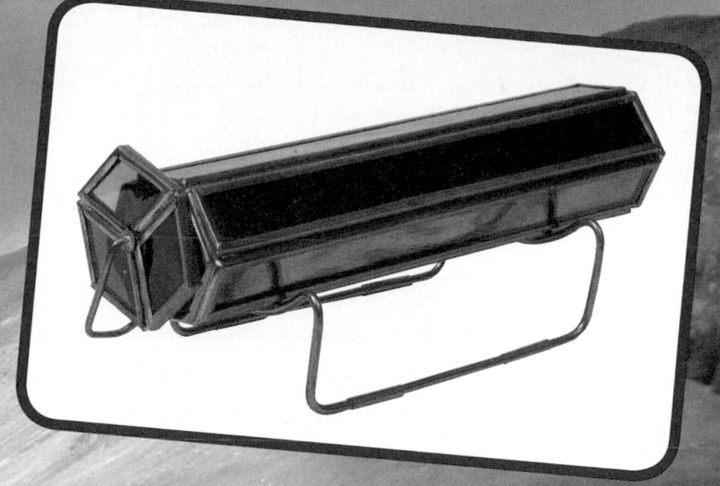

My Big Ideas Projects

A.3 Art Project
Building a Kaleidoscope

A.4 Science Project
Classifying Animals

A.1 Literature Project

Share Your Work at... My.BigIdeasMath.com

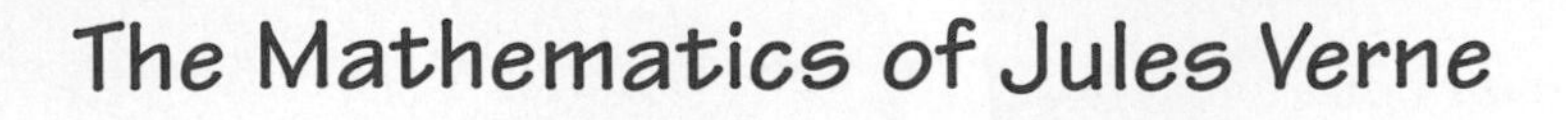

The Mathematics of Jules Verne

1 Project Overview

Jules Verne (1828–1905) was a famous French science fiction writer. He wrote about space, air, and underwater travel before aircraft and submarines were commonplace, and before any means of space travel had been devised.

For example, in his 1865 novel *From the Earth to the Moon*, he wrote about three astronauts who were launched from Florida and recovered through a splash landing. The first actual moon landing wasn't until 1969.

Essential Question How does the knowledge of mathematics influence science fiction writing?

Read one of Jules Verne's science fiction novels. Then write a book report about some of the mathematics used in the novel.

Sample: A league is an old measure of distance. It is approximately equal to 4 kilometers. You can convert 20,000 leagues to miles as follows.

$$20{,}000 \text{ leagues} \cdot \frac{4 \text{ km}}{1 \text{ league}} \cdot \frac{1 \text{ mile}}{1.6 \text{ km}} = 50{,}000 \text{ miles}$$

2 Things to Include

- Describe the major events in the plot.
- Write a brief paragraph describing the setting of the story.
- List and identify the main characters. Explain the contribution of each character to the story.
- Explain the major conflict in the story.
- Describe at least four examples of mathematics used in the story.
- Which of Jules Verne's scientific predictions have come true since he wrote the novel?

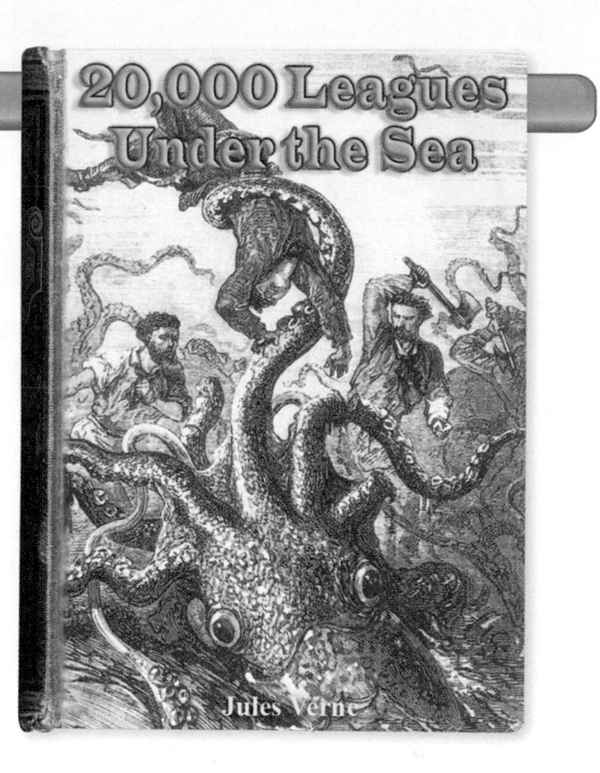

Jules Verne (1828–1905)

3 Things to Remember

- You can download one of Jules Verne's novels at *BigIdeasMath.com*.
- Add your own illustrations to your project.
- Organize your report in a folder, and think of a title for your report.

A.2 History Project

Mathematics in Ancient Greece

1 Getting Started

The ancient Greek period began around 1100 B.C. and lasted until the Roman conquest of Greece in 146 B.C.

The civilization of the ancient Greeks influenced the languages, politics, educational systems, philosophy, science, mathematics, and arts of Western Civilization. It was a primary force in the birth of the Renaissance in Europe between the 14th and 17th centuries.

Corinthian Helmet

Essential Question How do you use mathematical knowledge that was originally discovered by the Greeks?

Sample: Ancient Greek symbols for the numbers from 1 through 10 are shown in the table.

|	||	|||	||||	Γ	Γ|	Γ||	Γ|||	Γ||||	△
1	2	3	4	5	6	7	8	9	10

These same symbols were used to write the numbers between 11 and 39. Here are some examples.

△Γ||| = 18 △△△Γ = 35 △△|||| = 24

Alexander the Great

Parthenon

2 Things to Include

- Describe at least one contribution that each of the following people made to mathematics.

  Pythagoras (c. 570 B.C.–c. 490 B.C.)

  Aristotle (c. 384 B.C.–c. 322 B.C.)

  Euclid (c. 300 B.C.)

  Archimedes (c. 287 B.C.–c. 212 B.C.)

  Eratosthenes (c. 276 B.C.–c. 194 B.C.)

- Which of the people listed above was the teacher of Alexander the Great? What subjects did Alexander the Great study when he was in school?
- How did the ancient Greeks represent fractions?
- Describe how the ancient Greeks used mathematics. How does this compare with the ways in which mathematics is used today?

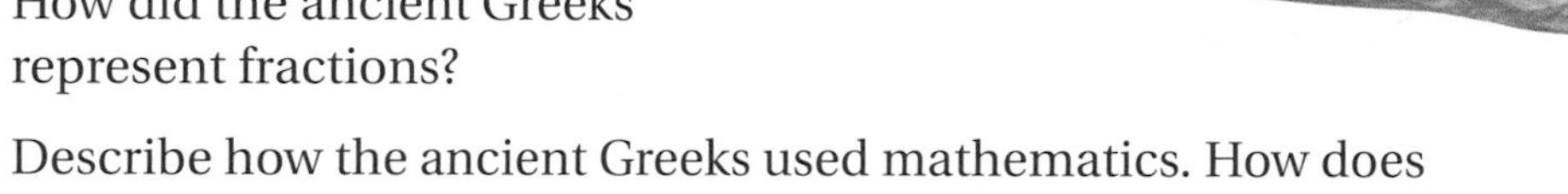

Α α	alpha	Ν ν	nu
Β β	beta	Ξ ξ	xi
Γ γ	gamma	Ο ο	omicron
Δ δ	delta	Π π	pi
Ε ε	epsilon	Ρ ρ	rho
Ζ ζ	zeta	Σ σ	sigma
Η η	eta	Τ τ	tau
Θ θ	theta	Υ υ	upsilon
Ι ι	iota	Φ φ	phi
Κ κ	kappa	Χ χ	chi
Λ λ	lambda	Ψ ψ	psi
Μ μ	mu	Ω ω	omega

3 Things to Remember

- Add your own illustrations to your project.
- Try to include as many different math concepts as possible. Your goal is to include at least one concept from each of the chapters you studied this year.
- Organize your report in a folder, and think of a title for your report.

Greek Pottery

Trireme Greek Warship

A.3 Art Project

Building a Kaleidoscope

1 Getting Started

A kaleidoscope is a tube of mirrors containing loose colored beads, pebbles, or other small colored objects. You look in one end and light enters the other end, reflecting off the mirrors.

Mirrors set at 60°

Essential Question How does the knowledge of mathematics help you create a kaleidoscope?

If the angle between the mirrors is 45°, you see 8 duplicate images. If the angle is 60°, you see 6 duplicate images. If the angle is 90°, you see 4 duplicate images. As the tube is rotated, the colored objects tumble, creating various patterns.

Write a report about kaleidoscopes. Discuss the mathematics you need to know in order to build a kaleidoscope.

Sample: A kaleidoscope whose mirrors meet at 60° angles has reflective symmetry and rotational symmetry.

Reflect

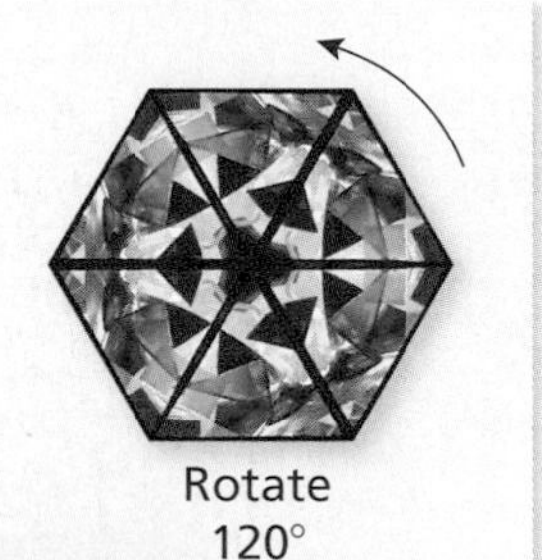
Rotate 120°

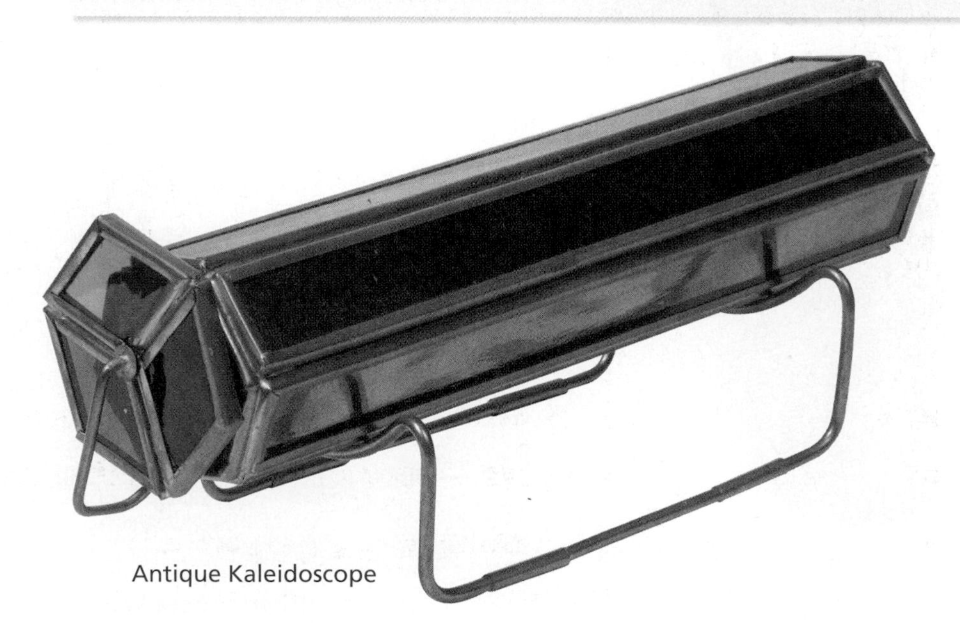
Antique Kaleidoscope

2 Things to Include

- How does the angle at which the mirrors meet affect the number of duplicate images that you see?
- What angles can you use other than 45°, 60°, and 90°? Explain your reasoning.
- Research the history of kaleidoscopes. Can you find examples of kaleidoscopes being used before they were patented by David Brewster in 1816?
- Make your own kaleidoscope.
- Describe the mathematics you used to create your kaleidoscope.

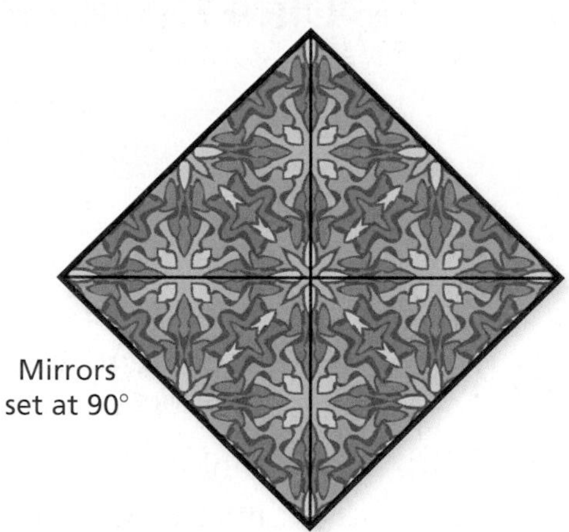

Mirrors set at 90°

Mirrors set at 60°

3 Things to Think About

- Add your own drawings and pattern creations to your project.
- Organize your report in a folder, and think of a title for your report.

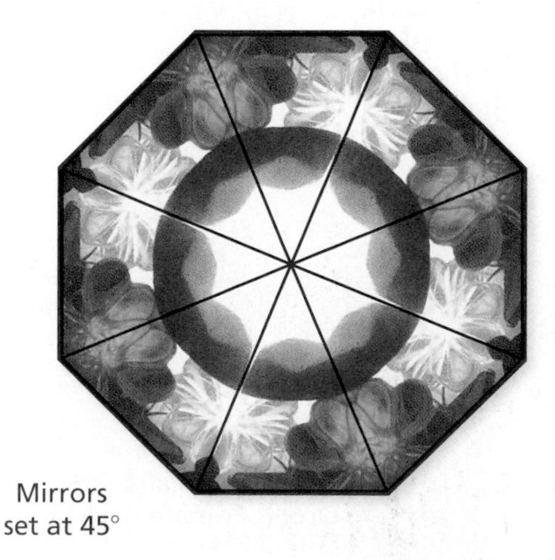

Mirrors set at 45°

Giant Kaleidoscope, San Diego harbor

A.4 Science Project

Classifying Animals

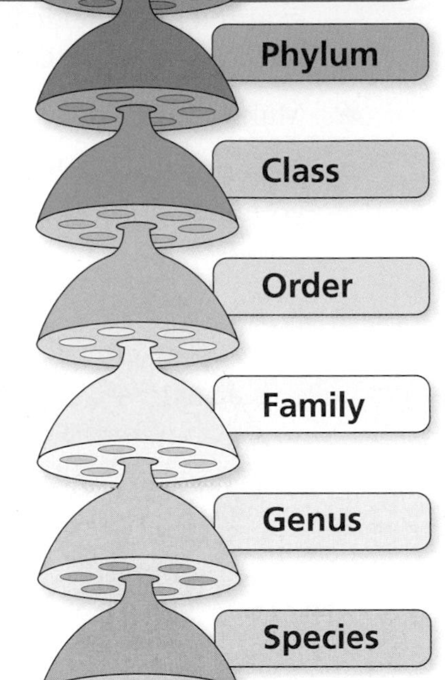

1 Getting Started

Biologists classify animals by placing them in phylums, or groups, with similar characteristics. Latin names, such as Chordata (having a spinal cord) or Arthropoda (having jointed limbs and rigid bodies) are used to describe these groups.

Biological classification is difficult, and scientists are still developing a complete system. There are seven main ranks of life on Earth; kingdom, phylum, class, order, family, genus, and species. However, scientists usually use more than these seven ranks to classify organisms.

Essential Question How does the classification of living organisms help you understand the similarities and differences of animals?

Write a report about how animals are classified. Choose several different animals and list the phylum, class, and order of each animal.

Wasp
Phylum: Arthropoda
Class: Insecta
Order: Hymenoptera (membranous wing)

Sample: A bat is classified as an animal in the phylum Chordata, class Mammalia, and order Chiroptera. *Chiroptera* is a Greek word meaning "hand-wing."

Bat
Phylum: Chordata
Class: Mammalia
Order: Chiroptera (hand-wing)

Monkey
Phylum: Chordata
Class: Mammalia
Order: Primate (large brain)

Kangaroo
Phylum: Chordata
Class: Mammalia
Order: Diprotodontia (two front teeth)

2 Things to Include

- List the different classes of phylum Chordata. Have you seen a member of each class?
- List the different classes of phylum Arthropoda. Have you seen a member of each class?
- Show how you can use graphic organizers to help classify animals. Which types of graphic organizers seem to be most helpful? Explain your reasoning.
- Summarize the number of species in each phlyum in an organized way. Be sure to include fractions, decimals, and percents.

Parrot
Phylum: Chordata
Class: Aves
Order: Psittaciformes
(strong, curved bill)

Frog
Phylum: Chordata
Class: Amphibia
Order: Anura (no tail)

Spider
Phylum: Arthropoda
Class: Arachnida
Order: Araneae (spider)

3 Things to Remember

- Organize your report in a folder, and think of a title for your report.

Lobster
Phylum: Arthropoda
Class: Malacostraca
Order: Decopada (ten footed)

Crocodile
Phylum: Chordata
Class: Reptilia
Order: Crocodilia
(pebble-worm)

Cougar
Phylum: Chordata
Class: Mammalia
Order: Carnivora (meat eater)

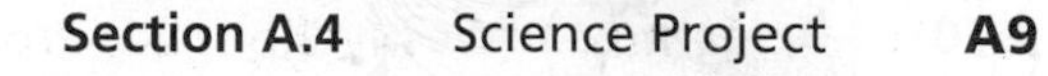

Selected Answers

Section 1.1 Integers and Absolute Value
(pages 6 and 7)

1. 9, -1, 15

3. -6; All of the other expressions are equal to 6.

5. 6 **7.** 10 **9.** 13 **11.** 12 **13.** 8 **15.** 18

17. 45 **19.** 125 **21.** $|-4| < 7$ **23.** $|-4| > -6$ **25.** $|5| = |-5|$

27. Because $|-5| = 5$, the statement is incorrect. $|-5| > 4$

29. $-8, 5$

31. $-7, -6, |5|, |-6|, 8$

33. $-17, |-11|, |20|, 21, |-34|$

35. -4

37. **a.** MATE

M A T E
-8 -6 -4 -2 0 2 4

b. TEAM

T E A M
-2 0 2 4 6 8 10

39. $n \geq 0$

41. The number closer to 0 is the greater integer.

43. **a.** Player 3 **b.** Player 2 **c.** Player 1

45. false; The absolute value of zero is zero, which is neither positive nor negative.

47. 144

49. 3170

Hmmm.

Section 1.2 Adding Integers
(pages 12 and 13)

1. Change the sign of the integer.

3. positive; 20 has the greater absolute value and is positive.

5. negative; The common sign is a negative sign.

7. false; A positive integer and its absolute value are equal, not opposites.

9. -10 **11.** 7 **13.** 0 **15.** 10

17. -7 **19.** -11 **21.** -4 **23.** -34

25. -10 and -10 are not opposites. $-10 + (-10) = -20$

27. \$48 **29.** -27 **31.** 21 **33.** -85

35. Use the Associate Property to add 13 and -13 first. -8

37. *Sample answer:* Use the Commutative Property to switch the last two terms. -12

39. *Sample answer:* Use the Commutative Property to switch the last two terms. 11

41. -13

43. *Sample answer:* $-26 + 1$; $-12 + (-13)$

45. $b = 2$

47. $6 + (-3) + 8$

49. Find the number in each row or column that already has two numbers in it before guessing.

51. 8

53. 183

Hint

Section 1.3 Subtracting Integers
(pages 18 and 19)

1. You add the integer's opposite.
3. What is 3 less than -2?; -5; 5
5. C
7. B
9. 13
11. -5
13. -10
15. 3
17. 17
19. 1
21. -22
23. -20
25. $-3 - 9$
27. 6
29. 9
31. 7
33. $m = 14$
35. $c = 15$
37. 2
39. 3
41. *Sample answer:* $x = -2, y = -1$; $x = -3, y = -2$
43. sometimes; It's positive only if the first integer is greater.
45. always; It's always positive because the first integer is always greater.
47. all values of a and b
49. when a and b have the same sign and $|a| > |b|$ or $|a| = |b|$, or $b = 0$
51. -45
53. 468
55. 2378

Section 1.4 Multiplying Integers
(pages 26 and 27)

1. **a.** They are the same. **b.** They are different.
3. negative; different signs
5. negative; different signs
7. false; The product of the first two negative integers is positive. The product of the positive result and the third negative integer is negative.
9. -21
11. 12
13. 27
15. 12
17. 0
19. -30
21. 78
23. 121
25. $-240{,}000$
27. 54
29. -105
31. 0
33. -1
35. -36
37. 54
39. The answer should be negative. $-10^2 = -(10 \cdot 10) = -100$
41. 32
43. -7500, 37,500
45. -12
47. **a.**

Month	Price of Skates
June	$165 = \$165$
July	$165 + (-12) = \$153$
August	$165 + 2(-12) = \$141$
September	$165 + 3(-12) = \$129$

b. The price drops $12 every month.

c. no; yes; In August you have $135 but the cost is $141. In September you have $153 and the cost is only $129.

49. 3
51. 14
53. D

Section 1.5 Dividing Integers
(pages 32 and 33)

1. They have the same sign. They have different signs. The dividend is zero.

3. *Sample answer:* −4, 2
5. negative
7. negative
9. −3
11. 3
13. 0
15. −6
17. 7
19. −10
21. undefined
23. 12
25. The quotient should be 0. $0 \div (-5) = 0$
27. 15 pages
29. −8
31. 65
33. 5
35. 4
37. −400 ft/min
39. 5
41. *Sample answer:* −20, −15, −10, −5, 0; Start with −10, then pair −15 with −5 and −20 with 0.
43. 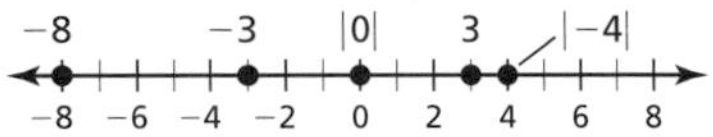

45. B

Section 1.6 The Coordinate Plane
(pages 38 and 39)

1. 4
3. (2, −2) is in Quadrant IV, (−2, 2) is in Quadrant II.
5. (3, 1)
7. (−2, 4)
9. (2, −2)
11. (−4, 2)
13. (4, 0)

15–25. See graph below.

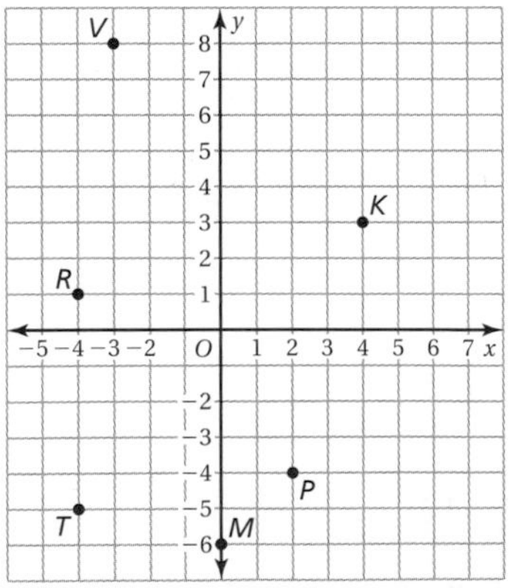

15. Quadrant I
17. *y*-axis
19. Quadrant IV
21. Quadrant II
23. Quadrant III
25. Quadrant II
27. The numbers are reversed. To plot (4, 5), start at (0, 0) and move 4 units right and 5 units up.
29. (−2, 1)
31. sometimes; It is true only for (0, 0).
33. always; The *x*-coordinate of a point in Quadrant II is negative, and so is the *y*-coordinate of a point in Quadrant IV.
35. Flamingo Café

37. Because the rainforest is in Quadrant IV, the x-coordinate of the point will be positive and the y-coordinate of the point will be negative.

39. (2, 2)

41–43. See graph below.

41. Quadrant II

43. x-axis

45. $-\frac{16}{2} < -\frac{12}{3}$

47. $3.45 > 3\frac{3}{8}$

Section 2.1 Rational Numbers
(pages 54 and 55)

1. A number is rational if it can be written as $\frac{a}{b}$ where a and b are integers and $b \neq 0$.

3. rational numbers, integers

5. rational numbers, integers, whole numbers

7. repeating

9. terminating

11. 0.875

13. $-0.\overline{7}$

15. $1.8\overline{3}$

17. $-5.58\overline{3}$

19. The bar should be over both digits to the right of the decimal point. $-\frac{7}{11} = -0.\overline{63}$

21. $\frac{9}{20}$

23. $-\frac{39}{125}$

25. $-1\frac{16}{25}$

27. $-12\frac{81}{200}$

29. $-2.5, -1.1, -\frac{4}{5}, 0.8, \frac{9}{5}$

31. $-\frac{9}{4}, -0.75, -\frac{6}{10}, \frac{5}{3}, 2.1$

33. $-2.4, -2.25, -\frac{11}{5}, \frac{15}{10}, 1.6$

35. spotted turtle

37. $-1.82 < -1.81$

39. $-4\frac{6}{10} > -4.65$

41. $-2\frac{13}{16} < -2\frac{11}{14}$

43. Michelle

45. No; The base of the skating pool is at -10 feet, which is deeper than $-9\frac{5}{6}$ feet.

47. **a.** when a is negative

b. when a and b have the same sign, $a \neq 0 \neq b$

49. $\frac{7}{30}$

50. 21.15

Selected Answers

Section 2.2 Adding and Subtracting Rational Numbers
(pages 60 and 61)

1. Because $|-8.46| > |5.31|$, subtract $|5.31|$ from $|-8.46|$ and the sign is negative.

3. What is 3.9 less than -4.8?; -8.7; -0.9

5. $-\frac{5}{14}$

7. $2\frac{3}{10}$

9. -0.9

11. 1.844

13. $1\frac{1}{2}$

15. $\frac{1}{18}$

17. $-18\frac{13}{24}$

19. -2.6

21. 14.963

23. $\frac{3}{8} - \frac{5}{6} = -\frac{11}{24}$

25. $\frac{1}{18}$

27. $-3\frac{9}{10}$

29. No, the cook needs $\frac{1}{12}$ cup more.

31–33. Subtract the least number from the greatest number.

Hint

35. $-\frac{n}{4}$

37. $-\frac{b}{24}$

39. 35.88

41. $8\frac{2}{3}$

43. C

Section 2.3 Multiplying and Dividing Rational Numbers
(pages 66 and 67)

1. The same rules for signs of integers are applied to rational numbers.

3. $-\frac{1}{3}$

5. $-\frac{3}{7}$

7. negative

9. positive

11. $-\frac{2}{3}$

13. $-\frac{1}{100}$

15. $2\frac{5}{14}$

17. $3.\overline{63}$

19. -6

21. -2.5875

23. $\frac{1}{3}$

25. $2\frac{1}{2}$

27. $-4\frac{17}{27}$

29. 0.025

31. 47.43

33. -0.064

35. The wrong fraction was inverted.

$$-\frac{1}{4} \div \frac{3}{2} = -\frac{1}{4} \times \frac{2}{3}$$
$$= -\frac{2}{12}$$
$$= -\frac{1}{6}$$

37. 8 packages

39. 1.3

41. $-4\frac{14}{15}$

43. $-1\frac{11}{36}$

45. $191\frac{11}{12}$ yd

47. How many spaces are between the boards?

49. a. -2, 4, -8, 16, -32, 64

b. When -2 is raised to an odd power, the product is negative. When -2 is raised to an even power, the product is positive.

c. negative

51. -5.4

53. $-8\frac{5}{18}$

Lesson 2.3b Number Properties *(pages 67A and 67B)*

1. $2 + 3 + (-2) = 2 + (-2) + 3$ Comm. Prop. of Add.
   $= 0 + 3$ Additive Inverse Property
   $= 3$ Addition Prop. of Zero

3. $4 \cdot 19 \cdot \frac{1}{2} = 4 \cdot \frac{1}{2} \cdot 19$ Comm. Prop. of Mult.
   $= 2 \cdot 19$ Multiply 4 and $\frac{1}{2}$.
   $= 38$ Multiply 2 and 19.

5. $5\left(\frac{7}{8} \cdot \frac{2}{5}\right) = 5\left(\frac{2}{5} \cdot \frac{7}{8}\right)$ Comm. Prop. of Mult.
   $= \left(5 \cdot \frac{2}{5}\right) \cdot \frac{7}{8}$ Assoc. Prop. of Mult.
   $= 2 \cdot \frac{7}{8}$ Multiply 5 and $\frac{2}{5}$.
   $= \frac{7}{4}$ Multiply 2 and $\frac{7}{8}$.

7. 0

9. *Sample answer:* Find a map, Lose a compass, Lose a compass

Section 2.4 Solving Equations Using Addition or Subtraction *(pages 74 and 75)*

1. Subtraction Property of Equality
3. No, $m = -8$ not -2 in the first equation.
5. $a = 19$
7. $k = -20$
9. $c = 3.6$
11. $q = -\frac{1}{6}$
13. $g = -10$
15. $y = -2.08$
17. $q = -\frac{7}{18}$
19. $w = -1\frac{13}{24}$
21. The 8 should have been subtracted rather than added.

$$\begin{array}{r} x + 8 = 10 \\ \underline{-8 \quad -8} \\ x = 2 \end{array}$$

23. $c + 10 = 3;\ c = -7$
25. $p - 6 = -14;\ p = -8$
27. $P + 2.54 = 1.38;\ -\$1.16$ million
29. $x + 8 = 12;\ 4$ cm
31. $x + 22.7 = 34.6;\ 11.9$ ft
33. Because your first jump is higher, your second jump went a farther distance than your first jump.
35. $m + 30.3 + 40.8 = 180;\ 108.9°$
37. -9
39. $6, -6$
41. -56
43. -9
45. B

Selected Answers

Section 2.5 Solving Equations Using Multiplication or Division *(pages 80 and 81)*

1. Multiplication is the inverse operation of division, so it can undo division.

3. dividing by 5

5. multiplying by -8

7. $h = 5$

9. $n = -14$

11. $m = -2$

13. $x = -8$

15. $p = -8$

17. $n = 8$

19. $g = -16$

21. $f = 6\frac{3}{4}$

23. They should divide by -4.2.

$$-4.2x = 21$$

$$\frac{-4.2x}{-4.2} = \frac{21}{-4.2}$$

$$x = -5$$

25. $\frac{2}{5}x = \frac{3}{20}$; $x = \frac{3}{8}$

27. $\frac{x}{-1.5} = 21$; $x = -31.5$

29. $\frac{x}{30} = 12\frac{3}{5}$; 378 ft

31–33. Sample answers are given.

31. **a.** $-2x = 4.4$ **b.** $\frac{x}{1.1} = -2$

33. **a.** $4x = -5$ **b.** $\frac{x}{5} = -\frac{1}{4}$

35. $-1.26n = -10.08$; 8 days

37. -50 ft

39. $-5, 5$

41. -7

43. 12

45. B

Lesson 2.5b Algebraic Expressions *(pages 81A and 81B)*

1. Terms: y, 10, $-\frac{3}{2}y$

Like terms: y and $-\frac{3}{2}y$

3. Terms: 7, $4p$, -5, p, $2q$

Like terms: 7 and -5, $4p$ and p

5. $-\frac{3}{8}b$

7. $3q + 2$

9. $7g + 3$

Section 2.6 Solving Two-Step Equations *(pages 86 and 87)*

1. Eliminate the constants on the side with the variable. Then solve for the variable using either division or multiplication.

3. D

5. A

7. $b = -3$

9. $t = -4$

11. $g = 4.22$

13. $p = 3\frac{1}{2}$

15. $h = -3.5$

17. $y = -6.4$

19. Each side should be divided by -3, not 3.

$$-3x + 2 = -7$$

$$-3x = -9$$

$$\frac{-3x}{-3} = \frac{-9}{-3}$$

$$x = 3$$

21. $a = 1\frac{1}{3}$

23. $b = 13\frac{1}{2}$

25. $v = -\frac{1}{30}$

27. $2.5 + 2.25x = 9.25$; 3 games

29. $v = -5$

31. $d = -12$

33. $m = -9$

35. *Sample answer:* You travel halfway up a ladder. Then you climb down two feet and are 8 feet above the ground. How long is the ladder? $x = 20$

37. the initial fee

39. Find the number of insects remaining and then find the number of insects you caught.

41. decrease the length by 10 cm; $2(25 + x) + 2(12) = 54$

43. $-6\frac{2}{3}$

45. 6.2

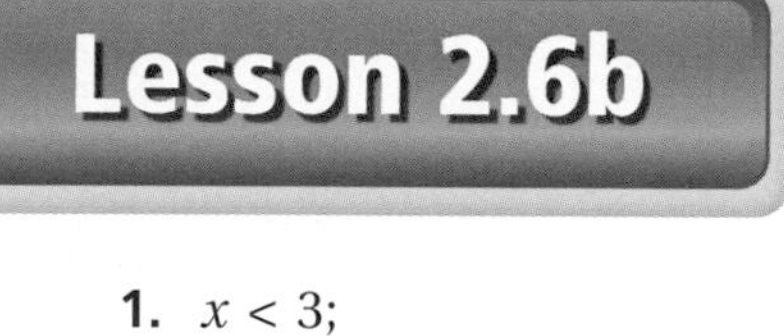

Lesson 2.6b Solving Inequalities
(pages 87A–87D)

1. $x < 3$;

0 1 2 3 4 5 6

3. $r > 0.7$;

5. $z \leq \frac{1}{5}$;

7. $b \geq -40$;

−70 −60 −50 −40 −30 −20 −10

9. $m \leq \frac{4}{3}$;

11. $x \geq 25$;

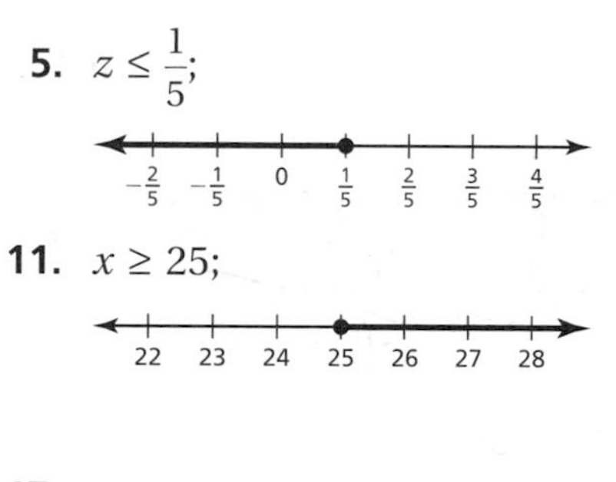

13. $j > -10.5$;

−10.5
−13 −12 −11 −10 −9 −8 −7

15. $k < 12$;

17. $z \leq -20$;

−50 −40 −30 −20 −10 0 10

19. $n < 3$;

0 1 2 3 4 5 6

21. $c \geq 18$;

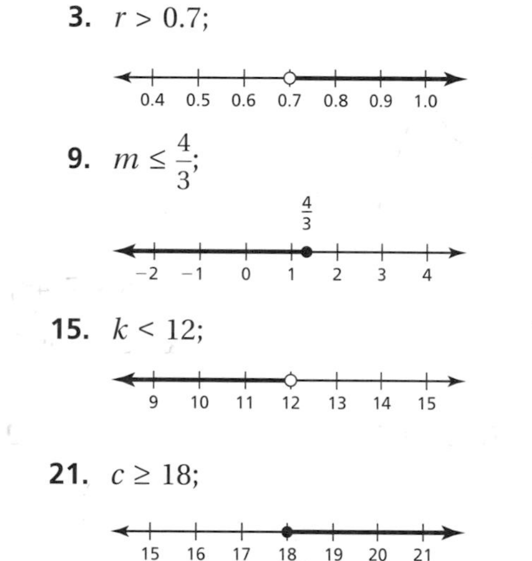

Section 3.1 Ratios and Rates
(pages 102 and 103)

1. It has a denominator of 1.

3. *Sample answer:* A basketball player runs 10 ft down the court in 2 sec.

5. \$0.10 per fl oz

7. \$72

9. 840 MB

11. $\frac{5}{9}$

13. $\frac{7}{3}$

15. $\frac{4}{3}$

17. 60 mi/h

19. \$2.40 per lb

21. 54 words per min

23. 90 calories per serving

25. 4.5 servings per package

27. 4.8 MB per min

Section 3.1 Ratios and Rates *(continued)*

(pages 102 and 103)

29. **a.** It costs $122 for 4 tickets.

b. $30.50 per ticket

c. $305

31. The 9-pack is the best buy at $2.55 per container.

33. Try searching for "fire hydrant colors."

35–37.

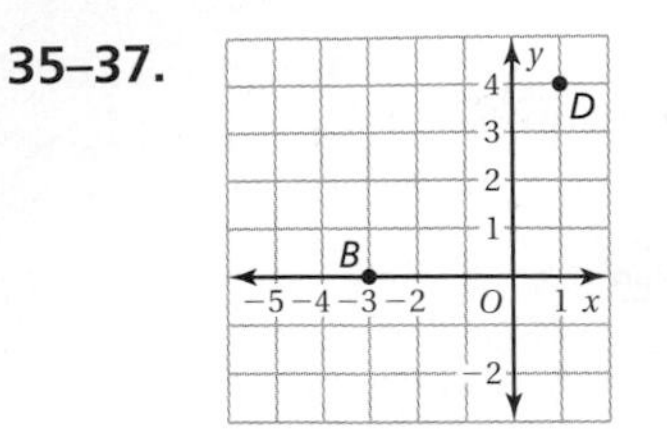

Section 3.2 Slope

(pages 108 and 109)

1. yes; Slope is the rate of change of a line.

3. 5; A ramp with a slope of 5 increases 5 units vertically for every 1 unit horizontally. A ramp with a slope of $\frac{1}{5}$ increases 1 unit vertically for every 5 units horizontally.

5. $\frac{3}{2}$

7. 1

9. $\frac{4}{5}$

11. **13.**

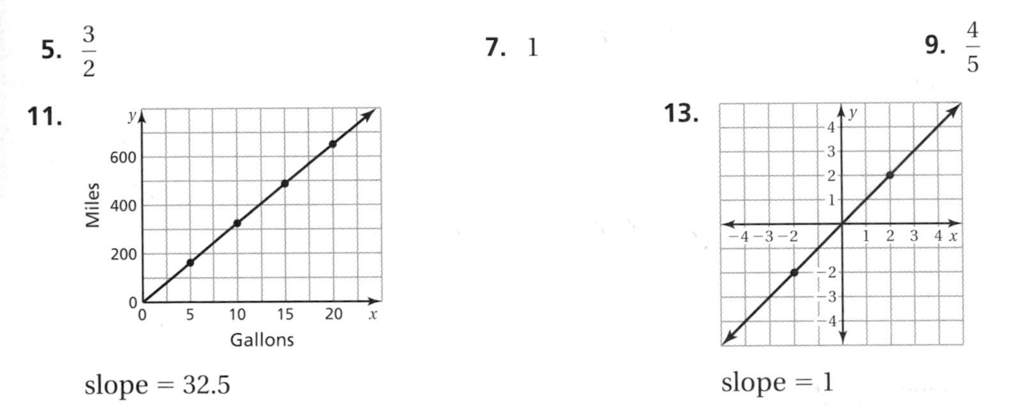

slope = 32.5

slope = 1

15. The change in y should be in the numerator. The change in x should be in the denominator.

Slope $= \frac{5}{4}$

17. **a.**

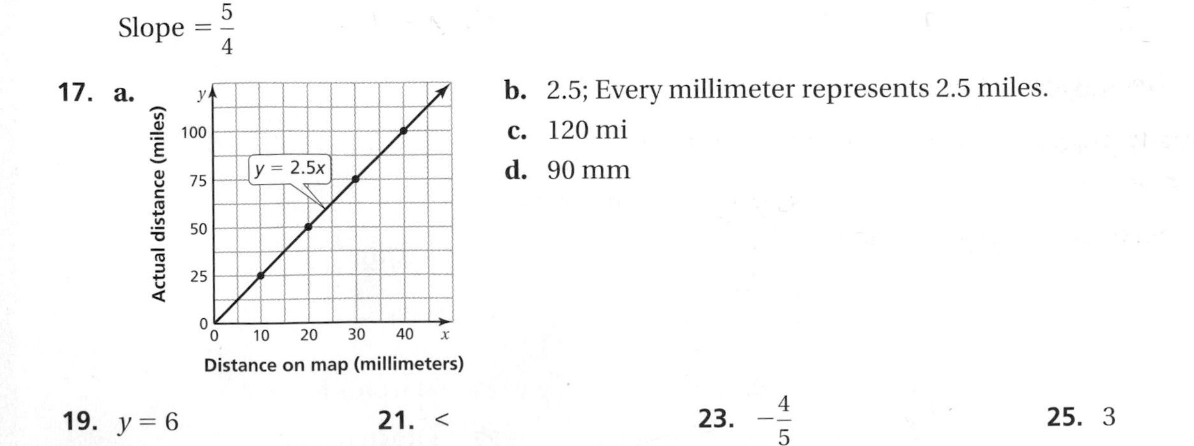

b. 2.5; Every millimeter represents 2.5 miles.

c. 120 mi

d. 90 mm

19. $y = 6$

21. $<$

23. $-\frac{4}{5}$

25. 3

Section 3.3 Proportions
(pages 114 and 115)

1. Both ratios are equal.
3. *Sample answer:* $\frac{6}{10}, \frac{12}{20}$
5. yes
7. no
9. yes
11. no
13. yes
15. yes
17. yes
19. no
21. yes
23. yes; Both can do 45 sit-ups per minute.
25. yes
27. yes
29. yes; They are both $\frac{4}{5}$.
31. **a.** Pitcher 3

    b. Pitcher 2 and Pitcher 4
33. **a.** no

    b. *Sample answer:* If the collection has 50 quarters and 30 dimes, when 10 of each coin are added, the new ratio of quarters to dimes is 3 : 2.
35. -13
37. -18
39. D

Section 3.4 Writing Proportions
(pages 120 and 121)

1. You can use the columns or the rows of the table to write a proportion.
3. *Sample answer:* $\frac{x}{12} = \frac{5}{6}$; $x = 10$
5. $\frac{x}{50} = \frac{78}{100}$
7. $\frac{x}{150} = \frac{96}{100}$
9. $\frac{n \text{ winners}}{85 \text{ entries}} = \frac{34 \text{ winners}}{170 \text{ entries}}$
11. $\frac{100 \text{ meters}}{x \text{ seconds}} = \frac{200 \text{ meters}}{22.4 \text{ seconds}}$
13. $\frac{\$24}{3 \text{ shirts}} = \frac{c}{7 \text{ shirts}}$
15. $\frac{5 \text{ 6th grade swimmers}}{16 \text{ swimmers}} = \frac{s \text{ 6th grade swimmers}}{80 \text{ swimmers}}$
17. $y = 16$
19. $c = 24$
21. $g = 14$
23. $\frac{1}{200} = \frac{19.5}{x}$; Dimensions for the model are in the numerators and the corresponding dimensions for the actual space shuttle are in the denominators.
25. Draw a diagram of the given information.
27. $x = 9$
29. $x = 140$

Hint

Selected Answers

Section 3.5 Solving Proportions
(pages 126 and 127)

1. mental math; Multiplication Property of Equality; Cross Products Property

3. yes; Both cross products give the equation $3x = 60$.

5. $h = 80$
7. $n = 15$
9. $y = 7\frac{1}{3}$
11. $k = 5.6$
13. $n = 10$
15. $d = 5.76$
17. $m = 20$
19. $d = 15$
21. $k = 5.4$
23. 108 pens
25. $x = 1.5$
27. $k = 4$
29. \$769.50
31. **a.** 16 mo **b.** 40 mo
33. Make a table to solve the problem.

35. 2; $\frac{1/2}{1/4} = \frac{1}{2} \times \frac{4}{1} = 2$
37. 6400
39. 7920

Section 3.6 Converting Measures Between Systems
(pages 134 and 135)

1. To convert between measurements, multiply by the ratio of the given relationship such that the desired unit is in the numerator, or set up and solve a proportion using the given relationship as one of the ratios.

3. Find the number of inches in 5 cm; 5 cm $\approx$ 1.97 in.; 5 in. $\approx$ 12.7 cm

5. $>$
7. $>$
9. $<$
11. 9.5
13. 21.08
15. 3.13
17. 121.92
19. 64.96
21. 2.38
23. about 584 km
25. $>$
27. $>$
29. $<$
31. 72
33. 4.72
35. 77,400 kg

37. A kilometer is shorter than a mile. So, the given speed when converted should be greater than 110.

39. about 3.7 gal

41.

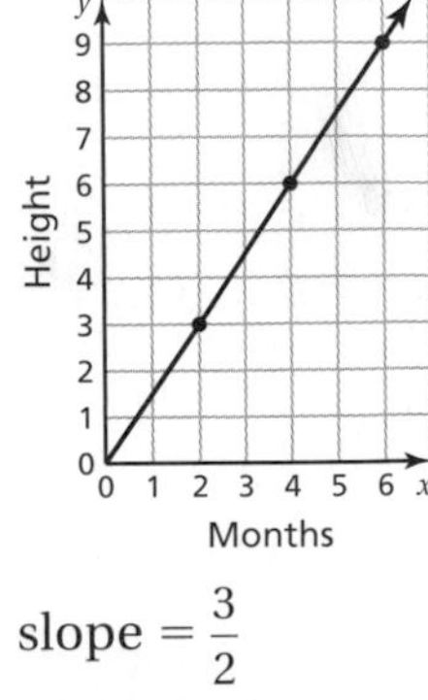

slope $= \frac{3}{2}$

43. C

Section 3.7 Direct Variation
(pages 140 and 141)

1. As one quantity increases, the other quantity increases.
3. the second graph; The points do not lie on a line.
5. no; The line does not pass through the origin.
7. yes; The points lie on a line that passes through the origin.
9. no; The line does not pass through the origin.
11. yes; The line passes through the origin.
13. yes; The line passes through the origin.
15. yes; The equation can be written as $y = kx$.
17. no; The equation cannot be written as $y = kx$.
19. yes; The equation can be written as $y = kx$.
21. no; The equation cannot be written as $y = kx$.
23. yes
25. $y = 5x$
27. $y = 24x$
29. $y = \frac{9}{8}x$
31. You can draw the ramp on a coordinate plane and write a direct variation equation.
33. no
35. Every graph of direct variation is a line; however, not all lines show direct variation because the line must pass through the origin.
37. $y = -60$
39. $d = -59\frac{1}{2}$

Lesson 3.7b Proportional Relationships
(pages 141A and 141B)

1. (0, 0): You earn \$0 for working 0 hours.
   (1, 15): You earn \$15 for working 1 hour; unit rate: $\frac{\$15}{1\text{ h}}$
   (4, 60): You earn \$60 for working 4 hours; unit rate: $\frac{\$60}{4\text{ h}} = \frac{\$15}{1\text{ h}}$
3. $y = 1.5$

Section 3.8 Inverse Variation
(pages 146 and 147)

1. As x increases, y decreases.
3. *Sample answer:* The wingspan of a bird varies inversely with its wing beat frequency.
5. inverse variation; The equation can be written as $y = \frac{k}{x}$.
7. direct variation; The equation can be written as $y = kx$.
9. neither; The equation cannot be written as $y = kx$ or $y = \frac{k}{x}$.

Section 3.8 Inverse Variation *(continued)*

(pages 146 and 147)

11. direct variation; The equation can be written as $y = kx$.

13. inverse variation; The equation can be written as $y = \frac{k}{x}$.

15. neither; The equation cannot be written as $y = kx$ or $y = \frac{k}{x}$.

17.

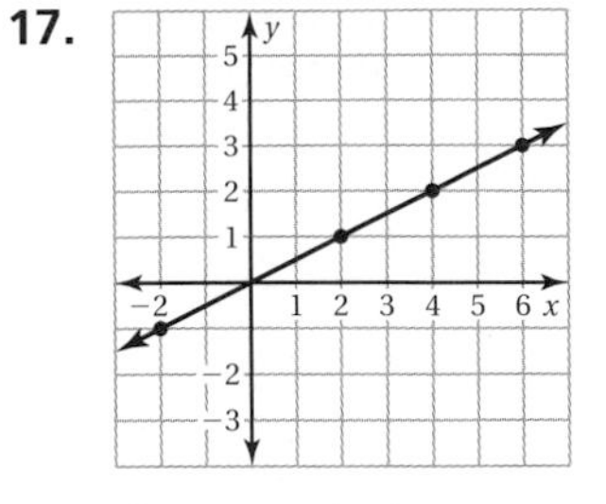

direct variation

19.

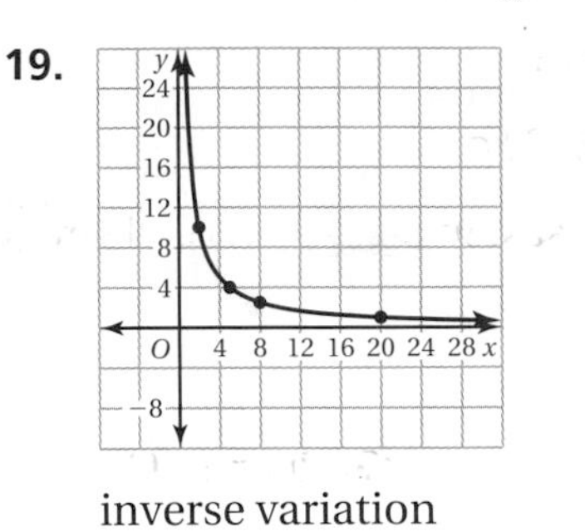

inverse variation

21. inverse variation; The equation can be written as $y = \frac{k}{x}$.

23. $y = \frac{4}{x}$

25. a. yes; $t = \frac{12}{s}$ **b.** 3 h

27. decreases **29.** 88 **31.** 63 **33.** yes **35.** yes **37.** B

Section 4.1 The Percent Equation

(pages 162 and 163)

1. A part of the whole is equal to a percent times the whole.

3. 55 is 20% of what number?; 275; 11

5. 37.5% **7.** 84 **9.** 64

11. $45 = p \cdot 60$; 75% **13.** $a = 0.32 \cdot 25$; 8

15. $12 = 0.005 \cdot w$; 2400 **17.** $102 = 1.2 \cdot w$; 85

19. 30 represents the part of the whole.

$30 = 0.6 \cdot w$

$50 = w$

21. $5400 **23.** 26 years old **25.** 56 signers

27. If the percent is less than 100%, the percent of a number is less than the number. If the percent is equal to 100%, the percent of a number will equal the number. If the percent is greater than 100%, the percent of a number is greater than the number.

29. Remember when writing a proportion that either the units are the same on each side of the proportion, or the numerators have the same units and the denominators have the same units.

31. 92% **33.** 0.88 **35.** 0.36

Section 4.2 Percents of Increase and Decrease (pages 168 and 169)

1. If the original amount decreases, the percent of change is a percent of decrease. If the original amount increases, the percent of change is a percent of increase.

3. The new amount is now 0.

5. decrease; 66.7%

7. increase; 225%

9. decrease; 12.5%

11. decrease; 37.5%

13. 10 m

15. 37 points

17. 153 students

19. 42.16 kg

21. They should have subtracted 10 in the last step because 25 is decreased by 40%.
   40% of $25 = 0.4 \cdot 25 = 10$
   So, $25 - 10 = 15$.

23. increase; 100%

25. increase; 133.3%

27. Increasing 20 to 40 is the same as increasing 20 by 20. So, it is a 100% increase. Decreasing 40 to 20 is the same as decreasing 40 by one-half of 40. So, it is a 50% decrease.

29. **a.** 100% increase

    b. 300% increase

31. less than; *Sample answer:* Let x represent the number. A 10% increase is equal to $x + 0.1x$, or $1.1x$. A 10% decrease of this new number is equal to $1.1x - 0.1(1.1x)$, or $0.99x$. Because $0.99x < x$, the result is less than the original number.

33. 10 girls

35. 35%

37. 56.25

Section 4.3 Discounts and Markups (pages 176 and 177)

1. *Sample answer:* Multiply the original price by $100\% - 25\% = 75\%$ to find the sale price.

3. **a.** 6% tax on a discounted price; The discounted price is less, so the tax is less.

   b. 30% markup on a $30 shirt; 30% of $30 is less than $30.

5. $35.70

7. $76.16

9. $53.33

11. $450

13. $172.40

15. 20%

17. no; Only the amount of markup should be in the numerator, $\frac{105 - 60}{60} = 0.75$.
    So, the markup is 75%.

19. $36

21. "Multiply $45.85 by 0.1" and "Multiply $45.85 by 0.9, then subtract from $45.85." Both will give the sale price of $4.59. The first method is easier because it is only one step.

23. no; $31.08

25. $30

27. 180

29. C

Section 4.4 Simple Interest
(pages 182 and 183)

1. I = simple interest, P = principal, r = annual interest rate (in decimal form), t = time (in years)

3. You have to change 6% to a decimal and 8 months to years.

5. a. \$300 **b.** \$1800

7. a. \$292.50 **b.** \$2092.50

9. a. \$308.20 **b.** \$1983.20

11. a. \$1722.24 **b.** \$6922.24

13. 3% **15.** 4% **17.** 2 yr **19.** 1.5 yr **21.** \$1440 **23.** 2 yr

25. \$2720 **27.** \$6700.80 **29.** \$8500 **31.** 5.25% **33.** 4 yr

35. 12.5 yr; Substitute \$2000 for P and I, 0.08 for r, and solve for t.

37. Year 1 = \$520; Year 2 = \$540.80; Year 3 = \$562.43

39. $n = 5$ **41.** $z = 9$

Section 5.1 Identifying Similar Figures
(pages 198 and 199)

1. They have the same measure.

3. *Sample answer:* A photograph of size 3 in. × 5 in. and another photograph of size 6 in. × 10 in.

5. $\angle A$ and $\angle W$, $\angle B$ and $\angle X$, $\angle C$ and $\angle Y$, $\angle D$ and $\angle Z$;
Side AB and Side WX, Side BC and Side XY, Side CD and Side YZ, Side AD and Side WZ

7.

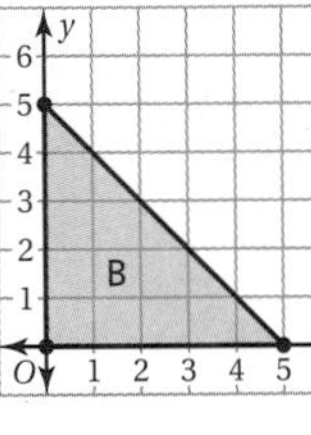

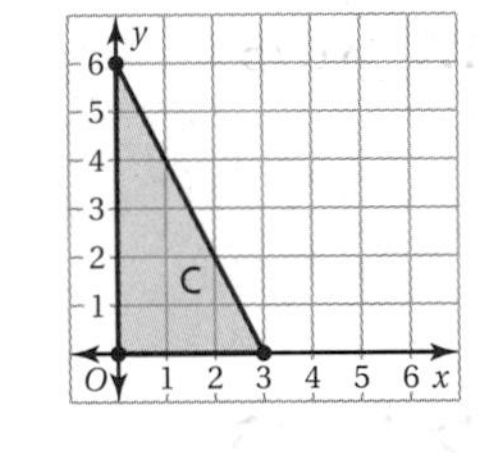

A and B; Corresponding side lengths are proportional and corresponding angles have the same measure.

9. similar; Corresponding angles have the same measure. Because $\frac{4}{6} = \frac{6}{9} = \frac{8}{12}$, the corresponding side lengths are proportional.

11. no **13.** 48° **15.** 42°

17. Simplify the ratios of length to width for each photo to see if any of the photos are similar.

19. yes; One could be a trapezoid and the other could be a parallelogram.

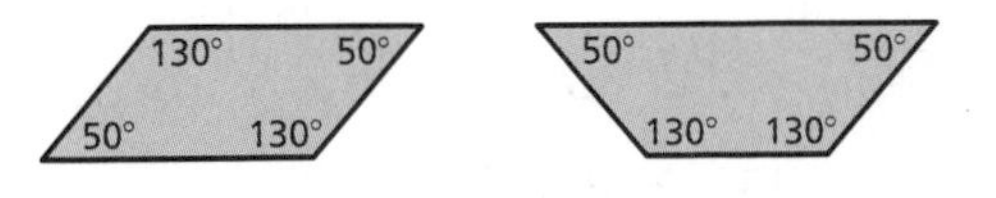

21. a. yes

b. yes; This is true for all similar triangles because the height of a triangle is a dimension of the triangle like the side lengths.

23. $\frac{16}{81}$ **25.** $\frac{49}{16}$ **27.** B

Section 5.2

Perimeters and Areas of Similar Figures
(pages 204 and 205)

1. The ratio of the perimeters is equal to the ratio of the corresponding side lengths.

3. 120 in.^{2}; Because the ratio of the corresponding side lengths is $\frac{1}{2}$, the ratio of the areas is equal to $\left(\frac{1}{2}\right)^2$. To find the area, solve the proportion $\frac{30}{x} = \frac{1}{4}$.

5. $\frac{5}{8}$; $\frac{25}{64}$

7. $\frac{14}{9}$; $\frac{196}{81}$

9. perimeter triples

11. Area is 16 times larger.

13. 45 in.

15. false; $\frac{\text{Area of } \triangle ABC}{\text{Area of } \triangle DEF} = \left(\frac{AB}{DE}\right)^2$

17. 39,900%; The ratio of the corresponding lengths is $\frac{6 \text{ in.}}{120 \text{ in.}} = \frac{1}{20}$. So, the ratio of the areas is $\frac{1}{400}$ and the area of the actual merry-go-round is 180,000 square inches. The percent of increase is $\frac{180{,}000 - 450}{450} = 399 = 39{,}900\%$.

19. $\frac{3}{4}$

21. 25% increase

23. 42.7% decrease

Section 5.3

Finding Unknown Measures in Similar Figures
(pages 210 and 211)

1. You can set up a proportion and solve for the unknown measure.

3. 15

5. 14.4

7. 8.4

9. 35 ft

11. 108 yd

13. 3 times

15. 12.5 bottles

17. 31.75

19. 3.88

21. 41.63

Section 5.4

Scale Drawings
(pages 216 and 217)

1. A scale is the ratio that compares the measurements of the drawing or model with the actual measurements. A scale factor is a scale without any units.

3. Convert one of the lengths into the same units as the other length. Then, form the scale and simplify.

5. 10 ft by 10 ft

7. 112.5%

9. 50 mi

11. 110 mi

13. 15 in.

15. 21.6 yd

17. The 5 cm should be in the numerator.

$$\frac{1 \text{ cm}}{20 \text{ m}} = \frac{5 \text{ cm}}{x \text{ m}}$$

$$x = 100 \text{ m}$$

19. 2.4 cm; 1 cm : 10 mm

Section 5.4 Scale Drawings *(continued)*

(pages 216 and 217)

21. **a.** *Answer should include, but is not limited to:* Make sure words and picture match the product.

b. Answers will vary.

23. Find the size of the object that would represent the model of the Sun.

25–27.

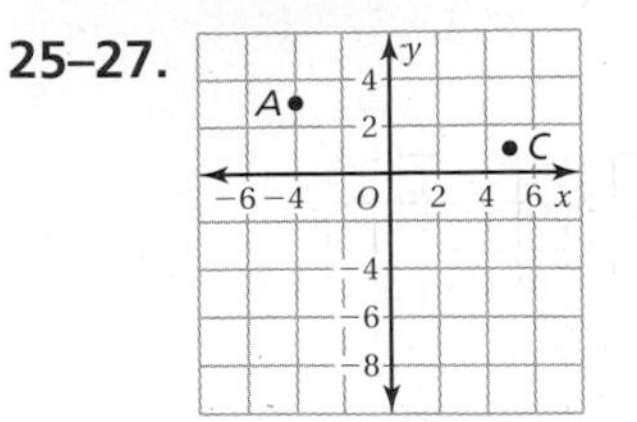

29. C

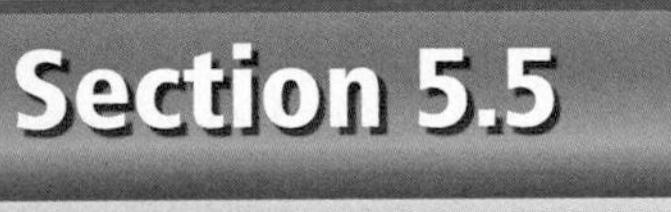

Lesson 5.4b Scale Drawings

(pages 217A and 217B)

1. 15 ft^2

3. 3 ft^2

5.

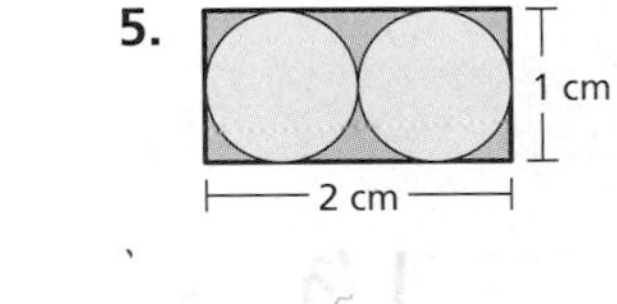

Section 5.5 Translations

(pages 224 and 225)

1. A

3. yes; Translate the letters T and O to the end.

5. no

7. yes

9. no

11. $A'(-3, 0)$, $B'(0, -1)$, $C'(1, -4)$, $D'(-3, -5)$

13.

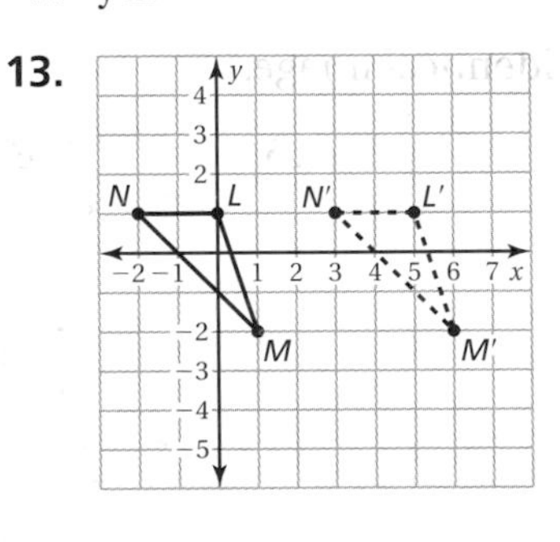

15.

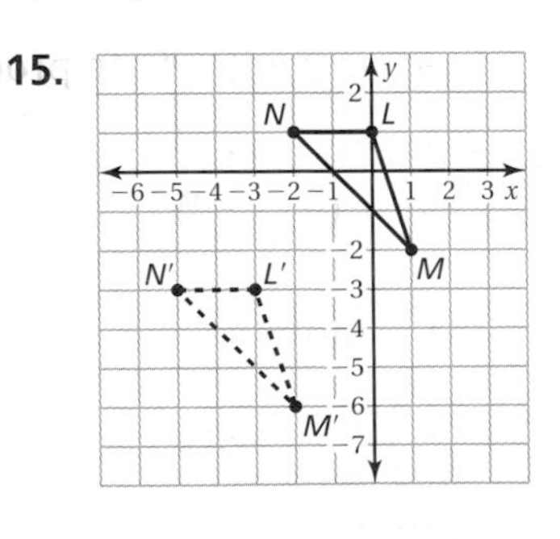

17. 2 units left and 2 units up

19. 6 units right and 3 units down

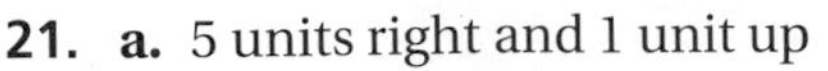

21. **a.** 5 units right and 1 unit up

b. no; It would hit the island.

c. 4 units right and 4 units up

23. If you are doing more than 10 moves and have not moved the knight to g5, you might want to start over.

25. no

27. yes

Section 5.6 Reflections
(pages 230 and 231)

1. The third one because it is not a reflection.
3. Quadrant IV
5. yes
7. no
9. no
11. $M'(-2, -1), N'(0, -3), P'(2, -2)$
13. $D'(-2, 1), E'(0, 2), F'(1, 5), G'(-1, 4)$
15. $T'(-1, -1), U'(-4, 2), V'(-6, -2)$
17. $J'(-2, 2), K'(-7, 4), L'(-9, -2), M'(-3, -1)$
19. x-axis
21. y-axis
23.

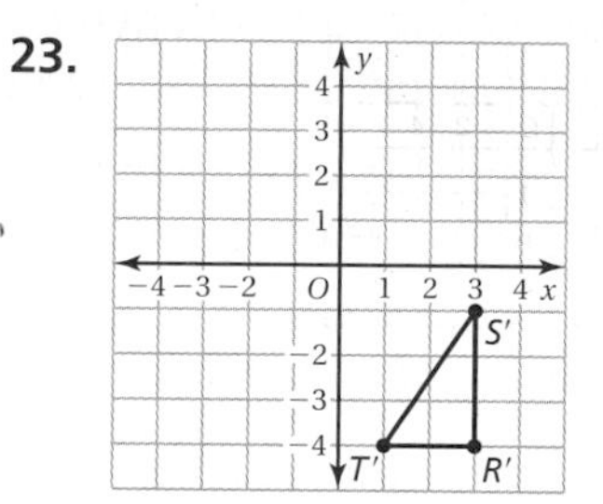

25. the first one; The left side of the face is a mirror image of the right side.
27.

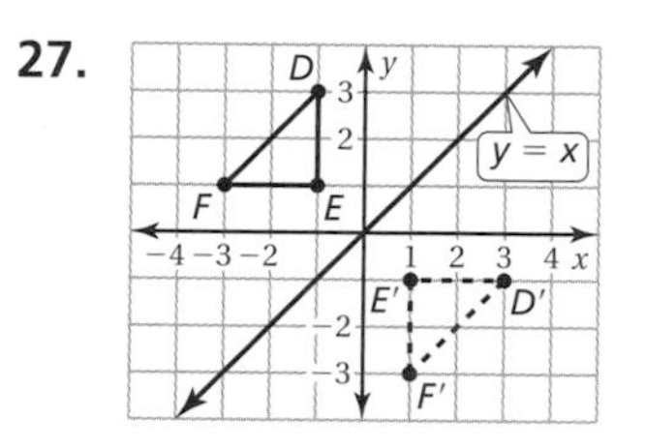

The x-coordinate and y-coordinate for each point are switched in the image.

29. straight
31. acute

Section 5.7 Rotations
(pages 236 and 237)

1. **a.** reflection **b.** rotation **c.** translation
3. Quadrant I
5. Quadrant III
7. No
9. yes; 180° clockwise or counterclockwise
11. It only needs to rotate 90° to produce an identical image.
13. $A'(-1, -4), B'(-4, -3), C'(-4, -1), D'(-1, -2)$

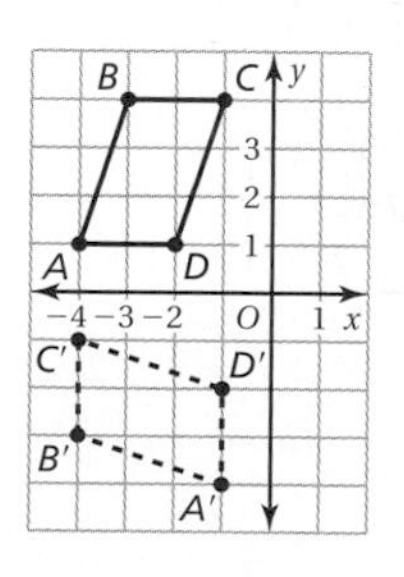

15. $A'(0, 1), B'(-1, -2), C'(-3, -2), D'(-2, 1)$

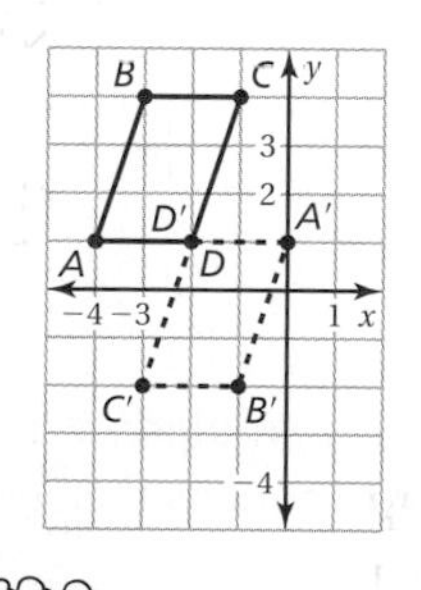

17. because both ways will produce the same image
19. Use Guess, Check, and Revise to solve this problem.
21. triangular prism
23. C

Hint

Section 6.1 Drawing 3-Dimensional Figures
(pages 254 and 255)

1. Prisms and cylinders both have two parallel, identical bases. The bases of a cylinder are circles. The bases of a prism are polygons. A prism has lateral faces that are parallelograms or rectangles. A cylinder has one smooth, round lateral surface.

3. *Sample answer:* Prisms: A cereal box is a rectangular prism. A pup tent with parallel triangular bases at the front and back is a triangular prism.

Pyramids: The Egyptian pyramids are rectangular pyramids. A house roof forms a pyramid if it has lateral faces that are triangles that meet at a common vertex.

Cylinders: Some examples of cylinders are a soup can, a tuna fish can, and a new, unsharpened, round pencil.

Cones: some examples of cones are a traffic cone, an ice cream sugar cone, a party hat, and the sharpened end of a pencil.

5. base: circle; solid: cylinder

7. front: side: top:

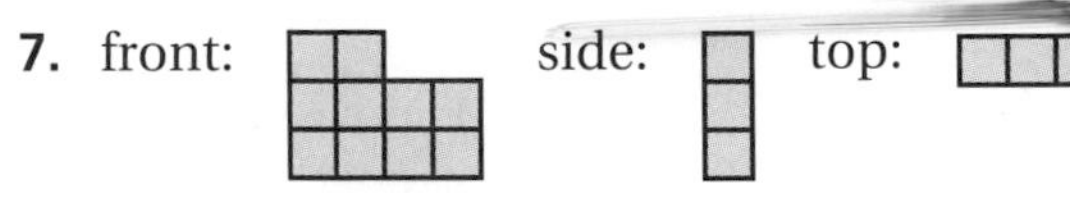

surface area: 34 units^2; volume: 10 units^3

9. front: side: top:

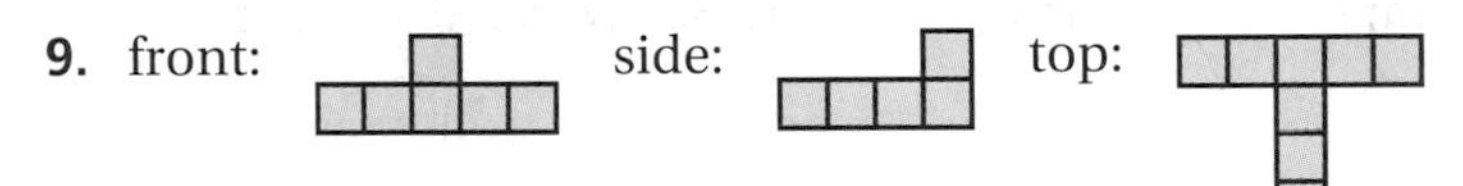

surface area: 38 units^2; volume: 9 units^3

11.

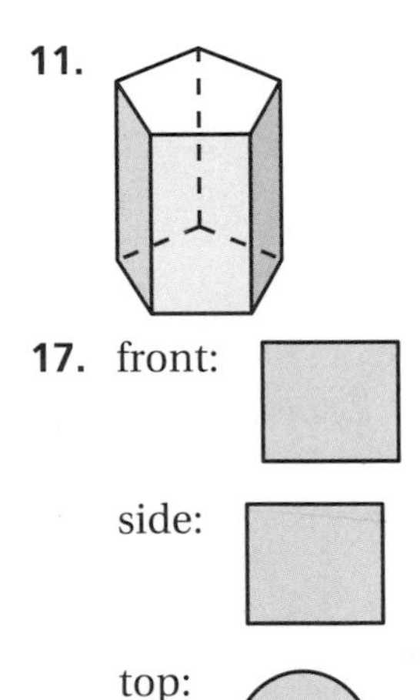

13.

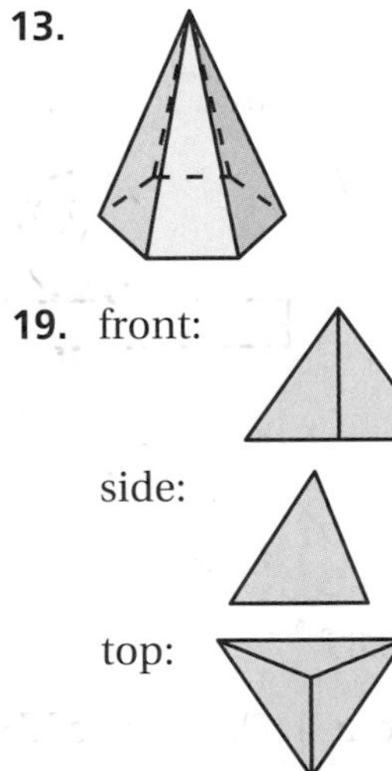

15.

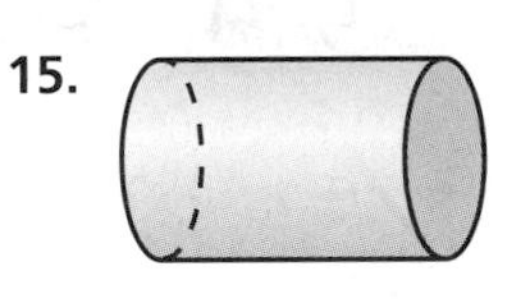

17. front:

side:

top:

19. front:

side:

top:

21. front:

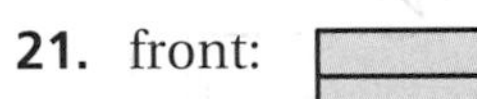

side:

top:

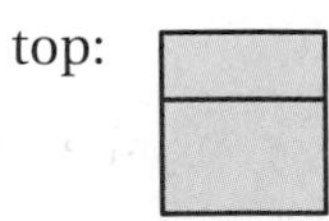

23. The Washington Monument is an *obelisk*. It consists of a pyramid sitting on top of a solid that tapers as it rises.

25.

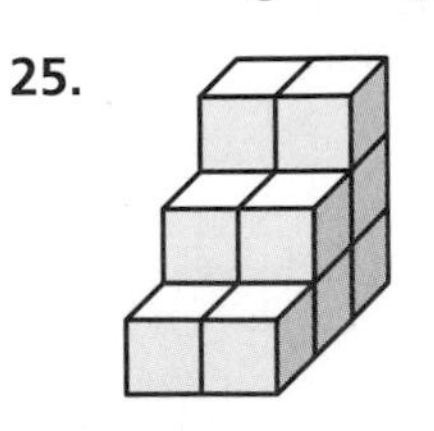

27. Use cubes to create solids that are possible.

29. 28 m^2

31. 15 ft^2

Section 6.2 Surface Areas of Prisms
(pages 260 and 261)

1. *Sample answer:* You want to paint a large toy chest in the form of a rectangular prism, and in order to know how much paint to buy, you need to know the surface area.

3. 18 cm^2

5. 108 cm^2

7. 3 cm, 3 cm, 4 cm, 4 cm, 5 cm, 5 cm

72 cm^2

9. 130 ft^2

11. 76 yd^2

13. 136 m^2

15. 448 in.^2; The surface area of the box is 448 square inches, so that is the least amount of paper needed to cover the box.

17. 156 in.^2

19. 83 ft^2

21. 2 qt

23. $S = 2B + Ph$

25. 48 units

27. C

Lesson 6.2b Circles
(pages 261A and 261B)

1. 18 in.

3. $C \approx 440$ cm; $A \approx 15{,}400\ \text{cm}^2$

5. $C \approx 31.4$ in.; $A \approx 78.5\ \text{in.}^2$

Section 6.3 Surface Areas of Cylinders
(pages 266 and 267)

1. $2\pi rh$

3. $36\pi \approx 113.0\ \text{cm}^2$

5. 3 ft, 2 ft, 3 ft

$30\pi \approx 94.2\ \text{ft}^2$

7. 7 ft, 5 ft, 7 ft

$168\pi \approx 527.5\ \text{ft}^2$

9. $156\pi \approx 489.8\ \text{ft}^2$

11. $120\pi \approx 376.8\ \text{ft}^2$

13. $28\pi \approx 87.9\ \text{m}^2$

15. The error is that only the lateral surface area is found. The areas of the bases should be added;

$$\begin{aligned} S &= 2\pi r^2 + 2\pi rh \\ &= 2\pi(6)^2 + 2\pi(6)(11) \\ &= 72\pi + 132\pi \\ &= 204\pi\ \text{ft}^2 \end{aligned}$$

17. The surface area of the cylinder with the height of 8.5 inches is greater than the surface area of the cylinder with the height of 11 inches.

19. After removing the wedge, is there any new surface area added?

21. 117

23. 56.52

Selected Answers

Section 6.4 Surface Areas of Pyramids
(pages 274 and 275)

1. the triangle and the hexagon
3. Knowing the slant height helps because it represents the height of the triangle that makes up each lateral face. So, the slant height helps you to find the area of each lateral face.
5. 178.3 mm^2
7. 144 ft^2
9. 170.1 yd^2
11. 1240.4 mm^2
13. 6 m

15. Determine how long the fabric needs to be so you can cut the fabric most efficiently.
17. 124 cm^2
19. $A \approx 452.16$ units^2; $C \approx 75.36$ units
21. $A \approx 572.265$ units^2; $C \approx 84.78$ units

Section 6.5 Surface Areas of Cones
(pages 280 and 281)

1. no; The base of a cone is a circle. A circle is not a polygon.
3. $\ell > r$
5. $36\pi \approx 113.0$ m^2
7. $119\pi \approx 373.7$ ft^2
9. $64\pi \approx 201.0$ yd^2
11. 15 cm
13. $130\pi \approx 408.2$ in.^{2}
15. $360\pi \approx 1130.4$ in.^{2}; $2.5\pi \approx 7.85$ ft^2
17. $96\pi \approx 301.44$ ft^2; $\frac{32}{3}\pi \approx 33.49\overline{3}$ yd^2
19. 12%
21. the lateral surface area
23. 45 in.^{2}
25. 16 ft^2

Section 6.6 Surface Areas of Composite Solids
(pages 286 and 287)

1. *Sample answer:*

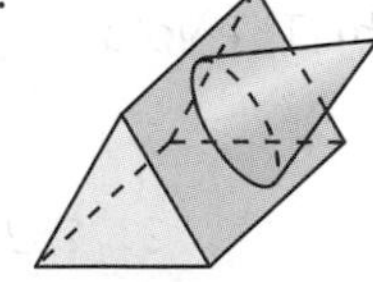

3. three cylinders
5. rectangular prism, half of a cylinder
7. cones; $104\pi \approx 326.6$ m^2
9. trapezoidal prism, rectangular prism; 152 cm^2
11. two rectangular prisms; 308 ft^2
13. 63.4%
15. $144\pi \approx 452.2$ in.^{2}
17. $806\pi \approx 2530.84$ mm^2
19. 10 ft^2
21. 47.5 in.^{2}

Section 7.1 Volumes of Prisms
(pages 302 and 303)

1. cubic units

3. *Sample answers:* Volume because you want to make sure the product will fit inside the package. Surface area because of the cost of packaging.

5. 288 cm^3

7. 160 yd^3

9. 420 mm^3

11. 645 mm^3

13. The area of the base is wrong.
$V = \frac{1}{2}(7)(5) \cdot 10 = 175\text{ cm}^3$

15. 225 $in.^3$

17. 7200 ft^3

19. 1728 $in.^3$

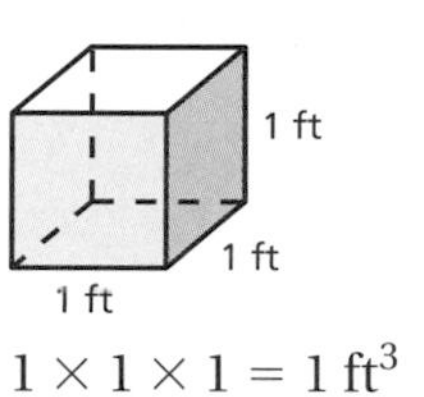

$1 \times 1 \times 1 = 1\text{ ft}^3$

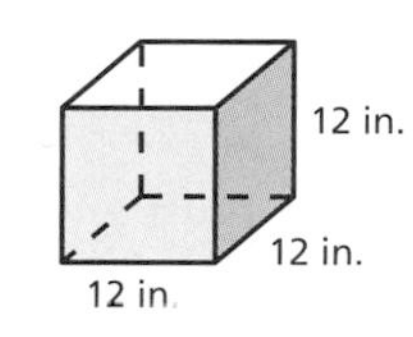

$12 \times 12 \times 12 = 1728\text{ in.}^3$

21. 20 cm

23. You can write the volume in cubic inches and use prime factorization to find the dimensions.

25. reflection

27. rotation

Section 7.2 Volumes of Cylinders
(pages 308 and 309)

1. How much does it take to cover the cylinder?; $170\pi \approx 533.8\text{ cm}^2$; $300\pi \approx 942\text{ cm}^3$

3. $486\pi \approx 1526.0\text{ ft}^3$

5. $245\pi \approx 769.3\text{ ft}^3$

7. $90\pi \approx 282.6\text{ mm}^3$

9. $63\pi \approx 197.8\text{ in.}^3$

11. $256\pi \approx 803.8\text{ cm}^3$

13. $\frac{125}{8\pi} \approx 5\text{ ft}$

15. $\frac{240}{\pi} \approx 76\text{ cm}$

Hint

17. Divide the volume of one round bale by the volume of one square bale.

19. $8325 - 729\pi \approx 6036\text{ m}^3$

21. $a = 0.5 \cdot 200$; 100

23. D

Section 7.3 Volumes of Pyramids
(pages 314 and 315)

1. The volume of a pyramid is $\frac{1}{3}$ times the area of the base times the height. The volume of a prism is the area of the base times the height.

3. 3 times greater

5. 20 mm^3

7. 80 $in.^3$

9. 252 mm^3

11. 700 mm^3

13. 30 $in.^2$

15. 7.5 ft

17. 12,000 $in.^3$; The volume of one paperweight is 12 cubic inches. So, 12 cubic inches of glass is needed to make one paperweight. So it takes $12 \times 1000 = 12{,}000$ cubic inches to make 1000 paperweights.

19. *Sample answer:* 5 ft by 4 ft

21. 28

23. 60

25. B

Section 7.4 Volumes of Cones
(pages 320 and 321)

1. The height of a cone is the distance from the vertex to the center of the base.

3. Divide by 3.

5. $9\pi \approx 28.3 \text{ m}^3$

7. $\frac{2\pi}{3} \approx 2.1 \text{ ft}^3$

9. $27\pi \approx 84.8 \text{ yd}^3$

11. $\frac{125\pi}{6} \approx 65.4 \text{ in.}^3$

13. The diameter was used instead of the radius.

    $V = \frac{1}{3}(\pi)(3)^2(8) = 24\pi \text{ m}^3$

15. 1.5 ft

17. $\frac{40}{3\pi} \approx 4.2$ in.

19. 24.1 min

21. $3y$

23. 315 m^3

25. $152\pi \approx 477.28 \text{ ft}^3$

Section 7.5 Volumes of Composite Solids
(pages 328 and 329)

1. A composite solid is a solid that is made up of more than one solid.

3. In Example 2, you had to subtract the volume of the cylinder-shaped hole from the volume of the entire cylinder. In Example 1, you had to find the volumes of the square prism and the square pyramid and add them together.

5. $125 + 16\pi \approx 175.2 \text{ in.}^3$

7. 220 cm^3

9. 173.3 ft^3

11. $216 - 24\pi \approx 140.6 \text{ m}^3$

13. **a.** *Sample answer:* 80% **b.** *Sample answer:* $100\pi \approx 314 \text{ in.}^3$

15. 13.875 in.^3; The volume of the hexagonal prism is 10.5(0.75) and the volume of the hexagonal pyramid is $\frac{1}{3}(6)(3)$.

17. $\frac{25}{9}$

19. B

Section 7.6 Surface Areas and Volumes of Similar Solids
(pages 335–337)

1. Similar solids are solids of the same type that have proportional corresponding linear measures.

3. **a.** $\frac{4}{9}$ **b.** $\frac{8}{27}$

5. no

7. no

9. $b = 18$ m; $c = 19.5$ m; $h = 9$ m

11. 1012.5 in.^2

13. $13{,}564.8 \text{ ft}^3$

15. 673.75 cm^2

17. a. yes; Because all circles are similar, the slant height and the circumference of the base of the cones are proportional.

b. no; because the ratio of the volumes of similar solids is equal to the cube of the ratio of their corresponding linear measures

19. Choose two variables, one to represent the surface area of the smallest doll and one to represent the volume of the smallest doll. Use these variables to find the surface areas and volumes of the other dolls.

21. 1

23. C

Section 8.1 Stem-and-Leaf Plots
(pages 352 and 353)

1. 3 is the stem; 4 is the leaf

3. From the leaves, you can see where most of the data lies and whether there are many values that are low or high.

5. 4; 42

7. no; There is no 2 as a leaf for the stem 3.

9. **Hours Online**

Stem	Leaf
0	0 2 6 8
1	2 2 4 5 7 8
2	1 4

Key: 2 | 1 = 21 hours

11. **Points Scored**

Stem	Leaf
3	8
4	2 2 3 3 5
5	0 1 6 8 8
6	
7	0 1 1 5

Key: 3 | 8 = 38 points

13. **Weights**

Stem	Leaf
0	8
1	2 5 7 8
2	4 4
3	1

Key: 2 | 4 = 24 pounds

Most of the weights are in the middle.

15. **Minutes in Line**

Stem	Leaf
1	6 9
2	0 2 6 7 9
3	1 1 6 8
4	0

Key: 4 | 0 = 4.0 minutes

17. mean: 56.6; median: 53; modes: 41, 43, 63; range: 56

19. 97; It increases the mean.

21. *Sample answer:* Points by a basketball player in his first 8 games

Points

Stem	Leaf
2	1 3 4
3	2 4
4	0 1 5

Key: 3 | 2 = 32 points

23.

25.

27. B

Selected Answers

Section 8.2 Histograms
(pages 358 and 359)

1. The *Test Scores* graph is a histogram because the number of students (frequency) achieving the test scores are shown in intervals of the same size (20).

3. No bar is shown on that interval.

5. flat

7. 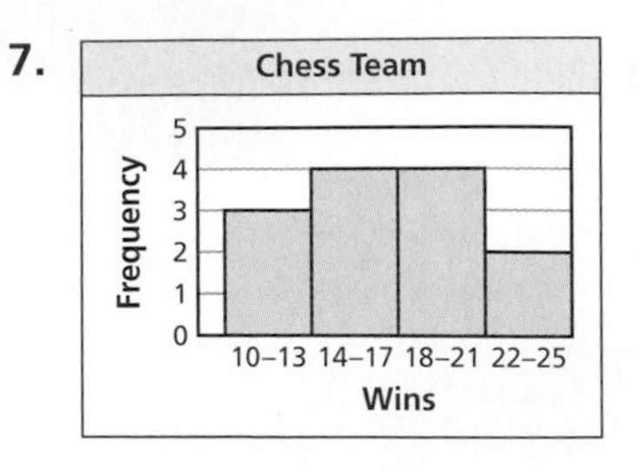

9. **a.** 4–5

   b. 20 students

   c. 85%

11. Pennsylvania; You can see from the intervals and frequencies that Pennsylvania counties are greater in area, which makes up for it having fewer counties.

13. Don't use a smaller interval because the distribution will appear flat.

15. 27

17. 51.2

Section 8.3 Circle Graphs
(pages 366 and 367)

1. Multiply the decimal form of each percent by 360° to find the angle measure for each section.

3. $\frac{1}{2}$ does not belong because it does not represent an entire circle.

5. orange

7. 20 students

9. 54°

11. 10.8°

13. 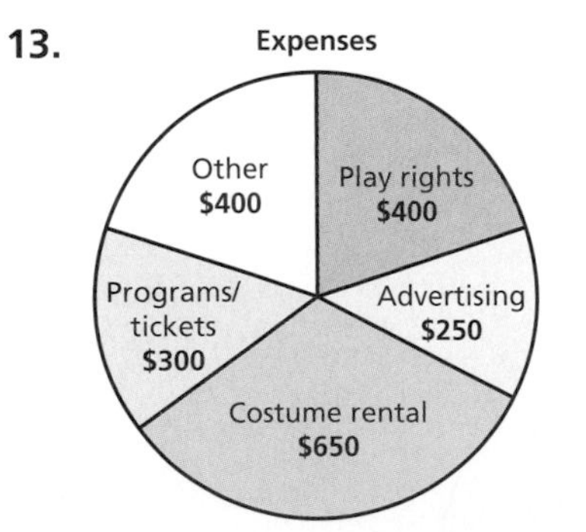

15. no; The sum of the percentages is greater than 100%. This would occur when students like more than one of these activities.

17. *Sample answer:* Knowledge of percentages, proportions, and degrees of a circle. How to convert from one form to another.

19. $x = 40$

21. $w = 1.5$

Section 8.4 Samples and Populations
(pages 372 and 373)

1. Samples are easier to obtain.

3. *Sample answer:* The results may be similar for middle school students, but not for children in first grade. Children in first grade probably do not know all of these nuts.

5. *Sample answer:* You could send a survey home with your classmates and have them ask one of their parents what their favorite nut is.

7. Population: All quarters in circulation
   Sample: 150 quarters

9. Population: All books in library
   Sample: 10 books

11. **a.** Population: All students at your school
    Sample: First 15 students at band class

    b. no; Your sample includes 15 students arriving at band class, and students who take band class play a musical instrument.

13. Sample A because it is representative of the population.

15. A population because there are few enough students in your homeroom to not make the surveying difficult.

17. 1260 students

19. Use the survey results to find the number of students in the school that plan to attend college.

21. 31.25%

23. $81.\overline{81}\%$

Lesson 8.4b Comparing Populations
(pages 373A and 373B)

1. **a.** *Sample answer:* randomly asking students at lunch
   b. *Sample answer:* randomly asking students on the football team

3. In general, boys are taller than girls.

Section 9.1 Introduction to Probability
(pages 388 and 389)

1. event; It is a collection of several outcomes.

3. *Sample answer:* flipping a coin and getting both heads and tails; rolling a number cube and getting a number between 1 and 6

5. no; They both have the same number of forward outcomes.

7. 6   9. 6, 7, 8, 9   11. 1, 2

13. **a.** 2 ways   **b.** blue, blue   15. **a.** 2 ways   **b.** purple, purple

17. **a.** 6 ways   **b.** yellow, green, blue, blue, purple, purple

19. There are 7 marbles that are *not* purple, even though there are only 4 colors. Choosing *not* purple could be red, red, red, blue, blue, green, or yellow.

21. false; five   23. false; red

25. no; More sections on a spinner does not necessarily mean you are more likely to spin red. It depends on the size of the sections of the spinner.

27. Do the number of outcomes increase, decrease, or stay the same?

29. 30   31. $-3\frac{1}{2}$

Section 9.2 Theoretical Probability
(pages 394 and 395)

1. There is a 50% chance you will get a favorable outcome.

3. Spinner 4; The other three spinners are fair.

5. $\frac{1}{6}$ or about 16.7%   7. $\frac{1}{2}$ or 50%   9. 0 or 0%   11. 9 chips

13. not fair, your friend   15. $\frac{1}{44}$ or about 2.3%

17. **a.** $\frac{4}{9}$ or about 44.4%   **b.** 5 males   19. There are 2 combinations for each.

21. $\frac{1}{4}$ 23. $-\frac{21}{40}$ 25. C

Section 9.3 Experimental Probability
(pages 402 and 403)

1. Perform an experiment several times. Count how often the event occurs and divide by the number of trials.

3. $\frac{2}{5}$ or 40%   5. $\frac{2}{5}$ or 40%   7. $\frac{7}{50}$ or 14%   9. $\frac{21}{25}$ or 84%   11. $\frac{17}{50}$ or 34%

13. The theoretical probability was found, not the experimental probability. $P(4) = \frac{11}{50}$

15. 45 tiles

17. 25

19. The experimental probability of 60% is close to the theoretical probability of 50%.

21. Make a list of all the possible ways to get each sum.

23. *Sample answer:* Roll two number cubes 50 times and find each product. Record how many times the product is at least 12. Divide this number by 50 to find the experimental probability.

25. $x = 5$

27. $x = 24$

Section 9.4 Independent and Dependent Events

(pages 409–411)

1. Draw a tree diagram or multiply $P(A)$ by $P(B)$.

3. *Sample answer:* independent events: a traffic jam and a sunny day; dependent events: temperatures below freezing and ice

5. independent; The outcome of the first roll does not affect the outcome of the second roll.

7. independent; You replace the marble, so the probability doesn't change.

9. dependent; There is one less person to choose from on the second draw.

11. $\frac{2}{9}$

13. $\frac{2}{9}$

15. $\frac{1}{20}$ or 5%

17. $\frac{3}{20}$ or 15%

19. $\frac{1}{42}$ or about 2.4%

21. $\frac{1}{21}$ or about 4.8%

23. $\frac{4}{21}$ or about 19%

25. $\frac{1}{2520}$ or about 0.04%

27. a. $\frac{1}{100}$ or 1%

b. It increases the probability that your choice is correct to $\frac{1}{25}$ or 4%, because each digit could be 0, 2, 4, 6, or 8.

29. a. $\frac{1}{9}$ or about 11.1%

b. It increases the probability that your guesses are correct to $\frac{1}{4}$ or 25%, because you are only choosing between 2 choices for each question.

31. $\frac{16}{25}$ or 64%

33. 1 : 5; 5 : 1

35. 1 : 35; 35 : 1

37. $n = -10.8$

39. B

Section 10.1 Solving Multi-Step Equations

(pages 426 and 427)

1. Add $6x$ and $3x$, subtract 2 from each side, and divide each side by 9.

3. 313 K

5. $K = C + 273$

7. $x = 3$

9. $m = -4$

11. $y = 3$

13. 15

15. a. \$7.25 **b.** \$1.75

Section 10.1 Solving Multi-Step Equations *(continued)* *(pages 426 and 427)*

17. The question asks for the sale price of the skirt, not the original price of the skirt.

19. $x = 12$; *Sample answer:* The left side has more steps, but does not deal with fractions like the right side does.

21. 12

Section 10.2 Solving Equations with Variables on Both Sides *(pages 432 and 433)*

1. Subtract $3x$ from each side and divide each side by 2.

3. Add $3x$ to each side so the constant is by itself.

5. $c = -3$ **7.** $a = -\frac{1}{2}$ **9.** $z = -2$ **11.** $p = -8\frac{1}{2}$ **13.** $d = -2$

15. $h = 2\frac{1}{2}$ **17.** $k = -6$ **19.** $f = -2\frac{2}{7}$ **21.** $b = 1$

23. They did not distribute the negative sign on the right side.

$$2(v - 5) = -(3v + 5)$$
$$2v - 10 = -3v - 5$$
$$5v = 5$$
$$v = 1$$

25. 27 **27.** 18

29. $5n - 3 = 8 - 6n$; $n = 1$

31. Use $d = rt$ and the fact that the distance is the same to write an equation with variables on both sides.

33. perimeter of $\triangle ABC$: 27;

perimeter of $\triangle DEF$: 45

35–37.

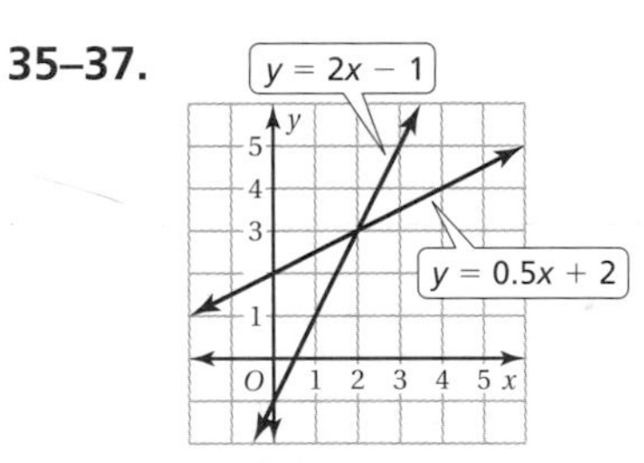

Lesson 10.2b Solutions of Linear Equations *(pages 433A and 433B)*

1. no solution **3.** $x = \frac{1}{3}$ **5.** no solution

7. no; There is no solution to the equation stating the areas are equal, $x + 1 = x$.

9. no solution **11.** infinitely many solutions **13.** $x = 2$

15. no solution **17.** infinitely many solutions **19.** $x = \frac{15}{16}$

Section 10.3 Solving Equations Using Tables and Graphs (pages 438 and 439)

1. $3x - 2 = x + 2; x = 2$

3. *Sample answer:* A table because graphs have to be accurate to find the point of intersection.

5. $h = 2$ 7. $g = 5$ 9. $a = -2$ 11. $p = -3$

13. $x = -3$ 15. $x = -2$ 17. 3 months 19. $x = 16; P = 24$ units

21. *Sample answer:* From the table, you can see that the solution is between 2 and 3. Then, you can use *Guess, Check, and Revise* to find the solution, which is 2.2.

23. Use $d = rt$ and the fact that both of you will have travelled the same distance when your friend catches you to write an equation with variables on both sides.

Hint

25. triangle: 27 square units; rectangle: 54 square units

27. $\frac{1}{2}$ 29. B

Slope of a Line (pages 446 and 447)

1. *Sample answer:* The rise is the difference of the distance from the ground to the peak of the roof and the distance from the ground to the lowest point on the roof. The run is the horizontal distance from the peak to the lowest point on the roof. The slope is the ratio of the rise to the run.

3. *Sample answer:* (5, −8), (−1, −4)

	Two Points	Change in y	Change in x	Slope of Line
5.	(−10, 4), (5, −20)	−24	15	$-\frac{8}{5}$
7.	(1, 9), (7, 6)	−3	6	$-\frac{1}{2}$

9. $-\frac{5}{2}$

11.
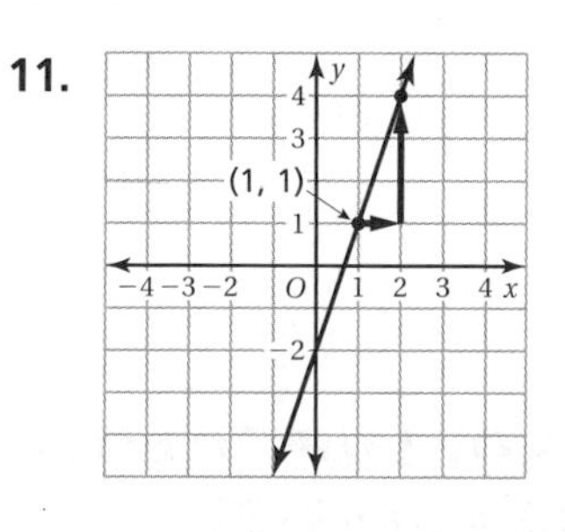

13.
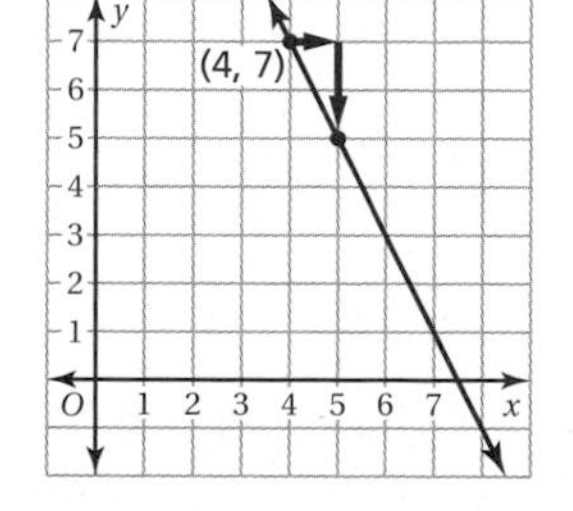

15. (−3, −7)

17. a. Daytona: $\frac{21}{34}$; Talladega: $\frac{13}{20}$

b. Talladega Superspeedway

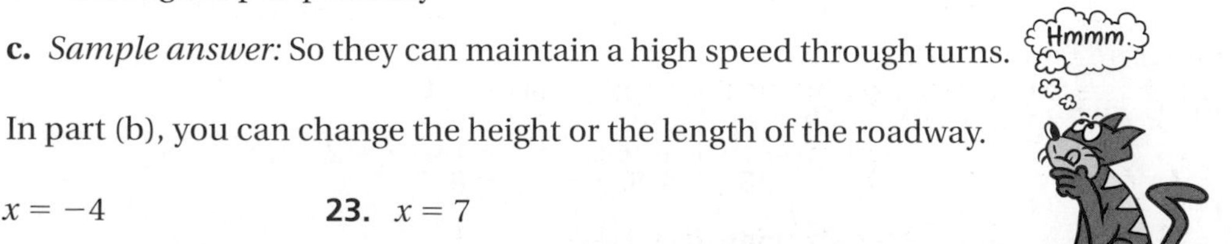
c. *Sample answer:* So they can maintain a high speed through turns.

Hmmm.

19. In part (b), you can change the height or the length of the roadway.

21. $x = -4$ 23. $x = 7$

Selected Answers

Section 10.5 Linear Functions
(pages 452 and 453)

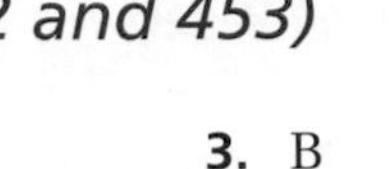

1. the y-coordinate of the point where the line crosses the y-axis

3. B

5. A

7. slope: -2; y-intercept: 6

9. slope: -5; y-intercept: 3

11. slope: $\frac{5}{2}$; y-intercept: -9

13.
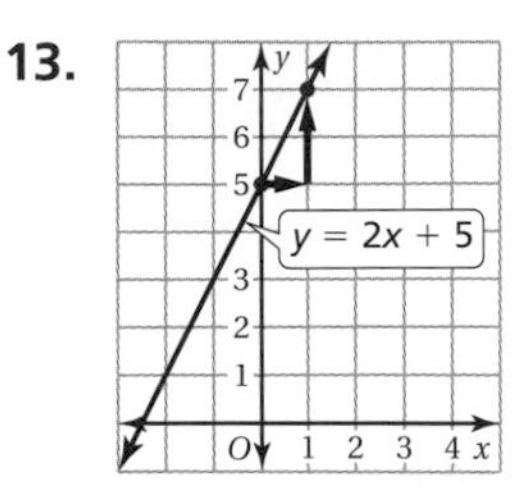

15.
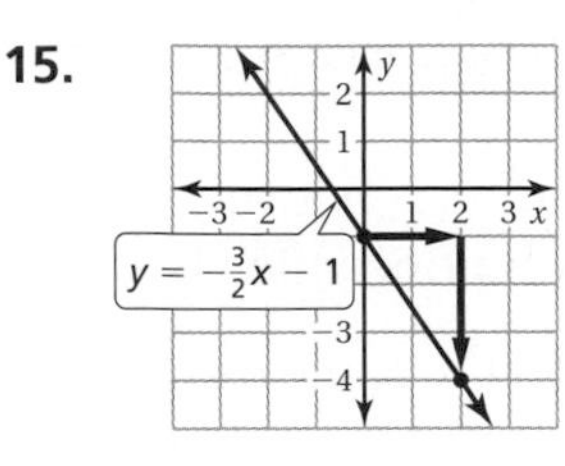

17.
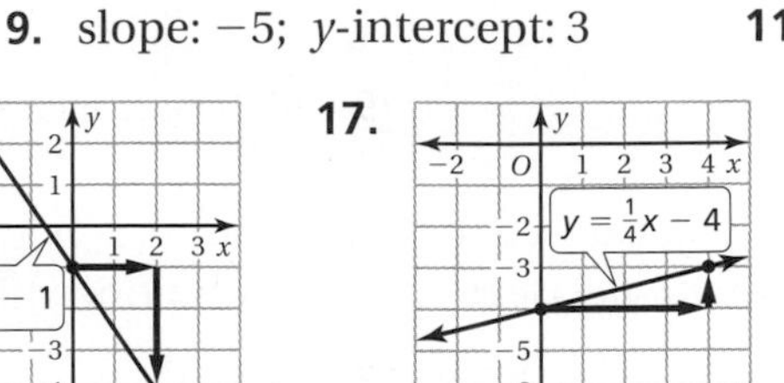

19. y = 6x − 5

21. 5y − 4x = −15

23.
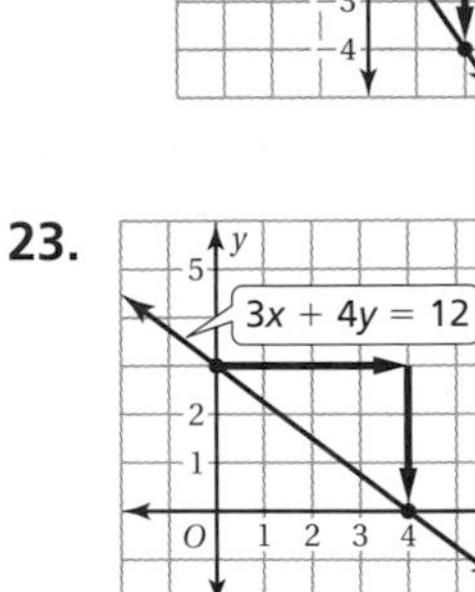

25. the number of cartoons the cartoonist has already completed

27. $y = -3x + 2$

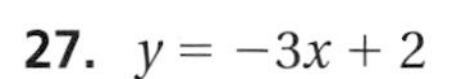

29. **a.** slope: 2; y-intercept: 7

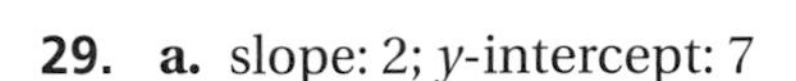

b. y = 2x + 7

c. no; The value of y is 5 when $x = -1$. The value of x cannot be negative because the side length of a rectangle is never negative.

31. slope: $-\frac{A}{B}$, y-intercept: $\frac{C}{B}$

33. $v = \frac{1}{2}$

35. A

Section 11.1 Classifying Angles
(pages 468 and 469)

1. The sum of the measures of two complementary angles is 90°. The sum of the measures of two supplementary angles is 180°.

3. sometimes; Either x or y may be obtuse.

5. never; Because x and y must both be less than 90° and greater than 0°.

7. complementary

9. supplementary

11. neither

13. 128

15. Vertical angles are congruent. The value of x is 35.

17. 37

19. 20

21. **a.** $\angle CBD$ and $\angle DBE$; $\angle ABF$ and $\angle FBE$

b. $\angle ABE$ and $\angle CBE$; $\angle ABD$ and $\angle CBD$; $\angle CBF$ and $\angle ABF$

23. 54°

25. $x = 8, y = 7, z = 8$

27. 29.3

29. C

Section 11.2 Angles and Sides of Triangles
(pages 474 and 475)

1. An equilateral triangle has three congruent sides. An isosceles triangle has at least two congruent sides. So, an equilateral triangle is a specific type of isosceles triangle.

3. right isosceles triangle

5. obtuse isosceles triangle

7. 94; obtuse triangle

9. 67.5; acute isosceles triangle

11. 24; obtuse isosceles triangle

13. **a.** 70 **b.** acute isosceles triangle

15. no; 39.5°

17. yes

19. If two angle measures of a triangle were greater than or equal to 90°, the sum of those two angle measures would be greater than or equal to 180°. The sum of the three angle measures would be greater than 180°, which is not possible.

21. $x + 2x + 2x + 8 + 5 = 48$; 7

23. $4x - 4 + 3\pi = 25.42$ or $2x - 4 = 6$; 5

Section 11.3 Angles of Polygons
(pages 481–483)

1. 

3. What is the measure of an angle of a regular pentagon?; 108°; 540°

5. 1260°

7. 720°

9. 1080°

11. no; The angle measures given add up to 535°, but the sum of the angle measures of a pentagon is 540°.

13. 135

15. 140°

17. 140°

19. The sum of the angle measures should have been divided by the number of angles, 20. $3240° \div 20 = 162°$; The measure of each angle is 162°.

21. 24 sides

23. convex; No line segment connecting two vertices lies outside the polygon.

25. no; All of the angles would not be congruent.

27. 135°

29. 120°

31. You can determine if it is a linear function by writing an equation or by graphing the points.

33. 9

35. 3

37. D

Selected Answers

Section 11.4 Using Similar Triangles
(pages 490 and 491)

1. Write a proportion that uses the missing measurement because the ratios of corresponding side lengths are equal.

3. Student should draw a triangle with the same angle measures as the textbook. The ratio of the corresponding side lengths, $\frac{\text{student's triangle length}}{\text{book's triangle length}}$, should be greater than one.

5. yes; The triangles have the same angle measures, 107°, 39°, and 34°.

7. no; The triangles do not have the same angle measures.

9. The numerators of the fractions should be from the same triangle.

$$\frac{18}{16} = \frac{x}{8}$$
$$16x = 144$$
$$x = 9$$

11. 65

13. no; Each side increases by 50%, so each side is multiplied by a factor of $\frac{3}{2}$. The area is $\frac{3}{2}\left(\frac{3}{2}\right) = \frac{9}{4}$ or 225% of the original area, which is a 125% increase.

15. When two triangles are similar, the ratios of corresponding sides are equal.

17. $24 = p \cdot 96$; 25%

19. $a = 0.18 \cdot 250$; 45

Section 11.5 Parallel Lines and Transversals
(pages 497–499)

1. *Sample answer:*

3. m and n

5. 8

7. $\angle 1 = 107°$, $\angle 2 = 73°$

9. $\angle 5 = 49°$, $\angle 6 = 131°$

11. 60°; Corresponding angles are congruent.

13. $\angle 1$, $\angle 3$, $\angle 5$, and $\angle 7$ are congruent. $\angle 2$, $\angle 4$, $\angle 6$, and $\angle 8$ are congruent.

15. $\angle 6 = 61°$; $\angle 6$ and the given angle are vertical angles.
$\angle 5 = 119°$ and $\angle 7 = 119°$; $\angle 5$ and $\angle 7$ are supplementary to the given angle.
$\angle 1 = 61°$; $\angle 1$ and the given angle are corresponding angles.
$\angle 3 = 61°$; $\angle 1$ and $\angle 3$ are vertical angles.
$\angle 2 = 119°$ and $\angle 4 = 119°$; $\angle 2$ and $\angle 4$ are supplementary to $\angle 1$.

17. $\angle 2 = 90°$; $\angle 2$ and the given angle are vertical angles.
$\angle 1 = 90°$ and $\angle 3 = 90°$; $\angle 1$ and $\angle 3$ are supplementary to the given angle.
$\angle 4 = 90°$; $\angle 4$ and the given angle are corresponding angles.
$\angle 6 = 90°$; $\angle 4$ and $\angle 6$ are vertical angles.
$\angle 5 = 90°$ and $\angle 7 = 90°$; $\angle 5$ and $\angle 7$ are supplementary to $\angle 4$.

19. 132°; *Sample answer:* ∠2 and ∠4 are alternate interior angles and ∠4 and ∠3 are supplementary.

21. 120°; *Sample answer:* ∠6 and ∠8 are alternate exterior angles.

23. 61.3°; *Sample answer:* ∠3 and ∠1 are alternate interior angles and ∠1 and ∠2 are supplementary.

25. They are all right angles because perpendicular lines form 90° angles.

27. 130

29. **a.** no; They look like they are spreading apart. **b.** Check students' work.

31. 13

33. 51

35. B

Section 12.1 Finding Square Roots
(pages 514 and 515)

1. no; There is no integer whose square is 26.

3. $\sqrt{256}$ represents the positive square root because there is not a – or a ± in front.

5. 1.3 km

7. 3 and −3

9. 2 and −2

11. 25

13. $\frac{1}{31}$ and $-\frac{1}{31}$

15. 2.2 and −2.2

17. The positive and negative square roots should have been given.

$$\pm\sqrt{\frac{1}{4}} = \frac{1}{2} \text{ and } -\frac{1}{2}$$

19. 9

21. 25

23. 40

25. because a negative radius does not make sense

27. =

29. 9 ft

31. 8 m/sec

33. 2.5 ft

35. 25

37. 144

39. B

Section 12.2 The Pythagorean Theorem
(pages 520 and 521)

1. The hypotenuse is the longest side and the legs are the other two sides.

3. 24 cm

5. 9 in.

7. 12 ft

9. The length of the hypotenuse was substituted for the wrong variable.

$$\begin{aligned} a^2 + b^2 &= c^2 \\ 7^2 + b^2 &= 25^2 \\ 49 + b^2 &= 625 \\ b^2 &= 576 \\ b &= 24 \end{aligned}$$

Selected Answers

Section 12.2 The Pythagorean Theorem *(continued)* *(pages 520 and 521)*

11. 16 cm

13. 10 ft

15. 8.4 cm

17. **a.** *Sample answer:*

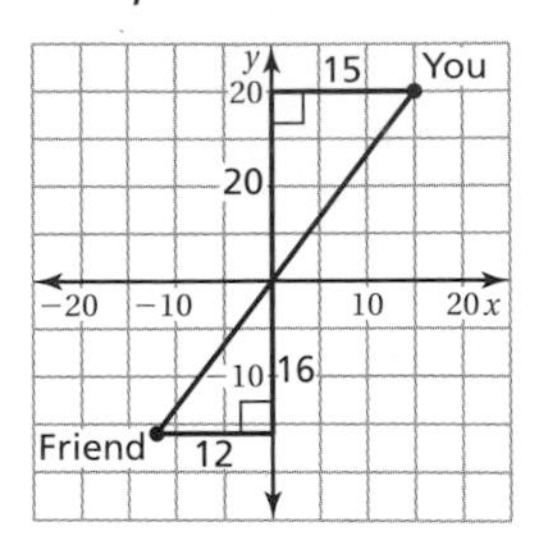

b. 45 ft

19. 6 and -6

21. 13

23. C

Section 12.3 Approximating Square Roots *(pages 529–531)*

1. A rational number can be written as the ratio of two integers. An irrational number cannot be written as the ratio of two integers.

3. all rational and irrational numbers; *Sample answer:* $-2, \frac{1}{8}, \sqrt{7}$

5. yes

7. no

9. rational; $3.\overline{6}$ is a repeating decimal.

11. irrational; 7 is not a perfect square.

13. rational; $-3\frac{8}{9}$ can be written as the ratio of two integers.

15. 144 is a perfect square. So, $\sqrt{144}$ is rational.

17. **a.** natural number

b. irrational number

c. irrational number

19. 26

21. -10

23. -13

25. 10; 10 is to the right of $\sqrt{20}$.

27. $\sqrt{133}$; $\sqrt{133}$ is to the right of $10\frac{3}{4}$.

29. -0.25; -0.25 is to the right of $-\sqrt{0.25}$.

31. 8 ft

33. *Sample answer:* $a = 82$, $b = 97$

35. 1.1

37. 30.1 m/sec

39. Falling objects do not fall at a linear rate. Their speed increases with each second they are falling.

41. $-3x + 3y$

43. $40k - 9$

Section 12.4 Simplifying Square Roots
(pages 536 and 537)

1. *Sample answer:* The square root is like a variable. So, you add or subtract the number in front to simplify.

3. about 1.62; yes   5. about 1.11; no   7. $\frac{\sqrt{7}+1}{3}$

9. $6\sqrt{3}$ 11. $2\sqrt{5}$ 13. $-7.7\sqrt{15}$

15. You do not add the radicands. $4\sqrt{5} + 3\sqrt{5} = 7\sqrt{5}$

17. $10\sqrt{2}$ 19. $4\sqrt{3}$ 21. $\frac{\sqrt{23}}{8}$ 23. $\frac{\sqrt{17}}{7}$

25. $10\sqrt{2}$ in.   27. $6\sqrt{6}$ 29. 210 ft^3

31. **a.** $88\sqrt{2}$ ft   **b.** 680 ft^2

Hint

33. Remember to take the square root of each side when solving for r.

35. 24 in.

37. C

Lesson 12.4b Cube Roots
(pages 537A and 537B)

1. -1 3. -6 5. 0   7. 9

9. $11\sqrt[3]{5}$ 11. $14\sqrt[3]{10}$ 13. -50

15. 1.9; 1.9 is to the right of $\sqrt[3]{5}$ on a number line.

17. $\sqrt[3]{100}$; $\sqrt[3]{100}$ is to the right of 4.25 on a number line.

19. $\sqrt[3]{-7}$; $\sqrt[3]{-7}$ is to the right of -2 on a number line.

21. 384 in.^{2}

Section 12.5 Using the Pythagorean Theorem
(pages 542 and 543)

1. *Sample answer:* You can plot a point at the origin and then draw lengths that represent the legs. Then, you can use the Pythagorean Theorem to find the hypotenuse of the triangle.

3. 27.7 m   5. 11.3 yd   7. 7.2 units   9. 27.5 ft   11. 15.1 m

13. yes   15. no   17. yes   19. 12.8 ft

21. **a.** *Sample answer:* 5 in., 7 in., 3 in.

    b. *Sample answer:* $BC \approx 8.6$ in.; $AB \approx 9.1$ in.

    c. Check students' work.

23. mean: 13; median: 12.5; mode: 12   25. mean: 58; median: 59; mode: 59

Section 13.1 Exponents
(pages 558 and 559)

1. An exponent describes the number of times the base is used as a factor. A power is the entire expression (base and exponent). A power tells you the value of the factor and the number of factors. No, the two cannot be used interchangeably.

3. 3^4 5. $\left(-\frac{1}{2}\right)^3$ 7. π^3x^4 9. 8^4b^3

11. 25 13. 1 15. $\frac{1}{144}$

17. The exponent 3 describes how many times the base 6 should be used as a factor. Three should not appear as a factor in the product. $6^3 = 6 \cdot 6 \cdot 6 = 216$

19. $-\left(\frac{1}{4}\right)^4$ 21. 29 23. 5 25. 66

27.

h	1	2	3	4	5
$2^h - 1$	1	3	7	15	31
2^{h-1}	1	2	4	8	16

$2^h - 1$; The option $2^h - 1$ pays you more money when $h > 1$.

29. Remember to add the black keys when finding how many notes you travel.

31. Associative Property of Multiplication

33. B

Section 13.2 Product of Powers Property
(pages 564 and 565)

1. When multiplying powers with the same base

3. 3^4 5. $(-4)^{12}$ 7. h^7

9. $\left(-\frac{5}{7}\right)^{17}$ 11. 5^{12} 13. 3.8^{12}

15. The bases should not be multiplied. $5^2 \cdot 5^9 = 5^{2+9} = 5^{11}$

17. $216g^3$ 19. $\frac{1}{25}k^2$ 21. $r^{12}t^{12}$

23. no; $3^2 + 3^3 = 9 + 27 = 36$ and $3^5 = 243$

25. 496 27. 78,125

29. **a.** $16\pi \approx 50.24$ in.^{3}

    b. $192\pi \approx 602.88$ in.^{3} Squaring each of the dimensions causes the volume to be 12 times larger.

31. Use the Commutative and Associative Properties of Multiplication to group the powers.

33. 4 35. 3 37. B

Section 13.3 Quotient of Powers Property
(pages 570 and 571)

1. To divide powers means to divide out the common factors of the numerator and denominator. To divide powers with the same base, write the power with the common base and an exponent found by subtracting the exponent in the denominator from the exponent in the numerator.

3. 6^6

5. $(-3)^3$

7. 5^6

9. $(-17)^3$

11. $(-6.4)^2$

13. b^{13}

15. You should subtract the exponents instead of dividing them. $\frac{6^{15}}{6^5} = 6^{15-5} = 6^{10}$

17. 2^9

19. π^8

21. k^{14}

23. $64x$

25. $125a^3b^2$

27. x^7y^6

29. You are checking to see if there is a constant rate of change in the prices, not if it is a linear function.

31. 10^{13} galaxies

33. -9

35. 61

37. B

Section 13.4 Zero and Negative Exponents
(pages 578 and 579)

1. no; Any nonzero base raised to the zero power is always 1.

3. $5^{-5}, 5^0, 5^4$

5.

n	4	3	2	1
$\frac{5^n}{5^2}$	$5^2 = 25$	$5^1 = 5$	$5^0 = 1$	$5^{-1} = \frac{1}{5}$

7. One-fifth of 5^1; $5^0 = \frac{1}{5}(5^1) = 1$

9. $\frac{1}{36}$

11. $\frac{1}{16}$

13. 1

15. $\frac{1}{125}$

17. The negative sign goes with the exponent, not the base. $(4)^{-3} = \frac{1}{4^3} = \frac{1}{64}$

19. 2^0; 10^0

21. $\frac{a^7}{64}$

23. $5b$

25. 12

27. $\frac{w^6}{9}$

29. 10,000 micrometers

31. 1,000,000 micrometers

33. Convert the blood donation to cubic millimeters before answering the parts.

35. If $a = 0$, then $0^n = 0$. Because you can not divide by 0, the expression $\frac{1}{0}$ is undefined.

37. 10^3

39. D

Section 13.5 Reading Scientific Notation
(pages 584 and 585)

1. Scientific notation uses a factor of at least one but less than 10 multiplied by a power of 10. A number in standard form is written out with all the zeros and place values included.

3. 0.00015 m

5. 20,000 mm^3

7. yes; The factor is at least 1 and less than 10. The power of 10 has an integer exponent.

9. no; The factor is greater than 10.

11. yes; The factor is at least 1 and less than 10. The power of 10 has an integer exponent.

13. no; The factor is less than 1.

15. 70,000,000

17. 500

19. 0.000044

21. 1,660,000,000

23. 9,725,000

25. **a.** 810,000,000 platelets
b. 1,350,000,000,000 platelets

27. **a.** Bellatrix
b. Betelgeuse

29. 5×10^{12} km^2

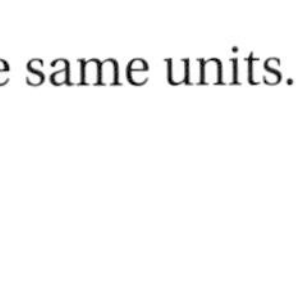

31. Be sure to convert some of the speeds so that they all have the same units.

33. 10^7

35. $\frac{1}{10^{16}}$

Section 13.6 Writing Scientific Notation
(pages 590 and 591)

1. If the number is greater than or equal to 10, the exponent will be positive. If the number is less than 1 and greater than 0, the exponent will be negative.

3. 2.1×10^{-3}

5. 3.21×10^{8}

7. 4×10^{-5}

9. 4.56×10^{10}

11. 8.4×10^{5}

13. 72.5 is not less than 10. The decimal point needs to move one more place to the left.
7.25×10^{7}

15. 9×10^{-10}

17. 1.6×10^{8}

19. 2.88×10^{-7}

21. 4.01×10^{7} m

23. 5.612×10^{14} cm^2

25. 9.75×10^{9} N•m per sec

27. *Answer should include, but is not limited to:* Make sure calculations using scientific notation are done correctly.

29. **a.** 2.65×10^{8} **b.** 2.2×10^{-4}

31. 200

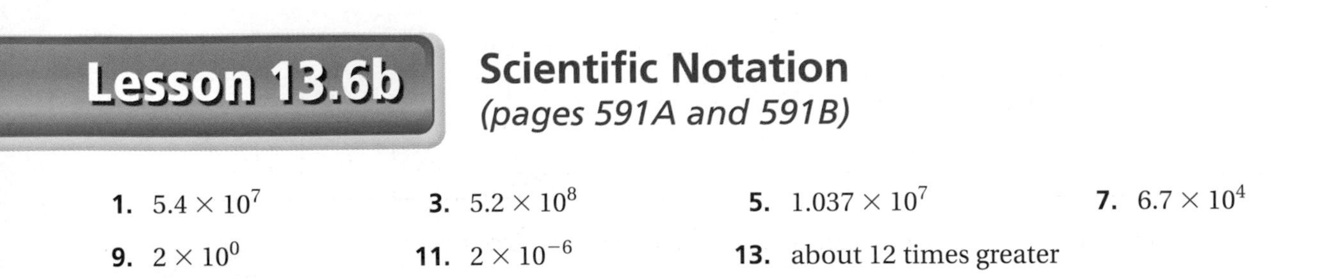

Lesson 13.6b Scientific Notation
(pages 591A and 591B)

1. 5.4×10^{7}

3. 5.2×10^{8}

5. 1.037×10^{7}

7. 6.7×10^{4}

9. 2×10^{0}

11. 2×10^{-6}

13. about 12 times greater

Topic 1

Angles
(pages AT2 and AT3)

1. complementary **3.** neither **5.** vertical; 90 **7.** 76

Topic 2

Geometry
(pages AT4–AT7)

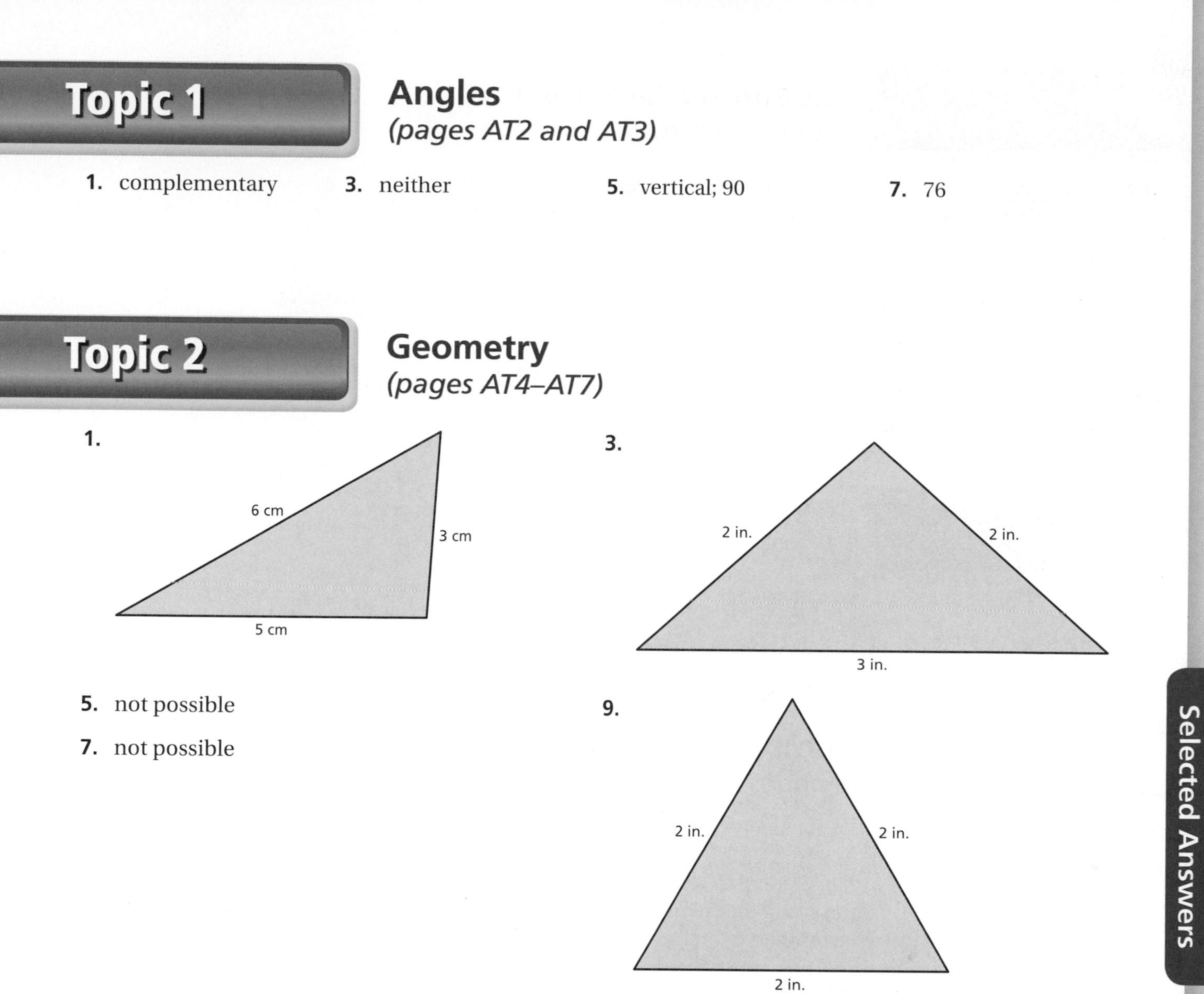

5. not possible

7. not possible

11. Side lengths in Exercise 5:

Side length	2 cm	4 cm	1 cm
Sum of other two side lengths	5 cm	3 cm	6 cm

Side lengths in Exercise 6:

Side length	6 cm	8 cm	10 cm
Sum of other two side lengths	18 cm	16 cm	14 cm

Side lengths in Exercise 7:

Side length	1 in.	2 in.	1 in.
Sum of other two side lengths	3 in.	2 in.	3 in.

Side lengths in Exercise 8:

Side length	5 cm	7 cm	4 cm
Sum of other two side lengths	11 cm	9 cm	12 cm

Side lengths in Exercise 9:

Side length	2 in.	2 in.	2 in.
Sum of other two side lengths	4 in.	4 in.	4 in.

Side lengths in Exercise 10:

Side length	1 in.	5 in.	3 in.
Sum of other two side lengths	8 in.	4 in.	6 in.

The sum of any two side lengths must be greater than the remaining side length.

Topic 2 Geometry *(continued)* *(pages AT4–AT7)*

13. 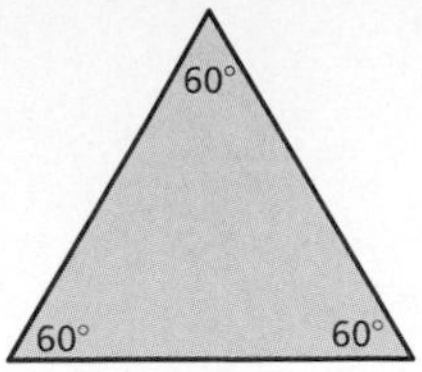

15. not possible

17. 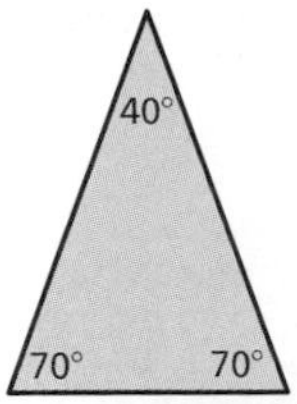

19. triangle

21. rectangle

23. triangle

25. The intersection is the shape of the base.

Topic 3 Converting Units of Measure *(pages AT8 and AT9)*

1. 176

2. 112.5

5. 0.55

7. 864 in.^{2}

Topic 4 Comparing Rates *(pages AT10 and AT11)*

1. a. fingernails

b. 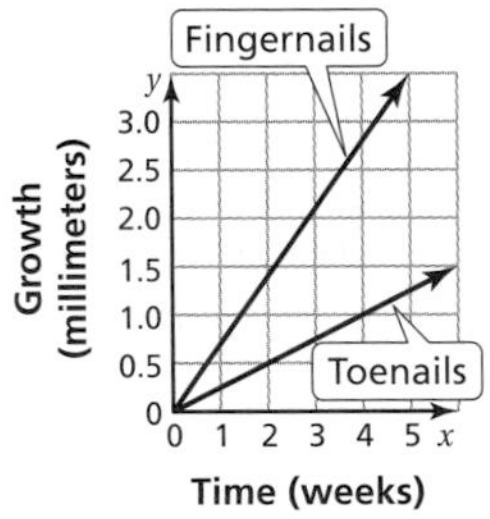

The graph that represents fingernails is steeper than the graph that represents toenails. So, fingernails grow faster than toenails.

Topic 5 Transformations *(pages AT12–AT15)*

1. 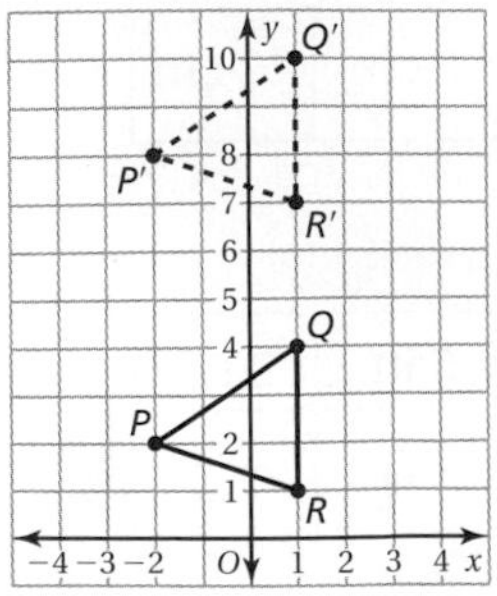

$P'(-2, 8)$, $Q'(1, 10)$, $R'(1, 7)$

3. 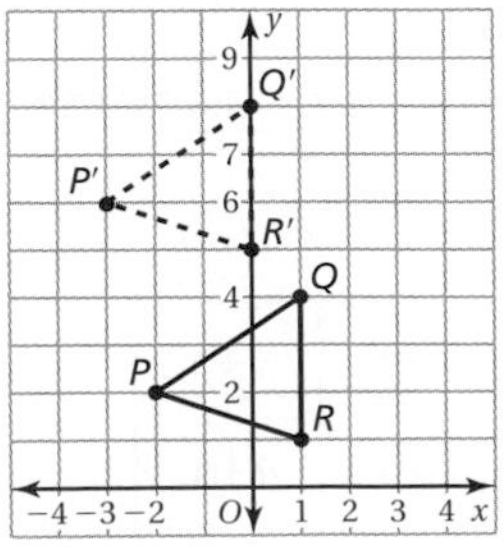

$P'(-3, 6)$, $Q'(0, 8)$, $R'(0, 5)$

5. a. side AB and side CD, side AD and side BC

b. *Sample answer:*

c. yes; *Sample answer:* A translation creates a congruent figure, so the sides remain parallel.

7. $L'(3, -1), M'(3, -4), N'(7, -4), P'(7, -1)$

9. $H'(6, -7), I'(6, -2), J'(3, -3), K'(3, -8)$

11. a. yes; *Sample answer:* The image is also a rectangle, so each angle measure is 90°.

b. yes; *Sample answer:* The image is congruent to the original, so side CD is the same length as side $C'D'$.

13. $L'(-1, -1), M'(-2, -4), N'(-4, -4), P'(-5, -1)$

15.

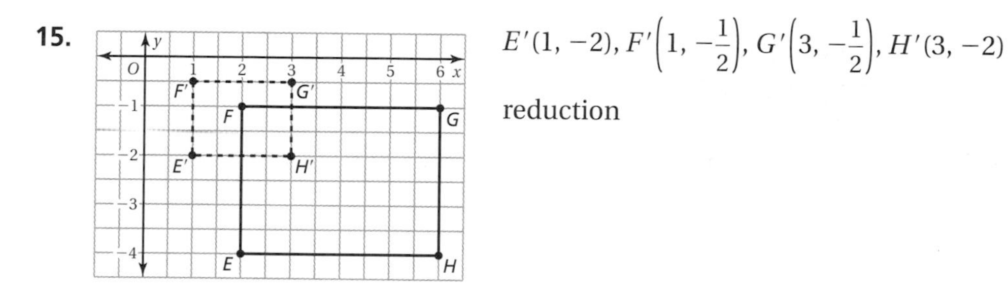

$E'(1, -2), F'\left(1, -\frac{1}{2}\right), G'\left(3, -\frac{1}{2}\right), H'(3, -2)$

reduction

17. a. yes; Triangle JKL is a 90° counterclockwise rotation about the origin of triangle XYZ.

b. yes; *Sample answer:* You can create triangle PQR by rotating triangle XYZ 90° counterclockwise about the origin and then dilating the image using a scale factor of 2.

Topic 6

Volumes of Spheres
(pages AT16 and AT17)

1. $2304\pi \approx 7234.6 \text{ cm}^3$

3. $\frac{1}{6}\pi \approx 0.5 \text{ in.}^3$

5. $\frac{10{,}624}{3}\pi \approx 11{,}120 \text{ ft}^3$

7. $54\pi \approx 170 \text{ cm}^3$

Topic 7

Triangles and Slope
(pages AT18 and AT19)

1. similar; Corresponding leg lengths are proportional.

3. The ratios are equal; *Sample answer:* Using the similar triangles in the Key Idea:

$$\frac{AB}{DE} = \frac{AC}{DF}$$

$$AB \cdot DF = DE \cdot AC$$

$$\frac{AB}{AC} = \frac{DE}{DF}$$

5. yes; The ratios of the corresponding leg lengths in the right triangles are proportional.

Selected Answers

Key Vocabulary Index

Mathematical terms are best understood when you see them used and defined *in context.* This index lists where you will find key vocabulary. A full glossary is available in your Record and Practice Journal and at *BigIdeasMath.com.*

Student Index

This student-friendly index will help you find vocabulary, key ideas, and concepts. It is easily accessible and designed to be a reference for you whether you are looking for a definition, real-life application, or help with avoiding common errors.

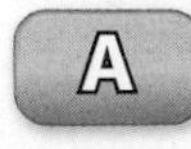

C

D

E

F

T

U

V

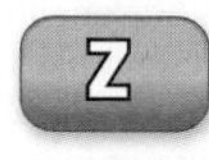

Student Index

Photo Credits

Front matter

iv Big Ideas Learning, LLC; **viii** *top* ©iStockphoto.com/Angel Rodriguez, ©iStockphoto.com/Ann Marie Kurtz; *bottom* Evok20/ Shutterstock.com; **ix** *top* ©iStockphoto.com/Stefan Klein; *bottom* Apollofoto/Shutterstock.com; **x** *top* ©iStockphoto.com/Angel Rodriguez, ©iStockphoto.com/Ann Marie Kurtz; *bottom* ©iStockphoto.com/ daaronj; **xi** *top* ©iStockphoto.com/Michael Flippo, ©iStockphoto.com/ Ann Marie Kurtz; *bottom* Apollofoto/Shutterstock.com; **xii** *top* ©iStockphoto.com/Alistair Cotton; *bottom* ©iStockphoto.com/ronen; **xiii** *top* ©iStockphoto.com/Lisa Thornberg, ©iStockphoto.com/Ann Marie Kurtz; *bottom* Heather Prosch-Jensen/Shutterstock.com; **xiv** *top* ©iStockphoto.com/Ron Sumners, ©iStockphoto.com/Ann Marie Kurtz; *bottom* Peter Close/Shutterstock.com; **xv** *top* ©iStockphoto.com/ Jonathan Larsen; *bottom* ©iStockphoto.com/ranplett; **xvi** *top* Varina and Jay Patel,©iStockphoto.com/Ann Marie Kurtz; *bottom* ©iStockphoto.com/Jane Norton; **xvii** *top* ©iStockphoto.com/Ann Marie Kurtz, infografick/Shutterstock.com; *bottom* wavebreakermedia ltd/ Shutterstock.com; **xviii** *top* ©iStockphoto.com/Ann Marie Kurtz, Kasiap/Shutterstock.com; *bottom* Odua Images/Shutterstock.com; **xix** *top* ©iStockphoto.com/Ann Marie Kurtz, Alexander Chaikin/ Shutterstock.com; *bottom* Edyta Pawlowska/Shutterstock.com; **xx** *top* ©iStockphoto.com/Ann Marie Kurtz, ©iStockphoto.com/pagadesign; *bottom* Sinisa Bobic/Shutterstock.com; **xxi** *top* stephan kerkhofs/ Shutterstock.com, Cigdem Sean Cooper/Shutterstock.com, ©iStockphoto.com/Andreas Gradin; *bottom* ©iStockphoto.com/ Thomas Perkins; **xxii** ©iStockphoto.com/kate_sept2004

Chapter 1

1 ©iStockphoto.com/Angel Rodriguez, ©iStockphoto.com/Ann Marie Kurtz; **21** ©iStockphoto.com/Ronnie Comeau; **34** *top* Estero High School; *bottom* Zephyrhills High School; **40** ©iStockphoto.com/ Rich Legg; **44** ©iStockphoto.com/susaro

Chapter 2

48 ©iStockphoto.com/Stefan Klein; **50** ©iStockphoto.com/shantell; **69** ©iStockphoto.com/Franky Sze; **75** ©iStockphoto.com/fotoVoyager

Chapter 3

96 ©iStockphoto.com/Angel Rodriguez, ©iStockphoto.com/Ann Marie Kurtz; **114** ©iStockphoto.com/Michael Kemter; **115** ©iStockphoto.com/ Vika Valter; **121** NASA; **122** Jean Thompson; **134** ©iStockphoto.com/ Eric Isselée; **136** Baldwin Online; Children's Literature Project at www.mainlesson.com; **137** ©iStockphoto.com/Brian Pamphilon; **139** Big Ideas Learning, LLC; **147** ©iStockphoto.com/Iain King; **148** ©iStockphoto.com/bluestocking

Chapter 4

156 ©iStockphoto.com/Michael Flippo, ©iStockphoto.com/ Ann Marie Kurtz; **166** ©iStockphoto.com/Robert Hadfield; **168** ©iStockphoto.com/TheSupe87; **175** ©iStockphoto.com/Wolfgang Amri; **176** ©iStockphoto.com/Albert Smirnov; **177** ©iStockphoto.com/ Lori Sparkia; **181** ©iStockphoto.com/Anne de Haas; **183** *top right* Big Ideas Learning, LLC; *center left* ©iStockphoto.com/Rui Matos; **184** ©iStockphoto.com/Michael Fernahl; **186** ©iStockphoto.com/ Ted Johns

Chapter 5

192 ©iStockphoto.com/Alistair Cotton; **194** *top* ©iStockphoto.com/ Viatcheslav Dusaleev; *bottom left* ©iStockphoto.com/Jason Mooy; *bottom center* ©iStockphoto.com/Andres Peiro Palmer; *bottom right* ©iStockphoto.com/Felix Möckel; **203** ©iStockphoto.com/Dennis Guyitt; **212** ©iStockphoto.com/Chris Schmidt; **213** ©iStockphoto.com/ Lisa F. Young; **216** ©iStockphoto.com/Dan Moore; **217** ©iStockphoto.com/Aldo Murillo; **217A** ©MesserWoland/ Wikipedia Commons; **224** ©iStockphoto.com/Er Ten Hong; **225** ©iStockphoto.com/Sergey Galushko, ©iStockphoto.com/Tryfonov Ievgenii; **226** ©iStockphoto.com/Ingmar Wesemann; **231** *center right* ©iStockphoto.com/asiseeit; *center left* ©iStockphoto.com/Hazlan Abdul Hakim; **237** ©iStockphoto.com/Maksim Shmeljov

Chapter 6

248 ©iStockphoto.com/Lisa Thornberg, ©iStockphoto.com/Ann Marie Kurtz; **251** *bottom left www.cartoonstock.com*; *bottom right* M.C. Escher's "Ascending and Descending" ©2010 The M.C. Escher Company-Holland. All rights reserved. *www.mcescher.com*; **254** *bottom left* ©iStockphoto.com/Rich Koele; *bottom center* ©iStockphoto.com/titelio; *bottom right* ©iStockphoto.com/David Spieth; **255** *top right* ©iStockphoto.com/Hedda Gjerpen; *center* ©iStockphoto.com/Rafal Zdeb; **266** ©iStockphoto.com/Tomasz Pietryszek; **267** *center left* Newcastle Drum Centre; *center right* ©iStockphoto.com/scol22; **270** *bottom row 1 left* ©iStockphoto.com/ Luke Daniek; *bottom row 1 right* ©iStockphoto.com/Jeff Whyte; *bottom row 2 left* ©2006 Michael Mattox/Image from BigStockPhoto.com; *bottom row 2 right* ©iStockphoto.com/Hedda Gjerpen; **271** ©iStockphoto.com/Josh Webb; **275** ©iStockphoto.com/ Robert Simon; **279** ©iStockphoto.com/Elena Schweitzer; **280** ©iStockphoto.com/Gordon Warlow; **282** ©iStockphoto.com/AVTG; **284** Courtesy NPS/Linda Chandler

Chapter 7

296 ©iStockphoto.com/Ron Sumners, ©iStockphoto.com/Ann Marie Kurtz; **303** *top left* ©iStockphoto.com/David Franklin; *top right* ©2008 Ruslan Kokarev. Image from BigStockPhoto.com; *center right* ©iStockphoto.com/Ebru Baraz, ©iStockphoto.com/Lev Mel; **304** ©iStockphoto.com/Jill Chen; **307** ©iStockphoto.com/camilla wisbauer; **309** *top left and center* ©iStockphoto.com/Achim Prill; *top right* ©iStockphoto.com/Marie-france Bélanger; *center left* ©iStockphoto.com/Matthew Dixon; *center right* ©iStockphoto.com/ Nilgun Bostanci; **310** *bottom left* ©iStockphoto.com/Jiri Vatka; *bottom right* ©iStockphoto.com/BostjanT; **314** Image © Courtesy of the Museum of Science, Boston, ©iStockphoto.com/ranplett; **315** *top left* ©iStockphoto.com/James Kingman; *top right* ©iStockphoto.com/Yails; **321** ©iStockphoto.com/Stefano Tiraboschi; **328** ©iStockphoto.com/ Gary Alvis; **329** *top left* ©iStockphoto.com/Jan Tyler; *center left* ©iStockphoto.com/Ben Greer; *center right* ©iStockphoto.com/Roberta Casaliggi; **336** GreenLight Collectibles, Ford Images; **337** *top left* ©iStockphoto.com/Daniel Cardiff; *center right* ©iStockphoto.com/ pomortzeff; *bottom left* ©iStockphoto.com/Ivana Starcevic

Chapter 8

346 ©iStockphoto.com/Jonathan Larsen; **349** ©iStockphoto.com/ Sharon Kaasa; **351** ©iStockphoto.com/Pekka Nikonen; **352** ©iStockphoto.com/Mehmet Salih Guler; **353** ©iStockphoto.com/ Rich Legg; **354** ©iStockphoto.com/Kirsty Pargeter; **357** *top* ©iStockphoto.com/susaro; *bottom* Arman Zhenikeyev/ Shutterstock.com; **359** ©iStockphoto.com/Eric Isselée; **361** ©iStockphoto.com/Vincent Chien Chow Chine; **365** Triff/ Shutterstock.com; **368** ©iStockphoto.com/iofoto; **369** ©iStockphoto.com/Andrew Manley; **370** *top left* ©iStockphoto.com/ Shannon Keegan; *top right* ©iStockphoto.com/Lorelyn Medina; **371** ©iStockphoto.com/Philip Lange; **372** *top* Big Ideas Learning, LLC; *Exercise 7 left* ©iStockphoto.com/kledge; *Exercise 7 right* ©iStockphoto.com/Stefan Klein; *Exercise 8 left* ©iStockphoto.com/ Antagain; *Exercise 8 right* ©iStockphoto.com/Borut Trdina; *Exercise 9* ©iStockphoto.com/Alex Slobodkin; **373** ©iStockphoto.com/7nuit, ©iStockphoto.com/Rodrigo Blanco, ©iStockphoto.com/blaneyphoto; **374** ©iStockphoto.com/machinim

Chapter 9

382 Varina and Jay Patel/Shutterstock.com, ©iStockphoto.com/Ann Marie Kurtz; **386** *top* ©iStockphoto.com/sweetym; *center left* Big Ideas Learning, LLC; **387** ©iStockphoto.com/sweetym; **388** ©iStockphoto.com/Joe_Potato, ©iStockphoto.com/sweetym; **389** *top left* ©iStockphoto.com/Jennifer Morgan; *top center* United States Golden Dollar Coin obverse featuring Sacagawea ©1999 United States Mint. All Rights Reserved. Used with permission.; *top right* United States coin image from the United States Mint.; **392** ©iStockphoto.com/Eric Ferguson; **395** ©iStockphoto.com/Elena Schweitzer; **397** *top* ©iStockphoto.com/Doug Cannell; *center* ©iStockphoto.com/Pawel Kuniec; **400** ©iStockphoto.com/Frank Van den Bergh; **404** ©iStockphoto.com/Joe_Potato, ©iStockphoto.com/sweetym; **405** ©iStockphoto.com/Joe_Potato, ©iStockphoto.com/sweetym; **406** ©United States Mint/Wikipedia Commons; **407** ©United States Mint/Wikipedia Commons; **408** ©iStockphoto.com/Tim Inman; **409** ©iStockphoto.com/Oleg Kozlov; **410** ©iStockphoto.com/Wojciech Krusinski; **411** ©iStockphoto.com/Andrey Armyagov; **412** *center left* ©iStockphoto.com/sweetym; *center right* ©iStockphoto.com/Alex Slobodkin; **415** ©iStockphoto.com/sweetym; **416** *top* ©iStockphoto.com/Frank Ramspott; *center* ©iStockphoto.com/7nuit

Chapter 10

420 infografick/Shutterstock.com, ©iStockphoto.com/Ann Marie Kurtz; **422** *top* ©iStockphoto.com/Joe Belanger; *bottom* ©iStockphoto.com/dra_schwartz; **423** ©iStockphoto.com/Jani Bryson; **427** ©iStockphoto.com/Robert Kelsey; **428** ©iStockphoto.com/Bill Lloyd; **429** ©iStockphoto.com/Jose Manuel Gelpi Diaz; **432** ©iStockphoto.com/Chris Schmidt; **434** ©iStockphoto.com/Yuriy Panyokov; **435** ©Matthew Gilson, used with permission of National Federation Independent Business, ©2006; All rights reserved.; **437** ©iStockphoto.com/Tinik; **438** ©iStockphoto.com/Tammy Bryngelson; **439** archana bhartia/Shutterstock.com; **441** ©iStockphoto.com/Roberto A Sanchez; **446** ©iStockphoto.com/Lev Mel; **447** *skier* ©iStockphoto.com/Ben Blankenburg; **448** ©iStockphoto.com/Stockphoto4u; **452** ©iStockphoto.com/Mark Stay; **453** ©iStockphoto.com/Frank Herzog; **454** ©iStockphoto.com/Nick Busey, ©iStockphoto.com/Jeff Clow; **460** ©iStockphoto.com/sweetym

Chapter 11

462 ©iStockphoto.com/Ann Marie Kurtz, Kasiap/Shutterstock.com; **469** ©iStockphoto.com/Jorgen Jacobsen; **471** Linda K. Bahner-Spiritartist.com; **474** *Exercise 9* ©iStockphoto.com/Chih-Feng Chen; *Exercise 10* ©iStockphoto.com/Andreas Gradin; *Exercise 11* ©iStockphoto.com/Jim Lopes; **475** ©iStockphoto.com/Zoran Kolundzija; **478** *bottom left* ©iStockphoto.com/Evgeny Terentev; *bottom right* ©iStockphoto.com/Vadym Volodin; **479** NASA; **482** ©Evelyn Peyton; **483** *top right* ©iStockphoto.com/Terraxplorer; *top left* ©iStockphoto.com/Lora Clark; *center right* ©iStockphoto.com/Jennifer Morgan; **492** Estate Craft Homes, Inc.

Chapter 12

508 ©iStockphoto.com/Ann Marie Kurtz, Alexander Chaikin; **513** Perfectblue97; **514** ©iStockphoto.com/Benjamin Lazare; **515** *top right* ©iStockphoto.com/iShootPhotos, LLC; *center left* ©iStockphoto.com/Jill Chen; **516** ©Oxford Science Archive/Heritage Images/Imagestate; **520** ©iStockphoto.com/Melissa Carroll; **521** *top left* ©iStockphoto.com/Alex Slobodkin, ©iStockphoto.com/Sebastian Duda; *top right* ©iStockphoto.com/Cathy Keifer; **523** *bottom left* ©iStockphoto.com/Yvan Dubé; *bottom right* ©iStockphoto.com/MACIEJ NOSKOWSKI; **524** ©iStockphoto.com/Kais Tolmats; **528** *top left* ©iStockphoto.com/Don Bayley; *center right* ©iStockphoto.com/iLexx; **531** ©iStockphoto.com/Marcio Silva; **533** Luminis/Shutterstock.com; **536** *Exercise 3* Joshua Haviv/Shutterstock.com; *Exercise 4* ©iStockphoto.com/William D Fergus McNeill; *Exercise 5* ©iStockphoto.com/Fotografie K.J. Schraa; **537** *top left* ©iStockphoto.com/parema; *center right* ©iStockphoto.com/Jet Chen Tan; **547** *center right* Hasan Shaheed/Shutterstock.com; *Exercise 18* ©iStockphoto.com/JenDen2005; *Exercise 19* Orla/Shutterstock.com; **548** *Exercise 14* MisterElements/Shutterstock.com; *Exercise 15* CD Lenzen/Shutterstock.com

Chapter 13

552 ©iStockphoto.com/pagadesign, ©iStockphoto.com/Ann Marie Kurtz; **554** ©iStockphoto.com/Franck Boston; **555** *a.* ©iStockphoto.com/Manfred Konrad; *b.* NASA/JPL-Caltech/R.Hurt(SSC); *c.* and *d.* NASA; *bottom right* Stevyn Colgan; **557** ©iStockphoto.com/Philippa Banks; **558** ©iStockphoto.com/clotilde hulin; **559** ©iStockphoto.com/Boris Yankov; **560** ©iStockphoto.com/John Tomaselli; **564** ©iStockphoto.com/VIKTORIA KULISH; **565** *top right* ©Paul Tessier; *center left* ©iStockphoto.com/subjug, ©iStockphoto.com/Valerie Loiseleux, ©iStockphoto.com/Linda Steward; **570** ©iStockphoto.com/Petrovich; **571** *top right* Dash/Shutterstock.com; *center left* NASA/JPL-Caltech/L.Cieza (UT Austin); **573** ©iStockphoto.com/Dan Moore; **577** ©iStockphoto.com/Aliaksandr Autayeu; **578** EugeneF/Shutterstock.com; **579** ©iStockphoto.com/Nancy Louie; **580** ©iStockphoto.com/Kais Tolmats; **581** *top* ©iStockphoto.com/Sebastian Kaulitzki; *center* ©iStockphoto.com/Henrik Jonsson; *bottom* 7art-screensavers.com; **583** *left* ©iStockphoto.com/Mark Stay; *center* ©iStockphoto.com/Frank Wright; *right* ©iStockphoto.com/Evgeniy Ivanov; *bottom left* ©iStockphoto.com/Oliver Sun Kim; **584** ©iStockphoto.com/Christian Jasiuk; **585** Microgen/Shutterstock.com; **586** *a.* ©iStockphoto.com/trigga; *b.* ©iStockphoto.com/subjug; *c.* ©iStockphoto.com/camilla wisbauer; *d.* ©iStockphoto.com/Joe Belanger; *e.* ©iStockphoto.com/thumb; *f.* ©iStockphoto.com/David Freund; **587** NASA; **588** *left* Google logo © Google Inc., 2011. Reprinted with permission.; *right* YouTube logo © Google Inc., 2011. Reprinted with permission.; **589** Elaine Barker/Shutterstock.com; **590** *bottom left* ©iStockphoto.com/Max Delson Martins Santos; *bottom right* ©iStockphoto.com/Jan Rysavy; **591** *top right* BORTEL Pavel/Shutterstock.com; *bottom left* ©iStockphoto.com/Boris Yankov; **591B** *center left* Sebastian Kaulitzki/Shutterstock.com; *center right* ©iStockphoto.com/Jan Rysavy; **596** *center left* Eric Isselée; *bottom right* ©iStockphoto.com/Eric Holsinger

Additional Topics

AT1 stephan kerhofs/Shutterstock.com, Cigdem Sean Cooper/Shutterstock.com, ©iStockphoto.com/Andreas Gradin; **AT3** Elena Elisseeva/Shutterstock.com; **AT8** Steve MacAulay Photography; **AT9** *top right* Benedict Gibson; *bottom right* Frantz Petion; **AT16** *top left* ©iStockphoto.com/Yury Kosourov; *center left* adziohiciek/Shutterstock.com; *Exercise 1* Carlos Caetano/Shutterstock.com; *Exercise 2* Samokhin/Shutterstock.com; *Exercise 3* cobalt88/Shutterstock.com; **AT17** *top left* Donald Joski/Shutterstock.com; *bottom left* bonchan/Shutterstock.com; **A0** ©iStockphoto.com/Björn Kindler; *center left* ©iStockphoto.com/Michael Mattner; *center right* AKaiser

Appendix A

A1 *top* ©iStockphoto.com/toddmedia; *bottom left* ©iStockphoto.com/Eric Isselée; *bottom right* ©iStockphoto.com/Loretta Hostettler; **A2** sgame/Shutterstock.com; **A3** *top* AKaiser/Shutterstock.com; *bottom left* Photononstop/SuperStock, ©iStockphoto.com/PhantomOfThe Opera; *bottom right* ©iStockphoto.com/Michael Mattner; **A4** ©iStockphoto.com/Duncan Walker; **A5** *top* ©iStockphoto.com/DNY59; *bottom left* ©iStockphoto.com/Boris Katsman; **A6** *top and center* ©iStockphoto.com/Loretta Hostettler; *bottom* ©iStockphoto.com/toddmedia; **A7** *top right* L M Bilder; *center left* ©iStockphoto.com/Clayton Hansen; *bottom left* Bill Whittaker; **A8** *wasp* ©iStockphoto.com/arlindo71; *bat* ©iStockphoto.com/Alexei Zaycev; *monkey and kangaroo* ©iStockphoto.com/Eric Isselée; **A9** *parrot* ©iStockphoto.com/Eric Isselée; *frog* ©iStockphoto.com/BrandonAlms; *spider* ©iStockphoto.com/arlindo71; *cougar* ©iStockphoto.com/Eric Isselée; *lobster* ©iStockphoto.com/Joan Vincent Canto Roig; *crocodile* ©iStockphoto.com/Mehmet Salih Guler

Cartoon illustrations Tyler Stout
Cover image Lechner & Benson Design

K

Counting and Cardinality	– Count to 100 by Ones and Tens; Compare Numbers
Operations and Algebraic Thinking	– Understand and Model Addition and Subtraction
Number and Operations in Base Ten	– Work with Numbers 11–19 to Gain Foundations for Place Value
Measurement and Data	– Describe and Compare Measurable Attributes; Classify Objects into Categories
Geometry	– Identify and Describe Shapes

1

Operations and Algebraic Thinking	– Represent and Solve Addition and Subtraction Problems
Number and Operations in Base Ten	– Understand Place Value for Two-Digit Numbers; Use Place Value and Properties to Add and Subtract
Measurement and Data	– Measure Lengths Indirectly; Write and Tell Time; Represent and Interpret Data
Geometry	– Draw Shapes; Partition Circles and Rectangles into Two and Four Equal Shares

2

Operations and Algebraic Thinking	– Solving One- and Two-Step Problems Involving Addition and Subtraction; Build a Foundation for Multiplication
Number and Operations in Base Ten	– Understand Place Value for Three-Digit Numbers; Use Place Value and Properties to Add and Subtract
Measurement and Data	– Measure and Estimate Lengths in Standard Units; Work with Time and Money
Geometry	– Draw and Identify Shapes; Partition Circles and Rectangles into Two, Three, and Four Equal Shares

3

Domain	Cluster
Operations and Algebraic Thinking	– Represent and Solve Problems Involving Multiplication and Division; Solve Two-Step Problems Involving Four Operations
Number and Operations in Base Ten	– Round Whole Numbers; Add, Subtract, and Multiply Multi-Digit Whole Numbers
Number and Operations — Fractions	– Understand Fractions as Numbers
Measurement and Data	– Solve Time, Liquid Volume, and Mass Problems; Understand Perimeter and Area
Geometry	– Reason with Shapes and Their Attributes

4

Domain	Cluster
Operations and Algebraic Thinking	– Use the Four Operations with Whole Numbers to Solve Problems; Understand Factors and Multiples
Number and Operations in Base Ten	– Generalize Place Value Understanding; Perform Multi-Digit Arithmetic
Number and Operations — Fractions	– Build Fractions from Unit Fractions; Understand Decimal Notation for Fractions
Measurement and Data	– Convert Measurements; Understand and Measure Angles
Geometry	– Draw and Identify Lines and Angles; Classify Shapes

5

Domain	Cluster
Operations and Algebraic Thinking	– Write and Interpret Numerical Expressions
Number and Operations in Base Ten	– Perform Operations with Multi-Digit Numbers and Decimals to Hundredths
Number and Operations — Fractions	– Add, Subtract, Multiply, and Divide Fractions
Measurement and Data	– Convert Measurements within a Measurement System, Understand Volume
Geometry	– Graph Points in the First Quadrant of the Coordinate Plane; Classify Two-Dimensional Figures

Mathematics Reference Sheet

Conversions

U.S. Customary
1 foot = 12 inches
1 yard = 3 feet
1 mile = 5280 feet
1 acre ≈ 43,560 square feet
1 cup = 8 fluid ounces
1 pint = 2 cups
1 quart = 2 pints
1 gallon = 4 quarts
1 gallon = 231 cubic inches
1 pound = 16 ounces
1 ton = 2000 pounds
1 cubic foot ≈ 7.5 gallons

Metric
1 centimeter = 10 millimeters
1 meter = 100 centimeters
1 kilometer = 1000 meters
1 liter = 1000 milliliters
1 kiloliter = 1000 liters
1 milliliter = 1 cubic centimeter
1 liter = 1000 cubic centimeters
1 cubic millimeter = 0.001 milliliter
1 gram = 1000 milligrams
1 kilogram = 1000 grams

Time
1 minute = 60 seconds
1 hour = 60 minutes
1 hour = 3600 seconds
1 year = 52 weeks

Temperature

$C = \frac{5}{9}(F - 32)$

$F = \frac{9}{5}C + 32$

U.S. Customary to Metric
1 inch ≈ 2.54 centimeters
1 foot ≈ 0.3 meter
1 mile ≈ 1.6 kilometers
1 quart ≈ 0.95 liter
1 gallon ≈ 3.79 liters
1 cup ≈ 237 milliliters
1 pound ≈ 0.45 kilogram
1 ounce ≈ 28.3 grams
1 gallon ≈ 3785 cubic centimeters

Metric to U.S. Customary
1 centimeter ≈ 0.39 inch
1 meter ≈ 3.28 feet
1 kilometer ≈ 0.6 mile
1 liter ≈ 1.06 quarts
1 liter ≈ 0.26 gallon
1 kilogram ≈ 2.2 pounds
1 gram ≈ 0.035 ounce
1 cubic meter ≈ 264 gallon

Rules of Exponents

Product of Powers Property: $a^m \cdot a^n = a^{m+n}$

Quotient of Powers Property: $\frac{a^m}{a^n} = a^{m-n}$, where $a \neq 0$

Zero Exponents: $a^0 = 1$, where $a \neq 0$

Negative Exponents: $a^{-n} = \frac{1}{a^n}$, where $a \neq 0$

Slope-Intercept Form

$y = mx + b$

slope (m), y-intercept (b)

Pythagorean Theorem

$a^2 + b^2 = c^2$

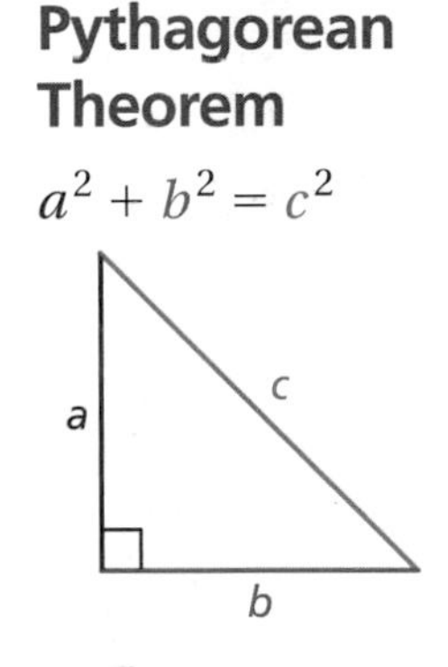

Surface Area and Volume

Prism

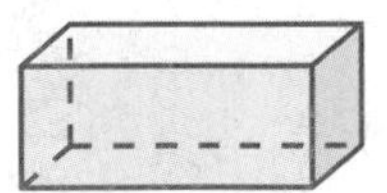

S = areas of bases + areas of lateral faces

$V = Bh$

Cylinder

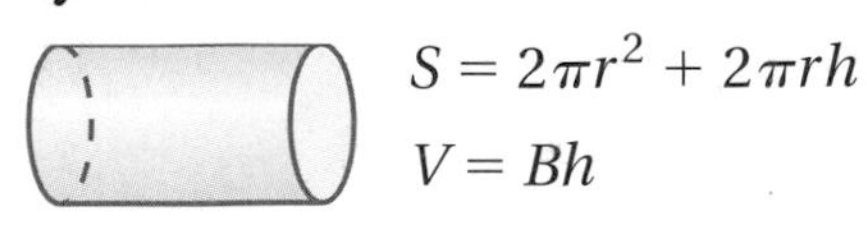

$S = 2\pi r^2 + 2\pi rh$

$V = Bh$

Pyramid

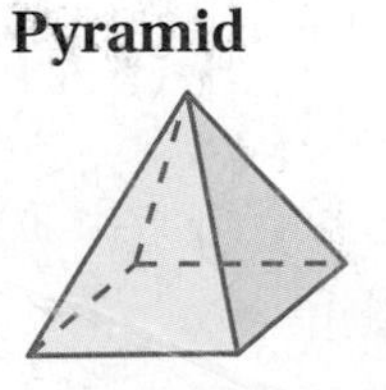

S = area of base + areas of lateral faces

$V = \frac{1}{3}Bh$

Cone

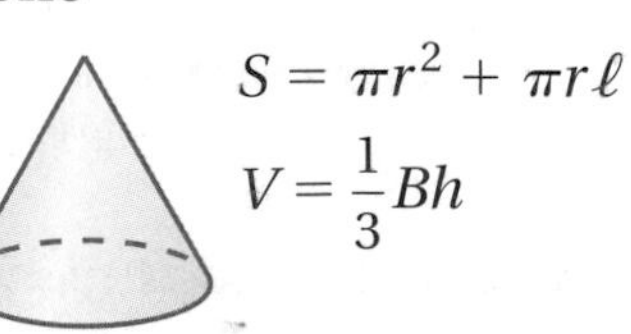

$S = \pi r^2 + \pi r\ell$

$V = \frac{1}{3}Bh$

Sphere

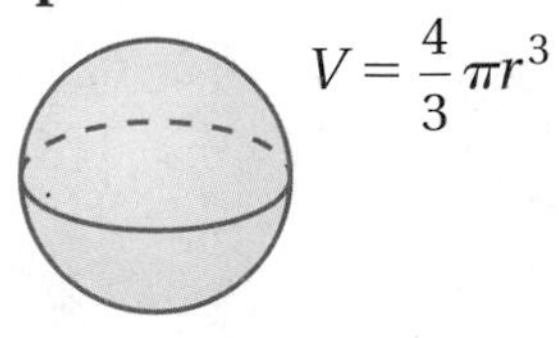

$V = \frac{4}{3}\pi r^3$

Circumference and Area of a Circle

$C = \pi d$ or $C = 2\pi r$

$A = \pi r^2$

$\pi \approx \frac{22}{7}$, or 3.14

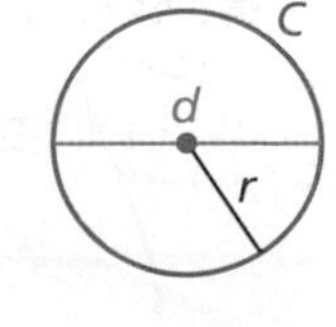